Slope

If $P_1(x_1, y_1)$ and $P_2(x_2, y_2)$ are two different points on a nonvertical line, then the slope m of the line is given by the formula

$$m = \frac{y_2 - y_1}{x_2 - x_1}.$$

Straight Line: Point-Slope Form

If a line contains the point $P(x_1, y_1)$ and has slope m, then the point slope form of the equation of a line is

$$y - y_1 = m(x - x_1).$$

Straight Line: Slope-Intercept Form

If a line has a y-intercept b and slope m, then the slope-intercept form of the equation of a line is $y = mx + b$.

Distance Formula

The distance between any two points $A(x_1, y_1)$ and $B(x_2, y_2)$ is given by the formula

$$d(A, B) = \sqrt{(x_2 - x_1)^2 + (y_2 - y_1)^2}.$$

Log/Exp Principle

For $b > 0$ and $b \neq 1$, $b^A = C$ is equivalent to $\log_b C = A$.

Rules of Logarithms

$$\log_b 1 = 0$$

$$\log_b b = 1$$

$$\log_b (xy) = \log_b x + \log_b y$$

$$\log_b (x^t) = t \log_b x$$

$$\log_b \left(\frac{x}{y}\right) = \log_b x - \log_b y$$

$$\log_b \left(\frac{1}{x}\right) = -\log_b x$$

Rules for Transformations

Transformation	Change in Equation	Effect on Graph
Horizontal translation	Replace x with $x - h$	If $h > 0$, graph moves $\lvert h \rvert$ units to the right. If $h < 0$, graph moves $\lvert h \rvert$ units to the left.
Vertical translation	Replace y with $y - k$	If $k > 0$, graph moves $\lvert k \rvert$ units up. If $k < 0$, graph moves $\lvert k \rvert$ units down.
Reflection across y-axis	Replace x with $-x$	Graph is reflected across y-axis.
Reflection across x-axis	Replace y with $-y$	Graph is reflected across x-axis.
Vertical change in shape	$y = f(x)$ becomes $y = b \cdot f(x)$	If $b > 1$, graph is stretched away from x-axis. If $0 < b < 1$, graph is shrunk down toward x-axis.
Horizontal change in shape	$y = f(x)$ becomes $y = f(ax)$	If $a > 1$, graph is shrunk toward y-axis. If $0 < a < 1$, graph is stretched away from y-axis.

College Algebra and Trigonometry

College Algebra and Trigonometry

SECOND EDITION

Timothy J. Kelly
Hamilton College

John T. Anderson
Hamilton College

Richard H. Balomenos

HOUGHTON MIFFLIN COMPANY BOSTON
Toronto Dallas Geneva, Illinois Palo Alto Princeton

Sponsoring Editor: Maureen O'Connor
Senior Development Editor: Tony Palermino
Senior Project Editor: Jean Andon
Design Coordinator: Martha Drury
Cover Designer: Mark Caleb
Production Coordinator: Frances Sharperson
Manufacturing Coordinator: Holly Schuster
Marketing Manager: Michael Ginley

Cover photograph by Keller and Pete Associates/Arlington, Massachusetts

Printed in the U.S.A.
Library of Congress Catalog Card Number: 91-71981

ISBN Numbers:
Text: 0-395-43215-4
Instructor's Edition: 0-395-60142-8
Solutions Manual: 0-395-60143-6
Test Bank: 0-395-60145-2
Student Solutions Manual: 0-395-60144-4

ABCDEFGHIJ-H-954321

Contents

Preface

Our increasingly technical and complex world continues to generate issues and problems for which the tools of algebra and trigonometry provide the basis for solutions. At the same time, our students are changing. Students today have more diverse interests and backgrounds, and they possess a stronger visual orientation.

In the second edition, we have sought to strike a balance between the demands of the subject matter and the needs of the students. Our text provides a sound and precise presentation of algebra and trigonometry that students can learn from. We have employed graphs liberally to visually represent functions and other difficult concepts. In addition, we have carefully selected examples that illustrate the concepts, and we have annotated the solutions to anticipate the kind of problems that students of diverse backgrounds encounter.

Care and attention to detail are the hallmarks of the second edition. Exercises, problem solving, applications, and end-of-chapter reviews have been revised extensively. We have written a broad array of exercises from simple drill problems that model examples to problems that will challenge the best students. To reinforce the importance of problem solving as an analytical tool, we have used this carefully-developed strategy throughout the text in many different settings. Furthermore, we have introduced a unique applications case study in each chapter, called *Analyzing Data,* that makes use of real data to show how the techniques in the chapter can be applied to solve actual problems. What follows is a detailed explanation of how our book strives to meet today's classroom needs.

Content and Pedagogy

Functions and Graphs Perhaps the single most important concept in algebra and trigonometry is the notion of function. In exploring the properties and applications of functions, we have fully exploited the nature of their graphs throughout the text. Moreover, we have consistently made use of the notions of transformations and symmetry to unify the treatment of functions.

Annotated Examples As an aid to student understanding, we have added annotations to the solutions of worked examples that explain how steps were obtained. Our intent is to simulate, whenever feasible, the explanation that an instructor would supply in presenting problems at the blackboard. Students and instructors who used the first edition of the text found this feature extraordinarily useful.

Exercise Sets We have carefully graded the exercises by level of difficulty. To help in the assignment of homework, each even-odd pairing is the same level of difficulty. We have ordered the exercises so they gradually become more difficult. Each exercise set ends with a collection of *Superset* exercises that extend and challenge students' understanding. Answers to *all* Superset problems appear in the Appendix, together with answers to the odd-numbered problems.

Problem Solving and Applications Mathematics students typically have trouble with word problems. We have devoted considerable time in Chapters 1 and 2 to developing this skill, and we have introduced a method for systematically solving word problems in Section 2.6. We then have used this problem-solving strategy in every subsequent chapter. Moreover, we have added dozens of new application problems in business and economics, biology, medicine, earth science, physics, chemistry, sociology, and environmental science.

Analyzing Data In our information society, students will be called upon to interpret and analyze quantitative data. Even the popular media increasingly have become reliant on basic statistics and graphs to communicate information. In response to this need, we have included in each chapter a carefully selected case study that connects the material presented in that chapter with an issue or problem that is solved by working with *real data*. Our own students have benefited greatly from these problems; the case studies served to motivate interest in the phenomenon under study and generated insightful questions *about mathematics.*

Calculator/Graphing Calculator We have included in the exercise sets several hundred problems which require the use of a hand-held calculator. In applications, computational complexities frequently require the use of a calculator, so we feel the experience is valuable for students using this text. We have also included calculator exercises that provide students the opportunity to discover, through computation, patterns that reinforce concepts. Our students have raised many good questions about the underlying mathematics following work on such problems. We also have included, at the end of Chapters 3 to 10, a set of *Graphing Calculator* problems that exploit graphing as a powerful problem-solving tool and as a means of building conceptual understanding through visualization. These problems can be solved using a graphing calculator, or with the *Math Assistant* graphing software available free-of-charge to users of this text.

End-of-Chapter Features We have concluded each chapter with an extensive array of features:

- *Chapter Review* is a one- to two-page summary of the essential concepts in the chapter that assists students as they review the chapter.
- *Review Exercises* are a comprehensive set of problems that cover all the major techniques and concepts presented in the chapter.
- *Chapter Test* examines students' understanding of the chapter. We have included complete solutions to all Chapter Test questions in the Appendix in the back of the book. This feature will be valuable to students needing help diagnosing the exact nature of their errors.

Changes in the Second Edition

Students, instructors, and reviewers have applauded the clarity and precision of the first edition. In refining the content and presentation for the second edition, we have attempted to maintain the same writing style that first edition readers found so appealing. However, we have made several content revisions and reorganizations for this edition.

In Chapter 1, we have added a discussion of scientific notation and significant digits, since applications requiring calculator use begin in that chapter.

We have moved the introduction to complex numbers to Chapter 2, so that complex numbers can be used in solving quadratic and higher-degree polynomial equations. We also have added scientific and business applications to Section 2.6, Problem Solving.

We have brought forward to Chapter 3 the discussion of inverse functions so that the treatment of composition of functions is now more complete. In Chapter 3, we have included two new sections. The first is Section 3.8, Interpreting Graphs. This section introduces the student to an applied skill that both mathematicians and mathematics educators have been calling for in recent years. Although it is an optional section, reviewers have hailed it as a very important contribution to the state of the art in precalculus texts, and so we heartily recommend its inclusion in your course. We have also included a new section on Variation, which, though also optional, presents an interesting topic for those interested in applications.

In Chapter 4, we have expanded the treatment of polynomial functions to include the major theorems (Factor Theorem, Remainder Theorem, Rational Root Theorem, etc.). In addition, we have included an optional section on the Bisection Method as an introduction to numerical approximation of zeros of polynomial functions. Those reviewers who are applied mathematicians have convinced us of the importance of this methodology.

Chapter 5 presents an introduction to exponential and logarithmic functions. We have moved the discussion of linear interpolation to an appendix, since most users rely on hand-held calculators. Also, we have included an entire section on applications to business, economics, biology, earth science, and physics.

We have streamlined our approach to Trigonometry by consolidating the material into Chapters 6, 7, and 8. This reorganization has allowed us to interject applications periodically throughout the treatment, by which we hope to offer the student a blend of both theory and practice.

We have expanded Chapter 9, Systems of Equations, to include more matrix algebra and a treatment of inverse matrices and their use in solving systems of equations.

Chapter 10 presents topics in algebra and probability that an instructor can choose from, depending on the students' interests and plans for future courses.

Supporting Materials

- *Instructor's Annotated Edition* contains answers to every exercise in the text.
- *Solutions Manual* contains complete, detailed solutions to every exercise in the text.
- *Student Solutions Manual* contains solutions to all odd-numbered exercises in the text.
- The *Computerized Test Generator* offers the instructor the capability of constructing multiple-choice or free-response tests from a 2000 item test bank, stored on a computer disk. The generator is available for both IBM and Macintosh users.
- *Printed Test Bank* is a printed version of the test questions in the Computerized Test Generator. Users who do not have access to a computer may select questions for use in a test prepared by hand.
- The *Math Assistant* software package contains a collection of programs that graph functions, solve equations, solve linear programming problems, and perform matrix calculations.
- *The Graphing Workbook for College Students* contains over 500 exercises suitable for graphing calculators or the Math Assistant.
- *The College Algebra and Trigonometry Video Library* contains twenty-four 30-minute reviews of the basic topics covered in the text. This is ideal for students who need out-of-class explanations that can be stopped, reviewed, and used as needed to gain mastery.

Acknowledgments

An expression of gratitude is due to the following reviewers who contributed to the shaping of the second edition:

Jasper Adams, State University of Texas
Michael Bennett, Bronx Community College
Frank Cerretto, Stockton State College
Pat Cook, Weatherford College
Richard Davis, Community College of Allegheny County
John DeCoursey, Vincennes University
Joel Greenstein, New York Technical College
Allen Hesse, Rochester Community College
Ira Kalantari, Western Illinois University
William Martin, Pima Community College, East Campus
Cecil McBride, Louisiana Tech University
Sharon Taylor Riley, Wayne County Community College
Gail K. Rogers, Blinn College
Michael Schramm, LeMoyne College
Joel Siegel, Sierra College
Charlotte Simpson, Bee County College
Rose Marie Smith, Texas Woman's University
Sam J. Tinsley, Richland Community College
Eric Wakkuri, Oregon Institute of Technology
Betsy Whitman, Florida Agricultural and Mechanical University
Jon Wilkin, Northern Virginia Community College, Alexandria Campus
Judith Willoughby, Minneapolis Community College
Paula J. Wojcik, Michigan State University

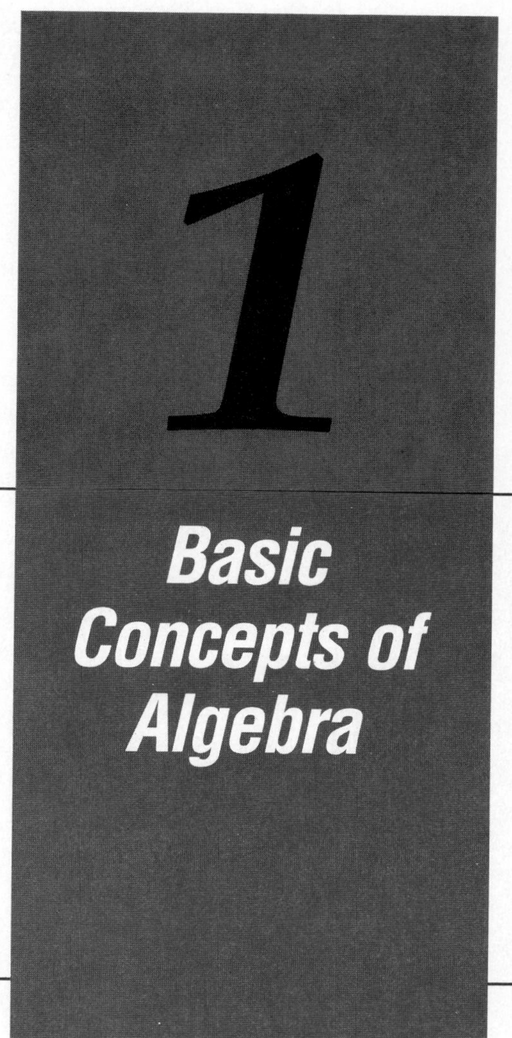

When information is quantitative in nature, it is frequently presented in a table or a graph. This is made quite clear by simply turning the pages of USA Today, *or* The Wall Street Journal. *But frequently it is possible to get even more information from tables or charts than is apparent. At the end of this chapter you will consider a simple type of chart, called a frequency histogram, which was presented in a research study of grading practices in a calculus course. You will see that by using information from this histogram, you can determine the average of the original set of data, without ever seeing the original data. Of course, to do this, you will need to evaluate an algebraic expression.*

1.1 Statements and Variables

In mathematics, as in everyday language, we use statements to communicate ideas. In algebra, our statements are about numbers and can be used to solve a wide variety of "real-life" problems. The skill in using algebra lies in translating such problems into statements about numbers.

We begin our review of algebra with two kinds of problems which may be familiar to you. They represent two different types of translation problems: describing a specific unknown value, and making generalizations about many values.

Problem 1. Five less than twice a certain number is one more than the number. What is the number?

Here, we are given information about some unknown number, referred to in the problem as a "certain number." Even though we do not know yet what the value of the number is, we can give it a name: x. We call x a **variable.** We can use a variable to talk about a number, whose value is not known to us yet.

English Sentence	Five less than twice a certain number	is	one more than the number
Algebraic Equation	$2x - 5$	$=$	$x + 1$

Recall that when two **expressions,** such as $2x - 5$ and $x + 1$ represent the same value, they can be written on opposite sides of an equals sign. The resulting statement is called an **equation.** The equation

$$2x - 5 = x + 1$$

summarizes, in algebraic form, the information about the unknown number x. The equation is then "solved for x." We will review methods for solving equations in Chapter 2. Our purpose in stating Problem 1 is to suggest one important use of algebra:

To write an equation that summarizes the information about a specific number whose value is not known yet. By solving the equation, we discover the value of this number.

Now consider a slightly different type of problem.

Problem 2. Think of a number. Double it. Add ten. Now take half of that result. Subtract your original number. Your answer is five.

To help understand the reason this "number trick" works, let us see what happens to a number that undergoes the process. Here we let the variable x stand for any number.

- Think of a number x
- Double it .... $2x$
- Add 10 .... $2x + 10$
- Take half .... $\frac{1}{2}(2x + 10)$
- Subtract your original number .... $\frac{1}{2}(2x + 10) - x$
- Your answer is 5 .... $\frac{1}{2}(2x + 10) - x = 5$

The reason the trick works is that the equation

$$\frac{1}{2}(2x + 10) - x = 5$$

is always true: no matter what number is chosen as the value of x, the expression $\frac{1}{2}(2x + 10) - x$ is always equal to 5. (Try it for some values.) Problem 2 suggests a second important use of algebra:

To state an **identity,** i.e. a statement that is true for all numbers.

Let us now summarize some important terminology.

Word	Meaning	Examples
Variable	A symbol (usually a letter) used to represent a number	x is the variable in $3x - 1$; x and y are the variables in $3x + 7y$.
Expression	A combination of numbers, variables, and operation symbols ($+, -, \times, \div, \sqrt{\ }$)	$4 + 2$, $5x - 13$, $\frac{x+y}{5}$, $5x^2 - xyz$
Equation	A statement that two expressions are equal	$2x + 1 = 11$, $x + y = 1$

Note that if two expressions, such as $7 + 2$ and $8 + 10$, do not represent the same number, then we can write $7 + 2 \neq 8 + 10$, where the symbol "$\neq$" is read "is not equal to."

EXAMPLE 1

Translate each of the following sentences into equations.

(a) Three times a certain number is eight less than the number.

(b) Three times a certain number is eight less the number.

(c) The product of two consecutive integers is four more than the smaller integer.

(d) The product of two numbers is five more than their average.

(e) The sum of two numbers is the same, regardless of the order in which they are added.

(f) One hundred more than a certain number is the same as one hundred less than the number.

Solution

(a) $3x = x - 8$

(b) $3x = 8 - x$ Notice that a small difference in wording between sentences (a) and (b) produces two very different equations.

(c) $n(n + 1) = n + 4$ Since the two integers are consecutive, we can call the smaller one n and the larger one $n + 1$, thereby using only one variable.

(d) $xy = \frac{1}{2}(x + y) + 5$ Since there is only one relationship between the two numbers, we must use two variables.

(e) $x + y = y + x$ This is an identity; it is true for any two numbers represented by x and y.

(f) $x + 100 = x - 100$ The equation faithfully represents the English sentence given. Of course, there can be no such number x. If there were, then a pay raise of \$100 and a pay cut of \$100 would mean the same thing.

As you saw in parts (a) and (b) of Example 1, simply changing one word or its position in a sentence can significantly change both the meaning of the sentence, and its algebraic translation. We explore this further in the next example, where we demonstrate some equations involving percents.

EXAMPLE 2

Translate each of the following sentences into an equation with two variables.

(a) x is 15% of y.

(b) x is 15% larger than y.

(c) x is 15% smaller than y.

(d) The median sales price for a new single-family home in 1987 had risen by roughly 70% over the price in 1977. (Let x represent the price in 1987, and y represent the price in 1977.)

Solution

(a) $x = 0.15y$ Remember: 15% means 15 hundredths, or 0.15; "15% of y" indicates multiplication: write $0.15 \cdot y$ or simply $0.15y$.

(b) $x = 1.15y$ The sentence can be rephrased as, "x is the same as y plus 15% of y." That is, x is 115% of y.

(c) $x = 0.85y$ The sentence can be rephrased as, "x is the same as y minus 15% of y." That is, x is 85% of y.

(d) $x = 1.7y$ The given sentence means that x is 170% of y.

To form the algebraic translation of some sentence, we must do more than simply replace words with variables and numbers. We must also make sure to translate the sense of the sentence. To do this successfully, we sometimes begin by asking ourselves, "What quantities are being compared in this sentence, and what are their relative sizes?" For example, consider this problem.

> Problem 3. Using C as the number of cars, and T as the number of trucks, translate the following sentence into an equation: The manufacturer produced seven times as many cars as trucks in 1991.

When comparing the number of cars and trucks, we first form a mental image of the equation.

$$C \quad ? \quad T$$

Because the sense of the equation is that there are more cars than trucks, we think

$$\underset{\textit{larger}}{C} \quad ? \quad \underset{\textit{smaller}}{T}$$

Finally, the words "seven times" suggest that we should multiply some quantity by 7. Clearly, we multiply the smaller quantity T by 7, to produce the larger quantity C. Thus the desired equation is

$$C = 7T.$$

The sentence that we have just translated is representative of many that are frequently translated incorrectly. The common error in this case would be to write "$7C = T$" since the words "seven times" appear closer to the word "cars" than to the word "trucks" in the problem statement. Clearly this equation would not capture the sense of the sentence.

Remember that you always can substitute specific values to determine whether the sense of the sentence has been correctly represented. For example, in Problem 3 above, suppose we take $C = 14$ and $T = 2$. We certainly have seven times as many cars as trucks, and we also have that $C = 7T$. Thus the sentence and the equation are consistent.

EXERCISE SET 1.1

In Exercises 1–8, translate the given statement into an equation with one variable.

1. Eight times a certain number is forty-seven.
2. Seven less than a number is fourteen.
3. Five more than a number is twice the number.
4. Ten more than a number is five times the number.
5. One hundred twenty percent of a number is four more than the number.
6. Seventy-five percent of a number is twelve less than the number.
7. The sum of two consecutive integers is three times the smaller integer.
8. The product of two consecutive integers is seven more than the square of the smaller integer.

In Exercises 9–22, translate the given statement into an equation with two variables.

9. The sum of two numbers is eleven.
10. The product of two numbers is fifty-three.
11. Twice a number is fifteen less than another number.

12. Five times a number is ten more than another number.
13. Six more than one number is eight less than twice another number.
14. Fifteen more than one number is two more than twice another number.
15. The average of two numbers is nineteen more than their product.
16. The average of two numbers is eleven less than their product.
17. One value is two less than 5% of another value.
18. One value is five more than 63% of another value.
19. One number is 2% larger than another number.
20. One number is 59% smaller than another number.
21. Six more than one number is 30% smaller than another number.
22. Seven less than one number is 6% larger than another number.

In Exercises 23–26, let n = the number of nuts, and let b = the number of bolts. In each case, write an equation relating n and b.

23. There are twice as many nuts as bolts in the toolbox.
24. There are a third as many bolts as nuts.
25. There are three less bolts than nuts.
26. There are seven more nuts than bolts.

In Exercises 27–38, translate the given phrase into an algebraic expression.

27. The value in cents of D dimes.
28. The value in cents of N nickels.
29. The number of feet in X inches.
30. The number of inches in Y yards.
31. The cost of an item selling for d dollars per dozen.
32. The cost of S square yards of carpeting selling for $19.95 per square yard.
33. The increase in the perimeter of a rectangle if its length of L feet is increased by 20% and width of W feet is increased by 15%.
34. The decrease in the circumference of a circle if its diameter of d cm is shortened by 30%.
35. The distance travelled if you walk n hours at 3 mph, then run m hours at 7 mph.
36. The distance travelled if you walk 5 hours at W mph, then swim 3 hours at S mph.
37. The total annual interest on two accounts, one containing n dollars at 5% annual interest, and the other containing m dollars at 9%.
38. The total cost of two items, originally priced at M dollars and N dollars, and now marked 20% off and 25% off respectively.

Superset

In Exercises 39–46, write an equation that represents the problem.

39. Twenty-two people were riding on a bus. At one stop no one got off, but several people got on the bus. Then there were thirty-six people on the bus. How many got on? (Let x represent the number of people who got on.)
40. A student sold 18 paperbacks to a used book dealer and had 32 left. How many did the student have originally? (Let p represent the number of paperbacks that the student had originally.)
41. Your cost for renting a car last week was $198.00. The charge quoted at the time of rental was a flat rate of $99.00 for the week, plus 20 cents per mile. How many miles did you drive the rental car? (Let m represent the number of miles that the rental car was driven.)
42. The length and width of a rectangular patio are each doubled, and as a result, the area of the patio is four times larger. If the final area is now 82.9 square meters, what was the original area? (Let a represent the original area.)
43. You paid $44 for a pair of shoes that were on sale for 20% off their original price. What was the original price of the shoes? (Let P be the original price.)
44. A student sold a used lounge chair for 5% more than she paid for it last year. Her profit was two dollars and forty-five cents. How much did she pay for the chair last year? (Let C represent the cost to the student last year.)
45. Write an expression that describes the percent increase in the cost of an item that had been N dollars, and is now selling for $N + 3$ dollars.
46. The average of four numbers is 17. Suppose that two of the numbers are M and P. Find an expression for the average of the other two numbers in terms of M and P.

1.2 The Real Numbers

Suppose you sold subscriptions for a newspaper and had been promised a commission of 50¢ for each subscription sold. Your paycheck stub lists your total commission as \$87.30. If we let n represent the number of subscriptions sold, then the equation

$$87.30 = 0.50n$$

models the situation. The solution of this equation is 174.6, since replacing the variable n with 174.6 produces the true statement

$$87.30 = 0.50(174.6).$$

But this solution makes no sense for the given problem. You could not have sold 174.6 subscriptions, since the number of subscriptions must be a whole number. Therefore, we conclude that 174.6 is not a meaningful replacement for the variable n, and the amount listed (\$87.30) is in error.

When writing an equation to solve a problem, you should ask yourself, "What kinds of numbers make sense for the variable in this problem?" We now review terminology that may help you answer this question.

Set	Description
Integers	$\{\ldots, -3, -2, -1, 0, 1, 2, 3, \ldots\}$ The set of positive integers is $\{1, 2, 3, \ldots\}$ The set of nonnegative integers is $\{0, 1, 2, 3, \ldots\}$ and is sometimes referred to as the **Whole Numbers.**
Rational Numbers	A rational number is any number that can be represented as a fraction $\frac{p}{q}$, where p and q are both integers, and $q \neq 0$ $\left(\text{for example, } \frac{17}{3}, -\frac{1}{3}\right)$. Every integer p is a rational number since it can be written as the fraction $\frac{p}{1}$.
Irrational Numbers	Numbers such as $\sqrt{2}$, $-\sqrt[3]{5}$, π, and $\frac{2}{\sqrt{3}}$, which are not rational numbers, are called irrational.
Real Numbers	All rational and irrational numbers, taken together, form the set of real numbers.

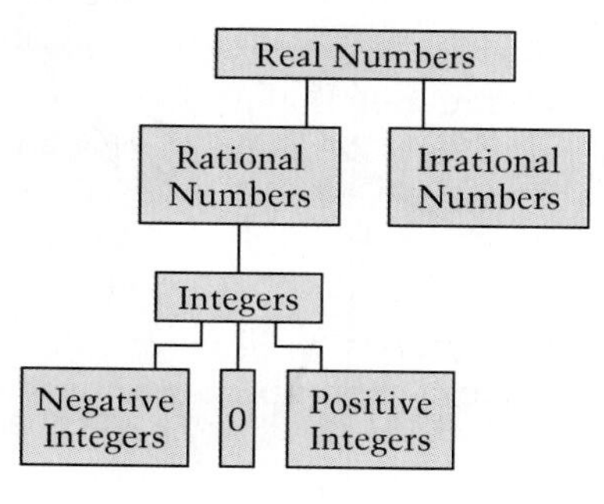

Figure 1.1

Every real number can be written in decimal form. One way of distinguishing a rational number from an irrational number is by looking at the number's decimal expansion.

In decimal form, *a rational number will either terminate or repeat a certain block of digits over and over.*

Terminating: $\frac{3}{8} = 0.375, \quad -\frac{7}{2} = -3.5, \quad \frac{1013}{500} = 2.026$

Repeating: $\frac{18}{11} = 1.63636363\ldots = 1.\overline{63}$ The bar over 63 means that it repeats indefinitely.

In decimal form, *an irrational number will neither terminate nor repeat.*

$\sqrt{7} = 2.6457513\ldots$ The decimal expansion continues, but without any pattern.

Careful! When using a calculator, don't forget that certain displayed values are not exact—they are only approximations. For example, the calculator will display 2.6457513 for $\sqrt{7}$ and 1.6363636 for $\frac{18}{11}$. But neither $\sqrt{7}$ nor $\frac{18}{11}$ can be expressed as a terminating decimal.

EXAMPLE 1

Classify each of the numbers below as one or more of the following: whole number, integer, rational number, irrational number, real number.
(a) -8 (b) 0 (c) $-\sqrt{11}$ (d) $0.958\overline{37}$ (e) 3.14

Solution

(a) integer, rational, real (b) whole number, integer, rational, real
(c) irrational, real (d) rational, real
(e) rational, real (Note: 3.14 is not π.)

Not every expression represents a real number. For example, $\sqrt{-1}$ is not a real number, because -1 has no real-valued square root; that is, there is no real number whose product with itself is equal to -1. We define a special symbol for $\sqrt{-1}$:

$$\sqrt{-1} = i.$$

With this definition we can write the "nonreal" number $\sqrt{-4}$ as $2i$, since $\sqrt{-4} = \sqrt{4} \cdot \sqrt{-1} = 2i$. Numbers involving i are called **complex numbers.** We shall discuss complex numbers in Chapter 2.

Properties of Real Numbers

We now list several important generalizations about the behavior of real numbers. We call these generalizations the Properties of Real Numbers. Unless otherwise noted, each variable used in the following table may represent any real number.

Properties of Real Numbers

Associative Property of Addition $(a + b) + c = a + (b + c)$

Associative Property of Multiplication $(ab)c = a(bc)$

Commutative Property of Addition $a + b = b + a$

Commutative Property of Multiplication $ab = ba$

Additive Identity Property There exists a unique number 0 such that

$$a + 0 = 0 + a = a$$

Multiplicative Identity Property There exists a unique number 1 such that

$$a \cdot 1 = 1 \cdot a = a$$

Additive Inverse Property For each real number a, there exists a unique real number $-a$ such that

$$a + (-a) = (-a) + a = 0$$

Multiplicative Inverse Property For each real number a not equal to 0, there exists a unique real number $\frac{1}{a}$ such that

$$a \cdot \frac{1}{a} = \frac{1}{a} \cdot a = 1$$

Distributive Property of Multiplication over Addition

$$a(b + c) = ab + ac$$
$$(b + c)a = ba + ca$$

EXAMPLE 2

Determine whether each statement is true or false.

(a) $3(8 + 4) = 24 + 12 = 36$

(b) $3(8 \cdot 4) = 24 \cdot 12 = 288$

(c) $16x + x = (16 + 1)x = 17x$

(d) $(4 + x) \cdot 3 = 4 + 3x$

(e) The additive inverse of -5 is 5.

(f) The multiplicative inverse of -5 is $-\frac{1}{5}$.

Solution

(a) True. This is an application of the Distributive Property.

(b) False. Don't be tricked into seeing this problem as an application of the Distributive Property. The parentheses simply tell you what to multiply first: $3(8 \cdot 4) = 3(32) = 96$.

(c) True. Remember that x and $1x$ mean the same thing.

(d) False. The Distributive Property requires that each term within the parentheses be multiplied by 3: $(4 + x) \cdot 3 = 12 + 3x$.

(e) True. This fact is sometimes represented as $-(-5) = 5$.

(f) True. There are three ways to write a negative fraction: $\frac{-1}{5}, \frac{1}{-5}$, or $-\frac{1}{5}$; the most common is $-\frac{1}{5}$.

Careful! Don't assume that a variable without a negative sign in front of it represents a positive number. A variable, such as x, might stand for a positive number, a negative number, or zero. Likewise, don't assume that an expression such as $-2x$ must be negative. For example, if x represents -3, then $-2x$ represents 6 (i.e., $-2x = -2(-3) = +6$).

We conclude this section with some number substitution problems, so that you might review the arithmetic of real numbers. Recall the **Order of Operations** convention for the use of grouping symbols and the four operations of arithmetic:

1. **Grouping Symbols** Parentheses ( ), brackets [ ], and fraction bars can be thought of as grouping symbols. In any expression, you should perform the operations in the innermost parentheses (or brackets) first, and work your way outward until all parentheses and brackets have been removed. Perform the operations in the numerator and denominator of a fraction separately.
2. **Arithmetic Operations** If no grouping symbols are present, we will agree that multiplications and divisions will be performed first, in the order in which they occur, followed by any remaining additions and subtractions, also in the order in which they occur.

Consider the expression $5 \cdot 10 + 2$. By virtue of the Order of Operations convention, the multiplication is done first, followed by the addition:

$$5 \cdot 10 + 2 = 50 + 2 = 52.$$

On the other hand, consider the expression $5 \cdot (10 + 2)$. Here the parentheses signify that the addition is to be done first, followed by the multiplication. Thus,

$$5 \cdot (10 + 2) = 5 \cdot 12 = 60.$$

Clearly, $5 \cdot 10 + 2 \neq 5 \cdot (10 + 2)$. Thus the order of operations does make a difference.

EXAMPLE 3

Evaluate each expression, given that $x = 5$, $y = -8$, and $z = 10$.

(a) $-(-z)$

(b) $-[x + (-y)]$

(c) $\dfrac{xz + x(-y)}{x - xy}$

(d) $([(x + y)z] + x) + yz$

Solution

(a) $-(-z) = -(-10) = 10$

(b) $-[x + (-y)] = -[5 + 8] = -13$

(c) $\dfrac{xz + x(-y)}{x - xy} = \dfrac{5(10) + 5(8)}{5 - (5)(-8)} = \dfrac{50 + 40}{5 - (-40)} = \dfrac{90}{45} = 2$

(d) $$\begin{aligned}([(x + y)z] + x) + yz &= ([(5 + (-8))10] + 5) + (-8)(10) \\ &= ([(-3)(10)] + 5) + (-8)(10) \\ &= ([-30] + 5) + (-80) \\ &= (-25) + (-80) = -105\end{aligned}$$

EXERCISE SET 1.2

In Exercises 1–8, classify the number as one or more of the following: whole number, integer, rational number, irrational number, real number.

1. $\frac{3}{5}$ **2.** -5 **3.** 0

4. $\frac{22}{7}$ **5.** $\sqrt{7}$ **6.** 3.715

7. $14.\overline{36}$ **8.** $\sqrt[3]{8}$

In Exercises 9–18, determine whether the statement is true or false.

9. $\sqrt{3} + 2 = 2 + \sqrt{3}$ **10.** $7\pi = \pi \cdot 7$

11. $3 \cdot 0 \cdot 8 = 24$ **12.** $3(6 - 6) = 0$

13. $(2 + 7)8 = 72$ **14.** $8(7 - 2) = 54$

15. $9x - x = 8$ **16.** $3x - 7x = 4x$

17. $-(3 \cdot 5) = (-3)(-5) = 15$

18. $2(3 \cdot 4) = (2 \cdot 3)(2 \cdot 4) = 48$

In Exercises 19–30, determine the additive and multiplicative inverses.

19. 5 **20.** -3 **21.** $-\frac{2}{3}$

22. -1 **23.** 0 **24.** $\frac{5}{2}$

25. π **26.** $\sqrt{2}$ **27.** $-x$

28. x **29.** $\frac{1}{x}$ **30.** $-xy$

In Exercises 31–40, evaluate the expression.

31. $6 + 3 \cdot 5$ **32.** $7 \div 4 - 2$

33. $8(4 + 3) \div 4$ **34.** $8 \cdot 4 + 3 \div 4$

35. $-3(2 - 5) + 1$ **36.** $(8 - 12)(-2) - 1$

37. $\dfrac{8 - 2(1 - 3)}{3 \cdot 2 - 6 \div 3}$ **38.** $\dfrac{1 - \frac{3}{8}}{\frac{2}{3} + 3}$

39. $\dfrac{2\left(3 - \frac{2}{7}\right)}{4 - \frac{5}{8}}$ **40.** $\dfrac{2(1 - 7)(-3)}{5 \cdot 4 - 1}$

In Exercises 41–48 evaluate the expression, then round your answer to the nearest tenth.

41. $(-3.5) - (7.1)(-4.6)$ 42. $5(-6.2) - 3(-5.4)$

43. $8.7 \div (2 - 9.3) - 5$

44. $(6.1 - 4.7) \div [5(8.1 - 9.7)]$

45. $\dfrac{3.7 - 6.25}{(0.89)(-10.3)}$ 46. $\dfrac{4(16.021 - 21.06)}{9.7 - 3.85}$

47. $\dfrac{6(5.01 - 8) - 10.2}{0.914 - 3(2.8 \div 7)}$

48. $\dfrac{0.003[85.6 \div (-0.01)]}{(0.4 - 7)(6 - 3.01) - 0.2}$

In Exercises 49–52, evaluate the expression, then round your answer to the nearest integer.

49. $[3697(618 - 9721)] \div 3178$

50. $80{,}000 \div (6291 + 899) - 3520$

51. $\dfrac{15{,}928 + 16(615 \div 12)}{1 - 3(3007 \div 48) - 5608}$

52. $\dfrac{87{,}600 - 3(729 \div 0.01)}{6281(1 - 0.007)}$

In Exercises 53–64, evaluate the expression, given that $a = 2$, $b = -10$, and $c = 5$.

53. $-b$ 54. $-a$ 55. $-(a)$

56. $-(-b)$ 57. $-(c - a)$ 58. $-(8 - c)$

59. $2ac - \dfrac{a(-b)}{c}$ 60. $3ca + \dfrac{c(-b)}{-a}$

61. $\dfrac{c - \dfrac{ab}{4}}{a(b - c)}$ 62. $\dfrac{ab - c + 1}{\dfrac{1}{a} - \dfrac{1}{c}}$

63. $a(c - b) \div b$ 64. $a[b - (c + a)]$

In Exercises 65–68 evaluate the expression given that $x = 5.095$, $y = -0.035$ and $z = -10.050$, then round your answer to the nearest thousandth.

65. $xy - \dfrac{x}{z}$ 66. $x(y - 0.011z)$

67. $3.142 \div y\left(\dfrac{3z}{x}\right)$ 68. $x + y \cdot z$

Superset

In Exercises 69–76, translate the statement into an equation.

69. One-fourth of a certain number is five times its multiplicative inverse.

70. The sum of a number and ten is twenty-four times its multiplicative inverse.

71. Three times the sum of a number and its reciprocal is four more than the number.

72. Twice the sum of a number and its reciprocal is one less than the number.

73. The additive inverse of the sum of two numbers is five less than their product.

74. The additive inverse of the product of two numbers is six more than twice their sum.

75. The sum of one number and 54% of a second number is 83% of the product of the two numbers.

76. The multiplicative inverse of the product of two numbers is twenty percent of the average of the two numbers.

In Exercises 77–82, specify whether the statement is true or false. If false, illustrate this with an example.

77. $(2x)(2y) = 2xy$ 78. $3(x + 7) = 3x + 7$

79. $-(3 - x) = x - 3$

80. $5 - (x + 2) = 5 - x + 2$

81. $\dfrac{1}{2x} = \dfrac{1}{2}x$ 82. $\dfrac{1}{\left(\dfrac{x}{y}\right)} = \dfrac{y}{x}$

In Exercises 83–86, add parentheses and brackets to the expression so that the value of the resulting expression is 12.

83. $3 \cdot 7 + 2 - 5$ 84. $3 \cdot 5 - 10 - 7$

85. $12 - 3 \cdot 8 \div 6$ 86. $12 + 3 \cdot 8 \div 3$

In Exercises 87–90, name the property of real numbers used to get from the previous step.

87. $3(1020 - 7) + 2(11 - 1530)$
$= (3060 - 21) + (22 - 3060)$ (a) ________
$= [(-21) + 3060] + [(-3060) + 22]$ (b) ________
$= (-21) + [3060 + (-3060)] + 22$ (c) ________
$= (-21) + 0 + 22$ (d) ________
$= -21 + 22$ (e) ________
$= 1$

88. $\left(\dfrac{1}{3} \cdot 2\right) \cdot 3 = \left(2 \cdot \dfrac{1}{3}\right) \cdot 3$ (a) ________
$= 2\left(\dfrac{1}{3} \cdot 3\right)$ (b) ________
$= 2(1)$ (c) ________
$= 2$ (d) ________

89. $3(5 - 16x) + (12x)(4)$
$= (15 - 48x) + (12x)(4)$ (a) ________
$= (15 - 48x) + 12(x \cdot 4)$ (b) ________
$= (15 - 48x) + 12(4 \cdot x)$ (c) ________
$= (15 - 48x) + (12 \cdot 4)x$ (d) ________
$= (15 - 48x) + 48x$
$= 15 + [(-48x) + 48x]$ (e) ________
$= 15 + [(-48 + 48)x]$ (f) ________
$= 15 + 0 \cdot x$ (g) ________
$= 15 + 0$
$= 15$ (h) ________

90. $(3x + 4)7 - 5x = (21x + 28) - 5x$ (a) ________
$= 21x + (28 - 5x)$ (b) ________
$= 21x + [(-5x) + 28]$ (c) ________
$= [21x + (-5x)] + 28$ (d) ________
$= [21 + (-5)]x + 28$ (e) ________
$= 16x + 28$

91. In this exercise we provide the steps of an indirect proof (proof by contradiction) that $\sqrt{2}$ is not a rational number. Your task is to justify each step. We begin by assuming that $\sqrt{2}$ is rational, that is, for some integers a and b we have $\sqrt{2} = \frac{a}{b}$. Further, we assume that $\frac{a}{b}$ is in lowest terms (a and b have no common factor).

i) The integer n is divisible by 2 if and only if n^2 is divisible by 2. (*Hint:* Consider the two cases of $n = 2k$ and $n = 2k+1$ and compute n^2 in each case.)

ii) Rewriting the equation $\sqrt{2} = \frac{a}{b}$ as $2b^2 = a^2$, we can conclude that 2 divides a.

iii) Next, 2 divides b.

Thus, a and b have a common factor of 2, a contradiction. So, $\sqrt{2}$ is irrational.

92. Prove that $\sqrt{3}$ is irrational. (*Hint:* Every integer can be written as $3k$, $3k+1$ or $3k+2$.)

1.3
Absolute Value, Distance, and the Real Line

Suppose the variable a represents some nonzero real number.

> If a represents a positive number, then $-a$ is negative. (For example, if $a = 7$, then $-a = -7$.)
>
> If a represents a negative number, then $-a$ is positive. (For example, if $a = -10$, then $-a = -(-10) = 10$.)

Thus, for any nonzero real number a, one of the numbers, a or $-a$, must be positive. The positive one is referred to as the *absolute value* of a.

Definition of Absolute Value

The **absolute value** of a nonzero real number a is the positive number in the set $\{a, -a\}$. The absolute value of 0 is 0. The absolute value of any real number a is denoted $|a|$. Notice that $|a| = |-a|$.

Observe that the absolute value of a positive number is the number itself, whereas the absolute value of a negative number is the additive inverse of the number. We can use these two facts to write an **alternate definition of absolute value:**

$$|a| = \begin{cases} a, & \text{if } a \text{ represents a positive number or } 0, \\ -a, & \text{if } a \text{ represents a negative number.} \end{cases}$$

You should think carefully about the last line in the alternative definition: if a represents a negative number, then the symbol "$-a$" represents a positive number, even though there is a negative sign "out front."

EXAMPLE 1

Evaluate each of the following:

(a) $|5|$ (b) $|-10|$ (c) $-\left|-\frac{1}{2}\right|$ (d) $|-4| - |-6|$ (e) $|0|$

Solution (a) $|5| = 5$ (b) $|-10| = 10$ (c) $-\left|-\frac{1}{2}\right| = -\left(\frac{1}{2}\right) = -\frac{1}{2}$

(d) $|-4| - |-6| = (4) - (6) = -2$ (e) $|0| = 0$

The Real Number Line

It is often helpful to visualize the set of real numbers as labels for the points on a line. Suppose we agree to choose one point on the line and label it 0. We call this point the **origin.** Then, suppose we let each real number a correspond to a point lying $|a|$ units from zero. If a is positive, the point lies to the right of 0; if a is negative, the point lies to the left of 0.

The real number corresponding to a particular point on the line is called the **coordinate** of the point. Points A, B, C, and D in Figure 1.2 have coordinates $-\frac{7}{2}$, $-\sqrt{2}$, $\frac{3}{4}$, and π, respectively.

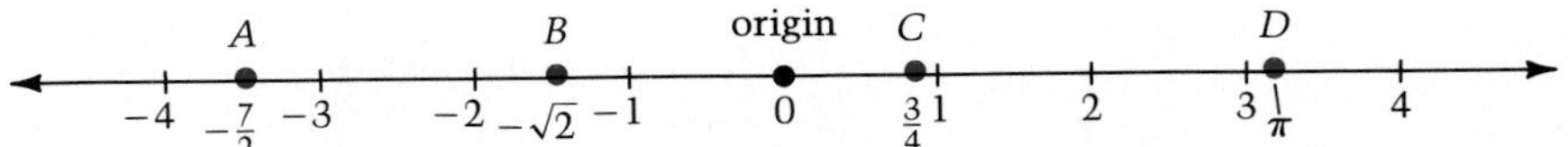

Figure 1.2 The Real Number Line.

To each real number there corresponds exactly one point on the real line, and to each point on the real line, there corresponds exactly one real number. Such a correspondence between the points on a line and the set of real numbers is called a **one-to-one correspondence.**

The relative position of two different points on the real line suggests a "greater than" or "less than" relationship between the two coordinates. For example, since 2 is to the right of -3, we say that 2 is greater than -3 and write $2 > -3$. Furthermore, since -3 is to the left of 2, we say that -3 is less than 2 and write $-3 < 2$. Relationships such as these are called **order relationships** or **inequalities.**

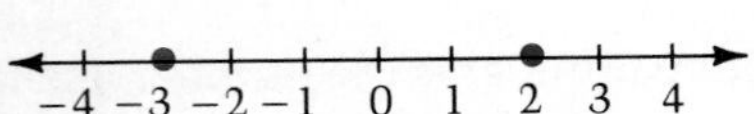

Figure 1.3 We can write either $-3 < 2$ or alternatively $2 > -3$.

EXAMPLE 2

In each case, use the given symbol and numbers to write an inequality:

(a) $<: 5, -10$ (b) $>: -1, -8$ (c) $<: \frac{1}{3}, 0.33$ (d) $>: \pi, 3.14$

Solution

(a) $-10 < 5$

(b) $-1 > -8$ Since -1 is to the right of -8 on the number line, it is greater than -8.

(c) $0.33 < \frac{1}{3}$

(d) $\pi > 3.14$ Since π can be approximated by 3.1416, we conclude π is greater than 3.14.

In Example 3, we use inequalities to translate sentences into algebraic statements.

EXAMPLE 3

Translate each of the following sentences into an inequality.

(a) Three times a certain number is less than seven.

(b) Eight less than a certain number is greater than twice the number.

(c) The product of two consecutive integers is less than their sum.

(d) The quarters in my pocket amount to more than five dollars.

(e) The sum of two consecutive integers is greater than the absolute value of the sum of the smaller integer and five.

Solution

(a) $3x < 7$ Here, x represents the "certain" number.

(b) $x - 8 > 2x$ Here, x represents the "certain" number.

(c) $n(n + 1) < 2n + 1$ Here, n represents the smaller integer, and $n + 1$ the larger. The sum is $n + (n + 1)$, which simplifies to $2n + 1$.

(d) $0.25q > 5$ Here, q represents the number of quarters; thus, $0.25q$ represents the value of those quarters in dollars.

(e) $2n + 1 > |n + 5|$ Here, n represents the smaller integer, and $n + 1$ represents its successor.

EXAMPLE 4

Graph each of the following sets of numbers on the real line.

(a) The set of integers between -5 and 1.

(b) The set of real numbers between -5 and 1.

(c) The set of real numbers less than or equal to 2.

(d) The set of real numbers whose absolute values are greater than 2.

Solution

(a)

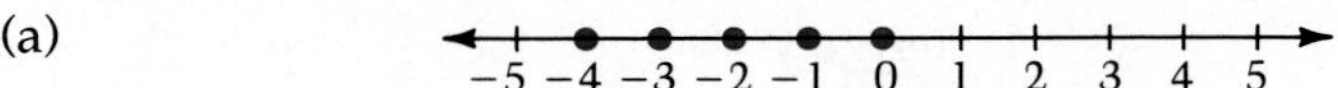

(b)

−5 −4 −3 −2 −1 0 1 2 3 4 5

When the endpoint of a set of real numbers is excluded, the dot is left open.

(c)

−5 −4 −3 −2 −1 0 1 2 3 4 5

When the endpoint of a set of real numbers is included, the dot is filled in.

(d)

−5 −4 −3 −2 −1 0 1 2 3 4 5

Figure 1.4

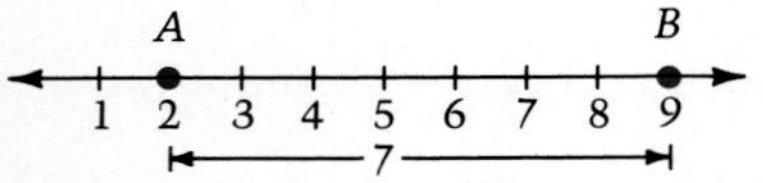

Figure 1.5 The distance between A and B is 7.

Let us now review the notion of distance on the real line. When someone asks you to estimate the distance between two cities, your answer will certainly not be a negative number. So you already know one of the most important characteristics of distance: it is a nonnegative quantity.

In Figure 1.5, the distance between points A and B is 7. To determine this, we find the absolute value of the difference between the coordinates 2 and 9.

$$|2 - 9| = |-7| = 7$$

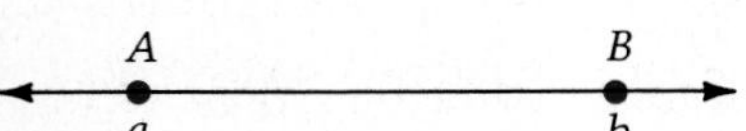

Figure 1.6 $d(A,B) = |a - b|$.

Definition of Distance

Suppose points A and B lie on the real number line and have coordinates a and b, respectively. The **distance** between points A and B, denoted $d(A, B)$, is given by the equation

$$d(A, B) = |a - b|.$$

When you calculate the distance between two points, it does not matter which coordinate you consider to be a and which you consider to be b. For the example above, the distance between A and B can be found from either of the following equations:

$$d(A, B) = |2 - 9| = |-7| = 7, \text{ or}$$
$$d(B, A) = |9 - 2| = |7| = 7.$$

In general, $d(A, B) = d(B, A)$, for any points A and B.

EXAMPLE 5

Using the information in Figure 1.7, find each distance:
(a) $d(C, O)$ (b) $d(A, C)$ (c) $d(C, A)$ (d) $d(A, A)$

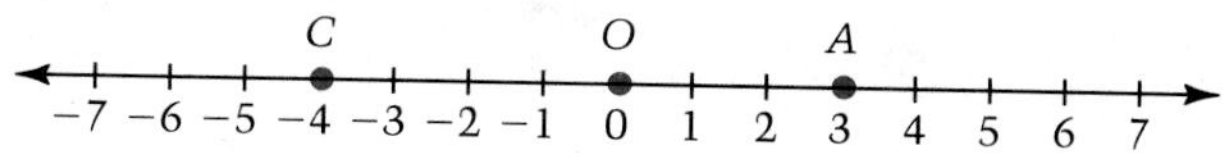

Figure 1.7

Solution

(a) $d(C, O) = |(-4) - (0)| = |-4| = 4$ The distance between the origin and a point is just the absolute value of the coordinate of the point.

(b) $d(A, C) = |3 - (-4)| = |7| = 7$

(c) $d(C, A) = 7$ Here, we use the result of part (b), together with the fact that $d(C, A) = d(A, C)$.

(d) $d(A, A) = |3 - 3| = |0| = 0$ The distance between a point and itself is 0.

EXERCISE SET 1.3

In Exercises 1–10, evaluate the expression.

1. $|-4|$ **2.** $\left|\frac{5}{9}\right|$ **3.** $-|0|$ **4.** $-|-\sqrt{3}|$

5. $-\left|-\frac{1}{5}\right|$ **6.** $-|3 + (-5)|$

7. $|10| + |-3|$ **8.** $|-5| - |6|$

9. $|-3|\,|-10|$ **10.** $\dfrac{|7 - (-2)|}{-5}$

In Exercises 11–20, use the given symbol and numbers to write an inequality.

11. $>: 0, -11$ **12.** $<: -20, -30$

13. $<: -0.05, -0.5$ **14.** $>: \frac{3}{5}, \frac{4}{7}$

15. $>: -\frac{5}{8}, -0.624$ **16.** $<: \frac{\pi}{2}, 1.577$

17. $<: |8.01 - 9.26|, \dfrac{6.8 + 5.7}{10.02}$

18. $>: \dfrac{6.7 - |3 - 7.5|}{16}, |0.28 - 5| \div 40$

19. $>: |8.14 - 9.011|, \dfrac{(6.07)(0.889)}{1 + |1 - 6.19|}$

20. $<: \left|5.908 - \dfrac{(16.1)(8.2)}{17.04}\right|, \dfrac{5 - 3(9.07)}{7.9 - 20}$

In Exercises 21–24, translate the sentence into an inequality.

21. Fifteen more than a certain number is less than -10.

22. Seven less than twice a certain number is greater than 20.

23. Five less than twice a certain number is nonnegative.

24. Six more than a certain number is positive.

In Exercises 25–30, graph the set of numbers on the real line.

25. $\left\{-1, -\frac{1}{2}, \sqrt{2}, 2\right\}$ **26.** $\{10, -30, 60\}$

27. The set of integers between -3 and 2.

28. The set of real numbers greater than or equal to 0.

29. The set of real numbers less than -2.

30. The set of integers greater than $-\frac{1}{2}$ and less than $2\frac{1}{2}$.

In Exercises 31–36, use the following graph to determine the distance.

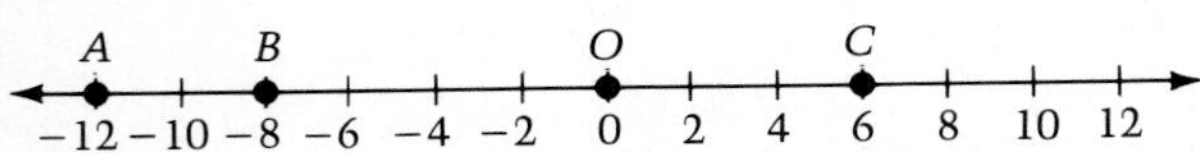

31. $d(O, A)$
32. $d(O, B)$
33. $d(C, A)$
34. $d(A, B)$
35. $d(A, C)$
36. $d(B, B)$

In Exercises 37–42, assume that points A, B, C, D, and E correspond to the real numbers 0.1, 0.01, -0.1, -0.01, and 0.005 respectively, on the real number line.

37. Determine the number corresponding to the point halfway between C and E.

38. Determine the number corresponding to the point halfway between D and E.

39. Determine the number corresponding to the point that is one-fourth of the way from point D to point A.

40. Determine the number corresponding to the point that is one-eighth of the way from point B to point E.

41. Which point (A, B, C, D, or E) is between the points corresponding to $\frac{1}{60} + \frac{1}{100}$ and $\frac{1}{60} - \frac{1}{100}$?

42. Which point (A, B, C, D, or E) is closest to the point corresponding to $\frac{111}{800} - \frac{57}{407}$?

Superset

43. For the following graph, $d(A, B) = 7$, $d(B, C) = 10$, and $d(O, C) = 5$. Determine the values of a, b, and c.

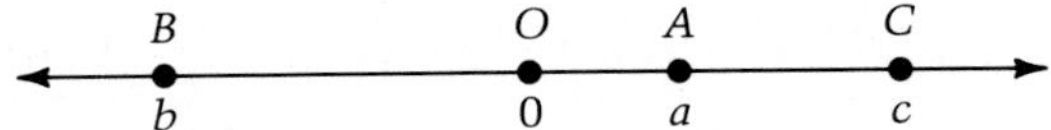

44. If a point A has coordinate a and a point B has coordinate b, show that $d(A, B) = d(B, A)$.

In Exercises 45–52, determine whether the statement is true or false. If false, illustrate this with an example.

45. $-x$ represents a negative number.

46. The absolute value of a number is always positive.

47. The sum of two negative numbers is always negative.

48. The product of two negative numbers is always negative.

49. Distance is never negative.

50. Distance is always positive.

51. $|x| = x$

52. $|x| > -x$

1.4
The Integers as Exponents

In this section we will review the arithmetic of expressions containing exponents. Integer exponents provide us with a streamlined way of representing repeated multiplication (by the same number). For example, consider the product $3 \times 3 \times 3 \times 3 \times 3$. As it is written here, it is said to be in **expanded form.** However, it can be rewritten in **exponential form** as 3^5. In the expression 3^5, 3 is called the **base,** and 5 is called the **exponent.** In general, for any real number a and any positive integer n, the symbol a^n represents the product of n factors, each equal to a.

$$\underbrace{a \cdot a \cdot a \cdot \cdots \cdot a \cdot a}_{n \text{ factors}} = a^n \qquad n \text{ is the exponent and } a \text{ is the base.}$$

The symbol a^n is called "the nth power of a", "a (raised) to the nth power," or simply "a to the n." If the exponent n is 1, it is usually omitted; thus a^1 is written a.

You probably encountered the first property of exponents by noticing a pattern like the one shown below.

$$\begin{array}{ccccc} (5 \cdot 5 \cdot 5) \cdot (5 \cdot 5 \cdot 5 \cdot 5) & = & 5 \cdot 5 \cdot 5 \cdot 5 \cdot 5 \cdot 5 \cdot 5 \\ 5^3 \quad \cdot \quad 5^4 & = & 5^7 \end{array}$$

We can generalize this result as follows:

If a is a real number, and if m and n are positive integers, then

$$a^m \cdot a^n = a^{m+n}.$$

In the equation above, we do not wish to restrict ourselves to positive exponents only, and so we must explore the meaning of the symbol a^n when the exponent n is 0 or a negative integer. Let us consider a^0 first. If we assume that the property $a^m \cdot a^n = a^{m+n}$ is true when m or n is zero, then it is correct to write

$$a^0 \cdot a^n = a^{0+n}.$$

Since the right side a^{0+n} can be written simply as a^n, we have

$$a^0 \cdot a^n = a^n.$$

Multiplying by a^0 doesn't change the value of a^n. So a^0 must be equal to 1.

Definition of Zero Exponent

$$a^0 = 1, \quad a \neq 0$$

We require that $a \neq 0$ to avoid the undefined expression 0^0.

Now suppose that n is a positive integer. Then a^{-n} has a negative exponent. If we assume that $a^m \cdot a^n = a^{m+n}$ holds true when the exponents are negative, then the following must be true:

$$a^{-n} \cdot a^n = a^{-n+n} = a^0 = 1.$$

Since the product of a^{-n} and a^n is 1, they must be reciprocals of one another. So we make the following definition.

Definition of Negative Exponent

$$a^{-n} = \frac{1}{a^n}, \quad a \neq 0$$

We require that $a \neq 0$ to avoid the undefined expression $\frac{1}{0}$.

EXAMPLE 1

Evaluate: (a) $(-2)^4$ (b) 5^{-3} (c) $\dfrac{1}{5^{-3}}$ (d) $(-3)(-3)^4$ (e) 10^{-5}

Solution

(a) $(-2)^4 = (-2)(-2)(-2)(-2) = 16$

Don't confuse $(-2)^4$ with -2^4. $-2^4 = -(2^4) = -(16) = -16$.

(b) $5^{-3} = \dfrac{1}{5^3} = \dfrac{1}{125}$

In (b) and (c) we have "moved" 5^{-3} from the numerator to the denominator or vice versa. Doing so requires that we change the sign of the exponent.

(c) $\dfrac{1}{5^{-3}} = 5^3 = 125$

$\left(5^{-3} = \dfrac{1}{5^3} \text{ and } \dfrac{1}{5^{-3}} = 5^3.\right)$

(d) $(-3)(-3)^4 = (-3)^1(-3)^4$
$= (-3)^5 = -243$

Remember: a negative number to an odd power is negative, and to an even power is positive.

(e) $10^{-5} = \dfrac{1}{10^5} = \dfrac{1}{100{,}000} = 0.00001$

To avoid having to repeat restrictions on variables over and over in our subsequent work, we will now make the following agreement.

Agreement

In the properties, examples, and exercises that follow, we will be using variables for bases and exponents. Let us agree that these variables have been restricted so as to avoid undefined terms of the form $\frac{1}{0}$, 0^0, and 0 to a negative power.

We now summarize the properties for integer exponents that you studied in your first course in algebra.

Properties of Integer Exponents

1. $a^m \cdot a^n = a^{m+n}$	$a^3 \cdot a^4 = a \cdot a \cdot a \cdot a \cdot a \cdot a \cdot a = a^7$
2. $(ab)^n = a^n b^n$	$(ab)^3 = (ab)(ab)(ab) = (aaa)(bbb) = a^3b^3$
3. $(a^m)^n = a^{mn}$	$(a^2)^3 = a^2 \cdot a^2 \cdot a^2 = (aa)(aa)(aa) = a^6$
4. $\dfrac{a^m}{a^n} = a^{m-n}$	$\dfrac{a^5}{a^2} = \dfrac{a \cdot a \cdot a \cdot a \cdot a}{a \cdot a} = a^3$
5. $\left(\dfrac{a}{b}\right)^n = \dfrac{a^n}{b^n}$	$\left(\dfrac{a}{b}\right)^3 = \dfrac{a}{b} \cdot \dfrac{a}{b} \cdot \dfrac{a}{b} = \dfrac{a \cdot a \cdot a}{b \cdot b \cdot b} = \dfrac{a^3}{b^3}$

Remember also: $a^0 = 1$ $\quad a = a^1$ $\quad a^{-n} = \dfrac{1}{a^n}$

EXAMPLE 2

Write each expression in the form a^n:

(a) $(2^3)^{-2}$ (b) $6^5\left(\frac{1}{6}\right)^5$ (c) $\frac{(3^2 \cdot 3^3)^4}{3^7}$ (d) $2^3 \cdot 4^5$ (e) $\frac{10^9 \times 10^7}{10^{-3}}$

Solution

(a) $(2^3)^{-2} = 2^{3(-2)} = 2^{-6}$ Property 3 is used.

(b) $6^5\left(\frac{1}{6}\right)^5 = \left(6 \cdot \frac{1}{6}\right)^5 = 1^5 = 1$ Property 2 is used.

(c) $\frac{(3^2 \cdot 3^3)^4}{3^7} = \frac{(3^5)^4}{3^7}$ $3^2 \cdot 3^3 = 3^5$ by Property 1.

$= \frac{3^{20}}{3^7}$ $(3^5)^4 = 3^{20}$ by Property 3.

$= 3^{13}$ $\frac{3^{20}}{3^7} = 3^{13}$ by Property 4.

(d) $2^3 \cdot 4^5 = 2^3(2^2)^5$ 4 is written as a power of 2.

$= 2^3 \cdot 2^{10}$ Now that the bases are the same (2), we can

$= 2^{13}$ use Property 1 and simply add the exponents.

(e) $\frac{10^9 \times 10^7}{10^{-3}} = \frac{10^{16}}{10^{-3}} = 10^{16-(-3)} = 10^{19}$

EXAMPLE 3

Simplify each expression so that x and y appear at most once, with no powers of powers, or negative exponents:

(a) $(3x^2y)(2xy)(4x^3y^2)$ (b) $(6xy^2)^3$ (c) $\left(\frac{4y}{x^2}\right)^3 x^6$

Solution

(a) $(3x^2y)(2xy)(4x^3y^2) = (3 \cdot 2 \cdot 4)(x^2 \cdot x \cdot x^3)(y \cdot y \cdot y^2)$ Group constants and like variables together.

$= 24x^6y^4$ Apply Property 1.

(b) $(6xy^2)^3 = 6^3x^3(y^2)^3 = 216x^3y^6$ When using Property 2 here, raise each factor, including the constant, to the third power.

(c) $\left(\frac{4y}{x^2}\right)^3 x^6 = \frac{(4y)^3}{(x^2)^3}x^6$ Property 5 is used.

$= (4y)^3x^{-6}x^6$ $\frac{1}{(x^2)^3} = \frac{1}{x^6} = x^{-6}$.

$= (4y)^3x^0$ Property 1 is used.

$= (4y)^3$ $x^0 = 1$.

$= 64y^3$ Property 3 is used.

One important result of the fact that $a^{-n} = \frac{1}{a^n}$ is that a fraction raised to a negative power $-n$ is equal to the reciprocal raised to the positive power n. For example,

$$\left(\frac{2}{3}\right)^{-5} = \frac{1}{\left(\frac{2}{3}\right)^5} = \frac{1^5}{\left(\frac{2}{3}\right)^5} = \left(\frac{1}{\frac{2}{3}}\right)^5 = \left(\frac{3}{2}\right)^5.$$

Careful! Remember that you can only use Property 1 if the bases are equal. Thus,

$$3^4 \cdot 9^5 = 3^4 \cdot (3^2)^5 = 3^4 \cdot 3^{10} = 3^{14},$$

and

$$3^4 \cdot 6^5 = 3^4 \cdot (3 \cdot 2)^5 = 3^4 \cdot 3^5 \cdot 2^5 = 3^9 \cdot 2^5.$$

In the second case, we cannot further rewrite $3^9 \cdot 2^5$ because the bases are not equal.

Scientific Notation

We can use exponential notation to write rather cumbersome expressions in a very compact form. There is a particular type of exponential form that is common when working with very large or very small numbers. For example, the amount of sugar consumed by Americans each year, roughly 18,250,000,000 pounds, can be written as 1.825×10^{10} lb, and the mass of an electron, approximately 0.000000000000000000000000000000911 kilograms, can be written as 9.11×10^{-31} kg. The numbers in these measurements, 1.825×10^{10} and 9.11×10^{-31}, are said to be written in scientific notation.

Definition of Scientific Notation

A positive number is in **scientific notation** when it is written as a product

$$A \times 10^n$$

where A is a number greater than or equal to 1, and less than 10, and n is an integer exponent.

To find the exponent necessary to write a number in scientific notation, simply count the number of places that the decimal point must be moved to produce the number A between 1 and 10. If the original number is greater than 1 (like the amount of sugar consumed), the exponent will be positive. If the original number is less than 1 (like the mass of an electron), the exponent will be negative.

EXAMPLE 4

Write each number in scientific notation: (a) 285,000 (b) 0.00079

Solution

(a) 2 8 5 0 0 0 . $= 2.85 \times 10^5$ To produce 2.85 (a number between 1 and 10) the decimal point must be moved 5 places to the **left.** Since 285,000 is greater than 1, the exponent is positive.

Thus, the given number in scientific notation is 2.85×10^5.

(b) 0 . 0 0 0 7 9 $= 7.9 \times 10^{-4}$ To produce 7.9 (a number between 1 and 10) the decimal point must be moved 4 places to the **right.** Since 0.00079 is less than 1, the exponent is negative.

Thus, the given number in scientific notation is 7.9×10^{-4}.

EXAMPLE 5

Write each number in decimal form (i.e., no exponents):
(a) 3.42×10^{-6} (b) 1.101×10^4

Solution

(a) $3.42 \times 10^{-6} = 0.00000342$ Since the exponent is negative, the value of the entire expression will be smaller than 3.42. Multiplying 3.42 by 10^{-6} has the same effect as moving the decimal point 6 places to the left.

(b) $1.101 \times 10^4 = 11{,}010.$ Since the exponent is positive, the value of the entire expression will be larger than 1.101. Multiplying 1.101 by 10^4 has the same effect as moving the decimal point 4 places to the right.

Many applications of mathematics involve taking measurements. Because some error is to be expected with any measuring device, we think of measurements as approximations. The number of **significant digits** in a measurement suggests the accuracy of the measurement. For example, if we report that the highest elevation in New York state is Mount Marcy at 5344 feet above sea level, we imply that the true measurement is somewhere between 5343.5 ft and 5344.5 ft, and that our measurement is accurate to the nearest foot. We also say that the measurement 5344 contains 4 significant digits, whereas a measurement accurate to the nearest tenth of a foot, say 5344.2, would contain 5 significant digits.

Consider a few more cases:

Value	Number of Significant Digits	Implied Accuracy: Measured to the Nearest . . .
5.28	3	hundredth
0.312	3	thousandth
0.001	1	thousandth
1562	4	one
6.0	2	tenth

There can be difficulties in working with significant digits. For example, consider a measurement of 200 ft. Was this the result of measuring to the nearest foot, in which case there are 3 significant digits, or to the nearest 10 feet, in which case there are 2 significant digits, or to the nearest 100 feet, in which case there is only one significant digit? The ambiguity is removed when the value is written in scientific notation $A \times 10^n$, and we agree that the number of significant digits is the number of digits in A.

Value	Number of Significant Digits	Value	Number of Significant Digits
2.00×10^{-6}	3	6.851×10^7	4
2.0×10^{11}	2	9.8×10^{-8}	2
2×10^{-5}	1	9.80×10^1	3

EXAMPLE 6

Compute the value of each expression. Write your answer in scientific notation, rounded to three significant digits.

(a) $(3.91 \times 10^{-12})(4.217 \times 10^{-8})$ (b) $58{,}500{,}000 \times 0.0841$

(c) $\dfrac{7.81 \times (6.624 \times 10^5)}{2.45 \times 10^8}$

Solution We have used a calculator to help with the computations here.

(a)
$$\begin{aligned}(3.91 \times 10^{-12})&(4.217 \times 10^{-8})\\ &= (3.91 \cdot 4.217) \times (10^{-12} \cdot 10^{-8})\\ &\approx 16.5 \times 10^{-20}\\ &= 1.65 \times 10^1 \times 10^{-20}\\ &= 1.65 \times 10^{-19}\end{aligned}$$

$(3.91) \cdot (4.217) = 16.48847$, which is rounded to 16.5. We write $\approx$ to indicate an approximation.

(b)
$$\begin{aligned}(58{,}500{,}000)(0.0841) &= (5.85 \times 10^7)(8.41 \times 10^{-2})\\ &= (5.85 \cdot 8.41) \times (10^7 \cdot 10^{-2})\\ &\approx 49.2 \times 10^5\\ &= 4.92 \times 10^6\end{aligned}$$

Think: $49.2 \times 10^5 = 4.92 \times 10^1 \times 10^5$

(c) $$\frac{7.81 \times (6.624 \times 10^5)}{2.45 \times 10^8} = \frac{(7.81 \cdot 6.624)}{2.45} \times \frac{10^5}{10^8}$$
$$\approx 21.1 \times 10^{-3}$$
$$= 2.11 \times 10^{-2}$$

EXERCISE SET 1.4

In Exercises 1–24, evaluate the expression.

1. $(-5)^4$ **2.** $(-3)^4$ **3.** -5^4

4. -3^4 **5.** 2^{-3} **6.** 3^{-2}

7. 0^8 **8.** 8^0 **9.** 0^{-7}

10. 0^0 **11.** 9^{-1} **12.** 9^{-2}

13. $\left(\frac{2}{3}\right)^{-1}$ **14.** $\left(\frac{5}{8}\right)^{-1}$ **15.** $\left(\frac{1}{2}\right)^{-3}$

16. $\left(\frac{1}{3}\right)^{-4}$ **17.** $\frac{1}{3^{-2}}$ **18.** $\frac{2}{4^{-3}}$

19. $\frac{3^2}{7^{-5}}$ **20.** $\frac{5^{-4}}{6^{-2}}$ **21.** 10^{-6}

22. $\left(\frac{1}{10}\right)^{-2}$ **23.** $\frac{10^2}{10^{-4}}$ **24.** $\frac{10^3}{10^{-3}}$

In Exercises 25–36, rewrite the expression with a positive exponent.

25. 6^{-2} **26.** $(-8)^{-3}$ **27.** $\left(\frac{3}{7}\right)^{-5}$

28. $\left(\frac{5}{8}\right)^{-3}$ **29.** a^{-2} **30.** b^{-10}

31. $\frac{1}{x^{-3}}$ **32.** $\frac{1}{y^{-7}}$ **33.** $\frac{1}{-3^{-4}}$

34. -2^{-3} **35.** x^{-1} **36.** $-y^{-1}$

In Exercises 37–48, rewrite the expression with a negative exponent.

37. $\frac{1}{8^3}$ **38.** $\frac{1}{(-10)^4}$ **39.** 4^5

40. 3^9 **41.** $\frac{1}{x^5}$ **42.** $\frac{1}{(-y)^7}$

43. $-\frac{1}{3^8}$ **44.** $-\frac{3}{4^3}$ **45.** $\left(\frac{2}{3}\right)^5$

46. $\left(\frac{3}{5}\right)^3$ **47.** $\frac{1}{y}$ **48.** $\frac{2}{-x}$

In Exercises 49–56, evaluate the expression, given that $x = 3, y = -5$, and $z = -4$.

49. $x^2 + y^2$ **50.** $x^2 - y^2$ **51.** $z^{-2}(x + y)^3$

52. $x(y - z)^2$ **53.** y^x **54.** x^0

55. $x^{-2} + y^{-1}$ **56.** $x^{-1} + y^{-2}$

In Exercises 57–60, evaluate the expression to the nearest hundredth. (**Careful:** Your calculator may not raise a negative number to a power. You may have to keep track of the sign of your calculation yourself.)

57. $(5.82)^2 - 6(3 - 4.71)^3$

58. $(3.75 - 6.8)^4 - (9.08)^2$

59. $\left[\frac{6.28 - 1.59}{(1.08)^2}\right]^3$

60. $(1.3)^{13} - (6.91 \div 5.75)^{13}$

In Exercises 61–68, write the expression in the form 3^n.

61. $3^3 \cdot 3^4$ **62.** $3^2 \cdot 3^5$

63. $3^2 \cdot 3^5 \cdot 3^{-7}$ **64.** $3^4 \cdot 3^{-3} \cdot 3^{-2}$

65. $(9^3)^5$ **66.** $[(27)^2]^{-3}$

67. $3^6 \cdot 9^4$ **68.** $3^5 \cdot \left(\frac{1}{9}\right)^2$

In Exercises 69–92, simplify the expression so that x, y, and z, appear at most once, and so that there are no negative exponents.

69. $(2x^2)(-2xy^3)(y^2)$ **70.** $(2x)(7yx)(x^2)$

71. $(3z^3)(4xyz)$ **72.** $(5x^2)(2xyz)$

73. $(3x^2)(5x^{-4}y)$ **74.** $(3y^{-2})(4x^{-1}y^5)$

75. $(6xy)(2z^{-5}y^{-1})$ **76.** $(2xyz)(x^{-1}yz^{-3})$

77. $(2x)^3$ **78.** $(3y)^2$

79. $(6xy^2z)^2$ **80.** $(4x^3yz)^2$

81. $(-5z^4)^3$ **82.** $(-6x^3y)^4$

83. $\frac{2xz^{-3}}{3yz^{-2}}$ **84.** $\frac{8y^{-2}z}{11yz^{-3}}$

85. $\frac{7x^{-2}y^{-5}}{10x^{-5}z^{-2}}$ 86. $\frac{5y^{-3}z^{2}}{6x^{-4}z^{-8}}$

87. $\left(\frac{x}{y}\right)^3\left(\frac{x^2y}{z}\right)^2$ 88. $\left(\frac{xy}{z}\right)\left(\frac{x}{yz^2}\right)^3$

89. $\left(\frac{2xy}{3z}\right)^{-3}$ 90. $\left(\frac{5x}{7yz}\right)^{-2}$

91. $\left(\frac{x^2}{3z^3}\right)^{-5}$ 92. $\left(\frac{y^3}{2x^2}\right)^{-4}$

In Exercises 93–105, write the number in scientific notation.

93. 0.0001 94. 71,900

95. 3,500,000,000 96. 0.036

97. 1600 98. 181,000

99. 0.00118 100. 1/200

101. 1/5000 102. $1/(2 \cdot 10^{-2})$

103. $2/(5 \cdot 10^{-3})$ 104. 0.006×10^8

105. 6000×10^{-7}

In Exercises 106–117, write the number in decimal form.

106. 6.41×10^{-3} 107. 6.41×10^5

108. 2.01×10^4 109. 9.9×10^{-5}

110. 8×10^{-6} 111. 3×10^8

112. 1.025×10^6 113. 5.555×10^9

114. 1.4×10^{-7} 115. 8.1×10^{-6}

116. 0.0034×10^{-6} 117. 849×10^{-7}

In Exercises 118–123, compute the value of each expression. Write your answer in scientific notation, rounded to three significant digits.

118. $(1.02 \times 10^{-3})(2.78 \times 10^8)$

119. $(6.334 \times 10^9)(5.55 \times 10^{-2})$

120. $2{,}850{,}000 \times 0.0729$

121. $660{,}000 \times 0.488$

122. $\frac{(6.12 \times 10^4)(8.02 \times 10^{-2})}{9.81 \times 10^3}$

123. $\frac{(1.56 \times 10^{12})(4.57 \times 10^{-8})}{7.48 \times 10^{-6}}$

Superset

In Exercises 124–133, determine whether the statement is true or false.

124. $3^2 \cdot 3^4 = 3^8$ 125. $(5^2)^2 = 5^4$

126. $(5^2)^3 = 5^5$ 127. $(-1)^0 = -1$

128. $(-6)^{-2} = \frac{1}{6^2}$ 129. $\frac{1}{(-3)^2} = (3)^{-2}$

130. $3^4 + 3^5 = 3^9$ 131. $2^3 \cdot 5^4 = 10^7$

132. $(-3)^4 = -3^4$ 133. $\frac{3^8}{3^4} = 3^2$

In Exercises 134–141, simplify the expression so that x, y, and z appear at most once, and so that there are no negative exponents.

134. $\frac{(2x^2y)^2}{xy^3z^2}$ 135. $\frac{3x^2y^2z}{(7x^3z^4)^3}$

136. $\frac{(4x^3z^{-1})^2}{(x^2yz^{-3})^{-1}}$ 137. $\frac{(y^{-5}z^2)^{-3}}{(3x^3y^4)^{-2}}$

138. $\frac{(xz^5)^{-4}(y^2z^{-2})^3}{(x^{-1})(x^2y^4)^3}$ 139. $\frac{(x^7y^3)^2(yz^{-1})^{-5}}{(xy^{-3})^{-1}(y^4z)^{-3}}$

140. $\frac{(x+y)^3(x+y)^{-2}}{(x+y)^4}$ 141. $\frac{(y-z)^{-5}(y-z)^{-6}}{(y-z)^{-7}}$

In Exercises 142–149, suppose that all you know is that $x^3 = 3$, $y^2 = 2$, and $z^4 = 4$. Use the Properties of Integer Exponents to evaluate each side of the inequality and decide whether the inequality is true or false.

142. $\frac{1}{x^6} < \frac{1}{y^6}$ 143. $z^8 > x^9y^{-2}$

144. $\left(\frac{xy}{z^2}\right)^6 > 1$ 145. $\left(\frac{2}{x^2}\right)^3 + \frac{1}{z^4+1} > 1$

146. $2x^{-3} - y^{-2} > z^{-4}$ 147. $2^{-4} + z^{-4} < \frac{2}{x^3y^2}$

148. $(x^{-1}yz)(xyz)^{-5} > 0.01$ 149. $\frac{1}{x^{-6}} > \frac{1}{y^{-4}} + \frac{1}{z^{-4}}$

In Exercises 150–157, determine which expression in the given pair represents the greater value.

150. $(1 - 0.004)^2, 1 - (0.004)^2$

151. $(1 + 0.02)^{-3}, 1 + (0.02)^{-3}$

152. $1 + 0.02, (1 + 0.02)^2$

153. $1 - 0.02, (1 - 0.02)^2$

154. $\frac{6.2 \times 10^{-8}}{9.1 \times 10^{-9}}, 6.9$ 155. $\frac{3.2 \times 10^{-40}}{7.5 \times 10^{-38}}, 0.004$

156. $\frac{1.8 \times 10^{56}}{1.1 \times 10^{60}}, 0.001$ 157. $\frac{5.4 \div 10^{-10}}{1.9 \div 10^{-12}}, \frac{1}{5}$

For Exercises 158–165, the necessary formulas from geometry are listed inside the front cover.

158. Determine the area of a circle whose radius is $2x^3$ cm.

159. Determine the volume of a sphere of radius $3a^2$ ft.

160. Determine the area of a square whose side measures $5p^{-1}$ in.

161. Determine the volume of a cube whose side measures $0.1y^3$ ft.

162. Determine the volume (in cubic centimeters) of a cylinder whose radius is n^2 cm and whose height is 1 m.

163. Determine the volume (in cubic feet) of a cylinder whose radius is .04 ft and whose height is 12 in.

164. Compare the volume of a sphere having radius 2^{-3} m, with the volume of a cylinder having radius 3^{-2} m and height $\frac{3}{2}$ m.

165. Which is greater, the area of a circle of radius $\left(\frac{3}{2}\right)^{-2}$ m or the surface area of a cube having side $\frac{1}{3}$ m?

In Exercises 166–171, a daily value has been given. Assuming that there are 365 days in a year, write the corresponding yearly value in scientific notation rounded to two significant digits, and then give your answer using the word million, billion, or trillion (10^{12}).*

166. In one day the United States government spends at least $2,000,000,000. How much does it spend in one year?

167. In one day roughly 4200 billion gallons of rain fall on the continental United States. How many gallons fall in one year?

168. In one day American drivers burn roughly 313 million gallons of fuel. How much do they burn in a year?

169. In one day Americans spend $2,500,000 washing their cars. How much do they spend in one year?

170. In one day Americans buy 50,000 new TV sets. How many do they buy in a year?

171. In one day Americans eat 47 million hot dogs. How many do they eat in a year?

172. A light-year is defined to be the distance that light will travel in one year. Assuming a 365-day year, and assuming that the speed of light is 186,000 miles per second, determine the number of miles in one light-year. Express your answer in scientific notation, rounded to three significant digits.

173. The sun is approximately 93 million miles from the earth. Using the appropriate information from the previous problem, determine the number of minutes it takes light to travel from the sun to the earth, rounded to two significant digits.

1.5
Polynomials: Addition, Subtraction, and Multiplication

A jet, beginning its takeoff, is at one end of a runway. If a represents its acceleration (in feet per second per second), and if it takes t seconds to take off, then the length of runway needed for takeoff is given by the expression

$$\frac{1}{2}at^2.$$

This expression is called a monomial. In general, a **monomial** is either a nonzero constant or the product of a nonzero constant and one or more variables, each of which is raised to a positive power. The constant is usually referred to as the **numerical coefficient** or simply the **coefficient** of the monomial. The **degree of the monomial** is the sum of the exponents of its variables. Examples of monomials are -3, $4x$, $16a^2bc^3$, and $-8x^2y^3$.

$-8x^2y^3$ The coefficient is -8; the degree is 5.

*Reprinted with permission from Tom Parker, *In One Day* (Boston: Houghton Mifflin Company, 1984)

The word **term** is often used to mean monomial. A **polynomial** is then defined as a sum of a finite number of terms. The **degree of a polynomial** is the highest degree of the terms in the polynomial. Examples of polynomials are $3x - 2$ and $8x^2 - 5xy + z^3$.

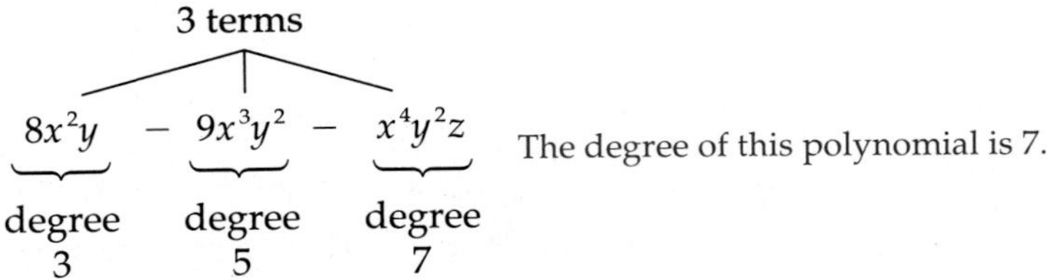

The degree of this polynomial is 7.

Careful! Not all variable expressions are polynomials. For example, $5x^2 + 4 + \frac{1}{x}$ is not a polynomial because the variable in the third term has a negative exponent $\left(\frac{1}{x} = x^{-1}\right)$.

Some polynomials have special names; a polynomial with two terms, such as $5x - 3$, is called a **binomial,** and a polynomial with three terms, such as $2x^2 - 7xy + 10y^2$, is called a **trinomial.** A nonzero number, such as 16, is called a **constant polynomial,** and has degree zero. The number 0 is called the **zero polynomial,** but its degree is not defined; it is said to have no degree.

A polynomial that contains only one variable is called a polynomial in that variable. Thus $y^5 - 6y^3 + y^2 - 1$ is called a **polynomial in *y*.** It is customary to write the term of highest degree first, followed by the term of next highest degree, and so forth. We say that the terms have been written in **descending order,** and understand that this refers to the degrees of the terms.

EXAMPLE 1

Select the appropriate expression to complete each sentence.

(a) The polynomial $5x^4 - 2x + 4y$ has ______ terms. [two/three/four]

(b) In the polynomial $3x^3 - 4x^2 - 7$, the coefficient of the x^2-term is ______. [−4/4]

(c) The degree of the monomial $5x^4y^2z$ is ______. [5/6/7/8]

(d) The degree of the polynomial $2xy - 3x^2y$ is ______. [2/3/5]

(e) The degree of the monomial 4 is ______. [0/1/not defined]

(f) The degree of the zero polynomial 0 is ______. [0/1/not defined]

Solution

(a) three (b) −4 (c) 7 (d) 3 (e) 0 (f) not defined

Terms having exactly the same variables with exactly the same exponents are called **like terms.**

$5x^2y \quad \sqrt{2}x^2y \quad -x^2y$ like terms

$7xy^2z \quad 7x^2yz$ unlike terms

Polynomials can be simplified by combining like terms. This is accomplished by using the Commutative and Distributive Properties.

EXAMPLE 2

Simplify the expression $5xy^3 - 6y^3 - 8xy^3$.

Solution

$$5xy^3 - 6y^3 - 8xy^3 = (5xy^3 - 8xy^3) - 6y^3$$ Like terms are grouped together.

$$= (5 - 8)xy^3 - 6y^3$$ The Distributive Property is used: $(ac - bc) = (a - b)c$.

$$= -3xy^3 - 6y^3$$

Polynomial Arithmetic

In Section 1.1 we observed that variables represent numbers, and in Section 1.2 we saw that real numbers behave according to the Properties of Real Numbers. Thus if x and y represent real numbers, then so do expressions such as $x - 3$ and $6xy - 5y^2$, and so they can be added, subtracted, multiplied, or divided like any other real numbers.

When adding two or more polynomials you simply combine like terms. To subtract a polynomial, you add its additive inverse. Be careful when finding the additive inverse of a polynomial.

$$-(3x^2 - 7xy + 18y^2) = -3x^2 + 7xy - 18y^2$$

The negative sign in front of the parentheses changes the sign of each term inside the parentheses.

EXAMPLE 3

Subtract $5x - 7y$ from $8x + 4y$.

Solution

$$(8x + 4y) - (5x - 7y) = 8x + 4y - 5x + 7y$$

$$= (8x - 5x) + (4y + 7y)$$ Like terms are grouped together.

$$= 3x + 11y$$

Addition and subtraction can be performed vertically. For example, to add $6x^3 - 3x + 7$ and $x^2 - 4x - 1$, we can write the two polynomials with like terms aligned vertically, making sure that we leave space for any omitted powers of x.

$$\begin{array}{rrrr} 6x^3 & & -\ 3x & +\ 7 \\ & x^2 & -\ 4x & -\ 1 \\ \hline 6x^3 & +\ x^2 & -\ 7x & +\ 6 \end{array} \quad \text{Align like terms.}$$

The simplest type of polynomial multiplication involves multiplying a polynomial by a monomial; a single application of the Distributive Property is required.

EXAMPLE 4

Multiply $3x^2$ by $5xy - 2x^3y^2$.

Solution

$$\begin{aligned} 3x^2(5xy - 2x^3y^2) &= (3x^2)(5xy) - (3x^2)(2x^3y^2) && \text{The Distributive Property is used.} \\ &= 15x^3y - 6x^5y^2 && \text{Property 1 of Exponents is used.} \end{aligned}$$

The general case of polynomial multiplication, where neither polynomial is a monomial, simply requires two applications of the Distributive Property.

EXAMPLE 5

Multiply $(5x + 2y)(3x^2 - 7y^2 + 4xy)$.

Solution

$$\begin{array}{ccccccc} (A + B) & \cdot & (\quad C \quad) & = A \cdot (\quad C \quad) & + & B \cdot (\quad C \quad) \\ \blacktriangledown & & \blacktriangledown & \blacktriangledown \quad \blacktriangledown & & \blacktriangledown \quad \blacktriangledown \\ (5x + 2y) & \cdot & (3x^2 - 7y^2 + 4xy) & = 5x(3x^2 - 7y^2 + 4xy) & + & 2y(3x^2 - 7y^2 + 4xy) \end{array}$$

The Distributive Property is used.

$$= 15x^3 - 35xy^2 + 20x^2y + 6x^2y - 14y^3 + 8xy^2$$

The Distributive Property is used again.

$$= 15x^3 + (20x^2y + 6x^2y) + (-35xy^2 + 8xy^2) - 14y^3$$

Like terms are grouped together.

$$= 15x^3 + 26x^2y - 27xy^2 - 14y^3$$

Multiplication can also be performed vertically. Notice the similarity to whole number multiplication when we multiply $(3x - 7)$ by $(2x + 4)$:

$$\begin{array}{rll} & 3x - 7 & \\ \times & 2x + 4 & \\ \hline & 12x - 28 & 4 \cdot (3x - 7) \\ 6x^2 - 14x & & 2x \cdot (3x - 7) \\ \hline 6x^2 - & 2x - 28 & \end{array}$$

Since products of binomials occur frequently in our work, they merit special attention. Suppose we wish to multiply two binomials, $(a + b)$ and $(c + d)$. Think of a and c as the first terms in the binomials, a and d as the outer terms, b and c as the inner terms, and b and d as the last terms.

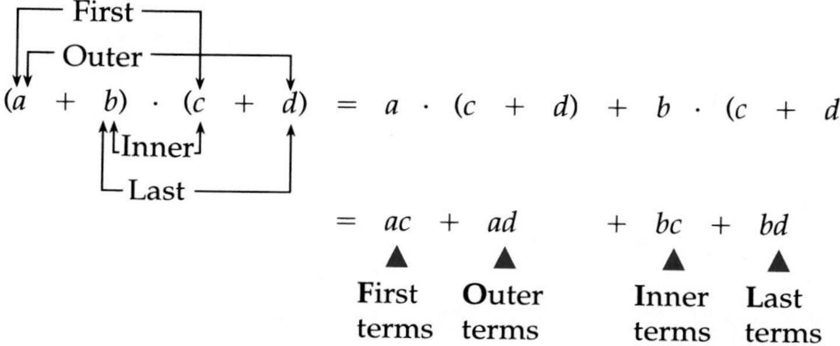

This procedure for finding the product of two binomials is sometimes called the **FOIL** method.

EXAMPLE 6

Multiply $(3x - 7)(4x + 3y)$.

Solution

$$(3x - 7)(4x + 3y) = \underset{\text{First}}{(3x)(4x)} + \underset{\text{Outer}}{(3x)(3y)} + \underset{\text{Inner}}{(-7)(4x)} + \underset{\text{Last}}{(-7)(3y)}$$

$$= 12x^2 + 9xy - 28x - 21y$$

Special Products

There are some special products of binomials that occur so frequently that we give them special names. In the derivations below, notice the use of the FOIL method.

Squares of Binomials

$\mathbf{(a + b)^2 = a^2 + 2ab + b^2}$

$$(a + b)^2 = (a + b)(a + b) = a^2 + ab + ba + b^2 = a^2 + 2ab + b^2$$

$\mathbf{(a - b)^2 = a^2 - 2ab + b^2}$

$$(a - b)^2 = (a - b)(a - b) = a^2 - ab - ba + b^2 = a^2 - 2ab + b^2$$

Difference of Two Squares

$$\mathbf{(a+b)(a-b)=a^2-b^2}$$

$$(a+b)(a-b)=a^2-ab+ba-b^2=a^2-b^2$$

Cubes of Binomials

$$\mathbf{(a+b)^3=a^3+3a^2b+3ab^2+b^3}$$

$$\begin{aligned}(a+b)^3=(a+b)^2(a+b)&=(a^2+2ab+b^2)(a+b)\\&=(a^3+2a^2b+ab^2)+(a^2b+2ab^2+b^3)\\&=a^3+3a^2b+3ab^2+b^3\end{aligned}$$

$$\mathbf{(a-b)^3=a^3-3a^2b+3ab^2-b^3}$$

$$\begin{aligned}(a-b)^3=(a-b)^2(a-b)&=(a^2-2ab+b^2)(a-b)\\&=(a^3-2a^2b+ab^2)+(-a^2b+2ab^2-b^3)\\&=a^3-3a^2b+3ab^2-b^3\end{aligned}$$

EXAMPLE 7

Determine $(6y+5x)^2$.

Solution

$$\begin{aligned}(6y+5x)^2&=(6y)^2+2(6y)(5x)+(5x)^2\\&=36y^2+60xy+25x^2\end{aligned}$$

Square the first term; double the product of the two terms; square the last term.

EXAMPLE 8

Perform the indicated operations.

(a) $(2x-7y)(2x+7y)$ (b) $(2x+5y)^3$

Solution

(a) Here we use the special product: $(a-b)(a+b)=a^2-b^2$.

$$\begin{aligned}(2x-7y)(2x+7y)&=(2x)^2-(7y)^2\\&=4x^2-49y^2\end{aligned}$$

(b) Here we use $(a+b)^3=a^3+3a^2b+3ab^2+b^3$ with $2x$ substituted for a, and $5y$ substituted for b.

$$\begin{aligned}(2x+5y)^3&=(2x)^3+3(2x)^2(5y)+3(2x)(5y)^2+(5y)^3\\&=8x^3+3\cdot4x^2\cdot5y+3\cdot2x\cdot25y^2+125y^3\\&=8x^3+60x^2y+150xy^2+125y^3\end{aligned}$$

EXERCISE SET 1.5

In Exercises 1–10, determine whether the given expression is or is not a polynomial. If it is a polynomial, state its number of terms and its degree.

1. $6x^2y^5z$

2. $10a^4b^4c - 1$

3. $5x^2 - 2x - \frac{1}{x}$

4. $x^3 - x^4 + 5xy$

5. $17xy^2 - 5x^3y + \frac{1}{3}xy$

6. $6x^2y - 17xyz + \sqrt{4}y^4$

7. $1 + 2xy + x^5 - 5x^2y^2$

8. $5x^2 - x - \frac{1}{x^2} + \frac{1}{x^3}$

9. 10

10. $7a^2b^2 - 5ac + 5abc^3 - 8$

In Exercises 11–20, add or subtract as indicated.

11. Add $2x^2 + 5x$ to $3x^3 - 7x^2 + 5x - 4$.

12. Add $-x^3 - 9x^2 + 5$ to $x^2 - 3x + 14$.

13. Subtract $u^2 - u + 1$ from $4u^4 - 2u^2 + 1$.

14. Subtract $x^3 + x^2 - 1$ from $4x^4 - 2x^2 + 1$.

15. $(x^4 - x^2 + 1) + (2x^2 - 8x + 5x^4 + 10)$

16. $(14a^2 - 6a - 5) + (7a^2 + 5a - 18)$

17. $(3v^2 - 8v + 14) - (5v^2 - 2v - 6)$

18. $(2y^3 - 3y + 4) - (y^3 - 4y^2 - 2)$

19. $-3(x - y) - (x^2 + y^2 - 3x - 3y)$

20. $-2(x + y) - (x^2 - y^2 + 2x - 2y)$

In Exercises 21–50, determine the indicated product.

21. $2x(5x^2 - 7x - 1)$

22. $-3a(a^2 + 9a - 1)$

23. $(-5)(ab - ba)$

24. $(-1)(uv - vu)$

25. $(3x - 2)(7x + 3)$

26. $(5x - 6)(2x + 7)$

27. $(4m - 3)\left(5m - \frac{1}{2}\right)$

28. $(9n - 4)\left(2n - \frac{2}{3}\right)$

29. $(5x)(6x + 5y)(3x + 2y)$

30. $(4u)(7u + 2v)(3u + 4v)$

31. $\left(b - \frac{1}{2}\right)\left(\frac{3}{2}b - \frac{1}{4}\right)$

32. $\left(n - \frac{1}{3}\right)\left(\frac{1}{2}n + \frac{2}{5}\right)$

33. $\left(3x + \frac{2}{5}\right)(x^2 + 3x - 4)$

34. $\left(2y - \frac{1}{3}\right)(y^2 - 5y + 2)$

35. $(4m + 3)(2m^3 - 3m^2 + 5m - 4)$

36. $(5x - 2)(x^4 - 2x^2 + x - 10)$

37. $(x^2 + x - 1)(3x^2 - 5x + 2)$

38. $(v^2 - v + 7)(v^2 + 3v - 2)$

39. $(3x^2 + 2y^2)(3x^2 - 2y^2)$

40. $(2m^2 + 3n^2)(2m^2 - 3n^2)$

41. $(3a + 7)^2$

42. $(4m - 3)^2$

43. $(5x + 3y)^2$

44. $(7a - 4b)^2$

45. $(x + 2)^3$

46. $(x - 2)^3$

47. $[(x - 2y) + 1]^2$

48. $[(3s + 2t) - 1]^2$

49. $(2x^2 + 5y^3)^3$

50. $(4x^3 - 3y^2)^3$

In Exercises 51–58, perform the indicated operations. Express the answer as a polynomial in x with terms in descending order.

51. $(x - 2)^2 - (x + 2)^2$

52. $(x + 5)^2 - (x + 2)^2$

53. $(x - 3)^2 (x + 3)^2$

54. $(x + 1)^2 (x - 1)^2$

55. $(2x + 3)^2 (x - 4)$

56. $(x + 7) (3x - 5)^2$

57. $[x - (a + b)]^2$

58. $[x + (a + b)]^3$

Superset

In Exercises 59–69, the answer should be written as a polynomial in simplified form.

59. The square below has a side with length $x + 2$. Find a polynomial that describes its area.

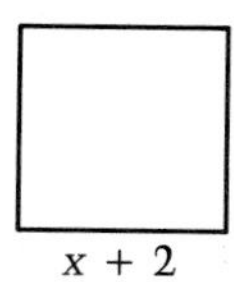

60. The square below has a side with length $2x + 1$. Find a polynomial that describes its area.

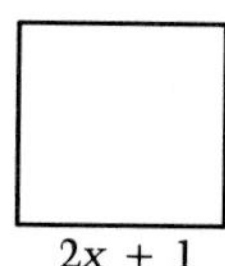

61. The triangle below has an altitude of length $2x$ and a base of length $x - 2$. Find a polynomial that describes its area.

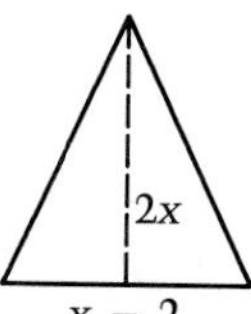

62. The circle below has a radius of length $x + 3$. Find a polynomial that describes its area.

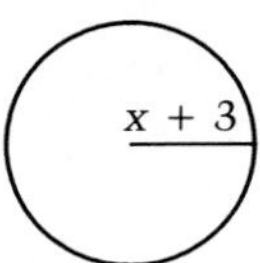

63. The sphere below has a radius of length $3x + 2$. Find a polynomial that describes its volume.

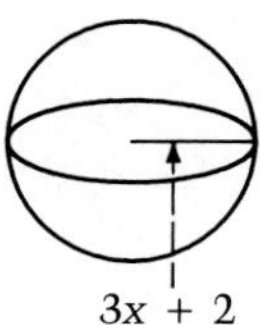

64. The cube below has an edge of length $x + 3$. Find a polynomial that describes its volume.

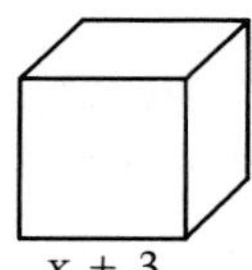

65. The height of the following box is $2x - 1$, and the area of the shaded top is $x^2 + 4$. Find a polynomial that describes the volume of the box.

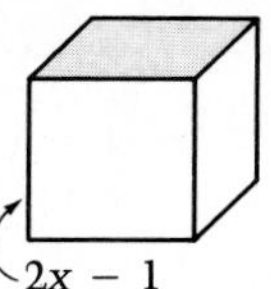

66. Find a polynomial that describes the surface area of the sphere in Exercise 63.

67. Find a polynomial that describes the surface area of the cube in Exercise 64.

68. Find a polynomial that describes the surface area of the box whose dimensions are $3x$, $2x - 1$ and $x + 4$.

69. Find a polynomial that describes the volume of the box in Exercise 68.

In Exercises 70–73, determine the coefficient of x^4 in the product, without computing the entire product.

70. $(6x^2 - 5)^3$

71. $(6x^2 - 5x)^3$

72. $3x(4x^2 - x)^2$

73. $2x^2(7 - x^2)^2$

74. Is it always true that $(x - 1)^2 = (1 - x)^2$? Explain your answer.

75. Is it always true that $(x - 1)^3 = (1 - x)^3$? Explain your answer.

76. Develop a formula for $(a + b + c)^2$.

77. Develop a formula for $(a + b)^4$.

78. Based on the result of Exercise 77, and your knowledge of formulas for $(a + b)^3$ and $(a - b)^3$, guess the formula for $(a - b)^4$. Verify by multiplication.

79. Multiply: $(A - B)(A^2 + AB + B^2)$.

80. Multiply: $(A + B)(A^2 - AB + B^2)$.

1.6 Polynomials: Factoring

In the last section we explored the process of multiplying two or more polynomials to produce a single polynomial. In this section we will examine techniques for reversing that process: that is, given a polynomial, we now want to rewrite it as the product of two (or more) polynomials. For example given an expression such as $2x^2y - 6xy^3$, we will rewrite it as $2xy(x - 3y^2)$.

Suppose a number a is the product of two numbers, b and c. We write

$a = bc$ and say that b and c are **factors** of a.

We could also say that a is **factored** into the product bc. We use the same terminology when we talk about monomials and polynomials.

$6 = 2 \cdot 3$	2 and 3 are factors of 6.
$3x^2yz = (3x^2)(yz)$	$3x^2$ and yz are factors of $3x^2yz$.
$x^2 - 5x + 6 = (x - 2)(x - 3)$	$(x - 2)$ and $(x - 3)$ are factors of $x^2 - 5x + 6$; or, $x^2 - 5x + 6$ is factored into the product $(x - 2)(x - 3)$.

Before we consider techniques for factoring polynomials, let us review the factorization of integers. Recall that a **prime number** is a positive integer greater than 1 whose only positive factors are 1 and itself. The first five prime numbers are

$$2, 3, 5, 7 \text{ and } 11.$$

The **greatest common factor (GCF)** of two integers is the largest factor that divides evenly into both integers. The **GCF** is found by first factoring each number into a product of powers of primes. For each of these primes, the highest power present in both factorizations will be a factor of the GCF.

$$360 = 2^3 \cdot 3^2 \cdot 5$$
$$84 = 2^2 \cdot 3 \cdot 7$$

The highest power of 2 that is a common factor of both 360 and 84 is 2^2; the highest power of 3 that is a common factor is 3^1; no other prime is a common factor. Thus the GCF of 360 and 84 is $2^2 \cdot 3 = 12$.

To determine the GCF of two or more monomials, we begin by looking at the factors of each monomial. We then form the product of those factors common to all the monomials.

EXAMPLE 1

Find the GCF of $-16x^2yz^3$, $12xyz^2$ and $20xz^2$.

Solution

$$-16x^2yz^3 = (-1)2^4 \cdot x^2 \cdot y \cdot z^3$$
$$12xyz^2 = 2^2 \cdot 3 \cdot x \cdot y \cdot z^2$$
$$20xz^2 = 2^2 \cdot 5 \cdot x \cdot z^2$$

2^2, x and z^2 are the common factors.

The GCF is $2^2 \cdot x \cdot z^2 = 4xz^2$.

In the remainder of this section we will suggest several strategies for factoring polynomials. We begin with the easiest.

Factoring Strategy 1 *Factor out a common factor*

The simplest type of polynomial factoring involves "factoring out" the GCF from each term of the polynomial. Essentially we are using the Distributive

Property to rewrite a polynomial $ab + ac$ as $a(b + c)$. Consider the polynomial $12x^2yz^2 - 18xy^2z^3$. Its terms are

$$12x^2yz^2 \quad \text{and} \quad -18xy^2z^3,$$

and the GCF of these terms is $6xyz^2$. In factored form, the polynomial will look like

$$6xyz^2(\qquad\qquad).$$

To find the expression inside the parentheses, we must determine "what's left" in each term of the polynomial after we factor out the GCF. In each case, "what's left" is the original term divided by the GCF.

Term	GCF	$\frac{\text{Term}}{\text{GCF}}$	= What's left
$12x^2yz^2$	$6xyz^2$	$\frac{12x^2yz^2}{6xyz^2}$	$= 2x$
$-18xy^2z^3$	$6xyz^2$	$\frac{-18xy^2z^3}{6xyz^2}$	$= -3yz$

Therefore,

$$12x^2yz^2 - 18xy^2z^3 = 6xyz^2(2x - 3yz),$$

where the expression on the right is called the **factored form** of the given polynomial.

EXAMPLE 2

Factor: (a) $16x^5 + 8x^3 - 6x^2$ (b) $14a^2bc^2 - 42ab^3c + 21ab^2c^4$ (c) $2n^3 + n^2$

Solution

(a) $16x^5 + 8x^3 - 6x^2 = 2x^2(8x^3 + 4x - 3)$ Check your work by multiplying the factors in your answer.

(b) $14a^2bc^2 - 42ab^3c + 21ab^2c^4 = 7abc(2ac - 6b^2 + 3bc^3)$

(c) $2n^3 + n^2 = n^2(2n + 1)$ Don't forget the factor "1" that's left in the second term.

Factoring Strategy 2 *Factor by observing a pattern*

In Section 1.5 we worked with several Special Products. We rely on these patterns here also, but this time we use them to factor polynomials rather than to multiply them. We add two new Special Products to our list: the difference of cubes and the sum of cubes. You should do the multiplication necessary to verify these identities. In each line of the following table, the polynomial on the left side of the equation is found by multiplying the factors on the right side.

	Polynomial		Factored Form
Difference of squares	$A^2 - B^2$	$=$	$(A + B)(A - B)$
Perfect trinomial square	$A^2 + 2AB + B^2$ $A^2 - 2AB + B^2$	$=$ $=$	$(A + B)^2$ $(A - B)^2$
Difference of cubes	$A^3 - B^3$	$=$	$(A - B)(A^2 + AB + B^2)$
Sum of cubes	$A^3 + B^3$	$=$	$(A + B)(A^2 - AB + B^2)$

We will use these patterns to factor these four types of polynomials. In practice, A and B may represent any real number or algebraic expression.

EXAMPLE 3

Factor each polynomial: (a) $x^2 + 6x + 9$ (b) $16x^2 - 81y^2$ (c) $16m^4 - n^4$

Solution

(a)
$$\begin{array}{ccccccc} x^2 & + & 6x & + & 9 & & \\ \blacktriangle & & \blacktriangle & & \blacktriangle & & \\ A^2 & + & 2AB & + & B^2 & = & (A + B)^2 \\ & & & & & & \blacktriangledown \quad \blacktriangledown \\ & & & & & & (x + 3)^2 \end{array}$$

This is a perfect trinomial square with $A = x$ and $B = 3$. In a perfect trinomial square, two of the terms are perfect squares of some expressions A and B, and the other term is twice the product of A and B.

Thus, $x^2 + 6x + 9 = (x + 3)^2$.

(b) $16x^2 - 81y^2 = (4x)^2 - (9y)^2$ — The polynomial has the form $A^2 - B^2$, where $A = 4x$ and $B = 9y$.

$= (4x + 9y)(4x - 9y)$ — The pattern $A^2 - B^2 = (A + B)(A - B)$ is used.

(c)
$$\begin{aligned} 16m^4 - n^4 &= (4m^2)^2 - (n^2)^2 \\ &= (4m^2 + n^2)(4m^2 - n^2) \\ &= (4m^2 + n^2)[(2m)^2 - (n)^2] \\ &= (4m^2 + n^2)(2m + n)(2m - n) \end{aligned}$$

We will use the Difference of Squares pattern twice in this problem.

EXAMPLE 4

Factor $27x^3 + 64y^3$.

Solution

The sum of cubes pattern is used: $A^3 + B^3 = (A + B)(A^2 - AB + B^2)$ where $A = 3x$ and $B = 4y$.

$$\begin{aligned} 27x^3 + 64y^3 &= (3x)^3 + (4y)^3 \\ &= (3x + 4y)(9x^2 - 12xy + 16y^2) \end{aligned}$$

Careful! The term $12xy$ in the second factor is AB, not $2AB$.

EXAMPLE 5

Factor $(2x + y)^3 - 27y^3$.

Solution

$$(A)^3 - (B)^3 = (A - B)[(A)^2 + (A)(B) + (B)^2]$$

$$\begin{aligned}(2x + y)^3 - (3y)^3 &= (2x + y - 3y)[(2x + y)^2 + (2x + y)(3y) + (3y)^2] \\ &= (2x - 2y)[(4x^2 + 4xy + y^2) + (6xy + 3y^2) + (9y^2)] \\ &= 2(x - y)(4x^2 + 10xy + 13y^2)\end{aligned}$$

In order to factor completely, we have factored out the constant 2.

Factoring Strategy 3 ***Factor by grouping***

Sometimes a polynomial containing four or more terms can be simplified by grouping pairs (or triples) of terms, and then factoring each grouping. This technique is demonstrated in the following example.

EXAMPLE 6

Factor $x^3 - 2x^2 + 4x - 8$.

Solution

$x^3 - 2x^2 + 4x - 8 = (x^3 - 2x^2) + (4x - 8)$ Two groupings are formed.

$= x^2(x - 2) + 4(x - 2)$ A monomial is factored out of each grouping to produce the common factor $(x - 2)$ in each.

$= (x^2 + 4)(x - 2)$ $(x - 2)$ is factored out of each term.

In Example 6, we could have grouped the terms as $(x^3 + 4x) - (2x^2 + 8)$. In general, there is always more than one way to form the groupings.

Factoring Strategy 4 ***Factor a trinomial by trial and error***

The product of $3x + 5$ and $2x + 3$ is the trinomial $6x^2 + 19x + 15$. Suppose instead, we begin with the trinomial and need to find its two factors. How do we go about finding them? We must factor $6x^2 + 19x + 15$ into a product of the form $(ax + b)(cx + d)$, which equals $acx^2 + adx + bcx + bd$, that is, $acx^2 + (ad + bc)x + bd$.

We will restrict our "trial and error" search to factors whose coefficients and constants are integers. Factors involving fractions are generally found using strategies that we will study later.

$$\begin{array}{lllllll} acx^2 & + & (ad + bc)x & + & bd & = & (ax + b)(cx + d) \\ 6x^2 & + & 19x & + & 15 & = & (\ ? \ )(\ ? \ ) \end{array}$$

$ac = 6$	$bd = 15$
$3 \times 2 = 6$	$3 \times 5 = 15$
$1 \times 6 = 6$	$1 \times 15 = 15$

First clue: $ac = 6$. Thus, the factorization has one of the following forms:

$$(3x + b)(2x + d) \quad \text{or} \quad (x + b)(6x + d).$$

Second clue: $bd = 15$, a positive number. Thus, b and d have the same sign. The coefficient of the x-term ($19x$) is positive; therefore, we can conclude that b and d are positive. The possible factorizations are shown below; each one produces an x^2-term of $6x^2$, and a constant term of 15.

	Middle term		Middle term
$(3x + 15)(2x + 1)$	$33x$	$(x + 15)(6x + 1)$	$91x$
$(3x + 1)(2x + 15)$	$47x$	$(x + 1)(6x + 15)$	$21x$
$(3x + 5)(2x + 3)$	$19x$	$(x + 5)(6x + 3)$	$33x$
$(3x + 3)(2x + 5)$	$21x$	$(x + 3)(6x + 5)$	$23x$

Third clue: The middle term is $19x$. Remember that this is found by adding the product of the inner terms to the product of the outer terms. The middle term of each possible factorization is shown above in color. Since only $(3x + 5)(2x + 3)$ has a middle term of $19x$, we conclude that $6x^2 + 19x + 15 = (3x + 5)(2x + 3)$.

EXAMPLE 7

Factor $2y^2 + 7y - 4$.

Solution

Trial and error: we want the form $(ay + b)(cy + d)$.
Since $ac = 2$, the form is $(2y + b)(y + d)$.
Since $bd = -4$, b and d have different signs. The possible factorizations are:

$$\begin{array}{lll} (2y + 2)(y - 2) & (2y + 4)(y - 1) & (2y + 1)(y - 4) \\ (2y - 2)(y + 2) & (2y - 4)(y + 1) & (2y - 1)(y + 4) \end{array}$$

Only $(2y - 1)(y + 4)$ produces the middle term $7y$.
Thus, $2y^2 + 7y - 4 = (2y - 1)(y + 4)$.

EXAMPLE 8

Factor $x^2 + x + 4$.

Solution Trial and error: The possible factorizations using integer coefficients and constants are $(x + 4)(x + 1)$ and $(x + 2)(x + 2)$. But neither produces the correct middle term. Thus, $x^2 + x + 4$ cannot be factored using the strategies of this section.

When a polynomial cannot be factored into a product of nonconstant polynomials with coefficients and constants in some specified set, the polynomial is said to be **irreducible over the set.** Thus, $x^2 + x + 4$ is irreducible over the set of integers, as are $2x + 5$ and $x^2 - 3$.

EXAMPLE 9

Factor $15y^3 + 27y^2 - 6y$ completely.

Solution

$$\begin{aligned} 15y^3 + 27y^2 - 6y &= 3y(5y^2 + 9y - 2) \\ &= 3y(5y - 1)(y + 2) \end{aligned}$$

To begin, factor out a common monomial. Then factor the remaining binomial by trial and error.

EXAMPLE 10

Factor $3x^3 + 2x^2 - 27x - 18$ completely.

Solution

$$\begin{aligned} 3x^3 + 2x^2 - 27x - 18 &= (3x^3 + 2x^2) - (27x + 18) \\ &= x^2(3x + 2) - 9(3x + 2) \end{aligned}$$

To begin, factor by grouping.

$$\begin{aligned} &= (x^2 - 9)(3x + 2) \\ &= (x + 3)(x - 3)(3x + 2) \end{aligned}$$

Now factor the difference of squares.

EXAMPLE 11

Factor $10x^2z^2 - 105z^2 + 5xz^2$ completely.

Solution

$$10x^2z^2 - 105z^2 + 5xz^2 = 5z^2(2x^2 - 21 + x)$$

Factor out a common monomial.

$$\begin{aligned} &= 5z^2(2x^2 + x - 21) \\ &= 5z^2(2x + 7)(x - 3) \end{aligned}$$

The terms of the trinomial have been reordered; now factor by trial and error.

EXERCISE SET 1.6

In Exercises 1–6, find the greatest common factor of the given monomials.

1. $5a^2b, 10ab, 15ab^2$
2. $6uv^2, 12u^2v, 18uv^2$
3. $4x^4y^3, 6x^2y^5, 8x^3y^4$
4. $10m^4n^5, 20m^5n^2, 35m^3n^4$
5. $30pqr, 60p^2, 75q^2, 30r^2$
6. $18x^2y^2z^3, 24xy^2, 12x^2z^2, 78xyz$

In Exercises 7–74, factor completely. If a polynomial cannot be factored, then state this fact.

7. $5m - m^2$
8. $3m^2 - 15m$
9. $4v^3 - 5v^2 - 6v$
10. $3v^3 - 7v^2 - 6v$
11. $x^2 - 100$
12. $x^2 - 64$
13. $4t^2 - 0.25$
14. $0.16t^2 - 49$
15. $p^2 - q^2$
16. $q^2 - 4p^2$
17. $4m^2 + 9n^2$
18. $25m^2 + 36n^2$
19. $8v^2 - 18u^2$
20. $18v^2 - 50u^2$
21. $s^3 - 8$
22. $s^3 - 27$
23. $1 - 64a^3$
24. $1 - 125a^3$
25. $8x^3 + 27y^3$
26. $64x^3 + 27y^3$
27. $0.001u^3v + v^4$
28. $v^2u^3 + 0.008v^5$
29. $m^2 + 14m + 49$
30. $m^2 + 10m + 25$
31. $z^2 - 14z + 49$
32. $z^2 - 10z + 25$
33. $4b^2 + 12b + 9$
34. $16b^2 + 40b + 25$
35. $3s^2 - 18s + 27$
36. $2s^2 - 8s + 8$
37. $2x^2 + xy - 2xy - y^2$
38. $3x^2 - xy + 3xy - y^2$
39. $p^3 - 2p^2 + p - 2$
40. $p^3 - 3p^2 + 2p - 6$
41. $t^2 + 7t + 12$
42. $t^2 + 7t + 10$
43. $a^5 - 0.027a^2b^3$
44. $b^4 - 125a^3b$
45. $8x^3 - 27y^3$
46. $64x^3 - 27y^3$
47. $y^2 + 2y - 15$
48. $y^2 + 5y - 14$
49. $s^2 - 2s + 2$
50. $s^2 + s + 1$
51. $p^2 - p - 6$
52. $p^2 - p - 12$
53. $2d^2 - 5d - 3$
54. $4d^2 - 7d + 3$
55. $m^2 + 15m + 36$
56. $m^2 + 16m + 28$
57. $4v^2 + 4v - 3$
58. $6v^2 + v - 2$
59. $15y^2 - 14y - 8$
60. $12y^2 - 8y - 15$
61. $6a^2 - 5ab - 4b^2$
62. $4a^2 - 4ab - 35b^2$
63. $t^4 - 8t^2 - 9$
64. $t^4 - 3t^2 - 4$
65. $n^4 - 5n^3 - 14n^2$
66. $n^5 - 2n^4 - 15n^3$
67. $18z^3 - 12z^2 + 2z$
68. $12z^4 - 12z^3 + 3z^2$
69. $64u^6 - v^6$
70. $8u^6 - 125v^3$
71. $a^2 - (4 - 4b + b^2)$
72. $a^2 - (9 - 6b + b^2)$
73. $16d^2 - s^4$
74. $9s^4 - d^2$

Superset

In Exercises 75–90, factor completely. Assume *all variable exponents represent positive integers.*

75. $x^n - 8x^{n+1}$
76. $5y^m + y^{2m}$
77. $x^{2m} + 2x^my^m + y^{2m}$
78. $y^{2n} - 2y^nz^n + z^{2n}$
79. $x^{2r} - 9y^{2s}$
80. $4x^{2m} - y^{2n}$
81. $36z^{2t} - 12z^ty^s + y^{2s}$
82. $y^{2r} - 10y^rx^m + 25x^{2m}$
83. $x^{2n+1} - xy^{2n}$
84. $y^{2n+3} - yx^{2n+2}$
85. $x^{3s} - y^{3t}$
86. $x^{4r} - y^{4t}$
87. $x^{4n+5} - xy^{4m+4}$
88. $x^{3m+7} - xy^{3m+6}$
89. $(a + b)^2 - 2(a + b)c + c^2$
90. $(a + b)^2 - 2(a^2 - b^2) + (a - b)^2$
91. The area of a square is given by the expression $x^2 + 8x + 16$. Describe the length of each side of the square in terms of x.
92. The area of a square is given by the expression $4x^2 - 12x + 9$. Describe the length of each side of the square in terms of x.
93. The area of the rectangle below is given by the expression $2c^2 - c - 15$. Describe the length of the rectangle in terms of c.

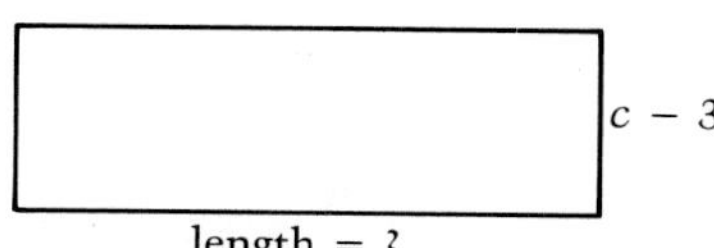

94. The area of the triangle below is given by the expression $t^2 - 0.5t - 3$. Express the length of the indicated base in terms of t.

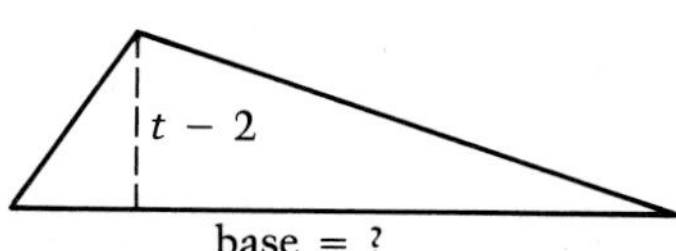

95. The volume of the box below is given by the expression $8a^3 - 27$. Find the area of the (shaded) top in terms of a.

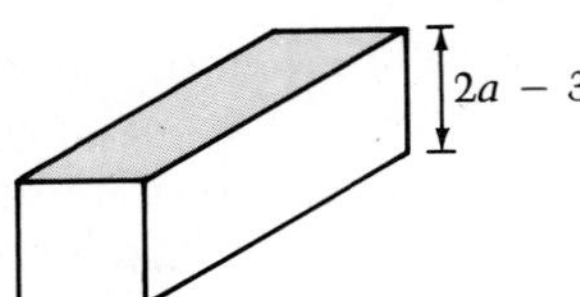

96. The area of the rectangle below is given by the expression $x^2 + 7x + 12$. Determine the area of the shaded region.

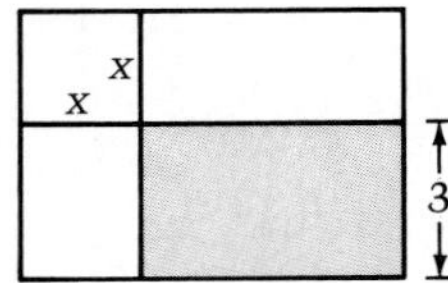

97. In the following rectangle, describe the area of the shaded region in terms of x.

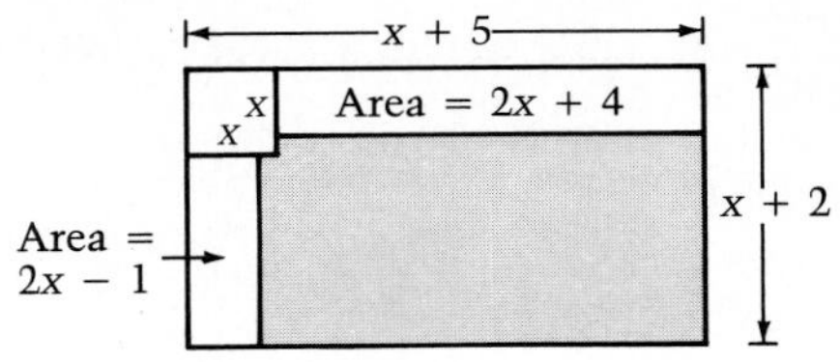

98. Draw a square with sides equal to $x + y$. Use the square to show that

$$(x + y)(x + y) \neq x^2 + y^2.$$

In Exercises 99–102, use one of the factoring patterns in this section to verify the given statement, without raising any number to a power.

99. $100^2 - 99^2 = 199$ **100.** $1000^2 - 999^2 = 1999$

101. $1000^2 - 990^2 = 19{,}990$ **102.** $100^2 - 90^2 = 1900$

Notice that $17 = 15 + 2$ and $13 = 15 - 2$. This means we can visualize the product $17 \cdot 13$ as the difference of two squares: $(15 + 2)(15 - 2) = 15^2 - 2^2$. Thus we can write $17 \cdot 13 = 15^2 - 2^2 = 225 - 4 = 221$. In Exercises 103–106, use this method to compute the value of the given product.

103. $15 \cdot 11$ **104.** $13 \cdot 9$ **105.** $21 \cdot 19$ **106.** $31 \cdot 29$

1.7
Radicals and Rational Exponents

The equation $3^4 = 81$ certainly tells us something about 81, namely, that "81 is 3 to the fourth power." But what does it tell us about 3? The equation says that 3 is a "fourth root of 81," and we write this in radical notation as

$$3 = \sqrt[4]{81}.$$

For $\sqrt[n]{b}$, n is the **index,** b is the **radicand,** and $\sqrt{}$ is the **radical sign.**

In general, if n is a positive integer, and a and b are real numbers, then the equation

$$a^n = b$$

tells us that a is an ***n*th root of *b*.** We indicate this by writing

$$a = \sqrt[n]{b}.$$

When the index is 2, it is usually not written. Thus $\sqrt{7}$ means $\sqrt[2]{7}$.

When the index n is even, a positive number b will have two real nth roots, one positive and the other negative. The positive root is called the *principal nth root*. For example, since $5^2 = 25$ and $(-5)^2 = 25$, both 5 and -5 are square roots of 25. But 5, the positive of the two, is called the principal square root of 25. **Thus, when n is even, the symbol $\sqrt[n]{\ }$ denotes the principal (positive) nth root.**

When the index n is odd, there is no ambiguity: a real number has only one real nth root, so the symbol $\sqrt[n]{\ }$ will denote that value. For example, $\sqrt[3]{-1000} = -10$, because -10 is the (only) real number that raised to the third power is -1000.

Real Number Values of $\sqrt[n]{b}$

	n is even	n is odd
$b = 0$	$\sqrt[n]{0} = 0$	$\sqrt[n]{0} = 0$
$b > 0$	Two real nth roots, but $\sqrt[n]{b}$ denotes only the positive one.	One real nth root.
$b < 0$	$\sqrt[n]{b}$ is not a real number.	One real nth root.

For example,

$$\sqrt{9} = 3 \quad (\text{not } \pm 3)$$
$$\sqrt[3]{8} = 2$$
$$\sqrt[3]{-8} = -2$$
$$\sqrt[4]{-8} \text{ is not a real number.}$$

EXAMPLE 1

Evaluate, if possible: (a) $\sqrt[5]{-32}$ (b) $\sqrt[3]{\frac{27}{8}}$ (c) $\sqrt{0}$ (d) $\sqrt{16}$ (e) $\sqrt{-16}$ (f) $-\sqrt[3]{-8}$ (g) The principal fourth root of $\frac{16}{81}$.

Solution

(a) $\sqrt[5]{-32} = -2$, since $(-2)^5 = -32$.

(b) $\sqrt[3]{\frac{27}{8}} = \frac{3}{2}$, since $\left(\frac{3}{2}\right)^3 = \frac{27}{8}$.

(c) $\sqrt{0} = 0$.

(d) $\sqrt{16} = 4$, since 4 is positive and $4^2 = 16$.

(e) $\sqrt{-16}$ is not a real number.

(f) $-\sqrt[3]{-8} = -(-2) = 2$

(g) The principal fourth root of $\frac{16}{81}$ is $\frac{2}{3}$, since $\frac{2}{3}$ is positive and $\left(\frac{2}{3}\right)^4 = \frac{16}{81}$.

Study the following statements:

$$\sqrt{3^2} = 3 \qquad \text{True, since } \sqrt{3^2} = \sqrt{9} = 3.$$

$$\sqrt{(-3)^2} = 3 \qquad \text{True, since } \sqrt{(-3)^2} = \sqrt{9} = 3.$$

$$\sqrt[5]{3^5} = 3 \qquad \text{True, since } \sqrt[5]{3^5} = \sqrt[5]{243} = 3.$$

$$\sqrt[5]{(-3)^5} = -3 \qquad \text{True, since } \sqrt[5]{(-3)^5} = \sqrt[5]{-243} = -3.$$

The first two statements suggest that when n is even, $\sqrt[n]{a^n}$ is never negative, no matter what a is. Thus, we use $|a|$ to define $\sqrt[n]{a^n}$ for even n. The last two statements imply that when n is odd, the value of $\sqrt[n]{a^n}$ is just a.

Property 1 of Radicals

If a is any real number, then

$$\sqrt[n]{a^n} = |a|, \quad \text{when } n \text{ is an even positive integer;}$$

$$\sqrt[n]{a^n} = a, \quad \text{when } n \text{ is an odd positive integer.}$$

In the next example, we use this Property of Radicals in a variety of situations involving numbers and variables.

EXAMPLE 2

Simplify each of the following: (a) $\sqrt[6]{(-4)^6}$ (b) $\sqrt[3]{2^3}$ (c) $\sqrt[3]{(-2)^3}$
(d) $\sqrt[4]{x^4}$ (e) $\sqrt[5]{x^5}$

Solution

(a) $\sqrt[6]{(-4)^6} = |-4| = 4$ (b) $\sqrt[3]{2^3} = 2$ (c) $\sqrt[3]{(-2)^3} = -2$
(d) $\sqrt[4]{x^4} = |x|$ (e) $\sqrt[5]{x^5} = x$

From this point on we shall assume that the variables in radical expressions have been restricted so that the expressions represent real numbers. For example, when you see $\sqrt{y}$, you may assume that the radicand y is not negative.

Other Properties of Radicals

For all real numbers a and b and positive integers n and m such that the radical expressions represent real numbers, the following properties apply.

2. $\sqrt[n]{a^m} = (\sqrt[n]{a})^m$

3. $\sqrt[n]{a} \cdot \sqrt[n]{b} = \sqrt[n]{ab}$

4. $\dfrac{\sqrt[n]{a}}{\sqrt[n]{b}} = \sqrt[n]{\dfrac{a}{b}}, \quad b \neq 0$

5. $\sqrt[n]{\sqrt[m]{a}} = \sqrt[nm]{a}$

When simplifying a radical expression we generally have two goals in mind: the final radicand should contain no exponents greater than or equal to the index, and there should be no fraction under the radical sign or radical in the denominator. For example, $\sqrt[3]{4^5}$ and $2/\sqrt{5}$ are not simplified. The process of rewriting an expression such as $2/\sqrt{5}$ so that it contains no radical in the denominator is called **rationalizing the denominator.** To do this, we multiply numerator and denominator by the same radical expression, chosen in such a way that the denominator in the product will contain no radical.

For example, to rationalize the denominator in the expression $2/\sqrt{5}$, we multiply by 1, expressed as $\sqrt{5}/\sqrt{5}$:

$$\frac{2}{\sqrt{5}} \times \frac{\sqrt{5}}{\sqrt{5}} = \frac{2 \times \sqrt{5}}{\sqrt{5} \times \sqrt{5}} = \frac{2\sqrt{5}}{(\sqrt{5})^2} = \frac{2\sqrt{5}}{5}.$$

Further examples are shown in Example 3.

EXAMPLE 3

Simplify each of the following:

(a) $\sqrt[3]{48}$ (b) $\sqrt[3]{(125)^2}$ (c) $\sqrt{8x^2y^5}$ (d) $\sqrt[3]{\dfrac{7}{100}}$ (e) $\sqrt{\dfrac{3y}{2z^5}}$ (f) $\sqrt[3]{\sqrt{128}}$

Solution

(a) $\sqrt[3]{48} = \sqrt[3]{2^3 \cdot 6}$ The index is 3; we look for perfect cubes in the radicand.

$= \sqrt[3]{2^3} \cdot \sqrt[3]{6}$ Property 3 of Radicals is used.

$= 2\sqrt[3]{6}$ Property 1 of Radicals is used.

(b) $\sqrt[3]{(125)^2} = (\sqrt[3]{125})^2$ Property 2 of Radicals is used. Note that we could have
$= (\sqrt[3]{5^3})^2$ squared 125, then found the cube root, but Property 2 spared us that unnecessary computation.

$= 5^2$ Property 1 of Radicals is used.

$= 25$

(c) $\sqrt{8x^2y^5} = \sqrt{2^2 \cdot 2 \cdot x^2 \cdot (y^2)^2 \cdot y}$ The index is 2; we look for perfect squares in the radicand.

$= \sqrt{2^2} \cdot \sqrt{2} \cdot \sqrt{x^2} \cdot \sqrt{(y^2)^2} \cdot \sqrt{y}$ Property 3 of Radicals is used.

$= 2 \cdot \sqrt{2} \cdot |x| \cdot y^2 \cdot \sqrt{y}$ Property 1 of Radicals is used.

$= 2|x|y^2\sqrt{2y}$ Property 3 of Radicals is used.

(d) $\sqrt[3]{\dfrac{7}{100}} = \sqrt[3]{\dfrac{7}{100} \cdot \dfrac{10}{10}}$ The index is 3. To simplify, we want a perfect cube in the denominator. Multiply by $\dfrac{10}{10}$ under the radical sign to

$= \dfrac{\sqrt[3]{70}}{\sqrt[3]{10^3}}$ get a denominator of $1000 = 10^3$.

$= \dfrac{\sqrt[3]{70}}{10}$ Note that the denominator is now a rational number.

(e) $\sqrt{\dfrac{3y}{2z^5}} = \sqrt{\dfrac{3y}{2z^5} \cdot \dfrac{2z}{2z}}$ The index is 2. We want perfect squares in the denominator: $2z^5 \cdot 2z = 2^2 \cdot z^6 = 2^2 \cdot (z^3)^2$.

$= \sqrt{\dfrac{6 \cdot yz}{2^2 \cdot (z^3)^2}}$

$= \dfrac{\sqrt{6yz}}{2|z^3|}$ z could be negative.

(f) $\sqrt[3]{\sqrt{128}} = \sqrt[6]{128}$ Property 5 of Radicals is used; the indices are multiplied.

$= \sqrt[6]{2^6 \cdot 2}$ Since the index is 6, we look for perfect 6th powers in the radicand.

$= \sqrt[6]{2^6} \cdot \sqrt[6]{2}$ Property 3 of Radicals is used.

$= 2\sqrt[6]{2}$ Property 1 of Radicals is used.

Let us agree that two radical terms will be called **like radicals,** provided they have the same index and same radicand. ($5\sqrt[3]{x}$ and $-2\sqrt[3]{x}$ are like radicals; $6\sqrt{y}$ and $6\sqrt[3]{y}$ are not.) To simplify sums and differences of radical expressions, we use the Properties of Radicals to simplify individual terms first, and then combine like radicals.

EXAMPLE 4

Simplify each of the following.

(a) $4\sqrt[3]{16} - 5\sqrt[3]{2} + 4\sqrt[3]{6}$ (b) $\sqrt{50x^2y^3} - 3\sqrt{8y}$

Solution

(a) $4\sqrt[3]{16} - 5\sqrt[3]{2} + 4\sqrt[3]{6} = 4\sqrt[3]{2^3 \cdot 2} - 5\sqrt[3]{2} + 4\sqrt[3]{6}$ Simplify individual terms.

$= 4 \cdot 2\sqrt[3]{2} - 5\sqrt[3]{2} + 4\sqrt[3]{6}$

$= (8\sqrt[3]{2} - 5\sqrt[3]{2}) + 4\sqrt[3]{6}$ Like terms are combined.

$= 3\sqrt[3]{2} + 4\sqrt[3]{6}$

(b) $\sqrt{50x^2y^3} - 3\sqrt{8y} = \sqrt{5^2 \cdot 2 \cdot x^2 \cdot y^2 \cdot y} - 3\sqrt{2^2 \cdot 2y}$ y cannot be negative; otherwise $\sqrt{50x^2y^3}$ and $3\sqrt{8y}$ would not be defined. Thus, $|y|$ is not necessary.

$= 5|x|y\sqrt{2y} - 3 \cdot 2\sqrt{2y}$

$= (5|x|y - 6)\sqrt{2y}$

Careful! An expression such as $\sqrt{7} - \sqrt{2}$ cannot be simplified any further. Don't make the mistake of writing:

$$\sqrt{7} - \sqrt{2} = \sqrt{5}. \qquad \text{False.}$$

EXAMPLE 5

Multiply each of the following:

(a) $\sqrt{5x} \cdot \sqrt{10xy}$ (b) $(5 + \sqrt{6x})(5 - \sqrt{6x})$.

Solution

(a) $\sqrt{5x} \cdot \sqrt{10xy} = \sqrt{5^2 \cdot 2 \cdot x^2 \cdot y}$ Property 3 of Radicals is used.

$= 5x\sqrt{2y}$ x cannot be negative; thus $|x|$ is unnecessary.

(b) $(5 + \sqrt{6x})(5 - \sqrt{6x}) = 5^2 - (\sqrt{6x})^2$ $(a + b)(a - b) = a^2 - b^2$

$= 25 - 6x$

Rational Exponents

In Section 1.4 we discussed exponential expressions of the form a^n, where the exponent n could be any integer. We also reviewed the Properties of Integer Exponents. We now wish to examine exponential expressions such as $2^{1/3}$ and $3^{4/5}$, where the exponents are rational numbers. We want the same properties established for integer exponents to be valid for rational exponents. We will see that this requirement helps us to settle on a way of defining a rational exponent.

Consider the expression $2^{1/3}$. One property of exponents requires that $(a^n)^m = a^{nm}$. Applying this in raising $2^{1/3}$ to the third power, we have

$$(2^{1/3})^3 = 2^{(1/3)(3)} = 2^1 = 2$$

Since $2^{1/3}$ raised to the third power is 2, we conclude that $2^{1/3}$ is the cube root of 2. Thus $2^{1/3} = \sqrt[3]{2}$. This argument generalizes for any positive integer n, and any nonnegative real number a.

$$(a^{1/n})^n = a^{(1/n)(n)} = a^1 = a, \qquad \text{and thus}$$
$$a^{1/n} = \sqrt[n]{a}.$$

Now what does an expression like $3^{4/5}$ mean? Again, one of the properties of exponents would require that

$$3^{4/5} = 3^{(1/5)(4)} = (3^{1/5})^4$$

while our definition of $3^{1/5}$ allows us to write

$$(3^{1/5})^4 = \left(\sqrt[5]{3}\right)^4 = \sqrt[5]{3^4}. \quad \text{Property 2 of Radicals is used.}$$

Thus we conclude that $3^{4/5}$ and $\sqrt[5]{3^4}$ must mean the same thing. This argument generalizes for any positive integers n and m, and any nonnegative real number a:

$$a^{m/n} = a^{(1/n)(m)} = (a^{1/n})^m = (\sqrt[n]{a})^m = \sqrt[n]{a^m}.$$

Note how our work with roots and radicals has helped us to give meaning to rational exponents. We summarize our results in the following definition.

Definition of Rational Exponent

For any nonnegative real number a, and for any positive integers m and n,

$$a^{1/n} = \sqrt[n]{a} \quad \text{and}$$
$$a^{m/n} = \sqrt[n]{a^m} = \left(\sqrt[n]{a}\right)^m.$$

In addition,

$$a^{-m/n} = \frac{1}{a^{m/n}}, \quad \text{provided } a \neq 0.$$

EXAMPLE 6

Evaluate each of the following: (a) $4^{3/2}$ (b) $\left(\dfrac{8}{27}\right)^{5/3}$ (c) $8^{-2/3}$

(d) $7^{4/5} \cdot 7^{2/5}$ (e) $3^{2/3} \cdot 3^{-2/3}$ (f) $\sqrt[5]{x} \cdot \sqrt{x}$

Solution

(a) $4^{3/2} = \left(\sqrt{4}\right)^3 = 2^3 = 8$

(b) $\left(\frac{8}{27}\right)^{5/3} = \left(\sqrt[3]{\frac{8}{27}}\right)^5 = \left(\frac{2}{3}\right)^5 = \frac{32}{243}$

(c) $8^{-2/3} = \frac{1}{8^{2/3}} = \frac{1}{(\sqrt[3]{8})^2} = \frac{1}{2^2} = \frac{1}{4}$

(d) $7^{4/5} \cdot 7^{2/5} = 7^{4/5+2/5} = 7^{6/5} = \sqrt[5]{7^6} = \sqrt[5]{7^5 \cdot 7} = 7\sqrt[5]{7}$

(e) $3^{2/3} \cdot 3^{-2/3} = 3^{2/3+(-2/3)} = 3^0 = 1$

(f) $\sqrt[5]{x} \cdot \sqrt{x} = x^{1/5} \cdot x^{1/2}$

$= x^{1/5+1/2}$

$= x^{7/10} = \sqrt[10]{x^7}$

Since the bases are equal, we can use Property 1 of Exponents.

EXAMPLE 7

Write each of the following expressions in the form $cx^r y^q$, where c is a real number, and r and q are rational numbers.

(a) $(3x^{-1/5}y^{1/4})^2$ (b) $\frac{5x^{1/2}y^{3/4}}{2x^{-1/3}y^{2/3}}$

Solution

(a) $(3x^{-1/5}y^{1/4})^2 = (3)^2(x^{-1/5})^2(y^{1/4})^2$ Property 2 of Exponents is used.

$= 9x^{-2/5}y^{1/2}$ Property 3 of Exponents is used.

(b) $\frac{5x^{1/2}y^{3/4}}{2x^{-1/3}y^{2/3}} = \frac{5}{2}x^{1/2}x^{1/3}y^{3/4}y^{-2/3}$ $\frac{1}{a^r} = a^{-r}$

$= \frac{5}{2}x^{(1/2+1/3)}y^{(3/4-2/3)}$ Property 4 of Exponents is used.

$= \frac{5}{2}x^{5/6}y^{1/12}$

EXERCISE SET 1.7

In Exercises 1–12, evaluate the expression. If the expression does not represent a real number, then state this fact.

1. $\sqrt[3]{1000}$ **2.** $\sqrt[4]{16}$ **3.** $\sqrt{0.0016}$

4. $\sqrt[3]{0.008}$ **5.** $-\sqrt{\frac{36}{49}}$ **6.** $-\sqrt{\frac{81}{25}}$

7. $-\sqrt[3]{-125}$ **8.** $\sqrt[3]{-64}$ **9.** $\sqrt[4]{-81}$

10. $\sqrt{-1}$ **11.** $\sqrt[3]{-0.027}$ **12.** $\sqrt[3]{-0.125}$

In Exercises 13–30, simplify the expression. Remember, "simplifying" includes rationalizing the denominator when appropriate.

13. $\sqrt[3]{40}$ **14.** $\sqrt[5]{96}$ **15.** $-\sqrt{27}$

16. $-\sqrt{50}$ **17.** $\sqrt{\frac{9}{10}}$ **18.** $\sqrt{\frac{4}{7}}$

19. $\sqrt[3]{\frac{-25}{9}}$
20. $\sqrt[3]{\frac{-9}{16}}$
21. $\frac{8}{\sqrt[3]{2}}$
22. $\frac{8}{\sqrt[3]{4}}$
23. $\sqrt[3]{\sqrt{729}}$
24. $\sqrt{\sqrt{243}}$
25. $\sqrt[3]{8x^4}$
26. $\sqrt[3]{27y^4}$
27. $\sqrt{\frac{x}{y}}$
28. $\sqrt[4]{\frac{y}{x^3}}$
29. $\sqrt[3]{\frac{27w^3}{2z}}$
30. $\sqrt[3]{\frac{8z^3}{5w}}$

In Exercises 31–46, perform the indicated operations and simplify.

31. $\sqrt{12} - \sqrt{3}$
32. $8\sqrt{48} - \sqrt{27}$
33. $2\sqrt{5} - \frac{5}{2}\sqrt{45}$
34. $\sqrt{2} + \frac{1}{2}\sqrt{18}$
35. $\sqrt{8} \cdot \sqrt{10}$
36. $\sqrt{6} \cdot \sqrt{12}$
37. $\sqrt[3]{5} \cdot \sqrt[3]{150}$
38. $\sqrt[3]{21} \cdot \sqrt[3]{18}$
39. $\frac{\sqrt{12}}{\sqrt{18}}$
40. $\frac{\sqrt{10}}{\sqrt{20}}$
41. $\sqrt{3}(\sqrt{6} - 2)$
42. $\sqrt{2}(5 - \sqrt{6})$
43. $(2 + \sqrt{5})(2 - \sqrt{5})$
44. $(3 - \sqrt{7})(3 + \sqrt{7})$
45. $(\sqrt{2} + \sqrt{7})^2$
46. $(\sqrt{2} + \sqrt{3})^2$

In Exercises 47–54, perform the indicated operations and simplify. *Assume that all variables represent positive real numbers.*

47. $\sqrt[3]{6pq^6} \cdot \sqrt[3]{12p^4}$
48. $\sqrt[3]{36p^5q} \cdot \sqrt[3]{144pq^2}$
49. $\sqrt{8ab^5} \cdot \sqrt{10a^3b}$
50. $\sqrt{8a^3b} \cdot \sqrt{6b^4a}$
51. $(2 + \sqrt{x})^2$
52. $(3 - \sqrt{x})^2$
53. $(3\sqrt{x} + \sqrt{y})(3\sqrt{x} - \sqrt{y})$
54. $(\sqrt{x} - 2\sqrt{y})(\sqrt{x} + 2\sqrt{y})$

In Exercises 55–66, evaluate the expression.

55. $100^{2/3}$
56. $25^{3/2}$
57. $\left(\frac{25}{36}\right)^{-3/2}$
58. $\left(\frac{9}{16}\right)^{-3/2}$
59. $3^{2/3} \cdot 3^{5/3}$
60. $9^{3/5} \cdot 9^{4/5}$
61. $2^{4/5} \cdot 4^{3/5}$
62. $3^{5/2} \cdot 27^{-1/2}$
63. $\sqrt[3]{25\sqrt[4]{125}}$
64. $\sqrt[4]{27\sqrt[3]{9}}$
65. $(5\sqrt[3]{4})^2$
66. $(2\sqrt[3]{9})^2$

In Exercises 67–84, simplify the expression. Write your answer in the form cx^ry^q where c is a real number, and r and q are rational numbers. *Assume that x and y represent positive real numbers.*

67. $(2x^{1/2})(5x^{1/3})$
68. $(3x^{1/4})(4x^{1/5})$
69. $(10xy^{1/2})(-2x^{1/3}y^{3/2})$
70. $(6x^{1/3}y^{1/4})(-7x^{2/3}y^{5/4})$
71. $(8x^{-1/2}y^{1/4})^3$
72. $(-2x^{3/5}y^8)^2$
73. $(16x^{1/2}y^{4/5})^{3/4}$
74. $(8x^{1/4}y^{3/8})^{1/3}$
75. $\frac{3x^{3/5}y^{1/3}}{6x^{1/10}y^{-5/6}}$
76. $\frac{14x^{3/4}y^{1/2}}{18x^{-1/2}y^{3/8}}$
77. $\left(\frac{2}{3x^2}\right)^{1/3}$
78. $\left(\frac{5}{2x^3}\right)^{1/4}$
79. $\sqrt[4]{\sqrt{x}}$
80. $\sqrt{\sqrt[3]{y}}$
81. $\sqrt[3]{x^2\sqrt[3]{y}}$
82. $\sqrt{y^3\sqrt[3]{x}}$
83. $\sqrt[3]{8y\sqrt[4]{xy^3}}$
84. $\sqrt[4]{16x\sqrt[3]{xy^2}}$

Superset

In Exercises 85–90 determine which expression in the given pair represents the greater value.

85. $1 - 0.02, \sqrt{1 - 0.02}$
86. $1 + 0.02, \sqrt{1 + 0.02}$
87. $\sqrt[8]{900}, \sqrt[9]{800}$
88. $\sqrt[3]{16.82}, \sqrt{7}$
89. $\sqrt{0.8}, (0.8)^2$
90. $\sqrt[8]{8}, \sqrt[4]{4}$

In Exercises 91–110 specify whether the statement is true or false. If false, illustrate this with an example.

91. $\sqrt{x^2} = x$
92. $\sqrt[3]{x^3} = x$
93. $\sqrt{x + y} = \sqrt{x} + \sqrt{y}$
94. $\sqrt{x^2 + y^2} = x + y$
95. $\frac{1}{\sqrt{x} + \sqrt{y}} = \frac{1}{\sqrt{x}} + \frac{1}{\sqrt{y}}$
96. $\frac{1}{\sqrt{xy}} = \frac{1}{\sqrt{x}} \cdot \frac{1}{\sqrt{y}}$
97. $x^{1/n} = \frac{1}{x^n}$
98. $\sqrt[3]{x} \cdot \sqrt{x} = \sqrt[6]{x}$
99. $\sqrt{x^2 - 4} = x - 2$
100. $\sqrt{x^2 + 2x + 1} = x + 1$
101. How does doubling a number affect its square root?
102. How does tripling a number affect its square root?
103. What would you do to a number in order to double its cube root?
104. What would you do to a number in order to triple its cube root?
105. If a number is increased ten-fold, is its square root more than doubled? More than tripled? More than quadrupled?

106. Verify that $\sqrt{8 + 2\sqrt{15}} = \sqrt{5} + \sqrt{3}$. (*Hint:* square both sides.)

107. Is $\sqrt{5 - 2\sqrt{6}} = \sqrt{2} - \sqrt{3}$? Explain.

108. Is $\sqrt{2} \cdot \sqrt{10 - 2\sqrt{21}} = \sqrt{14} - \sqrt{6}$? Explain.

109. Verify that $\sqrt{\pi} \cdot \sqrt{3 + 2\sqrt{2}} = \sqrt{\pi} + \sqrt{2\pi}$.

110. Determine the area of the shaded region in the figure below, given that figure *ABCD* is a square with area 2 square feet, and figure *EFGA* is a square with area 3 square feet. Explain your solution carefully.

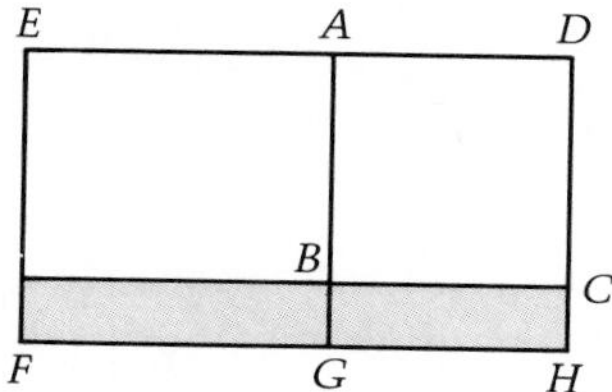

111. Using a calculator, determine $\sqrt{2}$. Take the result and press the square root key several more times. What pattern do you notice in the displayed values?

112. Perform the same procedure described in Exercise 111 with the value $\sqrt{3}$. Do you notice the same pattern?

113. Using a calculator, determine 2^2. Take the result and press the x^2 key several more times. Are the displayed values getting closer and closer to any specific value?

114. Take successive square roots of $\frac{1}{2}$ and successive squares of $\frac{1}{2}$. Do you notice any patterns?

1.8
Fractional Expressions

In Section 1.5 we reviewed sums, differences, products, and powers of polynomials. We now turn our attention to quotients of polynomials, which are commonly referred to as **rational expressions.** Examples of rational expressions are

$$\frac{14}{15}, \quad \frac{3xy^2 - 6x^3}{xy^4}, \quad \text{and} \quad \frac{8x^2 - 3x + 40}{x - 1}.$$

The arithmetic of rational expressions is patterned precisely after the arithmetic of fractions.

To begin our work with rational expressions, recall the basic principle for reducing fractions: for any real number a, and any nonzero real numbers b and c,

$$\frac{ac}{bc} = \frac{a}{b} \cdot \frac{c}{c} = \frac{a}{b}$$

The number c is called a **common factor.** If c is the greatest common factor of the numerator and denominator, then the reduced form $\frac{a}{b}$ is said to be in

lowest terms. To reduce a rational expression, we begin by factoring numerator and denominator, and then simply cancel any pair of common factors whose ratio equals 1.

$$\frac{x^2 + 4x + 3}{x^2 - 9} = \underbrace{\frac{(x + 1)\cancel{(x + 3)}}{(x - 3)\cancel{(x + 3)}}}_{\text{factor and cancel}} = \frac{(x + 1)}{(x - 3)}$$

It is important to note that the expression on the left is not defined when $x = 3$ or when $x = -3$, since both of these x-values produce a zero denominator. Thus to be precise we should write

$$\frac{x^2 + 4x + 3}{x^2 - 9} = \frac{x + 1}{x - 3}, \quad \text{provided } x \neq 3 \text{ and } x \neq -3.$$

In the examples and exercises that follow, we assume that numbers producing zero denominators have been excluded as values of the variables.

EXAMPLE 1

Simplify $\dfrac{4 - 2x}{x^3 - 8}$.

Solution

$$\frac{4 - 2x}{x^3 - 8} = \frac{(-2)(-2 + x)}{(x - 2)(x^2 + 2x + 4)}$$ Numerator and denominator are factored completely.

$$= \frac{(-2)\cancel{(x - 2)}}{\cancel{(x - 2)}(x^2 + 2x + 4)}$$ Observe that $-2 + x = x - 2$, and cancel common factors.

$$= \frac{-2}{x^2 + 2x + 4}$$ The given expression is now simplified, i.e., reduced to lowest terms.

We review the arithmetic of rational expressions by considering multiplication and division first. Recall the principles for multiplying and dividing fractions.

$$\frac{a}{b} \cdot \frac{c}{d} = \frac{ac}{bd} \qquad (b \text{ and } d \text{ are nonzero})$$

$$\frac{a}{b} \div \frac{c}{d} = \frac{a}{b} \cdot \frac{d}{c} = \frac{ad}{bc} \qquad (b, c, \text{ and } d \text{ are nonzero})$$

These principles are also true when a, b, c, and d are replaced with polynomials.

EXAMPLE 2

Perform the indicated operations and then simplify.

(a) $\dfrac{x}{2x+5} \cdot \dfrac{2x^2+3x-5}{x^2+3x}$ (b) $\dfrac{xy-3x}{x+2} \div \dfrac{y^2-9}{3x^2-x-14}$

Solution

(a) $\dfrac{x}{2x+5} \cdot \dfrac{2x^2+3x-5}{x^2+3x} = \dfrac{\cancel{x} \cdot \cancel{(2x+5)}(x-1)}{\cancel{(2x+5)} \cdot \cancel{x} \cdot (x+3)} = \dfrac{x-1}{x+3}$

(b) $\dfrac{xy-3x}{x+2} \div \dfrac{y^2-9}{3x^2-x-14} = \dfrac{xy-3x}{x+2} \cdot \dfrac{3x^2-x-14}{y^2-9}$ The second expression is inverted.

$$= \frac{x\cancel{(y-3)}(3x-7)\cancel{(x+2)}}{\cancel{(x+2)}(y+3)\cancel{(y-3)}}$$

$$= \frac{x(3x-7)}{y+3}$$

The crucial step to remember when adding or subtracting fractions is that you must first make the denominators the same. Recall the basic principles:

$$\frac{a}{c} + \frac{b}{c} = \frac{a+b}{c} \quad (c \neq 0),$$

$$\frac{a}{c} - \frac{b}{c} = \frac{a-b}{c} \quad (c \neq 0).$$

We call c a **common denominator.**

Generally it is easiest to add or subtract rational expressions when they have been written with a **least common denominator (LCD).** The LCD is that polynomial of least degree which is a multiple of each denominator. To find the LCD, we begin by factoring each denominator completely, and then construct the product of the greatest power of each factor present. The technique is illustrated in the next example.

EXAMPLE 3

Combine $\frac{2}{x} - \frac{3}{x-1} + \frac{x+3}{x^2-1}$.

Solution

x:	x
$x-1$:	$x-1$
x^2-1:	$x-1 \quad x+1$
LCD :	$x \cdot x-1 \cdot x+1$

Begin by factoring each of the denominators. The LCD is the product of the greatest power of each factor present in any of the factorizations.

$$\frac{2}{x} - \frac{3}{x-1} + \frac{x+3}{x^2-1}$$

Each term is multiplied by a form of 1 so that the denominator is the LCD.

$$= \frac{2}{x} \cdot \left[\frac{(x-1)(x+1)}{(x-1)(x+1)}\right] - \frac{3}{x-1}\left[\frac{x(x+1)}{x(x+1)}\right] + \frac{x+3}{x^2-1}\left[\frac{x}{x}\right]$$

$$= \frac{2(x^2-1)}{x(x^2-1)} - \frac{3x^2+3x}{x(x^2-1)} + \frac{x^2+3x}{x(x^2-1)}$$

$$= \frac{2x^2 - 2 - 3x^2 - 3x + x^2 + 3x}{x(x^2-1)}$$

$$= \frac{-2}{x(x^2-1)}$$

Up to this point, this section has been concerned with rational expressions, that is, quotients of polynomials. A **radical fractional expression** contains radicals in its numerator or denominator. Simplifying these expressions requires that denominators be rationalized. To do this we sometimes use a conjugate. Expressions such as $\sqrt{x} + \sqrt{2}$ and $\sqrt{x} - \sqrt{2}$ are called **conjugates.** The product of conjugates will not contain any radicals, and thus produces a rational number. The technique is illustrated in the following example.

EXAMPLE 4

Simplify $\frac{1}{\sqrt{x}+\sqrt{2}}$.

Solution

$$\frac{1}{\sqrt{x}+\sqrt{2}} = \frac{1}{\sqrt{x}+\sqrt{2}} \cdot \frac{\sqrt{x}-\sqrt{2}}{\sqrt{x}-\sqrt{2}} = \frac{\sqrt{x}-\sqrt{2}}{(\sqrt{x})^2-(\sqrt{2})^2} = \frac{\sqrt{x}-\sqrt{2}}{x-2}$$

The last type of fractional expression to be considered is called a **complex fractional expression.** In such an expression, the numerator or denominator itself contains a fraction.

EXAMPLE 5

Simplify the expression $\dfrac{1 - \dfrac{2}{x}}{1 - \dfrac{3}{x} + \dfrac{2}{x^2}}$.

Solution

$$\frac{1 - \dfrac{2}{x}}{1 - \dfrac{3}{x} + \dfrac{2}{x^2}} = \frac{1\left(\dfrac{x}{x}\right) - \dfrac{2}{x}}{1\left(\dfrac{x^2}{x^2}\right) - \dfrac{3}{x}\left(\dfrac{x}{x}\right) + \dfrac{2}{x^2}}$$

Begin by combining terms in the numerator and in the denominator.

$$= \frac{\dfrac{x}{x} - \dfrac{2}{x}}{\dfrac{x^2}{x^2} - \dfrac{3x}{x^2} + \dfrac{2}{x^2}}$$

$$= \frac{\dfrac{x - 2}{x}}{\dfrac{x^2 - 3x + 2}{x^2}}$$

The problem is now a fraction divided by a fraction; we shall invert and multiply.

$$= \frac{x - 2}{x} \cdot \frac{x^2}{x^2 - 3x + 2}$$

We now factor and cancel as before.

$$= \frac{\cancel{x - 2}}{\cancel{x}} \cdot \frac{\overset{x}{\cancel{x^2}}}{(\cancel{x - 2})(x - 1)}$$

$$= \frac{x}{x - 1}$$

There is an alternative approach that can be used in simplifying complex fractional expressions. We illustrate it using the problem of Example 5.

$$\frac{1-\frac{2}{x}}{1-\frac{3}{x}+\frac{2}{x^2}} = \frac{1-\frac{2}{x}}{1-\frac{3}{x}+\frac{2}{x^2}} \times \frac{x^2}{x^2}$$

Begin by finding the LCD of all rational expressions in the complex fraction. The LCD for $1, \frac{2}{x}, \frac{3}{x}$, and $\frac{2}{x^2}$ is x^2.

$$= \frac{(1)x^2 - \frac{2}{x}(x^2)}{1(x^2) - \frac{3}{x}(x^2) + \frac{2}{x^2}(x^2)}$$

The numerator and denominator are multiplied by the LCD, x^2.

$$= \frac{x^2 - 2x}{x^2 - 3x + 2}$$

Note that the fractions have been cleared from the numerator and the denominator. To simplify, factor and cancel.

$$= \frac{x(x-2)}{(x-2)(x-1)}$$

$$= \frac{x}{x-1}$$

EXERCISE SET 1.8

In Exercises 1–16, simplify the expression.

1. $\dfrac{(3-x)^2}{x-3}$
2. $\dfrac{(y-2)^2}{2-y}$
3. $\dfrac{2m^2-18}{m+3}$
4. $\dfrac{5m^2-45}{m-3}$
5. $\dfrac{v^3-4v}{v^2-v-6}$
6. $\dfrac{u^4-9u^2}{u^2+7u+12}$
7. $\dfrac{x^3-27}{x^2-9}$
8. $\dfrac{x^3+64}{x^2-16}$
9. $\dfrac{2a^2+8a+8}{3a^2+4a-4}$
10. $\dfrac{a^2+6a-27}{3a^2-27a+54}$
11. $\dfrac{2x+4-x^3-2x^2}{3x^2+12x+12}$
12. $\dfrac{5x^2+10x+5}{x^3+x^2-x-1}$
13. $\dfrac{-v^2+30v-225}{4v^2-58v-30}$
14. $\dfrac{6-v-v^2}{4v^2-10v-66}$
15. $(m^2n^2+2m^2n-24m^2)(n+6)^{-1}$
16. $(m^2n^2-2n^2m-15n^2)(m+3)^{-1}$

In Exercises 17–34, perform the indicated operation.

17. $\left(-\dfrac{5a}{b^2}\right)\left(\dfrac{3b}{a^3}\right)$
18. $\left(\dfrac{9a}{b^3}\right)\left(-\dfrac{2b^2}{a^3}\right)$
19. $\left(\dfrac{x-1}{x}\right)\left(\dfrac{x+1}{x}\right)$
20. $\left(\dfrac{3+y}{3-y}\right)\left(\dfrac{3+y}{y}\right)$
21. $\dfrac{5xyz^2}{18x^2}\left(\dfrac{yz}{3x}\right)^{-2}$
22. $\dfrac{7x^2yz}{12z^3}\left(\dfrac{xy}{2z}\right)^{-3}$
23. $(3m^2)(2n)^{-3}(8n)(2m)^{-2}$
24. $(4m^3)(3n)^{-2}(5n^2)(2m)^{-1}$
25. $\left(\dfrac{3v^2-12}{5v-15}\right)\left(\dfrac{2v^2-18}{v^2+5v+6}\right)$
26. $\left(\dfrac{v^2+10v+25}{v^2+7v+10}\right)\left(\dfrac{5-v}{2v^2-50}\right)$
27. $\dfrac{a}{a-b} \div \dfrac{b^2}{b-a}$
28. $\dfrac{a^2}{b-a} \div \dfrac{b}{a-b}$
29. $\dfrac{4m^2-4m-15}{4m+10} \div \dfrac{4m+6}{4m^2-25}$
30. $\dfrac{9m^2-18m-16}{6m+16} \div \dfrac{6m+4}{9m^2-64}$
31. $\dfrac{3}{v+2} - \dfrac{4}{v-2}$
32. $\dfrac{2}{v+3} - \dfrac{5}{v-3}$
33. $\dfrac{x^2-10}{x^2-2x-15} - \dfrac{5x}{2x^2-3x-35}$
34. $\dfrac{x+3}{2x^2+5x-12} - \dfrac{x-1}{2x^2+7x-4}$

In Exercises 35–64, simplify the expression.

35. $\dfrac{11}{6-\sqrt{3}}$

36. $\dfrac{18}{4+\sqrt{7}}$

37. $\dfrac{2-\sqrt{3}}{3+\sqrt{6}}$

38. $\dfrac{3+\sqrt{5}}{4-\sqrt{10}}$

39. $\dfrac{3}{\sqrt{2}+2\sqrt{5}}$

40. $\dfrac{2}{3\sqrt{7}-\sqrt{6}}$

41. $\dfrac{x}{\sqrt{x}+1}$

42. $\dfrac{y}{7-\sqrt{y}}$

43. $\dfrac{\sqrt{w}+3}{\sqrt{w}-2}$

44. $\dfrac{\sqrt{w}-4}{\sqrt{w}+3}$

45. $\dfrac{\sqrt{x}}{\sqrt{x}-\sqrt{y}}$

46. $\dfrac{\sqrt{y}}{\sqrt{x}+\sqrt{y}}$

47. $\dfrac{\sqrt{18}-\sqrt{t}}{\sqrt{2}-\sqrt{t}}$

48. $\dfrac{\sqrt{3}-\sqrt{t}}{2\sqrt{3}-\sqrt{t}}$

49. $\dfrac{3}{\sqrt{x+y}-2}$

50. $\dfrac{2x}{1-\sqrt{x+y}}$

51. $\dfrac{x}{\sqrt{x}-2\sqrt{y}}$

52. $\dfrac{x-2y}{2\sqrt{x}-\sqrt{y}}$

53. $(a+2+\sqrt{b})^{-1}$

54. $(\sqrt{a}+\sqrt{b})^{-2}$

55. $\dfrac{u-3u^{-2}}{u+2u^{-1}}$

56. $\dfrac{u-2u^{-2}}{u+5u^{-1}}$

57. $\dfrac{1-\dfrac{1}{v+1}}{1+\dfrac{1}{v-1}}$

58. $\dfrac{1+\dfrac{1}{v+1}}{1-\dfrac{1}{v-1}}$

59. $\dfrac{1-\dfrac{4}{x}-\dfrac{5}{x^2}}{1+\dfrac{4}{x}+\dfrac{3}{x^2}}$

60. $\dfrac{\dfrac{1}{x}+1-\dfrac{6}{x^2}}{\dfrac{2}{x}+1-\dfrac{3}{x^2}}$

61. $\dfrac{1+\dfrac{2}{x}}{\dfrac{2}{x}-\dfrac{1}{x+1}}$

62. $\dfrac{\dfrac{1}{x(x-4)}+\dfrac{1}{x}}{\dfrac{3}{x}-\dfrac{2}{x-1}}$

63. $\dfrac{\dfrac{1}{m+2}-\dfrac{m+1}{m}}{\dfrac{m}{m+2}+\dfrac{1}{m}}$

64. $\dfrac{\dfrac{n}{1+n}+\dfrac{1-n}{n}}{\dfrac{n}{1+n}-\dfrac{1-n}{n}}$

In Exercises 65–70, evaluate the expression to the nearest hundredth.

65. $\left(\dfrac{5.15-7.28}{3.01}\right)\div(6.37-2.89)$

66. $1\div\left(\dfrac{4.95-6.82}{9.15}\right)$

67. $\dfrac{\sqrt{20.5}-20.5}{\sqrt{10.5}-10.5}$

68. $\dfrac{\sqrt{1.05}-1.05}{\sqrt{0.95}-0.95}$

69. $\dfrac{1-\dfrac{1}{1+0.2}}{1+\dfrac{1}{1-0.2}}$

70. $1+\dfrac{1}{1+\dfrac{1}{1+0.03}}$

Superset

In Exercises 71–76, specify whether the statement is true or false. If false, illustrate this with an example.

71. $\dfrac{1}{x+y}=\dfrac{1}{x}+\dfrac{1}{y}$

72. $\dfrac{3}{4}+\dfrac{x}{y}=\dfrac{3+x}{4+y}$

73. $(x+y)^{-2}=x^{-2}+y^{-2}$

74. $\left(\dfrac{x}{y}+\dfrac{2}{3}\right)^{-1}=\dfrac{y}{x}+\dfrac{3}{2}$

75. $x^{-2}-y^{-2}=(x^{-1}-y^{-1})(x^{-1}+y^{-1})$

76. $(x\div y)\div z=x\div(y\div z)$

In Exercises 77–80, determine the area of the shaded region. Write your answer as a rational expression in x.

77.

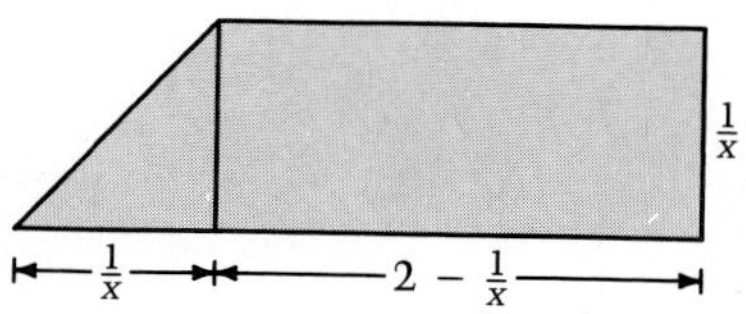

78.

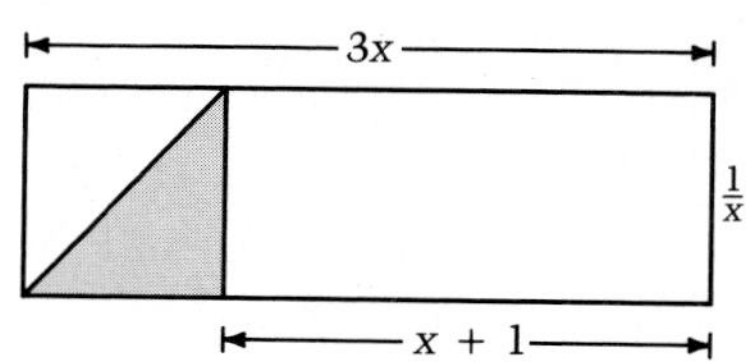

79.

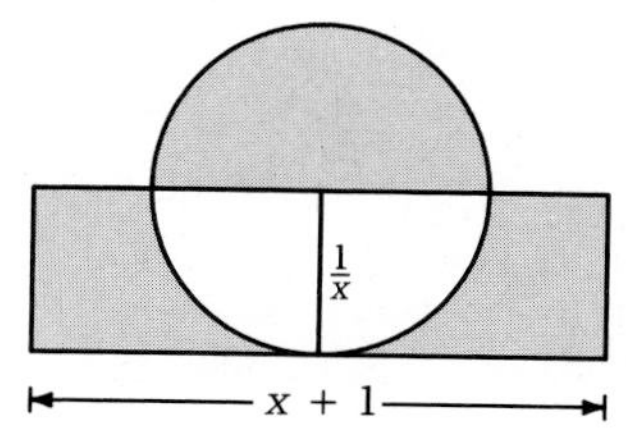

80.

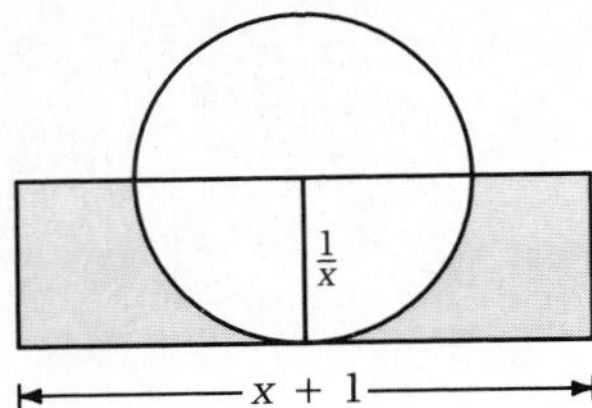

In calculus it is sometimes necessary to rewrite a quotient involving radicals so that the numerator is rational. In Exercises 81–84, rationalize the numerator.

81. $\dfrac{1+\sqrt{x}}{x}$ $\left(\textit{Hint:} \text{ multiply by } \dfrac{1-\sqrt{x}}{1-\sqrt{x}}.\right)$

82. $\dfrac{1-\sqrt{y}}{\sqrt{y}}$

83. $\dfrac{\sqrt{x+2}-\sqrt{x}}{2}$

84. $\dfrac{\sqrt{x+2}-\sqrt{x}}{\sqrt{x+2}+\sqrt{x}}$

The **weighted mean** of a set of values, $x_1, x_2, \ldots, x_m$, takes into account the number of times each value occurs, $n_1, n_2, \ldots, n_m$, respectively, and is given by the expression

$$\frac{x_1n_1 + x_2n_2 + \cdots + x_mn_m}{n_1 + n_2 + \cdots + n_m}$$

In Exercises 85–87, you will need to determine a weighted mean.

85. The seniors at a certain high school took the Advanced Placement Exam in Computer Science. Possible scores are 1, 2, 3, 4, and 5. The chart below shows how many students received each grade.

Grade	1	2	3	4	5
Students	8	10	24	19	5

Determine the average score of the students at that school. (Round to the nearest tenth.)

86. A person has invested \$5000 at 5.5%, \$12,000 at 7%, \$6000 at 9.5% and \$3000 at 12.5%. What is the average return (rate of profit) on these investments? (Round to the nearest tenth of a percent.)

87. A corporation has four levels of middle management positions. The mean salary is
\$21,500 for its 40 Level I managers,
\$26,800 for its 65 Level II managers,
\$33,000 for its 28 Level III managers, and
\$42,600 for its 15 Level IV managers.
What is the average salary paid to the corporation's middle management employees? (Round to the nearest dollar.)

88. Add up the first ten terms in the infinite sum $\frac{1}{2} + \frac{1}{4} + \frac{1}{8} + \frac{1}{16} + \frac{1}{32} + \cdots$ (the denominators are successive powers of 2). What pattern do you notice as you add more and more terms?

89. Add up the fifteen terms $1 + \frac{1}{2} + \frac{1}{3} + \frac{1}{4} + \cdots + \frac{1}{14} + \frac{1}{15}$. Do you notice any limiting pattern here, such as the one observed in Exercise 88?

Analyzing Data

The Case of the Generous Graders

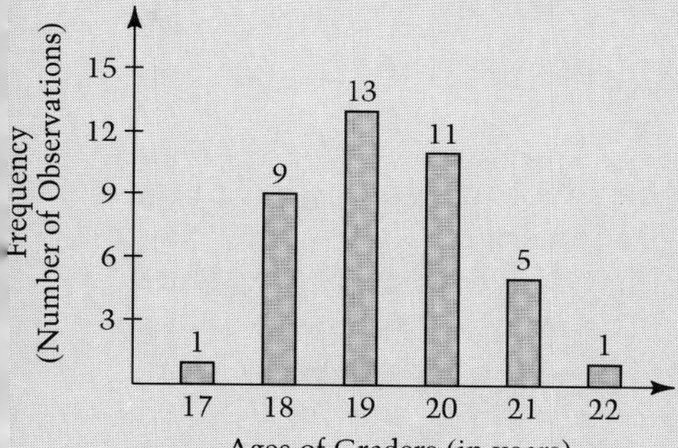

We would like now to consider a study that investigated grading practices in a calculus course at a large state university [Kelly, 1986]. A bit of explanation is necessary. For this particular course, all testing was conducted in a testing center, not in class. When a student completed a test, it would be graded immediately, by a specially trained undergraduate who was either a mathematics major or minor. The student would be present while the test was being graded. One question that the study addressed was whether grades received in this "face-to-face" grading situation would be different if the student had not been present during the grading process.

In addition to various background information on the exam graders themselves, the report presents the frequency histogram shown at the left. From this we see that one of the graders was 17 years old, nine were 18 years old, etc. When data are presented this way, you can determine the average value by computing a weighted mean. If we symbolize the values (in this case, ages) by the subscripted variables $x_1, x_2, x_3, \ldots, x_n$, and the corresponding frequencies by $f_1, f_2, f_3, \ldots, f_n$, then the average of the values (the weighted mean) is given by the expression

$$\frac{x_1 f_1 + x_2 f_2 + x_3 f_3 + \cdots + x_n f_n}{f_1 + f_2 + f_3 + \cdots + f_n}$$

$$= \frac{17(1) + 18(9) + 19(13) + 20(11) + 21(5) + 22(1)}{40}$$

$$\approx 19.3$$

By simply looking at the histogram, you might have guessed that the average would be somewhere around 19, since that is where the "visual center" of the histogram seems to be. Notice that the weighted mean takes into account the fact that some values occur more frequently than others and so should be "weighted" more heavily in the computation of the average.

Back to the study. Forty-one papers were randomly selected from those already graded, and, after being cleared of all grader markings, were photocopied and graded again, this time by different graders, who graded the exams without the students being present. The average test grade was almost four points lower if graded without the student present, a difference that was shown to be significant in the statistical tests used to analyze the data.

Source: Kelly, T.J. Effects of Field-Dependence/Independence and Gender on Patterns of Achievement and Grading in a First-Semester Calculus Course. *DAI.* Vol. 46A, 2953; Apr. 1986.

Chapter Review

1.1 Statements and Variables (pp. 1–6)

A *variable* is a symbol used to represent a number. (p. 3) An *expression* is a combination of numbers and variables (e.g., $3x^2 + 2yz$). An *equation* is a statement that two expressions have the same value. (p. 3)

1.2 The Real Numbers (pp. 7–13)

The set of *real numbers* is composed of *rational numbers* (e.g., $\frac{2}{3}$ and 8) and *irrational numbers* (e.g., $\sqrt{2}$ and π). The set of rational numbers is composed of *integers* and *noninteger fractions.* (p. 7) The set of nonnegative integers is sometimes referred to as *whole numbers.* (p. 7) The Properties of Real Numbers (p. 9) and the Order of Operations convention (p. 10) guide us in evaluating expressions involving real numbers.

1.3 Absolute Value, Distance, and the Real Number Line (pp. 13–18)

The *absolute value* of a nonzero real number a, denoted $|a|$, is the positive number in the set $\{a, -a\}$. The absolute value of zero is zero. (p. 13)

The set of real numbers can be placed in a one-to-one correspondence with the points on a line, called the *real line.* The number associated with a given point is the *coordinate* of the point. (p. 14) If points A and B have coordinates a and b respectively, then the *distance* between A and B, denoted $d(A, B)$, is equal to $|a - b|$. (p. 16)

1.4 The Integers as Exponents (pp. 18–27)

$$a^m \cdot a^n = a^{m+n}, \qquad (ab)^n = a^n b^n, \qquad (a^m)^n = a^{mn},$$

$$\frac{a^m}{a^n} = a^{m-n}, \qquad \left(\frac{a}{b}\right)^n = \frac{a^n}{b^n}.$$

In addition, if $a \neq 0$, then $a^0 = 1$ and $\dfrac{1}{a^n} = a^{-n}$. (p. 19)

A positive number is in *scientific notation* when it is written as a product $A \times 10^n$ where A is a number greater than or equal to 1, and less than 10, and n is an integer exponent. (p. 22) When a value is written in scientific notation $A \times 10^n$, the number of *significant digits* is the number of digits in A. (p. 23)

1.5 Polynomials: Addition, Subtraction, and Multiplication (pp. 27–34)

A *term,* or *monomial,* is either a constant or a product of constants and variables. Its *degree* is the sum of the exponents of its variables. The constant multiplier is called the *coefficient. Like terms* have the same variables with the same exponents. (p. 29)

A *polynomial* is a sum of terms, and can be simplified by combining like terms. (p. 29) A polynomial with two terms is called a *binomial*, and one with three terms, a *trinomial*. The FOIL method is used to find the product of two binomials. (p. 31)

Squares and cubes of polynomials are given by the following formulas: (p. 31)

$$(a + b)^2 = a^2 + 2ab + b^2$$
$$(a - b)^2 = a^2 - 2ab + b^2$$
$$(a + b)^3 = a^3 + 3a^2b + 3ab^2 + b^3$$
$$(a - b)^3 = a^3 - 3a^2b + 3ab^2 - b^3$$

1.6 Polynomials: Factoring (pp. 34–42)

To factor a polynomial, use one or more of the following strategies:

1. Factor out a common factor. (p. 35)
2. Observe a pattern: (p. 36)
$$A^2 + 2AB + B^2 = (A + B)^2$$
$$A^2 - 2AB + B^2 = (A - B)^2$$
$$A^2 - B^2 = (A + B)(A - B)$$
$$A^3 + B^3 = (A + B)(A^2 - AB + B^2)$$
$$A^3 - B^3 = (A - B)(A^2 + AB + B^2)$$
3. Grouping. (p. 38)
4. Trial and Error. (p. 38)

1.7 Radicals and Rational Exponents (pp. 42–51)

If n is even and $b > 0$, then $\sqrt[n]{b}$ refers to the *positive (principal) nth root*.

If n is even and $b < 0$, then $\sqrt[n]{b}$ is not a real number.

If n is odd, then $\sqrt[n]{b}$ refers to the one real nth root, regardless of the sign of b. (p. 43)

If a is nonnegative, the principal nth root of a can be written $a^{1/n}$. (p. 48)

For any nonnegative real number a, and for any positive integers m and n, $a^{1/n} = \sqrt[n]{a}$ and $a^{m/n} = \sqrt[n]{a^m} = \left(\sqrt[n]{a}\right)^m$. (p. 48)

1.8 Fractional Expressions (pp. 51–58)

A *rational expression* is a quotient of two polynomials. (p. 51)

The arithmetic of fractional expressions containing variables follows the same rules as the arithmetic of fractions. Reducing fractional expressions requires factoring numerator and denominator first. Adding fractional expressions requires common denominators. (p. 53)

Review Exercises

In Exercises 1–4, translate the given statement into an equation.

1. Three less than twice a certain number is fifty.
2. The average of two numbers is twice the smaller number.
3. One number is 50% larger than another.
4. The sum of three consecutive integers is fifteen. (Use only one variable.)

In Exercises 5 and 6, let p = the number of passenger jets, and c = the number of cargo jets, then write an equation relating p and c.

5. There are twice as many cargo jets as passenger jets.
6. There are ten less passenger jets than cargo jets.

In Exercises 7–10, classify the number as one or more of the following: whole number, integer, rational number, irrational number, real number.

7. $-\frac{3}{5}$
8. $\sqrt{18}$
9. 2.11
10. 7

In Exercises 11–12, evaluate the expression.

11. $3 - 2(6 - 4)$
12. $\dfrac{2 - \frac{1}{4}}{1 + \frac{1}{4}}$

In Exercises 13–16, evaluate the expression given that $x = 2, y = -3$, and $z = -1$.

13. $-y + 2z$
14. $xy - 3xz$
15. $\dfrac{y - x}{z(x - 5)}$
16. $\dfrac{3x + 2y}{\frac{1}{z} + \frac{1}{x}}$

In Exercises 17–18, evaluate the expression.

17. $|-3| - |7|$
18. $\dfrac{-|4 + (-3)|}{|-7|\,|-10|}$

In Exercises 19–20, translate the sentence into an inequality.

19. The average of two numbers is less than 10.
20. The product of two numbers is positive.

In Exercises 21–24, graph the set of numbers on the real line.

21. $\left\{-4, 0, \frac{3}{2}\right\}$
22. $\{-\sqrt{2}, \sqrt{2}\}$
23. The set of integers between $-\pi$ and π.
24. The set of real numbers greater than or equal to -3.

In Exercises 25–26, use the graph below to determine the distance.

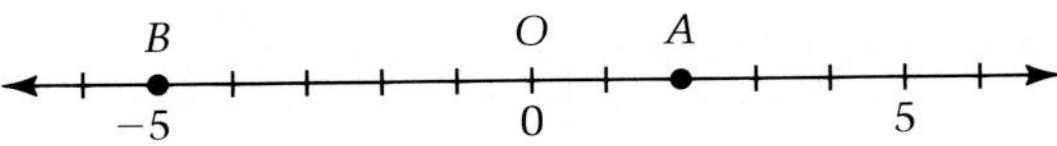

25. $d(O, B)$
26. $d(B, A)$

In Exercises 27–30, evaluate the given expression.

27. -2^3
28. 5^{-2}
29. $\left(\frac{3}{4}\right)^{-3}$
30. $\dfrac{3}{4^{-2}}$

In Exercises 31–32, evaluate the expression if $x = -2$ and $y = 3$.

31. $3x^2 - 2x + y$
32. $x(y - 5)^2$

In Exercises 33–34, rewrite the expression with a positive exponent.

33. $\left(\frac{b}{c}\right)^{-2}$
34. $\dfrac{1}{c^{-1}}$

In Exercises 35–38, simplify the expression so that x, y, and z appear at most once, and so that there are no negative exponents.

35. $(2x^3)(xy)^2(6x^{-1})$
36. $(4xy^2z^3)^3$
37. $\left(\frac{x^2}{6y}\right)^{-2}$
38. $\dfrac{(3xyz)(xy^{-1}z^2)^2}{2x^{-4}y}$

In Exercises 39–50, perform the indicated operations.

39. $(3x^2 + 4x - 7) + (-6x^2 - 5x + 1)$
40. $(8x^2 - x^3) - (1 + x - x^2 + x^3)$
41. $(y^3 - 16y^2 + 1) - (3 + y + 8y^2)$
42. $(3y^3 - 8y - 1) + (-4y^3 + 6y^2 + 3)$
43. $(2x)(x^2 - 6)$
44. $(-3x^2)(1 - x)$
45. $(x - 1)(7x + 3)$
46. $(2x - 5)(x + 3)$
47. $(x + 4)(3x^2 - 7x + 1)$
48. $(x - 2)(4x^3 + 3x^2 - 2)$
49. $(5n - 1)^2$
50. $(3 - 2n)^2$

In Exercises 51–64 factor completely. If a polynomial cannot be factored, state this fact.

51. $2xy^2 + 6x^2yz - 4yz^2$
52. $8m^2n^2 - 2m^2 - 2n^2$

53. $z^2 + 3z - 18$

54. $p^2 + 3p - 4$

55. $2x^2 - 5x - 12$

56. $3d^2 + 5d - 2$

57. $3x^2 - 2xy - 5y^2$

58. $12t^2 - 29t - 8$

59. $x^2 - 49$

60. $4x^2 + 100$

61. $2m^3 - 16$

62. $1 + 1000x^6$

63. $x^2 - y^2 - 2y - 1$

64. $9m^2 - n^2 + 2n - 1$

In Exercises 65–68 evaluate the expression. If the expression is not a real number, state this fact.

65. $\sqrt[3]{-1000}$

66. $\sqrt{.04}$

67. $-\sqrt{\dfrac{36}{5}}$

68. $\sqrt[4]{-16}$

In Exercises 69–72, simplify. Assume that all variables represent positive real numbers.

69. $\sqrt{18x^3y^2z}$

70. $\sqrt[3]{\dfrac{3}{5x^2}}$

71. $\sqrt[3]{\dfrac{125x^3}{2y}}$

72. $\sqrt[4]{\dfrac{64m^5}{27n^3}}$

In Exercises 73–78 perform the indicated operations and simplify. Assume that all variables represent positive real numbers.

73. $3\sqrt{18} - 5\sqrt{27} + \sqrt{8}$

74. $\sqrt[3]{15}\,\sqrt[3]{50}$

75. $(8 - 2\sqrt{x})^2$

76. $(\sqrt{x} - 5)^2$

77. $(2\sqrt{x} + \sqrt{y})(5\sqrt{x} - \sqrt{y})$

78. $(1 - \sqrt{x})^3$

In Exercises 79–82, simplify the expression. Write your answer in the form cx^ry^q where c is a real number, and r and q are rational numbers. *Assume that x and y represent positive real numbers.*

79. $(2x^{1/3}y^{1/5})(-7x^{1/3}y^{4/5})$

80. $\left(3x^{1/2}y^{2/3}\right)^{3/4}$

81. $\left(\dfrac{3}{5x^2}\right)^{1/5}$

82. $\sqrt[4]{16x^3y^2}$

In Exercises 83–90, simplify the expression.

83. $\dfrac{6x^2 - 23x + 7}{4x^2 - 49}$

84. $\dfrac{x^3 - 8}{x^2 - 4}$

85. $\dfrac{3}{1 - \sqrt{2}}$

86. $\dfrac{\sqrt{2}}{1 + \sqrt{5}}$

87. $\dfrac{x}{2 - 3\sqrt{x}}$

88. $\dfrac{3x}{2 - \sqrt{x + y}}$

89. $\dfrac{1 + \dfrac{1}{x + 1}}{1 + \dfrac{3}{x - 1}}$

90. $\dfrac{x - x^{-1}}{1 - x^{-2}}$

In Exercises 91–96, perform the indicated operations and simplify.

91. $\dfrac{1}{y + 2} + \dfrac{y + 9}{3y^2 + 5y - 2}$

92. $\dfrac{1}{1 - x^2} - \dfrac{1}{x^2 - 2x - 3}$

93. $\left[\dfrac{x^2 - m^2}{(x + m)^2}\right]\left[\dfrac{x + m}{x^3 - m^3}\right]$

94. $\left(\dfrac{64 - y^6}{2 - 3y + y^2}\right)\left(\dfrac{3 - 2y - y^2}{y^2 + 5y + 6}\right)$

95. $\dfrac{n^2 - 9}{n^3 - n^2 - 12n} \div \dfrac{n}{n - 3}$

96. $\dfrac{x^2}{27 - x^3} \div \left(\dfrac{4 - 6x}{3x^2 - 11x + 6}\right)$

97. Determine whether each of the following statements is true or false. If false, illustrate this with an example.
(a) The sum of a positive and a negative number is never positive.
(b) For any real number x, $|x| = x$.
(c) For any real number x, $\sqrt{x^2} = x$.

98. For the graph below the following statements are true:

$d(A, C) = 7 \qquad d(A, C) = d(B, D)$

$d(B, C) = 3 \qquad d(B, O) = d(O, C)$

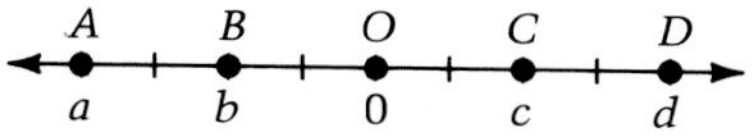

Determine $d(O, D)$.

99. In the figure below a circle of radius $x - 1$ has been inscribed in a square. Find a polynomial to describe the shaded area.

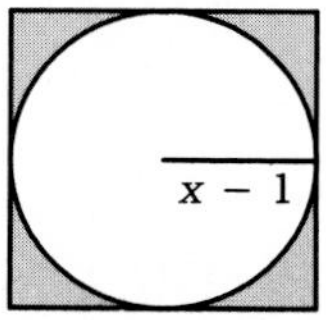

Chapter 1 Test

In Problems 1 and 2, translate the given sentence into an equation.

1. Thirteen more than twice a certain number is fifty-six.
2. The sum of three consecutive integers is 50% larger than the largest of the three integers.
3. Classify each number as one or more of the following: whole number, integer, rational number, irrational number, real number.

   (a) -5 (b) 0 (c) $\sqrt[3]{11}$

In Problems 4 and 5, evaluate the expression.

4. $-|-3| - |4|$
5. $\sqrt{0.0004}$
6. Use the graph below to determine the following distances: (a) $d(A, O)$ (b) $d(B, O)$.

A O B

-12 -9 -6 -3 0 3 6 9 12

7. Write each of the following numbers in scientific notation:

   (a) 0.0023×10^{-4} (b) $\dfrac{1}{5 \times 10^{-3}}$

In Problems 8–10, simplify the expression so that x, y, and z appear at the most once, with no powers of powers or negative exponents.

8. $(5x^2)(y^3)(3x^{-5})$
9. $(3xy^2z)^4$
10. $\left(\dfrac{2x}{5y^2}\right)^{-3}$

In Problems 11–15, perform the indicated operations.

11. $(u^2 + 3u - 7) - (u^3 - 2u^2 + u)$
12. $3x(x^2 - 7x + 4)$
13. $(8m - 2)(3m + 5)$
14. $(9z - 2)^2$
15. $\left(\dfrac{10u + 10v}{u^3 + 8v^2}\right)\left(\dfrac{u^2 + 3uv + 2v^2}{u^2 - 5uv - 6v^2}\right)$
16. Determine the greatest common factor of the following terms: $12x^2yz$, $8x^2z^3$, and $2x^5yz^2$.

In Problems 17 and 18, factor completely. If the polynomial cannot be factored, state this fact.

17. $6x^2y - 2xy^2 + 7x^2y^2$
18. $6t^2 - 5t - 4$

In Problems 19 and 20, simplify. Assume that all variables represent positive real numbers.

19. $\sqrt{24x^2y^2z}$
20. $\sqrt[4]{\dfrac{u^3}{32v^5}}$

In Problems 21 and 22, simplify. Write your answer in the form cx^ry^q.

21. $25x^{1/2}\left(2y^{1/3}\right)^2$
22. $\dfrac{3x^{2/3}}{8x^{1/9}y^{-1/2}}$

In Problems 23 and 24, simplify the expression.

23. $\dfrac{m^3 - 9m}{3m^2 + 8m - 3}$
24. $\dfrac{2 + \dfrac{1}{u - 5}}{\dfrac{19}{u - 5} - 12}$

2

Equations and Inequalities in One Variable

Most applications of mathematics involve equations of some sort. Whether we are looking over the shoulder of a chemist or a psychologist, we will not need to look far before seeing an equation or formula that models some process or describes some relationship. At the end of this chapter we consider a formula that will help us to solve a mystery involving two cities. These two cities have almost identical average daily high temperatures, but most people who have visited the cities would agree that they have vastly different climates. One of the keys to understanding how the cities are different lies in a formula that quantifies the extent to which the values in a given set vary from one another.

2.1 Introduction to Equations

An **equation** is a statement that two expressions represent the same number. The equation $2(5 + 7) = (-6)(-4)$ is true, since both expressions represent the same number, namely, 24. Equations containing variables, such as

$$2x - 1 = 5, \qquad x^2 = 3x - 4, \qquad \text{or} \qquad \frac{1}{x+1} - 3 = \frac{1}{x-1}$$

may be true for some values of the variables, and false for others. Those values of the variables that make an equation true are called **solutions** or **roots** of the equation, and are said to "satisfy" the equation. Thus 3 is a solution of the equation $2x - 1 = 5$, but 7 is not.

EXAMPLE 1

Determine whether $\frac{1}{3}$ is a solution of $7x - 3 = x + 1$.

Solution

$$7x - 3 = x + 1 \quad \text{Replace } x \text{ with } \tfrac{1}{3}.$$

$$7\left(\frac{1}{3}\right) - 3 = \frac{1}{3} + 1$$

$$-\frac{2}{3} = \frac{4}{3} \quad \text{False.}$$

Since the last equation is false, $\frac{1}{3}$ is not a solution.

EXAMPLE 2

Determine whether -2 and 3 are solutions of $x^2 - x - 6 = 0$.

Solution

$$x^2 - x - 6 = 0 \qquad\qquad x^2 - x - 6 = 0$$

$$(-2)^2 - (-2) - 6 = 0 \qquad\qquad (3)^2 - (3) - 6 = 0$$

$$4 + 2 - 6 = 0 \qquad\qquad 9 - 3 - 6 = 0$$

$$0 = 0 \quad \text{True} \qquad\qquad 0 = 0 \quad \text{True}$$

Both -2 and 3 are solutions of the equation $x^2 - x - 6 = 0$.

To "solve" an equation means to find *all* its solutions. The set of all solutions is called the **solution set.** Two equations with the same solution set are called **equivalent equations.** For example, the equations $9x - 17 = 10$ and $9x = 27$ are equivalent since each has $\{3\}$ as its solution set. The strategy used to solve an equation is to transform it into simpler equivalent equations, until the solution set is obvious. The following properties will help you to accomplish this.

Addition Property for Equivalent Equations

Adding the same real number (or variable expression) to both sides of an equation produces an equivalent equation.

Multiplication Property for Equivalent Equations

Multiplying both sides of an equation by the same nonzero real number (or variable expression) produces an equivalent equation.

EXAMPLE 3

Solve $2(x - 30) = 30 - (x - 9)$.

Solution

$$\begin{aligned} 2(x - 30) &= 30 - (x - 9) \\ 2x - 60 &= 30 - x + 9 \\ 2x - 60 &= 39 - x \end{aligned}$$

Begin by simplifying. Rewrite the equation without parentheses and collect like terms.

$$\begin{aligned} 2x + x - 60 &= 39 - x + x \\ 3x - 60 &= 39 \\ 3x - 60 + 60 &= 39 + 60 \\ 3x &= 99 \end{aligned}$$

The Addition Property is used to get a single x-term on one side and a constant term on the other.

$$\frac{1}{3} \cdot 3x = \frac{1}{3} \cdot 99$$

The Multiplication Property is used to get the coefficient of x equal to 1.

$$x = 33$$

The equation $x = 33$ is equivalent to the original equation.

Replacing x with 33 in the original equation produces a true statement. Thus, 33 is a solution, and we say that {33} is the solution set.

Definition of Linear Equation

A **linear equation** is an equation containing a first degree monomial, but no monomial of higher degree. A linear equation written in the form $ax + b = 0$, where a and b represent constants and $a \neq 0$, is called the **standard form of a linear equation in x.**

Any linear equation that can be written in the standard form $ax + b = 0$, where $a \neq 0$, can be transformed into the equivalent equation $x = -\frac{b}{a}$:

$$\begin{aligned} ax + b &= 0 \\ ax &= -b \\ x &= -\frac{b}{a} \end{aligned}$$

Therefore we conclude that such an equation has exactly one solution, namely, $-\frac{b}{a}$. We saw such an equation in Example 3. Certain linear equations, however, cannot be written in standard form, as shown in Example 4.

EXAMPLE 4

Solve for x: (a) $2x + 4 = 2(5 + x)$ (b) $2(3 - 6x) = 3(2 - 4x)$.

Solution

(a)
$$\begin{aligned} 2x + 4 &= 2(5 + x) \\ 2x + 4 &= 10 + 2x \\ 2x + (-2x) + 4 &= 10 + 2x + (-2x) \\ 4 &= 10 \end{aligned}$$

Since the final equation is never true, no matter what the value of x is, the original equation has no solutions.

(b)
$$\begin{aligned} 2(3 - 6x) &= 3(2 - 4x) \\ 6 - 12x &= 6 - 12x \\ 6 - 12x + 12x &= 6 - 12x + 12x \\ 6 &= 6 \end{aligned}$$

Since the final equation is always true, no matter what the value of x is, the original equation is an identity. It has the set of all real numbers as its solution set.

Frequently in applied problems, there is more than one letter in an equation. For example, a formula for the relationship between the Fahrenheit and Celsius temperature scales is given by

$$F = \frac{9}{5}C + 32$$

where F is the temperature in degrees Fahrenheit and C is the temperature in degrees Celsius. As written above, the equation has been "solved for F," and provides an efficient form for converting values from Celsius to Fahrenheit. However, we could "solve for C," to get $C = \frac{5}{9}(F - 32)$, a form useful for converting from Fahrenheit to Celsius.

$$\begin{aligned} F &= \frac{9}{5}C + 32 \\ F - 32 &= \frac{9}{5}C \\ \frac{5}{9}(F - 32) &= C \end{aligned}$$

Thus we "solve for a particular variable" by isolating it on one side of the equation, with constants and other variables appearing on the other side.

Though several letters may appear in an equation, some may actually represent constants. In practice, letters from the beginning of the alphabet (for example: a, b, c, d) frequently represent constants, while letters from the end of the alphabet (for example, s, t, u, v, x, y, z) usually represent variables.

EXAMPLE 5

Solve for y: $Ay - B = Cy + 7C$.

Solution $Ay - B = Cy + 7C$ — Begin by getting all y-terms on the same side of the equation.

$$Ay + (-Cy) - B = Cy + (-Cy) + 7C$$

$$Ay - Cy - B = 7C$$

Now get all constants on the right side.

$$Ay - Cy - B + B = 7C + B$$

$$Ay - Cy = 7C + B$$

Use the Distributive Property to factor out y from each term on the left side of the equation.

$$(A - C)y = 7C + B$$

Divide both sides by the coefficient of y.

$$y = \frac{7C + B}{A - C}$$

Of course, $A \neq C$.

One final note regarding linear equations. Suppose we have an equation such as

$$3x^2 + 2x + 5 = 3x^2 + 10.$$

By adding $-3x^2$ to both sides, the linear equation $2x + 5 = 10$ is produced, and can be solved by the methods of this section. An equation that does not look like a linear equation can sometimes be simplified to reveal a linear equation (perhaps with some restrictions on the values of possible solutions). We offer an example of such an equation in Example 6. There you will see an equation with fractional expressions involving variables. We will call this a **fractional equation.** An example of a fractional equation is

$$\frac{5}{x - 3} - 1 = 4.$$

Careful! When solving fractional equations, as in the next example, we multiply both sides by a variable expression. This method can have an undesirable side effect: it can introduce solutions that do not satisfy the original equation. Such numbers are called **extraneous roots.** Thus, whenever the sides of an equation have been multiplied by a variable expression, the solutions *must* be checked in the original equation.

EXAMPLE 6

Solve $\dfrac{x}{x+1} - 1 = \dfrac{2}{x^2-1}$.

Solution $\dfrac{x}{x+1} - 1 = \dfrac{2}{x^2-1}$ Begin by multiplying both sides by the least common denominator (LCD) of all fractions in the problem.

$$\left(\frac{x}{x+1} - 1\right)(x^2-1) = \left(\frac{2}{x^2-1}\right)(x^2-1)$$ Both sides are multiplied by the LCD $(x^2 - 1)$.

$$\frac{x\overset{(x-1)}{\cancel{(x^2-1)}}}{\cancel{x+1}} - 1(x^2-1) = \frac{2\cancel{(x^2-1)}}{\cancel{x^2-1}}$$ Think of $x^2 - 1$ as $(x+1)(x-1)$ and cancel.

$$x(x-1) - 1(x^2-1) = 2$$
$$x^2 - x - x^2 + 1 = 2$$
$$-x = 1$$
$$x = -1$$ This equation may not be equivalent to the given equation. The value *must* be checked.

Must check: $\dfrac{x}{x+1} - 1 = \dfrac{2}{x^2-1}$ Start with the original equation.

$$\frac{-1}{-1+1} - 1 = \frac{2}{(-1)^2 - 1}$$ Replace x with -1, then determine whether the resulting equation is true.

$$\frac{-1}{0} - 1 = \frac{2}{0}$$ Since -1 produces a zero denominator, it cannot be a solution.

The given equation has no solution.

Absolute Value Equations

Suppose $|A| = 19$. There are only two real numbers having an absolute value of 19, namely 19 and -19. Thus, either

$$A = 19 \quad \textbf{or} \quad A = -19.$$

If A represents a variable expression, then we can draw the same conclusions about A. Suppose $|1 - 4x| = 19$. Then either

$$1 - 4x = 19 \quad \textbf{or} \quad 1 - 4x = -19$$
$$-4x = 18 \qquad\qquad -4x = -20$$
$$x = -\frac{9}{2} \qquad\qquad x = 5$$

Thus, $-\frac{9}{2}$ and 5 are solutions of the equation $|1 - 4x| = 19$.

The technique used to solve the previous equation can be used even when the variable appears on both sides of the equation. For instance, if $|3x - 2| = 7x + 5$, we solve for x by first writing the following:

$$3x - 2 = 7x + 5 \quad \textbf{or} \quad 3x - 2 = -(7x + 5).$$

Absolute Value Equations

To solve the absolute value equation $|A| = B$, you must solve the two equations in the statement "$A = B$ **or** $A = -B$." You *must* check solutions.

EXAMPLE 7

Solve $|2x - 3| = 5 - 8x$.

Solution $|2x - 3| = 5 - 8x$ To solve, begin by writing two equations.

$$\begin{aligned} 2x - 3 &= 5 - 8x & \textbf{or} \quad 2x - 3 &= -(5 - 8x) \\ 10x - 3 &= 5 & 2x - 3 &= -5 + 8x \\ 10x &= 8 & -3 &= -5 + 6x \\ x &= \frac{4}{5} & 2 &= 6x \\ & & \frac{1}{3} &= x \end{aligned}$$

Both values, $\frac{4}{5}$ and $\frac{1}{3}$, *must* be checked.

Must Check:

$$|2x - 3| = 5 - 8x \qquad\qquad |2x - 3| = 5 - 8x$$

$$\left|2\left(\frac{4}{5}\right) - 3\right| = 5 - 8\left(\frac{4}{5}\right) \qquad\qquad \left|2\left(\frac{1}{3}\right) - 3\right| = 5 - 8\left(\frac{1}{3}\right)$$

$$\left|-\frac{7}{5}\right| = -\frac{7}{5} \quad \text{False.} \qquad\qquad \left|-\frac{7}{3}\right| = \frac{7}{3} \quad \text{True.}$$

The only solution is $\frac{1}{3}$.

EXERCISE SET 2.1

In Exercises 1–8, determine whether the values 2 and/or -2 are solutions.

1. $5x + 7 = 3(x + 1)$
2. $5 - y = 3(y - 1)$
3. $2u^2 - 3 = u$
4. $5t + 3 = t^2 - 2$
5. $|3s - 4| = s$
6. $|7x + 10| = -2x$
7. $\dfrac{y^2 - 4}{y + 2} + 6 = 3y$
8. $\dfrac{u + 2}{u - 2} = 2 + u$

In Exercises 9–24, solve the given equation.

9. $5y + 7 = -13$
10. $8 - 3z = -1$
11. $3 = 5 - 7u$
12. $16 + 3v = 5$
13. $9m - 4 = 15 - 2m$
14. $1 - 8n = 3 + 2n$
15. $5(s + 2) = 6s - 4$
16. $9t - 1 = 9(1 - t)$
17. $9t - 1 = 9(t - 1)$
18. $1 - 9(t - 1) = 9t$

19. $8s + 3(s - 4) = 10s$

20. $3(s - 4) - 5s = 2(6 - s)$

21. $(6y - 5)(y + 1) = (3y + 1)(2y) - 1$

22. $x(2x - 1) = (5 + x)(2x) + 1$

23. $(4x - 3)(x + 4) = (2x - 3)^2$

24. $(x - 2)(9x - 5) = (3x - 1)^2$

In Exercises 25–38, solve for the variable x.

25. $3x - b = 7$

26. $a - 2x = 9$

27. $2 - abx = c$

28. $a^2 + b^2x = c$

29. $5x + a = mx$

30. $3x = b - mx$

31. $2(a - x) = -2(x - a)$

32. $a(x + 2) = 2b(x + 1)$

33. $ax + 2 = 2x(b - 1)$

34. $2(ax - 2) + a(6 - 2x) = 2(3a - 2)$

35. $a(x - 1) - x(b + 2) = 0$

36. $2ax(b + 1) - ab(x - 2) = 0$

37. $x(x - 1) + x^2 - 1 = 2x^2$

38. $1 - (x^2 - 2x) + 3x^2 = 2x^2 - 1$

In Exercises 39–50, solve for the indicated variable in terms of the other variables.

39. Solve $C = \frac{5}{9}(F - 32)$ for F.

40. Solve $A = \frac{x + y + z}{3}$ for z.

41. Solve $P = 2(L + W)$ for W.

42. Solve $a = P(1 + rt)$ for t.

43. Solve $s = \frac{1}{2}at^2$ for a.

44. Solve $V = \frac{1}{3}\pi r^2 h$ for h.

45. Solve $P = 2wh + 2lh + 2lw$ for h.

46. Solve $\frac{1}{R} = \frac{1}{R_1} + \frac{1}{R_2}$ for R_1.

47. Solve $s = s_0 + v_0 t + \frac{1}{2}at^2$ for v_0.

48. Solve $S = \frac{a}{1 - r}$ for r.

49. Solve $t = a + (n - 1)d$ for n.

50. Solve $x = ta^2 + tb^2 + a^2b^2$ for t.

In Exercises 51–72, solve the given equation.

51. $\frac{8}{x} = \frac{4}{5}$

52. $\frac{2}{3} = \frac{t}{12}$

53. $\frac{2}{s + 3} = \frac{9}{10}$

54. $\frac{6}{1 - v} = \frac{3}{4}$

55. $\frac{w}{w - 1} = \frac{3}{5}$

56. $\frac{2}{5m + 1} = \frac{3}{m}$

57. $\frac{3}{n} + \frac{3}{4n} = 6$

58. $\frac{1}{u} = 7 - \frac{3}{2u}$

59. $\frac{1}{5x} + \frac{3}{2x} = \frac{17}{10x}$

60. $\frac{1}{5x} - \frac{3}{2x} + \frac{1}{x} = \frac{1}{5}$

61. $\frac{1}{x} - \frac{3}{2x} + \frac{4}{3x} = \frac{1}{2}$

62. $\frac{4}{3x} - \frac{1}{2x} = \frac{5}{6x}$

63. $\frac{x}{x - 2} = \frac{x + 2}{x + 4}$

64. $\frac{w + 1}{w + 2} = \frac{w - 3}{w - 1}$

65. $\frac{y}{y + 2} = 1 - \frac{1}{y + 3}$

66. $\frac{2x}{x + 3} = 3 - \frac{x}{x + 9}$

67. $\frac{x}{x - 4} - 2 = \frac{4}{x - 4}$

68. $\frac{v}{v - 3} = 2 + \frac{3}{v - 3}$

69. $\frac{6t}{t + 1} + \frac{12}{t^2 + 4t + 3} = \frac{6t - 1}{t + 3}$

70. $\frac{5x}{x + 2} = \frac{30}{x^2 + 7x + 10} + \frac{5x}{x + 5}$

71. $\frac{2p}{p - 3} = \frac{p}{p + 4} + \frac{24 + p^2}{p^2 + p - 12}$

72. $\frac{5m^2 + 1}{2m^2 + m - 1} - \frac{m}{2m - 1} = \frac{2m}{m + 1}$

In Exercises 73–84, solve the given equation.

73. $|3 + 2x| = 8$

74. $|4x - 5| = 15$

75. $|8 - 3y| = 7$

76. $|12 - 7p| = 31$

77. $|6 - 5x| = 11$

78. $|-4s + 1| = 11$

79. $|2 - 3v| = 7v - 3$

80. $|3t - 4| = 2t - 5$

81. $|9u - 2| = 3u - 8$

82. $|2n + 5| = 4n - 5$

83. $|3 - 2w| = 1 - 2w$

84. $|5m - 4| = 6 - 5m$

In Exercises 85–90, solve the given equation. Round solutions to the nearest tenth.

85. $8.2x - 6.1 = 3.8(x - 5)$

86. $5(2.7 - 9x) = 16.3 - 20.5x$

87. $3.95x + 16.33(x - 7) = 52.18 - 0.05x$

88. $2.25x - 5(16.8 - x) = x(83.04 - 16.91)$

89. $\frac{1}{x - 3.1} = \frac{6.03}{9.28}$

90. $\frac{8.75}{x + 2.05} = \frac{10.25}{11 - x}$

In Exercises 91–92, solve the given equation.

91. $|5x - 8.25| = 0.005$ 92. $|2x - 9.01| = 0.0001$

Superset

In Exercises 93–96, determine the value of a so that the solution set is the set of all real numbers.

93. $3(x + 4) = x - 2(a - x)$

94. $5x - 3(x - a) = x - (a - x)$

95. $\dfrac{1}{a} + \dfrac{4ax + 1}{2a} + \dfrac{1}{3a} = \dfrac{60x + 11}{30}$

96. $\dfrac{3 - 2x}{a} - \dfrac{3 + 4x}{2a} + 6 = \dfrac{27 - 8x}{4}$

In Exercises 97–106, solve the given equation. (*Hint:* To solve $|A| = |B|$, solve: "$A = B$ or $A = -B$", then check solutions.)

97. $|2 - 3v| = |7v - 3|$ **98.** $|3t - 4| = |2t - 5|$

99. $|9u - 2| = |3u - 8|$ **100.** $|2n + 5| = |4n - 5|$

101. $|4x - 1| = |4x + 7|$ **102.** $|8 + 3x| = |5x - 2|$

103. $|5 - 3x| = |3x + 8|$ **104.** $|2x + 3| = |2x - 5|$

105. $|6 - 3x| = |5x + 2|$ **106.** $|5x + 1| = |7 - 5x|$

107. To solve an absolute value equation of the form $|A| = B$ or $|A| = |B|$, we must solve the two equations in the statement "$A = B$ or $A = -B$." Formulate a similar statement for an equation of the form $|A| + |B| = C$, where A and B are expressions involving variables and C is some nonnegative constant. Then, solve the equation $|x + 2| + |2x - 5| = 7$.

108. (See Exercise 107.) Solve the equation: $|4x - 1| + |3x + 6| = 10$.

In each of the figures in Exercises 109–110, the shaded area is 10 square inches. Write an equation describing this fact in terms of the given variables.

109.

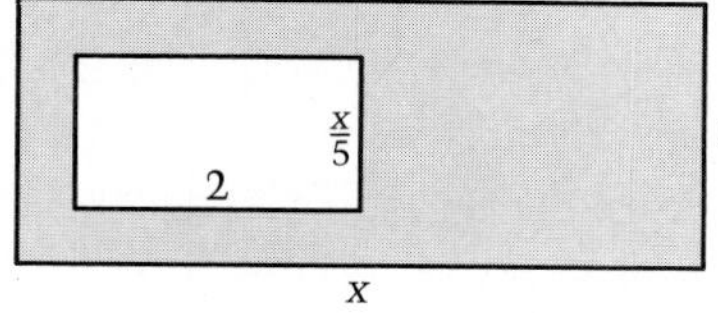

110.

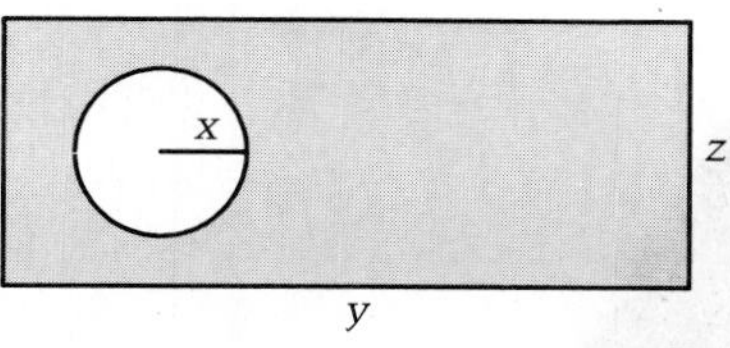

111. In the figure below, the area of the shaded region is 60 square meters. Find the dimensions of the larger rectangle. (Find x and $x + 2$.)

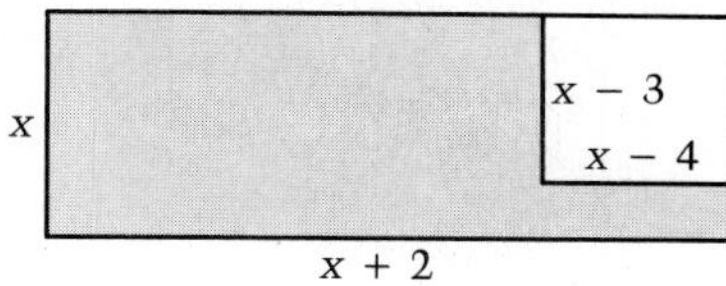

112. In the figure below, the area of the shaded region is 18 square meters. Find the height of the parallelogram (i.e., what is the value of x)?

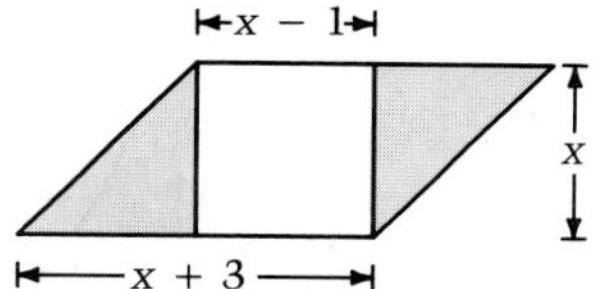

113. A student's grade G in a certain course is determined by the average of four test scores. The student scored 78, 86, and 81 respectively on the first three tests.

(a) Write an equation for G in terms of the student's score S on the fourth test.

(b) Solve the equation in part (a) for S and use this result to determine how well the student will have to do on the fourth test to have an average of 85 for the course.

114. A college book store marks up by 20% the price it pays publishers for textbooks.

(a) Write an equation for the selling price P of a textbook for which the book store pays the publisher S dollars.

(b) Solve the equation in part (a) for S and use this result to determine how much the store paid for a textbook it is selling for $45.

115. In a certain community the school tax T that each household must pay each year is $200 plus 5% of the assessed value V of their home.

(a) Write an equation for T in terms of V and use this formula to determine the tax bill of a household whose home is assessed at $65,000.

(b) Solve the equation in part (a) for V and use this result to determine the assessed value of the house whose owners have a tax bill of $2000.

116. The weekly income I of a salesperson in a local car dealership is \$180 plus a 1% commission on his total sales S for the week.
(a) Write an equation for I in terms of S and use this formula to determine the income of a salesperson in a week when his total sales were \$87,500.
(b) Solve the equation in part (a) for S and use this result to determine how much in sales the salesperson would need to have a weekly income of \$500.

117. The weekly take-home pay T of a worker at a factory is determined by deducting 30% of the worker's salary S to cover state, federal, and social security taxes. In addition, the worker has \$40 deducted to cover a payroll savings plan and union dues.
(a) Write an equation for T in terms of S and use this formula to determine the take-home pay of a worker who works a 40-hour week at \$15 per hour.
(b) Solve the equation in part (a) for S and use this result to determine the number of hours the worker worked in a week when his take-home pay was \$317.

118. The first place prize P in a regional curling tournament is set at 25% of the balance remaining after the promoter has deducted \$4500 for expenses from entry fees and gate receipts totaling X dollars.
(a) Write an equation for P in terms of X.
(b) Solve the equation in part (a) for X and use this result to determine the promoter's income from entry fees and gate receipts if the first place prize was \$820.

2.2 Introduction to Complex Numbers

All of the equations that we met in the last section had solutions that were real numbers. So at this point, it might seem that the set of real numbers is sufficient as a source of solutions for *all* equations. But in Section 2.3 we will consider quadratic equations—equations which have a term of second degree, such as $x^2 + 3x - 4 = 7$. We will see that some rather simple-looking quadratic equations don't have real number solutions. Perhaps the simplest is the equation

$$x^2 = -1$$

which has no real number solution, since there is no real number whose square is -1. By defining the number i, which *is* a solution of this equation, we will ultimately be in a position to solve any quadratic equation.

Definition of i

The number i has the property that

$$i^2 = -1.$$

For this reason, i can be considered a square root of -1. Thus, we write

$$i = \sqrt{-1}.$$

With this definition for $\sqrt{-1}$, we can define square roots of other negative numbers. Observe that

$$-2 = (-1)(2) = i^2(\sqrt{2})^2 = (i \cdot \sqrt{2})^2.$$

Thus, $i\sqrt{2}$ is a square root of -2. (Note that when i is multiplied by a radical, i is generally written first to avoid confusing a number like $\sqrt{2}i$ with $\sqrt{2i}$.) We can use the number i to form square roots of any negative number. If p is a positive real number, then $\sqrt{-p} = \sqrt{-1}\sqrt{p} = i\sqrt{p}$.

$$\sqrt{-8} = i\sqrt{8} = 2i\sqrt{2}$$
$$-\sqrt{-16} = -i\sqrt{16} = -4i$$

Using the number i and the set of all real numbers, we can form a set that allows us to do much more than simply form square roots of negative real numbers. As you will see in Chapter 4, the Fundamental Theorem of Algebra tells us that there is no need for any further invention of numbers. The set of *complex numbers*, as defined below, is large enough for solving any algebraic problem.

Definition of Complex Numbers

If a and b are real numbers, then any number of the form

$$a + bi$$

is called a **complex number.** The number a is called the **real part** of the complex number, and b is called the **imaginary part.**

Since every real number a can be written in the form $a + 0i$, every real number is a complex number. Any complex number of the form $a + bi$, where the imaginary part b is not zero, is called an **imaginary number.** Thus $2 + 3i$ is a complex number that is imaginary. Finally, since the number $7i$ can be written $0 + 7i$, it is a complex number. Since its imaginary part is not zero, it is also an imaginary number. Any number of the form bi, where $b \neq 0$, is called a **pure imaginary number.** Thus, $7i$ is a pure imaginary number.

EXAMPLE 1

Determine the real part and the imaginary part of each complex number.

(a) $4 + 7i$ (b) $\sqrt{3}$ (c) $-2 + i$ (d) $8 - 6i$ (e) $5i$ (f) 0
(g) -1

Solution

(a) The real part is 4; the imaginary part is 7.

(b) Since $\sqrt{3}$ can be written $\sqrt{3} + 0i$, the real part is $\sqrt{3}$ and the imaginary part is 0.

(c) Since $-2 + i$ can be written $-2 + 1i$, the real part is -2 and the imaginary part is 1.

(d) Since $8 - 6i$ can be written $8 + (-6)i$, the real part is 8 and the imaginary part is -6.

(e) Since $5i$ can be written $0 + 5i$, the real part is 0 and the imaginary part is 5.

(f) Since 0 can be written $0 + 0i$, the real part is 0 and the imaginary part is 0.

(g) Since -1 can be written $-1 + 0i$, the real part is -1 and the imaginary part is 0.

Thus, the set of complex numbers is formed by joining the set of real numbers and the set of imaginary numbers as shown in Figure 2.1. It can be shown that the Properties of Real Numbers and the rules governing exponents are also true for complex numbers.

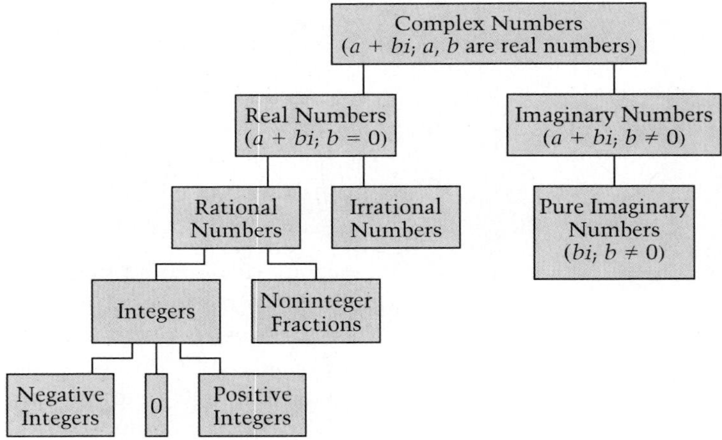

Figure 2.1

A few computational notes are in order. Since $i^2 = -1$, it follows that $i^4 = 1$, a fact that is quite useful when simplifying powers of i. Notice the following pattern:

$$\begin{aligned} i^1 &= i & i^5 &= i^4(i) = (1)(i) = i \\ i^2 &= -1 & i^6 &= i^4(i^2) = (1)(-1) = -1 \\ i^3 &= i^2(i) = -i & i^7 &= i^4(i^3) = (1)(-i) = -i \\ i^4 &= (i^2)(i^2) = 1 & i^8 &= i^4(i^4) = (1)(1) = 1 \end{aligned}$$

It follows that

$$i^4, i^8, i^{12}, i^{16}, \ldots$$

are all equal to 1. In general, for any positive integer n, $i^{4n} = 1$. With this information, we can simplify a number like i^{31} very easily. Since 28 is the greatest multiple of 4 less than 31, we can write the following:

$$i^{31} = i^{28}i^{3} = (1)(-i) = -i.$$

Careful! When simplifying an expression containing the square root of a negative number, you should always begin by rewriting the expression in terms of i. For example,

$$\sqrt{-4} \cdot \sqrt{-9} = (2i)(3i) = i^2 \cdot 6 = -6.$$

Failure to perform the above step may lead to serious errors. Note that $\sqrt{-4} \cdot \sqrt{-9} \neq \sqrt{(-4)(-9)}$. If we assumed that this statement were true, then we would obtain the following incorrect statement:

$$\sqrt{-4} \cdot \sqrt{-9} = \sqrt{(-4)(-9)} = \sqrt{36} = 6. \quad \text{False!}$$

Thus, $\sqrt{a} \cdot \sqrt{b} = \sqrt{ab}$ only if a and b are nonnegative real numbers.

EXAMPLE 2

Simplify each expression: (a) i^{64} (b) i^{39} (c) $\sqrt{-4} \cdot \sqrt{-8}$.

Solution

(a) $i^{64} = 1$ Since 64 is a multiple of 4, $i^{64} = 1$.

(b) $i^{39} = i^{36+3} = i^{36}i^{3} = (1)(-i) = -i$ 36 is the largest multiple of 4 in the exponent 39. Write 39 as $36 + 3$, then use Property 1 of Exponents.

(c)
$$\begin{aligned}\sqrt{-4} \cdot \sqrt{-8} &= i\sqrt{4} \cdot i\sqrt{8} \\ &= 2i \cdot 2i\sqrt{2} \\ &= 4i^2\sqrt{2} \\ &= -4\sqrt{2}\end{aligned}$$

As you might expect, for two complex numbers to be equal, their real parts must be equal, and their imaginary parts must be equal.

Equality of Complex Numbers

Two complex numbers $a + bi$ and $c + di$ are equal if and only if $a = c$ and $b = d$.

The sum of two complex numbers is a complex number whose real part is the sum of the real parts of the two numbers, and whose imaginary part is the sum of their imaginary parts. Their difference is found in a similar way.

Addition & Subtraction of Complex Numbers

If $a + bi$ and $c + di$ are complex numbers, then

$$(a + bi) + (c + di) = (a + c) + (b + d)i,$$

and

$$(a + bi) - (c + di) = (a - c) + (b - d)i.$$

We find the product of two complex numbers the same way we find the product of two binomials: we use the FOIL method.

$$\begin{aligned}(a + bi)(c + di) &= ac + adi + bci + bidi \\ &= ac + bdi^2 + (ad + bc)i \\ &= ac - bd + (ad + bc)i \qquad \text{Since } i^2 = -1.\end{aligned}$$

Multiplication of Complex Numbers

If $a + bi$ and $c + di$ are complex numbers, then

$$(a + bi)(c + di) = (ac - bd) + (ad + bc)i$$

In the following examples, we shall practice the arithmetic of complex numbers. Note that final answers are given in **standard form** $a + bi$. We will agree that if $a = 0$, the pure imaginary answer may be given in the form bi, and if $b = 0$, the answer may be stated as a real number.

EXAMPLE 3

Add or subtract. Write your answer in standard form.

(a) $(3 + 3i) + (7 - 5i)$ (b) $(6 - 2i) + 2i$ (c) $(10 + 7i) - (10 - 4i)$

Solution

(a) $$\begin{aligned}(3 + 3i) + (7 - 5i) &= (3 + 7) + (3i - 5i) \qquad \text{Group like terms and combine.} \\ &= 10 + (-2i) \\ &= 10 - 2i\end{aligned}$$

(b) $(6 - 2i) + 2i = 6 + (-2i + 2i) = 6$

(c) $$\begin{aligned}(10 + 7i) - (10 - 4i) &= (10 - 10) + (7i + 4i) \\ &= 11i\end{aligned}$$

EXAMPLE 4

Multiply. Write your answer in standard form.

(a) $-6i(4 + 9i)$ (b) $(2 + i)(3 - 4i)$

Solution

(a) $-6i(4 + 9i) = -24i - 54i^2$ Replace i^2 with -1.

$= -24i - 54(-1)$

$= 54 - 24i$

(b) $(2 + i)(3 - 4i) = 2 \cdot 3 + 2(-4i) + 3i + i(-4i)$

$= 6 - 8i + 3i - 4i^2$ Replace i^2 with -1.

$= 6 + 4 + [(-8i) + 3i]$

$= 10 - 5i$

Suppose you wanted to solve the following equation for z:

$$(3 - 4i)z = 2 + 5i$$

Your first impulse might be to multiply both sides by the reciprocal of $3 - 4i$, namely $\frac{1}{3-4i}$. And your impulse would be correct. But how can we rewrite this reciprocal in standard form $a + bi$? To do this we simply multiply the numerator and denominator by $3 + 4i$, as is shown in Example 5. This number $3 + 4i$ is called the **complex conjugate** of $3 - 4i$. In general, the complex conjugate of $a + bi$ is $a - bi$, and the complex conjugate of $a - bi$ is $a + bi$. Since $(a + bi)(a - bi) = a^2 - b^2i^2 = a^2 + b^2$, the product of a complex number and its conjugate is a real number.

Property of Complex Conjugates

$$(a + bi)(a - bi) = a^2 + b^2$$

EXAMPLE 5

Write the reciprocal of $3 - 4i$ in standard form $a + bi$.

Solution

The reciprocal of $3 - 4i$ is $\dfrac{1}{3 - 4i}$.

$$\frac{1}{3 - 4i} = \frac{1}{3 - 4i} \cdot \frac{3 + 4i}{3 + 4i}$$

Multiply the numerator and denominator by $3 + 4i$, the conjugate of the denominator $3 - 4i$.

$$= \frac{3 + 4i}{9 - 16i^2}$$

$$= \frac{3 + 4i}{9 + 16}$$

$$= \frac{3}{25} + \frac{4}{25}i$$

We can use the complex conjugate in a similar way when we divide complex numbers.

EXAMPLE 6

Simplify the quotient $\dfrac{2 - 3i}{-3 + i}$.

Solution

$$\frac{2 - 3i}{-3 + i} = \frac{2 - 3i}{-3 + i} \cdot \frac{-3 - i}{-3 - i}$$ Multiply the numerator and denominator by the conjugate of the denominator $-3 + i$.

$$= \frac{-6 - 2i + 9i + 3i^2}{9 - i^2}$$ We used the FOIL method to find the product in the numerator.

$$= \frac{-6 - 2i + 9i - 3}{9 - (-1)}$$ We used the fact that $i^2 = -1$.

$$= -\frac{9}{10} + \frac{7}{10}i$$

A bit of notation is in order. If we let the variable z represent the complex number $a + bi$, it is customary to let $\bar{z}$ represent its complex conjugate $a - bi$; that is,

$$\overline{a + bi} = a - bi.$$

For example, if $z = 3 + 7i$, then $\bar{z} = 3 - 7i$, and if $w = 4 - 3i$, then $\bar{w} = 4 + 3i$. We have already stated the most useful property of complex conjugates, namely, that if $z = a + bi$, then $z \cdot \bar{z} = a^2 + b^2$, a real number. But other properties can be established. We summarize them here.

Properties of Complex Conjugates

If $\bar{z}$ and $\bar{w}$ are complex conjugates of the complex numbers z and w, respectively, then

1. $\overline{z + w} = \bar{z} + \bar{w}$
2. $\overline{z - w} = \bar{z} - \bar{w}$
3. $\overline{z \cdot w} = \bar{z} \cdot \bar{w}$
4. $\overline{\left(\dfrac{z}{w}\right)} = \dfrac{\bar{z}}{\bar{w}}$, for $w \neq 0$
5. $\overline{z^n} = (\bar{z})^n$, for every positive integer n
6. $\bar{z} = z$, if z is a real number

EXERCISE SET 2.2

In Exercises 1–12, use the number i to rewrite the expression.

1. $\sqrt{-6}$
2. $\sqrt{-21}$
3. $\sqrt{-25}$
4. $\sqrt{-81}$
5. $\sqrt{-98}$
6. $\sqrt{-75}$
7. $-\sqrt{-9}$
8. $-\sqrt{-36}$
9. $-\sqrt{-20}$
10. $-\sqrt{-18}$
11. $-\sqrt{-63}$
12. $-\sqrt{-32}$

In Exercises 13–20, determine the real part and the imaginary part of the complex number.

13. $\frac{1}{2} + 6i$
14. $-7 - \frac{3}{5}i$
15. $6i$
16. 14
17. $-\sqrt{3}$
18. i
19. $\frac{1}{4} - i\sqrt{10}$
20. $-\sqrt{6} + \frac{2}{7}i$

In Exercises 21–60, simplify the expression.

21. i^{32}
22. i^{18}
23. $(-i)^{25}$
24. $(-i)^{51}$
25. $(2i)^3 \cdot i^{12}$
26. $(4i)^3 \cdot i^{50}$
27. $\sqrt{-7} \cdot \sqrt{-28}$
28. $\sqrt{-12} \cdot \sqrt{-3}$
29. $\sqrt{6} \cdot \sqrt{-2}$
30. $\sqrt{-5} \cdot \sqrt{10}$
31. $\sqrt{-2} \cdot \sqrt{-3} \cdot \sqrt{-30}$
32. $\sqrt{-5} \cdot \sqrt{-15} \cdot \sqrt{-6}$
33. $(2 + 3i) + (-4 + 5i)$
34. $(5 - 3i) + (8 + 2i)$
35. $(3 - 4i) - (-5 - 6i)$
36. $(1 + 6i) + (-1 + 6i)$
37. $(8 + 7i) + (6 - 7i)$
38. $(-12 - 4i) - (6 - i)$
39. $4(-6 + 8i)$
40. $-2(5 - 9i)$
41. $3i(2 - i)$
42. $-4i(1 + 8i)$
43. $(5 + 2i)(3 + i)$
44. $(1 + 4i)(1 + 2i)$
45. $(-7 + 2i)(2 - 5i)$
46. $(9 - 3i)(-2 + i)$
47. $(-1 - i)(8 + 2i)$
48. $(2 + 6i)(-3 - 3i)$
49. $(5 + 4i)(5 - 4i)$
50. $(2 + 3i)(2 - 3i)$
51. $(1 + i\sqrt{3})(1 - i\sqrt{3})$
52. $(\sqrt{7} - i)(\sqrt{7} + i)$
53. $(2 + 3i)^2$
54. $(3 + 4i)^2$
55. $(-5 + i)^2$
56. $(6 - i)^2$
57. $(-1 - 2i)^2 \cdot i$
58. $(2i)(2 + 5i)^2$
59. $(3i)^3(2 - i)$
60. $(-7 + 3i)(-2i)^3$

In Exercises 61–68, write the reciprocal of the complex number in the form $a + bi$.

61. $2 + 3i$
62. $1 + 4i$
63. $5 - 2i$
64. $3 - 7i$
65. $-\frac{1}{7}i$
66. $\frac{2}{3}i$
67. $\sqrt{3} - 2i$
68. $-\sqrt{11} + 5i$

In Exercises 69–82, simplify the quotient.

69. $\frac{6 + 2i}{2 - i}$
70. $\frac{8 + i}{2 - 3i}$
71. $\frac{2 + 3i}{1 + 4i}$
72. $\frac{1 + 4i}{3 + 2i}$
73. $\frac{5 - i}{2 + 5i}$
74. $\frac{3 - 4i}{6 + i}$
75. $\frac{34}{5 - 3i}$
76. $\frac{-53}{7 - 2i}$
77. $\frac{-2i}{5 + 4i}$
78. $\frac{10i}{3 + i}$
79. $\frac{3 - 7i}{-7i}$
80. $\frac{-9 + 8i}{3i}$
81. $\frac{1}{-6 - 2i}$
82. $\frac{1}{-2 - 4i}$

In Exercises 83–88, verify the following for the given values of the variables:

(a) $\overline{z + w} = \overline{z} + \overline{w}$
(b) $\overline{z - w} = \overline{z} - \overline{w}$
(c) $\overline{z \cdot w} = \overline{z} \cdot \overline{w}$
(d) $\overline{\left(\frac{z}{w}\right)} = \frac{\overline{z}}{\overline{w}}$

83. $z = 1 + 3i$ and $w = 2 - 4i$
84. $z = 5 + 2i$ and $w = 2 + 3i$
85. $z = 12 - 3i$ and $w = 5 + 9i$
86. $z = -11 + i$ and $w = 2 - 7i$
87. $z = -6 - 5i$ and $w = -1 - 4i$
88. $z = 9 - 9i$ and $w = -8 + 3i$

Superset

In Exercises 89–96, find real numbers x and y that satisfy the equation.

89. $2 + 3yi = -x + 9i$
90. $-4x + 7i = -8 - 7yi$
91. $8x - 4i = 6 + 2yi$
92. $9 + 5yi = 6x - 7i$
93. $-7 + i = (x - y) + (2x + y)i$
94. $(2x + y) - 8i = 5 + (x - 3y)i$
95. $(x - y) + 4i = 8 - (x - 2y)i$
96. $3 - 2i = (x + y) - (3x - y)i$

In Exercises 97–100, simplify the expression.

97. $i + i^2 + i^3 + \cdots + i^{48}$
98. $i + i^2 + i^3 + \cdots + i^{18}$
99. $i + i^2 + i^3 + \cdots + i^{29}$
100. $i + i^2 + i^3 + \cdots + i^{75}$

In Exercises 101–106, determine whether the statement is true or false. If false, illustrate this with an example.

101. If a is a positive real number and b is a negative real number, then $\sqrt{a} \cdot \sqrt{b} = \sqrt{ab}$.

102. If a and b are real numbers, then $\sqrt{a} \cdot \sqrt{b} = \sqrt{ab}$.

103. The product of two imaginary numbers is always imaginary.

104. The product of two imaginary numbers is always a real number.

105. If a and b are real numbers, then $(a + bi)(a - bi)$ is always a real number.

106. If a and b are real numbers, then $(a + bi)^2$ is always imaginary.

In Exercises 107–118, simplify the expression.

107. $\dfrac{1 + i}{(1 - 2i)^2}$ **108.** $\dfrac{1 - i}{(1 + 2i)^2}$

109. $\dfrac{5 - i}{(1 + i)(2 - 3i)}$ **110.** $\dfrac{-5 - i}{(1 - i)(2 + 3i)}$

111. $\dfrac{-1 + 8i}{(2 - i)(3 + 2i)}$ **112.** $\dfrac{14 + 2i}{(2 + i)(1 + 3i)}$

113. $\dfrac{4 - 3i}{(3 + 2i)(-1 + i)}$ **114.** $\dfrac{5 - i}{(3 - 2i)(1 + i)}$

115. $\dfrac{1}{2 + i} + \dfrac{-3}{1 - 3i}$ **116.** $\dfrac{-4}{1 + 3i} + \dfrac{2}{2 - i}$

117. $\dfrac{2 - 3i}{1 + i} - \dfrac{6 + 2i}{2 - i}$ **118.** $\dfrac{5 + 2i}{3 - i} - \dfrac{2 - 4i}{1 + i}$

In Exercises 119–126, find all complex numbers z that satisfy the given equation.

119. $(3 + 2i)z + i = 1 + 4i$

120. $(2 - 4i)z + 3i = -2 + 5i$

121. $(3 - 2i) + 5z = 3iz + (4 - 3i)$

122. $2iz - (8 - 5i) = z + (-10 + 7i)$

123. $2z + 5\bar{z} = 15 - 9i$ **124.** $7z - 3\bar{z} = -4 + 20i$

125. $z^2 = 8i$ **126.** $z^2 = -18i$

127. Let $z = a + bi$ and $w = c + di$. Show that $\overline{z - w} = \bar{z} - \bar{w}$.

128. Let $z = a + bi$ and $w = c + di$, where $w \neq 0$. Show that $\overline{\left(\dfrac{z}{w}\right)} = \dfrac{\bar{z}}{\bar{w}}$.

129. Let $z = a + bi$. Show that $\bar{\bar{z}} = z$.

130. Let $z = a + bi$. Show that $z \cdot \bar{z}$ is a real number.

131. Let $z = a + bi$. Show that $\frac{1}{2}(z + \bar{z}) = a$.

132. Let $z = a + bi$. Show that $\frac{1}{2}i(\bar{z} - z) = b$.

2.3 Quadratic Equations

All the problems in Section 2.1 involved solving equations that could ultimately be written in the form $ax + b = 0$. We now consider methods for solving equations of the form $ax^2 + bx + c = 0$.

Definition of Quadratic Equation

An equation containing a second degree monomial, but containing no monomial of higher degree, is called a **quadratic equation**. The form $ax^2 + bx + c = 0$, where a, b, and c are constants, and $a \neq 0$, is called the **standard form of a quadratic equation in x.**

The key to solving quadratic equations is the Property of Zero Products. This property says that if the product of two or more factors is zero, then at least one of the factors must be zero.

Property of Zero Products

If A and B are real numbers (or expressions representing real numbers) and if $A \cdot B = 0$, then $A = 0$, or $B = 0$, or both.

For example, to solve $(x - 2)(3x - 7) = 0$, we solve the two equations $x - 2 = 0$ and $3x - 7 = 0$. Note that in order to apply the Property of Zero Products, one side of the given equation must be factored and the other side must be 0.

EXAMPLE 1

Solve $2x^2 + x = 15$.

Solution

$$2x^2 + x = 15$$

$$2x^2 + x - 15 = 0 \qquad \text{All nonzero terms are on the left side.}$$

$$(2x - 5)(x + 3) = 0 \qquad \text{The nonzero side is factored.}$$

$2x - 5 = 0$	$x + 3 = 0$
$x = \frac{5}{2}$	$x = -3$

The Property of Zero Products is applied.

The solutions are $\frac{5}{2}$ and -3.

EXAMPLE 2

Solve $3x^2 - 30x + 75 = 0$.

Solution

$$3x^2 - 30x + 75 = 0$$

$$3(x^2 - 10x + 25) = 0 \qquad \text{3 is a common factor.}$$

$$(x - 5)(x - 5) = 0$$

Both sides are divided by 3, and $x^2 - 10x + 25$ is factored.

$x - 5 = 0$	
$x = 5$	

Since the factors are the same, we only need to solve $x - 5 = 0$ once.

The solution is 5.

In Example 2, the solution 5 is called a *root of multiplicity* 2, since the factor that produced it appeared twice in the factorization. In general, if a linear factor $ax + b$ appears exactly n times in the factorization, then $-\frac{b}{a}$ (the solution of $ax + b = 0$) is called a **root of multiplicity n.**

EXAMPLE 3

Solve $x^2 = c$ for x.

Solution

$$x^2 = c$$

$$x^2 - c = 0$$

$$x^2 - (\sqrt{c})^2 = 0 \qquad c \text{ is rewritten as } (\sqrt{c})^2\text{. The resulting difference of squares can be factored.}$$

$$(x - \sqrt{c})(x + \sqrt{c}) = 0$$

$x - \sqrt{c} = 0$	$x + \sqrt{c} = 0$	The Property of Zero Products is applied.
$x = \sqrt{c}$	$x = -\sqrt{c}$	

The solutions are $\sqrt{c}$ and $-\sqrt{c}$. This can also be written $\pm\sqrt{c}$.

Observe that in Example 3, it makes no difference whether c is positive or negative. If c is positive, say $c = 10$, then $x = \sqrt{10}$ or $x = -\sqrt{10}$. If c is negative, say $c = -10$, then $x = i\sqrt{10}$ or $x = -i\sqrt{10}$.

Property of Square Roots
If $A^2 = c$, where A represents a variable expression, then $A = \pm\sqrt{c}$.

EXAMPLE 4

Solve $(x - 4)^2 = 7$.

Solution

$$(x - 4)^2 = 7 \qquad \text{The equation has the form } A^2 = 7\text{, where } A \text{ is } x - 4.$$

$$x - 4 = \pm\sqrt{7} \qquad \text{By the Property of Square Roots, } A = \pm\sqrt{7}.$$

$$x = 4 \pm \sqrt{7}$$

The solutions are $4 + \sqrt{7}$ and $4 - \sqrt{7}$, which we write as $4 \pm \sqrt{7}$.

Completing the Square

Solving $x^2 + 4x - 9 = 0$ presents a problem since the trinomial cannot be factored easily. Example 5 demonstrates a method called **completing the square** which allows us to rewrite the equation in the form $(x + a)^2 = c$. An equation of this form can then be solved by the Property of Square Roots.

EXAMPLE 5

Solve $x^2 + 4x - 9 = 0$.

Solution

$x^2 + 4x - 9 = 0$ — We want an expression of the form $(x + a)^2$ on the left.

$x^2 + 4x + \square = 9 + \square$ — Rewrite so that only terms in x are on the left. The symbol $\square$ refers to the number to be added to each side.

$x^2 + 4x + 4 = 9 + 4$ — Take half the coefficient of x and square it. $\left(\frac{1}{2} \cdot 4\right)^2 = (2)^2 = 4$. Add 4 to both sides.

$(x + 2)^2 = 13$ — The left side is now a perfect trinomial square. Thus, we can use the Property of Square Roots.

$$x + 2 = \pm\sqrt{13}$$

$$x = -2 \pm \sqrt{13}$$

The solutions are $-2 \pm \sqrt{13}$.

Completing the Square

An expression of the form $x^2 + bx$ becomes a perfect trinomial square when $\left(\frac{b}{2}\right)^2$ is added, that is,

$$x^2 + bx + \left(\frac{b}{2}\right)^2 = \left(x + \frac{b}{2}\right)^2.$$

To apply the method of completing the square to the equation $3x^2 + 6x - 4 = 0$, an equation whose x^2-coefficient is not 1, we must first multiply both sides of this equation by $\frac{1}{3}$ so that the coefficient of x^2 is 1. The resulting equivalent equation, $x^2 + 2x - \frac{4}{3} = 0$, can then be solved by using the technique of Example 5. We now use the method of completing the square to solve a general quadratic equation.

EXAMPLE 6

Solve for x: $ax^2 + bx + c = 0$. Assume that $a \neq 0$.

Solution

$$ax^2 + bx + c = 0$$

Multiply both sides by $\frac{1}{a}$ to make the coefficient of x^2 equal to 1. Then complete the square.

$$x^2 + \frac{b}{a}x + \frac{c}{a} = 0$$

$$x^2 + \frac{b}{a}x + \square = -\frac{c}{a} + \square$$

The coefficient of x is $\frac{b}{a}$. Thus, $\left(\frac{1}{2} \cdot \frac{b}{a}\right)^2$ must be added to both sides to complete the square.

$$x^2 + \frac{b}{a}x + \left(\frac{b}{2a}\right)^2 = -\frac{c}{a} + \left(\frac{b}{2a}\right)^2$$

$$\left(x + \frac{b}{2a}\right)^2 = -\frac{c}{a} \cdot \frac{4a}{4a} + \frac{b^2}{4a^2}$$

Combine fractions.

$$\left(x + \frac{b}{2a}\right)^2 = \frac{b^2 - 4ac}{4a^2}$$

Apply the Property of Square Roots.

$$x + \frac{b}{2a} = \pm\sqrt{\frac{b^2 - 4ac}{4a^2}}$$

$$x = -\frac{b}{2a} \pm \frac{\sqrt{b^2 - 4ac}}{|2a|}$$

$$x = \frac{-b \pm \sqrt{b^2 - 4ac}}{2a}$$

For any $a \neq 0$, the expression $\pm\frac{1}{|2a|}$ represents $\pm\frac{1}{2a}$. Thus we drop the absolute value symbols.

The solutions are $\dfrac{-b \pm \sqrt{b^2 - 4ac}}{2a}$.

In Example 6 we have shown that we can determine the solutions of a quadratic equation by substituting its coefficients and constant term into a formula. We have proven the following important result.

The Quadratic Formula

Given an equation $ax^2 + bx + c = 0$, with $a \neq 0$, its solutions can be determined by the formula

$$x = \frac{-b \pm \sqrt{b^2 - 4ac}}{2a}.$$

EXAMPLE 7

Use the quadratic formula to solve $3x^2 + 2x - 4 = 0$.

Solution

$$3x^2 + 2x - 4 = 0$$

$$x = \frac{-b \pm \sqrt{b^2 - 4ac}}{2a}$$ Write the quadratic formula. Here $a = 3$, $b = 2$, and $c = -4$.

$$x = \frac{-2 \pm \sqrt{2^2 - 4(3)(-4)}}{2(3)}$$

$$x = \frac{-2 \pm \sqrt{52}}{6}$$

$$x = \frac{-2 \pm 2\sqrt{13}}{6} = \frac{-1 \pm \sqrt{13}}{3}$$

The solutions are $\frac{-1 \pm \sqrt{13}}{3}$.

EXAMPLE 8

Use the quadratic formula to solve $4 = 12y - 9y^2$.

Solution

$$4 = 12y - 9y^2$$

$$9y^2 - 12y + 4 = 0$$ To determine the values of a, b, and c, first rewrite the equation in standard form.

$$y = \frac{-b \pm \sqrt{b^2 - 4ac}}{2a}$$ In this problem $a = 9$, $b = -12$, and $c = 4$.

$$y = \frac{-(-12) \pm \sqrt{(-12)^2 - 4(9)(4)}}{2(9)}$$

$$y = \frac{12 \pm \sqrt{0}}{18} = \frac{12}{18} = \frac{2}{3}$$ There is only one root, a root of multiplicity 2.

The solution is $\frac{2}{3}$.

EXAMPLE 9

Use the quadratic formula to solve $x^2 + x = -2$.

Solution

$$x^2 + x + 2 = 0$$ Write the equation in standard form.

$$x = \frac{-1 \pm \sqrt{1^2 - 4(1)(2)}}{2(1)}$$ In this problem $a = 1, b = 1$, and $c = 2$.

$$x = \frac{-1 \pm \sqrt{-7}}{2}$$ The radicand is negative; thus the solutions are imaginary.

$$x = \frac{-1 \pm i\sqrt{7}}{2}$$

The solutions are $\frac{-1 + i\sqrt{7}}{2}$ and $\frac{-1 - i\sqrt{7}}{2}$.

The last three examples demonstrate the three different situations which can arise when we solve a quadratic equation. For real numbers a, b, and c, the expression $b^2 - 4ac$, called the **discriminant,** can be used to determine the nature of the solutions.

Discriminant	Nature of Solutions
$b^2 - 4ac > 0$	two different real number solutions
$b^2 - 4ac = 0$	one real number solution (of multiplicity 2) sometimes called a **double root.**
$b^2 - 4ac < 0$	two imaginary number solutions, which are conjugates of one another

EXAMPLE 10

Use the discriminant to determine the nature of the solutions of each equation.

(a) $4x^2 - 20x = -25$ (b) $x^2 + x = -3$

Solution

(a) $4x^2 - 20x + 25 = 0$ Write the equation in standard form.

$$b^2 - 4ac = (-20)^2 - 4(4)(25)$$ The discriminant is $b^2 - 4ac$; here, $a = 4, b = -20$, and $c = 25$.

$$= 400 - 400 = 0$$

Since the discriminant is 0, this quadratic equation has one real root (of multiplicity 2).

(b) $x^2 + x + 3 = 0$ The equation is written in standard form.

$$b^2 - 4ac = (1)^2 - 4(1)(3)$$ Here, $a = 1, b = 1$, and $c = 3$.

$$= 1 - 12 = -11$$

Since the discriminant (-11) is negative, the solutions of this quadratic are imaginary.

Figure 2.2

EXAMPLE 11

When an object is dropped from a height s_0 feet above the ground, its height s (in feet) after t seconds is given by the formula

$$s = s_0 - 16t^2.$$

Suppose a rock is dropped off a cliff 320 ft above a body of water. How long will it take for the rock to hit the water?

Solution

$s = s_0 - 16t^2$ — Here, $s_0 = 320$ ft, the height from which the rock was dropped.

$s = 320 - 16t^2$

$0 = 320 - 16t^2$ — We want to find the time t when the rock hits the water, that is, when the height s is 0.

$16t^2 = 320$ — Now solve for t.

$t^2 = 20$

$t = \pm\sqrt{20} = \pm 2\sqrt{5}$ — Since the negative value for t (time) makes no sense, only the positive value is a solution.

The rock takes $2\sqrt{5}$ seconds (roughly 4.5 seconds) to hit the water.

EXERCISE SET 2.3

In Exercises 1–32, solve the equation by factoring.

1. $x^2 + 9x + 8 = 0$
2. $y^2 - 5y + 6 = 0$
3. $s^2 - 3s - 10 = 0$
4. $t^2 + 4t - 21 = 0$
5. $u^2 - 8u + 16 = 0$
6. $m^2 + 12m + 36 = 0$
7. $n^2 - 64 = 0$
8. $p^2 + 100 = 0$
9. $3v^2 - 18v + 15 = 0$
10. $2q^2 + 16q - 18 = 0$
11. $5r^2 + 18r - 8 = 0$
12. $25z^2 - 9 = 0$
13. $16u^2 + 49 = 0$
14. $9x^2 - 6x + 1 = 0$
15. $y^2 + 10y = 0$
16. $2m^2 - m = 0$
17. $4s^2 + 4s + 1 = 0$
18. $7t^2 - 13t + 6 = 0$
19. $15n^2 - 37n - 8 = 0$
20. $6x^2 + 29x + 20 = 0$
21. $7 - z^2 = 16$
22. $y^2 = y + 12$
23. $w^2 + 10w = 24$
24. $u^2 + 14u = -45$
25. $x^2 - 4 = 3x$
26. $4p^2 + 9 = 17$
27. $3q - q^2 = -54$
28. $-60 - s^2 = 19s$
29. $n(12n - 35) = 3$
30. $r(4r - 23) = -15$
31. $(3w - 1)^2 = 6w - 3$
32. $27v(1 - 3v) = 9v + 1$

In Exercises 33–44, solve the equation by completing the square.

33. $x^2 + 4x - 21 = 0$
34. $u^2 + 10u = 24$
35. $s^2 - 14s = 15$
36. $v^2 - 16v - 36 = 0$
37. $y^2 + 6y + 11 = 0$
38. $x^2 - 8x + 21 = 0$
39. $t^2 - 5t = -6$
40. $w^2 + 11w + 30 = 0$
41. $4z^2 + 16z + 15 = 0$
42. $3p^2 - 20p = 32$
43. $5m^2 - 4m - 3 = 0$
44. $3n^2 + 6n = 1$

In Exercises 45–76, use the quadratic formula to solve the equation.

45. $x^2 + 5x + 4 = 0$
46. $y^2 + 11y + 18 = 0$
47. $v^2 - 8v + 16 = 0$
48. $w^2 + 4w - 60 = 0$
49. $z^2 + z + 10 = 0$
50. $p^2 + p - 10 = 0$
51. $q^2 - 64 = 0$
52. $s^2 + 64 = 0$
53. $v^2 - 8v + 13 = 0$
54. $4t^2 + 4t + 1 = 0$
55. $3x^2 - 7x - 6 = 0$
56. $3y^2 - 7y + 6 = 0$
57. $u^2 - 11 = 0$
58. $p^2 + 11p = 0$

59. $s^2 + 3s - 5 = 0$
60. $5t^2 - 11t + 2 = 0$
61. $36q^2 + 12q + 1 = 0$
62. $w^2 + 7 = 0$
63. $-x^2 + 2x - 2 = 0$
64. $-x^2 - 2x + 2 = 0$
65. $2n^2 = 9n$
66. $7 - 14t + 5t^2 = 0$
67. $3q^2 - 8q = 2$
68. $9 - 16z^2 = 0$
69. $x(4x - 3) = -1$
70. $(u + 8)^2 = 57 - u^2$
71. $8r - r^2 - 14 = 0$
72. $s^2 - 2s = 24$
73. $5y^2 - 8y + 1 = 0$
74. $5t^2 - 6t + 3 = 0$
75. $2m^2 + 12m + 6 = 0$
76. $16s^2 + 9 = 24s$

In Exercises 77–90, use the discriminant to determine the nature of the solutions without solving.

77. $x^2 - 5x + 3 = 0$
78. $25x^2 + 9 = 30x$
79. $x^2 + x + 1 = 0$
80. $x^2 + x - 1 = 0$
81. $3x^2 - 6x + 5 = 8$
82. $2x^2 - 7x + 7 = 3$
83. $9x^2 = 6x - 1$
84. $7x - 3 = 9x^2$
85. $3.7x^2 - 4.5x + 1.4 = 0$
86. $8.7x^2 + 0.08x - 0.3 = 0$
87. $0.2x^2 + 1.6x + 3.2 = 0$
88. $18.7 + 5.5x^2 = 19.2x$
89. $1897x^2 + 1650x + 374 = 0$
90. $508x^2 + 482x + 110 = 0$

In Exercises 91–96, use the quadratic formula to approximate the solutions of the given equation to the nearest tenth.

91. $4x^2 + 7x + 2 = 0$
92. $5x^2 + 4x - 4 = 0$
93. $2x^2 + 2x - 5 = 0$
94. $x^2 - 5x + 1 = 0$
95. $6.2x^2 - 8.1x + 2.5 = 0$
96. $16.85x^2 - 1.5x - 20 = 0$

97. When an object is propelled upward from a height of s_0 feet above the ground, with a velocity of v_0 ft/sec, its height s (in feet) after t seconds is given by the formula

$$s = s_0 + v_0t - 16t^2.$$

If an object is shot upward with a velocity of 88 ft/sec from the top of a 500 ft tall building, after how many seconds will it be 620 ft above the ground? Explain the reason you get two answers.

98. Approximately how many seconds after it is shot upward will the object in Exercise 97 hit the ground?

99. If an object is thrown downward from a height of s_0 meters above the ground with a speed of v_0 m/sec, its height s (in meters) after t seconds is given by the formula

$$s = s_0 - v_0t - 9.8t^2.$$

How long does it take an object thrown downward with a speed of 7 m/sec from the top of a 30 meter tall building to hit the ground?

100. Derive the general solution for the situation described in Exercise 99 by taking $s = 0$ and solving for t in terms of s_0 and v_0.

Superset

In Exercises 101–106, for what real number value(s) of a will the equation have one real root of multiplicity 2. If no such value exists, then state this. (*Hint:* the discriminant must be zero.)

101. $2x^2 - ax - 1 = 0$
102. $x^2 - ax - 7 = 11$
103. $x^2 + ax + 1 = 0$
104. $2x^2 + ax + a = 0$
105. $x^2 + (a - 3)x + a^2 = 0$
106. $ax^2 + 2ax + a = 0$

In Exercises 107–112, use the quadratic formula to solve the equation for the specified value.

107. $3 + Vr = r^2 + 2r$, solve for r.
108. $5s^2 + 1 = As + B$, solve for s.
109. $\dfrac{A}{x - A} + \dfrac{1}{x + 1} = 1$, solve for x.
110. $\dfrac{y + 2}{y} - my = 3$, solve for y.
111. $x^2 + 2xy + 3x + y^2 = 0$, solve for x.
112. $x^2 + 2xy + 3x + y^2 = 0$, solve for y.

In each of the figures in Exercises 113–116 the area of the shaded region is 10 square feet. Write an equation to describe this fact, and then solve for x. (*Remember:* x is positive.)

113.

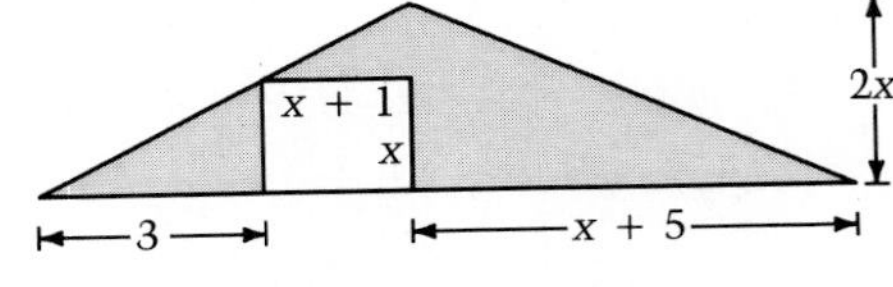

114.

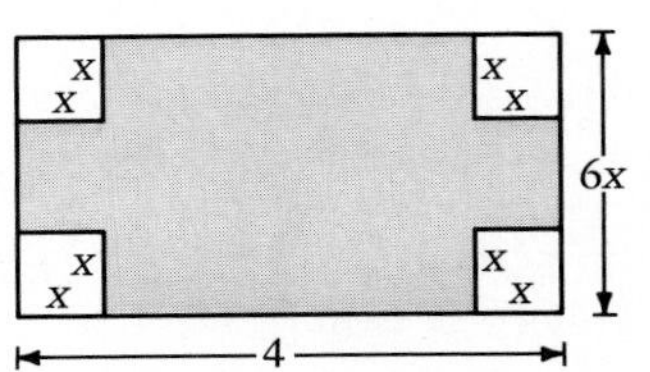

115.

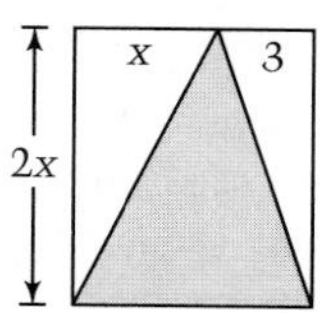

116.

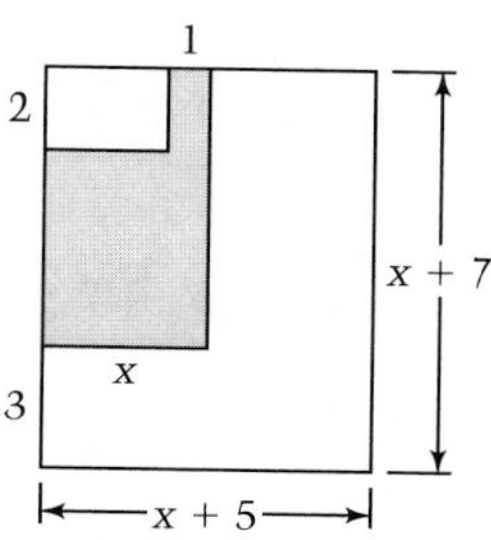

117. For the general quadratic equation $ax^2 + bx + c = 0$ show that

(a) the sum of the roots is $-\frac{b}{a}$, and

(b) the product of the roots is $\frac{c}{a}$.

118. (See Exercise 117.) Determine m and p so that the sum of the roots of the equation $mx^2 + 8x + p = 0$ is 3 and the product of the roots is 6.

119. The sum S of the first n positive integers is given by the formula $S = \frac{1}{2}n(n + 1)$. Use this fact to determine the value of n for which $1 + 2 + 3 + \cdots + n = 171$.

120. The number D of diagonals in an n-sided convex polygon is given by the formula $D = \frac{1}{2}n(n - 3)$. Determine the number of sides of a polygon having 35 diagonals.

2.4 Nonlinear Equations

We now consider a variety of situations where we can use the Property of Zero Products and the quadratic formula to solve equations.

EXAMPLE 1

Solve $x^5 - 5x^4 = 4x^3$.

Solution

$$x^5 - 5x^4 = 4x^3$$

$$x^3(x^2 - 5x - 4) = 0$$

Begin by rewriting all nonzero terms on one side. Then apply the Property of Zero Products.

$x^3 = 0$	$x^2 - 5x - 4 = 0$
	$x = \dfrac{-b \pm \sqrt{b^2 - 4ac}}{2a}$
$x = 0$	$x = \dfrac{5 \pm \sqrt{41}}{2}$

$a = 1, b = -5, c = -4.$

Zero is a root of multiplicity 3 since zero is the solution of the equation $x^3 = 0$.

The solutions are 0, $\dfrac{5 + \sqrt{41}}{2}$, and $\dfrac{5 - \sqrt{41}}{2}$.

The equation given in Example 1 is called a polynomial equation since each side of the equation is a polynomial. The **standard form** of a polynomial equation is obtained by collecting all the nonzero terms so that they appear on the same side of the equation. The degree of the resulting polynomial determines the degree of the equation. The equation in Example 1 is of fifth degree. Observe that it has one root of multiplicity 3, and two other roots, each of multiplicity 1. It is no accident that the sum of these multiplicities is 5, the degree of the equation. By the Fundamental Theorem of Algebra (Chapter 4), an nth degree polynomial equation has precisely n solutions (if we count multiplicities). Thus, for the third degree equation in the next example, we can expect to find three solutions (or perhaps two solutions, where one has multiplicity 1, and the other has multiplicity 2).

EXAMPLE 2

Solve $3x^3 + 2x^2 - 12x - 8 = 0$.

Solution

$$3x^3 + 2x^2 - 12x - 8 = 0$$

There is no common factor. The four terms suggest factoring by grouping.

$$x^2(3x + 2) - 4(3x + 2) = 0$$

$$(x^2 - 4)(3x + 2) = 0$$

The Property of Zero Products is now applied.

$x^2 - 4 = 0$	$3x + 2 = 0$
$x = \pm 2$	$x = -\frac{2}{3}$

The solutions are 2, -2, and $-\frac{2}{3}$.

When solving fractional equations, as in the next example, we multiply both sides by a variable expression. Recall that this method can introduce solutions that do not satisfy the original equation. Such numbers are called extraneous roots. Thus, whenever the sides of an equation have been multiplied by a variable expression, the solutions *must* be checked in the original equation.

EXAMPLE 3

Solve $\dfrac{x + 7}{x - 1} + \dfrac{x - 1}{x + 1} = \dfrac{4}{x^2 - 1}$.

Solution

$$\left(\frac{x+7}{x-1}+\frac{x-1}{x+1}\right)(x^2-1)=\left(\frac{4}{x^2-1}\right)(x^2-1)$$

Begin by multiplying both sides by the LCD (x^2-1).

$$\frac{x+7}{\cancel{x-1}}\cdot\overset{(x+1)}{\cancel{(x^2-1)}}+\frac{x-1}{\cancel{x+1}}\overset{(x-1)}{\cancel{(x^2-1)}}=\frac{4}{\cancel{x^2-1}}\cdot\cancel{(x^2-1)}$$

$$(x+7)(x+1)+(x-1)(x-1)=4$$

$$x^2+8x+7+x^2-2x+1=4$$

$$2x^2+6x+4=0$$

$$2(x+2)(x+1)=0$$

The Property of Zero Products is applied.

$x+2=0$	$x+1=0$
$x=-2$	$x=-1$

Must Check:

$$\mathbf{x=-2:}\quad \frac{-2+7}{-2-1}+\frac{-2-1}{-2+1}=\frac{4}{(-2)^2-1}$$

$$\frac{5}{-3}+3=\frac{4}{3}$$

True.

$$\mathbf{x=-1:}\quad \frac{-1+7}{-1-1}+\frac{-1-1}{-1+1}=\frac{4}{(-1)^2-1}$$

$$\frac{6}{-2}+\frac{-2}{0}=\frac{4}{0}$$

$\frac{-2}{0}$ and $\frac{4}{0}$ are not defined.

The only solution is -2.

Equations Quadratic in Form

An equation is said to be **quadratic in form** if replacing an expression with a single variable makes the equation quadratic in the new variable. Consider the equation $x^6+2x^3-3=0$.

$$x^6+2x^3-3=0$$

$$(x^3)^2+2(x^3)-3=0 \quad \text{Replace } x^3 \text{ with } u.$$

$$u^2+2u-3=0 \quad \text{The result is quadratic in } u.$$

As another example, consider $y^{2/3}-y^{1/3}+10=0$. It can be rewritten as $(y^{1/3})^2-(y^{1/3})+10=0$. Replacing $y^{1/3}$ with u produces $u^2-u+10=0$ which is quadratic in u.

EXAMPLE 4

Solve $x^4 + 3x^2 - 10 = 0$.

Solution

$$x^4 + 3x^2 - 10 = 0$$
$$(x^2)^2 + 3x^2 - 10 = 0$$
$$u^2 + 3u - 10 = 0 \qquad x^2 \text{ is replaced by } u.$$
$$(u + 5)(u - 2) = 0 \qquad \text{The Property of Zero Products is now applied.}$$

$u + 5 = 0$	$u - 2 = 0$	
$u = -5$	$u = 2$	
$x^2 = -5$	$x^2 = 2$	Remember that u represents x^2.
$x = \pm\sqrt{-5}$	$x = \pm\sqrt{2}$	
$x = \pm i\sqrt{5}$		

The solutions are $\pm i\sqrt{5}$ and $\pm\sqrt{2}$.

EXAMPLE 5

Solve $5p^{-2} + 4p^{-1} - 1 = 0$.

Solution

$$5p^{-2} + 4p^{-1} - 1 = 0$$
$$5(p^{-1})^2 + 4(p^{-1}) - 1 = 0$$
$$5u^2 + 4u - 1 = 0 \qquad p^{-1} \text{ is replaced by } u.$$
$$(5u - 1)(u + 1) = 0 \qquad \text{The Property of Zero Products is applied.}$$

$5u - 1 = 0$	$u + 1 = 0$	
$u = \frac{1}{5}$	$u = -1$	
$p^{-1} = \frac{1}{5}$	$p^{-1} = -1$	Remember that u represents p^{-1}.
$\frac{1}{p} = \frac{1}{5}$	$\frac{1}{p} = -1$	Recall that $p^{-1} = \frac{1}{p}$.
$p = 5$	$p = -1$	

The solutions are -1 and 5.

Radical Equations

In order to solve a radical equation such as $\sqrt{x - 1} = x - 3$, it is useful to begin by rewriting it in a form that does not contain radicals. We can accomplish this by squaring both sides of the equation. However, this technique may introduce extraneous roots.

We now apply this technique to $\sqrt{x - 1} + 3 = x$.

$$\sqrt{x - 1} + 3 = x$$

$$\sqrt{x - 1} = x - 3 \qquad \text{The radical term is isolated.}$$

$$x - 1 = x^2 - 6x + 9 \qquad \text{Squaring eliminates the radical.}$$

$$0 = x^2 - 7x + 10 \qquad \text{One side must be zero to use the Property of Zero Products.}$$

$$0 = (x - 5)(x - 2)$$

$x - 5 = 0$	$x - 2 = 0$
$x = 5$	$x = 2$

When you check the numbers 5 and 2, you find that 5 does satisfy the original equation, but 2 does not. Thus 5 is the only solution. (2 is an extraneous root.)

If both sides of an equation are raised to any positive integral power, then the solution set of the resulting equation will contain the solutions of the original equation. However, it may also contain extraneous roots. Thus you *must* always check solutions found after raising both sides of an equation to the same power.

Power Property for Equations

If $A = B$ and if n is a positive integer, then the solution set of the equation $A^n = B^n$ contains all the solutions of the equation $A = B$, but may also contain extraneous roots. Solutions *must* be checked.

EXAMPLE 6

Solve $\sqrt[3]{x^2 - 1} = 2$.

Solution

$$\sqrt[3]{x^2 - 1} = 2 \qquad \text{Cube both sides to eliminate the radical term.}$$

$$(\sqrt[3]{x^2 - 1})^3 = 2^3$$

$$x^2 - 1 = 8$$

$$x^2 = 9$$

$$x = \pm 3$$

Must Check:

$$x = 3: \quad \sqrt[3]{3^2 - 1} = 2$$
$$\sqrt[3]{8} = 2 \qquad \text{True.}$$

$$x = -3: \quad \sqrt[3]{(-3)^2 - 1} = 2$$
$$\sqrt[3]{8} = 2 \qquad \text{True.}$$

The solutions are ± 3.

EXAMPLE 7

Solve $1 + \sqrt{y + 4} = \sqrt{3y + 1}$.

Solution

$$1 + \sqrt{y + 4} = \sqrt{3y + 1} \qquad \text{Use the Power Property.}$$
$$(1 + \sqrt{y + 4})^2 = (\sqrt{3y + 1})^2$$
$$1 + 2\sqrt{y + 4} + y + 4 = 3y + 1 \qquad \text{Isolate the radical term.}$$
$$2\sqrt{y + 4} = 2y - 4$$
$$\sqrt{y + 4} = y - 2$$
$$(\sqrt{y + 4})^2 = (y - 2)^2 \qquad \text{Use the Power Property again.}$$
$$y + 4 = y^2 - 4y + 4$$
$$0 = y^2 - 5y$$
$$0 = y(y - 5) \qquad \text{The Property of Zero Products is applied.}$$

$y = 0$	$y - 5 = 0$
	$y = 5$

Must Check:

$$y = 0: \quad 1 + \sqrt{y + 4} = \sqrt{3y + 1}$$
$$1 + \sqrt{0 + 4} = \sqrt{3 \cdot 0 + 1}$$
$$1 + 2 = 1$$
$$3 = 1 \qquad \text{False.}$$

$$y = 5: \quad 1 + \sqrt{y + 4} = \sqrt{3y + 1}$$
$$1 + \sqrt{5 + 4} = \sqrt{3 \cdot 5 + 1}$$
$$1 + 3 = 4$$
$$4 = 4 \qquad \text{True.}$$

The only solution is 5.

EXERCISE SET 2.4

In Exercises 1–26, solve the equation.

1. $3s^3 + 8s = 14s^2$
2. $5m^6 + 4m^5 = 12m^4$
3. $v^4 + v^3 + v^2 = 0$
4. $w^4 = w^3 + w^2$
5. $2t^5 + 4t^3 = 7t^4$
6. $n^8 + 8n^6 = 5n^7$
7. $2p^3 - p^2 - 18p + 9 = 0$
8. $s^3 + 3s^2 - 25s - 75 = 0$
9. $3t^3 - 2t^2 + 48t - 32 = 0$
10. $8v^3 - 3v^2 - 32v + 12 = 0$
11. $r^4 - r^3 + 27r - 27 = 0$
12. $x^4 + 7x^3 - 8x - 56 = 0$
13. $5u^5 - 20u^4 - 4u^3 + 16u^2 = 0$
14. $7z^4 - 6z^3 + 28z^2 - 24z = 0$
15. $\dfrac{x+5}{x^2 - 2x + 1} = 2$
16. $\dfrac{2x^2 - 2}{x^2 - 4x + 11} = 3$
17. $\dfrac{1}{x} - \dfrac{2}{3} = \dfrac{1}{3x^2}$
18. $\dfrac{7}{x^2} - 1 = \dfrac{9}{4x}$
19. $\dfrac{3}{x-1} + \dfrac{4}{x+1} = 1$
20. $\dfrac{3}{x+5} = 1 - \dfrac{4}{x-5}$
21. $\dfrac{1}{x-2} = 1 + \dfrac{2}{x^2 - 2x}$
22. $\dfrac{5}{x+2} + \dfrac{10}{x^2 + 2x} = -2$
23. $\dfrac{3x}{x+2} - \dfrac{x}{x-1} = \dfrac{18}{x^2 + x - 2}$
24. $\dfrac{2x}{x+3} = \dfrac{x}{x-2} - \dfrac{13}{x^2 + x - 6}$
25. $\dfrac{6x}{2x-3} - \dfrac{2}{x} = 2$
26. $\dfrac{3x+2}{x-1} + \dfrac{2x}{x+1} = \dfrac{7x+3}{x^2 - 1}$

In Exercises 27–40, use a u-substitution to solve the equation.

27. $x^4 - 5x^2 - 14 = 0$
28. $x^4 - 11x^2 + 24 = 0$
29. $y^4 + 12 = 7y^2$
30. $y^4 = 2y^2 + 15$
31. $(t+2)^4 - 8(t+2) = 0$
32. $27(p-1) - (p-1)^4 = 0$
33. $x^4 = 81$
34. $x^4 = 16$
35. $3p^{-2} - 4p^{-1} = 0$
36. $\dfrac{6}{p^2} - \dfrac{5}{p} = -1$
37. $(p-1)^2 + 4(p-1) + 3 = 0$
38. $(t+4)^2 - 10(t+4) + 21 = 0$
39. $5x^{2/3} - 11x^{1/3} + 2 = 0$
40. $x^{1/2} - 8x^{1/4} + 15 = 0$

In Exercises 41–50, solve the radical equation.

41. $\sqrt[3]{5x+7} = 3$
42. $\sqrt[3]{3x-1} = 8$
43. $x = 5 + \sqrt{x+7}$
44. $3 + \sqrt{5-x} = x$
45. $\sqrt{x} + \sqrt{x-3} = 3$
46. $\sqrt{2x} + \sqrt{3-x} = 3$
47. $\sqrt{2x-5} - \sqrt{x-3} = 1$
48. $\sqrt{3x+1} - \sqrt{x-1} = 2$
49. $\sqrt{3x+1} = 2 + \sqrt{1-x}$
50. $\sqrt{2x-2} = 1 + \sqrt{x+6}$

Superset

In Exercises 51–56, solve the equation.

51. $\sqrt{1-\sqrt{x}} = 1 - \sqrt{x}$
52. $\sqrt{3-\sqrt{3-x}} = \sqrt{x}$
53. $1 + \dfrac{1}{\sqrt{x}+1} = \dfrac{1}{\sqrt{x}-1}$
54. $\dfrac{4}{3} - \dfrac{1}{\sqrt{x}+1} = \dfrac{1}{\sqrt{x}-1}$
55. $\dfrac{12}{\sqrt{x-9}} - \sqrt{x} = \dfrac{\sqrt{x-9}}{-2}$
56. $\dfrac{2}{1-x} = \dfrac{2}{1-\sqrt{x}} - \dfrac{1}{1+\sqrt{x}}$
57. In the equation

$$T = 2\pi\sqrt{\frac{x}{32}}$$

T is the time in seconds that it takes for one complete swing of a pendulum x ft long. Find the approximate length of a certain pendulum, if a complete swing takes 3.1 seconds. (Round to the nearest tenth and use 3.14 for π.)

58. The equation $V = \sqrt{64h}$ gives the speed V (in ft/sec), of an object right before it strikes the ground, after it has been dropped from a height of h ft. From what height must an object have been dropped if its speed is 30 mph when it strikes the ground?

In Exercises 59–62, solve the equation by first making a u-substitution.

59. $x - 3\sqrt{x} + 2 = 0$

60. $2\sqrt{x} - 7\sqrt[4]{x} + 6 = 0$

61. $\sqrt[6]{x} + 2\sqrt[3]{x} - 1 = 0$

62. $(\sqrt{x} + 1)^2 + \sqrt{x} - 1 = 0$
(*Hint:* Let $u = \sqrt{x} + 1$. You will need to rewrite the equation before making the substitution.)

2.5
Inequalities

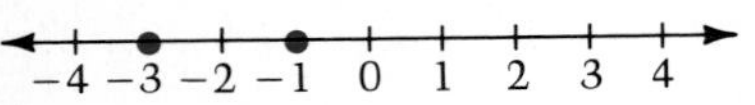

Figure 2.3 Recall that $-3 < -1$ and $-1 > -3$ represent the same order relationship.

In Chapter 1 we saw that the relative position of two different points on the real number line can be used to write an inequality involving the coordinates of the points. In this section we will consider inequalities that contain variables, such as

$$3x - 2 < 7 \quad \text{or} \quad |x - 7| > 3.$$

There are four types of inequalities corresponding to the symbols

$>$ (meaning: "is greater than"),
$<$ (meaning: "is less than"),
$\geq$ (meaning: "is greater than or equal to"),
$\leq$ (meaning: "is less than or equal to").

To *solve* an inequality means to determine its **solution set,** the set of numbers that make the inequality a true statement.

Consider the linear inequality $3x - 2 < 7$. We test -5 and 4 to determine whether either is a solution.

$$\begin{aligned} 3x - 2 &< 7 \\ 3(-5) - 2 &< 7 \\ -17 &< 7 \quad \text{True.} \end{aligned} \qquad \begin{aligned} 3x - 2 &< 7 \\ 3(4) - 2 &< 7 \\ 10 &< 7 \quad \text{False.} \end{aligned}$$

The number -5 is a solution since replacing x with -5 produces a true statement. However, 4 is not a solution.

Two inequalities having the same solution set are called **equivalent inequalities.** The strategy for solving an inequality is to transform it into an equivalent inequality whose solution set is obvious. The following properties suggest ways to produce equivalent inequalities.

Addition Property for Equivalent Inequalities

Adding the same real number (or variable expression) to both sides of an inequality produces an equivalent inequality.

Positive Multiplication Property for Equivalent Inequalities

Multiplying both sides of an inequality by the same positive real number (or variable expression) produces an equivalent inequality.

Negative Multiplication Property for Equivalent Inequalities

Multiplying both sides of an inequality by the same negative real number (or variable expression) and reversing the direction of the inequality produces an equivalent inequality.

$x + 3 < 4$ is equivalent to $x + 3 + (\mathbf{-3}) < 4 + (\mathbf{-3})$

$2x > 7$ is equivalent to $\mathbf{3} \cdot 2x > \mathbf{3} \cdot 7$

$8y > 5$ is equivalent to $\mathbf{-6} \cdot 8y < \mathbf{-6} \cdot 5$

EXAMPLE 1

Solve $13x - 17 > 4x + 1$ and graph the solution set.

Solution

$$13x - 17 > 4x + 1$$
$$13x + (-4x) - 17 > 4x + (-4x) + 1$$
$$9x - 17 > 1$$

Use the Addition Property to get a single x-term on one side and a constant on the other.

$$9x - 17 + 17 > 1 + 17$$
$$9x > 18$$
$$\frac{1}{9} \cdot 9x > \frac{1}{9} \cdot 18$$
$$x > 2$$

Now use the Positive Multiplication Property to make the x-coefficient 1.

The solution set is the set of all real numbers greater than 2.

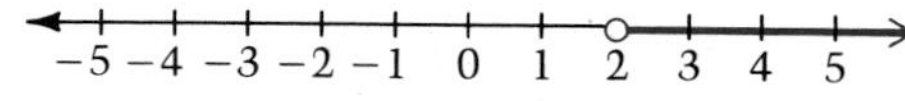

Figure 2.4

Note: A good way to check your work is to select one value in your solution set and verify, by substitution, that it is a solution. Then select one value not in your solution set and verify, by substitution, that it is not a solution. Although this method of checking is not foolproof, it will catch most common mistakes.

EXAMPLE 2

Solve $2 - 3x \geq 17$, and graph the solution set.

Solution

$$2 - 3x \geq 17 \qquad \text{Use the Addition Property.}$$
$$-3x \geq 15$$
$$-\frac{1}{3}(-3x) \leq -\frac{1}{3}(15) \qquad \text{Use the Negative Multiplication Property. Remember to reverse the direction of the inequality.}$$
$$x \leq -5$$

The solution set is the set of all real numbers less than or equal to -5.

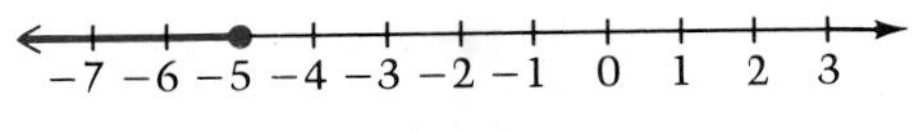

Figure 2.5

In addition to graphing, *set notation* and *interval notation* provide two more ways of describing the solution set of an inequality.

In **set notation** the solution sets for Examples 1 and 2 are written:

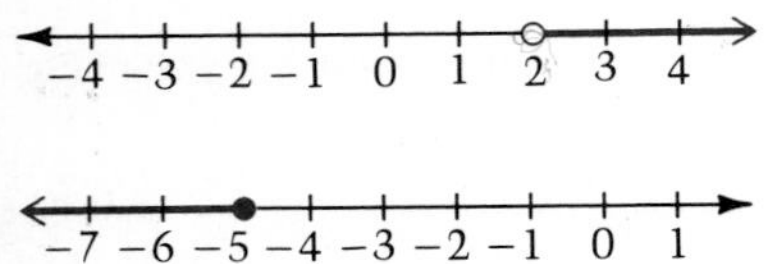

Figure 2.6

$\{x|x > 2\}$ This is read "The set of all real numbers x such that x is greater than 2."

$\{x|x \leq -5\}$ This is read "The set of all real numbers x such that x is less than or equal to -5."

In **interval notation** the solution set in Example 1 is written $(2, \infty)$, while the solution set in Example 2 is written $(-\infty, -5]$. The symbols $-\infty$ and ∞ are not real numbers. The ∞ in (a, ∞) denotes that the interval is unbounded and contains all real numbers greater than a; the $-\infty$ in $(-\infty, a)$ denotes that the interval is unbounded and contains all real numbers less than a.

In the intervals in Table 2.1, a and b are called **endpoints.** If endpoints are included, as in $[a, b]$, the interval is called a **closed interval.** If the endpoints are not included, as in (a, b) or (a, ∞), the interval is called an **open interval.** Intervals such as $[a, b)$ or $(-\infty, a]$, which include only one endpoint, are called **half-open intervals.** Notice that a square bracket is used to indicate that an endpoint is included, and a parenthesis is used when the endpoint is not included.

In some circumstances we are given that a certain expression has values between two numbers. For example, suppose we know that $-5 < 7 - 3x < 9$. In this case, we can solve for x by applying the Properties for Equivalent Inequalities to all three parts of this inequality. We demonstrate this process in Example 3.

Table 2.1

Interval Notation	Type of Interval	Set Notation	Graph
(a, b)	open	$\{x \mid a < x < b\}$	a b
$[a, b)$	half-open	$\{x \mid a \leq x < b\}$	a b
$(a, b]$	half-open	$\{x \mid a < x \leq b\}$	a b
$[a, b]$	closed	$\{x \mid a \leq x \leq b\}$	a b
(a, ∞)	open	$\{x \mid x > a\}$	a
$[a, \infty)$	half-open	$\{x \mid x \geq a\}$	a
$(-\infty, a)$	open	$\{x \mid x < a\}$	a
$(-\infty, a]$	half-open	$\{x \mid x \leq a\}$	a
$(-\infty, \infty)$	open	$\{x \mid x$ is a real number$\}$	

EXAMPLE 3

Solve $-5 < 7 - 3x < 9$.

Solution $$-5 < 7 - 3x < 9$$

$$(-7) - 5 < (-7) + 7 - 3x < (-7) + 9$$ Add (-7) to each part of the inequality.

$$-12 < -3x < 2$$

$$\left(-\frac{1}{3}\right)(-12) > \left(-\frac{1}{3}\right)(-3x) > \left(-\frac{1}{3}\right)2$$ Multiply each part by $-\frac{1}{3}$, and reverse the direction of each inequality symbol.

$$4 > x > -\frac{2}{3}$$

The solution set is $\{x \mid -\frac{2}{3} < x < 4\}$, which is $\left(-\frac{2}{3}, 4\right)$ in interval notation.

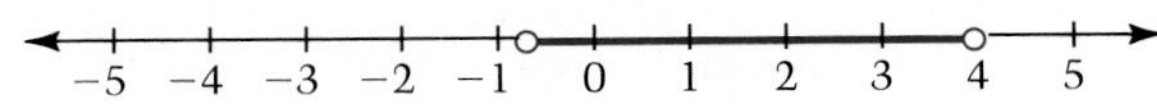

Figure 2.7

Two simple inequalities, such as

$$x > 5 \quad \text{or} \quad x \leq -4,$$

can be used to produce a **compound inequality.** There are two types of compound inequalities:

$\{x \mid x > 7$ **or** $x < -3\}$ This set is formed by combining all numbers greater than 7 or less than -3 into a single set.

$\{x \mid 2 \leq x < 9\}$
$\{x \mid x \geq 2$ **and** $x < 9\}$ This set is formed by taking those numbers common to both the sets $\{x \mid x \geq 2\}$ and $\{x \mid x < 9\}$. Note the two ways of representing this set.

EXAMPLE 4

Graph each of the following sets:
(a) $\{x \mid x < 3 \text{ and } x \geq -5\}$
(b) $\{x \mid x > 2 \text{ or } x < -2\}$
(c) $\{x \mid -10 < x < 5\}$

Solution

(a) −6 −5 −4 −3 −2 −1 0 1 2 3 4 5 "All numbers greater than or equal to -5 **and** less than 3."

(b) −4 −3 −2 −1 0 1 2 3 4 "All numbers greater than 2 **or** less than -2."

(c) −20 −15 −10 −5 0 5 10 15 20 "All numbers greater than -10 **and** less than 5."

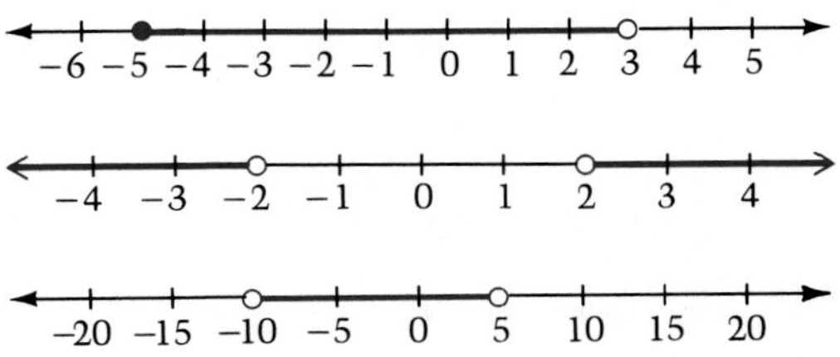

Figure 2.8

Notice that the solution set in Example 4(b) consists of the numbers in either of the intervals $(-\infty, -2)$ or $(2, \infty)$. We can represent this set by writing $(-\infty, -2) \cup (2, \infty)$, where the symbol $\cup$ signifies that we have a *union* of the two intervals.

Definition of the Union of Two Sets ($\cup$)

If A and B are two sets, then the **union** of sets A and B, written $A \cup B$, is the set of all elements which belong to A **or** B.

Absolute Value Inequalities

Suppose you wish to solve the inequality $|x| > 2$. Recall that we can think of $|x|$ as the distance between x and 0 on the real number line. The inequality $|x| > 2$ says that the distance between x and 0 is greater than 2. Since x can be any number more than 2 units away from 0, we conclude that x can be any number satisfying either of the inequalities: $x < -2$ **or** $x > 2$. Figure 2.9 represents the solution set of $|x| > 2$. It also represents the compound inequality $x > 2$ **or** $x < -2$.

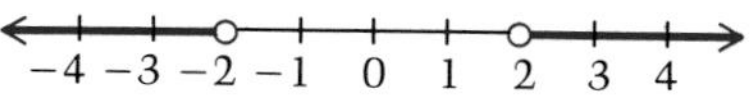

Figure 2.9

On the other hand, the inequality $|x| < 2$ says that the distance between x and 0 is less than 2. Thus x can be any number less than two units away from 0 on the real number line, that is $-2 < x < 2$. Stated another way, x can be any number simultaneously satisfying both of the inequalities: $x > -2$ **and** $x < 2$. Figure 2.10 represents the solution set of $|x| < 2$. It also represents the compound inequality $-2 < x < 2$.

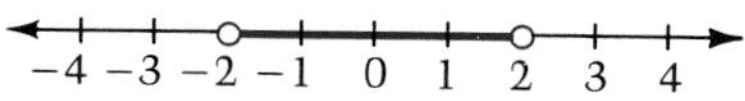

Figure 2.10

Theorem

If A is a variable expression and c is a positive real number, then

1. The solution set of the inequality $|A| > c$ is found by solving two inequalities: $A > c$ **or** $A < -c$.
2. The solution set of the inequality $|A| < c$ is found by solving the compound inequality $-c < A < c$.

In the above theorem the symbols $>$ and $<$ can be replaced by $\geq$ and $\leq$, respectively.

EXAMPLE 5

Solve $|5 - 3x| \le 8$ and describe the solution set in interval notation.

Solution

$$|5 - 3x| \le 8$$ Part 2 of the above theorem applies.

$$-8 \le 5 - 3x \le 8$$

$$(-5) - 8 \le (-5) + 5 - 3x \le (-5) + 8$$

$$-13 \le -3x \le 3$$

$$\left(-\frac{1}{3}\right)(-13) \ge \left(-\frac{1}{3}\right)(-3x) \ge \left(-\frac{1}{3}\right)(3)$$

$$\frac{13}{3} \ge x \ge -1$$

The solution set is $\{x \mid -1 \le x \le \frac{13}{3}\}$, which is $[-1, \frac{13}{3}]$ in interval notation.

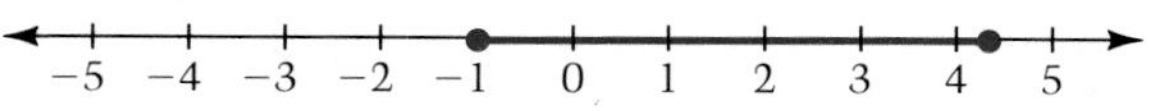

Figure 2.11

EXAMPLE 6

Solve $|2x - 3| \ge 7$ and sketch the graph.

Solution

$$|2x - 3| \ge 7$$

$$2x - 3 \ge 7 \quad \textbf{or} \quad 2x - 3 \le -7$$ Part 1 of the above Theorem applies.

$$2x \ge 10 \qquad 2x \le -4$$

$$x \ge 5 \qquad x \le -2$$

The solution set is $\{x \mid x \le -2$ **or** $x \ge 5\}$, which is $(-\infty, -2] \cup [5, \infty)$ in interval notation.

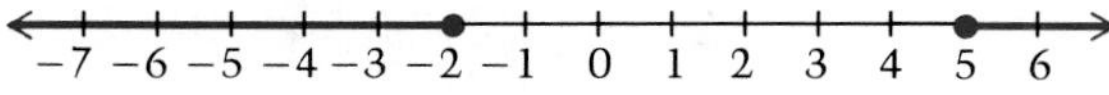

Figure 2.12

Quadratic Inequalities

In applied mathematics we frequently need to solve inequalities involving a quadratic expression. An inequality that contains a quadratic expression is called a **quadratic inequality.** To solve such an inequality, we shall use the techniques we learned for factoring polynomials as well as the simple rules about multiplying positive and negative numbers.

Suppose we wish to solve the inequality $x^2 - 4x - 12 < 0$. We begin by considering the associated equality $x^2 - 4x - 12 = 0$, which can be written in factored form as

$$(x - 6)(x + 2) = 0.$$

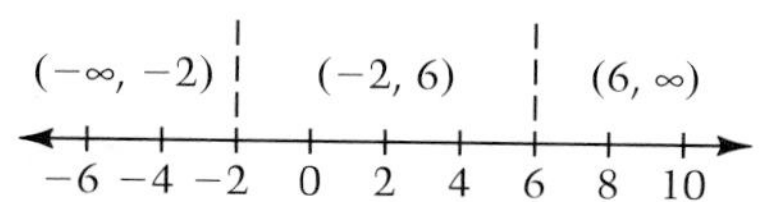

Figure 2.13

The two solutions of this equation, $x = 6$ and $x = -2$, separate the real number line into three distinct intervals: $(-\infty, -2)$, $(-2, 6)$, and $(6, \infty)$ as shown in Figure 2.13. Solving the inequality $x^2 - 4x - 12 < 0$, is equivalent to determining the interval(s) where $x^2 - 4x - 12$ is negative. We can do this by finding the intervals where the factors $(x - 6)$ and $(x + 2)$ are positive or negative.

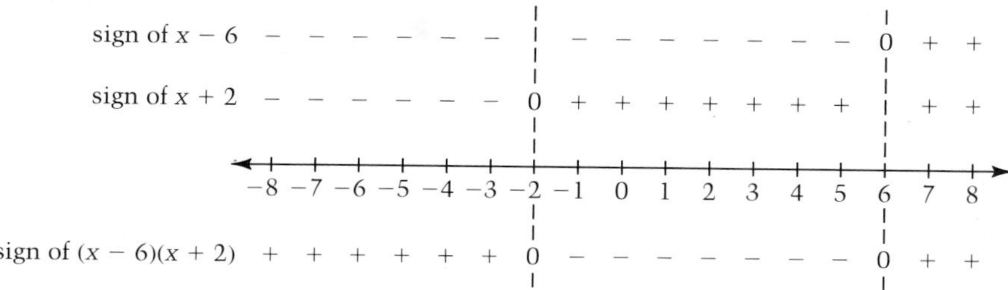

Figure 2.14 A sign chart for $(x - 6)(x + 2)$.

When both factors have the same sign, the product is positive; when the factors have different signs, the product is negative. Thus we can conclude that $(x - 6)(x + 2) = x^2 - 4x - 12$ is negative for values of x between -2 and 6, so the solution of the inequality $x^2 - 4x - 12 < 0$ is the interval $(-2, 6)$.

You might have observed in Figure 2.14 that the quadratic $x^2 - 4x - 12$ is always positive or always negative throughout each of the three intervals determined by the values $x = -2$ and $x = 6$. We can generalize this useful result.

Theorem

For a given quadratic $ax^2 + bx + c$, the real number solutions of the associated equality, $ax^2 + bx + c = 0$, break the real number line into intervals. Within each interval, $ax^2 + bx + c$ is either always positive (> 0) or always negative (< 0).

Because of this result, we need only test one value in each interval to determine whether the quadratic is positive or negative throughout the interval. Using these **test values** is a faster method for solving quadratic inequalities than the method illustrated in Figure 2.14.

EXAMPLE 7

Solve the inequality $x^2 + 4x > 21$. Describe the solution set in interval notation.

Solution

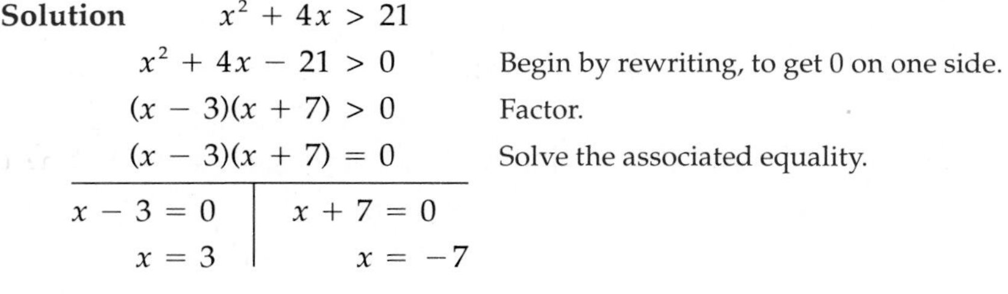

$$x^2 + 4x > 21$$

$$x^2 + 4x - 21 > 0 \qquad \text{Begin by rewriting, to get 0 on one side.}$$

$$(x - 3)(x + 7) > 0 \qquad \text{Factor.}$$

$$(x - 3)(x + 7) = 0 \qquad \text{Solve the associated equality.}$$

$x - 3 = 0$	$x + 7 = 0$
$x = 3$	$x = -7$

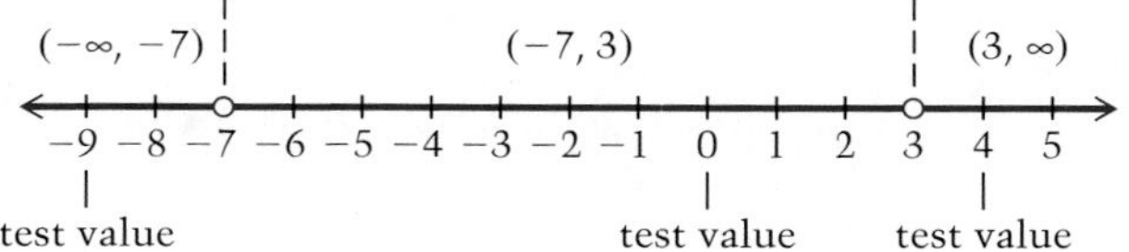

The solutions -7 and 3 break the real line into three intervals. We must select a test value from each interval.

Figure 2.15

Interval	Test Value	Value of $x^2 + 4x - 21$	Sign of $x^2 + 4x - 21$
$(-\infty, -7)$	-9	$(-9)^2 + 4(-9) - 21 = +24$	$+$
$(-7, 3)$	0	$(0)^2 + 4(0) - 21 = -21$	$-$
$(3, \infty)$	4	$(4)^2 + 4(4) - 21 = +11$	$+$

Since $x^2 + 4x - 21$ is positive on the intervals $(-\infty, -7)$ and $(3, \infty)$, we know that $x^2 + 4x - 21 > 0$ there. Thus the solution of the given inequality $x^2 + 4x > 21$ in interval notation is $(-\infty, -7) \cup (3, \infty)$.

If, in Example 7, we had been given the inequality $x^2 + 4x \geq 21$, then we would also have included the solutions of the equation $x^2 + 4x = 21$ in our solution set. In that case, the solution set would have been $(-\infty, -7] \cup [3, \infty)$.

Rational Inequalities

A slight variation of the procedure illustrated in Example 7 can be used to solve inequalities containing rational expressions, such as

$$\frac{4 - 3x}{2x + 1} > 0.$$

Here, we use the fact that a rational expression can change its sign only at points where the numerator or denominator is zero. Thus we use those values to break up the real number line into intervals. The rational expression is then always positive or always negative within each of those intervals.

EXAMPLE 8

Solve the inequality $\dfrac{6}{2x-5} \le 2$. Describe the solution set in interval notation.

Solution

$$\frac{6}{2x-5} \le 2$$ Begin by rewriting so that one side of the inequality is zero.

$$\frac{6}{2x-5} - 2 \le 0$$ Now combine terms to form a single rational expression.

$$\frac{6}{2x-5} - \frac{2(2x-5)}{2x-5} \le 0$$

$$\frac{16-4x}{2x-5} \le 0$$

The numerator is 0 when $x = 4$, and the denominator is 0 when $x = \frac{5}{2}$. We use these two values to split the real number line up into the three intervals, $\left(-\infty, \frac{5}{2}\right)$, $\left(\frac{5}{2}, 4\right)$, and $(4, \infty)$.

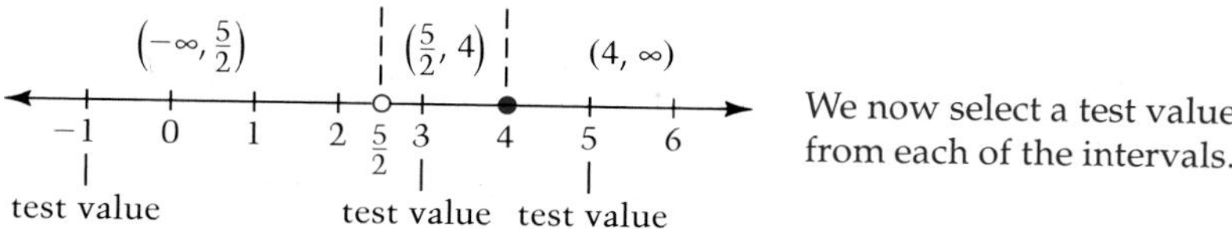

We now select a test value from each of the intervals.

Figure 2.16

Interval	Test Value	Value of $\dfrac{16-4x}{2x-5}$	Sign of $\dfrac{16-4x}{2x-5}$
$\left(-\infty, \frac{5}{2}\right)$	-1	$\dfrac{16-4(-1)}{2(-1)-5} = -\dfrac{20}{7}$	$-$
$\left(\frac{5}{2}, 4\right)$	3	$\dfrac{16-4(3)}{2(3)-5} = +4$	$+$
$(4, \infty)$	5	$\dfrac{16-4(5)}{2(5)-5} = -\dfrac{4}{5}$	$-$

Since $\dfrac{16-4x}{2x-5}$ is less than 0 on the intervals $\left(-\infty, \frac{5}{2}\right)$ or $(4, \infty)$, and equal to 0 when $x = 4$, the solution of the given inequality is $\left(-\infty, \frac{5}{2}\right) \cup [4, \infty)$.

EXERCISE SET 2.5

In Exercises 1–8, describe the interval in two ways: use set notation, and sketch the graph.

1. $(3, 8)$
2. $[3, 10]$
3. $(-2.5, 6]$
4. $[-4, 5.5)$
5. $(-\infty, -2)$
6. $(-2, \infty)$
7. $[-\sqrt{3}, \sqrt{3}]$
8. $(1, \pi)$

In Exercises 9–20, solve the inequality. Indicate the solution set (a) in set notation, (b) in interval notation, and (c) with a graph. (If there are no solutions, then state this fact.)

9. $3x - 8 > 4$
10. $9m + 2 < 7$
11. $2 \geq 3v + 5$
12. $9 > 2w - 6$
13. $8x - 7 > 5 - 2x$
14. $4u + 1 \leq 6u + 4$
15. $\frac{1}{4} - \frac{3}{4}x \geq 5x - 3$
16. $7 - \frac{10}{3}m > \frac{14}{5}m$
17. $5(2s + 1) < s + 3(3s + 1)$
18. $1 - 3(4 - w) \geq 6w - 3(w + 7)$
19. $6y - 2(y + 3) > 5(y - 2) - y$
20. $5z - 2(z - 1) < 3z - 1$

In Exercises 21–30, graph the set, then describe the set in interval notation.

21. $\{x | x \leq 10 \text{ and } x \geq -7\}$
22. $\{x | x > 3 \text{ and } x < 12\}$
23. $\left\{x \middle| x \geq \frac{7}{2} \text{ or } x \leq -\frac{8}{5}\right\}$
24. $\{x | x < 3 \text{ or } x > \pi\}$
25. $\left\{x \middle| x \geq \frac{\sqrt{3}}{2} \text{ or } x < -\sqrt{2}\right\}$
26. $\left\{x \middle| x < -\frac{\sqrt{3}}{2} \text{ or } x \geq \sqrt{2}\right\}$
27. $\{x | -3 < x \leq 3\}$
28. $\{x | -\sqrt{5} \leq x \leq 2\sqrt{5}\}$
29. $\{x | -6 < x < \sqrt{7}\}$
30. $\{x | -5 \leq x < -2\}$

In Exercises 31–36, graph the given set.

31. $[0, 4) \cup (5, 8]$
32. $(-1, 1) \cup [0, 2]$
33. $[-1, 0] \cup [0, 1]$
34. $[-10, -3] \cup [-2, 0]$
35. $(-\infty, 0) \cup (-1, \infty)$
36. $(-\infty, 5] \cup (5, \infty)$

In Exercises 37–50, solve the inequality. Indicate the solution set (a) in interval notation, and (b) with a graph. (If there are no solutions, then state this fact.)

37. $|x| > 1$
38. $|s| < 5$
39. $2|v| \leq 7$
40. $3|w| \geq 10$
41. $|y| > -1$
42. $|z| < -1$
43. $|3x - 5| > 10$
44. $|5v - 7| \leq 18$
45. $|8 - 3u| \leq 13$
46. $|3 - 10s| > 12$
47. $|8r - 1| \leq -3$
48. $|7 - 9t| \geq -2$
49. $|s - 7| \leq 0$
50. $|s - 7| < 0$

In Exercises 51–56, use interval notation to describe the solution set of the given inequality. Unless otherwise specified, round endpoint values to the nearest tenth.

51. $15.8x - 3(x + 2.7) < 46.3$
52. $5.6(3.5 - x) \geq 1 - 2.8x$
53. $|3x - 7| < 0.001$ (round to nearest thousandth)
54. $|8x - 6.2| < 0.001$ (round to nearest ten-thousandth)
55. $|1 - 5.4x| \geq 3.7$
56. $|9.8 - 0.7x| \leq 19.6$

In Exercises 57–74, solve the inequality. State your answer in interval notation.

57. $x^2 + 3x + 15 > 15$
58. $x^2 + 11x + 2 < 2$
59. $z^2 - 9z - 20 < -20$
60. $y^2 - 6y - 5 \geq -5$
61. $x^2 - 7x + 4 \geq 12$
62. $v^2 - 13v > -36$
63. $2m^2 - 4m \leq -1$
64. $t^2 + 6t + 3 < 0$
65. $x^3 - 9x < 0$
66. $y^3 - 36y \geq 0$
67. $\frac{10}{x - 1} - 3 < 2$
68. $\frac{6}{x + 4} - 2 \geq 1$
69. $\frac{6}{6 - x} - 1 < -3$
70. $\frac{2}{3 - x} + 5 > 4$
71. $\frac{8}{3x - 1} + 3 \leq 7$
72. $\frac{40}{4x + 3} - 7 > 1$
73. $\frac{-2}{15 - \frac{1}{2}x} \geq \frac{1}{10}$
74. $\frac{-3}{2x - 9} \leq \frac{1}{3}$

Superset

In Exercises 75–78, determine whether the statement is true or false. If false, illustrate this with an example.

75. $a < b$ always implies that $ac < bc$.

76. $a < b$ always implies that $\frac{1}{a} < \frac{1}{b}$, provided that neither a nor b is zero.

77. $a < b$ always implies that $\frac{a}{b} < 1$, provided $b \neq 0$.

78. $a < b$ always implies that $a^2 < b^2$.

In Exercises 79–82, determine an interval of values for x, so that the perimeter of the figure will be between 16.9 and 17.1 inches.

79.

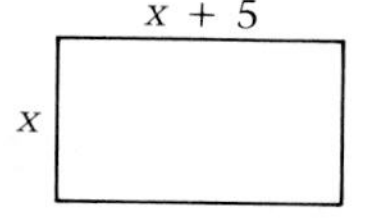

80.

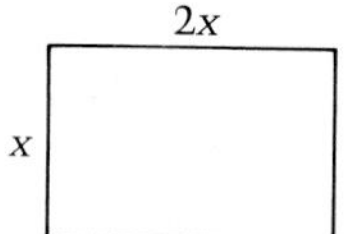

81.

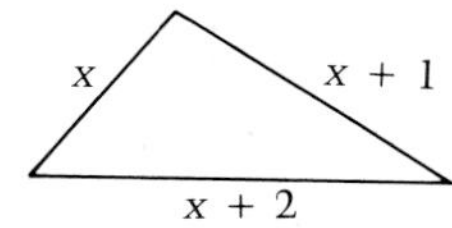

82.

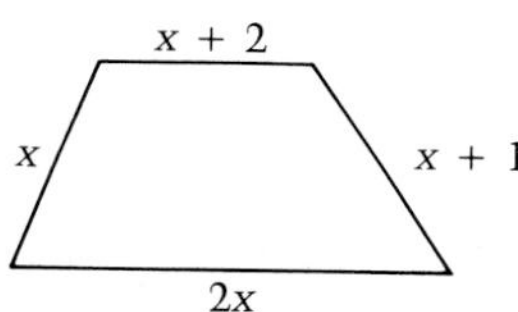

For Exercises 83 and 84, use the formula relating the Fahrenheit and Celsius temperature scales: $F = \frac{9}{5}C + 32$.

83. Suppose temperature readings range over the interval $(-30°, 30°)$ on the Celsius scale. What is the corresponding range on the Fahrenheit scale?

84. Suppose temperature readings range over the interval $(0°, 88°)$ on the Fahrenheit scale. What is the corresponding range on the Celsius scale?

For Exercises 85 and 86, use the following information: When an object is propelled upward from ground level with an initial velocity of v_0 ft/sec, its height s (in feet) above the ground t seconds later is given by the formula $s = v_0 t - 16t^2$.

85. If the initial velocity is 128 ft/sec, during what time interval will the object be more than 192 ft above ground?

86. If the initial velocity is 128 ft/sec, during what time intervals will the object be at most 112 ft above ground?

In Exercises 87–92, solve the inequality. State your answer in interval notation.

87. $|x^2 - 3| \leq 1$

88. $|x^2 - 5| > 4$

89. $|x^2 + 2x| < 1$

90. $|x^2 - 2x| > 3$

91. $|x^2 - 2x - 4| > 4$

92. $|x^2 + 4x + 1| \leq 1$

2.6 Problem Solving

Solving problems is very much like playing a game. In order to play the game, you must first learn the moves. Only after the moves have become second nature can you hope to be proficient in selecting the moves that will win the game.

The goal in problem solving is, generally, to determine some unknown quantity. The moves that lead you toward this goal are referred to as **strategies.** We are mainly concerned with problems that can be represented by **algebraic models.** For us that means equations or inequalities. We will discuss some strategies that will guide you in producing such models. Just like the moves in a game, this list of strategies should become second nature to you. They should serve as a mental checklist each time you approach a problem-solving situation.

Although the goal in solving a specific problem may be to find a certain value, the heart of problem solving is the *search for a useful strategy.* At this point we restrict ourselves to three problem-solving strategies.

Three Problem-Solving Strategies

- **Translate** the problem directly into an equation or an inequality.
- **Draw a sketch** showing how the various details of the problem are related.
- **Make a chart** that relates the known and unknown quantities in the problem.

These three strategies will serve as a guide in solving problems in this section. The strategies can be used singly or in combination to solve a problem.

Translate

It is often possible to translate a problem directly into an algebraic equation or an inequality which then can be solved. Consider the following problem.

Problem 1. A certain number is 15 more than twice its additive inverse. What is the number?

In *analyzing* this problem you should ask yourself what is it that you must find. A careful reading of the problem shows that you must determine the value of some unknown number. We can **translate** to produce an equation for use in finding this number.

Now you need to *organize* your work by introducing a variable so that you can create a model:

Let $x =$ the "certain number."

The *model* for the problem is the following equation:

$x = 15 + 2(-x)$. "certain number" is 15 more than twice its additive inverse."

The *answer,* 5, is found by solving the equation.

Draw a Sketch

The purpose of a sketch is to help you visualize the information in a problem. The sketch may suggest an equation or an inequality that models the problem. This strategy is particularly useful in problems involving geometric figures, where the sketch may remind you of an appropriate formula.

Problem 2. The length and width of a rectangular patio have been increased by 3 ft and 6 ft respectively. The enlarged patio is a square. If the area of the old patio was 130 ft^2 (square feet), what is the area of the new patio?

In *analyzing* this problem you should recognize that you must find the dimensions of the new patio, which depend on the dimensions of the old one. Your first impulse might be to draw a **sketch.** If so, your impulse is a good one.

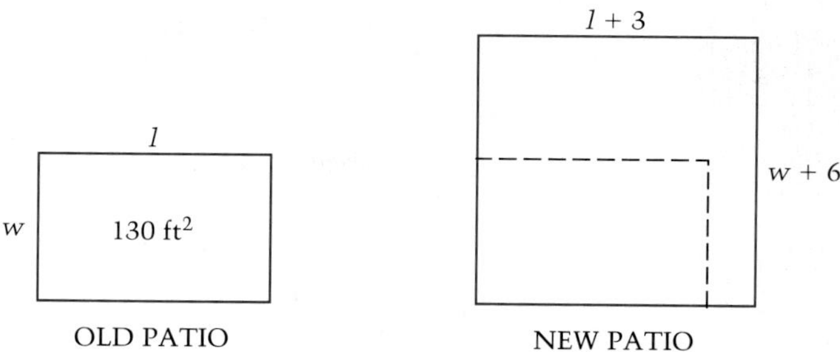

Figure 2.17

To *organize* your thinking about the problem, define the variables. Let l = length of the old patio and let w = width of the old patio.

From the sketches we can conclude that a *model* for the problem involves the following two equations.

$l \cdot w = 130$ Recall $A = lw$, where $A = 130 \text{ ft}^2$.

$l + 3 = w + 6$ Since the new patio is a square, the sides are equal.

We now combine the two equations to get a *simpler model* consisting of one equation in one variable. Since the second equation can be rewritten as

$$w = l - 3,$$

we can substitute $l - 3$ for w in the first equation to get a model requiring only one variable:

$$l(l - 3) = 130.$$

Solving this equation produces the result $l = 13$ or $l = -10$. The latter value is discarded since length cannot be negative. Since $l = 13$, the new patio has sides equal to $l + 3 = 16$, that is, each side measures 16 ft. Thus we have the *answer*: The area of the new patio is 256 ft^2.

Make a Chart

Sometimes the information in a problem is too complicated for simple translation, or is not easily represented in a sketch. A chart may help organize the information in a way that suggests an equation or inequality. Such is the case in the next problem.

Problem 3. At 5 P.M. a jet leaves an airport and flies due west at an average speed of 520 mph. At 6 P.M. a second jet leaves the same airport and flies due west at 650 mph. When does the second jet catch the first?

In *analyzing* this problem you should attempt to visualize what is happening: a second jet takes off an hour after the first and ultimately catches up. At that point in time the two jets will have covered the same distance. Since we must find the time it takes to do this, we use the formula $d = rt$, where d is the distance traveled, r is the rate of speed, and t is the time.

Often when a formula is used, a **chart** is useful to *organize* information. Let t = the time in hours for Jet 1 to cover the distance. Then $t - 1$ = the time in hours for Jet 2 to cover the same distance *in one less hour.*

	distance	=	**rate**	×	**time**
Jet 1	$520t$	=	520	×	t
Jet 2	$650(t - 1)$	=	650	×	$t - 1$

At the moment when Jet 2 catches up, both planes have covered the same distance, so it makes sense to equate the two different expressions for the same distance. This produces the *model:*

$$520t = 650(t - 1).$$

Solving the equation reveals that $t = 5$, and so the *answer* is that Jet 2 catches Jet 1 at 10 P.M., five hours after Jet 1 has taken off.

This kind of problem serves to remind us that the equation that models a given problem is often found by equating two different expressions for the same quantity. This useful idea will be used may times in this section.

You might have noticed that in each problem we emphasized the words ANALYZE, ORGANIZE, MODEL, and ANSWER. These terms represent four important *stages* in problem solving. Notice how these stages help to guide us through the problem-solving process.

EXAMPLE 1

The sum of two numbers is 15, and their product is 44. Find the two numbers.

Solution

- ANALYZE We must find two numbers that satisfy two conditions. Let us try translating the information in the problem directly into an equation.
- ORGANIZE Let the two numbers be represented by x and y. We ultimately want to find one equation in one variable. The problem presents two pieces of information which translate into the following equations.

Translate $x + y = 15$ "The sum of two numbers is 15 . . ."

$xy = 44$ ". . . their product is 44."

Solving the first equation for y shows that $y = 15 - x$. Now replace y with $15 - x$ in the second equation $xy = 44$.

- MODEL

$$\begin{aligned} x(15 - x) &= 44 \\ 15x - x^2 - 44 &= 0 \\ x^2 - 15x + 44 &= 0 \\ (x - 11)(x - 4) &= 0 \end{aligned}$$

The expressions $x(15 - x)$ and 44 both represent the product.

$x - 11 = 0$	$x - 4 = 0$
$x = 11$	$x = 4$

- ANSWER If $x = 11$, then since $x + y = 15$, we have $y = 4$. Thus, the numbers are 11 and 4. (If $x = 4$, then $y = 11$, which is the same answer.)

EXAMPLE 2

Two bikers have a pair of walkie-talkies that work up to a distance of 40 mi. One biker leaves from the town line traveling at an average rate of 16 mph. One hour later the other biker leaves from the same place traveling 21 mph in the opposite direction. How long will the second biker ride before the walkie-talkies fail?

Solution

- ANALYZE The unknown is the *time* it takes the two bikers to be 40 mi apart. Complicated data, such as in this problem, is best organized in a chart.
- ORGANIZE Let t = the time that the *second* biker rides before the walkie-talkies fail. Then $t + 1$ = the time that the *first* biker rides before the walkie-talkies fail.

A chart helps us find a model for the distances each biker travels.

Make a chart

	d	$= r \times t$
first biker	$16(t+1)$	$= 16 \times (t+1)$
second biker	$21t$	$= 21 \times t$

A sketch helps us visualize this information.

Draw a sketch

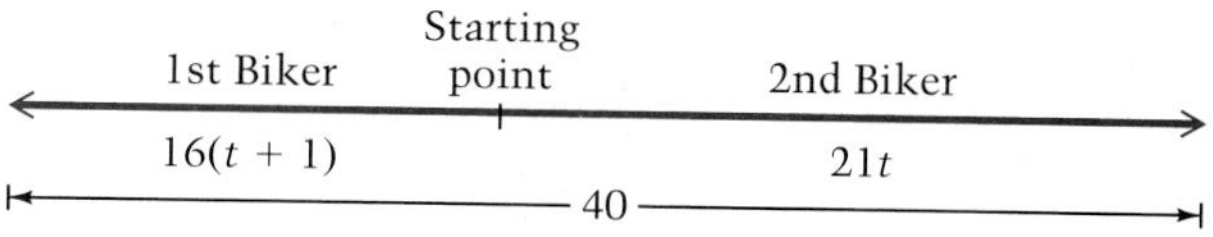

Figure 2.18

The distance at which the walkie-talkies stop working is 40 mi. From the sketch we see the total distance between the bikers is described by $16(t + 1) + 21t$. Thus, we want to equate these two quantities.

- MODEL

$$16(t+1) + 21t = 40$$
$$t = \frac{24}{37}$$

- ANSWER The second biker travels for $\frac{24}{37}$ hr, or approximately 39 min before the walkie-talkies stop working.

EXAMPLE 3

A man 6 ft tall is walking toward a street light that is 21 ft tall. What is the length of the man's shadow on the ground when he is 10 ft away from the light?

Solution

- ANALYZE You are to find the length of the man's shadow when he is 10 ft away from the light. Draw a sketch to help visualize the situation.
- ORGANIZE **Draw a sketch**

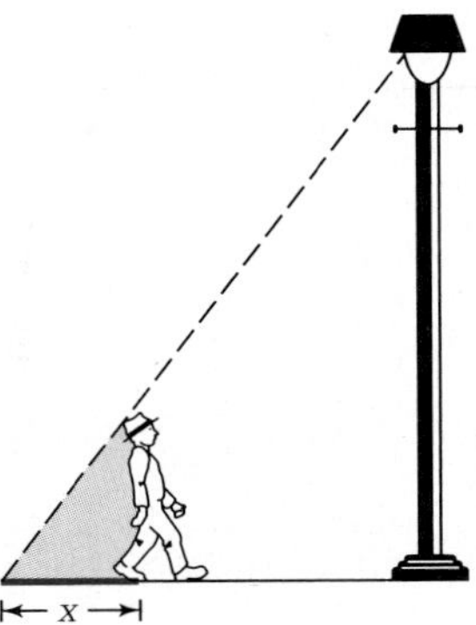

Figure 2.19

Let x = the length of the man's shadow.

The sketch suggests two similar triangles, one whose height is the man's and another whose height is the street light's.

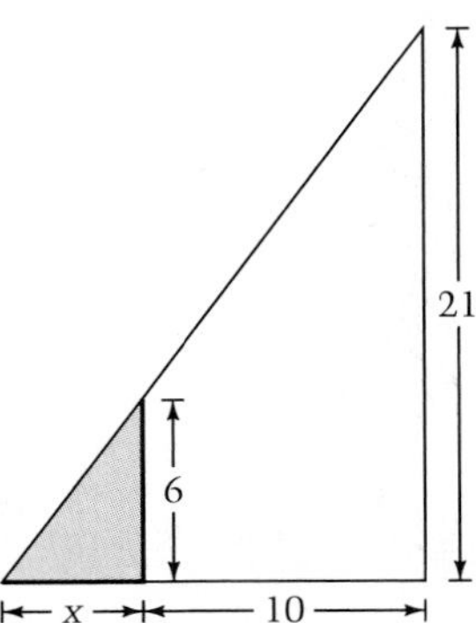

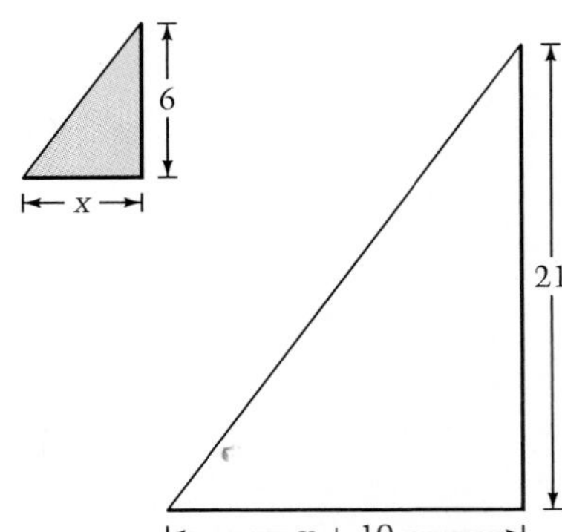

Figure 2.20

Use the ratios of the corresponding sides to set up a proportion. To set up these ratios correctly, it is often helpful to draw the similar triangles separately, as shown above on the right.

- MODEL

$$\frac{6}{x} = \frac{21}{x + 10}$$

$$6(x + 10) = 21\,x$$ Two steps have been combined by cross multiplying.

$$4 = x$$

- ANSWER The man's shadow is 4 ft long when he is 10 ft away from the light.

The final example in this section is representative of a class of problems referred to as *work problems.* Central to solving such problems is determining a work rate. For example, if a person can do a certain job, say frame a picture, in x hours, then the person's work rate is

$$\frac{1 \text{ job}}{x \text{ hr}} = \frac{1}{x} \text{ job per hour.}$$

Note that the work rate $\frac{1}{x}$ also represents the part of the job that can be done in one hour. Thus, if there are two workers doing a job, and

x_1 = the time it takes worker 1 to do the job **alone**
x_2 = the time it takes worker 2 to do the job **alone**
x = the time it takes for the workers to do the job **together,** then

the part of the job done by worker 1 in one hour	+	the part of the job done by worker 2 in one hour	=	the part of the job done by the workers together in one hour
$\frac{1}{x_1}$	+	$\frac{1}{x_2}$	=	$\frac{1}{x}$

EXAMPLE 4

When working together, Pat and Chris can install carpeting in an average-size room in three hours. Working alone, Pat is able to finish the job one hour faster than Chris. How long does it take Chris to do the job working alone?

Solution

- ANALYZE We need to find the time it takes Chris to do the job alone.
- ORGANIZE Introduce variables:

Let x_P = time it takes for Pat to finish the job alone.
Let x_C = time it takes for Chris to finish the job alone.

Thus $x_C = x_P + 1$ ("Pat can do the job 1 hr faster than Chris.")

3 hr = time it takes for Pat and Chris to finish the job together.

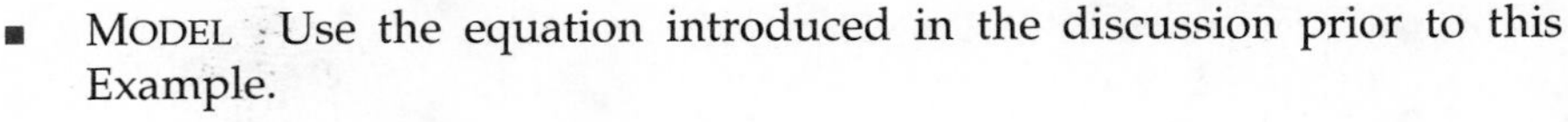

■ MODEL Use the equation introduced in the discussion prior to this Example.

$$\frac{1}{x_P} + \frac{1}{x_P + 1} = \frac{1}{3}$$

To solve, multiply both sides of the equation by the LCD, $3x_P(x_P + 1)$.

$$[3x_P(x_P + 1)]\left(\frac{1}{x_P} + \frac{1}{x_P + 1}\right) = \frac{1}{3}[3x_P(x_P + 1)]$$

$$3(x_P + 1) + 3x_P = x_P(x_P + 1)$$

$$3x_P + 3 + 3x_P = x_P^2 + x_P$$

$$x_P^2 - 5x_P - 3 = 0$$

$$x_P = \frac{-b \pm \sqrt{b^2 - 4ac}}{2a}$$

$$x_P = \frac{5 \pm \sqrt{25 - 4(1)(-3)}}{2} = \frac{5}{2} \pm \frac{\sqrt{37}}{2}$$

$$x_P = 5.5 \text{ hr or } -0.5 \text{ hr}$$

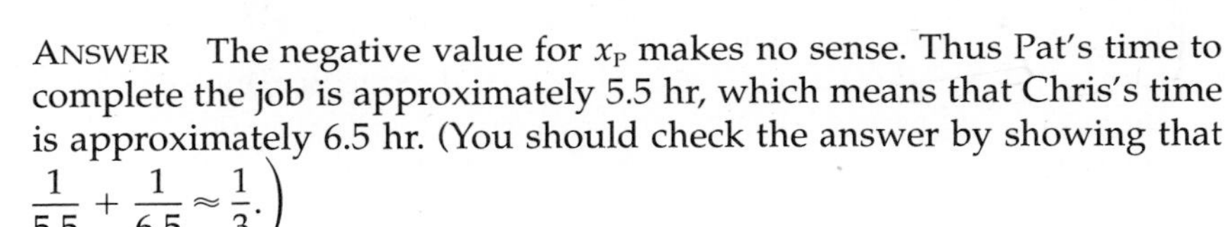

■ ANSWER The negative value for x_P makes no sense. Thus Pat's time to complete the job is approximately 5.5 hr, which means that Chris's time is approximately 6.5 hr. (You should check the answer by showing that $\frac{1}{5.5} + \frac{1}{6.5} \approx \frac{1}{3}$.)

EXERCISE 2.6

1. A number is increased by 5, and the result is twice the number. Find the original number.
2. Six more than a certain positive number is five more than twice the reciprocal of the number. What is the number?
3. The sum of three consecutive integers is 36. What are the three integers?
4. The sum of three consecutive even integers is twice the value of the smallest. What are the three integers?
5. A used car dealer recently purchased 3 cars. The station wagon costs $3000 less than the sports car, but is twice the cost of the subcompact. The total cost of the three cars is $32,760. Find the cost of each car.
6. Five years ago a tree was planted and measured 4 ft less than its present height. It is predicted that seven years from now it will be 16.5 ft high, triple its height at planting. How tall is the tree now?
7. To enclose a rectangular garden, 120 ft of fencing are used. If the garden must have an area of 900 ft^2, what are the dimensions of the garden?
8. A square piece of aluminum is used to manufacture a baking pan by cutting off from each corner a square 2 in. on a side. The flaps are then folded up to form the sides of the pan. How large a piece of aluminum should be used to produce a pan whose volume is 162 in.^{3}?
9. An investor wishes to invest a total of $10,000 in two accounts for one year. The first account offers an 8% annual percentage rate, and the second offers 12%. How much should be invested in each account so that the total interest is $920.
10. The entire stock of jeans at Hutton's Department Store would produce revenues of $4000, since there are 100 pairs and the price of each pair is $40. The manager decides to offer a 10% discount and later increases the discount to 25%. How many pairs should be sold at each discount to assure a total income of $3,378?

$\frac{1}{60}$ / min . $\frac{1}{30}$ / m

$t\left(\frac{3}{60}\right) = 1$

20 min

Jean Morris

4. Satz

Glanstle

11. When a small plane flew against a 40 mph wind from Portland to Watertown, the trip took $2\frac{1}{2}$ hr. On the return trip the plane flew with the wind, and the trip took $1\frac{1}{2}$ hr. On both legs of the trip the pilot flew at the maximum speed. What is the plane's maximum speed in calm air?

12. A train leaves Detroit heading for Chicago on the westbound track at an average speed of 60 mph. One hour later a train leaves Chicago on the eastbound track for Detroit at a speed of 45 mph. If Detroit and Chicago are 270 mi apart, how long has the second train been traveling when the trains meet? How far from Chicago will they be when they meet?

13. A camper's motorboat has a top speed of 6 mph in still water. The camper sets out downstream traveling at full speed. The motor breaks down, and the camper rows back to the dock at a rate of 2.5 mph. If the return trip takes the camper three times as long as the trip downstream, how fast is the current flowing?

14. A farmer prepares feed for the pigs by mixing two brands of feed. Feed A provides 140 grams of a certain nutrient per ounce, and Feed B provides 110 grams per ounce. Each pig is fed 30 ounces of feed. How should the feed be mixed so that each pig receives 3,660 grams of the nutrient?

15. An 18 ft light pole casts a 15 ft shadow, and simultaneously a 40 ft telephone pole casts a 12 ft shadow up a wall. How far is the telephone pole from the wall?

16. A rectangular corral is built beside a barn, using the barn as one side of the corral. If 61 ft of fencing are used for the three sides of the corral, what are the dimensions of the corral if its area must be 450 ft^2.

17. Admission prices at a local theatre for its Saturday matinees are $4 for adults and $1.50 for children under 12. If the theatre sold 253 tickets for a particular showing with total proceeds of $557, how many adults attended that matinee?

18. To determine the sticker price for new automobiles, a certain car dealership marks up the base price as set by the manufacturer by 15% but only marks up the "options package" by 5% as an incentive to potential buyers. The dealer is selling a specific model car with the options package for $12,884. The dealer paid the manufacturer $11,280 for this car with options. What was the profit realized by the dealer on the options package for this car?

19. Facing bankruptcy, a store owner must divest himself of his business. He sells off his inventory at a 40% loss but is able to sell his building at a 10% profit so that his total loss is $15,000. Given that the total value of the store owner's inventory and building was $157,500, what was the value of his inventory?

20. A high school sports boosters club sold pizzas to raise funds. The cost to the club for each pizza sold was $4. The club sold the pizzas at $7 each for single sales and offered a special of three pizzas for $17 for those wishing to buy more than one pizza. The club sold 302 pizzas altogether and realized a profit of $646. How many specials did the club sell?

21. When Joe mows the lawn, it takes an hour. Joe's mom can mow the lawn in 30 minutes. How long would it take Joe and his mom to mow the lawn together?

22. Margy takes twice as long as Maryann to go over the daily sales receipts at the end of a day's business. If they can complete the task together in an hour and twenty minutes, how long does it take each woman if she works by herself?

23. Jules and Jim work at a car wash. They are responsible for "finish work": cleaning the inside of the car, and wiping the car dry. When they perform the task individually, it takes Jules a minute more than Jim. Working together, they take only five minutes. How long does it take each of them individually?

24. A man 6′3″ tall is walking toward a street light that is $24\frac{1}{2}$ ft tall. How far away from the light is the man when his shadow is 5 ft long?

25. Three thousand feet of fencing are to be used to enclose a rectangular lot. If the lot is to have an area of five hundred thousand square feet, what should the dimensions be?

26. Two people leave their respective apartments at the same time to drive to a concert. Each must drive 42.5 mi to the concert. If, on the trip, one person averages 8 mph faster than the other, and arrives at the concert 8.5 min before the other, what were the average speeds of the two drivers?

Superset

27. A rectangular piece of sheet metal measuring 9 in. by 30 in. is cut along a line parallel to its shorter side in order to form two similar but unequal rectangular pieces. What is the area of the smaller piece?

28. A park ranger's motorboat can make an average of 7 mph in still water. The ranger sets out for a trip upstream but runs out of gas. He can row his boat at the rate of 3 mph in still water. The ranger rows back to his starting point and discovers that the round trip lasted 6

hours. If the current was flowing at 1 mph, how far upstream did the ranger go before running out of gas?

29. An olympian in training sets out upstream in a kayak for a strenuous workout and is able to maintain a constant speed of 7 mph. Unfortunately the olympian strains a shoulder muscle and decides it is best to drift back to the starting point with the current. If on this trip the olympian spent five times as long drifting as was spent paddling, how fast can the olympian paddle the kayak in still water?

30. A 50 foot tall telephone pole is standing 20 feet away from a 12 foot high flat-roofed building. The shadow cast by the telephone pole extends across the ground, up the side of the building, and 15 feet out along the flat roof of the building. How long a shadow would this telephone pole cast at this time of day if the building were not there?

31. A tourist walked 12 miles from Somerset to St. George's, had lunch, and returned by scenic tour bus over the same route. The entire excursion lasted 6 hours. The constant speed of the bus is four times the tourist's walking rate. How much time might the tourist have spent for lunch if the bus traveled at a constant rate between 15 and 18 mph?

32. One mile upstream from his starting point a canoeist passed a log floating with the current. After paddling upstream for one more hour, the canoeist turned and paddled back to his starting point, arriving at the same time as the log. How fast was the current flowing? What was the constant (still water) rate at which the canoeist was paddling the canoe?

33. A group of students rowed 6 mi downsteam and 6 mi back in 1 hr 47 min. If the current flowed at a rate of 4.9 mph, what was the rowing rate of the students in still water?

34. An isosceles triangle whose base is 10 and congruent sides are 13, is cut by a line parallel to the base to produce an isosceles triangle and a trapezoid. The area of the trapezoid is twice that of the triangle, that is, the area of the trapezoid is two-thirds that of the original triangle. What is the altitude of the trapezoid?

35. In Exercise 34 take R as a real number between 0 and 1 and suppose the ratio of the area of the trapezoid to that of the original triangle is R. What is the altitude of the trapezoid?

Exercises 36–39 deal with what are commonly called "mixture problems." Such problems can usually be modeled effectively by an equation that focuses on one of the constituents in the mixture. The general form of such an equation can be represented by

amount of given substance at the outset	+	amount of given substance added or subtracted	=	amount of given substance at the end

For example, consider the situation described by the passage: "an unknown amount (X) of acid is added to 18 ounces of a 65% solution of acid in water to produce a concentration that is 80% acid." If we focus on the acid, then the model is

$$0.65(18) + X = 0.80(18 + X)$$

Equivalently, we could write an equation about water to model this situation:

$$0.35(18) + 0 = 0.20(18 + X)$$

Only one of these equivalent equations is needed to model the situation and either equation could be solved to determine X.

36. Attempting to dilute 20 ounces of a 70% solution of acid in water, a chemist adds too much water. To obtain a 60% concentration the chemist must add an additional 4 ounces of acid. How much water did the chemist add?

37. Acid is added to 16 ounces of a 70% solution of acid in water in an attempt to derive an 80% solution. However, too much acid was added. How many ounces of acid were added if it was necessary to add 0.5 ounces of water to the resulting solution to produce that 80% solution?

38. Water is added to 30 ounces of a 75% solution of acid in water in an attempt to derive a 60% solution. After one-half of the resulting solution has been used up, it is discovered that the solution is too weak. The 60% solution is then attained by adding 5 ounces of acid to the remaining solution. How much water was added originally?

39. Water is added to 30 ounces of a 75% solution of acid in water in an attempt to derive a 60% solution. After R% of the resulting solution has been used up, it is discovered that the solution is too weak. The 60% solution is then attained by adding 5 ounces of acid to the remaining solution. How much water was added originally?

Analyzing Data

The Case of the Deceptive Means

In Chapter 1, we considered the mean (average) of a set of data. The mean provides us with a single value to represent the entire set of data—it tells us where the "middle" of the data lies. As a representative value it does a fair job, but it hardly tells the whole story behind the data. It does not tell us, for instance, how different from one another the values in the data set are.

Consider two American cities which, for the moment, will go unnamed. City A has a mean daily high temperature of 62.4° F and City B has a mean daily high temperature of 63.7° F. You might think that since these means are so close in value, the climates of the two cities would be quite similar. But upon closer inspection we see that this is not the case. Below we have listed the normal daily high temperatures for the two cities by *month.*

	Jan	Feb	Mar	Apr	May	Jun	Jul	Aug	Sep	Oct	Nov	Dec
City A	56°	59°	60°	61°	63°	64°	64°	65°	69°	68°	63°	57°
City B	39°	41°	51°	63°	73°	82°	86°	85°	79°	67°	55°	43°

$$s = \sqrt{\frac{\sum_{i=1}^{n} (x_i - \bar{x})^2}{n - 1}}$$

The symbol $\sum_{i=1}^{n}$ indicates a sum: add up n terms of the form $(x_i - \bar{x})^2$, one term for each value x_i in the data set.

Notice that the twelve normal high temperatures for City A do not vary nearly as much as those for City B. We rely on the *standard deviation* as a quantitative measure of the variability among the values in a data set. The formula for the standard deviation s of the values $x_1, x_2, \ldots, x_n$, is given at the left. The symbol $\bar{x}$ represents the mean of the set of values, and n is the number of values in the set.

We compute the standard deviation for each set of temperatures:

$$s_A = \sqrt{\frac{(56 - 62.4)^2 + (59 - 62.4)^2 + (60 - 62.4)^2 + \cdots + (57 - 62.4)^2}{12 - 1}} = 4.0$$

$$s_B = \sqrt{\frac{(39 - 63.7)^2 + (41 - 63.7)^2 + (51 - 63.7)^2 + \cdots + (43 - 63.7)^2}{12 - 1}} = 17.6$$

Clearly the standard deviation for City A, 4.0, is much smaller than that of City B, 17.6. This indicates less variability throughout the year in the normal daily high temperature in City A than in City B. Although the average high temperatures are close in value for the two cities, the difference in standard deviations suggests that the climates are quite different.

In case you have not guessed, City A is San Francisco and City B is Philadelphia.

Source: John W. Wright, ed., *The Universal Almanac, 1990.* Andrews & McMeel, New York, 1989.

Chapter Review

2.1 Introduction to Equations (pp. 65–74)

An *equation* is a statement that two expressions represent the same number (e.g., $6x + 4 = -12 + 4x$). Values of the variables that make an equation true are the *solutions* or *roots.* (p. 65) Equations with the same solution set are called *equivalent equations.* (p. 66)

Addition Property for Equivalent Equations: Adding the same real number (or variable expression) to both sides of an equation produces an equivalent equation. (p. 66)

Multiplication Property for Equivalent Equations: Multiplying both sides of an equation by the same nonzero real number (or variable expression) produces an equivalent equation. (p. 66)

A *linear equation* is an equation containing a first degree monomial, but no monomial of higher degree. A linear equation written in the form $ax + b = 0$, where a and b are constants and $a \neq 0$ is said to be in *standard form.* (p. 67)

A *fractional equation* is an equation containing fractional expressions involving variables. You *must* check solutions of these types of equations. (p. 69) An *absolute value equation* is an equation containing absolute values of terms involving variables. To solve an absolute value equation of the form $|A| = B$, you must solve the two equations in the statement "$A = B$ **or** $A = -B$." You *must* check solutions. (p. 69)

2.2 Introduction to Complex Numbers (pp. 74–82)

The number i has the property that $i^2 = -1$. For this reason, i can be considered a square root of -1. Thus, we write $i = \sqrt{-1}$. (p. 74) If p is a positive real number, then $\sqrt{-p} = i\sqrt{p}$.

If a and b are real numbers, then any number of the form $a + bi$ is called a *complex number.* The number a is called the *real part* of the complex number, and b is called the *imaginary part.* (p. 75) Any complex number of the form $a + bi$, where the imaginary part b is not zero, is called an *imaginary number.* Any number of the form bi, where b is not zero, is called a *pure imaginary number.* (p. 75)

Two complex numbers $a + bi$ and $c + di$ are equal if and only if $a = c$ and $b = d$. (p. 77)

If $a + bi$ and $c + di$ are complex numbers, then

$$(a + bi) + (c + di) = (a + c) + (b + d)i,$$

$$(a + bi) - (c + di) = (a - c) + (b - d)i, \text{ and}$$

$$(a + bi)(c + di) = (ac - bd) + (ad + bc)i. \text{ (p. 78)}$$

In general, for real numbers a and b, the complex conjugate of $a + bi$ is $a - bi$. Since $(a + bi)(a - bi) = a^2 + b^2$, the product of a complex number and its conjugate is a real number. (p. 79)

Let z be a complex number $a + bi$. Then $\bar{z}$ represents the complex conjugate $a - bi$. (p. 80)

If $\bar{z}$ and $\bar{w}$ are complex conjugates of the complex numbers z and w, respectively, then:

1. $\overline{z + w} = \bar{z} + \bar{w}$ 2. $\overline{z - w} = \bar{z} - \bar{w}$ 3. $\overline{z \cdot w} = \bar{z} \cdot \bar{w}$

4. $\overline{\left(\dfrac{z}{w}\right)} = \dfrac{\bar{z}}{\bar{w}}$, for $w \neq 0$ 5. $\overline{z^n} = (\bar{z})^n$, for every positive integer n

6. $\bar{z} = z$, if z is a real number

2.3 Quadratic Equations (pp. 82–91)

A *quadratic equation* is any equation containing a second degree monomial, but no monomial of higher degree. A quadratic equation written in the form $ax^2 + bx + c = 0$, where a, b, and c are constants, and $a \neq 0$, is said to be in *standard form*. (p. 82)

The *Property of Zero Products* is a basic tool in solving quadratic equations. (p. 82) The method for *completing the square* can be used to solve quadratic equations that cannot be factored. (p. 85)

The *quadratic formula* given below is used to solve quadratic equations. (p. 86)

$$x = \frac{-b \pm \sqrt{b^2 - 4ac}}{2a}$$

The *discriminant* $b^2 - 4ac$ can be used to determine the nature of the solutions of a quadratic equation.

2.4 Nonlinear Equations (pp. 91–98)

Many higher degree equations and fractional equations can be solved using techniques such as grouping and the Property of Zero Products. (p. 92) An equation is *quadratic in form* if replacing an expression with a single variable makes the equation quadratic.

Radical equations can be solved using the *Power Property for Equations*. You *must* check solutions when using this property. (p. 95)

2.5 Inequalities (pp. 98–109)

Use the Addition Property for Equivalent Inequalities to solve inequalities like $x + 3 > 4$. Use the Positive Multiplication Property for Equivalent Inequalities to solve $2x > -7$. Use the Negative Multiplication Property for

Equivalent Inequalities to solve $-6y < 5$; when using this property, remember to reverse the direction of the inequality. (p. 99)

The solution set of an inequality can be described in four ways: (1) interval notation, for example, $(a, b]$; (2) set notation, $\{x|a < x \le b\}$; (3) with a graph; and (4) in words. (p. 100)

To solve $|A| > c$ we solve the compound inequality $A > c$ **or** $A < -c$. To solve $|A| < c$ we solve the compound inequality $-c < A < c$. (p. 103) To solve a quadratic or rational inequality, use the associated equality in choosing test values to determine intervals where the expression is positive and intervals where it is negative. (p. 105)

2.6 Problem Solving (pp. 109–118)

The essential aim of problem solving is to find a useful strategy from the *Problem-Solving Strategies.* (p. 110) The four stages of problem solving are Analyze, Organize, Model, and Answer. (p. 110)

Review Exercises

In Exercises 1–10, solve the given equation.

1. $3x + 18(x - 1) = 2 - 3(x - 2)$

2. $5 - 4(v + 1) = 2(3 - v) + 17$

3. $7(5 - 6x) - 9 = 12 - 4(x - 1)$

4. $9 + 2(7x - 5) = 1 - 3(4 - 5x)$

5. $1 - \frac{m}{2} = 3$

6. $2 = \frac{x}{3} - 4$

7. $\frac{x}{x - 1} = \frac{3}{8}$

8. $\frac{3}{2 - m} = 0.5$

9. $\frac{2}{2 - m} + \frac{1}{4} = \frac{m}{m - 2}$

10. $\frac{2x}{3 - 3x} = \frac{x + 1}{x - 1}$

In Exercises 11–14, solve for the specified variable.

11. Solve for m.
$y - y_0 = mx - mx_0$

12. Solve for m.
$K = \frac{1}{2}mv^2$

13. Solve for t.
$v = v_0 - gt$

14. Solve for r.
$A = \pi s(R + r)$

In Exercises 15–20, solve the given equation.

15. $|3 - 6x| = 15$

16. $|7x - 6| = 8$

17. $|2y - 7| = 1 - y$

18. $|3v + 2| = 8 - v$

19. $|3(w - 5)| = 2 + w$

20. $|8(u - 3)| = u + 4$

In Exercises 21–48, perform the indicated operation, then write the answer in standard form.

21. i^{51}

22. i^{65}

23. $(-i)^{10}$

24. $(-i)^{35}$

25. $(3i)^4 \cdot i^{44}$

26. $(-2i)^7 \cdot i^{18}$

27. $\sqrt{-5} \cdot \sqrt{-3} \cdot \sqrt{6}$

28. $\sqrt{-8} \cdot \sqrt{12} \cdot \sqrt{75}$

29. $(1 - 3i) + (-2 + i)$

30. $(4 - 3i) - (6 - 2i)$

31. $5i(4 - 2i)$

32. $-6i(2 - 4i)$

33. $2i - i(8 + 3i)$

34. $2i(-3 - 5i) + 6i$

35. $(1 - 4i)(3 + 2i)$

36. $(3 + 5i)(7 - 4i)$

37. $(4 + 7i)(4 - 7i)$

38. $(2 - 9i)(2 + 9i)$

39. $(1 + i\sqrt{5})(1 - i\sqrt{5})$

40. $(\sqrt{11} - 2i)(\sqrt{11} + 2i)$

41. $(5 - 4i)^2$

42. $(3 + 8i)^2$

43. $\frac{1}{2 - 5i}$

44. $\frac{1}{3 + 11i}$

45. $\frac{6 - 2i}{2 + i}$

46. $\frac{3 + 4i}{6 - i}$

47. $\frac{2i}{3 - 5i}$

48. $\frac{3}{3 - 6i}$

In Exercises 49–64, solve the equation.

49. $x(1 - x) = 2$

50. $y(2y + 1) = 6$

51. $(2x + 1)(x - 7) = 0$

52. $3x^2 - 22x = 7$

53. $4x^2 - 19x = 5$

54. $x^2 - 2x + 3 = 0$

55. $x^2 + x + 3 = 0$

56. $(2x + 1)(x - 7) = 1$

57. $2x^2 - 5x + 1 = 0$

58. $x^2 + 3x + 1 = 0$

59. $7x^2 + 2x - 3 = 0$

60. $9 - x - 10x^2 = 0$

61. $2x^2 - 6x + 5 = 0$

62. $5x^2 = -2x - 3$

63. $x^2 + 5 = 0$

64. $x^2 - 5 = 0$

In Exercises 65–68, use the discriminant to determine the nature of the solutions without solving.

65. $x^2 - 10x + 1 = 0$

66. $3y^2 - 6y + 5 = 0$

67. $2p^2 - p = -1$

68. $8 = 2x - x^2$

In Exercises 69–82, solve each equation.

69. $2x^3 - 3x^2 - 5x = 0$

70. $m^3 - 10m^2 + 25m = 0$

71. $x^4 - 7x^2 + 10 = 0$

72. $9y^{-2} - 30y^{-1} + 25 = 0$

73. $x^{2/3} - 6x^{1/3} - 16 = 0$

74. $\dfrac{2}{x + 2} - \dfrac{3}{x - 3} = 0$

75. $\dfrac{x}{2x - 1} + \dfrac{2}{2x^2 - x} = 3$

76. $\dfrac{x + 4}{5x + 1} - \dfrac{10 - 2x^2}{15x^2 - 7x - 2} = \dfrac{x}{3x - 2}$

77. $\dfrac{1}{x} - \dfrac{x}{1 + x} = \dfrac{x - 3}{1 - x}$

78. $\dfrac{2x}{x^2 - 4} - \dfrac{1}{x + 2} = \dfrac{x}{x - 2}$

79. $\sqrt{y + 6} - 9 = -2y$

80. $\sqrt{2m + 3} - \sqrt{m + 1} = 1$

81. $\sqrt{x - 2} = 1 - \sqrt{13 - 2x}$

82. $\sqrt[3]{2x + 1} = 3$

In Exercises 83–92, solve the inequality. Write the answer in interval notation and sketch a graph.

83. $6x - 3 < 2 - x$

84. $-5x - 1 \geq x - 8$

85. $-2(3x + 4) \leq 6x - 4$

86. $2 - 3x > 1 - 3(x + 5)$

87. $-3 \leq 3 - 2x \leq 7$

88. $8 > 9x - 1 \geq -19$

89. $|3 - 2x| \leq 7$

90. $|1 + 3x| > 8$

91. $|4x - 7| \geq 9$

92. $|5 - 3x| < 6$

93. $y^2 - 6y < 7$

94. $m^2 - 8m \geq 4$

95. $v^2 - 8v > -9$

96. $3y^2 + 19y + 26 \leq 0$

97. $\dfrac{3}{2x - 1} < 1$

98. $\dfrac{x}{x - 3} \geq 2$

99. $\dfrac{1}{x + 1} > \dfrac{1}{x - 1}$

100. $x + 2 \leq \dfrac{x^2}{x - 3}$

101. The sum of three consecutive integers is 51. What are the three integers?

102. The sum of a number, the negative of the reciprocal of the number, and the square of the number equals 1. What number or numbers have this property? (*Hint:* You will need to factor by grouping.)

103. An investor wishes to invest a total of \$50,000 in two accounts for one year. The first account offers an 8% annual rate, while the second account offers $10\frac{1}{2}$%. How much should be placed in each account so that the total interest is \$5000?

104. An open box is to be made from a 16-inch by 20-inch piece of cardboard, by cutting out squares of equal size from the four corners, and then bending up the sides. What size squares should be cut out, if the bottom of the box is to have an area of 165 in.^{2}?

105. Dry Rock is 90 miles from West Branch. The two towns are connected by a straight highway. Suppose that at 1 P.M. a truck leaves Dry Rock, headed toward West Branch and travels at a speed of 40 miles per hour. Thirty minutes later a car leaves Dry Rock, traveling on the same road at a speed of 55 miles per hour. When does the car catch up to the truck, and how far from West Branch does it occur?

106. A man who is 5′9″ tall is walking toward a street light that is 23 ft tall. What is the length of the man's shadow on the ground when he is 12 ft away from the light?

107. Working together, Glen and Jerry take 8 hours to install a fence around an average size backyard. If Glen works three times as fast as Jerry, how long would it take each of the workers to install such a fence alone?

108. Is it possible for the sum of three consecutive integers to be 52? Explain.

109. Determine the value for a so that the equation

$$\frac{x}{a} - \frac{x}{2a} - \frac{x}{3a} = x$$

will be satisfied by all real numbers x.

110. Determine all values for c so that the equation

$$y^2 + y + c = 0$$

will have only one solution (of multiplicity 2).

111. In the figure below, the area of the shaded region is 12 square inches. Find the dimensions of the outer rectangle.

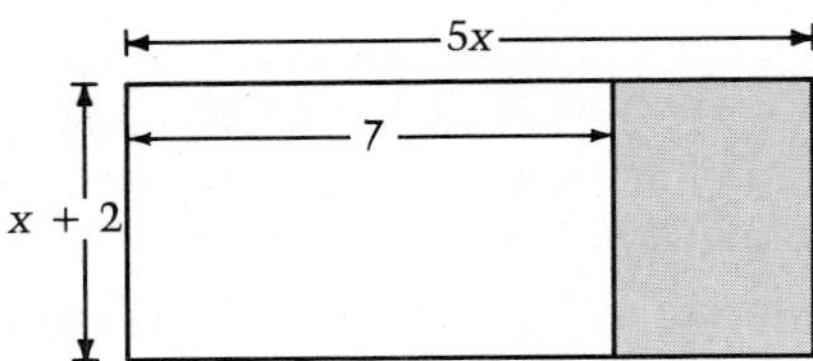

In Exercises 112–115, solve the given equation.

112. $\sqrt{x} - \sqrt{x-2} = \dfrac{1}{\sqrt{x-2}}$

113. $\sqrt{x} + 1 = 2\sqrt{x-5}$

114. $|x - 3| = 3 - x$

115. $|x| - 3 = 3 - x$

116. Determine whether the statement is true or false. If false, illustrate with an example.
(a) If $a < 0$ and $b > 0$, then $ab < b$.
(b) For all real numbers a and b, $|a + b| = |a| + |b|$.
(c) For all real numbers x, $\sqrt[3]{x^3} = x$.

117. A camper can row a boat at a rate of 7 mph in still water. After rowing upstream for 20 minutes and then downstream for 15 minutes, the camper is back at the starting point. Find the rate of the current, and the total distance that the camper rowed.

118. A chemistry student has 12 ml of a solution containing a 25% concentration of acid. How many ml of pure acid must be added to raise the concentration of acid to 50%?

119. There are two pipes, *A* and *B*, that can be used to fill a tank. If only pipe *A* is open, the tank fills in 40 minutes. If only pipe *B* is open, the tank fills in 30 minutes. There is also a drain pipe that can drain the entire tank in 20 minutes. If both pipes *A* and *B* are open and filling the tank, while the drain pipe is also open, how long will it take for the tank to fill?

Chapter 2 Test

In problems 1–4, solve the equation for x.

1. $(4x - 1)x = (2x + 3)(2x - 1)$

2. $ax - 1 = b(x - 4)$

3. $\dfrac{x}{x+2} = 2 - \dfrac{1}{x+2}$

4. $|6x - 1| = 8x + 5$

In problems 5–8, simplify the expression.

5. $(-i)^{27}$

6. $\sqrt{-3}\cdot\sqrt{-21}\cdot\sqrt{-7}$

7. $(9 - 5i) + (-3 - 4i)$

8. $(6 - i)(7 + 3i)$

9. Simplify the quotient $\dfrac{9+i}{2-3i}$.

10. Solve by factoring: $3v^2 + 11v - 20 = 0$.

11. Solve by completing the square: $2x^2 - 6x + 1 = 0$.

12. Solve by using the quadratic formula: $x^2 - 3x + 3 = 0$.

In problems 13–15, solve the equation.

13. $\dfrac{2}{y+5} = \dfrac{5}{6y+3} + \dfrac{3}{2y+1}$

14. $10x^{-2} + 4x^{-4} = 6$

15. $\sqrt{2y-1} = 2 + \sqrt{y-4}$

16. Sketch the graph of the following set:
$[-3.5, 2.5) \cup [3.5, \infty)$

In problems 17–20, solve the inequality, and describe the solution set in interval notation.

17. $3 + 2m > 15 + 3m$

18. $|3 - 2x| > 8$

19. $x^2 + 9x \leq -18$

20. $\dfrac{11}{x+3} \leq 2$

21. A small airplane leaves an airport, heading north at 190 mph. Two hours later a large jet leaves the airport, and heads in the same direction at 550 mph. How far away from the airport will the jet catch up to the small plane?

3

Graphs and Functions

In this chapter we turn to the business of studying functions. Much of the activity that goes on in scientific research is directed toward the discovery of functional relationships that exist between variables (e.g., body weight vs. daily caloric intake; sales revenues vs. advertising costs; pressure vs. temperature). At the end of this chapter you will consider a very powerful tool that relies on an understanding of linear functions. It is called a line of best fit, or sometimes, a least squares regression line. Its use is illustrated in a study of the smoking behaviors of California residents over the last twenty years. The case underscores the crucial role that a simple straight line can play in shedding light on a complicated phenomenon.

3.1 Equations and Graphs

An equation in x and y can be viewed as a way of associating y-values with x-values. For example, the equation $y = 5x + 1$ associates the y-value 11 with the x-value 2. This pair of numbers is denoted (2, 11), and is called an **ordered pair.** Each value in the pair is called a **coordinate.** Plotting an ordered pair requires two real number lines, one for each coordinate.

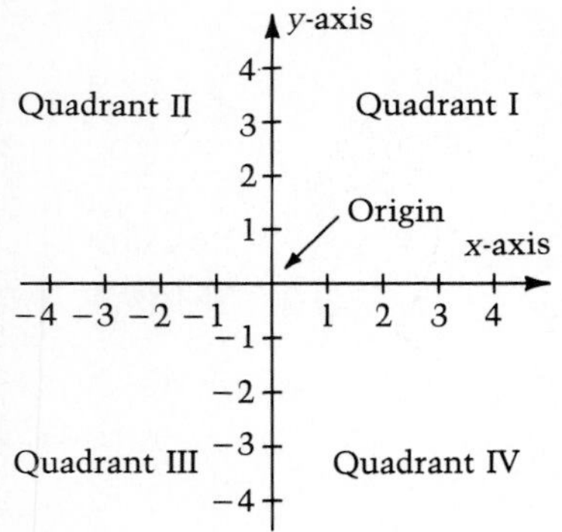

Figure 3.1 The entire system is called the ***xy*-plane** or the **coordinate plane,** and the point where the axes meet is the **origin.**

The two lines, called **axes,** separate the plane into four **quadrants** (Figure 3.1). Since we commonly use x and y as variables, we usually call the horizontal axis the ***x*-axis** and the vertical axis the ***y*-axis.**

Each ordered pair corresponds to a single point in the plane. Plotting a point requires two moves: one from the origin in the horizontal direction, followed by one in the vertical direction.

Suppose we wish to plot the point with coordinates $(3, -4)$. Start at the origin and move 3 units to the right. Then move 4 units downward. Plot the point by drawing a heavy dot (Figure 3.2).

The **graph of an equation** consists of those ordered pairs determined by the equation. For example, one of the ordered pairs determined by the equation $y = 5x + 1$ is $(2, 11)$, and so the point $(2, 11)$ lies on the graph of this equation. To find such points, we substitute a value for one of the variables, and solve for the other. After we have plotted a few such points, we try to decide on the shape of the graph, and then "connect the dots" accordingly. We will begin with a few simple cases where the shape of the graph can be anticipated from the form of the equation. The connection between an equation's graph and its algebraic form is a major theme of this book.

We first consider the straight line graph.

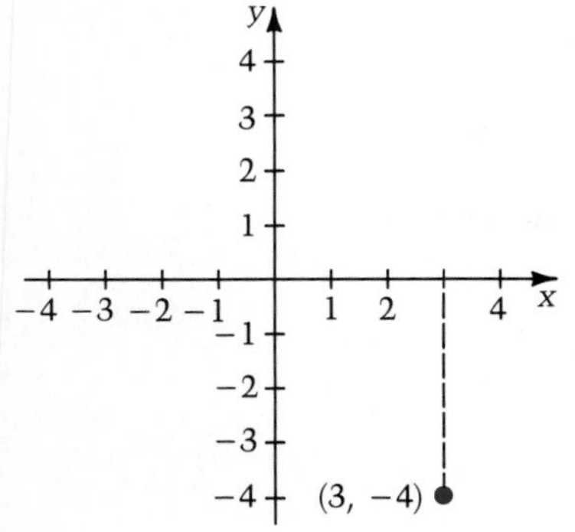

Figure 3.2

Straight Line

An equation of the form $y = mx + b$, where m and b are real numbers, has a **straight line** as its graph. The graph is completely determined by two points.

EXAMPLE 1

Graph the equation $y = 2x - 1$.

Solution

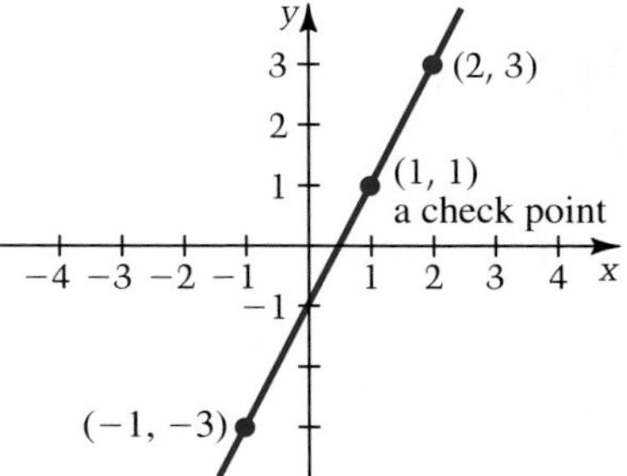

Figure 3.3

To begin, find two ordered pairs that satisfy the equation. Substituting $x = -1$, we get $y = 2(-1) - 1 = -3$. So one ordered pair is $(-1, -3)$. Substituting $x = 2$, we get a second ordered pair, $(2, 3)$.

Plot and label the points, and draw the line through them.

Check: Find another point determined by the equation. It should lie on the line. Here our check point is $(1, 1)$.

We now examine the graphs of equations containing quadratic expressions.

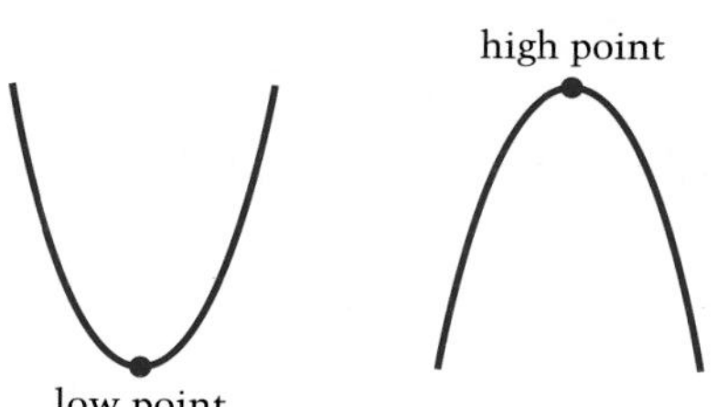

Figure 3.4

Parabola

An equation of the form $y = ax^2 + bx + c$, where a, b, and c are real numbers and $a \neq 0$, has a **parabola** as its graph. If $a > 0$, the parabola opens upward and has a low point; if $a < 0$, the parabola opens downward and has a high point.

Sometimes it is helpful to use a table to display ordered pairs that satisfy an equation. The x and y entries on each line of the table are then the coordinates of a point on the graph. We have constructed such a table in the next example.

EXAMPLE 2

Graph the equation $y = -x^2 + 4x + 1$.

Solution

To begin, find ordered pairs determined by the equation and display them in a table.

x	y	(x, y)
-1	-4	$(-1, -4)$
0	1	$(0, 1)$
1	4	$(1, 4)$
2	5	$(2, 5)$
3	4	$(3, 4)$
4	1	$(4, 1)$
5	-4	$(5, -4)$

$y = -(-1)^2 + 4(-1) + 1 = -4$

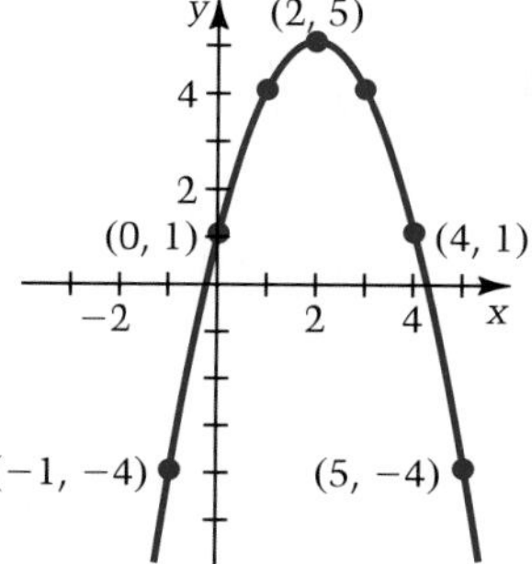

Figure 3.5

Plot and label these points, and draw a parabola through them.

Next we consider the graphs of equations containing absolute value expressions.

V-shaped Graph

An equation of the form $y = a|x| + k$, where a and k are real numbers and $a \neq 0$, has a **V-shaped graph.** The high or low point on such a graph occurs at the "corner" $(0, k)$. It is a low point on the graph if $a > 0$, and a high point if $a < 0$.

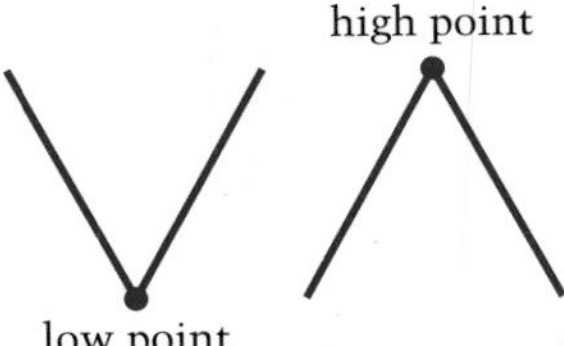

Figure 3.6

EXAMPLE 3

Graph the equation $y = -2|x| + 1$.

Solution

There is a corner point at (0, 1). It is a high point since a is negative (-2).

x	y	(x, y)
-2	-3	$(-2, -3)$
-1	-1	$(-1, -1)$
0	1	(0, 1)
1	-1	$(1, -1)$
2	-3	$(2, -3)$

$y = -2|-2| + 1 = -3$

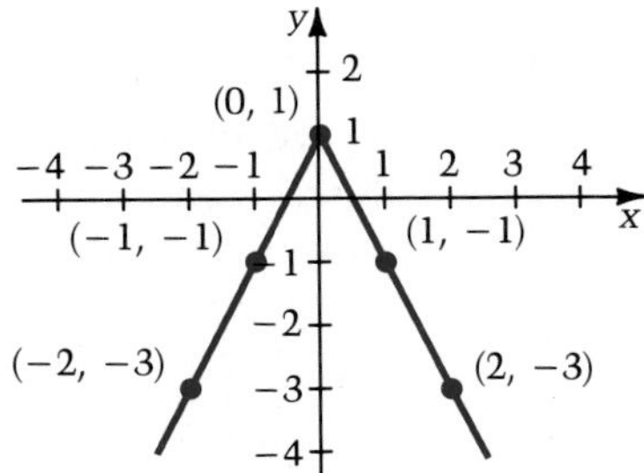

Figure 3.7

Distance in a Coordinate Plane

Recall the Pythagorean Theorem.

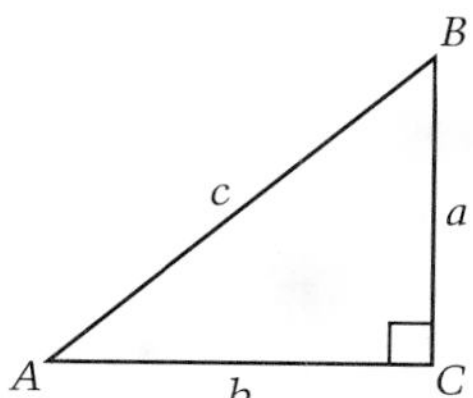

Figure 3.8 A right triangle with right angle at vertex C.

The Pythagorean Theorem

In a right triangle, where c is the length of the hypotenuse (the side opposite the right angle), and a and b are the lengths of the other two sides,

$$c^2 = a^2 + b^2.$$

Given any two points P and Q in the xy-plane, the Pythagorean Theorem can be used to determine the distance between them, symbolized $d(P, Q)$. In Figure 3.9(a), we consider the distance between points $P(1, 2)$ and $Q(4, 6)$. In Figure 3.9(b), we compute the distance between two general points $P_1(x_1, y_1)$ and $P_2(x_2, y_2)$.

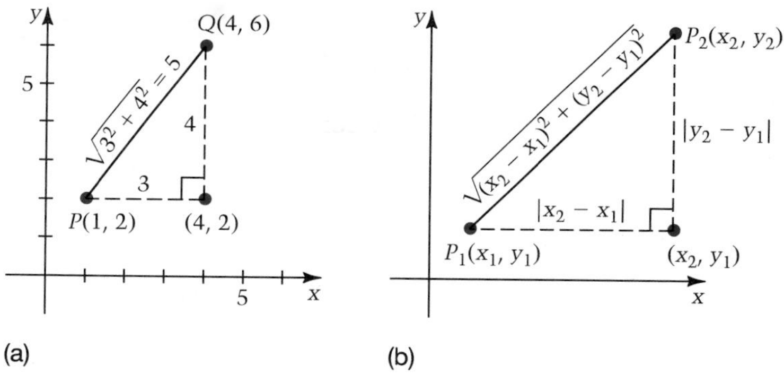

Figure 3.9 (a) $d(P,Q) = \sqrt{3^2 + 4^2} = \sqrt{25} = 5$.
(b) $d(P_1,P_2) = \sqrt{(|x_2 - x_1|)^2 + (|y_2 - y_1|)^2} = \sqrt{(x_2 - x_1)^2 + (y_2 - y_1)^2}$.
(Absolute values are unnecessary since the differences are squared.)

We conclude from Figure 3.9(b) the following general formula for finding the distance between two points.

The Distance Formula

The distance between two points $P_1(x_1, y_1)$ and $P_2(x_2, y_2)$ is given by the formula

$$d(P_1, P_2) = \sqrt{(x_2 - x_1)^2 + (y_2 - y_1)^2}.$$

EXAMPLE 4

Determine the distance d between the points $(2, -5)$ and $(6, 7)$.

Solution

$$\begin{aligned} d &= \sqrt{(x_2 - x_1)^2 + (y_2 - y_1)^2} \\ &= \sqrt{(6 - 2)^2 + (7 - (-5))^2} \\ &= \sqrt{4^2 + 12^2} \\ &= \sqrt{160} = 4\sqrt{10} \approx 12.6 \end{aligned}$$

Begin by writing the Distance Formula. Let $(x_1, y_1) = (2, -5)$ and $(x_2, y_2) = (6, 7)$.

It can be shown that the coordinates of the midpoint M of a line segment joining P_1 and P_2 are found by averaging the x coordinates of P_1 and P_2, and the y-coordinates of P_1 and P_2.

The Midpoint Formula

Suppose a line segment has endpoints $P_1(x_1, y_1)$ and $P_2(x_2, y_2)$. Then the midpoint M of the segment joining P_1 and P_2 has coordinates

$$\left(\frac{x_1 + x_2}{2}, \frac{y_1 + y_2}{2}\right).$$

EXAMPLE 5

Determine the midpoint of the line segment joining the points $(-1, 7)$ and $(4, -3)$.

Solution

$$\frac{x_1 + x_2}{2} = \frac{-1 + 4}{2} = \frac{3}{2}$$ Determine the average of the x-values.

$$\frac{y_1 + y_2}{2} = \frac{7 + (-3)}{2} = 2$$ Determine the average of the y-values.

The midpoint is $\left(\frac{3}{2}, 2\right)$.

Symmetry in a Coordinate Plane

When you look at the reflection of an object in a mirror, the reflection appears to be the same distance behind the mirror as the object is in front of the mirror. The notion of reflection has a useful application to graphing.

Figure 3.10

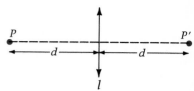

Suppose the mirror in Figure 3.10 was turned so that all you could see is the edge l, and suppose the object P was simply a point. Point P', the **reflection of point P across line l,** is found by drawing a dotted line through P, perpendicular to l, and then plotting P' the same distance behind l as P is in front of l.

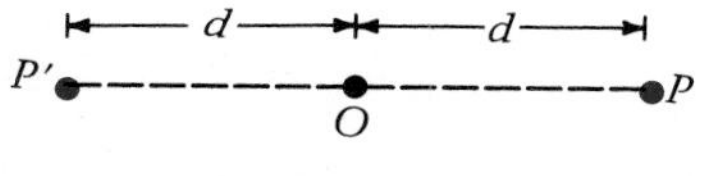

Figure 3.11

A point can also be reflected through a point. The point P', the **reflection of P through O,** is found by drawing a line through P and O. P' is the same distance from O as P is from O (Figure 3.11). It is often useful to reflect points through the origin (0, 0).

EXAMPLE 6

Plot and label the reflection of the point (2, 3) across the x-axis, across the y-axis, and through the origin.

Solution

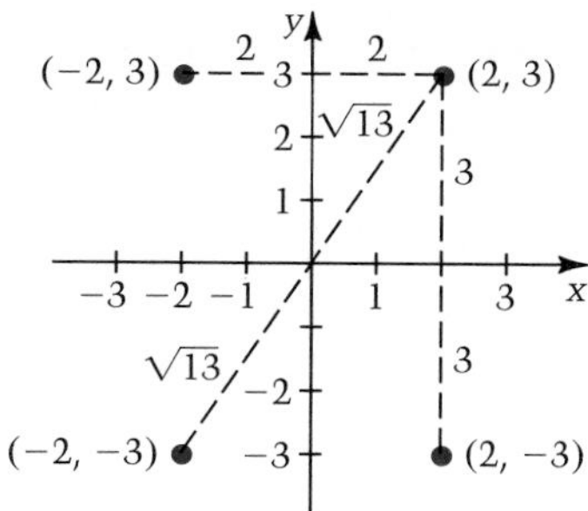

Figure 3.12

To reflect (2, 3) across the y-axis, change the sign of the x-coordinate. The reflection is $(-2, 3)$.

To reflect (2, 3) across the x-axis, change the sign of the y-coordinate. The reflection is $(2, -3)$.

To reflect (2, 3) through the origin, change the sign of both the x- and y-coordinate. The reflection is $(-2, -3)$.

In some cases, when graphing equations, you can find part of the graph by plotting points, and then reflect that part across the x-axis, or y-axis. We now use the notion of reflection to define an important property of graphs, called **symmetry.** There are three types of symmetry to consider. (Assume that we are graphing equations in the variables x and y.)

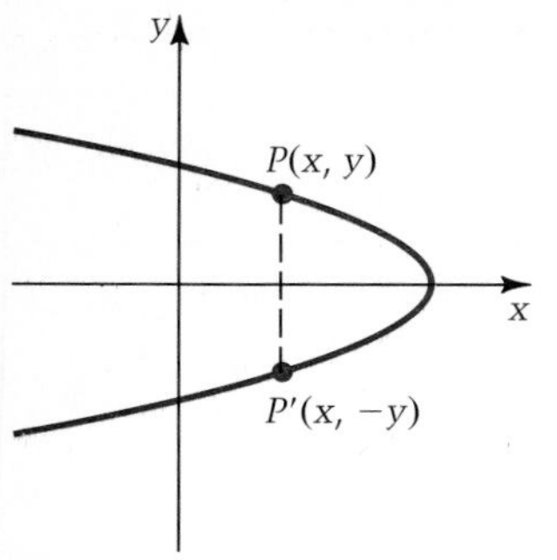

Figure 3.13 A graph symmetric with respect to the x-axis.

Symmetry with Respect to the x-axis

A graph is **symmetric with respect to the x-axis,** if for each point P on the graph, the reflection of P across the x-axis is also on the graph.

> **Test for x-axis Symmetry**
>
> The graph of an equation is symmetric with respect to the x-axis, if replacing y with $-y$ produces an equation equivalent to the original equation.

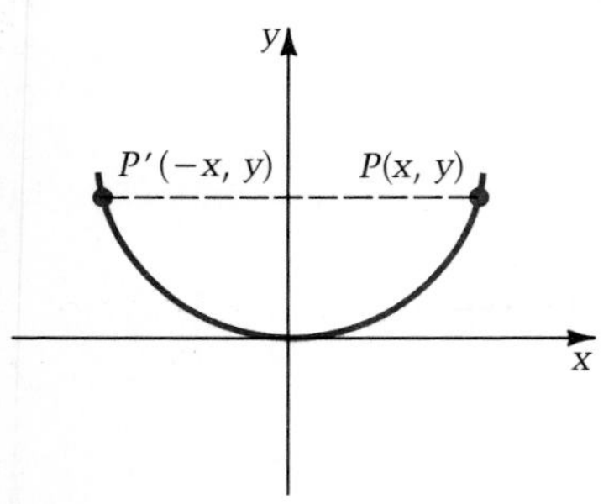

Figure 3.14 A graph symmetric with respect to the y-axis.

Symmetry with Respect to the y-axis

A graph is **symmetric with respect to the y-axis,** if for each point P on the graph, the reflection of P across the y-axis is also on the graph.

> **Test for y-axis Symmetry**
>
> The graph of an equation is symmetric with respect to the y-axis, if replacing x with $-x$ produces an equation equivalent to the original equation.

Symmetry with Respect to the Origin

A graph is **symmetric with respect to the origin,** if for each point P on the graph, the reflection of P through the origin is also on the graph.

> **Test for Symmetry with Respect to the Origin**
>
> The graph of an equation is symmetric with respect to the origin, if replacing x with $-x$, *and* y with $-y$ produces an equation equivalent to the original equation.

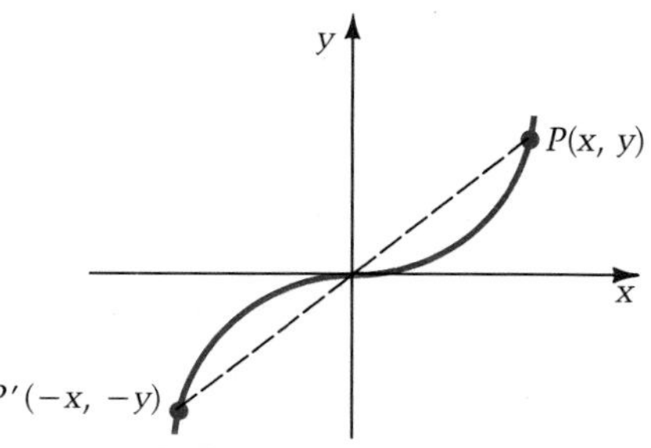

Figure 3.15 A graph symmetric with respect to the origin.

EXAMPLE 7

Is the graph of the equation $x = y^2 - 1$ symmetric with respect to the x-axis?

Solution

$x = (-y)^2 - 1$	Replace y with $-y$.
$x = y^2 - 1$	Since $(-y)^2 = (-y)(-y) = y^2$, the equation is equivalent to the original equation.

The graph *is* symmetric with respect to the x-axis.

EXAMPLE 8

Is the graph of the equation $y = 4x^2 + 9$ symmetric with respect to the y-axis?

Solution

$y = 4(-x)^2 + 9$	Replace x with $-x$.
$y = 4x^2 + 9$	The resulting equation is equivalent to the original equation since $(-x)^2 = x^2$.

The graph *is* symmetric with respect to the y-axis.

EXAMPLE 9

Test $y = x^5 - 2x$ for each of the three types of symmetry.

Solution

1. The test for symmetry with respect to the x-axis produces the equation $-y = x^5 - 2x$ so that $y = -x^5 + 2x$, which is *not* equivalent to the original equation. The graph is not symmetric with respect to the x-axis.
2. The test for symmetry with respect to the y-axis produces the equation $y = (-x)^5 - 2(-x)$ so that $y = -x^5 + 2x$, which is *not* equivalent to the original equation. The graph is not symmetric with respect to the y-axis.
3. Test for symmetry with respect to the origin.

$-y = (-x)^5 - 2(-x)$	Replace x with $-x$ and y with $-y$.
$-y = -x^5 + 2x$	
$y = x^5 - 2x$	The resulting equation is equivalent to the original equation.

The graph *is* symmetric with respect to the origin.

When graphing, you should test for symmetry before actually determining ordered pairs. Finding ordered pairs can be very time consuming. If symmetry exists, you can save time by first determining some points on one part of the graph, and then using symmetry to complete the graph.

EXAMPLE 10

Sketch the graph of the equation $y = x^3 - 4x$.

Solution First we test for symmetry. Replacing x with $-x$ and y with $-y$ produces $-y = (-x)^3 - 4(-x)$, which is equivalent to the given equation. Thus the graph is symmetric with respect to the origin.

x	y	(x, y)
0	0	(0, 0)
1	−3	(1, −3)
2	0	(2, 0)
3	15	(3, 15)

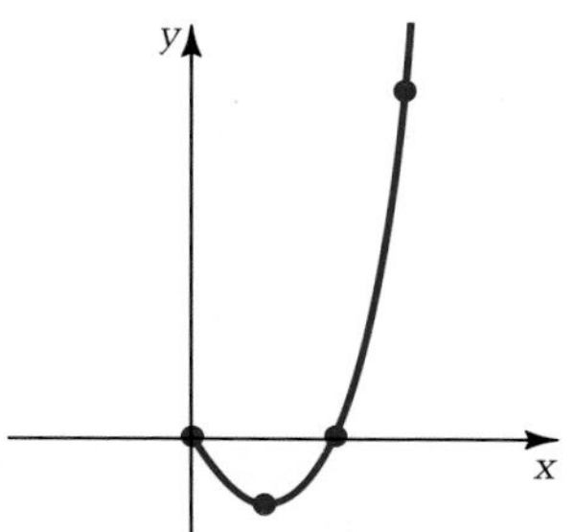

Figure 3.16 Reflect this curve through the origin.

Plot the points listed in the table, draw a curve through them, then reflect this curve through the origin to get the complete graph, as shown in Figure 3.17.

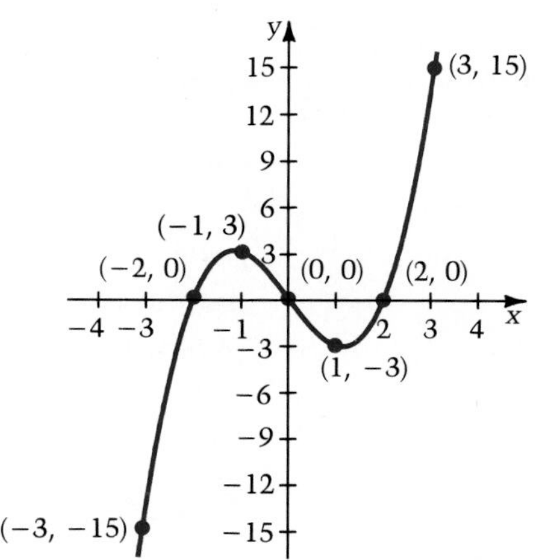

Figure 3.17

EXERCISE SET 3.1

In Exercises 1–22, graph the equation.

1. $y = 2x$

2. $y = \frac{1}{3}x$

3. $y = 3 - \frac{x}{2}$

4. $y = -5x + 1$

5. $y = x^2$

6. $y = x^2 - 3$

7. $y = x^2 - 1$

8. $y = x^2 + 1$

9. $y = 2 - x^2$

10. $y = 1 - x^2$

11. $y = 3x^2 - 1$

12. $y = 2x^2 + 3$

13. $2y = x^2 - 2x - 1$

14. $3y = x^2 - 8x + 12$

15. $y = |x|$

16. $y = |x| + 5$

17. $y = 3|x| + 2$

18. $y = 2|x| - 4$

19. $y = \frac{1}{3}|x|$

20. $y = -\frac{1}{2}|x|$

21. $y = 3 - 2|x|$

22. $y = 5 - 3|x|$

In Exercises 23–32, determine the distance $d(P, Q)$ and the midpoint of the segment PQ.

23. $P(0, 0), Q(-3, 4)$

24. $P(1, -1), Q(5, 2)$

25. $P(4, -5), Q(-1, -7)$

26. $P(8, 6), Q(0, 0)$

27. $P(8, -3), Q(6, -3)$

28. $P(7, 10), Q(7, -2)$

29. $P(-2, 3), Q(4, -1)$

30. $P(4, 5), Q(5, 8)$

31. $P(\sqrt{2}, -\sqrt{3}), Q(2\sqrt{2}, \sqrt{3})$

32. $P(-\sqrt{5}, \sqrt{3}), Q(2\sqrt{5}, 2\sqrt{3})$

In Exercises 33–38, plot and label the reflection of the point (a) across the x-axis, (b) across the y-axis, and (c) through the origin.

33. $(3, -2)$

34. $(-2, 1)$

35. $(4, 3)$

36. $(-4, -8)$

37. $(0, 5)$

38. $(-3, 0)$

In Exercises 39–50, test the graph of the given equation for each of the three types of symmetry.

39. $y = x^2 - 2$

40. $y^2 = 3 - x$

41. $y^2 = x^2$

42. $y^2 = x^4$

43. $y = (x^2 - 1)^2$

44. $y = x^2 - 4x^4$

45. $x^3 - y^5 = 1$

46. $x^4 - y^2 = 7$

47. $x^2 - y^2 - 4 = 0$

48. $y = 3x - x^2$

49. $xy = 2$

50. $x^2y^3 = -8$

In Exercises 51–56, reflect the given graph (a) across the x-axis, (b) across the y-axis, and (c) through the origin.

51.

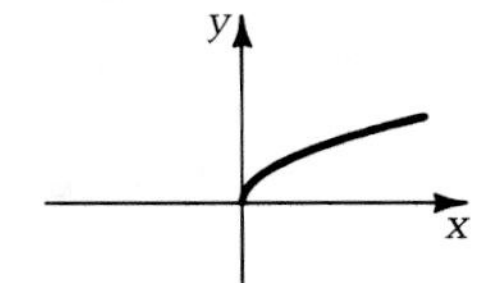

52.

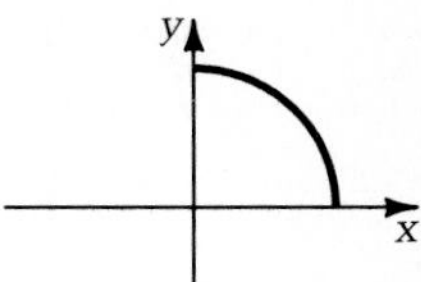

53.

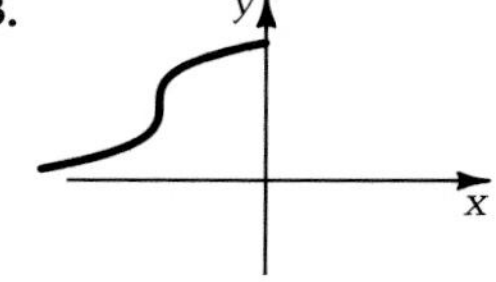

54.

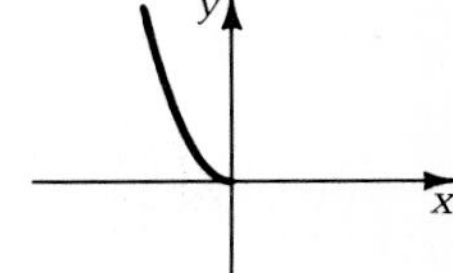

55.

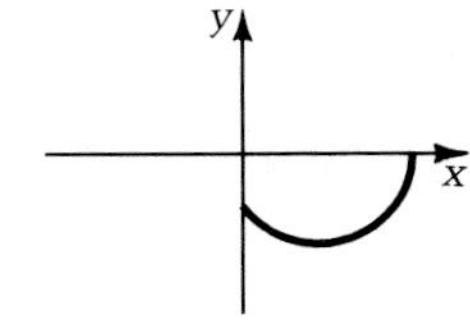

56.

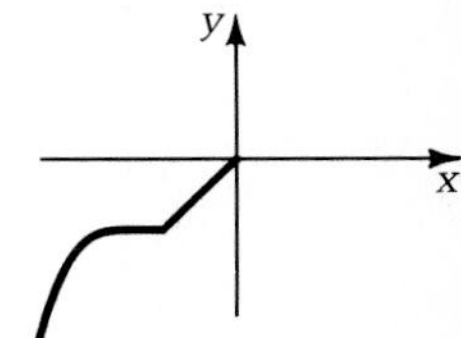

In Exercises 57–64, use tests for symmetry before graphing the equation.

57. $y^2 = x - 3$

58. $x = 25 - y^2$

59. $3y = x^3 - x$

60. $2x^3 - 3x + y = 0$

61. $y^2 = 4x^2$

62. $2x = y^3 - y$

63. $y^3 - 2y + x = 0$

64. $x^2 = 9y^2$

Superset

In Exercises 65–70, determine the coordinates of the reflection of the point across the line.

65. $(3, 2)$ across the line $x = 4$.

66. $(2, 7)$ across the line $x = 6$.

67. $(4, -3)$ across the line $x = -1$.

68. $(-2, -5)$ across the line $x = -3$.

69. $(3, 2)$ across the line $y = 5$.

70. $(2, 7)$ across the line $y = -4$.

Given three points P, Q, and R, there are three distances that can be computed: $d(P, Q)$, $d(Q, R)$, and $d(P, R)$. If the sum of two of those distances equals the third, then the three points lie on the same line, that is, they are **collinear.** In Exercises 71–76, use this fact to determine whether the three given points are collinear.

71. $(-1, 4), (1, 2), (3, 0)$

72. $(0, 5), (-3, 8), (5, 2)$

73. $(0, -7), (3, 11), (-1, 0)$

74. $(-2, 15), (0, 7), (1, 3)$

75. $(1, -1), (0, -3), (-2, -7)$

76. $(-1, 5), (0, 3), (1, 0)$

Three points P, Q, and R are the vertices of a right triangle, provided the three distances $d(P, Q)$, $d(Q, R)$, and $d(P, R)$, satisfy the Pythagorean Theorem. In Exercises 77–80, use this fact to determine whether the three given points are vertices of a right triangle.

77. $(0, 0), (-1, -1), (-2, 2)$

78. $(0, 0), (1, -3), (3, -1)$

79. $(0, 2), (-6, -1), (-10, 7)$

80. $(3, -2), (0, 4), (-2, 3)$

In Exercises 81–92, graph the equation.

81. $x = |y|$

82. $|x| = |y|$

83. $|x| + |y| = 1$

84. $|x| - |y| = 1$

85. $x^2 + y^2 = 25$

86. $x^2 = y^3$

87. $x^4 = y^2$

88. $x^2 + 4y^2 = 16$

89. $x^3 = y^2$

90. $x^2 + y^2 = 100$

91. $4x^2 + y^2 = 36$

92. $x^2 = y^6$

93. (a) Suppose the graph of an equation is symmetric with respect to both the x-axis and the y-axis. Explain why such a graph must also be symmetric with respect to the origin.

(b) By means of a graph, provide an example to show that the converse of the situation in part (a) need not hold, namely, if a graph is symmetric with respect to the origin, then the graph need not be symmetric with respect to either axis.

94. (a) This problem generalizes Exercise 93. Explain why the graph of an equation that displays any two of the three types of symmetry must also display the third type of symmetry.

(b) From part (a) conclude that a graph will display none, precisely one, or all three types of symmetry. By means of graphs, provide examples for each possibility.

3.2
Introduction to Functions

We often can describe a quantity by telling how it is related to, or dependent upon, some other quantity. For example, it is said that a child's rate of growth is dependent upon the quality and quantity of food consumed, and that happiness in life is dependent upon the degree of satisfaction you get from your job.

In mathematics we can sometimes express the relationship between things by using the word *function.* Our examples may be restated as

growth rate is a function of nutrition, and
happiness is a function of job satisfaction.

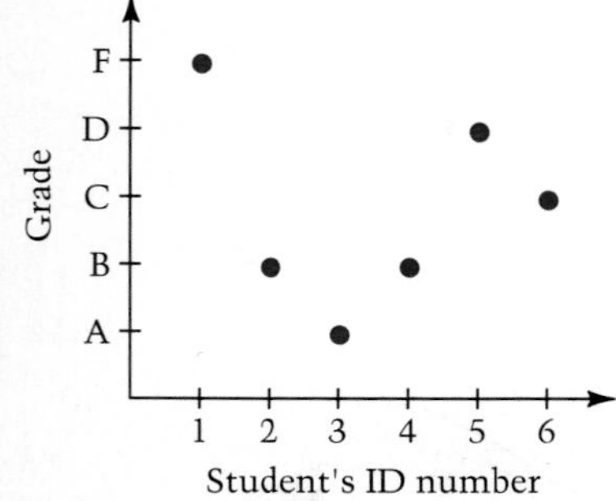

Figure 3.18

The concept of function is one of the most important in mathematics. We shall explore this concept in detail in this chapter.

The graph in Figure 3.18 shows the grades of six students. Suppose the teacher wishes to store the grades on a computer. Then it might be necessary to code the information as a list of pairs, where the first entry of the pair is the student's identification number and the second entry is the grade. This information can be written as the ordered pairs (1, F), (4, B), (2, B), (5, D), (3, A), and (6, C). The set of these ordered pairs is an example of a *function*.

Definition

Whenever a process pairs each member of a first set with exactly one member of a second set, the resulting set of ordered pairs is called a **function.** Thus, a function contains no two different ordered pairs having the same first coordinate. The set of all first coordinates is called the **domain** of the function and the set of all second coordinates is called the **range.**

The function described above is the set of ordered pairs

$$\{(1, F), (2, B), (3, A), (4, B), (5, D), (6, C)\}.$$

The domain of the function is the set

$$\{1, 2, 3, 4, 5, 6\}$$

and the range of the function is the set

$$\{A, B, C, D, F\}.$$

EXAMPLE 1

Use a set of ordered pairs to describe the function represented by the caloric graph shown below. Determine the domain and range of the function.

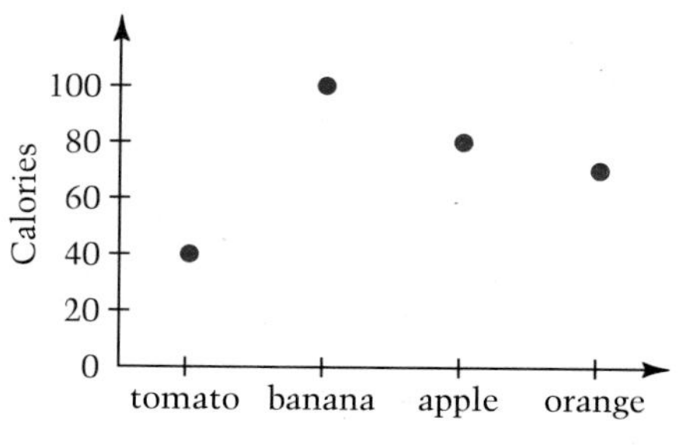

Figure 3.19

As in the graph of students' grades, remember to select the first coordinate from the horizontal scale and the second coordinate from the vertical scale.

Solution The function is the following set of ordered pairs: {(tomato, 40), (banana, 100), (apple, 80), (orange, 70)}. The domain is {tomato, banana, apple, orange}, and the range is {40, 70, 80, 100}.

Another example of a function is the relationship between crop yield and soil fertility. As corn grows, it leaches nitrogen and other nutrients from the soil. In order to maintain the same level of production from year to year, the nitrogen must be replenished, and fertilizers are used to replace the nitrogen. The table below shows the pounds per acre of nitrogen needed to maintain the fertility of the soil for various levels of corn production. This relationship establishes a function given by the following ordered pairs: {(12, 108), (20, 180), (25, 220), (30, 270)}.

Corn Production (tons per acre)	**Nitrogen Fertilizer (lbs per acre)**	
12	108	(12, 108)
20	180	(20, 180)
25	220	(25, 220)
30	270	(30, 270)

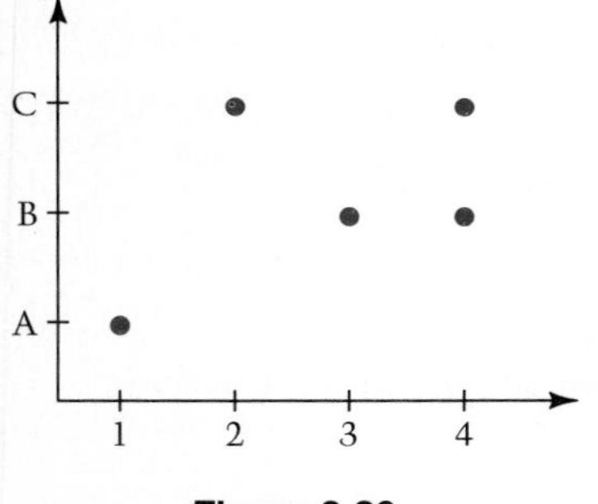

Figure 3.20

Not every set of ordered pairs is a function. The ordered pairs represented in the graph in Figure 3.20 do not form a function, since there are two ordered pairs with the same first coordinate, namely (4, B) and (4, C). In general, we refer to any set of ordered pairs as a **relation;** if no two different ordered pairs have the same first coordinate, we say that the relation is a function.

At some time you have probably already checked an "ideal weight" chart to determine your "perfect" weight. There are formulas which describe ideal weight as a function of height. For a male over age 25 with a medium frame, the formula might be $w = 4h - 121$, where w represents weight in pounds and h represents height in inches. The formula determines a function since to each value of height there is assigned exactly one ideal weight.

Given a value of height h, the ideal weight is found by *evaluating the function,* that is, by replacing h with the given height and then finding the corresponding value of w. The chart below lists some ordered pairs for this function: (62, 127), (68, 151), and (76, 183). The second coordinates are called **function values,** because they are found by evaluating the function.

$w = 4h - 121$		
h (in.)	**w (lb)**	**(h, w)**
62	127	(62, 127)
68	151	(68, 151)
76	183	(76, 183)

EXAMPLE 2

The formula for converting the temperature in degrees Fahrenheit (F) to degrees Celsius (C) is given by the function $C = \frac{5}{9}(F - 32)$. Evaluate the function for $F = 5, 14, 32$, and 50, and display the results in a table.

Solution

For $F = 5$, we get $C = \frac{5}{9} \cdot (5 - 32) = -15$

$F = 14$ $\quad C = \frac{5}{9} \cdot (14 - 32) = -10$

$F = 32$ $\quad C = \frac{5}{9} \cdot (32 - 32) = 0$

$F = 50$ $\quad C = \frac{5}{9} \cdot (50 - 32) = 10$

F	**C**	**(F, C)**
5	−15	(5, −15)
14	−10	(14, −10)
32	0	(32, 0)
50	10	(50, 10)

EXAMPLE 3

A certain function is described by the formula $y = 7x^2 + 3x - 1$. Evaluate the function for $x = -5, -1, 0$, and 5, and display the results as ordered pairs.

Solution

$$
\begin{aligned}
& & y &= 7 \cdot x^2 + 3 \cdot x - 1 \\
x &= -5 & y &= 7 \cdot (-5)^2 + 3 \cdot (-5) - 1 = 159 \\
x &= -1 & y &= 7 \cdot (-1)^2 + 3 \cdot (-1) - 1 = 3 \\
x &= 0 & y &= 7 \cdot (0)^2 + 3 \cdot (0) - 1 = -1 \\
x &= 5 & y &= 7 \cdot (5)^2 + 3 \cdot (5) - 1 = 189
\end{aligned}
$$

The resulting ordered pairs are $(-5, 159)$, $(-1, 3)$, $(0, -1)$, and $(5, 189)$.

In the two previous examples, we have seen that some functions can be described by equations involving two variables. In Example 2, C is the **dependent variable,** and F is the **independent variable.** C is dependent upon F because we first select a value for F and then use it to find C. We call F the independent variable because we are free to choose any domain value for it.

We can use letters to name functions. For example,

$$f = \{(-1, 2), (5, 3), (11, 12), (15, 25)\}.$$

In this case, we use $f(x)$ to denote the second coordinate in the ordered pair whose first coordinate is x. Thus, we write

$$f(-1) = 2, \quad f(5) = 3, \quad f(11) = 12, \quad f(15) = 25,$$

with $f(-1)$ read as "f of -1." Since $f(-1) = 2$, we say that 2 is the "function value at -1."

The equation $f(x) = 2x^2 + 7x - 4$ determines the same set of ordered pairs (and therefore the same function) as the equation $y = 2x^2 + 7x - 4$. You will see y and $f(x)$ used interchangeably, and you should become comfortable with both these usages.

EXAMPLE 4

For the function $g(x) = 2x^2 - 4x + 11$ find: (a) $g(3)$ (b) $[g(2)]^3$ (c) $\dfrac{g(1)}{g(-1)}$ (d) $g(b^3)$ (e) $g(3 + h)$.

Solution

(a) $g(3) = 2(3)^2 - 4(3) + 11 = 17$ — Replace x with 3.

(b) $[g(2)]^3 = [2(2)^2 - 4(2) + 11]^3 = 11^3 = 1331$ — Compute $g(2)$ first, then cube.

(c) $$\frac{g(1)}{g(-1)} = \frac{2(1)^2 - 4(1) + 11}{2(-1)^2 - 4(-1) + 11} = \frac{9}{17}$$

(d) $g(b^3) = 2(b^3)^2 - 4b^3 + 11 = 2b^6 - 4b^3 + 11$ — Replace x with b^3.

(e) $$\begin{aligned} g(3 + h) &= 2(3 + h)^2 - 4(3 + h) + 11 \\ &= 2(9 + 6h + h^2) - 12 - 4h + 11 \\ &= 18 + 12h + 2h^2 - 12 - 4h + 11 \\ &= 2h^2 + 8h + 17 \end{aligned}$$ — Replace x with $3 + h$.

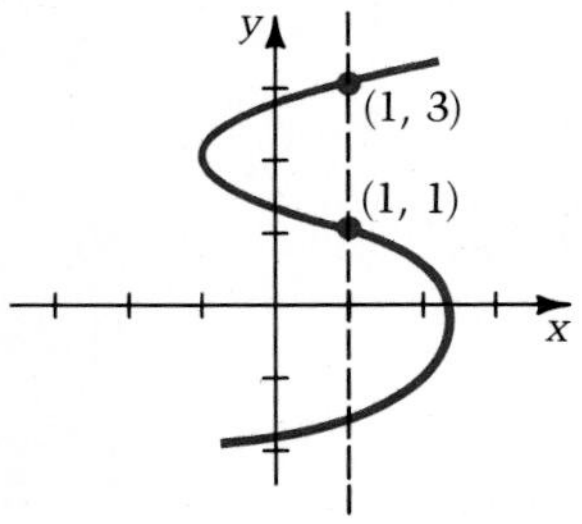

Figure 3.21

Recall that a set of ordered pairs is a function if no two different ordered pairs have the same first coordinate. The ordered pairs (1, 1) and (1, 3) have the same first coordinate; thus the curve in Figure 3.21 is not the graph of a function. The fact that such points lie on the same vertical line suggests a quick way of determining whether a graph represents a function.

The Vertical Line Test

For any graph in the xy-plane, if some vertical line can be drawn so that it intersects the graph more than once, then the graph does not represent a function of x. If no such line exists, the graph does represent a function of x.

In Figure 3.22, graphs (a) and (b) represent functions of x, while graphs (c) and (d) do not. For (c) and (d) we have found vertical lines (shown as

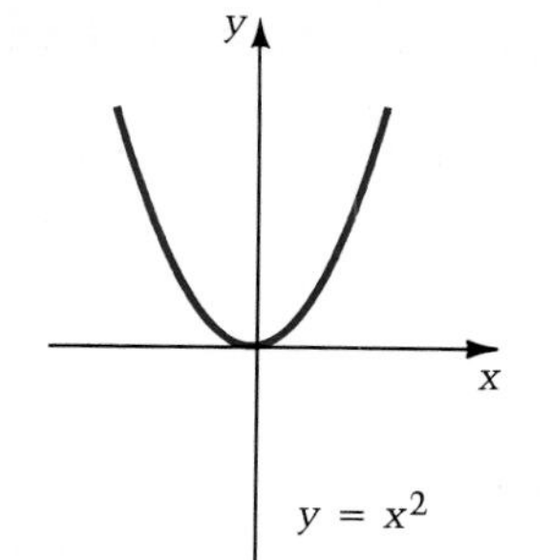

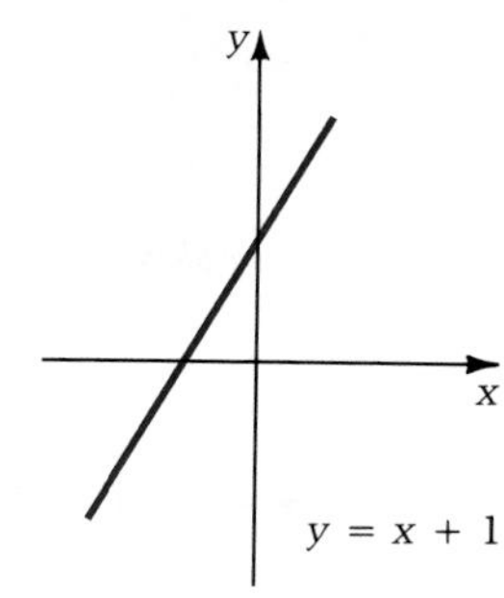

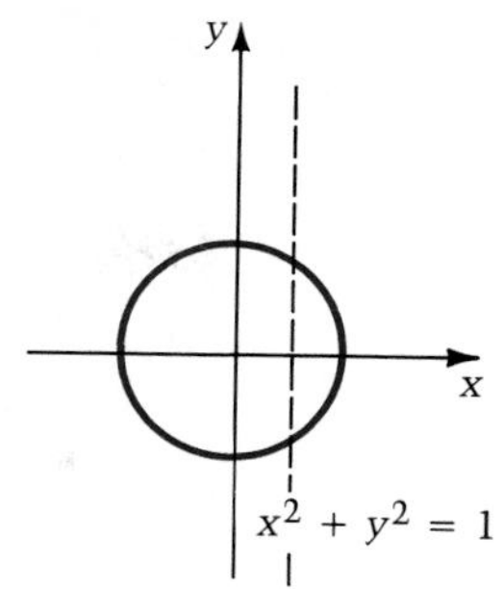

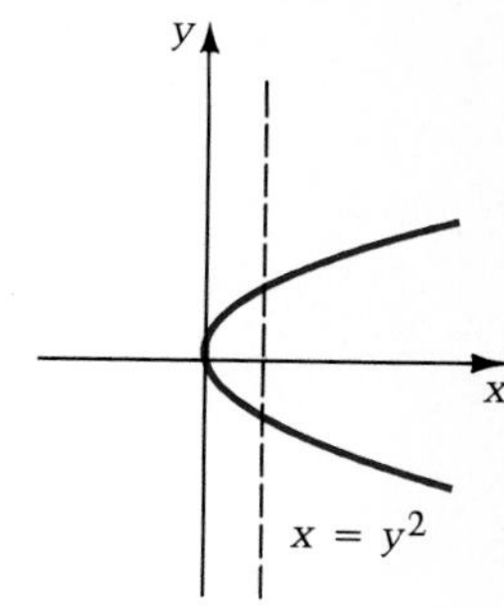

Figure 3.22

dashed lines) which intersect the graph in more than one point. In (a) and (b), no such vertical lines can be found.

Thus far we have discussed functions whose domains are the entire set of real numbers: x can be replaced by any real number. However, it is sometimes necessary to restrict the domain of a function in order to exclude values that produce undefined expressions.

$y = \sqrt{x}$ Since the square root of a negative number is not a real number, the domain of this function is the set of all nonnegative real numbers.

$y = \dfrac{1}{x}$ Since division by zero is undefined, the domain of this function is the set of all real numbers except zero.

Even though we have defined complex numbers in Chapter 2, we will restrict our attention to functions whose domains and ranges consist solely of real numbers.

EXAMPLE 5

Determine the domain of each function.

(a) $f(x) = \dfrac{2}{5 - 3x}$ (b) $H(x) = \dfrac{1}{x^2 + 2x - 15}$ (c) $Q(x) = \sqrt{25 - x^2}$

Solution

(a) $f(x) = \dfrac{2}{5 - 3x}$ Since the denominator cannot be 0, we solve $5 - 3x = 0$, to determine the value that is excluded from the domain.

$5 - 3x = 0$

$x = \dfrac{5}{3}$ The denominator is 0 when $x = \dfrac{5}{3}$.

The domain is the set of all real numbers except $\frac{5}{3}$; in interval notation: $\left(-\infty, \frac{5}{3}\right) \cup \left(\frac{5}{3}, \infty\right)$.

(b) $H(x) = \dfrac{1}{x^2 + 2x - 15}$ Since the denominator cannot be 0, we solve $x^2 + 2x - 15 = 0$, to determine the values that are excluded from the domain.

$$x^2 + 2x - 15 = 0$$
$$(x + 5)(x - 3) = 0$$

$x + 5 = 0$	$x - 3 = 0$
$x = -5$	$x = 3$

The denominator is 0 when $x = -5$ or $x = 3$.

The domain is the set of all real numbers except -5 and 3; in interval notation: $(-\infty, -5) \cup (-5, 3) \cup (3, \infty)$.

(c) $Q(x) = \sqrt{25 - x^2}$ Since the radicand must be nonnegative, we solve $25 - x^2 \geq 0$ to find the domain.

$$25 - x^2 \geq 0$$
$$(5 - x)(5 + x) \geq 0$$

Recall the method of using test values to solve quadratic inequalities.

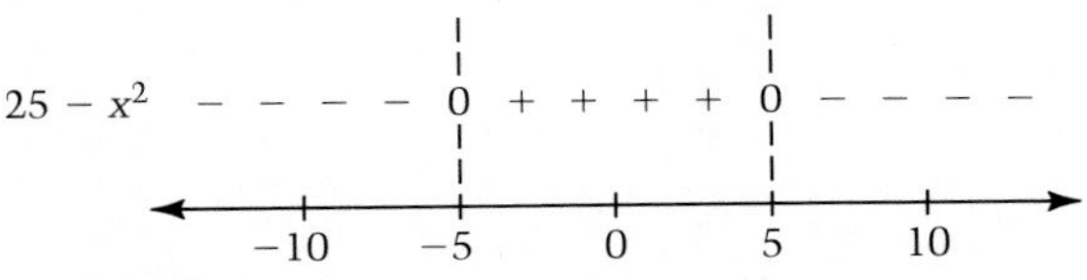

Figure 3.23

The domain is the set of all real numbers between -5 and 5 inclusive; in interval notation: $[-5, 5]$.

A function is called a **piecewise-defined function** if different formulas for $f(x)$ are valid for different parts of the domain. We consider such a function in the next example.

EXAMPLE 6

For the function $f(x)$, determine $f(-3), f(0), f(5), f(13)$.

$$f(x) = \begin{cases} x^2 + 1 & \text{if } x \leq 0 \\ x + 3 & \text{if } 0 < x \leq 5 \\ 7 & \text{if } x > 5 \end{cases}$$

Solution

$f(-3) = (-3)^2 + 1 = 10$ Since $-3 \leq 0$, the first piece, $x^2 + 1$, is valid.

$f(0) = (0)^2 + 1 = 1$ Since $0 \leq 0$, the first piece is valid.

$f(5) = (5) + 3 = 8$ Since $0 < 5 \leq 5$, the second piece, $x + 3$, is valid.

$f(13) = 7$ Since $13 > 5$, the third piece, 7, is valid.

Thus $f(-3) = 10, f(0) = 1, f(5) = 8, f(13) = 7$.

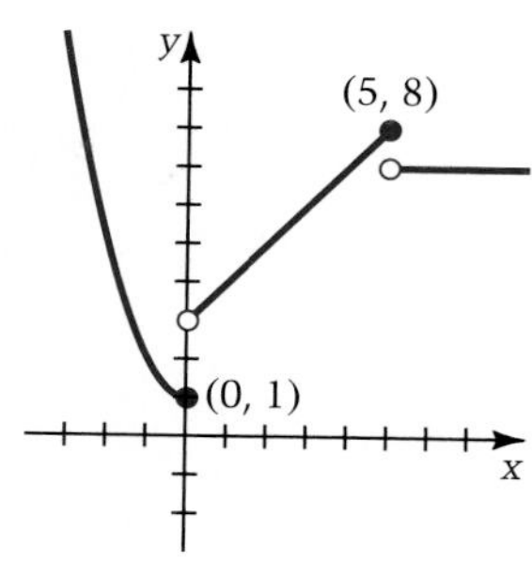

Figure 3.24

EXERCISE SET 3.2

In Exercises 1–4, describe the function with a set of ordered pairs. State the domain and range.

1.

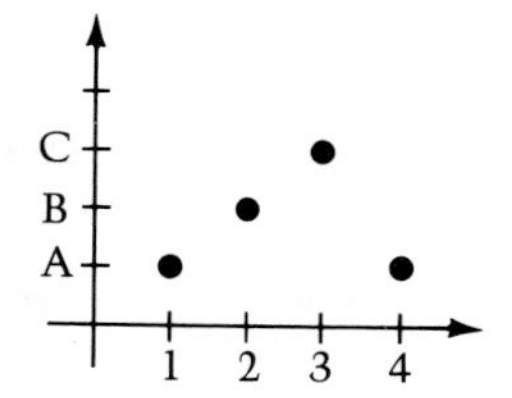

2.

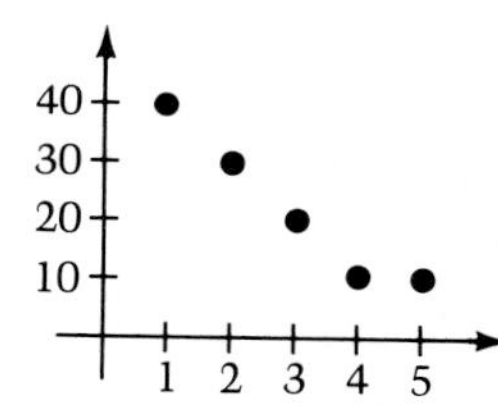

3.

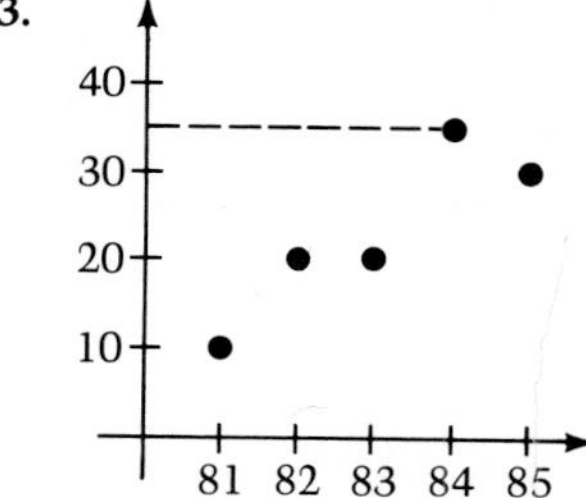

4.

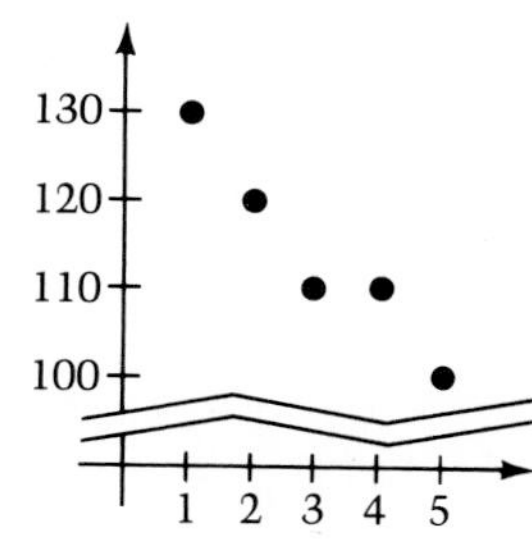

In Exercises 5–8, evaluate the function for the given values and display the results in a table.

5. $t = 2h - 1, h = -1, 0, 3, 4$

6. $x = -16t^2 + 2t, t = -2, -1, 0, 2$

7. $y = x^2 - 3x + 1, x = -5, -2, 1, 4$

8. $y = 5 - 4x - x^2, x = -10, -5, 5, 10$

In Exercises 9–14, determine $f(x)$ for the x-values 1, 1.2, 1.4, 1.6, 1.8, and 2.

9. $f(x) = 6x - 5$

10. $f(x) = 0.02x - 3.76$

11. $f(x) = x^2 - 5x + 3.66$

12. $f(x) = 3.5x^2 - 0.05x + 1$

13. $f(x) = \sqrt{\dfrac{x - 0.08}{2}}$ (round to nearest thousandth)

14. $f(x) = \sqrt{\dfrac{16 - 5.2x}{0.05}}$ (round to nearest thousandth)

In Exercises 15–30, find $f(-2)$, $f(0)$, $f(5)$, $f(c^2)$, $[f(c)]^2$, and $f(2 + h)$ for the given function. If the expression is not defined or not a real number, then state this fact.

15. $f(x) = 8 - 3x$

16. $f(x) = \frac{1}{2}x + 4$

17. $f(x) = 2|x - 1|$

18. $f(x) = 3|x - 2|$

19. $f(x) = 2x^2 - 1$

20. $f(x) = 6 - 2x^2$

21. $f(x) = 10$

22. $f(x) = -7$

23. $f(x) = x^2 - 5x + 6$

24. $f(x) = x^2 + 5x - 6$

25. $f(x) = \sqrt{1 - 3x}$

26. $f(x) = \sqrt{3x - 1}$

27. $f(x) = \sqrt{1 + x^2}$

28. $f(x) = \sqrt{2x^2 + 3}$

29. $f(x) = \dfrac{1}{2x + 4}$

30. $f(x) = \dfrac{1}{x^2 - 4}$

In Exercises 31–36, let $f(x) = \sqrt{x + 1}$, and $g(x) = \dfrac{1}{x - 6}$. Compute the given expression.

31. $f(3) - g(2)$

32. $[f(1)] \cdot [g(5)]$

33. $[f(0)] \cdot [g(1)]$

34. $f(1) - g(\sqrt{2})$

35. $\dfrac{2 - g(7)}{f(0)}$

36. $\dfrac{1 - g(7)}{f(1)}$

In Exercises 37–44, determine whether the given graph represents a function of x.

37.

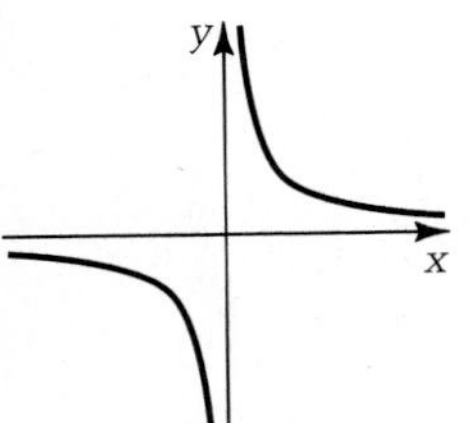

38.

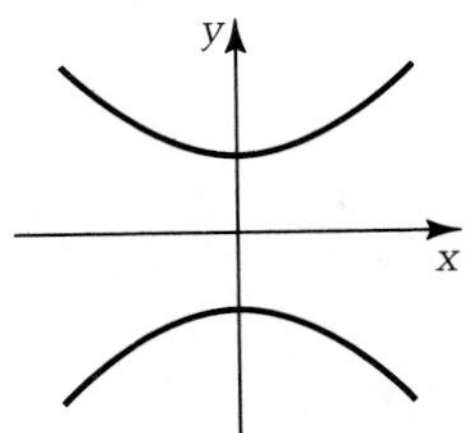

39.

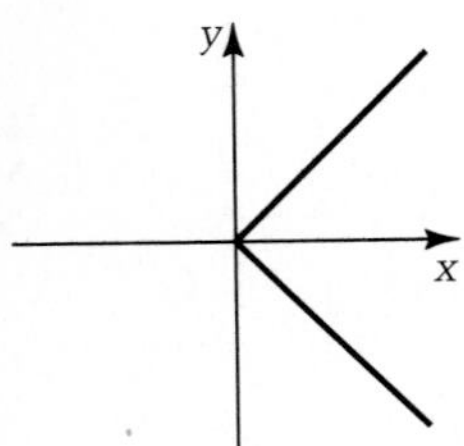

40.

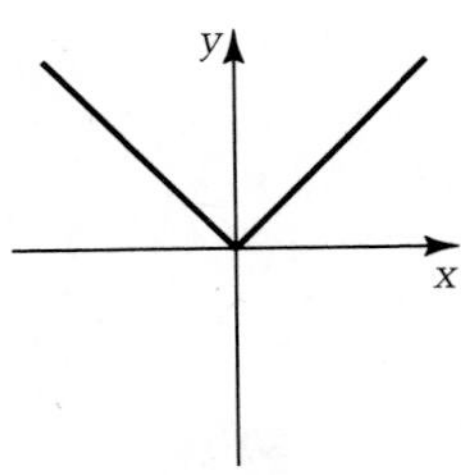

41.

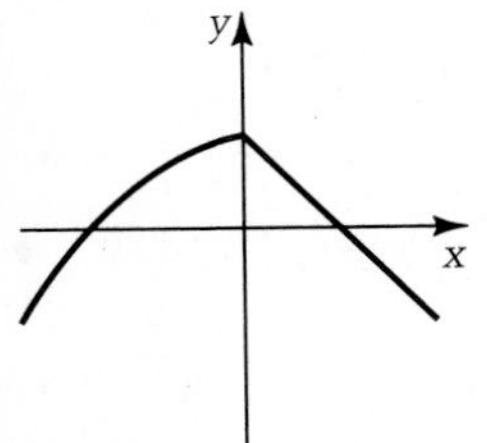

42.

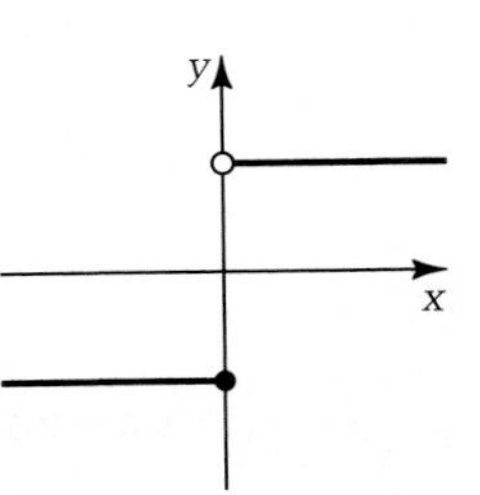

43.

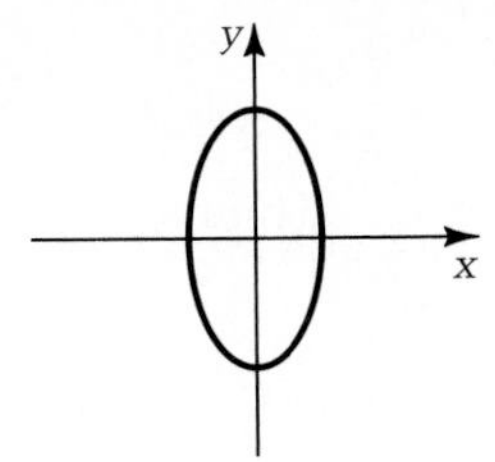

44.

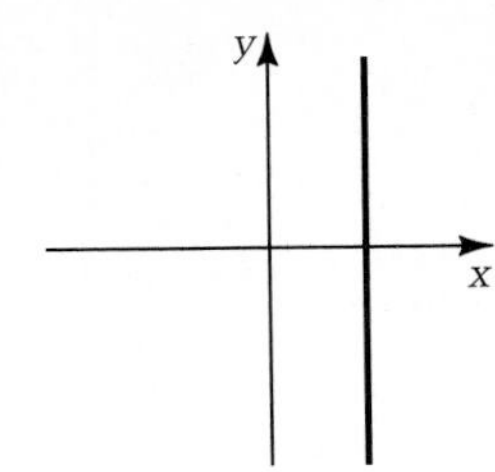

In Exercises 45–72, determine the domain of the given function. Write your answer in interval notation.

45. $f(x) = 2x - 4$

46. $g(x) = 1 - 3x$

47. $h(x) = 17$

48. $G(x) = -\sqrt{2}$

49. $H(x) = x^2 + \sqrt{2}x$

50. $F(x) = 1 - 4x^2$

51. $f(x) = 1 - |x|$

52. $g(x) = |1 - x|$

53. $f(x) = \sqrt{x + 3}$

54. $g(x) = \sqrt{x - 1}$

55. $y = \sqrt{4 - x}$

56. $y = \sqrt{2 + x}$

57. $y = \frac{1}{3}\sqrt{2x - 9}$

58. $y = \sqrt{\dfrac{3x - 15}{5}}$

59. $y = \dfrac{1}{x + 4}$

60. $y = \dfrac{1}{x - 8}$

61. $g(x) = \dfrac{1}{x - 7}$

62. $f(x) = \dfrac{1}{x + 14}$

63. $F(x) = \dfrac{1}{x^2 - 9}$

64. $G(x) = \dfrac{1}{x^2 + 5x + 6}$

65. $P(x) = \dfrac{1}{x^2 + 1}$

66. $y = \sqrt{x^2 + 4}$

67. $y = \sqrt{x^2 - 4}$

68. $y = \sqrt{3 - x^2}$

69. $f(x) = \sqrt{1 - x^2}$

70. $y = \sqrt{9 - x^2}$

71. $g(x) = \dfrac{1}{\sqrt{3 - 2x}}$

72. $h(x) = \dfrac{1}{\sqrt{2 - x^2}}$

In Exercises 73–76, evaluate $f(-2), f(-1), f(0), f(1)$, and $f(2)$.

73. $f(x) = \begin{cases} x, & \text{if } x < 0 \\ 2, & \text{if } x = 0 \\ x^2, & \text{if } x > 0 \end{cases}$

74. $f(x) = \begin{cases} 3, & \text{if } x \leq -2 \\ |x|, & \text{if } -2 < x < 2 \\ x^2 - 1, & \text{if } 2 \leq x \end{cases}$

75. $f(x) = \begin{cases} x + 3, & \text{if } x \leq -1 \\ x^2 + 1, & \text{if } -1 < x \leq 2 \\ 4 - x, & \text{if } 2 < x \end{cases}$

76. $f(x) = \begin{cases} |x+4|, & \text{if } x < -3 \\ 1, & \text{if } -3 \le x < 1 \\ 2x - 3, & \text{if } 1 \le x \end{cases}$

Superset

77. If $F(x) = x^2 + 1$, find (a) $F(3 + F(1))$, (b) $F(3) + F(1)$, and (c) $F(t^2 - F(t - 1))$.

78. Suppose $f(x) = \sqrt{x}$. Show that

$$\frac{f(3+h) - f((3)}{h} = \frac{1}{f(3+h) + f(3)}$$

In Exercises 79–86, evaluate the expression

$$\frac{f(x+h) - f(x)}{h}$$

for the given function. Simplify your answer as much as possible.

79. $f(x) = 2x + 5$

80. $f(x) = 3 - 4x$

81. $f(x) = 7$

82. $f(x) = x^2 - 2x - 3$

83. $f(x) = 3x - x^2$

84. $f(x) = x^3$

85. $f(x) = \dfrac{1}{x+1}$

86. $f(x) = \dfrac{1}{x-2}$

87. Suppose f is a function and for all real numbers a, b, $f(a + b) = f(a) \cdot f(b)$.
   (a) If $f(0) = 0$, what can be said about f?
   (b) What if $f(0) \neq 0$?
   (c) If $f(1) = c \neq 0$, what can be said about $f(n)$ where n is an integer?

3.3
Transformations and Graphing

In this chapter we have been concerned with functions defined by equations. Certain changes in an equation produce simple changes in its graph. These changes in the graph are called **transformations.** There are three general types of transformations: **translations, reflections,** and **changes in shape.**

Type I. Translations

A translation occurs when a graph is moved vertically or horizontally. The shape and size of the graph remain the same; only its position changes.

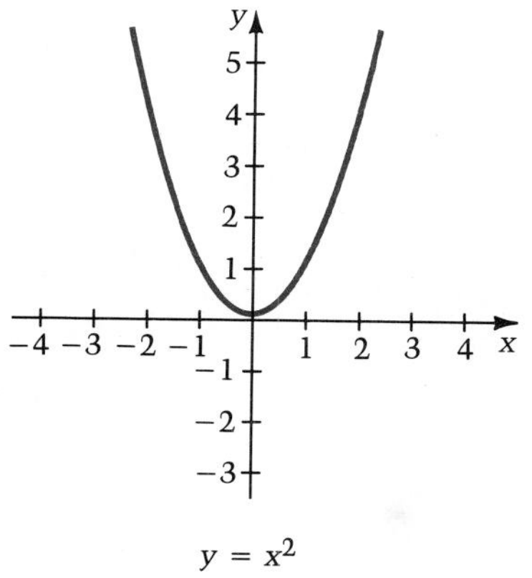

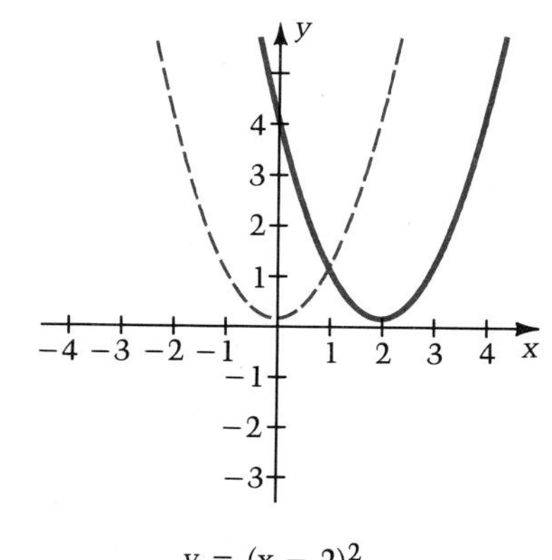

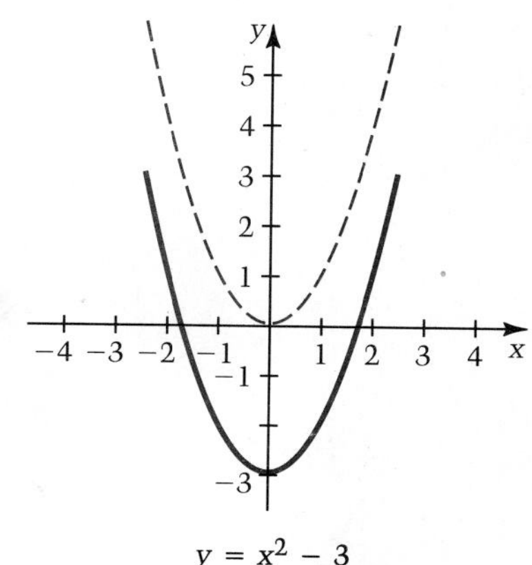

Figure 3.25 The graph of $y = (x - 2)^2$ is a translation of $y = x^2$ two units to the right; $y = x^2 - 3$ is a translation of $y = x^2$ three units down.

Type II. Reflections

The second type of transformation involves the reflection of a graph across the x-axis or y-axis.

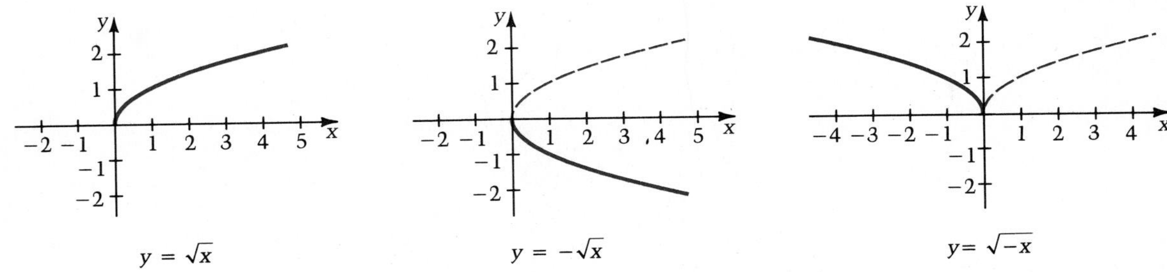

Figure 3.26 The graph of $y = -\sqrt{x}$ is a reflection of $y = \sqrt{x}$ across the x-axis; $y = \sqrt{-x}$ is a reflection of $y = \sqrt{x}$ across the y-axis.

Type III. Changes in Shape

The third type of transformation involves a change in the shape of a graph. A change in shape is also called a **stretching** or **shrinking.**

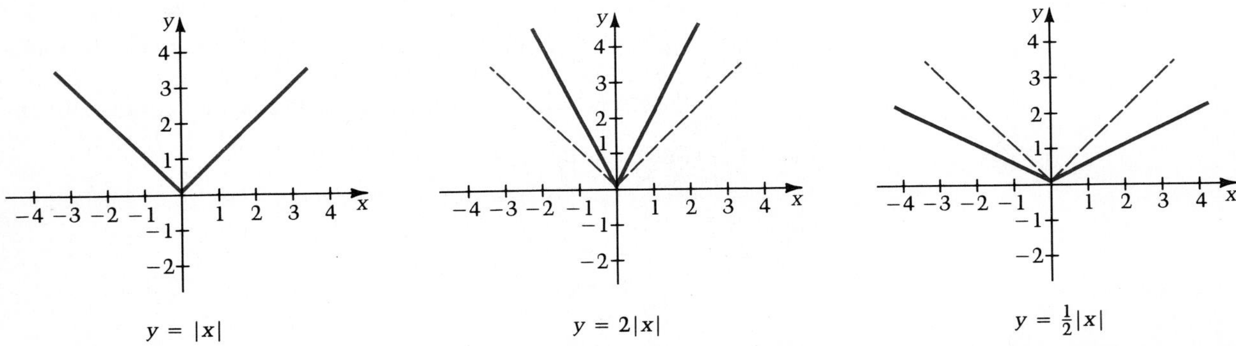

Figure 3.27 The graph of $y = 2|x|$ is a stretching of $y = |x|$ away from the x-axis; $y = \frac{1}{2}|x|$ is a shrinking of $y = |x|$ toward the x-axis.

Translations

Now that we have summarized the three general types of transformations, let us consider translations in detail. Compare the tables of ordered pairs for the functions on the facing page.

Notice that by adding 2 to each x-coordinate in the ordered pairs for $y = x^2$, you produce the ordered pairs for $y = (x - 2)^2$. But, adding 2 to each x-coordinate has the effect of moving each point 2 units to the right (positive direction). This suggests that the graph of $y = (x - 2)^2$ will be formed by

$y = x^2$			$y = (x - 2)^2$			$y = (x + 1)^2$		
x	y	(x, y)	x	y	(x, y)	x	y	(x, y)
-2	4	$(-2, 4)$	0	4	$(0, 4)$	-3	4	$(-3, 4)$
-1	1	$(-1, 1)$	1	1	$(1, 1)$	-2	1	$(-2, 1)$
0	0	$(0, 0)$	2	0	$(2, 0)$	-1	0	$(-1, 0)$
1	1	$(1, 1)$	3	1	$(3, 1)$	0	1	$(0, 1)$
2	4	$(2, 4)$	4	4	$(4, 4)$	1	4	$(1, 4)$

Add 2 to x-values to keep y-values the same

Add -1 to x-values to keep y-values the same

translating the graph $y = x^2$ two units in the positive x-direction. This is shown in Figure 3.28.

In a similar way, adding -1 to each x-coordinate in the ordered pairs for $y = x^2$ produces ordered pairs for $y = (x + 1)^2$. The graph of $y = (x + 1)^2$ is found by translating the graph $y = x^2$ one unit in the negative x-direction as shown in Figure 3.29. In this case, think of $(x + 1)^2$ as $(x - (-1))^2$.

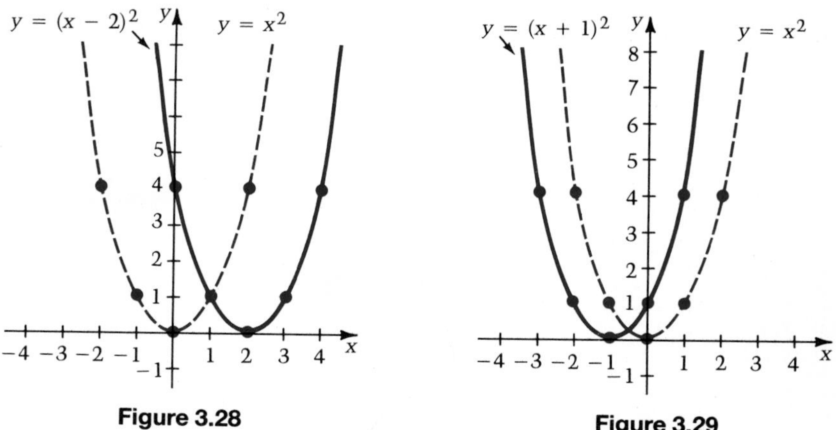

Figure 3.28

Figure 3.29

Horizontal Translations

Suppose h is a real number. Replacing x with $x - h$ in an equation translates the original graph horizontally. If $h > 0$, the graph moves $|h|$ units to the right; if $h < 0$, it moves $|h|$ units to the left.

$$y = (x - 2)^2 \qquad y = (x + 3)^2$$

$$h = 2 \qquad h = -3$$

h is *subtracted* from x

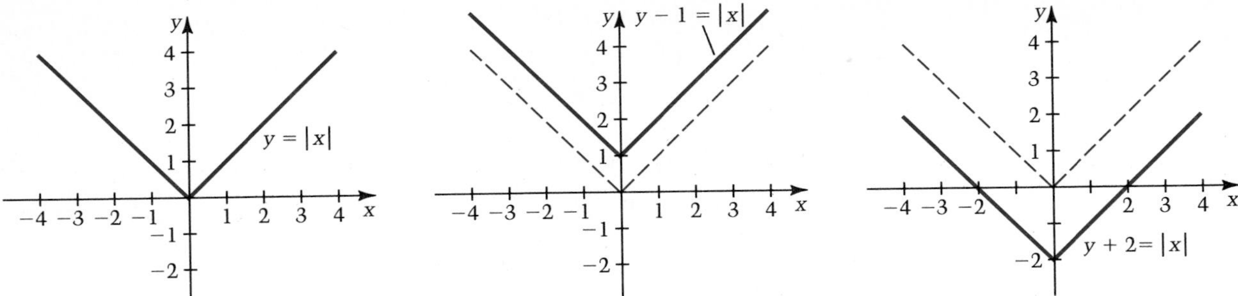

Figure 3.30 Replacing y with $y - 1$ in $y = |x|$ produces a translation 1 unit up. Replacing y with $y + 2$ produces a translation 2 units down.

We now examine changes in an equation that produce vertical translations, as shown in Figure 3.30.

> **Vertical Translations**
>
> Suppose k is a real number. Replacing y with $y - k$ in an equation translates the original graph vertically. If $k > 0$, the graph moves $|k|$ units up; if $k < 0$, the graph moves $|k|$ units down.

EXAMPLE 1

Graph each function by translating $y = \frac{1}{x}$: (a) $y = \frac{1}{x+3}$ (b) $y - 2 = \frac{1}{x}$

Solution

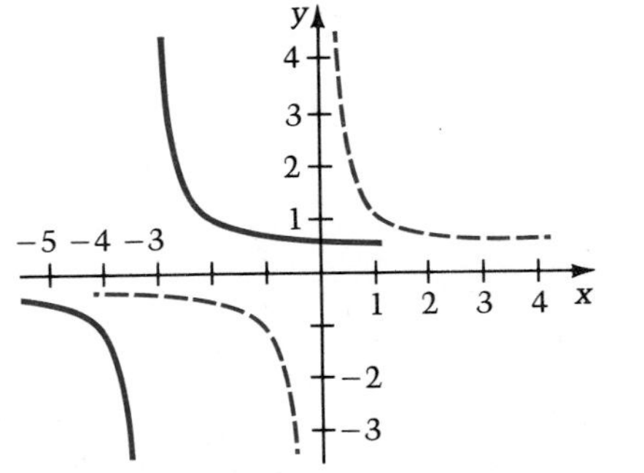

(a) $y = \frac{1}{x}$ is translated 3 units to the left.

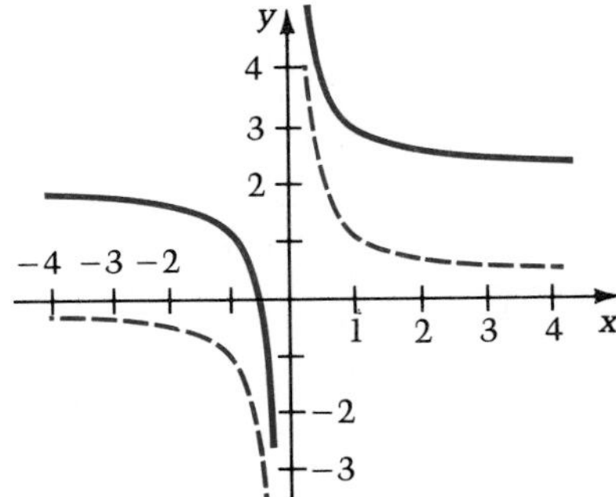

(b) $y = \frac{1}{x}$ is translated 2 units up.

Figure 3.31

Sometimes we discuss a function f without stating a specific equation. It is customary to use a general equation such as $y = f(x)$ to represent f.

EXAMPLE 2

Using the graph of $y = f(x)$ given in Figure 3.32, sketch the graphs of the following translations.

(a) $y = f(x - 1)$ (b) $y = f(x) - 1$

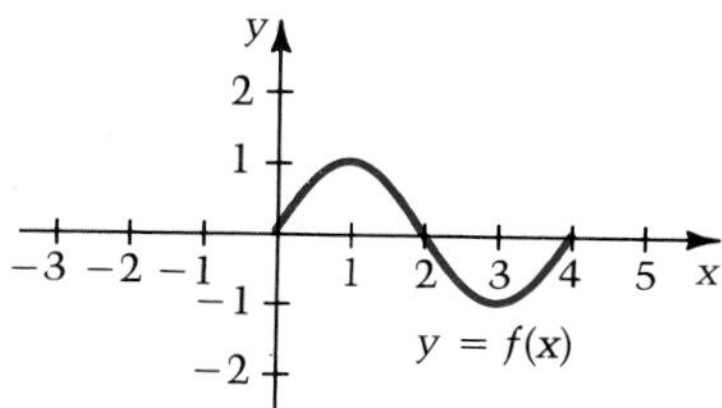

Figure 3.32

Solution

(a) Replacing x with $x - 1$ produces $y = f(x - 1)$. Thus, $y = f(x)$ is translated 1 unit to the right.

(b) Replacing y with $y + 1$ produces $y + 1 = f(x)$ so that $y = f(x) - 1$. Thus, $y = f(x)$ is translated 1 unit down.

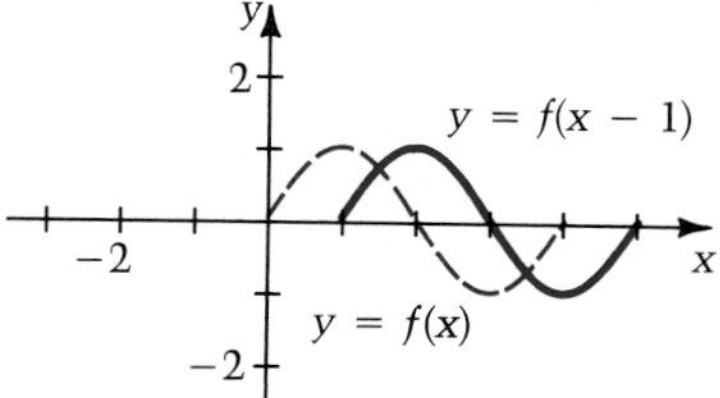

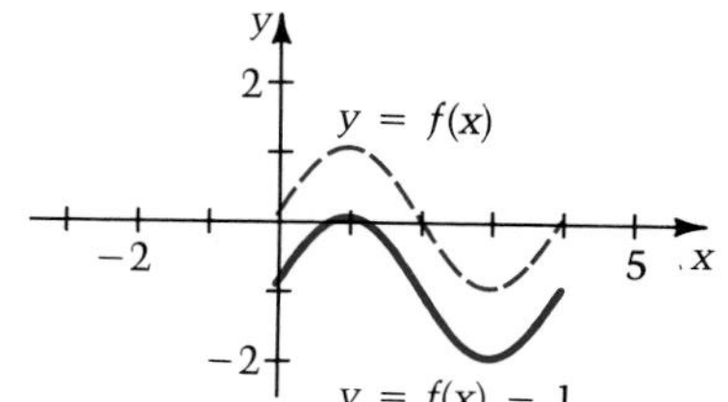

Figure 3.33

Reflections

Next, we examine graphs obtained by using reflections across the x-axis or y-axis. Compare the tables of ordered pairs for the functions on the next page.

$y = \sqrt{x}$			$y = -\sqrt{x}$			$y = \sqrt{-x}$		
x	y	(x, y)	x	y	(x, y)	x	y	(x, y)
0	0	(0, 0)	0	0	(0, 0)	0	0	(0, 0)
1	1	(1, 1)	1	−1	(1, −1)	−1	1	(−1, 1)
4	2	(4, 2)	4	−2	(4, −2)	−4	2	(−4, 2)
9	3	(9, 3)	9	−3	(9, −3)	−9	3	(−9, 3)

(a)

(b)

(a) Each y-coordinate is replaced by its negative, which causes a reflection of $y = \sqrt{x}$ across the x-axis (Figure 3.34). (b) Here each x-coordinate is replaced by its negative, thus reflecting the original function across the y-axis (Figure 3.35).

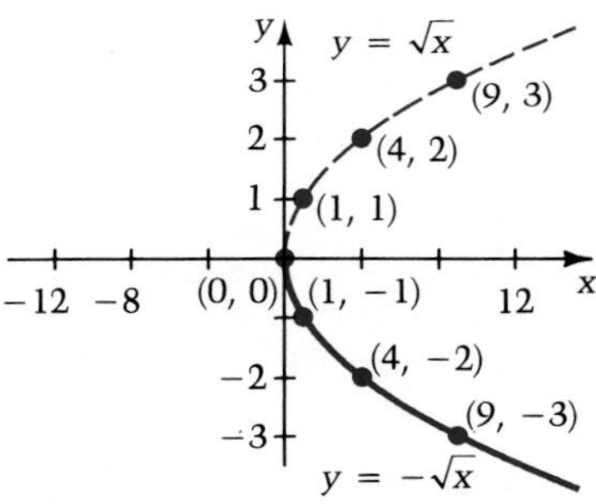

Figure 3.34

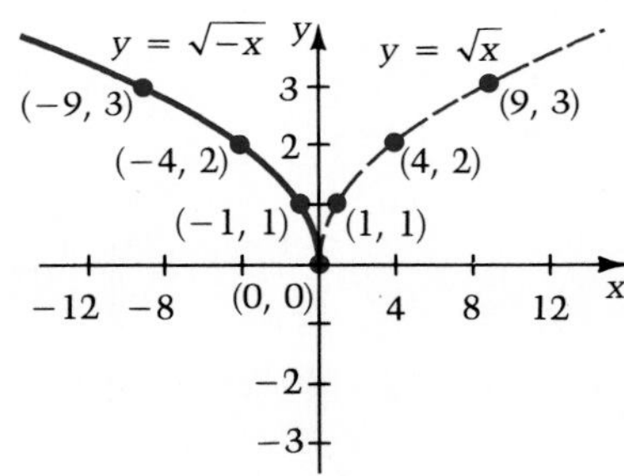

Figure 3.35

Reflecting Across the x-axis

Replacing y with $-y$ in an equation reflects the graph of the equation across the x-axis. If a function is denoted by $y = f(x)$, then the reflection of its graph across the x-axis has the equation $-y = f(x)$, that is, $y = -f(x)$.

Reflecting Across the y-axis

Replacing x with $-x$ in an equation reflects the graph of the equation across the y-axis. If a function is denoted by $y = f(x)$, then the reflection of its graph across the y-axis has the equation $y = f(-x)$.

EXAMPLE 3

Graph $y = |2 - x|$ by reflecting the graph of $y = |x + 2|$.

Solution

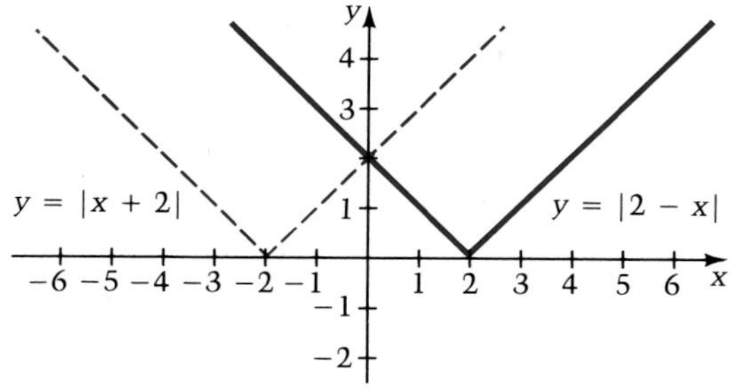

Figure 3.36

$y = |x + 2|$ is a horizontal translation of $y = |x|$ two units to the left.

$y = |2 - x|$ is obtained by replacing x by $-x$ in $y = |x + 2|$. This means we reflect $y = |x + 2|$ across the y-axis.

A bit of terminology is in order at this point. If the graph of a function f is symmetric with respect to the y-axis, then f is called an *even* function. However if the graph is symmetric with respect to the origin, then the function is called an *odd* function. In algebraic terms this terminology is defined as follows.

Definition of an Odd or Even Function

A function f is **even** if for all x in the domain of f, $f(-x) = f(x)$. A function f is **odd** if for all x in the domain of f, $f(-x) = -f(x)$.

For example, if $h(x) = x^4 - 3x^2 + 7$, then since

$$h(-x) = (-x)^4 - 3(-x)^2 + 7 = x^4 - 3x^2 + 7 = h(x),$$

the function h is even, and we conclude that its graph is symmetric with respect to the y-axis. On the other hand, if $g(x) = 2x^3 - 4x$, then since

$$g(-x) = 2(-x)^3 - 4(-x) = -2x^3 + 4x = -(2x^3 - 4x) = -g(x),$$

the function g is odd, and so its graph is symmetric with respect to the origin.

Changes in Shape

We now consider changes in the equation of a function that produce certain changes in the shape of a graph. These transformations are called stretchings and shrinkings. Compare the tables of ordered pairs for the following functions:

$y = |x|$

x	y	(x, y)
-1	1	$(-1, 1)$
0	0	$(0, 0)$
2	2	$(2, 2)$

$y = 3|x|$

x	y	(x, y)
-1	3	$(-1, 3)$
0	0	$(0, 0)$
2	6	$(2, 6)$

$y = \frac{1}{2}|x|$

x	y	(x, y)
-1	$\frac{1}{2}$	$\left(-1, \frac{1}{2}\right)$
0	0	$(0, 0)$
2	1	$(2, 1)$

(a) Each y-coordinate of $y = |x|$ has been multiplied (stretched) by 3.
(b) Each y-coordinate of $y = |x|$ has been multiplied (shrunk) by $\frac{1}{2}$.

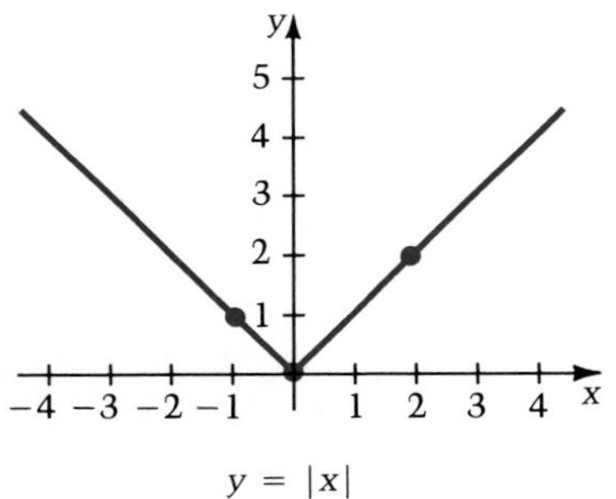

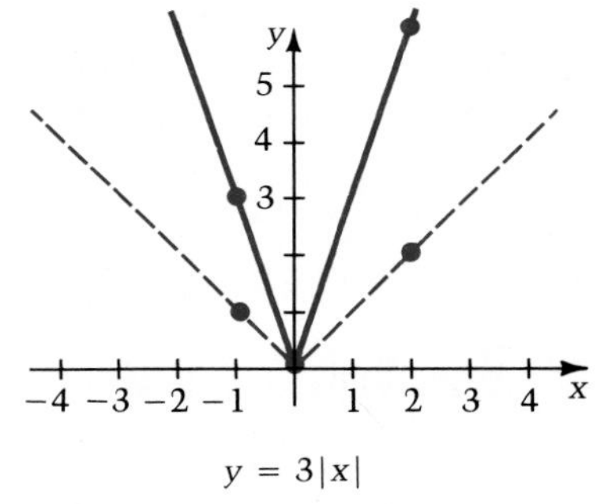

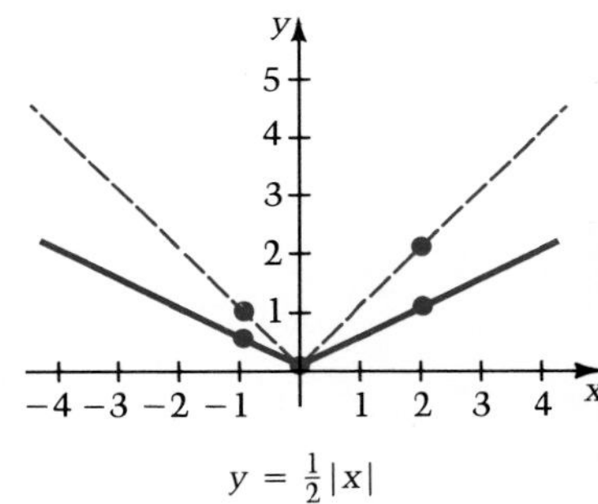

Figure 3.37

The graphs in Figure 3.37 demonstrate the effect of a vertical stretching or shrinking. Notice that in the graph of

$$y = 3|x|,$$

the y-values "increase faster" than the y-values in $y = |x|$. This accounts for the stretch away from the x-axis. The y-values of the graph $y = \frac{1}{2}|x|$ "increase more slowly" than those of $y = |x|$. This accounts for the shrinking toward the x-axis. Such changes in y-values always produce vertical stretching or shrinking. *Points having y-coordinate zero will not be affected by vertical stretching or shrinking.* Thus, these transformations leave the origin fixed.

Vertical Changes in Shape

Suppose that b is a positive real number. If $b > 1$, then the graph of $y = b \cdot f(x)$ is found by stretching the graph of $y = f(x)$ away from the x-axis. If $b < 1$, then $y = b \cdot f(x)$ is found by shrinking $y = f(x)$ down toward the x-axis. In both cases, points having y-coordinate zero are left fixed.

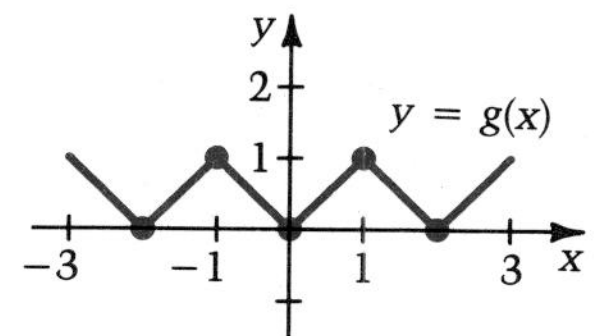

Figure 3.38

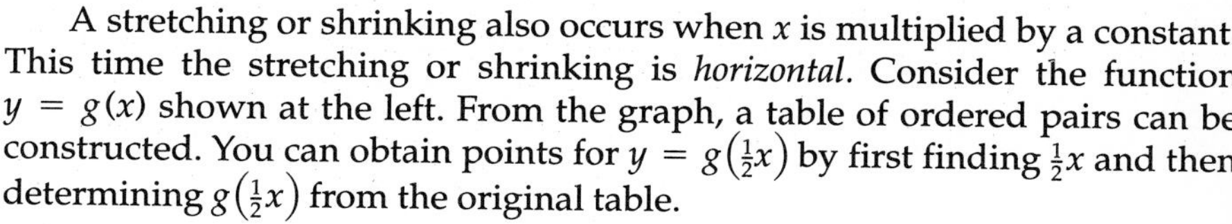

A stretching or shrinking also occurs when x is multiplied by a constant. This time the stretching or shrinking is *horizontal*. Consider the function $y = g(x)$ shown at the left. From the graph, a table of ordered pairs can be constructed. You can obtain points for $y = g\left(\frac{1}{2}x\right)$ by first finding $\frac{1}{2}x$ and then determining $g\left(\frac{1}{2}x\right)$ from the original table.

x	$g(x)$	$(x, g(x))$
-2	0	$(-2, 0)$
-1	1	$(-1, 1)$
0	0	$(0, 0)$
1	1	$(1, 1)$
2	0	$(2, 0)$

x	$\frac{1}{2}x$	$g\left(\frac{1}{2}x\right)$	$\left(x, g\left(\frac{1}{2}x\right)\right)$
-4	-2	0	$(-4, 0)$
-2	-1	1	$(-2, 1)$
0	0	0	$(0, 0)$
2	1	1	$(2, 1)$
4	2	0	$(4, 0)$

Replacing x with $\frac{1}{2}x$ in $y = g(x)$ stretches the graph horizontally. Any point on the graph having x-coordinate zero will not be moved. Thus, in this case, the origin will remain fixed (Figure 3.39).

Substituting $2x$ for x produces a shrinking of $y = g(x)$. In this case, each original y-value is paired with half the original x-value. Thus, the graph shrinks toward the y-axis (Figure 3.40).

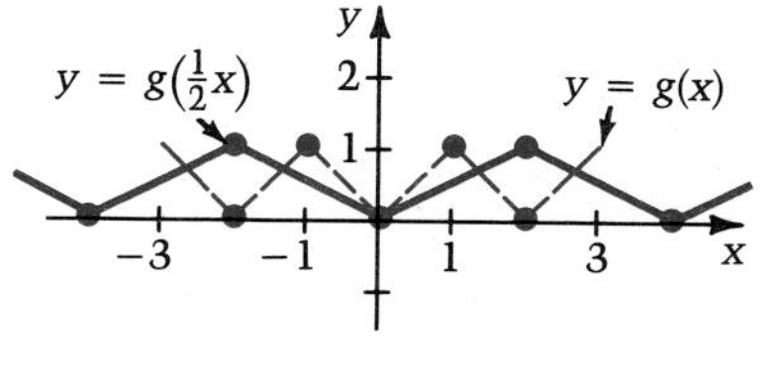

Figure 3.39

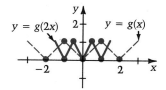

Figure 3.40

Horizontal Changes in Shape

Suppose a is a positive real number. If $a > 1$, then the graph of $y = f(ax)$ is found by shrinking $y = f(x)$ toward the y-axis. If $a < 1$, then the graph of $y = f(ax)$ is found by stretching $y = f(x)$ away from the y-axis. In both cases, points having an x-coordinate of zero are fixed.

Example 4 demonstrates a vertical stretching and a horizontal shrinking.

EXAMPLE 4

Using the graph of $y = f(x)$, sketch the following graphs.

(a) $y = 3f(x)$ (b) $y = f(2x)$

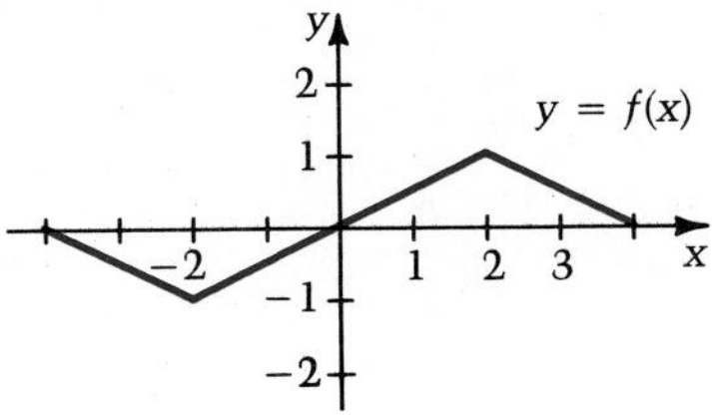

Figure 3.41

Solution

(a) Since $b > 1$, this is a stretching of $y = f(x)$ away from the x-axis. Since $b = 3$, each original y-coordinate is tripled.

(b) Since $a > 1$, we have a shrinking of $y = f(x)$ toward the y-axis. Since $a = 2$, for each original y-coordinate the original x-coordinate is halved.

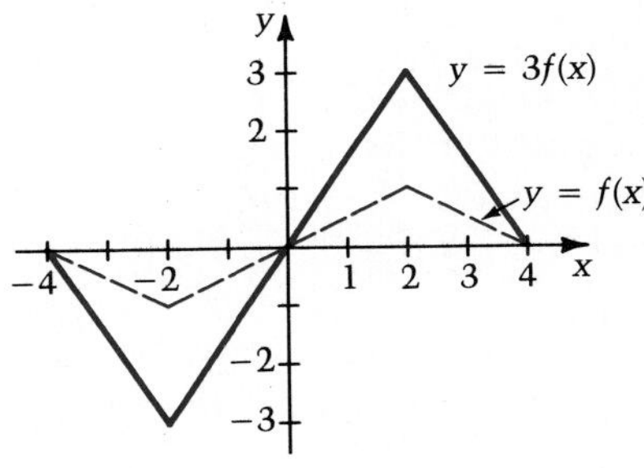

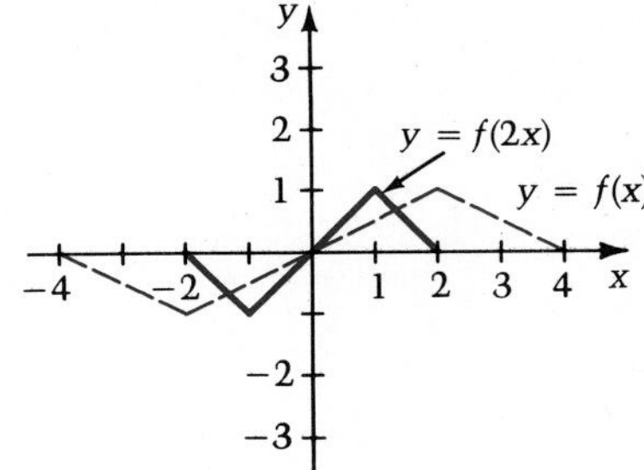

Figure 3.42

You have seen that there are three ways of transforming a graph: by translating it, by reflecting it, or by changing its shape (stretching or shrinking). Very often you can produce a desired graph by applying more than one transformation to a known graph, as shown in the next example.

EXAMPLE 5

Sketch the following graphs:

(a) $y = |x - 1| - 2$ (b) $y = -3(x + 1)^2$.

Solution

(a)

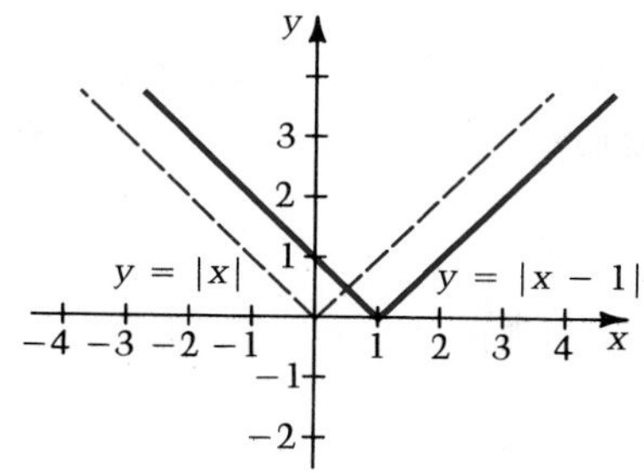

Step 1: We start with the graph of $y = |x|$. The $x - 1$ suggests translating $y = |x|$ one unit to the right to obtain $y = |x - 1|$.

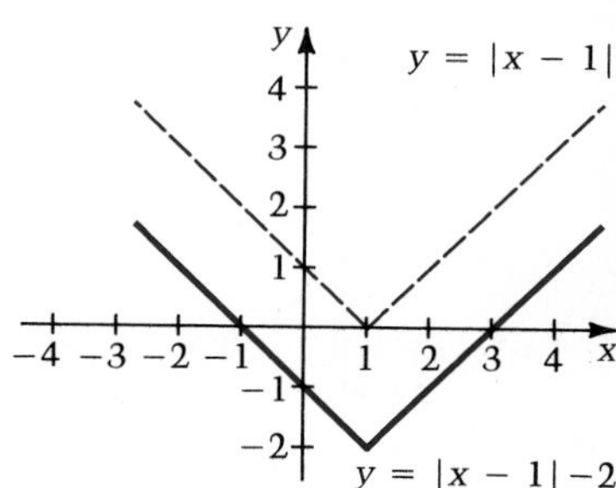

Step 2: Replacing y with $y + 2$ in the equation $y = |x - 1|$ yields $y = |x - 1| - 2$. Thus we have a translation of $y = |x - 1|$ two units down. This is the desired graph.

(b)
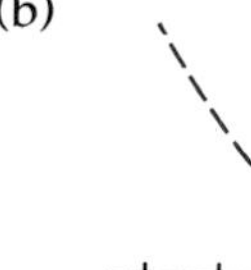

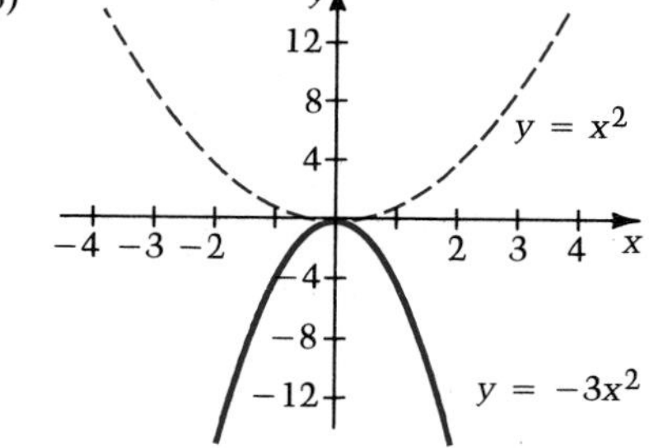

Step 1: We start with the graph of $y = x^2$. The minus sign suggests replacing y with $-y$ to get $y = -x^2$, a reflection across the x-axis. The 3 suggests stretching $y = -x^2$ away from the x-axis to get $y = -3x^2$.

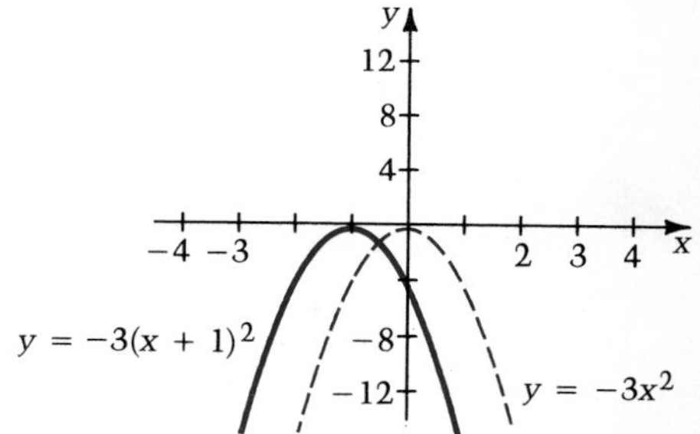

Step 2: The $(x + 1)$ suggests a translation of $y = -3x^2$ one unit to the left to get the graph of $y = -3(x + 1)^2$. This is the desired graph.

Figure 3.43

EXERCISE SET 3.3

In Exercises 1–14, graph the function by translating the graph of $y = x^2$, $y = |x|$ or $y = \frac{1}{x}$.

1. $y = (x - 2)^2$
2. $y = (x - 4)^2$
3. $y = \left(x + \frac{3}{2}\right)^2$
4. $y = \left(x - \frac{5}{2}\right)^2$
5. $y - 2 = x^2$
6. $y = x^2 - 4$
7. $y = |x + 4|$
8. $y = |x - 3|$
9. $y - 3 = |x|$
10. $y = |x| - 3$
11. $y = \dfrac{1}{x - 4}$
12. $y = \dfrac{1}{x - 2}$
13. $y = \dfrac{1}{x} - 4$
14. $y - 3 = \dfrac{1}{x}$

In Exercises 15–20, graph the function by reflecting the graph of $y = (x - 1)^2$, $y = |x - 1|$, or $y = \sqrt{x - 1}$ across the x- or y-axis.

15. $y = -\sqrt{x - 1}$ **16.** $y = |-x - 1|$

17. $y = (-x - 1)^2$ **18.** $y = \sqrt{-x - 1}$

19. $|x - 1| + y = 0$ **20.** $y = -(x - 1)^2$

In Exercises 21–26, graph the function by transforming the graph of $y = f(x)$ below.

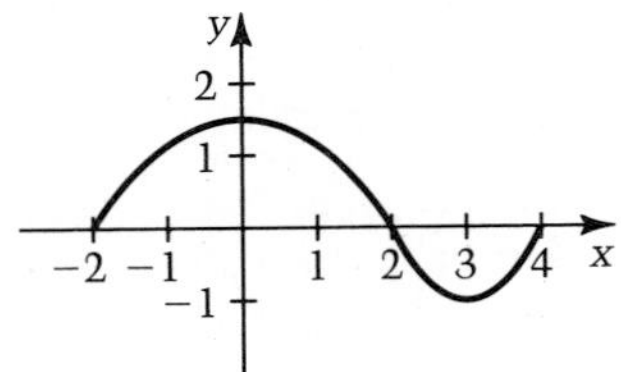

21. $y = f(2x)$ **22.** $y = f\left(\frac{1}{2}x\right)$ **23.** $y = f\left(\frac{1}{4}x\right)$

24. $y = f(4x)$ **25.** $y = 2f(x)$ **26.** $y = 4f(x)$

In Exercises 27–38, use transformations to graph the given function.

27. $y = (x - 2)^2 + 1$ **28.** $y = (x + 3)^2 - 1$

29. $y = 3(x - 2)^2 + 1$ **30.** $y = -2(x + 1)^2 - 3$

31. $y = -5|x|$ **32.** $y = -\frac{1}{2}|x|$

33. $y = 4|x + 1| - 4$ **34.** $y = 4|x - 2| + 3$

35. $y = \frac{1}{x + 3} + 2$ **36.** $y = \frac{1}{x + 4} + 3$

37. $y = 2 - \frac{1}{x - 3}$ **38.** $y = 3 - \frac{1}{x + 4}$

In Exercises 39–42, sketch the graph of the given function by transforming the graph used in Exercises 21–26.

39. $y = 2f(x - 3)$ **40.** $y = -2f(x - 1)$

41. $y = -\frac{1}{3}f(x + 2)$ **42.** $y = f\left(\frac{1}{3}x + 2\right)$

In Exercises 43–48, classify each function as even, odd or neither.

43. $f(x) = x$ **44.** $g(x) = x^2 + 1$

45. $h(x) = \sqrt{x}$ **46.** $F(x) = |x|$

47. $j(x) = (x - 1)(x + 1)$ **48.** $j(x) = x^5 - x^3 - 6x$

Superset

49. Consider the graphs below:

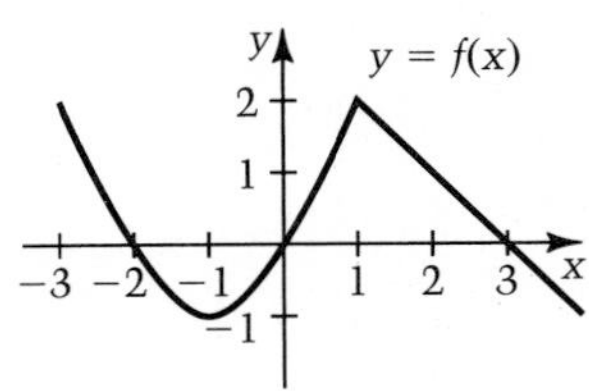

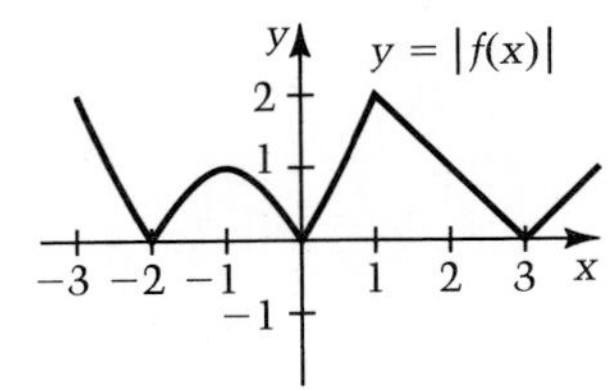

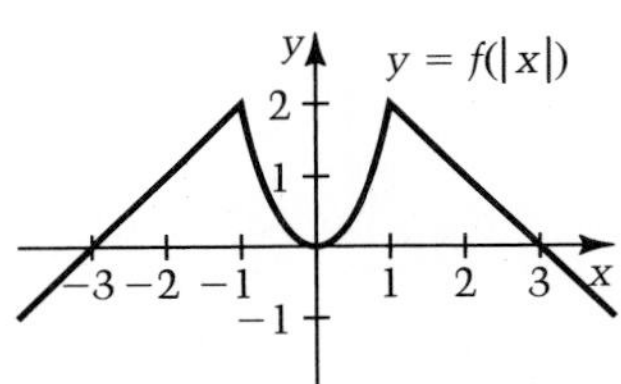

Describe a procedure for obtaining the graphs of $y = |f(x)|$ and $y = f(|x|)$ from the graph $y = f(x)$.

In Exercises 50–60, graph the function. (*Hint:* see Exercise 49.)

50. $y = |\sqrt{x} - 4|$ **51.** $y = |x^2 - 4|$

52. $y = |x^3 - 1|$ **53.** $y = \sqrt{|x|}$

54. $y = \sqrt{|x - 2|}$ **55.** $y = |2 - |x||$

56. $y = \frac{1}{|x - 1|}$ **57.** $y = \frac{1}{|x| - 1}$

58. $y = \frac{1}{|x| + 2}$ **59.** $y = \frac{-3}{|x + 2|}$

60. $y = |2|x| - 5|$

In Exercises 61–64, graph the following functions by transforming the given graph of $y = f(x)$.

(a) $y = |f(x) - 1|$ (b) $y = -2 \cdot f(|x|)$

(c) $y = |f(x - 1)|$ (d) $y = f\left(-\frac{1}{2}x\right)$

61.

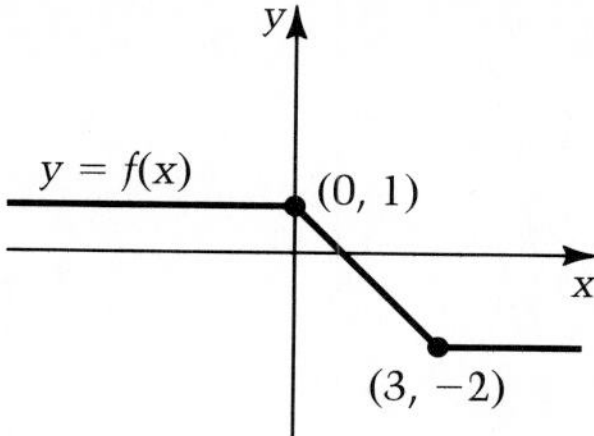

63.

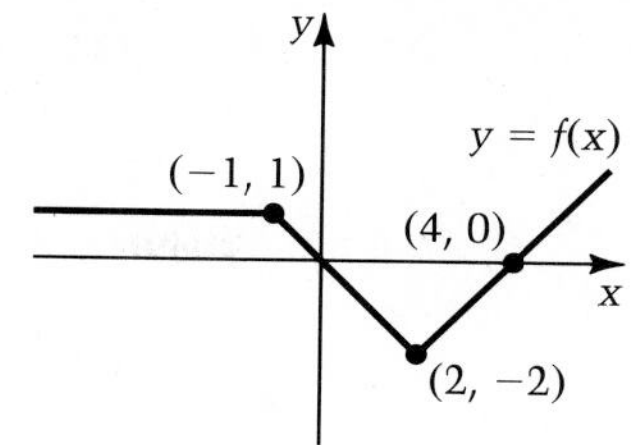

62.

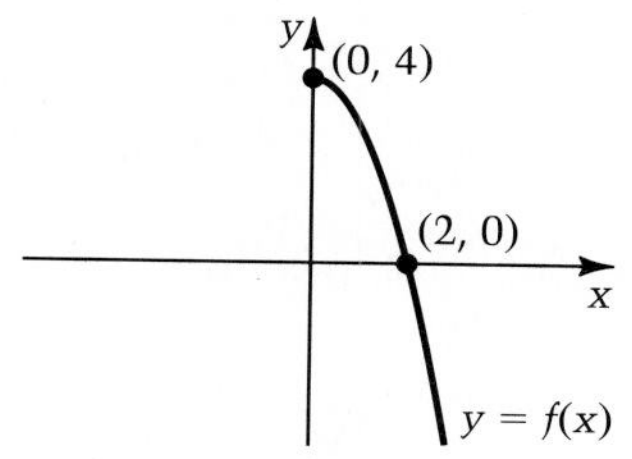

64.

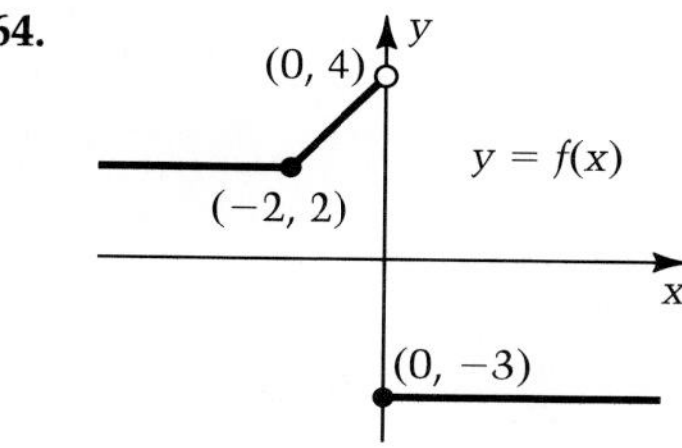

3.4
The Linear Function

A function described by an equation of the form

$$y = mx + b$$

is called a **linear function.** In Section 3.1 we saw that these functions have straight lines as their graphs. Given any two points on a line, we can determine an important characteristic of the line, its *slope*. A line's slope is a number that measures the slant (or inclination) of the line.

Let us consider the linear function

$$y = \frac{1}{2}x + \frac{3}{2},$$

and a table of some ordered pairs determined by the function.

Δx	x	y	Δy
	1	2	
2			1
	3	3	
4			2
	7	5	
14			7
	21	12	

No matter which two ordered pairs you choose, the difference between the y-values is always half the difference between the corresponding x-values.

Δx = Difference between x-values

Δy = Difference between y-values

Based on this data, we might hypothesize that

$$\text{difference in } y\text{-values} = \frac{1}{2}(\text{difference in } x\text{-values}),$$

that is,

$$\frac{\text{difference in } y\text{-values}}{\text{difference in } x\text{-values}} = \frac{1}{2}.$$

This ratio is called the slope of the line $y = \frac{1}{2}x + \frac{3}{2}$.

Definition of Slope of a Line

If $P_1(x_1, y_1)$ and $P_2(x_2, y_2)$ are two different points on a nonvertical line, then the **slope** m of the line is given by the formula

$$m = \frac{y_2 - y_1}{x_2 - x_1}.$$

When the equation of a line is in the form $y = mx + b$, the slope of the line is m, the coefficient of x.

The slope of a line can be found geometrically. For the points in Figure 3.44, we calculate the ratio of the change in vertical direction (difference in y-values) to the change in horizontal direction (difference in x-values).

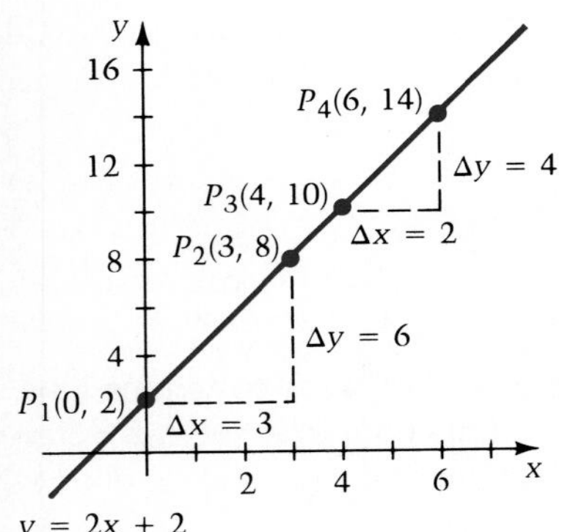

Figure 3.44 $y = 2x + 2$

For points P_3 and P_4, we calculate

$$\frac{\text{vertical change}}{\text{horizontal change}} = \frac{\Delta y}{\Delta x} = \frac{14 - 10}{6 - 4} = \frac{4}{2} = 2.$$

For points P_1 and P_2, we calculate

$$\frac{\text{vertical change}}{\text{horizontal change}} = \frac{\Delta y}{\Delta x} = \frac{8 - 2}{3 - 0} = \frac{6}{3} = 2.$$

Often the difference in y-values is called the **rise,** and the difference in x-values is called the **run.** Thus, the slope can be thought of as "the rise over the run."

EXAMPLE 1

Find the slope of the line determined by the following ordered pairs, then sketch the line. (a) $(3, 5)$ and $(1, -1)$ (b) $(-2, 3)$ and $(1, -2)$

Solution

(a) $m = \dfrac{-1 - 5}{1 - 3} = \dfrac{-6}{-2} = 3$ (b) $m = \dfrac{3 - (-2)}{-2 - 1} = \dfrac{5}{-3} = -\dfrac{5}{3}$

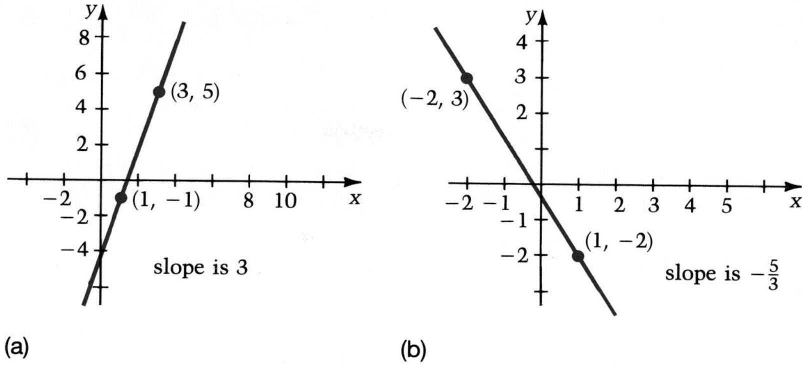

Figure 3.45

A positive slope means the line slants up from left to right (see Figure 3.45(a)). A negative slope means the line slants down from left to right (Figure 3.45(b)).

$$\frac{y_2 - y_1}{x_2 - x_1}$$

coordinates of P_2 — coordinates of P_1

Careful! When setting up the ratio to calculate the slope, make sure that the coordinates of the same ordered pair are aligned. Failure to do so will produce a value for the slope that has the wrong sign.

The graphs below show that the absolute value of the slope is a measure of the steepness of a line: the greater the absolute value, the steeper the line.

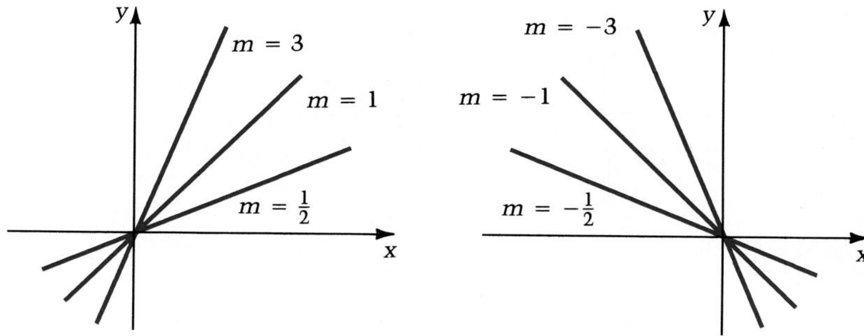

Figure 3.46

Horizontal Lines

Since every point on a horizontal line has the same y-coordinate, the equation of a horizontal line has the form $y = b$, where b is a real number. The slope of a horizontal line is 0. Note that the equation of the x-axis is $y = 0$.

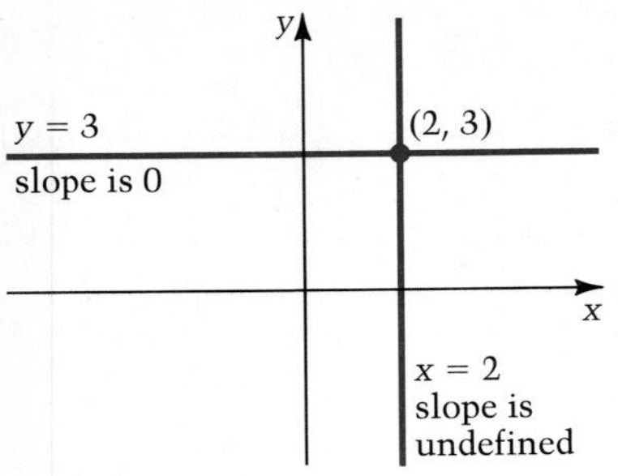

Figure 3.47

Vertical Lines

Since every point on a vertical line has the same x-coordinate, the equation of a vertical line has the form $x = c$, where c is a real number. The slope of a vertical line is not defined because the difference between the x-values of any two points on a vertical line is always zero, and division by zero is undefined. Note that the equation of the y-axis is $x = 0$.

Suppose we know that the point (2, 1) lies on a line whose slope is $\frac{3}{4}$. How can we find an equation of this line? Thinking of (2, 1) as (x_1, y_1) and (x, y) as (x_2, y_2), we can use the definition of slope to find an equation of the line.

$$\frac{3}{4} = \frac{y - 1}{x - 2} \qquad \text{Remember } m = \frac{y_2 - y_1}{x_2 - x_1}.$$

$$y - 1 = \frac{3}{4}(x - 2) \qquad \text{This is known as the point-slope form.}$$

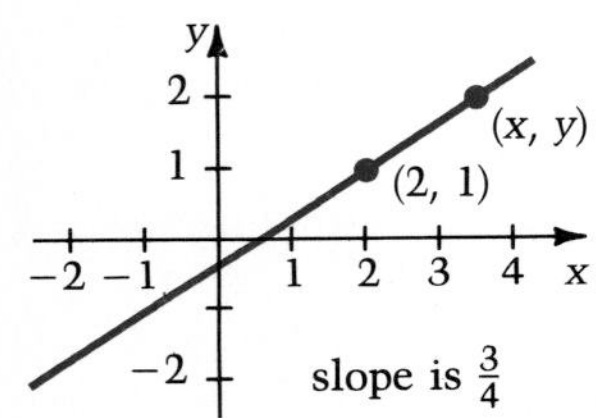

Figure 3.48

Straight line: point-slope form

If a line contains the point (x_1, y_1) and has slope m, then the **point-slope form** of the equation of the line is

$$y - y_1 = m(x - x_1).$$

EXAMPLE 2

(a) Write the point-slope form of the equation of the line containing (3, −1) and having slope 5. (b) Find an equation of the line containing (−2, 4) and having slope −1. Write it in the form $y = mx + b$.

Solution

(a)
$$y - y_1 = m(x - x_1)$$
$$y - (-1) = 5(x - 3)$$

(b)
$$y - 4 = -1(x - (-2))$$
$$y - 4 = -(x + 2)$$
$$y = -x - 2 + 4$$
$$y = -x + 2$$

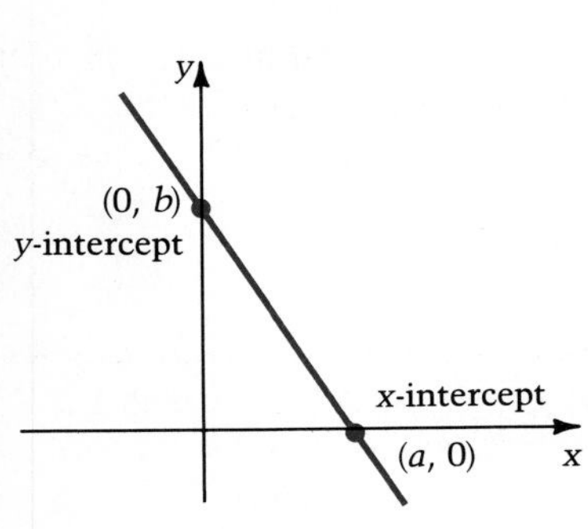

Figure 3.49

An important point on a line is the point where the line intersects the y-axis. The line graphed in Figure 3.49 crosses the y-axis at the point $(0, b)$. The number b is called the ***y*-intercept.** If we know the line has slope m, we can use the point-slope equation to write $y - b = m(x - 0)$ or simply, $y = mx + b$.

Straight line: slope-intercept form

If a line has a y-intercept b and slope m, then the **slope-intercept form** of the equation of the line is $y = mx + b$.

In Figure 3.49, the line crosses the x-axis at $(a, 0)$. The number a is called the **x-intercept.**

EXAMPLE 3

Write an equation of the line having y-intercept -1 and slope $\frac{3}{5}$. Sketch the graph.

Solution

$$y = mx + b$$

$$y = \frac{3}{5}x + (-1)$$

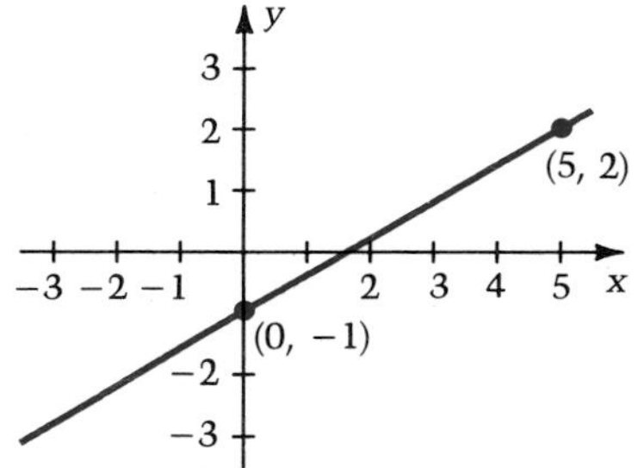

Figure 3.50

In order to sketch the line, we needed two points. We already knew $(0, -1)$ was a point on the line. To find a second point, we chose a value for x and found y. When $x = 5, y = 2$.

Given two points on a line, you can find an equation of the line by first calculating the slope and then using either the *point-slope form* or the *slope-intercept form.* Usually it is easier to use the point-slope form. This is how we proceed in the next example.

EXAMPLE 4

Determine an equation of the line containing the points $(-1, 3)$ and $(5, 2)$.

Solution $\quad m = \dfrac{3 - 2}{-1 - 5} = -\dfrac{1}{6}$ First find the slope.

$$y - y_1 = m(x - x_1)$$ Point-slope form is used.

$$y - 2 = -\frac{1}{6}(x - 5)$$ $m = -\frac{1}{6}$. We chose (5, 2) for (x_1, y_1).

$$y = -\frac{1}{6}x + \frac{5}{6} + 2$$

$$y = -\frac{1}{6}x + \frac{17}{6}$$

The equation found in Example 4 can be rewritten in the form $\frac{1}{6}x + y - \frac{17}{6} = 0$, known as the *general form* of the equation of this line.

Straight line: general form

An equation of the form $Ax + By + C = 0$, where A, B and C are real numbers, with A and B not both zero, is called the **general form** of the equation of the line.

Of the three forms of an equation of a line: the point-slope form, the slope-intercept form and the general form, the slope-intercept form is best suited for comparing lines, because the slope can be read directly from the equation. In addition, knowing the slope is essential to finding an equation of a line parallel or perpendicular to a given line.

Parallel and Perpendicular Lines

If two nonvertical lines are parallel, then their slopes are equal.

If two nonvertical lines are perpendicular, then their slopes are negative reciprocals. That is, if m_1 is the slope of a line, then $-\frac{1}{m_1}$ is the slope of the line perpendicular to it.

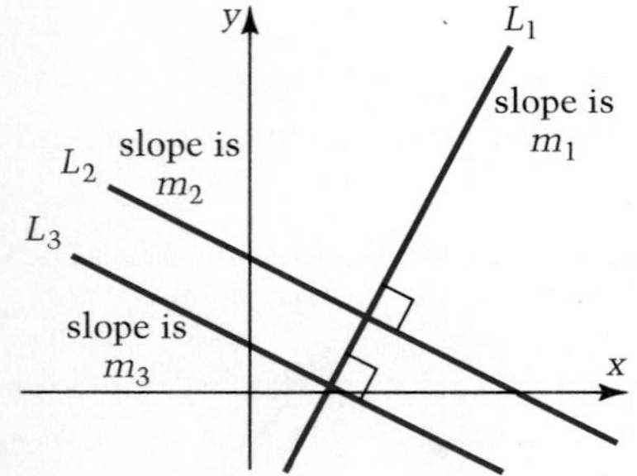

Figure 3.51 $m_2 = m_3$, $m_2 = -\frac{1}{m_1}$ and $m_3 = -\frac{1}{m_1}$

EXAMPLE 5

Line $\mathcal{M}$ is described by the equation $2x + 3y - 5 = 0$. Find an equation for the line $\mathcal{P}$ perpendicular to $\mathcal{M}$ which contains the point $(0, -3)$.

Solution

$$2x + 3y - 5 = 0$$ First, rewrite the equation for $\mathcal{M}$ in slope-intercept form.

$$y = -\frac{2}{3}x + \frac{5}{3}$$ We can now conclude $\mathcal{M}$ has slope $-\frac{2}{3}$.

Since $\mathcal{P}$ is perpendicular to $\mathcal{M}$, the slope of $\mathcal{P}$ is the negative reciprocal of $-\frac{2}{3}$, that is, $\frac{3}{2}$.

$y = \frac{3}{2}x + (-3)$ We use the slope-intercept form $y = mx + b$ because we know the y-intercept is -3.

$$\frac{3}{2}x - y - 3 = 0$$

Suppose you intend to purchase a certain product that sells for 79 cents per pound. A cost function describing this situation is given by

$$C(x) = 0.79x$$

where $C(x) =$ the total cost of your purchase, and $x =$ the number of pounds purchased. The graph of this cost function is a straight line passing through the origin (0, 0) and having slope 0.79. The graph is shown in Figure 3.52 on the left. If, in addition to the per-pound cost of the item, there was a flat packaging fee of $1.50, the cost function would be $C(x) = 1.50 + 0.79x$. The graph of this function is shown in Figure 3.52 on the right.

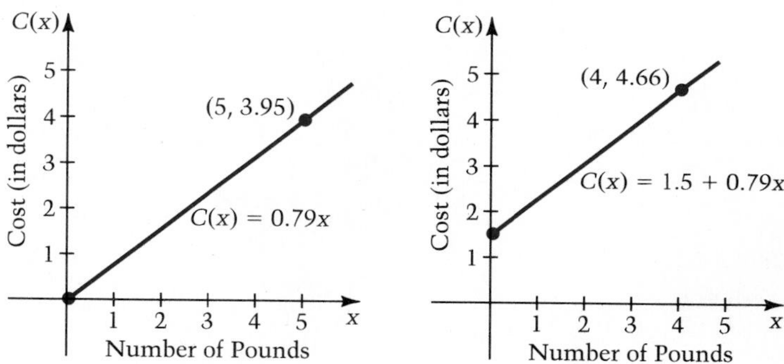

Figure 3.52

Notice that the slopes are the same for both of these lines, namely 0.79, the per pound cost of the product. However, the y-intercepts are different: 0.00 in the graph on the left and 1.50 in the graph on the right.

EXAMPLE 6

An overnight-delivery service promises to deliver any document anywhere in the continental United States by 9 A.M. next morning, provided it is ready for pick-up by 9 P.M. the night before. The cost for the service is a flat fee of $10, plus 2 cents per mile, up to a maximum of $40 per document. Graph the cost as a function of distance in miles.

Solution Let $C(x)$ represent the total cost (in dollars) of sending a document, where x represents the distance (in miles) that the document must travel. The cost function is given by

$$C(x) = 10 + 0.02x,$$

as long as $C(x)$ does not exceed a maximum of \$40. This maximum cost of \$40 is achieved when $40 = 10 + 0.02x$, that is, when $x = 1500$ miles.

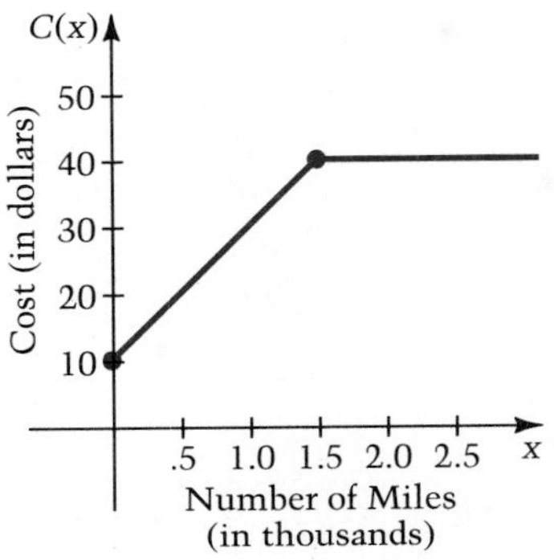

Figure 3.53

EXERCISE SET 3.4

In Exercises 1–14, find the slope of the line determined by the ordered pairs; then sketch the line.

1. $(3, 8)$ and $(4, 13)$
2. $(5, 2)$ and $(7, 10)$
3. $(8, -1)$ and $(-4, 3)$
4. $(-2, 7)$ and $(-7, 4)$
5. $(-5, -6)$ and $(3, -8)$
6. $(8, -5)$ and $(-10, 10)$
7. $(5, 2)$ and $(5, -7)$
8. $(2, 5)$ and $(-7, 5)$
9. $(3, 0)$ and $(1, 0)$
10. $(0, 8)$ and $(0, 11)$
11. $\left(\frac{1}{2}, \frac{1}{8}\right)$ and $\left(\frac{3}{2}, \frac{1}{4}\right)$
12. $\left(\frac{1}{3}, \frac{1}{4}\right)$ and $\left(\frac{2}{3}, \frac{3}{8}\right)$
13. $(0.1, 0.01)$ and $(1, 1)$
14. $(0.2, 1.3)$ and $(1.8, 1.5)$

For Exercises 15–28, determine an equation of the line passing through the pairs of points given in Exercises 1–14. Write your answer in (a) point-slope form, (b) slope-intercept form, and (c) general form.

In Exercises 29–46, determine an equation of the line satisfying the stated conditions. Write your answer in (a) point-slope form, (b) slope-intercept form, and (c) general form.

29. slope is -2; passes through $(3, 5)$
30. slope is 3; passes through $(5, -1)$
31. slope is $-\frac{2}{3}$; passes through $(1, 1)$
32. slope is $\frac{1}{2}$; passes through $(-1, 1)$
33. slope is 0; passes through $(2, -2)$
34. slope is not defined; passes through $(2, -2)$
35. y-intercept is 3; slope is $\frac{1}{2}$
36. y-intercept is -2; slope is $\frac{3}{4}$
37. y-intercept is 0; slope is -2
38. y-intercept is -2; slope is 0
39. line is parallel to the graph of $3x - 7y + 8 = 0$ and passes through $(1, -2)$
40. line is parallel to the graph of $2x + y - 10 = 0$ and passes through the origin

41. line is parallel to the y-axis and passes through $(0, 8)$
42. line is parallel to the x-axis and passes through $(-2, -8)$
43. line is perpendicular to the graph of $y = 6x - 10$ and passes through $(5, -2)$
44. line is perpendicular to the graph of $3x - 5y + 7 = 0$ and passes through $(-1, 3)$
45. line is perpendicular to the graph of $x = 2$ and passes through $(5, -10)$
46. line is perpendicular to the graph of $y = -7$ and passes through $(5, -10)$.

In Exercises 47–54, determine the slope-intercept form ($y = mx + b$) of the line that is described. Round the values of m and b to the nearest hundredth after all computations are performed.

47. line passes through $(3.41, 8.22)$ and $(-1.73, 4.56)$
48. line passes through $(-7.80, -5.49)$ and $(5, 2.11)$
49. line passes through $(-1, 5.37)$ and $(2.12, 1.48)$
50. line passes through $(0, 0.82)$ and $(1.03, -0.08)$
51. line passes through $(5.71, 3.08)$ and is parallel to the line $y = -0.89x + 1.7$
52. line passes through $(6.88, -0.21)$ and is parallel to the line $4.80x - 7.68y + 8 = 0$
53. line passes through $(0.53, -0.72)$ and is perpendicular to the line $1.20x + 6.30y - 8.95 = 0$
54. line passes through $(-1.78, -6.81)$ and is perpendicular to the line $6.30x - 2.8y - 5 = 0$

In Exercises 55–62, (a) carefully graph the given linear function over the interval $0 \le x \le 5$; (b) use your graph to estimate $f(2.5)$; (c) use your calculator to determine $f(2.5)$ exactly.

55. $f(x) = 2x - 7$
56. $f(x) = 5 - 3x$
57. $f(x) = -2.2x + 7.4$
58. $f(x) = 9.6x - 13.8$
59. $f(x) = 0.2x + 5.1$
60. $f(x) = 0.3x - 1.2$
61. $f(x) = \pi - \sqrt{2}x$
62. $f(x) = \sqrt{2}x - \pi$

63. A newly formed local rock group decides to produce and market its own music video. The one-time expenses incurred by the group for items such as the rental of a recording studio total \$28,770. For each copy of the tape that the group sells, its expenses for items such as blank tape and shipping will be \$4.45. The group plans to sell the video at \$14.95 a copy. Write a function that expresses the profit P as a function of the number of copies x of the music video that they sell. Use this function to determine how many copies of the tape must be sold for the group to 'break even'. (Of course, if P is negative it represents a loss.)
64. A simple technique used by wildlife management organizations to determine the population of a given species in a certain site, such as the number of bass in a lake, is known as "capture-tag-recapture." According to this technique, some fish are caught, in some manner marked or tagged, and then released back into the lake. After this first sample has had sufficient time to mix with the rest of the fish population in the lake, a second sample of approximately the same number are caught. The basic assumption is that the ratio of the number of tagged fish caught this second time to the size of the second sample is roughly the same as the ratio of the number of fish tagged originally to the entire population. Assume that 200 fish are caught and tagged in the first sample, while the second sample contained 240 fish, and of these x had been previously tagged. Write a function that gives the estimate using this "capture-tag-recapture" technique for the fish population P as a function of x.
65. A dog kennel feeds its animals a mixture of two brands of food. Brand A is 15% protein and Brand B is 25% protein. Express the percentage P of protein provided by a 10-ounce serving as a function of the number of ounces x of Brand A used for this serving.
66. A market research department has determined that its company will sell 13,400 units of a certain product if the price is \$8.80 per unit and will sell 6600 units if the price is \$11.30 per unit. Determine a linear function $y = f(x)$ compatible with these results, so that the function can predict how many units y the company will sell if the price is x dollars per unit.
67. An all-day parking garage in a certain metropolitan area has a capacity of 400 full-size cars for which the weekly parking fee is \$78 each. The garage charges only \$51 per week for certain subcompact and other small cars since it can park two of them in the space required to park and maneuver one full-size car. The weekly fixed expenses for the parking garage owners for items such as mortgage, taxes, salaries, insurance, and maintenance total \$27,400. Express the profit P for the owners as a function of the number x of full-sized cars in the garage during a week when the garage is filled to capacity.
68. A toy manufacturer has spent \$287,500 on the design, field-testing and advertising for a new board game and the production of the first 10,000 copies of the game.

Each subsequent copy of the game will cost the manufacturer \$4.70 to produce and distribute. Assuming the manufacturer sells at least 15,000 copies of the game, determine the profit P (a loss, if negative) if x copies of the game are sold at \$19.50 each. Use this function to determine how many copies of the game must be sold for the manufacturer to 'break even.'

Superset

69. Show that the three points (1, 3), (5, 1), and (6, 3) are vertices of a right triangle. (Use slopes to show that two of the sides are perpendicular.)

70. Show that the points (1, 4), (2, 7), and (−3, −8) lie on the same line. (Use slopes.)

71. An ant starts at the point (0, 4) and walks 4 units along the line $y = 3x + 4$ in the first quadrant before stopping. Did the ant pass through either of the points (1, 5) or (1, 7)? Explain.

72. Another ant starts at the origin and walks 3 units up a course parallel to the line $y = 3x + 4$ before stopping.
(a) Use an equation to describe the line along which this ant is walking.
(b) Did the ant pass through the point (1, 3)? Explain.
(c) How far above the x-axis was the ant when it stopped?

73. A third ant starts at the point (1, 2) when $t = 0$ and follows a path on the xy-plane such that its coordinates at any time t are given by the equations

$$x = 3t + 1$$
$$y = 2 - t.$$

By solving each equation for t and equating these expressions in x and y, determine a function of x that describes the ant's path. What is the domain of the function?

74. Follow the procedure outlined in Exercise 73 to determine an equation in slope-intercept form that describes the graph given by each of the following pairs of equations.

(a) $x = 2t - 3$, $y = t + 4$
(b) $x = t$, $y = 5 - 6t$
(c) $x = 3t + 2$, $y = 2 - 3t$
(d) $x = t^2 - 3$, $y = t^2 + 4$

75. A teacher has given a test in which the highest grade was 75, and the lowest grade was 40. Determine a linear function, $y = f(x)$, that can be used to distribute the test scores so that the grade of 40 becomes 60 and 75 becomes 100.

76. We say that y *varies directly* with x provided there exists some nonzero constant k such that $y = kx$. Suppose V varies directly with t, and when $t = \frac{1}{2}$, $V = \pi$. For what value of t is V equal to 1?

77. Three of the four vertices of a parallelogram are (1, 0), (3, 4), and (2, 6). Determine the coordinates of the fourth vertex of the parallelogram. (*Hint:* there are three ways to position the fourth vertex.)

78. Three of the four vertices of a parallelogram are (−2, 1), (1, −1) and (0, 5). Determine the coordinates of the fourth vertex of the parallelogram.

3.5 The Quadratic Function

We have seen that the graph of $f(x) = a(x - h)^2 + k$ is a parabola produced by a series of transformations of $f(x) = x^2$. Let us review this method by graphing $f(x) = -2(x - 1)^2 + 3$ (Figure 3.54).

Let us examine the graph of

$$f(x) = -2(x - 1)^2 + 3$$

in detail (Figure 3.55). It is a parabola that opens downward because the coefficient of $(x - 1)^2$ is negative. The graph has its highest point at (1, 3); thus

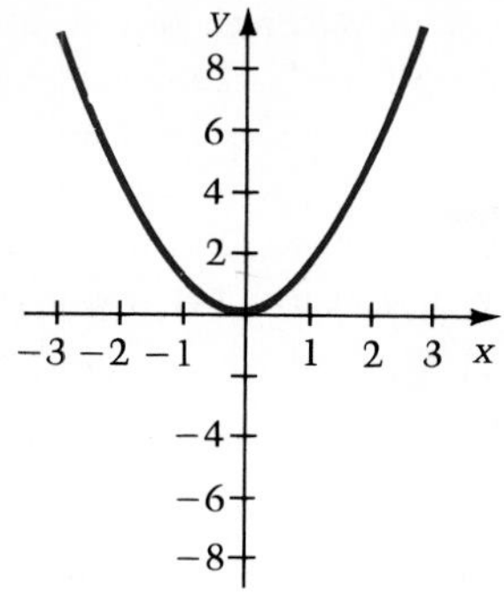

(**a**) $f(x) = x^2$

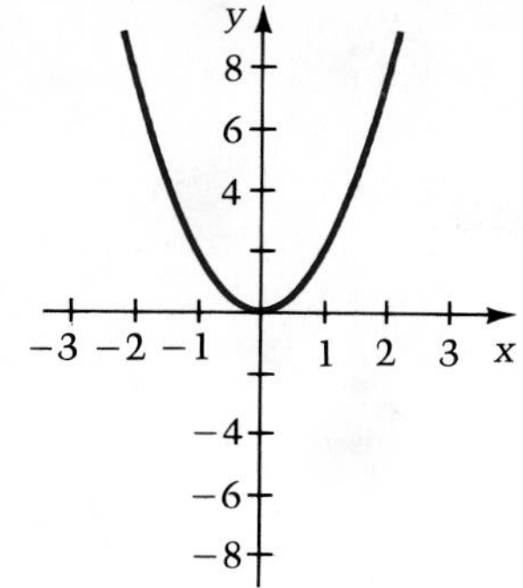

(**b**) $f(x) = 2x^2$
A vertical stretch of (**a**)

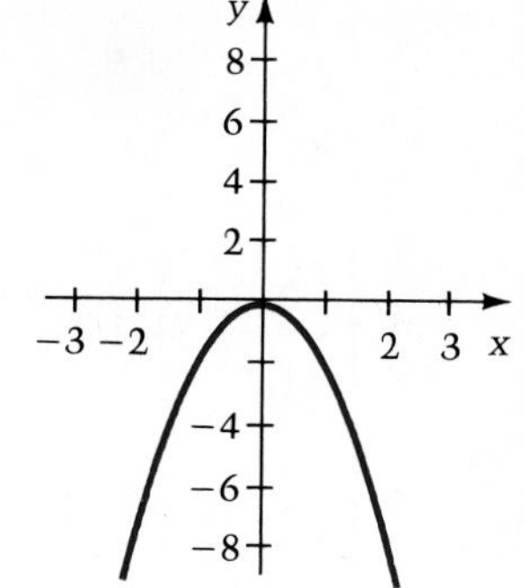

(**c**) $f(x) = -2x^2$
A reflection of (**b**) across the x-axis

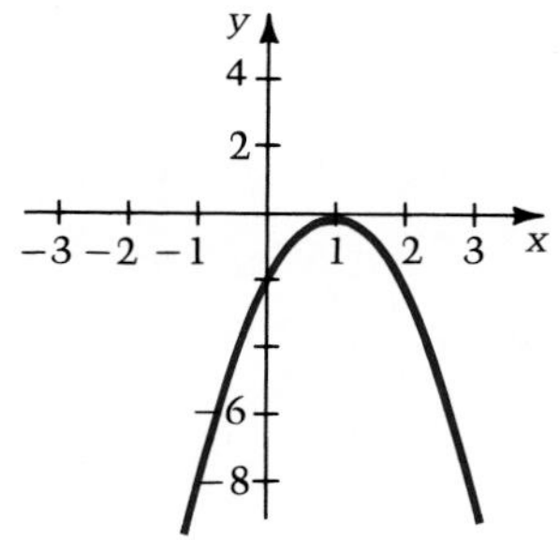

(**d**) $f(x) = -2\,(x - 1)^2$
a horizontal translation of (**c**)

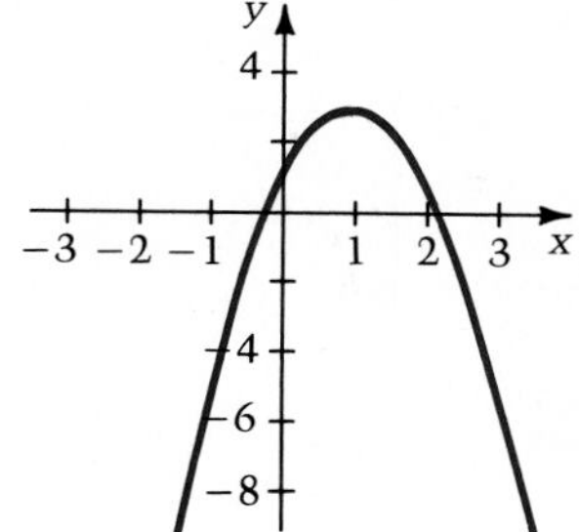

(**e**) $f(x) = -2(x - 1)^2 + 3$
a vertical translation of (**d**)

Figure 3.54

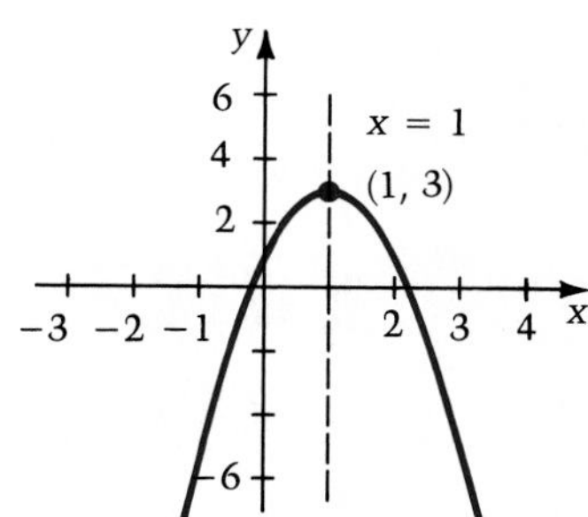

Figure 3.55 $y = -2(x - 1)^2 + 3.$

the greatest function value is 3, and there is no least function value. The graph is symmetric with respect to the line $x = 1$.

Now let us consider the graph of the function $g(x) = 2(x + 3)^2 - 2$ (Figure 3.56). Again, we have a parabola, but this graph opens upward because

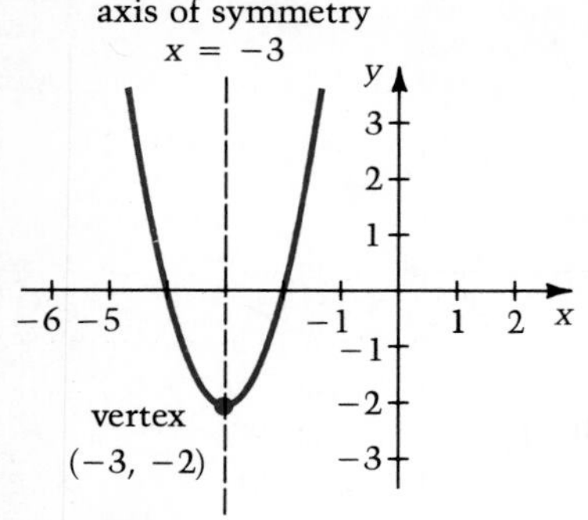

Figure 3.56 $g(x) = 2(x + 3)^2 - 2$.

the coefficient of $(x + 3)^2$ is positive. The parabola has its lowest point at $(-3, -2)$; therefore, the least function value is -2, and there is no greatest function value. The graph is symmetric with respect to the line $x = -3$.

In general, parabolas have either a high or low point, called the **vertex.** Passing through the vertex is a line of symmetry called the **axis of symmetry.** This information is easily obtained from the **vertex form of a parabola,** which is an equation of the form

$$f(x) = a(x - h)^2 + k$$ A negative sign must precede h and a positive sign must precede k.

From this equation we can deduce some features of the parabola:

The graph opens $\begin{cases} \text{upward if } a > 0, \\ \text{downward if } a < 0. \end{cases}$

The graph is $\begin{cases} \text{wider than } f(x) = x^2 \text{ if } |a| < 1, \\ \text{narrower than } f(x) = x^2 \text{ if } |a| > 1. \end{cases}$

The vertex (h, k) is $\begin{cases} \text{a high point if the parabola opens downward,} \\ \text{a low point if the parabola opens upward.} \end{cases}$

The axis of symmetry is the line $x = h$.

If the vertex (h, k) is a high point, then k is the *maximum* function value; if (h, k) is a low point, then k is the *minimum* function value.

EXAMPLE 1

Write each function in vertex form. Find the vertex, the axis of symmetry and the direction in which the parabola opens.

(a) $y = -3(x - 2)^2 - 1$ (b) $y = \frac{1}{2}(x + 2)^2 - 3$

Solution

(a) The vertex form is $y = -3(x - 2)^2 + (-1)$.
The vertex is $(2, -1)$. Here, $h = 2, k = -1$.
The axis of symmetry is $x = 2$. The axis is $x = h$.
The parabola opens downward. Here, $a = -3$; that is, $a < 0$.

(b) The vertex form is $y = \frac{1}{2}(x - (-2))^2 + (-3)$.
The vertex is $(-2, -3)$. Here, $h = -2, k = -3$.
The axis of symmetry is $x = -2$.
The parabola opens upward. Here, $a = \frac{1}{2}$; that is, $a > 0$.

The vertex form $y = a(x - h)^2 + k$ is one way to write an equation of a parabola. Expanding $(x - h)^2$ and simplifying yields an equivalent equation known as the *general form of the quadratic function:*

$$
\begin{aligned}
f(x) &= a(x - h)^2 + k \\
&= a(x^2 - 2hx + h^2) + k \\
&= ax^2 - 2ahx + ah^2 + k \quad \text{Replace } -2ah \text{ with } b \text{ and } (ah^2 + k) \text{ with } c. \\
f(x) &= ax^2 + bx + c
\end{aligned}
$$

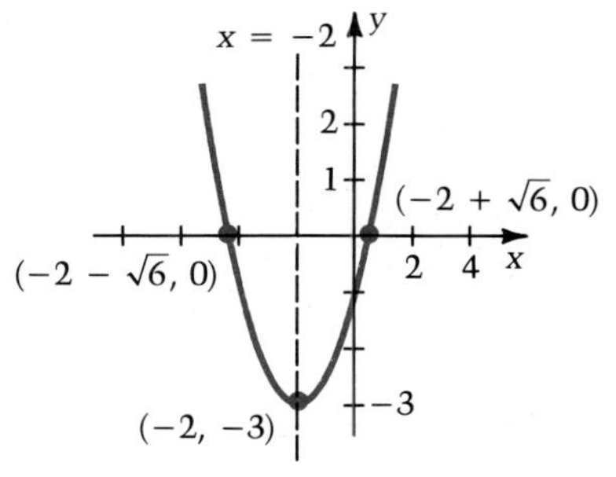

Figure 3.57 $y = \frac{1}{2}(x + 2)^2 - 3$. The x-intercepts are $-2 \pm \sqrt{6}$.

Definition of a Quadratic Function

A function of the form $f(x) = ax^2 + bx + c$, with a, b, and c real numbers and $a \neq 0$, is called a **quadratic function.** The graph of a quadratic function is a parabola. We call $f(x) = ax^2 + bx + c$ the **general form of the quadratic function.**

When graphing a function, it is useful to determine the ***x*-intercepts,** that is, the values of x where $f(x) = 0$. Since these are the values of x where $y = 0$, the x-intercepts tell us where the graph intersects the x-axis. For example, the graph of the function $y = \frac{1}{2}(x + 2)^2 - 3$ has two x-intercepts.

To find the x-intercepts of a parabola, solve $ax^2 + bx + c = 0$.

EXAMPLE 2

Find the x-intercepts of each of the following quadratic functions.

(a) $f(x) = 2x^2 - x - 6$ (b) $g(x) = x^2 - x - 10$

Solution

(a) Set the function equal to 0, then factor and apply the Principle of Zero Products.

$$2x^2 - x - 6 = 0$$
$$(2x + 3)(x - 2) = 0$$

$2x + 3 = 0$	$x - 2 = 0$
$x = -\dfrac{3}{2}$	$x = 2$

The x-intercepts are $x = -\dfrac{3}{2}$ and 2.

(b) Set the function equal to 0, then use the quadratic formula.

$$x^2 - x - 10 = 0$$

$$x = \frac{-(-1) \pm \sqrt{(-1)^2 - 4(1)(-10)}}{2(1)} = \frac{1 \pm \sqrt{41}}{2}$$

The x-intercepts are $x = \dfrac{1 + \sqrt{41}}{2}$ and $\dfrac{1 - \sqrt{41}}{2}$.

Suppose we need to graph the function $f(x) = 2x^2 - 4x - 6$. We know the graph is a parabola. It would be easier to sketch the graph if the equation were in vertex form. A method called **completing the square,** which we used in Chapter 2 to factor quadratic expressions, can be used to convert a quadratic equation from general form to vertex form.

$f(x) = 2x^2 - 4x - 6$ $= 2(x^2 - 2x + \square) - 6$	First, factor the coefficient of x^2 out of the x^2- and x-terms. A space is left for the number that makes the expression in parentheses a perfect square.
$= 2(x^2 - 2x + 1) - 6 - 2 \cdot 1$ ▲ the square of $\left(\frac{1}{2} \cdot (-2)\right)$	In the space write the square of half the coefficient of the x-term. Since we have added $2 \cdot 1$, we must subtract $2 \cdot 1$.
$= 2(x - 1)^2 - 8$	The vertex is $(1, -8)$.

EXAMPLE 3

Graph the function $y = 2x^2 - 4x - 6$. Label the vertex, the axis of symmetry, the x-intercepts, and two other points.

Solution We have just found that the vertex form of this parabola is $f(x) = 2(x - 1)^2 - 8$. Thus, the vertex is $(1, -8)$ and the axis of symmetry is $x = 1$. Since $a = 2$, that is, $a > 0$, the parabola opens upward. To find the x-intercepts, set $2x^2 - 4x - 6$ equal to zero. Solving $2x^2 - 4x - 6 = 0$, we get $x = -1$ and $x = 3$. The x-intercepts are -1 and 3.

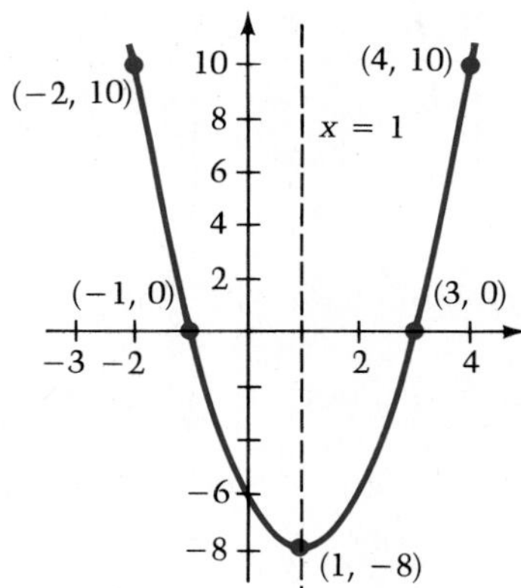

Figure 3.58

Finally, we find two more points:

$x = -2 \qquad y = 2(-2)^2 - 4(-2) - 6 = 10$

$x = 4 \qquad y = 2(4)^2 - 4(4) - 6 = 10$

EXAMPLE 4

Graph the function $y = 4x^2 - 16x + 7$. Label the vertex, the axis of symmetry, the x-intercepts, and two other points. Is there a maximum or minimum function value?

Solution

(1) First, find the x-intercepts.

$$x = \frac{-(-16) \pm \sqrt{(-16)^2 - 4(4)(7)}}{2(4)}$$ Use the quadratic formula.

$$x = \frac{7}{2} \text{ and } x = \frac{1}{2}$$ The x-intercepts are $\frac{7}{2}$ and $\frac{1}{2}$.

(2) Next, find the vertex form.

$y = 4(x^2 - 4x + \square) + 7$ Complete the square.

$y = 4(x^2 - 4x + (-2)^2) + 7 - 4(-2)^2$

$y = 4(x - 2)^2 - 9$ Thus, $h = 2$ and $k = -9$.

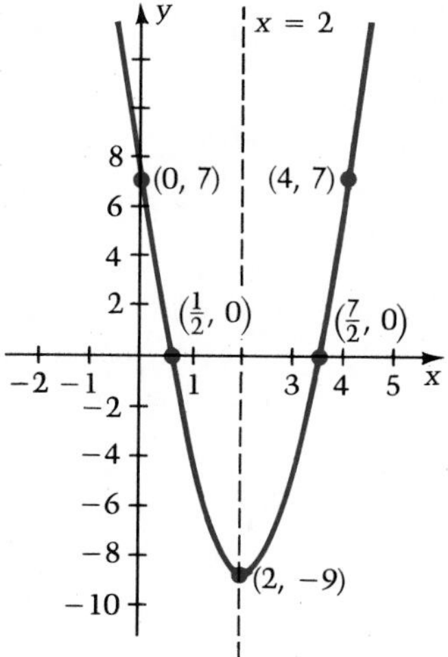

Figure 3.59

The vertex is $(2, -9)$ and the axis of symmetry is $x = 2$.

Since $a = 4$, that is, $a > 0$, the parabola opens upward.

Finally, we find two more points: Select points where $x = 0$ and 4.

$x = 0$ $\quad y = 4(0)^2 - 16(0) + 7 = 7$

$x = 4$ $\quad y = 4(4)^2 - 16(4) + 7 = 7$

The two other points are (0, 7) and (4, 7).

There is no maximum function value (y-value). The minimum function value is -9, the y-coordinate of the vertex.

EXAMPLE 5

You have gone into the video tape rental business. Profits are described by the quadratic function

$$P(x) = -4000 + 5000x - 1000x^2,$$

where $P(x)$ = monthly profits in dollars, and
x = number (in thousands) of video tapes rented per month.

(a) Determine $P(0)$. Explain what this number represents.

(b) Determine monthly profits for monthly rentals of 1000, 2000, 3000, 4000, and 5000 (that is, for $x = 1, 2, 3, 4$, and 5).

(c) Graph the parabola described by the above equation. How many rentals are necessary in a month in order to maximize profits? What is the maximum profit possible?

Solution

(a) $P(0) = -4000 + 5000(0) - 1000(0)^2 = -4000 + 0 + 0 = -4000.$

This means that if no video tapes are rented ($x = 0$), the business is $4000 "in the hole"; that is, the business owes $4000.

(b)

x	1	2	3	4	5
$P(x)$	0	2000	2000	0	-4000

(c) $P(x) = -1000(x^2 - 5x + 6.25) - 4000 + 6250 = -1000(x - 2.5)^2 + 2250$

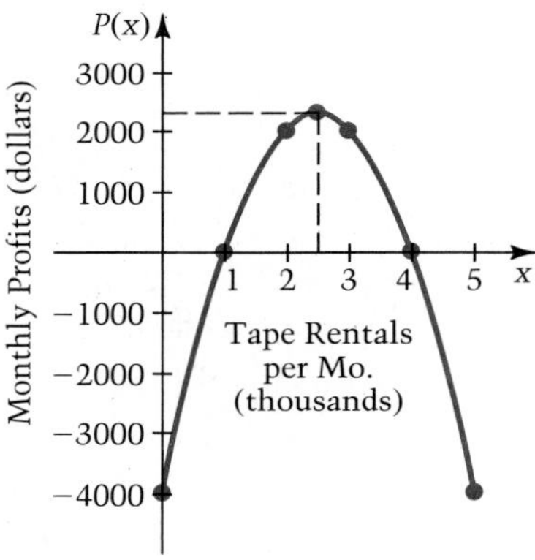

Figure 3.60

The vertex (2.5, 2250) is a high point. Thus, maximum profits occur when there are 2500 rentals, producing profits of $2250.

EXAMPLE 6

At noon a jumbo jet is passing over Chicago, flying due south at 600 mph. At the same time, a small plane is 120 mi south of the jet, flying due east at 200 mph. At what time are the two aircraft closest? How close are they at this time?

Solution

- ANALYZE We must determine the minimum distance between the two aircraft. The information in the problem is best organized in a sketch.

■ ORGANIZE Let t = the time elapsed (in hours) since noon.
Let $d(t)$ = the distance between the two aircraft at time t.

Draw a sketch

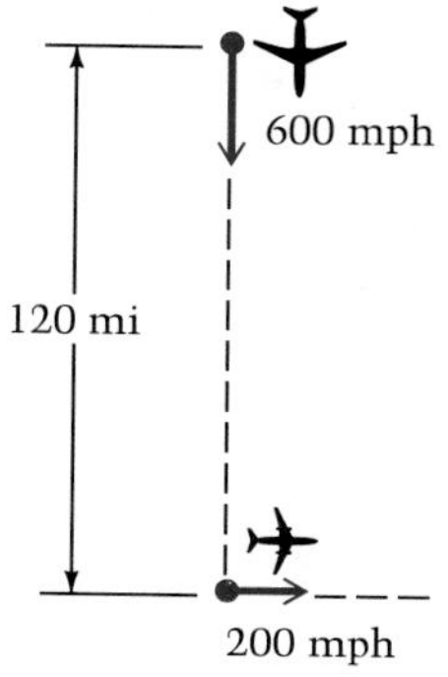

Situation at noon ($t = 0$).

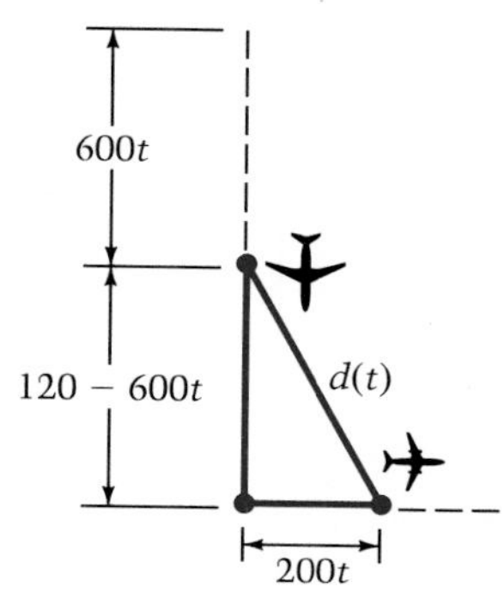

Situation at time t.

Note that at time t, the jet has traveled due south $600t$ miles and the small plane has traveled due east $200t$ miles.

■ MODEL By the Pythagorean Theorem, the distance between the two aircraft at time t is given by the function $d(t) = \sqrt{(120 - 600t)^2 + (200t)^2}$. The distance $d(t)$ is a minimum at the same time that the square of the distance, $[d(t)]^2$, is a minimum. Because $[d(t)]^2$ yields a simpler expression, we will work with it.

$$\begin{aligned}[d(t)]^2 &= (120 - 600t)^2 + (200t)^2 \\ &= 14400 - 144000t + 360000t^2 + 40000t^2 \\ &= 14400 - 144000t + 400000t^2 \\ &= 400000\left(t^2 - \frac{9}{25}t + \square\right) + 14400 && \text{Complete the square.} \\ &= 400000\left(t^2 - \frac{9}{25}t + \left(-\frac{9}{50}\right)^2\right) + 14400 - 12960 \\ &= 400000\left(t - \frac{9}{50}\right)^2 + 1440\end{aligned}$$

The function's graph is a parabola which opens upward. Thus the minimum value of $[d(t)]^2$ occurs at the vertex $\left(\frac{9}{50}, 1440\right)$, that is, when $t = \frac{9}{50}$ hr. This minimum value of $[d(t)]^2$ is 1440 mi^2.

■ ANSWER The two aircraft are closest $\frac{9}{50}(60) \approx 11$ minutes after noon, that is, approximately 12:11 P.M. The minimum distance is $\sqrt{1440} \approx 38$ mi.

EXERCISE SET 3.5

In Exercises 1–18, write the given equation in vertex form. Find the vertex, the axis of symmetry, and the direction in which the parabola opens.

1. $y = 3(x - 1)^2 + 4$
2. $y = -2(x + 3)^2 - 4$
3. $y = -2(x + 6)^2$
4. $y = 5(x - 7)^2$
5. $y = 2(x - 3)^2$
6. $y = -3(x + 7)^2$
7. $y = -(x + 5)^2 - 1$
8. $y = -(x - 2)^2 + 3$
9. $y + 7 = 3(x + 4)^2 + 5$
10. $y - 8 = (x - 3)^2 + 2$
11. $3y = -(x + 1)^2 - 5$
12. $-2y = -(x - 4)^2 - 7$
13. $3(y + 1) = (x - 9)^2$
14. $4(y - 6) = (x + 2)^2$
15. $7y = -2(x + 6)^2$
16. $10y = 3(x + 1)^2$
17. $y = x^2 + 4$
18. $2y = 3x^2 - 8$

For Exercises 19–36, find the x-intercepts of the parabolas in Exercises 1–18.

In Exercises 37–50, graph the given function. Label the vertex, the x-intercepts, the axis of symmetry, and two other points.

37. $f(x) = 5x^2 - 3$
38. $g(x) = 4 - 2x^2$
39. $y = x^2 - 4x - 5$
40. $y = x^2 + 6x + 6$
41. $y = 4x - x^2 - 4$
42. $y = 2 - x^2 - x$
43. $y = x^2 - 6x + 9$
44. $y = x^2 + 4x + 4$
45. $f(x) = 4x^2 - 20x + 25$
46. $f(x) = 9x^2 + 48x + 64$
47. $g(x) = 4x^2 - 16x + 15$
48. $f(x) = 4x^2 + 10x + 21$
49. $y = -2x^2 + 20x - 54$
50. $y = 3x^2 - 12x + 20$

In Exercises 51–54, determine the maximum or minimum function value. State whether this value is a maximum or minimum.

51. $y = -3(x + 1)^2$
52. $y = 5(x - 2)^2$
53. $y = x^2 - 4x - 5$
54. $y = 4x - x^2$

55. The height $h(t)$ of a projectile shot vertically upward from ground level at time $t = 0$ with an initial velocity of 352 ft/sec can be described by a quadratic function of time t:

$$h(t) = 352t - 16t^2.$$

(a) By finding the y-coordinate of the vertex, determine the maximum height of the projectile.
(b) When does the projectile achieve its maximum height?
(c) When does the projectile return to the ground?

56. The sum of two numbers is 12.
(a) Describe the two numbers using the variable x.
(b) Describe the product of the two numbers as a function of x.
(c) Determine the two numbers that make the product as large as possible.

In Exercises 57–64, (a) carefully graph the given quadratic function over the interval $0 \le x \le 5$; (b) use your graph to estimate $f(2.5)$; (c) use your calculator to determine $f(2.5)$ exactly.

57. $f(x) = 4 - 0.2x^2$
58. $f(x) = 3 + 0.3x^2$
59. $f(x) = x^2 - 3x + 4$
60. $f(x) = x^2 - x - 3$
61. $f(x) = 0.4x^2 - 1.1x$
62. $f(x) = 10.3x - 2.1x^2$
63. $f(x) = 0.3x^2 - x - 0.43$
64. $f(x) = 0.2x^2 - 0.08x - 3.92$

The height $h(t)$ of an object shot vertically upward (from ground level) with a velocity of v_0 ft/sec can be described by a function of time t: $h(t) = v_0t - 16t^2$. In Exercises 65–68, determine the time at which the object reaches its maximum height, and the maximum height achieved. (After computations, round answers to the nearest tenth.)

65. $v_0 = 80.5$ ft/sec
66. $v_0 = 22.6$ ft/sec
67. $v_0 = 50.2$ ft/sec
68. $v_0 = 197.4$ ft/sec

Superset

69. A small mining operation can produce T tons of coal each day at a total cost of $8T^2 + 25T + 2000$ dollars. The price at which the coal can be sold is $1185 - 12T$ dollars per ton.
(a) Express the profit P of the company as a function of T.
(b) How many tons of coal should the company produce to maximize its daily profit?

70. A local grocer has found that she can sell $528 - 150x$ boxes of cereal each week, if the price is set at x dollars per box.
(a) Express the weekly revenue R of the grocer's cereal sales as a function of x.
(b) Determine the price which will maximize the grocer's revenue from cereal sales.

71. The perimeter of a rectangle is 60 ft.
(a) Describe the area of the rectangle as a function of one variable.

(b) Determine the dimensions of the rectangle that produce the largest area, and state the value of this maximum area.

72. The owner of a stable wishes to fence in a rectangular region, and intends to use 100 feet of fencing that was given to him by a friend. The fencing is needed for only three sides, since the fourth side will be along the barn. Write a function that describes the area of the region in terms of x. Determine the maximum possible area, and the dimensions that produce this area.

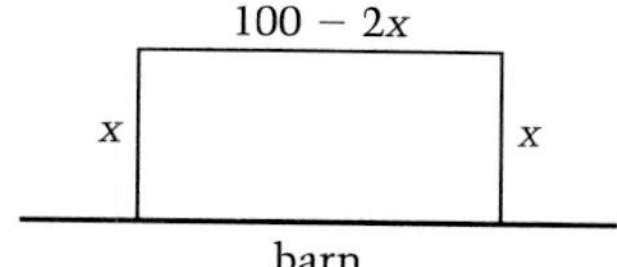

73. The market research division of a clothing company predicts that the company will sell $11{,}800 - 200x$ swimsuits in its new line this season if the swimsuits are priced at x dollars each. Production and distribution costs to the company are \$16.20 for each swimsuit sold.
 (a) Express the profit P of the company on its swimsuit sales as a function of x.
 (b) What price should the company set for the swimsuits to maximize profit?

74. (See Exercise 73.) Express the revenue R of the company as a function of x and show that the price set to maximize revenues is different than the price set to maximize profit.

75. The City Museum can ship most of its smaller artifacts in crates that are 12 inches deep. Their shipper restricts them to crates whose length, width and depth add up to 96 inches or less. The museum designs its crates so that it just meets this restriction.
 (a) Express the volume V (in cubic inches) of these 12 inch deep crates as a function of their width x.
 (b) Determine the length and width of the crate of greatest volume.

76. (a) Express the surface area S (all six sides) of the crates described in Exercise 75 as a function of x.
 (b) Determine the length and width of the crate of greatest surface area.

77. Determine b such that the quadratic function $f(x) = x^2 - bx + 6$ has exactly one x-intercept.

78. The quadratic function $h(x) = 2x^2 - 3x + C$ has 5 as one of its x-intercepts. Determine the value of C and the other x-intercept.

79. We say that y varies directly with x^2 provided there exists some nonzero constant k such that $y = kx^2$. Suppose d varies directly with t^2 and that d is 5 when $t = 3$. Find d when $t = 5$.

3.6
Composition and the Algebra of Functions

Just as real numbers can be added, subtracted, multiplied or divided, so too can functions. The resulting function is determined by the effect it has on a domain value. For example, given two functions f and g, the *sum function* $f + g$ takes a domain value x and produces the range value $f(x) + g(x)$. That is, if $f(x) = x^2 - 7$ and $g(x) = 5 + \sqrt{x}$, then

$$(f + g)(9) = f(9) + g(9) = (9^2 - 7) + (5 + \sqrt{9}) = 81 - 7 + 5 + 3 = 82,$$

and in general, the sum function $f + g$ is given by

$$(f + g)(x) = f(x) + g(x) = (x^2 - 7) + (5 + \sqrt{x}) = x^2 + \sqrt{x} - 2,$$

with domain $[0, \infty)$.

Operations on Functions

If f and g are functions, then for any real number x that is common to both the domain of f *and* the domain of g,

the **sum function** $f + g$ is given by $(f + g)(x) = f(x) + g(x)$;
the **difference function** $f - g$ is given by $(f - g)(x) = f(x) - g(x)$;
the **product function** fg is given by $(fg)(x) = f(x) \cdot g(x)$;
the **quotient function** $\frac{f}{g}$ is given by $\left(\frac{f}{g}\right)(x) = \frac{f(x)}{g(x)}$, provided $g(x) \neq 0$.

Note that in each case, the domain of the resulting function consists of those real numbers that are common to both the domain of f *and* the domain of g, and that values of x such that $g(x) = 0$ are excluded from the domain of the quotient function $\frac{f}{g}$.

EXAMPLE 1

If $f(x) = 2x^2 - 1$ and $g(x) = 3x - \sqrt{x}$, determine each of the following:

(a) $(f + g)(4)$ (b) $(f - g)(0)$ (c) $(fg)\left(\frac{1}{4}\right)$ (d) $\left(\frac{f}{g}\right)(2)$

(e) $(f - g)(x)$ (f) $\left(\frac{f}{g}\right)(x)$

Solution

(a) $(f + g)(4) = f(4) + g(4) = [2(4)^2 - 1] + [3(4) - \sqrt{4}] = 41$

(b) $(f - g)(0) = f(0) - g(0) = [2(0)^2 - 1] - [3(0) - \sqrt{0}] = -1$

(c) $(fg)\left(\frac{1}{4}\right) = f\left(\frac{1}{4}\right) \cdot g\left(\frac{1}{4}\right) = \left[2\left(\frac{1}{4}\right)^2 - 1\right] \cdot \left[3\left(\frac{1}{4}\right) - \sqrt{\frac{1}{4}}\right] = -\frac{7}{32}$

(d) $\left(\frac{f}{g}\right)(2) = \frac{f(2)}{g(2)} = \frac{2(2)^2 - 1}{3(2) - \sqrt{2}} = \frac{7}{6 - \sqrt{2}}$

(e) $(f - g)(x) = f(x) - g(x) = (2x^2 - 1) - (3x - \sqrt{x}) = 2x^2 - 3x + \sqrt{x} - 1$

The domain of $f - g$ is the set of all real numbers in the interval $[0, \infty)$.

(f) $\left(\frac{f}{g}\right)(x) = \frac{f(x)}{g(x)} = \frac{2x^2 - 1}{3x - \sqrt{x}}$

Since the denominator is 0 when $x = 0$ or when $x = \frac{1}{9}$, the domain of the quotient function $\frac{f}{g}$ is the set of all positive real numbers except $\frac{1}{9}$. That is, the domain is $\left(0, \frac{1}{9}\right) \cup \left(\frac{1}{9}, \infty\right)$.

Composite Functions

A function may be viewed as a number processor that takes a domain value as input and produces a range value as output. For example, we can treat the functions $f(x) = x^2 + 4$ and $g(x) = 3x - 1$ as processors.

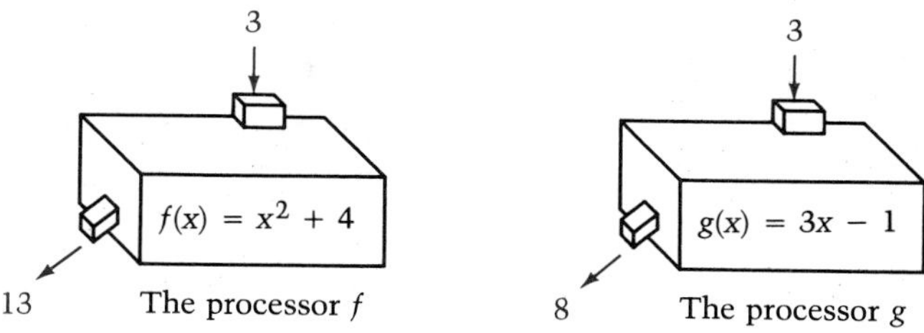

Figure 3.61

The function f takes an x-value from the domain as input, squares it and adds four to produce the function value $y = f(x)$. The function g takes an x-value, multiplies it by three and subtracts one to produce the function value $y = g(x)$.

The functions g and f can be combined to generate the *composite function* $f \circ g$, read "f circle g." This is illustrated below.

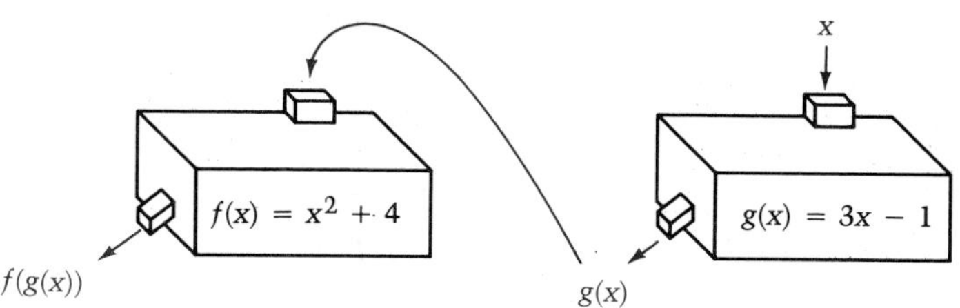

Figure 3.62

The function $f \circ g$ is the composition of g followed by f. First, x is processed by g and then the output $g(x)$ becomes the input for f.

Definition of Composite Function

Let f and g be functions with the range of g contained in the domain of f. For each x in the domain of g, the **composite function** $f \circ g$ is defined as

$$(f \circ g)(x) = f(g(x)).$$

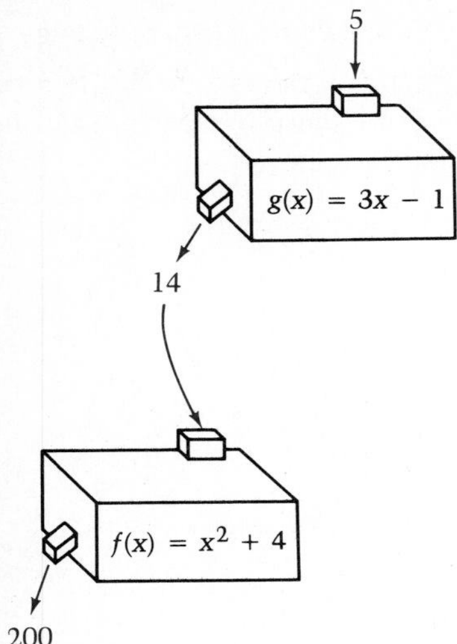

Figure 3.63 $(f \circ g)(5) = 200$.

How does the $f \circ g$ processor work? The domain value x is first processed by g to produce $g(x)$. Then $g(x)$ is processed by f to obtain $f(g(x))$. For example, for the functions f and g shown at the left, we can find $(f \circ g)(5)$ as follows.

$$\begin{aligned} (f \circ g)(5) &= f(g(5)) && \text{First, we evaluate } g(5). \\ &= f(14) && g(5) = 3(5) - 1 = 14. \\ &= 200 \end{aligned}$$

We can also compute $(g \circ f)(5)$.

$$\begin{aligned} (g \circ f)(5) &= g(f(5)) && \text{First, we evaluate } f(5). \\ &= g(29) && f(5) = 5^2 + 4 = 29. \\ &= 86 \end{aligned}$$

Notice that $(f \circ g)(5) = 200$, but $(g \circ f)(5) = 86$. In general, $f \circ g$ and $g \circ f$ are *not* the same function. The order of the functions in a composition is important.

In the definition of $f \circ g$, it is necessary that the range of g be contained in the domain of f so that every output of g is acceptable input for f. For example, suppose $f(x) = \sqrt{x}$ and $g(x) = x - 10$. To determine $(f \circ g)(3)$, we write $f(g(3)) = f(-7) = \sqrt{-7}$. But the f processor, $\sqrt{}$, cannot accept -7 as input. We say that $(f \circ g)(3)$ is not defined. (Remember, we are restricting our work to functions whose domains and ranges consist solely of real numbers.)

EXAMPLE 2

Suppose $f(x) = 2x^2 - 3$, $g(x) = |x| - 5$, and $h(x) = \sqrt{x}$. Evaluate:

(a) $(f \circ g \circ h)(4)$ (b) $(h \circ f)(1)$.

Solution

(a)
$$\begin{aligned} (f \circ g \circ h)(4) &= f(g(h(4))) && \text{First, we evaluate } h(4). \\ &= f(g(2)) && h(4) = \sqrt{4} = 2. \\ &= f(-3) && g(2) = |2| - 5 = -3. \\ &= 15 && f(-3) = 2(-3)^2 - 3 = 15. \end{aligned}$$

(b)
$$\begin{aligned} (h \circ f)(1) &= h(f(1)) && \text{First evaluate } f(1). \\ &= h(-1) && f(1) = 2(1)^2 - 3 = -1. \\ &= \sqrt{-1} \end{aligned}$$

$(h \circ f)(1)$ is not defined because $f(1) = -1$ is not in the domain of h.

Suppose we have two functions f and g that are described by the equations $f(x) = 3x^2 + 4x$ and $g(x) = 2x - 1$. To describe $f \circ g$, we determine $(f \circ g)(x)$.

$$\begin{aligned}(f \circ g)(x) &= f(g(x)) \\ &= f(2x - 1) \\ &= 3(2x - 1)^2 + 4(2x - 1) \qquad f(\square) = 3(\square)^2 + 4(\square) \\ &= 3(4x^2 - 4x + 1) + 4(2x - 1) \\ &= 12x^2 - 4x - 1\end{aligned}$$

On the other hand,

$$\begin{aligned}(g \circ f)(x) &= g(f(x)) = g(3x^2 + 4x) \\ &= 2(3x^2 + 4x) - 1 \\ &= 6x^2 + 8x - 1\end{aligned}$$

so that $f \circ g$ and $g \circ f$ are not the same function.

EXAMPLE 3

If $f(x) = 2x + 3$, $g(x) = x^2 - 1$ and $h(x) = \frac{1}{x}$, determine

(a) $(h \circ g)(x)$ (b) $(f \circ g)(x)$.

Solution

(a) $(h \circ g)(x) = h(g(x)) = h(x^2 - 1) = \dfrac{1}{x^2 - 1}$

(b) $(f \circ g)(x) = f(g(x)) = f(x^2 - 1) = 2(x^2 - 1) + 3 = 2x^2 + 1$

EXAMPLE 4

Suppose $f(x) = 2x + 5$ and $g(x) = \sqrt{x}$. Determine the domain of $(g \circ f)(x)$.

Solution Since $(g \circ f)(x) = g(f(x))$, we need to find those values of x for which $f(x)$ is in the domain of g. We know the domain of g is the set of nonnegative real numbers; thus we need to determine values of x such that $f(x)$ is nonnegative.

$$\begin{aligned}2x + 5 &\geq 0 \qquad \text{Recall that } f(x) = 2x + 5. \\ 2x &\geq -5 \\ x &\geq -\frac{5}{2}\end{aligned}$$

Thus, for all $x \geq -\frac{5}{2}$, $f(x)$ is in the domain of g, and $(g \circ f)(x)$ is defined. The domain of $g \circ f$ is $\left[-\frac{5}{2}, \infty\right)$.

Given the equations for two functions f and g, we can determine the equation for the composite function $f \circ g$. Let us now consider the reverse situation. Suppose we know that $F(x) = (x - 7)^2$. How might we describe $F(x)$ as a composition?

To answer this question, notice that there are two processes operating when we evaluate $F(6)$.

$$\begin{aligned} F(6) &= (6 - 7)^2 && \text{First process: subtract 7.} \\ &= (-1)^2 && \text{Second process: square.} \\ &= 1 \end{aligned}$$

Each process can be represented as a function.

$f(x) = x - 7$ First function: the function that subtracts 7 from a number

$g(x) = x^2$ Second function: the function that squares a number

Thus we can write $F(x)$ as the composition of the two functions f and g.

$$F(x) = (g \circ f)(x)$$

(f: first function; g: second function)

When we build the composite function $F(x)$, the first function is placed closest to x because it will process x first. That is, a composition is built from right to left.

To find simple functions used to build a composite function, you must discover a sequence of simple processes that produce the composite function.

EXAMPLE 5

Express $G(x) = x^3 + 15$ as a composite function.

Solution Since G cubes and then adds 15, we may write

$$G(x) = (g \circ f)(x)$$

where $f(x) = x^3$ and $g(x) = x + 15$.

EXAMPLE 6

Consider the following simple functions.

$$f(x) = x + 2 \qquad g(x) = |x| \qquad h(x) = x^2 \qquad k(x) = \frac{1}{x}$$

Write each of the following functions as a composition of two or more of these simple functions.

(a) $F(x) = |x| + 2$ (b) $G(x) = |x + 2|$ (c) $H(x) = \dfrac{1}{x^2 + 2}$

Solution

(a) To determine $F(x)$, first take the absolute value of x: $g(x) = |x|$
Then add 2: $f(g(x)) = |x| + 2$
Thus, $F(x) = (f \circ g)(x)$

(b) To determine $G(x)$, first add 2: $f(x) = x + 2$
Then take the absolute value: $g(f(x)) = |x + 2|$
Thus, $G(x) = (g \circ f)(x)$.

(c) To determine $H(x)$, first form the square: $h(x) = x^2$
Then add 2: $f(h(x)) = x^2 + 2$
Finally, take the reciprocal: $k(f(h(x))) = \dfrac{1}{x^2 + 2}$
Thus, $H(x) = (k \circ f \circ h)(x)$.

EXERCISE SET 3.6

In Exercises 1–12, evaluate the given expression for $f(x) = x^2 - 4$ and $g(x) = x + 4$.

1. $(f + g)(-2)$
2. $(f - g)(-3)$
3. $(fg)(0)$
4. $(fg)(-1)$
5. $\left(\dfrac{f}{g}\right)(-4)$
6. $\left(\dfrac{f}{g}\right)(0)$
7. $\left(\dfrac{g}{f}\right)(3)$
8. $\left(\dfrac{g}{f}\right)(2)$
9. $1 - (fg)(1)$
10. $f(1) - (fg)(1)$
11. $a^2 - (f + g)(a)$
12. $2a^2 + (f - g)(a)$

In Exercises 13–24, for the given pair of functions find (a) $(f + g)(x)$, (b) $(f - g)(x)$, (c) $(fg)(x)$, (d) $\left(\dfrac{f}{g}\right)(x)$, and state the domain of each.

13. $f(x) = 3x + 1;\ g(x) = 2x - 7$
14. $f(x) = 5x - 2;\ g(x) = 2x + 3$
15. $f(x) = 5x - 1;\ g(x) = 1 - x$
16. $f(x) = 3x + 2;\ g(x) = 1 - 2x$
17. $f(x) = x^2;\ g(x) = x + 5$
18. $f(x) = x^2 + 2;\ g(x) = 2 - x$
19. $f(x) = \sqrt{x};\ g(x) = x - 1$
20. $f(x) = \dfrac{1}{x};\ g(x) = x^2$
21. $f(x) = x^2 - 3x + 2;\ g(x) = x^2 + 1$
22. $f(x) = x^2 - x - 1;\ g(x) = x^2 - 4$
23. $f(x) = \dfrac{x - 2}{x + 2};\ g(x) = \dfrac{x}{x + 2}$
24. $f(x) = \dfrac{x}{x - 3};\ g(x) = x^2 - 9$

In Exercises 25–30, evaluate the given expression for $f(x) = 2x - 1$ and $g(x) = |x|$.

25. $(f \circ g)(-2)$
26. $(g \circ f)(0)$
27. $(g \circ f)(-2)$
28. $(f \circ g)(0)$
29. $(f \circ g)\left(\dfrac{1}{3}\right)$
30. $(g \circ f)\left(\dfrac{1}{3}\right)$

In Exercises 31–36, evaluate the given expression for $f(x) = 3$ and $g(x) = 10x - 7$.

31. $(g \circ f)(5)$
32. $(f \circ g)(5)$
33. $(g \circ f)(-5)$
34. $(f \circ g)(-5)$
35. $(f \circ g)(0)$
36. $(g \circ f)(0)$

In Exercises 37–50, evaluate the given expression for $f(x) = \dfrac{1}{x - 4}$, $g(x) = 4x^2 - 8$, and $h(x) = \sqrt{x}$.

37. $(g \circ f)(-6)$
38. $(g \circ f)(-2)$

39. $(g \circ f)(0)$

40. $(f \circ g)(0)$

41. $(f \circ h)(16)$

42. $(f \circ g)\left(\frac{1}{2}\right)$

43. $(h \circ g)(\sqrt{2})$

44. $(f \circ g)(\sqrt{3})$

45. $(f \circ f)(2)$

46. $(g \circ g)(-1)$

47. $(g \circ h)(4)$

48. $(h \circ g)(1)$

49. $(h \circ g)(0)$

50. $(g \circ h)(0)$

In Exercises 51–60, find an equation that describes $(f \circ g)(x)$ and state the domain of the composite function.

51. $f(x) = x^2 + 1;\ g(x) = x - 3$

52. $f(x) = x + 4;\ g(x) = 2x^2 - 1$

53. $f(x) = |x|;\ g(x) = \dfrac{1}{x - 3}$

54. $f(x) = \dfrac{1}{x + 3};\ g(x) = |x + 1|$

55. $f(x) = 1 - \sqrt{x};\ g(x) = x - 5$

56. $f(x) = \sqrt{x - 1};\ g(x) = 5 - x$

57. $f(x) = \dfrac{1}{x + 4};\ g(x) = x^2 + 5$

58. $f(x) = \dfrac{1}{x - 1};\ g(x) = x^2 + 5$

59. $f(x) = \dfrac{1}{x};\ g(x) = \sqrt{x}$

60. $f(x) = \dfrac{1}{x + 1};\ g(x) = \sqrt{x - 3}$

In Exercises 61–68, write the given function $y = F(x)$ as a composition of two or more of the following:

$$f(x) = x + 2,\ j(x) = |x|,\ g(x) = x^2,$$
$$k(x) = \frac{3}{x},\ h(x) = \sqrt{x}.$$

61. $F(x) = |x| + 2$

62. $F(x) = \dfrac{3}{|x|}$

63. $F(x) = \sqrt{\dfrac{3}{x}}$

64. $F(x) = \sqrt{x + 2}$

65. $F(x) = \dfrac{3}{|x| + 2}$

66. $F(x) = \sqrt{\dfrac{3}{x + 2}}$

67. $F(x) = \dfrac{3}{x + 4}$

68. $F(x) = x^4 + 2$

Superset

69. Suppose $y = 3x - 2$, $x = 1 - z$, and $z = 5t + 2$. Write y as a function of t.

70. Suppose $x = 3t + 2$, $z = 6 - t$, and $z = 2y + 1$. Write y as a function of x.

71. A physiologist monitored a person moving a heavy object. The subject exerted a force for a period of 60 seconds which caused an increase in pulse rate. Functional relationships observed in this test are shown below. Use the graphs to determine how much time elapsed before the pulse rate reached 120.

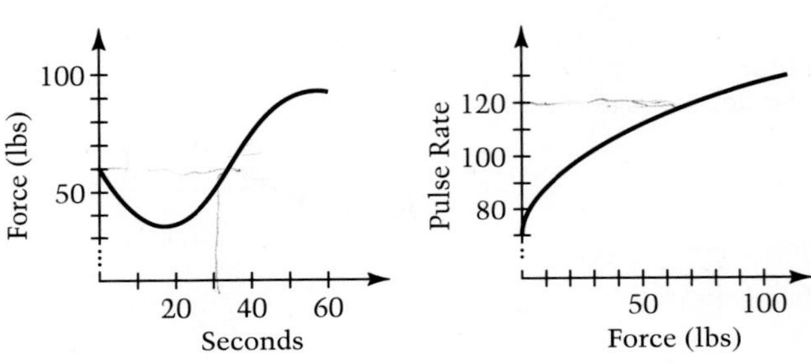

In Exercises 72–76, the functions f and g are given with restricted domains. In each case, find an equation for $(f \circ g)(x)$, and state the domain of the composite function.

72. $f(x) = 3x - 1$; domain: $[-1, 1]$
$g(x) = x - 5$; domain: $[0, 5]$

73. $f(x) = 2x + 1$; domain: $[0, 2]$
$g(x) = x + 1$; domain: $[-2, 2]$

74. $f(x) = \sqrt{x}$; domain: $[0, \infty)$
$g(x) = x - 4$; domain: $[0, 10]$

75. $f(x) = \sqrt{x}$; domain: $[0, \infty)$
$g(x) = x + 3$; domain: $[-5, 0]$

76. $f(x) = x^2$; domain: $[-10, 10]$
$g(x) = \sqrt{-x}$; domain: $(-\infty, 0]$

77. Show that the composition of two odd functions is an odd function.

78. Show that the composition of an odd function and an even function is an even function.

3.7 Inverse Functions

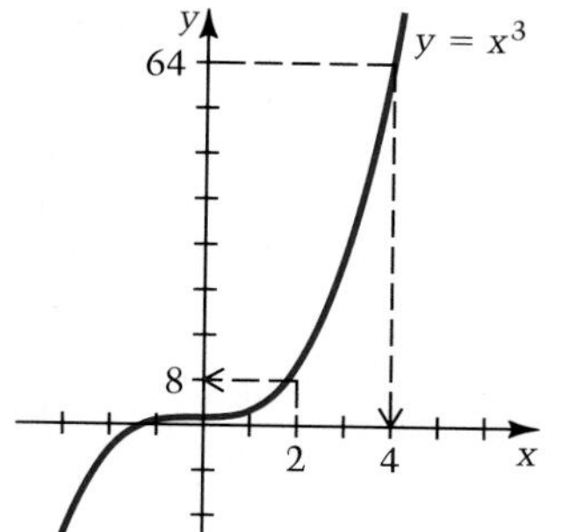

Figure 3.64 "Given $y = 64$, find x." The answer is unique.

We now consider two questions that can be asked about a function $y = f(x)$. The first is straightforward: given a value of x, what is the corresponding value of y? To answer this, we simply evaluate the function. For example,

$$\text{if } y = x^3 \text{ and } x = 2, \text{ then } y = 8.$$

The second question reverses the first: given a value of y, what is the corresponding value of x? Sometimes this question has a unique answer. For example (see Figure 3.64),

$$\text{for the function } y = x^3\text{:} \quad \text{if } y = 64, \text{ then } x = 4.$$

Sometimes, however, the answer is not unique. For example (see Figure 3.65),

$$\text{for the function } y = x^2\text{:} \quad \text{if } y = 16, \text{ then } x = 4 \text{ or } x = -4.$$

We use the term **one-to-one** to describe those functions which always have a unique answer to the question "given y, find x." Thus, the function in Figure 3.64 is one-to-one, while the function in Figure 3.65 is not.

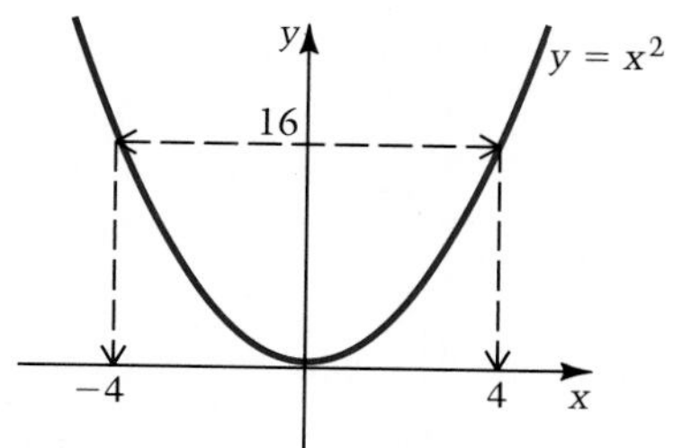

Figure 3.65 "Given $y = 16$, find x." The answer is not unique.

Definition of a One-to-One Function

A function $y = f(x)$ is said to be **one-to-one** if each y-value in the range of f corresponds to precisely one x-value in the domain of f; that is, for any numbers a and b in the domain of f,

$$f(a) = f(b) \text{ implies } a = b.$$

EXAMPLE 1

Determine whether the function is one-to-one.

(a) $f(x) = 7x + 2$ (b) $y = x^2 - 6x + 8$

Solution

(a) $7a + 2 = 7b + 2$ Assume $f(a) = f(b)$ then simplify.

$$7a = 7b$$

$$a = b$$

Since the assumption that $f(a) = f(b)$ leads to the statement that $a = b$, we conclude that f is one-to-one.

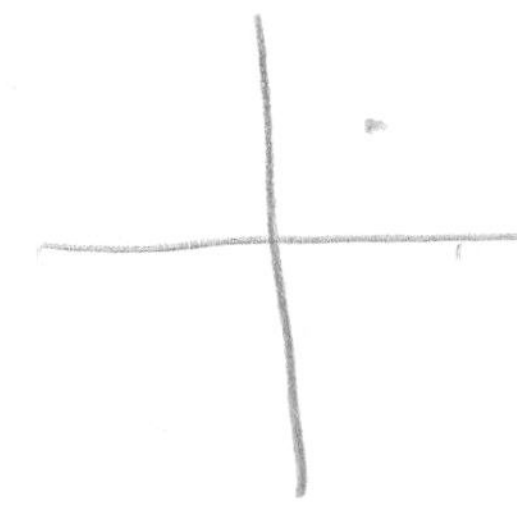

(b) Here we will find a y-value which is associated with two different x-values. Let us choose $y = 3$.

$$y = x^2 - 6x + 8$$
$$3 = x^2 - 6x + 8 \quad \text{Replace } y \text{ with 3, and solve for } x.$$
$$0 = x^2 - 6x + 5$$
$$0 = (x - 5)(x - 1)$$

$x - 5 = 0$	$x - 1 = 0$
$x = 5$	$x = 1$

We have found that when $x = 5$ or when $x = 1$, the same y-value, 3, is produced. Thus the function is not one-to-one.

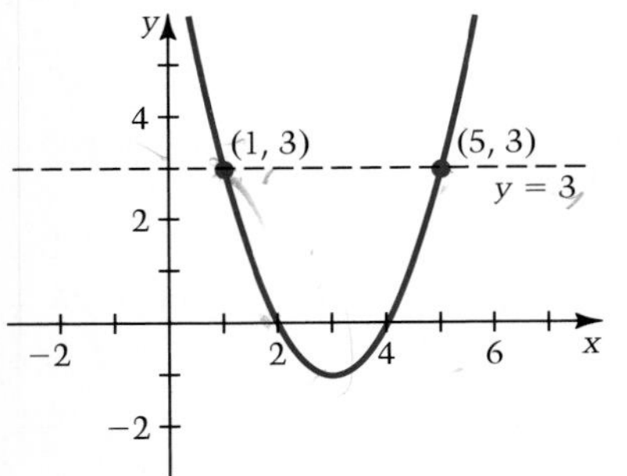

Figure 3.66

As we saw in part (b) of Example 1, we can show that a function is *not* one-to-one by finding two different domain values that produce the same range value. In that example we found that the points (1, 3) and (5, 3) both lie on the graph of the function $y = x^2 - 6x + 8$, as illustrated in Figure 3.66. Notice that these points lie on the horizontal line $y = 3$. Whenever two points have different x-coordinates, but the same y-coordinate, they will determine a horizontal line. This fact leads us to a graphical test to determine whether a function is one-to-one.

The Horizontal Line Test

For the graph of a function f in the xy-plane, if a horizontal line can be found that intersects the graph of the function more than once, then the function is **not** one-to-one. If no such line exists, then the function **is** one-to-one.

Notice in Figure 3.67(a) that every horizontal line will intersect the graph at only one point, but in Figure 3.67(b) *not every* horizontal line intersects the graph more than once; there need be only one such line for the function not to be one-to-one.

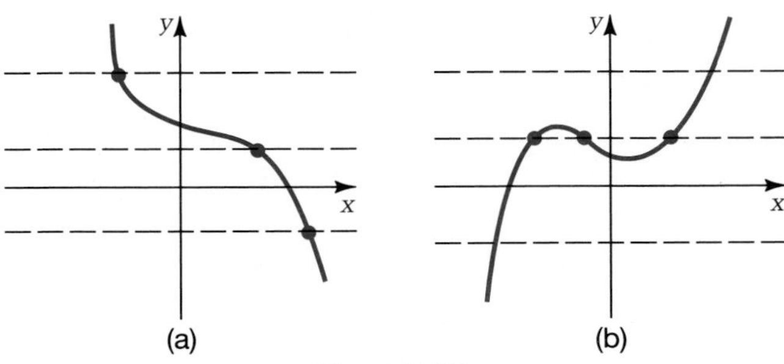

Figure 3.67

This horizontal line test provides an easy way for deciding whether a function is one-to-one. It is useful when you are certain what the graph of a function looks like. For example, from your work in Section 3 of this chapter, you are familiar with the graph of $y = \frac{1}{x}$. It is clearly one-to-one by the horizontal line test, as is any horizontal or vertical translation of that graph. Thus any function of the form $y = \frac{1}{x - h}$ or $y = \frac{1}{x} + k$ is also one-to-one. By the same token, since $y = x^2$ fails the horizontal line test, neither it nor any function of the form $y = (x - h)^2$ or $y = x^2 + k$ is one-to-one.

Inverse Functions

If a function f is one-to-one, then there exists an *inverse function*, denoted f^{-1}, such that $f(f^{-1}(x)) = x$, and $(f^{-1}(f(x)) = x$. That is, the inverse function f^{-1} undoes the process f. For example, if f processes 3 and produces 10, then f^{-1} processes 10 and produces 3:

$$\text{if } f(3) = 10, \text{ then } f^{-1}(10) = 3.$$

The function f^{-1} interchanges the inputs and outputs of the function f: the domain of f becomes the range of f^{-1} and the range of f becomes the domain of f^{-1}.

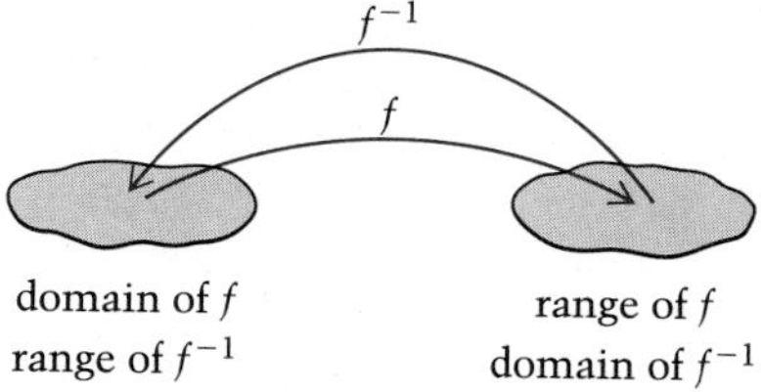

Figure 3.68

Definition of Inverse Function

If a function f is one-to-one, then there exists a unique function f^{-1}, called the **inverse function of *f*,** such that

$$(f^{-1} \circ f)(x) = x \quad \text{for any } x \text{ in the domain of } f,$$
$$(f \circ f^{-1})(x) = x \quad \text{for any } x \text{ in the domain of } f^{-1}.$$

Caution The "−1" that appears in the notation f^{-1} for the inverse function of f should **not** be thought of as a negative exponent; f^{-1} is not $\frac{1}{f}$. Rather, the "−1" is part of the name for the inverse function.

EXAMPLE 2

The function $f(x) = 3x - 2$ is one-to-one. Find its inverse function, $f^{-1}(x)$.

Solution

$f(x) = 3x - 2$ We begin by writing the function in terms of x and y.

$y = 3x - 2$ We now solve for x in terms of y.

$y + 2 = 3x$

$x = \dfrac{y+2}{3}$ We will now interchange the roles of x and y to produce the inverse function.

$y = \dfrac{x+2}{3}$

$f^{-1}(x) = \dfrac{x+2}{3}$ y is replaced with $f^{-1}(x)$ to signify the inverse function.

We can check that the equation for f^{-1} is correct by verifying that $(f^{-1} \circ f)(x) = x$ and that $(f \circ f^{-1})(x) = x$:

$$(f^{-1} \circ f)(x) = f^{-1}(f(x)) = f^{-1}(3x - 2) = \frac{(3x - 2) + 2}{3} = \frac{3x}{3} = x,$$

$$(f \circ f^{-1})(x) = f(f^{-1}(x)) = f\left(\frac{x+2}{3}\right) = 3\left(\frac{x+2}{3}\right) - 2 = (x + 2) - 2 = x.$$

Note that the domains and ranges of f and f^{-1} are $(-\infty, \infty)$.

EXAMPLE 3

The function $g(x) = x^3 - 5$ is one-to-one. Find its inverse function, $g^{-1}(x)$.

Solution

$g(x) = x^3 - 5$

$y = x^3 - 5$

$y + 5 = x^3$ We now solve for x in terms of y.

$x = \sqrt[3]{y + 5}$ Interchanging the roles of x and y will produce the inverse function.

$y = \sqrt[3]{x + 5}$

$g^{-1}(x) = \sqrt[3]{x + 5}$ y is replaced with $g^{-1}(x)$ to signify the inverse function.

Note that the domains and ranges of g and g^{-1} are $(-\infty, \infty)$.

We now summarize the method for finding the equation of an inverse function.

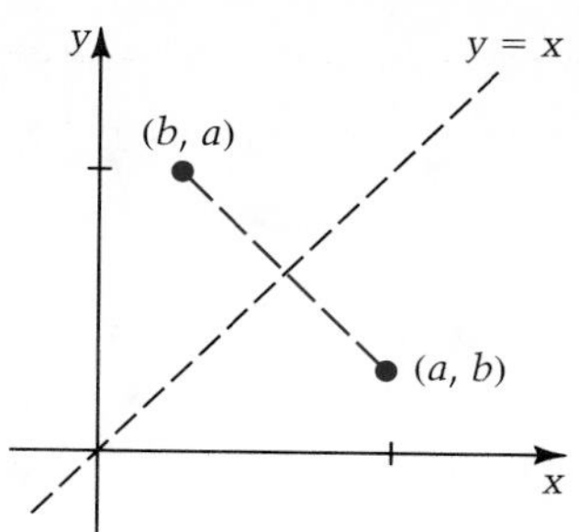

Figure 3.69

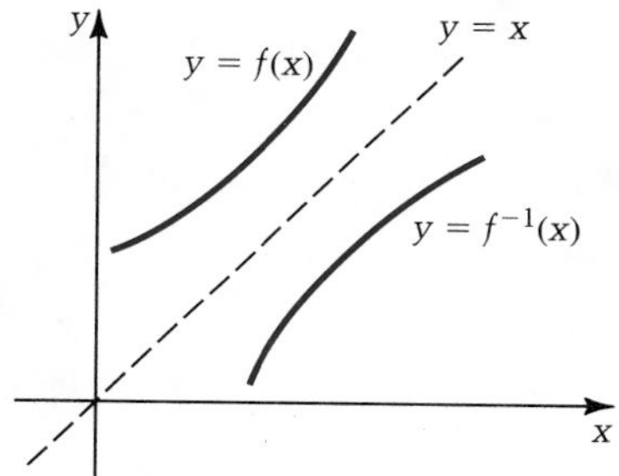

Figure 3.70

Finding an Equation for an Inverse Function

1. Verify that $y = f(x)$ is one-to-one.
2. Solve for x in terms of y.
3. Interchange x and y, and replace y with $f^{-1}(x)$.
4. Verify that the domain of f is the range of f^{-1}, and that the domain of f^{-1} is the range of f.

Since f^{-1} interchanges the roles of domain and range for f, it follows that for every point (a, b) on the graph of f, the point (b, a) is on the graph of f^{-1}. In Figure 3.69, notice the graphical relationship that exists between points of the form (a, b) and (b, a): they are reflections of one another across the line $y = x$. Consequently, for *every* point on the graph of f, its mirror image across the line $y = x$ is to be found on the graph of f^{-1}. As a result, the graphs of f and f^{-1} are reflections of each other across the line $y = x$.

Graphical Property of f and f^{-1}

The graphs of f and f^{-1} are reflections of one another across the line $y = x$.

EXAMPLE 4

Consider $f(x) = x^2 - 4$ with domain $[1, 3]$. Determine $f^{-1}(x)$, and sketch its graph.

Solution $f(x) = x^2 - 4$; domain $[1, 3]$.

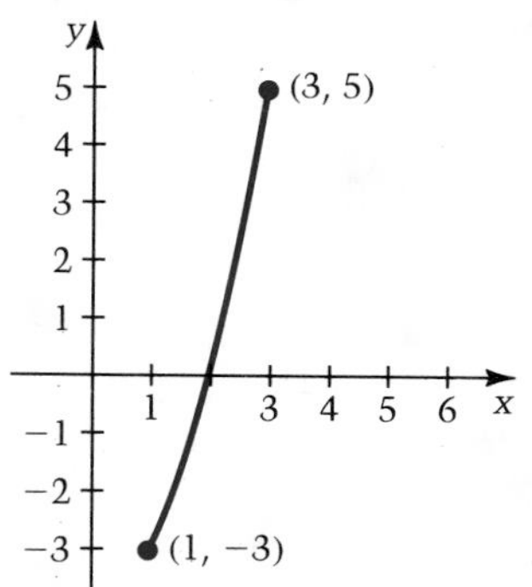

Figure 3.71

The function is one-to-one, since the graph, shown in color at the left, clearly satisfies the Horizontal Line Test.

$$y = x^2 - 4$$

Write the equation for f in terms of x and y, then solve for x in terms of y.

$$y + 4 = x^2$$

$$\pm\sqrt{y+4} = x$$
$$\sqrt{y+4} = x$$

Notice from the graph, that since $1 \le x \le 3$, x is always positive, so only the + sign in the "±" symbol is appropriate.

$$\sqrt{x+4} = y$$
$$f^{-1}(x) = \sqrt{x+4}$$

Now interchange x and y, and replace y with $f^{-1}(x)$.

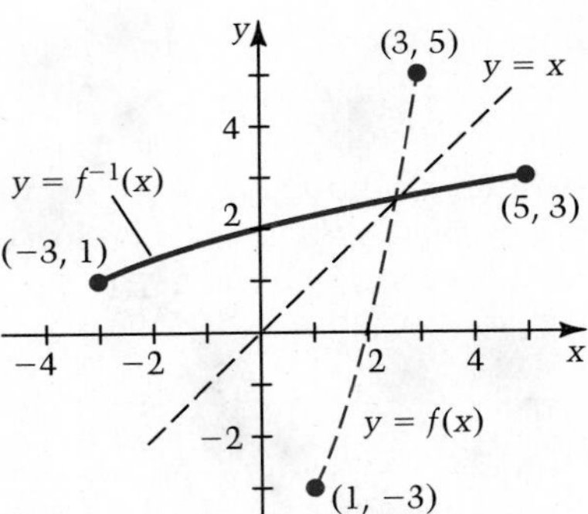

Figure 3.72

Domain of f^{-1} = range of $f = [-3, 5]$.
Domain of f = range of $f^{-1} = [1, 3]$.

EXERCISE SET 3.7

In Exercises 1–10, determine whether the function is one-to-one.

1. $f(x) = 7 - 4x$
2. $f(x) = 5 - 2x^3$
3. $f(x) = x^3 + 2$
4. $f(x) = |3 - x|$
5. $f(x) = |2x - 1|$
6. $f(x) = 3x + 1$
7. $f(x) = x^2 + 4x - 21$
8. $f(x) = \dfrac{x-3}{x+2}$
9. $f(x) = \dfrac{x+2}{x-1}$
10. $f(x) = x^2 - 2x - 15$

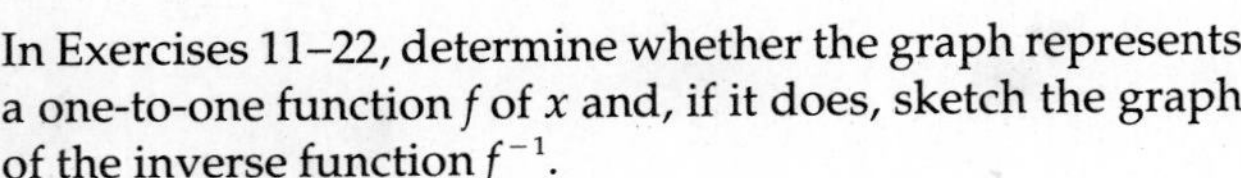

In Exercises 11–22, determine whether the graph represents a one-to-one function f of x and, if it does, sketch the graph of the inverse function f^{-1}.

11.

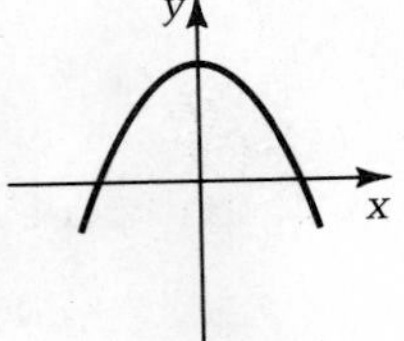

12.

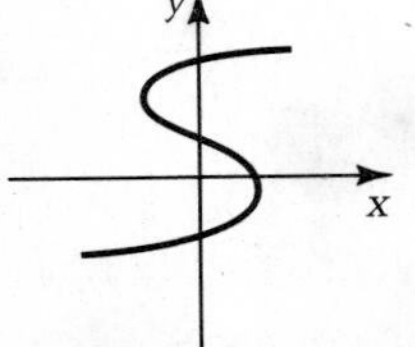

13.

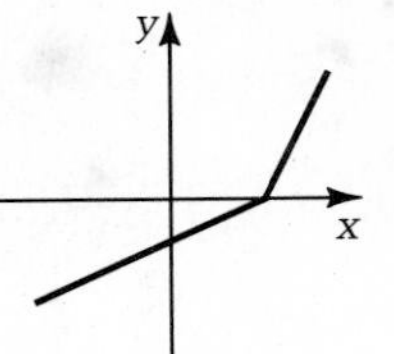

14.

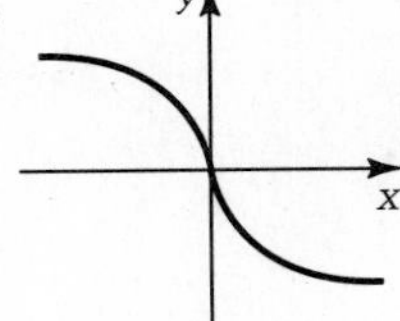

15.

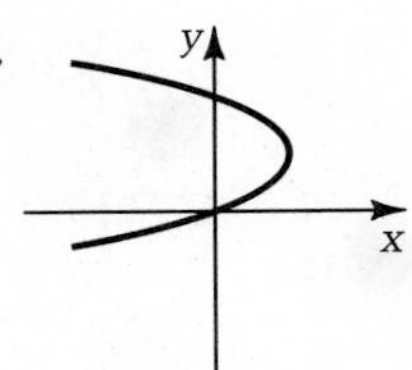

16.

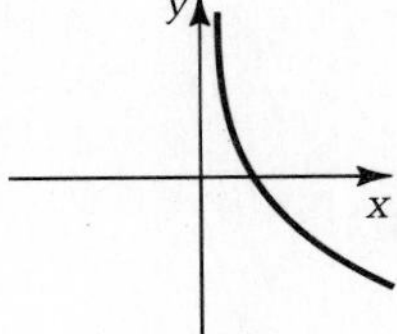

17.

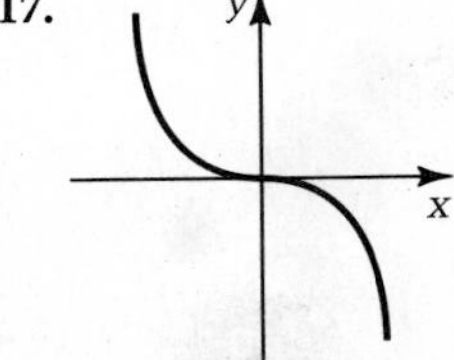

18.

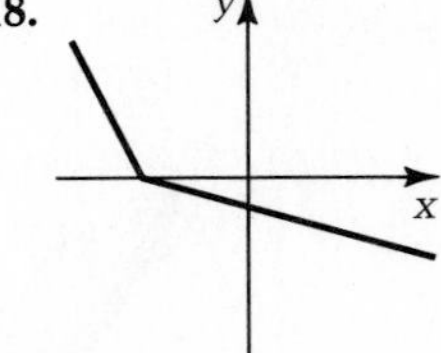

19. 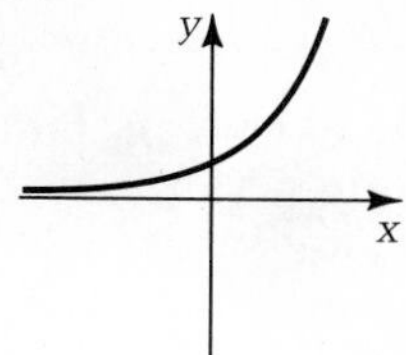

20. 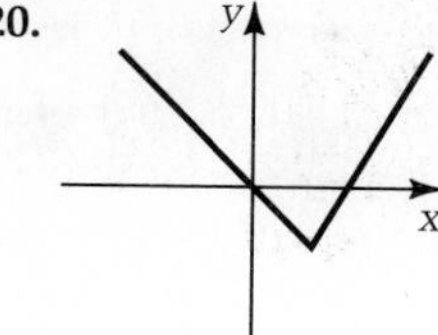

21. 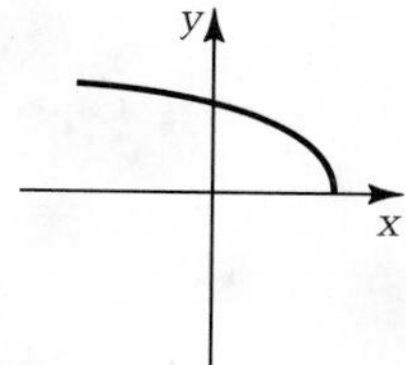

22. 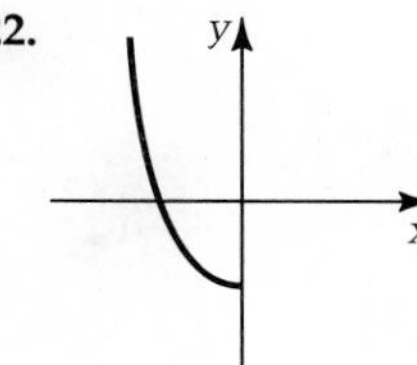

In Exercises 23–36, the given function $y = f((x)$ is one-to-one. Find an equation for the inverse function f^{-1} and specify the domain and range for f^{-1}.

23. $f(x) = 6 - 2x, 1 \le x \le 3$
24. $f(x) = \sqrt{3 + x}, -3 \le x$
25. $f(x) = x^2, 0 \le x \le 3$
26. $f(x) = 7 + 4x, -2 \le x \le 1$
27. $f(x) = \sqrt{4 - x}, x \le 4$
28. $f(x) = (x - 2)^2, -1 \le x \le 2$
29. $f(x) = 8(x - 1)^3, 0 < x < 3$
30. $f(x) = 16x^4, -1 \le x \le 0$
31. $f(x) = 4 - \sqrt{x}, 1 < x \le 9$
32. $f(x) = 3 + x^{2/3}, -1 \le x < 0$
33. $f(x) = x^{2/3} - 1, 1 \le x < 8$
34. $f(x) = \sqrt{x} - 2, 4 < x \le 9$
35. $f(x) = 7 - x^2, -3 \le x \le -1$
36. $f(x) = (3 - x)^3, 1 < x < 3$

In Exercises 37–46, determine whether the given function $y = f(x)$ has an inverse. If not, explain why not. If f^{-1} exists, find an equation, the domain, and the range for the inverse function.

37. $f(x) = |x - 2|, -3 \le x \le 1$
38. $f(x) = |x + 1|, -2 \le x \le 1$
39. $f(x) = |3 - 2x|, 1 \le x \le 4$
40. $f(x) = |8 - 3x|, 0 \le x \le 2$
41. $f(x) = \dfrac{x - 2}{2x + 1}, 0 \le x < 3$
42. $f(x) = \dfrac{3x - 2}{2x - 1}, 1 < x \le 4$
43. $f(x) = x^2 - 4x, -1 < x \le 4$
44. $f(x) = 6x - x^2, 0 \le x < 2$
45. $f(x) = x^2 + 8x + 7, -2 \le x \le 0$
46. $f(x) = x^2 + 2x - 3, -2 \le x \le 3$

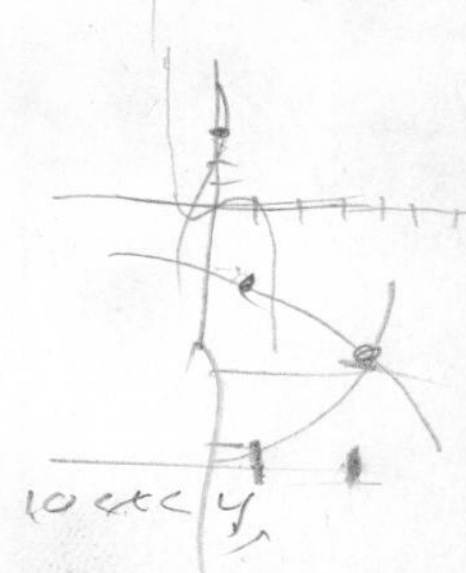

Superset

In Exercises 47–52, determine those real values of A for which the given function $y = f(x)$ is one-to-one.

47. $f(x) = x^2 + Ax + 1, -1 \le x \le 4$
48. $f(x) = x^2 - 4x + 3, -3 \le x \le A$
49. $f(x) = x^2 + 6x + 5, A \le x \le 2$
50. $f(x) = x^2 + Ax + 3, -2 \le x \le 6$
51. $f(x) = |3x + A|, -2 \le x \le 1$
52. $f(x) = |2x - 5|, 1 \le x \le A$
53. Determine those open intervals (a, b) on which the function $f(x) = \dfrac{1}{x}$ is its own inverse.
54. Other than the functions found in Exercise 53, are there any functions which are their own inverses on some open interval?

3.8 Interpreting Graphs

In this chapter you have learned a great deal about functions and how to graph them. At this point we will consider how graphs can be used to answer questions about the phenomenon that the function models. Your intuition for this skill will grow through experience with specific graphs, and so we will devote most of this section to examples. Of course we will point out important generalizations along the way.

EXAMPLE 1

Temperature (in degrees Fahrenheit) has been measured at a rocket launch site over a 24-hour period (from midnight Sunday to midnight Monday). The graph of temperature as a function of time is shown in Figure 3.73.

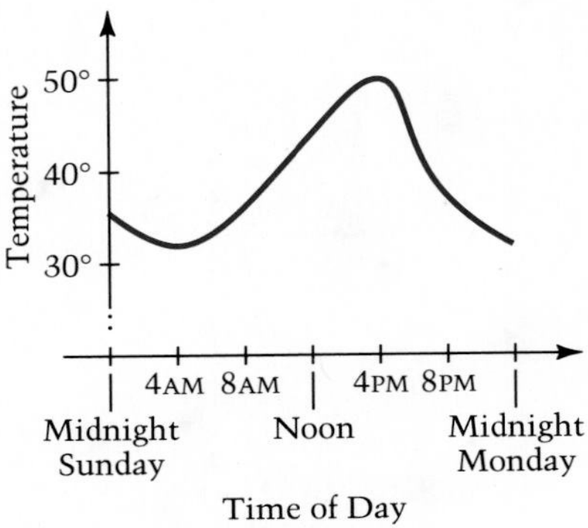

Figure 3.73

(a) What was the (approximate) temperature at noon?

(b) What was the temperature range (difference between the high and low temperatures) over the 24-hour period?

(c) At what time(s) over the 24-hour period was the temperature 40°?

(d) Over what time interval(s) was the temperature less than 40°?

(e) Over what time interval(s) was the temperature increasing? Over what time interval(s) was it decreasing?

Solution

(a) Find the point on the graph whose domain value is "noon" (look directly above "noon" on the time axis). The corresponding function value is the temperature directly to the left of this point. Thus the temperature at noon was (approximately) 44°. (See Figure 3.74(a).)

(b) We determine the high point and the low point on the graph, and find the corresponding temperatures.

$$\text{high temperature} \approx 50^\circ$$
$$\text{low temperature} \approx 32^\circ$$

The temperature range for the day was $50^\circ - 32^\circ = 18^\circ$ (Figure 3.74(b)).

(c) By drawing a horizontal line at the 40° mark, we find the points on the graph with a function value of 40°. The associated times (domain values) are found by looking directly below these points on the time axis. The

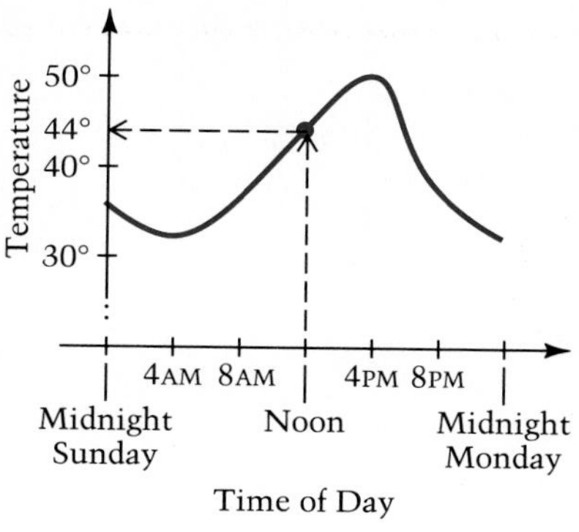

(a)

Temperature
50°
40°
32°
30°
high point
low point
4AM 8AM 4PM 8PM
Midnight Sunday
Noon
Midnight Monday
Time of Day

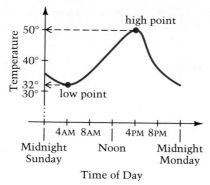

(b)

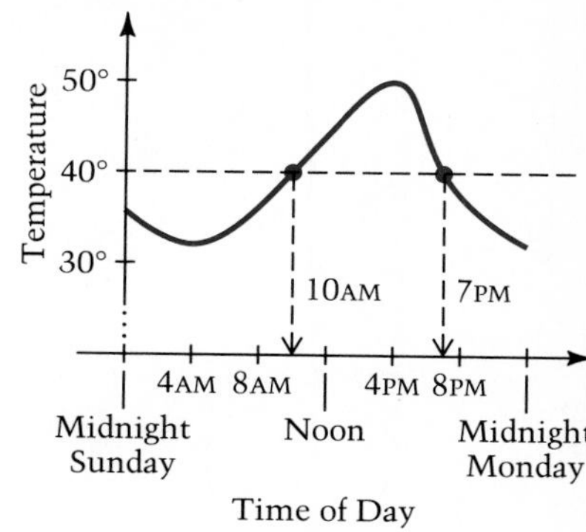

(c)

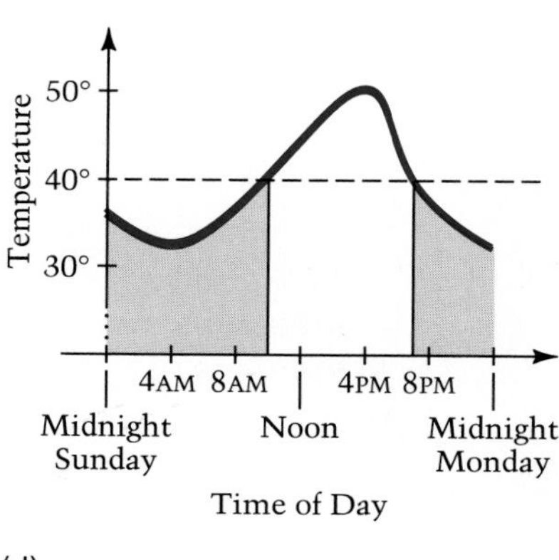

(d)

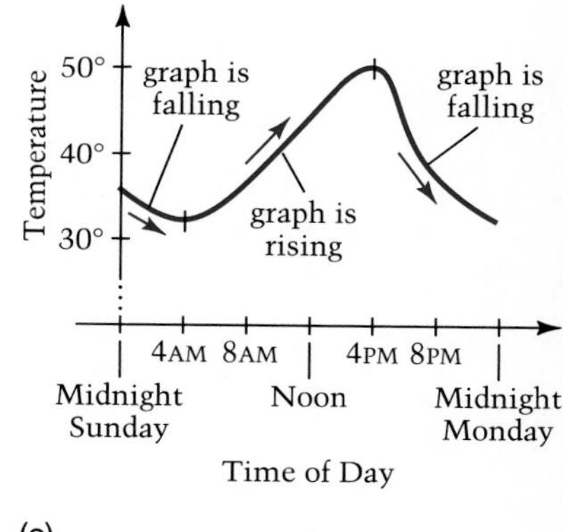

(e)

Figure 3.74

temperature hit 40° twice: first at approximately 10 A.M., then again at approximately 7 P.M. (Figure 3.74(c)).

(d) By observing when the graph is below the 40° line, we find the desired intervals. The temperature was below 40° from midnight Sunday till 10 A.M., and then again from 7 P.M. till midnight Monday (Figure 3.74(d)).

(e) We determine the interval when the temperature is increasing by observing those periods when the graph is rising as we follow it from left to right. Clearly the temperature is increasing on the interval from 4 A.M. to 4 P.M.. The temperature is decreasing on those intervals when the graph is falling as we follow it from left to right. Thus the temperature is decreasing on the intervals from midnight Sunday to 4 A.M., and then from 4 P.M. to midnight Monday (Figure 3.74(e)).

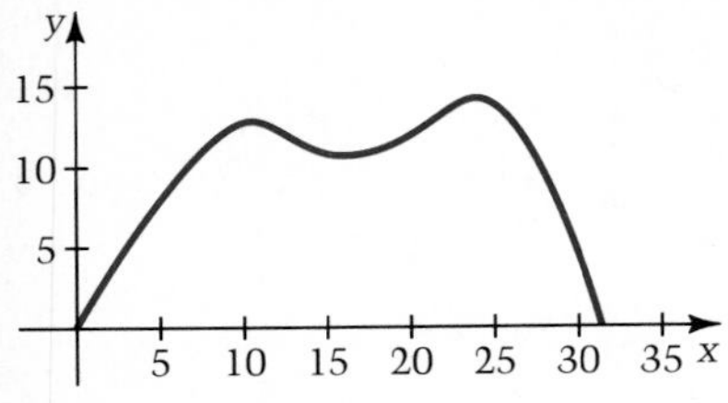

Figure 3.75

In Example 1 you observed that the temperature was increasing on those intervals when the graph was "heading upward" as we follow it from left to right. Similarly, the temperature was decreasing on those intervals when the graph was "heading downward." Note in Figure 3.75 that the function values (y-values) are increasing over both intervals (0, 10) and (16, 24). However, because the increase is "steeper" over the interval (0, 10), we say that the increase is more rapid there than over (16, 24). For similar reasons we say that the y-values are decreasing more rapidly over the interval (24, 31) than over (10, 16). We explore this notion further in the next example.

EXAMPLE 2

Suppose that we follow the motion of a car on a relatively straight road over a 30-minute period. The graph below represents this motion. Note that values on the d-axis represent distance from the starting point, and values on the t-axis represent elapsed time.

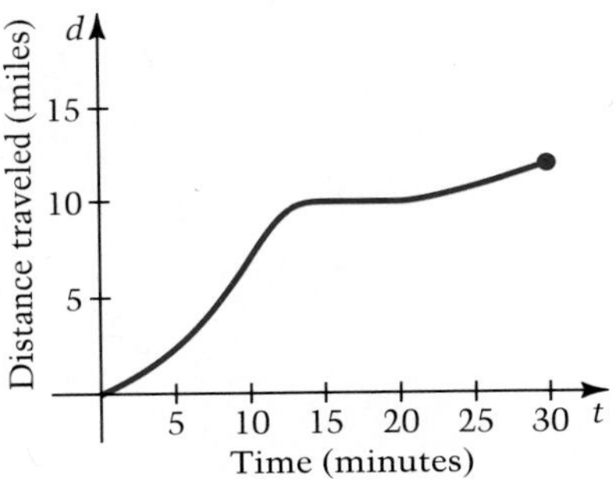

Figure 3.76

(a) How far did the car travel during the first 10 minutes?

(b) How far did the car travel during the last 10 minutes?

(c) What was the average velocity (in miles per hour) over the entire 30-minute interval? Over the first 10 minutes? Over the last 10 minutes?

(d) What was the car doing when $t = 15$?

Solution

(a) When $t = 10$, we see that $d = 7$. So the distance traveled was 7 miles.

(b) The last 10 minutes occur between $t = 20$ and $t = 30$. So, the car traveled $12 - 10 = 2$ miles during the last 10 minutes. Perhaps the road was bumpy, or the traffic was heavy.

t	d
20	10
30	12

(c) During the 30-minute interval the car traveled 12 miles. Recall that

$$\text{distance} = \text{velocity} \cdot \text{time},$$

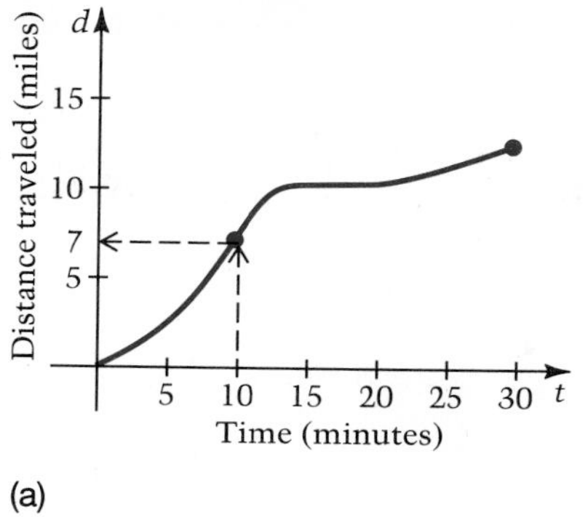

(a)

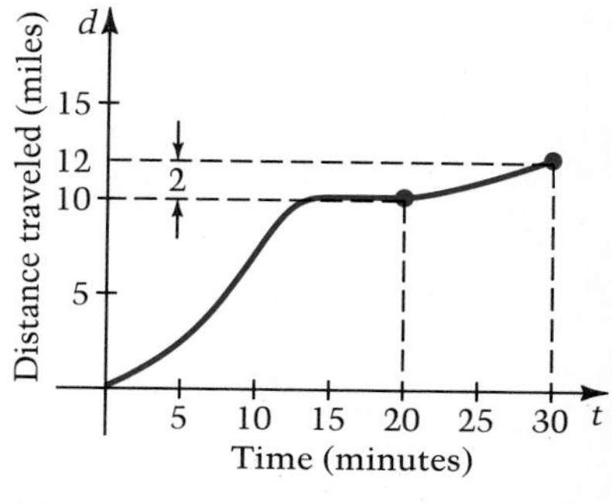

(b)

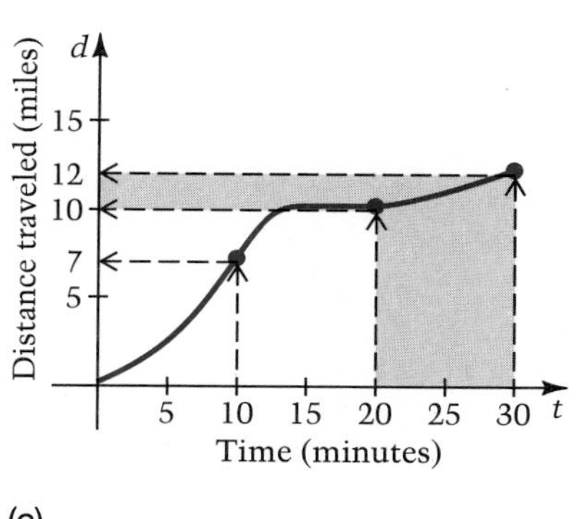

(c)

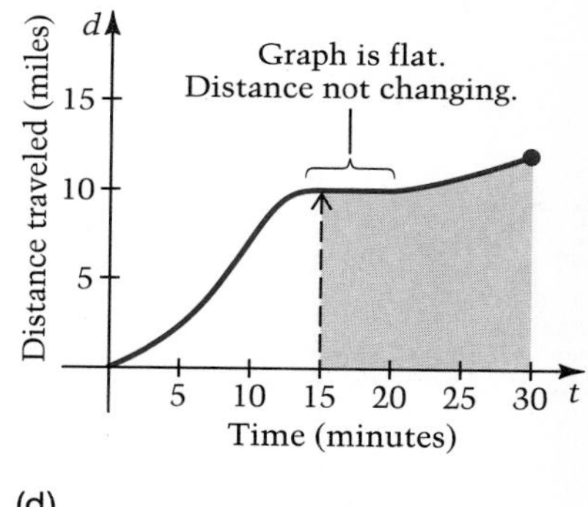

(d)

Figure 3.77

where "velocity" refers to the average velocity over the time period. So,

$$\text{velocity} = \frac{\text{distance}}{\text{time}} = \frac{12 \text{ mi}}{0.5 \text{ hr}} = 24 \text{ mph.}$$

Over the first 10 minutes, the car traveled approximately 7 miles, so

$$\text{velocity} = \frac{\text{distance}}{\text{time}} = \frac{7 \text{ mi}}{\frac{1}{6} \text{ hr}} = 42 \text{ mph.}$$

Over the last 10 minutes the car traveled approximately 2 miles, so

$$\text{velocity} = \frac{\text{distance}}{\text{time}} = \frac{2 \text{ mi}}{\frac{1}{6} \text{ hr}} = 12 \text{ mph.}$$

(d) When $t = 15$ min, the car is not moving. It appears that over the 6 minute interval from approximately $t = 14$ to $t = 20$ the car was stopped. Perhaps there was road construction, diminished visibility, or an accident.

In parts (c) and (d) of Example 2, we computed velocities. Notice that the average velocity was greater over the first 10 minutes than over the last 10 minutes. This can be determined by comparing the steepness of the curve over those intervals. The steeper the increase in a distance/time graph, the greater the velocity. This follows because

$$\text{velocity} = \frac{\text{distance}}{\text{time}}$$

and it is precisely the steepness that measures this ratio of y-values to x-values. For example, in Figure 3.78, car A is traveling faster than car B at time t_1. This is clear because the graph for car A is rising more steeply at time t_1 than is the graph for car B at that time.

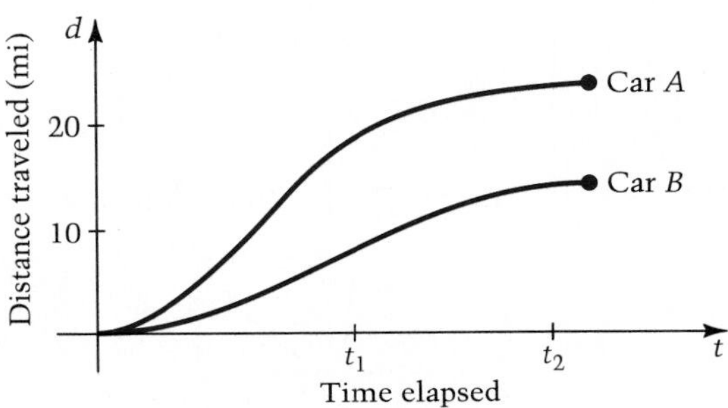

Figure 3.78

To compare velocities it sometimes helps to draw line segments that approximate the curve at the points of interest. The line segments have the advantage that they have slopes which are easy to compare. For example, in Figure

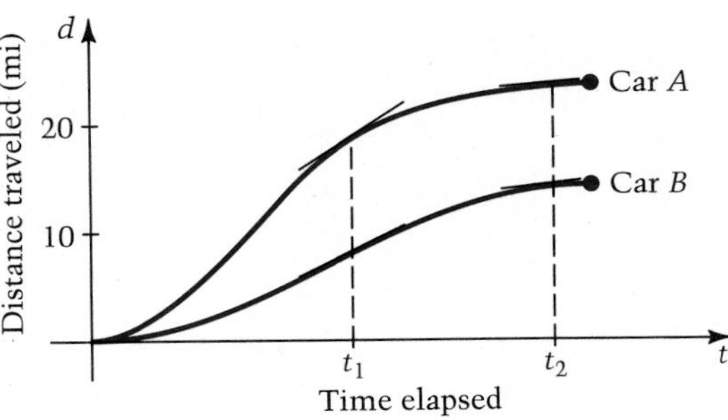

Figure 3.79

3.79 it is clear that the segment for car A at time t_1 has a steeper slope than the segment for car B at time t_1. On the other hand, at time t_2, the two segments look almost parallel, signifying that the two cars have roughly the same velocities at that time. However, note that car A's higher speed during the first part of the trip has accounted for a ten mile lead by time t_2, as represented by the vertical distance between the two graphs at that time.

EXERCISE SET 3.8

1. The graph below shows the depth (in inches) of snow on the ground (during a recent snowstorm) as a function of the time of day for a 24-hour period from noon on Monday until noon on Tuesday.
(a) How much snow was on the ground at noon on Monday before this storm began?
(b) How much snow was on the ground by midnight?
(c) At what time (approximately) did the storm end?
(d) How much snow fell during the storm?
(e) At what time (approximately) was the snow accumulating at the quickest rate?

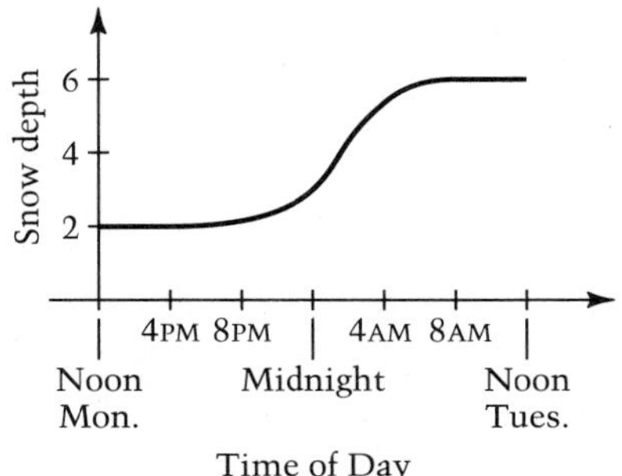

2. The local agricultural co-op maintains a record of rainfall during a growing season that lasts 70 days. The graph below shows the cumulative rainfall (in inches) for last season as a function of time t (in days) since the season began.
(a) During which of the 10 weeks (days 0–7, 7–14, 14–21, etc) did the most rain fall?
(b) During which week did virtually no rain fall?
(c) After how many days (approximately) was the cumulative rainfall for the season about 3 inches? How many days (approximately) did it then take for the next 3 inches of rain to fall?
(d) On which day (approximately) did the rain fall most heavily?

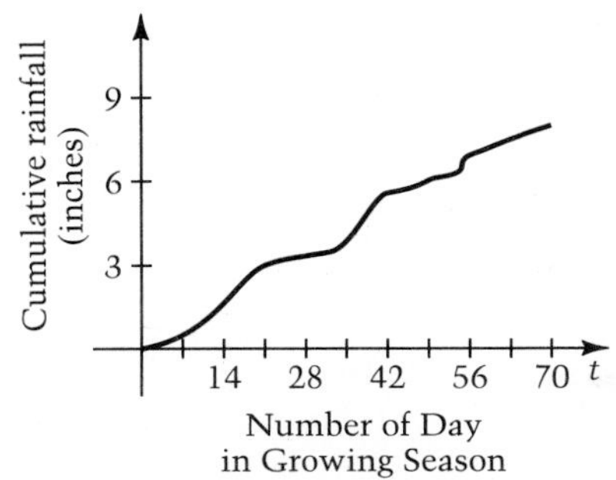

3. Fuel economy tests on a new automobile engine determined the number of gallons of gasoline the engine will consume when operated for an hour at constant speed. The graph below reflects the results of these tests with the number of gallons consumed as a function of the rate (mph) at which the engine was operated.
(a) How many gallons will the engine consume in an hour if the engine is run at 30 mph.
(b) How many gallons will the engine consume in 30 minutes if the engine is run at 40 mph?
(c) At what constant rate is the engine run if it consumes 2 gallons in an hour?
(d) How many gallons of gasoline would this engine consume on a 100-mile trip if a constant speed of 40 mph was maintained for the entire trip?
(e) For a 100 mile trip would it be more economical

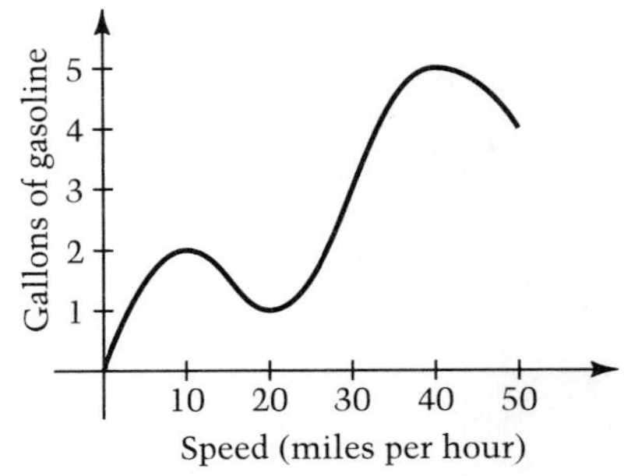

(fewer gallons consumed) to maintain a constant speed of 20 mph or of 50 mph? of 30 mph or of 50 mph?

4. The graph below shows the altitude (in feet) of a hot air balloon during its 2-hour flight as a function of the elapsed time t (in minutes) since the balloon was launched.
   (a) What was the altitude of the balloon 30 minutes after it was launched?
   (b) During which time interval(s) (approximately) was the balloon more than 500 feet above the ground?
   (c) How many minutes (approximately) after it was launched did the balloon attain its maximum altitude?
   (d) During which time interval(s) (approximately) was the balloon descending?
   (e) Approximately how many minutes into its flight was the balloon ascending at the fastest rate of the entire flight?

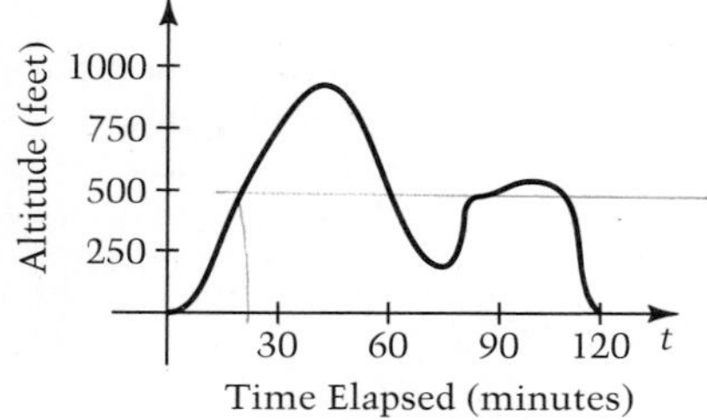

5. Monitoring equipment at certain unstaffed meterological stations are activated whenever wind velocities reach 40 mph. The graph below shows the recorded average wind velocities (rather than occasional gusts) during a 30-hour period as a hurricane passed through the vicinity of one of these stations. The graph shows wind velocity (in mph) as a function of elapsed time t (in hours) since the first winds of 40 mph were detected.
   (a) When (approximate value of t) did the station first record winds of 80 mph?

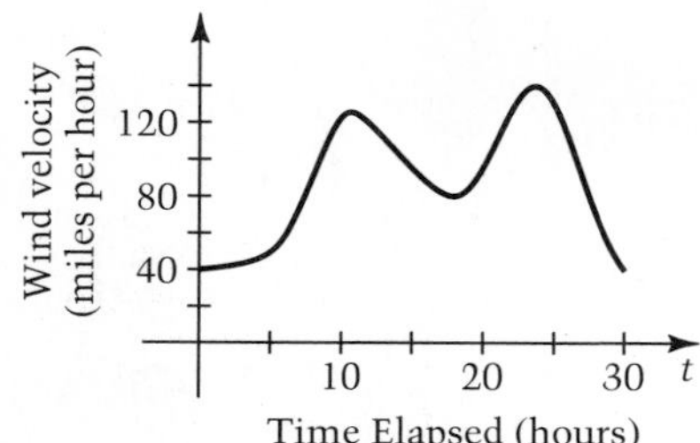

   (b) During which time interval(s) (approximately) did the station measure winds with velocity in excess of 100 mph?
   (c) During which time interval(s) (approximately) was wind velocity increasing?
   (d) How many hours after the station began recording data on this storm were the peak winds (maximum velocity) recorded?

6. The graph below shows the number of calories that a person burned as a function of elapsed time t (in minutes) during a 30-minute workout on a stationary bicycle. At various times the person adjusted the resistance of the bicycle and, thus altered the rate at which calories were expended.
   (a) How many calories had the person expended after the first 10 minutes?
   (b) How many minutes (approximately) passed before the person had expended 32 calories?
   (c) How many more minutes (approximately) did it take for the person to expend another 32 calories?
   (d) During which time interval was the workout the most beneficial (greatest rate of calories expended per minute)?
   (e) During which time interval did the person apparently cheat by not pedaling?

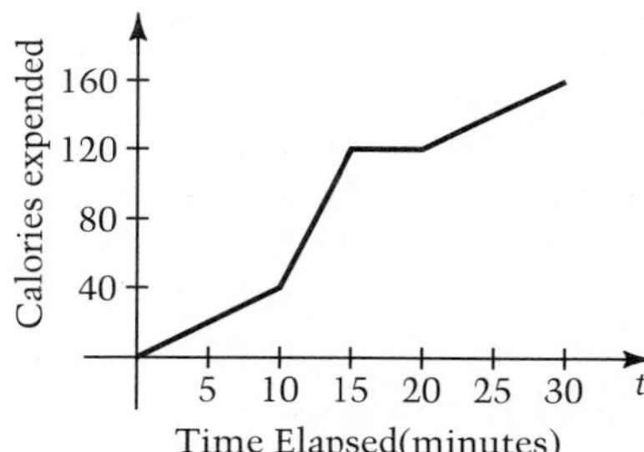

7. The following graph shows the weight (in pounds) of a newborn who was brought to a pediatrician once every 3 months for the first 18 months following birth.
   (a) What was the child's weight at birth?
   (b) How many months old (approximately) was the child before the weight was at least 15 pounds?
   (c) During which 3-month span (0–3, 3–6, 6–9, etc) did the child have the greatest weight gain?
   (d) Was there any 3-month span (0–3, 3–6, etc) when the child's weight increased by at least 50%?
   (e) The child gained 2 pounds during each of the fourth and fifth 3-month spans. Was the percentage

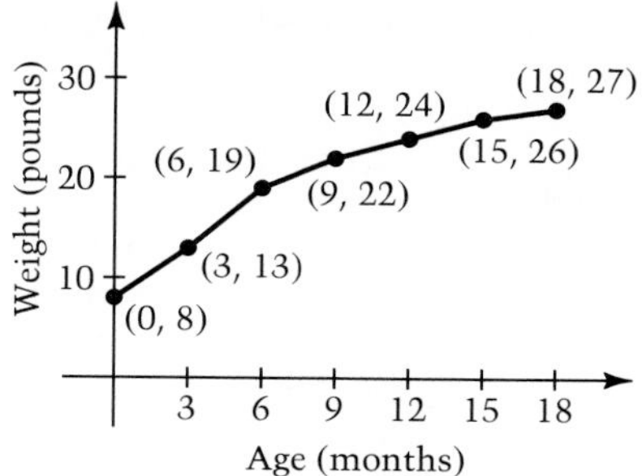

gain in weight during the 9–12 month span the same as, more than, or less than the percentage gain in weight during the 12–15 month span?

8. The graph below shows the height (actually 'length' for infants) in inches for the child whose weights were displayed in Exercise 7.
(a) What was the child's length (height) at birth?
(b) How many months old (approximately) was the child before the length was at least 30 inches?
(c) During which 3-month span (0–3, 3–6, 6–9, etc) did the child's length increase the most?
(d) Was there a 6-month span (0–6, 3–9, 6–12, 9–15, or 12–18) during which the child's length did not increase by at least 10%?
(e) During the first three months of life, this child's length increased as much as it did during the last 9 months recorded on the graph. Was the percentage gain in the child's length between the 9th and 18th months the same as, more than or less than the percentage gain in length during the first three months of life?

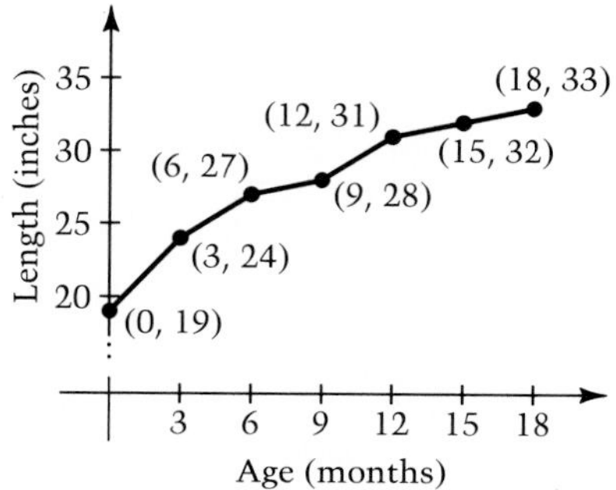

9. A weight is suspended at the end of a spring. The weight is pulled down and then released causing the spring to contract and expand as the weight bounces up and down. The graph below shows the length of the spring (in centimeters) as a function of the elapsed time t (in seconds) after the weight is released.
(a) What was the length of the spring when the weight was released?
(b) For the 20 seconds of motion shown by the graph, at what times (approximately) was the spring 15 centimeters long?
(c) During which time intervals was the length of the spring decreasing (weight moving upward)?
(d) At which instants was the spring shortest (minimum length)?

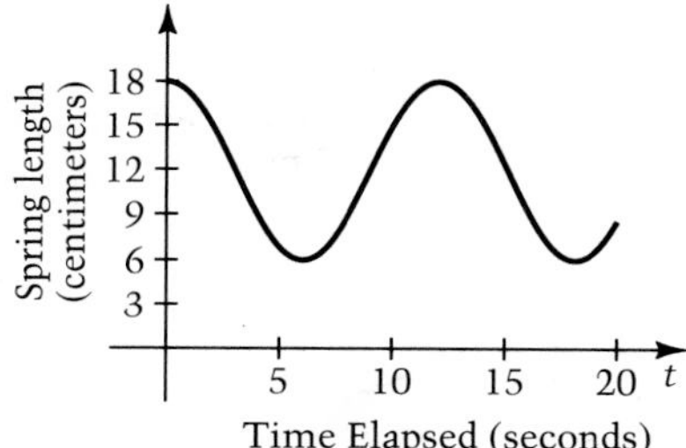

10. The graph below shows the official time for sunset in a part of the US that does not employ Daylight Savings Time. Each mark on the horizontal axis corresponds to the first day of a month.
(a) During which month did the longest day of the year occur?
(b) Within 10 days, estimate the first day of the year that the sun set after 7 P.M.
(c) Assuming 30 days per month, for how many days (within approximately 10) did the sun set earlier than 6 P.M.?
(d) Within 10 minutes, estimate the difference between the earliest and the latest that the sun set during the year.

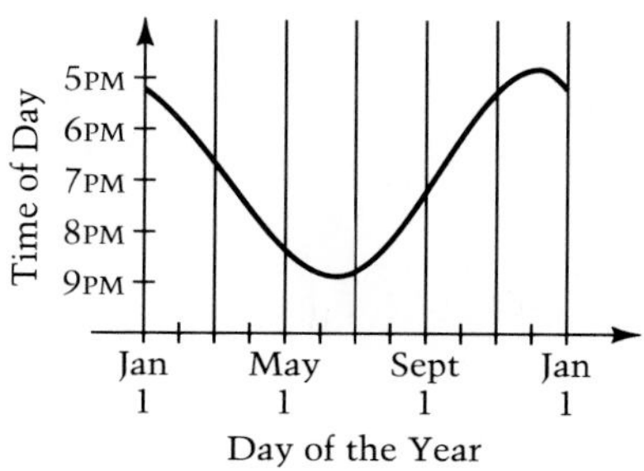

11. The graph below shows the distance (in miles) a marathon runner has run as a function of elapsed time t (in hours).

(a) What is the total distance that the individual ran?
(b) How far did the runner run during the second hour?
(c) What was the average velocity (in mph) of the runner during the first two and one-half hours?
(d) During which 30-minute interval (0–30, 30–60, etc) did the runner maintain the greatest speed?
(e) For how many hours (approximately) did the runner actually run?

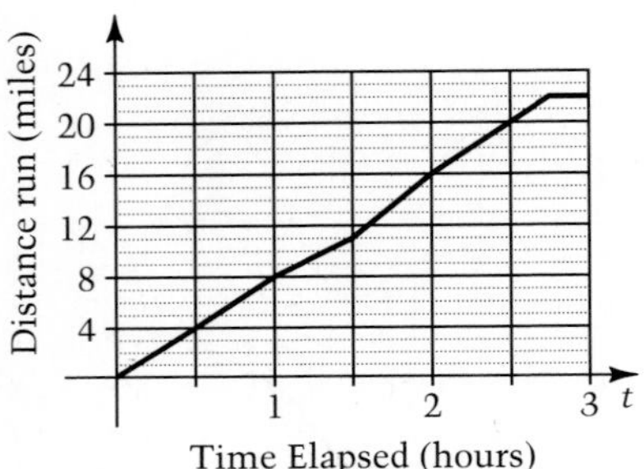

12. Eight years ago you purchased and planted a Hybrid Poplar seedling from the Michigan Bulb Co. Each year on the tree's anniversary you measured the height of the tree (in feet). Your data is represented in the graph below as a function of time t (in years).

(a) What was the height of the tree (seedling) when you planted it?
(b) How much taller was the tree after the first year?
(c) It was claimed that the tree would grow 4′–6′ for each of the first five years. Identify the year(s) when this did not happen.
(d) It was claimed that the tree would reach a height of 20′–30′ in 5 years. Did this happen?
(e) Were there two successive years during which the tree grew by the same amount during each year?

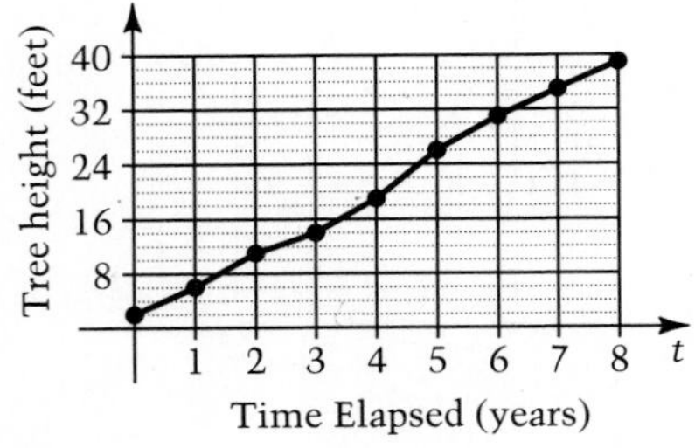

For Exercises 13–16 see the graph below. It is approximately 500 miles from Boston, MA to Buffalo, NY. One afternoon three planes (A, B, and C) begin the flight from Boston to Buffalo, leaving Boston at different times and arriving in Buffalo before midnight that night. The graph shows the distance between the Boston airport and each of the three planes as a function of the time of day from noon until midnight.

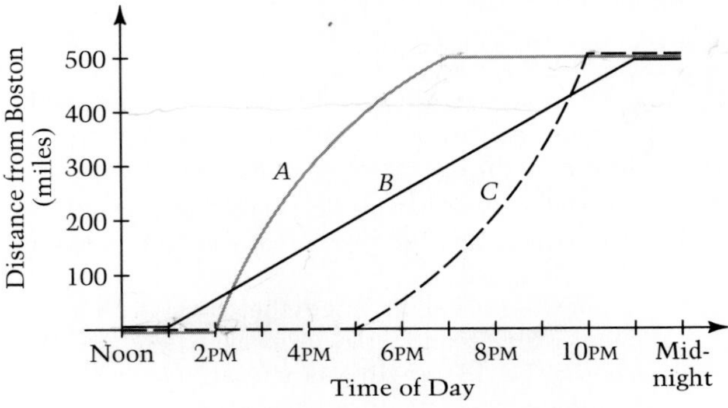

13. (a) Which plane left Boston first?
(b) Which plane's airspeed increased throughout its flight?
(c) At what time of day was plane B approximately 400 miles from Buffalo?
(d) At what time did plane C pass plane B?
(e) During which time interval(s) was precisely one plane airborne?

14. (a) Which plane landed in Buffalo first?
(b) Which plane's airspeed decreased throughout its flight?
(c) At what time of day was plane A approximately 300 miles from Boston?
(d) At what time did plane A pass plane B?
(e) During which time interval(s) were precisely two planes airborne?

15. (a) Which plane(s) reached Buffalo by 10 P.M.?
(b) Which plane flew at a constant speed throughout its flight?
(c) At what time of day was plane B approximately 200 miles from Boston?
(d) Which plane(s) were airborne for at least 6 hours?
(e) During which time interval(s) was plane C closer to Boston than plane B?

16. (a) Which plane(s) were still on the ground in Boston at 4 P.M.?

(b) All three planes were airborne at 6 P.M. At that time which plane was flying the fastest?
(c) At what time of day was plane C approximately 400 miles from Buffalo?
(d) Which plane had the shortest flying time?
(e) During which time interval was plane A closer to Buffalo than plane B?

17. An investor invested \$1000 in each of three stocks at the same time. About a month later the investor began receiving daily reports from her broker on the status of each investment. The graph below shows the dollar value of each investment (A, B, and C) as a function of time t (in days) beginning with the broker's first report and continuing for the next 100 market days.
(a) Which investment had decreased in value the most by the time of the broker's first report?
(b) Which investment fluctuated in value the most wildly throughout the 100 days?
(c) Which investment grew in value at a constant rate throughout the 100 days?
(d) Which investment showed the greatest change in value during the last 50 days?
(e) Which investment gained the most in value during the last 50 days?

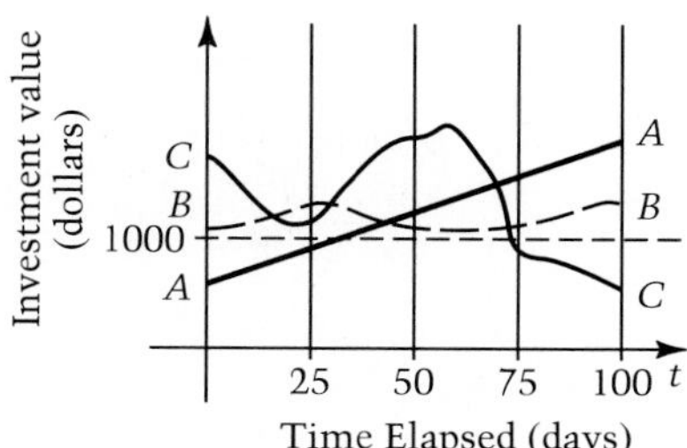

18. The accompanying graph shows the temperature and the relative humidity during a particular 24-hour period as a function of the time of day from midnight to midnight. The vertical axis measures temperature in degrees Fahrenheit and relative humidity in percent using the same numerical scale.
(a) What was the highest temperature attained?
(b) During which time interval(s) did both the temperature and the relative humidity increase?
(c) For how many total hours during this 24-hour period did the temperature decrease?
(d) What was the difference between the highest and the lowest readings of relative humidity?
(e) Assuming an increase in relative humidity coupled

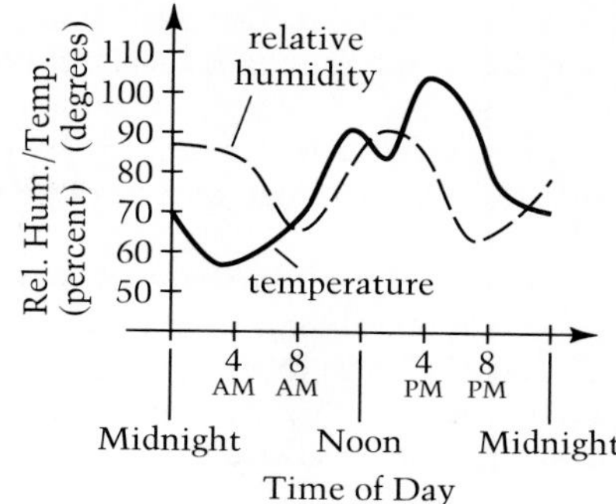

with a decrease in temperature correspond to a likely period of rain (or showers), during which time interval(s) did it likely rain?

Superset

19. One of the factors that affects the rate at which a given volume of water will evaporate is the surface area of the water that is exposed to the air. Thus, a given volume of water will evaporate more quickly from a flat open dish than from a tall, thin cylindrical container. Each of three containers is filled at the same time with 8 ounces of water which is then allowed to evaporate. The graphs below show the volume of water (in ounces)

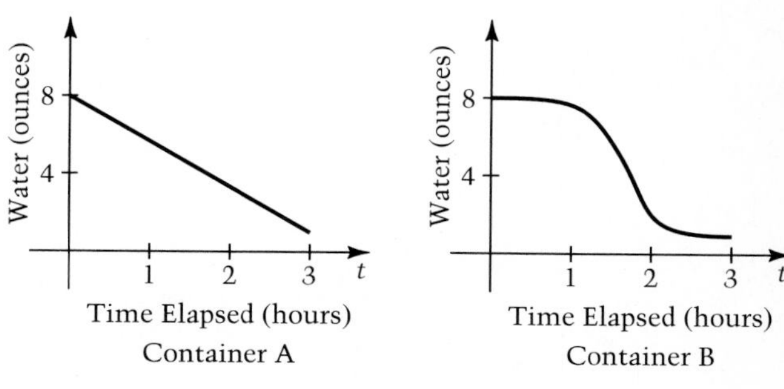

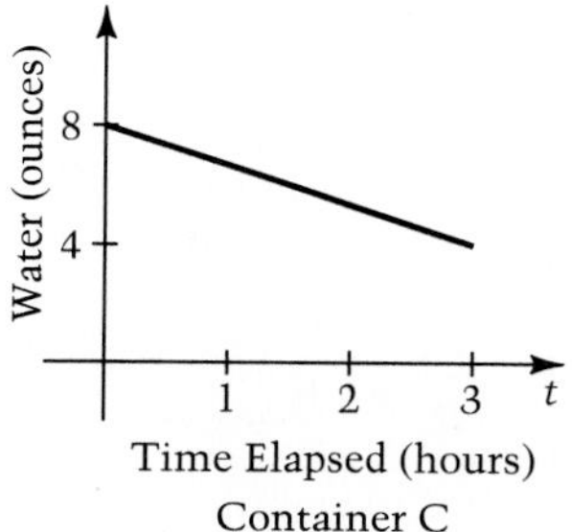

remaining in each of the three containers as a function of the elapsed time t (in hours) since the containers were filled. One of the containers is not a circular cylinder, the other two are. One of the circular cylinders is shorter than the other one, and thus has a greater radius than the other one.

(a) Which container is not a circular cylinder?
(b) Which container is the shorter circular cylinder?
(c) Which container has the most water left in it after 3 hours?
(d) From which container has the most water evaporated after 1 hour?

In Exercises 20–28, a common situation is described as a function of time beginning at a certain instant. From among the graphs given below, select the graph which best represents the function described.

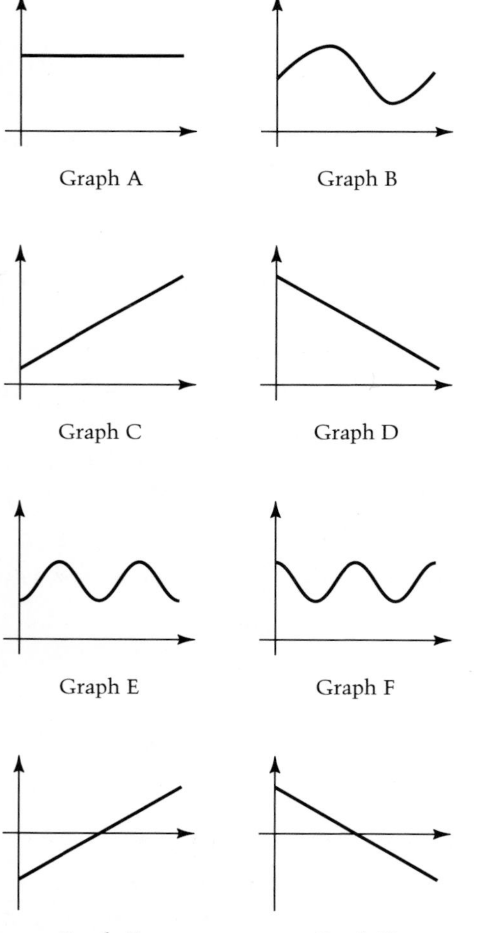

20. The height above the ground of a child on a playground swing between successive pushes.

21. The height of a missile launched from a submerged submarine since the time of the launch.

22. The temperature of a can of soda since the can was removed from the refrigerator and left standing on a counter.

23. The volume of air in a person's lungs as the person inhales and exhales twice.

24. The height of the grass in a lawn since the lawn was last mowed.

25. The height of the tides at an ocean beach during a 12-hour period.

26. The amount of money in a checking account that begins in good standing and is then overdrawn.

27. The distance between the sun and the earth during any 24-hour period.

28. The height above the ground of a fly ball in baseball since the ball was struck by the bat.

In Exercises 29–36, the distance of a highwire performer from one end of the highwire ($D = 0$) is represented by the given graph as a function of time during the one minute that it takes the performer to go from one end of the highwire to the other end ($D = 40$). Note that the performer may pause one or more times to execute a trick or to turn and change direction. Sketch a graph that represents the velocity of the performer along the highwire as a function of time as the performer moves from one end ($D = 0$) to the other end ($D = 40$) of the highwire.

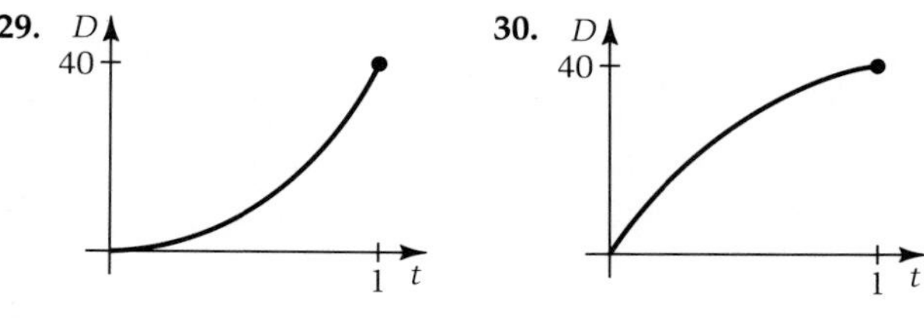

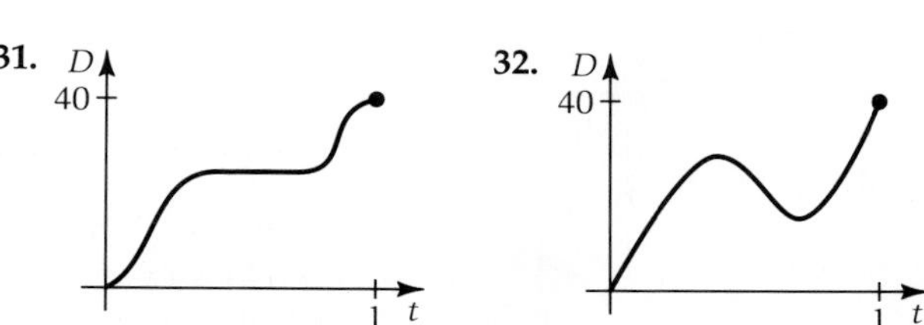

33. 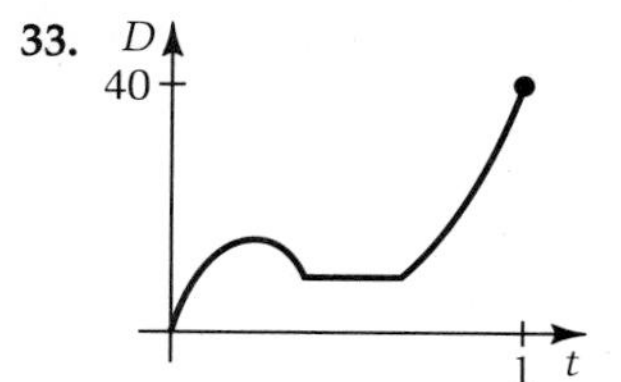

34. 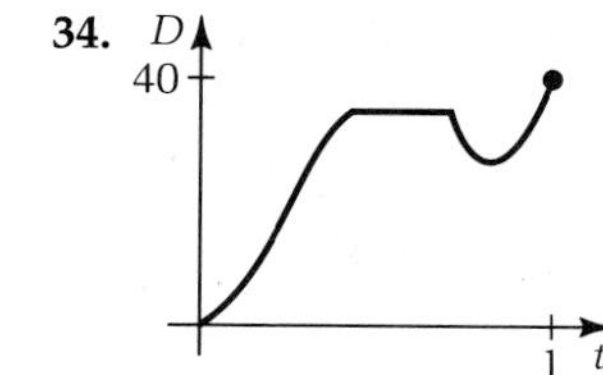

35. 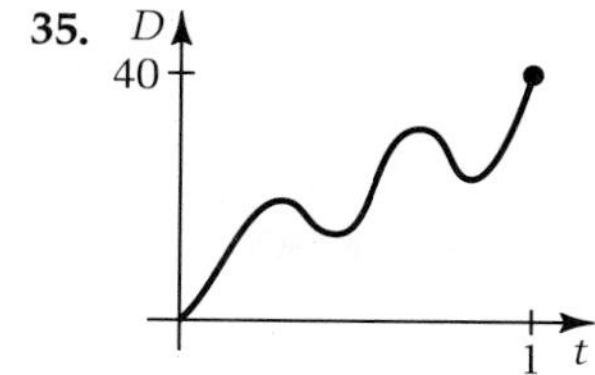

36. 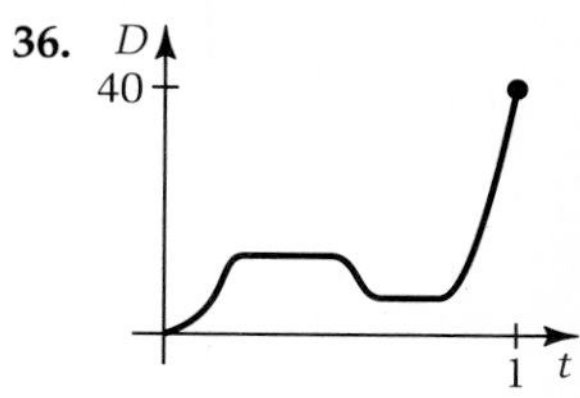

3.9 Variation

Sometimes the mathematical models that represent situations or processes in the world can be quite complex. However, a surprisingly large number of situations can be modeled by the simple though powerful notion of variation (or proportionality). Consider the following:

1. The pressure in your ears while you swim underwater is a function of the depth at which you are swimming.
2. The intensity of light on your desk top is a function of the square of the distance between your desk and the light source.
3. The distance you continue to travel after applying your foot to the brake pedal is a function of the square of the speed at which you are traveling.

All of these functional relationships are based on the fact that as one quantity varies, some other quantity varies in a predictable way.

The first type of variation we consider is called *direct variation*.

Definition of Direct Variation

If x and y denote two quantities, we say that y is **directly proportional** to x, or **y varies directly as x,** if there is a nonzero constant k such that

$$y = kx.$$

We refer to k as the **constant of proportionality.**

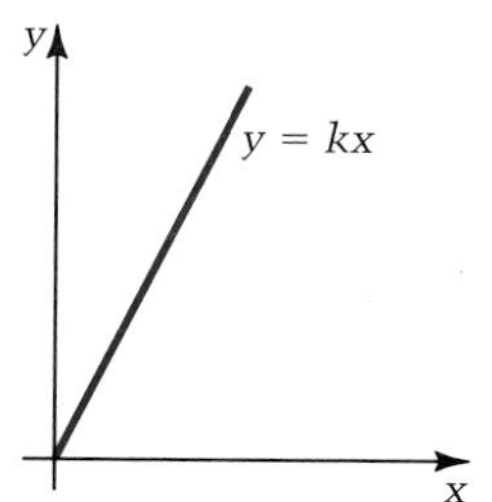

Figure 3.80 The graph of a direct proportion ($k > 0$).

As shown in Figure 3.80, the graph of a direct proportion is a straight line that passes through the origin. Thus, in a given application we can determine the constant of proportionality k by simply knowing one specific pair of nonzero values for x and y.

EXAMPLE 1

The pressure in your ears when you swim underwater is directly proportional to the depth at which you are swimming. At 20 feet the pressure is measured to be approximately 8.6 pounds per square inch (psi). Determine the equation that describes pressure in terms of depth, and use it to find the pressure at 125 feet.

Solution Let p = the pressure (in psi) and let d = the depth (in feet).

$p = kd$ — p is directly proportional to d.

$8.6 = k \cdot 20$ — The given pair of values, $p = 8.6$ and $d = 20$, is substituted to find k.

$\frac{8.6}{20} = k$

$k = 0.43$

$p = 0.43d$ — This is the equation describing pressure in terms of depth.

$p = (0.43)(125)$ — 125 is substituted for d to find the pressure at 125 feet.

$p = 53.75$

The desired equation is $p = 0.43d$, and the pressure at 125 feet is 53.75 psi.

There are three important notes regarding direct variation:

1. If two quantities are directly proportional, then as one quantity takes on larger values, the other quantity also takes on larger values.
2. If y is directly proportional to x, then the *quotient* $\frac{y}{x}$ is always equal to the constant k.
3. The statement "y is proportional to x" is a shorthand way of indicating that y is directly proportional to x.

There are many situations where a quantity y varies directly with a power of x. That is

$$y = kx^n,$$

where n is any positive real number. For example, if we ignore air resistance, the distance d an object falls (from rest) varies directly as the square of the amount of time t it falls:

$$d = kt^2.$$

If we know that an object fell 144 feet in 3 seconds, we can use that information to determine k:

$$144 = k(3)^2$$

$$\frac{144}{9} = k$$

$$16 = k$$

Thus the equation describing such motion is

$$d = 16t^2.$$

Inverse Variation

A second type of variation is referred to as *inverse variation*.

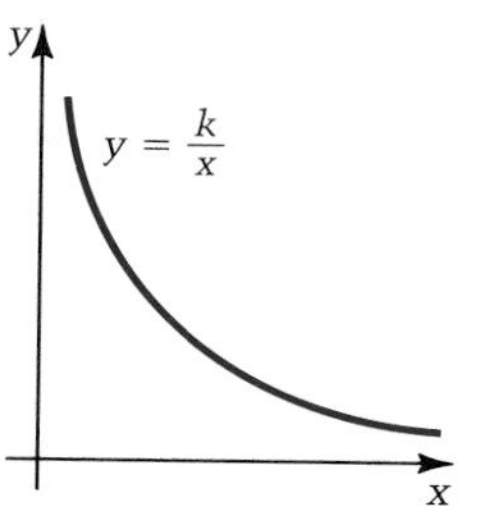

Figure 3.81 The graph of an inverse proportion ($k > 0$).

Definition of Inverse Variation

If x and y denote two quantities, we say that y is **inversely proportional** to x, or y **varies inversely** as x, if there is a nonzero constant k such that

$$y = \frac{k}{x}.$$

Figure 3.81 shows that with an inverse proportion, as one quantity gets larger, the other quantity gets smaller. Also note that the product xy is always equal to the constant k.

Just as with direct variation, we can talk about a quantity y varying inversely as a power of x, and describe this relationship as

$$y = \frac{k}{x^n}.$$

EXAMPLE 2

Suppose y is inversely proportional to x^3. If $y = 12$ when $x = 2$, determine the value of y when $x = 4$.

$y = \frac{k}{x^3}$ — y is inversely proportional to x^3.

$12 = \frac{k}{8}$ — The given pair of values, $y = 12$ and $x = 2$, is substituted to determine k.

$k = 96$

$y = \frac{96}{x^3}$ — Equation of proportionality is rewritten using value for k.

$y = \frac{96}{(4)^3}$ — $x = 4$ is substituted.

$= \frac{96}{64} = \frac{3}{2}$

When $x = 4$, the value of y is $\frac{3}{2}$.

Sometimes a statement of variation might combine the two types of proportionality defined above. Accordingly, we refer to this possibility as **combined variation.** The term **joint variation** is usually reserved for the situation where one variable depends on the product of two or more expressions.

EXAMPLE 3

Express each statement as an equation using k as the constant of proportionality.

(a) y varies jointly as x^2 and z.

(b) r is directly proportional to s^3 and inversely proportional to t.

(c) m is proportional to a and b and varies inversely as c^4.

(d) y varies jointly as A and the square of d, and varies inversely as the cube of s.

Solution (a) $y = kx^2z$ (b) $r = \dfrac{ks^3}{t}$ (c) $m = \dfrac{kab}{c^4}$ (d) $y = \dfrac{kAd^2}{s^3}$

EXERCISE SET 3.9

In Exercises 1–10, express the statement as an equation using k as the constant of proportionality.

1. y is proportional to x^2.
2. x varies inversely as y^3.
3. F is inversely proportional to $R^{3/2}$.
4. D is proportional to C^2.
5. m varies jointly as p and q^2.
6. n varies directly as p^2 and inversely as q and s.
7. a is directly proportional to b^2 and inversely proportional to c^3.
8. z is inversely proportional to x and $t^{1/3}$.
9. y varies directly as x^2 and z^3 and inversely as w.
10. p varies jointly as r^2 and s.

In Exercises 11–18, determine the value of the constant of proportionality. Write the equation of proportionality; then find the specified value.

11. y varies inversely as x. If $y = 8$ when $x = 3$, find y when $x = 5$.
12. x is directly proportional to y^2. If $x = 9$ when $y = 2$, find x when $y = 3$.
13. m varies jointly as p and q^2. If $m = 36$ when $p = 2$ and $q = 3$, find m when $p = 3$ and $q = 2$.
14. z varies inversely as w^2 and x. If $z = 6$ when $w = 2$ and $x = 3$, find z when $w = 3$ and $x = 2$.
15. a is directly proportional to b^2 and inversely proportional to c. If $a = 48$ when $b = 8$ and $c = 2$, find a when $b = 4$ and $c = 9$.
16. m is directly proportional to p and inversely proportional to n^2. If $m = 3$ when $p = 36$ and $n = 2$, find m when $p = 12$ and $n = 4$.
17. x is proportional to y^3 and varies inversely as w and z^2. If $x = 9$ when $y = 18$, $w = 12$ and $z = 6$, find x when $y = 6$, $w = 4$ and $z = 9$.
18. w varies directly as x^2 and inversely as z and y. If $w = 2$ when $x = 6$, $z = 3$ and $y = 4$, find w when $x = 4$, $z = 2$ and $y = 3$.

In Exercises 19–22, we work with the result of the Italian astronomer Galileo (1564–1642), that says the distance s through which an object "free falls" from rest is directly proportional to the square of the time t it falls. For an object in free fall we assume the gravitational pull of the earth is constant and there is no air resistance.

19. If an object falls 400 feet in 5 seconds, how far will it fall in 10 seconds?

20. If an object falls 490 meters in 10 seconds, how far will it fall in 20 seconds?

21. If an object falls 1960 centimeters in 2 seconds, how far will it fall in 3 seconds?

22. If an object falls 768 inches in 2 seconds, how far will it fall in 3 seconds?

In Exercises 23–26, we work with Hooke's law for a spring: the distance s a spring is stretched beyond its natural length is proportional to the force F applied.

23. If a force of 15 pounds stretches a spring 2 inches beyond its natural length, what force is required to stretch the spring 6 inches beyond its natural length.

24. If a force of 10 pounds stretches a spring 8 inches beyond its natural length, what force is required to stretch the spring one foot beyond its natural length?

25. If a force of 12 pounds stretches a spring 5 inches beyond its natural length, how far beyond its natural length will the spring be stretched by a force of 80 ounces?

26. If a force of 18 pounds stretches a spring 10 inches beyond its natural length, how far beyond its natural length will a spring be stretched by a force of 15 pounds?

Superset

In Exercises 27–32, the table of values is determined by an equation of proportionality of the form $y = kx^a$, where k is the constant of proportionality and a is an integer (positive or negative). Determine the equation.

27.

x	3	6	9	12	15
y	2	4	6	8	10

28.

x	2	4	6	8	10
y	2	8	18	32	50

29.

x	-2	-1	1	2	3
y	16	2	-2	-16	-54

30.

x	-2	-1	1	2	3
y	9	18	-18	-9	-6

31.

x	-1	1	2	3	4
y	-12	12	6	4	3

32.

x	-2	-1	1	2	3
y	-10	-5	5	10	15

In Exercises 33–38, the table of values is determined by an equation of proportionality of the form $z = kx^a y^b$, where k is the constant of proportionality and a and b are integers (positive or negative). Determine the equation.

33.

x	1	2	1	1	3	2
y	2	1	3	4	2	4
z	4	4	6	8	12	16

34.

x	-2	2	2	-3	2	3
y	2	-2	2	-2	3	3
z	4	4	-4	-6	-6	-9

35.

x	1	2	2	1	3	2
y	2	1	2	3	1	3
z	6	12	24	9	27	36

36.

x	1	2	1	3	2	3
y	2	1	3	1	3	2
z	3	12	2	18	4	9

37.

x	1	2	1	4	2	4
y	2	1	4	1	4	2
z	4	16	2	32	4	16

38.

x	1	2	2	1	3	2
y	2	1	2	3	1	3
z	16	8	32	36	12	72

In Exercises 39–46, determine the value of the constant of proportionality to the nearest tenth; write the equation of proportionality, and find the specified value to the nearest tenth.

39. D is proportional to C^2. If $D = 13.4$ when $C = 2.8$, find D when $C = 5.9$.

40. a is directly proportional to b^2 and inversely proportional to c^3. If $a = 17.8$ when $b = 8.3$ and $c = 1.4$, find a when $b = 0.9$ and $c = 1.3$.

41. p varies jointly as r^2 and s. If $p = 0.7$ when $r = 0.5$ and $s = 6.3$, find p when $r = 1.5$ and $s = 2.1$.

42. F is inversely proportional to $R^{3/2}$. If $F = 72.3$ when $R = 10$, find F when $R = 5$.

43. z is inversely proportional to x and $t^{1/3}$. If $z = 4.0$ when $x = 5.1$ and $t = 9$, find z when $x = 3.4$ and $t = 72$.

44. m varies jointly as p and q^2. If $m = 24.9$ when $p = -0.7$ and $q = 4.1$, find m when $p = 0.3$ and $q = -8.5$.

45. n varies directly as p^2 and inversely as q and s. If $n = 9.2$ when $p = -2.1, q = -4.5$ and $s = 0.3$, find n when $p = 8.6, q = -1.4$ and $s = -1.8$.

46. y varies directly as x^2 and z^3 and inversely as w. If $y = 18$ when $x = 7, z = 3$ and $w = 20$, find y when $x = 18, z = 6.2$ and $w = 174.6$.

47. Newton's Law of Gravitation states that the attractive force F between two objects varies jointly as the masses of the objects, m kg and M kg, and is inversely proportional to the square of the distance r between the objects. If the mass of one of the objects is halved, how much closer to one another would the objects have to be so that the attractive force is unchanged? Express your answer as a percentage to the nearest tenth.

48. The strength s of various horizontal beams with rectangular cross-sections all having the same width varies directly as the depth D of the beam and is inversely proportional to the distance x between the supports spanned by the beam. How much must the depth (thickness) of the beam be increased if the beam is to be twice as strong and span three times the distance? Express your answer as a percentage to the nearest tenth.

49. The illumination I provided by a light source is inversely proportional to the square of the distance d from the source. Suppose for a certain light source that $I = 48$ candela are provided at a distance of 300 centimeters. Determine to the nearest tenth of a centimeter the distance at which this light source will provide $I = 35$ candela.

50. In the early seventeenth century, the German astronomer-mathematician Johannes Kepler formulated what are now known as Kepler's Laws of Planetary Motion. The third of these three laws can be stated as follows: the square of the period of revolution (about the sun) of a planet is directly proportional to the cube of the planet's average distance from the sun, and the constant of proportionality is the same for all planets. On average the planet Venus is 0.72 times as far from the sun as the earth. Compute the length (in earth years) of a "Venus year" to the nearest hundredth.

51. The kinetic energy E of an object varies jointly as the mass m and the square of the velocity v of the object. Given that $E = 450$ ft-lb for a 4 pound object moving at 15 ft/sec, determine the velocity to the nearest tenth of a ft/sec for an object weighing 6 pounds with kinetic energy of 750 ft-lb.

52. The ideal gas law states that for a given quantity of gas the temperature T of the gas varies jointly as its pressure P and volume V. Compute the percentage increase in V (to the nearest tenth) if T is increased by 20% and P is increased 10%.

Analyzing Data

The Case of the Vanishing Smokers

Can a linear function save lives? Well, perhaps not directly, but it certainly can help in establishing that lives are being saved, or, at least, that behaviors are changing. Our statistics problem in this chapter deals with the apparent decrease in the number of smokers in the state of California over (roughly) the last twenty years. But before we consider the specifics of this case, let us discuss the method of fitting a linear function to real (observed) data.

Suppose you want to estimate the college algebra grades of a sample of students *before* they actually take the course. Furthermore, suppose that you *do* know the high school algebra grades of these students. How might you go about making these predictions? One way is to use data on students who have already taken the college algebra course. That is, plot one ordered pair (high school grade, college grade) for each student, and then construct a linear function that best fits the data, as shown in the figure below. The graph of such a function is called a **line of best fit** (or a **least squares regression line**), and is written $\hat{y} = a + bx$. With this equation, you can predict a student's college algebra grade ($\hat{y}$), by simply plugging in the student's high school algebra grade (x).

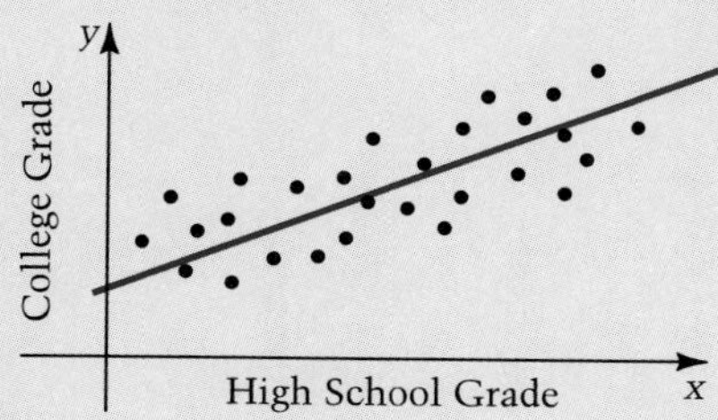

Determining a line of best fit is precisely what a group of researchers did in exploring changes over time, in the smoking habits of California residents [Pierce, et al., 1991]. Using data from National Health Surveys from 1974 through 1987, they constructed a line of best fit that could then be used to predict the percent of the California population that would still be smoking in 1989 and 1990. (See figure (a) on the next page.) However, in figure (b) you see (in color) the points corresponding to the actual percent of Californians still smoking in 1989 and 1990, according to data now available. It is clear that these points fall quite a bit below the predicted values. So the question

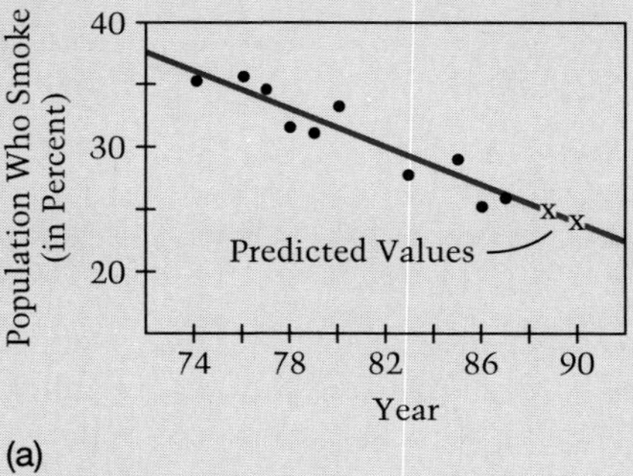

(a)

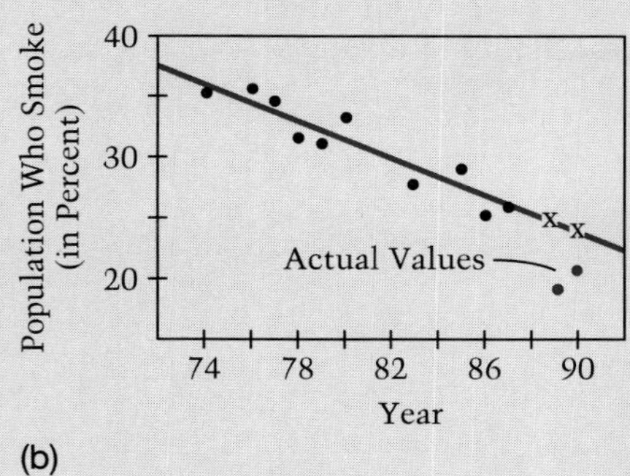

(b)

is, what happened in California to cause a more rapid decrease during 1989 and 1990 than had been observed in the previous 15 years?

In 1988, California passed Proposition 99 and, as a result, the state excise tax on cigarettes increased dramatically, thus providing funds ($294 million) for extensive anti-smoking campaigns. The researchers conclude that the significant drop in smoking in 1989 and 1990 may be attributed to Proposition 99, as evidenced in the figure.

Source: Pierce J. P., et al. Reducing Tobacco Consumption in California: Proposition 99 Seems to Work. *Journal of the American Medical Association,* Vol. 265, No. 10: 1257–1258; March 1991. Copyright 1991, American Medical Association.

Chapter Review

3.1 Equations and Graphs (pp. 125–136)

We graph ordered pairs of real numbers on two real number lines, called *axes.* The axes separate the plane into four *quadrants.* The horizontal axis is named for the independent variable (i.e., *x-axis*), and the vertical axis is named for the dependent variable (i.e., *y-axis*). The axes meet at the *origin,* and the entire system is the *xy-plane.* (p. 125)

Equation of a *straight line:* $y = mx + b$, where m and b are real numbers.

Equation of a *parabola:* $y = ax^2 + bx + c$, where a, b, and c are real numbers and $a \neq 0$.

Equation of a *V-shaped graph:* $y = a|x| + k$, where a and k are real numbers and $a \neq 0$. (p. 128)

A graph is

- *symmetric with respect to the x-axis,* if for each point P on the graph, the reflection of P across the x-axis is also on the graph. If the graph is described by an equation, then replacing y with $-y$ produces an equation equivalent to the original. (p. 132)
- *symmetric with respect to the y-axis,* if for each point P on the graph, the reflection of P across the y-axis is also on the graph. If the graph is described by an equation, then replacing x with $-x$ produces an equation equivalent to the original. (p. 132)
- *symmetric with respect to the origin,* if for each point P on the graph, the reflection of P through the origin is also on the graph. If the graph is described by an equation, then replacing x with $-x$ *and* y with $-y$ produces an equation equivalent to the original. (p. 132)

The *distance between two points* $P_1(x_1, y_1)$ and $P_2(x_2, y_2)$ is given by the formula $d(P_1, P_2) = \sqrt{(x_2 - x_1)^2 + (y_2 - y_1)^2}$. (p. 129)

The *midpoint* of the line segment joining points $P_1(x_1, y_1)$ and $P_2(x_2, y_2)$ has coordinates

$$\left(\frac{x_1 + x_2}{2}, \frac{y_1 + y_2}{2}\right). \quad \text{(p. 130)}$$

3.2 Introduction to Functions (pp. 136–145)

A *function* is a set of ordered pairs, with no two of them having the same first coordinate and different second coordinate. The set of all first coordinates is the *domain,* and the set of all second coordinates is the *range.* In general, any set of ordered pairs is a *relation.* (p. 137)

For the equation $y = 2x + 5$, we call x the *independent variable* and y the *dependent variable* because we are free to choose any value for x, but the value of y is dependent upon the value of x we choose. (p. 139)

The *Vertical Line Test:* For any graph in the xy-plane, if some vertical line can be drawn so that it intersects the graph more than once, then the graph does not represent a function of x. If no such line exists, the graph does represent a function of x. (p. 140)

3.3 Transformations and Graphing (pp. 145–157)

There are three general types of transformations: *translations, reflections,* and *changes in shape.* (pp. 145–146)

Transformation	Change in Equation	Effect on Graph
Horizontal translation	Replace x with $x - h$	If $h > 0$, graph moves $\|h\|$ units to the right. If $h < 0$, graph moves $\|h\|$ units to the left.
Vertical translation	Replace y with $y - k$	If $k > 0$, graph moves $\|k\|$ units up. If $k < 0$, graph moves $\|k\|$ units down.
Reflection across y-axis	Replace x with $-x$	Graph is reflected across y-axis.
Reflection across x-axis	Replace y with $-y$	Graph is reflected across x-axis.
Vertical change in shape	$y = f(x)$ becomes $y = b \cdot f(x)$	If $b > 1$, graph is stretched away from x-axis. If $0 < b < 1$, graph is shrunk toward x-axis.
Horizontal change in shape	$y = f(x)$ becomes $y = f(ax)$	If $a > 1$, graph is shrunk toward y-axis. If $0 < a < 1$, graph is stretched away from y-axis.

A function f is *even* if for all x in the domain of f, $f(-x) = f(x)$; f is *odd* if for all x in the domain, $f(-x) = -f(x)$. (p. 151)

3.4 The Linear Function (pp. 157–166)

An equation of the form

$$y = mx + b$$

describes a *linear function*. Its graph is a straight line. The *slope m* of any nonvertical line is given by the formula

$$m = \frac{y_2 - y_1}{x_2 - x_1},$$

where (x_1, y_1) and (x_2, y_2) are any two different points on the line. (p. 158)

There are three different types of straight line equations (pp. 160–161):

Point-slope form	*Slope-intercept form*	*General form*
$y - y_1 = m(x - x_1)$	$y = mx + b$	$Ax + By + C = 0$

If two nonvertical lines are parallel, then their slopes are equal. If two nonvertical lines are perpendicular, then their slopes are negative reciprocals of each other. (p. 162)

3.5 The Quadratic Function (pp. 166–175)

In general, parabolas have either a high or low point, called the *vertex*. The *axis of symmetry* is a line of symmetry that passes through the vertex. An equation of the form $f(x) = a(x - h)^2 + k$ is called the *vertex form of a parabola*, and (h, k) is the vertex. (p. 168) If the vertex is a high point, then k is the *maximum* function value; if (h, k) is a low point, then k is the *minimum* function value. (p. 168)

The equation $f(x) = ax^2 + bx + c$, where a, b, and c are real numbers and $a \neq 0$, is called the *general form of the quadratic function*. Its graph is a parabola. (p. 168) The method called *completing the square* can be used to convert a quadratic equation from general form into vertex form. (p. 170)

3.6 Composition and the Algebra of Functions (pp. 175–182)

If f and g are functions, then for any real number x that is common to both the domain of f *and* the domain of g,

the *sum function* $f + g$ is given by $(f + g)(x) = f(x) + g(x)$;

the *difference function* $f - g$ is given by $(f - g)(x) = f(x) - g(x)$;

the *product function* fg is given by $(fg)(x) = f(x) \cdot g(x)$;

the *quotient function* $\frac{f}{g}$ is given by $\left(\frac{f}{g}\right)(x) = \frac{f(x)}{g(x)}$, provided $g(x) \neq 0$.

Let f and g be functions, with the range of g contained in the domain of f. For each x in the domain of g, the *composite function* $f \circ g$ is defined as $(f \circ g)(x) = f(g(x))$. The order of the functions in a composition is important. In general, $f \circ g$ and $g \circ f$ are not the same function. (p. 177)

3.7 Inverse Functions (pp. 183–189)

A function f is *one–to–one* if each y-value in the range of f corresponds to exactly one x-value in the domain.

The *Horizontal Line Test:* For the graph of a function in the xy-plane, if a horizontal line can be found that intersects the graph of the function more than once, then the function is not one–to–one. If no such line exists, then the function is one–to–one. (p. 184)

If a function f is one–to–one, then there exists a unique *inverse function* f^{-1} such that $(f^{-1} \circ f)(x) = x$ for any x in the domain of f, and $(f \circ f^{-1})(x) = x$ for any x in the domain of f^{-1}. The domain of f becomes the range of f^{-1}, and the range of f becomes the domain of f^{-1}. (p. 185)

The graph of $y = f^{-1}(x)$ is the reflection of $y = f(x)$ across the line $y = x$. (p. 187)

3.8 Interpreting Graphs (pp. 189–201)

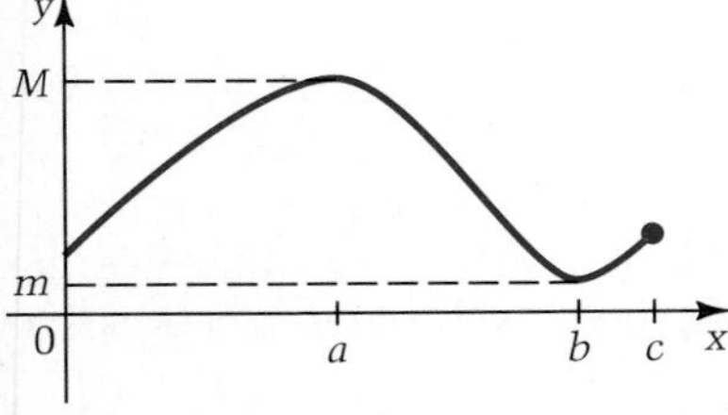

Figure 3.82

For the situation modeled by the graph in Figure 3.82, the *maximum value* of the dependent variable (y) is M; this maximum occurs when the independent variable (x) is a. Similarly, the *minimum value* of y is m, and occurs when x is b.

Where the graph rises, as it does on the interval $[b, c]$, we say that the quantity y is *increasing*. Similarly, where the graph falls, as it does on the interval $[a, b]$, we say that the quantity y is *decreasing*.

Using the slopes of approximating line segments, we can estimate how fast y is increasing or decreasing. (p. 194)

3.9 Variation (pp. 201–206)

If A and B represent variable expressions, we say that

- A is *directly proportional* to B (or, A *varies directly* as B), if there is a nonzero constant k such that $A = kB$;
- A is *inversely proportional* to B (or, A *varies inversely* as B), if there is a nonzero constant k such that $A = k/B$.

In either case, we refer to k as the constant of proportionality. (p. 201)

An equation of proportionality that combines several proportional statements, with one variable dependent on the product and/or quotient of two or more variables is called *combined variation.* (p. 204)

Review Exercises

In Exercises 1–4, test the graph of the given equation for each of the three types of symmetry.

1. $y = x^3 - x$ **2.** $y^2 = 5 - x^4$

3. $y^2 = x + 2$ **4.** $y = (2 - x^2)^3$

In Exercises 5–6, determine the distance $d(P, Q)$ and the midpoint of the segment PQ.

5. $P(-2, 3), Q(7, 1)$ **6.** $P(4, 2), Q(6, -5)$

In Exercises 7–10, find $f(-3), f(0), f(2), f(a)$ and $f(3 + h)$ for the given function. If the expression is not defined or not a real number, then state this fact.

7. $f(x) = \dfrac{1}{x - 2}$ **8.** $f(x) = 9 - x^2$

9. $f(x) = 4$ **10.** $f(x) = \sqrt{1 - x}$

In Exercises 11–14, determine whether the given graph represents a function of x

11.

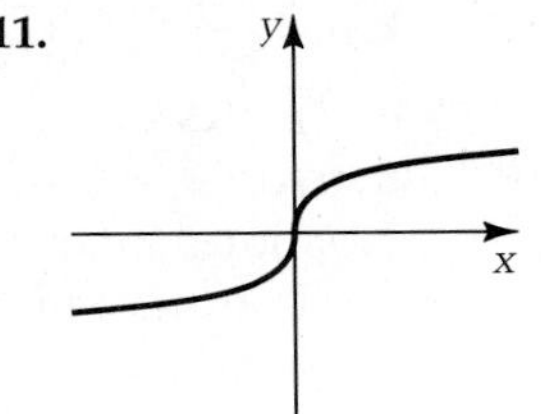

12.

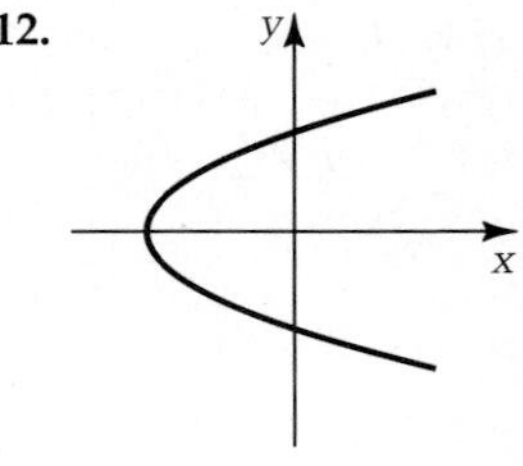

13. 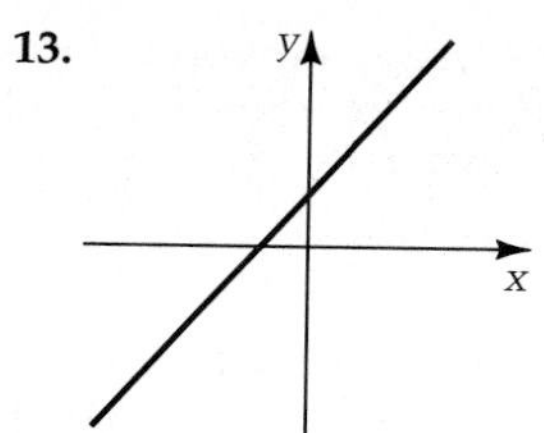

14. 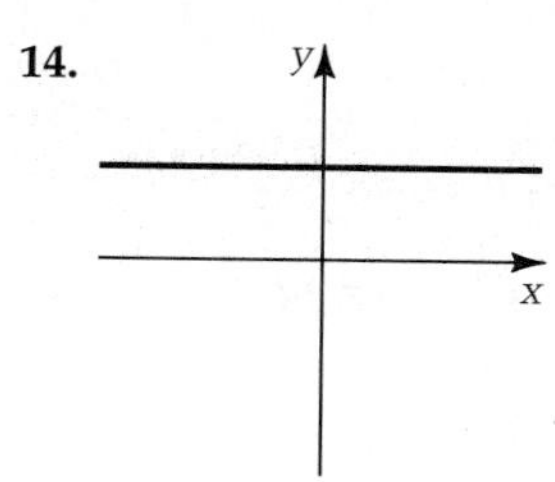

For Exercises 15–18, determine whether the given graph in Exercises 11–14 represents a function $y = f(x)$ that has an inverse.

In Exercises 19–22, determine the domain of the given function. Write your answer in interval notation.

19. $f(x) = \sqrt{4 - x^2}$

20. $f(x) = \dfrac{1}{\sqrt{9 - x^2}}$

21. $f(x) = \dfrac{1}{x^2 + 4}$

22. $f(x) = \dfrac{1}{x^2 + 5x + 4}$

In Exercises 23–28, use transformations to graph the given function.

23. $y = 2|x - 1|$

24. $y = 2 + \dfrac{1}{x - 3}$

25. $y = 1 - (x + 2)^2$

26. $y = -1 - 2x^2$

27. $y = \dfrac{1}{2 - x} - 1$

28. $y = 4 - |x + 1|$

In Exercises 29–36, determine an equation of the line satisfying the stated conditions. Write your answer in (a) point-slope form, (b) slope-intercept form, and (c) general form.

29. slope is -3, passes through $(-2, 1)$.

30. slope is 0, passes through $(3, -4)$.

31. slope is not defined, passes through $(3, -4)$.

32. y-intercept is -2, slope is $\frac{1}{2}$.

33. line is parallel to the graph of $2x - 3y + 7 = 0$ and passes through $(-1, -4)$.

34. line is perpendicular to the graph of $3x = 4y + 12$ and passes through $(2, 3)$.

35. line is perpendicular to the graph of $y = 4$ and passes through $(6, 2)$.

36. line is parallel to the x-axis and passes through $(3, 5)$.

In Exercises 37–42, write the given equation in vertex form and sketch the graph. Label the vertex, the axis of symmetry, and two other points.

37. $y = 2(x - 1)^2 - 3$

38. $y = 4 - (x + 2)^2$

39. $y = 7 + 4x - x^2$

40. $y = x^2 + 4x + 7$

41. $y = 2x^2 + 4x + 6$

42. $y = 10 - 8x - 2x^2$

In Exercises 43–54, evaluate the given expression for $f(x) = 2x + 1$, $g(x) = 5 + x^2$, and $h(x) = \sqrt{x}$.

43. $(g \circ h)(2)$

44. $(h \circ f)(4)$

45. $(h \circ g)(2)$

46. $(f \circ h)(4)$

47. $(g \circ g)(1)$

48. $(f \circ f)(2)$

49. $(f \circ g \circ h)(4)$

50. $(h \circ g \circ f)(-2)$

51. $(g \circ h \circ f)(4)$

52. $(f \circ h \circ g)(-2)$

53. $(f \circ f \circ f)(1)$

54. $(g \circ g \circ g)(0)$

In Exercises 55–58, determine whether the function is one–to–one.

55. $f(x) = 3 - 4x$

56. $f(x) = |2 + x|$

57. $f(x) = 4 - x^2$

58. $f(x) = x^3 - 8$

In Exercises 59–64, the given function $y = f(x)$ is one–to–one. Find an equation for the inverse function f^{-1} and specify the domain and range for f^{-1}.

59. $f(x) = 7 - 3x, 1 \le x < 4$

60. $f(x) = 3 - x^{1/3}, -1 < x \le 8$

61. $f(x) = 6 + 2\sqrt{x}, 1 < x \le 4$

62. $f(x) = (3 + x)^2, -1 \le x \le 2$

63. $f(x) = x^2 - 4x, -2 \le x \le 1$

64. $f(x) = |x - 4|, -2 \le x < 3$

65. The pulse rate (in beats per minute) of a certain individual was monitored continuously over a 30-minute span while the individual was in the recovery room following minor surgery that required a local anesthesia. See Figure 3.83.

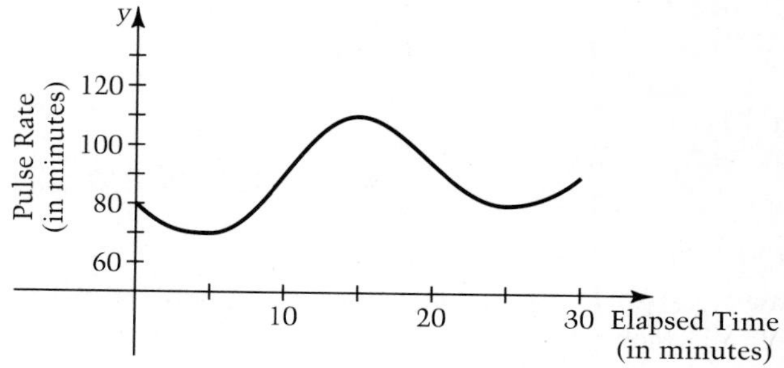

Figure 3.83

(a) What was the person's pulse rate at the start ($t = 0$)?
(b) How many minutes later did the pulse rate reach a maximum for the 30-minute span?
(c) What was the range in the pulse rate during the 30-minute span?
(d) Over what time intervals did the pulse rate increase?

66. The elevation (in floors) of a slow-moving freight elevator in a local hotel is recorded over a one-hour period. See Figure 3.84.
(a) On which floor was the elevator when the recording period began?
(b) How many times during the hour did the elevator stop at a floor for at least one minute?
(c) What was the longest time (approximately) that the elevator spent during one stop at a floor?
(d) How many times did the elevator pass the 5th floor?

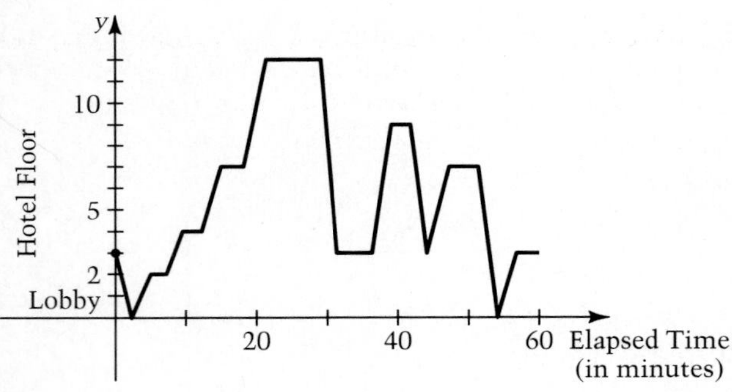

Figure 3.84

In Exercises 67–70, determine the value of the constant of proportionality, write the equation of proportionality, and find the specified value.

67. y is directly proportional to x^2. If $y = 4$ when $x = 3$, find y when $x = 2$.

68. m varies inversely as p. If $m = 7$ when $p = 3$, find m when $p = 4$.

69. z varies inversely as s and t^3. If $z = 8$ when $s = 6$ and $t = 2$, find z when $s = 2$ and $t = 6$.

70. p varies jointly as t^2 and y^3. If $p = 18$ when $t = 0.5$ and $y = 6$, find p when $t = 2$ and $y = 3$.

71. Sketch the graph of $\big||x| - |y|\big| = 4$.

72. The x- and y-coordinates of the position of an ant in the xy-plane at any time t are given by the equations

$$x = 2 + t^2 \quad \text{and} \quad y = -3 + 2t^2.$$

Solve the equations simultaneously to eliminate t and obtain a single equation in terms of x and y. Explain why the path of the ant's trip in the xy-plane is *not* a straight line. *Hint:* Consider what happens when $t = 0$.

73. The diagonals of a square intersect at the point (3, 1). One vertex of the square is the point (2, 5). Determine the coordinates of the three other vertices of the square.

74. The perimeter of a rectangle is 40. Determine the dimensions of the rectangle with the shortest diagonal.

75. Determine an equation for a parabola with a vertical axis of symmetry such that the points (3, 3), (5, 3) and (7, 7) lie on the parabola.

76. Suppose y is proportional to the sum of x and w. Further, suppose w is proportional to the product of x and y. If $y = 4$ when $x = 2$ and $y = 9$ when $x = 1$, determine y when $x = 5$.

Graphing Calculator Exercises

[*Note: Round all estimated values to the nearest hundredth.*]

In Exercises 1–6, solve the given equation over the set of real numbers. (*Hint:* To solve an equation of the form $h(x) = g(x)$, first rewrite the equation in the form $h(x) - g(x) = 0$; then sketch the function $f(x) = h(x) - g(x)$. The desired solutions are the x-intercepts of f. For example, to solve the equation $3x^2 - x = x + 1$, first rewrite it as $3x^2 - 2x - 1 = 0$, and then graph $f(x) = 3x^2 - 2x - 1$ and approximate its x-intercepts.)

1. $x^3 - x - 1 = 12$

2. $x^3 - 12x - 1 = 3$

3. $|x| = x^2 - 1$

4. $x^3 + x^2 = 10 - x$

5. $12 - x^3 = x^2 + 2x$

6. $x^4 - \sqrt{x} = 17$

7. Use the graph of $f(x) = 1 - x^2 + \sqrt{x}$ to determine x-values such that (a) $f(x) = 1$, (b) $f(x) = 0$, (c) $f(x) = -1$, (d) $f(x) = 5$.

8. For the function in Exercise 7, determine the maximum function value, and the x-value that produces it.

9. For the function in Exercise 7, determine the domain and range.

10. Use the graph of $f(x) = x^2 - 5x + 2\sqrt{x} - 1$ to determine x-values such that (a) $f(x) = 1$, (b) $f(x) = 0$, (c) $f(x) = -1$, (d) $f(x) = -5$.

11. For the function in Exercise 10, determine the minimum function value, and the x-value that produces it.

12. For the function in Exercise 10, determine the domain and range.

13. An open topped box is to be constructed from a rectangular piece of cardboard, 10 in. wide by 20 in. long, by cutting out squares of equal size from the four corners, and then folding up the sides. What is the maximum volume of such a box, and what size squares must be cut out to produce it? (*Hint:* Express the volume V as a function of x, the side of each square, then graph this function to find the maximum value of V.)

14. A wire of length 10 in. is to be cut into two pieces. One piece will be bent to form a circle, and the other will be bent to form a square.
   (a) How should the wire be cut so that the total area enclosed is a minimum? What is this minimum area?
   (b) How should the wire be cut so that the total area enclosed is a maximum? What is this maximum area?

Chapter Test

In Problems 1 and 2, sketch the graph of the function described by the given equation.

1. $f(x) = x^2 - 3x - 8$

2. $g(x) = 11 - 4|x|$

3. Reflect the graph shown below (a) across the x-axis, (b) across the y-axis, and (c) through the origin.

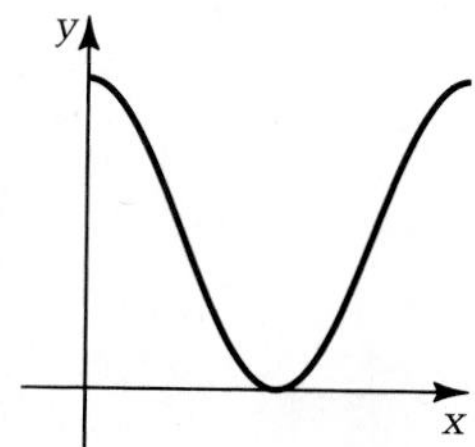

In Problems 4 and 5, test the graph of the given equation for each of the three types of symmetry.

4. $x^2 - y^2 + 8 = 0$

5. $x^3y = -5$

6. Determine $f(-2)$, $f(0)$, $f(h^2)$, and $f(b - 1)$ for the function $f(x) = -x^2 - 5x + 9$.

In Problems 7 and 8, sketch the graph of the function by translating the graph of $y = |x|$, or $y = \dfrac{1}{x}$.

7. $y + 2 = |x|$

8. $y = \dfrac{1}{x} + 5$

9. Sketch the graph of the function $y = -(x + 1)^2$, by reflecting the graph of $y = (x + 1)^2$.

10. Sketch the graph of the given function by transforming the graph of $y = f(x)$ shown below.

(a) $y = f(3x)$
(b) $y = 4f(x)$
(c) $y = \dfrac{1}{3}f(x)$

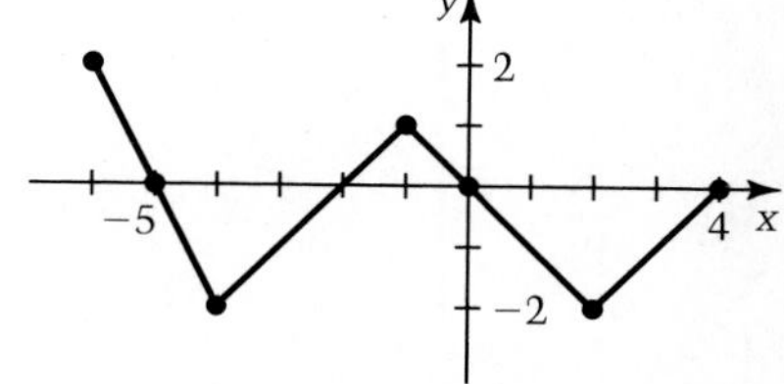

11. Find the slope of the line passing through the points $(3, -1)$ and $(5, 4)$. Then sketch the line.

12. A line is perpendicular to the graph of $y = -5$, and passes through the point $(6, -1)$. Write an equation of this line in (a) point-slope form, (b) slope-intercept form, and (c) general form.

13. Sketch the graph of the function $y = x^2 - 4x - 21$. Label the vertex, the x-intercepts, the axis of symmetry, and two additional points.

14. The graph below represents the volume V (in liters) of air in a person's lungs as a function of time t (in seconds). The person was told by a doctor to inhale and exhale completely several times.

(a) Was the person inhaling or exhaling when $t = 5$?
(b) How much time elapsed between successive breaths?
(c) What is the apparent capacity of the person's lungs?

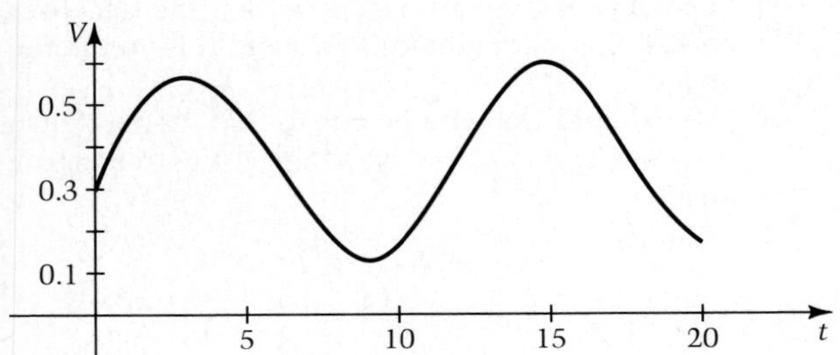

15. The quantity A varies directly as b^2 and inversely as c. Express this statement as an equation using k as the constant of proportionality.

16. The quantity m varies jointly as n and p^2. When $n = 3$ and $p = 2$, we know that $m = 6$. Find m when $n = 4$ and $p = 9$.

In Problems 17 and 18, evaluate the expression given that $f(x) = 3x + 3$, and $g(x) = |x| - 2$.

17. $(f \circ g)(-2)$

18. $(g \circ f)(0)$

19. If $f(x) = \dfrac{1}{2 - x}$, and $g(x) = \sqrt{x + 6}$, find an equation that describes $(f \circ g)(x)$ and state the domain of the composite function.

20. Write the function $F(x) = |x + 4|^2$ as a composition of the following: $f(x) = x^2$, $g(x) = x + 4$, $h(x) = |x|$.

21. The function $f(x) = \sqrt{x} + 3$, for $1 \le x < 9$, is one–to–one. Find the equation for the inverse function f^{-1}, specify the domain and range of f^{-1}, and sketch the graph of $y = f^{-1}(x)$.

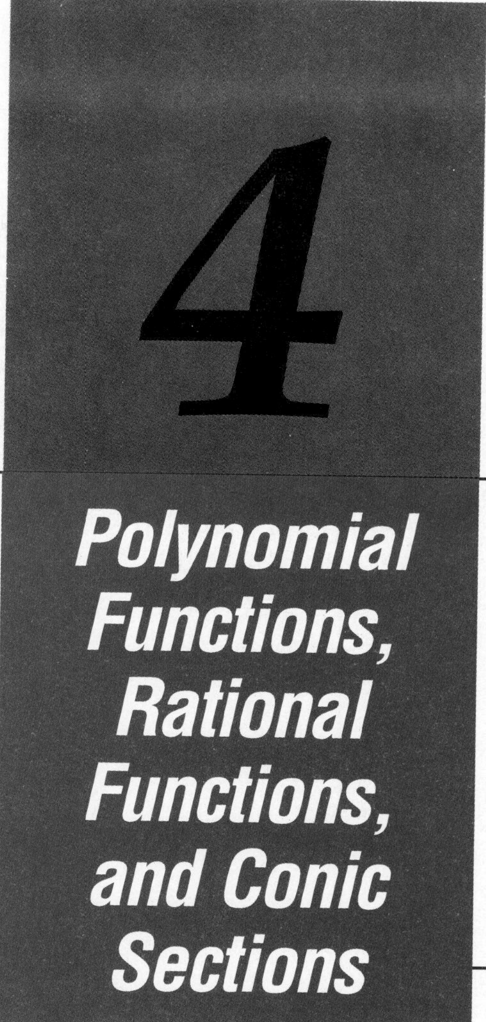

It would be quite convenient if the only models needed to represent processes and relationships between variables were linear (straight lines). In applying mathematics, however, one quickly finds that other models are needed to capture the complexity of certain phenomena. At the end of this chapter we consider the question of whether it's possible to lower one's blood pressure too much. *The researchers whose work is considered claim that it is, and that in order to understand this, one must be willing to accept the fact that a straight line, though simple, is not always an adequate model. Other models (such as higher degree polynomials) are often more appropriate and enlightening.*

4.1 Important Features of Polynomial Functions

In Chapter 3, we saw that the graph of a linear function, such as $f(x) = 2x - 7$, is a straight line, and the graph of a quadratic function, such as $g(x) = 3x^2 - 5x + 4$, is a parabola. Since $2x - 7$ and $3x^2 - 5x + 4$ are polynomials, the functions $f(x) = 2x - 7$ and $g(x) = 3x^2 - 5x + 4$ are called *polynomial functions.* Any function described by a polynomial is a polynomial function. Polynomial functions are classified by the degree of the polynomial.

Polynomial Function	Degree	Common Name
$f(x) = 2$	0	constant function
$g(x) = 2x - 7$	1	linear function
$y = 3x^2 - 5x + 4$	2	quadratic function
$y = -5x^3 + 4x^2 - 2x + 17$	3	cubic function
$p(x) = x^6 - 15x^2 + 10$	6	6th degree polynomial function
$q(x) = 4x^{11} + 2x^6 - 5x^2$	11	11th degree polynomial function

It is customary to write a polynomial so that the degrees of its terms are in descending order. A polynomial function written this way is said to be in **standard form.**

Definition

An ***n*th degree polynomial function in *x*** is a function that can be written in the form

$$f(x) = a_n x^n + a_{n-1}x^{n-1} + \cdots + a_2x^2 + a_1x + a_0$$

where n is a nonnegative integer, $a_n \neq 0$, and the coefficients $a_n, a_{n-1}, \ldots a_2, a_1, a_0$ are real numbers.

The term of highest degree of a polynomial function is called the **leading term,** and the coefficient of the leading term is called the **leading coefficient.** We refer to a_0 as the **constant term.**

EXAMPLE 1

Determine whether each of the following is a polynomial function. If so, write it in standard form and identify the leading term.

(a) $f(x) = 2x - 5x^3 + 4$ (b) $g(x) = x^2 + 3\sqrt{x}$

(c) $F(x) = 3x + \sqrt{2}x^4 - x^2$ (d) $G(x) = x^4 + 3x + \dfrac{2}{x}$

Solution

(a) f is a polynomial function. Its standard form is

$$f(x) = -5x^3 + 2x + 4,$$

and the leading term is $-5x^3$.

(b) g is not a polynomial function. The exponents of the terms of a polynomial function must be nonnegative *integers.* Since $3\sqrt{x} = 3x^{1/2}$, its exponent does not meet this requirement.

(c) F is a polynomial function. Its standard form is

$$F(x) = \sqrt{2}x^4 - x^2 + 3x,$$

and the leading term is $\sqrt{2}x^4$.

(d) G is not a polynomial function. The exponents of the terms of a polynomial function must be *nonnegative* integers. Since $\frac{2}{x} = 2x^{-1}$, its exponent does not meet this requirement.

In this and the next section we shall explore some important features of polynomial functions and their graphs. Our goal is to introduce a few general techniques to help us graph a polynomial function. We begin by considering polynomial functions described by monomials.

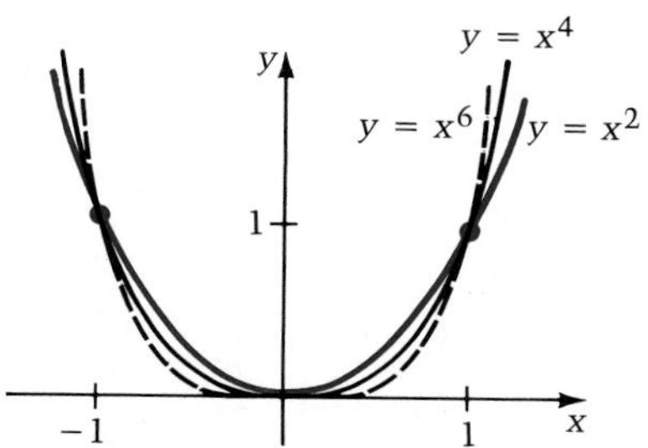

Figure 4.1 Graphs of the form $y = x^{\text{even}}$

$y = x^p$, where p is an even positive integer (Figure 4.1)

Recall that the graph of $y = x^2$ is a parabola. Although the graphs of $y = x^4$ and $y = x^6$ look like parabolas, they are not. They are, however, symmetric with respect to the y-axis and never drop below the x-axis. (Why?) As the exponents increase, the graphs become flatter between -1 and 1 and steeper outside this interval. In all cases, as the x-values get further from zero, the y-values increase more rapidly.

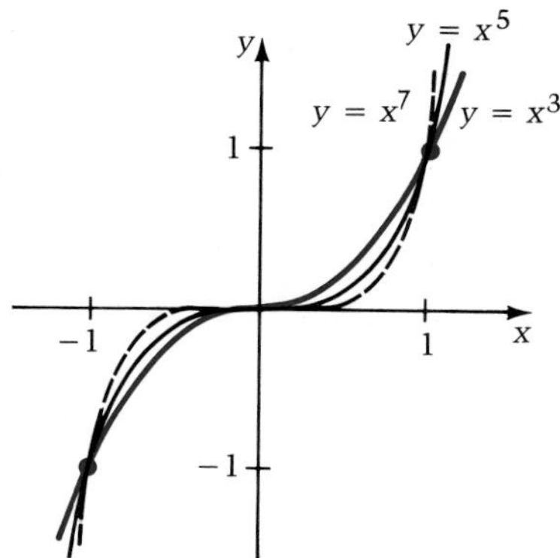

Figure 4.2 Graphs of the form $y = x^{\text{odd}}$

$y = x^q$, where q is an odd positive integer greater than 1 (Figure 4.2)

The graphs of $y = x^3$, $y = x^5$, and $y = x^7$ are all symmetric with respect to the origin. (Why?) As the exponents increase, the graphs become flatter between -1 and 1 and steeper outside this interval. In all cases, as the x-values get further to the left of zero, the y-values decrease more rapidly, and as the x-values get further to the right of zero, the y-values increase more rapidly.

From our work with symmetry in Chapter 3, we know that the graph of $y = -x^n$, where n is *any* positive integer, is the reflection of $y = x^n$ across the x-axis (Figure 4.3).

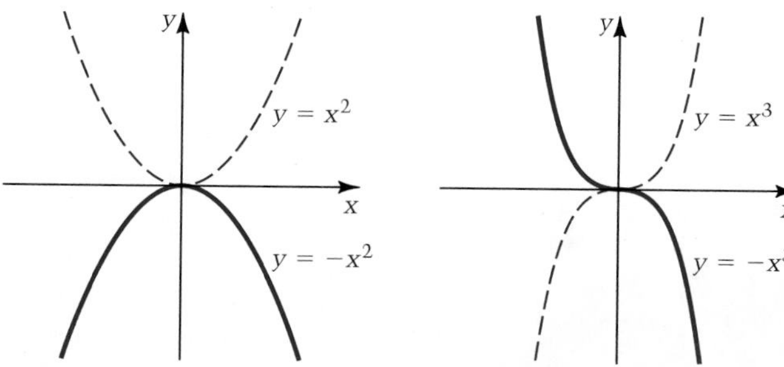

Figure 4.3

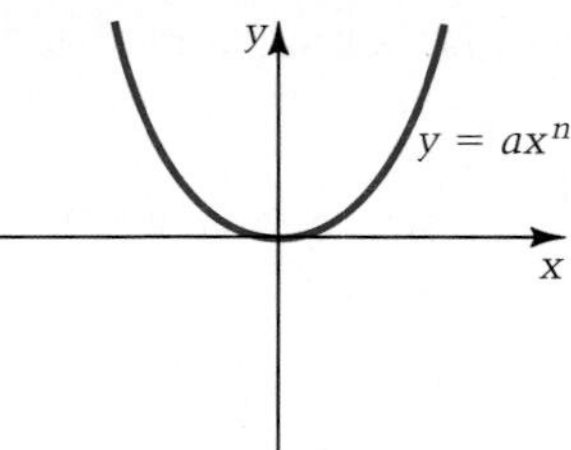

(a) Type I: *a* positive, *n* even

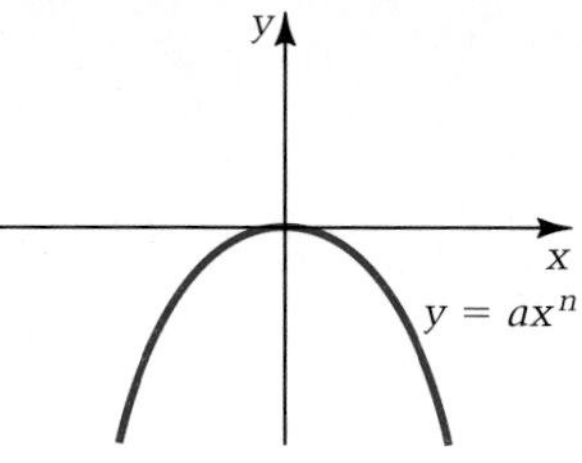

(b) Type II: *a* negative, *n* even

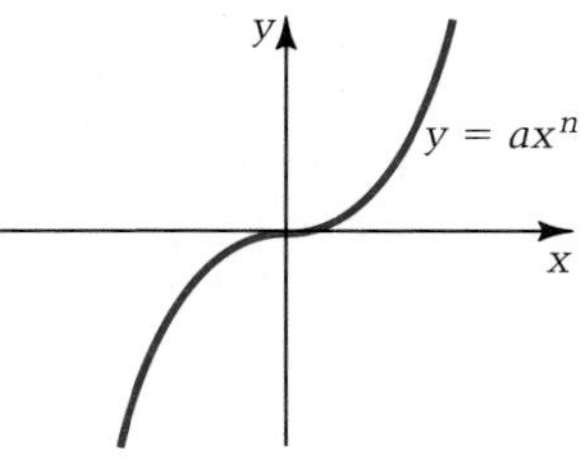

(c) Type III: *a* positive, *n* odd

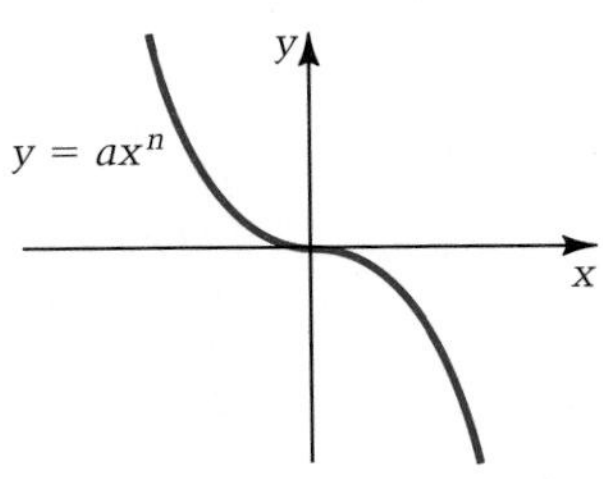

(d) Type IV: *a* negative, *n* odd

Figure 4.4

Thus, the graph of a polynomial function of the form $f(x) = ax^n$, where a is nonzero, and n is any positive integer greater than 1, will have one of the four shapes shown in Figure 4.4. The basic shape of the graph depends upon the sign of the coefficient a and whether n is odd or even.

EXAMPLE 2

Classify the graph of each polynomial function as one of the types above.
(a) $y = -3x^5$ (b) $f(x) = x^6$ (c) $g(x) = 5x^7$ (d) $y = -3x^4$

Solution

(a) Type IV, since $a < 0$ and n is odd. (b) Type I, since $a > 0$ and n is even.
(c) Type III, since $a > 0$ and n is odd. (d) Type II, since $a < 0$ and n is even.

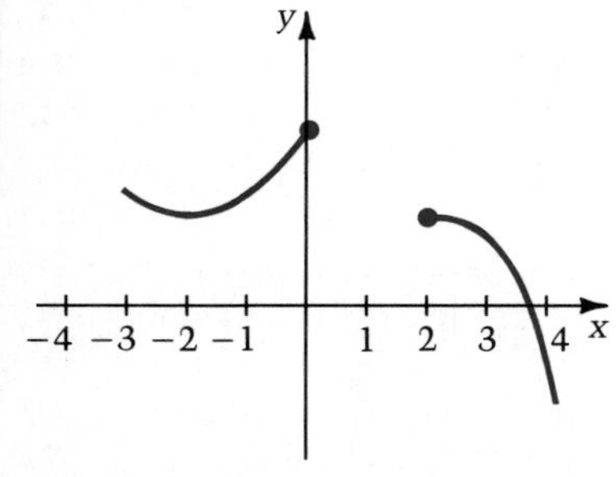

Figure 4.5 Not the graph of a polynomial function.

Next, we consider three important features of any polynomial function: its domain, the general shape of its graph, and its behavior for x-values to the extreme right or extreme left of 0.

I. Domain

The domain of any polynomial function is the set of all real numbers. The graph continues indefinitely to the right of 0 and to the left of 0 and has no gaps. The graph in Figure 4.5 could not be the graph of a polynomial function since it *is not defined for real numbers between 0 and 2.*

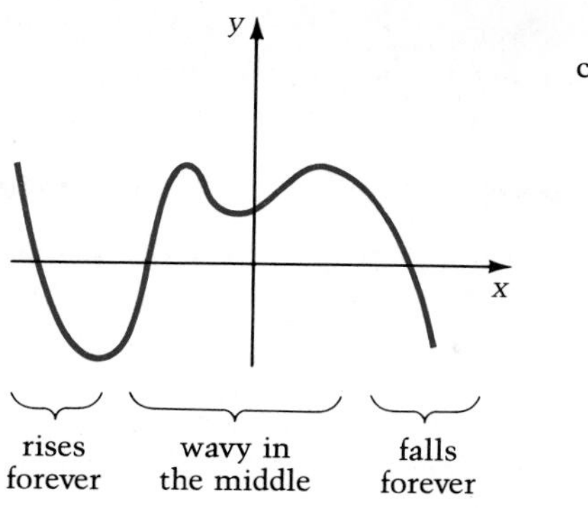

(a) The graph of a polynomial function.

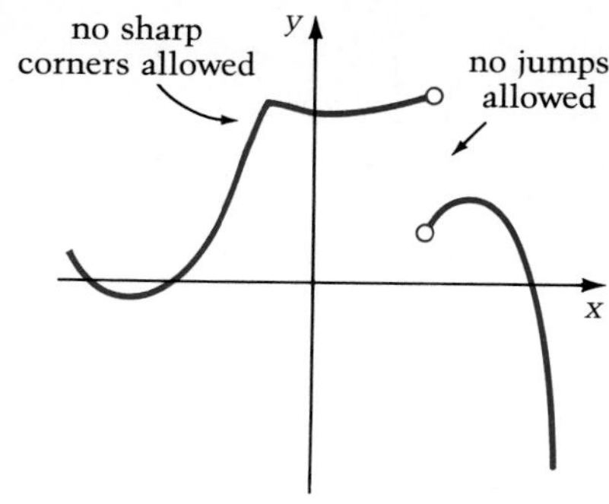

(b) Not the graph of a polynomial function.

Figure 4.6

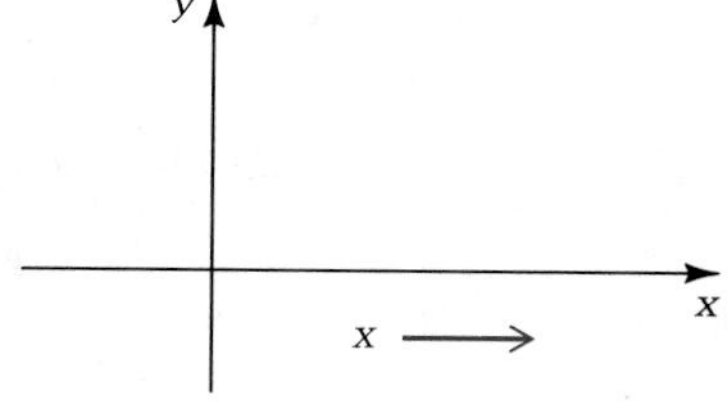

Figure 4.7 $x \to \infty$

II. General Shape

We already know how to graph first degree polynomial functions (linear functions) and second degree polynomial functions (quadratic functions). For degrees higher than 2, the graphs of polynomial functions can be characterized crudely by saying that inside the extremes they are composed of smooth "waves," and for extreme values of x, they either rise or fall indefinitely. Furthermore, the graph is always smooth and unbroken: there are no sharp corners, and no jumps or holes of any kind (Figure 4.6).

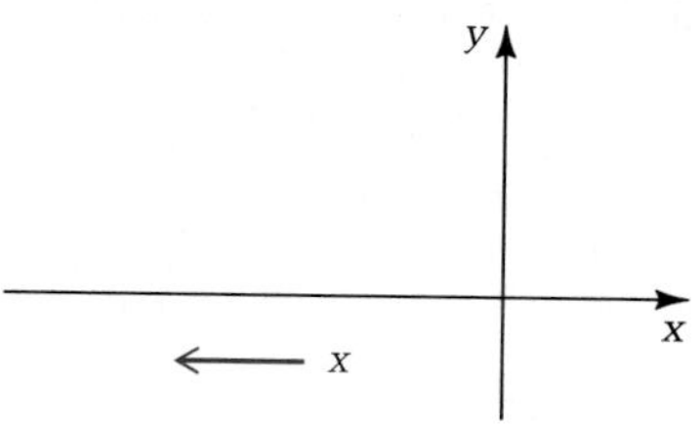

Figure 4.8 $x \to -\infty$

III. Behavior for Extreme Values of x

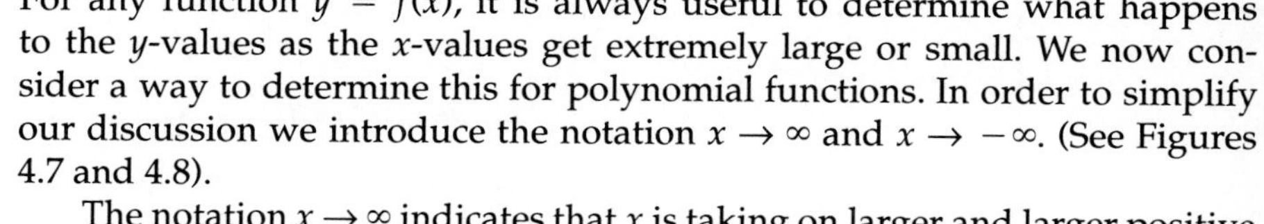

For any function $y = f(x)$, it is always useful to determine what happens to the y-values as the x-values get extremely large or small. We now consider a way to determine this for polynomial functions. In order to simplify our discussion we introduce the notation $x \to \infty$ and $x \to -\infty$. (See Figures 4.7 and 4.8).

The notation $x \to \infty$ indicates that x is taking on larger and larger positive values *without bound*. For example, suppose x takes on the values 10,000, 20,000, 30,000, etc. (successive multiples of 10,000). The x-values are getting larger and larger and will eventually be larger than any number you can imagine. We say that x is increasing without bound, and we write $x \to \infty$.

The notation $x \to -\infty$ indicates that x is taking on smaller and smaller negative values *without bound*. For example, if x takes on the values -10, -100, $-1{,}000$, $-10{,}000$, etc., then we say x is decreasing without bound, and we write $x \to -\infty$.

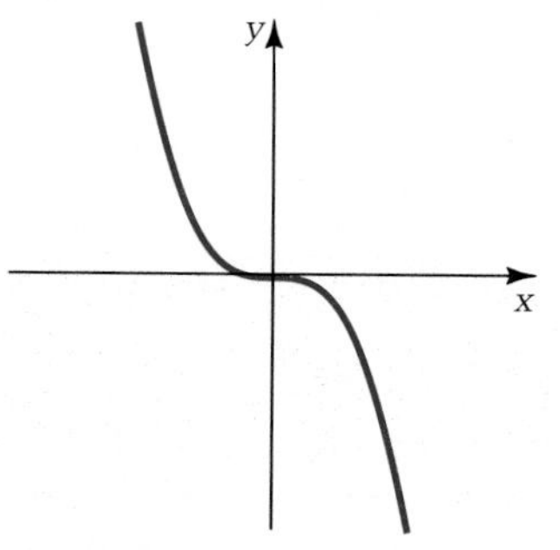

Figure 4.9 $y = -x^5$.

We use this notation to describe the behavior of the graph of a function to the extreme right ($x \to \infty$) or to the extreme left ($x \to -\infty$) of zero. For example, the graph of the function $y = -x^5$ falls without bound as $x \to \infty$, and it rises without bound as $x \to -\infty$. (See Figure 4.9.)

Saying that a graph falls without bound is equivalent to saying that the y-values are decreasing without bound. Thus, for the function $y = -x^5$, we can write

$$\text{as } x \to \infty, \quad y \to -\infty.$$

Similarly, if a graph rises without bound, this means that the y-values are increasing without bound. Thus, for the function $y = -x^5$ we can write

$$\text{as } x \to -\infty, \quad y \to \infty.$$

To determine the behavior of the graph of a polynomial function at the extreme right or left, you need only look at the leading term of the polynomial.

First Property of Polynomial Functions

Consider the polynomial function f, described by the equation

$$f(x) = a_n x^n + a_{n-1}x^{n-1} + \cdots + a_1 x + a_0.$$

As $x \to \infty$, and as $x \to -\infty$, the graph of f behaves like the graph of $y = a_n x^n$.

For example, knowing what the graph of $y = 2x^3$ looks like toward the extreme left and extreme right tells us what the graph of the function $y = 2x^3 + 4x^2 - 5x + 13$ looks like toward those extremes.

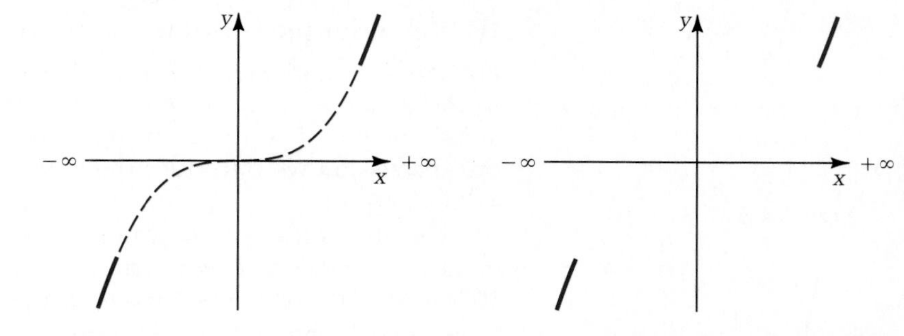

(a) $y = 2x^3$.

(b) $y = 2x^3 + 4x^2 - 5x + 13$ toward the extremes.

Figure 4.10

We have seen that there are only four possible ways that a polynomial function of the form $y = ax^n$ ($n > 1, a \neq 0$) can behave for extreme values of x. Therefore, by the First Property of Polynomial Functions, there are only four possible ways (Figure 4.11) that *any* polynomial function of positive degree can behave for extreme values of x.

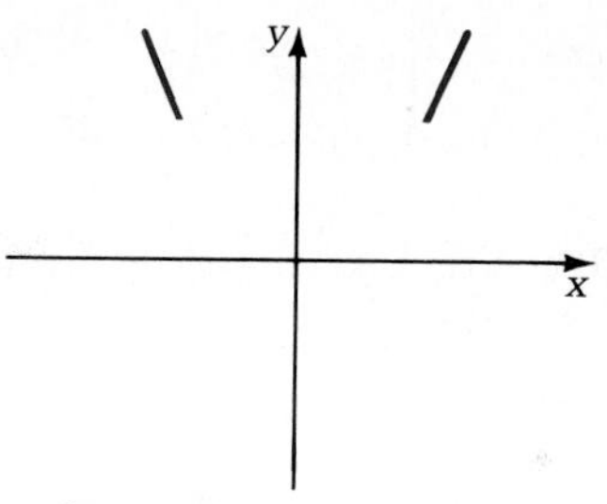

Leading coefficient is positive; degree of the polynomial is even.

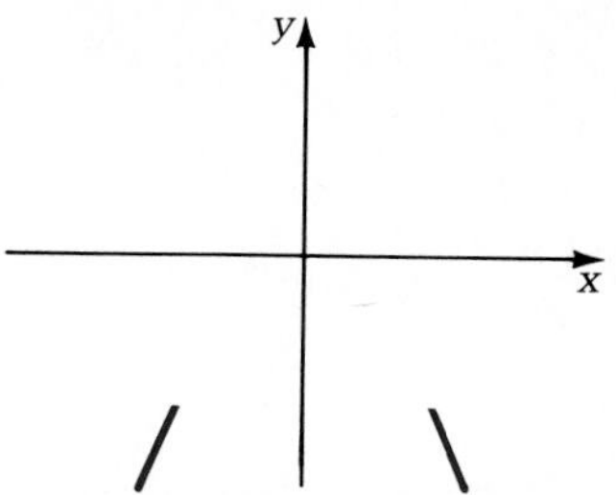

Leading coefficient is negative; degree of the polynomial is even.

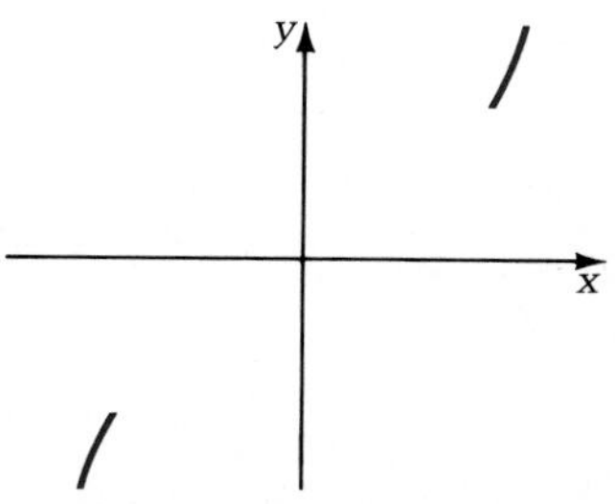

Leading coefficient is positive; degree of the polynomial is odd.

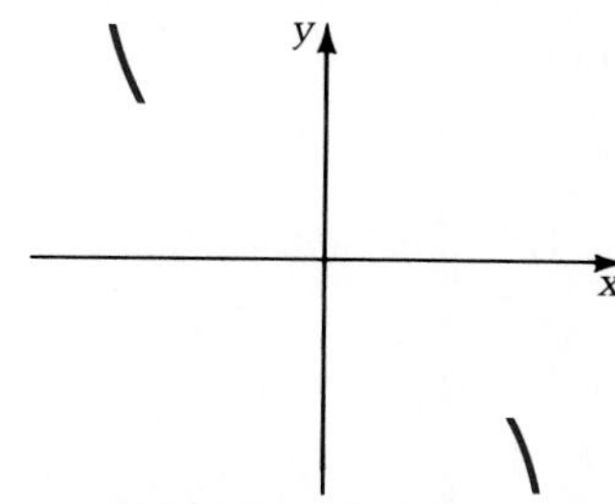

Leading coefficient is negative; degree of the polynomial is odd.

Figure 4.11

A useful point on the graph of $f(x) = a_nx^n + \cdots + a_1x + a_0$ is the *y-intercept*: $(0, a_0)$. Notice that the y-coordinate of this point is the constant term of the polynomial function.

EXAMPLE 3

For each polynomial function below, select the graph that best represents it.

(a) $f(x) = x^9 + 5x^2 + 2$
(b) $f(x) = 2x^8 - 3x^7 + 2$
(c) $f(x) = -x^8 + 5x^2 + 2$
(d) $f(x) = -3x^9 + 2x^8 + 2$ **(Careful!)**

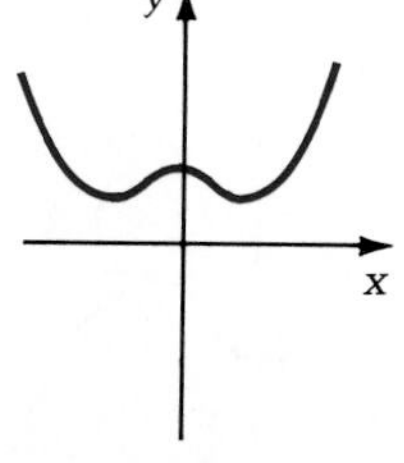

Graph No. 1

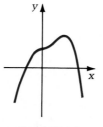

Graph No. 2

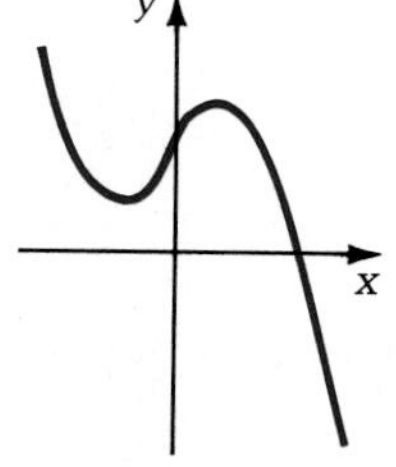

Graph No. 3

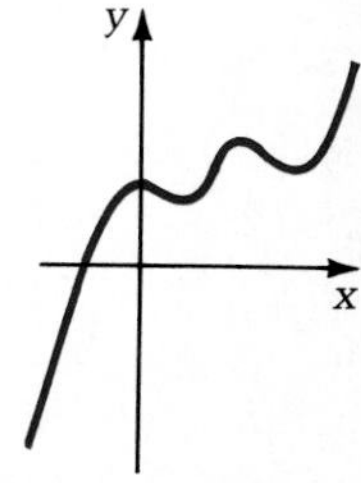

Graph No. 4

Figure 4.12

Solution

(a) Graph No. 4, since the leading term is x^9.

(b) Graph No. 1, since the leading term is $2x^8$.

(c) Graph No. 2, since the leading term is $-x^8$.

(d) Graph No. 3, since the leading term is $-3x^9$.

EXERCISE SET 4.1

In Exercises 1–12, determine whether the function is a polynomial function. If so, write it in standard form and identify its leading term.

1. $f(x) = 2 + 7x - x^3$
2. $g(x) = 2x^2 - 5x^4 + 9x - 4$
3. $F(x) = 3x^2 - 2x\sqrt{x} + x$
4. $f(x) = 6x - 10x^2 + 3x^3 - x$
5. $H(x) = 3x^3 + 4x^2 - 2x^5 + 5x^2 + 1$
6. $h(x) = x^2 - \dfrac{3}{x} + 5$
7. $g(x) = 8x^2 - 4x + 2 - x^{-1}$
8. $F(x) = 1 - x + x^2 - x^4$
9. $f(x) = (3x + 2)^2 + (x + 2) + 5$
10. $g(x) = 6 - 2x - (5x + 2)^2$
11. $k(x) = 7x + 3x^3 - \sqrt{5}x^4 - 3$
12. $G(x) = x^2 - x + x^{1/3}$

In Exercises 13–28, classify the graph of the polynomial function as one of the types listed on page 220.

13. $y = 2x^3$
14. $y = -3x^2$
15. $y = -4x^6$
16. $y = -3x^5$
17. $y = 5x^4$
18. $y = 10x^7$
19. $y = -x^2$
20. $y = \dfrac{7}{2}x^{10}$
21. $5y = x^3$
22. $-2y = x^9$
23. $-2y = 3x^7$
24. $\dfrac{1}{3}y = 4x^8$
25. $y + 2x^3 = 0$
26. $3x^4 + y = 0$
27. $x^2 - 2y = 0$
28. $3y - x^5 = 0$

In Exercises 29–32, select the graph that best represents the given polynomial function.

29. $y = x^3 + 2x^2 + 1$
30. $y = 1 - 2x^2 - x^3$
31. $y = 1 - 2x^4$
32. $y = 2x^3 + x^4 + 1$

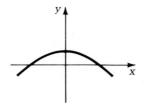

Graph No. 1

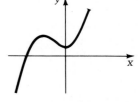

Graph No. 2

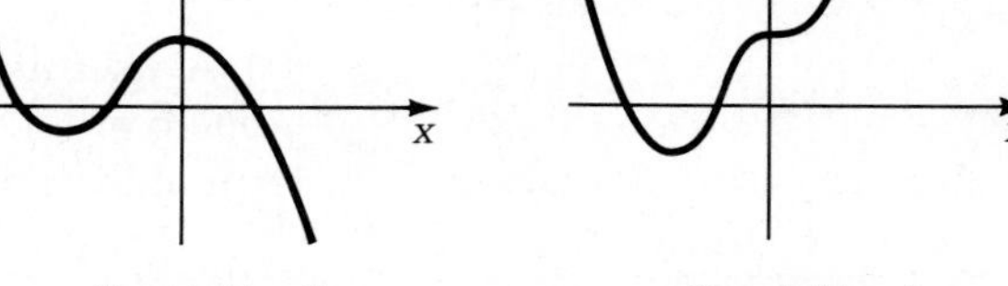

Graph No. 3

Graph No. 4

In Exercises 33–36, select the graph that best represents the given polynomial function.

33. $y = 2x^5 + 3x^2 - 2$
34. $y = 2x^4 - x^2 - 2$
35. $y = x^3 - x^4 - 2$
36. $y = 2x^2 - 3x^7 - 2$

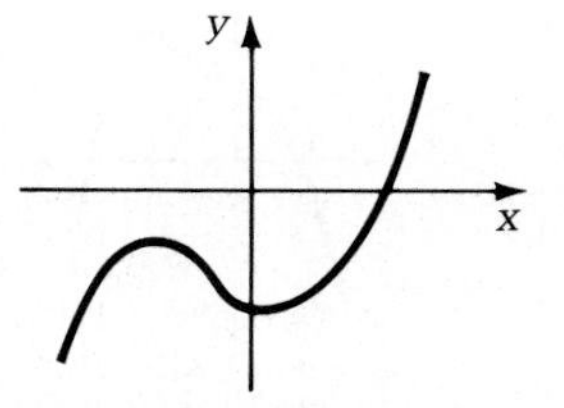

Graph No. 1

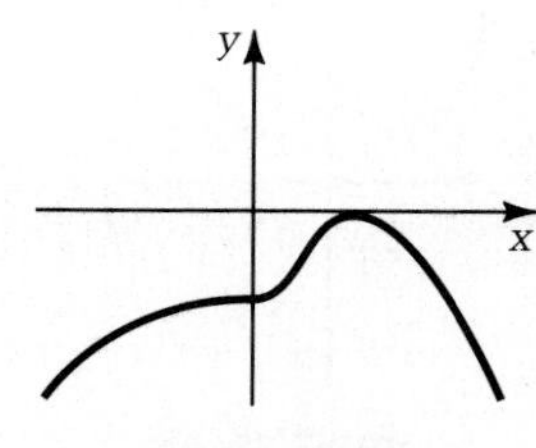

Graph No. 2

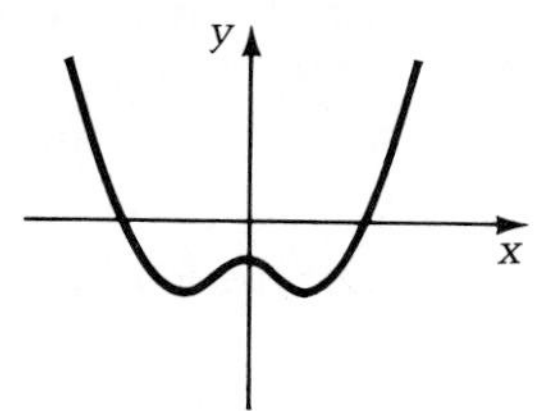

Graph No. 3

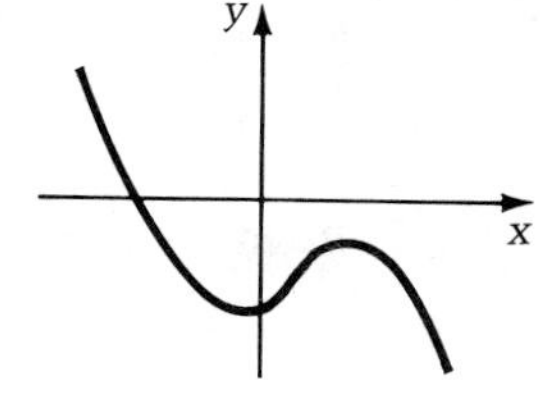

Graph No. 4

Superset

In Exercises 37–46, use your knowledge of symmetry and transformations (Sections 3.1 and 3.3) to graph the given polynomial function.

37. $f(x) = -x^3 + 7$

38. $g(x) = -x^4 - 1$

39. $h(x) = 2 + (x - 1)^3$

40. $k(x) = 3 + 2(x + 1)^3$

41. $A(x) = 1 - (x + 2)^4$

42. $B(x) = 2(x + 3)^4 - 20$

43. $F(x) = 2x^4 + 1$

44. $G(x) = 3x^3 + 2$

45. $f(x) = 2(x + 1)^4$

46. $g(x) = 3(x + 2)^3$

In Exercises 47–52, determine from its graph whether the function could be a polynomial function. Explain.

47.

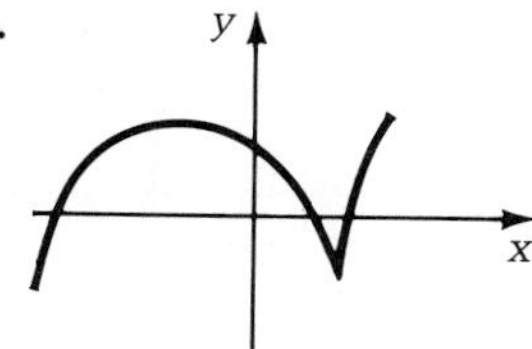

48.

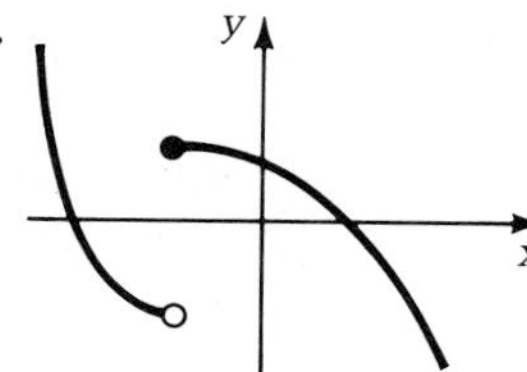

49.

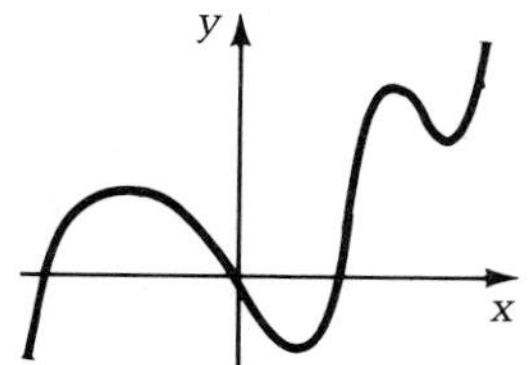

50.

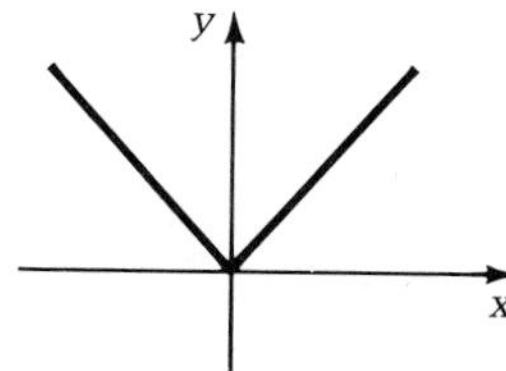

51.

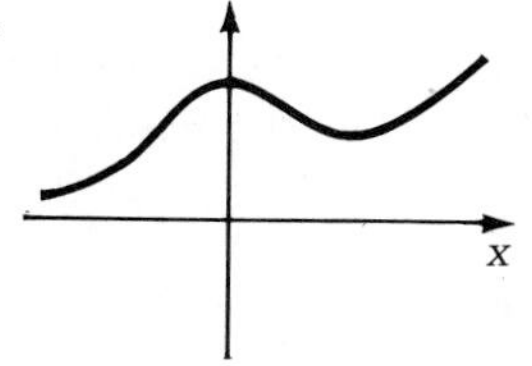

52.

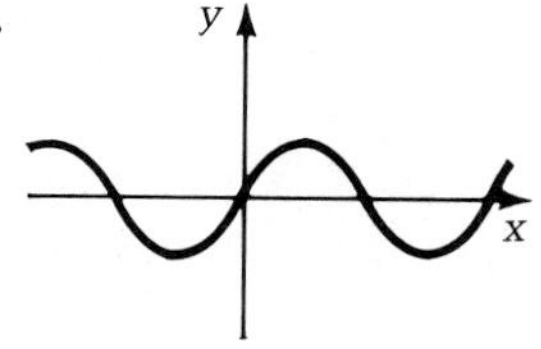

In Exercises 53–56, select the graph that best represents the given polynomial function. You should use the First Property of Polynomial Functions. (Also, determining the y-intercept might help.)

53. $f(x) = -x^3 + x^2 + x - 1$

Graph No. 1

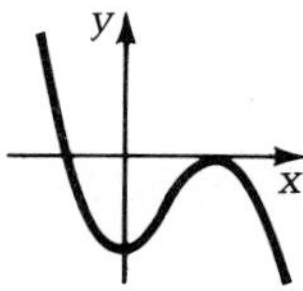

Graph No. 2

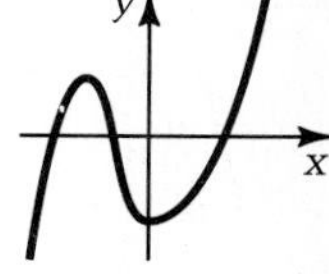

Graph No. 3

54. $f(x) = x^3 - 6x^2 + 9x + 5$

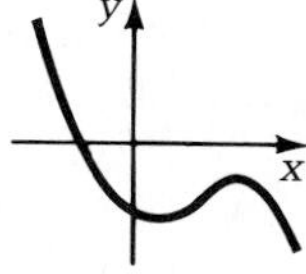

Graph No. 1

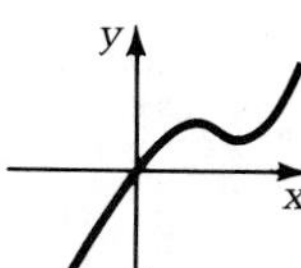

Graph No. 2

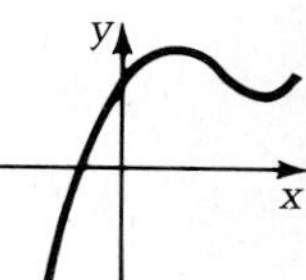

Graph No. 3

55. $y = x^4 - 3x^3 + 3x^2 - x$

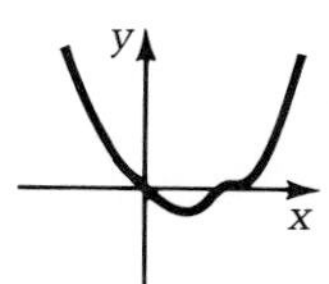

Graph No. 1

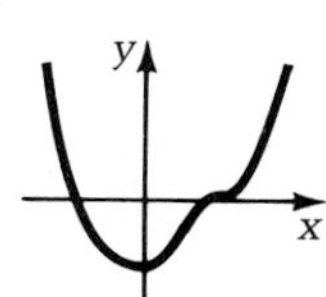

Graph No. 2

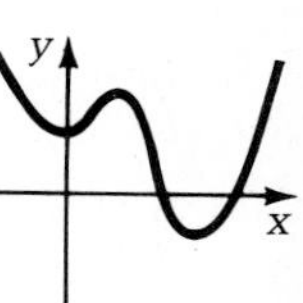

Graph No. 3

56. $f(x) = x^4 - x^3 - 4x^2 + 4x$

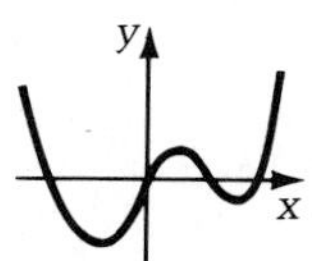

Graph No. 1

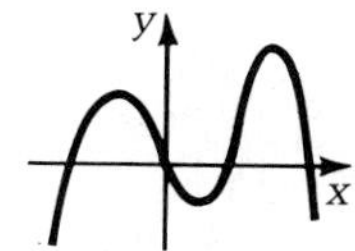

Graph No. 2

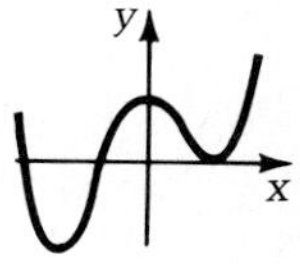

Graph No. 3

4.2
Graphing Polynomial Functions

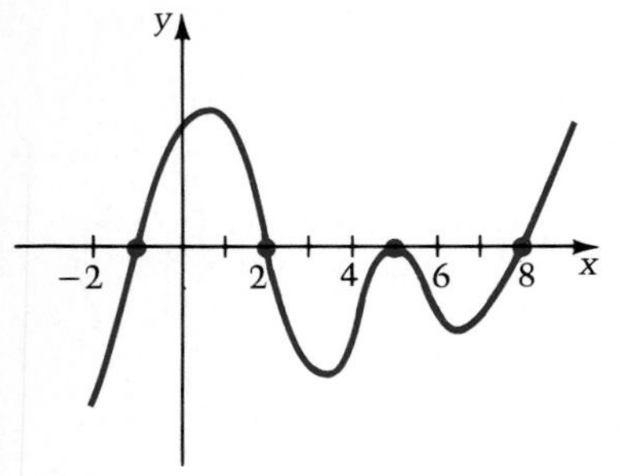

Figure 4.13

The function graphed in Figure 4.13 has x-intercepts at $-1, 2, 5$, and 8. These four x-values are called **zeros of the function,** since they produce function values (y-values) equal to 0. Because they are real numbers, they are called **real zeros** of the function. At a real zero of a function, the graph can behave in one of two ways: either the graph crosses the x-axis (this occurs at -1, 2, and 8 in Figure 4.13) or the graph touches the x-axis without crossing it. (This occurs at 5 in the figure.)

In Figure 4.14 (a), the quadratic function $y = x^2 + x + 3$ is graphed. It has no real zeros and thus will have no x-intercepts. However, since the complex numbers $-\frac{1}{2} + \frac{\sqrt{11}}{2}i$ and $-\frac{1}{2} - \frac{\sqrt{11}}{2}i$ are solutions of the equation $0 = x^2 + x + 3$, they produce function values equal to zero, and thus are called zeros of the function. Since they are nonreal complex numbers, we will refer to them as **imaginary zeros** of the function. (Recall that an imaginary number is a complex number of the form $a + bi$, where $b \neq 0$.) Imaginary zeros *do not* appear as x-intercepts of the graph. Taken together, the real zeros and the imaginary zeros are called the **complex zeros** of the function.

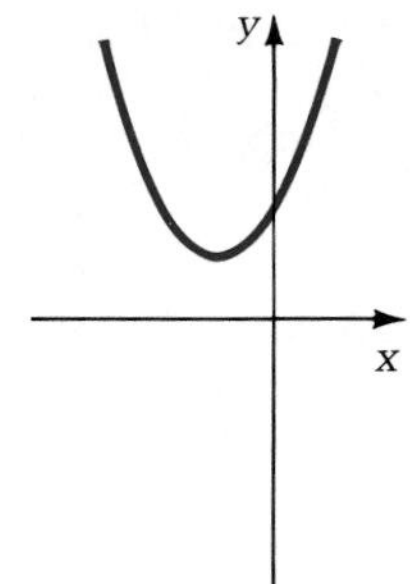

(a) $y = x^2 + x + 3$
No real zeros.
Two imaginary zeros.

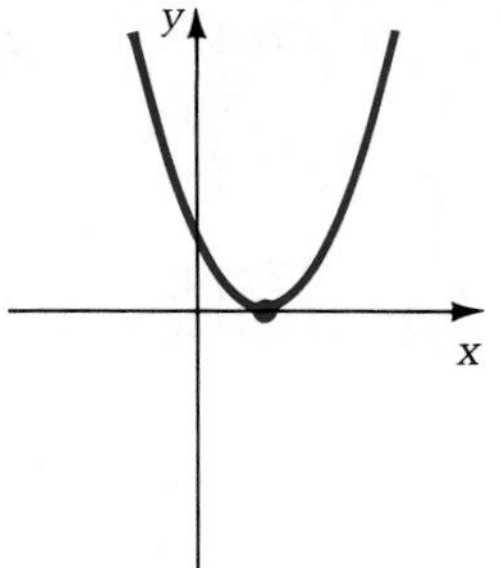

(b) $y = x^2 - 2x + 1$
1 real zero, $x = 1$, of multiplicity two.

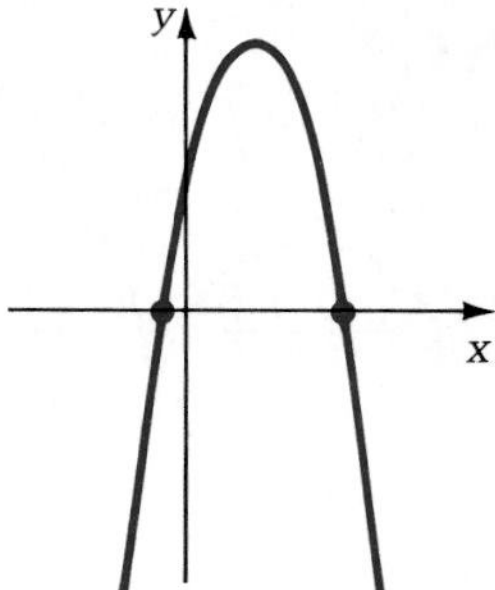

(c) $y = -x^2 + 4x + 5$
2 real zeros, $x = -1$ and $x = 5$.

Figure 4.14

It is clear from Figure 4.14 that a quadratic function $f(x) = ax^2 + bx + c$ can have 0, 1, or 2 real zeros, depending on whether the equation $0 = ax^2 + bx + c$ has 0, 1, or 2 real number solutions. Since a quadratic function has at most two real zeros, its graph cannot intersect the x-axis more than twice.

In general, an nth degree polynomial equation of the form

$$0 = a_n x^n + a_{n-1}x^{n-1} + \cdots + a_1 x + a_0$$

can have *at most n* real number solutions (also referred to as **roots**). Since the roots of this equation are precisely the zeros of the function

$$f(x) = a_n x^n + a_{n-1}x^{n-1} + \cdots + a_1 x + a_0$$

we can conclude that an nth degree polynomial function can have at most n real zeros (x-intercepts).

Second Property of Polynomial Functions

The graph of an nth degree polynomial function cannot intersect (cross or touch) the x-axis more than n times.

EXAMPLE 1

Select the graph from the four below that best represents the graph of $f(x) = 2x^4 - 9x^3 + 11x^2 - 4$.

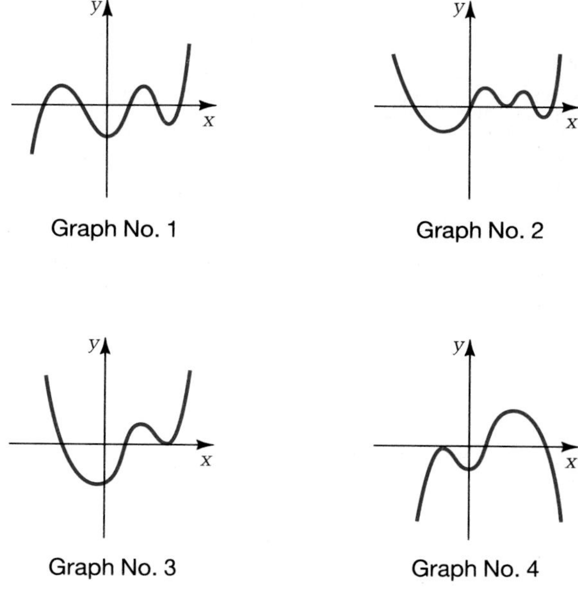

Figure 4.15

Solution By the First Property of Polynomial Functions, we know that the graph of the given function must look like the graph of $y = 2x^4$ for extreme values of x. This narrows the possibilities to Graphs 2 and 3. By the Second Property, there can be at most 4 x-intercepts, and so Graph 2 is disqualified. Thus, Graph 3 is the correct choice.

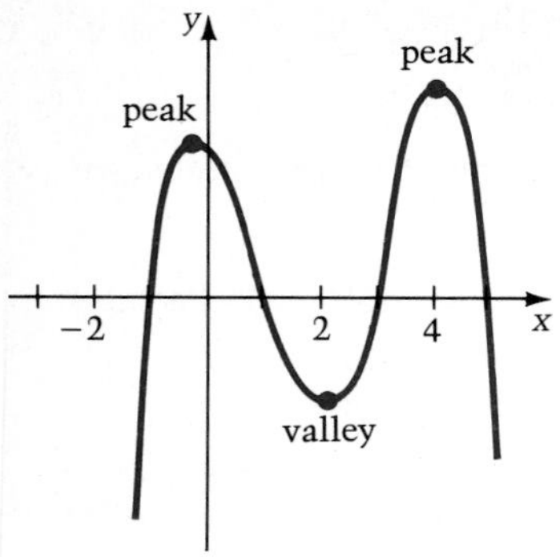

Figure 4.16

At the left is the graph of a fourth degree polynomial function having 4 zeros. To the left of the smallest zero, -1, and to the right of the largest zero, 5, the graph must rise or fall indefinitely. (In between -1 and 5 the graph contains waves.) The four zeros divide the x-axis into five intervals, and on three of these intervals the graph must have a peak or a valley. We shall use the term **relative maximum** instead of peak, and the term **relative minimum** instead of valley (plural: maxima and minima).

Figure 4.16 suggests a very useful property for graphing polynomial functions.

Third Property of Polynomial Functions

The graph of an nth degree polynomial function can have at most $n - 1$ relative maxima and minima.

EXAMPLE 2

Select the graph from the four below that best represents the graph of $f(x) = -2x^5 + x^4 - 3x^2 + x - 2$.

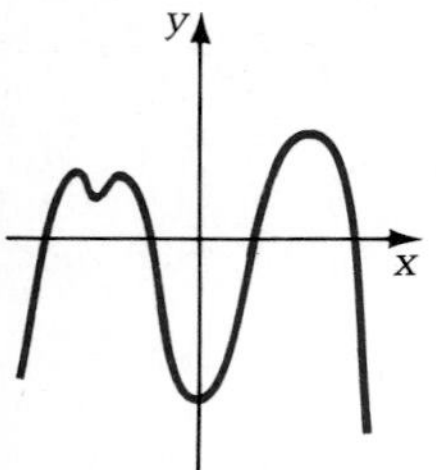

Graph No. 1

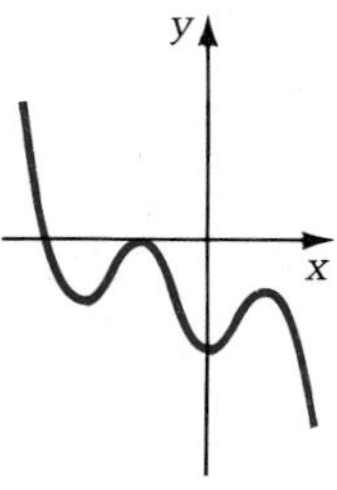

Graph No. 2

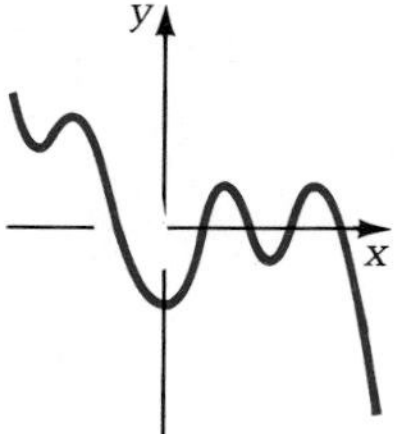

Graph No. 3

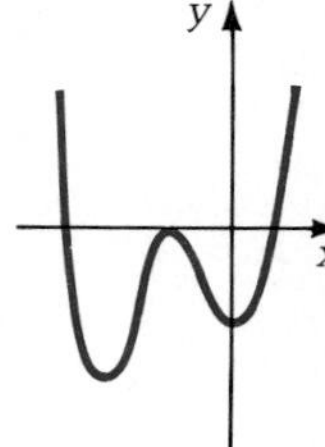

Graph No. 4

Figure 4.17

Solution By the Third Property of Polynomial Functions, we know that the graph of the given function has at most 4 relative maxima and minima. This narrows down the possibilities to Graphs 2 and 4. By the First Property we know that the graph must look like the graph of $y = -2x^5$ towards the extremes; therefore Graph 4 is disqualified. Thus, Graph 2 is the correct choice.

In order to determine the zeros of the polynomial function $y = f(x)$, it is necessary to solve the equation $f(x) = 0$. For example, for the quadratic function $y = x^2 - 7x - 8$, we follow a three step procedure to find the zeros.

$x^2 - 7x - 8 = 0$		The polynomial is set equal to 0.
$(x - 8)(x + 1) = 0$		The left side is factored.
$x - 8 = 0$	$x + 1 = 0$	The Property of Zero Products is applied. The
$x = 8$	$x = -1$	zeros are 8 and -1.

We can use the same method to find the zeros of any polynomial function.

EXAMPLE 3

Determine the zeros of the function $g(x) = x^4 + 5x^3 + 4x^2$.

Solution

$$x^4 + 5x^3 + 4x^2 = 0$$
$$x^2(x^2 + 5x + 4) = 0$$
$$x^2(x + 4)(x + 1) = 0$$

Factor the polynomial as completely as possible, and set each factor equal to zero.

The zeros are 0, -4, and -1. (Note, 0 is a zero of multiplicity 2, because x^2 is a factor.)

EXAMPLE 4

Determine the real zeros of the function $h(x) = (3x + 1)(x - 7)(x^2 + 2x + 2)$.

Solution

$$(3x + 1)(x - 7)(x^2 + 2x + 2) = 0$$

$3x + 1 = 0$	$x - 7 = 0$	$x^2 + 2x + 2 = 0$
$x = -\dfrac{1}{3}$	$x = 7$	The discriminant $b^2 - 4ac < 0$; thus there are no real solutions.

The real zeros are $-\dfrac{1}{3}$ and 7.

Determining the zeros of a polynomial function which is *not* in factored form can be difficult. (Techniques for finding zeros are presented in Sections 3, 4, 5, and 6 of this chapter.) For the moment, we shall restrict our attention to polynomial functions for which complete factorizations are relatively simple.

Let us agree that if c_1 and c_2 are real zeros of a polynomial function f, such that there are no other zeros between c_1 and c_2, then c_1 and c_2 will be called **adjacent zeros** of f.

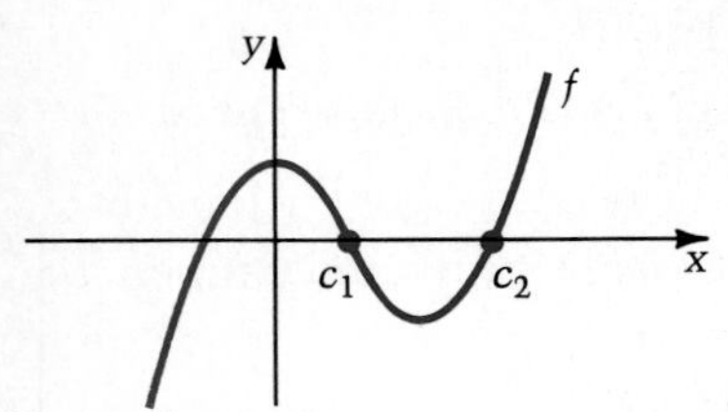

Figure 4.18 c_1 and c_2 are adjacent zeros of f.

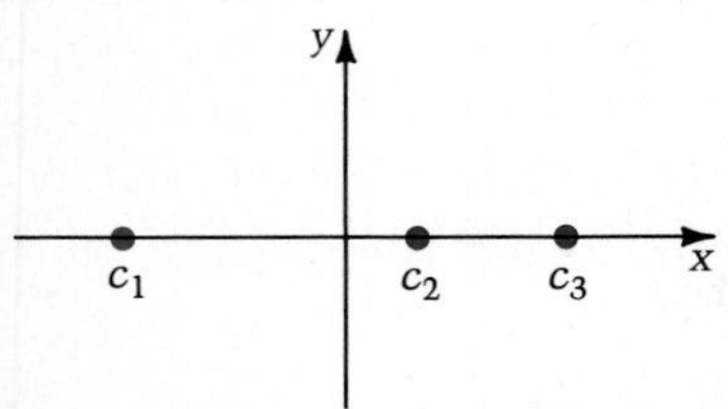

Figure 4.19 Zeros c_1, c_2, and c_3 break the x-axis into 4 intervals: $(-\infty, c_1)$, (c_1, c_2), (c_2, c_3), and (c_3, ∞).

Fourth Property of Polynomial Functions

If c_1 and c_2 are two adjacent zeros of a polynomial function f, then the graph of f lies either completely above or completely below the x-axis for all values between c_1 and c_2.

The Fourth Property is of great use in sketching the graph of a polynomial function. First, find all the real zeros of the function, and order them from smallest to largest. Next, use these zeros to divide the x-axis into open intervals. Evaluate the function at some convenient test value in each interval. If the function is positive at the test value, then the graph lies completely above the x-axis for all x-values in that interval. If the function is negative at the test value, then the graph lies completely below the x-axis for all x-values in that interval.

We are now in a position to outline a general procedure for sketching the graph of a polynomial function.

Step 1. Find the real zeros of the function.

Step 2. Apply the Fourth Property (select test values).

Step 3. Use the results of Step 2 to begin the sketch.

Step 4. Determine the leading term and apply the First, Second, and Third Properties.

EXAMPLE 5

Sketch the graph of the function $f(x) = (x - 1)(x + 2)^2(x - 2)$.

Solution

Step 1. To find the real zeros, solve the equation

$$(x - 1)(x + 2)^2(x - 2) = 0.$$

The zeros are -2, 1, and 2 in order from smallest to largest. Plot these zeros.

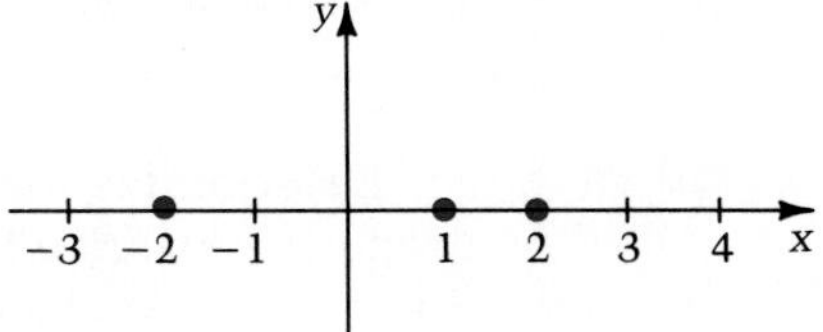

Figure 4.20

Step 2. To apply the Fourth Property, we must first determine the intervals having adjacent zeros as endpoints, and then evaluate the function at a convenient test value in each interval.

Interval	Test Value t	$f(t)$	
$(-\infty, -2)$	-3	20	$f(t) > 0$; graph lies above x-axis
$(-2, 1)$	0	8	$f(t) > 0$; graph lies above x-axis
$(1, 2)$	$\frac{3}{2}$	$-\frac{49}{16}$	$f(t) < 0$; graph lies below x-axis
$(2, \infty)$	3	50	$f(t) > 0$; graph lies above x-axis

Step 3. Using the results from Step 2, sketch the graph a little to the right and to the left of each zero. Since the graph lies above the x-axis both to the left *and* to the right of -2, it must touch but not cross the x-axis at -2. The graph lies above the x-axis to the left of 1 and below the x-axis to the right of 1, so it must cross at 1. Similar reasoning leads us to conclude that the graph is rising as it crosses at 2.

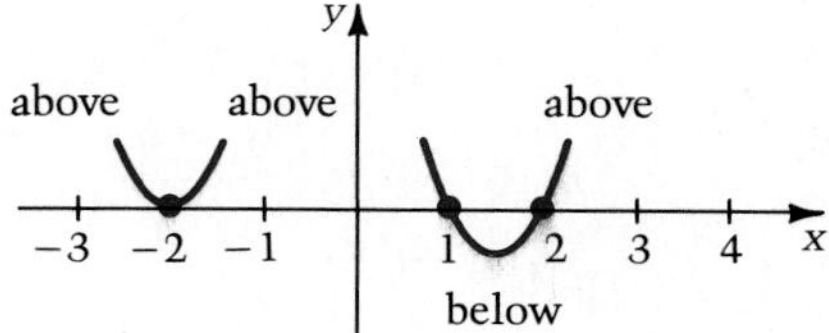

Figure 4.21

Step 4. Multiplying the leading terms from each of the three factors of the polynomial gives us the leading term (and the degree) of the polynomial: $x \cdot x^2 \cdot x = x^4$. Since the degree is 4, there are at most 3 relative maxima and minima (Third Property). Since we know one is located at $x = -2$, one must occur between -2 and 1, and another between 1 and 2.

Although we cannot determine exactly where the relative maxima and minima occur without the aid of calculus, we can complete the graph by plotting a few additional points, such as those chosen in Step 2. Notice that toward the extremes the graph behaves like $y = x^4$ (First Property).

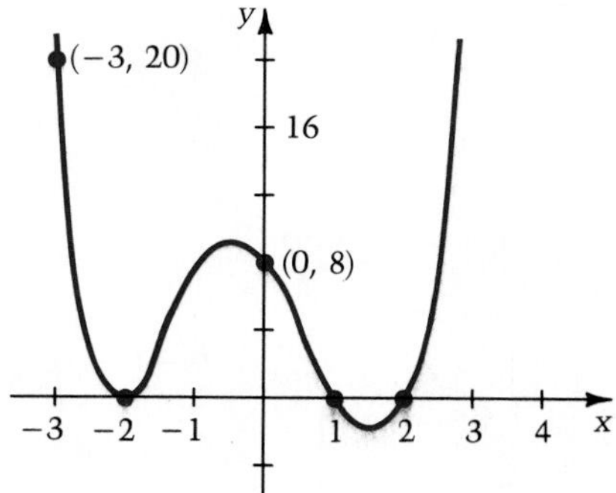

Figure 4.22

As the previous example illustrates, it is sometimes easier to graph a polynomial function if we first factor the polynomial. Recall that one factoring technique is u-substitution, a method that is applied to polynomials that are quadratic in form. In the next example, we shall use u-substitution to factor the polynomial.

EXAMPLE 6

Sketch the graph of $f(x) = (x - 3)(x^4 - 5x^2 + 4)$.

Solution

Step 1. Since $x - 3$ is a factor of the polynomial, we know that 3 is a zero. Notice that the other factor is quadratic in x^2. We can use u-substitution to factor it.

$$(x - 3)(x^4 - 5x^2 + 4) = 0$$

$x - 3 = 0$	$x^4 - 5x^2 + 4 = 0$	Let $u = x^2$.
	$u^2 - 5u + 4 = 0$	
	$(u - 4)(u - 1) = 0$	Thus, $u = 1$ or 4, i.e. $x^2 = 1$ or 4.

The real zeros of this function are $x = -2, -1, 1, 2$, and 3 from smallest to largest.

Step 2. Now we select a test value in each interval and apply the Fourth Property.

Interval	Test Value t	$f(t)$	
$(-\infty, -2)$	-3	-240	$f(t) < 0$; graph lies below x-axis
$(-2, -1)$	$-\frac{3}{2}$	$\frac{315}{32}$	$f(t) > 0$; graph lies above x-axis
$(-1, 1)$	0	-12	$f(t) < 0$; graph lies below x-axis
$(1, 2)$	$\frac{3}{2}$	$\frac{105}{32}$	$f(t) > 0$; graph lies above x-axis
$(2, 3)$	$\frac{5}{2}$	$-\frac{189}{32}$	$f(t) < 0$; graph lies below x-axis
$(3, \infty)$	4	180	$f(t) > 0$; graph lies above x-axis

Step 3.

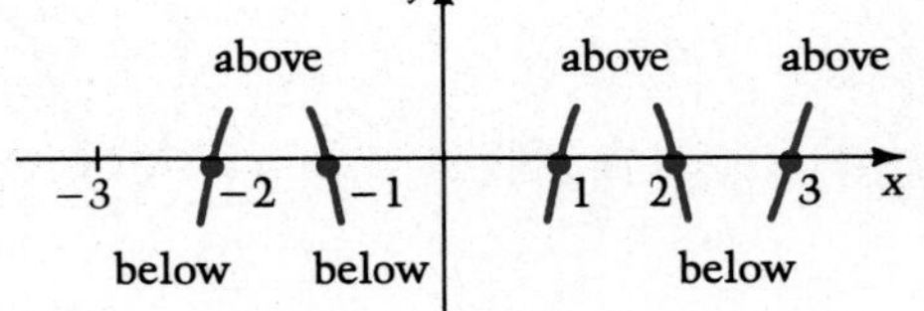

Sketch the graph a little to the right and to the left of each zero.

Figure 4.23

Step 4. Since the degree of the polynomial is 5, there can be at most 4 relative maxima and minima (Third Property). The relative maxima occur in the intervals $(-2, -1)$, and $(1, 2)$. The relative minima are located in the intervals $(-1, 1)$ and $(2, 3)$.

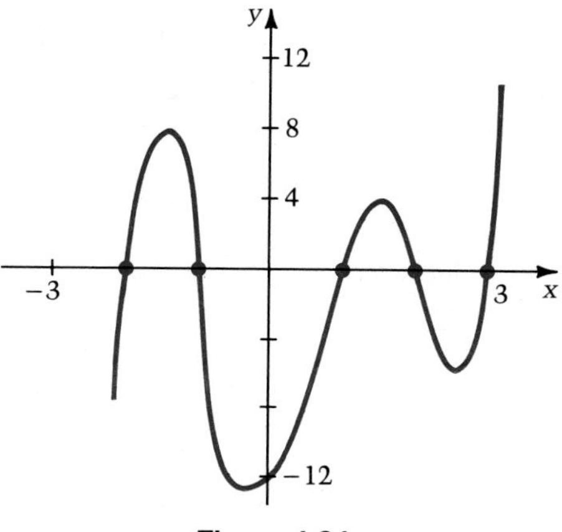

Figure 4.24

Thus far we have only worked examples of graphs for nth degree polynomial functions having precisely n real roots (counting multiplicities). It is natural to wonder what happens when some of the zeros are not real numbers. Of course, the Properties still apply but might not give us the complete picture. Consider the polynomial functions whose graphs are shown in Figure 4.25.

Notice that each function has one real zero and two imaginary zeros. Yet the two graphs are significantly different. One graph has no relative maxi-

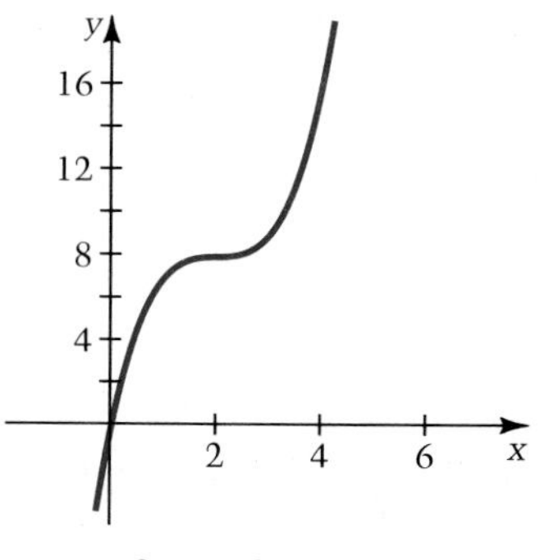

$f(x) = x^3 - 6x^2 + 12x$
zeros: $0, 3 + i\sqrt{3}, 3 - i\sqrt{3}$

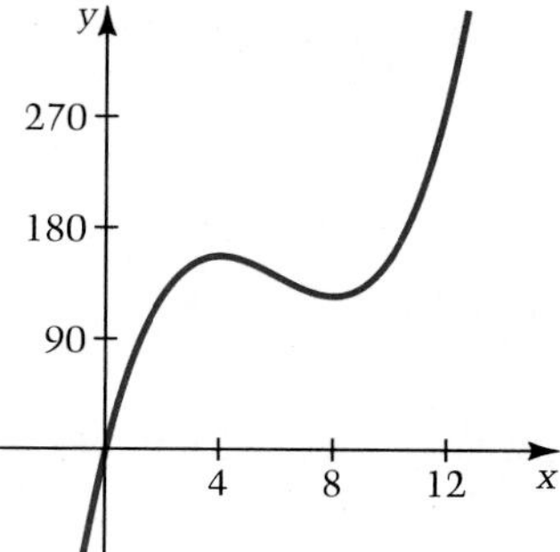

$g(x) = x^3 - 18x^2 + 96x$
zeros: $0, 9 + i\sqrt{15}, 9 - i\sqrt{15}$

Figure 4.25

mum or minimum, while the other graph has both a relative maximum and a relative minimum. To determine such distinctions we must wait for techniques that are covered in a calculus course. Until then, plotting many points will usually help in sketching a graph precisely, and is best accomplished with the assistance of a calculator or computer.

EXERCISE SET 4.2

In Exercises 1–4, use the First, Second, and Third Properties of Polynomial Functions to select the graph that best represents the function.

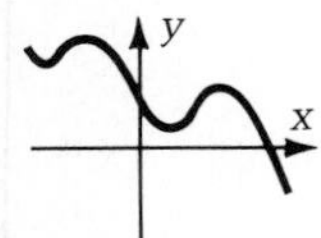

Graph No. 1

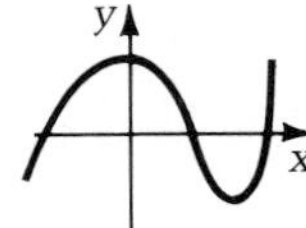

Graph No. 2

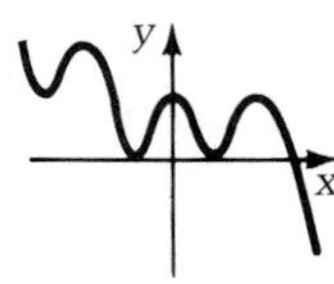

Graph No. 3

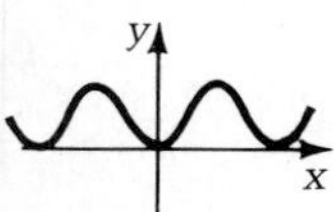

Graph No. 4

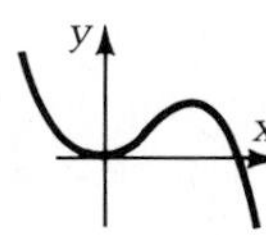

Graph No. 5

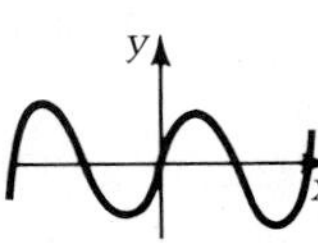

Graph No. 6

1. $y = 12x^2 - x^3 - x$ **2.** $y = x^6 - 6x^4 + 9x^2$

3. $y = x^5 - 5x^3 + 4x$

4. $y = -x^5 + 10x^3 - 25x + 20$

In Exercises 5–8, use the First, Second, and Third Properties of Polynomial Functions to select the graph that best represents the function.

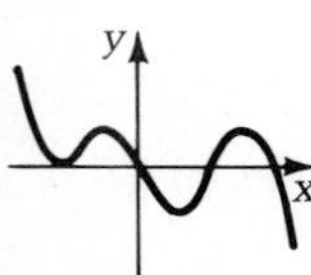

Graph No. 1

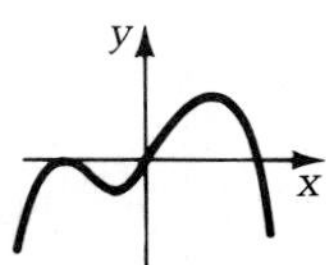

Graph No. 2

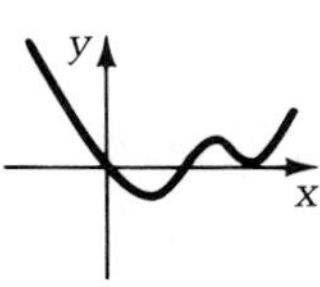

Graph No. 3

Graph No. 4

Graph No. 5

Graph No. 6

5. $y = x^3 - 2x^2$ **6.** $y = x^4 - x^3 - x^6 + x$

7. $f(x) = 2x^4 - 5x^3 + 4x^2 - x$

8. $f(x) = 2x^4 + x^3 - 1$

In Exercises 9–16, determine the real zeros of the given function.

9. $f(x) = (x + 3)(x - 5)(x + 7)^3$

10. $f(x) = -5x(x + 2)^2(3x - 1)$

11. $f(x) = x(3x - 2)^2(x^2 + 3x + 2)$

12. $f(x) = x^2(2 - x)^3(x^2 + 4x + 3)$

13. $f(x) = (x - 5)^2(2x + 5)^3(x^2 + 2x + 2)$

14. $f(x) = x(7 - 2x)^2(x^2 + x + 1)$

15. $f(x) = x^4 - 7x^2 + 12$ **16.** $f(x) = x^4 - 11x^2 + 18$

For Exercises 17–24, sketch the graphs of the functions in Exercises 9–16.

In Exercises 25–38, find the real zeros of the given function; then sketch the graph.

25. $y = x^3 - 3x^2$ **26.** $y = 5x^3 - x^4$

27. $y = (x^2 + 3x + 2)(x^2 - 3x - 4)$

28. $y = (x^2 - 2x - 3)(x^2 - 5x + 6)$

29. $y = (2x - 1 - x^2)(2x + 2 + x^2)$

30. $y = (2x - 1 - 2x^2)(6x - 9 - x^2)$

31. $y = x^7 - 3x^5 - 4x^3$ **32.** $y = 7x^4 - x^6 + 18x^2$

33. $y = (2 - x)^3(x + 3)^2$ **34.** $y = (x - 1)^4(x - 3)^2$

35. $y = (1 - 2x)^2(2 + x)^2$ **36.** $y = (4 + x)(5 - 2x)^3$

37. $y = (2 - x)(x + 4)^2(3x + 2)^2$

38. $y = (3 - x)(5 + x)(4 - 3x)^3$

Superset

In Exercises 39–46, give an example of a polynomial function, written as a product of linear factors, whose graph according to the First, Second, Third, and Fourth Properties resembles the given graph.

39.

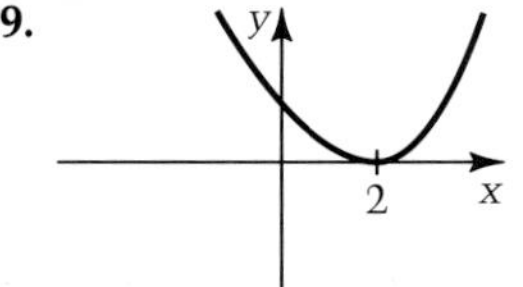

40.

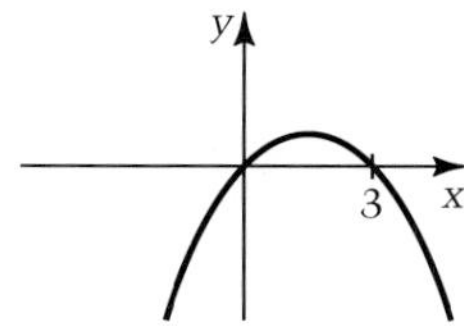

41.

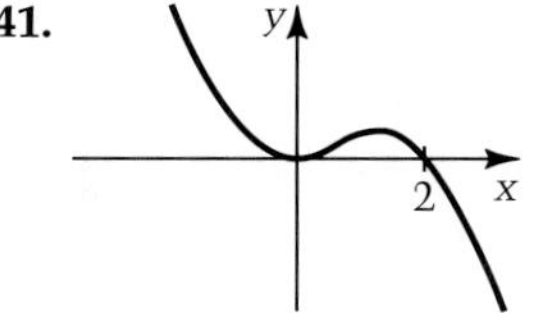

42.

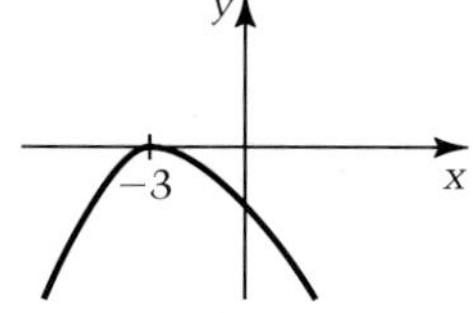

43.

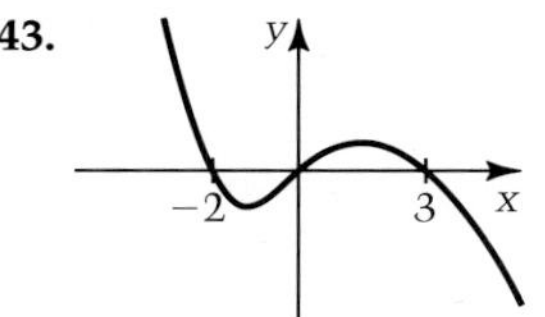

44.

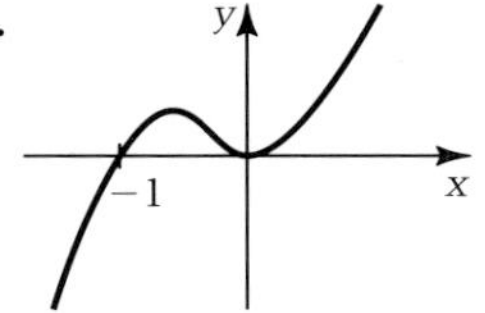

45.

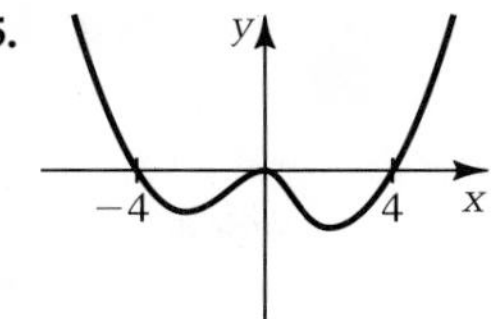

46.

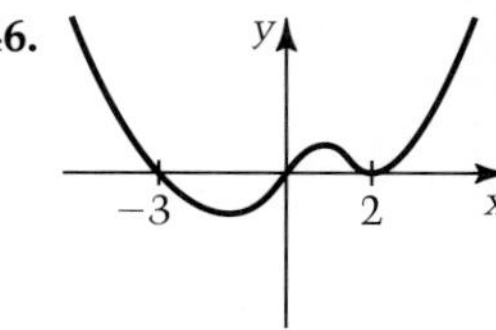

In Exercises 47–50, use the First, Second, Third, and Fourth Properties to determine the sign (positive or negative) of the lead coefficient, the smallest possible value of the degree $= n$ and the smallest possible number $= m$ of real zeros (counting multiplicities) for the polynomial function whose graph is given.

47.

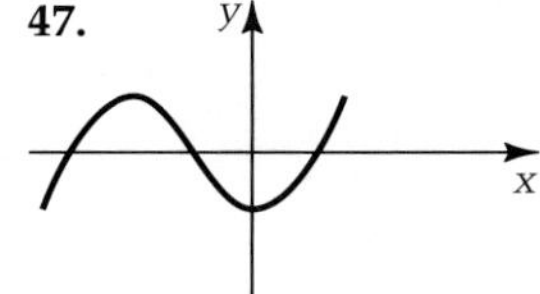

48.

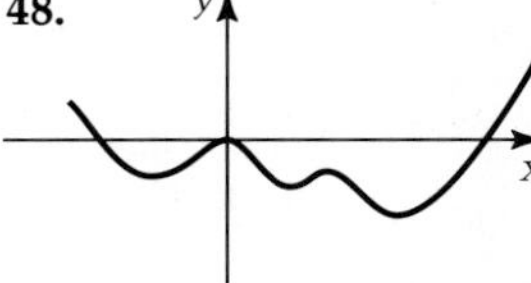

49.

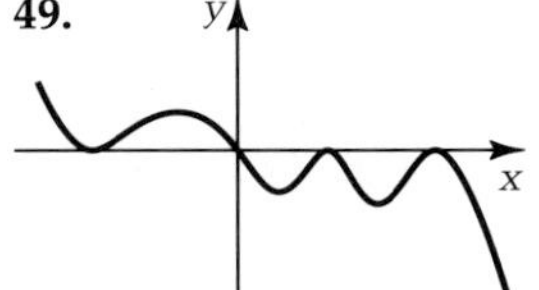

50.

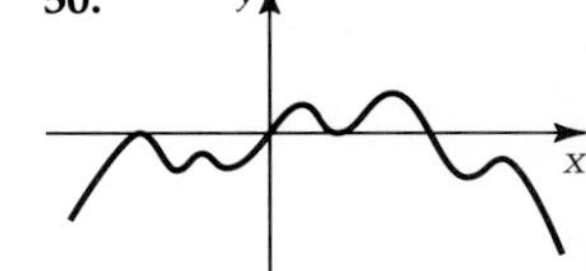

4.3
Division of Polynomials and the Factor Theorem

In this section we begin a rather careful study of techniques for determining the zeros of a polynomial function $f(x)$. We will discover that if a number c is a zero of $f(x)$, that is, $f(c) = 0$, then $x - c$ is a factor of the polynomial $f(x)$. We start our work by reviewing polynomial division.

The method for dividing one polynomial by another is quite similar to the long division process used with whole numbers. In fact, in both cases the names of the key elements are precisely the same.

$$
\begin{array}{rrl}
\text{divisor} & 37 & \leftarrow \text{quotient} \\
\hookrightarrow 7\,) & 260 & \leftarrow \text{dividend} \\
 & -21 & \\
\hline
 & 50 & \\
 & -49 & \\
\hline
 & 1 & \leftarrow \text{remainder}
\end{array}
$$

Thus, $260 = 7(37) + 1$.

Now we divide polynomials.

$$
\begin{array}{rrl}
\text{divisor} & x - 3 & \longleftarrow \text{quotient} \\
\hookrightarrow x - 2 \,\big) & x^2 - 5x + 8 & \longleftarrow \text{dividend} \\
& -(x^2 - 2x) & \\
\hline
& -3x + 8 & \\
& -(-3x + 6) & \\
\hline
& 2 & \longleftarrow \text{remainder}
\end{array}
$$

Thus, $x^2 - 5x + 8 = (x - 2)(x - 3) + 2$.

We will illustrate the technique for dividing polynomials in the following examples.

EXAMPLE 1

Divide $f(x) = 6x^4 + x^3 - 3x + 5$ by $d(x) = 3x^2 + 5x + 6$.

Solution The coefficient of x^2 in the dividend, $f(x)$, is zero, so we leave space for x^2-terms that will occur during the division process. The process itself consists of a two-step sequence of "multiplying and subtracting" that is repeated *until the degree of the remainder is less than the degree of the divisor, $d(x)$.*

The quotient of the leading terms in dividend and divisor gives us the first term in the quotient: $\dfrac{6x^4}{3x^2} = 2x^2$.

$$
\begin{array}{r l}
 & 2x^2 \\
3x^2 + 5x + 6 \,\big) & 6x^4 + x^3 - 3x + 5
\end{array}
$$

Now multiply $2x^2$ by the divisor, then subtract this product from the dividend.

$$
\begin{array}{r rrrrr}
 & 2x^2 & -\ 3x & +\ 1 & & \\
3x^2 + 5x + 6 \,\big) & 6x^4 & +\ x^3 & & -\ 3x & +\ 5 \\
 & -(6x^4 & +\ 10x^3 & +\ 12x^2) & & \\
\hline
 & & -\ 9x^3 & -\ 12x^2 & -\ 3x & +\ 5 \\
 & & -(-\ 9x^3 & -\ 15x^2 & -\ 18x) & \\
\hline
 & & & 3x^2 & +\ 15x & +\ 5 \\
 & & & -(3x^2 & +\ 5x & +\ 6) \\
\hline
 & & & & 10x & -\ 1
\end{array}
$$

$\dfrac{-9x^2}{3x^2} = -3x$; add $-3x$ to quotient; multiply divisor by $-3x$; subtract.

$\dfrac{3x^2}{3x^2} = 1$, add 1 in quotient; multiply divisor by 1; and subtract.

The quotient is $2x^2 - 3x + 1$, and the remainder is $10x - 1$. The result can be written as $6x^4 + x^3 - 3x + 5 = (3x^2 + 5x + 6)(2x^2 - 3x + 1) + (10x - 1)$.

EXAMPLE 2

Divide $f(x) = 2x^3 - 5x^2 - 11x - 4$ by $d(x) = x - 4$.

Solution

$$
\begin{array}{r|l}
 & 2x^2 + 3x + 1 \\
\hline
x - 4 & 2x^3 - 5x^2 - 11x - 4 \\
 & -(2x^3 - 8x^2) \\
\hline
 & 3x^2 - 11x - 4 \\
 & -(3x^2 - 12x) \\
\hline
 & x - 4 \\
 & -(x - 4) \\
\hline
 & 0
\end{array}
$$

$\frac{2x^3}{x} = 2x^2$. Place $2x^2$ in quotient; multiply divisor by $2x^2$; subtract.

$\frac{3x^2}{x} = 3x$. Add $3x$ to quotient; multiply divisor by $3x$; subtract.

$\frac{x}{x} = 1$. Add 1 to quotient; multiply divisor by 1; subtract.

The remainder is zero.

The quotient is $2x^2 + 3x + 1$, and the remainder is 0. The result can be written as $2x^3 - 5x^2 - 11x - 4 = (x - 4)(2x^2 + 3x + 1) + 0$.

Notice that in Example 2 the remainder was 0. This means that $x - 4$ *divided evenly* into $2x^3 - 5x^2 - 11x - 4$. In other words, $x - 4$ is a *factor* of $2x^3 - 5x^2 - 11x - 4$. Consequently, we can say that $2x^2 + 3x + 1$ is also a factor of $2x^3 - 5x^2 - 11x - 4$. In each of the previous two examples we wrote the dividend as a product of the divisor times the quotient, plus the remainder. We can always summarize the division of polynomials this way, as the following statement suggests.

The Division Algorithm

When a polynomial $f(x)$ is divided by a nonzero polynomial $d(x)$, there exist unique polynomials $q(x)$, the *quotient*, and $r(x)$, the *remainder*, such that

$$f(x) = d(x) \cdot q(x) + r(x),$$

where either $r(x) = 0$, or else the degree of $r(x)$ is less than the degree of $d(x)$.

In Example 2 we divided by a linear polynomial of the form $x - c$ (in the example $c = 4$). Dividing by such linear polynomials is of great use in finding the zeros of a polynomial function. For that reason we now consider a streamlined technique, called **synthetic division,** that can be used in those situations. On the next page we show a problem that employs the standard long division technique illustrated in Example 2. Beside it we show the same process, but with three simplifications: (1) the variables have been omitted;

(2) terms usually "brought down" for convenience of computation have been omitted; and (3) the first number in each of the products has been deleted since it invariably combines with the number above it to produce 0. We note that omitting the variables is legitimate since the coefficients alone are needed in completing the division process. Of course, we must agree to write the coefficients in order of decreasing powers of x.

$$\begin{array}{r|l}
 & \;2x^2 + 5x - 1 \leftarrow \text{quotient} \\ \hline
x-3 & 2x^3 - x^2 - 16x + 7 \\
 & -(2x^3 - 6x^2) \\ \cline{2-2}
 & \quad 5x^2 - 16x \\
 & \quad -(5x^2 - 15x) \\ \cline{2-2}
 & \qquad -x + 7 \\
 & \qquad -(-x + 3) \\ \cline{2-2}
 & \qquad\quad 4 \leftarrow \text{remainder}
\end{array}
\qquad
\begin{array}{r|rrrr}
\text{quotient} \rightarrow & 2 & 5 & -1 & \\ \hline
1-3 & 2 & -1 & -16 & 7 \\
 & & 6 & & \\ \cline{3-3}
 & & 5 & & \\
 & & & 15 & \\ \cline{4-4}
 & & & -1 & \\
 & & & & -3 \\ \cline{5-5}
\text{remainder} \rightarrow & & & & 4
\end{array}$$

We can further simplify the form on the right by moving all the numbers upward so as to occupy four lines. Since the divisor will always be of the form $x - c$, we need only write c (in this case 3). We show this further simplification below on the left. Notice that the bottom row contains all but the first coefficient of the quotient (in this case 2), and ends with the remainder. By placing this first coefficient in the bottom row, that row will contain all the coefficients of the quotient, followed by the remainder. Thus, the top row is unnecessary. This final simplification is shown on the right below.

$$\begin{array}{r|rrrr}
 & 2 & 5 & -1 & \\ \hline
3 & 2 & -1 & -16 & 7 \\
 & & 6 & 15 & -3 \\ \cline{3-5}
 & & 5 & -1 & 4
\end{array}
\qquad\qquad
\begin{array}{r|rrrr}
3 & 2 & -1 & -16 & 7 \\
 & & 6 & 15 & -3 \\ \cline{2-5}
 & 2 & 5 & -1 & 4
\end{array}$$

Left: 5, −1: coefficients of the quotient with first coefficient missing; 4: remainder

Right: 2, 5, −1: coefficients of the quotient; 4: remainder

In the following example, we describe the steps involved in using synthetic division directly. Note that this technique relies only on the value of c and the coefficients of the polynomial to be divided by $x - c$. Because we are dividing by $x - c$, the degree of the quotient will always be one less than the degree of the dividend.

EXAMPLE 3

Use synthetic division to divide $f(x) = 2x^3 - x^2 - 16x + 7$ by $d(x) = x - 3$.

Solution Begin by writing the top row: $c\rfloor$, from the divisor $x - c$ (in this case $c = 3$), followed by the coefficients of the dividend.

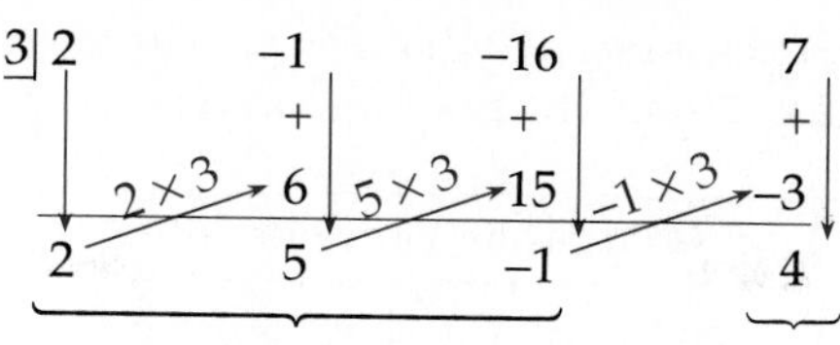

Bring down the first coefficient, 2. Then successively multiply by 3 and add to the next coefficient.

The degree of the quotient is 2, one less than that of the dividend.

quotient: $2x^2 + 5x - 1$ remainder

Thus, $f(x) = (x - 3)(2x^2 + 5x - 1) + 4$.

EXAMPLE 4

Use synthetic division to divide $f(x) = x^4 - 3x^3 + 2x - 1$ by $d(x) = x + 2$.

Solution Since $x + 2 = x - (-2)$, in this case $c = -2$. For a missing power of x, record the coefficient 0.

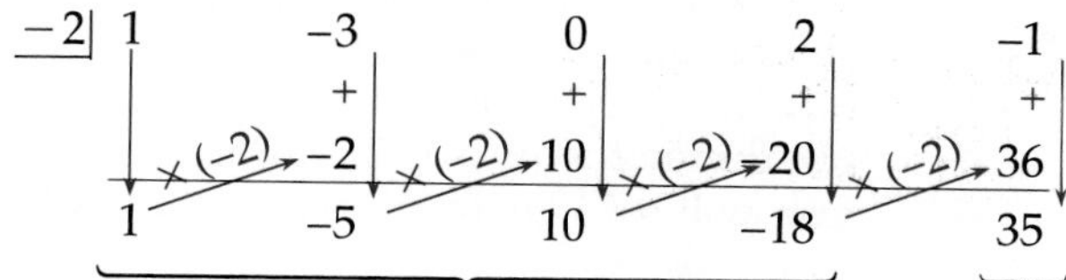

Since there is no x^2-term, its coefficient must be recorded as 0.

quotient: $x^3 - 5x^2 + 10x - 18$ remainder

The degree of the quotient is one less than that of the dividend.

Thus, $f(x) = (x + 2)(x^3 - 5x^2 + 10x - 18) + 35$.

The remainders obtained in the last two examples are not just any two numbers. To discover what these two numbers really are, we look at the Division Algorithm with the divisor $d(x) = x - c$:

$$f(x) = (x - c) \cdot q(x) + r(x).$$

The Division Algorithm tells us that either $r(x) = 0$, or the degree of $r(x)$ is less than the degree of $x - c$, which is 1. This means that either $r(x) = 0$, or the degree of $r(x)$ is 0. In either case $r(x)$ must be a constant, so we can write $f(x) = (x - c) \cdot q(x) + r$, where the remainder is some number. This statement is true for any value of x, and so it must be true in particular when $x = c$. Substituting c for x, we have

$$f(c) = (c - c) \cdot q(c) + r$$
$$f(c) = 0 \cdot q(c) + r$$
$$f(c) = r$$

This last line says that the remainder r is simply the number you get when you evaluate the polynomial $f(x)$ at c.

The Remainder Theorem

If a polynomial $f(x)$ is divided by $x - c$, then the remainder is equal to $f(c)$.

In Example 3, with $f(x) = 2x^3 - x^2 - 16x + 7$, we found that

$$f(x) = (x - 3)(2x^2 + 5x - 1) + 4.$$

The Remainder Theorem assures us that $f(3) = 4$, the remainder. Similarly, in the division that was illustrated in Example 4, we concluded that with $f(x) = x^4 - 3x^3 + 2x - 1$, we have

$$f(x) = (x + 2)(x^3 - 5x^2 + 10x - 18) + 35,$$

so that by the Remainder Theorem, $f(-2) = 35$. Note that in this last case, since the divisor was $x + 2 = x - (-2)$, the value of c was -2.

Thus to compute $f(c)$ for some polynomial $f(x)$ and some number c, we can divide $f(x)$ by $x - c$. The remainder is then $f(c)$. Of course synthetic division will help us to do such division very efficiently.

EXAMPLE 5

Let $f(x) = x^3 - 7x + 2$. Evaluate $f(-3)$.

Solution We use synthetic division to divide $f(x)$ by $x - (-3) = x + 3$.

-3	1	0	-7	2	The coefficient of x^2 is 0.
		-3	9	-6	
	1	-3	2	-4	
	quotient			remainder	

By the Remainder Theorem, $f(-3) = -4$.

For the polynomial $f(x)$ and the linear factor $x - c$, we have seen that

$$f(x) = (x - c) \cdot q(x) + r.$$

Certainly if $x - c$ is a factor of $f(x)$, then the remainder r must be 0. But, by the Remainder Theorem, this means that $f(c) = 0$, and so c is a zero of $f(x)$. Conversely, suppose c is a zero of $f(x)$. This means that $f(c) = 0$, but since $f(c) = r$ by the Remainder Theorem, we can say that $r = 0$. Thus $x - c$ must be a factor of $f(x)$. We have just proved the following theorem, which will be central to our efforts in finding zeros of polynomial functions.

The Factor Theorem

The number c is a zero of a polynomial function $y = f(x)$ if and only if $x - c$ is one of the factors of $f(x)$.

The Factor Theorem says that if c is one of the zeros of a polynomial function, then $x - c$ must be one of the factors of the polynomial. Knowing one of the factors of an expression is often the key to factoring the expression completely.

EXAMPLE 6

Use synthetic division to determine whether $x - 1$ or $x + 3$ is a factor of $x^3 - x^2 - 8x + 12$.

Solution First test $x - 1$, by dividing synthetically by 1.

$$\begin{array}{r|rrrr} 1 & 1 & -1 & -8 & 12 \\ & & 1 & 0 & -8 \\ \hline & 1 & 0 & -8 & 4 \end{array} \quad \leftarrow \text{remainder} = f(1) = 4$$

Since $f(1)$ *is not* equal to 0, 1 is not a zero of the polynomial, and so $x - 1$ *is not* a factor.

Next test $x + 3 = x - (-3)$, by dividing synthetically by -3.

$$\begin{array}{r|rrrr} -3 & 1 & -1 & -8 & 12 \\ & & -3 & 12 & -12 \\ \hline & 1 & -4 & 4 & 0 \end{array} \quad \leftarrow \text{remainder} = f(-3) = 0$$

Since $f(-3)$ *is equal* to 0, -3 is a zero of the polynomial, and so $x + 3$ *is* a factor.

Now consider a related problem. Suppose we are trying to graph some polynomial function $f(x)$, and we need to find all the real number zeros, since these will correspond to the x-intercepts of the graph. Furthermore, suppose we know that some real number c is one of the zeros. The following example demonstrates that we can first divide $f(x)$ by $x - c$, and then concentrate on factoring the quotient, which may be easier to factor since it is of a lesser degree.

EXAMPLE 7

Find all the real zeros of the function $f(x) = 2x^3 + 9x^2 - 32x + 21$ given that one of the zeros is -7.

Solution Since -7 is a zero of $f(x)$, we know that $x - (-7) = x + 7$ is a factor of $f(x)$.

$$\begin{array}{r|rrrr} -7 & 2 & 9 & -32 & 21 \\ & & -14 & 35 & -21 \\ \hline & 2 & -5 & 3 & 0 \end{array}$$

We use synthetic division to divide $f(x)$ by $x + 7$. We want to write $f(x) = (x + 7) \cdot q(x)$, where $q(x)$ is the quotient.

We conclude that $f(x) = (x + 7)(2x^2 - 5x + 3)$. To find the zeros, set the polynomial equal to zero, factor the polynomial, then solve.

$$(x + 7)(2x^2 - 5x + 3) = 0$$
$$(x + 7)(2x - 3)(x - 1) = 0$$

$x + 7 = 0$	$2x - 3 = 0$	$x - 1 = 0$
$x = -7$	$2x = 3$	$x = 1$
	$x = \frac{3}{2}$	

Don't forget! The Property of Zero Products is the main strategy for solving polynomial equations.

The real zeros are $-7, \frac{3}{2}$, and 1.

EXERCISE SET 4.3

In Exercises 1–12, perform the indicated division.

1. Divide $x^3 - 2x^2 + 5x + 6$ by $x + 1$.
2. Divide $x^3 - 3x^2 + 4x + 1$ by $x - 2$.
3. Divide $x^4 - 3x^3 + 2x - 1$ by $x - 3$.
4. Divide $x^4 - 7x^2 + 4x - 6$ by $x + 3$.
5. Divide $x^4 + 2x^3 - 5x + 2$ by $x - 1$.
6. Divide $x^4 - 5x^2 + 3x + 7$ by $x + 2$.
7. Divide $x^3 - 3x^2 + 4x$ by $x^2 + 2$.
8. Divide $x^3 + 5x - 2$ by $x^2 + 3$.
9. Divide $x^4 + 13$ by $x^2 - 2x + 2$.
10. Divide $x^4 + 5$ by $x^2 + 2x + 2$.
11. Divide $x^6 - 1$ by $x^2 + x + 1$.
12. Divide $x^5 + x^2$ by $x^2 - x + 1$.

In Exercises 13–18, use synthetic division to determine which, if any, of the linear polynomials $x - 1, x + 2, x - 2$, and $x + 3$ is a factor of the given polynomial.

13. $x^4 + 4x^3 - 4x^2 - 28x - 21$
14. $x^4 - 13x^2 + 12x$
15. $x^5 + 2x^4 - 19x^3 - 8x^2 + 60x$
16. $x^4 + 4x^3 - 2x^2 - 12x + 9$
17. $x^4 - x^3 - 9x^2 + 11x + 6$
18. $x^5 - 2x^4 - 2x^3 + 4x^2 + x - 2$

In Exercises 19–24, use the Remainder Theorem to verify the indicated function values.

19. $f(x) = 2x^3 - 3x^2 + 6x - 7; f(-1) = -18, f(1) = -2$
20. $g(x) = x^4 - 3x^3 + 4x^2 - x - 1; g(-2) = 57, g(2) = 5$

21. $G(x) = 10 - 4x + x^2 - 3x^3 + x^5$; $G(-2) = 14$, $G(1) = 5$

22. $F(x) = 1 - x + 8x^3 - 10x^5$; $F(-1) = 4$, $F(2) = -257$

23. $g(x) = 3 - 2x^2 + x^4 - 3x^5 - x$; $g(-1) = 6$, $g(1) = -2$

24. $f(x) = 4x^3 - x^4 + 3 + 7x + 5x^2$; $f(-1) = -4$, $f(1) = 18$

In Exercises 25–32, divide $f(x)$ by $d(x)$ and write your answer in the Division Algorithm form:

$$f(x) = g(x) \cdot d(x) + r(x).$$

25. $f(x) = x^2 + 6x + 2$, $d(x) = x + 1$

26. $f(x) = x^2 + 4x + 3$, $d(x) = x - 3$

27. $f(x) = x^3 + 8$, $d(x) = x - 2$

28. $f(x) = x^3 - 27$, $d(x) = x + 3$

29. $f(x) = x^3 - 6x^2 + 11x - 6$, $d(x) = x - 2$

30. $f(x) = x^4 - 10x^2 + 5$, $d(x) = x - 3$

31. $f(x) = x^4 - 5x^2 + 10$, $d(x) = x + 2$

32. $f(x) = x^3 + 3x^2 + 2x$, $d(x) = x + 2$

In Exercises 33–40, a polynomial function and one or more of its zeros are given. Find all the zeros of the polynomial function.

33. $f(x) = x^2 - 3x - 28$; -4

34. $f(x) = x^2 - 11x + 30$; 6

35. $f(x) = 12x^2 - 11x + 2$; $\frac{2}{3}$

36. $f(x) = 10x^2 + 7x - 12$; $-\frac{3}{2}$

37. $f(x) = x^3 + 8x^2 + 14x + 4$; -2

38. $f(x) = x^3 + 7x^2 + 13x + 3$; -3

39. $f(x) = x^3 + 7x^2 + 2x - 40$; 2 and -4

40. $f(x) = x^3 - 5x^2 - 12x + 36$; 2 and -3

Superset

41. If c is a positive real number, for which positive integers n is
    (a) $x - c$ a factor of $x^n - c^n$?
    (b) $x - c$ a factor of $x^n + c^n$?
    (c) $x + c$ a factor of $x^n - c^n$?
    (d) $x + c$ a factor of $x^n + c^n$?

42. Determine k so that $x + 2$ is a factor of $x^3 + kx^2 - 7x + 3$.

43. Determine k so that $x - k$ is a factor of $x^2 + kx - 5$.

44. Determine A and B so that $x - 1$ and $x + 2$ are factors of $x^3 + Ax^2 - 3x + B$.

4.4
Real Zeros of Polynomial Functions

In this section we concern ourselves with a few techniques that help us to narrow down the list of possibilities in our search for zeros of polynomial functions. Recall that the number c is a *zero* of a polynomial $f(x)$ provided $f(c) = 0$, and if c is a real number, then we refer to it as a *real zero*. We will begin with a theorem that tells us what rational numbers are possible zeros of a polynomial function with integer coefficients. The proof of this theorem is outlined in the exercise set.

The Rational Root Theorem

If the polynomial function

$$f(x) = a_nx^n + a_{n-1}x^{n-1} + \cdots + a_2x^2 + a_1x + a_0$$

has integer coefficients, and if $\frac{p}{q}$ is a rational zero of $f(x)$ in lowest terms, then p must be a divisor of a_0, the constant term, and q must be a divisor of a_n, the leading coefficient.

For example, if $f(x) = 2x^3 - x^2 - 8x + 3$, we have

divisors of the constant term, 3: $+1, -1, +3, -3$
divisors of the leading coefficient, 2: $+1, -1, +2, -2.$

By forming all possible quotients having a divisor of the constant term in the numerator and a divisor of the leading coefficient in the denominator, we have a list of all

possible rational zeros of $f(x)$: $\pm 1, \pm \frac{1}{2}, \pm 3, \pm \frac{3}{2}.$

It bears emphasizing that the Rational Root Theorem not only gives us this list of eight rational numbers as possible zeros of $y = f(x)$, but it tells us that no other rational number can be a zero. This is the power of the theorem: it tells us exactly which rational numbers to test.

EXAMPLE 1

Find all the real zeros of the function $f(x) = 3x^3 + \frac{5}{2}x^2 - 1.$

Solution To apply the Rational Root Theorem the coefficients of the polynomial must be integers. Thus, we work with the equation

$6x^3 + 5x^2 - 2 = 0.$ $3x^3 + \frac{5}{2}x^2 - 1 = 0$ is multiplied by 2 to 'clear fractions'.

Any roots of this equation will be zeros of the given function. The possible rational roots are:

$\pm 1, \pm 2, \pm \frac{1}{2}, \pm \frac{1}{3}, \pm \frac{2}{3}, \pm \frac{1}{6}.$ factors of constant, -2: $\pm 1, \pm 2$
factors of lead coefficient, 6: $\pm 1, \pm 2, \pm 3, \pm 6$

We use synthetic division to test some of these possible roots. Remember, c is a root if and only if the remainder is 0.

$$\begin{array}{r|rrrr} 1 & 6 & 5 & 0 & -2 \\ & & 6 & 11 & 11 \\ \hline & 6 & 11 & 11 & 9 \neq 0 \end{array}$$

1 is not a root

$$\begin{array}{r|rrrr} -1 & 6 & 5 & 0 & -2 \\ & & -6 & 1 & -1 \\ \hline & 6 & -1 & 1 & -3 \neq 0 \end{array}$$

-1 is not a root

$$\begin{array}{r|rrrr} \frac{1}{2} & 6 & 5 & 0 & -2 \\ & & 3 & 4 & 2 \\ \hline & 6 & 8 & 4 & 0 \end{array}$$

$\frac{1}{2}$ is a root.

From the synthetic division by $\frac{1}{2}$, we know that $\left(x - \frac{1}{2}\right)$ is a factor and

$$6x^3 + 5x^2 - 2 = \left(x - \frac{1}{2}\right)(6x^2 + 8x + 4).$$

But the quadratic $6x^2 + 8x + 4$ does not have real roots since its discriminant is negative: $b^2 - 4ac = (8)^2 - 4(6)(4) = 64 - 96 < 0$. Thus, the function $f(x) = 3x^3 + \frac{5}{2}x^2 - 1$ has precisely one real zero: $\frac{1}{2}$.

EXAMPLE 2

Find all the real zeros of $f(x) = x^4 - 4x^3 + x^2 + 12x - 12$.

Solution The possible rational roots of $f(x) = 0$ are

$\pm 1, \pm 2, \pm 3, \pm 4, \pm 6, \pm 12$. The constant is -12; leading coefficient is 1.

We use synthetic division to test these possibilities.

$$\begin{array}{r|rrrrr} 2 & 1 & -4 & 1 & 12 & -12 \\ & & 2 & -4 & -6 & 12 \\ \hline & 1 & -2 & -3 & 6 & 0 \end{array}$$

The remainder is 0, so 2 is a root, and $x - 2$ is a factor of $f(x)$.

Thus, $f(x) = (x - 2)(x^3 - 2x^2 - 3x + 6)$. By the Rational Root Theorem the possible rational roots of $x^3 - 2x^2 - 3x + 6 = 0$ are

$\pm 1, \pm 2, \pm 3, \pm 6$. The constant is 6; the leading coefficient is 1.

Again we use synthetic division to test the possibilities.

$$\begin{array}{r|rrrr} 2 & 1 & -2 & -3 & 6 \\ & & 2 & 0 & -6 \\ \hline & 1 & 0 & -3 & 0 \end{array}$$

2 is a root, so $x - 2$ is a factor of $x^3 - 2x^2 - 3x + 6$.

Thus,

$$\begin{aligned} f(x) &= (x - 2)(x - 2)(x^2 - 3) \\ &= (x - 2)^2(x - \sqrt{3})(x + \sqrt{3}). \end{aligned}$$

The real zeros are 2 (multiplicity 2), $\sqrt{3}$, and $-\sqrt{3}$.

Descartes' Rule of Signs

We will now consider a result that allows us to determine the *number* of positive real zeros and negative real zeros that a polynomial function can have. Before we do that, we must discuss a bit of terminology. Suppose the terms of a polynomial are written in decreasing powers of x. (Missing powers of x

can be ignored.) If two successive terms in the polynomial have different signs, we say that there is a **variation in sign.**

$$7x^6 - 4x^5 - 8x^4 - 9x^2 + 2x - 5 \qquad \text{3 variations in sign.}$$

① ② ③

We use this notion of variation in sign in the following theorem.

Descartes' Rule of Signs

Let $f(x)$ be a polynomial of degree $n > 0$, with real coefficients. Let P be the number of variations in sign of $f(x)$; let N be the number of variations in sign of $f(-x)$. Then,

(a) the number of positive real roots of $f(x) = 0$ is either P, or P minus some positive even integer;

(b) the number of negative real roots of $f(x) = 0$ is either N, or N minus some positive even integer.

EXAMPLE 3

Determine the possible number of positive and negative real zeros for the polynomial function $f(x) = x^5 + 4x^3 - x^2 + 7$.

Solution First, we compute the variations in sign for $f(x)$ and for $f(-x)$.

$f(x) = x^5 + 4x^3 - x^2 + 7$ ① ② — $P = 2$; thus, $f(x) = 0$ has either 0 or 2 positive real roots.

$f(-x) = -x^5 - 4x^3 - x^2 + 7$ ① — $N = 1$; thus, $f(x) = 0$ has precisely one negative real root.

Thus, $f(x)$ has either two or no positive real zeros, and has precisely one negative real zero.

Clearly the function

$$f(x) = (x - 4)^2(x - 5) = x^3 - 13x^2 + 56x - 80$$

has zeros 4 and 5. According to Descartes' Rule, there are either three or one positive real zeros. Since the factor $x - 4$ is squared, the 4 counts as two zeros (a zero of multiplicity 2), then 5 is the third zero. So remember that when using Descartes' Rule, a zero of multiplicity k counts as k zeros.

Note that the function

$$f(x) = x^4 + 5x^3 + 2x^2 + 8x + 11$$

can have no positive real zeros since there are no variations in sign in the polynomial. However, since $f(-x) = x^4 - 5x^3 + 2x^2 - 8x + 11$ has four variations in sign, there may be 4, or 2, or no negative real zeros.

Bounds for the Real Zeros

If all the positive real zeros of a polynomial are less than or equal to the real number b, then we say that b is an **upper bound** for the real zeros of $f(x)$. Similarly, if all the negative real zeros are greater than the real number a, then we say that a is a **lower bound** for the real zeros of $f(x)$. We conclude this section by considering a theorem which sometimes allows us to put upper and lower bounds on the real zeros of a polynomial function.

The Bounds Theorem

Suppose $f(x)$ is a polynomial with real coefficients, and with leading coefficient positive.

(a) If b is a positive real number, and if each entry in the third row of the synthetic division of $f(x)$ by $x - b$ is positive or zero, then all of the real zeros of $f(x)$ are less than or equal to b.

(b) If a is a negative real number, and if the numbers in the third row of the synthetic division of $f(x)$ by $x - a$ alternate between nonnegative and nonpositive numbers, then all of the real zeros of f are greater than or equal to a.

Thus, the real zeros of $f(x)$ lie in the interval $[a, b]$.

EXAMPLE 4

For the function $f(x) = 4x^5 - 2x^4 + 22x^3 - 11x^2 - 12x + 6$, show that the real zeros of f lie in the interval $[-2, 1]$.

Solution We apply the Bounds Theorem as follows.

(a) We use synthetic division to divide $f(x)$ by $x - 1$.

1	4	−2	22	−11	−12	6	
		4	2	24	13	1	
	4	2	24	13	1	7	Third row.

Since each entry in the third row of numbers is positive (or it could be zero), the real zeros of $f(x)$ are less than or equal to 1.

(b) We use synthetic division to divide $f(x)$ by $x - (-2)$.

$$\begin{array}{r|rrrrrr} -2 & 4 & -2 & 22 & -11 & -12 & 6 \\ & & -8 & 20 & -84 & 190 & -356 \\ \hline & 4 & -10 & 42 & -95 & 178 & -350 \end{array}$$ Third row.

Since the numbers in the third row alternate in sign, the real zeros of $f(x)$ are greater than or equal to -2.

Thus, the real zeros of f lie in the interval $[-2, 1]$.

EXAMPLE 5

Find all the real zeros of $f(x) = 4x^5 - 2x^4 + 22x^3 - 11x^2 - 12x + 6$

Solution By the Rational Root Theorem the only possible rational roots are

$$\pm 1, \pm 2, \pm 3, \pm 4, \pm 6, \pm\frac{1}{2}, \pm\frac{3}{2}, \pm\frac{1}{4}, \pm\frac{3}{4}.$$ $a_0 = 6, a_n = 4.$

Because of our work in Example 4 we can discard possibilities from this list that are *not* in the interval $[-2, 1]$. This leaves us with $\pm 1, -2, \pm\frac{1}{2}, -\frac{3}{2}, \pm\frac{1}{4}, \pm\frac{3}{4}$ as the only possible rational roots.

$$\begin{array}{r|rrrrrr} \frac{1}{2} & 4 & -2 & 22 & -11 & -12 & 6 \\ & & 2 & 0 & 11 & 0 & -6 \\ \hline & 4 & 0 & 22 & 0 & -12 & 0 \end{array}$$ The possible root $\frac{1}{2}$ is tested.

Thus, $x - \frac{1}{2}$ is a factor and we can write

$$\begin{aligned} f(x) &= \left(x - \frac{1}{2}\right)(4x^4 + 22x^2 - 12) \\ &= (2x - 1)(2x^4 + 11x^2 - 6) \end{aligned}$$

The second factor is quadratic in x^2.

$$\begin{aligned} 2x^4 + 11x^2 - 6 &= 2u^2 + 11u - 6 \\ &= (2u - 1)(u + 6) \\ &= (2x^2 - 1)(x^2 + 6) \end{aligned}$$ $u = x^2$ is used.

Thus, the complete factorization is given by

$$f(x) = (2x - 1)(\sqrt{2}x - 1)(\sqrt{2}x + 1)(x^2 + 6).$$

The real zeros of f are $\frac{1}{2}, \frac{1}{\sqrt{2}}$, and $-\frac{1}{\sqrt{2}}$.

EXAMPLE 6

A rectangular piece of sheet metal measuring 36 cm by 28 cm is to be used in manufacturing an open top box. The process requires that squares of equal size be cut from the four corners, and then the flaps can be folded up to form the sides of the box. If the volume of the box is required to be 1980 cm^3, what size squares should be cut from each corner?

Solution

- ANALYZE We must determine the size (dimensions) of a square to be cut from each corner of a rectangular piece of sheet metal, so that the constructed box has a volume of 1980 cm^3. The information in this problem is best organized with a sketch.
- ORGANIZE Let x = the length of each side of the square.

Draw a sketch

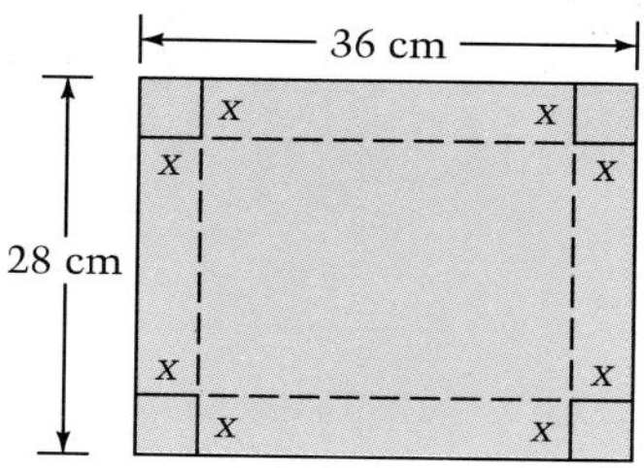

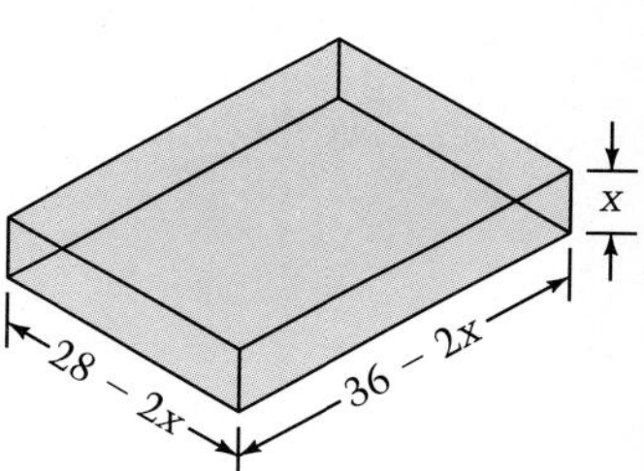

The box is constructed by cutting out the squares and folding up the flaps.

- MODEL The volume V of the constructed box is given by the formula

$$V = \text{length} \times \text{width} \times \text{height}.$$

Since volume must be 1980 cm^3, we write the following equation, then solve for x.

$$1980 = (36 - 2x)(28 - 2x)(x)$$
$$1980 = 1008x - 128x^2 + 4x^3$$
$$0 = 4x^3 - 128x^2 + 1008x - 1980 \quad \text{Divide through by 4 to simplify the equation.}$$
$$0 = x^3 - 32x^2 + 252x - 495$$

By the Rational Root Theorem, the possible rational roots are the factors of 495: ± 1, ± 3, ± 5, ± 9, ± 11, etc. Testing the first few of these values, we observe that 3 is a root:

$$\begin{array}{r|rrrr} 3 & 1 & -32 & 252 & -495 \\ & & 3 & -87 & 495 \\ \hline & 1 & -29 & 165 & 0 \end{array}$$

From the third line in the synthetic division we can conclude that $x^3 - 32x^2 + 252x - 495 = (x - 3)(x^2 - 29x + 165)$. With the polynomial thus factored, we can solve the equation.

$$0 = x^3 - 32x^2 + 252x - 495$$
$$0 = (x - 3)(x^2 - 29x + 165)$$

$x = 3$	$x = \dfrac{-(-29) \pm \sqrt{(-29)^2 - 4(1)(165)}}{2}$
	$x \approx 21.2$ or 7.8

Of the possible values of x, 21.2 is disallowed since it would yield negative values for the sides of the box.

■ ANSWER If each of the cut out squares measures 3 cm on a side, the resulting box will have volume (exactly) 1980 cm^3; or, if each of the cut out squares measures (approximately) 7.8 cm on a side, then the resulting volume will be (approximately) 1980 cm^3. The two values of x yield very different types of boxes; the x-value that works best for the manufacturer would depend on how the box will be marketed to the consumer.

EXERCISE SET 4.4

In Exercises 1–6, use the Rational Root Theorem to form a list of possible rational zeros of the polynomial function.

1. $f(x) = 2x^3 - 5x^2 + 7x + 6$

2. $f(x) = x^4 + \frac{1}{2}x^3 - \frac{1}{6}x^2 + 4x - \frac{1}{4}$

3. $f(x) = 5x - 8x^4 - 1$

4. $f(x) = 8x^3 + 5x^2 - 4$

5. $f(x) = \frac{3}{2}x^2 + 3x^5 - \frac{2}{3}x - 2$

6. $f(x) = 6x^5 - 3x^4 + 1$

In Exercises 7–14, without solving for x use Descartes' Rule of Signs to determine the possible number of positive and negative real zeros.

7. $f(x) = x^7 - 3x^4 + 2x^3 + x + 5$

8. $f(x) = x^5 + 2x^4 - 3x + 4$

9. $f(x) = x^6 + 4x^4 + 6x^3 - 7$

10. $f(x) = x^4 - x^3 - x + 2$

11. $f(x) = 7x - 3x^2 - x^3 + 5$

12. $f(x) = x^4 + x^2 + 7x - 5$

13. $f(x) = x^6 - 2x^5 - 3x^3 + 4$

14. $f(x) = 9x^3 - 4x^4 - x^7 + 9$

In Exercises 15–20, use the Bounds Theorem to determine integers $a < 0$ and $b > 0$ so that all of the real zeros of f lie in the interval $[a, b]$.

15. $f(x) = x^3 - 5x^2 + 5x + 2$

16. $f(x) = x^3 - 6x - 5$

17. $f(x) = x^5 + 2x^2 - 3$

18. $f(x) = x^4 + 3x^3 - x^2 - 2x - 4$

19. $f(x) = 6x^3 + 13x^2 + 2x - 5$

20. $f(x) = 12x^3 + x^2 - 47x - 36$

In Exercises 21–36, find the real zeros of the function.

21. $f(x) = x^4 - 6x^3 + 9x^2 + 4x - 12$

22. $f(x) = x^4 + 7x^3 + 7x^2 - 21x - 30$

23. $f(x) = x^4 + x^3 - 17x^2 - 5x + 60$

24. $f(x) = 3x^5 + 20x^4 + 17x^3 + 28x^2 - 12x$

25. $f(x) = x^5 + 3x^4 + 4x^3 + 4x^2 + 3x + 1$

26. $f(x) = 12x^3 - 17x^2 - 13x - 2$

27. $f(x) = x^4 + 6x^3 - 29x - 6$

28. $f(x) = x^5 - 6x^4 + 15x^3 - 26x^2 + 36x - 24$

29. $f(x) = 6x^3 + x^2 - 4x + 1$

30. $f(x) = x^4 - 9x^2 - 4x + 12$
31. $f(x) = 2x^4 + 11x^3 + 15x^2 + 13x + 4$
32. $f(x) = 10x^4 - 51x^3 + 38x^2 + 24x$
33. $f(x) = 6x^4 + 35x^3 + 34x^2 - 40x$
34. $f(x) = x^4 - 8x^3 + 20x^2 - 17x + 4$
35. $f(x) = 12x^4 - 4x^3 - 9x^2 + 1$
36. $f(x) = 12x^4 - 4x^3 - 9x^2 + 6x - 1$

Superset

37. In this exercise we sketch the steps that could be followed to prove the Rational Root Theorem. Suppose $\frac{p}{q}$ is a rational number in lowest terms, and it is known to be a solution of the equation $f(x) = 0$, where $f(x) = a_nx^n + a_{n-1}x^{n-1} + \cdots + a_1x + a_0$ and each of the coefficients $a_n, a_{n-1}, \ldots, a_1, a_0$ is an integer. Write out $f\left(\frac{p}{q}\right) = 0$ and multiply this equation by q^n to "clear fractions". Conclude that p divides a_0 and that q divides a_n.
38. Suppose that the zeros of the quadratic function $f(x) = 2x^2 + bx + c$ are integers and that b and c are integers. Prove that b and c are even integers.
39. Suppose the quadratic function $f(x) = x^2 + bx + 1$ has distinct integer zeros and that b is an integer. What are the possible values of b?
40. Give an example of a polynomial function to which the Rational Root Theorem cannot be applied.
41. Give an example of a cubic function having four terms and no positive real zeros.
42. Suppose that $\frac{2}{3}$ and $-\frac{5}{14}$ are known to be zeros of the function $f(x) = Ax^3 + 3x^2 + Bx + C$, where A, B, and C are integers. What else can you say about the values of A and B?

In Exercises 43–46, write an equation for a quadratic function f whose zeros and y-intercept are shown in the graph.

43.
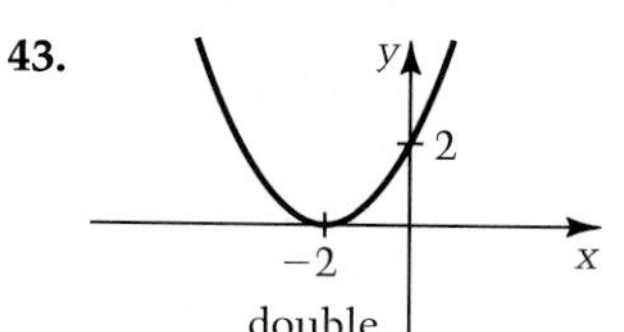

44.
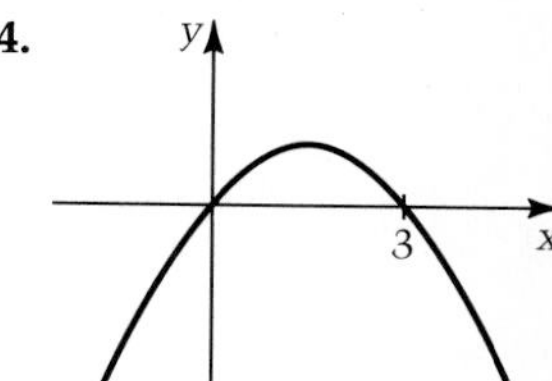

45.
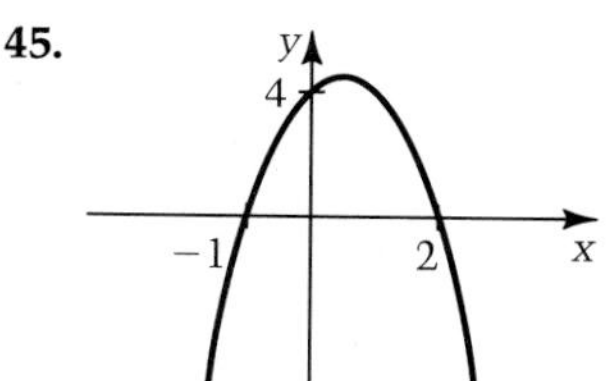

46.
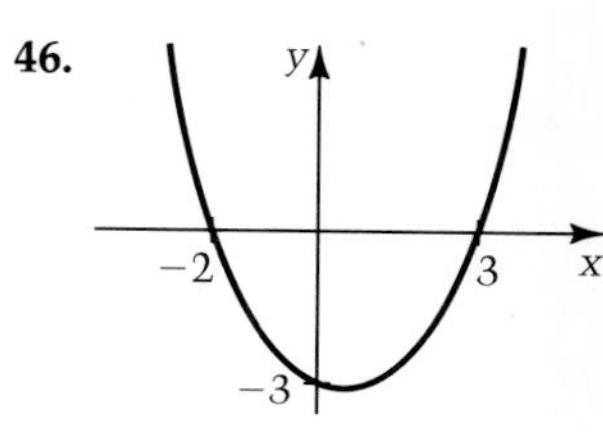

In Exercises 47–50, write an equation for a cubic function f whose zeros and y-intercept are shown in the graph.

47.
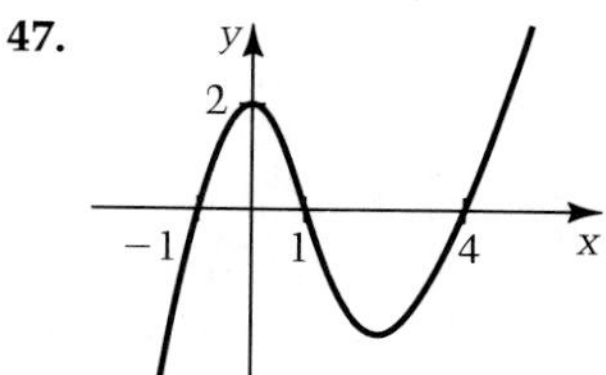

48.
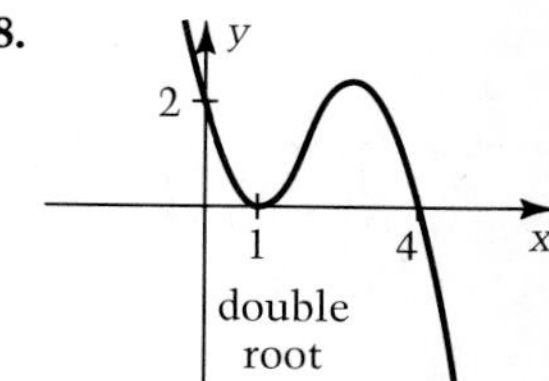

49.
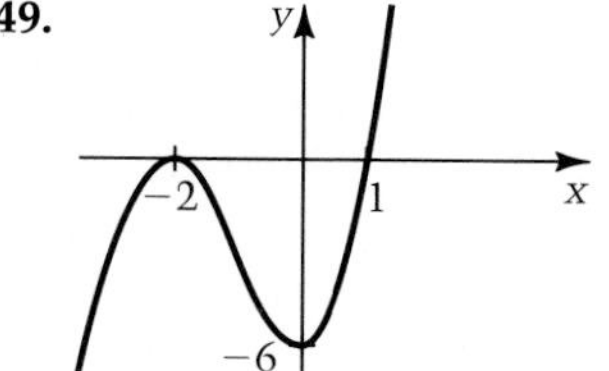

50.
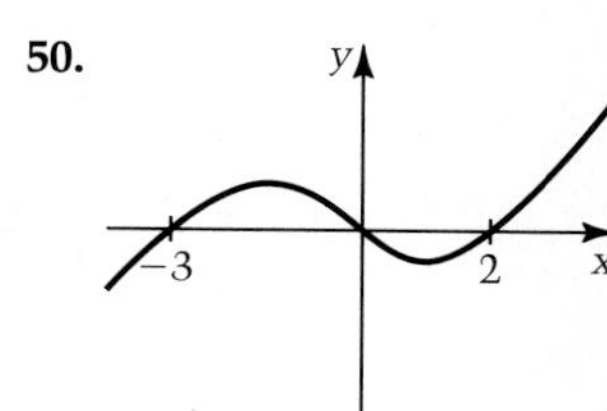

4.5
Complex Zeros of Polynomial Functions

Recall that the set of complex numbers consists of real numbers (numbers of the form $a + bi$, where $b = 0$) and imaginary numbers (numbers of the form $a + bi$, where $b \neq 0$.) Hence imaginary numbers are the "nonreal" complex numbers. For this reason, we can use the phrases "imaginary number" and "nonreal complex number" interchangeably.

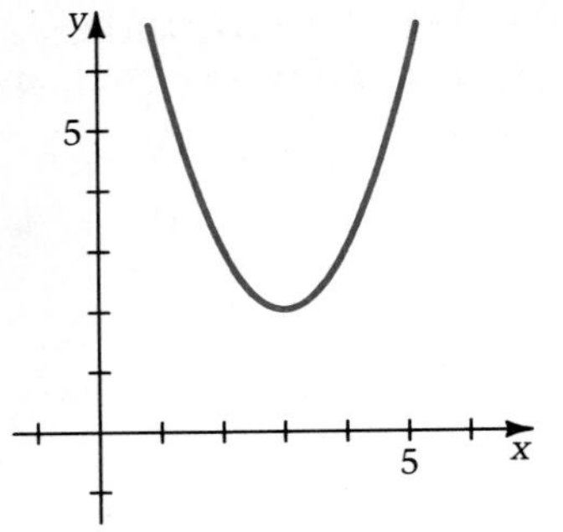

Figure 4.26 $y = x^2 - 6x + 11$

One of the reasons for defining the complex numbers was to provide a set of numbers over which *any* quadratic equation can be solved. Consider the graph of the function $y = x^2 - 6x + 11$ shown in Figure 4.26. The function has no real zeros. Since the x-intercepts of the graph of a function correspond to the real zeros of the function, the parabola in the figure never intersects the x-axis. However the function does have *imaginary zeros*. To determine these zeros, we solve the equation $x^2 - 6x + 11 = 0$, by using the quadratic formula.

$$x = \frac{-(-6) \pm \sqrt{36 - 44}}{2} = \frac{6 \pm \sqrt{-8}}{2} = \frac{6 \pm 2i\sqrt{2}}{2} = 3 \pm i\sqrt{2}.$$

The two roots of the equation are the zeros of the function. They are imaginary numbers, that is, nonreal complex numbers. They are not represented anywhere on the graph.

It is quite clear from our work with quadratic equations that exactly one of the following must occur:

1. The equation has two distinct real roots (the discriminant $b^2 - 4ac$ is positive.)
2. The equation has one real root of multiplicity two (the discriminant is 0.)
3. The equation has two imaginary roots (the discriminant is negative.)

Thus, if we count multiplicities, every quadratic equation has exactly two roots. (Remember, counting multiplicities means that for an equation such as $x^2 - 6x + 9 = (x - 3)(x - 3) = 0$, the root 3 would be counted as two roots, since the factor $x - 3$ appears twice.)

The natural question then is whether every third degree polynomial equation has exactly three roots (counting multiplicities), and, in general, whether every nth degree polynomial has exactly n roots (counting multiplicities).

A first step in arriving at an answer to this question is provided by the *Fundamental Theorem of Algebra*, a result proved by the great German mathematician Karl Friedrich Gauss (1777–1855) in his doctoral dissertation in 1799.

The Fundamental Theorem of Algebra

If $f(x)$ is a polynomial of degree greater than 0, then $f(x)$ has at least one complex zero.

It is important to mention at this point that the Remainder Theorem and the Factor Theorem are true not only for real numbers, but for all complex numbers as well. Thus the Remainder Theorem can be rephrased to say: *If a* *polynomial $f(x)$ is divided by $x - c$, where c is any complex number, then the remainder is equal to $f(c)$.* The Factor Theorem says: *The complex number c is a zero of the polynomial function $f(x)$ if and only if $x - c$ is a factor of $f(x)$.*

Now suppose $f(x)$ is a polynomial of degree n.

$$f(x) = a_n x^n + a_{n-1}x^{n-1} + \cdots + a_2x^2 + a_1x + a_0$$

By the Fundamental Theorem of Algebra, $f(x)$ has at least one zero, call it c_1. Hence by the Factor Theorem we know that $x - c_1$ is a factor of $f(x)$, and so

$$f(x) = (x - c_1) \cdot q_1(x)$$

where $q_1(x)$ is a polynomial of degree $n - 1$. If that degree is greater than 0, then by the Fundamental Theorem of Algebra again $q_1(x)$ has a zero c_2, and so by the Factor Theorem we can write

$$f(x) = (x - c_1) \cdot (x - c_2) \cdot q_2(x)$$

where $q_2(x)$ is of degree $n - 2$. This factoring process is performed n times, producing n linear factors, with $q_n(x)$ equal to the leading coefficient a_n. That is,

$$f(x) = a_n(x - c_1) \cdot (x - c_2) \cdots (x - c_n).$$

(Though some of the c's may be the same, we can say that each of them is a zero of $f(x)$.) We can summarize this argument in the following theorem.

Linear Factorization Theorem

If $f(x)$ is a polynomial of degree $n > 0$, then $f(x)$ can be factored into a product of n linear factors,

$$f(x) = a_n(x - c_1) \cdot (x - c_2) \cdots (x - c_n),$$

where $c_1, c_2, \ldots, c_n$ are complex numbers, and a_n is the leading coefficient of $f(x)$.

EXAMPLE 1

Determine a polynomial $f(x)$ of degree 3 with zeros 2, 1, and -1, such that $f(0) = 3$.

Solution

$f(x) = a(x - 2) \cdot (x - 1) \cdot [x - (-1)]$

$f(x) = a(x - 2) \cdot (x - 1) \cdot (x + 1)$

By the Linear Factorization Theorem, $f(x)$ has a linear factor for each zero. We must find a, the leading coefficient.

Find a:

$3 = a(0 - 2) \cdot (0 - 1) \cdot (0 + 1)$ Since $f(0) = 3$, replace x with 0, and $f(x)$ with 3, then solve for a.

$3 = a(-2) \cdot (-1) \cdot (1)$

$3 = 2a$

$\frac{3}{2} = a$

$f(x) = \frac{3}{2}(x - 2)(x - 1)(x + 1)$ This is $f(x)$ in factored form.

$f(x) = \frac{3}{2}x^3 - 3x^2 - \frac{3}{2}x + 3$ This is $f(x)$ in standard form.

EXAMPLE 2

Determine a polynomial $f(x)$ of degree 4 with -3 a zero of multiplicity 2, imaginary zeros i and $-i$, and $f(-2) = 10$.

Solution

$f(x) = a(x + 3)^2(x - i)[x - (-i)]$ Use the Linear Factorization Theorem.

Find a:

$10 = a(-2 + 3)^2(-2 - i)(-2 + i)$ Since $f(-2) = 10$, replace x with -2, and $f(x)$ with 10, then solve for a.

$10 = a(1)^2(4 - i^2)$ Recall: $i^2 = -1$.

$10 = 5a$

$2 = a$

$f(x) = 2(x + 3)^2(x - i)(x + i)$ This is $f(x)$ in factored form.

$f(x) = 2(x + 3)^2(x^2 + 1)$ $(x - i)(x + i) = x^2 - i^2 = x^2 + 1$.

$f(x) = 2x^4 + 12x^3 + 20x^2 + 12x + 18$ This is $f(x)$ in standard form.

By virtue of the Linear Factorization Theorem, an nth degree polynomial $f(x)$ can be written as

$$f(x) = a_n(x - c_1) \cdot (x - c_2) \cdots (x - c_n)$$

where the c's are complex (real or imaginary) zeros of $f(x)$. It may be the case that the c's are not all distinct. If c is a zero of multiplicity k, then the factor $x - c$ will appear in the factorization k times, and is written $(x - c)^k$. This suggests that if we count a zero of multiplicity k as k zeros, our earlier conjecture, that an nth degree polynomial function has n zeros, is indeed true!

Theorem

Let $f(x) = a_n x^n + a_{n-1}x^{n-1} + \cdots + a_2 x^2 + a_1 x + a_0$ be a polynomial function of degree $n > 0$. The function has precisely n zeros, and the equation $f(x) = 0$ has precisely n roots, provided a root/zero of multiplicity k is counted k times.

EXAMPLE 3

Determine all the zeros of the function $f(x) = x^3 - 6x^2 + 21x - 26$.

Solution Since $f(x)$ is a third degree polynomial, we expect to find three zeros (counting multiplicities) according to the previous theorem.

$$f(x) = x^3 - 6x^2 + 21x - 26.$$ Use Descartes' Rule of Signs.

① ② ③

Since there are three variations in sign, there are either three positive real roots, or one positive real root.

divisors of the constant, -26: $\pm 1, \pm 2, \pm 13, \pm 26$
divisors of the leading coefficient, 1: ± 1
possible rational roots: $\pm 1, \pm 2, \pm 13, \pm 26$

Use the Rational Root Theorem.

$$\begin{array}{r|rrrr} 2 & 1 & -6 & 21 & -26 \\ & & 2 & -8 & 26 \\ \hline & 1 & -4 & 13 & 0 \end{array}$$

Test the possible roots using synthetic division. 2 is a zero since division by $x - 2$ produces a 0 remainder.

From the synthetic division we know that

$$f(x) = (x - 2)(x^2 - 4x + 13)$$
$$(x - 2)(x^2 - 4x + 13) = 0$$

To find the other zeros, set $f(x)$ equal to 0, then solve.

$x - 2 = 0$	$x^2 - 4x + 13 = 0$
$x = 2$	$x = \dfrac{4 \pm \sqrt{16 - 52}}{2}$
	$x = 2 \pm 3i$

The three zeros are 2, $2 - 3i$, and $2 + 3i$. Each zero is of multiplicity one.

Recall that we refer to the imaginary numbers $a + bi$ and $a - bi$ as complex conjugates of one another. It is no coincidence that the imaginary zeros we found in Example 3, namely, $2 + 3i$ and $2 - 3i$, are conjugates. In general, we have the following result, whose proof is discussed in the exercises.

Conjugate Roots Theorem

Suppose $f(x) = a_nx^n + a_{n-1}x^{n-1} + \cdots + a_1x + a_0$ is a polynomial function of degree $n > 0$ with real coefficients. If $a + bi$ is an imaginary root of $f(x) = 0$, then the conjugate $a - bi$ is also a root of $f(x) = 0$. That is, the imaginary zeros of a polynomial function with real coefficients occur in conjugate pairs.

Because of the Conjugate Roots Theorem, we can be certain that a polynomial of odd degree with real coefficients will have at least one real zero, since the nonreal complex zeros occur in pairs. Consequently, a polynomial of odd degree, having real coefficients, always has at least one linear factor of the form $x - r$, where r is real, and the graph of a polynomial function of odd degree *must* pass through the x-axis at least once. On the other hand, if the degree of the polynomial is even, there may be no real zeros, and hence no x-intercepts for the graph.

EXAMPLE 4

Determine a polynomial of lowest degree having real coefficients, and having zeros 5 and $3 - i$.

Solution Since the polynomial is to have real coefficients, the imaginary zeros must occur in conjugate pairs; since $3 - i$ is a zero, $3 + i$ must be a zero also.

$$f(x) = a(x - 5)[x - (3 - i)][x - (3 + i)] \quad \text{Use the Linear Factorization Theorem.}$$

$$= a(x - 5)[x^2 - (3 - i)x - (3 + i)x + (3 - i)(3 + i)]$$

$$= a(x - 5)(x^2 - 6x + 10)$$

There is not enough information to determine a uniquely. Any nonzero value of a, say $a = 1$, will produce a polynomial that satisfies the given conditions. Thus $f(x) = (x - 5)(x^2 - 6x + 10)$ satisfies the given conditions. In standard form this polynomial is $f(x) = x^3 - 11x^2 + 40x - 50$.

If, in Example 4, an additional piece of information about the function had been given, e.g., $f(0) = 10$, then the leading coefficient a could have been determined.

EXAMPLE 5

Determine all the zeros of the polynomial function

$$f(x) = x^5 + 3x^4 + 2x^3 + 14x^2 + 29x + 15,$$

given that one of the zeros is $1 + 2i$.

Solution Using the Conjugate Roots Theorem, we can conclude that $1 + 2i$ and its conjugate $1 - 2i$ are zeros of $f(x)$. Thus the product of the corresponding factors,

$$[x - (1 + 2i)][x - (1 - 2i)] = x^2 - 2x + 5$$

divides evenly into $f(x)$; that is, $f(x) = (x^2 - 2x + 5) \cdot q(x)$ for some third degree polynomial $q(x)$. To determine $q(x)$, we can use long division to obtain $f(x) = (x^2 - 2x + 5)(x^3 + 5x^2 + 7x + 3)$. (You should verify this result by performing the division.) We know the zeros that arise from the quadratic factor; we must determine the zeros that arise from the cubic factor.

$q(x) = x^3 + 5x^2 + 7x + 3$ — Use the Rational Root Theorem.
divisors of the constant, 3: $\pm 1, \pm 3$
divisors of the lead coefficient, 1: ± 1
possible rational roots: $\pm 1, \pm 3$.

There are no variations in sign in $q(x)$. — Use Descartes' Rule of Signs.

Therefore there can be no positive real roots, so we test only -1, and/or -3.

$$\begin{array}{r|rrrr} -1 & 1 & 5 & 7 & 3 \\ & & -1 & -4 & -3 \\ \hline & 1 & 4 & 3 & 0 \end{array}$$

Test -1 by synthetic division. Since the remainder is 0, -1 is a zero and $x + 1$ is a factor.

Thus,

$$\begin{aligned} q(x) &= (x + 1)(x^2 + 4x + 3) && \text{Now factor by trial and error.} \\ &= (x + 1)(x + 3)(x + 1) \end{aligned}$$

Putting it all together, we have:

$$\begin{aligned} f(x) &= (x^2 - 2x + 5)(x^3 + 5x^2 + 7x + 3) && \text{From the long division.} \\ &= (x^2 - 2x + 5)(x + 1)^2(x + 3) && \text{From synthetic division and factoring above.} \end{aligned}$$

Thus the complete list of zeros is $1 - 2i, 1 + 2i, -1$ (a zero of multiplicity 2), and -3. We thus have found five zeros, counting multiplicities. This was to be expected since $f(x)$ is a fifth degree polynomial.

EXERCISE SET 4.5

In Exercises 1–8, determine a polynomial $f(x)$ of the given degree having the given zeros and y-intercept.

1. second degree; zeros: $i, -2i$; $f(0) = 3$
2. second degree; zeros: $1, \sqrt{2}$; $f(0) = 2$
3. second degree; zeros: $3, i$; $f(0) = 12$
4. second degree; zeros: $i\sqrt{2}, -3i\sqrt{2}$; $f(0) = -3$
5. third degree; zeros: $\frac{1}{2}, \frac{1}{3}, -1$; $f(0) = -2$

6. third degree; zeros: $\frac{1}{4}, -\frac{2}{5}, 6$; $f(0) = -6$

7. third degree; zeros: $1, -5i, i\sqrt{2}$; $f(0) = -10$

8. third degree; zeros: $-2, -2i, -\frac{1}{3}$; $f(0) = 12$

In Exercises 9–20, determine all the zeros of the function.

9. $f(x) = x^2 + 3x + 7$
10. $f(x) = x^2 - 5x + 7$
11. $f(x) = 9x^2 + 64$
12. $f(x) = 4x^2 + 25$
13. $f(x) = 4x - 5x^2 - 2$
14. $f(x) = 3x - 1 - 9x^2$
15. $f(x) = (x - 3)(x^2 - 6x + 11)$
16. $f(x) = (x + 8)(x^2 - 2x + 7)$
17. $f(x) = (4x^2 + x + 1)(x^2 - x + 1)$
18. $f(x) = (4x^2 - 12x + 11)(x^2 + 12)$
19. $f(x) = x^3 + 1$
20. $f(x) = x^3 - 8$

In Exercises 21–28, determine a polynomial function f of least degree having real coefficients and having the given zeros. Leave your answer in factored form.

21. $4 + 3i$
22. $3 - 2i$
23. 4 and $-5i$
24. 4 and $2 - i\sqrt{5}$
25. $-3i$ (multiplicity 2)
26. $2 + i$ and $1 - i\sqrt{3}$
27. $-2, 1 + 2i$ and $i\sqrt{3}$
28. -1 and $2 + i$ (multiplicity 2)

In Exercises 29–34, a polynomial function and one or more of its zeros are given. Determine all the zeros of the function.

29. $f(x) = x^3 - 4x^2 + 5x - 6$; 3
30. $f(x) = x^3 + 8x^2 + 20x + 25$; -5
31. $f(x) = x^4 + 5x^3 + 11x^2 + 13x + 6$; -2 and -1
32. $f(x) = x^4 - x^2 + 4x - 4$; 1 and -2
33. $f(x) = x^4 - 6x^3 + 13x^2 - 24x + 36$; $2i$
34. $f(x) = x^4 - 4x^3 + 16x^2 - 24x + 20$; $1 + 3i$

Superset

35. Let $f(x) = a_nx^n + a_{n-1}x^{n-1} + \cdots + a_1x + a_0$ be a polynomial function of degree $n > 0$ with real coefficients. Assume that the imaginary number z is a zero of f, that is, $f(z) = 0$. Use the Properties of Complex Conjugates (p. 80) to revise the expression $\overline{f(z)}$ and thereby show that if $f(z) = 0$, then $f(\overline{z}) = 0$. This proves the Conjugate Roots Theorem.

36. Prove that a polynomial equation of odd degree with real coefficients has at least one real root.

In Exercises 37–42, find all the solutions of the equation. (*Hint:* Start by looking for rational solutions.)

37. $x^4 + x^3 + 2x^2 + 4x - 8 = 0$
38. $x^4 - 5x^3 + 8x^2 - 10x + 12 = 0$
39. $2x^4 + x^3 - 2x^2 - 4x - 3 = 0$
40. $3x^4 - x^3 + 3x - 1 = 0$
41. $2x^4 - 5x^3 - x^2 - 5x - 3 = 0$
42. $3x^4 + 5x^3 + 4x^2 + 10x - 4 = 0$

In Exercises 43–50, we build upon the following observations

$$(x - c_1)(x - c_2) = x^2 - (c_1 + c_2)x + c_1c_2 \text{ and}$$
$$(x - c_1)(x - c_2)(x - c_3) = x^3 - (c_1 + c_2 + c_3)x^2 + (c_1c_2 + c_2c_3 + c_1c_3)x - c_1c_2c_3$$

43. Write out a similar result for a polynomial that is the product of four linear factors. In general, for a polynomial of the form $f(x) = x^n + a_{n-1}x^{n-1} + \cdots + a_1x + a_0$ what is the significance of a_0? of a_{n-1}?

44. Determine A and B given that two of the roots of the equation $x^3 + 3x^2 + Ax + B = 0$ are 2 and -3.

45. Determine A and B given that two of the roots of the equation $x^3 + Ax^2 + Bx + 6 = 0$ are -1 and 3.

46. The roots of the equation
$$x^4 + Ax^3 + Bx^2 + Cx + D = 0$$
are 2, 3, -1 and 4. Determine the values of A, B, C, and D without computing the product
$$(x - 2)(x - 3)(x + 1)(x - 4)$$
directly.

47. What is the coefficient of x^{n-1} in the expansion of $(x - 1)^n$?

48. What is the coefficient of x^{n-1} in the expansion of $(x + 2)^n$?

49. What is the coefficient of x^9 in the expansion of $(x - 2)^5(x + 2)^5$?

50. What is the coefficient of x^9 in the expansion of $(x + 2)^4(x - 2)^6$?

4.6
Approximating Real Zeros of Polynomial Functions

As we have already seen, finding the zeros of a polynomial function $f(x)$ reduces to finding the solutions or roots of the equation $f(x) = 0$. Linear polynomial equations, such as $3x - 7 = 0$, are simple to solve; likewise, the quadratic formula assures us that any quadratic (second degree polynomial) equation can also be solved without difficulty. There are formulas for solving cubic (third degree polynomial) and quartic (fourth degree polynomial) equations, but these formulas are rather cumbersome and are rarely used.

In the early nineteenth century the Norwegian mathematician Niels Henrik Abel (1802–1829) proved that there is no formula for solving fifth (or higher) degree polynomial equations. Add to this difficulty the fact that techniques such as the Rational Root Theorem and Descartes' Rule of Signs are frequently not sufficient for finding all the zeros of a polynomial function $f(x)$, and we are led to the following realization: the only effective approach for finding the zeros in many cases involves numerical methods, that is, methods whereby we *approximate* the zeros.

In this section we will consider a simple but extremely effective procedure for approximating the real zeros of a polynomial function $f(x)$, or equivalently, for approximating the real roots of the polynomial equation $f(x) = 0$. This procedure is known as the **Bisection Method,** and is based on an important theorem that we state here without proof.

The Intermediate Value Theorem for Polynomial Functions

Let $y = f(x)$ be a polynomial function such that $f(a) \neq f(b)$ for $a < b$. Then for every real number C between $f(a)$ and $f(b)$, there exists at least one value c between a and b, such that $f(c) = C$.

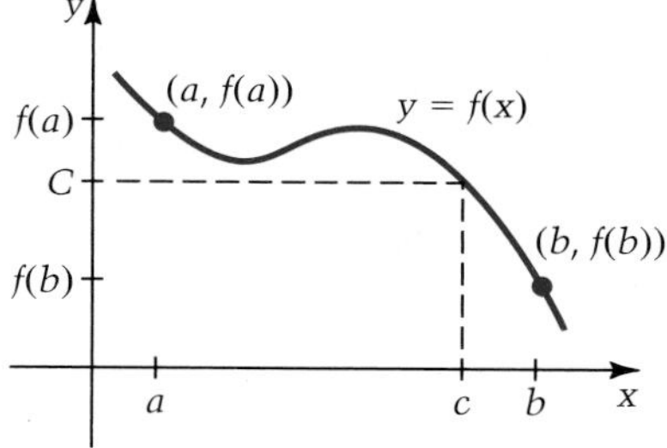

Figure 4.27

The theorem is illustrated in Figure 4.27. For our purposes, this theorem is quite useful when we know that $f(a)$ and $f(b)$ have different signs, for then

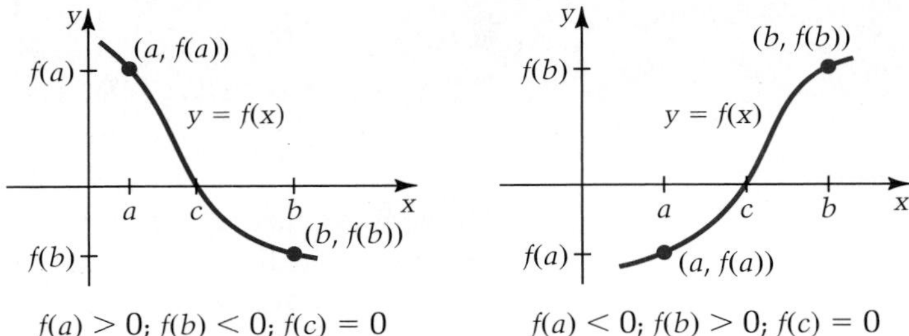

Figure 4.28

we can conclude that there must be some number c between a and b, such that $f(c) = 0$. That is, if $f(a)$ and $f(b)$ have different signs, then the function has a zero c somewhere between a and b, and the graph of f will cross the x-axis at c. (See Figure 4.28)

Our goal is to approximate the zero c. To do this, we consider the midpoint of the interval $[a, b]$ as our first approximation: $m_1 = \dfrac{a + b}{2}$. One of the following three things must happen:

(a) m_1 turns out to be a zero of the function, that is, $f(m_1) = 0$, so $c = m_1$, and we are finished. (See Figure 4.29a.)

(b) $f(a)$ and $f(m_1)$ have different signs, so the zero c is somewhere between a and m_1. (See Figure 4.29b.)

(c) $f(m_1)$ and $f(b)$ have different signs, so the zero c is between m_1 and b. (See Figure 4.29c.)

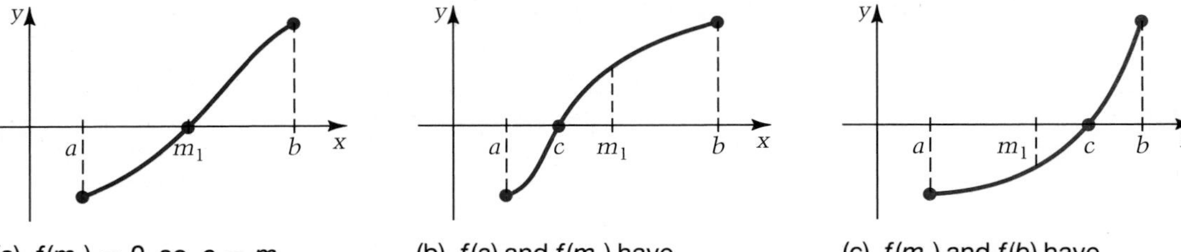

(a) $f(m_1) = 0$ so $c = m_1$.

(b) $f(a)$ and $f(m_1)$ have different signs so $a < c < m_1$.

(c) $f(m_1)$ and $f(b)$ have different signs so $m_1 < c < b$.

Figure 4.29

Note that in cases (b) and (c), $f(m_1)$ has a sign opposite to that of either $f(a)$ or $f(b)$, and so we have located the zero c either in the interval $[a, m_1]$, the "left half" of the interval $[a, b]$, or in $[m_1, b]$, the "right half" of the interval $[a, b]$. For argument's sake, let's assume that the zero c is in $[a, m_1]$. See Figure 4.30. We then find the midpoint of $[a, m_1]$; call it m_2. After evaluating the

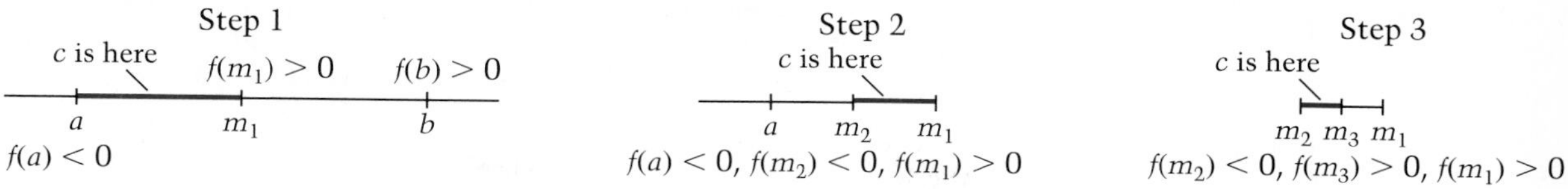

Figure 4.30

function at m_2, we arrive at one of the three possibilities as before: either m_2 is the desired zero c, or c lies in one of the "quarter intervals" $[a, m_2]$ or $[m_2, m_1]$. We continue this process, cutting the interval in half at each stage (that is, *bisecting* it), and thus get closer and closer to the zero c.

EXAMPLE 1

Use the Bisection method to determine the positive zero of $f(x) = x^2 - 2$, accurate to the nearest hundredth. (Note that here we are approximating the positive root of the equation $x^2 - 2 = 0$, that is, $x^2 = 2$. Thus we are approximating $\sqrt{2}$ to the nearest hundredth.)

Solution Since $f(1) = -1 < 0$, and $f(2) = 2 > 0$, we know that the desired zero is between 1 and 2. So we begin by finding the midpoint of the interval $[1, 2]$: $m_1 = \dfrac{1 + 2}{2} = 1.5$; now compute: $f(1.5) = 0.25 > 0$.

Step	L Left Endpoint	m_1 Midpoint	R Right Endpoint	$f(L)$	$f(m_1)$	$f(R)$
1	1 ⟷ (endpoints for next step)	1.5	2	−	+	+

From this we see that since a change of sign occurs between $f(1)$ and $f(1.5)$, the zero lies in the interval $[1, 1.5]$; we've indicated this in the chart by screening the sign change. Thus we take 1 and 1.5 as the endpoints for step 2, with $m_2 = \dfrac{1 + 1.5}{2} = 1.25$ as the step 2 midpoint. Then we compute: $f(1.25) = -0.4375 < 0$.

n Step	L Left Endpoint	m_n Midpoint	R Right Endpoint	$f(L)$	$f(m_n)$	$f(R)$
1	1	1.5	2	−	+	+
2	1	1.25 ⟷ (endpoints for next step)	1.5	−	−	+

A change of sign occurs between $f(1.25)$ and $f(1.5)$, so the zero lies in the interval [1.25, 1.5]. The midpoint of this interval is $m_3 = \frac{1.25 + 1.5}{2} = 1.375$, with $f(1.375) = -0.109375$. We summarize our work in the following table.

n Step	L Left Endpoint	m_n Midpoint	R Right Endpoint	$f(L)$	$f(m_n)$	$f(R)$
				Note interval of change		
1	1	1.5	2	−	+	+
2	1	1.25	1.5	−	−	+
3	1.25	1.375	1.5	−	−	+
4	1.375	1.4375	1.5	−	+	+
5	1.375	1.40625	1.4375	−	−	+
6	1.40625	1.421875	1.4375	−	+	+
7	1.40625	1.414063	1.421875	−	−	+
8	1.414063	1.417969	1.421875	−	+	+
9	1.414063	1.416016	1.417969	−	+	+
10	1.414063	1.415040	1.416016	−	+	+
11	1.414063	1.414551	1.415040	−	+	+

The zero is in the interval [1.414063, 1.414551]. Rounding either endpoint to the nearest hundredth yields 1.41. Thus we may stop at this point. The desired zero is approximately 1.41.

In general, the width of the interval at step n is given by the formula $\frac{b-a}{2^{n-1}}$ where $[a, b]$ is the original interval. If the nth midpoint m_n is used to estimate the zero, then the most that the error ($|\text{actual zero} - m_n|$) could be is half the nth interval width, that is,

$$\text{MAXIMUM ERROR} = \frac{1}{2}\left(\frac{b-a}{2^{n-1}}\right) = \frac{b-a}{2^n}$$

Thus, if $m_3 = 1.375$ is used to estimate the zero in Example 1, the maximum error possible is

$$\frac{b-a}{2^n} = \frac{2-1}{2^3} = \frac{1}{8} = 0.125$$

which is half the width of the interval at step 3.

In order to become familiar with the Bisection Method, you should use a calculator to replicate the table in the previous example. Doing this will reinforce the technique of generating endpoints for successive steps.

EXERCISE SET 4.6

Note: The accuracy of the estimates to be found in these exercises has not been specified as the accuracy will depend a great deal upon the calculating device that you use. A computer program was used to generate answers for the answer section accurate to six decimal places.

In Exercises 1–8, the function f has a zero between the given values of a and b. Use the Bisection Method to estimate this root.

1. $f(x) = x^3 - x^2 + x + 7; a = -2, b = -1$

2. $f(x) = 3x^3 + x - 21; a = 1, b = 2$

3. $f(x) = 16 - x^3 - x^4; a = 1, b = 2$

4. $f(x) = x^5 - 3x^4 - x + 10; a = 1, b = 2$

5. $f(x) = x^4 - 2x^2 - 5x - 10; a = 2, b = 3$

6. $f(x) = x^6 - x^3 + 2x - 18; a = -2, b = -1$

7. $f(x) = 20 + x - x^2 + x^3 - x^4; a = 2, b = 3$

8. $f(x) = 4x^4 - 3x^3 + 2x^2 - x; a = 0, b = 1$

In Exercises 9–16, find a polynomial function with integer coefficients for which the given value is a zero. Then use the Bisection Method with your function to approximate the value.

9. $\sqrt{3}$ **10.** $\sqrt{5}$ **11.** $3 - \sqrt{6}$

12. $-1 + \sqrt{7}$ **13.** $-\sqrt[3]{2}$ **14.** $\sqrt[3]{3}$

15. $1 + \sqrt[3]{3}$ **16.** $2 + \sqrt[3]{2}$

Superset

In Exercises 17–22, use the Bisection Method to estimate both zeros of the given quadratic function. Then use the quadratic formula to check your answers.

17. $f(x) = x^2 + 6x + 3$ **18.** $f(x) = x^2 - 3x - 1$

19. $f(x) = 3x^2 + x - 4$ **20.** $f(x) = 5x^2 + 2x - 3$

21. $f(x) = 1.2x^2 - 3.7x - 0.8$

22. $f(x) = 3.1x^2 + 7.6x + 1.9$

In Exercises 23–28, the given cubic function has three real zeros, one of which is rational. Estimate the three zeros as follows: (i) Use the Bisection Method to estimate one zero. The zero you find might not be rational. (ii) Use synthetic division to determine a quadratic equation (with approximate coefficients) whose roots are (approximately) the other two zeros. (iii) Use the quadratic formula to estimate these other two zeros.

23. $f(x) = 3x^3 + x^2 - 5x + 2$

24. $f(x) = 4x^3 - 16x^2 + 7x + 20$

25. $f(x) = 5x^3 + 23x^2 - 4$

26. $f(x) = 3x^3 + 20x^2 - 8$

27. $f(x) = 6x^3 + 35x^2 + 55x + 21$

28. $f(x) = 4x^3 - x^2 - 7x - 3$

4.7 Rational Functions

$$f(x) = \frac{x^2 - 3x + 7}{x^2 + 3x - 10}$$

$$g(x) = \frac{1}{x^2 - 1}$$

Examples of Rational Functions

Recall that a rational expression is a fraction in which both numerator and denominator are polynomials. Functions defined by rational expressions are called **rational functions.** Although rational functions are not as well behaved as polynomial functions, there are a few general techniques which will help us graph rational functions.

The simplest rational function is given by the equation

$$y = \frac{1}{x}.$$

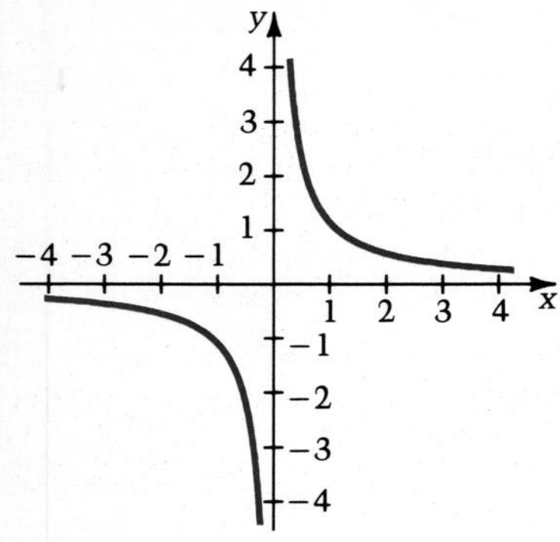

Figure 4.31 $y = \frac{1}{x}$.

Notice that in Figure 4.31, the graph gets closer and closer to the x-axis but never touches it. (The graph approaches the y-axis in the same fashion.) When a graph approaches a line in this manner, we say that the line is an **asymptote** of the graph.

If the line is vertical, it is called a **vertical asymptote;** if it is horizontal, it is called a **horizontal asymptote.**

Consider the transformations of the graph of $y = \frac{1}{x}$ shown in Figure 4.32. As is customary, asymptotes are indicated with dashed lines.

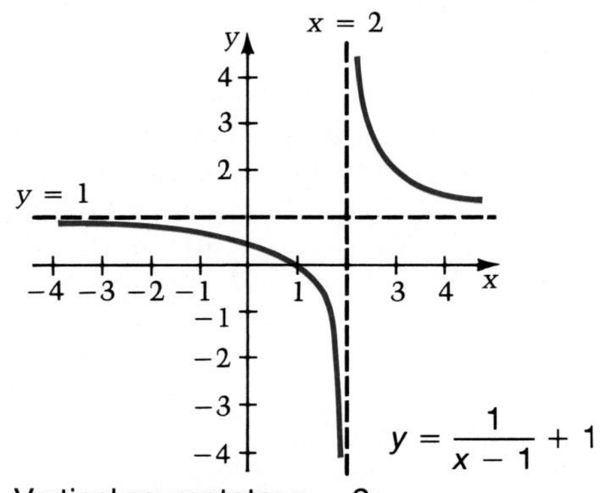

Vertical asymptote: $x = 2$; horizontal asymptote: $y = 1$.

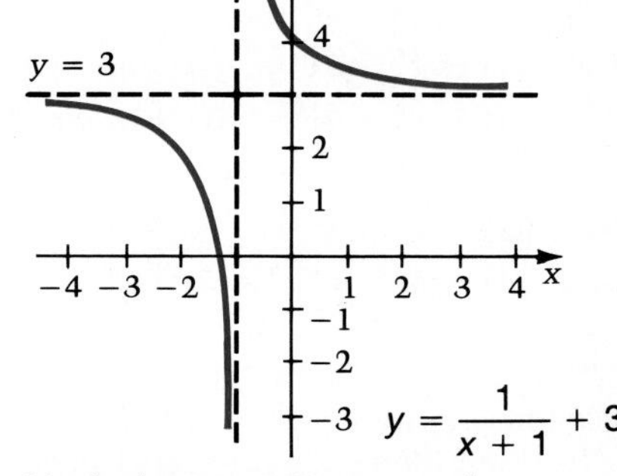

Vertical asymptote: $x = -1$; horizontal asymptote: $y = 3$.

Figure 4.32

Not all rational functions are transformations of $y = \frac{1}{x}$. In the following example, we consider some rational functions whose graphs do not resemble the graph of $y = \frac{1}{x}$.

EXAMPLE 1

Write the equations of any vertical or horizontal asymptotes of each graph.

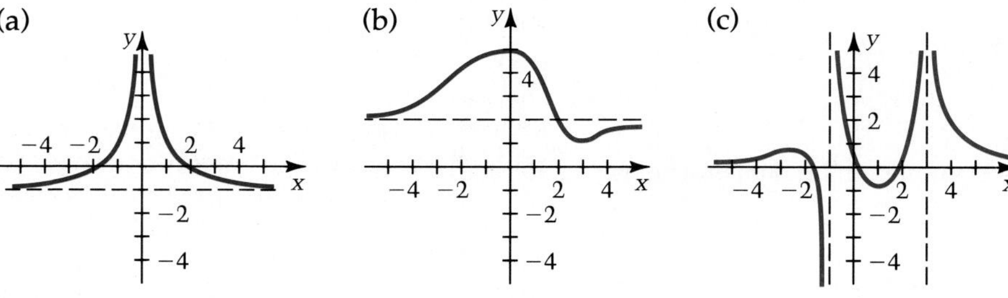

Figure 4.33

Solution

(a) Vertical asymptote: $x = 0$; horizontal asymptote: $y = -1$.
(b) Vertical asymptote: none; horizontal asymptote: $y = 2$.
(c) Vertical asymptotes: $x = -1$ and $x = 3$; horizontal asymptote: $y = 0$.

Notice that the graph in part (b) crosses the asymptote at $x = 2$, and eventually approaches it from below.

Remember that a vertical or horizontal asymptote is described by an equation either in the form $x = c$ or $y = d$, respectively (for real numbers c and d).

Asymptotes are characteristic features of the graphs of rational functions and are essential to sketching such graphs. We now consider methods for finding vertical and horizontal asymptotes, given only the equation of the function. In Figure 4.32(b), we saw that the line $x = -1$ is a vertical asymptote of the graph of the function

$$y = \frac{1}{x+1} + 3 = \frac{3x+4}{x+1}.$$

If we try to evaluate this function at $x = -1$, we find that (i) the numerator is nonzero, and (ii) the denominator is zero. This is true in general for all vertical asymptotes.

First Property of Rational Functions

If f is a rational function defined by

$$f(x) = \frac{g(x)}{h(x)}$$

then the line $x = c$ is a vertical asymptote of the graph of f provided $g(c) \neq 0$ and $h(c) = 0$.

EXAMPLE 2

Find all vertical asymptotes of the graph of each function.

(a) $y = \dfrac{3x^2}{x^2 - 1}$ (b) $y = \dfrac{x^3 - 2x^2}{x^2 - 3x + 2}$

Solution

(a)

$$x^2 - 1 = 0$$
$$(x - 1)(x + 1) = 0$$

Determine the values that make the denominator 0.

$x - 1 = 0$	$x + 1 = 0$
$x = 1$	$x = -1$

Check whether the numerator is nonzero for these values.

For $x = 1$, the numerator $3(1)^2 \neq 0$.
For $x = -1$, the numerator $3(-1)^2 \neq 0$.
Thus, the lines $x = 1$ and $x = -1$ are vertical asymptotes.

(b) $x^2 - 3x + 2 = 0$ — Determine the values that make the denominator 0.

$(x - 1)(x - 2) = 0$

$x - 1 = 0$	$x - 2 = 0$
$x = 1$	$x = 2$

Check whether the numerator is nonzero for these values.

For $x = 1$, the numerator $1^3 - 2(1)^2 \neq 0$.
For $x = 2$, the numerator $2^3 - 2(2)^2 = 0$.
Therefore, the line $x = 1$ is the only vertical asymptote.

The graph of a rational function can have at most one horizontal asymptote. To determine this asymptote, we need only look at the leading terms of the numerator and the denominator.

Second Property of Rational Functions

If

$$f(x) = \frac{a_n x^n + a_{n-1}x^{n-1} + \cdots + a_2x^2 + a_1x + a_0}{b_m x^m + b_{m-1}x^{m-1} + \cdots + b_2x^2 + b_1x + b_0}$$

is a rational function such that the numerator has degree n and the denominator has degree m, then the graph of f has

(i) the x-axis ($y = 0$) as a horizontal asymptote if $n < m$;

(ii) the line $y = \frac{a_n}{b_m}$ as a horizontal asymptote if $n = m$;

(iii) no horizontal asymptote if $n > m$.

EXAMPLE 3

For the graph of each function, determine the horizontal asymptote, if it exists.

(a) $y = \dfrac{3x^2}{7x^2 - 1}$ (b) $f(x) = \dfrac{x^3 - 2x^2}{x^2 - 3x + 2}$ (c) $g(x) = \dfrac{3x + 4}{7x^2 - 5}$

Solution

(a) The degree of the numerator and denominator are the same. By (ii) of the Second Property, the horizontal asymptote is $y = \frac{3}{7}$, the ratio of the leading coefficients.

(b) The degree of the numerator is greater than the degree of the denominator. Therefore, by (iii) of the Second Property, there is no horizontal asymptote.

(c) The degree of the numerator is less than the degree of the denominator. By (i) of the Second Property, the horizontal asymptote for the graph of g is the x-axis ($y = 0$).

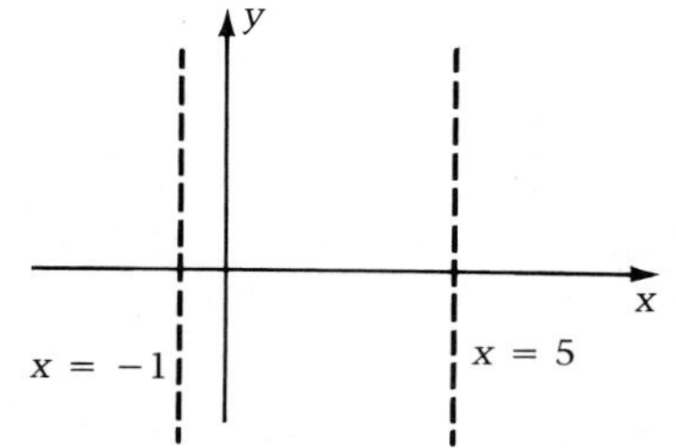

Figure 4.34 Asymptotes of the rational function

$$f(x) = \frac{x + 4}{(x + 1)(x - 5)}$$

To graph rational functions, we shall follow a procedure similar to that used to graph polynomial functions. In Section 4.2 we found that the key to graphing polynomial functions was to find the intervals where the graph lies below the x-axis and the intervals where the graph lies above the x-axis. This information, together with the vertical and horizontal asymptotes, will help us graph rational functions.

Suppose we wish to graph the rational function

$$f(x) = \frac{x + 4}{(x + 1)(x - 5)}.$$

Using the First and Second Properties of Rational Functions, we find that the graph has vertical asymptotes $x = -1$ and $x = 5$, and has a horizontal asymptote $y = 0$.

Once the asymptotes are found, we can proceed as we did in Section 4.2. We must now find the zeros of the function. (Recall that these values are the x-intercepts of the graph.) To find the zeros, we must solve the equation

$$\frac{x + 4}{(x + 1)(x - 5)} = 0.$$

Recall that a fraction is 0 only when its numerator is 0 and its denominator is not zero. Therefore, the solution of the equation is -4; that is, -4 is the only zero of the function. The function is undefined when the denominator is zero, that is, when

$$x = 5 \quad \text{and} \quad x = -1.$$

Recall that when graphing polynomial functions, we use the zeros to divide the x-axis into open intervals in order to determine where the function is positive and where it is negative. For rational functions, the situation is a little different.

Third Property of Rational Functions

Between any two successive x-values where a rational function f is either 0 or undefined, the graph of f lies either completely above or completely below the x-axis.

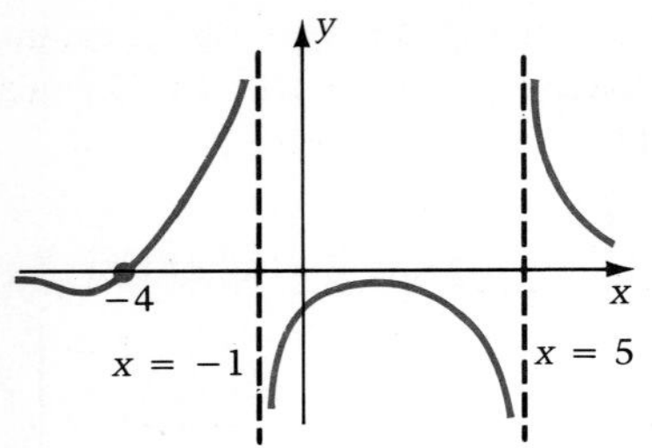

Figure 4.35 $f(x) = \dfrac{x + 4}{(x + 1)(x - 5)}$

Since f has a zero at -4, and is undefined at -1 and 5, by this Property we can conclude that on each of the intervals $(-\infty, -4)$, $(-4, -1)$, $(-1, 5)$ and $(5, \infty)$, the graph lies either completely above or completely below the x-axis. As with polynomial functions, we use test values to determine this.

The graph of the function f is shown in Figure 4.35. Since the graph is above the x-axis to the right of 5 and below the x-axis to the left of 5, we say informally that the graph jumps across the x-axis at 5. Notice that the graph also jumps across the x-axis at -1.

We now summarize the general procedure for graphing rational functions of the form

$$f(x) = \frac{g(x)}{h(x)}.$$

Step 1. Find the horizontal and vertical asymptotes.

Step 2. Find the zeros of the function and the values where the function is undefined.

Step 3. Apply the Third Property of Rational Functions (for each interval, select test values to determine whether the graph lies above the x-axis or below the x-axis.)

Step 4. Plot a few additional points and draw a smooth curve. Recall that a graph can cross a horizontal asymptote.

EXAMPLE 4

Sketch the graph of the rational function $f(x) = \dfrac{2x^2 - 10}{x^2 - 6x + 5}$.

Solution

Step 1. To find the horizontal asymptote, we use the Second Property. Since the degrees of the numerator and denominator are equal, the line $y = \frac{2}{1}$ (i.e., $y = 2$) is the horizontal asymptote. To determine the vertical asymptotes, rewrite the denominator in factored form:

$$f(x) = \frac{2(x^2 - 5)}{(x - 5)(x - 1)}.$$

The denominator is 0 when $x = 5$ or $x = 1$, and the numerator is not 0 for either of these values. Thus, the lines $x = 5$ and $x = 1$ are vertical asymptotes.

Step 2. To find the zeros of the function, we determine the values where the numerator is 0 but the denominator is not 0, namely $\sqrt{5}$ and $-\sqrt{5}$. In addition, the function is undefined when $x = 5$ or $x = 1$. Plot the zeros and graph the asymptotes.

Step 3. To apply the Third Property, we use the zeros and the x-values where the function is undefined to divide the x-axis into open intervals, as shown in Figure 4.36. Select a test value in each interval.

Interval	Test Value t	$f(t)$	
$(-\infty, -\sqrt{5})$	-3	$\frac{1}{4}$	$f(t) > 0$; graph lies above x-axis
$(-\sqrt{5}, 1)$	0	-2	$f(t) < 0$; graph lies below x-axis
$(1, \sqrt{5})$	2	$\frac{2}{3}$	$f(t) > 0$; graph lies above x-axis
$(\sqrt{5}, 5)$	3	-2	$f(t) < 0$; graph lies below x-axis
$(5, \infty)$	7	$7\frac{1}{3}$	$f(t) > 0$; graph lies above x-axis

Step 4. Plot a few additional points (use the test points from Step 3) and draw a smooth curve (Figure 4.37).

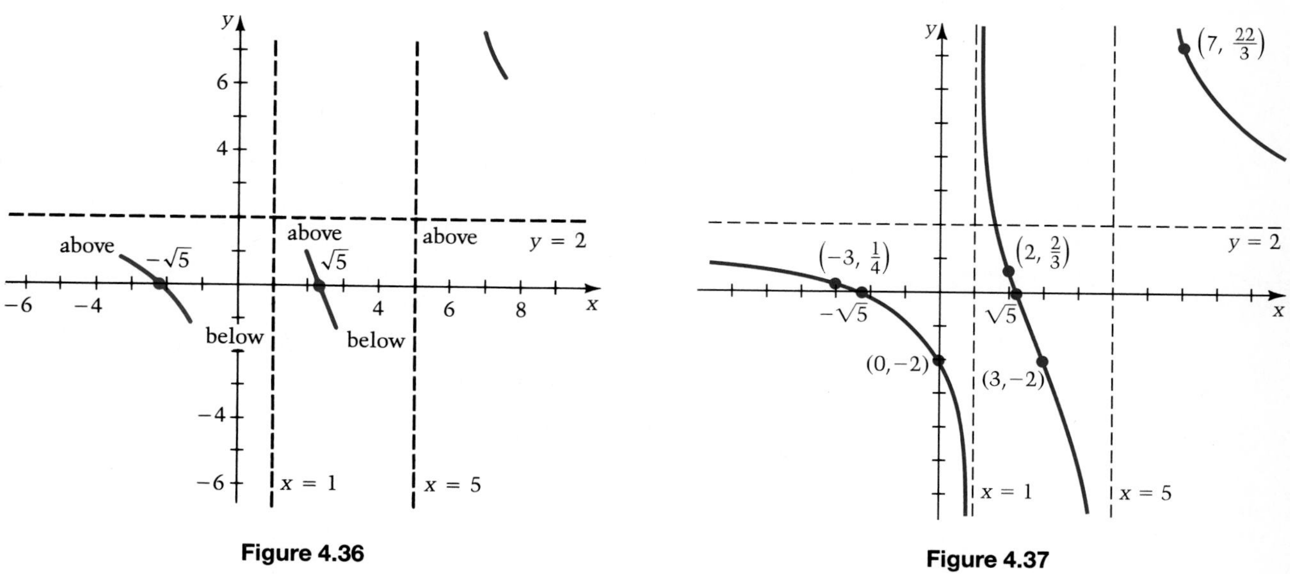

Figure 4.36

Figure 4.37

We have said that the zeros of a rational function occur where the numerator $g(x)$ is 0 and the denominator $h(x)$ is not 0. The vertical asymptotes occur

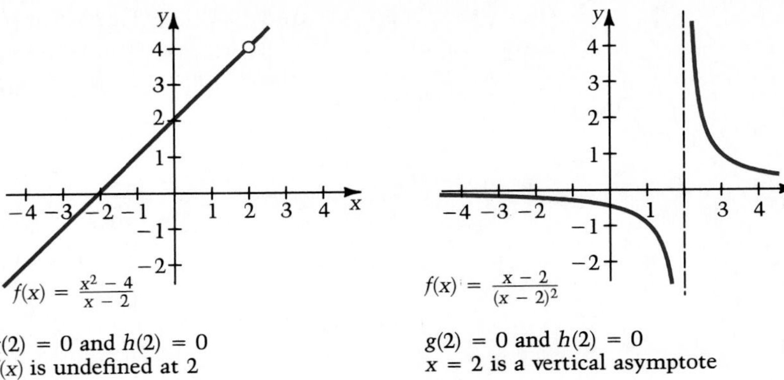

Figure 4.38

where $g(x)$ is not 0 and $h(x)$ is 0. We have not considered what might happen if $g(x) = 0$ and $h(x) = 0$. The graphs in Figure 4.38 illustrate what could occur.

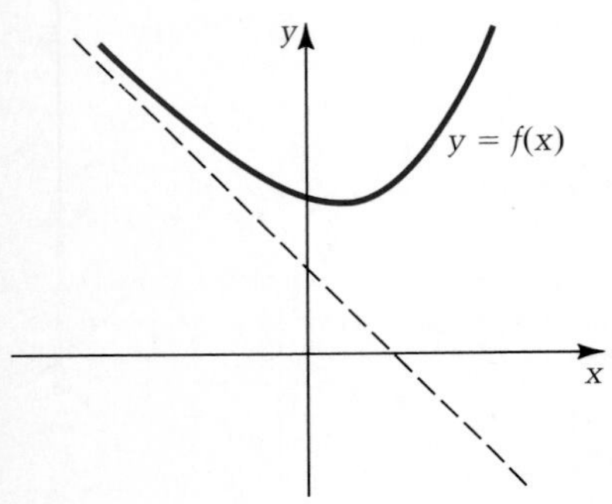

Figure 4.39 The dashed line is an oblique asymptote.

Oblique Asymptotes

Thus far the asymptotes we have considered have been vertical or horizontal lines. A line which is neither horizontal nor vertical is called an *oblique line.* Such a line can be an asymptote for a rational function, in which case it is referred to as an **oblique asymptote.** If for a rational function $f(x) = \frac{g(x)}{h(x)}$ the degree of the numerator is one more than the degree of the denominator, then the graph of $f(x)$ will have an oblique asymptote. Dividing $g(x)$ by $h(x)$ produces a quotient of the form $mx + b$, from which we can conclude that the line $y = mx + b$ is the oblique asymptote.

EXAMPLE 5

Sketch the graph of $f(x) = \frac{x^2 + x - 2}{x - 2}$.

Solution

Step 1. Since the degree of the numerator is one more than the degree of the denominator, there is no horizontal asymptote, but there will be an oblique asymptote. We determine its equation by dividing.

$$\begin{array}{r} x + 3 \\ x - 2 \overline{)\, x^2 + x - 2} \\ -(x^2 - 2x) \\ \hline 3x - 2 \\ -(3x - 6) \\ \hline 4 \end{array}$$

$$\frac{x^2 + x - 2}{x - 2} = x + 3 + \frac{4}{x - 2}.$$

As $x \to \infty, \frac{4}{x - 2} \to 0$, so the line $y = x + 3$ is an oblique asymptote.

Step 2. The zeros of the function are the solutions of the equation $x^2 + x - 2 = 0$, namely -2 and 1.

Steps 3 and 4. We use the zeros, the vertical asymptote, the oblique asymptote, and a few other points to sketch the graph.

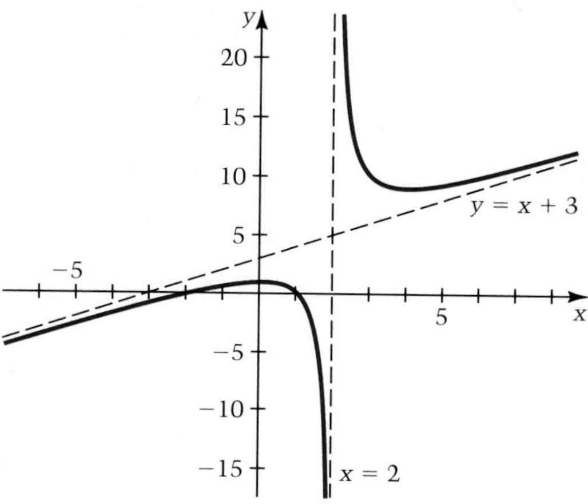

Figure 4.40

EXERCISE SET 4.7

In Exercises 1–4, write the equations of any vertical or horizontal asymptotes.

1.
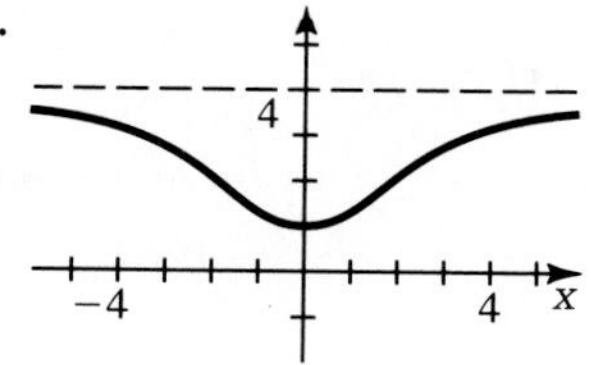

2.
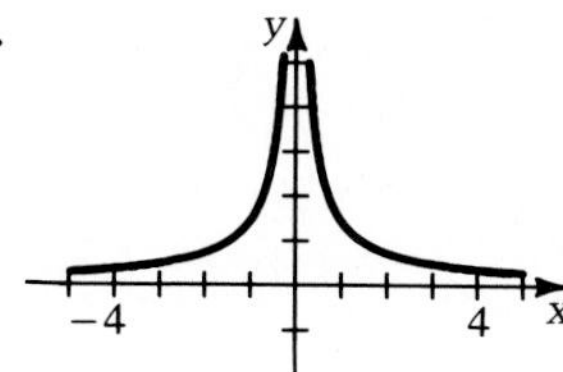

3.
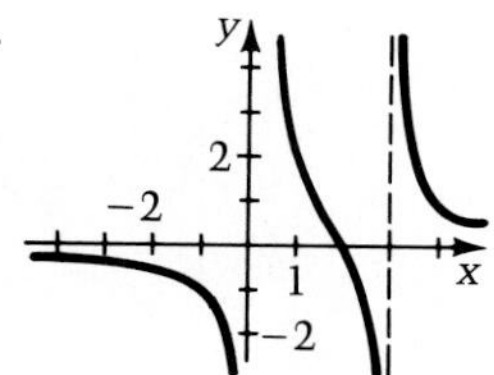

4.
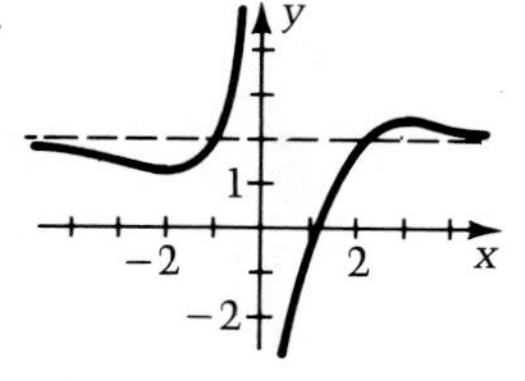

In Exercises 5–14, write the equations of any horizontal or vertical asymptotes.

5. $y = \dfrac{1}{x + 9}$

6. $y = \dfrac{1}{x - 3}$

7. $y = \dfrac{2x + 9}{x^2 + 3x + 2}$

8. $y = \dfrac{x - 6}{x^2 - 9}$

9. $y = \dfrac{x^2 - x - 2}{(2x - 1)^2}$

10. $y = \dfrac{x^2 - 1}{2x^2 - 8}$

11. $y = \dfrac{10x^2 + 2x + 1}{5x^2 + 6x + 1}$

12. $y = \dfrac{x^3 - 8}{(3x - 1)^3}$

13. $y = \dfrac{2x^4 - 7x^2 + 4}{(2x - 1)(x + 1)^2}$

14. $y = \dfrac{x^4 + 3x^2 + 4}{(x - 3)^2(x + 5)}$

In Exercises 15–26, graph the rational function.

15. $y = \dfrac{1}{x + 1}$

16. $y = \dfrac{1}{x - 2}$

17. $y = \dfrac{x - 1}{x - 2}$

18. $y = \dfrac{2x - 1}{x - 1}$

19. $y = \dfrac{4}{x^2 - 2x - 3}$

20. $y = \dfrac{2}{x^2 + x - 6}$

21. $y = \dfrac{1}{x^2 + 4x + 4}$

22. $y = \dfrac{3}{x^2 + 6x + 9}$

23. $y = \dfrac{3x^2}{x^2 - 4}$

24. $y = \dfrac{2x^2}{x^2 - 16}$

25. $y = \dfrac{x^2 - 9}{2x^2}$

26. $y = \dfrac{2x^2 - 8}{x^2}$

In Exercises 27–32, sketch the graph of the given function and show the oblique asymptote.

27. $f(x) = \dfrac{x^2 - 1}{x}$

28. $y = \dfrac{x^2 - 3x + 1}{x}$

29. $f(x) = \dfrac{x^2 + 2x + 1}{x}$

30. $f(x) = \dfrac{2x^2 - 3}{x}$

31. $f(x) = \dfrac{3x^2 - 5x - 1}{x - 2}$

32. $y = \dfrac{2x^2 - 7x + 4}{x - 3}$

Superset

In Exercises 33–38, sketch the graph of the function and use that graph to sketch the graph of $y = \dfrac{1}{f(x)}$.

33. $f(x) = x + 2$

34. $f(x) = x - 3$

35. $f(x) = x^2 - 4$

36. $f(x) = 1 - x^2$

37. $f(x) = |9 - x^2|$

38. $f(x) = |x^2 - 4|$

4.8
Conic Sections I: Parabola and Circle

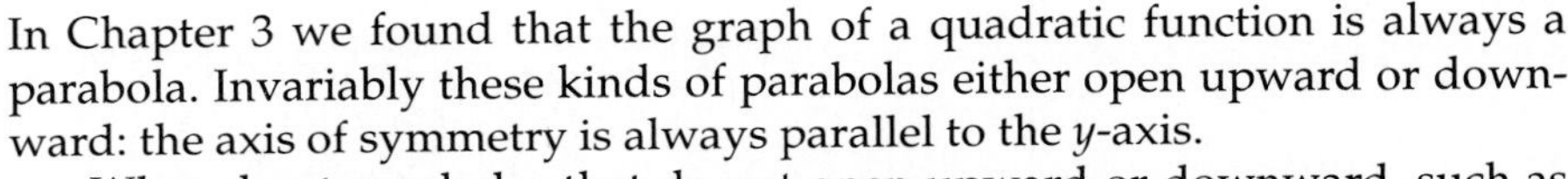

In Chapter 3 we found that the graph of a quadratic function is always a parabola. Invariably these kinds of parabolas either open upward or downward: the axis of symmetry is always parallel to the y-axis.

Figure 4.41

What about parabolas that do not open upward or downward, such as those at the left? Clearly these are not the graphs of functions because they fail the vertical line test. Recall that we refer to such sets of points as *relations*.

In this section, we begin our study of a special group of relations called **conic sections.** This group consists of the parabola, the circle, the ellipse, and the hyperbola. They are called conic sections because each can be formed by cutting a double cone with a plane as shown in Figure 4.42.

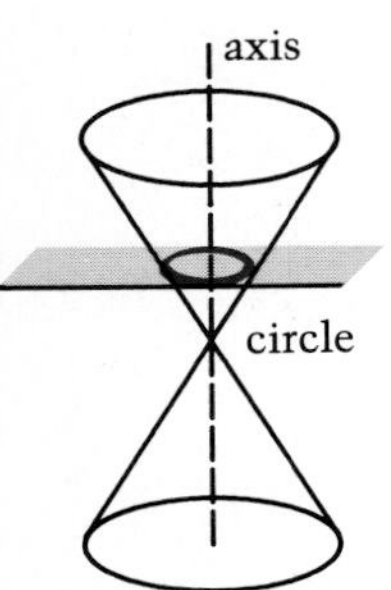

The plane is perpendicular to the axis of the cone.

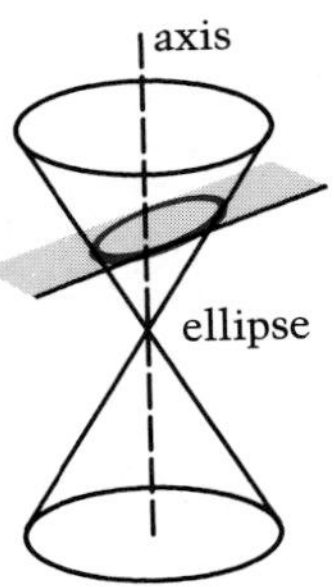

The plane is tilted.

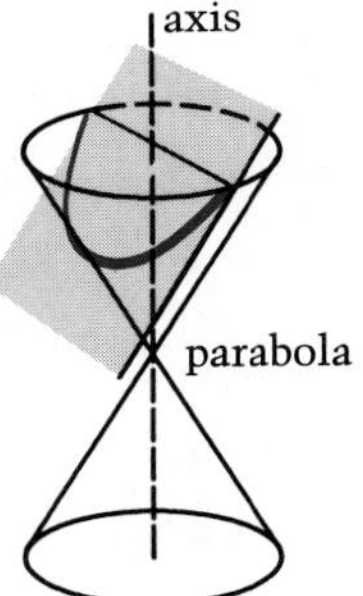

The plane is parallel to one side of the cone.

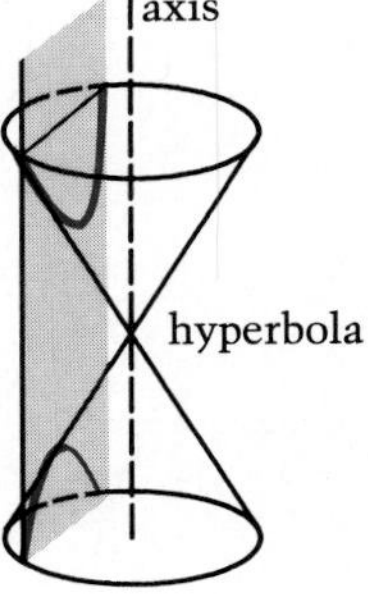

The plane is parallel to the axis of the cone.

Figure 4.42

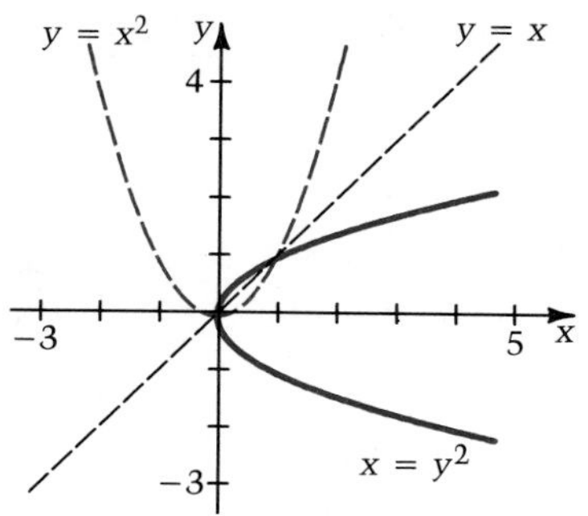

Figure 4.43

We shall begin by considering parabolas. The parabola described by the equation $y = x^2$ is shown in Figure 4.43. Interchanging x and y in this equation produces the equation $x = y^2$, which also describes a parabola. When x and y are interchanged in an equation, the resulting graph is the reflection of the original graph across the line $y = x$. Although $x = y^2$ does not describe a function of x, the rules of symmetry and transformations still apply. We summarize these rules in the following table.

Change in Equation	Change in the Graph
x is replaced with $x - h$	If $h > 0$, graph moves $\lvert h \rvert$ units to the right. If $h < 0$, graph moves $\lvert h \rvert$ units to the left.
y is replaced with $y - k$	If $k > 0$, graph moves $\lvert k \rvert$ units up. If $k < 0$, graph moves $\lvert k \rvert$ units down.
x is replaced with $-x$	Graph is reflected across the y-axis.
y is replaced with $-y$	Graph is reflected across the x-axis.

For example, two transformations of the graph of

$$x = y^2$$

are shown below. Replacing x with $-\frac{1}{3}x$ reflects the graph of $x = y^2$ across the y-axis and stretches the graph horizontally (the new function is $-\frac{1}{3}x = y^2$ or $x = -3y^2$). This is displayed in Figure 4.44(a). Replacing y with $y - (-3)$ and x with $x - 2$ translates the original graph 3 units down and 2 units to the right (the new function is $x = (y + 3)^2 + 2$). This is shown in Figure 4.44(b).

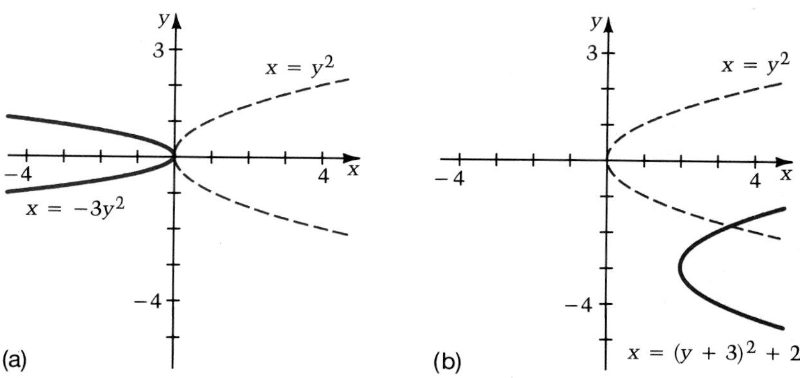

Figure 4.44

Recall that it is easier to sketch the graph of a quadratic function if the equation is written in vertex form: $f(x) = a(x - h)^2 + k$. For example, the graph of $f(x) = -2(x - 5)^2 + 7$ has its vertex at (5, 7), the axis of symmetry is the line $x = 5$, and the parabola opens downward since a (in this case -2)

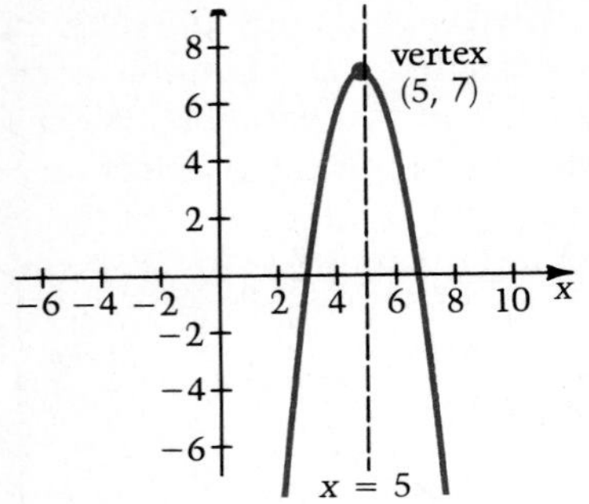

Figure 4.45

is negative (Figure 4.45). There is an analogous vertex form of the equation of a parabola opening to the right or the left.

Parabola: Vertex Form

The graph of an equation of the form $x = a(y - k)^2 + h$ is a parabola whose vertex is (h, k) and whose axis of symmetry is the line $y = k$. If $a > 0$, the parabola opens to the right; if $a < 0$ the parabola opens to the left.

EXAMPLE 1

Sketch the graph of the parabola $x = 2(y + 3)^2 - 4$. Determine the vertex, the axis of symmetry, and the x- and y-intercepts.

Solution $x = 2[y - (-3)]^2 + (-4)$ The equation is written in vertex form: $x = a(y - k)^2 + h$.

Since $a > 0$, the parabola opens to the right. The vertex is $(-4, -3)$, and the axis of symmetry is the line $y = -3$. The y-intercepts occur when $x = 0$.

$$0 = 2(y + 3)^2 - 4$$
$$4 = 2(y + 3)^2$$
$$2 = (y + 3)^2$$

Thus, the y-intercepts are $-3 \pm \sqrt{2}$ or approximately -1.6 and -4.4. The x-intercept occurs when $y = 0$, that is, $x = 2(0 + 3)^2 - 4 = 14$.

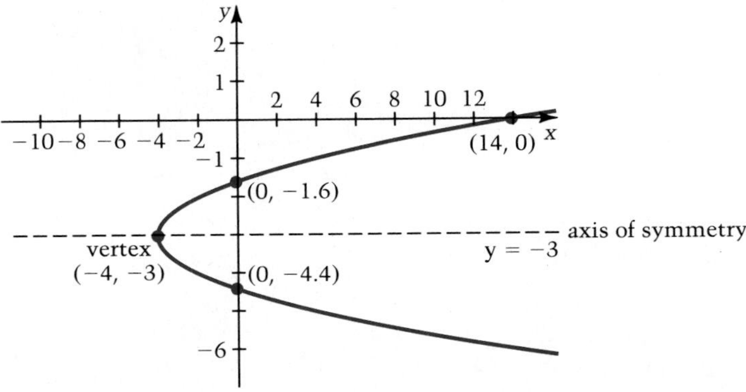

Figure 4.46

In Chapter 3 we saw that the method of completing the square is used to rewrite a quadratic equation in vertex form. This method is used in the following example.

EXAMPLE 2

Sketch the graph of the parabola $x = -3y^2 + 6y + 5$. Determine the vertex, the axis of symmetry, and the x- and y-intercepts.

Solution

$x = -3(y^2 - 2y \quad\quad) + 5$	Begin completing the square by factoring the coefficient of y^2 out of the variable terms.
$x = -3(y^2 - 2y + 1) + 5 + 3$ $x = -3(y - 1)^2 + 8$	Take half of the coefficient of the y-term and square it. Add this inside the parentheses. In effect we added -3; thus, we must also add 3.

The vertex is (8, 1). Since $a < 0$ the parabola opens to the left. To find the y-intercepts, use the quadratic formula to find solutions of the equation $0 = -3y^2 + 6y + 5$. The y-intercepts are approximately 2.6 and -0.6. The x-intercept occurs when $y = 0$. Since $x = -3(0 - 1)^2 + 8 = 5$, the x-intercept is 5.

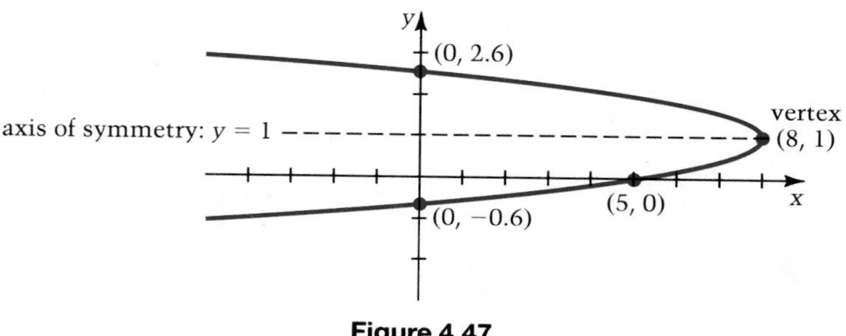

Figure 4.47

The Circle

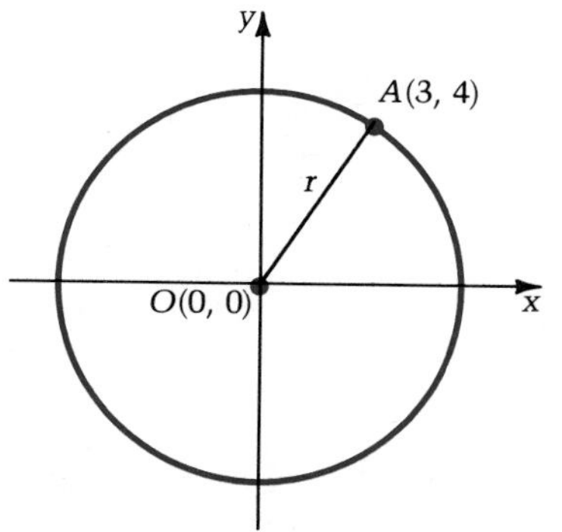

Figure 4.48

Suppose we are given that a circle has its center at $O(0, 0)$ and contains the point $A(3, 4)$ (Figure 4.48). How can we determine r, the radius of the circle, that is, the distance between $O(0, 0)$ and $A(3, 4)$?

To determine this distance, recall the Distance Formula.

The Distance Formula

The distance between any two points $P_1(x_1, y_1)$ and $P_2(x_2, y_2)$ is given by the formula

$$d(P_1, P_2) = \sqrt{(x_2 - x_1)^2 + (y_2 - y_1)^2}.$$

Using the Distance Formula, we conclude that for the circle in Figure 4.48

$$r = d(O, A) = \sqrt{(3-0)^2 + (4-0)^2} = \sqrt{9+16} = \sqrt{25} = 5.$$

If a circle has its center at $O(0, 0)$, we can use the Distance Formula to describe the distance between any point $A(x, y)$ on the circle and the center $O(0, 0)$: $d(O, A) = \sqrt{(x-0)^2 + (y-0)^2} = \sqrt{x^2 + y^2}$. Since this distance is the radius, we can write $\sqrt{x^2 + y^2} = r$ so that

$$x^2 + y^2 = r^2.$$

Replacing x with $x - h$ and y with $y - k$ in the equation $x^2 + y^2 = r^2$ translates the original circle so that its new center is at (h, k). The radius remains unchanged.

Circle: Center/Radius Form

An equation of the form

$$(x-h)^2 + (y-k)^2 = r^2 \quad (r > 0)$$

describes a circle with center at (h, k) and radius r.

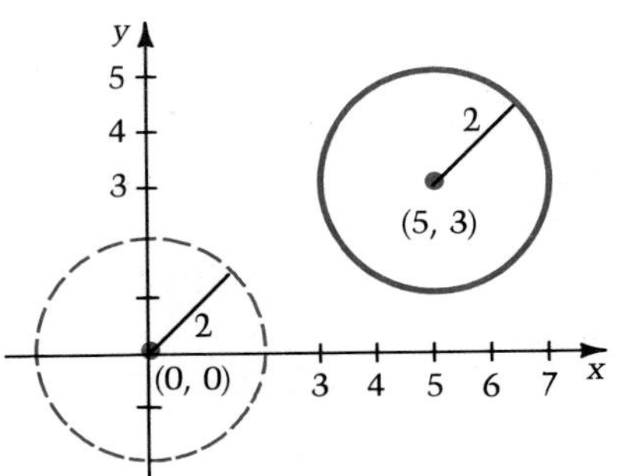

The graph of the relation $(x-5)^2 + (y-3)^2 = 4$ is a circle with center at (5, 3) and radius 2. It is found by translating the graph of $x^2 + y^2 = 4$ five units to the right and three units up.

Figure 4.49

Suppose the equation $Ax^2 + Bx + Ay^2 + Cy + D = 0$ describes a circle. To graph this, rewrite the equation in center/radius form.

EXAMPLE 4

Sketch the graph of the circle $4x^2 - 8x + 4y^2 + 16y + 11 = 0$. Determine the center and radius.

Solution

$(4x^2 - 8x \qquad) + (4y^2 + 16y \qquad) = -11$ Prepare to complete the square.

$$4(x^2 - 2x \qquad) + 4(y^2 + 4y \qquad) = -11$$

$$4(x^2 - 2x + 1) + 4(y^2 + 4y + 4) = -11 + 4 \cdot 1 + 4 \cdot 4$$ Complete the square in x and y.

$$(x - 1)^2 + (y + 2)^2 = \frac{9}{4}$$

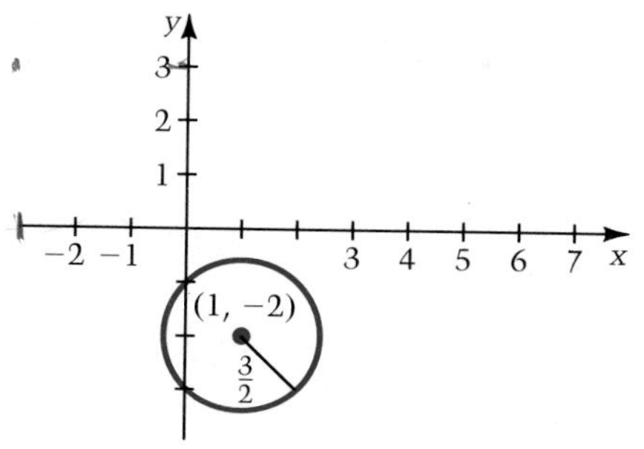

Figure 4.50

The graph is a circle of radius $\frac{3}{2}$ with center at $(1, -2)$.

EXERCISE SET 4.8

In Exercises 1–12, sketch the graph of the given parabola. Determine the vertex, axis of symmetry, and the x- and y-intercepts.

1. $y^2 = 4x$ **2.** $2y^2 = -10x$

3. $9y^2 = 3x$ **4.** $4y^2 = x$

5. $x = -2(y - 3)^2 + 5$ **6.** $x = 3(y + 1)^2 - 2$

7. $x = 2(y + 1)^2 + \frac{3}{2}$ **8.** $x = -\frac{1}{2}(y - 2)^2 + 4$

9. $x = -y^2 + 6y - 13$ **10.** $x = 3 - 4y - y^2$

11. $y^2 + 2y + 3x - 8 = 0$ **12.** $y^2 - 6y - 2x - 9 = 0$

In Exercises 13–24, sketch the graph of the given circle. Determine the center and radius.

13. $x^2 + y^2 = 16$ **14.** $x^2 + y^2 = 25$

15. $(x + 1)^2 + (y - 3)^2 = 9$ **16.** $(x - 3)^2 + (y + 4)^2 = 1$

17. $x^2 + y^2 - 4x = 0$ **18.** $x^2 + y^2 - 6y + 5 = 0$

19. $x^2 + 4x + y^2 - 2y - 11 = 0$

20. $x^2 - 3x + y^2 - 5y - \frac{1}{2} = 0$

21. $4x^2 + 4y^2 - 4x + 4y - 10 = 0$

22. $9x^2 + 9y^2 - 6x + 18y - 8 = 0$

23. $25x^2 + 25y^2 - 10x - 150y + 1 = 0$

24. $9x^2 + 9y^2 - 3x - 6y - 1 = 0$

The unit circle is a circle centered at (0, 0) with radius equal to 1. Therefore, the unit circle is the set of all points in the plane that lie at a distance of exactly 1 unit from the origin. In Exercises 25–32, use the Distance Formula to determine whether the given point lies inside or outside the unit circle.

25. (0.65, 0.85) 26. (−0.25, −0.75)

27. (−0.6, −0.7) 28. (−0.9, −0.1)

29. $\left(\frac{\sqrt{3}}{3}, -\frac{\sqrt{3}}{3}\right)$ 30. $\left(-\frac{\sqrt{6}}{3}, -\frac{\sqrt{6}}{4}\right)$

31. $\left(\frac{\sqrt{2}}{2}, \frac{\sqrt{3}}{2}\right)$ 32. $\left(-\frac{1}{\pi}, \frac{\pi}{4}\right)$

Superset

33. A circle of radius 3 has its center in the second quadrant and is tangent to both coordinate axes. Write an equation for the circle.

34. The line tangent to a circle is perpendicular to the radius at the point where the radius and tangent meet. Write an equation of the line tangent to the circle $x^2 + y^2 + 6x - 4y - 12 = 0$ at the point $(1, 5)$.

35. Prove that if one side of a triangle inscribed in a circle is a diameter of that circle, then the triangle is a right triangle. (*Hint:* Let the circle be $x^2 + y^2 = 1$ and let the vertices be $(-1, 0)$, $(1, 0)$ and $(x, \sqrt{1 - x^2})$ with $0 \le x < 1$.)

36. Write an equation for each circle which is tangent to both coordinate axes and passes through the point $(2, 4)$.

37. Show that the collection of all points which are twice as far from the origin as they are from the point $(4, 0)$ is a circle.

38. Show that the collection of all points equidistant from two given points is the line which is the perpendicular bisector of the line segment joining the two points. (*Hint:* Take (a, b) and $(-a, -b)$ as the two points.)

A parabola can be defined as the set of all points equidistant from a point, called the *focus* of the parabola, and a line, called the *directrix* of the parabola. In the diagram we have taken the directrix as a horizontal line. The condition $d(P, F) = d(P, l)$ gives us

$$\sqrt{(x - a)^2 + (y - b)^2} = |y - c|.$$

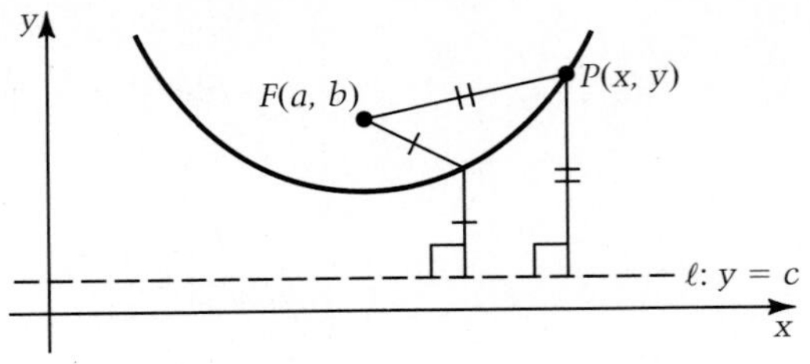

In Exercises 39–46, use this condition to derive the equation in vertex form of the parabola with the given focus and directrix. Identify the vertex and axis of symmetry for the parabola.

39. focus: $(0, 2)$; directrix: $y = 0$

40. focus: $(3, 0)$; directrix: $x = 0$

41. focus: $(0, -2)$; directrix: $y = 0$

42. focus: $(-3, 0)$; directrix: $x = 0$

43. focus: $(1, 2)$; directrix: $x = -3$

44. focus: $(3, 4)$; directrix: $y = -2$

45. focus: $(c, 0)$; directrix: $x = -c$

46. focus: $(0, c)$; directrix: $y = -c$

4.9
Conic Sections II: Ellipse and Hyperbola

Suppose we stretched a circle horizontally. This distorted circle, called an **ellipse,** would look like Figure 4.51. The ellipse has its **center** at the origin. The points $(a, 0)$, $(-a, 0)$, $(0, b)$, and $(0, -b)$ are the **vertices.**

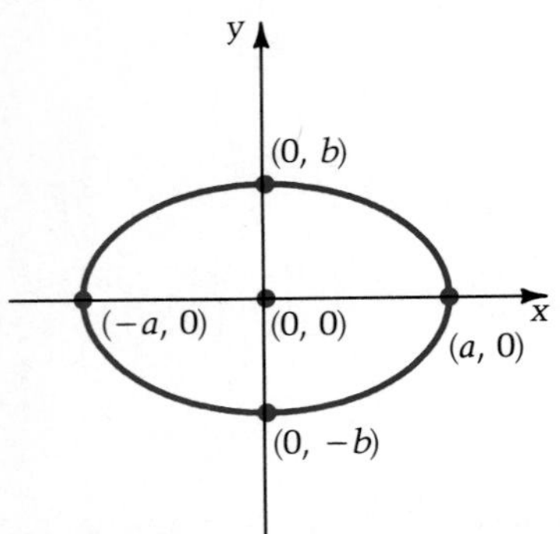

Figure 4.51

Ellipse Centered at the Origin

An equation of the form

$$\frac{x^2}{a^2} + \frac{y^2}{b^2} = 1$$

describes an ellipse with center at the origin and x-intercepts $\pm a$ and y-intercepts $\pm b$. The vertices are $(a, 0)$, $(-a, 0)$, $(0, b)$, and $(0, -b)$. Notice that the right side of the equation is $+1$.

EXAMPLE 1

Sketch the graph of each ellipse. Determine the center, x- and y-intercepts, and vertices.

(a) $\dfrac{x^2}{16} + \dfrac{y^2}{9} = 1$

(b) $100x^2 + 25y^2 = 100$

Solution

(a) The center is at the origin. The x-intercepts are ± 4, and the y-intercepts are ± 3. The vertices are $(4, 0)$, $(-4, 0)$, $(0, 3)$, and $(0, -3)$.

(b) Divide both sides by 100 to get $\dfrac{x^2}{1} + \dfrac{y^2}{4} = 1$. The center is at the origin, and the x-intercepts are ± 1; the y-intercepts are ± 2. The vertices are $(1, 0)$, $(-1, 0)$, $(0, 2)$, and $(0, -2)$.

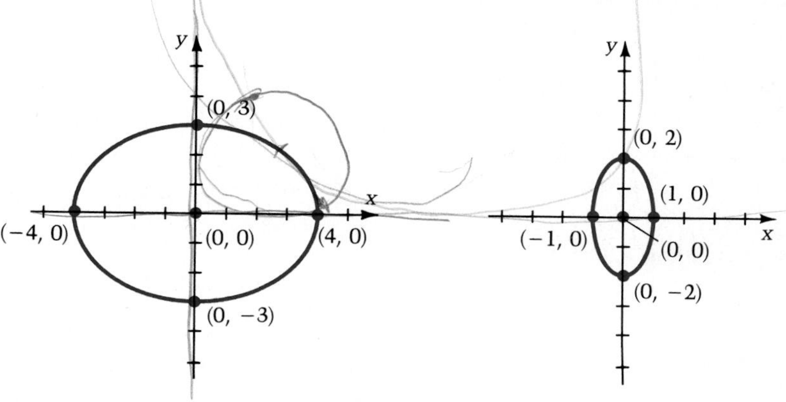

Figure 4.52

Suppose we translate the graph of

$$\frac{x^2}{a^2} + \frac{y^2}{b^2} = 1$$

so that its center is at (h, k). As you might expect, the equation of the new ellipse is

$$\frac{(x - h)^2}{a^2} + \frac{(y - k)^2}{b^2} = 1.$$

The point (h, k) corresponds to the old center $(0, 0)$. By adding a or $-a$ to the x-coordinate of the new center, and adding b or $-b$ to the y-coordinate of the new center, we can find the new vertices. See Figure 4.53 on the next page.

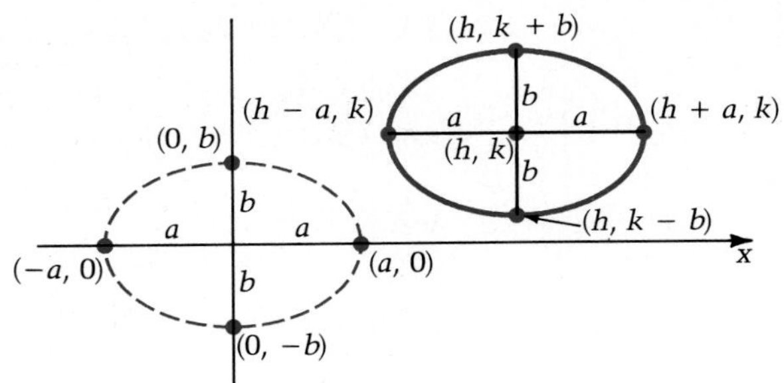

Figure 4.53

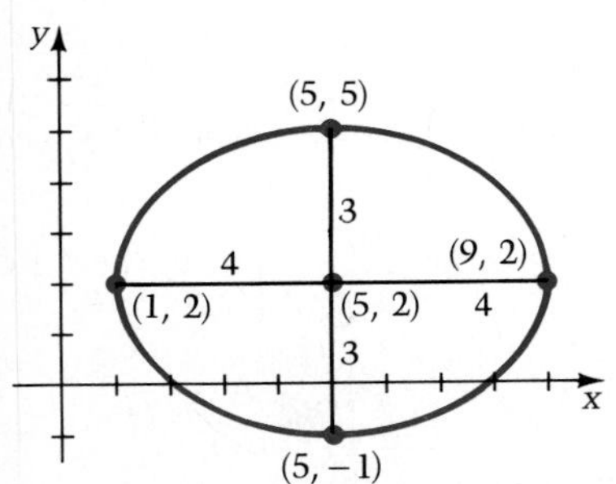

Figure 4.54

For example, the equation

$$\frac{(x-5)^2}{16} + \frac{(y-2)^2}{9} = 1$$

describes an ellipse with center at (5, 2). Since $a = 4$ and $b = 3$, the vertices are $(5 + 4, 2)$, $(5 - 4, 2)$, $(5, 2 + 3)$, and $(5, 2 - 3)$ (i.e., (9, 2), (1, 2), (5, 5), and (5, −1)).

If you are given an equation of an ellipse in the form

$$Ax^2 + Bx + Cy^2 + Dy + E = 0,$$

and you wish to sketch its graph, begin by completing the square (twice) to obtain an equation of the form

$$\frac{(x-h)^2}{a^2} + \frac{(y-k)^2}{b^2} = 1.$$

EXAMPLE 2

Sketch the graph of the ellipse $9x^2 + 18x + 4y^2 - 40y + 73 = 0$. Determine the center and vertices.

Solution

$$(9x^2 + 18x \quad) + (4y^2 - 40y \quad) = -73$$

Complete the square for x and y.

$$9(x^2 + 2x \quad) + 4(y^2 - 10y \quad) = -73$$

$$9(x^2 + 2x + 1) + 4(y^2 - 10y + 25) = -73 + 9 + 100$$

$$9(x+1)^2 + 4(y-5)^2 = 36$$

Now divide by 36 to get the proper form.

$$\frac{(x+1)^2}{4} + \frac{(y-5)^2}{9} = 1$$

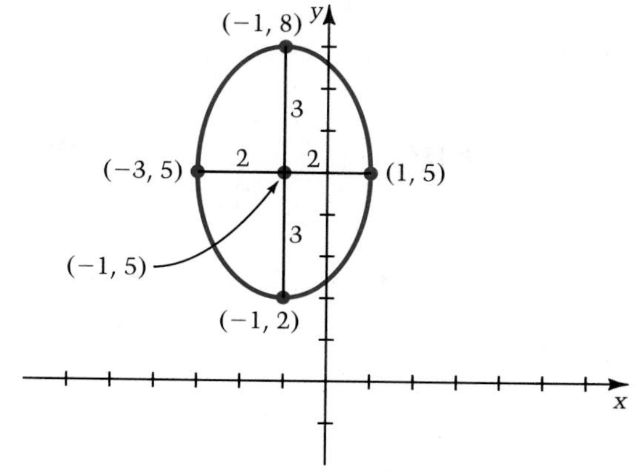

The ellipse has its center at $(-1, 5)$; $a = 2$ and $b = 3$. The vertices are $(-3, 5)$, $(1, 5)$, $(-1, 8)$, and $(-1, 2)$.

Figure 4.55

The Hyperbola

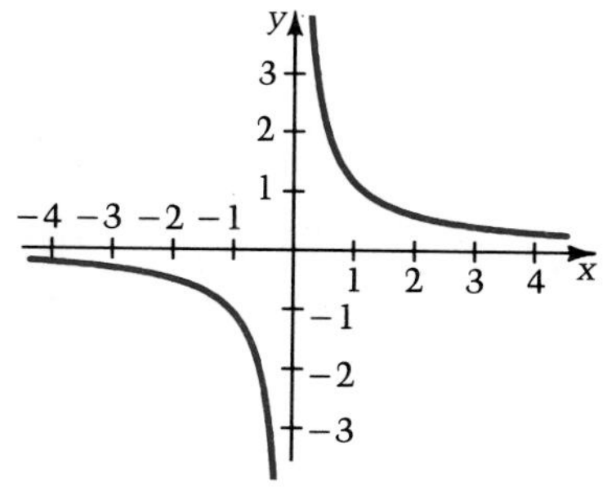

Figure 4.56 $y = \frac{1}{x}$.

Consider the equations

$$\frac{x^2}{a^2} - \frac{y^2}{b^2} = 1 \quad \text{and} \quad \frac{y^2}{b^2} - \frac{x^2}{a^2} = 1.$$

Although these equations differ only slightly from the equation of an ellipse (notice the minus sign), their graphs differ greatly. The graphs of the above equations are **hyperbolas.** The graph of $y = \frac{1}{x}$, which we considered in Section 4.7, is a hyperbola having the x-axis as a horizontal asymptote and the y-axis as a vertical asymptote.

In this section we shall study hyperbolas like those in Figure 4.57 that open in the x-direction or the y-direction.

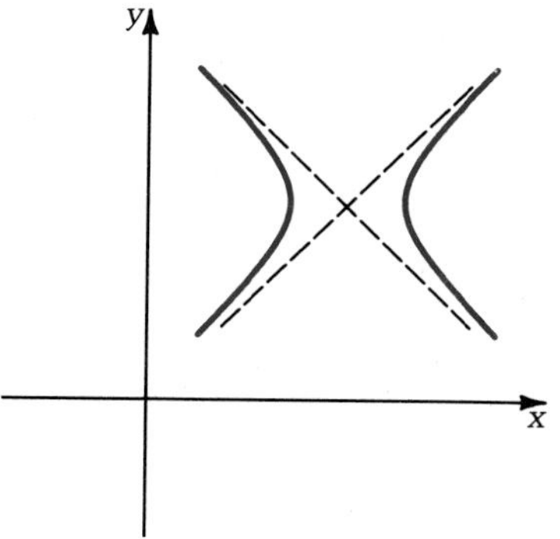

(a) Hyperbola opening in x-direction.

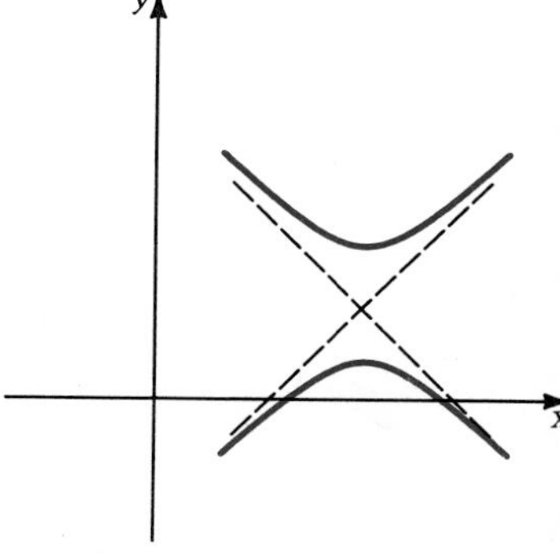

(b) Hyperbola opening in y-direction.

Figure 4.57

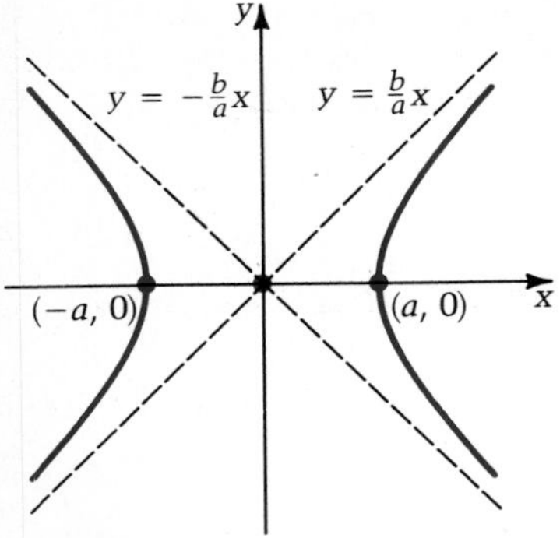

Figure 4.58

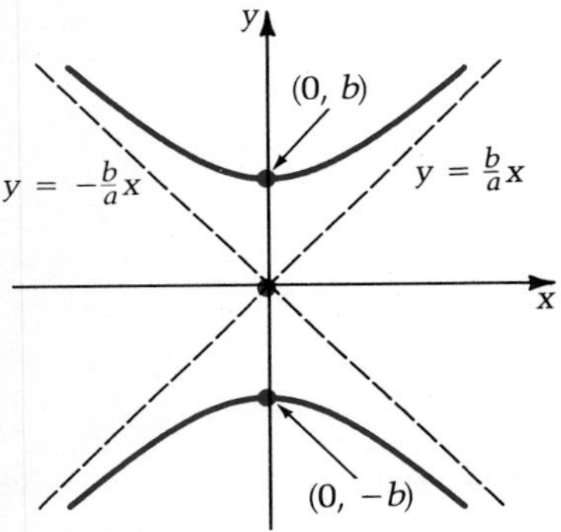

Figure 4.59

We begin by considering hyperbolas whose asymptotes intersect at the origin. We say that such hyperbolas have their centers at the origin.

Hyperbolas Centered at Origin

The graph of an equation of the form

$$\frac{x^2}{a^2} - \frac{y^2}{b^2} = 1$$

is a **hyperbola** that opens on the x-axis. The vertices are $(a, 0)$ and $(-a, 0)$, and the asymptotes are the lines $y = \frac{b}{a}x$ and $y = -\frac{b}{a}x$. There are no y-intercepts. The center of the hyperbola is at $(0, 0)$.

The graph of an equation of the form

$$\frac{y^2}{b^2} - \frac{x^2}{a^2} = 1$$

is a **hyperbola** that opens on the y-axis. The vertices are $(0, b)$ and $(0, -b)$, and the asymptotes are the lines $y = \frac{b}{a}x$ and $y = -\frac{b}{a}x$. There are no x-intercepts. The center of the hyperbola is at $(0, 0)$.

Notice that the right side of the equation in either of the above forms is $+1$. If the x^2-term is positive, the hyperbola opens on the x-axis, and if the y^2-term is positive, the hyperbola opens on the y-axis. The two cases are illustrated in Figure 4.60.

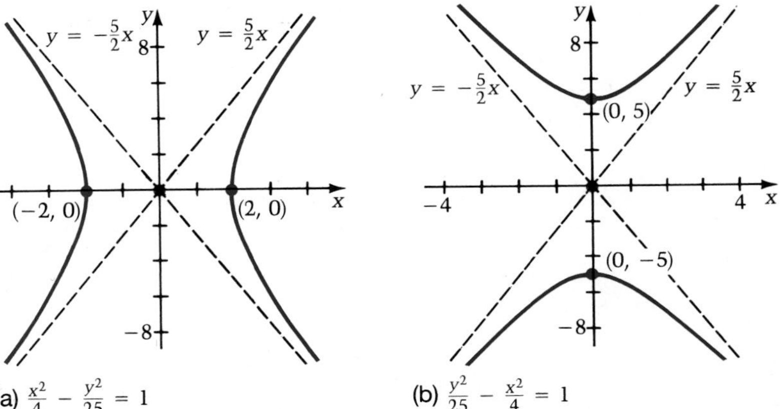

Figure 4.60

To determine the asymptotes, plot the points (a, b), $(-a, b)$, $(a, -b)$, and $(-a, -b)$. These four points determine the corners of a rectangle. The diagonals of the rectangle, when extended, form the two asymptotes.

EXAMPLE 3

Sketch the graph of the hyperbola $\frac{x^2}{16} - \frac{y^2}{9} = 1$. Show the center, vertices, and asymptotes.

Solution Since the x^2-term is positive, this hyperbola opens on the x-axis with vertices $(4, 0)$ and $(-4, 0)$. Since $a = 4$ and $b = 3$, we can find the asymptotes by drawing the rectangle with corners $(4, 3)$, $(-4, 3)$, $(4, -3)$, and $(-4, -3)$, and then using the diagonals to graph the asymptotes.

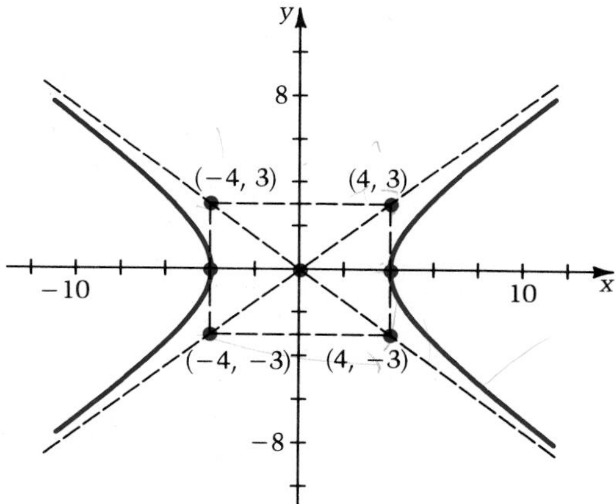

Figure 4.61

The two equations

$$\frac{(x-h)^2}{a^2} - \frac{(y-k)^2}{b^2} = 1 \quad \text{and} \quad \frac{(y-k)^2}{b^2} - \frac{(x-h)^2}{a^2} = 1$$

describe hyperbolas with centers at (h, k). If the $(x - h)^2$-term is positive, the hyperbola opens in the x-direction with vertices $(h + a, k)$ and $(h - a, k)$. If the $(y - k)^2$-term is positive, the hyperbola opens in the y-direction with vertices $(h, k + b)$ and $(h, k - b)$. The four corners of the rectangle that determine the asymptotes are $(h \pm a, k \pm b)$.

EXAMPLE 4

Sketch the graph of the hyperbola $9x^2 - 16y^2 + 18x - 64y + 521 = 0$. Show the center, vertices and asymptotes.

Solution

$$(9x^2 + 18x \quad\;) + (-16y^2 - 64y \quad\;) = -521$$

Complete the square for x and y.

$$9(x^2 + 2x \quad\;) - 16(y^2 + 4y \quad\;) = -521$$

$$9(x^2 + 2x + 1) - 16(y^2 + 4y + 4) = -521 + 9 - 64$$

$$9(x + 1)^2 - 16(y + 2)^2 = -576$$

Divide by -576 to get the proper form.

$$\frac{(y + 2)^2}{36} - \frac{(x + 1)^2}{64} = 1$$

Since the $(y + 2)^2$-term is positive, the hyperbola opens in the y-direction. The center of the hyperbola is at $(-1, -2)$. Since $a = 8$ and $b = 6$, the corners of the rectangle used to find the asymptotes are $(-1 \pm 8, -2 \pm 6)$, that is, $(7, -8)$, $(7, 4)$, $(-9, -8)$, and $(-9, 4)$. The vertices are $(-1, -2 + 6)$ and $(-1, -2 - 6)$ or simply $(-1, 4)$ and $(-1, -8)$.

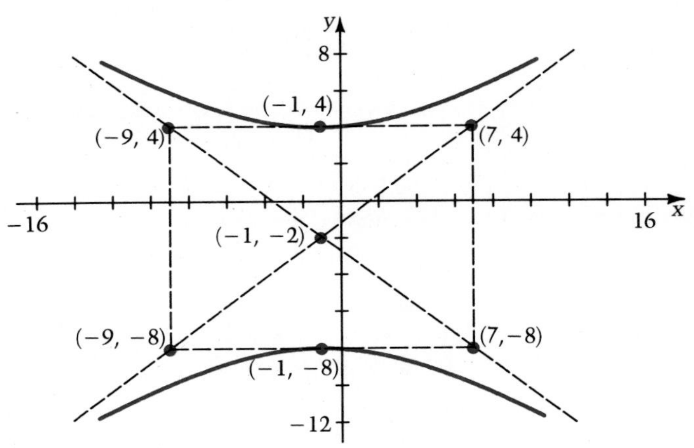

Figure 4.62

EXERCISE SET 4.9

In Exercises 1–14, sketch the graph of the given ellipse. Determine the center and vertices.

1. $\dfrac{x^2}{9} + \dfrac{y^2}{16} = 1$

2. $\dfrac{x^2}{25} + \dfrac{y^2}{4} = 1$

3. $\dfrac{(x - 2)^2}{4} + \dfrac{(y + 3)^2}{100} = 1$

4. $\dfrac{(x + 1)^2}{36} + \dfrac{(y - 1)^2}{16} = 1$

5. $\dfrac{(x - 5)^2}{10} + \dfrac{y^2}{16} = 1$

6. $\dfrac{x^2}{3} + (y - 2)^2 = 1$

7. $\dfrac{25x^2}{121} + \dfrac{y^2}{16} = 1$

8. $x^2 + \dfrac{16y^2}{81} = 1$

9. $16(x - 1)^2 + 9(y - 2)^2 = 144$

10. $25(x + 3)^2 + (y + 6)^2 = 100$

11. $8x^2 + y^2 + 48x + 56 = 0$

12. $36x^2 + y^2 - 36x - 27 = 0$

13. $x^2 + 9y^2 + 2x - 18y + 1 = 0$

14. $9x^2 + 4y^2 - 36x + 8y + 4 = 0$

In Exercises 15–28, sketch the graph of the given hyperbola. Determine the center, vertices, and asymptotes.

15. $\dfrac{x^2}{16} - \dfrac{y^2}{9} = 1$

16. $\dfrac{y^2}{9} - \dfrac{x^2}{16} = 1$

17. $\dfrac{x^2}{2} - \dfrac{y^2}{10} = 1$

18. $\dfrac{y^2}{20} - \dfrac{x^2}{12} = 1$

19. $\dfrac{(y - 2)^2}{25} - \dfrac{(x - 1)^2}{4} = 1$

20. $\dfrac{(x - 3)^2}{36} - \dfrac{y^2}{16} = 1$

21. $\dfrac{16(x - 2)^2}{25} - \dfrac{64(y - 3)^2}{49} = 1$

22. $36\left(x - \dfrac{1}{2}\right)^2 - \dfrac{9y^2}{4} = 1$

23. $4x^2 - 25y^2 - 100 = 0$

24. $y^2 - 10x^2 - 10 = 0$

25. $16y^2 - x^2 - 48y - 28 = 0$

26. $16x^2 - y^2 - 56x - 207 = 0$

27. $9y^2 - 4x^2 + 18y - 16x - 43 = 0$

28. $x^2 - 8y^2 - 4x - 32y - 36 = 0$

Superset

In Exercises 29–36, identify the graph of each equation as a parabola, circle, ellipse, or hyperbola; then sketch the graph.

29. $x^2 + y^2 - 4x + 2y - 4 = 0$

30. $x^2 + 10x + 2y^2 + 23 = 0$

31. $x = 3y^2 - 12y + 11$

32. $2y^2 - 3x^2 - 24y - 30x - 9 = 0$

33. $4x^2 + 3y^2 + 24x - 6y + 27 = 0$

34. $x^2 + y^2 + 8x + 2y + 13 = 0$

35. $4x^2 - 3y^2 - 8x - 12y - 20 = 0$

36. $x = 4y^2 + 8y + 1$

In Exercises 37–44, graph the given relation, and state whether or not it is a function. Begin by squaring both sides. Remember that the graph of the resulting conic section must be restricted in order to obtain the graph described by the given equation.

37. $y = \sqrt{9 - x^2}$

38. $x = -\sqrt{9 - y^2}$

39. $x = \sqrt{4 - y^2}$

40. $y = -\sqrt{4 - x^2}$

41. $y = -\sqrt{x - 2}$

42. $y = \sqrt{2 - x}$

43. $x = \sqrt{y^2 - 1}$

44. $x = -\sqrt{y^2 + 4}$

An ellipse can be defined as the set of all points the sum of whose distances from two points, called the *foci* of the ellipse, is a positive constant K:

$$d(P, F_1) + d(P, F_2) = K.$$

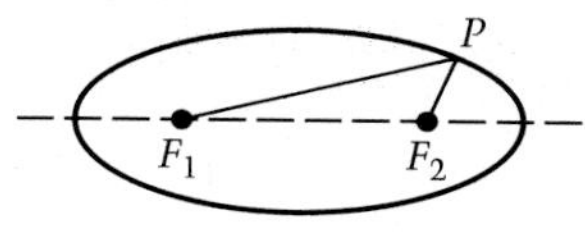

Notice by symmetry that the line through the foci passes through two of the vertices of the ellipse and, in fact, the distance between these vertices is the constant K. For example, with $F_1(2, 3)$, $F_2(8, 3)$, $P(x, y)$ and $K = 10$ we have

$$\sqrt{(x - 2)^2 + (y - 3)^2} + \sqrt{(x - 8)^2 + (y - 3)^2} = 10.$$

To simplify this equation, we transfer one square root to the other side of the equation and square each side of the equation. The resulting equation will have a term with a square root. We rewrite this equation so the square root term appears on one side and all other terms are on the other side of the equation. We then square each side of this equation to produce an equation with no more square roots. We then complete the squares for x and for y, as usual, to find the center and vertices of the ellipse.

In Exercises 45–50, use the distance condition described above to derive an equation for the ellipse with given foci and constant K. Identify the center and vertices for the ellipse.

45. foci: $(0, 0)$ and $(0, 8)$; $K = 10$

46. foci: $(0, 0)$ and $(6, 0)$; $K = 10$

47. foci: $(-2, 0)$ and $(8, 0)$; $K = 26$

48. foci: $(1, -3)$ and $(1, 7)$; $K = 26$

49. foci: $(c, 0)$ and $(-c, 0)$; $K = 2a$

50. foci: $(0, c)$ and $(0, -c)$; $K = 2b$

A hyperbola can be defined as the set of all points the difference of whose distances from two points, called the *foci* of the hyperbola, is a positive constant K:

$$|d(P, F_1) - d(P, F_2)| = K$$

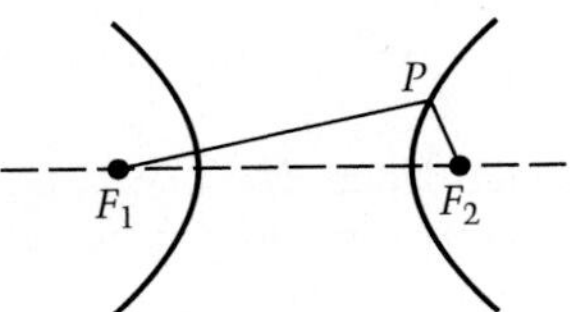

This equation involves two square roots and is simplified in the same way that the comparable equation for an ellipse is simplified.

In Exercises 51–56, use the distance condition described above to derive an equation for the hyperbola with given foci and constant K. Identify the center, vertices and asymptotes for the hyperbola.

51. foci: $(0, 0)$ and $(10, 0)$; $K = 8$

52. foci: $(0, 0)$ and $(10, 0)$; $K = 6$

53. foci: $(1, -7)$ and $(1, 19)$; $K = 10$

54. foci: $(-10, 3)$ and $(16, 3)$; $K = 24$

55. foci: $(c, 0)$ and $(-c, 0)$; $K = 2a$

56. foci: $(0, c)$ and $(0, -c)$; $K = 2b$

The Case of the Diminishing Returns

In the Analyzing Data feature in Chapter 3, we considered the straight line that best fit a set of data. Though a straight line offers great simplicity, it often fails to capture important "twists and turns" in the data. These twists and turns can be crucial to a complete understanding of the phenomenon under study, and can sometimes be a matter of life and death.

Our problem in this chapter focuses on the question, addressed recently by a group of researchers [Farnett, et al., 1991], of whether it is wise to follow the old adage: "the lower the blood pressure, the better." High blood pressure (hypertension) is known to be a major cause of heart attack, kidney disease, and stroke. So it seems reasonable to expect that a simple straight-line model might fit the relationship between heart attack and high blood pressure, with low blood pressure being associated with a low incidence of heart attack and high blood pressure being associated with a high incidence of heart attack. But the researchers found that, at least in patients being treated for high blood pressure, this is not always the case.

Farnett and her colleagues reviewed a large body of research that dealt with this problem. They found that many studies had *assumed* a straight line relationship and so detecting a higher degree of polynomial relationship was virtually impossible.

One well-done study that they reviewed [Cooper, et al., 1988] presented data as shown in Figure 4.63 on the following page. In the figure, values on the horizontal axis correspond to *decreases* in blood pressure; note that *negative values of a decrease* correspond to an *increase*. So " − 10.00" on the horizontal axis corresponds to an increase of 10 "points" in the blood pressure reading. The vertical axis represents the death rate over a 5-year period for the 10,000 patients in the study. You can see that the death rate is highest when the reduction in blood pressure is 45 "points," and lowest somewhere between 10 to 20 "points." The death rate appears to be about the same, roughly 0.025, when there is an increase of 15 points or a decrease of 30 points. So it seems apparent that although a moderate reduction in blood pressure minimized the risk of death, reducing the blood pressure too much may not be the healthiest path for these patients.

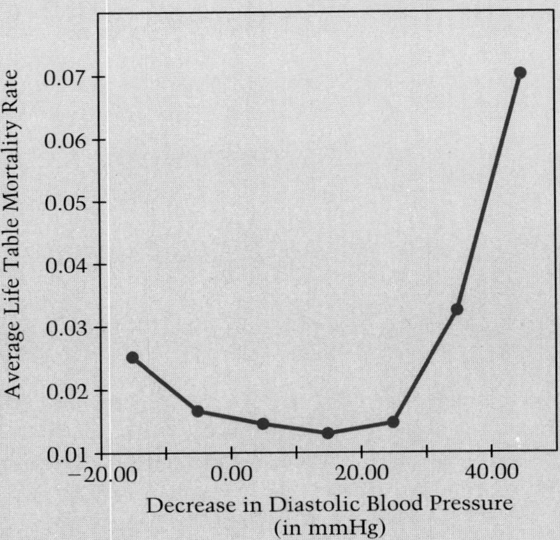

Figure 4.63

The researchers call this graph a J-shaped curve. You might have guessed that the data presented here can be fit much better by a quadratic or cubic curve than a straight line. Though a straight line fit would be the simplest, it simply wouldn't tell the whole story.

Sources: Farnett, L., et al. The J-Curve Phenomenon and the Treatment of Hypertension. *Journal of the American Medical Association,* Vol. 265, No. 4: 489–495, January 1991.

Cooper, S., et al. The Relation between Degree of Blood Pressure Reduction and Mortality among Hypertensives in the Hypertension Detection and Follow-up Program. *American Journal of Epidemiology,* Vol. 127: 387–402, 1988. Reprinted by permission.

Chapter Review

4.1 Important Features of Polynomial Functions (pp. 217–225)

An *nth degree polynomial function* in x is a function that has the form $f(x) = a_nx^n + a_{n-1}x^{n-1} + \cdots + a_1x + a_0$, where $a_n \neq 0$, and n is a nonnegative integer. (p. 218) The graph of a polynomial function of the form $y = ax^n$ (where n is a positive integer greater than 1) has one of the following shapes. (p. 220)

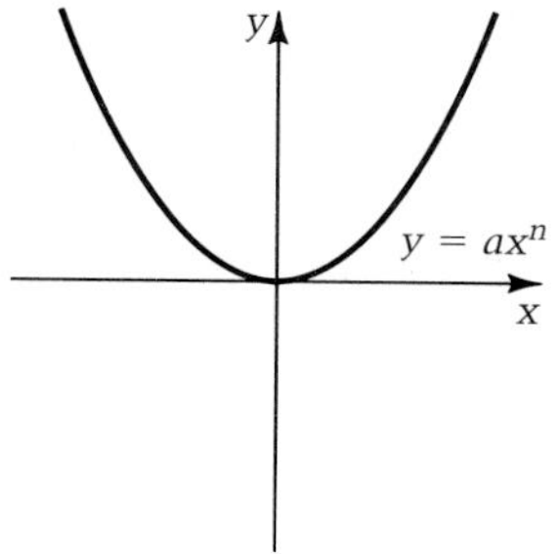

Type I: *a* positive, *n* even

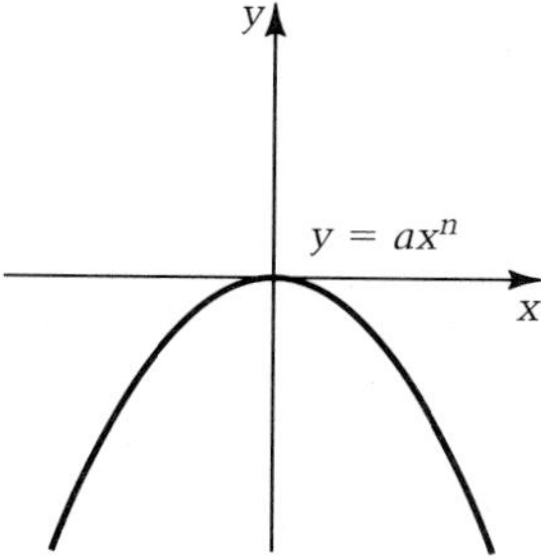

Type II: *a* negative, *n* even

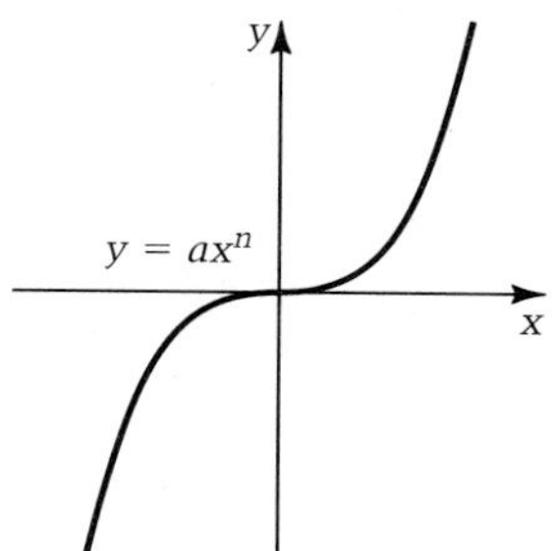

Type III: *a* positive, *n* odd

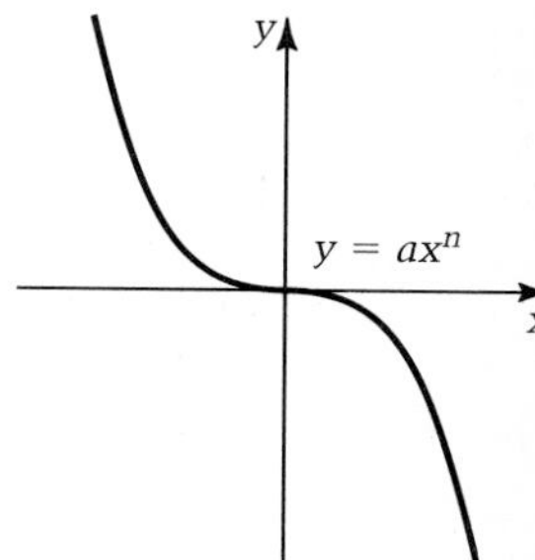

Type IV: *a* negative, *n* odd

First Property of Polynomial Functions: As $x \to \infty$ and as $x \to -\infty$, the graph of a polynomial function described by the equation

$$f(x) = a_nx^n + a_{n-1}x^{n-1} + \cdots + a_1x + a_0$$

behaves like the graph of $g(x) = a_nx^n$. (p. 222)

4.2 Graphing Polynomial Functions (pp. 226–235)

An nth degree polynomial function $f(x) = a_nx^n + a_{n-1}x^{n-1} + \cdots + a_1x + a_0$ has at most n real zeros (x-intercepts). (p. 227)

Second Property of Polynomial Functions: The graph of an nth degree polynomial function cannot intersect (cross or touch) the x-axis more than n times. (p. 227)

Third Property of Polynomial Functions: The graph of an nth degree polynomial function can have at most $n - 1$ relative maxima and minima (peaks and valleys). (p. 228)

Fourth Property of Polynomial Functions: If c_1 and c_2 are adjacent zeros of polynomial function f, then the graph of f lies either completely above or completely below the x-axis for all values between c_1 and c_2. (p. 230)

4.3 Division of Polynomials and the Factor Theorem (pp. 235–243)

Dividing one polynomial by another is similar to the long division process used with whole numbers. (p. 235)

The *Division Algorithm:* When a polynomial $f(x)$ is divided by a nonzero polynomial $d(x)$, there exist unique polynomials $q(x)$, the *quotient*, and $r(x)$, the *remainder*, such that $f(x) = d(x) \cdot q(x) + r(x)$, where either $r(x) = 0$, or else the degree of $r(x)$ is less than the degree of $d(x)$. (p. 237)

Synthetic Division is a streamlined technique for dividing a polynomial by a linear polynomial of the form $x - c$. (p. 237)

The *Remainder Theorem:* If a polynomial $f(x)$ is divided by $x - c$, then the remainder is equal to $f(c)$. (p. 240)

The *Factor Theorem:* The number c is a zero of a polynomial function $y = f(x)$ if and only if $x - c$ is one of the factors of $f(x)$. (p. 241)

4.4 Real Zeros of Polynomial Functions (pp. 243–251)

The *Rational Root Theorem:* If the polynomial function

$$f(x) = a_n x^n + a_{n-1}x^{n-1} + \cdots + a_2x^2 + a_1x + a_0$$

has integer coefficients, and if $\frac{p}{q}$ is a rational zero of $f(x)$ in lowest terms, then p must be a divisor of a_0, the constant term, and q must be a divisor of a_n, the leading coefficient. (p. 243)

Descartes' Rule of Signs provides a way to determine for a given polynomial function the number of positive or negative real zeros that are possible. (p. 246)

The *Bounds Theorem* gives us a way of determining for a given polynomial function an interval $[a, b]$ in which all of the real zeros must lie. (p. 247)

4.5 Complex Zeros of Polynomial Functions (pp. 251–258)

The complex zeros of a polynomial function with real coefficients come in *conjugate pairs:* $a + bi$ and $a - bi$. (p. 256)

The *Fundamental Theorem of Algebra* states that a polynomial of degree greater than 0 has at least one complex zero. (p. 252)

As a consequence of this theorem, we can show that if $f(x)$ is an nth degree polynomial with $n > 0$, then the function $y = f(x)$ has precisely n zeros, and the equation $f(x) = 0$ has precisely n roots, provided a root/zero of multiplicity k is counted k times. (p. 255)

4.6 Approximating Real Zeros of Polynomial Functions (pp. 259–263)

The *Bisection Method* provides an effective way of approximating the real zeros of a polynomial function $f(x)$, to any desired degree of accuracy. The method is based on the simple observation that if $f(a)$ and $f(b)$ have different signs, then the function has a zero at some value of x between a and b.

4.7 Rational Functions (pp. 263–272)

A *rational function* is a function of the form $f(x) = \dfrac{g(x)}{h(x)}$, where $g(x)$ and $h(x)$ are polynomials. (p. 263) If a graph gets closer and closer to a certain line, then the line is an *asymptote* of the graph. (p. 264)

First Property of Rational Functions: The line $x = c$ is a *vertical asymptote* of the graph of the rational function f defined above provided $g(c) \neq 0$ and $h(c) = 0$. (p. 265)

Second Property of Rational Functions: If f is a rational function whose numerator has degree n and whose denominator has degree m, then (a) if $n < m$, then the x-axis is a *horizontal asymptote;* (b) if $n = m$, then the line $y = c$, where c is the quotient of the leading coefficients, is a horizontal asymptote; or (c) if $n > m$, there is no horizontal asymptote. (p. 266)

Third Property of Rational Functions: Between any two successive x-values where a rational function f is either 0 or not defined, the graph of f lies either completely above or completely below the x-axis. (p. 267)

4.8 Conic Sections I: Parabola and Circle (pp. 272–278)

The equation of a *parabola* that opens to the right or left:

$$x = a(y - k)^2 + h. \qquad \text{(p. 274)}$$

The equation of a *circle* with center at (h, k) and radius r:

$$(x - h)^2 + (y - k)^2 = r^2. \qquad \text{(p. 276)}$$

4.9 Conic Sections II: Ellipse and Hyperbola (pp. 278–286)

The equation of an *ellipse* with center at (h, k):

$$\frac{(x-h)^2}{a^2} + \frac{(y-k)^2}{b^2} = 1. \qquad \text{(p. 279)}$$

The equation of a *hyperbola* with center at (h, k) is given by one of the forms

$$\frac{(x-h)^2}{a^2} - \frac{(y-k)^2}{b^2} = 1 \quad \text{or} \quad \frac{(y-k)^2}{b^2} - \frac{(x-h)^2}{a^2} = 1. \qquad \text{(p. 283)}$$

Review Exercises

In Exercises 1–4, select the graph that best represents the function.

1. $y = 4x - x^3$
2. $y = 4x^2 - x^4$
3. $y = x^4 - 4x^3$
4. $y = x^2 - 4x$

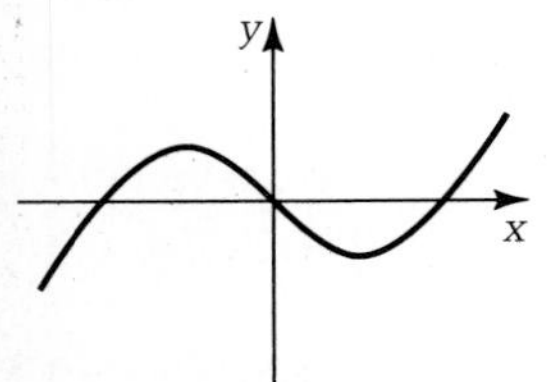

Graph No. 1

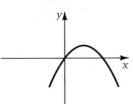

Graph No. 2

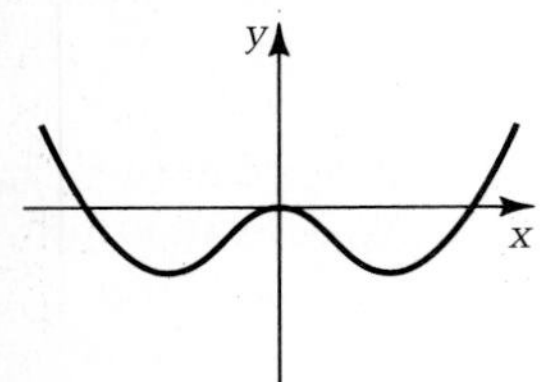

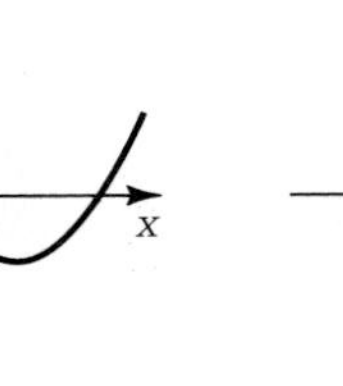

Graph No. 3

Graph No. 4

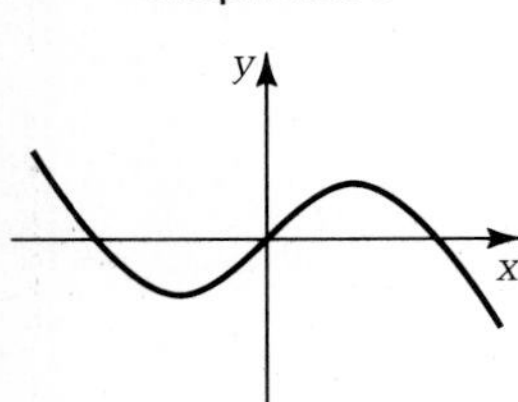

Graph No. 5

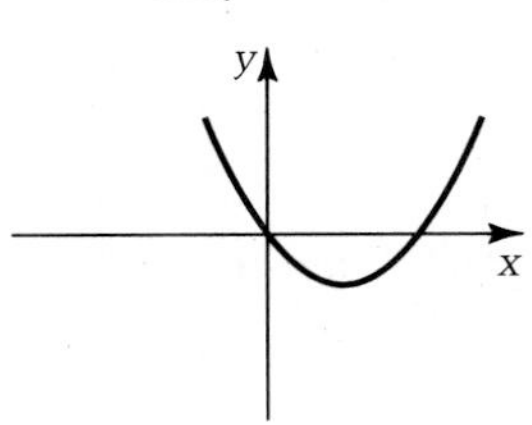

Graph No. 6

In Exercises 5–16, find the real zeros of the given function and then sketch the graph.

5. $y = x^5 - 9x^3$
6. $y = 16 - x^4$
7. $y = 4x^2 + 3x^3 - x^4$
8. $y = x^7 - 5x^5 + 4x^3$
9. $y = (x - 2)^2(x + 3)^2$
10. $y = (3 - x)(x - 1)^3$
11. $y = (x^2 - 3x + 2)(x^2 - 4x - 5)$
12. $y = (9x - x^2)(x^2 + x + 3)$
13. $y = (x^2 + 4x + 1)(4x - x^3)$
14. $y = (x^2 + 3x)^2(x^2 + 4x + 2)$
15. $y = (x^2 + x + 1)(8 - 2x - x^2)$
16. $y = (9x - x^3)(x + 1)^2$

In Exercises 17–20, use synthetic division to determine which, if any, of the linear polynomials $x + 1$, $x - 1$, $x + 2$ and $x - 2$ is a factor of the given polynomial.

17. $x^4 - 4x^2 + x + 2$
18. $x^4 + 3x^3 - 3x^2 - 12x - 4$
19. $x^5 - 2x^4 + 4x^3 - 5x + 2$
20. $x^5 - x^4 - 2x^3 - 3x + 6$

In Exercises 21–24, use the Remainder Theorem to verify the indicated function values.

21. $f(x) = 3x^3 - 4x^2 + 5x + 8$; $f(2) = 26$, $f(-1) = -4$
22. $g(x) = x^4 - 3x^2 - 2x + 9$; $g(-2) = 17$, $g(1) = 5$

23. $G(x) = 7 - 3x + 2x^3 - x^6;\ G(-2) = -67,\ G(1) = 5$

24. $F(x) = 4 - x^2 + 5x^3 - x^5;\ F(2) = 8,\ F(-1) = -1$

In Exercises 25–28, use the Rational Root Theorem to form a list of possible rational zeros of the polynomial function.

25. $f(x) = 3x^3 - 5x^2 + 2x + 6$

26. $f(x) = x^2 - 3x^4 + 2 - 4x$

27. $f(x) = x^3 - \frac{1}{2}x^5 + \frac{3}{4} + 5x$

28. $f(x) = x^4 - \frac{5}{6}x + 1$

In Exercises 29–32, without solving for x, use Descartes' Rule of Signs to determine the possible number of positive and negative real zeros.

29. $f(x) = x^5 - 2x^4 + 3x^3 + x + 6$

30. $f(x) = x^6 - x^3 - 3x + 5$

31. $f(x) = 5x^5 - 2x^6 - x^9 + 13$

32. $f(x) = 4x - 5x^4 - x^5 + 3$

In Exercises 33–36, determine a polynomial function of the given degree with real coefficients, whose roots include the value(s) listed.

33. third degree; $-2, 3 - i$

34. fourth degree; $2 + i$ (multiplicity 2)

35. fourth degree; $1 + i, 4 - i\sqrt{3}$

36. third degree; $3, -2 + i\sqrt{2}$

In Exercises 37–48, determine all the zeros of the given function.

37. $f(x) = x^2 + 5x + 7$

38. $f(x) = 9x^2 + 25$

39. $f(x) = 4x^2 - 3x - 2x^3$

40. $f(x) = 5x - 4x^2 + 2x^3$

41. $f(x) = x^3 - 1$

42. $f(x) = x^2 + 6x + 10$

43. $f(x) = 16x^2 + 9$

44. $f(x) = 27 - x^3$

45. $f(x) = x^4 + 3x^3 - 7x^2 - 27x - 18$

46. $f(x) = x^4 - 9x^3 + 19x^2 + 9x - 20$

47. $f(x) = x^4 + 3x^3 - 9x^2 + 3x - 10$, given that one of the zeros is i.

48. $f(x) = x^4 - 4x^3 - 4x^2 + 36x - 45$, given that one of the zeros is $2 + i$.

In Exercises 49–52, the function f has a zero between the given values of a and b. Use the Bisection Method to estimate this zero.

49. $f(x) = x^3 - 3x + 4;\ a = -3,\ b = -2$

50. $f(x) = 2x^3 - 3x - 12;\ a = 2,\ b = 3$

51. $f(x) = 3x^2 - x^4;\ a = 1,\ b = 2$

52. $f(x) = 5x - x^3;\ a = -3,\ b = -2$

In Exercises 53–58, write the equations of any horizontal or vertical asymptotes for the graph of the given function.

53. $y = \dfrac{3x - 4}{x^2 + 5x + 4}$

54. $y = \dfrac{(3x - 1)^3}{x(x^2 - 5x - 6)}$

55. $y = \dfrac{x^3 - 8}{x(2x - 3)^2}$

56. $y = \dfrac{(2x^2 - 1)^2}{x^3 - 9x}$

57. $y = \dfrac{x^4 + 5x^2 + 4}{(x - 2)(3x - 1)^2}$

58. $y = \dfrac{7x - 5}{4x - 3 - x^2}$

In Exercises 59–62, graph the rational function.

59. $y = \dfrac{1}{x - 3}$

60. $y = \dfrac{2}{5 - x}$

61. $y = \dfrac{(x - 2)^3}{x^2}$

62. $y = \dfrac{4x^2 - 9}{x}$

In Exercises 63–78, determine whether the equation represents a parabola, circle, ellipse or hyperbola, and then sketch the graph. Your graph should identify the following: the vertex and axis of symmetry for a parabola, the center and radius for a circle, the center and vertices for an ellipse, the center, vertices and asymptotes for a hyperbola.

63. $x^2 - 4x + y - 7 = 0$

64. $x^2 + y^2 + 6x - 7 = 0$

65. $x^2 + 4y^2 - 36 = 0$

66. $y^2 + 2y - x + 3 = 0$

67. $x^2 + y^2 - 4x + 6y + 8 = 0$

68. $9x^2 + y^2 - 144 = 0$

69. $x^2 - 4y^2 - 36 = 0$

70. $9x^2 - y^2 - 144 = 0$

71. $y^2 - 6y + x = 0$

72. $4x^2 + 9y^2 - 8x + 18y - 131 = 0$

73. $x^2 + y^2 - 2y + 4x + 3 = 0$

74. $9y^2 - x^2 - 6x - 45 = 0$

75. $4x^2 - 9y^2 - 8x + 18y - 149 = 0$

76. $x^2 + 6x - y + 5 = 0$

77. $4x^2 + y^2 + 16x - 48 = 0$

78. $x^2 + y^2 - 2y - 1 = 0$

79. Sketch the graph and give an equation for a polynomial function that has 6 real zeros but only 3 relative maxima or minima. (*Hint:* The zeros need not be distinct.)

80. What does it mean geometrically if when we apply synthetic division to a polynomial function f for each of three different x-values, the remainder is the same in each case?

81. What is the coefficient of x^{10} in the expansion of $(x - 2)^4(x + 3)^5(x - 1)^2$? (*Hint:* See the comment preceding Exercises 43–50 in Section 4.5.)

82. Suppose the four distinct roots of a fourth degree polynomial equation with real coefficients are pure imaginary. Show that the coefficients of both x and x^3 are zero.

83. Write an equation for a rational function whose only asymptotes are the lines $x = 2$, $x = -1$ and $y = 4$.

84. The sum and the product of the zeros of a quadratic function with real coefficients are each equal to P. What can be said about this common value P if the zeros are not real?

85. Find the points on the ellipse $x^2 - 2x + 4y^2 = 27$ which are 4 units away from the center of the ellipse.

86. In applying the Bisection Method to estimate a root of the equation $f(x) = 0$ on the interval $[a, b]$, a student has found incorrectly that $f(a) < 0$ and $f(b) > 0$. In fact, $f(x) > 0$ for all x in the interval, and there is no real zero of f on the interval $[a, b]$. What will the student find when applying the method under the false impression that $f(a) < 0$?

Graphing Calculator Exercises

[*Note: Round all estimated values to the nearest hundredth.*]

In Exercises 1–4, use a graph to determine the real zeros of the given function.

1. $f(x) = x^3 - 2x^2 + 12$ **2.** $f(x) = x^3 - 12x^2 + 12$

3. $f(x) = x^4 - 12x^2 + 12$ **4.** $f(x) = x^4 - 12x^3 + 12$

In Exercises 5–12, solve the given equation over the set of real numbers. *Hint:* Recall that to solve the equation $h(x) = g(x)$, sketch the function $f(x) = h(x) - g(x)$ and approximate the x-intercepts.)

5. $x^5 - 3x^4 + \frac{1}{x} = 10$ **6.** $3 - 10x^3 = 2x - \frac{1}{x^2}$

7. $x^4 - 3x^2 - 2x = 16$

8. $2x^4 + 2x^3 - 10 = x^3 - 9$

9. $x^6 - 8x^4 + x = x^3 - 1$ **10.** $x^8 + x^5 - 10x = 1$

11. $\frac{1}{x^2 - 3} = x + 12$ **12.** $x - \frac{1}{x^3 + 1} = x^2$

In Exercises 13–16 solve the given inequality. Write your answers in interval form. (*Hint:* To solve an inequality of the form $h(x) < g(x)$, first rewrite the inequality in the form $h(x) - g(x) < 0$, and sketch the function $f(x) = h(x) - g(x)$. The desired intervals are those where the graph falls below the x-axis. Similarly, to solve $h(x) > g(x)$, determine the intervals where the graph of $f(x) = h(x) - g(x)$ lies above the x-axis. For example, to solve $x^2 - 1 < 2x + 4$, determine the interval(s) where $f(x) = x^2 - 2x - 5$ lies below the x-axis.)

13. $x^2 - 1 < 2x + 4$ **14.** $x^3 - 16x^2 > x - 20$

15. $x^2 - \frac{1}{x} > \frac{1}{x^2}$ **16.** $\frac{1}{x^2 - 1} < x + 2$

17. Determine the interval(s) where the graph of $f(x) = x^2 + 1$ lies above the graph of $g(x) = 2x^3 - x - 8$. (*Hint:* Solve the inequality $x^2 + 1 > 2x^3 - x - 8$.)

18. Determine the interval(s) where the graph of $f(x) = \frac{1}{x^2} + 3$ lies below the graph of $g(x) = x^3 + 1$.

Circles, ellipses, and certain parabolas and hyperbolas are not functions and so cannot be sketched directly by a function grapher. However, a sketch *can* be produced. For example, the unit circle $x^2 + y^2 = 1$ can be drawn by sketching the two functions $y = \sqrt{1 - x^2}$ and $y = -\sqrt{1 - x^2}$ on the same set of axes. In Exercises 19–24, sketch the equation by using this method.

19. $x^2 + y^2 = 1$ **20.** $x^2 + y^2 = 5$ **21.** $x - y^2 = 1$

22. $x^2 - y^2 = 3$ **23.** $x^2 - 2y^2 = 5$ **24.** $x^2 + 2y^2 = 5$

Chapter 4 Test

In Problems 1–4, select the graph from the four below that best represents the given function.

1. $f(x) = x^3 - x^2 - x + 1$
2. $f(x) = 1 - x^3 + x^2 + x$
3. $f(x) = 1 - x^4 + 4x^2$
4. $f(x) = x^4 - 4x^2 + 4$

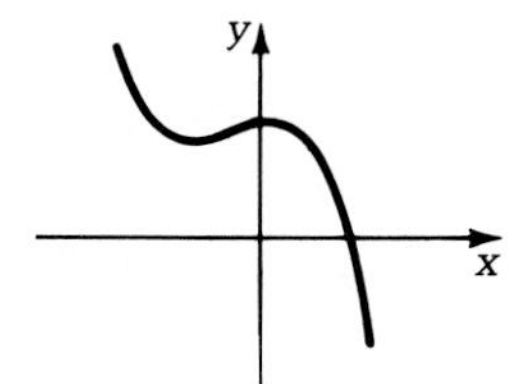

Graph No. 1

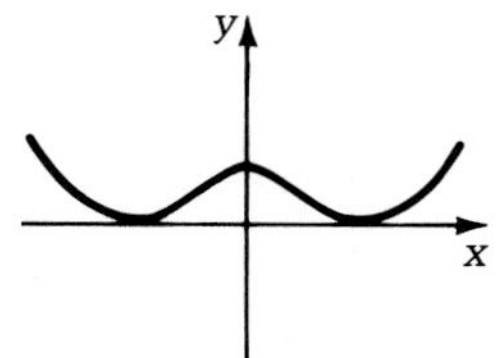

Graph No. 2

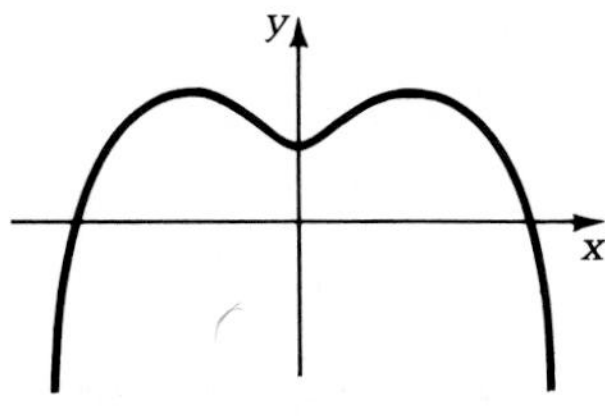

Graph No. 3

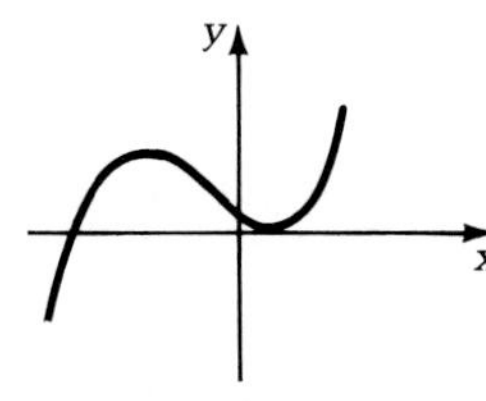

Graph No. 4

In Problems 5 and 6, use the First, Second, and Third Properties of Polynomial Functions to select the graph from the four below that best represents the given function.

5. $f(x) = x^3 - 5x^2$
6. $f(x) = x^3 + 4x^2 - 11x + 6$

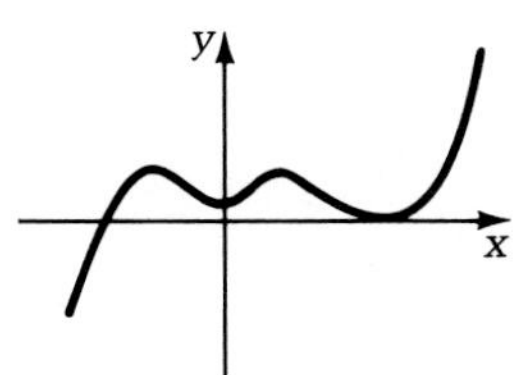

Graph No. 1

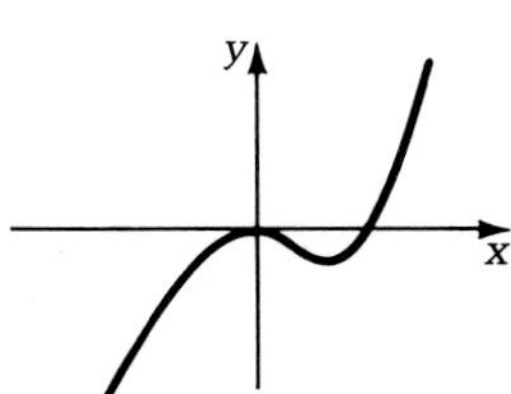

Graph No. 2

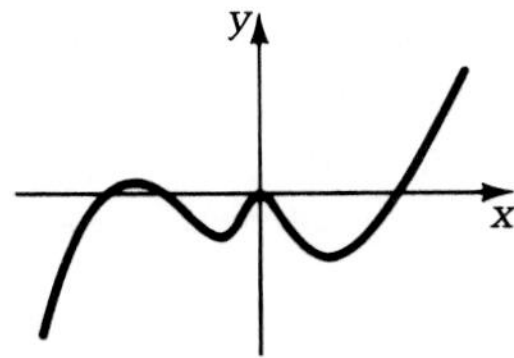

Graph No. 3

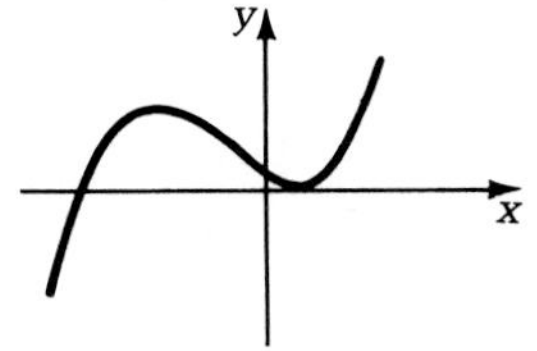

Graph No. 4

7. Find the zeros of $f(x) = x(x - 2)(x + 4)^4$, then sketch the graph.
8. Use the Remainder Theorem and synthetic division to determine the value of $f(-2)$ for the function given by $f(x) = x^5 - 3x^2 + 4x - 7$.
9. Without solving for x, use Descartes' Rule of Signs to determine the possible number of positive and negative real zeros of the function $f(x) = x^4 - 2x^3 - 3x + 5$.

In Problems 10 and 11, find all zeros of the given function.

10. $f(x) = 4x^3 - 12x^2 + 9x - 2$
11. $f(x) = 4x^4 + 4x^3 + 5x^2 + 8x - 6$
12. Determine a polynomial function of the third degree with real coefficients, and whose roots include 2 and $2 - 3i$.
13. Determine an interval of the form $[n, n+1]$, where n is an integer such that the function $f(x) = x^3 - 2x - 27$ has a zero that could be approximated by the Bisection Method.

In Problems 14 and 15, write the equations of any horizontal or vertical asymptotes of the graph of the given function.

14. $y = \dfrac{3x^2 - 12}{6x^2 - 5x + 1}$
15. $f(x) = \dfrac{5x}{2x^2 - 18}$
16. Sketch the graph of the rational function
$$f(x) = \frac{1}{x^2 - 4x - 5}.$$
17. Sketch the parabola $y^2 - 6y - x + 5 = 0$. Determine the vertex, the axis of symmetry and the x- and y-intercepts.
18. Sketch the circle $x^2 + 2x + y^2 - 6y + 7 = 0$. Determine the center and the radius.
19. Sketch the ellipse $\dfrac{(x-5)^2}{25} + \dfrac{y^2}{4} = 1$. Determine the center and the vertices.
20. Sketch the hyperbola $\dfrac{(y+2)^2}{24} - \dfrac{(x-1)^2}{4} = 1$. Determine the center, vertices, and asymptotes.

5

Exponential and Logarithmic Functions

In this chapter we study exponential functions, such as $f(x) = 2^x$ and $g(x) = e^x$, and their inverses, which are called logarithmic functions. These functions enjoy a broad range of applications, from modeling the growth of bacteria to measuring the strength of an earthquake. At the end of this chapter, we will consider the explosive growth in the number of VCRs to be found in households in recent years. We will attempt to model this growth, which has been called **exponential growth** *by many. But is it really exponential?*

5.1 Exponential Functions

Our work with the functions in this chapter will rely upon an understanding of the properties of exponents. These properties are listed on the next page. Although we have defined b^x for rational exponents only, the expression b^x makes sense even when x is not a rational number.

Properties of Real Number Exponents

Property 1 $b^0 = 1$ **Property 2** $b^1 = b$

Property 3 $b^x b^y = b^{x+y}$ **Property 4** $(b^x)^t = b^{xt}$

Property 5 $\dfrac{b^x}{b^y} = b^{x-y}$ **Property 6** $b^{-x} = \dfrac{1}{b^x}$

where the base b is positive, and the exponents x, y, and t represent any real numbers.

EXAMPLE 1

Simplify the following expressions: (a) $2(2^x)^3$ (b) $\dfrac{3^{x+1}}{3^2}$ (c) $4^{x+5} \cdot 8^{1-x}$

Solution

(a) $2(2^x)^3 = 2(2^{3x}) = 2^{3x+1}$ Properties 4 and 3 of Real Number Exponents are used. Note that $(2^x)^3$ is not the same as 2^{x+3}, nor 2^{x^3}.

(b) $\dfrac{3^{x+1}}{3^2} = 3^{(x+1)-2} = 3^{x-1}$ Property 5 of Real Number Exponents is used.

(c) $4^{x+5} \cdot 8^{1-x} = (2^2)^{x+5}(2^3)^{1-x}$ Begin by rewriting each factor with the same base.

$= 2^{2x+10}2^{3-3x}$

$= 2^{-x+13}$

We now consider functions that are described by equations containing exponential expressions. These functions are used to solve problems involving population growth, compound interest, and radioactive decay.

Definition of Exponential Function

An **exponential function** is a function of the form

$$f(x) = b^x,$$

where the base b is any positive real number except 1.

To sketch the graph of the exponential function $f(x) = 2^x$, we begin with a table of values.

x	-4	-3	-2	-1	0	1	2	3
$f(x)$	$\frac{1}{16}$	$\frac{1}{8}$	$\frac{1}{4}$	$\frac{1}{2}$	1	2	4	8

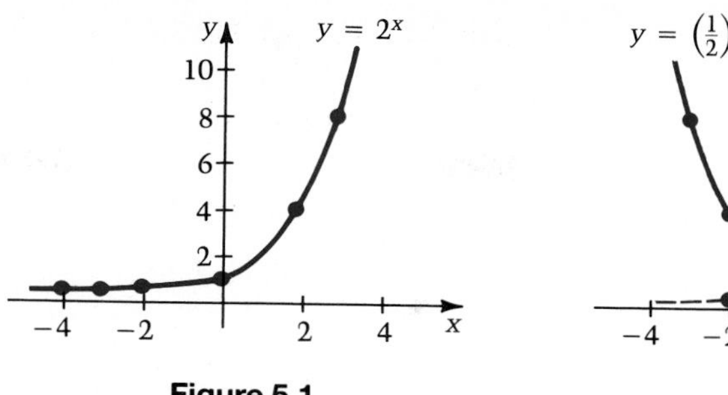

Figure 5.1

Figure 5.2

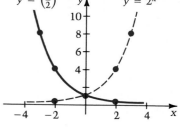

We plot these points and draw a smooth curve as shown in Figure 5.1. Notice that as $x \to -\infty$, the corresponding values of $f(x)$ remain positive but get very close to zero. Thus, the x-axis is a horizontal asymptote.

Next we sketch the graph of $f(x) = (\frac{1}{2})^x$. By Property 6 of exponents we know that

$$\left(\frac{1}{2}\right)^x = \frac{1}{2^x} = 2^{-x}.$$

How does the graph of $y = 2^{-x}$ compare with the graph of $y = 2^x$? Recall that in general, replacing x with $-x$ in an equation reflects a graph across the y-axis. Thus, $f(x) = (\frac{1}{2})^x$ (that is, $y = 2^{-x}$) is the reflection of $y = 2^x$ across the y-axis (Figure 5.2). We generalize these graphing results in Figure 5.3 for any positive base b not equal to 1.

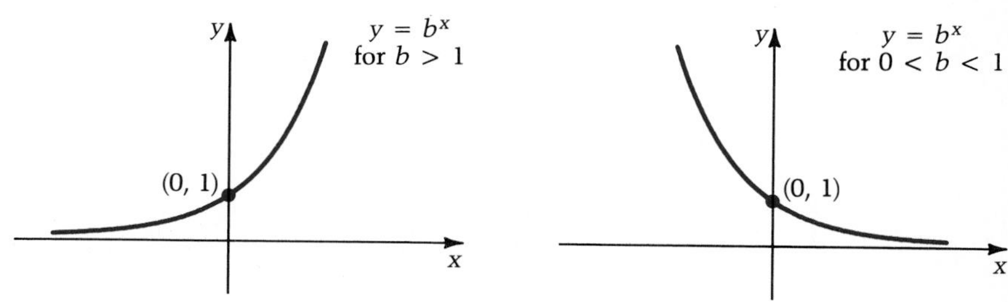

Figure 5.3

These graphs illustrate five important features of the function $f(x) = b^x$, where $b > 0$ and $b \neq 1$:

- the domain is the set of all real numbers;
- the range is the set of all positive real numbers;
- $b^0 = 1$;
- the x-axis is a horizontal asymptote of the graph;
- the function is one-to-one.

In the next example, notice the following:

For $x > 0$, larger values of the base b produce larger values of b^x.
For $x < 0$, larger values of the base b produce smaller values of b^x.

EXAMPLE 2

Sketch the graphs of each set of equations on a single set of coordinate axes.

(a) $y = 2^x, y = 3^x$, and $y = 10^x$ (b) $y = \left(\frac{1}{2}\right)^x$ and $y = \left(\frac{1}{3}\right)^x$

Solution

(a)

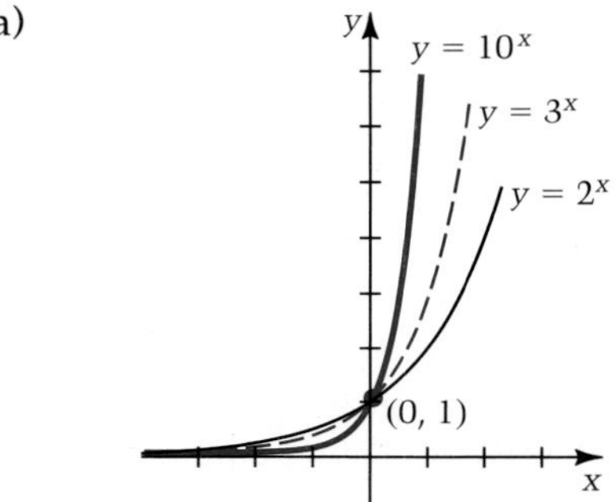

(b)

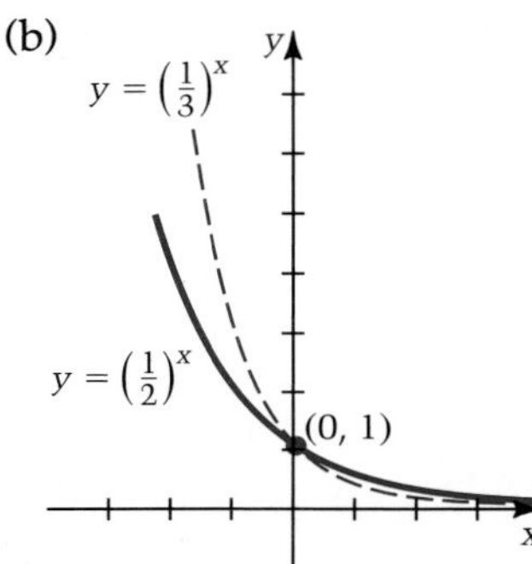

Figure 5.4

We can use the graphing techniques of Section 3.3 to sketch the graphs of more complicated exponential functions.

EXAMPLE 3

Sketch the graph of: (a) $y = 2^x - 4$ (b) $y = 1 + \left(\frac{1}{3}\right)^{x+2}$.

Solution

(a)

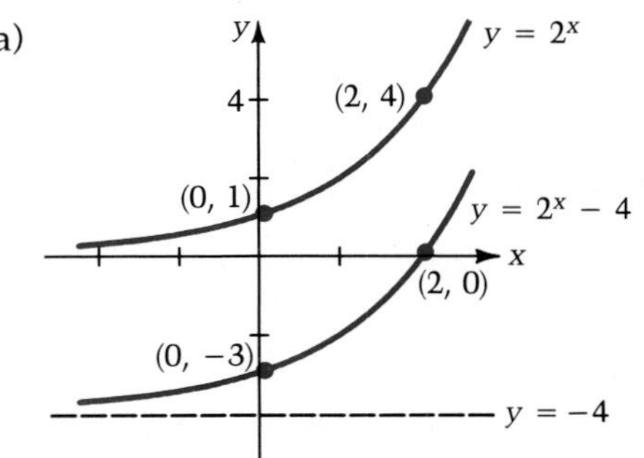

Figure 5.5

Recall that the graph of $y = f(x) - 4$ is a translation of $y = f(x)$ four units down. The graph of $y = 2^x$ is lowered four units to obtain $y = 2^x - 4$.

The line $y = -4$ is a horizontal asymptote.

(b)

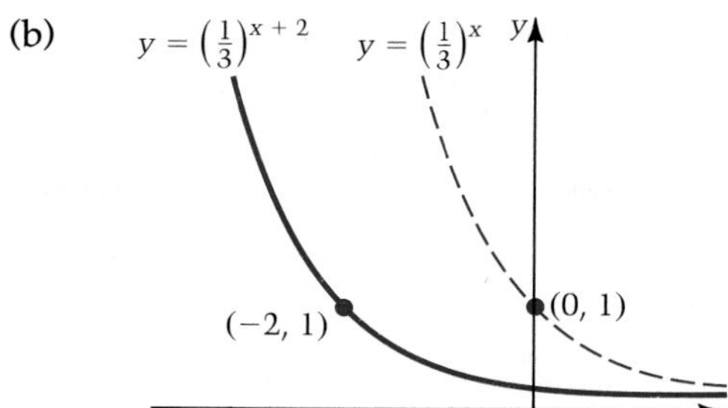

Step 1: We start with the graph of $y = \left(\frac{1}{3}\right)^x$. The $x + 2$ suggests translating this graph two units to the left to obtain the graph of $y = \left(\frac{1}{3}\right)^{x+2}$.

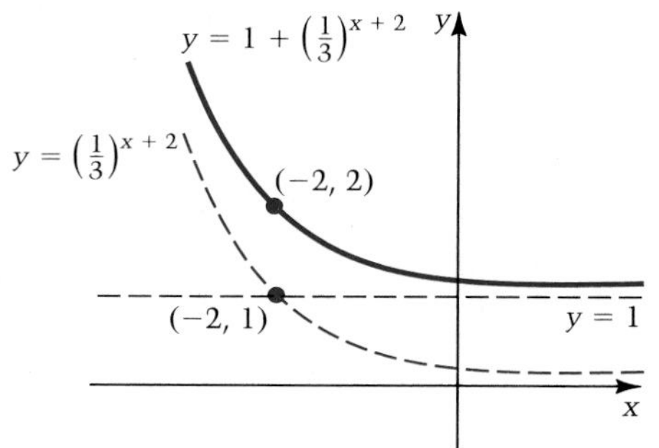

Step 2: Replacing y with $y - 1$ in $y = \left(\frac{1}{3}\right)^{x+2}$ yields $y = 1 + \left(\frac{1}{3}\right)^{x+2}$. Thus, we have a translation of $y = \left(\frac{1}{3}\right)^{x+2}$ one unit up. This is the desired graph. The horizontal asymptote is the line $y = 1$.

Figure 5.6

Before we can use the Properties of Real Number Exponents to simplify and solve exponential equations, we must first rewrite exponential expressions so that they have the same base. Then, we can write a simpler equivalent equation by using the fact that $f(x) = b^x$ is a one-to-one function, that is, if $b > 0$ and $b \neq 1$, then

$$b^a = b^c \text{ is equivalent to } a = c.$$

EXAMPLE 4

Solve the following equation for x: $2^{6-x} = \frac{1}{4}(8^x)$.

Solution

$$2^{6-x} = \frac{1}{4}(8^x)$$ Begin by writing each term with the same base.

$$2^{6-x} = \frac{1}{2^2}(2^3)^x$$ To simplify, use Properties 3, 4, and 6 of Real Number Exponents.

$$2^{6-x} = 2^{-2} \cdot 2^{3x}$$

$$2^{6-x} = 2^{-2+3x}$$ Since the bases are equal, the exponents are equal.

$$6 - x = -2 + 3x$$

$$x = 2$$

EXERCISE SET 5.1

In Exercises 1–14, simplify the expression.

1. $(2^x)^{-3}$ **2.** $(3^x)^{-2}$ **3.** $5^{x-7} \cdot 5^{x+3}$

4. $7^{4-x} \cdot 7^{x-9}$ **5.** $2^{x+2} \cdot 4^{-x}$ **6.** $3^{x-2} \cdot 9^x$

7. $\frac{3^{x+2}}{3^{4-x}}$ **8.** $\frac{2^{1-x}}{2^{5+x}}$ **9.** $4^{x+1} \cdot 8^{2-x}$

10. $9^{2-3x} \cdot 27^{x-4}$ **11.** $\frac{27^{2x-1}}{9^{x+2}}$ **12.** $\frac{8^{x-1}}{4^{3+2x}}$

13. $2^x + 4^x$ **14.** $9^x - 3^x$

In Exercises 15–26, sketch the graph of the equation.

15. $y = 4^x$ **16.** $y = 5^x$ **17.** $y = 3^{-x}$

18. $y = 5^{-x}$ **19.** $y = 3^x - 1$ **20.** $y = 2^x + 5$

21. $y = 3^{x-1}$ **22.** $y = 2^{x+5}$

23. $y = \left(\frac{1}{2}\right)^x - 1$ **24.** $y = 1 - \left(\frac{1}{3}\right)^x$

25. $y = -\left(\frac{1}{3}\right)^{x-2}$ **26.** $y = -\left(\frac{1}{2}\right)^{x+1}$

In Exercises 27–38, solve for x.

27. $2^{x+1} = 16$ **28.** $3^{7-x} = 81$ **29.** $8^x = 4$

30. $9^x = 27$ **31.** $3^x = 9^{x+4}$ **32.** $2^x = 8^{5-x}$

33. $3^x = 9(3^{5-x})$ **34.** $2^x = 2^{3x}(4^2)$ **35.** $2^x = -4$

36. $3^{x-2} = 0$ **37.** $4^x = \frac{1}{2}(8^{x+1})$ **38.** $9^x = \frac{1}{27}(3^{1-x})$

Superset

39. The carbon isotope ^{14}C has a half-life of 5760 years. (Thus, if there were N of the ^{14}C atoms present at a certain time, there would be $\frac{1}{2}N$ left 5760 years later.) If there are 10 mg of ^{14}C present at time $t = 0$, then the amount $f(t)$ present after t years is given by

$$f(t) = 10\left(\frac{1}{2}\right)^{t/5760}.$$

Determine the amount of ^{14}C present after:
(a) 1000 years
(b) 10,000 years
(c) 50,000 years.

40. A certain type of bacteria doubles its population size every hour. The number N of bacteria present t hours after we begin observing a certain colony is given by the formula

$$N = (100)2^t$$

Determine the number of bacteria present after:
(a) one hour
(b) $3\frac{1}{2}$ hours
(c) 1 day.

In Exercises 41–46, solve for x.

41. $2^{x^2} = 4$ **42.** $2^{x^3} = \frac{1}{2}$

43. $2^x(2^x - 1) = 0$ **44.** $3^x(9 - 3^x) = 0$

45. $2^{2x} - 5 \cdot 2^x + 4 = 0$ **46.** $3^{2x} - 12 \cdot 3^x + 27 = 0$

47. Carefully graph $y = 2^x$ and use your graph to estimate the value of:
(a) $2^{\sqrt{2}}$ (b) 2^{π} (c) $2^{-\sqrt{3}}$

48. Carefully graph $y = 10^x$ and use your graph to estimate the value of x when:
(a) $y = 50$ (b) $y = 75$ (c) $y = 5$

49. Suppose some quantity Q is related to another quantity k by the relationship $Q = 3^k$. What happens to Q if the value of k is doubled? tripled? increased tenfold?

50. Does the equation $3^x = x$ have any solutions? (*Hint:* graph $y = 3^x$ and $y = x$ on the same set of axes.)

In Exercises 51–58, sketch the graph of the equation.

51. $y = 2^{|x|}$ **52.** $y = |2^x|$ **53.** $y = 2^{x^2}$

54. $y = 2^{1/x}$ 55. $y = -2^{x^2}$ 56. $y = 2^{-|x|}$

57. $y = 2^x + 2^{-x}$ 58. $y = 2^x - 2^{-x}$

59. What must be true of the function f if the graph of $y = 2^{f(x)}$ is symmetric with respect to the y-axis?

60. Explain why there are no functions f for which the graph of $y = 2^{f(x)}$ is symmetric with respect to the origin.

61. Derive Property 5 of Real Number Exponents from the other five properties.

62. Derive Property 6 of Real Number Exponents from the other five properties.

63. Use the y^x key on your calculator to determine 2^x for the x-values 3.10, 3.11, 3.12, 3.13, 3.14, and 3.15. Which of these six values most closely approximate 2^π?

64. Which value is greater, 3^π or π^3?

In Exercises 65–74, estimate the quantity to the nearest hundredth.

65. $2^{\sqrt{2}}$ 66. $2^{\sqrt{3}}$ 67. $3^{\sqrt{2}+\sqrt{3}}$

68. $3^{1-\sqrt{2}}$ 69. $\pi^{2-\sqrt{2}}$ 70. $\pi^{3+\sqrt{3}}$

71. $(2.72)^{-0.25}$ 72. $(2.72)^{-3.14}$ 73. $\sqrt{2}^{\sqrt{2}}$

74. π^π

In Exercises 75–78, an equation and three values are given. Determine which of the three values best approximates a solution of the equation.

75. $3^x = 1 + x$; -0.41, -0.17, 1.15

76. $2^{x-1} = 1 - x$; -1.15, 0.35, 1.5

77. $2^{x-2} = \sqrt{x-1}$; 0.5, 1.0, 1.5

78. $2^{x+1} = x^2 - x + 1$; -0.8, -0.4, 0.4

5.2 Logarithmic Functions

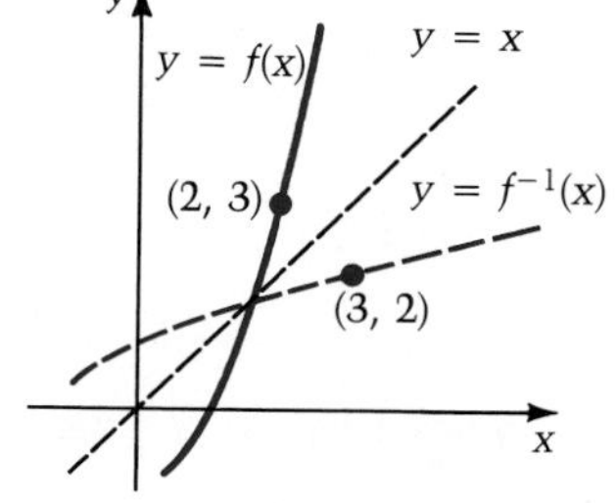

Figure 5.7

Since the graph of the exponential function $y = b^x$ passes the Horizontal Line Test, we can conclude that this function is one-to-one.

Recall that any one-to-one function has an inverse. Part of the procedure we follow to find the inverse of a one-to-one function $y = f(x)$ is to interchange x and y. Because of this interchange, the graphs of $y = f(x)$ and $y = f^{-1}(x)$ are reflections of one another across the line $y = x$, as shown in Figure 5.7.

The graphs of $y = b^x$ and its inverse $x = b^y$ appear in Figure 5.8, one for the case where $b > 1$, and one for the case where $0 < b < 1$.

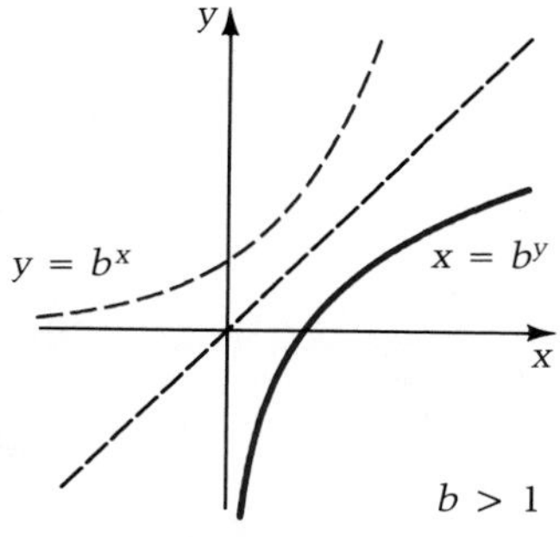

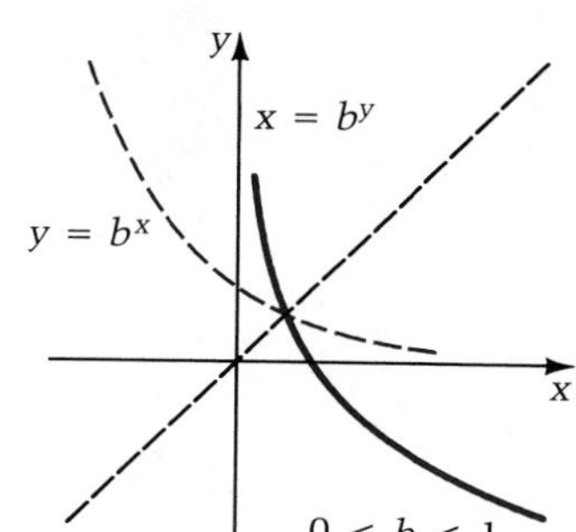

Figure 5.8

In order to describe these inverse functions in the usual form $y = f^{-1}(x)$, it is necessary to write the equation $x = b^y$ with y in terms of x. We can describe y in words:

y is the power to which you raise b in order to obtain x.

We define *logarithm* as an abbreviation for this description of y.

Definition of Logarithm

For $b > 0$ and $b \neq 1$, the function $f(x) = b^x$ has an inverse function, denoted $f^{-1}(x) = \log_b x$. This notation is an abbreviation for the **logarithm of x to the base b.** The domain of the logarithmic function is $(0, \infty)$ and the range is $(-\infty, \infty)$.

Notice that we have taken y in the equation $x = b^y$ and defined it as a logarithm. Therefore, *a logarithm is an exponent*, and every exponential statement can be rewritten as an equivalent statement about logarithms. This equivalence is summarized below.

Log/Exp Principle

For $b > 0$ and $b \neq 1$, $b^A = C$ is equivalent to $\log_b C = A$.

EXAMPLE 1

Rewrite the following as logarithmic equations: (a) $x^3 = 64$ (b) $9^x = 3$.

Solution

(a) $x^3 = 64$ Use the Log/Exp Principle.

$\log_x 64 = 3$

(b) $9^x = 3$ Use the Log/Exp Principle.

$\log_9 3 = x$

EXAMPLE 2

Solve each of the following equations for x.

(a) $\log_3 x = 2$ (b) $\log_8 x = \dfrac{1}{3}$ (c) $\log_x 64 = 3$ (d) $\log_9 3 = x$

Solution

(a) $\log_3 x = 2$

$3^2 = x$

$x = 9$

(b) $\log_8 x = \frac{1}{3}$

$8^{1/3} = x$

$x = 2$

(c) $\log_x 64 = 3$

$x^3 = 64$

$x = 64^{1/3}$

$x = 4$

(d) $\log_9 3 = x$

$9^x = 3$

$x = \frac{1}{2}$

Graphs of Logarithmic Functions

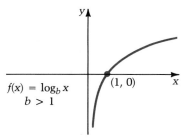

You should keep in mind that a logarithm is just an exponent. When we say $\log_3 9 = 2$, we are saying that 2 is the exponent to which we raise 3 in order to get 9. That is, $3^2 = 9$.

The two general types of logarithmic graphs ($y = \log_b x$) are shown in Figure 5.9. By remembering these graphs, we can recall five important features of the logarithmic function:

- the domain is $(0, \infty)$;
- the range is $(-\infty, \infty)$;
- $\log_b 1 = 0$;
- the y-axis is a vertical asymptote of the graph;
- the function is one-to-one.

Since the logarithmic function is one-to-one, no two x-values correspond to the same y-value. Thus,

$$\text{if } \log_b A = \log_b C, \text{ then } A = C.$$

This fact will be useful when we solve logarithmic equations.

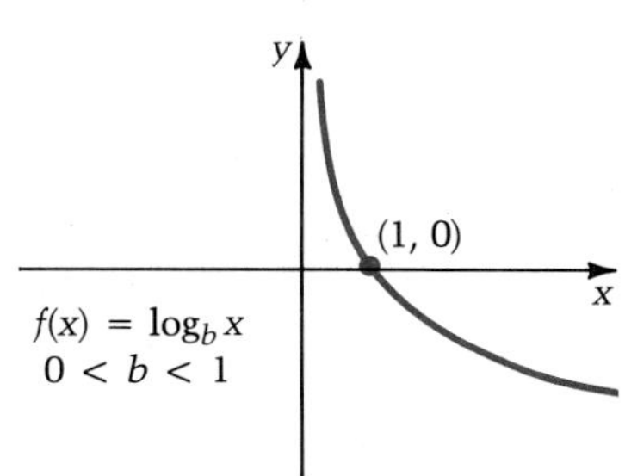

Figure 5.9

EXAMPLE 3

Sketch the graph of: (a) $y = \log_2 x$ (b) $y = \log_2(x - 1)$.

Solution

(a)

x	$\frac{1}{4}$	$\frac{1}{2}$	1	2	4
y	-2	-1	0	1	2

To compute a table of values, we use the Log/Exp Principle to write $y = \log_2 x$ as $x = 2^y$.

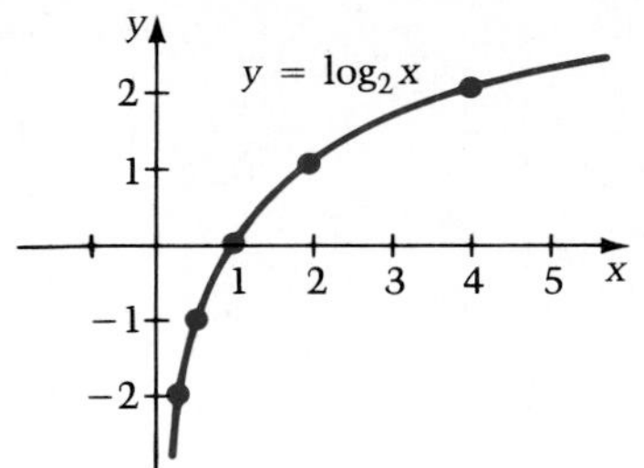

Plot the points and sketch the curve. The base for the logarithm in this example is $b = 2$. Note that this sketch is similar to the graph given in Figure 5.9 for the general case $b > 1$. The line $x = 0$ is a vertical asymptote.

(b)

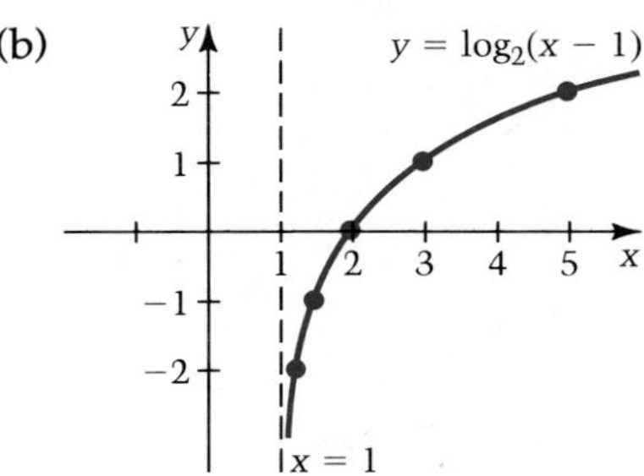

Recall that the graph of $y = f(x - 1)$ is a translation of $y = f(x)$ one unit to the right. Thus, the graph of $y = \log_2 x$ is translated one unit to the right to obtain the graph of $y = \log_2(x - 1)$. The line $x = 1$ is a vertical asymptote.

Figure 5.10

Properties of Logarithms

We use the following properties to simplify and evaluate logarithmic expressions. The proofs of these properties will be postponed until you have seen how they work.

Properties of Logarithms

Property 1 $\log_b 1 = 0$

Property 2 $\log_b b = 1$

Property 3 $\log_b (xy) = \log_b x + \log_b y$

Property 4 $\log_b (x^t) = t \log_b x$

Property 5 $\log_b \left(\dfrac{x}{y}\right) = \log_b x - \log_b y$

Property 6 $\log_b \left(\dfrac{1}{x}\right) = -\log_b x$

where x and y are positive real numbers, t is any real number, and base b is positive, but not equal to 1.

EXAMPLE 4

Given that $\log_2 x = 3.1$ and $\log_2 y = 1.6$, evaluate $\log_2 \frac{8x}{y^3}$.

Solution

$$\begin{aligned}\log_2 \frac{8x}{y^3} &= \log_2 8x - \log_2 y^3 && \text{Property 5 of Logarithms is used.}\\ &= \log_2 8 + \log_2 x - \log_2 y^3 && \text{Property 3 of Logarithms is used.}\\ &= \log_2 8 + \log_2 x - 3\log_2 y && \text{Property 4 of Logarithms is used.}\\ &= 3 + 3.1 - 3(1.6) && \log_2 8 = 3.\\ &= 1.3\end{aligned}$$

EXAMPLE 5

Given that $\log_b 2 = A$, $\log_b 3 = B$, and $\log_b 5 = C$, express $\log_b \left(\frac{2\sqrt{5}}{81}\right)$ in terms of A, B, and C.

Solution

$$\begin{aligned}\log_b \left(\frac{2\sqrt{5}}{81}\right) &= \log_b 2\sqrt{5} - \log_b 81 && \text{Property 5 of Logarithms is used.}\\ &= \log_b 2 + \log_b \sqrt{5} - \log_b 81 && \text{Property 3 of Logarithms is used.}\\ &= \log_b 2 + \log_b (5^{1/2}) - \log_b (3^4)\\ &= \log_b 2 + \frac{1}{2}\log_b 5 - 4\log_b 3 && \text{Property 4 of Logarithms is used.}\\ &= A + \frac{1}{2}C - 4B\end{aligned}$$

Sometimes we can use the Properties of Logarithms to rewrite an equation and then apply the graphing techniques of Section 3.3 on transformations to graph more complicated logarithmic equations.

EXAMPLE 6

Sketch the graph of $y = \log_2 \left[\frac{1}{(x+1)^3}\right]$.

Solution We begin by using Property 4 of Logarithms to rewrite the equation:

$$y = \log_2 \left[\frac{1}{(x+1)^3}\right] = \log_2 [(x+1)^{-3}] = -3 \log_2 (x+1).$$

To obtain the graph of $y = -3 \log_2 (x+1)$ we start with a graph known to us: $y = \log_2 x$, and carry out a succession of transformations that correspond to a succession of simple algebraic changes.

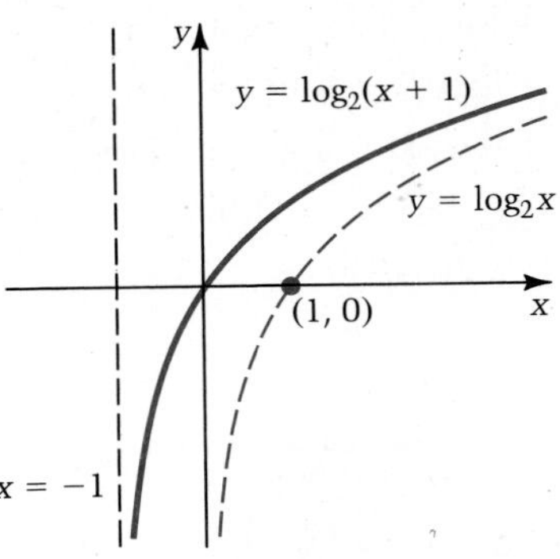

Step 1: We begin with the graph of $y = \log_2 x$. The $x + 1$ suggests translating this graph one unit to the left to obtain the graph of $y = \log_2 (x+1)$. The line $x = -1$ is the vertical asymptote.

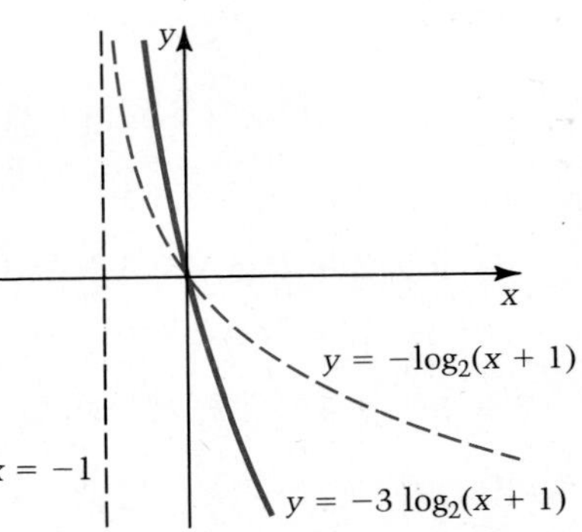

Step 2: The minus sign suggests replacing y by $-y$ to get $y = -\log_2 (x+1)$, a reflection across the x-axis. The 3 suggests stretching this graph away from the x-axis to get $y = -3 \log_2 (x+1)$. The line $x = -1$ is still the vertical asymptote.

Figure 5.11

In general, for a one-to-one function $y = f(x)$ and its inverse $y = f^{-1}(x)$ we know that

$$(f \circ f^{-1})(x) = f(f^{-1}(x)) = x \qquad \text{and} \qquad (f^{-1} \circ f)(x) = f^{-1}(f(x)) = x.$$

For the exponential and logarithmic functions these two results can be written as

$$b^{\log_b x} = x \quad (x > 0) \qquad \text{and} \qquad \log_b (b^x) = x \quad (\text{all real } x).$$

For example, for $x > 0$ we can simplify the expression $8^{\log_2 x}$ as follows:

$$8^{\log_2 x} = (2^3)^{\log_2 x} = 2^{3\log_2 x} = 2^{\log_2 (x^3)} = x^3.$$

We conclude by proving Properties 1, 3, and 4 of Logarithms. The proofs of Properties 2, 5, and 6 are left as exercises.

Proof of Property 1: $\log_b 1 = 0$

Property 1 of Real Number Exponents states that $b^0 = 1$

Then, by the Log/Exp Principle $\log_b 1 = 0$ ■

Proof of Property 3: $\log_b (xy) = \log_b x + \log_b y$

Begin by letting $r = \log_b x$ and $s = \log_b y$

By the Log/Exp Principle $x = b^r$ and $y = b^s$

By Property 3 of Real Number Exponents $xy = b^{r+s}$

By the Log/Exp Principle $\log_b (xy) = r + s$

Replacing r with $\log_b x$ and s with $\log_b y$, we obtain $\log_b (xy) = \log_b x + \log_b y$ ■

Proof of Property 4: $\log_b (x^t) = t \cdot \log_b x$

Begin by letting $r = \log_b x$

By the Log/Exp Principle $x = b^r$

By Property 4 of Real Number Exponents $x^t = b^{rt}$

By the Log/Exp Principle $\log_b (x^t) = rt$

Replacing r with $\log_b x$, we obtain $\log_b (x^t) = t \cdot \log_b x$ ■

EXERCISE SET 5.2

In Exercises 1–6, rewrite the given equation as a logarithmic equation.

1. $x^2 = 25$ **2.** $x^4 = 16$ **3.** $16^x = 4$

4. $27^x = 3$ **5.** $5^3 = x$ **6.** $10^5 = x$

In Exercises 7–20, use the Log/Exp Principle to solve for x.

7. $\log_2 x = 3$ **8.** $\log_4 x = 3$ **9.** $\log_8 x = \frac{1}{2}$

10. $\log_9 x = \frac{1}{2}$ **11.** $\log_x 4 = 2$ **12.** $\log_x 8 = 3$

13. $\log_8 4 = x$ **14.** $\log_9 27 = x$ **15.** $\log_x 3 = -2$

16. $\log_x 16 = -4$ **17.** $\log_x 32 = 5$ **18.** $\log_x 16 = 4$

19. $\log_4 8 = x$ **20.** $\log_{27} 9 = x$

In Exercises 21–34, sketch the graph of the equation.

21. $y = \log_3 x$ **22.** $y = \log_4 x$

23. $y = \log_{1/2} x$ **24.** $y = \log_{1/3} x$

25. $y = \log_2(x + 2)$ **26.** $y = \log_3(x - 3)$

27. $y = \log_3(-x)$ **28.** $y = \log_2(-x)$

29. $y = 1 + \log_3 x$ **30.** $y = -1 + \log_2 x$

31. $y = -2 + \log_{1/2} x$ **32.** $y = 3 + \log_{1/3} x$

33. $y = -\log_4 x$ **34.** $y = -\log_3 x$

In Exercises 35–50, use the three values $\log_2 x = 3.1$, $\log_2 y = 1.6$, and $\log_2 z = -2.7$ to evaluate the expression.

35. $\log_2(xyz)$ **36.** $\log_2\left(\frac{4}{xyz}\right)$ **37.** $\log_2\left(\frac{16x}{yz^3}\right)$

38. $\log_2\left(\frac{xy^2}{4z}\right)$ **39.** $\log_2\left(\frac{x^2y}{8}\right)$ **40.** $\log_2\left(\frac{y^2z}{32x}\right)$

41. $\log_2(\sqrt{xz})$ **42.** $\log_2(x\sqrt{y})$

43. $\log_2((x^2y)^{1/3})$ **44.** $\log_2\left(\left(\frac{x}{yz}\right)^{1/3}\right)$

45. $\log_2\left(\frac{x}{y}\right) - \frac{\log_2 x}{\log_2 y}$

46. $\log_2(xy) - (\log_2 x \cdot \log_2 y)$

47. $(\log_2 x)^2 - \log_2(x^2)$ **48.** $(\log_2 x)^3 - \log_2(x^3)$

49. $\log_2(\log_2(y^5))$ **50.** $\log_2 (10 \log_2(xz))$

In Exercises 51–68, write the expression in terms of A, B, and C given that $\log_b 2 = A$, $\log_b 3 = B$, and $\log_b 5 = C$.

51. $\log_b 15$ **52.** $\log_b 10$ **53.** $\log_b 16$

54. $\log_b 81$ **55.** $\log_b 600$ **56.** $\log_b 450$

57. $\log_b(5 - 2)$ **58.** $\log_b(2 + 3)$ **59.** $\log_b \frac{1}{3}$

60. $\log_b \frac{1}{2}$ **61.** $\log_b \frac{4}{5}$ **62.** $\log_b \frac{5}{8}$

63. $\log_b 0.3$ **64.** $\log_b 0.03$ **65.** $\log_b 1.2$

66. $\log_b 24.3$ **67.** $\log_b 0.081$ **68.** $\log_b 0.12$

Superset

In Exercises 69–84, sketch the graph of the equation.

69. $y = -\log_3(x - 2)$ **70.** $y = -\log_2(x + 3)$

71. $y = 2 - \log_2 x$ **72.** $y = 3 - \log_3 x$

73. $y = \log_3(x^2)$ **74.** $y = \log_2(x^3)$

75. $y = \log_2|x - 2|$ **76.** $y = \log_3|x + 1|$

77. $y = |1 + \log_3 x|$ **78.** $y = |-1 + \log_2 x|$

79. $y = \log_8(2^x)$ **80.** $y = \log_2(8^x)$ **81.** $y = 4^{\log_2 x}$

82. $y = 2^{\log_4 x}$ **83.** $y = 2^{\log_8 x}$ **84.** $y = 8^{\log_2 x}$

85. Prove Property 2 of Logarithms.

86. Prove Property 5 of Logarithms.

87. Prove Property 6 of Logarithms.

88. Prove the *Change of Base Formula for Logarithms:* $\log_b x = \dfrac{\log_B x}{\log_B b}$, where x is any positive real number, and b and B are positive real numbers with neither equal to 1. (*Hint:* set $s = \log_b x$ and $t = \log_B b$, and then get $x = B^{st}$.)

5.3 Solving Exponential and Logarithmic Equations

In this section we use the Properties of Logarithms to solve equations. We shall also see how logarithms can be used to carry out certain computations.

Our first two examples serve as a reminder of the central role played by the Log/Exp Principle: for $b > 0$ and $b \neq 1$,

$$b^A = C \quad \text{is equivalent to} \quad \log_b C = A.$$

Recall as well the following two facts that can be quite useful in simplifying equations:

$$\text{If } b^A = b^C\text{, then } A = C$$
$$\text{If } \log_b A = \log_b C\text{, then } A = C.$$

EXAMPLE 1

Solve for x: $\log_{1/4}(1/8) = x + 5$.

Solution

$$\log_{1/4}\left(\frac{1}{8}\right) = x + 5 \qquad \text{Use the Log/Exp Principle.}$$

$$\frac{1}{8} = \left(\frac{1}{4}\right)^{x+5}$$

$$2^{-3} = (2^{-2})^{x+5}$$

$$2^{-3} = 2^{-2x-10} \qquad \text{Bases are the same, so the exponents can be equated.}$$
$$-3 = -2x - 10$$
$$x = -\frac{7}{2}$$

EXAMPLE 2

Solve for x: $\log_5[\log_3(\log_2 x)] = 0$.

Solution By virtue of the Log/Exp Principle, the statement $\log_5[T] = 0$ tells us $T = 5^0$, that is, $T = 1$. Thus, in the given equation, the expression in brackets equals 1.

$$\log_3(\log_2 x) = 1$$
$$\log_2 x = 3 \qquad \text{The Log/Exp Principle is used.}$$
$$x = 2^3 \qquad \text{The Log/Exp Principle is used.}$$

Thus, $x = 8$.

The fundamental approach to solving any equation is to rewrite the equation in an equivalent, but simpler form, so the solution is easier to see. Sometimes, even though the steps we perform are legitimate, they don't produce an equivalent equation. In these cases we must check that any solutions obtained are actually solutions of the original equation. For example, when we square both sides of an equation, we must beware of extraneous roots. Since logarithms are defined for positive real numbers only, we must discard any solution that would require taking the logarithm of zero or a negative number when substituted back in the original equation. The next two examples illustrate this very point.

EXAMPLE 3

Solve for x: $2\log_5(3 - x) = \log_5(13 - 3x)$.

Solution

$$2\log_5(3 - x) = \log_5(13 - 3x) \qquad \text{Use Property 4 of Logarithms.}$$
$$\log_5[(3 - x)^2] = \log_5(13 - 3x)$$
$$(3 - x)^2 = 13 - 3x \qquad \text{If } \log_b A = \log_b C, \text{ then } A = C.$$
$$x^2 - 6x + 9 = 13 - 3x$$
$$x^2 - 3x - 4 = 0 \qquad \text{Use the Property of Zero Products.}$$
$$(x + 1)(x - 4) = 0$$

$x + 1 = 0$	$x - 4 = 0$	
$x = -1$	$x = 4$	Now we must check solutions.

Must Check:

$x = -1$	$x = 4$
$2\log_5 4 = \log_5(13 + 3)$	$2\log_5(-1) = \log_5(13 - 12)$
$\log_5 4^2 = \log_5 16$ True.	$\log_5(-1)$ is undefined.

The only solution is $x = -1$.

Notice in both this example and the next one, the decision about whether to discard a solution is not based on whether the value of x is positive, negative, or zero. Instead, we examine whether a particular value of x makes sense when substituted back into the original equation.

EXAMPLE 4

Solve for x: $\log_3(2 - x) + \log_3(3 - x) = \log_3 6$.

Solution

$$\log_3(2 - x) + \log_3(3 - x) = \log_3 6 \quad \text{Use Property 3 of Logarithms.}$$
$$\log_3[(2 - x)(3 - x)] = \log_3 6$$
$$(2 - x)(3 - x) = 6 \quad \text{If } \log_b A = \log_b C, \text{ then } A = C.$$
$$x^2 - 5x + 6 = 6$$
$$x^2 - 5x = 0 \quad \text{Factor and solve.}$$
$$x(x - 5) = 0$$
$$x = 0 \text{ or } 5$$

Must Check:

$x = 0$	$x = 5$
$\log_3 2 + \log_3 3 = \log_3 6$ True	$\log_3(-3) + \log_3(-2) = \log_3 6$
	False; $\log_3 t$ defined only for $t > 0$.

The only solution is $x = 0$.

Common Logarithms

From the early seventeenth century until just recently, 10 was commonly used as a base for logarithms, because the base for our system of enumeration is 10.

$$\log_{10}(0.01) = -2 \qquad \log_{10}(10) = 1$$
$$\log_{10}(0.1) = -1 \qquad \log_{10}(100) = 2$$
$$\log_{10}(1) = 0$$

By using scientific notation and the Properties of Logarithms, we need only a table of values of $\log_{10} x$ for values of x between 1 and 10 to compute the logarithm of any positive number. For example,

$$\begin{aligned}\log_{10}(178) &= \log_{10}(1.78 \times 10^2) = \log_{10}(1.78) + \log_{10}(10^2)\\ &= \log_{10}(1.78) + 2.\end{aligned}$$

Similarly,

$$\log_{10}(0.00178) = \log_{10}(1.78 \times 10^{-3}) = \log_{10}(1.78) - 3.$$

To complete either of these computations we only need to know that $\log_{10}(1.78)$ is approximately 0.2504. Today we use calculators rather than a table of values to access these approximate values, but the same principle underlies how calculators work and how tables are formed.

Base 10 logarithms are called *common logarithms.* Common logarithms were of great use historically in carrying out elaborate calculations. For example, by taking the common logarithm of each side of the equation $x = (17.23)(431.8)$ and using Property 3 of Logarithms, this multiplication problem could be converted to a problem in addition:

$$\log_{10} x = \log_{10}(17.23) + \log_{10}(431.8).$$

The underlying principle of the slide rule, used by engineers until the 1960s, was based precisely on this sort of conversion.

Natural Logarithms

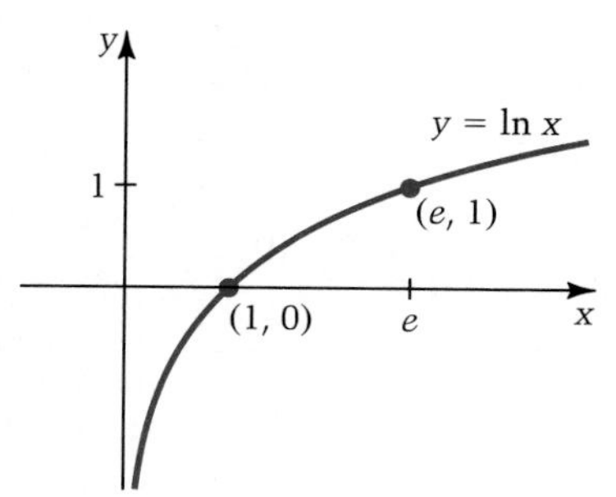

Figure 5.12

With the advent of computers and inexpensive calculators the use of logarithms for computations has been significantly reduced. Nowadays, logarithms are used primarily in those settings where they arise naturally as part of a mathematical model. We examine such models in the next section. It turns out that it is far more convenient to use a logarithm with a base e, where e represents a particular irrational number approximately equal to 2.718. (The value of e will be discussed in greater detail in the next section.) The function $y = \log_e x$ is usually written

$$y = \ln x$$

and is called the *natural logarithmic function.*

$$\begin{aligned}&\ln \frac{1}{e} = \ln e^{-1} = -1 \qquad && \ln e^2 = 2\\ &\ln 1 = 0 && \ln e^3 = 3\\ &\ln e = 1 &&\end{aligned}$$

We now illustrate the three simple types of computations that can arise from an exponential equation. That is, for the equation $B^A = C$, we may want to find a value for A, B, or C, given values for the other two quantities. In the solution of the first example below, we use the phrase "take the natural logarithm." This phrase refers to rewriting an equation $X = Y$ as $\ln X = \ln Y$. The two equations are equivalent provided X and Y are positive.

EXAMPLE 5

Solve for x: $2^x = 5$.

Solution

$$
\begin{aligned}
2^x &= 5 && \text{Take the natural logarithm of each side of the equation.}\\
\ln 2^x &= \ln 5 \\
x \ln 2 &= \ln 5 && \text{Property 4 of Logarithms is used.}\\
x &= \frac{\ln 5}{\ln 2}
\end{aligned}
$$

In the same way that $\frac{1}{7}$, $\sqrt{2}$ and π are names of real numbers for which we frequently use decimal approxmations (0.14, 1.41, and 3.14, respectively), the quantities ln 5 and ln 2 that appear in the example above should be thought of as the names of real numbers. Decimal approximations for these numbers and other values of the natural logarithm function are given in Table 3 at the back of the text. You can also obtain such decimal approximations by using the **ln** key on a calculator. By either technique, the solution to the equation in the previous example, $2^x = 5$, can be approximated as follows:

$$x = \frac{\ln 5}{\ln 2} \approx \frac{1.6094}{0.6931} \approx 2.32 \quad \text{(to the nearest hundredth).}$$

Notice that this approximation tells us that $2^{2.32} \approx 5$. We can make a simple check on the *reasonableness* of this answer: $2^2 = 4$ and $2^3 = 8$, therefore the solution of the equation $2^x = 5$ should be somewhere in between 2 and 3. Our solution, 2.32, certainly is.

Most errors in computations produce results that err by an order of magnitude (power of ten) and such errors can be uncovered by this kind of reasonableness test. You should get in the habit of asking if answers are reasonable, especially when you use a computer or calculator.

Previously in this chapter, we solved exponential equations such as $4^{x-1} = 8^{5x}$, where the two expressions could be rewritten with the same base, in this case as $2^{2(x-1)} = 2^{3(5x)}$. We are now in a position to solve a more general type of problem with the help of logarithms.

EXAMPLE 6

Solve for x: $2^{x-1} = 3^{1+4x}$.

Solution $2^{x-1} = 3^{1+4x}$ Begin by taking the natural logarithm of both sides.

$\ln(2^{x-1}) = \ln(3^{1+4x})$ Now use Property 4 of Logarithms.

$(x - 1)\ln 2 = (1 + 4x)\ln 3$

$x \ln 2 - \ln 2 = \ln 3 + 4x \ln 3$ Isolate x on one side of the equation, then solve.

$x(\ln 2 - 4 \ln 3) = \ln 3 + \ln 2$

$x = \dfrac{\ln 3 + \ln 2}{\ln 2 - 4 \ln 3}$ A calculator or table is needed at this point.

$x \approx -0.4841$

Before working our next example, we need to introduce some terminology. If $A = \ln C$, then by the Log/Exp Principle we can write

$$e^A = C.$$

(Remember that $\ln C$ means $\log_e C$.) Clearly A is a logarithm, since it is the exponent to which we raise e to get C. In this situation the number C has a name also:

C is the **antilogarithm of A, base e.**

Thus the antilogarithm of A, base e, is found by raising e to the power A. If you are using a calculator, this means that pressing the INV key, followed by the ln key will produce the same result as pressing the e^x key. Thus to determine $e^{0.78}$, we can use Table 2 in the back of the book, or we can use a calculator:

Enter *Press* INV *Press* ln .... *Display* 2.181472265

or, alternatively,

Enter *Press* e^x .... *Display* 2.181472265

In the next example you will use antilogarithms to evaluate $70^{1/5}$. Of course, if your calculator has a y^x key, you could use it to evaluate $70^{1/5}$ with a single "keystroke." However, after reading this example you might appreciate the steps your calculator actually goes through to produce the result.

EXAMPLE 7

Solve for x to four decimal places: $x = 70^{1/5}$.

Solution We begin with a reasonableness test. Since we are looking for x such that $x^5 = 70$, we think

$$2^5 = 32$$
$$x^5 = 70 \qquad x \text{ must be between 2 and 3.}$$
$$3^5 = 243$$

Now we begin solving.

$$x = 70^{1/5} \qquad \text{Take the natural logarithm of each side.}$$
$$\ln x = \ln(70^{1/5}) \qquad \text{Simplify, if possible.}$$
$$\ln x = \frac{1}{5}\ln 70 \qquad \text{Property 4 of Logarithms was used to simplify.}$$
$$\ln x \approx \frac{1}{5}(4.2485) \qquad \text{Table 3 or a calculator was used.}$$
$$\ln x \approx 0.8497 \qquad \text{Now take antilogarithm, base } e\text{, of each side.}$$
$$e^{\ln x} \approx e^{0.8497} \qquad \text{Recall: } b^{\log_b x} = x. \text{ So, } e^{\ln x} = x.$$
$$x \approx 2.3389 \qquad \text{Table 2 or a calculator was used.}$$

EXERCISE SET 5.3

In Exercises 1–28, solve the equation for x.

1. $4^{\log_2 x} = 5$
2. $3^{\log_9 x} = 2$
3. $2^{\log_4 x} = 5$
4. $9^{\log_3 x} = 2$
5. $\log_3(9^x) = -2$
6. $\log_4(2^x) = -3$
7. $\log_9(3^x) = -2$
8. $\log_2(4^x) = -3$
9. $\log_2(\log_3 x) = -1$
10. $\log_2(\log_5 x) = -1$
11. $\log_2(x^3) = \log_2(8x)$
12. $\log_3(x^4) = 2\log_3(5x)$
13. $\log_2(x^2 - 6) = \log_2(-x)$
14. $\log_2(7x) = \log_2 7 + \log_2 x$
15. $\log_3(5x) = \log_3 5 + \log_3 x$
16. $\log_3(x^2 - 8) = \log_3(-2x)$
17. $(3 - \log_2 x)\log_2 x = 0$
18. $(5 - \log_2 x)\log_2 x = 0$
19. $\ln(x + 1) + \ln(x + 4) = \ln(x + 9)$
20. $\ln(3 - x) + \ln(2 - x) = \ln(11 - x)$
21. $4\ln x = \ln(5x^2 - 4)$
22. $4\ln x = \ln(5x^2 - 6)$
23. $\log_x 4 = 2$
24. $\log_x 9 = 2$
25. $\log_2(\ln x) = 1$
26. $\log_3(\ln x) = 1$
27. $\ln 3 + \ln x = -1$
28. $\ln 5 + \ln x = 1$

In Exercises 29–42, sketch the graph of the equation. Label any asymptotes and intercepts.

29. $y = \ln(x - 4)$
30. $y = \ln(x + 4)$
31. $y = -2\ln x$
32. $y = 3\ln x$
33. $y = \ln(-2x)$
34. $y = \ln(3x)$
35. $y = \ln(e^x)$
36. $y = \ln(e^{-x})$
37. $y = \ln(e^2 x)$
38. $y = e^{2\ln x}$
39. $y = e^{3\ln x}$
40. $y = \ln\left(\frac{x}{e}\right)$
41. $y = \ln(e^{x^2})$
42. $y = \ln(e^{x^3})$

Superset

In Exercises 43–50, solve the equation for x.

43. $\log_x 1 = 0$
44. $\log_3(x^2) = 2\log_3 x$

45. $\log_4(x + 7) = \log_2(x + 1)$

46. $\log_2(x - 7) = \log_4(x - 1)$

47. $\log_3(3x) = \log_3 27 \cdot \log_3 x$

48. $\log_2(9x^2) = \log_2 16 \cdot \log_2 x$

49. $\ln(\ln x) = -1$

50. $\ln(\log_2 x) = 1$

In Exercises 51–58, sketch the graph of the equation. Label any asymptotes and intercepts.

51. $y = \ln \dfrac{1}{x - 2}$

52. $y = \ln \dfrac{1}{x + 4}$

53. $y = \ln \sqrt{x + 3}$

54. $y = \ln \sqrt{-x}$

55. $y = \ln|x + 3|$

56. $y = \ln|x - 2|$

57. $y = |\ln(x + 3)|$

58. $y = |\ln(x - 2)|$

In Exercises 59–68, determine whether the statement is true or false. If false, illustrate this with an example.

59. $\ln x + \ln y = \ln(x + y)$

60. $\ln x - \ln y = \ln(x - y)$

61. $(\ln x)(\ln y) = \ln(xy)$

62. $(\ln x)^2 = 2 \ln x$

63. $\ln(x^2) = 2 \ln x$

64. $x + \ln x^{-1} = x - \ln x$

65. $\ln(ex) = e \ln x$

66. $\ln e^x = x$

67. $\ln e^x = \ln x^e$

68. $\ln(e + x) = 1 + \ln x$

69. (a) Suppose x is an integer and $\log_{10} x = 6.18$. How many digits does x have?

(b) Suppose x is an integer and $\ln x = 6.18$. How many digits does x have? (*Hint:* By the Change of Base Formula (Exercise 88, Section 5.2), we know that $\log_{10} x = (\log_{10} e)(\ln x) \approx 0.4343 \ln x$.)

70. The product of the first n positive integers is denoted by $n!$ (read "n factorial"):

$$n! = 1 \cdot 2 \cdot 3 \cdot 4 \cdots (n - 2)(n - 1)n.$$

For example, $7! = 1 \cdot 2 \cdot 3 \cdot 4 \cdot 5 \cdot 6 \cdot 7 = 5040$. Use the fact that

$$\ln n! = \ln 1 + \ln 2 + \ln 3 + \cdots + \ln n$$

to estimate the number of digits in (a) 5! (b) 10! (c) 20!. (*Hint:* See Exercise 69.)

71. Complete the following table, where

$$f(x) = x - \frac{x^2}{2} + \frac{x^3}{3}.$$

x	0.05	0.10	0.15	0.20	0.25
$\ln(1 + x)$					
$f(x)$					

What do you notice?

72. Suppose some quantity Q is related to another quantity x by the relationship $Q = \log_2 x$. What happens to the value of Q if x is doubled? quadrupled? squared?

In Exercises 73–90, use natural logarithms to solve for x to the nearest hundredth.

73. $x^3 = 2$

74. $x^5 = 7$

75. $5^x = 23$

76. $3^x = 8$

77. $12^x = 5$

78. $15^x = 9$

79. $2^x = \dfrac{1}{3}$

80. $3^x = \dfrac{1}{4}$

81. $x = \left(1 + \dfrac{1}{10}\right)^{10}$

82. $x = \left(1 - \dfrac{1}{10}\right)^{10}$

83. $3^x = 4^{x-1}$

84. $2^x = 5^{x-2}$

85. $x^5 - 7x = 0, x > 1$

86. $x^8 - 3x = 0, x > 0$

87. $(x - 1)^5 = 0.0015, \quad x > 1$

88. $(x + 1)^8 = 37.92, \quad x > -1$

89. $x^2 = e$

90. $x^3 = e^2$

5.4 Applications

Exponential and logarithmic functions are used to solve a variety of applied problems. Interest on savings and loans, population growth, radioactive decay, and earthquake intensity are some of those subject areas employing exponential or logarithmic models.

t	$N(t)$
0	1
1	2
2	4
3	8
4	16
5	32

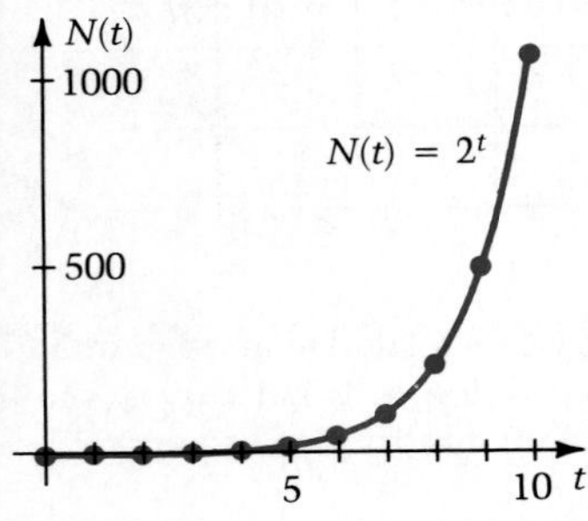

Figure 5.13

Consider the following hypothetical example of population growth. Suppose we are studying the growth of bacteria, and we begin with a single cell that splits after an hour of growth to form two cells. After another hour each of the two cells splits to form two more cells, making a total of four cells. This doubling process continues, and at the end of the third hour there is a total of eight cells. The data for the first 5 hours are shown in the table at the left (t is the number of hours elapsed, and $N(t)$ is the number of cells at time t). A graph for $0 \le t \le 10$ is shown in Figure 5.13.

As both the graph and the table suggest, the function that models this situation is $N(t) = 2^t$, and its domain is the set of positive integers. We show a smooth curve to indicate the exponential nature of the process. Note that in this situation, the initial population was 1 cell.

If there are B_0 cells initially, and if the population doubles every hour, then after t hours, the size of the population, $N(t)$, is given by the equation

$$N(t) = B_0 2^t.$$

EXAMPLE 1

The bacteria *E. coli* is found in many organisms, including humans. Suppose that, under certain conditions, the number of these bacteria present in an experimental colony is given by the equation $N(t) = B_0 2^t$. How many *E. coli* are present 6 hours after the start of the experiment, if there were 1,500,000 initially?

Solution $N(t) = B_0 2^t$

$N(6) = 1{,}500{,}000 \cdot 2^6$ Values are substituted for B_0 and t.

$N(6) = 96{,}000{,}000$

There are 96,000,000 bacteria present after 6 hours.

EXAMPLE 2

How long will it take the colony in Example 1 to grow to 20,000,000 bacteria?

Solution $N(t) = B_0 2^t$

$20{,}000{,}000 = 1{,}500{,}000 \cdot 2^t$ Values are substituted for $N(t)$ and B_0.

$13.3 \approx 2^t$

$\ln 13.3 \approx \ln 2^t$ We take the log of both sides and then use Property 4 of Logarithms to solve for t.

$\ln 13.3 \approx t \cdot \ln 2$

$$\frac{\ln 13.3}{\ln 2} \approx t$$

ln 13.3 ≈ 2.5878 and ln 2 ≈ 0.6931.

$$t \approx \frac{2.5878}{0.6931} \approx 3.7$$

There will be 20,000,000 bacteria in the colony after approximately 3.7 hours.

Exponential and logarithmic models are used to solve problems concerning investments and loans. If P dollars are invested at an interest rate i (expressed as a decimal), and the interest is compounded n times per year, then the amount, A, of money in the account after t years is

$$A = P\left(1 + \frac{i}{n}\right)^{nt}.$$

EXAMPLE 3

Suppose \$2000 is deposited in an account that advertises a 9.2% interest rate compounded quarterly. What will the value of the account be in 25 years?

Solution

$$A = P\left(1 + \frac{i}{n}\right)^{nt}$$

$$A = 2000\left(1 + \frac{0.092}{4}\right)^{4(25)}$$

$$A = 2000(1.023)^{100}$$

$$\ln A = \ln[2000(1.023)^{100}]$$

$$\ln A = \ln 2000 + 100 \ln 1.023$$

$$\ln A \approx 7.6009 + 100(0.0227)$$

$$\ln A \approx 9.8709$$

$$A \approx 1.94 \times 10^4$$

We shall use logarithms to perform this complicated computation. If you have a calculator with a y^x key, you can find $(1.023)^{100}$ directly.

We took the antilogarithm of both sides to find A.

The value of the account will be approximately \$19,400 in 25 years.

Let us perform an experiment to determine the effect of increasing the number of times interest is compounded each year. To make things simple, suppose that \$1.00 is invested for one year at an interest rate of 100% (i.e., $P = 1$, $t = 1$, and $i = 1$). In the following table we show the value of the investment after one year as we vary the number of times the interest is compounded. (It is customary to use a 360-day year when compounding interest.)

n	**Interest is compounded:**	$A = \left(1 + \frac{1}{n}\right)^n$
1	annually	2
2	semiannually	2.25
4	quarterly	≈ 2.441406
12	monthly	≈ 2.613035
52	weekly	≈ 2.692597
360	daily	≈ 2.714516
8,640	hourly	≈ 2.718125
518,400	each minute	≈ 2.718262
31,104,000	each second	≈ 2.718282

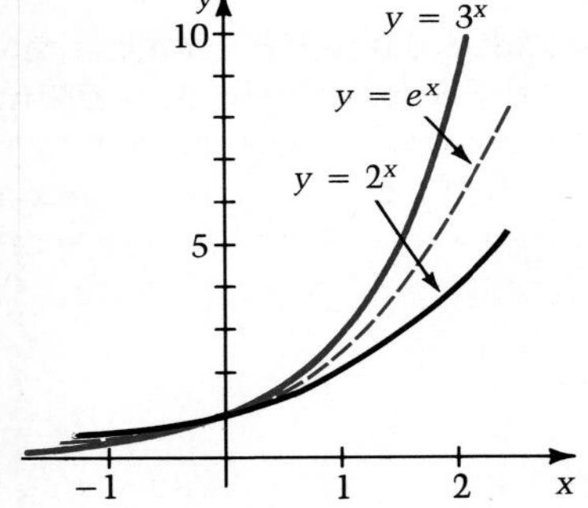

Figure 5.14

Notice that as n increases, the value of the investment approaches some constant value. This constant value is the irrational number e we spoke of in the last section.

$$e \approx 2.718282.$$

The function $y = e^x$ is called the **natural exponential function,** and the function $y = \log_e x$, usually written $y = \ln x$, is called the **natural logarithmic function.** As you might expect, since e is between 2 and 3, the graph of $y = e^x$ lies between the graphs of $y = 2^x$ and $y = 3^x$, as shown in Figure 5.14.

When interest is advertised as being "compounded continuously", the model used to determine the value of an account after t years is

$$A = Pe^{it}$$

where P is the amount initially invested and i is the annual interest rate.

EXAMPLE 4

An amount of $12,000 is placed in an account advertising an 8% annual interest rate, compounded continuously.

(a) How much is the account worth after 5 years have passed?

(b) How long is it before the account is worth twice the initial investment?

Solution

(a) $A = Pe^{it}$ Substitute values for P, i, and t.

$A = 12{,}000 \cdot e^{(0.08)(5)}$

$A \approx 12{,}000 \cdot 1.4918$ $e^{0.4} \approx 1.4918$

$A \approx 17{,}902$

After 5 years, the value of the account is approximately $17,902.

(b)
$$\begin{aligned}
24{,}000 &= 12{,}000 \cdot e^{0.08t} \\
2 &= e^{0.08t} \\
\ln 2 &= \ln e^{0.08t} \\
\ln 2 &= 0.08t \ln e && \text{Property 4 of Logarithms is used.} \\
\ln 2 &= (0.08t)(1) && \ln e = \log_e e = 1 \text{ by Property 2 of Logarithms.} \\
\frac{0.6931}{0.08} &\approx t \\
t &\approx 8.66
\end{aligned}$$

The amount in the account will double its value in roughly $8\frac{2}{3}$ years.

The previous example describes an **exponential growth** model because the value of the account is increasing. Of equal interest is the **exponential decay** model. The most common exponential decay problems involve radioactive substances.

Carbon-14 is a radioactive substance present in all living matter. While an organism is alive, the amount of Carbon-14 it contains remains constant. Once the organism dies, however, the number of Carbon-14 atoms begins to decrease. This process is called **radioactive decay.** By comparing the level of radiation in a fossil with the level present in a similar living sample, we can estimate how long ago the fossil was a living organism. Essential to such an investigation is the formula

$$y = y_0 e^{kt},$$

where y_0 is the amount of radioactive substance present initially, y is the amount present after t years, and k is the *decay constant.*

The rate of decay for a radioactive substance is usually expressed as its *half-life,* the time it takes for a given sample to reduce to half its size. We can find a formula for the decay constant k in terms of the half-life T by substituting $y = \frac{1}{2}y_0$ in the formula $y = y_0 e^{kt}$:

$$\begin{aligned}
\frac{1}{2}y_0 &= y_0 e^{kT} && \text{Assume } y = \frac{1}{2}y_0 \text{ at time } t = T. \\
\frac{1}{2} &= e^{kT} && \text{Take natural log of each side of equation and use Properties 4 and 6 of Logarithms.} \\
\ln\frac{1}{2} &= \ln(e^{kT}) \\
-\ln 2 &= kT \ln e \\
-\ln 2 &= kT && \ln e = 1. \\
k &= -\frac{\ln 2}{T} \approx -\frac{0.6931}{T}
\end{aligned}$$

EXAMPLE 5

Carbon-14 has a half-life of 5750 years. Suppose that 100 grams of Carbon-14 were present in an organism when it lived 3000 years ago. How much Carbon-14 would remain now?

Solution

$k \approx \dfrac{-0.6931}{5750}$ Begin by determining the decay constant.

$k \approx -0.00012$

$y \approx y_0 e^{(-0.00012)t}$ The value of k is substituted into the radioactive decay model.

$y \approx 100e^{(-0.00012)3000}$ Values of y_0 and t are substituted.

$y \approx 100e^{-0.36}$

$y \approx 100(0.6977)$ $e^{-0.36} \approx 0.6977$

$y \approx 69.77$

There would be approximately 69.8 grams of Carbon-14 present today.

EXAMPLE 6

The following equation describes the relationship between the energy E released by an earthquake and the earthquake's magnitude M on the Richter scale.

$$\log_{10} E = 11.4 + 1.5\,M \quad \text{(energy } E \text{ is measured in ergs).}$$

For example, the 1989 San Francisco earthquake registered $M = 7.1$ on the Richter scale and released $E = 1.12 \times 10^{22}$ ergs of energy.

John Paulos, in his book *Innumeracy,* claims that the total TNT equivalent of all existing nuclear weapons is 25,000 megatons. (1 megaton = 1 million tons, and one ton of TNT represents 4.2×10^6 ergs of energy.) What would the Richter scale magnitude be if the energy of all existing nuclear weapons were released?

Solution

- ANALYZE With the help of the given equation, a Richter scale magnitude M must be calculated for a specific value of E.
- ORGANIZE The energy equivalent E of all existing nuclear weapons must be computed in ergs.

$$\begin{aligned} E &= (25{,}000 \text{ megatons of TNT})\left(\frac{1 \text{ million tons}}{\text{megaton}}\right)\left(\frac{4.2 \times 10^6 \text{ ergs}}{\text{ton of TNT}}\right) \\ &= (2.5 \times 10^4)(10^6)(4.2 \times 10^6) \text{ ergs} \\ &= 10.5 \times 10^{16} \text{ ergs} \end{aligned}$$

- MODEL The model was given in the statement of the problem.

$$\begin{aligned} \log_{10}E &= 11.4 + 1.5\,M \\ \log_{10}(10.5 \times 10^{16}) &= 11.4 + 1.5\,M \\ 1.0212 + 16 &\approx 11.4 + 1.5\,M \\ \frac{1.0212 + 16 - 11.4}{1.5} &\approx M \\ 3.7 &\approx M \end{aligned}$$

- ANSWER The release of energy would rate a 3.7 on the Richter scale.

EXERCISE SET 5.4

For Exercises 1–4, refer to Example 1 (p. 318).

1. How many bacteria are present 7 hours after the start of the experiment, if there were 1350 initially?
2. How many bacteria are present 4 hours after the start of the experiment, if there were 58,000 initially?
3. How long will it take a colony of 5000 bacteria to triple in number?
4. How long ago were there less than 1000 bacteria in the colony, if there are 7500 present now?
5. If \$1000 is placed in an account that earns 8% compounded quarterly, what is the value of the account 10 years later?
6. If \$500 is placed in an account that earns 6% compounded semiannually, what is the value of the account 12 years later?
7. How much money must be placed in an account that earns 7% compounded semiannually, so that there will be \$2500 in the account 5 years from now?
8. How much money must be placed in an account that earns 6% compounded quarterly, so that there will be \$1000 in the account 8 years from now?
9. If you put 1¢ in your bank account today, 2¢ tomorrow, 4¢ the day after tomorrow, 8¢ the next day, and so on, in how many days will you have to deposit at least \$100?
10. (Refer to Exercise 9) Suppose instead, you deposit 1¢, 3¢, 9¢, 27¢, and so on. In how many days will you have to deposit at least \$100?
11. If \$1000 is placed in an account that earns 8% compounded continuously, what is the value of the account 10 years later?
12. If \$500 is placed in an account that earns 6% compounded continuously, what is the value of the account 12 years later?
13. Suppose \$1762 is placed in an account that earns 7% compounded continuously. How long will it be before the account is worth \$2500?
14. Suppose \$619 is placed in an account that earns 6% compounded continuously. How long will it be before the account is worth \$1000?
15. Suppose there were 150 gm of Carbon-14 in an organism when it lived roughly 10,000 years ago. How much Carbon-14 remains today?
16. Suppose there are 20 gm of Carbon-14 remaining in an organism known to have lived 2500 years ago. How many grams of Carbon-14 were in the organism when it lived?
17. Suppose we know that there were 180 gm of Carbon-14 in an organism when it lived and only 25 gm remain today? How long ago did the organism live?
18. Suppose there are 200 gm of Carbon-14 in an organism that dies today. In how many years will there be less than 75 gm left in the organism?
19. If inflation persists at 6% annually compounded continuously, how much will an item that costs \$10 today cost in 5 years?
20. The population of the United States tends to grow at a rate proportional to the size of the population. Thus, we may represent the population size by the exponential model $A(t) = 180e^{0.013t}$, where t is the number of years since 1960 and $A(t)$ is the population in millions. What will the population of the U.S. be in the year 2000?

Superset

21. How much money must be placed in an account that earns 6% compounded continuously in order that \$1000 may be withdrawn at the end of each year forever?

22. (Refer to Exercise 21) How much must be placed in the account if the interest is compounded quarterly?

23. The **effective annual rate** of interest in an account that holds $A(t)$ dollars after t years is defined by

$$\frac{[A(t) - A(0)]}{A(0)} \times 100\%.$$

What is the effective annual rate of interest on money invested at 6% compounded quarterly?

24. (Refer to Exercise 23) What is the effective annual rate of interest on money invested at 6% compounded continuously?

25. At what rate, compounded continuously, must a sum of money be invested so that the actual earnings are 8% per year?

26. (Refer to Exercise 20) In what year will the population of the United States exceed the current world population of 4 billion?

27. Suppose \$18,500 is placed in an account that advertises a 9.13% annual interest rate compounded continuously.
(a) How much is the account worth after $4\frac{1}{2}$ years?
(b) How long will it take for the account to be worth three times the initial investment?

28. If 86.2 gm of Carbon-14 were present in an organism when it was alive, and 38.7 gm are present in the remains today, how long ago did the organism live?

29. A new water treatment installation continuously reduces the percentage of pollutant in the water by $\frac{1}{50}$ per day ($k = -0.02$). How many days does it take to cut the pollutant in the water to less than 1% of what it was originally?

30. A certain species of bird is in danger of extinction. It is estimated that only 900 of these creatures are still alive. Estimates five years ago placed the population at 1200. Experts claim that once the population drops below 200, this bird's situation will be irreversible. In how many years will this happen? (Assume that the population changes at a rate proportional to its size.)

31. The Richter scale converts seismographic data into numbers that make it easier to refer to the strength of an earthquake. The magnitude M of an earthquake is related to a seismographic reading of x millimeters by the formula $x = x_0 10^M$, where $x_0 = 10^{-3}$ millimeters is the seismographic reading of a benchmark earthquake 100 km from its epicenter. The intensity of an earthquake is the ratio of x to x_0. Notice that for each increase by 1 in the Richter scale rating M, the intensity of an earthquake increases by a factor of 10.
If the Richter scale rating for a certain earthquake was 4.5, what would be the Richter scale rating for an earthquake that was twice as intense?

32. The Oakland quake of 1989 was rated by several authorities at 6.1 on the Richter scale. How many more times intense was the famous San Francisco quake of '06 that has been given a rating of 8.9 on the Richter scale? (See Exercise 31.)

33. Newton's Law of Cooling states that the temperature of an object will drop exponentially towards the temperature of the surrounding medium. Namely, if an object with temperature T_0 is placed in an environment with constant temperature C, then the temperature T of the object x minutes later is given by

$$T = C + (T_0 - C)e^{kx}.$$

The constant k is a negative number expressing the rate of cooling.
Three minutes after you serve yourself a bowl of soup, the temperature of the soup has dropped from 190°F to 165°F. The constant temperature of your dining area is 70°F. (This information can be used to determine the value of the constant k.) What was the temperature of the soup one minute after serving?

34. (See Exercise 33.) How many minutes must you wait after serving the soup before the temperature of the soup will be 135°F?

35. Sound intensity is measured on a logarithmic scale because of the extreme range in values. The unit of measure is called a **decibel.** If $I_0 \approx 10^{-16}$ watts per square centimeter is the power of sound just below the threshold of human hearing and I is the power (in watt/cm^2) of a certain sound, the decibel reading N for that sound is given by the formula

$$N = 10 \log_{10}\left(\frac{I}{I_0}\right) = \frac{10}{\ln 10}\ln\left(\frac{I}{I_0}\right).$$

Determine the decibel ratings of the following sounds: (a) normal conversation: 3.2×10^{-10} watt/cm^2, (b) heavy traffic: 7.4×10^{-8} watt/cm^2.

36. (See Exercise 35.) What is the power rating (watt/cm^2) of the sound of severe thunder that registers at 125 decibels?

Analyzing Data

The Case of the Ubiquitous VCR

In the Data Analysis features in Chapters 3 and 4, we discussed the technique of fitting straight lines or parabolas to data that have been observed. We are now in a position to model processes that exhibit exponential behavior. Consider the following table which shows the number of U.S. households with videocassette recorders (VCRs) for each of the years 1978–1987. We have also included the natural logarithm of the number of households for later use.

Year	Number of Households with VCRs (in ten thousands)	
x	y	$\ln y$
1978 (1)	20	3.0
1979 (2)	40	3.7
1980 (3)	84	4.4
1981 (4)	144	5.0
1982 (5)	253	5.5

Year	Number of Households with VCRs (in ten thousands)	
x	y	$\ln y$
1983 (6)	458	6.1
1984 (7)	888	6.8
1985 (8)	1760	7.5
1986 (9)	3092	8.0
1987 (10)	4256	8.4

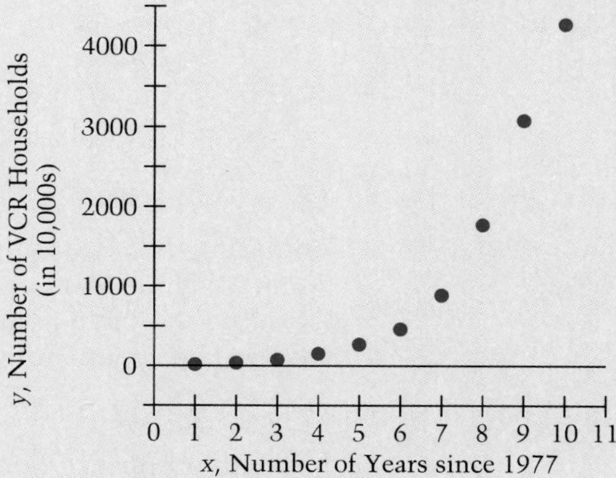

Figure 5.15

In Figure 5.15 we have plotted the ten (x, y) pairs from the table. Note that we have recoded the x-values 1 through 10, so the precise meaning of the x-value in that graph is "number of years since 1977." It is clear that the data, as shown in the graph, do not follow a straight line. Cases such as this one (i.e. where growth is apparent) are often best modeled by an exponential function $y = y_0 e^{kx}$.

In the table we also listed the value of ln y for each of the y-values. To determine the exponential function of best fit (we will need to determine values for y_0 and k), let's start by graphing ln y (instead of y) on the vertical axis, with x on the horizontal axis. See Figure 5.16.

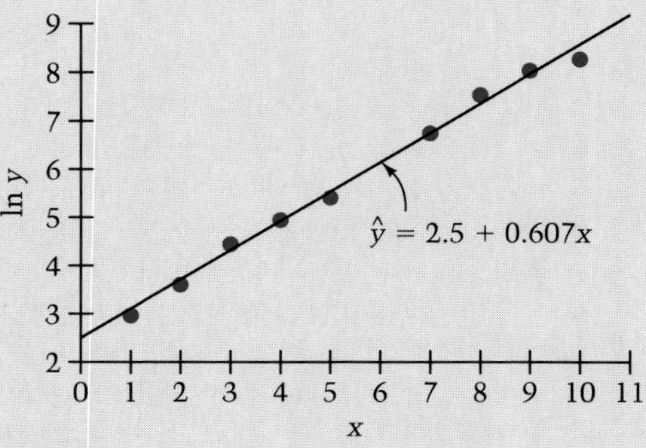

Figure 5.16

Notice that by replacing each y-value with ln y, the points exhibit straight line behavior, and so it makes sense to determine an equation for the line of best fit, which is drawn in the figure. We estimate that line to be $\hat{y} = 2.5 + 0.607x$. (Remember that $\hat{y}$ represents ln y.) Below we show how to use logarithms to rewrite the exponential function $y = y_0e^{kx}$, so that it can be compared with our line of best fit.

$$y = y_0e^{kx}$$

$$\ln y = \ln (y = y_0e^{kx})$$

$$\hat{y} = a + bx \qquad \ln y = \ln y_0 + \ln e^{kx}$$

$$\hat{y} = 2.5 + 0.607x \qquad \ln y = \ln y_0 + kx \ln e$$

$$\ln y = 2.5 + 0.607x \qquad \ln y = \ln y_0 + kx$$

Comparing the two forms, we conclude that $k = 0.607$, and, since $2.5 = \ln y_0$ we can conclude that $y_0 = e^{2.5} \approx 12.2$. Thus, the exponential model of best fit to the data shown in Figure 5.15 is $y = 12.2e^{0.607x}$. We can use our exponential model to predict the number of households with VCRs for 1988 ($x = 11$).

$$y = 12.2e^{0.607x} = 12.2e^{0.607(11)} \approx 9686 \quad \text{(in ten thousands)}$$

The *actual* number of households (in ten thousands) with VCRs in 1988 was 5139. The estimate based on our model is quite high. Our exponential model fits well up to 1987, but thereafter it overestimates the actual number of VCRs. In other words, the increase in the number of VCRs has begun to slow down, and so the graph will start leveling off. This phenomenon frequently follows periods of exponential growth, which generally cannot continue forever. Afterall, there are just so many households in the U.S.

Source: Wright, J. ed. *The Universal Almanac 1990*. Andrews & McMeel, New York, p. 281, 1989.

Chapter Review

5.1 Exponential Functions (pp. 297–303)

Suppose x, y, and t are any real numbers and b is positive. Then the *Properties of Real Number Exponents* are as follows (p. 298):

$$b^0 = 1 \qquad b^1 = b \qquad b^x b^y = b^{x+y} \qquad (b^x)^t = b^{xt} \qquad \frac{b^x}{b^y} = b^{x-y} \qquad b^{-x} = \frac{1}{b^x}$$

A function of the form $y = b^x$, where b is any positive constant except 1, is called an *exponential function*; b is called the *base*. (p. 298) Five important features of the exponential function $y = b^x$ are (p. 299):

- the domain is $(-\infty, \infty)$;
- the range is $(0, \infty)$;
- $b^0 = 1$;
- the x-axis is a horizontal asymptote of the graph;
- the function is one-to-one.

5.2 Logarithmic Functions (pp. 303–310)

For $b > 0$ with $b \neq 1$, the exponential function $f(x) = b^x$ has an inverse called the *logarithmic function with base b* and denoted $f^{-1}(x) = \log_b x$. Five important features of the logarithmic function are (p. 305):

- the domain is $(0, \infty)$;
- the range is $(-\infty, \infty)$;
- $\log_b 1 = 0$;
- the y-axis is a vertical asymptote of the graph;
- the function is one-to-one.

Log/Exp Principle: For $b > 0$ and $b \neq 1$, $b^A = C$ is equivalent to $\log_b C = A$. (p. 304)

Suppose x and y are positive real numbers, t is any real number, and base b is positive, but not equal to 1. The *Properties of Logarithms* are as follows (p. 306):

$$\log_b 1 = 0 \qquad \log_b b = 1 \qquad \log_b(xy) = \log_b x + \log_b y$$

$$\log_b(x^t) = t \log_b x \qquad \log_b\left(\frac{x}{y}\right) = \log_b x - \log_b y \qquad \log_b\left(\frac{1}{x}\right) = -\log_b x$$

Also remember that: $b^{\log_b x} = x\,(x > 0)$ and $\log_b b^x = x$. (p. 308)

5.3 Solving Exponential and Logarithmic Equations (pp. 310–317)

Exponential and logarithmic equations can often be solved by using the Log/Exp Principle, or one of the following (p. 310):

$$\text{If } b^A = b^C, \text{ then } A = C.$$

$$\text{If } \log_b A = \log_b C, \text{ then } A = C.$$

Base 10 logarithms are called *common logarithms.* (p. 312) The function $y = \log_e x$, where $e \approx 2.718$, is called the *natural logarithm function.* It is customary to write $\log_e x$ as $\ln x$. (p. 313)

If $e^A = C$, then C is called the *antilogarithm* of A, base e. (p. 315)

5.4 Applications (pp. 317–324)

Exponential and logarithmic functions serve as models for problems involving population growth, compound interest, and radioactive decay. (p. 317) The function $y = e^x$, where $e \approx 2.718$, is called the *natural exponential function.* (p. 320)

Review Exercises

In Exercises 1–6, simplify the expression.

1. $(5^x)^2$ **2.** $(3^{-x})^{-4}$ **3.** $2^{x+3} \cdot 8^{-2x}$

4. $3^{x-1} \cdot 9^{-x}$ **5.** $9^{x-2} \cdot 27^{3-2x}$ **6.** $4^{2-x} \cdot 8^{3x-2}$

In Exercises 7–12, solve for x.

7. $9^x = 243$ **8.** $8^{2-x} = 4^x$

9. $4^{x+1} = 2(8^{3-x})$ **10.** $9^{2-3x} = \frac{1}{3}(27^{2x-1})$

11. $3^{x+2} = 0$ **12.** $5^x = -\frac{1}{5}$

In Exercises 13–18, sketch the graph of the equation. Label any asymptotes and intercepts.

13. $y = 2^{x-1}$ **14.** $y = 3^{x+1}$ **15.** $y = 4 - 2^{-x}$

16. $y = 1 + 3^{-x}$ **17.** $y = -\left(\frac{1}{2}\right)^{x-2}$ **18.** $y = \frac{1}{8} - \frac{1}{4^x}$

In Exercises 19–24, rewrite the equation as a logarithmic equation.

19. $x^3 = 27$ **20.** $9^x = 27$ **21.** $3^x = 27$

22. $2^5 = x$ **23.** $5^2 = x$ **24.** $x^9 = 27$

In Exercises 25–32, use the Log/Exp Principle to solve for x.

25. $\log_x 9 = -2$ **26.** $\log_8 16 = x$ **27.** $\log_{16} 8 = x$

28. $\log_x 2 = -3$ **29.** $\log_x 3 = \frac{1}{2}$ **30.** $\log_x\left(\frac{1}{2}\right) = 3$

31. $\log_2\left(\frac{1}{x}\right) = -3$ **32.** $\log_3\left(\frac{1}{x}\right) = -2$

In Exercises 33–40, sketch the graph of the equation. Label any asymptotes and intercepts.

33. $y = \log_2(x + 1)$ **34.** $y = \log_3\left(\frac{1}{x}\right)$

35. $y = \log_4(-x)$ **36.** $y = \log_{1/4} x$

37. $y = -1 + \log_{1/2} x$ **38.** $y = \log_3(x + 3)$

39. $y = \log_2(4 - x)$ **40.** $y = 1 - \log_3 x$

In Exercises 41–46, use the three values $\log_2 x = 4.1$, $\log_2 y = 1.7$ and $\log_2 z = -3.4$ to evaluate the expression.

41. $\log_2(8z^3y)$ **42.** $\log_2(x^2y^{-3})$ **43.** $\log_2\left(\frac{xy}{2z}\right)$

44. $\log_2\left(8y\left(\frac{z}{x}\right)^{1/3}\right)$ **45.** $\log_2(z\sqrt{xy})$ **46.** $\log_2\left(\frac{\sqrt{z}}{16x}\right)$

In Exercises 47–52, write the expression in terms of A, B and C given that $\log_b 2 = A$, $\log_b 3 = B$ and $\log_b 5 = C$.

47. $\log_b 22.5$ **48.** $\log_b 0.27$ **49.** $\log_b\left(\frac{125}{12}\right)$

50. $\log_b 400$ **51.** $\log_b(40b^2)$ **52.** $\log_b\left(\frac{b^3}{15}\right)$

In Exercises 53–62, solve the equation for x.

53. $27^{\log_8 x} = 9$ **54.** $2^{\log_8 x} = 4$

55. $\log_3(\log_2 x) = -1$ **56.** $\log_3(\log_5 x) = 1$

57. $4 \ln x = \ln(4x^2 - 3)$ **58.** $\ln(x^2 - 6) = \ln x$

59. $\ln 4 + \ln x = -3$ **60.** $\ln(x + 1) = \ln 3 + \ln x$

61. $\ln(x + 2) + \ln(x - 2) = 0$

62. $\ln(\ln x) = 0$

In Exercises 63–70, sketch the graph of the equation. Label any asymptotes and intercepts.

63. $y = \ln(x - 1)$ **64.** $y = 2 \ln x$ **65.** $y = \ln(e^{2x})$

66. $y = \ln(e^3 x)$ **67.** $y = e^{\ln 2x}$ **68.** $y = e^{\ln(e^x)}$

69. $y = \ln\left(\frac{e^2}{x}\right)$ **70.** $y = 1 - \ln x$

71. If \$1000 is placed in an account that earns 6% compounded quarterly, what is the value of the account 12 years later?

72. If \$1000 is placed in an account that earns 6% compounded continuously, how much interest does the account earn in 5 years?

73. Suppose we know that there were 200 gm of Carbon-14 in an organism when it lived and only 20 gm remain today. How long ago did the organism live?

74. At what rate must a sum of money be invested in an account so that the value of the account will at least double in 8 years if the interest is compounded semiannually?

In Exercises 75–78, solve for x.

75. $2^{x^2 - x} = 4$ **76.** $(2^x)^x = 8$ **77.** $4^x + 8 = 6(2^x)$

78. $2^x(2^{2x} + 3) = 3 \cdot 2^{2x} + 1$

In Exercises 79–86, use natural logarithms to solve for x to the nearest hundredth.

79. $x^3 = 12$ 80. $x^2 = 7$ 81. $3^x = 13$

82. $2^x = \frac{1}{5}$ 83. $x^{6/5} = 0.48$ 84. $x^{2/5} = 27.1$

85. $2^x = 3^{x-1}$ 86. $5^{x+1} = 2^{2x-1}$

Graphing Calculator Exercises

[*Note: Round all estimated values to the nearest hundredth.*]

In Exercises 1–2, determine (a) the domain; (b) the range; (c) the x-intercepts; (d) the interval(s) where the graph lies above the x-axis; (e) the value(s) of x such that $f(x) = -1$; (f) the maximum function value, if any, and the corresponding x-value; (g) the minimum function value, if any, and the corresponding x-value.

1. $f(x) = 1 - x^2 - 2^x$ **2.** $f(x) = x^3 - 3^x$

In Exercises 3–10, use a graph to solve the given equation. (*Hint:* Recall that to solve the equation $h(x) = g(x)$, sketch the function $f(x) = h(x) - g(x)$ and approximate the x-intercepts.)

3. $3^x = -x$ **4.** $2^{-x} = x$ **5.** $x = \ln x$

6. $\frac{1}{x} = \ln x$ **7.** $2 + x = e^x$ **8.** $x = e^x$

9. $1 + x = e^x$ **10.** $e^{-x} = x$

11. Graph each of the following pairs of functions on the same set of axes for values of x in the interval $[-1, 1]$.

$y = e^x$ and $y = 1 + x$

$y = e^x$ and $y = 1 + x + \frac{x^2}{2}$

$y = e^x$ and $y = 1 + x + \frac{x^2}{2} + \frac{x^3}{6}$

What do you observe?

12. Graph each of the following pairs of functions on the same set of axes for values of x in the interval $(-1, 1]$.

$y = \ln(1 + x)$ and $y = x - \frac{x^2}{2}$

$y = \ln(1 + x)$ and $y = x - \frac{x^2}{2} + \frac{x^3}{3}$

$y = \ln(1 + x)$ and $y = x - \frac{x^2}{2} + \frac{x^3}{3} - \frac{x^4}{4}$

What do you observe?

Chapter Test

In Problems 1–2, simplify the expression.

1. $4^{x-1} \cdot 8^{2-3x}$ **2.** $\dfrac{9^{x+2}}{27^{2x}}$

In Problems 3–4, sketch the graph of the equation.

3. $y = 3^{x-2}$ **4.** $y = \left(\dfrac{1}{2}\right)^x - 4$

In Problems 5–6, solve for x.

5. $3^{x+1} = 0$ **6.** $4^{-2x} = 8^{5x-2}$

In Problems 7–9, rewrite the equation as a logarithmic equation.

7. $x^3 = 125$ **8.** $4^x = 64$ **9.** $7^4 = x$

In Problems 10–11, use the Log/Exp Principle to solve for x.

10. $\log_3 27 = x$ **11.** $\log_x 16 = -2$

In Problems 12–13, sketch the graph of the equation.

12. $y = \log_2(x - 3)$ **13.** $y = \log_{1/4} x$

In Problems 14–15, use the values $\log_2 x = 4.3$, $\log_2 y = 2.8$, and $\log_2 z = -3.1$ to evaluate the expression.

14. $\log_2\left(\dfrac{8x}{yz}\right)$ **15.** $\log_2(xy^2z^3)$

In Problems 16–23, solve for x.

16. $\log_4 2^x = -5$ **17.** $\log_3(\log_2 x) = 0$

18. $\log_2 x^3 = \log_2 9x$ **19.** $\log_2 8x = \log_2 8 + \log_2 x$

20. $\ln(x - 3) + \ln(x - 4) = \ln(7 - x)$

21. $2 \ln x = \ln(3x + 4)$

22. $\ln(e^2x) = 5$ **23.** $\ln\left(\dfrac{e}{x}\right) = 3$

In Problems 24–25, sketch the graph of the equation.

24. $y = \ln(-x)$ **25.** $y = \ln(x + 3)$

In Problems 26–27, use natural logarithms to solve for x to the nearest hundredth.

26. $x^2 = 3$ **27.** $2^x = 6$

28. The number of bacteria present in an experimental colony is given by the equation $N(t) = B_0 2^t$, where t is measured in hours. If there are 1,600,000 bacteria cells present initially, how long will it take the colony to grow to 2,400,000 bacteria?

29. If \$500 is placed in an account that earns 8% compounded quarterly, what is the value of the account 6 years later?

30. If \$500 is placed in an account that earns 8% compounded continuously, what is the value of the account 6 years later?

31. Suppose there were 80 gm of Carbon-14 in an organism when it lived 300,000 years ago. How much Carbon-14 remains today? (Use the formula $y = y_0 e^{kt}$ where $k = -0.00012$.)

6

Trigonometry: An Introduction

In this chapter we begin our study of the trigonometric functions by first considering angles in a right triangle, the way trigonometry arose historically. But the real power in applying trigonometric functions comes in those situations where the same pattern of values repeats itself over and over, in a consistent and predictable fashion. For example, one might think that there are no times of the day more popular than any other times when it comes to giving birth. Looking at birth-time data from one perspective appears to support this position. But, as we will show at the end of this chapter, a careful look over a twenty-four hour period presents a rather surprising (trigonometric) picture.

6.1 Triangles

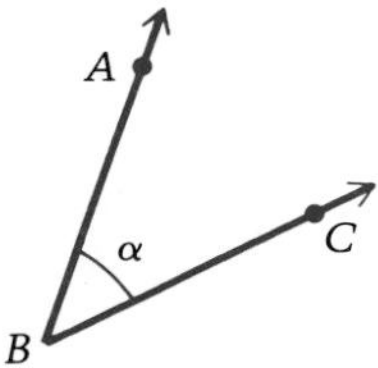

Figure 6.1

Trigonometry is a branch of mathematics that developed as a tool for solving problems involving triangles. In fact, the word "trigonometry" is derived from the Greek words for "measurement of triangles." It is appropriate, therefore, to begin our study of trigonometry by reviewing some basic information about triangles.

An **angle** is a figure consisting of two rays having the same endpoint. This common endpoint is called the **vertex.** For the angle in Figure 6.1, B is the vertex. This angle can be described in three ways: $\angle ABC$, $\angle B$, and α. The

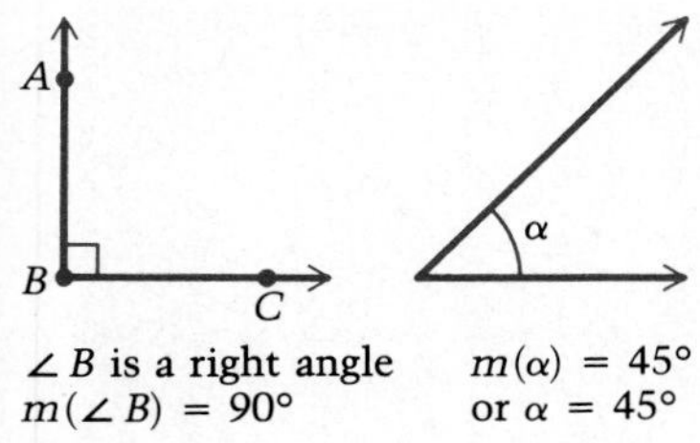

$\angle B$ is a right angle
$m(\angle B) = 90°$

$m(\alpha) = 45°$
or $\alpha = 45°$

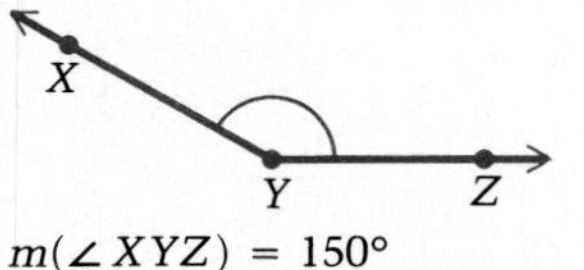

$m(\angle XYZ) = 150°$

Figure 6.2

symbol α is the Greek letter *alpha*. Other Greek letters often used in describing angles are θ (*theta*), φ (*phi*), and β (*beta*).

A common unit for measuring angles is the degree (°). Angles of various measures are shown in Figure 6.2. The symbols $m(\angle B)$ and $m(\alpha)$ are used to indicate the measures of angles B and α, respectively. However, the "m" is often omitted. Thus, you may see $m(\theta) = 45°$ or simply $\theta = 45°$.

An angle with measure 90° is called a **right angle.** The special symbol □, signifies a right angle when placed at a vertex as in Figure 6.2. An angle with measure between 0° and 90° is called **acute,** and an angle with measure between 90° and 180° is called **obtuse.**

Recall that a triangle has 3 angles and 3 sides. The triangle below is referred to as $\triangle ABC$. Points A, B, and C are called **vertices** of the triangle.

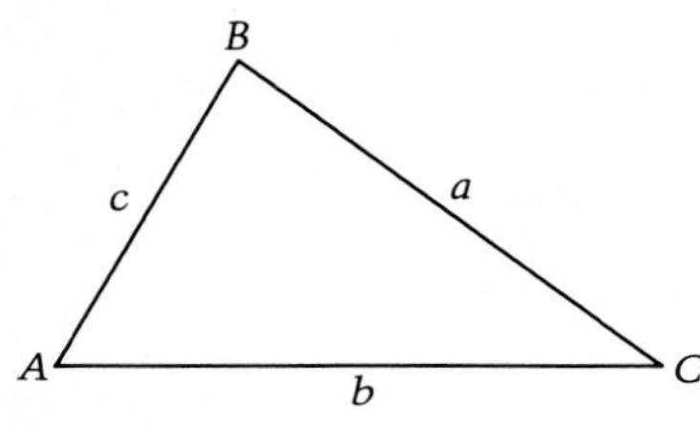

Figure 6.3

When we label a triangle, it is common to let a stand for the side opposite vertex A, b for the side opposite vertex B, and c for the side opposite vertex C.

Theorem
In any triangle, the sum of the measures of the angles is 180°.

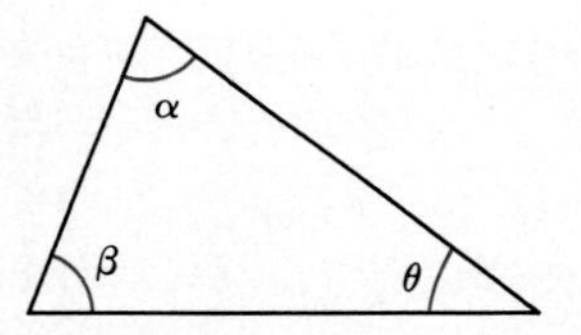

Figure 6.4 $\alpha + \beta + \theta = 180°$
$m(\alpha) + m(\beta) + m(\theta) = 180°$.

Certain types of triangles are of special importance to the development of trigonometry. An **isosceles** triangle is a triangle with two sides of equal length. In an isosceles triangle, the angles opposite the sides of equal length have the same measure.

EXAMPLE 1

In $\triangle ABC$, $a = 5$, $b = 5$, and $m(\angle A) = 65°$. Find $m(\angle C)$.

Solution

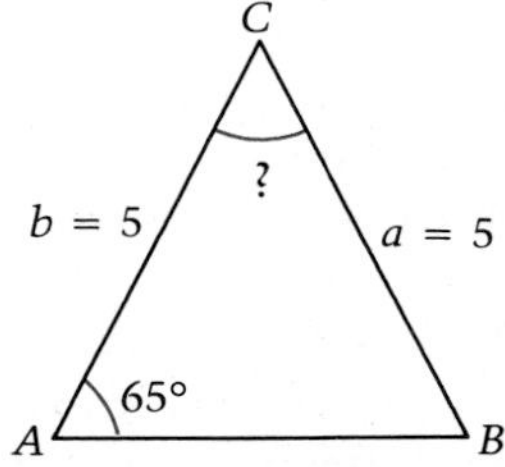

Figure 6.5

Begin by drawing a triangle and labeling its known parts.

Since $a = b$, the triangle is isosceles. Thus $m(\angle B) = m(\angle A) = 65°$.

$$\begin{aligned} m(\angle A) + m(\angle B) + m(\angle C) &= 180° \\ 65° + 65° + m(\angle C) &= 180° \\ m(\angle C) &= 180° - 65° - 65° \\ m(\angle C) &= 50° \end{aligned}$$

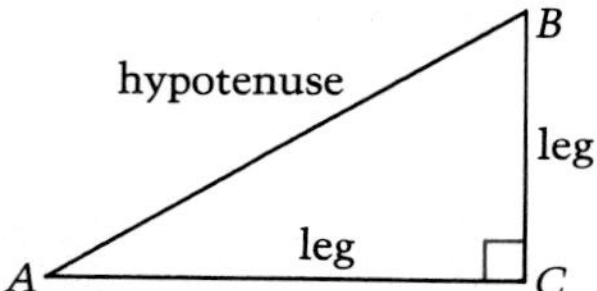

Figure 6.6 A right triangle.

A **right triangle** is a triangle in which one of the angles is a right angle. In a right triangle, it is common to label the right angle C. The longest side of a right triangle is opposite the right angle and is called the **hypotenuse.** The other two sides are called **legs.**

EXAMPLE 2

$\triangle ABC$ is an isosceles right triangle with $m(\angle C) = 90°$. Find $m(\angle A)$.

Solution

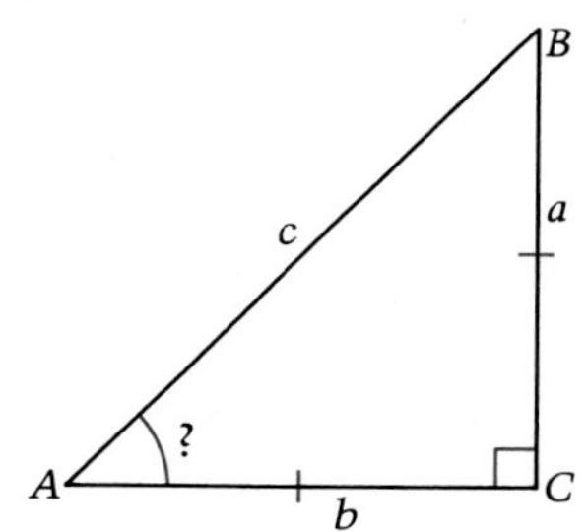

Figure 6.7

Begin by drawing a triangle and labeling its known parts.

Sides (or angles) having equal measure are denoted by the same hash marks. The hash marks in this figure indicate that $a = b$.

$m(\angle A) + m(\angle B) + m(\angle C) = 180°$	True for any triangle.
$m(\angle A) + m(\angle B) + 90° = 180°$	C is a right angle.
$m(\angle A) + m(\angle A) = 90°$	$m(\angle B) = m(\angle A)$ since $\triangle ABC$ is isosceles.
$m(\angle A) = 45°$	

One useful fact about a right triangle is that the square of the hypotenuse is equal to the sum of the squares of the legs. This fact is known as the Pythagorean Theorem.

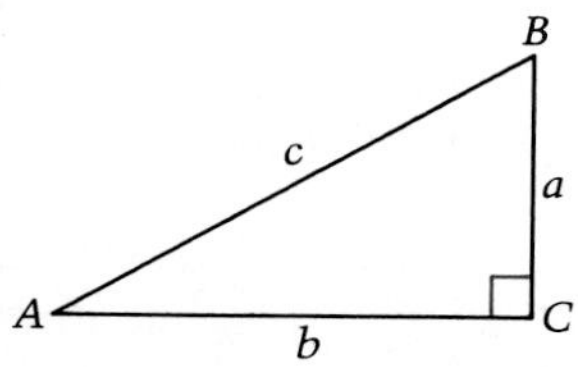

Figure 6.8 $c^2 = a^2 + b^2$

Pythagorean Theorem

If ABC is a right triangle, with a and b the lengths of the legs and c the length of the hypotenuse, then $c^2 = a^2 + b^2$.

A triangle whose three sides have the same length is called an **equilateral triangle.** In an equilateral triangle, each of the angles measures 60°. You should convince yourself that the following statement is true: every equilateral triangle is also isosceles, but not every isosceles triangle is equilateral.

EXAMPLE 3

Given an equilateral triangle with side of length 1, find the length of an altitude.

Solution

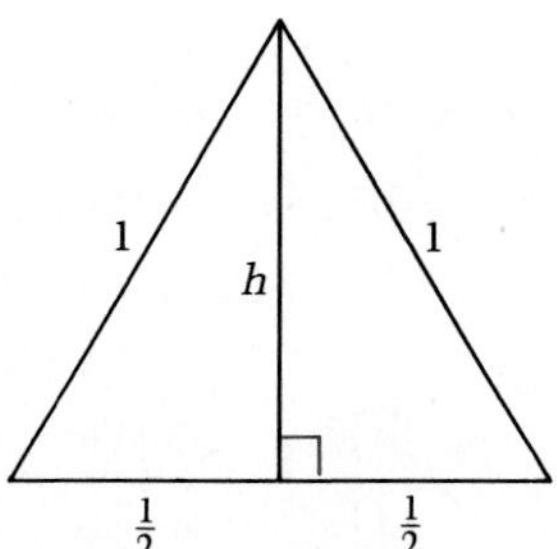

Figure 6.9

Begin by drawing an equilateral triangle with altitude h. The **altitude** is the perpendicular from a vertex to the opposite side. In an equilateral triangle, the altitude cuts the opposite side into two equal lengths.

$$1^2 = h^2 + \left(\frac{1}{2}\right)^2$$

Apply the Pythagorean Theorem to either of the two smaller triangles. They are congruent.

$$1 = h^2 + \frac{1}{4}$$

$$h^2 = \frac{3}{4}$$

The solutions of the equation $h^2 = \frac{3}{4}$ are $\pm \frac{\sqrt{3}}{2}$.

$$h = \sqrt{\frac{3}{4}} = \frac{\sqrt{3}}{2}$$

Since length is nonnegative, the value $-\frac{\sqrt{3}}{2}$ has been discarded.

The altitude is $\frac{\sqrt{3}}{2}$.

We call triangles **similar** if they have the same shape, even though they may have different sizes. This means that one of the triangles is a "magnification" or a "reduction" of the other.

In Figure 6.10, triangle ABC and triangle XYZ are similar. This is written $\triangle ABC \sim \triangle XYZ$. When we say that $\triangle ABC$ is similar to $\triangle XYZ$, we imply a special correspondence between the two triangles such that the measures of

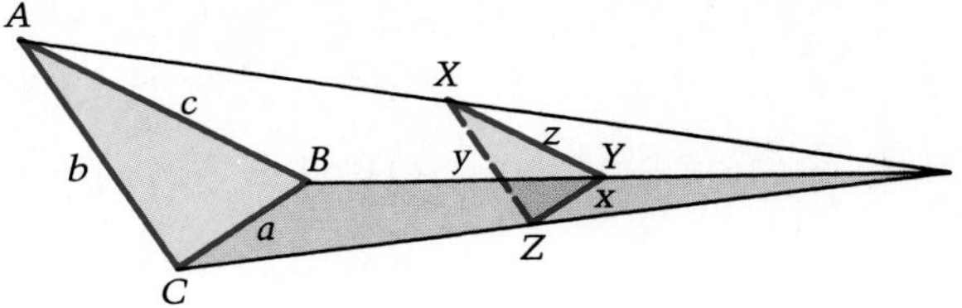

Figure 6.10

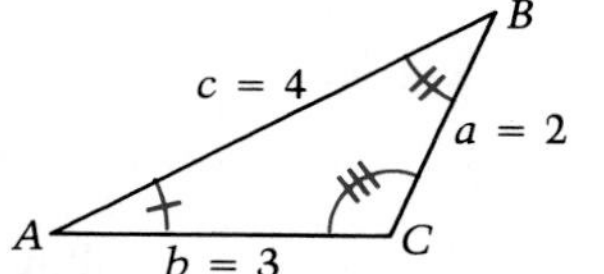

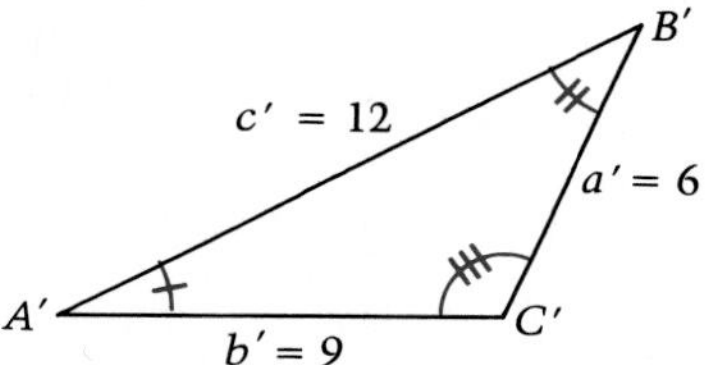

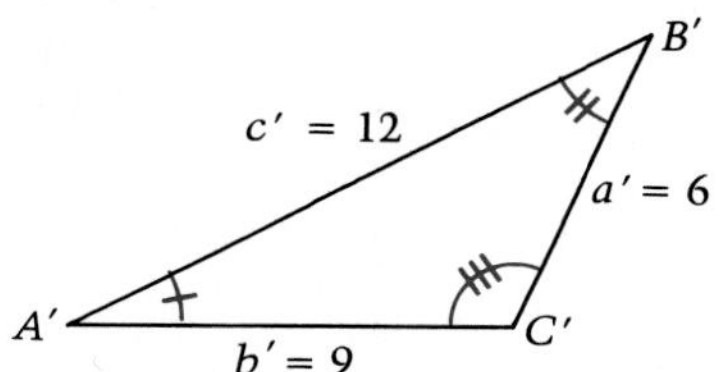

Figure 6.11

the corresponding angles are equal and the corresponding sides of the triangles are proportional. Thus,

$$m(\angle A) = m(\angle X),\ m(\angle B) = m(\angle Y),\ \text{and } m(\angle C) = m(\angle Z),$$

and

$$\frac{a}{x} = \frac{b}{y} = \frac{c}{z}.$$

In Figure 6.11, $\triangle A'B'C' \sim \triangle ABC$. Notice that

$$\frac{a}{a'} = \frac{2}{6} = \frac{1}{3} \qquad \frac{b}{b'} = \frac{3}{9} = \frac{1}{3} \qquad \frac{c}{c'} = \frac{4}{12} = \frac{1}{3}.$$

Thus, the ratio of the corresponding sides is $\frac{1}{3}$.

You may recall from geometry that to prove two triangles are similar, you must find two pairs of corresponding angles having the same measure.

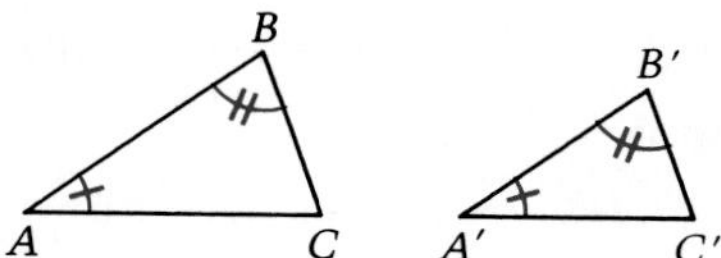

Figure 6.12 $\triangle ABC \sim \triangle A'B'C'$

Angle-Angle (A-A) Similarity Theorem

Two triangles are similar if two angles of one triangle have the same measure as two angles of the other triangle.

EXAMPLE 4

In the triangles below, $m(\angle A) = m(\angle X)$, $m(\angle B) = m(\angle Y)$, and the lengths of some sides are given. Find c.

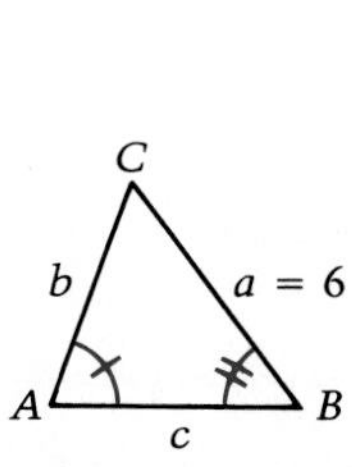

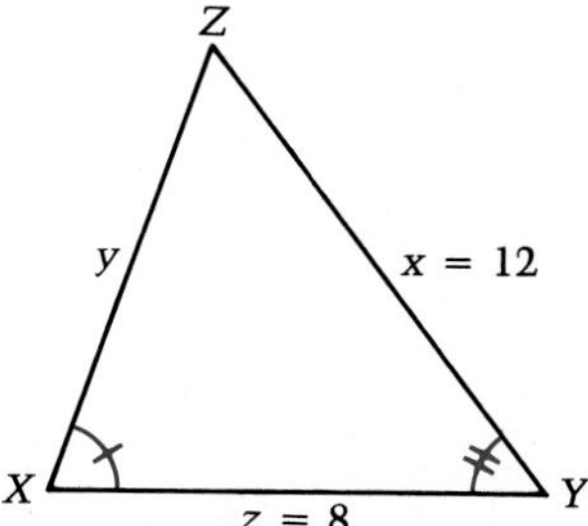

Figure 6.13

Solution

$\triangle ABC \sim \triangle XYZ$ The triangles are similar by the A-A Similarity Theorem.

$\frac{6}{12} = \frac{c}{8}$ By similarity, $\frac{a}{x} = \frac{c}{z}$.

$12c = 6 \cdot 8$

$c = 4$

EXAMPLE 5

In the right triangles below $m(\angle A) = m(\angle A')$, $a = 3$, $b = 4$, and $a' = 9$. Find c'.

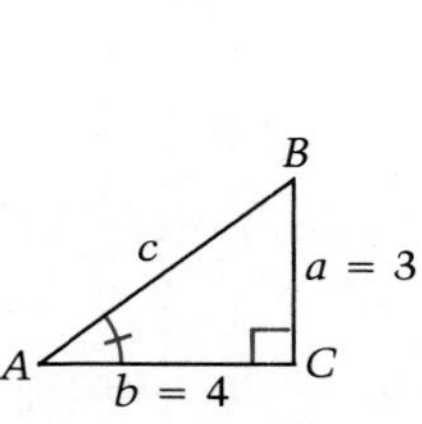

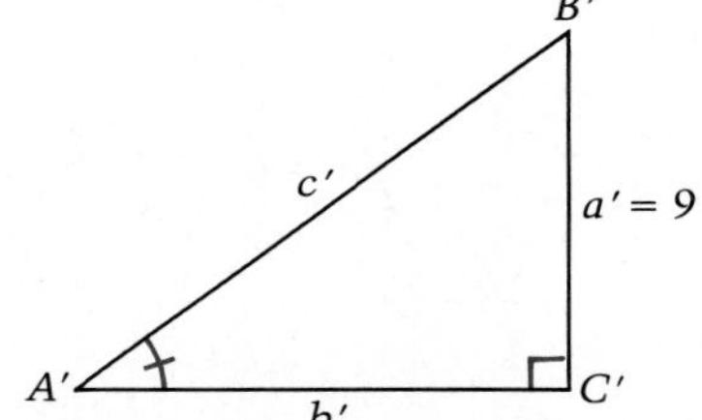

Figure 6.14

Solution

$\triangle ABC \sim \triangle A'B'C'$ The A-A Similarity Theorem is used.

$\frac{a}{a'} = \frac{b}{b'} = \frac{c}{c'}$ This proportion follows from similarity. To find c', we must first find c.

$c^2 = a^2 + b^2$

$c = \sqrt{a^2 + b^2} = \sqrt{3^2 + 4^2}$

$c = 5$ Now that c has been determined, we use the proportion to find c'.

$\frac{a}{a'} = \frac{c}{c'}$

$\frac{3}{9} = \frac{5}{c'}$

$c' = 15$

EXAMPLE 6

A building casts a shadow 84 ft long. At the same instant a nearby fence casts a shadow 7 ft long. If the fence is 10 ft high, what is the height of the building?

Solution

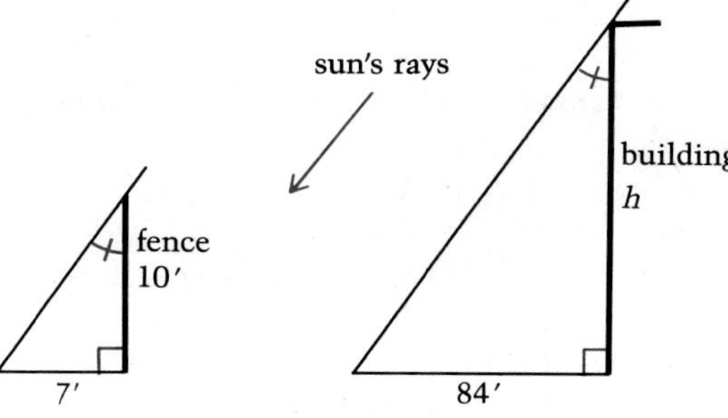

Draw a figure. If we assume that the building and fence are perpendicular to the ground, and the sun's rays are parallel, then the two triangles are similar.

Figure 6.15

$$\frac{h}{10} = \frac{84}{7}$$ Set up a proportion using corresponding sides of the two triangles.

$$7h = 10 \cdot 84$$

$$h = 120$$

The height of the building is 120 ft.

EXERCISE SET 6.1

In Exercises 1–6, use the given information about $\triangle ABC$ to solve for the missing part.

1. $m(\angle A) = 40°, m(\angle B) = 73°, m(\angle C) = ?$

2. $m(\angle A) = 58°, m(\angle C) = 16°, m(\angle B) = ?$

3. $a = b, m(\angle C) = 30°, m(\angle B) = ?$

4. $a = b, m(\angle C) = 82°, m(\angle B) = ?$

5. $m(\angle A) = m(\angle B), a = 18, b = ?$

6. $m(\angle C) = 90°, a = b, m(\angle A) = ?$

In Exercises 7–18, $\triangle ABC$ is a right triangle, c is the length of the hypotenuse, and a and b are the lengths of the legs. Find the length of the third side given the lengths of two sides.

7. $a = 6, b = 8$

8. $a = 12, c = 15$

9. $b = 15, c = 39$

10. $a = 15, c = 17$

11. $b = 5, c = 13$

12. $a = 12, c = 20$

13. $a = 2, b = 3$

14. $a = 5, b = 5$

15. $b = 1, c = 1\frac{1}{4}$

16. $a = \frac{6}{5}, b = \frac{8}{5}$

17. $a = 1.4, b = 4.8$

18. $a = 7.5, b = 4.0$

In Exercises 19–22, $\triangle ABC$ and $\triangle A'B'C'$ are two triangles such that $m(\angle A) = m(\angle A')$ and $m(\angle B) = m(\angle B')$. Given the lengths of some sides, find the length of the indicated side.

19. $a = 10, b = 8, a' = 8, b' = ?$

20. $a = 12, b = 9, a' = ?, b' = 6$

21. $a = 18, b = ?, a' = 9, b' = 5$

22. $a = ?, b = 27, a' = 6, b' = 12$

In Exercises 23–28, $\triangle ABC$ is a right triangle and PQ is perpendicular to AC (see the figure below). Find the length of the indicated side.

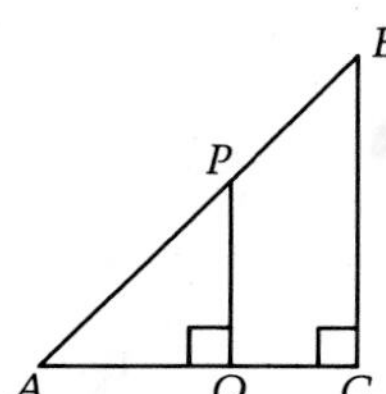

23. $AP = 8, AQ = 6, AC = 18, AB = ?$

24. $AQ = 10, PQ = 8, BC = 12, AC = ?$

25. $AP = 6, PQ = 4, AC = 18, BC = ?$

26. $AQ = 10, PQ = 5, AB = 30, BC = ?$

27. $\frac{PQ}{AQ} = \frac{1}{3}, AB = 18, BC = ?$

28. $\frac{PQ}{AP} = \frac{\sqrt{3}}{3}, AC = 10\sqrt{3}, BC = ?$

29. A 6 ft pole perpendicular to the ground casts a shadow 8 ft long at the same time that a telephone pole casts a shadow 56 ft long. What is the height of the telephone pole?

30. The top of a church spire casts a shadow 200 ft long. At the same time a nearby 8 ft wall casts a shadow 25 ft long. How high is the top of the spire?

Superset

31. A boat travels 8 mi due east and then 15 mi due north. How far is the boat from its starting point?

32. A 40 ft ladder is placed against a wall, with its foot 24 ft from the base of the wall. At what height does the ladder touch the wall?

33. A television set has a square picture screen with a 19 in. diagonal. What are the dimensions of the screen? What is its area?

In Exercises 34–41, $\triangle ABC$ is an equilateral triangle, BD is the altitude to side AC and the length of AB is 5. Find the following.

34. $m(\angle ADB)$
35. $m(\angle ABD)$
36. $m(\angle DBC)$
37. BD
38. AD
39. DC
40. area of $\triangle ABC$
41. area of $\triangle ABD$

42. Find the length of a side of a square having the same area as a rectangle with base of length 12 and diagonal of length 15.

43. Find the area of a right triangle with one leg of length 16 and hypotenuse of length 34.

44. Find the length of a diagonal of a rectangle with sides of lengths 32 and 24.

45. Find the length of a diagonal of a square with side of length 12.

46. Find the length of a side of an equilateral triangle with altitude of length h.

6.2 Trigonometric Ratios of Acute Angles

Figure 6.16 shows two right triangles ABC and $A'B'C'$ with

$$\alpha = m(\angle A) = m(\angle A').$$

Since each triangle has a right angle and each triangle has an acute angle with the same measure, the triangles are similar by the A-A Similarity Theorem. Because of the similarity we can write

$$\frac{a}{a'} = \frac{b}{b'}.$$

Multiplying each side of the above equation by $\frac{a'}{b}$ produces

$$\frac{a}{b} = \frac{a'}{b'}.$$

This last equation tells us that a ratio of sides of $\triangle ABC$ is equal to the ratio of corresponding sides in $\triangle A'B'C'$. These ratios are equal because the equality of $m(\angle A)$ and $m(\angle A')$ implies that the two right triangles are similar. Moreover, it is the value of α that determines the value of this ratio.

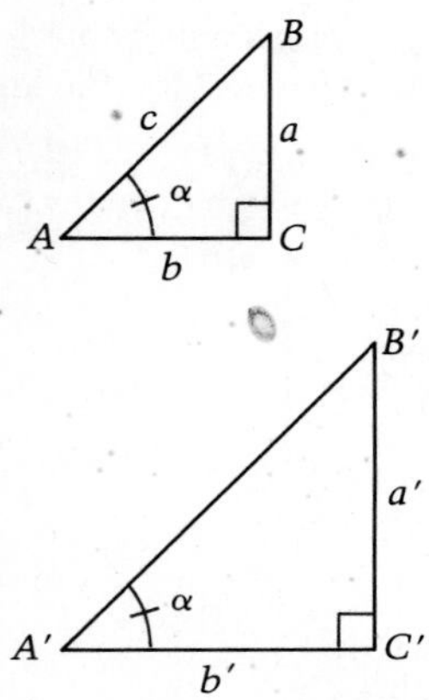

Figure 6.16

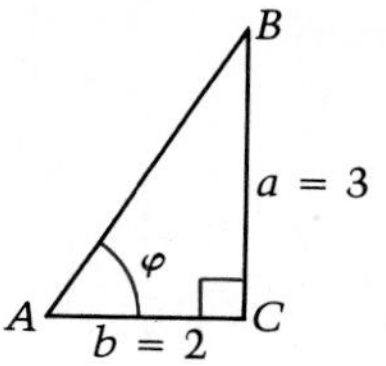

Figure 6.17

For example, suppose that in the right triangle at the left, $\angle A$ has measure φ. Notice that the ratio $\frac{a}{b}$ is $\frac{3}{2}$ in this triangle. Now consider the right triangles in Figure 6.18. Because the corresponding acute angle has measure φ, the corresponding ratio in each of these triangles is also $\frac{3}{2}$.

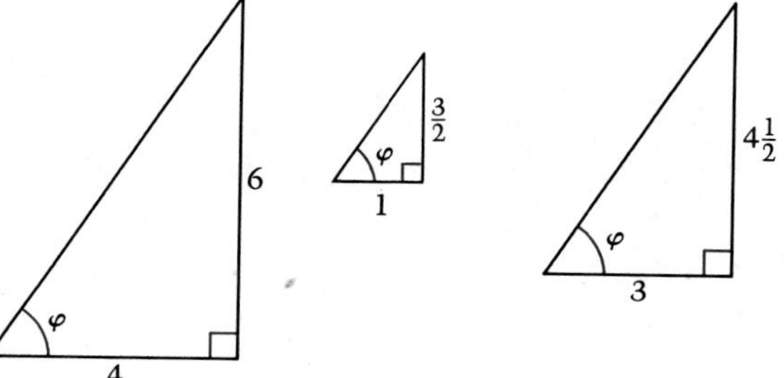

Figure 6.18

As we see in Figure 6.18, a particular acute angle (here φ) always produces the same ratio of opposite to adjacent sides (in this case the ratio is $\frac{3}{2}$). Therefore, we can think of the angle as a domain value for a function having the ratio as the function value.

Let us adopt a way of describing the sides of a right triangle in terms of an acute angle θ.

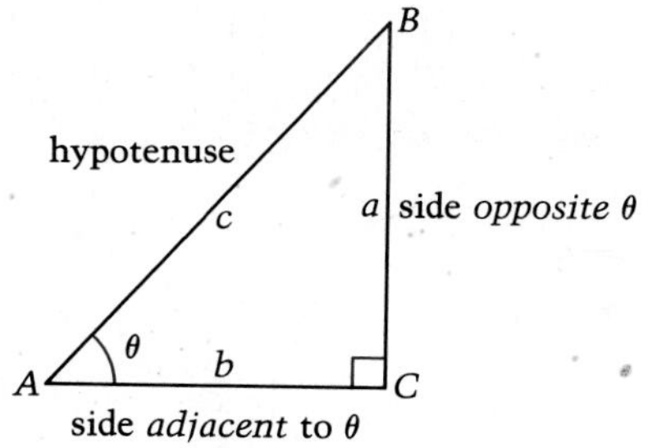

Figure 6.19

We can think of the ratio $\frac{a}{b}$ as $\frac{\text{length of side opposite } \theta}{\text{length of side adjacent to } \theta}$.

The opposite/adjacent ratio is called the **tangent** of the angle. For an acute angle θ we write

$$\text{tangent}(\theta) = \frac{\text{length of side opposite } \theta}{\text{length of side adjacent to } \theta}.$$

For any right triangle ABC, there are six possible ratios of sides that can be formed. Each ratio can be defined as a function of an acute angle θ.

Figure 6.20

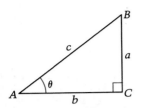

$$\sin\theta = \frac{a}{c} \qquad \csc\theta = \frac{c}{a}$$

$$\cos\theta = \frac{b}{c} \qquad \sec\theta = \frac{c}{b}$$

$$\tan\theta = \frac{a}{b} \qquad \cot\theta = \frac{b}{a}$$

Definition of the Trigonometric Functions of an Acute Angle

Let θ be an acute angle of a right triangle. The trigonometric functions of θ are defined by the following ratios:

$$\text{sine } \theta = \frac{\text{length of side } \textbf{opposite } \theta}{\text{length of } \textbf{hypotenuse}}, \quad \text{denoted } \sin\theta$$

$$\text{cosine } \theta = \frac{\text{length of side } \textbf{adjacent } \text{to } \theta}{\text{length of } \textbf{hypotenuse}}, \quad \text{denoted } \cos\theta$$

$$\text{tangent } \theta = \frac{\text{length of side } \textbf{opposite } \theta}{\text{length of side } \textbf{adjacent } \text{to } \theta}, \quad \text{denoted } \tan\theta$$

$$\text{cotangent } \theta = \frac{\text{length of side } \textbf{adjacent } \text{to } \theta}{\text{length of side } \textbf{opposite } \theta}, \quad \text{denoted } \cot\theta$$

$$\text{secant } \theta = \frac{\text{length of } \textbf{hypotenuse}}{\text{length of side } \textbf{adjacent } \text{to } \theta}, \quad \text{denoted } \sec\theta$$

$$\text{cosecant } \theta = \frac{\text{length of } \textbf{hypotenuse}}{\text{length of side } \textbf{opposite } \theta}, \quad \text{denoted } \csc\theta$$

EXAMPLE 1

Find the six trigonometric ratios of α in the right triangle in Figure 6.21.

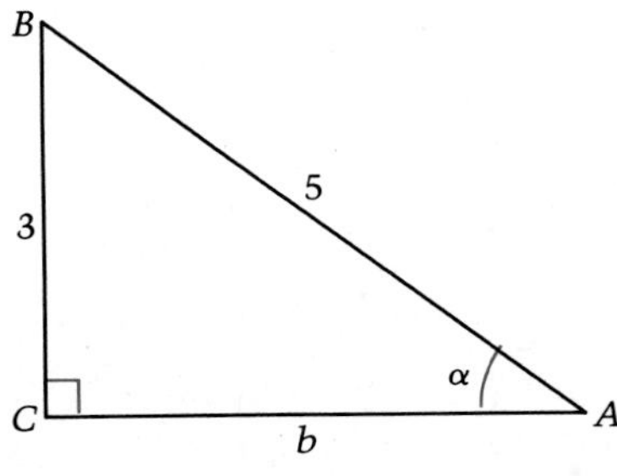

Figure 6.21

Solution

$$5^2 = 3^2 + b^2$$
$$b^2 = 25 - 9 = 16$$
$$b = 4$$

To determine the six trigonometric ratios of α, we need to know the lengths of the three sides of the triangle. The Pythagorean Theorem is used to find b.

$$\sin\alpha = \frac{\text{opposite}}{\text{hypotenuse}} = \frac{3}{5} \qquad \cos\alpha = \frac{\text{adjacent}}{\text{hypotenuse}} = \frac{4}{5} \qquad \tan\alpha = \frac{\text{opposite}}{\text{adjacent}} = \frac{3}{4}$$

$$\csc\alpha = \frac{\text{hypotenuse}}{\text{opposite}} = \frac{5}{3} \qquad \sec\alpha = \frac{\text{hypotenuse}}{\text{adjacent}} = \frac{5}{4} \qquad \cot\alpha = \frac{\text{adjacent}}{\text{opposite}} = \frac{4}{3}$$

Note that in Example 1, the function values of $\sin\alpha$ and $\csc\alpha$ are reciprocals of one another. This is also the case for the pair $\cos\alpha$ and $\sec\alpha$, and for the pair $\tan\alpha$ and $\cot\alpha$. Note also that we do not know the measure of α, even though we do know its six trigonometric function values.

EXAMPLE 2

Angles α and β are the two acute angles in the right triangle shown in Figure 6.22. Show that $\sin\alpha = \cos\beta$.

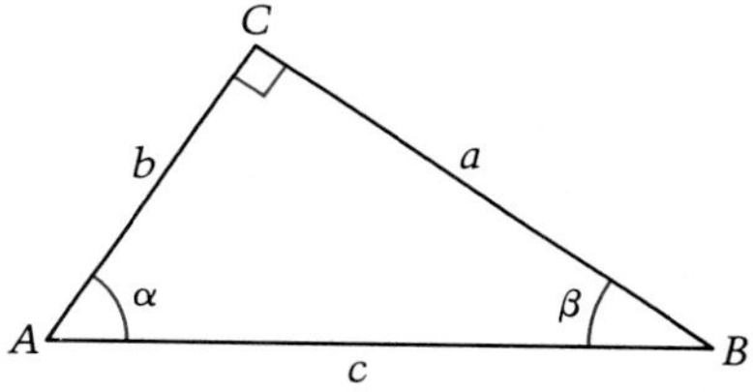

Figure 6.22

Solution

$$\sin\alpha = \frac{\text{side opposite } \alpha}{\text{hypotenuse}} = \frac{a}{c} \qquad \cos\beta = \frac{\text{side adjacent to } \beta}{\text{hypotenuse}} = \frac{a}{c}$$

Since $\sin\alpha$ and $\cos\beta$ each equal $\frac{a}{c}$, $\sin\alpha = \cos\beta$.

To determine the trigonometric ratios of an angle whose measure is 45°, consider the square $PQRS$ with sides of length 1. The diagonal of the square bisects the right angle P and forms the hypotenuse of an isosceles right triangle. Its length is $\sqrt{2}$ (by the Pythagorean Theorem). We can use $\triangle PRS$ to determine the six trigonometric ratios of 45°.

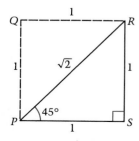

Figure 6.23

$$\sin 45^\circ = \frac{1}{\sqrt{2}} = \frac{\sqrt{2}}{2} \qquad \cos 45^\circ = \frac{1}{\sqrt{2}} = \frac{\sqrt{2}}{2} \qquad \tan 45^\circ = \frac{1}{1} = 1$$

$$\csc 45^\circ = \frac{\sqrt{2}}{1} = \sqrt{2} \qquad \sec 45^\circ = \frac{\sqrt{2}}{1} = \sqrt{2} \qquad \cot 45^\circ = \frac{1}{1} = 1$$

The trigonometric ratios of 30° and 60° can be found by using an equilateral triangle with sides of length 1. Recall that in an equilateral triangle the altitude from a vertex to the opposite side bisects both the angle and the side. Thus, we can use $\triangle ABC$ in Figure 6.24 to determine the six trigonometric ratios of 30° and 60°.

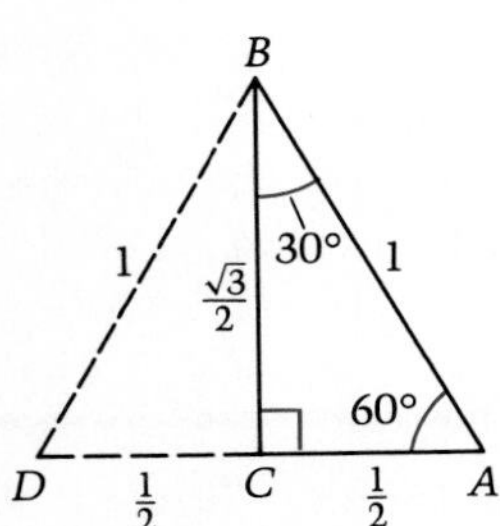

Figure 6.24

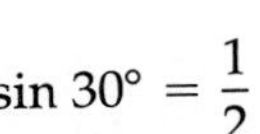

$$\sin 30° = \frac{1}{2} \qquad \cos 30° = \frac{\sqrt{3}}{2} \qquad \tan 30° = \frac{1}{\sqrt{3}} = \frac{\sqrt{3}}{3}$$

$$\csc 30° = 2 \qquad \sec 30° = \frac{2}{\sqrt{3}} = \frac{2\sqrt{3}}{3} \qquad \cot 30° = \sqrt{3}$$

$$\sin 60° = \frac{\sqrt{3}}{2} \qquad \cos 60° = \frac{1}{2} \qquad \tan 60° = \sqrt{3}$$

$$\csc 60° = \frac{2}{\sqrt{3}} = \frac{2\sqrt{3}}{3} \qquad \sec 60° = 2 \qquad \cot 60° = \frac{1}{\sqrt{3}} = \frac{\sqrt{3}}{3}$$

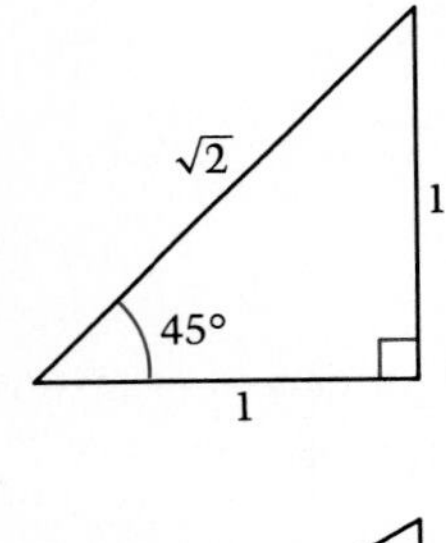

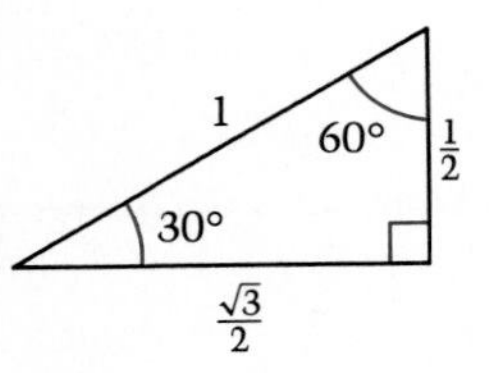

Figure 6.25

Trigonometric Function Values for Special Angles

$f(\theta)$ \ θ	**30°**	**45°**	**60°**
sin θ	$\frac{1}{2}$	$\frac{\sqrt{2}}{2}$	$\frac{\sqrt{3}}{2}$
cos θ	$\frac{\sqrt{3}}{2}$	$\frac{\sqrt{2}}{2}$	$\frac{1}{2}$
tan θ	$\frac{\sqrt{3}}{3}$	1	$\sqrt{3}$
cot θ	$\sqrt{3}$	1	$\frac{\sqrt{3}}{3}$
sec θ	$\frac{2\sqrt{3}}{3}$	$\sqrt{2}$	2
csc θ	2	$\sqrt{2}$	$\frac{2\sqrt{3}}{3}$

EXAMPLE 3

Given a right triangle FED with $m(\angle F) = 30°$ and hypotenuse of length 40, find the length of side f.

Solution

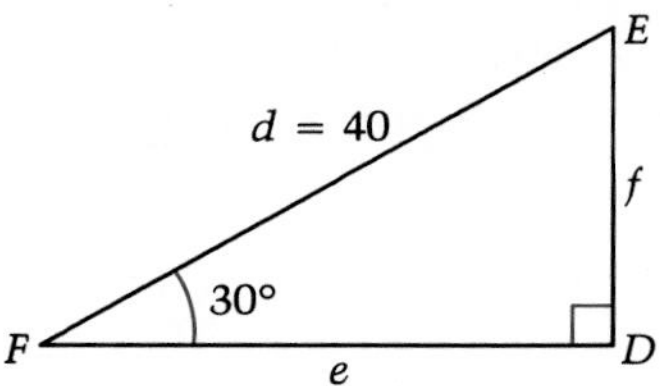

Figure 6.26

Begin by drawing a right triangle and labeling its parts.

To find f, we choose a trigonometric ratio involving f and some known side.

$$\sin 30° = \frac{f}{d} = \frac{1}{2}$$

$\sin\theta = \dfrac{\text{side opposite }\theta}{\text{hypotenuse}}$ and $\sin 30° = \dfrac{1}{2}$.

$$\frac{f}{40} = \frac{1}{2}$$

The known value of d is substituted.

$$f = 20$$

For an acute angle, if one trigonometric ratio is known, we can draw a right triangle and use it to find the other five ratios.

EXAMPLE 4

Given that $\sin\beta = \dfrac{5}{7}$ and β is an acute angle, find $\cos\beta$.

Solution

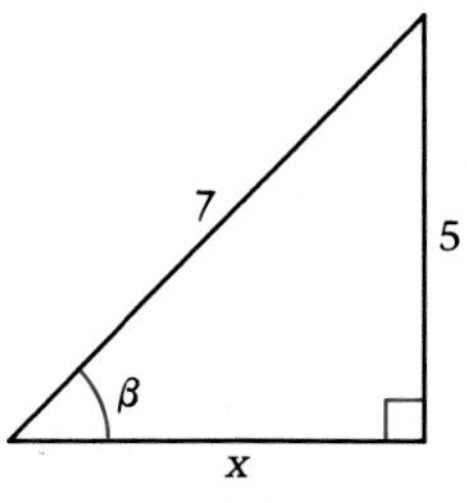

Figure 6.27

Since $\sin\beta$ is the ratio of the side opposite β to the hypotenuse, we begin by drawing a right triangle in which this ratio is $\frac{5}{7}$. The side opposite β is labeled 5 and the hypotenuse is labeled 7.

The side adjacent to β is labeled x and will be needed in determining $\cos\beta$.

$$7^2 = x^2 + 5^2$$

The Pythagorean Theorem is used.

$$x^2 = 7^2 - 5^2 = 24$$

$$x = \sqrt{24} = 2\sqrt{6}$$

Remember to choose the positive value only.

$$\cos\beta = \frac{2\sqrt{6}}{7}$$

$\cos\beta = \dfrac{\text{side adjacent to }\beta}{\text{hypotenuse}}$.

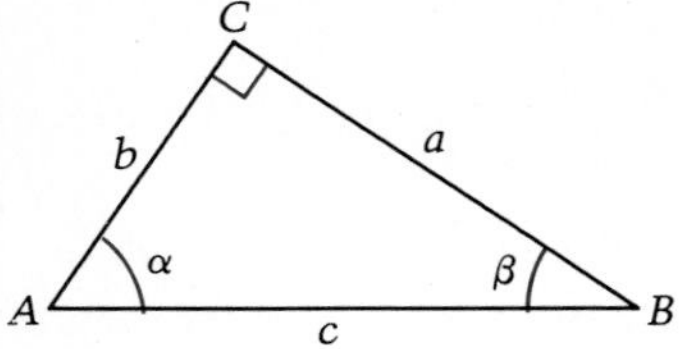

Figure 6.28 Angles α and β are complementary.

Two angles are **complementary** if the sum of their measures is 90°. In any right triangle, the acute angles are complementary. This is illustrated in Figure 6.28, which shows that $\alpha + \beta = 90°$. In Example 2, we discovered that $\sin \alpha = \cos \beta$, that is, sine of α = cosine of the complement of α:

$$\sin \alpha = \cos(90° - \alpha).$$

EXAMPLE 5

If α and β are complementary, and $\cos \alpha = \dfrac{\sqrt{11}}{6}$, find $\cos \beta$ and $\tan \beta$.

Solution

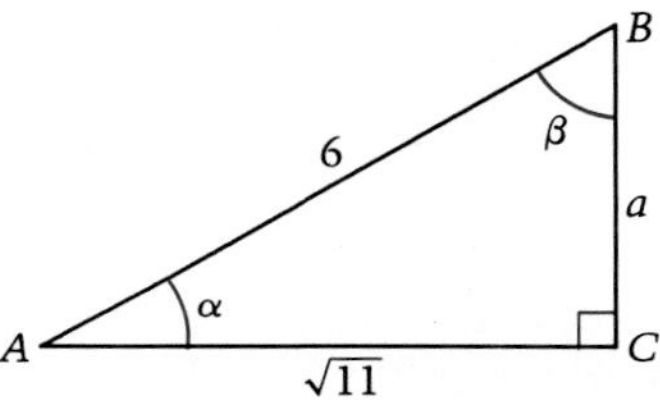

Figure 6.29

Since α and β are complementary, we can draw a right triangle having α and β as acute angles.

$\cos \alpha = \dfrac{\text{side adjacent to } \alpha}{\text{hypotenuse}} = \dfrac{\sqrt{11}}{6}$, so side AC is labeled $\sqrt{11}$ and AB is labeled 6.

Pythagorean Theorem: $6^2 = (\sqrt{11})^2 + a^2$ implies $a^2 = 36 - 11 = 25$. Thus, $a = 5$.

$$\cos \beta = \frac{\text{side adjacent to } \beta}{\text{hypotenuse}} = \frac{5}{6}$$

EXERCISE SET 6.2

In Exercises 1–16, use the given information about the right triangle ABC to find the six trigonometric ratios of α.

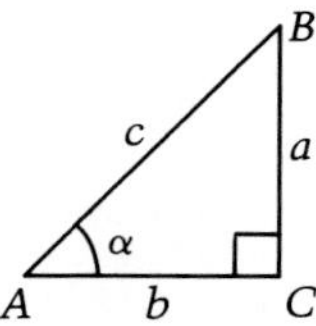

1. $a = 6, b = 8$
2. $a = 7, b = 24$
3. $a = 8, c = 17$
4. $b = 10, c = 26$
5. $c = 37, b = 12$
6. $a = 48, c = 50$
7. $a = 21, c = 35$
8. $c = 41, b = 9$
9. $b = 1, c = \sqrt{2}$
10. $a = 2, b = 2$
11. $a = 1, c = 3$
12. $c = 2, a = \sqrt{3}$
13. $b = 2, a = 2\sqrt{3}$
14. $c = 4, b = 1$
15. $a = 1, b = \sqrt{2}$
16. $b = \sqrt{5}, a = 1$

For Exercises 17–26, refer back to the right triangle used in Exercises 1–16. In each exercise, given the lengths of two sides, use the Pythagorean Theorem to find the length of the third side, then find the values of the six trigonometric ratios of α.

17. $a = 42.72 \quad b = 31.68$
18. $a = 34.35 \quad b = 63.45$
19. $a = 293.4 \quad c = 568.3$
20. $b = 349.6 \quad c = 638.2$

21. $a = 0.7184$ $b = 0.6319$
22. $a = 1.231$ $b = 1.893$
23. $b = 2462$ $c = 4091$
24. $a = 5000$ $c = 5663$
25. $a = 12.427$ $c = 22.645$
26. $b = 79.351$ $c = 83.381$

In Exercises 27–30, α and β are acute angles in right triangle ABC. Verify the given statement.

27. $\tan \alpha = \cot \beta$
28. $\csc \alpha = \sec \beta$
29. $\sin \beta = \cos \alpha$
30. $\tan \beta = \cot \alpha$

In Exercises 31–40, given an equilateral triangle $\triangle ABC$ with side of length s and altitude of length h, find the indicated part.

31. If $s = 6$, find h
32. If $s = 12$, find h
33. If $h = 6$, find s
34. If $h = 15$, find s
35. If $h = 5\sqrt{3}$, find the perimeter of $\triangle ABC$
36. If $h = 9$, find the area of $\triangle ABC$
37. If $s = 6.348$, find h.
38. If $h = 15.259$, find s.
39. If $h = 25.9808$, find the perimeter of $\triangle ABC$.
40. If $h = 9.4873$, find the area of $\triangle ABC$.

In Exercises 41–50, $\triangle ABC$ has right angle at C and sides of length, a, b, and c. Given the following information, find the indicated part.

41. $m(\angle A) = 30°$, $b = 21$, find c
42. $m(\angle B) = 30°$, $c = 58$, find a
43. $m(\angle B) = 45°$, $b = 24$, find c
44. $m(\angle A) = 45°$, $c = 50$, find b
45. $m(\angle B) = 60°$, $b = 15\sqrt{3}$, find a
46. $m(\angle B) = 60°$, $a = 24$, find b
47. $a = 12$, $b = 12$, find $m(\angle A)$
48. $b = 30$, $c = 60$, find $m(\angle A)$
49. $b = 24$, $a = 24\sqrt{3}$, find $m(\angle A)$
50. $a = 10\sqrt{3}$, $c = 20$, find $m(\angle A)$

In Exercises 51–60, α and β are complementary. Given the following information, find the indicated trigonometric ratios.

51. $\cos \alpha = \frac{3}{5}$, find $\sin \alpha$ and $\tan \beta$
52. $\sin \alpha = \frac{5}{13}$, find $\cos \alpha$ and $\cos \beta$
53. $\csc \alpha = \frac{17}{15}$, find $\tan \alpha$ and $\cos \beta$
54. $\sec \alpha = \frac{29}{21}$, find $\tan \beta$ and $\tan \alpha$
55. $\tan \alpha = \frac{3}{4}$, find $\sin \beta$ and $\cos \alpha$
56. $\tan \alpha = \frac{4}{3}$, find $\sin \alpha$ and $\sec \beta$
57. $\sin \alpha = \frac{7}{25}$, find $\cos \alpha$ and $\tan \alpha$
58. $\cos \alpha = \frac{35}{37}$, find $\tan \alpha$ and $\sin \beta$
59. $\cos \alpha = 0.9$, find $\cos \beta$ and $\sin \beta$
60. $\sin \alpha = 0.3$, find $\cos \beta$ and $\tan \alpha$

In Exercises 61–66, α and β are complementary. From the given information, find the indicated trigonometric ratios.

61. $\sin \alpha = 0.3846$, find $\cos \alpha$ and $\cos \beta$.
62. $\cos \alpha = 0.4286$, find $\sin \alpha$ and $\tan \beta$.
63. $\csc \alpha = 1.1333$, find $\tan \alpha$ and $\cos \beta$.
64. $\sec \alpha = 1.3810$, find $\tan \beta$ and $\tan \alpha$.
65. $\cos \alpha = 0.9459$, find $\tan \alpha$ and $\sin \beta$.
66. $\sin \alpha = 0.2800$, find $\cos \alpha$ and $\tan \alpha$.

Superset

In Exercises 67–70, use the trigonometric ratios of special angles to solve the problem.

67. Show that in an equilateral triangle with side of length s that the altitude is $\frac{\sqrt{3}}{2}s$.
68. If the diagonal of a square has length $6\sqrt{2}$, what is the area of the square?
69. In an isosceles right triangle with hypotenuse 12, what is the length of the altitude to the hypotenuse?
70. Find the lengths of the sides of $\triangle ABC$ if $m(\angle A) = 30°$, $m(\angle B) = 60°$, and the altitude from C has length 8.

In Exercises 71–72, find the area of the triangle.

71.

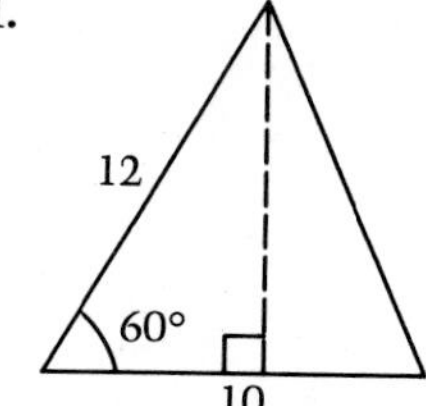

72.

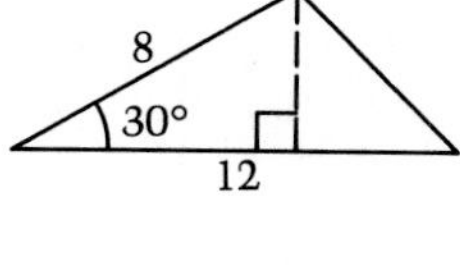

73. The local fire department's longest ladder measures 72 ft. If the angle between the ground and the ladder must be 60°, how high can the ladder reach? How far from a building should the foot of the ladder be?

74. One of the world's tallest flagpoles is 256 ft high. Guy wires are used to support the pole as shown in the diagram at the right. If the guy wires are anchored 60 ft from the foot of the flagpole, how long is each wire? How high up the pole are the wires fastened?

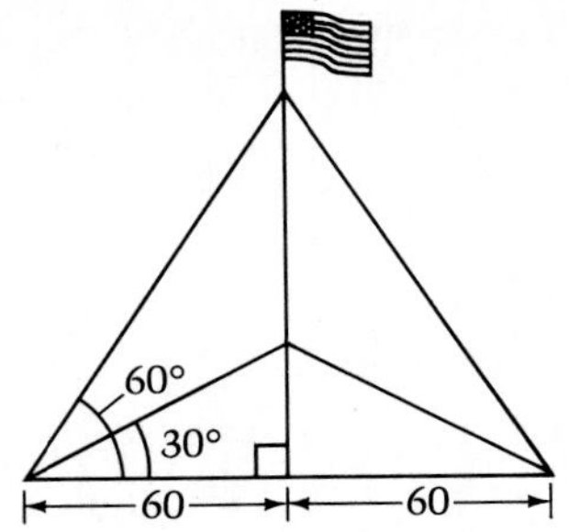

6.3 Angles of Rotation

Thus far we have defined the trigonometric functions for angles between 0° and 90°. In this section we shall extend our definitions to include *all* angles. In order to do this, it is useful to think of an angle as being formed by rotating a ray.

Figure 6.30

Between twelve midnight and 4 A.M. the hour hand of a clock rotates to produce the angle shown in the figure. The ray pointing to 12 is called the **initial side** of the angle, and the ray pointing to 4 is called the **terminal side.**

An angle is in **standard position** in the xy-plane, if its vertex is at the origin and its initial side lies along the positive x-axis. An angle in standard position is usually named by the quadrant in which the terminal side lies. Each of the angles below is in standard position. Notice that an angle may be the result of one or more complete rotations about the vertex, as shown in Figure 6.31(b).

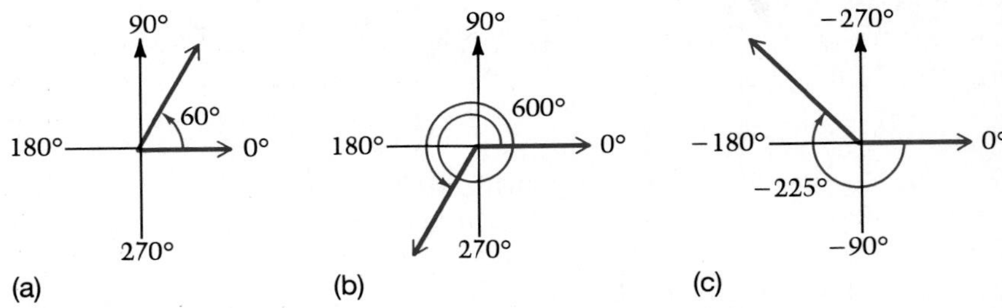

Figure 6.31 (a) A first quadrant angle. (b) A third quadrant angle. (c) A second quadrant angle.

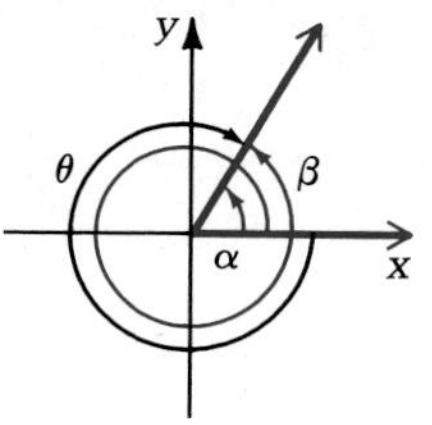

Figure 6.32

We agree that a counterclockwise rotation produces a positive angle as shown in Figure 6.31(a) and Figure 6.31(b). A clockwise rotation produces a negative angle as shown in Figure 6.31(c).

An angle whose terminal side lies on the x- or y-axis is called a **quadrantal angle.** When angles in standard position have the same terminal side, they are called **coterminal.** Angles α, β, and θ in Figure 6.32 are coterminal.

EXAMPLE 1

Sketch each of the following angles in standard position.

(a) 120° (b) 210° (c) −60° (d) 315°

Solution (a)

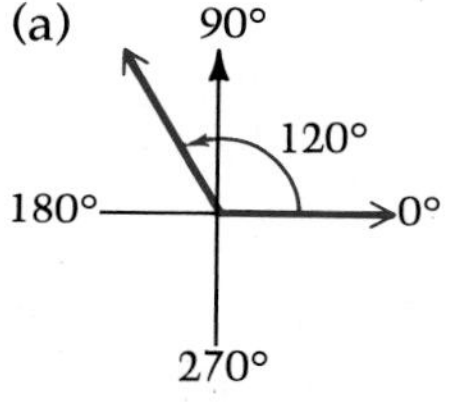

(b)

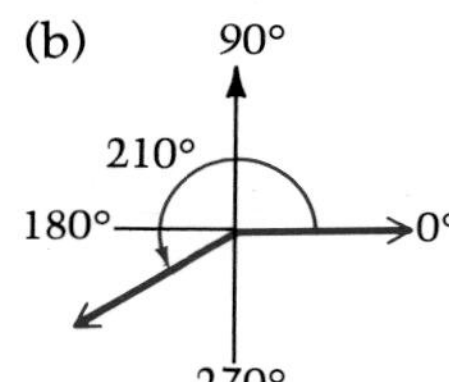

(c)

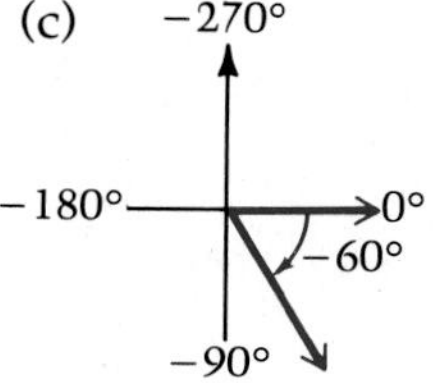

(d)

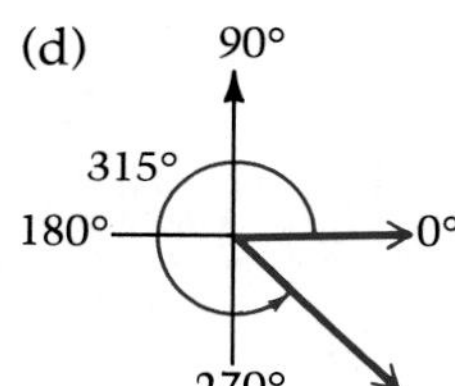

Figure 6.33

EXAMPLE 2

Sketch each of the following angles in standard position. Determine the measure of the angle between 0° and 360° that is coterminal with each angle.

(a) −30° (b) 398° (c) 810°

Solution

(a)

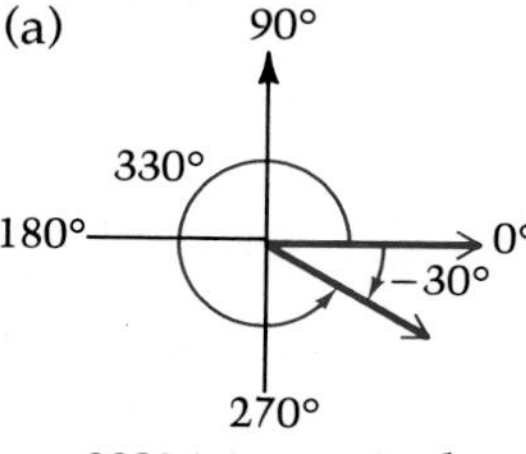

330° is coterminal with −30°

(b)

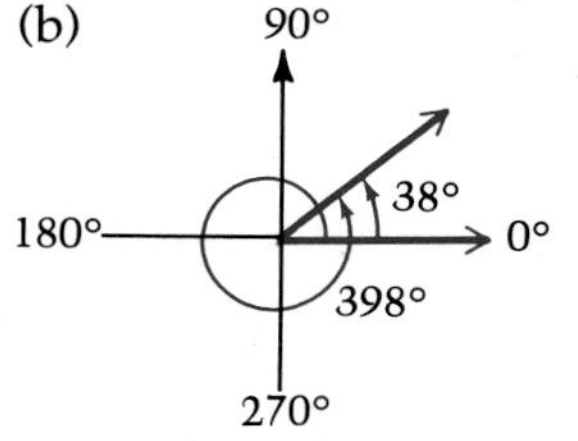

38° is coterminal with 398°

(c)

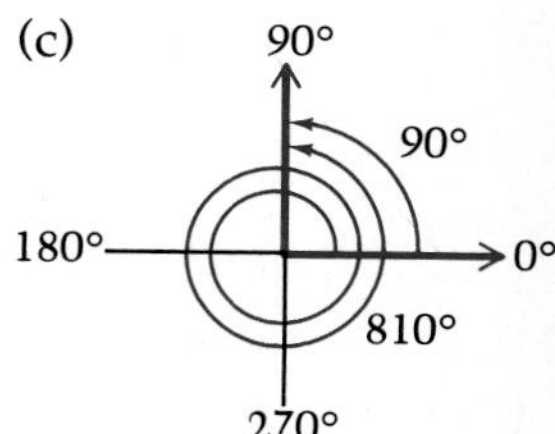

90° is coterminal with 810°

Figure 6.34

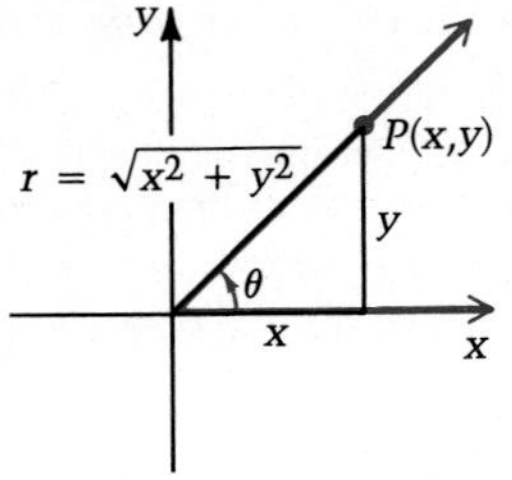

Figure 6.35

Suppose we have an acute angle θ in standard position, and we choose a point $P(x, y)$ on the terminal side of θ as shown in Figure 6.35. If a perpendicular is drawn from P to the x-axis, we can then define the trigonometric functions of θ in terms of the sides of the resulting right triangle.

$$\sin \theta = \frac{y}{r} \qquad \cos \theta = \frac{x}{r} \qquad \tan \theta = \frac{y}{x}$$

$$\csc \theta = \frac{r}{y} \qquad \sec \theta = \frac{r}{x} \qquad \cot \theta = \frac{x}{y}$$

Note that the trigonometric functions of θ have been defined in terms of the coordinates of a point on the terminal side.

This method suggests a way of defining the values of the trigonometric functions of an angle in any quadrant. We simply need to know the x- and y-coordinates of a point on the terminal side of the angle.

Definition of the Trigonometric Functions of Any Angle

Let θ be an angle in standard position with $P(x, y)$ a point on its terminal side. Then

$$\sin \theta = \frac{y}{r} \qquad \cos \theta = \frac{x}{r} \qquad \tan \theta = \frac{y}{x} \quad (x \neq 0)$$

$$\csc \theta = \frac{r}{y} \quad (y \neq 0) \qquad \sec \theta = \frac{r}{x} \quad (x \neq 0) \qquad \cot \theta = \frac{x}{y} \quad (y \neq 0)$$

where $r = \sqrt{x^2 + y^2}$ is the distance from P to the origin.

EXAMPLE 3

If $(-3, 4)$ is a point on the terminal side of an angle α in standard position, determine the values of the six trigonometric functions of α.

Solution Begin by plotting the point $(-3, 4)$ and sketching the angle α. To determine the trigonometric function values, we must first find the values of x, y, and r.

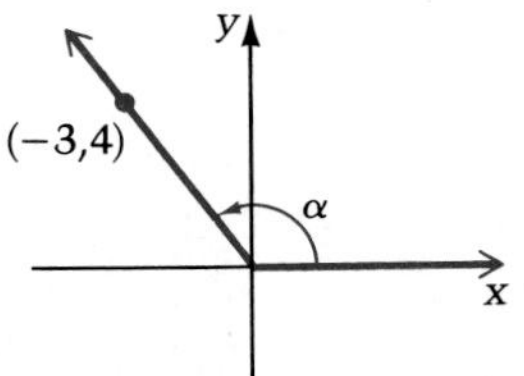

Figure 6.36

$$x = -3, y = 4, \text{ and } r = \sqrt{x^2 + y^2} = \sqrt{(-3)^2 + (4)^2} = \sqrt{25} = 5.$$

$$\sin \alpha = \frac{y}{r} = \frac{4}{5} \qquad \cos \alpha = \frac{x}{r} = -\frac{3}{5} \qquad \tan \alpha = \frac{y}{x} = -\frac{4}{3}$$

$$\csc \alpha = \frac{r}{y} = \frac{5}{4} \qquad \sec \alpha = \frac{r}{x} = -\frac{5}{3} \qquad \cot \alpha = \frac{x}{y} = -\frac{3}{4}$$

Thus far we have not computed the values of the trigonometric functions of 0°, 90°, or any of the other quadrantal angles. To determine these values, first select a point (x, y) on the terminal side of the angle, and then apply the definition of the trigonometric functions.

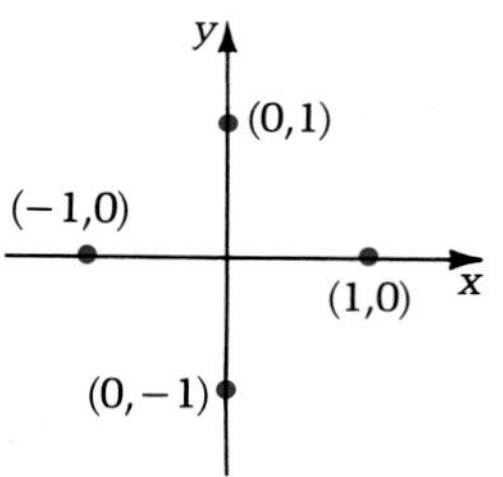

Figure 6.37

There are only four possible terminal sides for a quadrantal angle: the positive x-axis, the positive y-axis, the negative x-axis, and the negative y-axis.

Use the points plotted in Figure 6.37 to determine the values of the trigonometric functions of quadrantal angles.

For example, an angle of 90° has the positive y-axis as its terminal side. The point (0, 1) can be used to determine trigonometric function values of 90°: $x = 0, y = 1, r = \sqrt{0^2 + 1^2} = 1$.

$$\sin 90° = \frac{1}{1} = 1 \qquad \cos 90° = \frac{0}{1} = 0 \qquad \tan 90° = \frac{1}{0} \quad \text{undefined}$$

$$\csc 90° = \frac{1}{1} = 1 \qquad \sec 90° = \frac{1}{0} \quad \text{undefined} \qquad \cot 90° = \frac{0}{1} = 0$$

On the other hand, an angle of 0° has the positive x-axis as its terminal (and initial) side, so the point (1, 0) is used to determine the function values: $x = 1, y = 0, r = \sqrt{1^2 + 0^2} = 1$.

$$\sin 0° = \frac{0}{1} = 0 \qquad \cos 0° = \frac{1}{1} = 1 \qquad \tan 0° = \frac{0}{1} = 0$$

$$\csc 0° = \frac{1}{0} \quad \text{undefined} \qquad \sec 0° = \frac{1}{1} = 1 \qquad \cot 0° = \frac{1}{0} \quad \text{undefined}$$

EXAMPLE 4

Compute the following: (a) cos 270° (b) tan (−270°) (c) sec 540°.

Solution

(a) 270° has its terminal side on the negative y-axis, so use $(0, -1)$.

$$x = 0, y = -1, r = \sqrt{0^2 + (-1)^2} = 1; \cos 270° = \frac{x}{r} = \frac{0}{1} = 0$$

(b) $-270°$ has its terminal side on the positive y-axis, so use $(0, 1)$.

$$x = 0, y = 1, r = \sqrt{0^2 + 1^2} = 1; \tan(-270°) = \frac{y}{x} = \frac{1}{0} \quad \text{undefined}$$

(c) 540° has its terminal side on the negative x-axis, so use $(-1, 0)$.

$$x = -1, y = 0, r = \sqrt{(-1)^2 + 0^2} = 1; \sec 540° = \frac{r}{x} = \frac{1}{-1} = -1$$

We add the angles 0° and 90° to our list of special angles, and we expand the table given in the previous section to include these two angles.

Trigonometric Function Values for Special Angles

$f(\theta)$ \ θ	0°	30°	45°	60°	90°
sin θ	0	$\frac{1}{2}$	$\frac{\sqrt{2}}{2}$	$\frac{\sqrt{3}}{2}$	1
cos θ	1	$\frac{\sqrt{3}}{2}$	$\frac{\sqrt{2}}{2}$	$\frac{1}{2}$	0
tan θ	0	$\frac{\sqrt{3}}{3}$	1	$\sqrt{3}$	undefined
cot θ	undefined	$\sqrt{3}$	1	$\frac{\sqrt{3}}{3}$	0
sec θ	1	$\frac{2\sqrt{3}}{3}$	$\sqrt{2}$	2	undefined
csc θ	undefined	2	$\sqrt{2}$	$\frac{2\sqrt{3}}{3}$	1

EXERCISE SET 6.3

In Exercises 1–12, sketch the angle in standard position.

1. 60° **2.** 135° **3.** 330°
4. 255° **5.** 390° **6.** 480°
7. $-30°$ **8.** $-120°$ **9.** $-390°$
10. $-480°$ **11.** $-725°$ **12.** $-1142°$

In Exercises 13–20, determine the measure of the angle between 0° and 360° that is coterminal with the given angle.

13. 405° **14.** 485° **15.** 723°
16. 990° **17.** $-38°$ **18.** $-180°$
19. $-660°$ **20.** $-1689°$

In Exercises 21–34, determine a positive angle between 0° and 360° inclusive and a negative angle between 0° and −360° inclusive that are coterminal with the given angle.

21. 20° **22.** 102° **23.** 225°

24. 270° **25.** −410° **26.** −450°

27. −1351° **28.** 1300° **29.** 0°

30. 1080° **31.** 1575° **32.** −940°

33. $\frac{1^\circ}{2}$ **34.** $-\frac{1^\circ}{2}$

In Exercises 35–38, determine the values of the six trigonometric functions of α.

35.

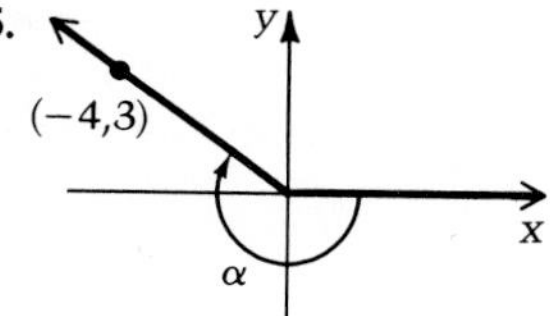

36.

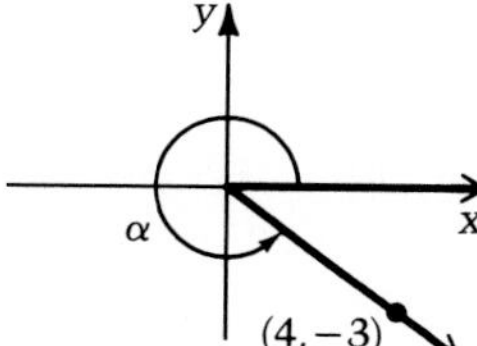

37.

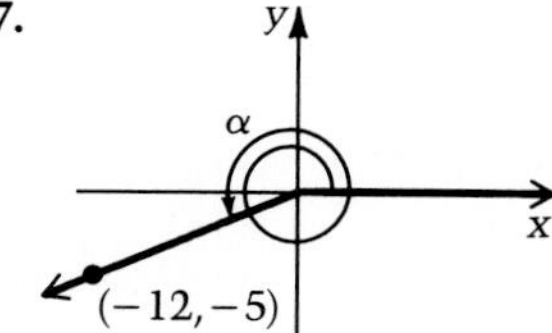

38.

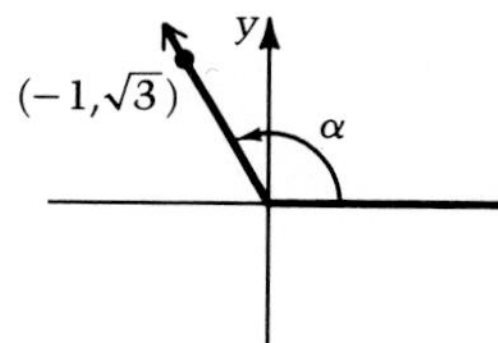

In Exercises 39–54, the given point is on the terminal side of an angle α. Determine the values of the six trigonometric functions of α.

39. $(-8, 6)$ **40.** $(-5, 12)$ **41.** $(1, 1)$

42. $(2, 6)$ **43.** $(3, -3)$ **44.** $(1, -2)$

45. $(-3, -9)$ **46.** $(-\sqrt{3}, -1)$ **47.** $(-\sqrt{5}, 2)$

48. $(-2\sqrt{3}, 4)$

49. $(-5\sqrt{7}, 3\sqrt{3})$ 50. $(-4\sqrt{5}, -7\sqrt{5})$

51. $(25.34, -72.67)$ 52. $(2.872, -6.449)$

53. $(-13.856, -5.657)$ 54. $(-11.180, 6.471)$

In Exercises 55–66, determine the value.

55. sin 90° **56.** cos 180° **57.** cot 180°

58. tan 270° **59.** sec 450° **60.** sin 810°

61. cos(−90°) **62.** csc(−630°) **63.** tan(−540°)

64. sec(−1440°) **65.** csc(−720°) **66.** cot(−900°)

Superset

In Exercises 67–70, determine the measure of angle φ.

67. φ is one-third of a complete revolution in a counterclockwise direction.

68. φ is one-third of a complete revolution in a clockwise direction.

69. φ is one and one quarter complete revolutions in a clockwise direction.

70. φ is three tenths of a complete revolution in a counterclockwise direction.

In Exercises 71–78, the value of one trigonometric function of φ is given. Assuming that φ is a second quadrant angle, find the values of the other five trigonometric function values of φ.

71. $\tan \varphi = -5$ **72.** $\sin \varphi = \frac{4}{5}$

73. $\csc \varphi = \frac{5}{3}$ **74.** $\cot \varphi = -\frac{3}{4}$

75. $\cot \varphi = -\frac{5}{3}$ **76.** $\tan \varphi = -1$

77. $\cos \varphi = -\frac{3}{5}$ **78.** $\sec \varphi = -\sqrt{3}$

In Exercises 79–86, the value of one trigonometric function of φ is given. If the terminal side of φ lies in the given quadrant, find the values of the other five trigonometric functions of φ.

79. third quadrant; $\sin \varphi = -\frac{2}{5}$

80. second quadrant; $\sin \varphi = \frac{3}{5}$

81. second quadrant; $\cos \varphi = -0.6$

82. fourth quadrant; $\cos \varphi = 0.7$

83. fourth quadrant; $\sec \varphi = 1.2$

84. third quadrant; $\tan \varphi = 1$

85. fourth quadrant; $\tan \varphi = -1$

86. second quadrant; $\csc \varphi = 1.5$

6.4 Identities and Tables

An **identity** is an equation that is true for all values of the variables for which both sides of the equation are defined. Several trigonometric identities are easily derived from the definitions of the trigonometric functions.

Reciprocal Identities

	Derivation
(1) $\sin\theta = \dfrac{1}{\csc\theta}$	$\sin\theta = \dfrac{y}{r} = \dfrac{1}{\frac{r}{y}} = \dfrac{1}{\csc\theta}$
(2) $\cos\theta = \dfrac{1}{\sec\theta}$	$\cos\theta = \dfrac{x}{r} = \dfrac{1}{\frac{r}{x}} = \dfrac{1}{\sec\theta}$
(3) $\tan\theta = \dfrac{1}{\cot\theta}$	$\tan\theta = \dfrac{y}{x} = \dfrac{1}{\frac{x}{y}} = \dfrac{1}{\cot\theta}$
(4) $\cot\theta = \dfrac{1}{\tan\theta}$	$\cot\theta = \dfrac{x}{y} = \dfrac{1}{\frac{y}{x}} = \dfrac{1}{\tan\theta}$
(5) $\sec\theta = \dfrac{1}{\cos\theta}$	$\sec\theta = \dfrac{r}{x} = \dfrac{1}{\frac{x}{r}} = \dfrac{1}{\cos\theta}$
(6) $\csc\theta = \dfrac{1}{\sin\theta}$	$\csc\theta = \dfrac{r}{y} = \dfrac{1}{\frac{y}{r}} = \dfrac{1}{\sin\theta}$

Quotient Identities

	Derivation
(7) $\dfrac{\sin\theta}{\cos\theta} = \tan\theta$	$\dfrac{\sin\theta}{\cos\theta} = \dfrac{\frac{y}{r}}{\frac{x}{r}} = \dfrac{y}{r}\cdot\dfrac{r}{x} = \dfrac{y}{x} = \tan\theta$
(8) $\dfrac{\cos\theta}{\sin\theta} = \cot\theta$	$\dfrac{\cos\theta}{\sin\theta} = \dfrac{\frac{x}{r}}{\frac{y}{r}} = \dfrac{x}{r}\cdot\dfrac{r}{y} = \dfrac{x}{y} = \cot\theta$

EXAMPLE 1

Given $\sin \beta = 0.324$ and $\cos \beta = 0.946$, find the other four trigonometric functions of β. Round answers to three decimal places.

Solution

$$\tan \beta = \frac{\sin \beta}{\cos \beta} = \frac{0.324}{0.946} \approx 0.342$$

0.342 is an approximation for $\tan \beta$, and we write $\tan \beta \approx 0.342$.

$$\cot \beta = \frac{1}{\tan \beta} = \frac{1}{0.342} \approx 2.924$$

If the identity $\frac{\cos \beta}{\sin \beta} = \cot \beta$ is used, the answer is 2.920. The difference is due to rounding error.

$$\sec \beta = \frac{1}{\cos \beta} = \frac{1}{0.946} \approx 1.057$$

$$\csc \beta = \frac{1}{\sin \beta} = \frac{1}{0.324} \approx 3.086$$

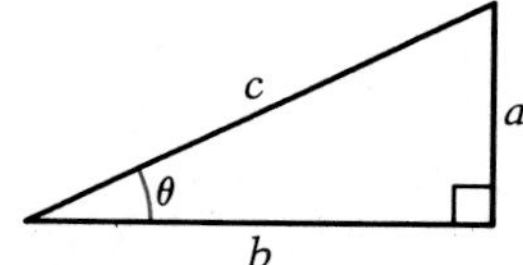

Figure 6.38

The next group of identities that we will consider are referred to as the Pythagorean Identities because their derivation depends on the Pythagorean Theorem. Using the right triangle at the left, we can make the following statements.

$$a^2 + b^2 = c^2$$ This is the Pythagorean Theorem.

$$\frac{a^2}{c^2} + \frac{b^2}{c^2} = \frac{c^2}{c^2}$$ Both sides are divided by c^2.

$$\left(\frac{a}{c}\right)^2 + \left(\frac{b}{c}\right)^2 = 1$$

$$(\sin \theta)^2 + (\cos \theta)^2 = 1$$ $\sin \theta = \frac{a}{c}$; $\cos \theta = \frac{b}{c}$.

It is common to write $(\sin \theta)^2$ as $\sin^2\theta$ and $(\cos \theta)^2$ as $\cos^2\theta$.

Pythagorean Identities

	Derivation
(9) $\sin^2\theta + \cos^2\theta = 1$	Shown above.
(10) $\tan^2\theta + 1 = \sec^2\theta$	Divide both sides of (9) by $\cos^2\theta$. Then use the Quotient and Reciprocal Identities.
(11) $1 + \cot^2\theta = \csc^2\theta$	Divide both sides of (9) by $\sin^2\theta$. Then use the Quotient and Reciprocal Identities.

EXAMPLE 2

In each of the following, θ is an acute angle.

(a) If $\sin\theta = \dfrac{1}{3}$, find $\cos\theta$. (b) If $\tan\theta = 1.60$, find $\sec\theta$.

Solution

(a)
$$\sin^2\theta + \cos^2\theta = 1$$
Identity (9).
$$\left(\frac{1}{3}\right)^2 + \cos^2\theta = 1$$
$$\cos^2\theta = 1 - \frac{1}{9}$$
$$\cos\theta = +\sqrt{\frac{8}{9}} = \frac{2\sqrt{2}}{3} \approx 0.94$$
For an acute angle all trigonometric function values are positive.

(b)
$$\tan^2\theta + 1 = \sec^2\theta$$
Identity (10).
$$(1.60)^2 + 1 = \sec^2\theta$$
$$\sec^2\theta = 3.56$$
$$\sec\theta = +\sqrt{3.56} \approx 1.89$$
As in part (a), select the positive root.

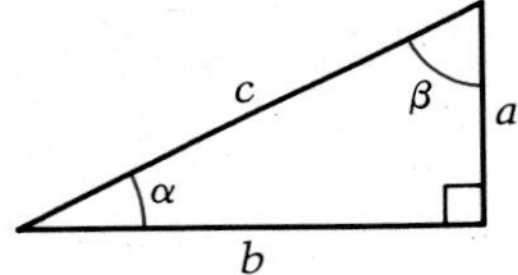

Figure 6.39 Since $\sin\alpha = \frac{a}{c}$ and $\cos\beta = \frac{a}{c}$, $\sin\alpha = \cos\beta$.

Recall that if α and β are complementary angles, then $\alpha + \beta = 90°$. Figure 6.39 shows that in this case $\sin\alpha = \cos\beta$. Since $\beta = 90° - \alpha$, we can write $\sin\alpha = \cos(90° - \alpha)$. This is called a **Cofunction Identity** since the sine and cosine are cofunctions of each other. The tangent and cotangent are also cofunctions of each other, as are the secant and cosecant. We now list the cofunction identities.

Figure 6.40 Since $\cos\alpha = \frac{b}{c}$ and $\sin(90° - \alpha) = \frac{b}{c}$, we conclude that $\cos\alpha = \sin(90° - \alpha)$.

Cofunction Identities

	Derivation
$\sin\alpha = \cos(90° - \alpha)$	See Figure 6.39.
$\cos\alpha = \sin(90° - \alpha)$	See Figure 6.40.
$\tan\alpha = \cot(90° - \alpha)$	$\tan\alpha = \dfrac{\sin\alpha}{\cos\alpha} = \dfrac{\cos(90° - \alpha)}{\sin(90° - \alpha)} = \cot(90° - \alpha)$
$\cot\alpha = \tan(90° - \alpha)$	$\cot\alpha = \dfrac{1}{\tan\alpha} = \dfrac{1}{\cot(90° - \alpha)} = \tan(90° - \alpha)$
$\sec\alpha = \csc(90° - \alpha)$	$\sec\alpha = \dfrac{1}{\cos\alpha} = \dfrac{1}{\cos(90° - \alpha)} = \csc(90° - \alpha)$
$\csc\alpha = \sec(90° - \alpha)$	$\csc\alpha = \dfrac{1}{\sin\alpha} = \dfrac{1}{\cos(90° - \alpha)} = \sec(90° - \alpha)$

In words, the cofunction identities say that the trigonometric function of an angle is equal to the cofunction of the complement of the angle.

EXAMPLE 3

Complete the following statements with an acute angle.

(a) $\sin 37° = \cos$ ____ (b) $\cot 52° = \tan$ ____ (c) $\csc 45° = \sec$ ____

Solution

(a) $\sin 37° = \cos(90° - 37°) = \cos 53°$

(b) $\cot 52° = \tan(90° - 52°) = \tan 38°$

(c) $\csc 45° = \sec(90° - 45°) = \sec 45°$

EXAMPLE 4

Given $\sin 70° \approx 0.940$ and $\cos 70° \approx 0.342$, find: (a) $\cot 20°$ (b) $\csc 20°$. Round each answer to three decimal places.

Solution

(a) $\cot 20° = \tan(90° - 20°) = \tan 70° = \dfrac{\sin 70°}{\cos 70°} \approx \dfrac{0.940}{0.342} \approx 2.749$

(b) $\csc 20° = \dfrac{1}{\sin 20°} = \dfrac{1}{\cos 70°} \approx \dfrac{1}{0.342} \approx 2.924$

We know the trigonometric function values of the special angles, but what about cos 37° or sin 72°. Such values are not as easy to compute, so we need to use a calculator or a table of values. (Certain polynomial functions can be used to generate very accurate approximate values. This is how calculators that evaluate trigonometric functions work and how tables of values are formed.) Table 4 in the Appendix lists values of the six trigonometric functions of angles between 0° and 90°. Although the values in this table are approximations, we agree to write $\sin 15° = 0.2588$ (from Table 4) even though the exact value of sin 15° is $\dfrac{\sqrt{2 - \sqrt{3}}}{2}$, and 0.2588 is only an approximation for $\dfrac{\sqrt{2 - \sqrt{3}}}{2}$. The same comments about approximations apply to calculator-generated values.

The degree measure of an angle can be written in two ways: decimal form and degree-minute-second form. For the same historical reasons that we take the measure of a circle to be 360 degrees, the fractional parts of a degree are sometimes measured in a "base 60" system, with 60 minutes in a degree ($1° = 60'$), and with 60 seconds in a minute ($1' = 60''$). For example, $42.3° = 42°18'$. The earliest applications of trigonometry used astronomical observations to facilitate navigation and to obtain accurate calendars for planning planting seasons and religious celebrations. The use of the degree-minute-second form persists today in certain settings, though the availability of calculators has made the decimal degree form more convenient to work with.

The next two examples demonstrate a technique for converting one form of degree measure to the other. A calculator will normally make it easier to perform such conversions.

EXAMPLE 5

Convert $41°22'35''$ to decimal form, rounded to the nearest hundredth.

Solution Since $1' = \left(\frac{1}{60}\right)^{\circ}$ and $1'' = \frac{1}{60}(1') = \left(\frac{1}{3600}\right)^{\circ}$, we begin by expressing the degree measure as a rational number of degrees.

$$\begin{aligned} 41°22'35'' &= \left[41 + 22\left(\frac{1}{60}\right) + 35\left(\frac{1}{3600}\right)\right]^{\circ} \\ &\approx [41 + 0.3667 + 0.0097]^{\circ} = 41.3764° \end{aligned}$$

Thus, $41°22'35'' \approx 41.38°$.

EXAMPLE 6

Convert $27.3711°$ to degree-minute-second form, rounded to the nearest second.

Solution The conversion takes two steps. First, we express the fractional part of the degree measure in minutes.

$$\begin{aligned} 0.3711° &= (0.3711)(1°) = (0.3711)(60') \qquad 1° = 60' \\ &= 22.266' \end{aligned}$$

Thus, $27.3711° = 27°22.266'$. Next, we express the decimal part of the minutes in seconds.

$$\begin{aligned} 0.266' &= (0.266)(1') = (0.266)(60'') \qquad 1' = 60'' \\ &= 15.96'' \qquad 15.96'' \approx 16'' \end{aligned}$$

Thus, $27.3711° \approx 27°22'16''$.

Some Comments About Calculators

Careful! A calculator will compute the trigonometric function values for domain values measured in degrees or radians. (We discuss radian measure in Chapter 7.) You can select between degrees and radians on calculators by means of a key (DRG on some calculators; DEG on some others). You should be careful that the calculator is in the proper mode for an intended calculation with trigonometric values.

To find the angle α to two decimal places such that $\sin \alpha = 0.5402$, we use the inverse key, INV .

ENTER .5402 PRESS INV PRESS sin

DISPLAY 32.69725475

Thus, $\alpha \approx 32.70°$. (If you did not get this answer, check to be sure you have your calculator in the degree mode.) The inverse key followed by the sine key "undoes" what the sine key "does." The INV key, when followed by the sin , cos , or tan key calculates the smallest angle in Quadrant I having the positive trigonometric function value entered.

To evaluate the cotangent, secant, and cosecant functions, it is necessary to use the Reciprocal Identities and the reciprocal key $\frac{1}{x}$ on your calculator. For example,

$$\csc 36.25° = \frac{1}{\sin 36.25°} = \frac{1}{0.5913} = 1.6912$$

To perform this computation using a calculator, take the following steps:

First, make sure your calculator is in degree mode, then

ENTER 36.25 PRESS sin PRESS $\frac{1}{x}$

DISPLAY 1.691161311

Use of Trigonometric Tables

We conclude with two examples on how to use Table 4 in the Appendix in the event that you will not be using a calculator for your work with trigonometric functions. At this point you should consult with your instructor regarding the use of the trigonometric tables. Your instructor may want you to study the method of interpolation outlined in the Appendix, particularly if you will be relying on the tables for computation, rather than on a calculator.

EXAMPLE 7

Use Table 4 to determine: (a) $\tan 39°20'$ (b) $\cos 62°40'$.

Solution

(a) $\tan 39°20' = 0.8195$ For angles between 0° and 45°, start at the *top left* of Table 4 and read down the extreme left columns until you find 39°20′. Then move across to the column having $\tan \alpha$ at the top.

(b) $\cos 62°40' = 0.4592$ For angles between 45° and 90°, start at the *bottom right* of Table 4 and read up the extreme right columns until you reach 62°40′. Then move across to the column having $\cos \alpha$ at the bottom.

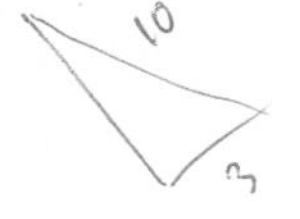

EXAMPLE 8

Use Table 4 to determine acute angle α if $\csc \alpha = 1.211$.

Solution $\alpha = 55°40'$ Read up the column in Table 4 having $\csc \alpha$ at the bottom, until you find 1.211. Move across to the right-hand column to find 55°40′.

EXERCISE SET 6.4

In Exercises 1–4, two approximate trigonometric function values of an angle are given. Find the other four trigonometric function values rounded to two decimal places.

1. $\sin 32° = 0.53$, $\cos 32° = 0.85$

2. $\tan 55° = 1.43$, $\sin 55° = 0.82$

3. $\tan 66° = 2.25$, $\sec 66° = 2.46$

4. $\tan 23° = 0.42$, $\cos 23° = 0.92$

In Exercises 5–10, complete the statement with an acute angle.

5. $\sin 54° = \cos$ ____ **6.** $\cos 12° = \sin$ ____

7. $\tan 38° = \cot$ ____ **8.** $\cot 81° = \tan$ ____

9. $\sec 45° = \csc$ ____ **10.** $\csc 71° = \sec$ ____

In Exercises 11–18, rewrite the expression as a trigonometric function of an angle between 45° and 90°.

11. $\cos 27°$ **12.** $\sec 12°$ **13.** $\cot 38°$

14. $\tan 41°$ **15.** $\csc 3°$ **16.** $\cos 43°$

17. $\sin 19°$ **18.** $\sin 23°$

In Exercises 19–24, approximate the function value given that $\sin 32° = 0.53$, $\cos 32° = 0.85$, $\sin 23° = 0.39$, and $\tan 23° = 0.42$. Round your answer to two decimal places.

19. $\sin 58°$ **20.** $\sec 58°$ **21.** $\sec 67°$

22. $\tan 58°$ **23.** $\tan 67°$ **24.** $\cot 67°$

In Exercises 25–32, use the Reciprocal and Pythagorean Identities to determine the values of the other five trigonometric functions of the acute angle α.

25. $\sin \alpha = \frac{1}{2}$ **26.** $\cos \alpha = \frac{3}{10}$

27. $\tan \alpha = \frac{3}{2}$ **28.** $\cot \alpha = \frac{1}{2}$

29. $\csc \alpha = 3$ **30.** $\sec \alpha = 10$

31. $\cos \alpha = \frac{4}{5}$ **32.** $\sin \alpha = \frac{7}{10}$

In Exercises 33–42, convert the given degree measure from degree-minute-second form to decimal form. Round your answer to the nearest thousandth.

33. 18°46′55″ **34.** 82°26′9″ **35.** 71°13′36″
36. 9°0′49″ **37.** 24°51′3″ **38.** 32°58′14″
39. 43°0′28″ **40.** 51°37′56″ **41.** 45°15′54″
42. 30°45′36″

In Exercises 43–52, convert the given degree measure from decimal form to degree-minute-second form. Round your answer to the nearest second.

43. 69.510° **44.** 38.791° **45.** 20.942°
46. 74.143° **47.** 8.372° **48.** 13.450°
49. 86.704° **50.** 2.619° **51.** 47.165°
52. 53.281°

In Exercises 53–64, use Table 4 in the Appendix to determine the value.

53. sin 31°10′ **54.** cos 43°50′ **55.** tan 21°40′
56. tan 49°20′ **57.** sec 56°50′ **58.** sec 10°30′
59. cot 48°10′ **60.** csc 15°40′ **61.** cos 52°20′
62. sin 80°10′ **63.** csc 45°30′ **64.** cot 78°50′

In Exercises 65–76, use Table 4 to determine the value of the acute angle α.

65. $\sin \alpha = 0.1132$ **66.** $\cos \alpha = 0.9563$
67. $\cot \alpha = 0.3443$ **68.** $\tan \alpha = 1.024$
69. $\cos \alpha = 0.4975$ **70.** $\csc \alpha = 2.542$
71. $\sec \alpha = 1.410$ **72.** $\sin \alpha = 0.9750$
73. $\tan \alpha = 0.2462$ **74.** $\sec \alpha = 8.834$
75. $\csc \alpha = 1.127$ **76.** $\cot \alpha = 1.483$

In Exercises 77–80, two approximate trigonometric function values of an angle are given. Use the Reciprocal and Quotient Identities to find the other four trigonometric function values rounded to four decimal places.

77. sin 32° = 0.5299, cos 32° = 0.8480
78. tan 66° = 2.2460, sec 66° = 2.4586
79. tan 55° = 1.4281, sin 55° = 0.8192
80. tan 23° = 0.4245, csc 23° = 2.5593

In Exercises 81–84, use the Reciprocal and Pythagorean Identities to find the values of the other five trigonometric functions of α rounded to four decimal places.

81. $\sin \alpha = 0.5124$ 82. $\cos \alpha = 0.3051$
83. $\cos \alpha = 0.8236$ 84. $\sin \alpha = 0.7645$

In Exercises 85–96, determine the value to four decimal places.

85. sin 31.2° 86. cos 43.8° 87. tan 21.66°
88. tan 49.33° 89. cos 52.25° 90. sin 80.46°
91. csc 56.3° 92. csc 10.5° 93. cot 48.2°
94. sec 15.7° 95. sec 45.5° 96. cot 78.9°

In Exercises 97–108, determine the acute angle α in degrees to two decimal places.

97. $\sin \alpha = 0.3486$ 98. $\cos \alpha = 0.1542$
99. $\cos \alpha = 0.7816$ 100. $\tan \alpha = 0.6230$
101. $\tan \alpha = 2.036$ 102. $\sin \alpha = 0.9108$
103. $\csc \alpha = 1.6239$ 104. $\sec \alpha = 1.0872$
105. $\sec \alpha = 1.3100$ 106. $\cot \alpha = 3.5712$
107. $\cot \alpha = 0.5774$ 108. $\csc \alpha = 1.1547$

In Exercises 109–120, determine the value rounded to four decimal places.

109. sin 39°33′ 110. tan 12°15′ 111. tan 31°23′
112. cos 19°3′ 113. cos 71°27′ 114. sin 63°11′
115. cot 82°18′ 116. cot 75°52′ 117. sec 87°47′
118. csc 45°36′ 119. csc 22°54′ 120. sec 37°42′

Superset

In Exercises 121–128, determine whether the statement is an identity.

121. $\cos \alpha \tan \alpha = \sin \alpha$ **122.** $\cos \alpha \sec \alpha = \cot \alpha$
123. $\sec \alpha = \tan \alpha \csc \alpha$ **124.** $\sin \alpha \cot \alpha = \cos \alpha$
125. $\dfrac{\tan \alpha}{\sin \alpha} = \sec \alpha$ **126.** $\dfrac{\csc \alpha}{\sec \alpha} = \cot \alpha$
127. $\dfrac{1}{\tan \alpha} = \dfrac{\cos \alpha}{\sin \alpha}$ **128.** $\cot \alpha = \dfrac{\sec \alpha}{\csc \alpha}$

In Exercises 129–132, evaluate the pair of expressions by using the values of the special angles.

129. $\sin(90° - 30°)$, $\sin 90° - \sin 30°$
130. $\sin(60° + 30°)$, $\sin 60° + \sin 30°$
131. $\cot 90° - \cot 30°$, $\cot(90° - 30°)$
132. $\cos(45° + 45°)$, $\cos 45° + \cos 45°$

In Exercises 133–136, verify that the statement is true.

133. $\sin(60° + 30°) = \sin 60° \cos 30° + \sin 30° \cos 60°$
134. $\sin 60° \cos 30° = \frac{1}{2}(\sin 90° + \sin 30°)$

135. $\cos 90^\circ = (\cos 45^\circ)(\cos 45^\circ) - (\sin 45^\circ)(\sin 45^\circ)$

136. $\tan 60^\circ = \dfrac{2 \tan 30^\circ}{1 - \tan^2 30^\circ}$

In Exercises 137–146, determine the value to four decimal places.

137. $\cos 408^\circ$ 138. $\tan 597^\circ$

139. $\tan 212^\circ$ 140. $\sin 320^\circ$

141. $\sin(-1134^\circ)$ 142. $\cos(-666^\circ)$

143. $\cos 196.2^\circ$ 144. $\sin 196.2^\circ$

145. $\csc(-287^\circ)$ 146. $\sec(-136^\circ)$

In Exercises 147–154, find α in degrees to two decimal places.

147. $\cos \alpha = -0.7145$, α in Quadrant II

148. $\sin \alpha = -0.7593$, α in Quadrant III

149. $\tan \alpha = -10.988$, α in Quadrant IV

150. $\tan \alpha = -4.915$, α in Quadrant II

151. $\sin \alpha = -0.6211$, α in Quadrant IV

152. $\cos \alpha = 0.9898$, α in Quadrant IV

153. $\cot \alpha = 3.420$, α in Quadrant III

154. $\csc \alpha = -1.070$, α in Quadrant III

6.5 Applications Involving Right Triangles

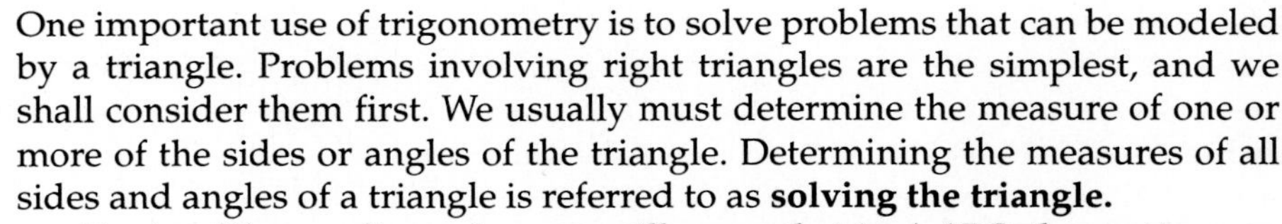

One important use of trigonometry is to solve problems that can be modeled by a triangle. Problems involving right triangles are the simplest, and we shall consider them first. We usually must determine the measure of one or more of the sides or angles of the triangle. Determining the measures of all sides and angles of a triangle is referred to as **solving the triangle.**

To simplify our discussion, we will agree that in $\triangle ABC$, the vertices are A, B, and C, and the sides opposite these vertices are a, b, and c, respectively (Figure 6.41). Also, we will agree that the statement "$A = 42^\circ$" will mean "the measure of the angle at vertex A is 42°."

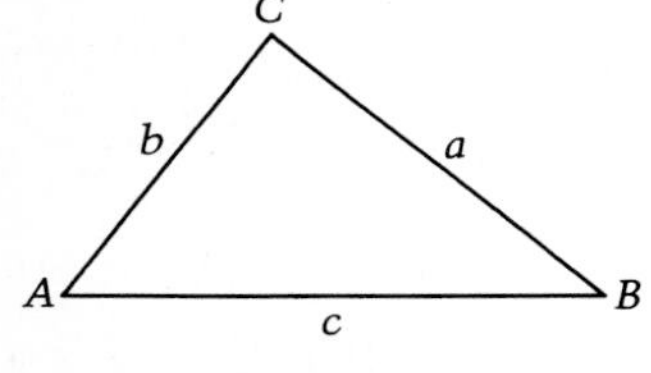

Figure 6.41

EXAMPLE 1

Solve the right triangle ABC, given that $C = 90^\circ$, $B = 40.2^\circ$, and $a = 10.3$.

Solution Begin by drawing a triangle and labeling its parts.

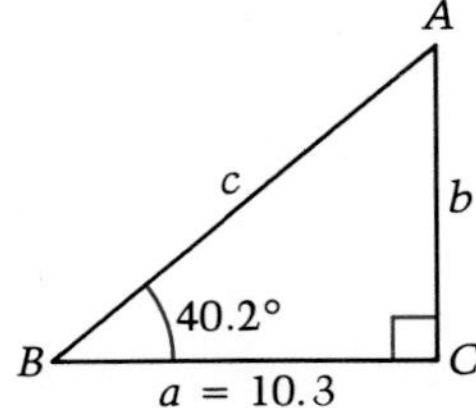

Figure 6.42

Find c: $\cos 40.2^\circ = \dfrac{10.3}{c}$ $\quad \cos B = \dfrac{a}{c}$.

$$c = \frac{10.3}{\cos 40.2^\circ} = \frac{10.3}{0.7638} \approx 13.5$$

We chose $\cos B = \frac{a}{c}$ because this equation involved the unknown c that we wished to find and two variables, B and a, that were known. Trying $\sin B = \frac{b}{c}$ would have been pointless because two variables in that equation were then unknown.

Find b: $\tan 40.2° = \frac{b}{10.3}$ $\tan B = \frac{b}{a}$ is used to find b because B and a are known.

$$b = (10.3)(\tan 40.2°)$$
$$b = (10.3)(0.8451) \approx 8.70$$

Find A: Since A and B are complementary angles, $A = 49.8°$.

We digress for a moment to discuss accuracy of answers to applied problems. Trigonometric function values from a calculator or from Table 4 are approximations. Also, measuring instruments are subject to some measurement error and thus they too produce approximations. So when we solve problems involving measurements of lengths and angles, our solutions are approximations. The question then is how should such solutions be "rounded"? To answer this we continue the discussion we began in Chapter 1 on the notion of *significant digits.*

The length 10.3 in Example 1 is said to have three significant digits. The number 0.061 has two significant digits, and the number 72,000, if it has been rounded to the nearest thousand, also has two significant digits. We apply the following rule of thumb:

> If a number N can be written in scientific notation as $N = A \cdot 10^n$ where $|A|$ is greater than or equal to 1 and less than 10, then the number of **significant digits** in N is the number of digits in A.

When solving problems involving angles and sides of triangles, you should round your calculations to be consistent with these standard rules.

Number of significant digits in length, or in angle measure in decimal degree form	Measure of angle in degree-minute-second form should be rounded to
1	the nearest multiple of 10°
2	the nearest degree
3	the nearest multiple of 10′
4	the nearest minute

When measurements are added or subtracted, the answer is rounded to *the least number of decimal places* of any of the measurements. When measurements are multiplied or divided, the answer is rounded to the *least number of significant digits* of any of the measurements. The power or root of a measurement is rounded to the same number of significant digits as is the measurement itself.

EXAMPLE 2

Solve $\triangle ABC$ given that it is a right triangle with $C = 90°$, $a = 16.5$, and $c = 30.2$.

Solution Draw a triangle and label its parts.

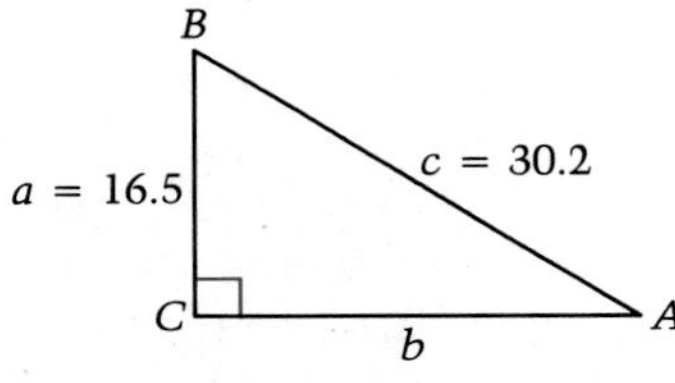

Figure 6.43

Find *b*:

$$(30.2)^2 = (16.5)^2 + b^2$$
$$b^2 \approx 912 - 272 = 640$$
$$b \approx 25.3$$

Find *A*:

$$\sin A = \frac{16.5}{30.2} \approx 0.546$$
$$A \approx 33°10'$$

Find *B*:

$$B \approx 56°50'$$

Note that A and B are rounded to the nearest 10′.

It should be clear from Examples 1 and 2 that a right triangle can be solved provided either one side and one of the acute angles are known, or two sides are known. Suppose you are given both acute angles. Is that enough information to solve the triangle? (The answer is "No.")

Some practical applications involve angles formed by the horizontal and the line of sight of an observer. If the line of sight is above the horizontal, the angle is called an **angle of elevation.** If the line of sight is below the horizontal, the angle is called an **angle of depression** (Figure 6.44).

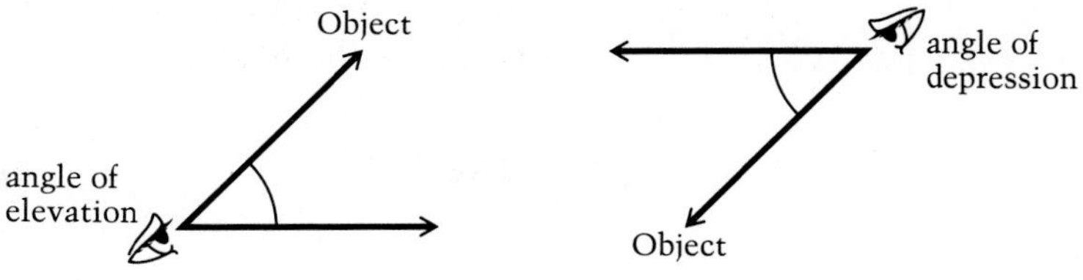

Figure 6.44

EXAMPLE 3

A hot air balloon is rising vertically in still air. An observer is standing on level ground, 100 feet away from the point of launch. At one instant the observer measures the angle of elevation of the balloon as 30°00′. One minute later, the angle of elevation is 76°10′. How far did the balloon travel during that minute?

Solution Draw a figure that models the problem. We wish to determine $b - a$, the distance traveled during the one-minute period.

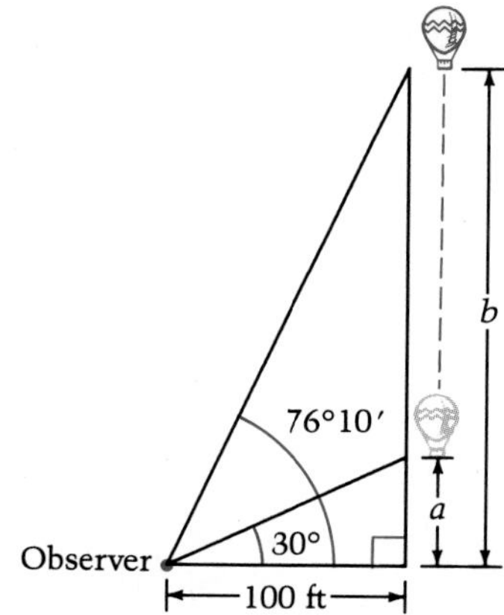

Figure 6.45

Find *a*:

$$\tan 30° = \frac{a}{100}$$
$$a = 100(\tan 30°)$$
$$= 100\left(\frac{\sqrt{3}}{3}\right)$$
$$\approx 57.7$$

Find *b*:

$$\tan 76°10' = \frac{b}{100}$$
$$b = 100(\tan 76°10')$$
$$= 100(4.061)$$
$$\approx 406$$

The balloon traveled $406 - 57.7 \approx 348$ feet during the minute.

Surveyors and navigators measure angles in terms of the north-south line. Two methods of measure are used: azimuth and bearing. The **azimuth** is the measure of an angle from due north in the clockwise direction.

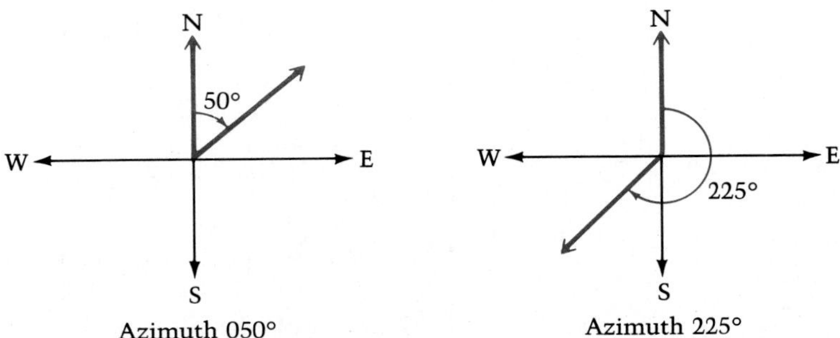

Figure 6.46

The **bearing** is the measure of the acute angle between the north-south line and the line representing the direction. It is described in terms of compass directions.

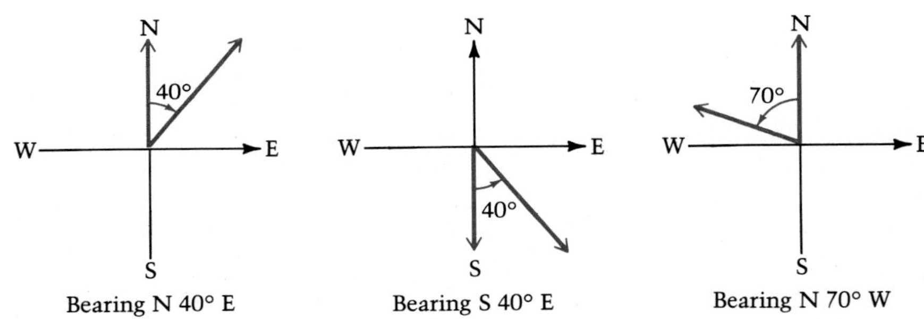

Figure 6.47

EXAMPLE 4

A plane flies from County Airport on a bearing of N 45° E for three hours, and then flies on a bearing of S 45° E for four hours. If the speed of the plane is 400 mph, and we ignore the effects of wind, what is the plane's distance and azimuth from the County Airport after the seven hours?

Solution We begin by drawing a figure. Note that $C = 90°$ so $\triangle ABC$ is a right triangle. We wish to determine side c and angle (azimuth) θ. Note that $\theta = A + 45°$.

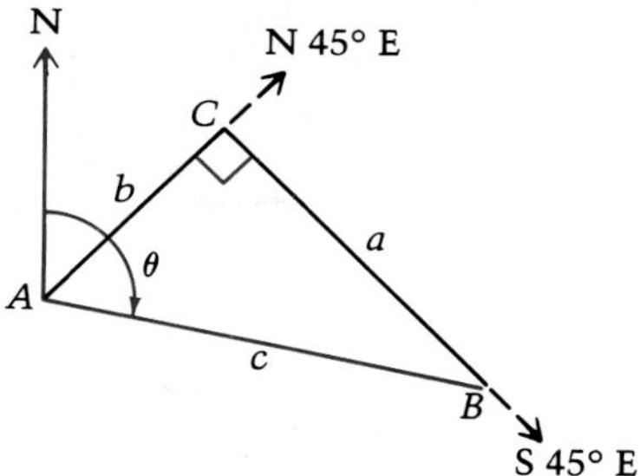

Figure 6.48

$$b = (\text{speed})(\text{time on N } 45^\circ \text{ E bearing})$$
$$= (400 \text{ mph})(3 \text{ hr}) = 1200 \text{ mi}$$
$$a = (\text{speed})(\text{time on S } 45^\circ \text{ E bearing})$$
$$= (400)(4 \text{ hr}) = 1600 \text{ mi}$$
$$c = \sqrt{1200^2 + 1600^2} = 2000$$
$$\tan A = \frac{\text{opposite}}{\text{adjacent}} = \frac{a}{b} = \frac{1600}{1200} = \frac{4}{3}$$
$$A \approx 53^\circ$$ *A is rounded to the nearest degree.*

Since $\theta = 45^\circ + A = 45^\circ + 53^\circ = 98^\circ$, we conclude that the azimuth of the plane is approximately 98°. The plane is at a distance of 2000 mi from County Airport.

EXAMPLE 5

You are traveling at a speed of 50 mph in the rightmost lane of a highway having three lanes in each direction. A van passes you ($\approx$ 25 ft to your left), and two seconds later you estimate that the angle between the direction of motion and your line of sight to the van is roughly 20°. How fast is the van traveling?

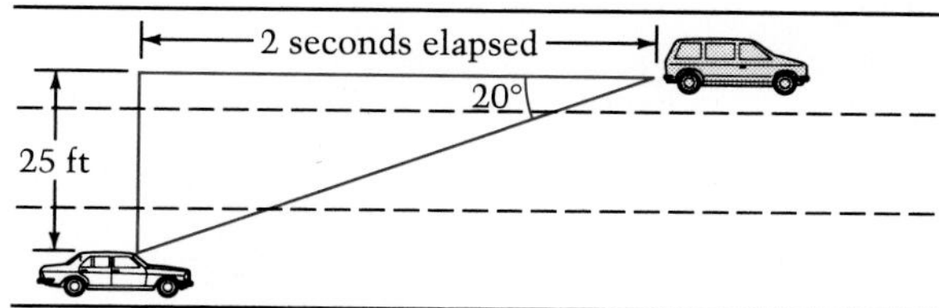

Solution

- ANALYZE We must determine the speed of the van. To organize our work we draw a right triangle that summarizes the information.

■ ORGANIZE Let x = distance traveled by van in 2 seconds.

Draw a sketch

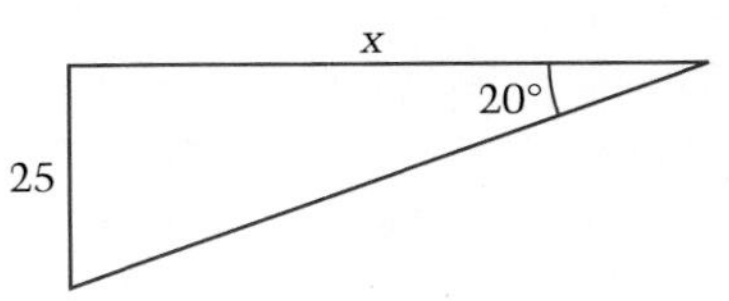

Ask yourself: which trig function involves the known side and the desired side?

■ MODEL

$$\tan 20^\circ = \frac{25}{x}$$

$$0.364 \approx \frac{25}{x}$$

$$x = \frac{25}{0.364}$$

$$x \approx 69$$

Thus, relative to you, the van is traveling at a rate of 69 feet in 2 seconds. We convert that rate to miles per hour.

$$\left(\frac{69 \text{ ft}}{2 \text{ sec}}\right)\left(\frac{1 \text{ mi}}{5280 \text{ ft}}\right)\left(\frac{3600 \text{ sec}}{1 \text{ hr}}\right) \approx 24 \text{ mph}$$

■ ANSWER The van is traveling 24 mph *faster* than you are. Therefore its speed is approximately 74 mph.

EXERCISE SET 6.5

In Exercises 1–10, some information about the right triangle ABC is given. Use only the Table of Special Angles to solve $\triangle ABC$. (Assume the right angle is at C.)

1. $B = 45^\circ, c = 64$
2. $A = 30^\circ, c = 48$
3. $A = 60^\circ, b = 44$
4. $B = 60^\circ, b = 12$
5. $a = 15, b = 15\sqrt{3}$
6. $b = 40, c = 80$
7. $A = 45^\circ, a = 44$
8. $B = 30^\circ, a = 24$
9. $a = 25\sqrt{2}, c = 50$
10. $b = 25, a = 25$

In Exercises 11–20, use the given information to solve the right triangle ABC. Be sure to adjust the number of significant digits in the answer.

11. $A = 36^\circ, a = 70$
12. $B = 72^\circ, a = 40$
13. $B = 12^\circ, a = 9$
14. $A = 53^\circ, c = 20$
15. $c = 12.3, A = 38^\circ$
16. $a = 42.5, A = 67^\circ$
17. $b = 0.244, A = 28^\circ 20'$
18. $A = 51^\circ 40', b = 7.85$
19. $c = 142.5, B = 71^\circ 20'$
20. $c = 6.782, B = 23^\circ 40'$

In Exercises 21–30, solve the right triangle ABC with right angle at C.

21. $a = 23.47, A = 32.8^\circ$
22. $a = 1.831, B = 58.2^\circ$
23. $b = 1293.0, A = 47.73^\circ$
24. $b = 2.828, A = 12.62^\circ$
25. $c = 19.800, A = 28.19^\circ$
26. $c = 505.07, B = 41.38^\circ$
27. $a = 2.3261, b = 2.4495$
28. $a = 84.853, c = 305.123$
29. $b = 0.9847, a = 0.2231$
30. $b = 398.24, c = 1763.66$

31. A 24 ft ladder is leaning against a building. If the ladder makes an angle of 60° with the ground, how far up the building does the ladder reach?

32. If an 18 ft ladder is placed against a building so that it reaches a window sill 9 ft off the ground, what is the acute angle between the ladder and the ground?

33. A jet is flying at an altitude of 2000 ft. The angle of depression to an aircraft carrier is 20°. How far is the jet from the carrier?

34. When the angle of elevation of the sun is 60°, a certain flagpole casts a shadow 30 ft long. How tall is this flagpole?

35. A 60 ft long ramp is inclined at an angle of 5° with the level ground. How high does the ramp rise above the ground?

36. A ramp for wheelchairs is to be built beside the main steps of the library. The total vertical rise of the steps is 3 ft and the ramp will be inclined at an angle of 12°. How long a ramp is needed?

37. A television crew 2600 ft from a launch pad is filming the launch of a space shuttle. What is the angle of elevation of the camera when the shuttle is 4000 ft directly above the pad?

38. From an 80 ft tall lighthouse on the coast, an overturned sailboat is sighted. If the angle of depression is 9°, how far is the boat from the lighthouse?

39. A 50 ft tall flagpole casts a shadow on level ground. What is the angle of elevation of the sun when the shadow is (a) 29 ft long? (b) 60.0 ft long? (c) 12.2 ft long? (d) 125.00 ft long?

40. The Charleston Light in Charleston, SC is one of the most powerful lighthouses in the Western Hemisphere. It is 163 ft high, and its light can be seen 19 mi out at sea. What is the distance of a small boat from the foot of the tower if the angle of depression of the boat from the tower is (a) 30°? (b) 9°32′?

Superset

41. From level ground, your angle of elevation to a distant cliff is 30°. By walking a distance of 2000 ft directly toward the foot of the cliff, your angle of elevation becomes 45°. What is the height of the cliff?

42. A 20 ft tall flagpole is mounted on the edge of the roof of a building. A six foot tall person standing level with the base of the building measures the angle of elevation to the top of the flagpole to be 65°. From the same spot, the angle of elevation to the foot of the flagpole is 60°. What is the height of the building?

43. A flagpole is mounted on the edge of the roof of a 90 foot tall building. A five foot tall person standing level with the base of the building measures the angle of elevation to the top of the flagpole to be 62°. From the same spot, the angle of elevation to the foot of the flagpole is 56°. How tall is the flagpole?

44. A video camera is to be installed in a bank to monitor the bank teller's counter, which is 4 feet high. The camera will be mounted on the wall at a height of 10 ft. The counter is 20 ft from the wall on which the camera is to be mounted. To aim the camera at the top of the counter, what should be the angle of depression of the camera?

45. A 52 ft long guy wire runs from level ground to the top of an antenna. The guy wire makes an angle of 68.6° with the antenna. How far out from the base of the antenna is the guy wire anchored?

46. A railroad track passes below a long, level highway bridge and at a right angle to it. At the same instant that a train traveling 60 mph passes under the bridge, a car traveling 45 mph on the bridge is directly above the train. If the highway bridge is 100 feet above the track, find the distance between the train and car 10 seconds later. (Note that 60 mph = 88 ft/s and 45 mph = 66 ft/s.)

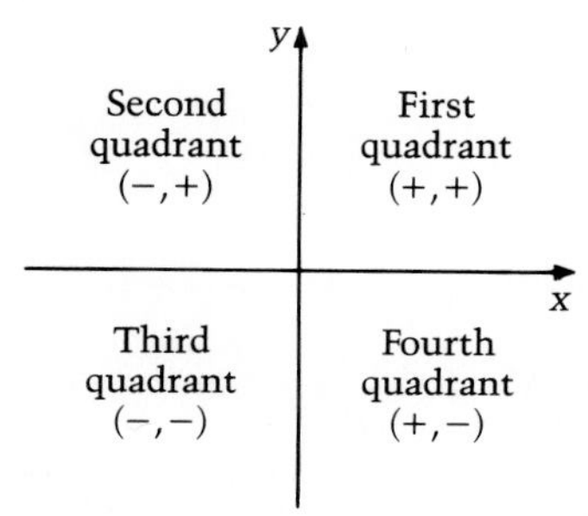

Figure 6.49 The signs of (x, y)

6.6 Reference Angles

We have defined the six trigonometric functions of an angle θ in terms of x, y, and r, where x and y are coordinates of a point on the terminal side of θ, and r is the distance between that point and the origin. Since r must be positive, the sign of a trigonometric function value, such as cos 210°, is determined by the signs of x and y. For example, since $\cos\theta = \dfrac{x}{r}$ and r is positive, the sign of $\cos\theta$ is determined by the sign of x, which is positive in the first and fourth quadrants and negative in the second and third quadrants.

The chart below summarizes information needed to determine the sign of a trigonometric function value.

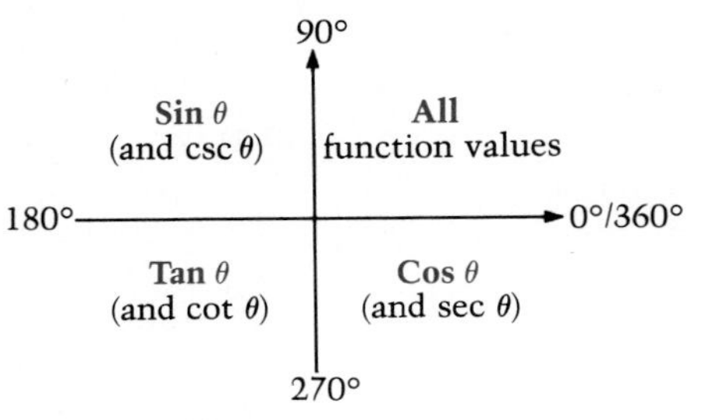

Figure 6.50 For each quadrant, the trigonometric functions which are positive.

The memory device ASTC (Always Study Trigonometry Carefully) is often used as a reminder of which trigonometric functions are positive in which quadrants.

positive functions	A	S	T	C
quadrant	1st	2nd	3rd	4th

EXAMPLE 1

Determine the sign of the following:

(a) sin 190° (b) tan 135° (c) cos(−50°)

Solution

(a) 190° is a third quadrant angle.

Think:

A	S	T	C
1	2	3	4

In the third quadrant, only tan and cot are positive.

sin 190° is negative.

(b) 135° is a second quadrant angle.

Think:

A	S	T	C
1	2	3	4

In the second quadrant, only sin and csc are positive.

tan 135° is negative.

(c) −50° is a fourth quadrant angle.

Think:

A	S	T	C
1	2	3	4

In the fourth quadrant, only cos and sec are positive.

cos (−50°) is positive.

Suppose you wish to evaluate cos 210°. You already know one part of the answer: the *sign* of cos 210° is negative. Thus,

$$\cos 210° = -\underline{\quad\quad}.$$

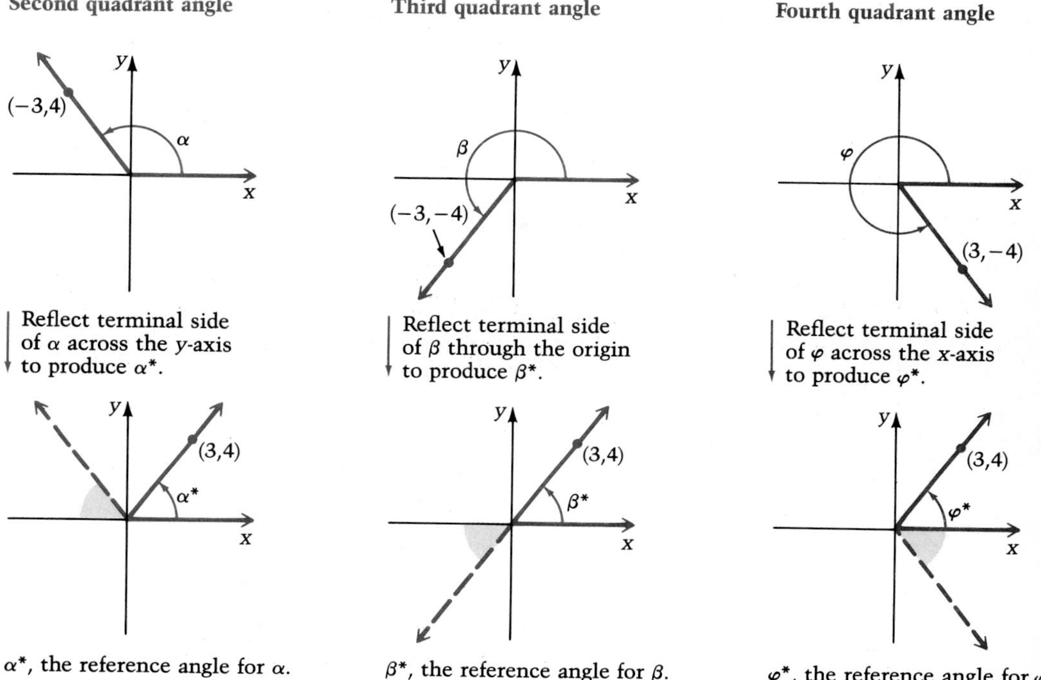
Second quadrant angle
Third quadrant angle
Fourth quadrant angle
(−3,4)
α
(−3,−4)
β
φ
(3,−4)
Reflect terminal side of α across the y-axis to produce α*.
Reflect terminal side of β through the origin to produce β*.
Reflect terminal side of φ across the x-axis to produce φ*.
(3,4)
α*
β*
φ*
α*, the reference angle for α.
β*, the reference angle for β.
φ*, the reference angle for φ.

Figure 6.51 An angle equal to the reference angle is shaded.

To complete the statement $\cos 210° = -$____, we use a *reference angle.* A **reference angle** for a given angle θ is a first quadrant angle whose trigonometric function values are numerically the same as the trigonometric function values of θ—only the signs may differ.

Figure 6.51 presents second, third, and fourth quadrant angles. In each case, the reference angle is found by performing the appropriate reflection of the terminal side necessary to make it a first quadrant angle. We denote the reference angle of any angle θ by attaching an asterisk (θ^*).

Figure 6.51 illustrates the following useful procedure.

Procedure for Finding a Reference Angle

Suppose θ is an angle between 0° and 360° inclusive and θ^* is the reference angle of θ.

If θ is a first quadrant angle, $\theta^* = \theta$.
If θ is a second quadrant angle, $\theta^* = 180° - \theta$.
If θ is a third quadrant angle, $\theta^* = \theta - 180°$.
If θ is a fourth quadrant angle, $\theta^* = 360° - \theta$.

If you are given an angle greater than 360° or less than 0°, add or subtract an appropriate multiple of 360° so that the measure of the resulting angle is between 0° and 360°. Then use the above procedure.

EXAMPLE 2

Determine the reference angle for each of the following angles.

(a) $\theta = 240°$ (b) $\theta = 1000°$ (c) $\theta = -200°$

Solution

(a) 240° is a third quadrant angle. Therefore $\theta^* = 240° - 180° = 60°$. The reference angle is 60°.

(b) 1000° is greater than 360°. Subtract $2 \cdot 360°$ from 1000° to get 280°. Now, 280° is a fourth quadrant angle. Thus, $\theta^* = 360° - 280° = 80°$. The reference angle is 80°.

(c) $-200°$ is less than 0°. Add $1 \cdot 360°$ to get 160°. Now, 160° is a second quadrant angle. Thus, $\theta^* = 180° - 160° = 20°$. The reference angle is 20°.

positive functions	A	S	T	C
quadrant	1st	2nd	3rd	4th

ASTC Chart

Because of the way we defined a reference angle, the trigonometric function values of a given angle are the same as the trigonometric function values of the reference angle, except maybe for the sign. We use the ASTC chart to determine the correct sign.

To evaluate an expression like tan 240°, we follow a three step procedure:

Step 1. Determine the sign of $\tan\theta$: tan 240° is positive

Step 2. Find the reference angle θ^*: $240° - 180° = 60°$

Step 3. Evaluate $\tan\theta^*$: $\tan 60° = \sqrt{3}$

Thus, $\tan 240° = +\tan 60° = \sqrt{3}$.

EXAMPLE 3

Evaluate (a) cos 120° (b) sin 280.3°.

Solution

(a) $\cos 120° = \text{sign} \cos \left(\text{reference angle of } 120°\right)$

$= -\cos \left(\text{reference angle of } 120°\right)$ — 120° is a second quadrant angle; cos is negative there.

$= -\cos 60°$ — $\theta^* = 180° - 120° = 60°$.

$= -\dfrac{1}{2}$ — Note that this problem can be done without a calculator.

(b) $\sin 280.3° = \text{sign} \ \sin \boxed{\text{reference angle of } 280.3°}$

$= -\sin \boxed{\text{reference angle of } 280.3°}$ 280.3° is a fourth quadrant angle; sin is negative there.

$= -\sin 79.7° = -0.9839$ $\theta^* = 360° - 280.3° = 79.7°$

EXERCISE SET 6.6

In Exercises 1–14, determine the sign of the quantity.

1. cot 200° 2. sec 175° 3. cos 315°
4. tan 285° 5. sin 457° 6. cot 703°
7. csc 1210° 8. sin 1568° 9. tan(−279°)
10. csc(−112°) 11. sec(−763°) 12. cos(−581°)
13. sin 196°36′ 14. tan(−185°42′)
15. cos(−300.6°) 16. sin 328.8°

In Exercises 17–32, determine the reference angle for the given angle.

17. 200° 18. 185° 19. 320°
20. 197° 21. 485° 22. 696°
23. −250° 24. −185° 25. −444°
26. −715° 27. −1445° 28. −1081°
29. 265°35′ 30. −95°58′ 31. 510.3°
32. −112.5°

In Exercises 33–66, evaluate the quantity.

33. cos 240° 34. csc 240° 35. tan 300°
36. sin 120° 37. cot 225° 38. sec 135°
39. sec 330° 40. tan 180° 41. csc 300°
42. sec 300° 43. tan 750° 44. cos 1290°
45. sin(−210°) 46. sin(−1290°) 47. tan 158°
48. sin 345° 49. cos 408° 50. cot 497°
51. cot 320° 52. tan 212° 53. cos 1258°
54. cos 485° 55. sec(−280°) 56. csc(−272°)
57. sin(−1134°) 58. tan(−666°)
59. csc(−138°) 60. sec(−295°)
61. sin 95°20′ 62. cot 245°40′
63. cot(−100°30′) 64. cos 250°10′
65. sin 310.7° 66. tan(−155.4°)

Superset

In Exercises 67–76, assume that the angle θ terminates in the given quadrant. Find the sign of the given trigonometric function value.

67. second quadrant; cos θ
68. third quadrant; sin θ
69. fourth quadrant; tan θ
70. fourth quadrant; sec θ
71. third quadrant; cos θ
72. fourth quadrant; cot θ
73. second quadrant; sec θ
74. second quadrant; tan θ
75. fourth quadrant; sin θ
76. second quadrant; csc θ

In Exercises 77–82, assume that the angle θ has measure between 180° and 270°. Find the sign of the given trigonometric function value.

77. $\sin \frac{\theta}{2}$ 78. $\sin 2\theta$ 79. $\sin(90° + \theta)$
80. $\sin(90° - \theta)$ 81. $\sin(180° - \theta)$ 82. $\sin(180° + \theta)$

6.7
Law of Sines

In Section 6.5 we used trigonometric functions to solve right triangles. In this section and the next, we shall derive the Law of Sines and the Law of Cosines, which allow us to solve triangles which are *not* right triangles. Triangles that are not right triangles are called **oblique.**

First, we shall consider the Law of Sines. We begin by considering any oblique triangle ABC. There are two possibilities. **Case 1:** $\triangle ABC$ is an *acute* triangle (all angles are less than 90°) as shown in Figure 6.52(a). **Case 2:** $\triangle ABC$ is an *obtuse* triangle (one of its angles is greater than 90°) as shown in Figure 6.52(b). We have drawn the altitude h from vertex B to side b (Case 1), or to the line containing side b (Case 2).

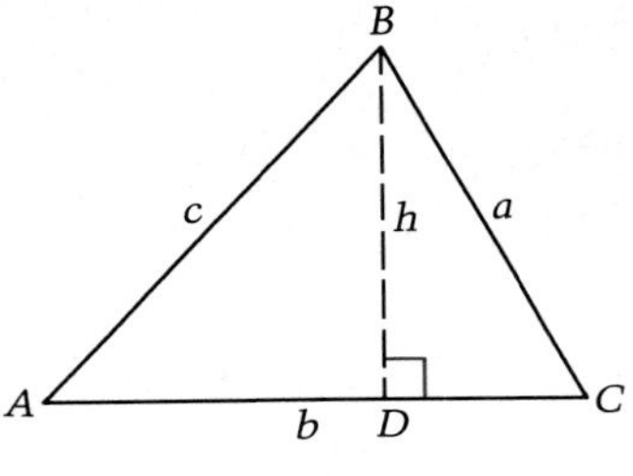

(**a**) Case 1: acute triangle

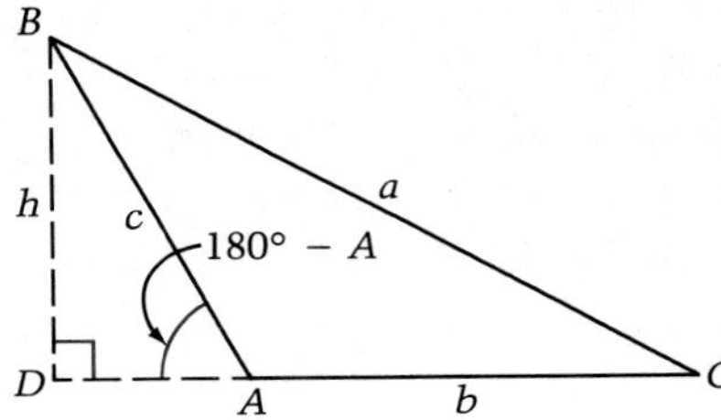

(**b**) Case 2: obtuse triangle

Figure 6.52

Note that in Case 2, A is a second quadrant angle, and so, by our work with reference angles, we know that

$$\sin(180° - A) = \sin A.$$

Thus, in both cases,

$$h = a \sin C, \quad \text{and} \quad h = c \sin A.$$

Equating the two expressions for h, we get $a \sin C = c \sin A$, or

$$\frac{\sin A}{a} = \frac{\sin C}{c}.$$

Had we drawn the altitude from vertex C to side c, we would have concluded

$$\frac{\sin A}{a} = \frac{\sin B}{b}.$$

We summarize our results as follows:

Law of Sines

In any triangle with angles A, B, and C, and opposite sides a, b, and c, respectively,

$$\frac{\sin A}{a} = \frac{\sin B}{b} = \frac{\sin C}{c}.$$

EXAMPLE 1

Given $\triangle ABC$ with $A = 50.1°$, $B = 70.6°$, and $c = 10.5$, solve the triangle.

Solution Begin by drawing a triangle and labeling its parts.

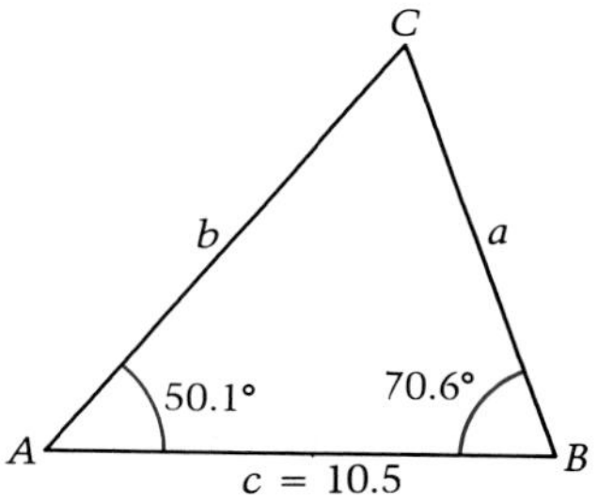

Figure 6.53

Find C: $C = 180° - 50.1° - 70.6° = 59.3°$

Find a: $\frac{\sin 50.1°}{a} = \frac{\sin 59.3°}{10.5}$ $\qquad \frac{\sin A}{a} = \frac{\sin C}{c}.$

$$a = \frac{(10.5)(\sin 50.1°)}{\sin 59.3°}$$

$$= \frac{(10.5)(0.7672)}{0.8599} \approx 9.37$$ Round to three significant digits.

Find b: $\frac{\sin 70.6°}{b} = \frac{\sin 59.3°}{10.5}$ $\qquad \frac{\sin B}{b} = \frac{\sin C}{c}.$

$$b = \frac{(10.5)(0.9432)}{0.8599} \approx 11.5$$ Round to three significant digits.

We use the Law of Sines to solve oblique triangles when we are given (1) one side and two angles, or (2) two sides and the angle opposite one of them. The first type of problem was treated in Example 1. The second type is more complicated.

Consider what might happen if we are given two sides, say a and b, and acute angle A. There are four situations that can occur. (In the figures below, h is the altitude from vertex C.)

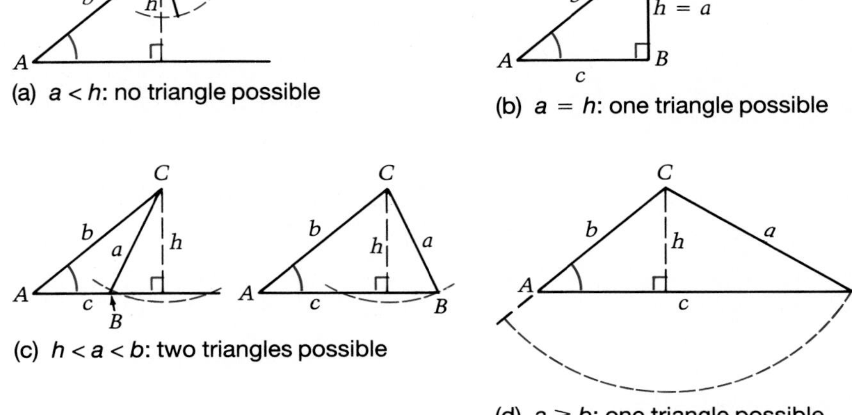

Figure 6.54

Example 2 illustrates case (c) above, namely the case where two triangles are possible. This is generally called *the ambiguous case.*

EXAMPLE 2

Solve $\triangle ABC$, given that $A = 24°30'$, $a = 6.00$, and $b = 12.2$.

Solution Before we can draw the triangle, we must determine another angle.

Find B: $$\frac{\sin B}{12.2} = \frac{\sin 24°30'}{6.00} \qquad \frac{\sin B}{b} = \frac{\sin A}{a}.$$

$$\sin B = \frac{0.4147}{6.00}(12.2)$$

$$\approx 0.8432$$

Recall that the sine function is positive for angles in the first or second quadrant. Since $\sin 57°30' \approx 0.8432$, the second quadrant angle having the same sine function value is $180° - 57°30' = 122°30'$. Thus, we have two cases, as shown in Figure 6.55:

$$B = 57°30' \quad \text{or} \quad 122°30'.$$

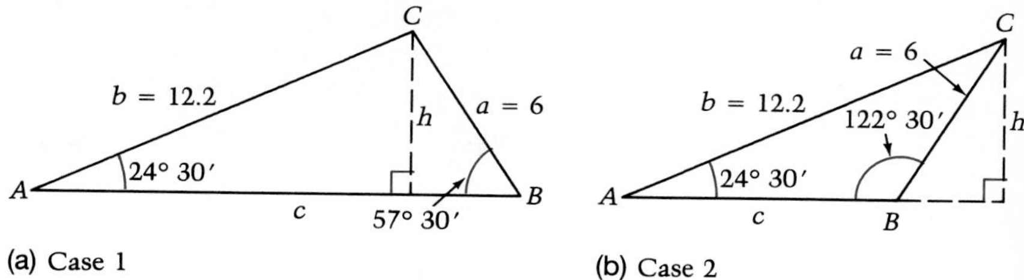

Figure 6.55

Case 1: $B = 57°30'$

Find C: $C = 180° - 24°30' - 57°30' = 98°00'$

Find c:

$$\frac{\sin 98°00'}{c} = \frac{\sin 24°30'}{6.00} \qquad \frac{\sin C}{c} = \frac{\sin A}{a}.$$

$$\frac{0.9903}{c} = \frac{0.4147}{6.00} \qquad \sin 98°00' = \sin(180° - 98°00') = \sin 82°00' = 0.9903$$

$$c = \left(\frac{6.00}{0.4147}\right)(0.9903)$$

$$\approx 14.3$$

Case 2: $B = 122°30'$

Find C: $C = 180° - 24°30' - 122°30' = 33°00'$

Find c:

$$\frac{\sin 33°00'}{c} = \frac{0.4147}{6.00} \qquad \frac{\sin C}{c} = \frac{\sin A}{a}.$$

$$c = \frac{6.00}{0.4147}(0.5446) \approx 7.88$$

In discussing the case where two sides and an opposite angle are given, we have not considered the case where the given angle is obtuse.

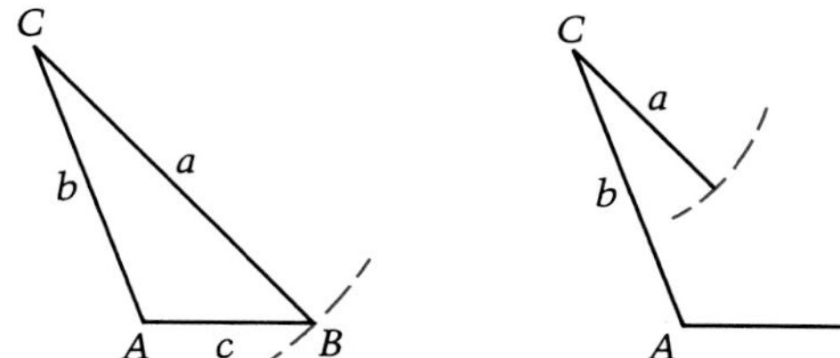

Figure 6.56

Clearly in this situation, a triangle cannot be formed if $a \le b$.

In practice, applying the Law of Sines will supply you with the information you need in order to decide which of the many possible cases can occur for the given data. For example, in the case of "no possible triangle," one step in the computation might yield that the sine of an unknown angle is greater than 1, which is impossible.

EXAMPLE 3

A forest fire is spotted by observers in two fire towers 12.0 miles apart. Tower B is on a bearing of S 12°10′ E from Tower A. If the bearing of the fire from Tower A is S 45°40′ W and from Tower B is N 75°20′ W, how far is the fire from Tower B?

Solution

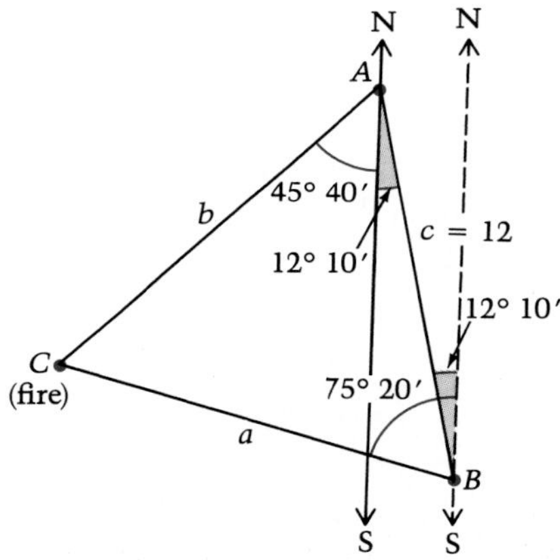

Draw a figure to represent the given data.

To determine B, use the fact that when parallel lines are cut by a transversal, alternate interior angles (shaded angles) are congruent. Thus

$$B = 75°20' - 12°10' = 63°10'.$$

Note that $A = 45°40' + 12°10' = 57°50'$.

Let us now simplify our diagram, showing the known parts of the triangle.

$$\begin{aligned} C &= 180° - A - B \\ &= 180° - 57°50' - 63°10' \\ &= 59°00'. \end{aligned}$$

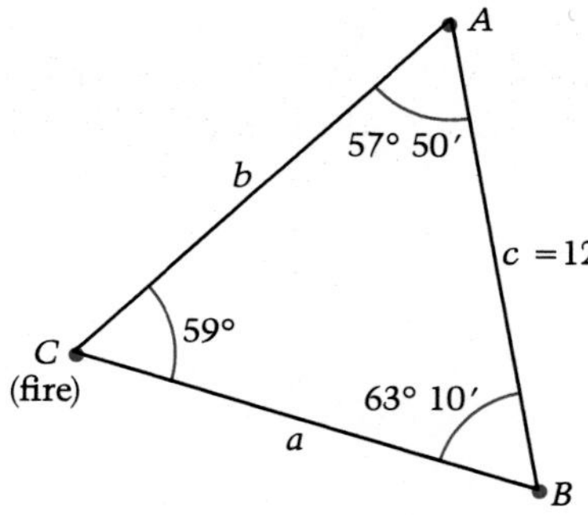

Figure 6.57

Find a: $$\frac{\sin 57°50'}{a} = \frac{\sin 59°00'}{12.0} \qquad \frac{\sin A}{a} = \frac{\sin C}{c}$$

$$a = \frac{12.0}{0.8572}(0.8465) \approx 11.9 \text{ mi}$$

The fire is approximately 11.9 mi from Tower B.

EXERCISE SET 6.7

In Exercises 1–10, the measures of the two angles and one side of $\triangle ABC$ are given. Solve the triangle.

1. $A = 30°, B = 45°, a = 10\sqrt{2}$
2. $A = 60°, B = 45°, b = 8\sqrt{6}$
3. $B = 30°, C = 135°, b = 4\sqrt{2}$
4. $B = 45°, C = 120°, c = 4\sqrt{6}$
5. $a = 24.0, A = 100°, B = 24°$
6. $b = 54.0, A = 76°, B = 41°$
7. $c = 42.0, B = 36°, C = 64°$
8. $a = 96.0, A = 105°, C = 21°$
9. $a = 0.7280, B = 71°30', C = 66°50'$
10. $b = 0.4980, A = 37°20', C = 92°10'$

In Exercises 11–20, solve triangle ABC.

11. $A = 101.7°, B = 19.3°, c = 118.5$
12. $A = 42.8°, C = 51.1°, b = 92.75$
13. $B = 75.6°, C = 33.2°, a = 4384.0$
14. $A = 48.6°, B = 94.1°, c = 282.84$
15. $a = 14.95, b = 17.43, B = 58.6°$
16. $b = 0.4731, a = 0.4347, B = 78.4°$
17. $a = 6241, c = 8233, C = 31.7°$
18. $a = 3.3569, c = 2.7973, A = 62.5°$
19. $b = 756.2, c = 541.6, B = 112.3°$
20. $b = 464.5, c = 637.3, C = 73.8°$

In Exercises 21–28, you are given measures for A, a, and b. State whether it is possible to have $\triangle ABC$. If it is, solve the triangle. (Determine both solutions if there is more than one.)

21. $A = 30°, a = 12, b = 24$
22. $A = 60°, a = 36, b = 24$
23. $A = 120°, a = 18, b = 24$
24. $A = 135°, a = 24, b = 24\sqrt{2}$
25. $A = 120°, a = 54.3, b = 48.8$
26. $A = 135°, b = 36.4, a = 44.7$
27. $a = 32.2, b = 36.4, A = 52°30'$
28. $b = 24.8, a = 20.6, A = 38°30'$

29. Points A and B are on opposite sides of the Grand Canyon. Point C is 200 yd from A, $m(\angle BAC) = 87°30'$, and $m(\angle ACB) = 67°12'$. What is the distance between A and B?

30. Two observers standing on shore $\frac{1}{2}$ mi apart at points A and B measure the angle to a sailboat at point C at the same time. If $m(\angle CAB) = 63.4°$ and $m(\angle CBA) = 56.6°$, find the distance from each observer to the boat.

31. Ship A is 485 m due east of ship B. A lighthouse 1600 m from ship A is on a bearing of N 17°18′ E from ship B. What is the bearing of the lighthouse from ship A?

32. Two observers 2 mi apart on level ground are in line with a spot directly below a hot air balloon. If the angle of elevation of the balloon for one observer is 68.9° and, at the same time, 26.4° for the other, what is the altitude of the balloon to the nearest tenth of a mile?

33. An observer on a ship spots a life raft at a bearing of N 75°24′ E while, at the same instant, a second observer on another ship takes a bearing of the life raft of N 15°54′ E. If the second ship is at a distance of 5.5 miles and a bearing of S 27°54′ E from the first ship, find the distance from each ship to the life raft.

34. An observer on a ship spots a life raft at a bearing of N 35°18′ W while, at the same instant, a second observer on another ship takes a bearing of the life raft of N 10°36′ E. If the second ship is 4.8 mi and on a bearing of S 55°42′ W from the first ship, how far is each ship from the life raft?

35. A wind storm caused a 64 ft tall telephone pole to lean due east at an angle of 63.8° to the ground. A 106 ft long guy wire is then attached to the top of the pole and anchored in the ground due west of the foot of the pole. How far from the foot of the pole is the guy wire anchored?

36. Two lighthouses are located 49 mi apart on the coast of Massachusetts, one near Salem and the other in Provincetown on Cape Cod. The Provincetown lighthouse is on a bearing of S 46° E from the Salem lighthouse. A fishing boat in distress radios an SOS signal. The bearing of the signal from the Salem station is S 55°12′ E and from Provincetown N 47°36′ E. How far is the ship from Provincetown?

Superset

37. Express x in terms of θ, φ and m.

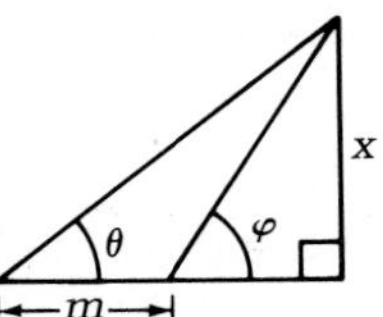

In Exercises 38–39, prove the given statement.

38. $\dfrac{a + b}{b} = \dfrac{\sin A + \sin B}{\sin B}$

39. $\dfrac{a - b}{a + b} = \dfrac{\sin A - \sin B}{\sin A + \sin B}$

40. Show that the area of triangle ABC is given by

$$\frac{a^2 \sin B \sin C}{2 \sin A}.$$

6.8 Law of Cosines

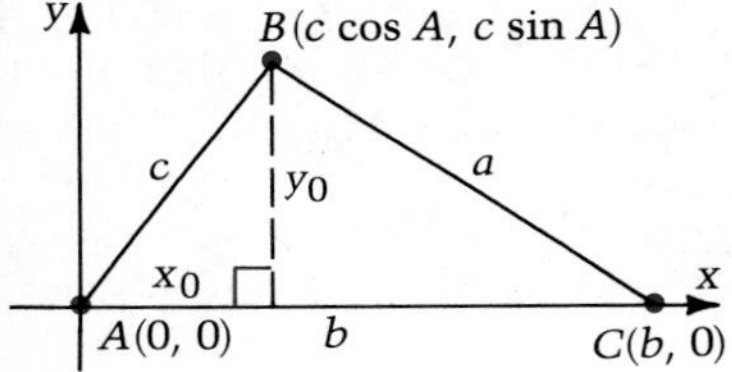

Figure 6.58 *B* has coordinates $x_0 = c \cos A$, $y_0 = c \sin A$.

In the last section, we considered the Law of Sines and how it is used to solve triangles when we are given (1) one side and two angles, or (2) two sides and an angle opposite one of them. We now derive a formula, called the Law of Cosines, which will be useful if we are given (3) two sides and the included angle, or (4) three sides. Suppose we have $\triangle ABC$ positioned so that side b lies on the x-axis (Figure 6.58). Then

$$x_0 = c \cos A, \quad \text{and} \quad y_0 = c \sin A. \qquad \cos A = \frac{x_0}{c} \text{ and } \sin A = \frac{y_0}{c}.$$

By the distance formula, we have

$$\begin{aligned} a^2 &= (c \cos A - b)^2 + (c \sin A - 0)^2 \\ &= (c \cos A)^2 - 2(c \cos A)(b) + b^2 + (c \sin A)^2 \\ &= c^2 \cos^2 A - 2bc \cos A + b^2 + c^2 \sin^2 A \\ &= b^2 + c^2(\sin^2 A + \cos^2 A) - 2bc \cos A \\ &= b^2 + c^2 - 2bc \cos A \qquad \sin^2 A + \cos^2 A = 1. \end{aligned}$$

Note that the formula gives us a means of determining the square of one side of a triangle, given the other two sides and their included angle. There are three forms of this formula, as stated below.

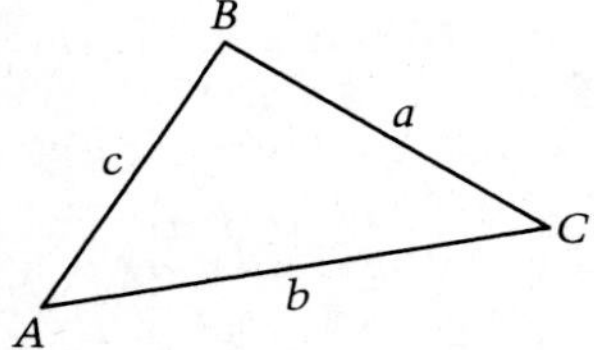

Figure 6.59

Law of Cosines

In any triangle with angles A, B, and C, and opposite sides a, b, and c, respectively,

$$\begin{aligned} c^2 &= a^2 + b^2 - 2ab \cos C, \\ a^2 &= b^2 + c^2 - 2bc \cos A, \\ b^2 &= a^2 + c^2 - 2ac \cos B. \end{aligned}$$

EXAMPLE 1

Given $\triangle ABC$ with $a = 4$, $b = 6$, and $C = 120°$, find c.

Solution Begin by drawing a triangle that represents the given data.

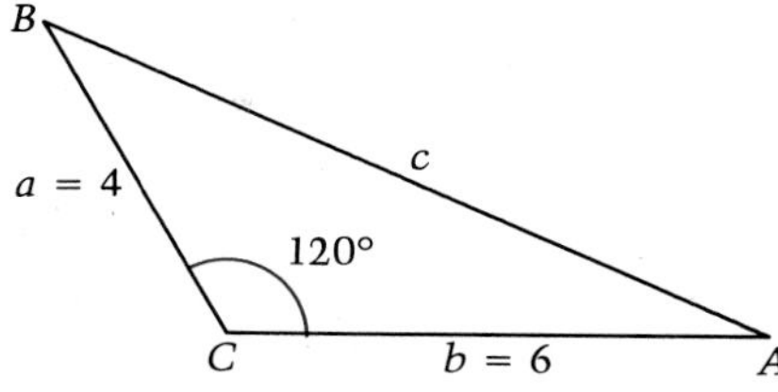

Figure 6.60

$$c^2 = a^2 + b^2 - 2ab \cos C$$
$$c^2 = 4^2 + 6^2 - 2(4)(6) \cos 120°$$
$$c^2 = 16 + 36 - 48\left(-\frac{1}{2}\right) \qquad \cos 120° = -\cos 60° = -\frac{1}{2}.$$
$$c^2 = 52 + 24 = 76$$
$$c = \sqrt{76} \approx 8.7$$

EXAMPLE 2

Given $\triangle ABC$ with $a = 6.00$, $b = 12.0$, and $c = 7.00$, find B.

Solution

$$b^2 = a^2 + c^2 - 2ac \cos B$$
$$12^2 = 6^2 + 7^2 - 2(6)(7) \cos B$$

The form of the Law of Cosines that involves angle B is used.

$$\cos B = \frac{144 - 36 - 49}{-2(6)(7)} \approx -0.7024$$

Since $\cos B$ is negative, B is the second quadrant angle whose reference angle has cosine function value 0.7024.

Since $\cos 45.4° \approx 0.7024$, we get $B \approx 180° - 45.4° = 134.6°$.

EXAMPLE 3

Highway 102 runs east-west and is intersected by Route 66, in a direction 20° north of due east. Car A is traveling along Highway 102 and is 4 miles east of the intersection. Car B is traveling eastbound on Route 66 and is 18 miles past the intersection. What is the distance between the two cars?

Solution

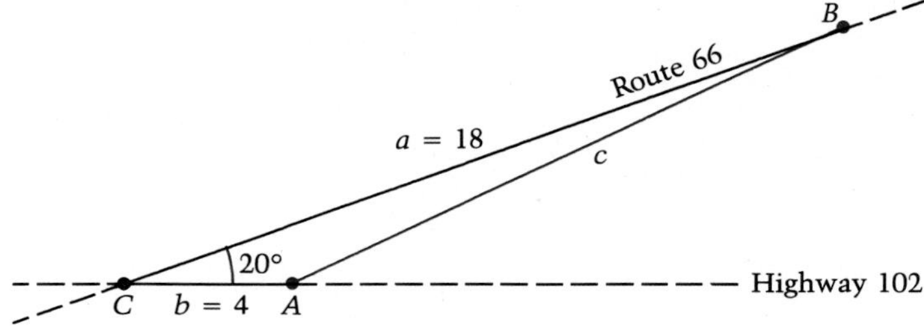

Figure 6.61

We wish to determine the distance c shown above.

$$c^2 = a^2 + b^2 - 2ab\cos C$$
$$c^2 = 18^2 + 4^2 - 2(4)(18)\cos 20°$$
$$c^2 = 324 + 16 - 144(0.9397) \approx 204.7$$
$$c \approx 14.3$$

The distance between the two cars is approximately 14 mi.

EXAMPLE 4

Points B and C are on opposite sides of a reservoir. If the distance from point A to B is known to be 1.25 mi, and from point A to C is 1.15 mi, what is the distance between B and C if $\angle BAC = 55°10'$?

Solution Draw a figure that represents the data.

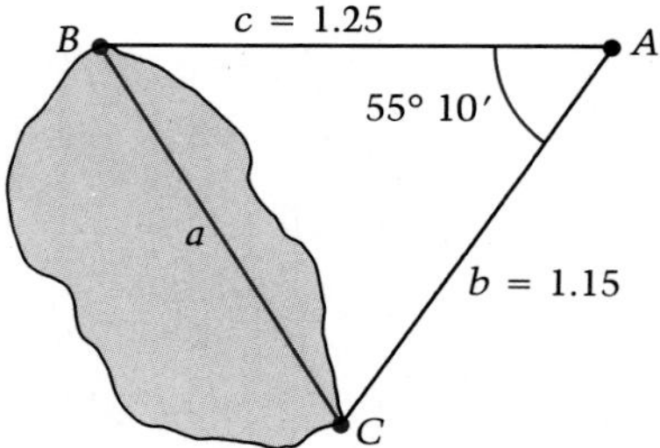

Figure 6.62

$$a^2 = (1.15)^2 + (1.25)^2 - 2(1.15)(1.25)(\cos 55°10') \approx 1.24$$

Law of Cosines is applied to $\triangle ABC$. The desired distance is a.

$$a \approx 1.11$$

The distance between B and C is approximately 1.11 mi.

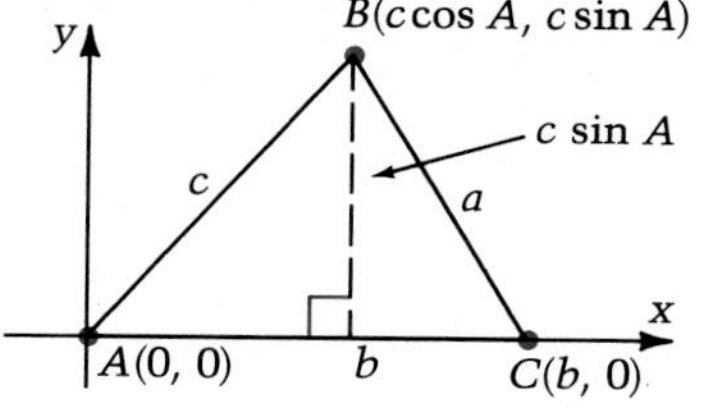

Figure 6.63

We now return to the figure we used to motivate the derivation of the Law of Cosines (see Figure 6.63). According to the formula for the area of a triangle, we can describe the area $\mathbb{A}$ of $\triangle ABC$ as

$$\mathbb{A} = \frac{1}{2}(\text{altitude})(\text{base}) = \frac{1}{2}(c \sin A)(b) = \frac{1}{2}bc \sin A.$$

In general, the area of a triangle is half the product of the lengths of any two sides times the sine of the included angle. For $\triangle ABC$, $\mathbb{A}$ can be expressed three ways:

$$\mathbb{A} = \frac{1}{2}ab \sin C, \qquad \mathbb{A} = \frac{1}{2}ac \sin B, \qquad \mathbb{A} = \frac{1}{2}bc \sin A.$$

EXAMPLE 5

Find the area of $\triangle ABC$ if $a = 13$ cm, $b = 10$ cm, and $C = 30°$.

Solution Draw a triangle that represents the data.

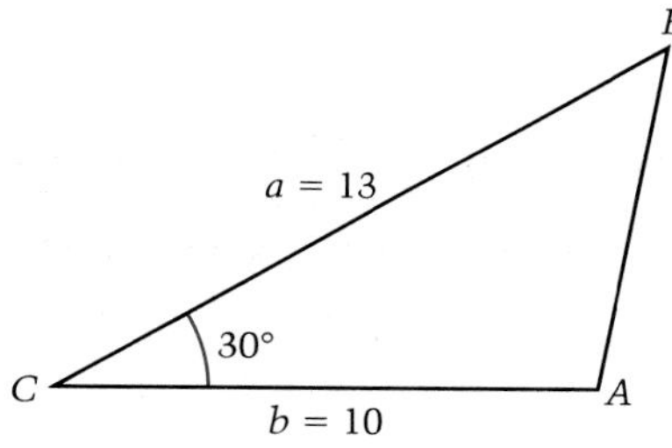

Figure 6.64

$$\begin{aligned} \text{Area} &= \frac{1}{2}(\text{product of two sides}) \cdot \sin(\text{included angle}) \\ &= \frac{1}{2}(ab)(\sin 30°) \\ &= \frac{1}{2}(13 \cdot 10)\left(\frac{1}{2}\right) = 32.5 \end{aligned}$$

The area of the triangle is 33 cm^2, rounded to two significant digits.

EXAMPLE 6

Find the area of $\triangle ABC$ if $a = 22$ in., $b = 16$ in., and $c = 18$ in.

Solution Draw a triangle that represents the data (see Figure 6.65). Since we are not given the lengths of two sides *and* the measure of an included angle, we first apply the Law of Cosines to determine one of the angles. We will find C.

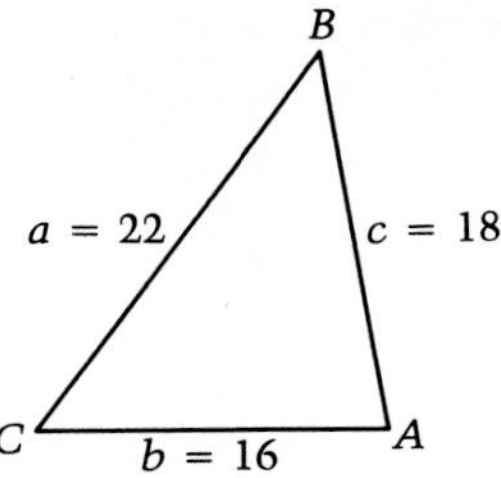

Figure 6.65

$$c^2 = a^2 + b^2 - 2ab\cos C$$
$$\cos C = \frac{c^2 - a^2 - b^2}{-2ab} = \frac{a^2 + b^2 - c^2}{2ab} = \frac{22^2 + 16^2 - 18^2}{2(22)(16)} \approx 0.5909$$
$$C \approx 54^\circ$$

Now that we have a and b and angle C, we can determine the area.

$$\text{Area} = \frac{1}{2}(ab)\sin 54^\circ = \frac{1}{2}(22 \cdot 16)(0.8090) \approx 140 \text{ in.}^2 \text{ (two significant digits)}$$

In the last example, it did not matter which of the three angles we chose in order to apply the formula for area; the result would have been the same. In addition, once we found that $\cos C \approx 0.5909$, we could have noted that since we need $\sin C$ in the formula for area, and since

$$\sin C = \sqrt{1 - \cos^2 C} \quad \text{The Pythagorean Identity}$$

we could have determined $\sin C$ (necessary for computing the area) without ever determining the measure of C.

EXERCISE SET 6.8

In Exercises 1–8, solve $\triangle ABC$, given two sides and the included angle.

1. $a = 12, b = 10, C = 60^\circ$
2. $a = 12, b = 10, C = 30^\circ$
3. $b = 20, c = 16, A = 120^\circ$
4. $b = 24, c = 30, A = 135^\circ$
5. $a = 112, c = 96, B = 54^\circ$
6. $a = 78, c = 125, B = 128^\circ$
7. $A = 147^\circ 40', b = 12.6, c = 16.3$
8. $B = 34^\circ 50', a = 22.3, c = 18.2$

In Exercises 9–16, solve $\triangle ABC$, given a, b, and c.

9. $a = 3.0, b = 4.0, c = 5.0$
10. $a = 5.0, b = 12.0, c = 13.0$
11. $a = 2.8, b = 2.7, c = 2.4$
12. $a = 3.6, b = 5.2, c = 4.8$
13. $a = 3.2, b = 4.8, c = 6.4$
14. $a = 9.0, b = 6.3, c = 5.4$
15. $a = 45.0, b = 30.0, c = 50.0$
16. $a = 30.0, b = 20.0, c = 35.0$

In Exercises 17–26, solve the triangle ABC.

17. $a = 5808, b = 7920, C = 109.3°$

18. $a = 39.12, b = 36.37, C = 80.7°$

19. $b = 9.591, c = 6.856, A = 51.5°$

20. $b = 207.36, c = 214.48, A = 98.4°$

21. $a = 0.9674, c = 0.8834, B = 61.6°$

22. $a = 8347.0, c = 6723.0, B = 68.71°$

23. $a = 2.2365, b = 3.4641, c = 3.1623$

24. $a = 46.90, b = 38.73, c = 33.17$

25. $a = 0.2500, b = 0.1887, c = 0.3448$

26. $a = 3218.8, b = 3555.3, c = 3806.7$

27. Points A and B are sighted from point C. If $C = 98°$, $AC = 128$ m, and $BC = 96$ m, how far apart are the points A and B?

28. Points A and B are sighted from point C. If $C = 36°$, $AC = 118$ ft, and $BC = 105$ ft, how far apart are the points A and B?

29. Two sides and the included angle of a parallelogram have measures 3.2, 4.8, and 54°20′, respectively. Find the lengths of the diagonals.

30. The lengths of two sides of a parallelogram are 24.6 in. and 38.2 in. The angle at one vertex has measure 108°40′. Find the lengths of the diagonals.

31. A bridge is supported by triangular braces. If the sides of each brace have lengths 63 ft, 46 ft, and 40 ft, find the measure of the angle opposite the 46 ft side.

32. The measures of two sides of a parallelogram are 28 in. and 42 in. If the longer diagonal has measure 58 in., find the measures of the angles at the vertices.

In Exercises 33–44, determine the area of $\triangle ABC$ using the given information.

33. $A = 60°, b = 12.6, c = 18.3$

34. $A = 45°, c = 23.7, b = 16.4$

35. $B = 37°12', a = 10.9, c = 15.8$

36. $B = 24°54', c = 10.5, a = 14.6$

37. $C = 112°, b = 44.6, a = 32.5$

38. $C = 118°, a = 18.7, b = 30.6$

39. $A = 13°30', b = 254, c = 261$

40. $A = 66°20', c = 0.231, b = 0.176$

41. $A = 41.8°, b = 35.49, c = 42.78$

42. $B = 69.6°, a = 479.6, c = 404.9$

43. $C = 105.1°, a = 3.784, b = 3.465$

44. $a = 8.485, b = 5.568, c = 7.616$

Superset

45. The lengths of two sides of a triangle are 12 in. and 16 in., and the area is 87.36 in^2. Solve the triangle.

46. Given $\triangle ABC$ with $a = 12, b = 16$, and $c = 10$, find the length of the altitude from vertex B.

47. Find the area of a triangle with vertices at the points with coordinates $(-5, 0)$, $(6, 3)$, and $(8, -5)$.

In Exercises 48–50, assume you are given an isosceles triangle ABC with $a = c$ and $B = \theta$.

48. Find the length of the base b in terms of a and θ.

49. Show that $b = 2a \sin \frac{1}{2}\theta$.

50. Prove that $b^2 = 2a^2(1 - \cos \theta)$.

51. Determine the perimeter of a regular octagon inscribed in a circle of radius 24.6 in.

52. Determine the perimeter of a regular 12-gon inscribed in a circle of radius 35 cm.

53. Find the area of the octagon in Exercise 51.

54. Two sides and the included angle of a parallelogram have measures 17.88, 21.91 and 54.4°, respectively. Find the area of the parallelogram.

Analyzing Data

The Case of the Tidal Tots

In this chapter we defined the trigonometric functions by forming ratios involving all possible pairs of sides in a right triangle: six possible ratios and six different trigonometric functions of an acute angle in a right triangle. However, we found that we could easily extend our definition of the trigonometric functions to any angle. For example, sin 295° and cos (−29°) make sense, even though neither 295° nor −29° is an acute angle in a right triangle.

One of the most important observations we could make at this point is that trigonometric function values repeat over and over in a consistent and predictable fashion. For example, $\sin 45° = \sin 415° = \sin 775° = \frac{\sqrt{2}}{2}$: adding or subtracting any multiple of 360° to angle θ produces the same sine function value. We call this behavior *periodic*, and consider it to be the primary characteristic of the trigonometric functions. What we will learn in the next chapter is that trigonometric functions can be defined so that domain values need not be angles but can be any real number. In practice this means that trigonometric functions are potential models for any process that repeats the same values periodically over time, like the tides. This brings us to our application.

It has long been held that the number of normal births that occur during the daytime hours (6 A.M. to 6 P.M.) is roughly equal to the number of births occurring during the nighttime hours (6 P.M. to 6 A.M.). As far back as 1936, large studies have supported this claim [Guthmann & Bienhuls, 1936]. But in 1956, Peter King suggested that the 24-hour day be broken up differently to uncover periodic birth patterns that might otherwise go undetected [King, 1956]. This data is summarized in the table below. Study it carefully.

Time Period	3 A.M.–11 P.M.	11 A.M.–3 P.M.	3 P.M.–11 P.M.	11 P.M.–3 A.M.
Avg. number of births hourly	1561	1375	1213	1375

Notice that we can think of the hours between 3 A.M. and 11 A.M. as "peak" hours, when the highest hourly birth rates occur, and the hours between 3 P.M. and 11 P.M. as "low" hours, when significantly fewer births occur.

Now consider the graph in Figure 6.66 which shows an even more detailed accounting of birthrate by hour. It is taken from Kendall's *Time-Series,* based on data from results reported by both Bliss [1958] and King [1956]. The vertical axis represents the square root of the number of births at four hospitals in the studies, and the horizontal axis represents the hour of the day, starting with midnight (24, far left). A smooth rising-then-falling curve has been fitted to the data points. (We will be seeing many curves similar to this as we continue our study of trigonometry.) Notice the striking cyclical pattern in the occurrence of births, with a clear peaking at around 6 A.M. and a low point at around 6 P.M. This somewhat mysterious pattern serves not only to generate many questions about the factors that stimulate the birth process, but also serves as a stunning example of the cyclical nature of trigonometric functions—a concept that we will explore in detail in the next two chapters.

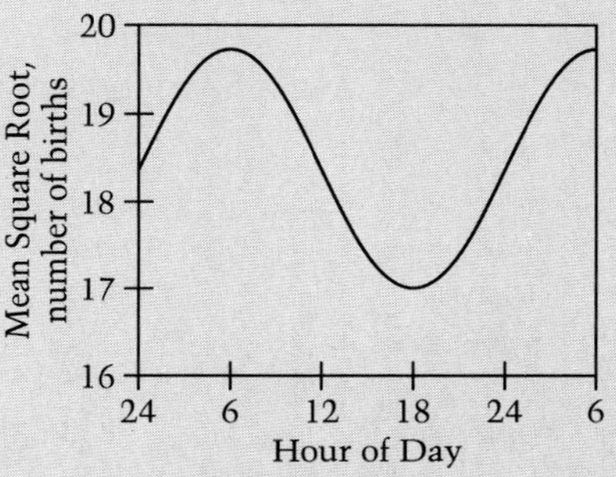

Figure 6.66

Sources: Guthmann, H. & Bienhuls, M. *Monatschr, Gerburtsh. und Gynak.*, **103,** 337 (1936).

King, P. Increased frequency of birth in the morning hours, *Science*, **123,** 985 (1956).

Kendall, M. *Time-Series*, Hafner Press, New York, p. 16, 1976. Reproduced by permission of Edward Arnold Publishers.

Bliss, C. Periodic Regression in Biology and Climatology, Bulletin No. 615, Connecticut Agricultural Experiment Station, New Haven (1958).

Chapter Review

6.1 Triangles (pp. 331–338)

An *isosceles triangle* is a triangle with two sides of equal length. In any isosceles triangle the angles opposite the two sides of equal length have the same measure. (p. 332) A *right triangle* is a triangle in which one of the angles is a right angle. (p. 333) An *equilateral triangle* is a triangle with three sides of equal length. Each angle of an equilateral triangle has measure 60°. (p. 334)

A-A Similarity Theorem: Two triangles are similar if two angles of one triangle have the same measure as two angles of the other triangle. (p. 335)

6.2 Trigonometric Ratios of Acute Angles (pp. 338–346)

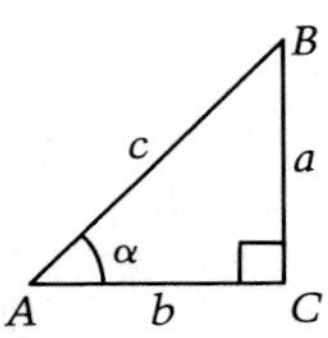

Let α be an acute angle in a right triangle ABC. Then the trigonometric functions of α are defined by the following ratios (p. 340):

$$\sin \alpha = \frac{a}{c} \qquad \cos \alpha = \frac{b}{c} \qquad \tan \alpha = \frac{a}{b}$$

$$\csc \alpha = \frac{c}{a} \qquad \sec \alpha = \frac{c}{b} \qquad \cot \alpha = \frac{b}{a}$$

6.3 Angles of Rotation (pp. 346–351)

An angle is in *standard position* in the xy-plane if its vertex is at the origin, and its initial side lies on the positive x-axis. The angle is *positive* if the rotation is counterclockwise and *negative* if the rotation is clockwise. (p. 346)

Let θ be an angle in standard position with $P(x, y)$ a point on its terminal side. Then

$$\sin \theta = \frac{y}{r} \qquad \cos \theta = \frac{x}{r} \qquad \tan \theta = \frac{y}{x} \quad (x \neq 0)$$

$$\csc \theta = \frac{r}{y} \quad (y \neq 0) \qquad \sec \theta = \frac{r}{x} \quad (x \neq 0) \qquad \cot \theta = \frac{x}{y} \quad (y \neq 0)$$

where $r = \sqrt{x^2 + y^2}$ is the distance from P to the origin. (p. 348)

6.4 Identities and Tables (pp. 352–360)

Reciprocal Identities

$$\sin \theta = \frac{1}{\csc \theta} \qquad \cos \theta = \frac{1}{\sec \theta} \qquad \tan \theta = \frac{1}{\cot \theta}$$

$$\csc \theta = \frac{1}{\sin \theta} \qquad \sec \theta = \frac{1}{\cos \theta} \qquad \cot \theta = \frac{1}{\tan \theta}$$

Quotient Identities

$$\frac{\sin \theta}{\cos \theta} = \tan \theta$$

$$\frac{\cos \theta}{\sin \theta} = \cot \theta$$

Cofunction Identities

$\sin\theta = \cos(90° - \theta)$ $\quad\cos\theta = \sin(90° - \theta)$

$\tan\theta = \cot(90° - \theta)$ $\quad\cot\theta = \tan(90° - \theta)$

$\sec\theta = \csc(90° - \theta)$ $\quad\csc\theta = \sec(90° - \theta)$

Pythagorean Identities

$\sin^2\theta + \cos^2\theta = 1$

$\tan^2\theta + 1 = \sec^2\theta$

$1 + \cot^2\theta = \csc^2\theta$

6.5 Applications Involving Right Triangles (pp. 360–367)

The process of determining the measures of all the sides and all the angles in a triangle is referred to as *solving the triangle.* A right triangle can be solved whenever one side and one of the acute angles are known, or whenever two sides are known. (p. 360)

6.6 Reference Angles (pp. 367–371)

The sign of a trigonometric function value depends on the quadrant in which the angle terminates. It is found by using the ASTC chart (p. 368):

positive functions	A	S	T	C
quadrant	1	2	3	4

A *reference angle* θ^* for a given angle θ is a first quadrant angle whose trigonometric function values are numerically the same as θ's. Only the signs may differ. (p. 369)

6.7 Law of Sines (pp. 372–378)

Law of Sines: In any triangle with angles A, B, C, and opposite sides a, b, and c respectively,

$$\frac{\sin A}{a} = \frac{\sin B}{b} = \frac{\sin C}{c}.$$

The Law of Sines is used to solve oblique triangles when given (1) one side and two angles, or (2) two sides and the angle opposite one of them. (p. 373)

6.8 Law of Cosines (pp. 378–383)

Law of Cosines: In any triangle with angles A, B, and C, and opposite sides a, b, and c, respectively,

$$a^2 = b^2 + c^2 - 2bc\cos A,$$
$$b^2 = a^2 + c^2 - 2ac\cos B,$$
$$c^2 = a^2 + b^2 - 2ab\cos C. \quad \text{(p. 378)}$$

Review Exercises

In Exercises 1–4, ABC is a right triangle and PQ is perpendicular to AC. Find the length of the indicated side.

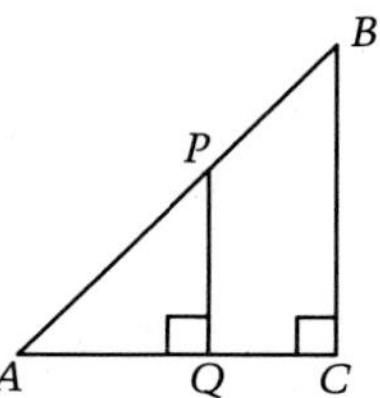

1. $AQ = 3, PQ = 5, BC = 20, AB = ?$
2. $AP = 4, AB = 10, AC = 5, PQ = ?$
3. $AQ = 6, QC = 3, AP = 9, AB = ?$
4. $AC = 10, QC = 4, PQ = 9, PB = ?$
5. Determine the area of an equilateral triangle whose altitude is 8.
6. Determine the area of an isosceles triangle whose base angles are 30° and whose base is 6.

In Exercises 7–10, use the given information about the right triangle ABC to find the six trigonometric ratios of α.

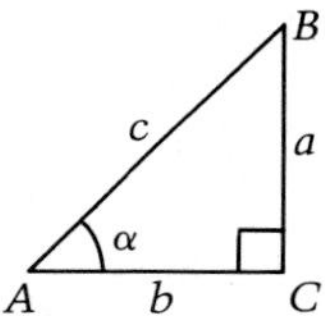

7. $a = 2, b = 3$
8. $a = 1, c = 2$
9. $b = 2, c = 4$
10. $a = 10, b = 2$

In Exercises 11–14, refer to the following figure.

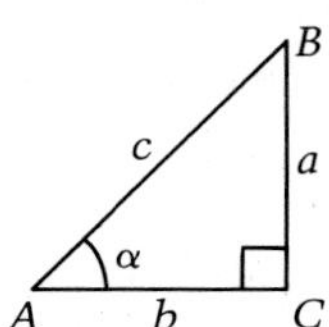

11. $m(\angle A) = 30°, b = 5$, find c.
12. $m(\angle B) = 45°, c = 20$, find a.
13. $a = 2\sqrt{3}, c = 4$, find $m(\angle A)$.
14. $a = \sqrt{2}, b = \sqrt{6}$, find $m(\angle A)$.

In Exercises 15–18, α and β are complementary angles. Use the given information to find the indicated trigonometric ratios.

15. $\sin \alpha = \frac{2}{5}$, find $\cos \alpha$ and $\tan \beta$.
16. $\tan \alpha = \frac{5}{3}$, find $\sec \alpha$ and $\cos \beta$.
17. $\csc \alpha = 3$, find $\cos \alpha$ and $\tan \beta$.
18. $\sec \alpha = 2$, find $\cot \alpha$ and $\sin \beta$.

In Exercises 19–22, the given point is on the terminal side of an angle α. Determine the values of the six trigonometric functions of α.

19. $(-5, 2)$
20. $(4, -1)$
21. $(0, -3)$
22. $(-2, 0)$

In Exercises 23–26, rewrite the expression as a trigonometric function of an angle between 45° and 90°.

23. sin 32°
24. sec 4°
25. cot 18°
26. csc 44°

In Exercises 27–30, convert the given degree measure from degree-minute-second form to decimal form or vice versa.

27. 23°45′36″
28. 47°27′18″
29. 73.265°
30. 11.805°

In Exercises 31–36, determine the value.

31. sin 35°10′
32. cot 57°30′
33. sec 8°40′
34. cos 65°20′
35. tan 78°50′
36. csc 17°20′

In Exercises 37–42, determine α to the nearest degree.

37. $\cos \alpha = 0.9135$
38. $\tan \alpha = 1.483$
39. $\csc \alpha = 1.589$
40. $\sin \alpha = 0.7431$
41. $\cot \alpha = 3.732$
42. $\sec \alpha = 2.366$

In Exercises 43–48, use the Reciprocal, Quotient, and Pythagorean Identities to find the values of the other five trigonometric functions of α rounded to four decimal places.

43. $\cos \alpha = 0.5614$
44. $\sin \alpha = 0.3452$
45. $\tan \alpha = 1.8602$
46. $\sec \alpha = 2.6113$
47. $\csc \alpha = 4.0871$
48. $\cot \alpha = 0.2547$

In Exercises 49–52 use the Reciprocal and Pythagorean Identities to determine the values of the other five trigonometric functions of the acute angle x.

49. $\sin x = \frac{3}{4}$
50. $\cos x = \frac{9}{10}$

51. $\tan x = \frac{4}{3}$

52. $\cot x = \frac{1}{3}$

In Exercises 53–56, determine the value to four decimal places.

53. sin 72.3°

54. tan 36.8°

55. sec 61.4°

56. csc 11.2°

In Exercises 57–60, determine the angle α in degrees to two decimal places.

57. $\cos \alpha = 0.8713$

58. $\cot \alpha = 0.6625$

59. $\csc \alpha = 1.6224$

60. $\sin \alpha = 0.0481$

61. A 32 ft ladder is leaning against a building. If the ladder makes an angle of 72° with the ground, how far up the building does the ladder reach?

62. A 48 ft long ramp is inclined at an angle of 8° with the level ground. How high does the ramp rise above the ground?

In Exercises 63–70, evaluate the quantity.

63. sin 855°

64. cos(−390°)

65. sec(−870°)

66. csc 1500°

67. cot 630°

68. tan 540°

69. cos(−810°)

70. sin(−630°)

In Exercises 71–78, solve the triangle.

71. $a = 14, A = 38°, B = 73°$

72. $a = 17, B = 46°, C = 81°$

73. $a = 11, b = 16, A = 40°$

74. $a = 18, b = 7.0, B = 23°$

75. $a = 7.0, b = 9.0, C = 26°$

76. $a = 10, b = 13, C = 100°$

77. $a = 12, b = 15, c = 23$

78. $a = 14, b = 19, c = 13$

In Exercises 79–80, find an expression in t for the other five trigonometric functions of α, given that α is an acute angle.

79. $\sin \alpha = t$

80. $\tan \alpha = t$

In Exercises 81–82, find an expression in t for the other five trigonometric functions of α, given that α is a second quadrant angle.

81. $\tan \alpha = t$

82. $\sec \alpha = t$

83. Determine to the nearest hundredth the perimeter of a regular 10-sided polygon inscribed in a circle of radius 1.

84. Determine to the nearest hundredth the area of a regular 12-sided polygon inscribed in a circle of radius 1.

85. Determine to the nearest tenth of a degree the measure of the largest angle in a triangle in which one side is twice another and the third side is the square root of the product of the other two sides.

86. Determine the length of a diagonal in a regular pentagon of side 1.

Graphing Calculator Exercises

[*Note: Round all estimated values to the nearest hundredth.*]

1. The hypotenuse of a right triangle is 10 in.
   (a) Express the area A of the triangle as a function of x, the length of one of the legs, then specify the domain of this function.
   (b) Use the graph of the function to determine the maximum area of such a right triangle, and the dimensions that produce this maximum area.
   (c) In theory, the triangle you found in part (b) is isosceles. What are its *exact* dimensions?

2. The right triangle shaded in the figure has one leg on the positive x-axis, and its hypotenuse is a radius of the unit circle $x^2 + y^2 = 1$.

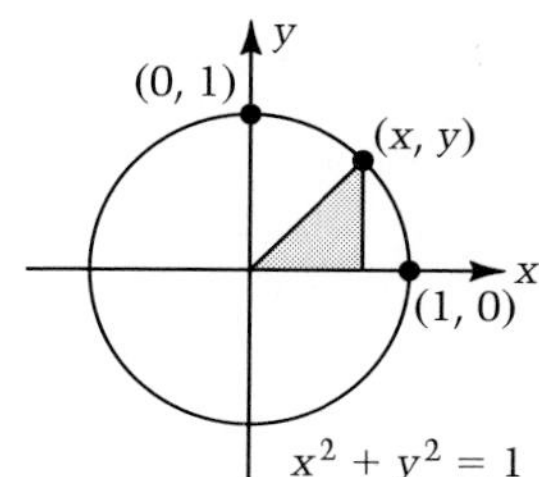

   (a) Express the area A of the triangle as a function of x, then specify the domain of this function.
   (b) Use the graph of the function to determine the maximum area of such a right triangle, and the dimensions that produce this maximum area.

(c) What do you think are the *exact* dimensions of the triangle in part (b)?

3. The triangle shaded in the figure has one leg on the x-axis, with opposite vertex $P(x, y)$ in the first quadrant, lying on the graph of $y = 20 - x^2$.

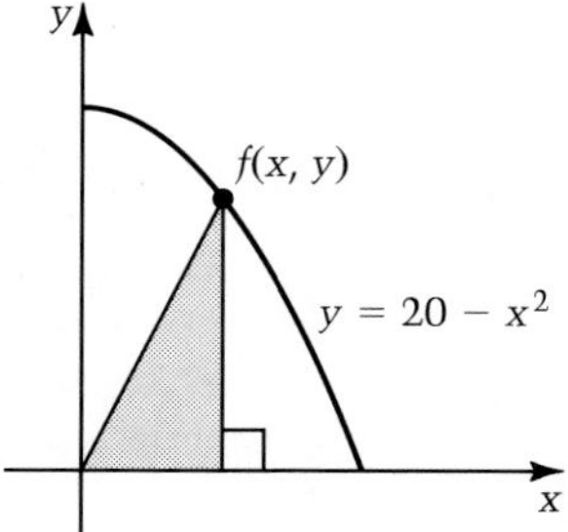

(a) Express the area A of the triangle as a function of x, then specify the domain of this function.
(b) Use the graph of the function found in part (a) to determine the maximum area of such a triangle, and the dimensions that produce this maximum area.

4. The perimeter of an isosceles triangle is 20 cm.
(a) Express the area A of the triangle as a function of the base x, then specify the domain of this function.

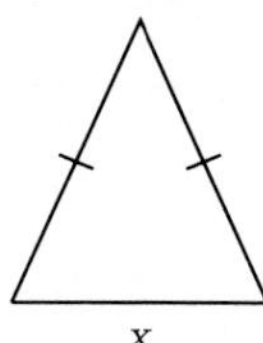

(b) Use the graph of the function to determine the maximum area of such a triangle, and the dimensions that produce this maximum area.

In Exercises 5–10, determine whether the given equation is an identity for θ between 0° and 360°. Make sure your function grapher is set for degree measurement. (*Hint:* One way of checking that an equation $h(x) = g(x)$ is an identity is to graph the functions $y = h(x)$ and $y = g(x)$ on the same set of axes, and observe that the graphs coincide.

5. $\sin \theta = \sin(180° - \theta)$

6. $\cos \theta = -\cos(-\theta)$

7. $\sin \theta = -\sin(-\theta)$

8. $\sin \theta = \sin(\theta - 180°)$

9. $\sin \theta + \cos \theta = 1$

10. $\sin \theta \cos \theta = \dfrac{\sin 2\theta}{2}$

Chapter Test

In Problems 1–2, assume that $\triangle ABC$ is a right triangle, c is the length of the hypotenuse and a and b are the lengths of the legs. Find the length of the third side given the lengths of two sides.

1. $a = 12, b = 16$

2. $a = 10, c = 26$

In Problems 3–4, assume that $\triangle ABC$ and $\triangle A'B'C'$ are two triangles with $m(\angle A) = m(\angle A')$ and $m(\angle B) = m(\angle B')$. Find the length of the indicated side.

3. $a = 2, b = 5, a' = 12, b' = ?$

4. $a = 7, b = 8, a' = ?, b' = 20$

In Problems 5–6, use the given information about the right triangle ABC to find the values of the six trigonometric ratios of α.

5.

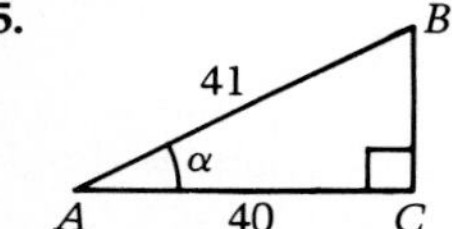

6.

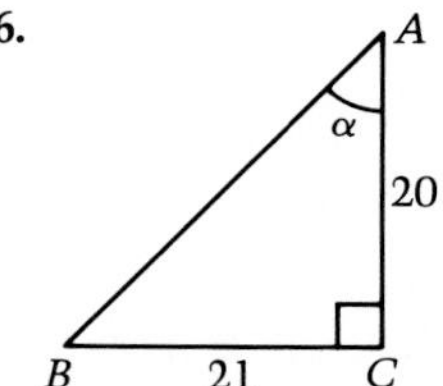

In Problems 7–8, $\triangle ABC$ has right angle at C and sides of length a, b, and c. Given the following information, find the indicated side.

7. $m(\angle A) = 30°, a = 24$, find c

8. $m(\angle A) = 45°, c = 18$, find a

In Problems 9–10, the angles α and β are complementary. Given the following information, find $\tan \alpha$ and $\tan \beta$.

9. $\sin \alpha = \dfrac{3}{7}$

10. $\cos \alpha = \dfrac{5}{7}$

In Problems 11–12, the given point is located on the terminal side of angle θ. Find $\sin\theta$ and $\tan\theta$.

11. $(-10, 24)$ **12.** $(-12, -9)$

In Problems 13–14, evaluate each quantity.

13. $\cos(-270°)$ **14.** $\sin 180°$

In Problems 15–16, given that $\sin 72° = 0.95$ and $\cos 72° = 0.31$, find the function value. Round your answer to two decimal places.

15. $\sec 18°$ **16.** $\cot 18°$

In Problems 17–18, use the Reciprocal and Pythagorean Identities to determine the values of the other five trigonometric functions of acute angle α.

17. $\tan\alpha = \frac{3}{2}$ **18.** $\sin\alpha = \frac{5}{6}$

In Problems 19–22, evaluate the quantity.

19. $\cos 42°23'$ **20.** $\sec 63°48'$

21. $\csc 78.4°$ **22.** $\sin 19.7°$

In Problems 23–25, use the given information to solve the right triangle ABC.

23. $B = 60°, b = 12$

24. $A = 33°42', b = 96.8$

25. $a = 36.3, b = 46.9$

26. What is the angle of elevation of the sun at the time that a 68 ft tall flagpole casts a shadow 25 ft long?

27. A 25 ft long ladder is leaning against a building. If the foot of the ladder is 6 ft from the base of the building (on level ground), what acute angle does the ladder make with the ground?

In Problems 28–29, evaluate the quantity without using a table or calculator.

28. $\tan 135°$ **29.** $\cot(-240°)$

30. Solve $\triangle ABC$ given that $A = 41°$, $B = 83°$, and $c = 44.6$

In Problems 31–32 you are given the measure of one angle and two sides. State whether it is possible to have $\triangle ABC$. If it is, solve the triangle. (Determine both solutions if there is more than one.)

31. $A = 24°18', c = 9.4, a = 6.2$

32. $B = 46°54', a = 12.4, b = 10.6$

33. A boat B is observed simultaneously by two Coast Guard stations S and T which are 550 yd apart. Angles STB and TSB are observed to be 132°48′ and 21°42′, respectively. Find the distance from the boat to station T.

34. Points A and B are 126 ft apart and on the same side of a river. A tree at point T, on the other side of the river, is observed, with angles ABT and BAT equal to 74°12′ and 42°36′, respectively. Find the distance from point T to point B.

In Problems 35–36, solve $\triangle ABC$, given the following information.

35. $a = 1.52, b = 2.31, C = 119°$

36. $a = 26.7, b = 34.2, c = 41.8$

7 Trigonometric Functions

Applications of trigonometric functions abound: seasonal business cycles, biological rhythms, and meteorological patterns all exhibit strong periodic, and thus trigonometric, behavior. At the end of this chapter we consider two examples of what is called **time-series data.** *In these examples we will see that from money to measles, trigonometric models provide us with a powerful tool for understanding cyclical processes.*

7.1 Radian Measure and the Unit Circle

Up to this point, our study of trigonometry has focused on angles and triangles. In this chapter we shall develop a way of defining the trigonometric functions so that their domains consist of real numbers, not just angle measurements. This allows us to apply trigonometric functions to a much wider variety of problems. Many natural phenomena such as the motion of planets, the ocean's tides, and the beat of a human heart, are *periodic,* that is, certain behavior is repeated at regular intervals. When defined as functions of real

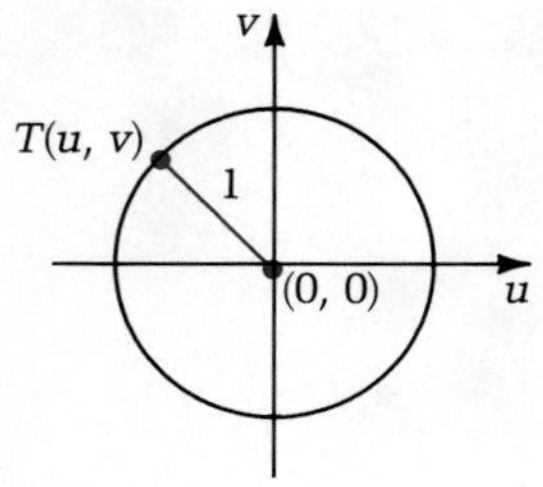

Figure 7.1 The unit circle in the *uv*-plane.

numbers, trigonometric functions provide an excellent means of describing periodic behavior in nature.

The circle of radius 1 with center at the origin is called the **unit circle.** In Figure 7.1 we have labeled the horizontal axis the u-axis and the vertical axis the v-axis. An equation for the unit circle is easily derived by means of the distance formula. Since any point $T(u, v)$ on the unit circle is at a distance 1 from the origin (0, 0), we have, by the distance formula,

$$\sqrt{(u - 0)^2 + (v - 0)^2} = 1.$$

Upon squaring both sides of this equation, we get

$$u^2 + v^2 = 1,$$

which is the common way of describing the unit circle in the uv-plane.

EXAMPLE 1

Determine whether each of the following points is on the unit circle.

(a) $\left(\frac{1}{2}, \frac{\sqrt{3}}{2}\right)$ (b) $\left(-\frac{\sqrt{2}}{2}, \frac{\sqrt{2}}{2}\right)$ (c) $\left(\frac{1}{3}, \frac{2}{3}\right)$

Solution

(a) $\left(\frac{1}{2}\right)^2 + \left(\frac{\sqrt{3}}{2}\right)^2 = \frac{1}{4} + \frac{3}{4} = 1$

To determine whether a point (u, v) is on the unit circle, check to see if its coordinates satisfy the equation $u^2 + v^2 = 1$.

Thus, $\left(\frac{1}{2}, \frac{\sqrt{3}}{2}\right)$ is on the unit circle.

(b) $\left(\frac{-\sqrt{2}}{2}\right)^2 + \left(\frac{\sqrt{2}}{2}\right)^2 = \frac{2}{4} + \frac{2}{4} = 1$

Thus, $\left(-\frac{\sqrt{2}}{2}, \frac{\sqrt{2}}{2}\right)$ is on the unit circle.

(c) $\left(\frac{1}{3}\right)^2 + \left(\frac{2}{3}\right)^2 = \frac{1}{9} + \frac{4}{9} = \frac{5}{9} \neq 1$

Thus, $\left(\frac{1}{3}, \frac{2}{3}\right)$ is not on the unit circle.

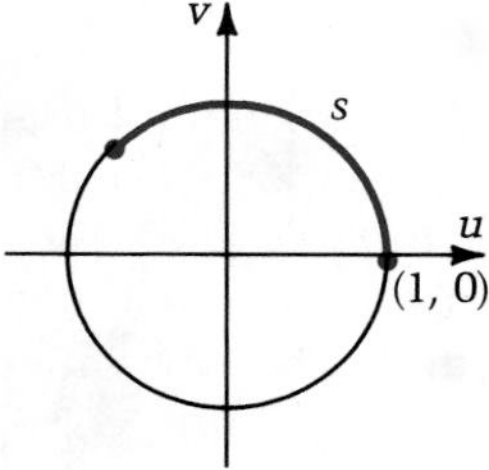

Figure 7.2 The standard arc *s*.

Let us define a **standard arc** s as an arc on the unit circle which starts at the point (1, 0) and travels s units counterclockwise if s is positive, and s units clockwise if s is negative. Since the circumference C of the unit circle is

$$C = 2\pi r = 2\pi(1) = 2\pi,$$

arcs whose lengths are rational multiples of π $\left(\text{e.g., } \frac{\pi}{2}, -\frac{3\pi}{4}, 5\pi\right)$ are easy to visualize.

EXAMPLE 2

Represent each of the following real numbers as a standard arc.

(a) π (b) $\dfrac{\pi}{2}$ (c) $\dfrac{\pi}{3}$ (d) $-\dfrac{3\pi}{2}$ (e) $-\dfrac{2\pi}{3}$ (f) $\dfrac{13\pi}{6}$

Solution

(a)

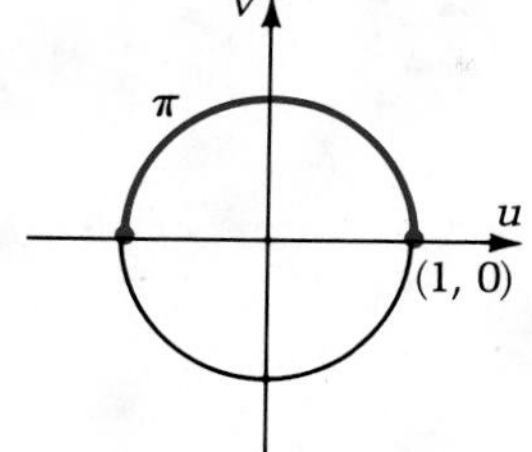

$\pi = \frac{1}{2}(2\pi)$, which is $\frac{1}{2}$ of the circumference.

(b)

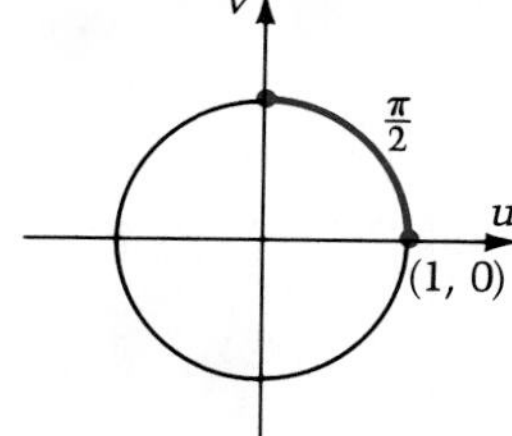

$\frac{\pi}{2} = \frac{1}{4}(2\pi)$, which is $\frac{1}{4}$ of the circumference.

(c)

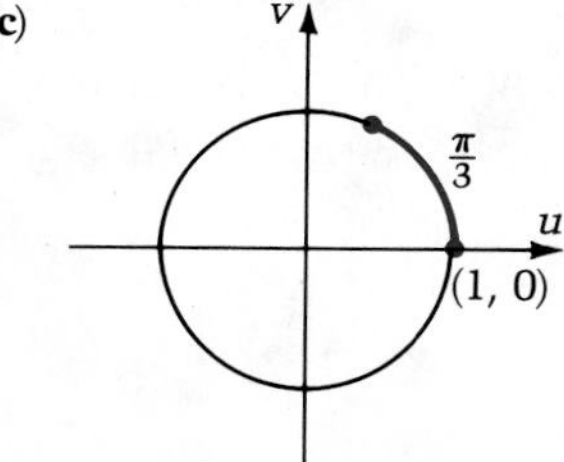

$\frac{\pi}{3} = \frac{1}{6}(2\pi)$, which is $\frac{1}{6}$ of the circumference.

(d)

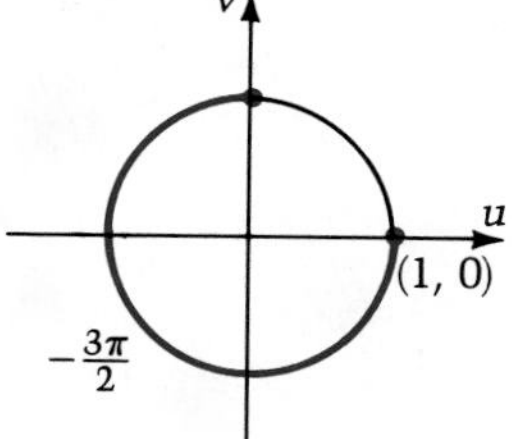

$\frac{3\pi}{2}$ is $\frac{3}{4}$ of the circumference. Since $-\frac{3\pi}{2} < 0$, the arc travels clockwise.

(e)

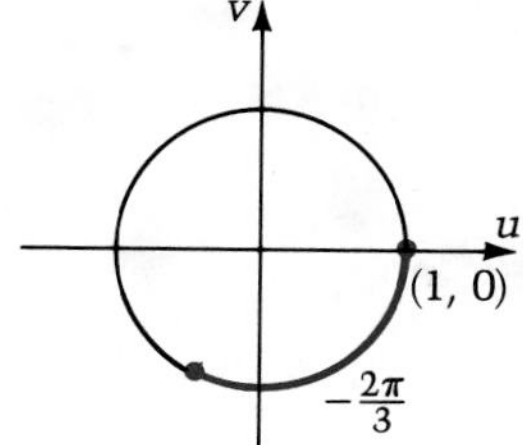

$\frac{2\pi}{3}$ is $\frac{1}{3}$ of the circumference. Since $-\frac{2\pi}{3} < 0$, the arc travels clockwise.

(f)

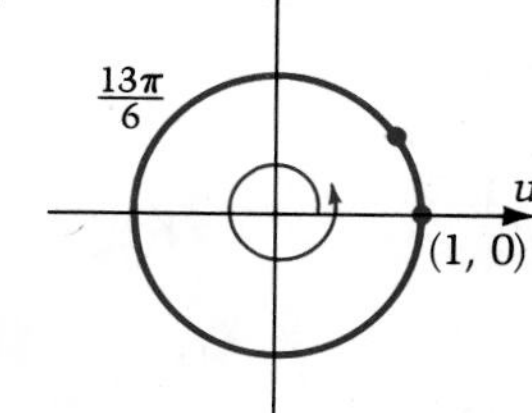

$\frac{13\pi}{6}$ is $1\frac{1}{12}$ of the circumference.

Figure 7.3

Radian Measure

In Chapter 6 we discussed the trigonometric functions of angles measured in degrees and used these functions to solve some interesting problems. However, to use trigonometric functions to solve other types of problems, we need to define them in such a way that their domains consist of real numbers (with no units attached).

Figure 7.4 shows a positive angle θ in standard position, and a circle of radius r with center at the origin. The angle θ is called a **central angle** since its vertex is the center of the circle. We say that angle θ "subtends" or "cuts" an arc of the circle. Let s be the length of this arc.

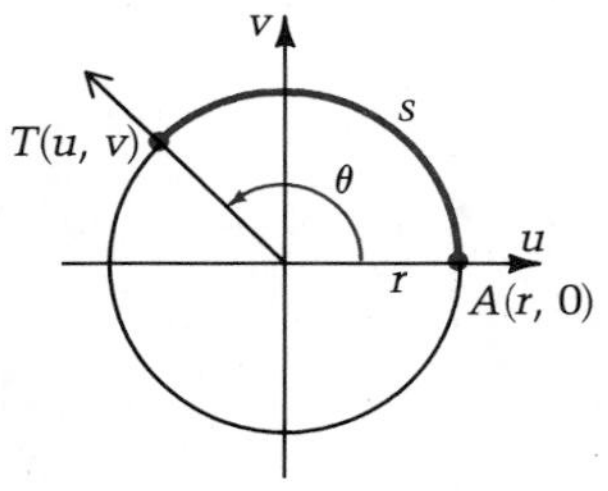

Figure 7.4

Notice that $A(r, 0)$ is the point where the initial side of angle θ intersects the circle. Furthermore, $T(u, v)$ is the point where the terminal side of angle θ intersects the circle, and s is the length of the arc cut by angle θ. We use s, r, and θ, as just described, to make the following definition.

Definition of Radian Measure

The **radian measure** of a positive central angle θ is defined as the ratio of the arclength s to the radius r:

$$\theta = \frac{s}{r}.$$

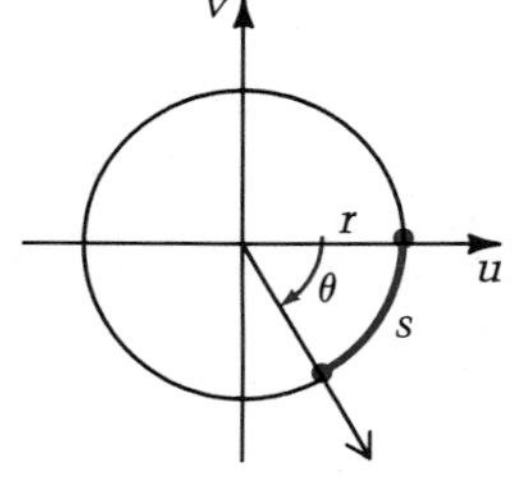

Figure 7.5 $\theta = \frac{s}{r}, s < 0$

Since both s and r have the same units of length, their ratio θ will have no units. However, we will say that the angle has a measure of θ "radians" to remind us that it was measured by using the radius. If θ is a negative angle, its radian measure is defined to be negative as Figure 7.5 suggests.

In a unit circle the radian measure of a positive central angle is equal to the length of the arc that it cuts. This is shown in Figure 7.6.

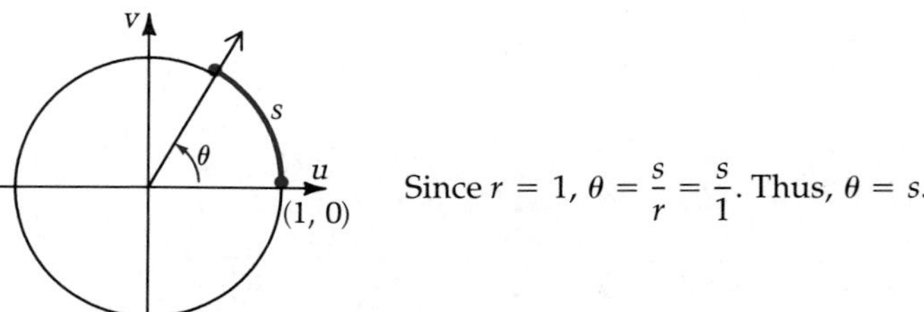

Figure 7.6

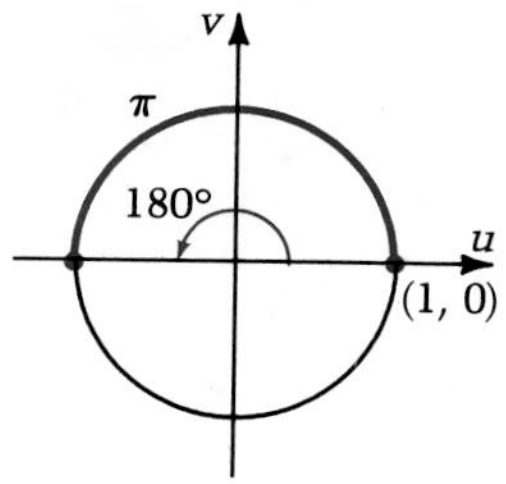

Figure 7.7

degrees	radians
0°	0
30°	$\frac{\pi}{6}$
45°	$\frac{\pi}{4}$
90°	$\frac{\pi}{2}$

Now consider a central angle of 180° in a unit circle as shown in Figure 7.7. Since an angle of 180° cuts an arc of length π (half the circumference), the above statement suggests that an angle of 180° has a radian measure of π. That is,

$$180° = \pi \text{ radians.}$$

We use this relationship in setting up a proportion for converting the degree measure of an angle to radian measure and vice versa.

Degree/Radian Conversion Formula

If d is the degree measure of an angle and α is its radian measure, then

$$\frac{d}{180°} = \frac{\alpha}{\pi}.$$

EXAMPLE 3

(a) Convert 165° to radians.

(b) Convert $-\frac{3\pi}{4}$ radians to degrees.

(c) Convert 3 radians to degrees.

(d) Convert 40.6° to radians.

Solution

(a) $\frac{165°}{180°} = \frac{\alpha}{\pi}$ Set up the proportion: $\frac{d}{180°} = \frac{\alpha}{\pi}$.

$$\alpha = \frac{165}{180}\pi = \frac{11}{12}\pi \text{ radians}$$

(b) $\frac{d}{180°} = \frac{-3\pi/4}{\pi}$

$$d = -\frac{3\pi}{4} \cdot \frac{1}{\pi} \cdot 180° = -135°$$

(c) $\frac{d}{180°} = \frac{3}{\pi}$

$$d = \frac{3 \times 180°}{\pi} \approx 172°$$ $\pi \approx 3.14$.

(d) $\frac{40.6°}{180°} = \frac{\alpha}{\pi}$

$$\alpha = \frac{40.6}{180} \cdot \pi \approx 0.708 \text{ radians}$$ $\pi \approx 3.14$.

Since the definition of radian measure depends on the radius and arc of a circle, a variety of problems involving circular arcs can be solved using radian measure.

EXAMPLE 4

Determine the length of an arc cut by a central angle of 45° in a circle of radius 2.15 yd.

Solution

$\theta = \frac{s}{r}$ To use this formula, 45° must first be converted to radians: $\frac{45°}{180°} = \frac{\theta}{\pi}$, thus $\theta = \frac{45}{180}\pi = \frac{\pi}{4}$.

$\frac{\pi}{4} = \frac{s}{2.15}$ $\frac{\pi}{4}$ is substituted for θ.

$s = \frac{\pi \cdot 2.15}{4}$

$s \approx 1.69$

The arc has length 1.69 yd (rounded to two decimal places).

EXAMPLE 5

A wheel of radius 80 cm rolls along the ground without slipping and rotates through an angle of 60°. How far does the wheel move?

Solution

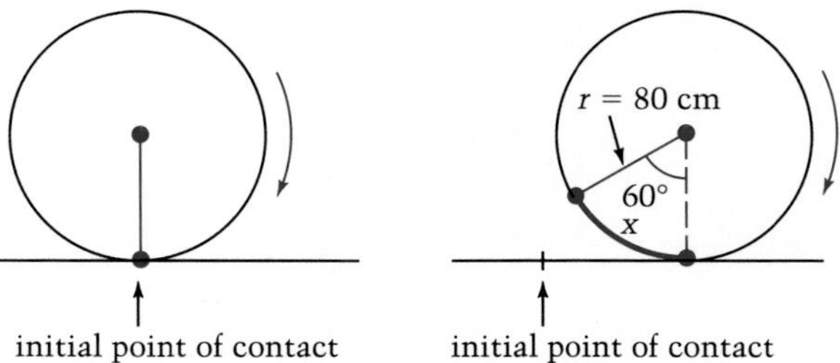

Figure 7.8

The distance that the wheel advances is equal to the length of the arc shown in the right-hand figure above. We use this fact to find s.

$\theta = \dfrac{s}{r}$ To use this formula, θ must be measured in radians. $\dfrac{60°}{180°} = \dfrac{\theta}{\pi}$, thus $\theta = \dfrac{60}{180}\pi = \dfrac{\pi}{3}$.

$$\frac{\pi}{3} = \frac{s}{80}$$

$$s = \frac{80\pi}{3}$$

The wheel has moved $\dfrac{80\pi}{3}$ cm ($\approx$ 84 cm) to the right.

EXERCISE SET 7.1

In Exercises 1–4, determine whether the point is on the unit circle.

1. $\left(-\dfrac{\sqrt{3}}{2}, -\dfrac{1}{2}\right)$ **2.** $\left(-\dfrac{\sqrt{3}}{3}, -\dfrac{1}{3}\right)$

3. $\left(\dfrac{\sqrt{2}}{2}, -\dfrac{1}{2}\right)$ **4.** $\left(-\dfrac{\sqrt{3}}{2}, \dfrac{\sqrt{2}}{2}\right)$

In Exercises 5–8, the point (x, y) is on the unit circle. You are given one of its coordinates and the quadrant in which the point lies. Determine an approximate value of the other coordinate to four decimal places.

5. $x = 0.3672$, Quadrant IV
6. $y = 0.7218$, Quadrant II
7. $y = -0.4516$, Quadrant III
8. $x = -0.6711$, Quadrant II

In Exercises 9–16, represent the real number as a standard arc.

9. $\dfrac{\pi}{4}$ **10.** $\dfrac{5\pi}{6}$ **11.** $\dfrac{2\pi}{3}$ **12.** $\dfrac{5\pi}{4}$

13. $-\dfrac{\pi}{4}$ **14.** $-\dfrac{5\pi}{6}$ **15.** $-\dfrac{13\pi}{4}$ **16.** $-\dfrac{11\pi}{3}$

In Exercises 17–38, convert the measure of the angle from degrees to radians or vice versa.

17. 60° **18.** 45° **19.** 315° **20.** 360°

21. 210° **22.** 330° **23.** $-30°$ **24.** $-15°$

25. 75° **26.** 300° **27.** $-540°$ **28.** $-200°$

29. $\dfrac{3\pi}{2}$ **30.** $\dfrac{\pi}{3}$ **31.** $\dfrac{5\pi}{6}$ **32.** $\dfrac{5\pi}{2}$

33. -3π **34.** 5π **35.** $-\dfrac{25\pi}{6}$ **36.** $-\dfrac{19\pi}{9}$

37. 1.5 **38.** 5

In Exercises 39–46, convert the measure of the angle from degrees to radians or vice versa. Approximate all answers to three decimal places.

39. 35.8° 40. 169.2° 41. 248°36′

42. $-312°48'$ 43. 12.5664 44. 15.7080

45. -5.4264 46. -18.1681

47. Determine the length of an arc cut by a central angle of 90° in a circle of radius 4 in.

48. Determine the length of an arc cut by a central angle of 30° in a circle of radius 2 cm.

49. A central angle of 1.5 radians cuts an arc of length 12 m. Find the radius of the circle.

50. A central angle of 2.5 radians cuts an arc of 25 ft. Find the radius of the circle.

51. What is the measure in radians of the angle through which the minute hand of a clock turns in 42 min?

52. What is the measure of the angle in degrees through which the minute hand of a clock turns between 1:30 P.M. and 2:20 P.M. of the same day?

Superset

53. Find the number of radians through which each of the hands of a clock move in (a) 12 hours, (b) in one hour, (c) in 30 min, (d) in 5 min.

54. After midnight, at what time are the hands of a clock first perpendicular to each other?

55. The minute hand of a clock is of length 5 cm and the hour hand is of length 3.6 cm. How far does the tip of the minute hand move in 12 hours? How far does the tip of the hour hand move in one hour?

56. The tires of an automobile are 24 in. in diameter. If the automobile backs up 20 ft, through what angle does each wheel turn?

57. (Refer to Exercise 56) Through what angle does each wheel turn if the automobile travels one mile?

58. Through what angle does a water wheel of radius 32 ft turn if it rotates at the rate of 2 mi per hour for 1 min?

59. Suppose θ is an angle in standard position. Determine the coordinates of the point of intersection of the terminal side of θ with the unit circle if

(a) $\theta = \frac{\pi}{6}$, (b) $\theta = \frac{\pi}{4}$, (c) $\theta = \frac{\pi}{3}$.

60. Between midnight and noon, how many times are the hands of a clock perpendicular to each other?

If P is the terminal point of a standard arc of length s, then

P is in Quadrant I when $0 < s < \frac{\pi}{2}$,

P is in Quadrant II when $\frac{\pi}{2} < s < \pi$,

P is in Quadrant III when $\pi < s < \frac{3\pi}{2}$,

P is in Quadrant IV when $\frac{3\pi}{2} < s < 2\pi$,

P is in Quadrant I when $2\pi < s < \frac{5\pi}{2}$, etc.

In Exercises 61–70, given that the real number is represented as a standard arc, determine the quadrant in which the standard arc terminates.

61. 25 62. 32 63. $7\sqrt{11}$ 64. $6\sqrt{3}$

65. -416 66. -128 67. $3\sqrt{5} + 21$

68. $2\sqrt{7} - 34$ 69. $\sqrt{\pi} + 24$ 70. $\pi^2 + 10$

7.2
Trigonometric Functions of Real Numbers

One of the central ideas in mathematics is the concept of function. Of special interest to us are those functions whose domains are sets of real numbers. Radian measure provides a means for defining the trigonometric functions so that their domains are sets of real numbers.

The transition from angle measurements to real numbers as domain values is made possible by noticing the following:

> *Every real number x can be associated uniquely with the central angle θ that cuts standard arc x on the unit circle.*

This result is illustrated by the following procedure. To determine the angle θ associated with the real number x, begin by representing x as a standard arc. This arc terminates at a unique point $T(u, v)$, and is cut by a unique central angle θ, having $T(u, v)$ on its terminal side. We have thus associated angle θ with the real number x.

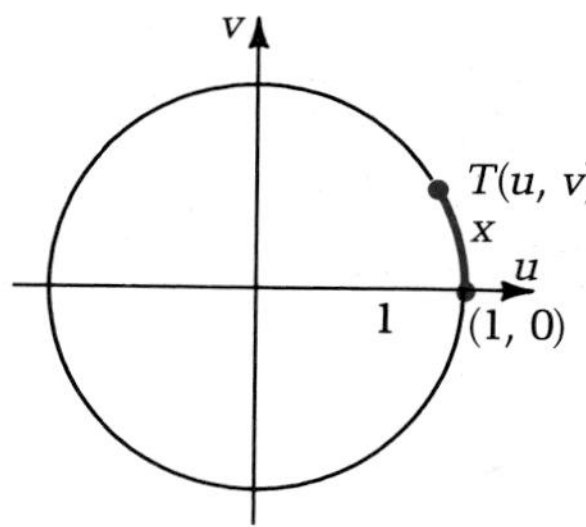

(a) Represent x as a standard arc.

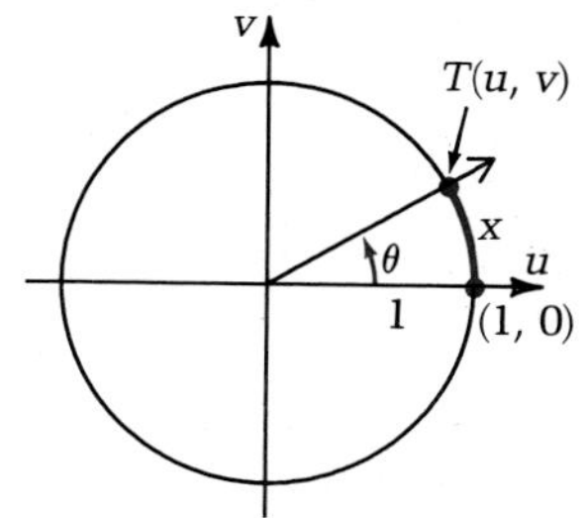

(b) θ is the unique angle that cuts the standard arc x.

Figure 7.9

Recall that we previously defined the trigonometric functions of θ in terms of the coordinates of the point $T(u, v)$. Since $r = 1$, we have

$$\theta = \frac{s}{r} = \frac{x}{1} = x.$$

Since $\theta = x$, it is reasonable to define the trigonometric functions of the real number x to be the same as the trigonometric functions of angle θ.

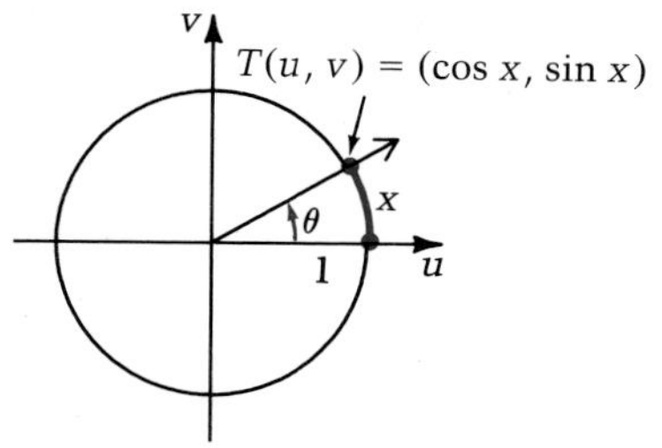

Figure 7.10

Definition of Trigonometric Functions of a Real Number

Let x be any real number, let θ be the central angle in standard position in the unit circle having radian measure x, and let $T(u, v)$ be the terminal point on standard arc x. Then

$\sin x = \sin \theta = v$ domain: all real numbers

$\cos x = \cos \theta = u$ domain: all real numbers

$\tan x = \tan \theta = \dfrac{v}{u}$ domain: all real numbers except odd multiples of $\dfrac{\pi}{2}$ (which make denominator $u = 0$).

$\cot x = \cot \theta = \dfrac{u}{v}$ domain: all real numbers except integral multiples of π (which make denominator $v = 0$).

$\sec x = \sec \theta = \dfrac{1}{u}$ domain: all real numbers except odd multiples of $\dfrac{\pi}{2}$.

$\csc x = \csc \theta = \dfrac{1}{v}$ domain: all real numbers except integral multiples of π.

EXAMPLE 1

Evaluate each of the following expressions.

(a) $\sin \dfrac{\pi}{2}$ (b) $\cos(-\pi)$ (c) $\tan \dfrac{3\pi}{2}$ (d) $\sin 0$

Solution

(a)

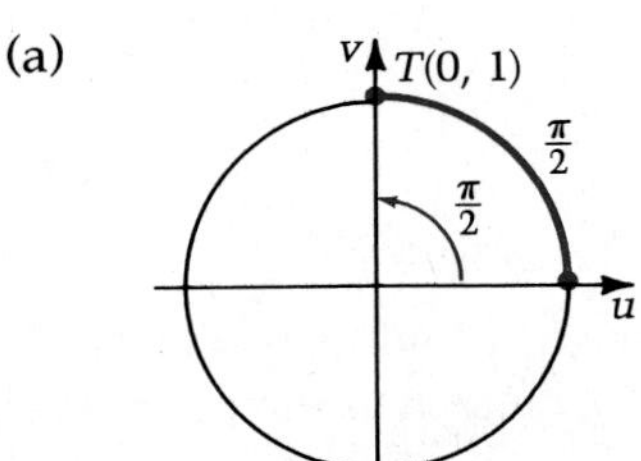

Draw a unit circle and standard arc $\dfrac{\pi}{2}$.
(Note: $\dfrac{\pi}{2}$ is $\dfrac{1}{4}$ of the circumference.)

$\sin \dfrac{\pi}{2}$ = second coordinate of $T(0, 1) = 1$.

Thus $\sin \dfrac{\pi}{2} = 1$.

(b)

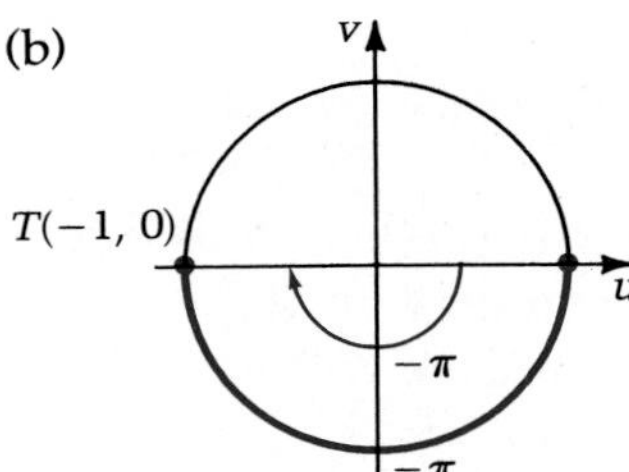

Draw a unit circle and standard arc $-\pi$.

$\cos(-\pi)$ = first coordinate of $T(-1, 0)$
$= -1$.

Thus $\cos(-\pi) = -1$.

(c)

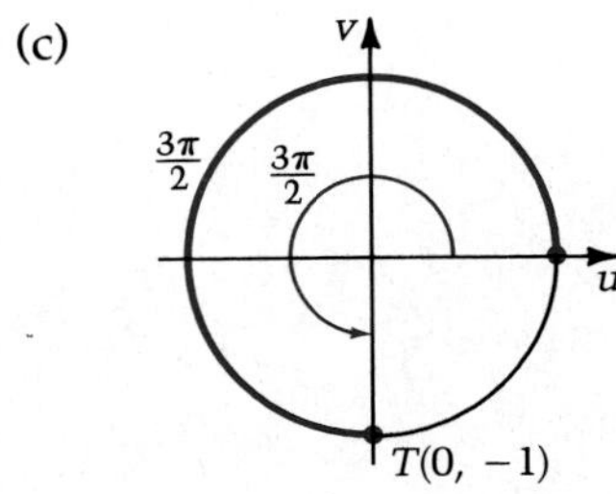

Draw a unit circle and standard arc $\dfrac{3\pi}{2}$.

$$\tan \frac{3\pi}{2} = \frac{\text{second coordinate of } T(0, -1)}{\text{first coordinate of } T(0, -1)}$$

$$= \frac{-1}{0} \text{ which is not defined.}$$

Thus $\tan \dfrac{3\pi}{2}$ is not defined.

Figure 7.11

(d) The point (1, 0) is the terminal point of standard arc 0 on the unit circle.

$$\sin 0 = \text{the second coordinate of } T(1, 0) = 0$$

Thus $\sin 0 = 0$.

In Example 1 we saw that trigonometric function values of integral multiples of $\frac{\pi}{2}$ are easily determined, since the corresponding arcs terminate at one of the four points (1, 0), (0, 1), (−1, 0), or (0, −1). Three other special arcs and their terminal points are shown on the unit circles below.

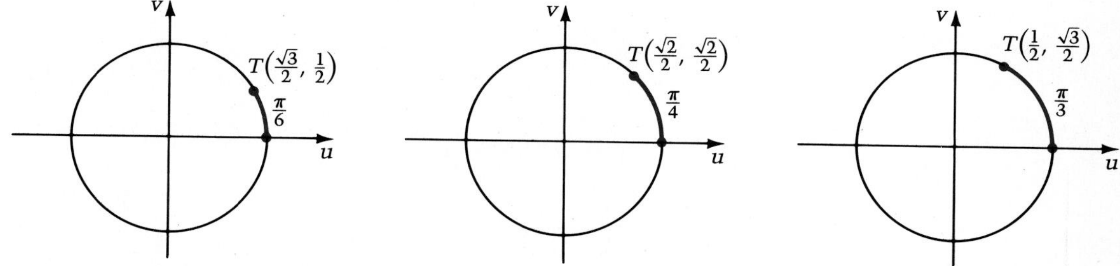

Figure 7.12

Knowing the coordinates of these terminal points allows us to determine trigonometric function values of $\frac{\pi}{6}$, $\frac{\pi}{4}$, and $\frac{\pi}{3}$. For example,

$$\sin\frac{\pi}{4} = \text{the second coordinate of } \left(\frac{\sqrt{2}}{2}, \frac{\sqrt{2}}{2}\right) = \frac{\sqrt{2}}{2},$$

$$\sec\frac{\pi}{3} = \frac{1}{\cos\frac{\pi}{3}} = \frac{1}{\text{the first coordinate of } \left(\frac{1}{2}, \frac{\sqrt{3}}{2}\right)} = \frac{1}{1/2} = 2.$$

In addition, we can determine trigonometric function values of some other real numbers by reflecting these three points through the origin or across either axis. This is demonstrated in the next example.

EXAMPLE 2

Evaluate each of the following expressions.

(a) $\cos\frac{5\pi}{6}$ (b) $\sin\left(-\frac{\pi}{4}\right)$ (c) $\tan\frac{4\pi}{3}$

Solution

(a)

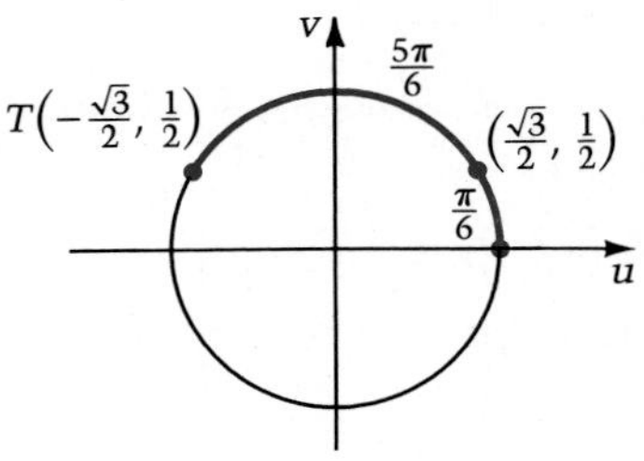

Draw the standard arc $\frac{5\pi}{6}$. The terminal point is found by reflecting the terminal point of $\frac{\pi}{6}$ across the vertical axis. Thus,

$$\cos\frac{5\pi}{6} = \text{first coordinate of } \left(-\frac{\sqrt{3}}{2}, \frac{1}{2}\right) = -\frac{\sqrt{3}}{2}.$$

Thus, $\cos\frac{5\pi}{6} = -\frac{\sqrt{3}}{2}$.

(b)

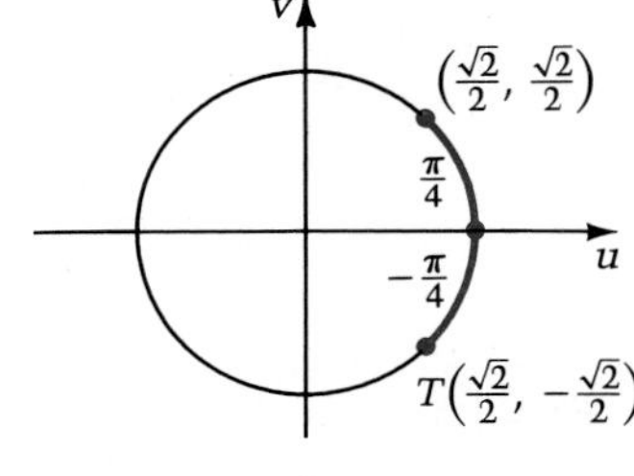

Draw the standard arc $-\frac{\pi}{4}$. The terminal point is found by reflecting the terminal point of $\frac{\pi}{4}$ across the horizontal axis. Thus, $\sin\left(-\frac{\pi}{4}\right) =$ the second coordinate of $\left(\frac{\sqrt{2}}{2}, -\frac{\sqrt{2}}{2}\right) = -\frac{\sqrt{2}}{2}$.

Thus, $\sin\left(-\frac{\pi}{4}\right) = -\frac{\sqrt{2}}{2}$.

(c)

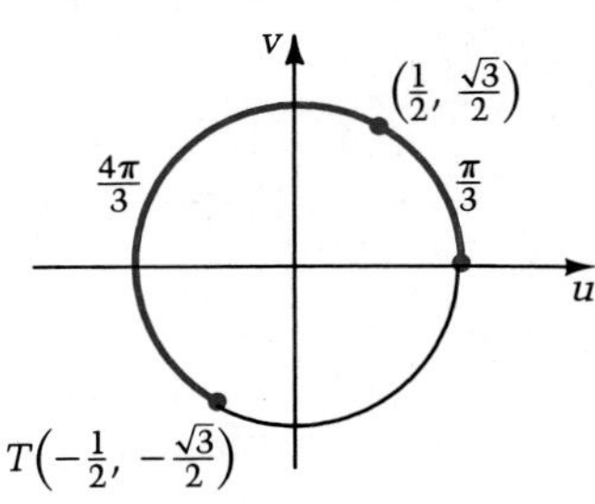

Figure 7.13

Draw the standard arc $\frac{4\pi}{3}$. The terminal point is found by reflecting the terminal point of $\frac{\pi}{3}$ through the origin. Thus,

$$\tan\frac{4\pi}{3} = \frac{\text{the second coordinate of } \left(-\frac{1}{2}, -\frac{\sqrt{3}}{2}\right)}{\text{the first coordinate of } \left(-\frac{1}{2}, -\frac{\sqrt{3}}{2}\right)}$$

$$= \frac{\frac{-\sqrt{3}}{2}}{-\frac{1}{2}} = \sqrt{3}$$

Thus, $\tan\frac{4\pi}{3} = \sqrt{3}$.

By virtue of the method used in the previous two examples, we can update the table of special angles from the previous chapter.

$f(x)$ \ x	0 (or 0°)	$\frac{\pi}{6}$ (or 30°)	$\frac{\pi}{4}$ (or 45°)	$\frac{\pi}{3}$ (or 60°)	$\frac{\pi}{2}$ (or 90°)
$\sin x$	0	$\frac{1}{2}$	$\frac{\sqrt{2}}{2}$	$\frac{\sqrt{3}}{2}$	1
$\cos x$	1	$\frac{\sqrt{3}}{2}$	$\frac{\sqrt{2}}{2}$	$\frac{1}{2}$	0
$\tan x$	0	$\frac{\sqrt{3}}{3}$	1	$\sqrt{3}$	undefined
$\cot x$	undefined	$\sqrt{3}$	1	$\frac{\sqrt{3}}{3}$	0
$\sec x$	1	$\frac{2\sqrt{3}}{3}$	$\sqrt{2}$	2	undefined
$\csc x$	undefined	2	$\sqrt{2}$	$\frac{2\sqrt{3}}{3}$	1

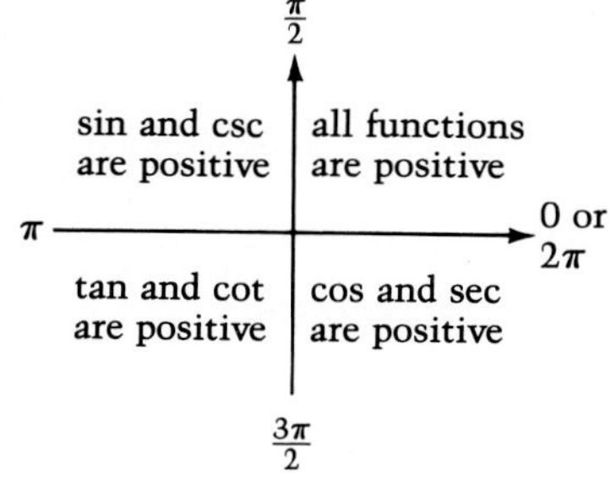

Figure 7.14

With a few simple changes, you can use the "reference angle" technique developed in the previous chapter to determine trigonometric function values of numbers outside the interval $\left[0, \frac{\pi}{2}\right]$. To find the trigonometric function value, attach the appropriate sign to the trigonometric function value of the reference value of x. The sign of your answer will depend on the quadrant in which the standard arc x terminates.

For any x between 0 and 2π, the reference value for x is determined as follows. (Recall that $180° = \pi$ radians.)

x	reference value for x
between 0 and $\frac{\pi}{2}$	x
between $\frac{\pi}{2}$ and π	$\pi - x$
between π and $\frac{3\pi}{2}$	$x - \pi$
between $\frac{3\pi}{2}$ and 2π	$2\pi - x$

To determine the trigonometric function value of a number outside the interval $[0, 2\pi)$, begin by adding or subtracting an appropriate multiple of 2π to produce a value between 0 and 2π.

EXAMPLE 3

Determine each of the following function values.

(a) $\cos \frac{7\pi}{6}$ (b) $\tan\left(-\frac{\pi}{4}\right)$ (c) $\csc \frac{20\pi}{3}$

Solution

(a) $\cos \frac{7\pi}{6} = \text{sign} \cos \left[\text{reference value for } \frac{7\pi}{6}\right]$

$= -\cos \left[\text{reference value for } \frac{7\pi}{6}\right]$ Standard arc $\frac{7\pi}{6}$ terminates in the third quadrant, where the cosine is negative.

$= -\cos \frac{\pi}{6} = -\frac{\sqrt{3}}{2}$ Reference value: $\frac{7\pi}{6} - \pi = \frac{\pi}{6}$.

(b) $\tan\left(-\frac{\pi}{4}\right) = \tan \frac{7\pi}{4}$ Add 2π to $-\frac{\pi}{4}$ to produce a value between 0 and 2π.

$= \text{sign} \tan \left[\text{reference value for } \frac{7\pi}{4}\right]$ Standard arc $\frac{7\pi}{4}$ terminates in the fourth quadrant, where the tangent is negative.

$= -\tan \frac{\pi}{4} = -1$ Reference value: $2\pi - \frac{7\pi}{4} = \frac{\pi}{4}$.

(c) $\csc \frac{20\pi}{3} = \csc \frac{2\pi}{3}$ Subtract $3 \cdot 2\pi$ from $\frac{20\pi}{3}$ to produce a value between 0 and 2π.

$= \text{sign} \csc \left[\text{reference value for } \frac{2\pi}{3}\right]$

$= +\csc \left[\text{reference value for } \frac{2\pi}{3}\right]$ Standard arc $\frac{2\pi}{3}$ terminates in the second quadrant, where the cosecant is positive.

$= +\csc \frac{\pi}{3} = \frac{2\sqrt{3}}{3}$ Reference value: $\pi - \frac{2\pi}{3} = \frac{\pi}{3}$.

To evaluate the trigonometric function values of a real number x that is not associated with a special arc, we can use a calculator or Table 4 in the Appendix. Remember that when you use your calculator, it should be set in the proper (degree/radian) mode for values of x. In Table 4, the column headed by the word "radians" contains the values of x.

EXAMPLE 4

Determine each of the following function values. Round your answer to two decimal places: (a) $\cos 3.3$ (b) $\sec \frac{21\pi}{5}$

Solution

(a) $\cos 3.3 = \boxed{\text{sign}} \cos \boxed{\begin{matrix}\text{reference value}\\ \text{for } 3.3\end{matrix}}$

cos 3.3 means "the cosine of 3.3 radians."

$= -\cos \boxed{\begin{matrix}\text{reference value}\\ \text{for } 3.3\end{matrix}}$

Since 3.3 is between π (≈ 3.14) and $\frac{3\pi}{2}$ (≈ 4.71), standard arc 3.3 terminates in the third quadrant, where cosine is negative.

$\approx -\cos 0.16$

Reference value: $3.3 - \pi \approx 0.16$.

≈ -0.99

(b) $\sec \frac{21\pi}{5} = \sec \frac{\pi}{5}$

Subtract $2 \cdot 2\pi$ from $\frac{21\pi}{5}$ to produce a value between 0 and 2π.

$= \boxed{\text{sign}} \sec \boxed{\begin{matrix}\text{reference value}\\ \text{for } \frac{\pi}{5}\end{matrix}}$

$= +\sec \boxed{\begin{matrix}\text{reference value}\\ \text{for } \frac{\pi}{5}\end{matrix}}$

Standard arc $\frac{\pi}{5}$ terminates in the first quadrant where all trigonometric functions are positive.

$= +\sec \frac{\pi}{5}$

Reference value: $\frac{\pi}{5}$

$\approx \sec 0.63$

$\frac{\pi}{5} \approx 0.63$

≈ 1.24

EXERCISE SET 7.2

In Exercises 1–10, determine the function value.

1. $\cot 0$ **2.** $\csc \frac{\pi}{2}$ **3.** $\tan 2\pi$ **4.** $\csc 2\pi$

5. $\sin \frac{\pi}{3}$ **6.** $\tan \frac{\pi}{4}$ **7.** $\csc \frac{\pi}{6}$ **8.** $\sec \frac{\pi}{6}$

9. $\cot \frac{\pi}{4}$ **10.** $\cos \frac{\pi}{3}$

In Exercises 11–30, determine the function value.

11. $\cos \frac{2\pi}{3}$ **12.** $\cot \frac{5\pi}{3}$ **13.** $\sec\left(-\frac{\pi}{6}\right)$

14. $\tan\left(-\frac{5\pi}{6}\right)$ **15.** $\tan \frac{3\pi}{4}$ **16.** $\csc \frac{5\pi}{4}$

17. $\sin\left(-\frac{3\pi}{4}\right)$ **18.** $\sec\left(-\frac{5\pi}{4}\right)$ **19.** $\cot\left(-\frac{7\pi}{6}\right)$

20. $\sin\left(-\frac{4\pi}{3}\right)$ **21.** $\csc\left(-\frac{7\pi}{4}\right)$ **22.** $\sec\left(-\frac{11\pi}{6}\right)$

23. $\cos\frac{5\pi}{2}$ **24.** $\sin 5\pi$ **25.** $\cot\frac{10\pi}{3}$

26. $\csc\frac{9\pi}{4}$ **27.** $\tan\left(-\frac{7\pi}{3}\right)$ **28.** $\csc\left(-\frac{11\pi}{4}\right)$

29. $\sec\frac{25\pi}{6}$ **30.** $\cos\left(-\frac{21\pi}{4}\right)$

In Exercises 31–38, determine the function value. Round to two decimal places.

31. $\sin 3.7$ 32. $\tan 5$ 33. $\sec 8.5$

34. $\cos 10.5$ 35. $\csc\frac{12\pi}{5}$ 36. $\sin\frac{5\pi}{7}$

37. $\tan\left(-\frac{\pi}{9}\right)$ 38. $\cot\left(-\frac{4\pi}{7}\right)$

Superset

In Exercises 39–44, verify that the given statement is true.

39. $\sin\frac{13\pi}{6}\sec\frac{5\pi}{3} = \tan\frac{9\pi}{4}$

40. $\tan\frac{4\pi}{3}\tan\frac{7\pi}{6} = \tan\frac{13\pi}{4}$

41. $\tan\frac{\pi}{3} = \dfrac{1-\cos\frac{4\pi}{3}}{\sin\frac{\pi}{3}}$

42. $\tan\frac{\pi}{6} = \dfrac{1-\cos\frac{\pi}{3}}{\sin\frac{\pi}{3}}$

43. $\cos\frac{5\pi}{6} = \cos\frac{\pi}{2}\cos\frac{\pi}{3} - \sin\frac{\pi}{2}\sin\frac{\pi}{3}$

44. $\sin\frac{\pi}{3} = 2\sin\frac{\pi}{6}\cos\frac{\pi}{6}$

In Exercises 45–50, solve for the smallest positive value of x.

45. $\cot x = -1$ **46.** $\csc x = -\sqrt{2}$

47. $\sec x = -\frac{2\sqrt{3}}{3}$ **48.** $\tan x = -\sqrt{3}$

49. $\sec(-x) = \sqrt{2}$ **50.** $\csc(-x) = -2$

7.3 The Trigonometric Functions: Basic Graphs and Properties

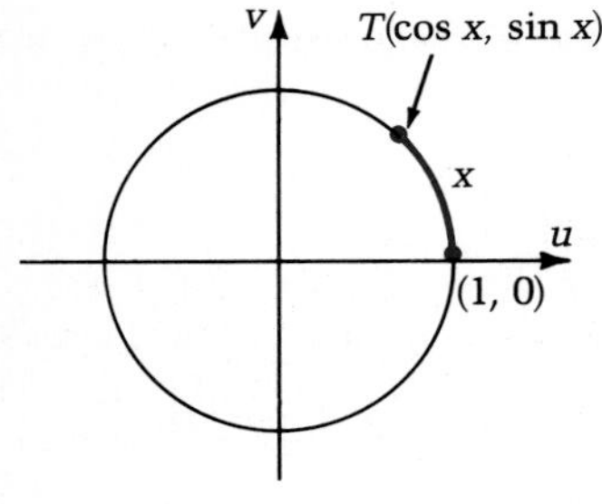

Figure 7.15

We have seen that every real number x uniquely determines a standard arc on the unit circle. The coordinates of the terminal point T of this arc are used to define the six trigonometric functions of the real number x. The first coordinate of T is $\cos x$ and the second coordinate is $\sin x$. There are many different standard arcs (and thus many different values of x) that are associated with the same terminal point T. Fortunately, these values of x occur in a predictable, or *periodic* way.

Figure 7.16 demonstrates that $\frac{\pi}{3}$, $\frac{7\pi}{3}$, and $\frac{13\pi}{3}$ all determine the same terminal point $\left(\frac{1}{2}, \frac{\sqrt{3}}{2}\right)$. Adding any positive or negative multiple of 2π to $\frac{\pi}{3}$ adds one or more complete revolutions to the arc. The resulting standard arc still has $\left(\frac{1}{2}, \frac{\sqrt{3}}{2}\right)$ as its terminal point. This means that $\sin\frac{\pi}{3}$, $\sin\frac{7\pi}{3}$, and

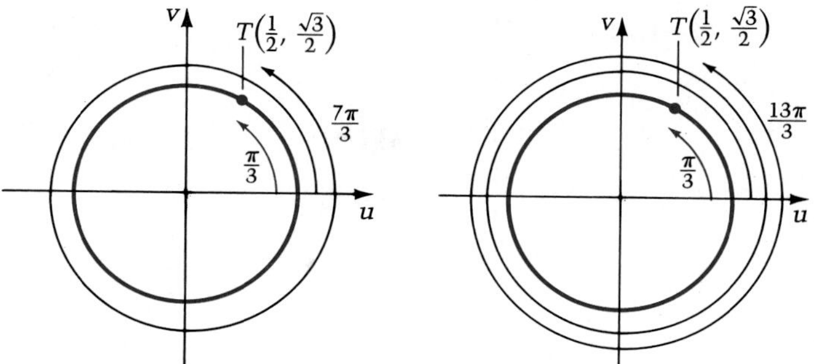

Figure 7.16

$\sin \dfrac{13\pi}{3}$ are the same: $\dfrac{\sqrt{3}}{2}$. Thus, the sine function repeats the same value each time the domain value changes by 2π. Functions exhibiting such repetitive behavior are called *periodic.*

Definition of Periodic Function

A function f is called a **periodic function** if there is a positive real number p such that

$$f(x) = f(x + p)$$

for all x in the domain of f; the smallest such positive number p is called the **period** of the function. The graph of the function over an interval of length p is called a **cycle** of the graph.

Both the sine and cosine functions are periodic with period 2π. For that reason we write

$$\sin x = \sin(x + n \cdot 2\pi), \text{ for any integer } n, \text{ and}$$
$$\cos x = \cos(x + n \cdot 2\pi), \text{ for any integer } n.$$

To graph the sine function, we shall first study its behavior for x-values from 0 to 2π, and thus produce one complete cycle of the graph. Then, since the function has period 2π, the pattern observed in the graph for values between 0 and 2π will be repeated over and over.

Recall that the sine function is defined by means of the second coordinate of a standard arc's terminal point T. Thus, values on the horizontal axis of the graph of the sine function correspond to standard arcs, and values on the vertical axis correspond to the second coordinates of the arcs' terminal points (see Figure 7.17).

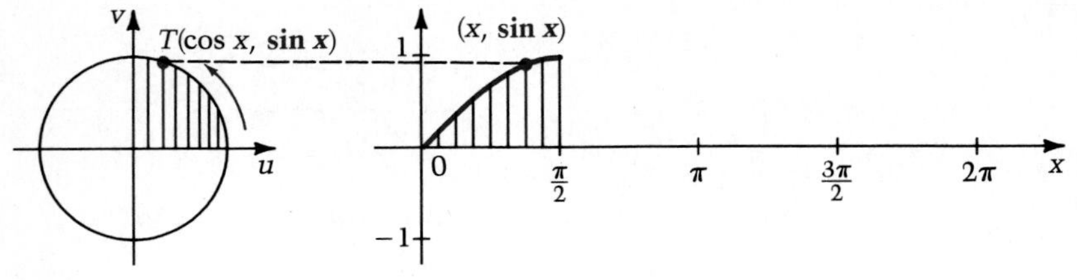

As x increases from 0 to $\frac{\pi}{2}$, second coordinates ("heights") of terminal points increase from 0 to 1.

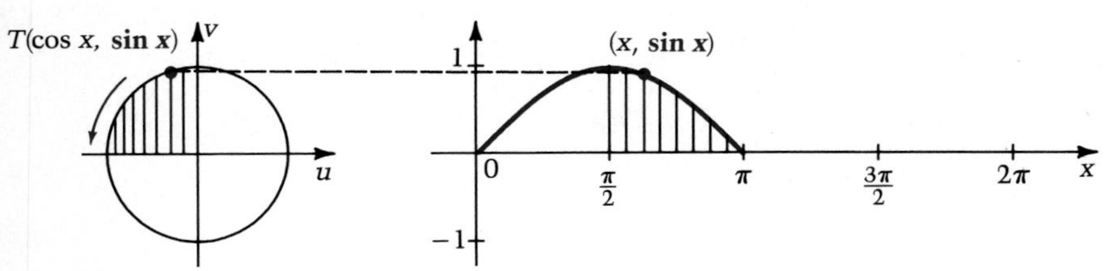

As x increases from $\frac{\pi}{2}$ to π, second coordinates decrease from 1 to 0.

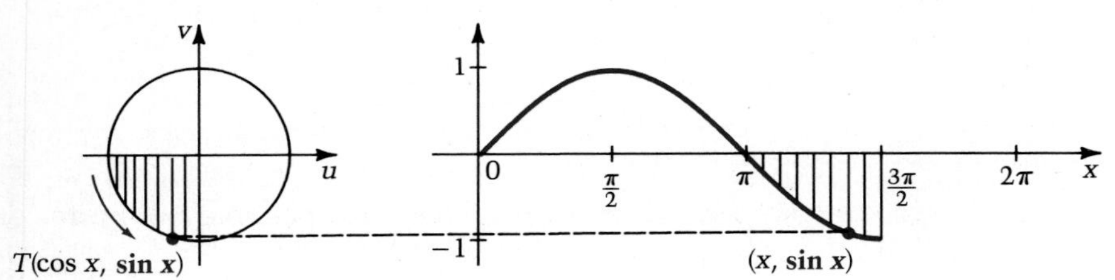

As x increases from π to $\frac{3\pi}{2}$, second coordinates decrease from 0 to -1.

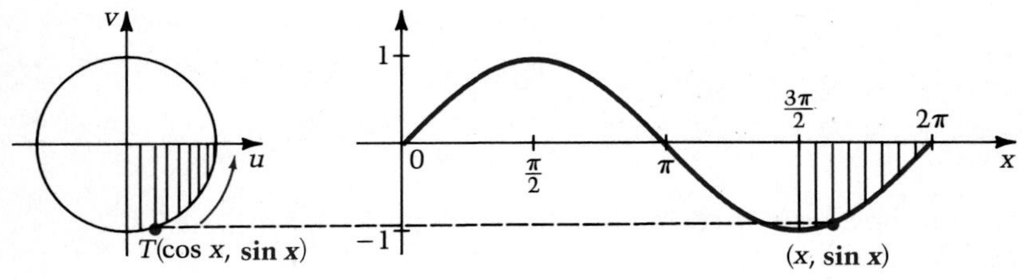

As x increases from $\frac{3\pi}{2}$ to 2π, second coordinates increase from -1 to 0.

Figure 7.17 Constructing one cycle of the graph of $y = \sin x$.

To get a more precise graph of the sine function, we construct a table of special values of x, plot the corresponding ordered pairs, and draw a smooth curve through the points. Since the sine function is periodic with period 2π, the portion of the graph between 0 and 2π will repeat every 2π units. The complete graph is shown in Figure 7.18. You should be able to sketch this graph quickly from memory.

x	0	$\frac{\pi}{6}$	$\frac{\pi}{4}$	$\frac{\pi}{3}$	$\frac{\pi}{2}$	$\frac{3\pi}{4}$	π	$\frac{5\pi}{4}$	$\frac{3\pi}{2}$	$\frac{7\pi}{4}$	2π
sin x	0	$\frac{1}{2}$	$\frac{\sqrt{2}}{2}$	$\frac{\sqrt{3}}{2}$	1	$\frac{\sqrt{2}}{2}$	0	$-\frac{\sqrt{2}}{2}$	-1	$-\frac{\sqrt{2}}{2}$	0

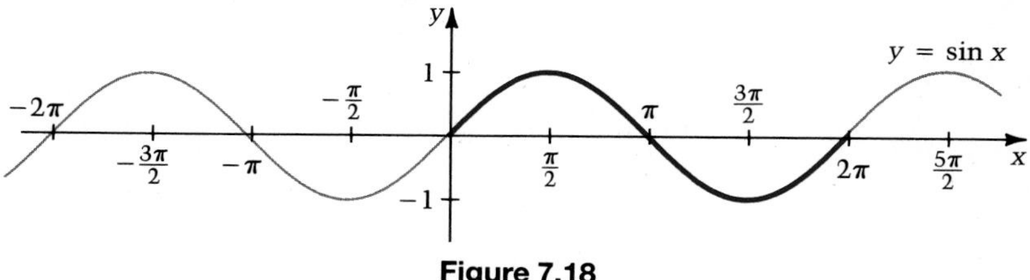

Figure 7.18

Properties of the Sine Function	
Domain	the set of all real numbers
Range	the set of real numbers between -1 and 1 inclusive
Period	2π
Symmetry	with respect to the origin; thus sine is an odd function, that is, $\sin(-x) = -\sin x$

We now turn our attention to the cosine function. Recall that the cosine is defined by means of the first coordinate of an arc's terminal point. Thus, values on the horizontal axis of the graph of the cosine function correspond to standard arcs, and values on the vertical axis correspond to the first coordinates of the arcs' terminal points. For example, to graph the cosine function for values of x between 0 and $\frac{\pi}{2}$, we notice that as standard arc x increases from 0 to $\frac{\pi}{2}$, first coordinates of the terminal points *decrease* from 1 to 0.

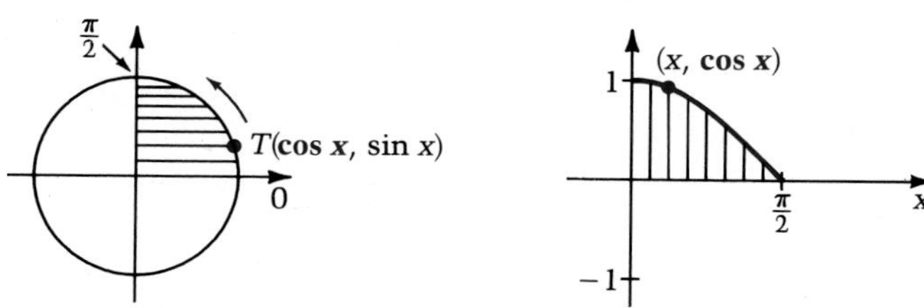

Figure 7.19 Constructing the graph of $y = \cos x$ for $0 \le x \le \frac{\pi}{2}$.

To get a more precise graph of the cosine function, we construct a table of special values of x, plot the corresponding ordered pairs, and draw a smooth curve through the points. We use the fact that the cosine function is periodic with period 2π to complete the graph. Note that the graph of the cosine function is symmetric with respect to the y-axis.

x	0	$\frac{\pi}{6}$	$\frac{\pi}{4}$	$\frac{\pi}{3}$	$\frac{\pi}{2}$	$\frac{3\pi}{4}$	π	$\frac{5\pi}{4}$	$\frac{3\pi}{2}$	$\frac{7\pi}{4}$	2π
$\cos x$	1	$\frac{\sqrt{3}}{2}$	$\frac{\sqrt{2}}{2}$	$\frac{1}{2}$	0	$-\frac{\sqrt{2}}{2}$	-1	$-\frac{\sqrt{2}}{2}$	0	$\frac{\sqrt{2}}{2}$	1

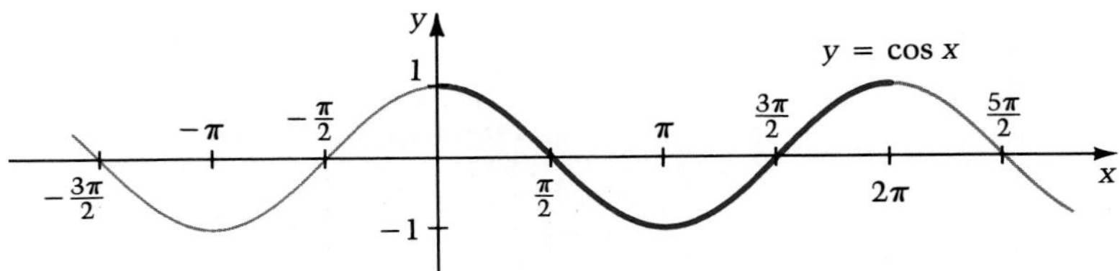

Figure 7.20

Properties of the Cosine Function

Domain the set of all real numbers

Range the set of real numbers between -1 and 1 inclusive

Period 2π

Symmetry with respect to the y-axis; thus cosine is an even function, that is, $\cos(-x) = \cos x$

EXAMPLE 1

(a) Graph $y = \sin x$ and $y = \sin(-x)$ on the same set of axes.

(b) Graph $y = \cos x$ and $y = \cos\left(x + \frac{\pi}{2}\right)$ on the same set of axes.

Solution

(a) To obtain the graph of $y = \sin(-x)$, reflect the graph of $y = \sin x$ across the y-axis.

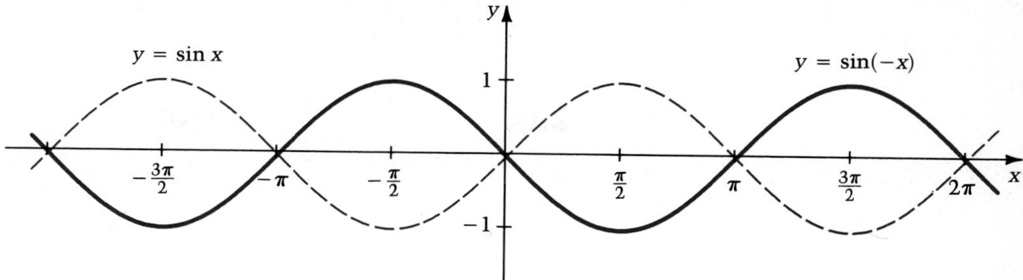

Figure 7.21

(b) To obtain the graph of $y = \cos\left(x + \frac{\pi}{2}\right)$, translate the graph of $y = \cos x$ $\frac{\pi}{2}$ units to the left.

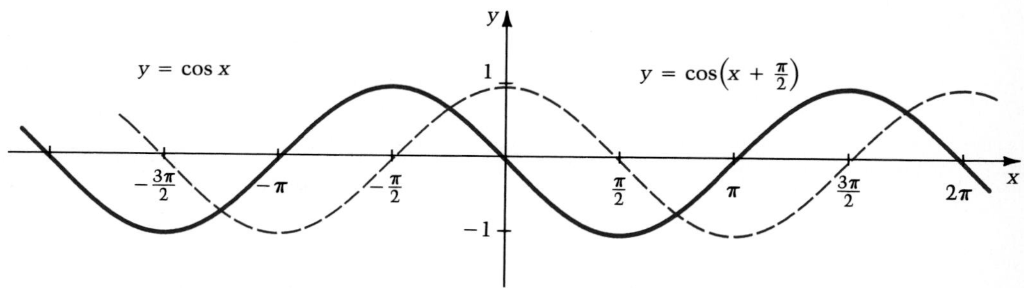

Figure 7.22

The graphs of the tangent and cotangent functions look very different from those of the sine and cosine functions. To get an idea of what the graph of $y = \tan x$ looks like, recall that

$$\tan x = \frac{\sin x}{\cos x},$$

and thus, the tangent function will not be defined when the denominator, $\cos x$, is 0. Since $\cos x$ is 0 for all x in the set

$$\left\{\ldots -\frac{5\pi}{2}, -\frac{3\pi}{2}, -\frac{\pi}{2}, \frac{\pi}{2}, \frac{3\pi}{2}, \frac{5\pi}{2}, \ldots\right\},$$

the tangent function is not defined for any of these values. We begin the graph of the tangent function by drawing a vertical asymptote at each of these values (see Figure 7.23).

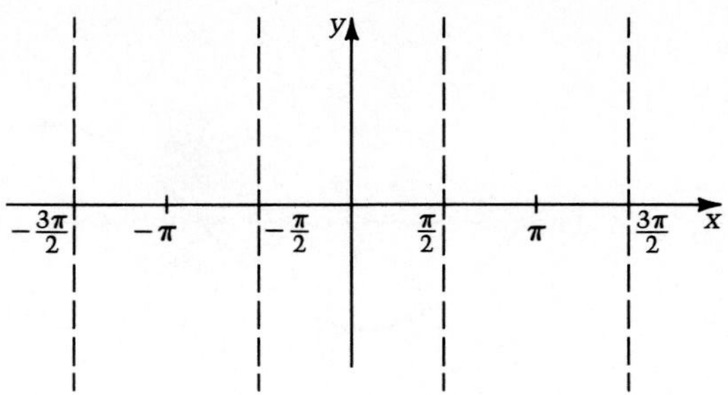

Figure 7.23

Notice that the tangent function is defined for all numbers between $-\frac{\pi}{2}$ and $\frac{\pi}{2}$. Let us consider what the graph will look like on that interval.

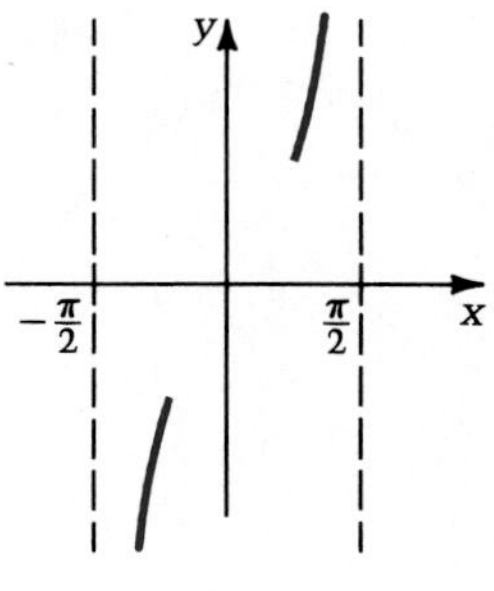

Figure 7.24

As x increases to $\frac{\pi}{2}$, $\sin x$ approaches 1, and $\cos x$ is nearing 0, so that $\tan x$, which equals $\frac{\sin x}{\cos x}$, becomes larger and larger.

A similar argument establishes that as x decreases towards $-\frac{\pi}{2}$, $\sin x$ approaches -1, $\cos x$ approaches 0, and so $\tan x$ takes on negative numbers with larger and larger absolute values.

To get a more complete graph of the tangent function for values between $-\frac{\pi}{2}$ and $\frac{\pi}{2}$, we construct a table of special values of x, plot the corresponding ordered pairs, and draw a smooth curve through the points (Figure 7.25). This curve is then repeated within each pair of adjacent asymptotes.

x	$-\frac{\pi}{2}$	$-\frac{\pi}{3}$	$-\frac{\pi}{4}$	$-\frac{\pi}{6}$	0	$\frac{\pi}{6}$	$\frac{\pi}{4}$	$\frac{\pi}{3}$	$\frac{\pi}{2}$
tan x	*	$-\sqrt{3}$	-1	$-\frac{\sqrt{3}}{3}$	0	$\frac{\sqrt{3}}{3}$	1	$\sqrt{3}$	*

*undefined

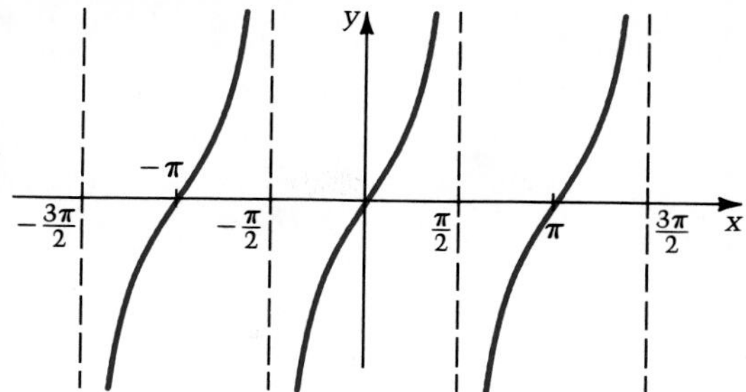

Figure 7.25 $y = \tan x$

Properties of the Tangent Function

Domain	all real numbers *except* $x = \frac{\pi}{2} + k\pi$, for any integer k
Range	all real numbers
Period	π
Symmetry	with respect to the origin; tangent is an odd function, thus $\tan(-x) = -\tan x$
Asymptotes	vertical asymptotes $x = \frac{\pi}{2} + k\pi$, for any integer k

The graph of $y = \tan x$ is periodic, but unlike the sine and cosine functions, the period is π. Thus, the portion of the graph between $-\frac{\pi}{2}$ and $\frac{\pi}{2}$ repeats itself over and over as suggested in Figure 7.25.

The graph of the cotangent function can be found by using a method similar to that used to derive the graph of $y = \tan x$. Recall that

$$\cot x = \frac{1}{\tan x}.$$

This fact suggests that the graph of $y = \cot x$ becomes infinite (has a vertical asymptote) where $\tan x$ is 0, and is zero where the graph of $y = \tan x$ has vertical asymptotes. The complete graph is shown in Figure 7.26. Like the tangent function, the cotangent function has period π. In addition, the graph of the cotangent function, like that of the tangent function, is symmetric with respect to the origin.

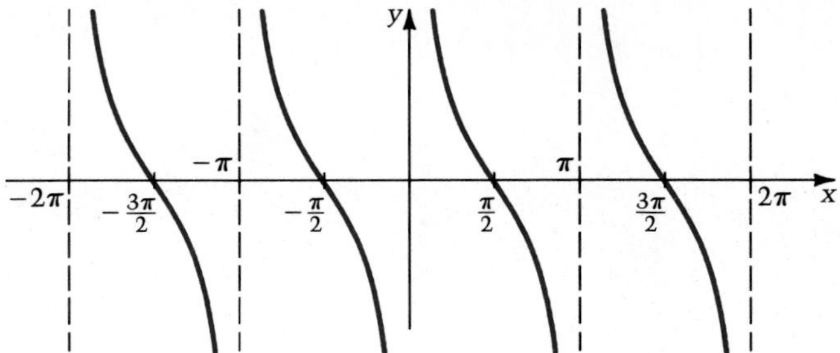

Figure 7.26 $y = \cot x$

Properties of the Cotangent Function	
Domain	all real numbers *except* $x = k\pi$, for any integer k
Range	all real numbers
Period	π
Symmetry	with respect to the origin; cotangent is an odd function, thus $\cot(-x) = -\cot x$
Asymptotes	vertical asymptotes $x = k\pi$, for any integer k

EXAMPLE 2

Verify that the graphs of the tangent and cotangent functions are symmetric with respect to the origin.

Solution Recall that to verify that the graph of a function $y = f(x)$ is symmetric with respect to the origin, we must show that when y is replaced with $-y$ and x is replaced with $-x$, the resulting equation is equivalent to $y = f(x)$.

$$y = \tan x = \frac{\sin x}{\cos x}$$

$$-y = \frac{\sin(-x)}{\cos(-x)} \qquad x \text{ is replaced with } -x \text{ and } y \text{ is replaced with } -y.$$

$$-y = \frac{-\sin x}{\cos x} \qquad \text{Recall that } \sin(-x) = -\sin x \text{ and } \cos(-x) = \cos x.$$

$$y = \frac{\sin x}{\cos x} \qquad \text{Both sides have been multiplied by } -1.$$

$$y = \tan x$$

Thus, the graph of $y = \tan x$ is symmetric with respect to the origin. Symmetry with respect to the origin for the graph of $y = \cot x$ is established in a similar manner.

The graph of the secant function can be easily produced by recalling that

$$\sec x = \frac{1}{\cos x}, \quad \text{for } \cos x \neq 0.$$

Since the secant function is not defined for values of x where $\cos x$ is 0, its domain does not include any value in the set

$$\left\{\ldots, -\frac{5\pi}{2}, -\frac{3\pi}{2}, -\frac{\pi}{2}, \frac{\pi}{2}, \frac{3\pi}{2}, \frac{5\pi}{2}, \ldots\right\},$$

and its graph has vertical asymptotes at these values. Since the secant function can be thought of as the reciprocal of the cosine function, it is helpful to graph the cosine function as a dashed curve, and then generate the graph of the secant function by viewing the y-values of the secant function as reciprocals of the y-values of the cosine function.

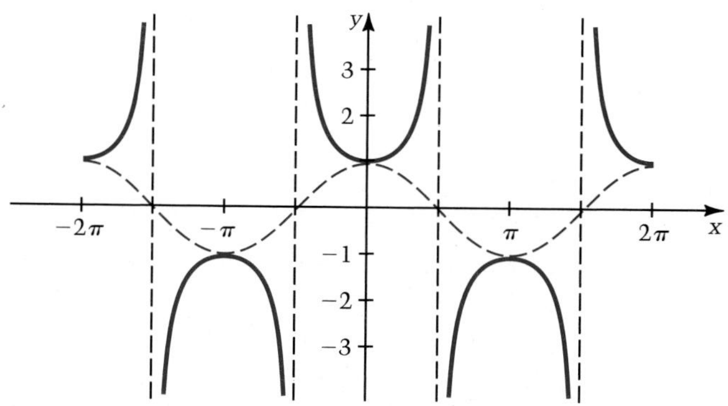

Figure 7.27 $y = \sec x$
($y = \cos x$ is shown as a dashed curve).

Notice that wherever the cosine function is 0, the secant function is undefined, and wherever the cosine function is 1, so is the secant function.

Furthermore, whenever the cosine function is positive (negative), the secant function is positive (negative).

Properties of the Secant Function	
Domain	all real numbers *except* $x = \frac{\pi}{2} + k\pi$, for any integer k
Range	all real numbers y such that $y \leq -1$ or $y \geq 1$
Period	2π
Symmetry	with respect to the y-axis (an even function, like its reciprocal, cosine)
Asymptotes	vertical asymptotes $x = \frac{\pi}{2} + k\pi$ (values where cosine is 0)

Note that the secant function has period 2π, the same as that of its reciprocal, the cosine function. Moreover, like the cosine function, the graph of the secant function is symmetric with respect to the y-axis.

The graph of the cosecant function is given in Figure 7.28. The cosecant function is the reciprocal of the sine function:

$$y = \frac{1}{\sin x}, \quad \text{for } \sin x \neq 0.$$

(The graph of the sine function is shown as a dashed curve.) Like the sine function, the cosecant function has period 2π, and is symmetric with respect to the origin.

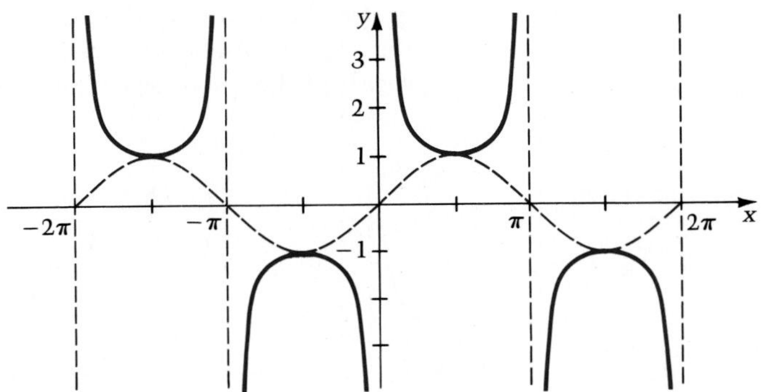

Figure 7.28 $y = \csc x$ ($y = \sin x$ is shown as a dashed curve).

Properties of the Cosecant Function	
Domain	all real numbers *except* $x = k\pi$, for any integer k
Range	all real numbers y such that $y \leq -1$ or $y \geq 1$
Period	2π
Symmetry	with respect to the origin (an odd function, like its reciprocal, sine)
Asymptotes	vertical asymptotes $x = k\pi$ (values where sine is 0)

EXAMPLE 3

Use the graphs of $y = \csc\left(x - \frac{\pi}{2}\right)$ and $y = -\sec x$ to determine whether $\csc\left(x - \frac{\pi}{2}\right) = -\sec x$ is an identity.

Solution

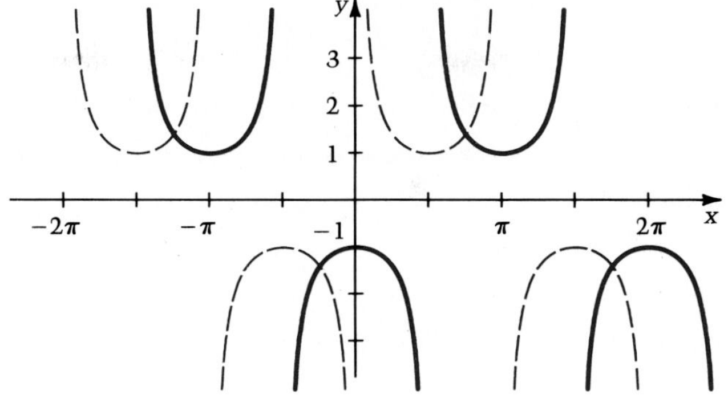

$y = \csc\left(x - \frac{\pi}{2}\right)$. The graph of $y = \csc x$ (dashed curve) is translated $\frac{\pi}{2}$ units to the right.

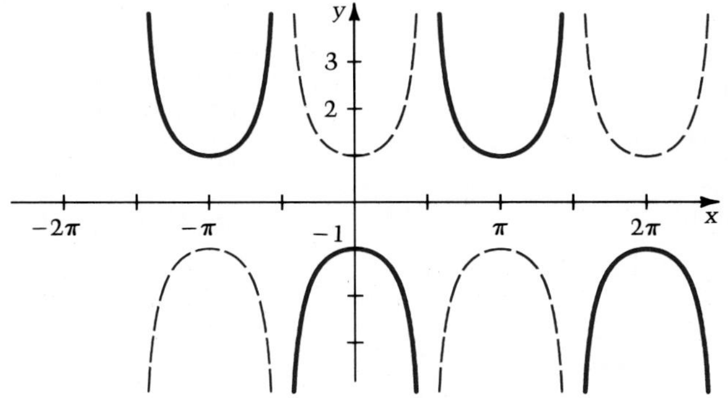

$y = -\sec x$. The graph of $y = \sec x$ (dashed curve) is reflected across the x-axis.

Figure 7.29

Since the graphs of $y = \csc\left(x - \frac{\pi}{2}\right)$ and $y = -\sec x$ are identical, the equation $\csc\left(x - \frac{\pi}{2}\right) = -\sec x$ is an identity.

EXAMPLE 4

Answer True or False.

(a) $\frac{7\pi}{2}$ is not in the domain of the tangent function.

(b) $\frac{3}{2}$ is in the range of the cosine function.

(c) The sine, cosine, and tangent functions all have the same period.

Solution

(a) True $\tan x = \dfrac{\sin x}{\cos x}$, and $\cos\dfrac{7\pi}{2} = 0$.

(b) False $-1 \le \cos x \le 1$ for all real x.

(c) False Tangent has period π; sine and cosine have period 2π.

EXERCISE SET 7.3

In Exercises 1–12, graph the pair of functions on the same axes.

1. $y = \sin x,\quad y = \sin(x + \pi)$

2. $y = \cos x,\quad y = \cos(x - \pi)$

3. $y = \sin x,\quad y = \sin\left(\dfrac{\pi}{2} + x\right)$

4. $y = \sin x,\quad y = \sin\left(x - \dfrac{\pi}{2}\right)$

5. $y = \cos x,\quad y = \cos(-x)$

6. $y = \sin x,\quad y = -\sin x$

7. $y = \sin x,\quad y = \sin\left(\dfrac{\pi}{4} - x\right)$

8. $y = \cos x,\quad y = \cos\left(-x - \dfrac{\pi}{4}\right)$

9. $y = \cos(-x),\quad y = -\cos(-x - \pi)$

10. $y = -\sin x,\quad y = -\sin(-x)$

11. $y = \sin x,\quad y = \sin(x - 2\pi)$

12. $y = \cos x,\quad y = \cos(x + 4\pi)$

In Exercises 13–22, graph the pair of functions on the same set of axes.

13. $y = \tan x,\quad y = \tan\left(x - \dfrac{\pi}{2}\right)$

14. $y = \sec x,\quad y = \sec\left(x + \dfrac{\pi}{3}\right)$

15. $y = \cot x,\quad y = \cot\left(x - \dfrac{\pi}{6}\right)$

16. $y = \tan x,\quad y = \tan(-x)$

17. $y = \csc x,\quad y = -\csc x$

18. $y = \cot x,\quad y = -\cot x$

19. $y = \sec x,\quad y = \sec\left(\dfrac{\pi}{4} - x\right)$

20. $y = \tan x,\quad y = \tan\left(\dfrac{\pi}{2} - x\right)$

21. $y = \csc(-x),\quad y = -\csc(-x)$

22. $y = \csc(-x),\quad y = \csc\left(-x - \dfrac{\pi}{2}\right)$

In Exercises 23–30, determine whether the statement is True or False.

23. $-\dfrac{\pi}{2}$ is not in the domain of $y = \cot x$.

24. 7π is in the domain of $y = \csc x$.

25. 2π is in the range of $y = \cot x$.

26. $x = 2\pi$ is an asymptote of $y = \cot x$.

27. Secant and sine have the same period.

28. All trigonometric functions except for tangent and cotangent have period 2π.

29. $\csc(-x) = \sin x$ is an identity.

30. $\cot(-x) = -\tan x$ is an identity.

Superset

In Exercises 31–38, sketch the graph of the function.

31. $y = |\sin x|$

32. $y = |\cot x|$

33. $y = \tan|x|$

34. $y = \sec|x|$

35. $y = \csc|x + \pi|$

36. $y = \sin|x - \pi|$

37. $y = \cos\left|\dfrac{\pi}{4} - x\right|$

38. $y = \csc\left|\dfrac{\pi}{3} + x\right|$

In Exercises 39–42, sketch the graphs of the pair of equations on the same axes.

39. $y = \sin x,\quad x = \sin y$

40. $y = \cos x,\quad x = \cos y$

41. $y = \tan x, \quad x = \tan y, -\frac{\pi}{2} < x < \frac{\pi}{2}$

42. $y = \cot x, \quad x = \cot y, 0 < x < \pi$

In Exercises 43–48, evaluate the given function for the x-values 1, -3, 21, 7.1416, -15.7080, and 235.

43. $f(x) = \sin(x + \pi)$

44. $f(x) = \cos(x - \pi)$

45. $f(x) = -\sin\left(x - \frac{\pi}{4}\right)$

46. $f(x) = -\cos\left(x + \frac{\pi}{4}\right)$

47. $f(x) = \tan\left(x - \frac{\pi}{2}\right)$

48. $f(x) = \cot\left(x - \frac{\pi}{6}\right)$

7.4 Transformations of the Trigonometric Functions

In this section we will rely heavily on our knowledge of transformations to develop an efficient way of graphing periodic functions. Essentially we will be concerned with translations, stretchings, and shrinkings of the basic trigonometric graphs.

For example, consider the function $y = 3 \sin x$. For this function, each y-coordinate is three times the corresponding y-coordinate of the function $y = \sin x$.

x	0	$\frac{\pi}{6}$	$\frac{\pi}{4}$	$\frac{\pi}{3}$	$\frac{\pi}{2}$	$\frac{2\pi}{3}$	$\frac{3\pi}{4}$	π	$\frac{3\pi}{2}$	2π
sin x	0	$\frac{1}{2}$	$\frac{\sqrt{2}}{2}$	$\frac{\sqrt{3}}{2}$	1	$\frac{\sqrt{3}}{2}$	$\frac{\sqrt{2}}{2}$	0	-1	0
3 sin x	0	$\frac{3}{2}$	$\frac{3\sqrt{2}}{2}$	$\frac{3\sqrt{3}}{2}$	3	$\frac{3\sqrt{3}}{2}$	$\frac{3\sqrt{2}}{2}$	0	-3	0

In Figure 7.30, the graph of $y = 3 \sin x$ is shown as the result of stretching the graph of $y = \sin x$ vertically (away from the x-axis). Recall that, in general, the graph of $y = k \cdot f(x)$ is found by vertically stretching or shrinking the graph of $y = f(x)$. If $k < 0$, the stretching or shrinking is accompanied by a reflection across the x-axis.

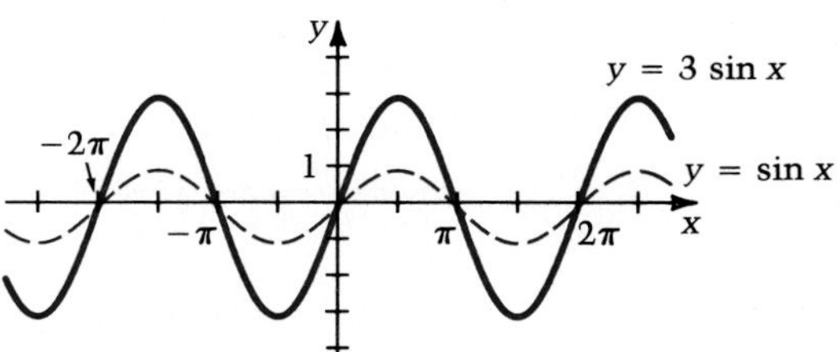

Figure 7.30

Notice that the range of the function $y = \sin x$ is the set of real numbers from -1 to 1 inclusive, whereas the range of $y = 3 \sin x$ is the set of real numbers from -3 to 3 inclusive. We call 3 the *amplitude* of $y = 3 \sin x$.

Definition of Amplitude

For functions of the form $y = a \sin x$ and $y = a \cos x$, the number $|a|$ is called the **amplitude.** The amplitude of a periodic function is one-half the difference between the maximum and minimum values of the function.

EXAMPLE 1

Sketch the graph of: (a) $y = \frac{1}{2} \cos x$ (b) $y = -2 \sin x$.

Solution

(a)

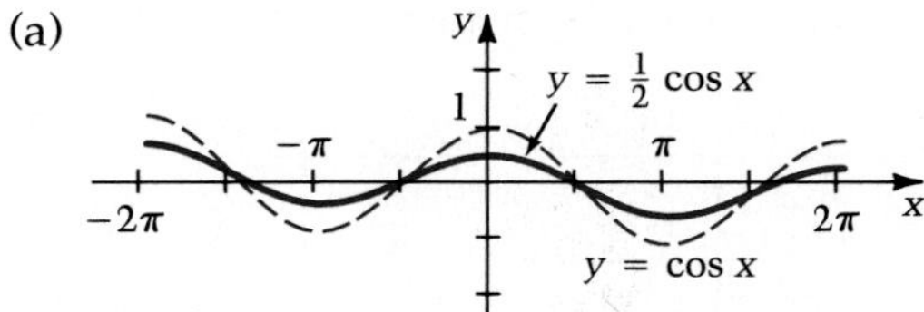

(b)

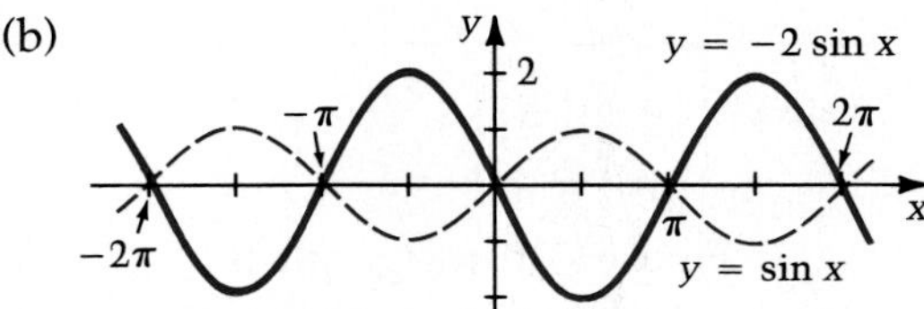

Figure 7.31

In (a), each y-coordinate for the graph of $y = \frac{1}{2} \cos x$ is $\frac{1}{2}$ the corresponding y-coordinate for the graph of $y = \cos x$. In (b), the graph of $y = -2 \sin x$ involves both a vertical stretching and a reflection across the x-axis.

Next we consider what happens to the graph of $y = \sin x$ when x is multiplied by 3. That is, we wish to determine how the graph of $y = \sin x$ is transformed to produce the graph of $y = \sin 3x$. Consider the following table of values.

x	0	$\frac{\pi}{6}$	$\frac{\pi}{4}$	$\frac{\pi}{3}$	$\frac{\pi}{2}$	$\frac{2\pi}{3}$	$\frac{3\pi}{4}$	π	$\frac{3\pi}{2}$	2π
$3x$	0	$\frac{\pi}{2}$	$\frac{3\pi}{4}$	π	$\frac{3\pi}{2}$	2π	$\frac{9\pi}{4}$	3π	$\frac{9\pi}{2}$	6π
$\sin 3x$	0	1	$\frac{\sqrt{2}}{2}$	0	-1	0	$\frac{\sqrt{2}}{2}$	0	1	0

Notice that as x takes on values from 0 to 2π, $3x$ takes on values from 0 to 6π. As a result, the graph of $y = \sin 3x$ completes *three* full cycles as x goes from 0 to 2π, with one full cycle as x goes from 0 to $\frac{2\pi}{3}$. Recall that the period is the length of the interval over which a periodic function makes one complete cycle. We therefore conclude that the period of $y = \sin 3x$ is $\frac{2\pi}{3}$, or one-third the period of $y = \sin x$. Graphs of

$$y = \sin 3x \quad \text{and} \quad y = \sin x$$

are shown in Figure 7.32.

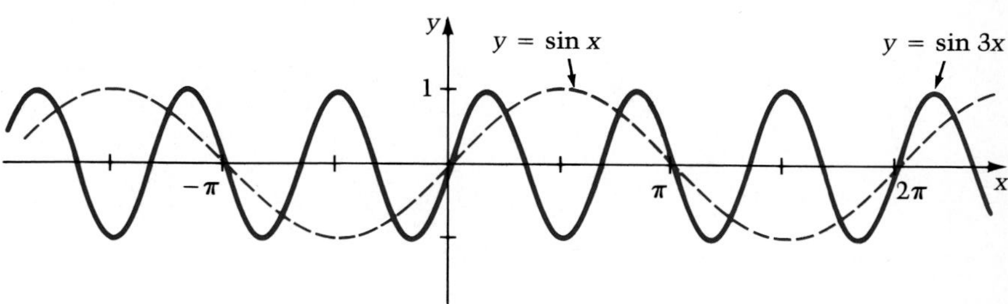

Figure 7.32

The graphs above suggest the following generalization.

> Functions of the form $y = \sin bx$ and $y = \cos bx$ have period equal to $\left|\frac{2\pi}{b}\right|$.

EXAMPLE 2

Sketch the graph of $y = \cos 2x$.

Solution

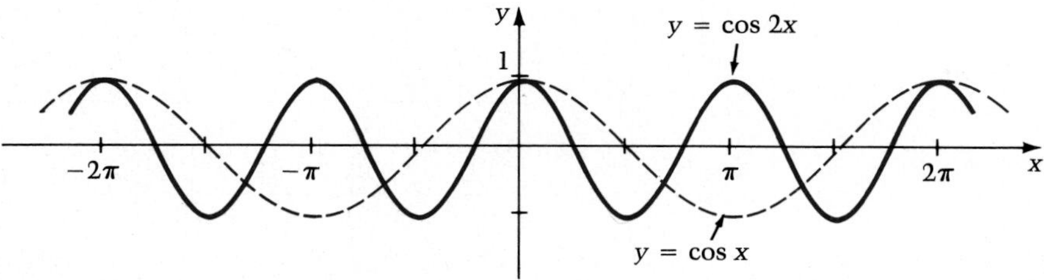

Figure 7.33

The period of $y = \cos 2x$ is $\left|\frac{2\pi}{2}\right| = \pi$, thus the given function completes one full cycle every π units.

Let us summarize our results for functions of the following two forms:

$$y = a \sin bx \quad \text{or} \quad y = a \cos bx.$$

- The amplitude $|a|$ indicates a vertical stretching of the basic sine or cosine curve if $|a| > 1$, and a vertical shrinking if $0 < |a| < 1$.
- The value $|b|$ indicates a horizontal stretching of the basic sine or cosine curve if $0 < |b| < 1$, and a horizontal shrinking if $|b| > 1$.

EXAMPLE 3

Sketch the graph of (a) $y = -2 \sin \frac{1}{3}x$ (b) $y = 3 \sin(-4x)$.

Solution

(a) We begin by sketching the graph of $y = \sin \frac{1}{3}x$ as a dashed curve. It is then stretched vertically and reflected across the x-axis to produce the graph of $y = -2 \sin \frac{1}{3}x$.

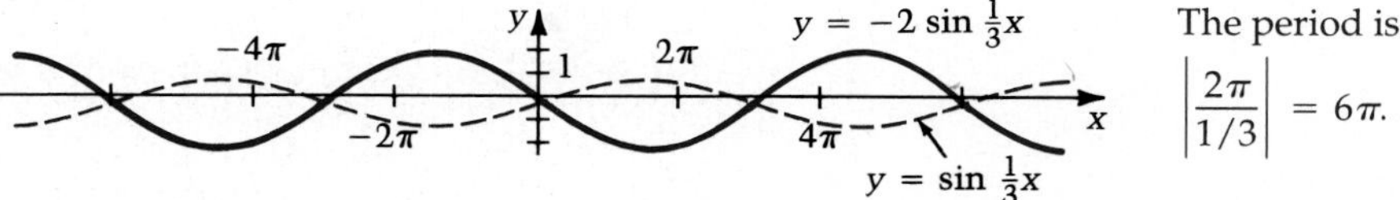

The period is $\left|\frac{2\pi}{1/3}\right| = 6\pi.$

Figure 7.34

(b) Recall that the graph of $y = f(-x)$ is the result of reflecting $y = f(x)$ across the y-axis. We begin by sketching the graph of $y = 3\sin(4x)$ as a dashed curve, then reflect it across the y-axis to produce $y = 3\sin(-4x)$.

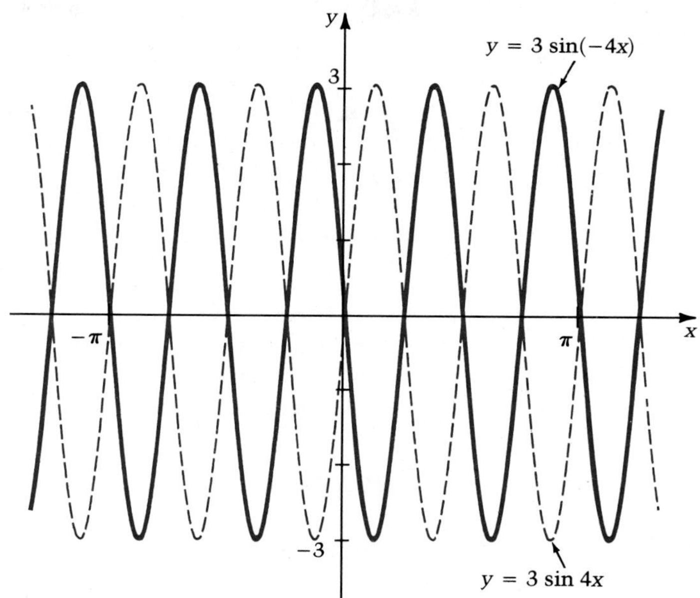

Figure 7.35

The period is $\left|\frac{2\pi}{-4}\right| = \frac{\pi}{2}$.

Consider the function $y = 1 + 5\sin 6\left(x - \frac{\pi}{4}\right)$. If this were simply the function $y = 5\sin 6x$, then we could use results already discussed in this section to conclude that the amplitude is equal to 5 and the period is equal to $\left|\frac{2\pi}{6}\right| = \frac{\pi}{3}$. As it turns out, the function $y = 1 + 5\sin 6\left(x - \frac{\pi}{4}\right)$ has the same period and amplitude as $y = 5\sin 6x$.

> The function $y = k + a\sin b(x - h)$ has the same amplitude $|a|$ and period $\left|\frac{2\pi}{b}\right|$ as the function $y = a\sin bx$.

The question then is how the graph of $y = 1 + 5\sin 6\left(x - \frac{\pi}{4}\right)$ differs from that of $y = 5\sin 6x$. Recall from Chapter 3 that the graph of the function

$y = k + f(x - h)$ can be produced by translating the graph of $y = f(x)$ horizontally h units and vertically k units.

> The graph of $y = k + a \sin b(x - h)$ is produced by translating the graph of $y = a \sin bx$:
>
> $$|h| \text{ units} \begin{cases} \text{to the right,} & \text{if } h > 0, \\ \text{to the left,} & \text{if } h < 0, \end{cases}$$
>
> $$|k| \text{ units} \begin{cases} \text{upward,} & \text{if } k > 0, \\ \text{downward,} & \text{if } k < 0. \end{cases}$$
>
> Similar statements hold for the cosine function.

The number h that determines the extent of horizontal translation is called the **phase shift.** When determining the phase shift, be careful to express the periodic function in precisely the form stated above. Example 4 shows how the facts stated above may be applied to the cosine function.

EXAMPLE 4

For the function $y = 3\cos\left(\frac{1}{2}x - \frac{\pi}{8}\right) - 1$, (a) determine the period, amplitude, and phase shift, and (b) sketch the graph.

Solution

(a) $y = -1 + 3\cos\frac{1}{2}\left(x - \frac{\pi}{4}\right)$ Begin by restating as $y = k + a\cos b(x - h)$.

The period is $\left|\frac{2\pi}{b}\right| = \frac{2\pi}{1/2} = 4\pi$, the amplitude is $|3| = 3$, and the phase shift is $\frac{\pi}{4}$.

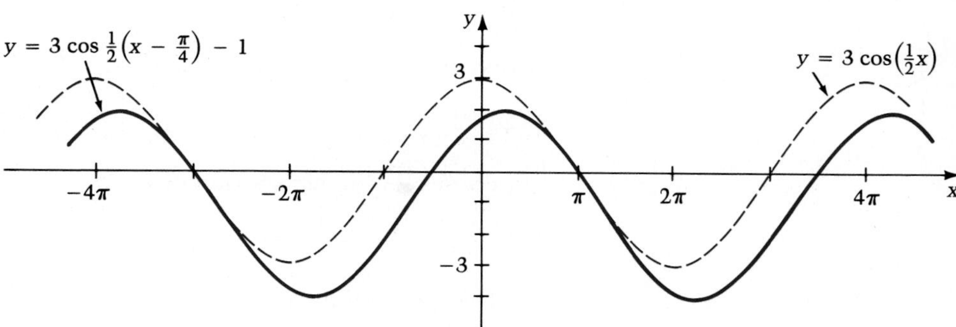

Figure 7.36

(b) As shown in Figure 7.36, we sketch the graph of $y = 3\cos\left(\frac{1}{2}x\right)$ as a dashed curve, then translate this graph $\frac{\pi}{4}$ units to the right and one unit downward to produce the graph of $y = 3\cos\frac{1}{2}\left(x - \frac{\pi}{4}\right) - 1$.

Suppose we wish to sketch the graph of the function $f(x) = \frac{1}{2}x + \sin x$. Taken individually, the two component functions $y = \frac{1}{2}x$ and $y = \sin x$ are familiar to us (a straight line and the sine curve, respectively). One method for graphing the function $f(x) = \frac{1}{2}x + \sin x$ is to shift each point on the graph of $y = \sin x$ by an amount equal to $\frac{1}{2}x$. Example 5 uses this method, which is called graphing by **addition of ordinates.**

EXAMPLE 5

Sketch the graph of $y = \dfrac{1}{2}x + \sin x$.

Solution

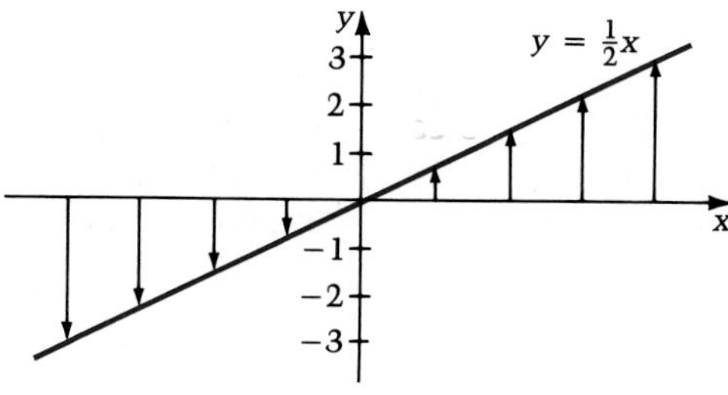

Sketch the graph of $y = \dfrac{1}{2}x$ and use arrows to represent y-coordinates for various values of x.

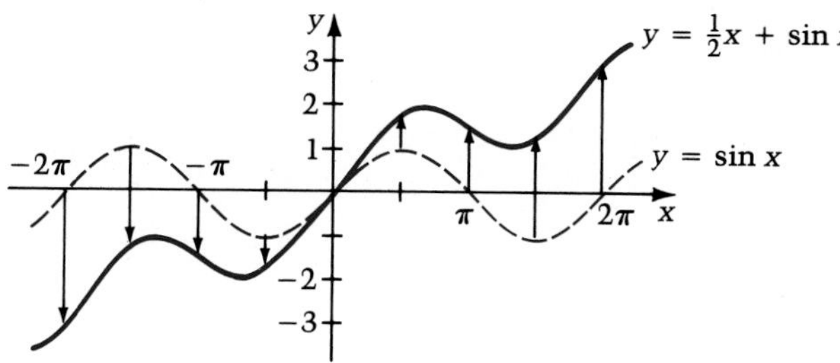

Sketch the graph of $y = \sin x$ as a dashed curve and use the arrows from the figure above to indicate the result of adding $\dfrac{1}{2}x$. Draw a smooth curve through the arrowheads.

Figure 7.37

EXERCISE SET 7.4

In Exercises 1–16, sketch the graph of the given function on the interval $[0, 2\pi]$.

1. $y = \frac{1}{3}\cos x$
2. $y = 4\sin x$
3. $y = -\frac{1}{2}\sin x$
4. $y = -\frac{3}{2}\cos x$
5. $y = -3\sin x$
6. $y = \sqrt{2}\sin x$
7. $y = \cos\frac{1}{2}x$
8. $y = \sin\left(-\frac{1}{3}x\right)$
9. $y = \cos(-3x)$
10. $y = \sin\frac{3}{2}x$
11. $y = 2\sin\frac{1}{2}x$
12. $y = \frac{1}{2}\cos 2x$
13. $y = -3\sin\frac{1}{3}x$
14. $y = -2\cos 3x$
15. $y = \sqrt{2}\cos(-2x)$
16. $y = \sqrt{3}\cos\left(-\frac{1}{2}x\right)$

In Exercises 17–30, determine the amplitude, period, and phase shift, and sketch the graph of the given function.

17. $y = 2\cos\left(x + \frac{\pi}{2}\right)$
18. $y = 3\cos\left(x - \frac{\pi}{2}\right)$
19. $y = 2\cos\frac{1}{2}\left(x - \frac{\pi}{3}\right)$
20. $y = \frac{1}{2}\sin 2(x - \pi)$
21. $y = 2\sin\left(\frac{1}{2}x + \frac{\pi}{4}\right)$
22. $y = 2\cos(2x + \pi)$
23. $y = \sqrt{3}\sin\left(-\frac{x}{2} - \frac{\pi}{2}\right)$
24. $y = 3\sin\left(-x + \frac{\pi}{2}\right)$
25. $y = -3\sin(2x - \pi)$
26. $y = -2\cos\left(\frac{x}{2} + \frac{\pi}{2}\right)$
27. $y = 3\cos\left(\frac{1}{2}x + \frac{\pi}{2}\right) - 2$
28. $y = -3\sin\left(2x + \frac{\pi}{2}\right) + 1$
29. $y = 2\sin\left(x - \frac{\pi}{2}\right) + 1$
30. $y = -2\cos\left(\frac{x}{2} + \frac{\pi}{2}\right) - 1$

In Exercises 31–38, sketch the graph of the given function.

31. $y = \sin x - 2x$
32. $y = 2x - \sin x$
33. $y = \frac{1}{2}x + \sin 2x$
34. $y = \frac{1}{2}x - \cos 2x$
35. $y = -3\cos 2x + x$
36. $y = -2\sin\frac{1}{2}x + 3x$
37. $y = -2\sin\frac{1}{2}\left(x - \frac{\pi}{2}\right) - 2x$
38. $y = -\frac{1}{2}\cos\left(x - \frac{\pi}{2}\right) + 4x$

Superset

39. Find b such that the period of the function $y = \frac{1}{2}\sin bx$ is $\frac{\pi}{4}$.
40. Find b so that the function $y = 3\cos b(x - \pi)$ has period 4π.

In Exercises 41–48, sketch the graph of the given function.

41. $y = |\sin 2x|$
42. $y = |3\sin x|$
43. $y = |-2\cos x|$
44. $y = \cos|-2x|$
45. $y = \sin x + \cos x$
46. $y = \sin x - \cos x$
47. $y = \sqrt{3}\sin x - \cos x$
48. $y = \frac{\sqrt{3}}{2}\sin x + \frac{1}{2}\cos x$

In Exercises 49–52, determine whether the statement is True or False.

49. The function $y = \sin x + \cos x$ is periodic with period 4π.
50. The maximum value of $4\sin\frac{1}{2}x$ is 2.
51. The maximum value of $-2\cos\frac{1}{2}x$ is 2.
52. The minimum value of $-\frac{3}{2}\cos\frac{2}{3}x$ is $\frac{3}{2}$.

In Exercises 53–56, evaluate the given function for the x-values 1, -2.5, 4.512, 24.67, -31.56, and 49.223.

53. $f(x) = 2\sin\left(\frac{1}{2}x + \frac{\pi}{4}\right)$
54. $f(x) = 2\cos(2x + \pi)$
55. $f(x) = \sqrt{3}\cos\left(-\frac{1}{2}x - \frac{\pi}{2}\right)$
56. $f(x) = \frac{3}{2}\sin\left(-x + \frac{\pi}{2}\right)$

7.5
The Inverse Trigonometric Functions

By the definition of a function, we are assured that to each value in the domain of a trigonometric function, there is assigned exactly one range value. For example, the sine function assigns to the domain value $\frac{\pi}{6}$ exactly one range value, namely $\frac{1}{2}$. Thus, if we know that $\sin\frac{\pi}{6} = y$, then y must be $\frac{1}{2}$. However, it is not true that if we know a particular y-value, say $\frac{1}{2}$, the corresponding x-value is unique. For example, if $\sin x = \frac{1}{2}$, then x can be any number in the set $\left\{\ldots, -\frac{11\pi}{6}, -\frac{7\pi}{6}, \frac{\pi}{6}, \frac{5\pi}{6}, \frac{13\pi}{6}, \ldots\right\}$.

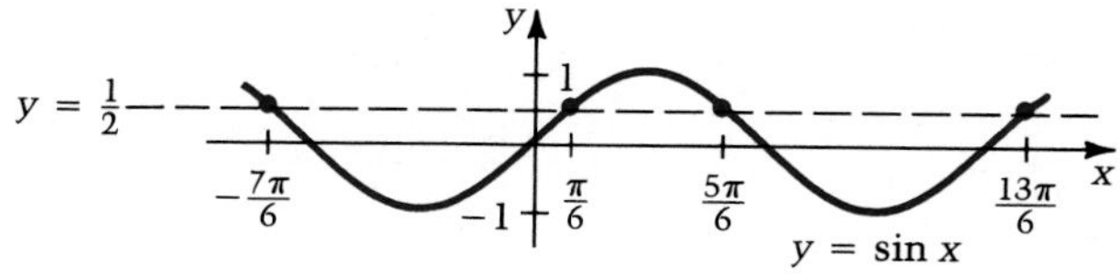

Figure 7.38 There are an infinite number of values of x such that $\sin x = \frac{1}{2}$.

In this section we would like to define an inverse sine function, that is, a function which "undoes" what the sine function "does." To be a function, this "inverse sine" must take a value like $\frac{1}{2}$ and produce *exactly one number* whose sine is $\frac{1}{2}$, for example, $\frac{\pi}{6}$. (If it produced more than one, it would not be a function!) The problem is that the sine function is not one-to-one, and thus cannot have an inverse function. To resolve this problem, we look at the sine function and restrict its domain to some interval on which it *is* one-to-one. In Figure 7.39(a), we have graphed the sine function for domain values

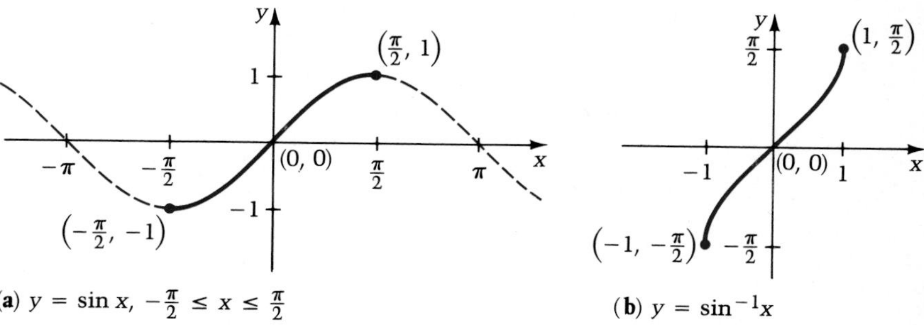

Figure 7.39 (a) The restricted sine function. (b) The inverse sine function.

in the interval $\left[-\frac{\pi}{2}, \frac{\pi}{2}\right]$. For these values, the sine function is one-to-one and thus has an inverse function. Recall that the graph of an inverse can be found by reflecting the original graph across the line $y = x$. The graph of the inverse sine function is shown in Figure 7.39(b). We describe this inverse with the equation $y = \sin^{-1} x$.

Careful! The symbol $\sin^{-1} x$ does not mean $\frac{1}{\sin x}$; as you know,

$$\frac{1}{\sin x} = \csc x.$$

Definition of the Inverse Sine Function

The inverse sine function is defined as follows:

$$y = \sin^{-1} x \quad \text{with domain } -1 \le x \le 1$$

$$\text{if and only if} \qquad \sin y = x \quad \text{and} \quad -\frac{\pi}{2} \le y \le \frac{\pi}{2}.$$

As a consequence of the definition of $y = \sin^{-1} x$, it is often useful to think of the expression $\sin^{-1} x$ as

$$\text{the angle between } -\frac{\pi}{2} \text{ and } \frac{\pi}{2} \text{ inclusive whose sine is } x.$$

The function name "arcsin" is sometimes used instead of $\sin^{-1}$ to refer to the inverse sine function. Thus, arcsin and $\sin^{-1}$ denote the same thing and can be used interchangeably.

EXAMPLE 1

Determine the value of each expression without using a calculator or Table 4.

(a) $\sin^{-1}\left(\frac{1}{2}\right)$ (b) $\arcsin\left(\frac{\sqrt{2}}{2}\right)$ (c) $\sin^{-1}\left(-\frac{\sqrt{3}}{2}\right)$

(d) $\tan\left(\arcsin\left(-\frac{1}{2}\right)\right)$ (e) $\sin^{-1}\left(\sin\frac{5\pi}{4}\right)$

Solution

(a) Think of $\sin^{-1}\left(\frac{1}{2}\right)$ as the angle between $-\frac{\pi}{2}$ and $\frac{\pi}{2}$ whose sine is $\frac{1}{2}$. Since

$\sin\left(\frac{\pi}{6}\right) = \frac{1}{2}$, and $\frac{\pi}{6}$ is between $-\frac{\pi}{2}$ and $\frac{\pi}{2}$, we conclude that

$$\sin^{-1}\left(\frac{1}{2}\right) = \frac{\pi}{6}.$$

(b) $\arcsin\left(\frac{\sqrt{2}}{2}\right)$ is the angle between $-\frac{\pi}{2}$ and $\frac{\pi}{2}$ whose sine is $\frac{\sqrt{2}}{2}$. Since $\sin\left(\frac{\pi}{4}\right) = \frac{\sqrt{2}}{2}$, and $\frac{\pi}{4}$ is between $-\frac{\pi}{2}$ and $\frac{\pi}{2}$,

$$\arcsin\left(\frac{\sqrt{2}}{2}\right) = \frac{\pi}{4}.$$

(c) $\sin^{-1}\left(-\frac{\sqrt{3}}{2}\right)$ is the angle between $-\frac{\pi}{2}$ and $\frac{\pi}{2}$ whose sine is $-\frac{\sqrt{3}}{2}$. Since $\sin\left(-\frac{\pi}{3}\right) = -\frac{\sqrt{3}}{2}$, and $-\frac{\pi}{3}$ is between $-\frac{\pi}{2}$ and $\frac{\pi}{2}$,

$$\sin^{-1}\left(-\frac{\sqrt{3}}{2}\right) = -\frac{\pi}{3}.$$

(d) First determine $\arcsin\left(-\frac{1}{2}\right)$: $\arcsin\left(-\frac{1}{2}\right)$ is the angle between $-\frac{\pi}{2}$ and $\frac{\pi}{2}$ whose sine is $-\frac{1}{2}$. Since $\sin\left(-\frac{\pi}{6}\right) = -\frac{1}{2}$,

$$\arcsin\left(-\frac{1}{2}\right) = -\frac{\pi}{6}.$$

Thus,

$$\begin{aligned} \tan\left(\arcsin\left(-\frac{1}{2}\right)\right) &= \tan\left(-\frac{\pi}{6}\right) \\ &= -\tan\left(\frac{\pi}{6}\right) && \text{Tangent is an odd function.} \\ &= -\frac{\sqrt{3}}{3}. \end{aligned}$$

(e) First determine $\sin\frac{5\pi}{4}$.

$$\sin\left(\frac{5\pi}{4}\right) = -\sin\left(\frac{\pi}{4}\right) = -\frac{\sqrt{2}}{2} \qquad \text{Use the reference angle } \frac{\pi}{4}.$$

Thus,

$$\begin{aligned} \sin^{-1}\left(\sin\frac{5\pi}{4}\right) &= \sin^{-1}\left(-\frac{\sqrt{2}}{2}\right) \\ &= -\frac{\pi}{4} \end{aligned}$$

$\sin^{-1}\left(-\frac{\sqrt{2}}{2}\right)$ is the angle between $-\frac{\pi}{2}$ and $\frac{\pi}{2}$ whose sine is $-\frac{\sqrt{2}}{2}$, namely, $-\frac{\pi}{4}$.

Recall that if f and f^{-1} are inverses of one another, then

$$f(f^{-1}(x)) = x \quad \text{for all } x \text{ in the domain of } f^{-1}, \text{ and}$$
$$f^{-1}(f(x)) = x \quad \text{for all } x \text{ in the domain of } f.$$

If we let $f(x) = \sin x$ and $f^{-1}(x) = \sin^{-1} x$, then those statements become

$$\sin(\sin^{-1} x) = x \quad \text{for all } x \text{ such that } -1 \leq x \leq 1, \text{ and}$$
$$\sin^{-1}(\sin x) = x \quad \text{for all } x \text{ such that } -\frac{\pi}{2} \leq x \leq \frac{\pi}{2}.$$

By virtue of the last equation, it is *not* true that $\sin^{-1}(\sin x) = x$ for all x in the domain of the sine function—Example 1(e) is a case in point. However, for all x between $-\frac{\pi}{2}$ and $\frac{\pi}{2}$ (the domain of the restricted sine function), it is true that $\sin^{-1}(\sin x) = x$.

In a manner similar to the case of $\sin^{-1} x$, we can restrict the domains of the other trigonometric functions so that inverse functions can be defined. By restricting $y = \cos x$ to values of x between 0 and π inclusive, the function becomes one-to-one and has an inverse cosine function.

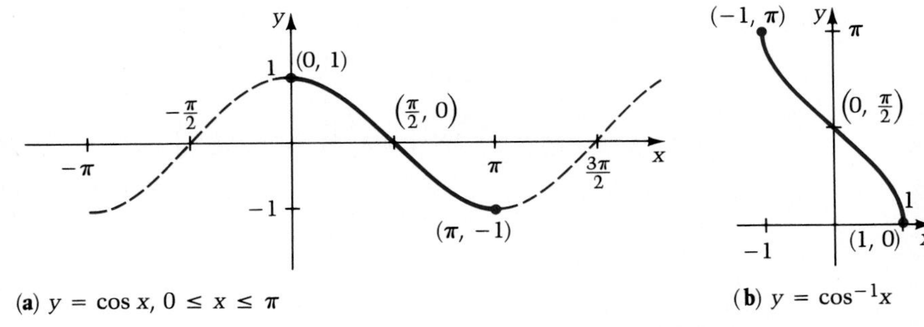

Figure 7.40 (a) The restricted cosine function. (b) The inverse cosine function.

Definition of the Inverse Cosine Function

The inverse cosine function is defined as follows:

$$y = \cos^{-1} x \quad \text{with domain } -1 \leq x \leq 1$$

if and only if $\quad \cos y = x \quad \text{and} \quad 0 \leq y \leq \pi.$

By restricting $y = \tan x$ to values of x between $-\frac{\pi}{2}$ and $\frac{\pi}{2}$, we can define the inverse function $y = \tan^{-1} x$.

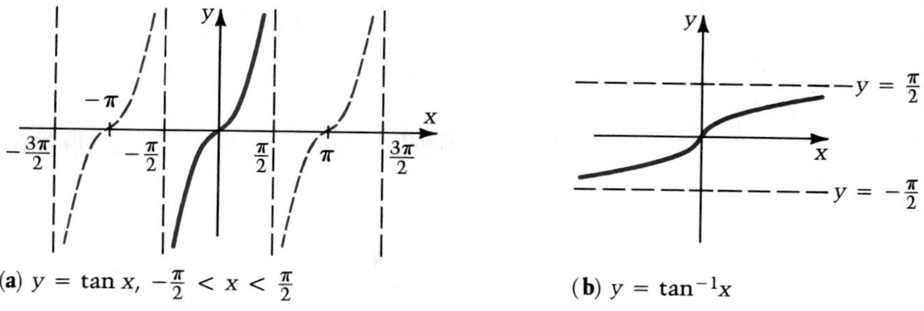

Figure 7.41 (a) The restricted tangent function. (b) The inverse tangent function.

Definition of the Inverse Tangent Function

The inverse tangent function is defined as follows:

$$y = \tan^{-1} x \quad \text{with domain } -\infty < x < \infty,$$

if and only if $\quad \tan y = x \quad \text{and} \quad -\frac{\pi}{2} < y < \frac{\pi}{2}.$

By virtue of the definitions of $\cos^{-1} x$ and $\tan^{-1} x$, we can think of $\cos^{-1} x$ as "the angle between 0 and π whose cosine is equal to x," and $\tan^{-1} x$ as "the angle between $-\frac{\pi}{2}$ and $\frac{\pi}{2}$ whose tangent is equal to x." The expressions arccos x and arctan x are used interchangeably with $\cos^{-1} x$ and $\tan^{-1} x$.

EXAMPLE 2

Evaluate each expression, without using a calculator or Table 4.

(a) arctan(1) (b) $\cos^{-1}\left(-\frac{\sqrt{3}}{2}\right)$ (c) $\tan^{-1}\left(\tan\frac{2\pi}{3}\right)$

Solution

(a) arctan(1) is the angle between $-\frac{\pi}{2}$ and $\frac{\pi}{2}$ whose tangent is 1. Since $\tan\left(\frac{\pi}{4}\right) = 1$, and $\frac{\pi}{4}$ is between $-\frac{\pi}{2}$ and $\frac{\pi}{2}$, arctan(1) $= \frac{\pi}{4}$.

(b) $\cos^{-1}\left(-\frac{\sqrt{3}}{2}\right)$ is the angle between 0 and π inclusive whose cosine is $-\frac{\sqrt{3}}{2}$. Since $\cos\left(\frac{\pi}{6}\right)$ is $\frac{\sqrt{3}}{2}$, we are looking for an angle that has $\frac{\pi}{6}$ as a reference angle, and whose cosine is negative. Since the angle must be between 0 and π, our only choice is $\frac{5\pi}{6}$. Thus, $\cos^{-1}\left(-\frac{\sqrt{3}}{2}\right) = \frac{5\pi}{6}$.

(c) We begin by determining $\tan\left(\frac{2\pi}{3}\right)$:

$$\tan\left(\frac{2\pi}{3}\right) = -\tan\left(\frac{\pi}{3}\right) = -\sqrt{3}.$$

Now determine $\tan^{-1}(-\sqrt{3})$: $\tan^{-1}(-\sqrt{3})$ is the angle between $-\frac{\pi}{2}$ and $\frac{\pi}{2}$ whose tangent is $-\sqrt{3}$. Since the tangent is negative, the angle is between $-\frac{\pi}{2}$ and 0, and since the value of the tangent is $-\sqrt{3}$, the reference angle is $\frac{\pi}{3}$. Thus $\tan^{-1}(-\sqrt{3}) = -\frac{\pi}{3}$, and we have $\tan^{-1}\left(\tan\left(\frac{2\pi}{3}\right)\right) = -\frac{\pi}{3}$.

Inverses can be defined for the cotangent, secant, and cosecant functions by suitably restricting the domains of these three trigonometric functions. We leave these problems for the exercise set.

Composite Trigonometric Expressions

Sometimes it is necessary to evaluate expressions such as $\tan(\arcsin(-\frac{3}{5}))$ or $\cos(\tan^{-1}(\frac{5}{12}))$. Such composite expressions can be evaluated without resorting to calculators, tables, or facts about special angles. These problems require that you recall three things:

- The ranges of the inverse trigonometric functions:

$$-\frac{\pi}{2} \le \sin^{-1} x \le \frac{\pi}{2}, \qquad 0 \le \cos^{-1} x \le \pi, \qquad -\frac{\pi}{2} < \tan^{-1} x < \frac{\pi}{2}.$$

- The ASTC memory device for determining the quadrant in which an angle in standard position will terminate.
- The trigonometric function values of an angle in standard position can be determined by the coordinates of any point on the terminal side of the angle.

EXAMPLE 3

Determine the exact values of the following:

(a) $\tan\left(\arcsin\left(-\frac{3}{5}\right)\right)$ (b) $\cos\left(\tan^{-1}\left(\frac{5}{12}\right)\right)$

Solution

(a) Begin by letting $t = \arcsin\left(-\frac{3}{5}\right)$. Then $\sin t = \sin\left(\arcsin\left(-\frac{3}{5}\right)\right) = -\frac{3}{5}$. Since $\sin t = -\frac{3}{5}$, a negative number, t must terminate in either the third or fourth quadrant (use ASTC). But, because t is in the range of the arcsin function, $-\frac{\pi}{2} \le t \le \frac{\pi}{2}$, t cannot terminate in the third quadrant. Thus, t must be a negative angle terminating in the fourth quadrant.

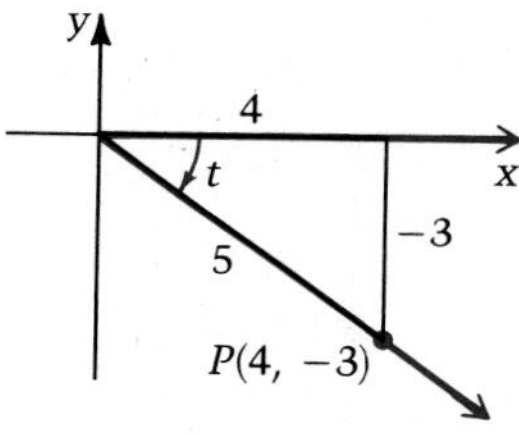

Figure 7.42

Draw angle t and label a point P on its terminal side: since $\sin t = \frac{y}{r} = \frac{-3}{5}$, let P have y-coordinate -3, and lie 5 units from the origin. To find the x-coordinate of P, use the Pythagorean Theorem:

$$x^2 + (-3)^2 = 5^2 \text{ so } x^2 = 16.$$

Since $x > 0$, we get $x = 4$.

Thus, $\tan\left(\arcsin\left(-\frac{3}{5}\right)\right) = \tan t = \frac{y}{x} = -\frac{3}{4}$.

(b) Begin by letting $t = \tan^{-1}\left(\frac{5}{12}\right)$. Then $\tan t = \tan\left(\tan^{-1}\left(\frac{5}{12}\right)\right) = \frac{5}{12}$. Since $\tan t$ is a positive number, t must terminate in either the first or third quadrant. The range of the inverse tangent function requires that $-\frac{\pi}{2} < t < \frac{\pi}{2}$. Thus, t is a first quadrant angle.

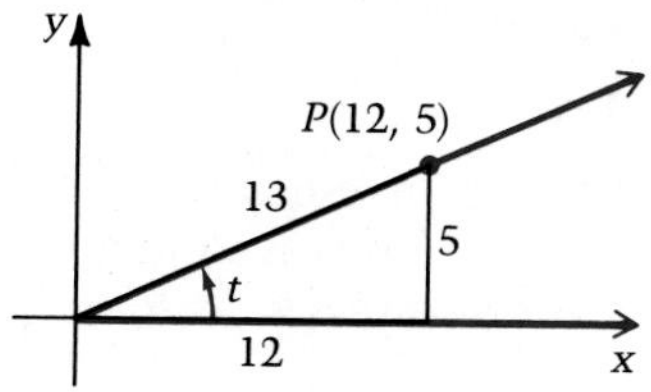

Figure 7.43

Draw t and a point P on its terminal side. Since $\tan t = \frac{y}{x} = \frac{5}{12}$, let P have coordinates (12, 5). By the Pythagorean Theorem, we have

$$r^2 = 12^2 + 5^2 = 169$$
$$r = 13.$$

Thus, $\cos\left(\tan^{-1}\left(\frac{5}{12}\right)\right) = \cos t = \frac{x}{r} = \frac{12}{13}$.

EXAMPLE 4

Write $\cos^2(\tan^{-1} z)$, for $z \geq 0$, as an expression involving no trigonometric functions.

Solution Let $t = \tan^{-1} z$. Then $\tan t = z$. Since $z \geq 0$, t must be between 0 and $\frac{\pi}{2}$.

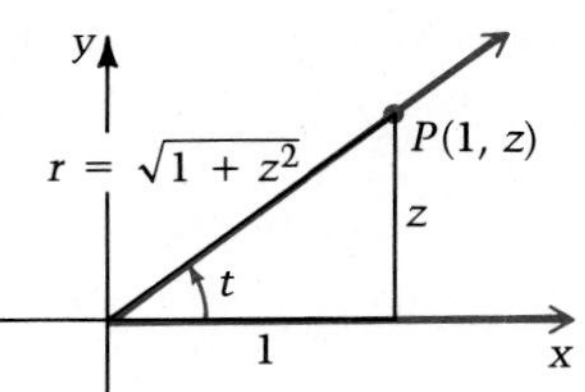

Figure 7.44

Since $\tan t = \frac{y}{x} = \frac{z}{1}$, draw angle t with $P(1, z)$ on its terminal side. By the Pythagorean Theorem,

$$r^2 = 1^2 + z^2$$
$$r = \sqrt{1 + z^2}.$$

Thus, $\cos^2(\tan^{-1} z) = (\cos t)^2 = \left(\frac{x}{r}\right)^2 = \left(\frac{1}{\sqrt{1 + z^2}}\right)^2 = \frac{1}{1 + z^2}.$

EXERCISE SET 7.5

In Exercises 1–24, evaluate the expression without using a calculator or Table 4.

1. $\sin^{-1} \frac{\sqrt{3}}{2}$
2. $\sin^{-1} 1$
3. $\cos^{-1}(-1)$
4. $\tan^{-1}(\sqrt{3})$
5. $\tan^{-1}\left(-\frac{\sqrt{3}}{3}\right)$
6. $\cos^{-1}\left(-\frac{\sqrt{2}}{2}\right)$
7. $\arctan(-1)$
8. $\tan^{-1}(-\sqrt{3})$
9. $\arcsin 0$
10. $\arccos 0$
11. $\sin\left(\cos^{-1}\left(\frac{\sqrt{3}}{2}\right)\right)$
12. $\tan\left(\cos^{-1}\left(\frac{\sqrt{3}}{2}\right)\right)$
13. $\tan\left(\cos^{-1}\left(-\frac{1}{2}\right)\right)$
14. $\sin\left(\cos^{-1}\left(-\frac{1}{2}\right)\right)$
15. $\cos(\sin^{-1}(-1))$
16. $\cos\left(\sin^{-1}\left(-\frac{\sqrt{2}}{2}\right)\right)$
17. $\sin(\tan^{-1}(-\sqrt{3}))$
18. $\cos(\tan^{-1}(-1))$
19. $\cos\left(\cos^{-1}\left(-\frac{\sqrt{2}}{2}\right)\right)$
20. $\cos\left(\sin^{-1}\left(-\frac{1}{2}\right)\right)$
21. $\tan^{-1}\left(\tan \frac{3\pi}{4}\right)$
22. $\cos^{-1}\left(\cos \frac{\pi}{3}\right)$
23. $\sin^{-1}\left(\cos\left(-\frac{\pi}{3}\right)\right)$
24. $\tan(\tan^{-1}(-\sqrt{3}))$

In Exercises 25–32, evaluate the expression. Round your answers to four decimal places, with angles expressed in radians.

25. $\sin^{-1}\left(\frac{\sqrt{5}}{12}\right)$
26. $\arcsin\left(\frac{\sqrt{7}}{3}\right)$
27. $\arctan(-\sqrt{11})$
28. $\tan^{-1}\left(\frac{\sqrt{17}}{\sqrt{3}}\right)$
29. $\sin(\arctan(3.1562))$
30. $\cos(\arcsin(-0.7246))$
31. $\tan^{-1}(\tan(36.7°))$
32. $\tan(\tan^{-1}(-24))$

In Exercises 33–42, use a right triangle to evaluate the given expression.

33. $\sin\left(\sin^{-1}\frac{4}{5}\right)$ **34.** $\cos\left(\sin^{-1}\frac{4}{5}\right)$

35. $\cos\left(\sin^{-1}\left(-\frac{3}{5}\right)\right)$ **36.** $\tan\left(\arccos\frac{4}{5}\right)$

37. $\cos\left(\tan^{-1}\left(-\frac{3}{4}\right)\right)$ **38.** $\sin\left(\tan^{-1}\frac{7}{24}\right)$

39. $\sin\left(\arccos\left(-\frac{12}{13}\right)\right)$ **40.** $\tan\left(\sin^{-1}\left(\frac{5}{13}\right)\right)$

41. $\tan\left(\arcsin\left(-\frac{2}{7}\right)\right)$ **42.** $\sin(\arctan 2)$

In Exercises 43–52, rewrite the expression without using trigonometric functions. Assume $z \geq 0$.

43. $\sin^2(\arctan z)$ **44.** $\cos(\tan^{-1} z)$

45. $\cos^2(\sin^{-1} z)$ **46.** $\tan^2(\sin^{-1} z)$

47. $\tan(\arcsin(-z))$ **48.** $\sin(\arcsin(-z))$

49. $\cot(\tan^{-1} z)$ **50.** $\sec^2(\arctan z)$

51. $\csc^2(\tan^{-1}(2z))$ **52.** $\cot(\sin^{-1}(3z))$

Superset

Inverse cotangent, secant, and cosecant functions can be defined as follows:

$y = \cot^{-1} x$ if and only if $\cot y = x$, where $-\infty < x < \infty$, and $0 < y < \pi$,

$y = \sec^{-1} x$ if and only if $\sec y = x$, where $x \leq -1$ or $x \geq 1$, and $0 \leq y \leq \pi$, $y \neq \frac{\pi}{2}$,

$y = \csc^{-1} x$ if and only if $\csc y = x$, where $x \leq -1$ or $x \geq 1$, and $-\frac{\pi}{2} \leq y \leq \frac{\pi}{2}$, $y \neq 0$.

Use these definitions in evaluating the expressions in Exercises 53–62, without using a calculator or Table 4.

53. $\operatorname{arcsec}(-2)$ **54.** $\operatorname{arccsc}(-\sqrt{2})$

55. $\sin\left(\csc^{-1}\left(-\frac{13}{5}\right)\right)$ **56.** $\cos\left(\sec^{-1}\left(-\frac{13}{12}\right)\right)$

57. $\cos\left(\sec^{-1}\frac{3}{\sqrt{5}}\right)$ **58.** $\cot\left(\sec^{-1}\frac{7}{3}\right)$

59. $\cot\left(\sin^{-1}\frac{2}{3}\right)$ **60.** $\sec\left(\sin^{-1}\left(-\frac{1}{2}\right)\right)$

61. $\sec\left(\cot^{-1}\left(-\frac{\sqrt{11}}{2}\right)\right)$ **62.** $\cot\left(\tan^{-1}\frac{\sqrt{6}}{4}\right)$

For Exercises 63–68, use a calculator and the definitions above to evaluate the expression.

63. $\arccos(0.9483)$ **64.** $\sin^{-1}(0.4226)$

65. $\tan^{-1}(7.115)$ **66.** $\operatorname{arccot}(-28.64)$

67. $\sin(\cot^{-1}(12.25))$ **68.** $\sin(\cos^{-1}(-0.2221))$

In Exercises 69–78, sketch the graph.

69. $y = 3\cos^{-1} x$ **70.** $y = \cos^{-1}\left(\frac{1}{3}x\right)$

71. $y = \sin^{-1}(2x)$ **72.** $y = \frac{1}{2}\sin^{-1} x$

73. $y = \sin^{-1}(x + 1)$ **74.** $y = \cos^{-1}(x - 2)$

75. $y = -\cos^{-1}(2x)$ **76.** $y = -2\tan^{-1} x$

77. $y = \frac{\pi}{2} - \sin^{-1} x$ **78.** $y = \frac{\pi}{2} - \cos^{-1} x$

7.6 Vectors in the Plane

The statement "The distance between points A and B is 12 miles" tells us only the distance between A and B. The statement "An object undergoes a displacement of 12 miles due east from point A to point B" tells us both the distance between A and B and the direction in which the object travels. Quantities like distance, which indicate a magnitude (size) but no direction, are called **scalar quantities.** Quantities like displacement, which indicate both magnitude and direction, are called **vector quantities.**

Since a vector quantity contains information about magnitude and direction, it is convenient to represent it by an arrow (a directed line segment) in

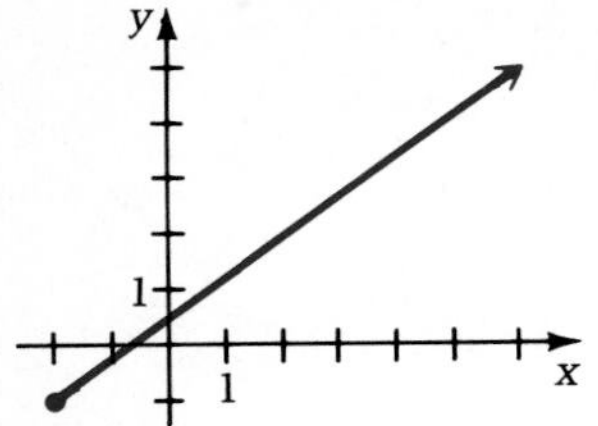

Figure 7.45

the xy-plane (Figure 7.45). The magnitude of the vector quantity is given by the length of the arrow. The direction of the vector quantity is given by the angle that the arrow makes with the horizontal in the direction of the positive x-axis.

EXAMPLE 1

Determine the magnitude and direction of the vector shown in Figure 7.46.

Solution Begin by drawing a right triangle with the arrow as its hypotenuse. Label the coordinates of the tail A and head B of the vector.

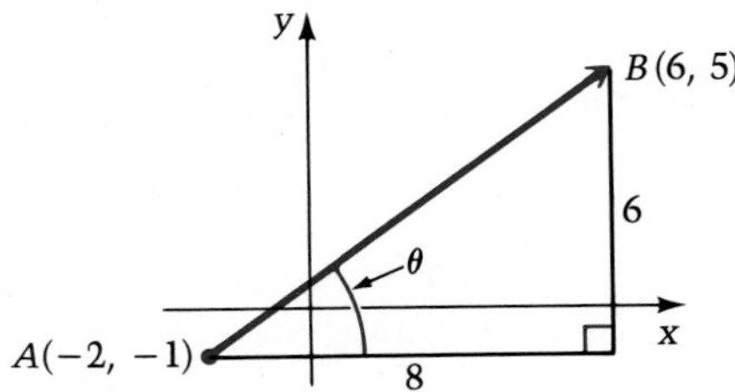

Figure 7.46

$$\begin{aligned} d &= \sqrt{(x_2 - x_1)^2 + (y_2 - y_1)^2} \\ &= \sqrt{(6 - (-2))^2 + (5 - (-1))^2} \\ &= \sqrt{64 + 36} = 10 \end{aligned}$$

The magnitude of the vector can be found by applying the distance formula to the coordinates of the tail $A(-2, -1)$ and head $B(6, 5)$ of the vector.

$$\tan\theta = \frac{\text{opposite}}{\text{adjacent}} = \frac{6}{8} = 0.75$$

$$\theta \approx 37°$$

The direction can be found by determining angle θ in the triangle.

The magnitude of the vector is 10, and it makes an angle of approximately 37° with the horizontal.

We usually use boldface letters to represent vector quantities (refer to Figure 7.47).

We will need to determine the sum of two vectors, and the product of a scalar (a real number) and a vector. The sum $\mathbf{a} + \mathbf{b}$ of two vectors $\mathbf{a}$ and $\mathbf{b}$ is found by first moving $\mathbf{b}$ without changing its magnitude or direction, so that the tail of $\mathbf{b}$ is placed at the head of $\mathbf{a}$. Then draw an arrow from the tail of $\mathbf{a}$ to the head of $\mathbf{b}$ (Figure 7.48).

The product $k\mathbf{a}$ of a scalar k and a vector $\mathbf{a}$ is represented by drawing the arrow whose length is k times the length of $\mathbf{a}$. If k is a positive number, $\mathbf{a}$ and $k\mathbf{a}$ have the same direction; if k is a negative number, $\mathbf{a}$ and $k\mathbf{a}$ have opposite

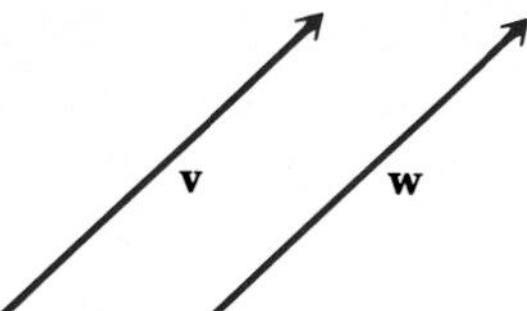

Vectors **v** and **w** are equal since they have the same magnitude and direction.

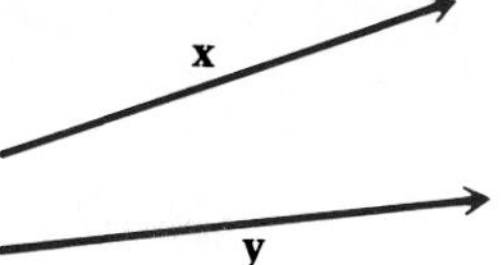

Vectors **x** and **y** have the same magnitude, but different directions.

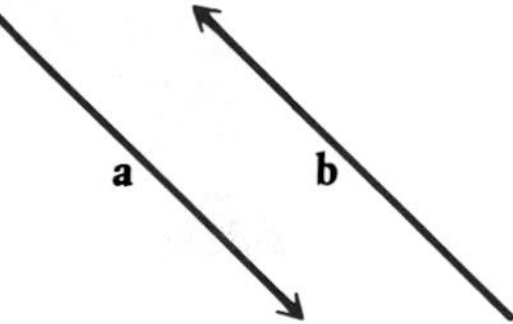

Vectors **a** and **b** are said to have the same magnitude but *opposite* directions.

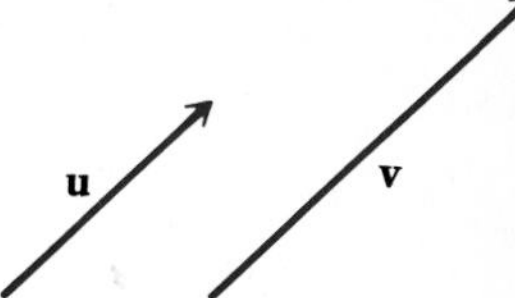

Vectors **u** and **v** have the same direction, but different magnitudes.

Figure 7.47

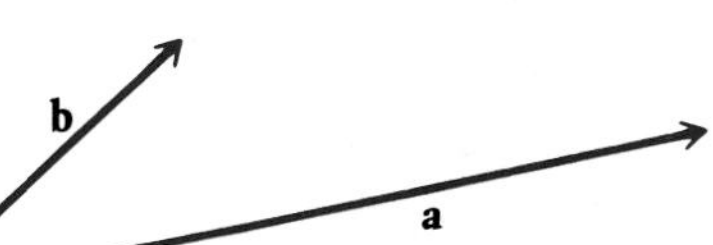

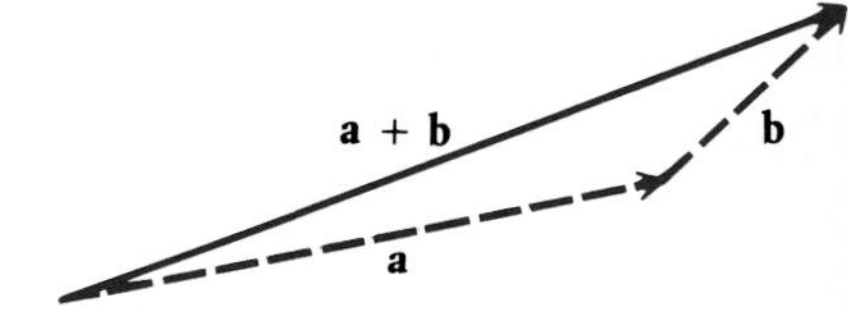

Figure 7.48 Vector addition

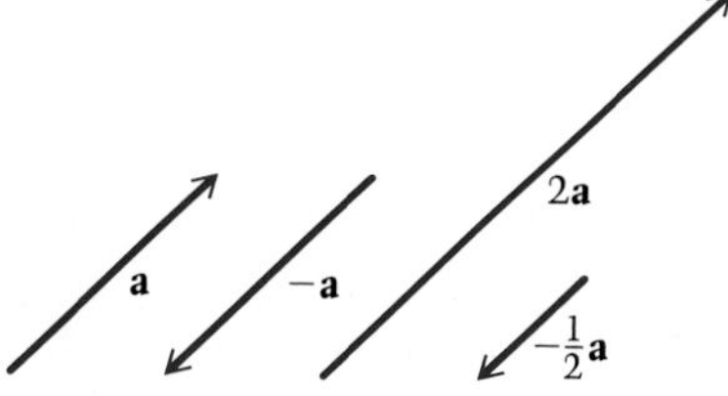

Figure 7.49 Parallel vectors

directions. Note that as a consequence of the way we have defined the sum of two vectors and the product of a scalar and a vector, we can represent the difference $\mathbf{a} - \mathbf{b}$ as $\mathbf{a} + (-1)\mathbf{b}$. Note also that two vectors are **parallel** if and only if one is a scalar multiple of the other. This is illustrated in Figure 7.49.

EXAMPLE 2

Vectors **a** and **b** are represented in Figure 7.50. Draw $2\mathbf{a} + 3\mathbf{b}$ and $2\mathbf{b} - \mathbf{a}$.

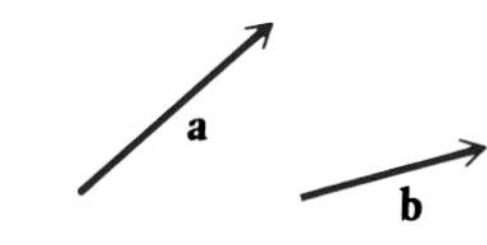

Figure 7.50

Solution

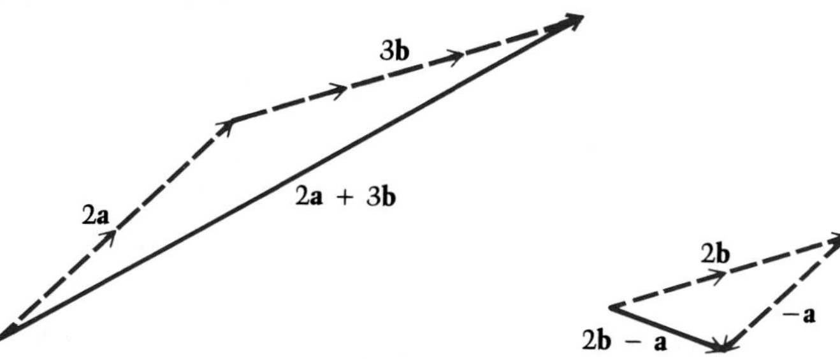

Figure 7.51

We have illustrated a method for adding (or subtracting) vectors by placing them in a tail-to-head arrangement. We can also add (or subtract) two vectors by placing them in a tail-to-tail arrangement. When positioned this way, the two vectors determine a parallelogram as shown in Figure 7.52. The sum **a** + **b** is found by drawing a vector from the point where the tails meet, along the diagonal of the parallelogram to the opposite vertex. The difference **a** − **b** is found by drawing a vector from the head of **b** to the head of **a**.

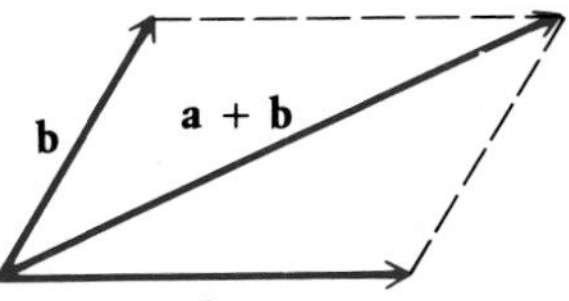

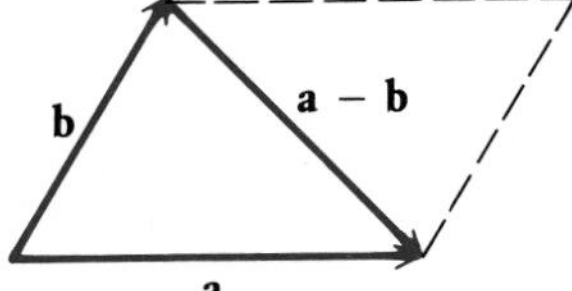

Figure 7.52

We have frequently referred to the tail and head of a vector. These points are alternately called the **initial point** and the **terminal point** of the vector, respectively. It is sometimes convenient to use these points when naming the vector. For example, if we know that vector **v** in Figure 7.53 has initial point A and terminal point B, we may refer to it as $\overrightarrow{AB}$.

Figure 7.53 $\mathbf{v} = \overrightarrow{AB}$

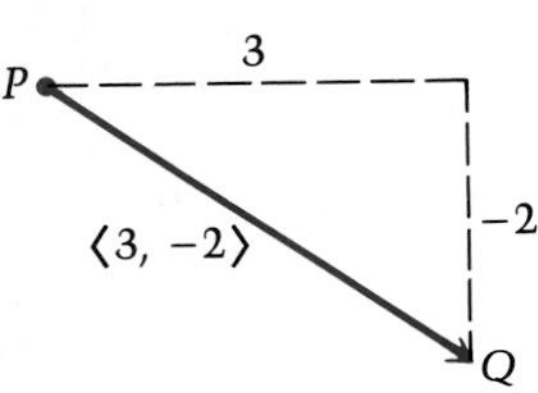

Figure 7.54

Vectors as Ordered Pairs

One of the most common ways of naming a vector is with an ordered pair of numbers. The first number of the ordered pair represents the change in x from the initial point to the terminal point of the vector, and the second number represents the change in y from the initial point to the terminal point. To distinguish ordered pairs representing vectors from ordered pairs representing points, we will use angular brackets $\langle\ ,\ \rangle$ when referring to vectors. For example, in Figure 7.54, we denote the vector $\overrightarrow{PQ}$ by $\langle 3, -2\rangle$. This indicates that the change in x from P to Q is 3 units in the positive x-direction and the change in y is 2 units in the negative y-direction.

In a natural way, the x-coordinates and y-coordinates of two points P_1 and P_2 can be used to determine the vector $\overrightarrow{P_1P_2}$.

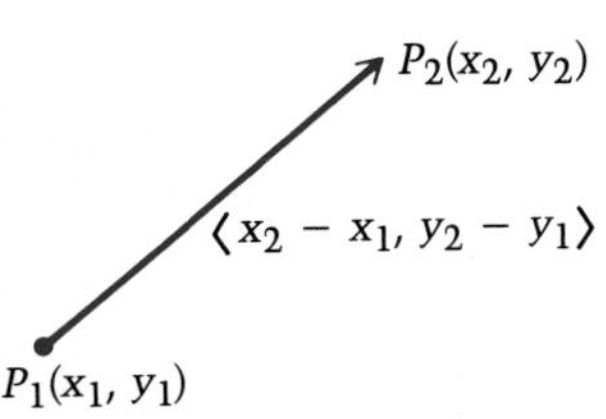

Figure 7.55

Definition of Scalar Components

Given two points, $P_1(x_1, y_1)$ and $P_2(x_2, y_2)$, the vector with initial point P_1 and terminal point P_2 is defined as follows:

$$\overrightarrow{P_1P_2} = \langle x_2 - x_1, y_2 - y_1\rangle.$$

The numbers $x_2 - x_1$ and $y_2 - y_1$ are called the **scalar components** of the vector $\overrightarrow{P_1P_2}$.

EXAMPLE 3

(a) Describe the vector from $A(-3, 2)$ to $B(9, -7)$ as an ordered pair.

(b) If $\mathbf{v} = \langle 3, -5\rangle$ is positioned in the plane so that its initial point is at $(1, 2)$, what are the coordinates of its terminal point?

Solution

(a) $\overrightarrow{AB} = \langle 9 - (-3), -7 - 2\rangle = \langle 12, -9\rangle$.

(Note: $\overrightarrow{BA} = \langle -3 - 9, 2 - (-7)\rangle = \langle -12, 9\rangle$, thus the vector $\overrightarrow{BA}$ is the same as the vector $-\overrightarrow{AB}$.

(b) The terminal point is found by adding 3 to the x-coordinate of the initial point and -5 to the y-coordinate of the initial point. Thus, the terminal point has coordinates $(1 + 3, 2 - 5) = (4, -3)$.

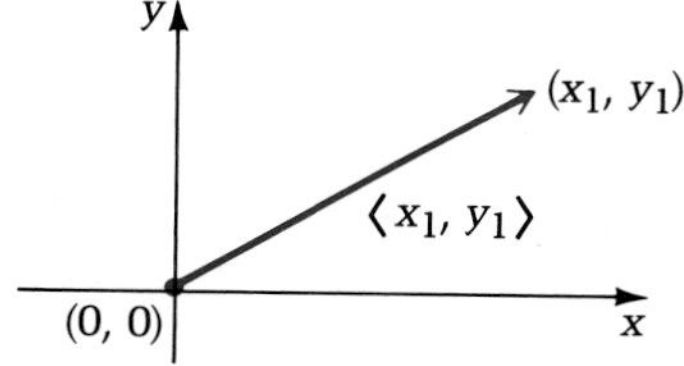

Figure 7.56

A vector that is positioned with its initial point at the origin $(0, 0)$ is called a **position vector**, or **radius vector.** In this case, if the vector is given by $\langle x_1, y_1\rangle$, then the terminal point is (x_1, y_1). That is, the scalar components of a position vector are the same as the coordinates of the terminal point. The vector $\langle 0, 0\rangle = \mathbf{0}$ is called the **zero vector.** We now formally define the notions of equality, addition, and scalar multiplication of vectors.

Definition of Vector Operations and Equality

Given two vectors $\mathbf{a} = \langle a_1, a_2\rangle$ and $\mathbf{b} = \langle b_1, b_2\rangle$, and a scalar k, we have the following definitions.

Equality of vectors	$\mathbf{a} = \mathbf{b}$ if and only if $a_1 = b_1$ and $a_2 = b_2$.
Addition of vectors	$\mathbf{a} + \mathbf{b} = \langle a_1 + b_1, a_2 + b_2\rangle$
Scalar Multiplication of a vector	$k\mathbf{a} = \langle ka_1, ka_2\rangle$

EXAMPLE 4

Suppose $\mathbf{v} = \langle 2, 3\rangle$, $\mathbf{w} = \langle 5, -4\rangle$ and $\mathbf{j} = \langle 0, 1\rangle$. Determine:
(a) $2\mathbf{v} + 3\mathbf{w}$ (b) $\mathbf{w} - 6\mathbf{j}$.

Solution

(a)
$$\begin{aligned} 2\mathbf{v} + 3\mathbf{w} &= 2\langle 2, 3\rangle + 3\langle 5, -4\rangle \\ &= \langle 4, 6\rangle + \langle 15, -12\rangle \\ &= \langle 4 + 15, 6 - 12\rangle \\ &= \langle 19, -6\rangle \end{aligned}$$

(b)
$$\begin{aligned} \mathbf{w} - 6\mathbf{j} &= \langle 5, -4\rangle + (-6)\langle 0, 1\rangle \\ &= \langle 5, -4\rangle + \langle 0, -6\rangle \\ &= \langle 5 + 0, -4 - 6\rangle \\ &= \langle 5, -10\rangle \end{aligned}$$

The length of a vector **v** is called the **norm** of **v** and is denoted $\|\mathbf{v}\|$. Using the Pythagorean Theorem and the ordered pair definition of a vector, we have the following definition.

Definition of Vector Norm

If $\mathbf{a} = \langle a_1, a_2 \rangle$, then $\|\mathbf{a}\|$ is the **norm** of **a**, defined $\|\mathbf{a}\| = \sqrt{a_1^2 + a_2^2}$.

Note that the norm of a vector is a scalar quantity—it measures the length of the vector. A vector is called a **unit vector** if it has length equal to 1. Given a vector **a**, we can determine the **unit vector in the direction of a,** denoted $\mathbf{u}_\mathbf{a}$, by multiplying vector **a** by the scalar $\frac{1}{\|\mathbf{a}\|}$. That is,

$$\mathbf{u}_\mathbf{a} = \frac{\mathbf{a}}{\|\mathbf{a}\|}$$

is the unit vector in the direction of **a.**

EXAMPLE 5

Suppose $\mathbf{v} = \langle 3, -4 \rangle$. Determine the following:

(a) $\|\mathbf{v}\|$ (b) $\mathbf{u}_\mathbf{v}$ (c) a vector of length 2 in the direction of **v**

Solution

(a) $\|\mathbf{v}\| = \sqrt{3^2 + (-4)^2} = \sqrt{9 + 16} = 5$

(b) $\mathbf{u}_\mathbf{v} = \frac{1}{\|\mathbf{v}\|}\mathbf{v} = \frac{1}{5}\langle 3, -4 \rangle = \left\langle \frac{3}{5}, -\frac{4}{5} \right\rangle$

Thus the unit vector in the direction of **v** is $\left\langle \frac{3}{5}, -\frac{4}{5} \right\rangle$.

(c) To determine a vector of length 2 in the direction of **v**, multiply the unit vector (in that direction) by 2.

$$2\mathbf{u}_\mathbf{v} = 2\left\langle \frac{3}{5}, -\frac{4}{5} \right\rangle = \left\langle \frac{6}{5}, -\frac{8}{5} \right\rangle$$

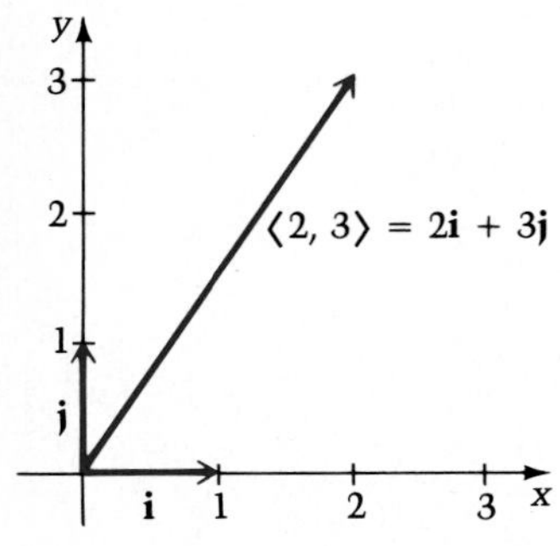

Figure 7.57

The unit vectors $\langle 1, 0 \rangle$ and $\langle 0, 1 \rangle$ occur so frequently that we give them the special names **i** and **j**, respectively. Thus, **i** is a vector of length 1 in the positive x-direction, and **j** is a vector of length 1 in the positive y-direction. These two vectors are referred to as the **unit coordinate vectors.** For any real numbers a_1 and a_2,

$$a_1\mathbf{i} + a_2\mathbf{j} = a_1\langle 1, 0 \rangle + a_2\langle 0, 1 \rangle = \langle a_1, 0 \rangle + \langle 0, a_2 \rangle = \langle a_1, a_2 \rangle.$$

Thus, any vector $\mathbf{a} = \langle a_1, a_2 \rangle$ can be written as $\mathbf{a} = a_1\mathbf{i} + a_2\mathbf{j}$, which is referred to as a **linear combination** of **i** and **j**.

EXAMPLE 6

Suppose $\mathbf{v} = 2\mathbf{i} + 3\mathbf{j}$ and $\mathbf{w} = \mathbf{i} - 4\mathbf{j}$. Determine the following:

(a) $5\mathbf{v} - \mathbf{w}$ (b) $\|\mathbf{v}\|$ (c) a unit vector in the direction of **v**

Solution

(a) $5\mathbf{v} - \mathbf{w} = 5(2\mathbf{i} + 3\mathbf{j}) - (\mathbf{i} - 4\mathbf{j}) = 10\mathbf{i} + 15\mathbf{j} - \mathbf{i} + 4\mathbf{j} = 9\mathbf{i} + 19\mathbf{j}$

(b) $\|\mathbf{v}\| = \|2\mathbf{i} + 3\mathbf{j}\| = \sqrt{2^2 + 3^2} = \sqrt{4 + 9} = \sqrt{13}$

(c) $\mathbf{u}_\mathbf{v} = \dfrac{1}{\|\mathbf{v}\|}\mathbf{v} = \dfrac{1}{\sqrt{13}}(2\mathbf{i} + 3\mathbf{j}) = \dfrac{2}{\sqrt{13}}\mathbf{i} + \dfrac{3}{\sqrt{13}}\mathbf{j} = \dfrac{2\sqrt{13}}{13}\mathbf{i} + \dfrac{3\sqrt{13}}{13}\mathbf{j}$

EXERCISE SET 7.6

In Exercises 1–8, determine the magnitude and direction of the vector.

1.

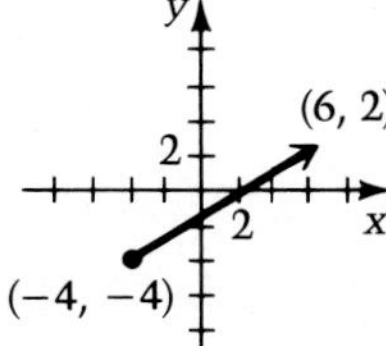

2.

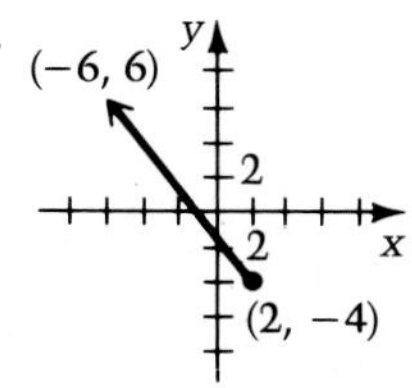

3.

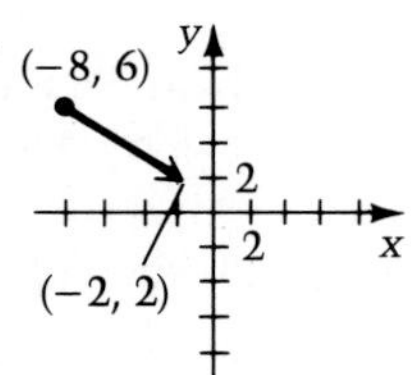

4.

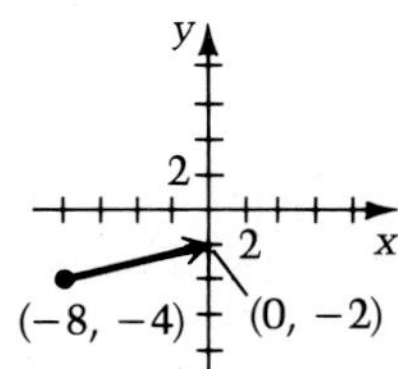

5.

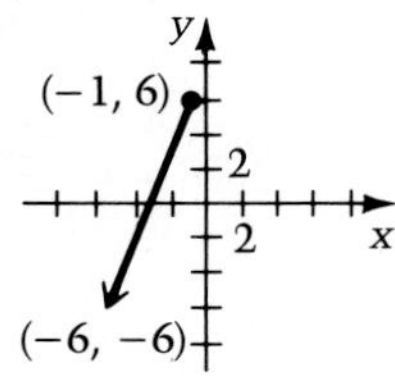

6.

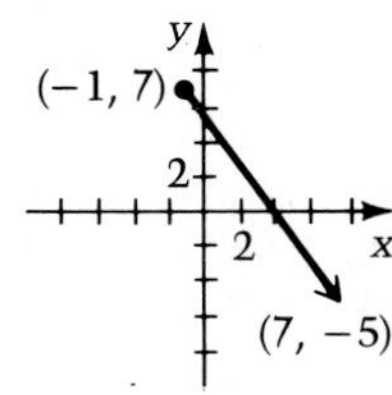

7.

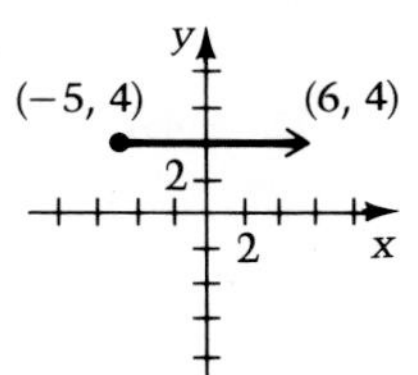

8.

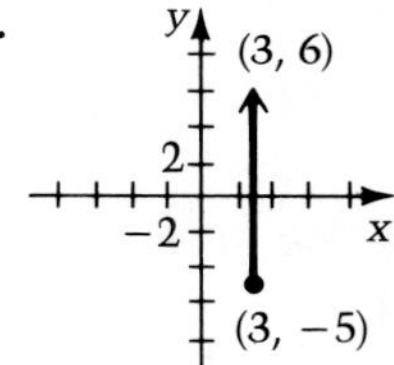

In Exercises 9–20, using the given vectors, draw the following.

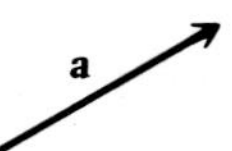

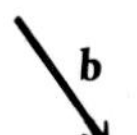

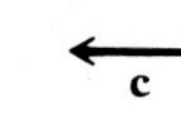

9. $\mathbf{a} + \mathbf{d}$ **10.** $\mathbf{b} + \mathbf{c}$ **11.** $\mathbf{a} - \mathbf{d}$

12. $\mathbf{b} - \mathbf{c}$ **13.** $2\mathbf{a} - 3\mathbf{b}$ **14.** $3\mathbf{d} - 4\mathbf{b}$

15. $\frac{1}{2}\mathbf{a} + \frac{1}{2}\mathbf{d}$ **16.** $2\mathbf{b} + 2\mathbf{c}$ **17.** $-2\mathbf{b}$

18. $-3\mathbf{c}$ **19.** $\mathbf{a} + 2\mathbf{b} - \mathbf{c}$ **20.** $2\mathbf{a} - \mathbf{c} - 2\mathbf{d}$

In Exercises 21–26, points *A* and *B* are given. Describe the vector from *A* to *B* as an ordered pair.

21. $A(-1, -3), B(4, 9)$ **22.** $A(2, -2), B(-4, 6)$

23. $A(-5, 9), B(-2, 5)$ **24.** $A(-10, -7), B(-2, -1)$

25. $A(1, 4), B(2, -4)$ **26.** $A\left(-2, 3\frac{1}{2}\right), B(-5, 4)$

In Exercises 27–30, points A and B are given. Describe the vector from A to B as an ordered pair and determine its magnitude and direction.

27. $A(2.37, -1.89)$, $B(-2.75, 3.45)$
28. $A(-2.64, -3.31)$, $B(3.73, 1.80)$
29. $A(-5.57, -2.48)$, $B(1.96, -3.73)$
30. $A(7.57, 3.09)$, $B(-3.78, -1.07)$

In Exercises 31–36, given the vector **v** and the coordinates of its initial point, find the coordinates of its terminal point.

31. $\mathbf{v} = \langle 3, -2\rangle, (-1, 4)$ **32.** $\mathbf{v} = \langle -2, 3\rangle, (-1, 4)$

33. $\mathbf{v} = \langle -4, -5\rangle, \left(1\frac{1}{2}, 3\frac{1}{2}\right)$ **34.** $\mathbf{v} = \langle -5, 2\rangle, \left(1\frac{1}{2}, 3\frac{1}{2}\right)$

35. $\mathbf{v} = \left\langle \frac{13}{2}, \frac{7}{2}\right\rangle, (-4, -1)$ **36.** $\mathbf{v} = \left\langle -\frac{3}{2}, -\frac{7}{2}\right\rangle, (1, -2)$

In Exercises 37–44, for the vectors $\mathbf{u} = \langle 3, 6\rangle$, $\mathbf{v} = \langle -8, 2\rangle$, and $\mathbf{w} = \langle 2, -1\rangle$, determine the following.

37. $3\mathbf{v} + 2\mathbf{w}$ **38.** $2\mathbf{v} + \mathbf{u}$ **39.** $\frac{1}{2}\mathbf{v} + 3\mathbf{u}$

40. $2\mathbf{w} + 3\mathbf{u}$ **41.** $-\frac{1}{2}\mathbf{v} + 3\mathbf{u}$ **42.** $-4\mathbf{w} + \frac{1}{3}\mathbf{u}$

43. $3\mathbf{v} + 2\mathbf{w} - \frac{1}{3}\mathbf{u}$ **44.** $\frac{1}{2}\mathbf{v} - 3\mathbf{w} + \frac{2}{3}\mathbf{u}$

In Exercises 45–52, for the given vector **v**, determine $\|\mathbf{v}\|$, $\mathbf{u}_\mathbf{v}$, and $-2\mathbf{u}_\mathbf{v}$.

45. $\langle -6, 8\rangle$ **46.** $\langle -4, -3\rangle$ **47.** $\langle 0, -4\rangle$

48. $\langle -6, 0\rangle$ **49.** $\langle 2\sqrt{2}, 2\rangle$ **50.** $\langle 3, \sqrt{7}\rangle$

51. $\left\langle \frac{8}{3}, -2\right\rangle$ **52.** $\left\langle -\frac{5}{2}, 6\right\rangle$

In Exercises 53–58, given vectors **v** and **w**, determine $2\mathbf{v} - 3\mathbf{w}$ as a linear combination of the unit vectors **i** and **j**. Determine $\|\mathbf{v}\|$ and $\mathbf{u}_\mathbf{v}$.

53. $\mathbf{v} = \mathbf{i} + \mathbf{j}, \mathbf{w} = \mathbf{i} - 2\mathbf{j}$

54. $\mathbf{v} = 2\mathbf{i} - \mathbf{j}, \mathbf{w} = \mathbf{i} - 2\mathbf{j}$

55. $\mathbf{v} = 3\mathbf{i} - 2\mathbf{j}, \mathbf{w} = \mathbf{i} + 4\mathbf{j}$

56. $\mathbf{v} = 2\mathbf{i} - 3\mathbf{j}, \mathbf{w} = -\mathbf{i} + 4\mathbf{j}$

57. $\mathbf{v} = \frac{1}{2}\mathbf{i} + \frac{5}{2}\mathbf{j}, \mathbf{w} = \frac{7}{2}\mathbf{i} - \frac{1}{2}\mathbf{j}$

58. $\mathbf{v} = \frac{2}{3}\mathbf{i} + 4\mathbf{j}, \mathbf{w} = \frac{4}{3}\mathbf{i} + \frac{1}{2}\mathbf{j}$

Superset

In Exercises 59–64, describe the radius vector as an ordered pair.

59.

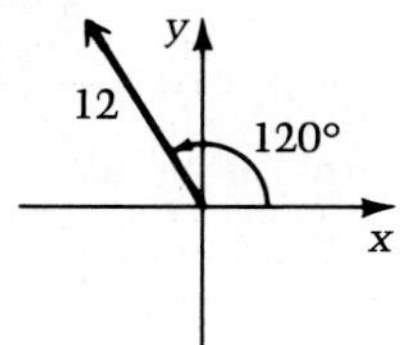

60.

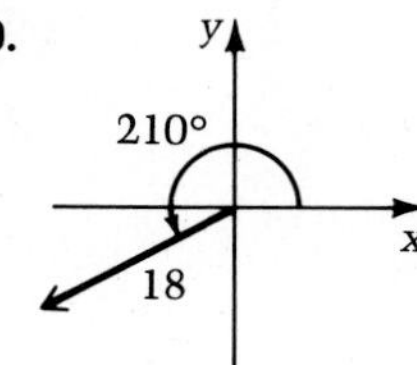

61.

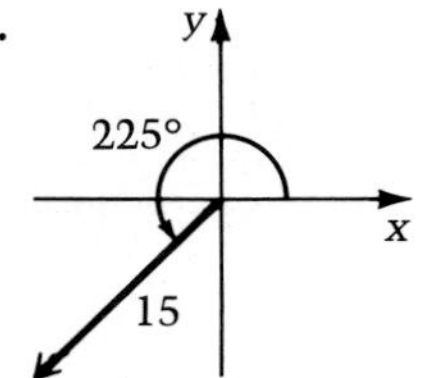

62.

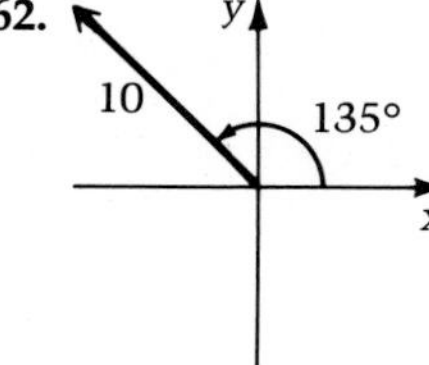

63.

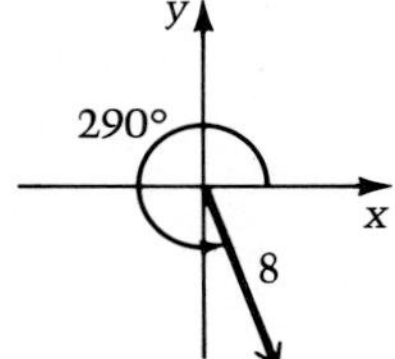

64.

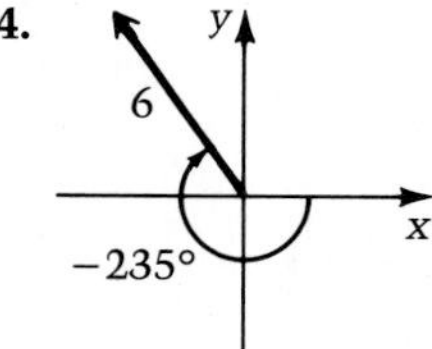

65. If $\mathbf{a} = 2\mathbf{i} + 3\mathbf{j}$, $\mathbf{b} = \mathbf{i} - 2\mathbf{j}$, and $\mathbf{c} = 4\mathbf{i} - \mathbf{j}$, find scalars r and s such that $\mathbf{c} = r\mathbf{a} - s\mathbf{b}$.

66. Find the angle between the radius vectors $\langle -2, 1\rangle$ and $\langle 3, 4\rangle$.

67. Determine a radius vector **v** which forms a right angle with the radius vector $\mathbf{u} = \langle 2, 3\rangle$ such that $\|\mathbf{v}\| = \|\mathbf{u}\|$. (*Hint:* There are two answers.)

68. Express the unit vector **i** as a linear combination of the vectors $\mathbf{u} = 2\mathbf{i} - 3\mathbf{j}$ and $\mathbf{v} = 5\mathbf{i} + 4\mathbf{j}$.

In Exercises 69–72, vector **a** has magnitude $\|\mathbf{a}\|$ and direction θ, vector **b** has magnitude $\|\mathbf{b}\|$ and direction φ. Express vector $\mathbf{c} = \mathbf{a} + \mathbf{b}$ as an ordered pair.

69. $\|\mathbf{a}\| = 9.89, \theta = 87°$; $\|\mathbf{b}\| = 2.79, \varphi = 73°$

70. $\|\mathbf{a}\| = 65.4, \theta = 42°$; $\|\mathbf{b}\| = 78.8, \varphi = 64°$

71. $\|\mathbf{a}\| = 979.8, \theta = 205.6°$; $\|\mathbf{b}\| = 346.4, \varphi = 81.4°$

72. $\|\mathbf{a}\| = 286.2, \theta = 101.2°$; $\|\mathbf{b}\| = 303.0, \varphi = 293.6°$

7.7
Vector Applications and the Dot Product

Often the quantities involved in applied problems are vectors, that is, quantities having both magnitude and direction. For example, the **velocity v** of an object is a vector quantity whose magnitude, $\|\mathbf{v}\|$, is the **speed** of the object and whose **direction** is the direction in which the object is moving.

In aviation, one frequently uses the concepts of *air speed* and *ground speed.* The **air speed** is the speed at which an airplane would fly in still air; the **ground speed** is the airplane's speed relative to the ground, after the effect of the wind has been accounted for. Thus, the (true) ground speed and true direction of an airplane are determined by forming the vector sum of the airplane's velocity vector and the wind's velocity vector.

EXAMPLE 1

An airplane's air speed is set at 300 km/h and its bearing is set at N 90° E (i.e., due east). A 50 km/h wind is blowing with a bearing S 60° E (i.e., in a direction 60° east of due south). Determine the ground speed and true bearing of the airplane.

Solution

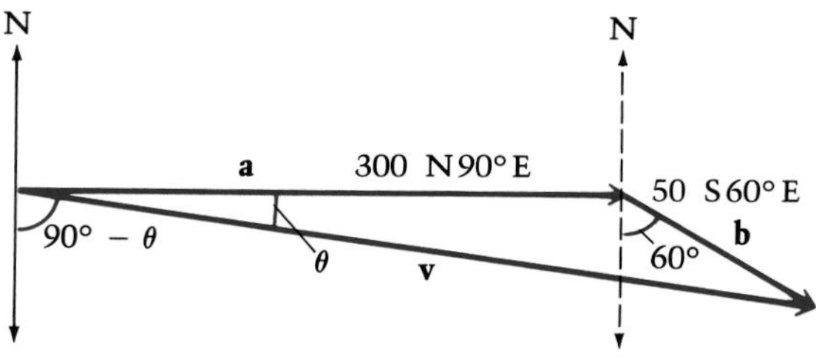

Figure 7.58

Draw a diagram. Let **a** be the airplane's velocity, and let **b** be the wind velocity. The ground velocity **v** is the vector sum of vectors **a** and **b**.

$$\|\mathbf{v}\|^2 = 300^2 + 50^2 - 2(300)(50)\cos 150^\circ$$
$$= 90{,}000 + 2500 - 30{,}000\left(-\frac{\sqrt{3}}{2}\right) \approx 118{,}481$$
$$\|\mathbf{v}\| = \sqrt{118{,}481} \approx 344 \text{ km/h}$$

The Law of Cosines is used to find $\|\mathbf{v}\|$, the ground speed. Note that the angle formed by **a** and **b** is $90^\circ + 60^\circ = 150^\circ$.

$$\frac{\sin\theta}{50} = \frac{\sin 150^\circ}{344}$$
$$\sin\theta = \frac{50}{344}(\sin 150^\circ) = \frac{50}{344}\left(\frac{1}{2}\right) \approx 0.0727$$
$$\theta \approx 4^\circ 10'$$

The Law of Sines is used to find the true direction which is $S(90^\circ - \theta)$ E.

The airplane's approximate ground speed is 344 km/h; its true bearing is S 85°50′ E.

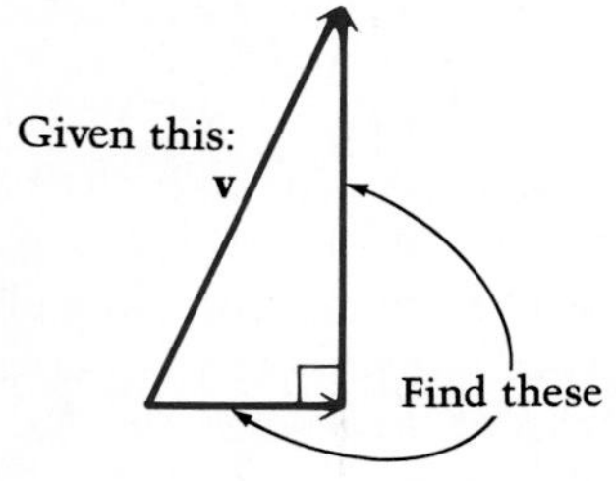

Figure 7.59

Our next application is a common problem in physics and engineering: given a vector **v**, find two other vectors whose vector sum is **v** (Figure 7.59). The two "other" vectors are usually perpendicular to each other, and are called **vector components of v.** The process of determining these two component vectors is referred to as **resolving v** into perpendicular components.

Of special importance is the resolution of a vector **v** into horizontal and vertical components. (We will refer to these vector components as $\mathbf{v}_x$ and $\mathbf{v}_y$, respectively, as shown in Figure 7.60.) To do this, we position **v** with its tail at the origin of the xy-plane, and draw perpendiculars from the head of **v** to the x-axis and to the y-axis.

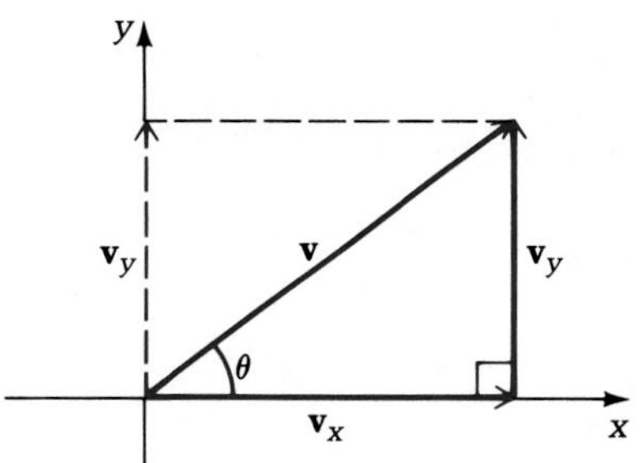

Figure 7.60

We move $\mathbf{v}_y$ to the position on the right in order to visualize it as one leg in a right triangle, with $\mathbf{v}_x$ as the other leg.

From Figure 7.60 above, it is clear that

$$\cos\theta = \frac{\text{adjacent}}{\text{hypotenuse}} = \frac{\|\mathbf{v}_x\|}{\|\mathbf{v}\|} \quad \text{and} \quad \sin\theta = \frac{\text{opposite}}{\text{hypotenuse}} = \frac{\|\mathbf{v}_y\|}{\|\mathbf{v}\|}.$$

Thus,

$$\|\mathbf{v}_x\| = \|\mathbf{v}\|\cos\theta \quad \text{and} \quad \|\mathbf{v}_y\| = \|\mathbf{v}\|\sin\theta.$$

Since $\mathbf{v}_x$ is a vector with magnitude $\|\mathbf{v}\|\cos\theta$ in the direction of **i**, and $\mathbf{v}_y$ is a vector with magnitude $\|\mathbf{v}\|\sin\theta$ in the direction of **j**, we write

$$\mathbf{v}_x = (\|\mathbf{v}\|\cos\theta)\mathbf{i} \quad \text{and} \quad \mathbf{v}_y = (\|\mathbf{v}\|\sin\theta)\mathbf{j}$$

where θ is the angle formed by **v** and the positive x-axis. Note that

$$\mathbf{v} = \mathbf{v}_x + \mathbf{v}_y = (\|\mathbf{v}\|\cos\theta)\mathbf{i} + (\|\mathbf{v}\|\sin\theta)\mathbf{j},$$

and so $\mathbf{v}_x$ and $\mathbf{v}_y$ are truly vector components of **v**.

EXAMPLE 2

Vector **v** has magnitude 8. Resolve **v** into horizontal and vertical components

(a) if it makes an angle of 60° with the positive x-axis;

(b) if it makes an angle of 135° with the positive x-axis.

Solution

(a)

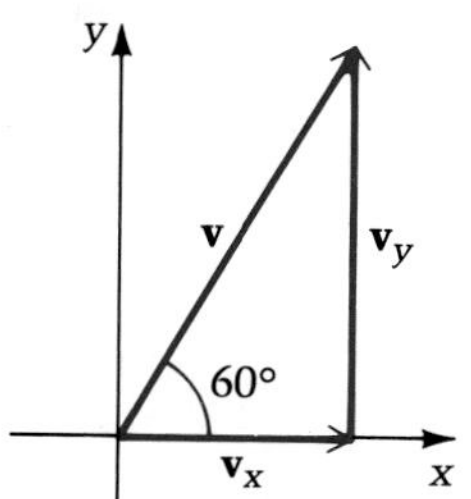

Figure 7.61

$$\mathbf{v}_x = (\|\mathbf{v}\| \cos \theta)\mathbf{i} = (8 \cdot \cos 60°)\mathbf{i} = (8)\left(\frac{1}{2}\right)\mathbf{i} = 4\mathbf{i}$$

$$\mathbf{v}_y = (\|\mathbf{v}\| \sin \theta)\mathbf{j} = (8 \cdot \sin 60°)\mathbf{j} = (8)\left(\frac{\sqrt{3}}{2}\right)\mathbf{j} = 4\sqrt{3}\mathbf{j}$$

$$\mathbf{v} = \mathbf{v}_x + \mathbf{v}_y = 4\mathbf{i} + 4\sqrt{3}\mathbf{j}.$$

(b)

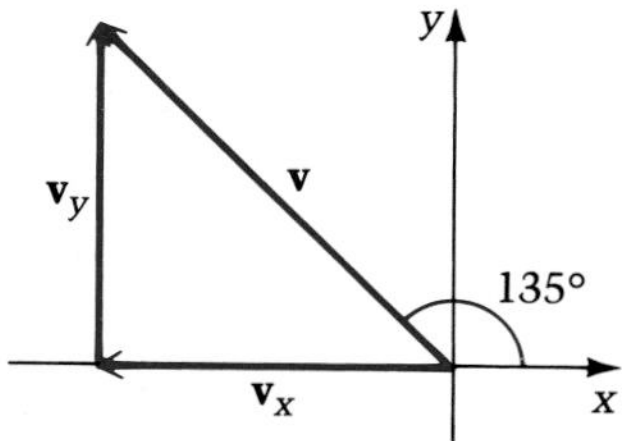

Figure 7.62

$$\mathbf{v}_x = (\|\mathbf{v}\| \cos \theta)\mathbf{i} = (8 \cdot \cos 135°)\mathbf{i} = \left[8 \cdot \left(-\frac{\sqrt{2}}{2}\right)\right]\mathbf{i} = -4\sqrt{2}\mathbf{i}$$

$$\mathbf{v}_y = (\|\mathbf{v}\| \sin \theta)\mathbf{j} = (8 \cdot \sin 135°)\mathbf{j} = \left(8 \cdot \frac{\sqrt{2}}{2}\right)\mathbf{j} = 4\sqrt{2}\mathbf{j}$$

$$\mathbf{v} = \mathbf{v}_x + \mathbf{v}_y = -4\sqrt{2}\mathbf{i} + 4\sqrt{2}\mathbf{j}.$$

To study an object under the influence of a force, it is useful to represent the force as a vector. We say that a force of one newton (N) is required to accelerate a mass of 1 kg at a rate of 1 m/s^2. One of the most common forces is **g**, the force that gravity exerts on an object (also called weight). If an object has a mass of M kg, then the magnitude of **g** is $9.8 \cdot M$ newtons.

When an object is not accelerating (either it is moving at a constant velocity or it is at rest), we say that **forces are in equilibrium.** This means that the vector sum of all forces is zero.

EXAMPLE 3

A 400 kg piano is being rolled down a ramp. The ramp makes an angle of 30° with the level ground below. If we neglect friction, what is the force required to hold the piano stationary on the ramp?

Solution

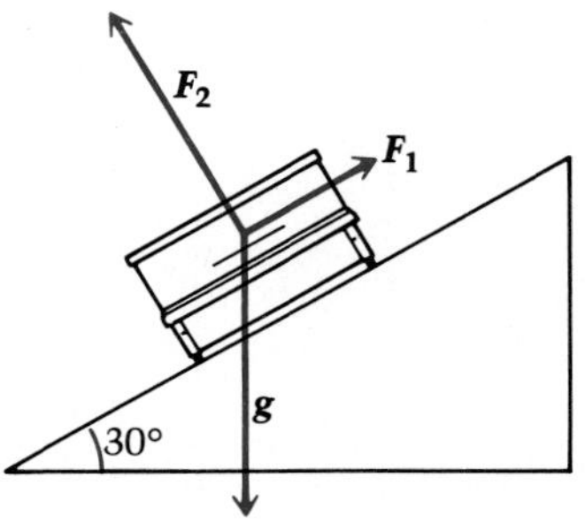

Begin by drawing a figure. Three forces act on the object:

g, the force of gravity;

$\mathbf{F}_1$, the force needed to hold the piano stationary ($\mathbf{F}_1$ is parallel to the ramp);

$\mathbf{F}_2$, a force perpendicular to the surface of the ramp that keeps the piano from crashing through the ramp.

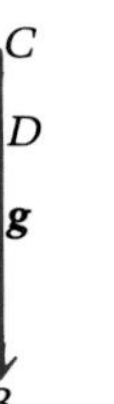

Figure 7.63

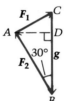

We draw a triangle representing the forces. Since there is no motion, the total vector sum is **0**. That is, $\mathbf{F}_1 + \mathbf{g} + \mathbf{F}_2 = \mathbf{0}$. Notice that $\mathbf{F}_1 + \mathbf{g} = -\mathbf{F}_2$.

The question asks us to determine $\mathbf{F}_1$. To determine $\angle B$ in the triangle above, we have drawn a perpendicular from A that meets side BC at point D. Given $\angle CAD = 30°$, and since $\angle CAB = 90°$, we conclude that $\angle DAB = 60°$. Thus, in right triangle ADB we have $\angle B = 30°$.

$$\sin 30° = \frac{\text{opposite}}{\text{hypotenuse}} = \frac{\|\mathbf{F}_1\|}{\|\mathbf{g}\|} = \frac{\|\mathbf{F}_1\|}{(9.8)(400)}$$

To determine $\mathbf{F}_1$, consider $\sin B$ in right triangle ABC.

$$\|\mathbf{F}_1\| = (\sin 30°)(9.8)(400) = \left(\frac{1}{2}\right)(3920) = 1960 \text{ N}$$

The force required to hold the piano on the ramp is 1960 N (roughly 441 lb).

EXAMPLE 4

An object is fired at an angle of 60° with the horizontal, and with a speed of 88 feet per second (that is, 60 mph). Assuming that the ground is level, and that air resistance can be neglected, how far down range does the object land?

Solution

- ANALYZE The object is fired at an angle of 60° with the ground, first rising, then falling, and ultimately returning to the ground. We must determine the horizontal distance it travels during this time.
- ORGANIZE Think of the initial velocity as a vector **v** that can be decomposed into a vertical component and a horizontal component. The *magnitudes* of these component vectors are shown in the figure below.

  Draw a sketch

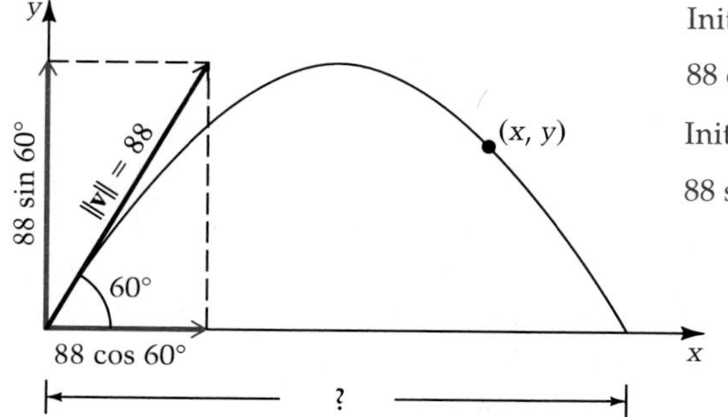

Initial horizontal speed:

$88 \cos 60° = 88 \cdot \frac{1}{2} = 44$ ft/sec

Initial vertical speed:

$88 \sin 60° = 88 \cdot \frac{\sqrt{3}}{2} = 44\sqrt{3}$ ft/sec

Let t = the time (in seconds) elapsed since the object was fired.
Let x = the x-coordinate of the object's position at time t.
Let y = the y-coordinate of the object's position at time t.

Note that both x and y are functions of t.

- MODEL We need formulas to describe both x and y at time t. The x-coordinate (horizontal position) depends entirely on the initial horizontal speed. The y-coordinate (vertical position, also called *height*) depends on the initial vertical speed, and also on the constant downward acceleration due to gravity.

  Horizontal Position $x = 44t$ (distance = speed × time)
  Vertical Position $y = 44\sqrt{3}\,t - 16t^2$ (See Exercise 97, p. 90)

We first determine the time t when the object's height is 0.

$$0 = 44\sqrt{3}\,t - 16t^2$$
$$0 = t\left(44\sqrt{3} - 16t\right)$$

$t = 0$	$44\sqrt{3} - 16t = 0$
	$t = \dfrac{44\sqrt{3}}{16} = \dfrac{11\sqrt{3}}{4}$

At time $t = 0$, the object is just leaving the ground. At time $t = \dfrac{11\sqrt{3}}{4}$, the object is just returning to the ground. To find the horizontal distance traveled we substitute this latter time, $t = \dfrac{11\sqrt{3}}{4}$, into the equation $x = 44t$.

$$x = 44\left(\frac{11\sqrt{3}}{4}\right) = 121\sqrt{3}\text{ ft}$$

■ ANSWER The object lands $121\sqrt{3}$ ft ($\approx$210 ft) down range from the point at which it was launched.

The Dot Product of Two Vectors

Up to this point, the only type of vector multiplication that we have discussed is the product of a scalar k and a vector **v** (the result k**v** is a vector). We now define a product of two vectors in such a way that the product is a scalar, not a vector.

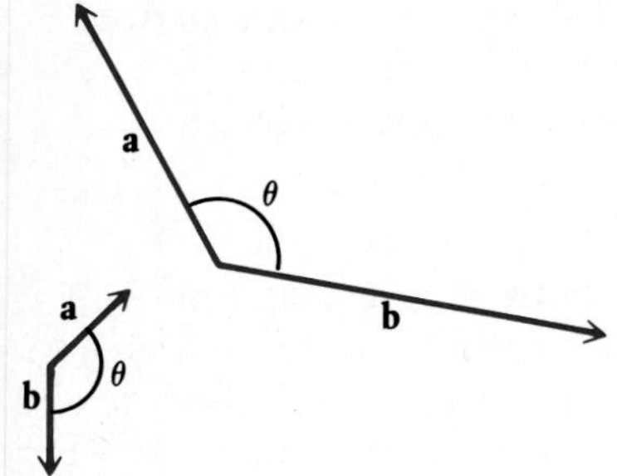

Figure 7.64 The angle between **a** and **b** is θ.

Definition of Dot Product

Let **a** and **b** be two nonzero vectors. The **dot product** of **a** and **b**, denoted **a** · **b**, is defined as

$$\mathbf{a} \cdot \mathbf{b} = \|\mathbf{a}\|\,\|\mathbf{b}\| \cos\theta$$

where θ is the angle between **a** and **b** such that $0° \le \theta \le 180°$.

There is an alternate definition of the dot product which involves the components of the two vectors **a** and **b**.

Alternate Definition of Dot Product

Let $\mathbf{a} = \langle a_1, a_2 \rangle$ and $\mathbf{b} = \langle b_1, b_2 \rangle$. Then

$$\mathbf{a} \cdot \mathbf{b} = a_1 b_1 + a_2 b_2.$$

Careful! The dot product of two vectors is a scalar, not a vector.

Taken together, the two definitions of dot product provide a way to determine the angle between two vectors. The technique is illustrated in Example 4. Note that when the angle between the two vectors is 90°, $\cos\theta = 0$, and so the dot product is 0. In addition, if the dot product of two nonzero vectors is 0, then $\theta = 90°$. Thus, if **a** and **b** are nonzero vectors,

$$\mathbf{a} \cdot \mathbf{b} = 0 \text{ if and only if } \mathbf{a} \text{ and } \mathbf{b} \text{ are perpendicular.}$$

EXAMPLE 5

Find $\mathbf{v} \cdot \mathbf{w}$ if $\mathbf{v} = \langle 3, 2 \rangle$ and $\mathbf{w} = \langle 4, -5 \rangle$. Determine the angle between **v** and **w**.

Solution Although a figure is not necessary, it does help in visualizing the problem.

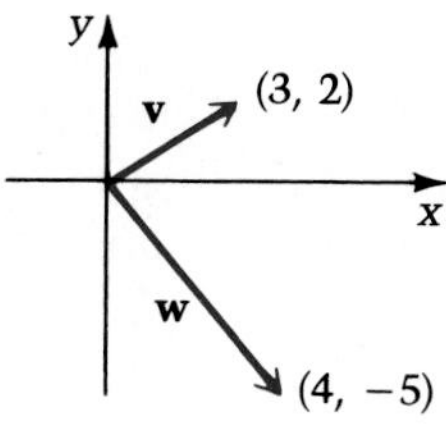

Figure 7.65

$$\mathbf{v} \cdot \mathbf{w} = \langle 3, 2 \rangle \cdot \langle 4, -5 \rangle = (3)(4) + (2)(-5) = 12 - 10 = 2$$

$$\mathbf{v} \cdot \mathbf{w} = \|\mathbf{v}\|\,\|\mathbf{w}\| \cos\theta$$

Use $\mathbf{v} \cdot \mathbf{w}$ to approximate θ, the angle between **v** and **w**.

$$2 = \|\langle 3, 2 \rangle\|\,\|\langle 4, -5 \rangle\| \cos\theta$$

$$2 = \left[\sqrt{3^2 + 2^2} \cdot \sqrt{4^2 + (-5)^2}\right] \cos\theta$$

$$\cos\theta = \frac{2}{\sqrt{13} \cdot \sqrt{41}} = \frac{2}{\sqrt{533}} \approx 0.0866$$

$$\theta \approx 85°$$

EXERCISE SET 7.7

In Exercises 1–4, use the fact that eastbound airplanes "hitch a ride" on the jet stream.

1. A jet's air speed is 450 mph and its bearing is N 80° E. If the jet stream is blowing due east at 120 mph, what is the jet's ground speed and true bearing?
2. A jet's air speed is 450 mph and its bearing is S 80° W. If the jet stream is blowing due east at 120 mph, what is the jet's ground speed and true bearing?
3. A jet's ground speed is 410 mph and its true bearing is due north. If the jet stream is blowing due east at 120 mph, what is the jet's air speed and bearing?
4. A jet's ground speed is 510 mph and its true bearing is N 40° E. If the jet stream is blowing due east at 120 mph, what is the jet's air speed and bearing?
5. A 62 kg weight is on a ramp which is inclined at 40°. What is the force needed to hold the weight stationary on the ramp?
6. A crate containing a 300 kg wood stove is being unloaded from a delivery truck by sliding it down an inclined plank which is at an angle of 32° with the horizontal. What is the force needed to hold the crate at rest?
7. A 500 lb boat is being lowered into the water down an inclined ramp which is at an angle of 25° with the horizontal. What is the force needed to hold the boat at rest?
8. A force of 42 lb is required to hold a 250 lb block of granite from sliding down an inclined ramp. At what angle is the ramp inclined?

In Exercises 9–16, determine the dot product of the given vectors. Find the angle between the vectors to the nearest degree.

9. $\langle 3, 1\rangle, \langle -2, 4\rangle$
10. $\langle 1, 1\rangle, \langle 3, -1\rangle$
11. $\langle 2, -1\rangle, \langle 2, -2\rangle$
12. $\langle -1, 2\rangle, \langle 4, 6\rangle$
13. $\langle 1, 9\rangle, \langle 9, -1\rangle$
14. $\langle 4, 3\rangle, \langle 3, -4\rangle$
15. $\langle -5, -3\rangle, \langle 4, -3\rangle$
16. $\langle -10, 2\rangle, \langle -3, -6\rangle$

Superset

In Exercises 17–22, two vectors are given. Determine whether they are parallel, perpendicular, or neither.

17. $\langle 2, -4\rangle, \langle 2, 1\rangle$
18. $\langle 1, -3\rangle, \langle 3, -1\rangle$
19. $\langle 1, 9\rangle, \left\langle \frac{3}{2}, \frac{1}{6}\right\rangle$
20. $\langle 8, 6\rangle, \left\langle \frac{1}{2}, -\frac{2}{3}\right\rangle$
21. $\langle -9, 6\rangle, \left\langle 2, -\frac{4}{3}\right\rangle$
22. $\left\langle -\frac{1}{3}, \frac{2}{3}\right\rangle, \langle 4, -8\rangle$

In physics, **work** is said to be done when a force applied to an object causes the object to move. In particular, if a constant force $\mathbf{F}$ is applied in the direction of motion, then the work W done by force $\mathbf{F}$ in moving the object a distance d is the scalar quantity $W = \|\mathbf{F}\| \cdot d$. However, if a constant force $\mathbf{F}$ is applied at an angle θ to the direction of the motion, then the work done by $\mathbf{F}$ in moving the object a distance d is

$$\begin{aligned} W &= \begin{pmatrix}\text{Component of force in}\\ \text{direction of motion}\end{pmatrix} \cdot (\text{Distance}) \\ &= (\|\mathbf{F}\| \cos\theta) \cdot d \end{aligned}$$

In Exercises 23–26, solve the work problem.

23. A wagon loaded with 180 lb of patio bricks is pulled 100 yd over level ground by a handle which makes an angle of 43° with the horizontal. Find the work done if a force of 22 lb is exerted in pulling the wagon.
24. A large crate is pushed across a level floor. Find the work done in moving the crate 20 ft if a force of 35 lb is applied at an angle of 18° with the horizontal.
25. A wagon is used to haul three small children a distance of a half mile over level ground. The handle used to pull the wagon makes an angle of 35° with the horizontal. Find the work done if a force of 25 lb is exerted on the handle.
26. A box is pulled by exerting a force of 16 lb on a rope attached to the box that makes an angle of 40° with the horizontal. Determine the work done in pulling the box 46 ft.

In Exercises 27–30, we consider a few of the theorems from geometry that can be proven using vector methods. In some instances you may need to use the fact that $\mathbf{a} \cdot \mathbf{a} = \|\mathbf{a}\|^2$, which follows directly from the definition of dot product. (Take $\mathbf{a} = \mathbf{b}$ so that $\theta = 0$.)

27. Prove that the diagonals of a parallelogram are perpendicular if and only if the parallelogram is a rhombus. (*Hint:* In the figure below, the diagonals are $\mathbf{a} + \mathbf{b}$ and $\mathbf{a} - \mathbf{b}$.)

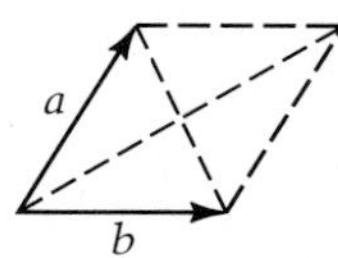

28. Prove that every angle inscribed in a semicircle is a right angle. (*Hint:* In the figure below, express **c** and **d** in terms of **a** and **b**.)

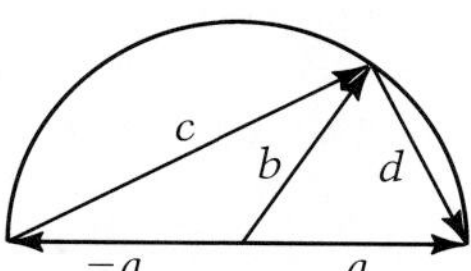

29. Prove that the line segment that joins the midpoints of two sides of a triangle is parallel to and half as long as the third side of the triangle. (*Hint:* In the figure below, you need to show that if $\mathbf{e} = \frac{1}{2}\mathbf{a}$ and $\mathbf{f} = \frac{1}{2}\mathbf{b}$, then $\mathbf{d} = \frac{1}{2}\mathbf{c}$.)

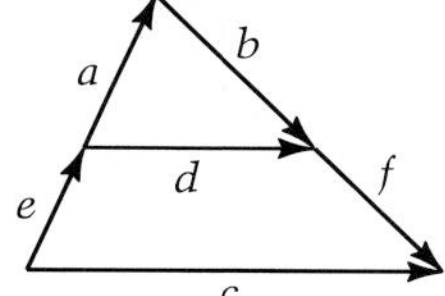

30. Prove that in a parallelogram the sum of the squares of the lengths of the diagonals is equal to the sum of the squares of the lengths of the sides.

In Exercises 31–34, the magnitude $\|\mathbf{a}\|$ and direction θ of the vector **a** are given. Find $\mathbf{a}_x$ and $\mathbf{a}_y$.

31. $\|\mathbf{a}\| = 9.89$, $\theta = 87°$ 32. $\|\mathbf{a}\| = 65.4$, $\theta = 42°$

33. $\|\mathbf{a}\| = 979.8$, $\theta = 205.6°$

34. $\|\mathbf{a}\| = 286.2$, $\theta = 101.2$

35. Along a stretch of the Mississippi River, the water flows directly south with a current of 5 mph. A person who can row a boat at 4 mph in still water attempts to row due west across the river. What is the actual direction and speed of the boat?

36. Two forces of 26 newtons and 40 newtons act on a point in the plane. If the angle between the forces is 73.6°, find the magnitude of the resultant force and its angle with the smaller of the two given forces.

37. An object weighing 100 lb is suspended between two trees by two ropes attached at the top of the object. The rope off to the left makes an angle of 58° with the vertical and the rope off to the right an angle of 27° with the vertical. Find the tension (magnitude of the force vector) along each rope.

Analyzing Data

The Case of the Time Series: Money & Measles

We need not look far to find applications involving trigonometric functions: seasonal business cycles (e.g., highs and lows in sales figures throughout the year), biological rhythms (e.g., electroencephalogram, electrocardiogram, breathing patterns), and meteorological patterns all bear striking testimony to the importance of periodic functions. We present two examples of such behavior.

First, in Figure 7.66 we see a graph taken from Chatfield's *The Analysis of Time Series: An Introduction*. The graph is based on data from a study by Chatfield and Prothero [1973], and shows sales figures for the years 1965 through 1971 for an engineering firm. Observations that are made periodically, over time, such as those used to generate this graph, are referred to as time-series data.

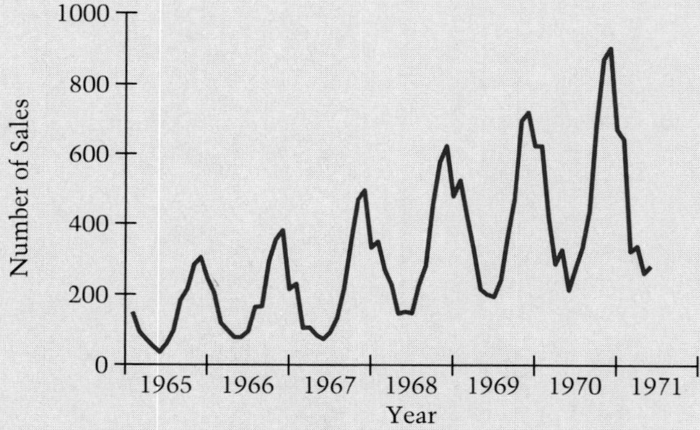

Figure 7.66 Sales of a certain engineering company in successive months.

As Chatfield points out, time-series data are useful in business both "to forecast future sales," and "to plan production." Notice in the figure that the sales cycle repeats itself with a minor variation each year. That variation is most likely quite acceptable to the company officials since the low point is rising steadily each year, and the high point seems to be rising even more dramatically. Clearly sales are increasing over time.

As a second example, consider the graphs shown in Figure 7.67 taken from Batschelet's *Introduction to Mathematics for Life Sciences*. The figure originally appeared in a research report by Bliss and Blevins [1959]. The graph in the upper box shows the seasonal pattern of outbreak of measles with a high point occurring around May. The solid dots represent the actual data points, and a trigonometric function has been fitted to the data.

The fitted function consists of a constant term and components of the form $a \sin bt$ and $c \cos dt$. Such a function is referred to as a *trigonometric polynomial*. The graph in the lower box shows sketches of the components. As Batschelet reminds us, *"Any continuous or reasonable discontinuous periodic function can be represented by a trigonometric polynomial with any degree of accuracy."*

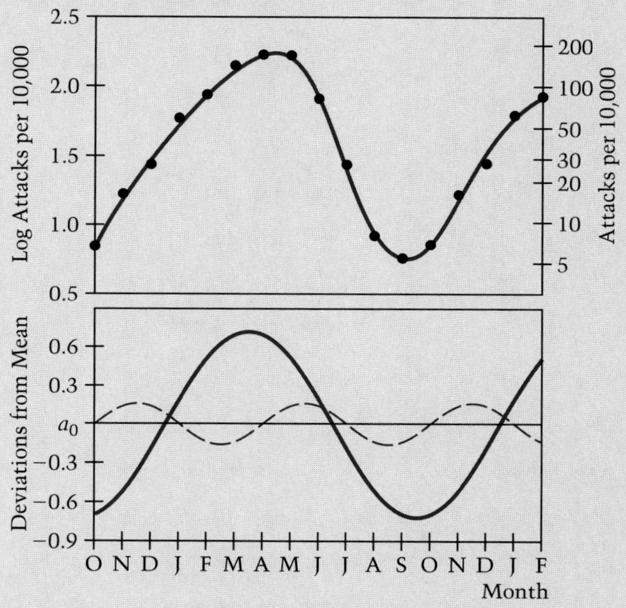

Figure 7.67

Sources: Chatfield, C. *The Analysis of Time Series: An Introduction*, 4th ed. Chapman & Hall, London, p. 3, 1989. Reprinted by permission.

Chatfield, C. and Prothero, D. L. Box-Jenkins seasonal forecasting: problems in a case-study, *J. R. Statistical Society*, A, **136,** 295–336 (1973).

Batschelet, E. *Introduction to Mathematics for Life Sciences*, Springer-Verlag, Heidelberg, p. 138, 1979. Reprinted by permission.

Bliss, C. I. and Blevins, D. L. The analysis of seasonal variation in measles. *Am. J. Hyg.*, **70,** 328–334 (1959).

Chapter Review

7.1 Radian Measure and the Unit Circle (pp. 393–400)

The unit circle is the circle of radius 1 with center at the origin. A point $T(u, v)$ is on the unit circle if and only if $u^2 + v^2 = 1$. (p. 394) A *standard arc s* is an arc on the unit circle which starts at (1, 0), has length $|s|$, and travels counterclockwise if $s > 0$ and clockwise if $s < 0$. (p. 394)

A *central angle* θ is an angle whose vertex is at the center of a circle. (p. 396) The *radian measure* of a positive central angle θ is defined as a ratio: $\theta = s/r$, where s is the arclength of the arc cut by θ, and r is the radius of the circle. (p. 396)

7.2 Trigonometric Functions of Real Numbers (pp. 400–408)

Every real number x can be associated uniquely with a central angle θ that cuts standard arc x on the unit circle. (p. 400) To define the trigonometric functions of any real number x, let $T(u, v)$ be the terminal point on the standard arc x. (p. 401)

$$\sin x = v \qquad \cos x = u \qquad \tan x = \frac{v}{u}$$

$$\csc x = \frac{1}{v} \qquad \sec x = \frac{1}{u} \qquad \cot x = \frac{u}{v}$$

To find the trigonometric function value of any real number x, we determine the sign by means of the ASTC chart, and then find the trigonometric function value of the reference value for x:

x between	0 and $\frac{\pi}{2}$	$\frac{\pi}{2}$ and π	π and $\frac{3\pi}{2}$	$\frac{3\pi}{2}$ and 2π
reference value	x	$\pi - x$	$x - \pi$	$2\pi - x$

If x is outside the interval $[0, 2\pi)$, begin by adding an appropriate integer multiple of 2π to produce a value between 0 and 2π, then use the procedure described above. (p. 405)

7.3 The Trigonometric Functions: Basic Graphs and Properties (pp. 408–421)

A function f is called a *periodic function* if there is a positive real number p such that $f(x) = f(x + p)$ for all x in the domain of f. The smallest such positive number p is called the *period* of the function. The graph of the function over an interval of length p is called a *cycle* of the graph. (p. 409)

7.4 Transformations of the Trigonometric Functions (pp. 421–428)

For functions described by equations of the form $y = k + a \sin b(x - h)$ or $y = k + a \cos b(x - h)$, the number $|a|$ is the *amplitude*, the number $\left|\frac{2\pi}{b}\right|$ is the *period*, and the number h is the *phase shift*. (pp. 422–426)

7.5 The Inverse Trigonometric Functions (pp. 429–437)

If each of the trigonometric functions is restricted to a suitable portion of its domain, the resulting function is one-to-one, and thus has an inverse function. The inverses of the sine, cosine, and tangent functions are

$y = \sin^{-1} x$ with domain $-1 \le x \le 1$ if and only if $\sin y = x$ and $-\frac{\pi}{2} \le y \le \frac{\pi}{2}$,

$y = \cos^{-1} x$ with domain $-1 \le x \le 1$ if and only if $\cos y = x$ and $0 \le y \le \pi$,

$y = \tan^{-1} x$ with domain $-\infty < x < \infty$ if and only if $\tan y = x$ and $-\frac{\pi}{2} < x < \frac{\pi}{2}$.

7.6 Vectors in the Plane (pp. 437–444)

A vector with initial point $P_1(x_1, y_1)$ and terminal point $P_2(x_2, y_2)$ is given by the ordered pair $\langle x_2 - x_1, y_2 - y_1\rangle$. The numbers $x_2 - x_1$ and $y_2 - y_1$ are called the *scalar components* of the vector. (p. 440)

The length of a vector **v** is called the *norm* of **v** and is denoted $\|\mathbf{v}\|$. If $\mathbf{a} = \langle a_1, a_2\rangle$, then $\|\mathbf{a}\| = \sqrt{a_1^2 + a_2^2}$. (p. 442)

7.7 Vector Applications and the Dot Product (pp. 445–453)

For a given vector **a**, the process of determining two component vectors **b** and **c** such that **b** is perpendicular to **c** and $\mathbf{b} + \mathbf{c} = \mathbf{a}$ is referred to as *resolving* **a** into perpendicular components. For a given vector **v**, the horizontal component vector is denoted $\mathbf{v}_x$ and the vertical component vector is denoted $\mathbf{v}_y$. (p. 446)

The *dot product* of two nonzero vectors $\mathbf{a} = \langle a_1, a_2\rangle$ and $\mathbf{b} = \langle b_1, b_2\rangle$ is denoted $\mathbf{a} \cdot \mathbf{b}$. The dot product can be defined in two equivalent ways: (1) $\mathbf{a} \cdot \mathbf{b} = \|\mathbf{a}\|\,\|\mathbf{b}\| \cos \theta$, where θ is the angle between **a** and **b** such that $0° \le \theta \le 180°$; (2) $\mathbf{a} \cdot \mathbf{b} = a_1b_1 + a_2b_2$. (p. 450)

Review Exercises

In Exercises 1–8, convert the measure of the angle from degrees to radians or vice versa.

1. $120°$
2. $-135°$
3. $225°$
4. $765°$
5. $\frac{3\pi}{4}$
6. $\frac{11\pi}{2}$
7. $-\frac{2\pi}{3}$
8. $-\frac{5\pi}{6}$

9. Determine the length of an arc cut by a central angle of 80° in a circle of radius 5.4 cm.

10. A central angle of 0.4 radians cuts an arc of length 14 inches. Find the radius of the circle.

In Exercises 11–22, determine the function value.

11. $\sin\frac{\pi}{4}$
12. $\cos\frac{5\pi}{4}$
13. $\cos\left(-\frac{\pi}{3}\right)$
14. $\tan\left(-\frac{2\pi}{3}\right)$
15. $\tan\frac{\pi}{6}$
16. $\sin\left(-\frac{5\pi}{6}\right)$
17. $\sec\pi$
18. $\cot\frac{\pi}{2}$
19. $\csc\left(-\frac{3\pi}{4}\right)$
20. $\sec\left(-\frac{7\pi}{6}\right)$
21. $\cot\frac{7\pi}{6}$
22. $\csc\frac{11\pi}{6}$

In Exercises 23–40, sketch the graph on the interval $[0, 2\pi]$.

23. $y = 1 - \sin x$
24. $y = 2 + \cos x$
25. $y = -1 + 2\cos x$
26. $y = 1 - 2\sin x$
27. $y = 3\sin\left(x - \frac{\pi}{4}\right)$
28. $y = -2\cos(x + \pi)$
29. $y = -2 + 3\cos(-x)$
30. $y = 1 + 2\sin(-x)$
31. $y = -1 + \sin(2x - \pi)$
32. $y = 2 - \cos(3x - \pi)$
33. $y = \tan 2x$
34. $y = \sec\frac{x}{2}$
35. $y = -\csc\frac{x}{3}$
36. $y = -\cot 3x$
37. $y = \cot(2x + \pi)$
38. $y = \tan\left(x + \frac{\pi}{2}\right)$
39. $y = 2\sec\left(x + \frac{\pi}{2}\right)$
40. $y = -3\csc\left(\frac{\pi}{2} - x\right)$

In Exercises 41–56, evaluate the expression.

41. $\arcsin\left(-\frac{1}{2}\right)$
42. $\arctan\frac{\sqrt{3}}{3}$
43. $\arccos 1$
44. $\sin^{-1}\left(-\frac{\sqrt{2}}{2}\right)$
45. $\tan^{-1} 0$
46. $\cos^{-1}\left(-\frac{1}{2}\right)$
47. $\sin^{-1}\left(\sin\frac{5\pi}{6}\right)$
48. $\cos^{-1}\left(\cos\left(-\frac{\pi}{4}\right)\right)$
49. $\cos^{-1}\left(\sin\frac{\pi}{3}\right)$
50. $\sin^{-1}\left(\cos\frac{\pi}{6}\right)$
51. $\cos\left(\sin^{-1}\frac{2}{3}\right)$
52. $\sec\left(\tan^{-1}\left(-\frac{2}{5}\right)\right)$
53. $\csc\left(\cos^{-1}\frac{1}{4}\right)$
54. $\cot\left(\arcsin\left(-\frac{4}{5}\right)\right)$
55. $\sin(\arctan 5)$
56. $\tan\left(\arccos\frac{1}{3}\right)$

In Exercises 57–58, rewrite the expression as an expression involving no trigonometric functions. Assume $z \geq 0$.

57. $\cot^2(\arcsin z)$
58. $\sec(\arctan z)$

In Exercises 59–77, use the vectors $\mathbf{u} = \langle 2, -3\rangle$, $\mathbf{v} = \langle -4, 1\rangle$, and $\mathbf{w} = \langle 6, 0\rangle$, to determine the following.

59. $2\mathbf{u} - \mathbf{v} + \mathbf{w}$
60. $3(\mathbf{v} - \mathbf{w}) - \mathbf{u}$
61. The terminal point of $\mathbf{u}$ given that its initial point is $(4, 1)$
62. The initial point of $\mathbf{v}$ given that its terminal point is $(1, 3)$
63. A vector parallel to and twice as long as $\mathbf{v}$
64. A vector one-third as long as $\mathbf{w}$ and in the opposite direction of $\mathbf{w}$
65. $\mathbf{u}_{\mathbf{v}}$
66. $\mathbf{u}_{\mathbf{w}}$
67. $\|2\mathbf{i} - \mathbf{v}\|$
68. $\|-\mathbf{j}\|$
69. A vector $\mathbf{x}$ so that $(\mathbf{u} + \mathbf{x}) \cdot (\mathbf{u} + \mathbf{x}) = 0$
70. Scalars c and k so that $c\mathbf{u} + k\mathbf{v} = \mathbf{w}$
71. The horizontal component of $\mathbf{u}$
72. The vertical component of $\mathbf{v}$
73. $\mathbf{u} \cdot \mathbf{v}$
74. $\mathbf{v} \cdot (\mathbf{w} + \mathbf{j})$
75. A real number c so that $\mathbf{v}$ and $\langle c, 2\rangle$ are perpendicular
76. A vector $\mathbf{x}$ of length 1 so that $\mathbf{u} \cdot \mathbf{x} = 0$
77. The angle (to the nearest degree) between $\mathbf{u}$ and $\mathbf{v}$

78. A jet's air speed is 420 mph and its bearing is N70°E. If the jet stream is blowing due east at 110 mph, what is the jet's ground speed and true bearing?

79. A 70 kg weight is on a ramp which is inclined at 20°. What is the force needed to hold the weight stationary on the ramp?

In Exercises 80–81, determine whether the statement is true or false.

80. $2 \sin \frac{\pi}{3} \cos \frac{\pi}{3} = \sin \frac{2\pi}{3}$

81. $\sin \frac{\pi}{6} = \sqrt{\frac{1 + \cos \frac{\pi}{3}}{2}}$

In Exercises 82–83, find a value of b such that the period p of the function is the given value.

82. $y = \sin bx, p = \frac{\pi}{3}$

83. $y = \cos bx, p = \frac{\pi}{2}$

84. An equilateral triangle is inscribed in a circle with a 6-inch radius. Find the length of the arc cut by one side of the triangle.

Graphing Calculator Exercises

[*Note: Round all estimated values to the nearest hundredth.*]

In Exercises 1–5, let the function $f(x) = \frac{\sin x}{x}$.

1. Use the graph of this function to help determine its domain.
2. What happens to the function values as x-values get close to 0?
3. Is there a minimum function value? If so, what is it?
4. Is there a minimum function value? If so, what is it?
5. Determine the x-intercepts.

In Exercises 6–11, let the function $f(x) = \frac{1 - \cos x}{x}$.

6. Use the graph of this function to help determine its domain.
7. What happens to the function values as x-values get close to 0?
8. What is the maximum function value, and what value of x produces it?
9. What is the minimum function value, and what value of x produces it?
10. Determine the x-intercepts.
11. Explain why the low points and high points of this graph get closer to the x-axis as $|x|$ gets larger.

12. Graph each of the following pairs of functions on the same set of axes, for values of x in the interval $[-\pi, \pi]$.

$y = \sin x$ and $y = x - \frac{x^3}{6}$

$y = \sin x$ and $y = x - \frac{x^3}{6} + \frac{x^5}{120}$

$y = \sin x$ and $y = x - \frac{x^3}{6} + \frac{x^5}{120} - \frac{x^7}{5040}$

What do you observe?

13. Graph each of the following pairs of functions on the same set of axes, for values of x in the interval of $[-\pi, \pi]$.

$y = \cos x$ and $y = 1 - \frac{x^2}{2}$

$y = \cos x$ and $y = 1 - \frac{x^2}{2} + \frac{x^4}{24}$

$y = \cos x$ and $y = 1 - \frac{x^2}{2} + \frac{x^4}{24} - \frac{x^6}{720}$

What do you observe?

Chapter Test

1. Determine whether the point $\left(\frac{\sqrt{2}}{6}, \frac{4}{6}\right)$ is on the unit circle.

In Problems 2–4, convert the measure of the given angle from degrees to radians.

2. $-135°$ 3. $240°$ 4. $-735°$

In Problems 5–7, convert the measure of the given angle from radians to degrees.

5. $\frac{9\pi}{2}$ 6. $\frac{16\pi}{3}$ 7. $-\frac{17\pi}{6}$

8. Determine the measure in radians and in degrees of the central angle which cuts an arc of length 12π cm in a circle with radius 8 cm.

In Problems 9–12, determine the given trigonometric function value.

9. $\sin\frac{5\pi}{4}$ 10. $\cos\left(-\frac{4\pi}{3}\right)$

11. $\sec\frac{19\pi}{2}$ 12. $\tan\left(-\frac{17\pi}{3}\right)$

In Problems 13–14, determine the trigonometric function value.

13. $\sin 4.3$ 14. $\cos 5.6$

15. Graph the following pair of functions on the same axes:

$$y = \sin x,\ y = -\sin\left(x + \frac{\pi}{2}\right).$$

16. Graph the following pair of functions on the same axes: $y = \tan x,\ y = \tan(x - \pi)$.

In Problems 17–20, determine whether the statement is true or false.

17. $\tan x$ is not defined for $x = -3\pi$.

18. $\cot x$ is not defined for $x = -\frac{3\pi}{2}$.

19. $\csc(-x) = -\frac{1}{\sin x}$ is an identity.

20. $\cos(-x) = -\cos x$ is an identity.

21. Determine the amplitude and period, then sketch the graph of the function $y = 3\cos\frac{1}{2}x$.

22. Determine the amplitude, period, and phase shift, then sketch the graph of the function $y = -4\sin(2x + \pi)$.

23. Sketch the graph of the function $y = -3\sin\frac{1}{2}x + x$.

In Problems 24–26, evaluate the expression without using a calculator or Table 4.

24. $\arcsin\left(-\frac{1}{2}\right)$ 25. $\cos^{-1}\frac{\sqrt{3}}{2}$ 26. $\sin^{-1}\left(\cot\frac{3\pi}{4}\right)$

In Problems 27–28, use a right triangle to evaluate the expression.

27. $\cos\left(\tan^{-1}\frac{3}{4}\right)$ 28. $\sin\left(\operatorname{arcsec}\frac{13}{5}\right)$

29. The vector **a** has initial point (0, 0) and terminal point (−3, 3). The vector **b** has initial point (0, 0) and terminal point (1, 1).
(a) Determine the magnitude and direction of vector **a**.
(b) Draw **a** − 2**b**.

In Problems 30–31, points A and B are given. Describe the vector $\overrightarrow{AB}$ as an ordered pair.

30. $A(-3, -2), B(2, 7)$ 31. $A(6, -3), B(-1, 3)$

In Problems 32–33, use the vectors $\mathbf{a} = \left\langle -\frac{1}{2}, \frac{7}{2}\right\rangle$ and $\mathbf{b} = \langle 6, -1\rangle$ to determine the following.

32. $\|\mathbf{a}\|$ and $2\mathbf{a} - \frac{1}{2}\mathbf{b}$ 33. $\|\mathbf{b}\|$ and $\mathbf{b} - 4\mathbf{a}$

In Problems 34–35, the magnitude $\|\mathbf{v}\|$ and the angle θ between the vector **v** and the positive x-axis are given. Find the horizontal and vertical components of **v**.

34. $\|\mathbf{v}\| = 8.6, \theta = 148°$ 35. $\|\mathbf{v}\| = 10.8, \theta = 206°$

36. A plane's air speed is 500 mph and its bearing is N 75°E. An 8 mph wind is blowing in the direction N 70°W. Determine the ground speed and true bearing of the plane.

37. Find the force required to hold a 120 lb crate stationary on a ramp inclined at 25°.

In Problems 38–39, determine the dot product of the given vectors. Find the angle between the vectors to the nearest degree.

38. $\langle 3, -1\rangle, \langle -2, -2\rangle$ 39. $\langle 4, 6\rangle, \langle -3, 2\rangle$

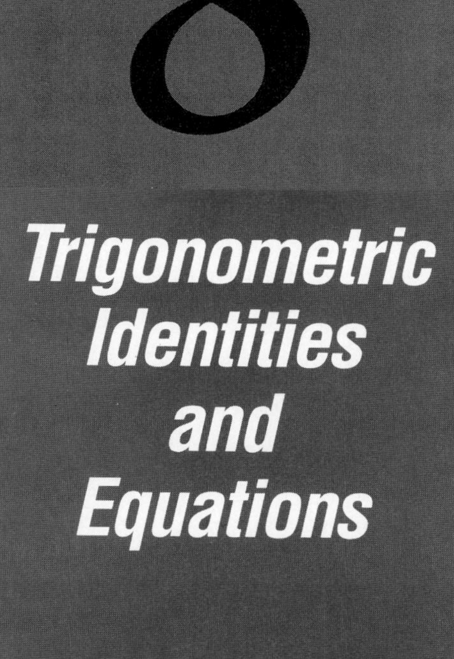

In this chapter we complete our discussion of the trigonometric functions. Because the basic trigonometric functions are defined by relying on arcs of the unit circle, they are sometimes referred to as **circular functions.** *There are some relatively profound results about circles that we simply take for granted (that is, as "given"). At the end of this chapter, with the help of some experimental data from a third grade class, we will use the knowledge gained in this course to construct, from scratch, a formula for the area of a circle.*

8.1 Basic Trigonometric Identities

We refer to equations such as

$$(x + 1)^2 = x^2 + 2x + 1 \quad \text{or} \quad \frac{x^2 - 1}{x - 1} = x + 1$$

as identities. An **identity** is an equation that is true for all values of x for which the expressions are defined. The equation

$$(x + 1)^2 = x^2 + 2x + 1$$

is true for all real numbers x. The equation

$$\frac{x^2 - 1}{x - 1} = x + 1$$

is true for $x \neq 1$. As long as $x \neq 1$, the expression on the left-hand side of the equation is defined, and may be simplified by factoring and canceling:

$$\frac{x^2 - 1}{x - 1} = \frac{(x - 1)(x + 1)}{x - 1} = x + 1.$$

We have already seen trigonometric identities such as

$$\tan \theta = \frac{\sin \theta}{\cos \theta}.$$

This identity is true for all θ such that $\cos \theta \neq 0$. It is one of the basic identities we derived earlier. Recall that we derived eleven basic identities using the definitions of the trigonometric functions and the properties of the unit circle. They are called the **basic identities** because only the trigonometric definitions are used to derive them.

We classified the eleven basic identities into three categories, shown at the top of page 463.

You should memorize the basic identities listed in each category. We now consider an example in which knowledge of the Basic Identities is essential in simplifying the given expression.

EXAMPLE 1

Show that the expression $\dfrac{1 + \tan^2 \theta}{\csc^2 \theta}$ may be simplified to $\tan^2 \theta$.

Solution

$$\frac{1 + \tan^2 \theta}{\csc^2 \theta} = \frac{\sec^2 \theta}{\csc^2 \theta} \quad \text{By identity (10), } \tan^2 \theta + 1 = \sec^2 \theta.$$

$$= \frac{\dfrac{1}{\cos^2 \theta}}{\dfrac{1}{\sin^2 \theta}} \quad \text{By (5), } \sec^2 \theta = \frac{1}{\cos^2 \theta}. \quad \text{By (6), } \csc^2 \theta = \frac{1}{\sin^2 \theta}.$$

$$= \frac{\sin^2 \theta}{\cos^2 \theta} \quad \text{Divide: multiply by the reciprocal of the denominator.}$$

$$= \tan^2 \theta \quad \text{By (7), } \frac{\sin \theta}{\cos \theta} = \tan \theta.$$

There are no standard steps to take to simplify a trigonometric expression. Simplifying trigonometric expressions is similar to factoring polynomials: by trial and error and by experience, you learn what will work in which situations. One useful technique is to begin by rewriting the entire expression in terms of sines and cosines.

Basic Trigonometric Identities

I. Reciprocal Identities

(1) $$\sin\theta = \frac{1}{\csc\theta}$$

(2) $$\cos\theta = \frac{1}{\sec\theta}$$

(3) $$\tan\theta = \frac{1}{\cot\theta}$$

(4) $$\cot\theta = \frac{1}{\tan\theta}$$

(5) $$\sec\theta = \frac{1}{\cos\theta}$$

(6) $$\csc\theta = \frac{1}{\sin\theta}$$

II. Quotient Identities

(7) $$\frac{\sin\theta}{\cos\theta} = \tan\theta$$

(8) $$\frac{\cos\theta}{\sin\theta} = \cot\theta$$

III. Pythagorean Identities

(9) $$\sin^2\theta + \cos^2\theta = 1$$

(10) $$\tan^2\theta + 1 = \sec^2\theta$$

(11) $$1 + \cot^2\theta = \csc^2\theta$$

EXAMPLE 2

Express $\left(1 - \frac{1}{\csc\theta}\right)^2 + \cos^2\theta$ in terms of $\sin\theta$.

Solution

$$\left(1 - \frac{1}{\csc\theta}\right)^2 + \cos^2\theta = (1 - \sin\theta)^2 + \cos^2\theta \qquad \text{By (1)}$$

$$= 1 - 2\sin\theta + \sin^2\theta + \cos^2\theta \qquad (1 - \sin\theta)^2 \text{ is expanded.}$$

$$= 1 - 2\sin\theta + 1 \qquad \text{By (9)}$$

$$= 2 - 2\sin\theta$$

We are now in a position to begin verifying new trigonometric identities. The primary strategy that we will use is to choose the expression on one side of the identity and to transform it so that it is identical to the expression on the other side. This "transformation" is made by using the rules of algebra and the Basic Identities. Before we consider some examples, we offer a few words of advice.

Guidelines for Verifying Identities

1. Memorize the Basic Identities. Verifying a new identity often means that you must run through a mental checklist of options that might prove effective in rewriting one of the expressions.
2. It is often best to proceed by first writing down the more complicated side of the identity, and then transforming it in a step-by-step fashion until it looks exactly like the other side of the identity.
3. As an aid in rewriting the more complicated side of the identity, it is often useful to rewrite this expression in terms of sines and cosines only.

Careful! One approach *not* to be used is to write down the entire identity as your first step, and "do the same thing to both sides," as you would in solving an equation. Working on both sides of an equation in this way can be done only when the equation is assumed to be true. In verifying an identity, you are trying to prove that the equation is true. Thus, you *may not* assume that it is already true.

EXAMPLE 3

Verify the identity: $\dfrac{1 - \cos^2 \theta}{\tan \theta} = \sin \theta \cos \theta$.

Solution

$$\frac{1 - \cos^2 \theta}{\tan \theta} = \frac{\sin^2 \theta}{\tan \theta} \qquad \text{By (9), } 1 - \cos^2 \theta = \sin^2 \theta.$$

$$= \frac{\sin^2 \theta}{\dfrac{\sin \theta}{\cos \theta}} \qquad \text{By (7)}$$

$$= \frac{\sin^2 \theta}{1} \cdot \frac{\cos \theta}{\sin \theta}$$

$$= \sin \theta \cos \theta$$

EXAMPLE 4

Verify the identity: $\dfrac{\tan x}{1 - \sec x} = -\dfrac{1 + \sec x}{\tan x}$.

Solution

$$\frac{\tan x}{1-\sec x} = \frac{\tan x}{1-\sec x}\cdot\frac{1+\sec x}{1+\sec x}$$ Notice that multiplying by $\frac{1+\sec x}{1+\sec x}$ transforms the denominator into $1-\sec^2 x$.

$$= \frac{(\tan x)(1+\sec x)}{1-\sec^2 x}$$

$$= \frac{(\tan x)(1+\sec x)}{-\tan^2 x}$$ By (10), $1-\sec^2 x = -\tan^2 x$.

$$= -\frac{1+\sec x}{\tan x}$$ $\tan x$ has been canceled from the numerator and denominator.

EXERCISE SET 8.1

In Exercises 1–8, show that the first expression may be rewritten as the second by using the Reciprocal and Quotient Identities.

1. $\sin\alpha\sec\alpha$; $\tan\alpha$

2. $\dfrac{\csc\alpha}{\sec\alpha}$; $\cot\alpha$

3. $\dfrac{\sin\varphi}{\csc\varphi}$; $\sin^2\varphi$

4. $\dfrac{\cos\varphi}{\sec\varphi}$; $\cos^2\varphi$

5. $\csc\beta$; $\sec\beta\cot\beta$

6. $\sin\beta$; $\dfrac{\tan\beta}{\sec\beta}$

7. $\csc\theta\cot\theta$; $\cos\theta\csc^2\theta$

8. $\csc^2\theta\tan\theta$; $\sec\theta\csc\theta$

In Exercises 9–18, rewrite the expression in terms of only the sine or only the tangent.

9. $\dfrac{\csc\alpha}{1+\cot^2\alpha}$

10. $\dfrac{\tan\alpha\sec\alpha}{1+\tan^2\alpha}$

11. $\cot\beta(\sec^2\beta-1)$

12. $\sec^2\alpha-2$

13. $\csc\beta-\cos\beta\cot\beta$

14. $\dfrac{\sec\beta-\cos\beta}{\tan\beta}$

15. $(1+\tan^2\varphi)\sin^2\varphi$

16. $\dfrac{\sec^2\varphi-1}{\sec^2\varphi}$

17. $\dfrac{\sin\alpha+\cos\alpha}{\cos\alpha}$

18. $\dfrac{\csc\alpha+\sec\alpha}{\csc\alpha}$

In Exercises 19–38, verify the identity by using the Basic Identities.

19. $\tan\alpha+\cot\alpha=\csc\alpha\sec\alpha$

20. $\dfrac{\sec\alpha}{\tan\alpha}-\dfrac{\tan\alpha}{\sec\alpha}=\cos\alpha\cot\alpha$

21. $(1+\cot\beta)^2-\csc^2\beta=2\cot\beta$

22. $(1+\csc^2\beta)-\cot^2\beta=2$

23. $\cos\alpha(\tan\alpha+\sec\alpha)=1+\sin\alpha$

24. $\sin\alpha(\cot\alpha+\csc\alpha)=\cos\alpha+1$

25. $\sec\beta+\csc\beta\cot\beta=\csc^2\beta\sec\beta$

26. $\sec\beta-\sin\beta\cot\beta=\tan\beta\sin\beta$

27. $\dfrac{\sin\gamma}{\cos\gamma+1}+\dfrac{\cos\gamma-1}{\sin\gamma}=0$

28. $\dfrac{\sin\gamma}{\csc\gamma}+\dfrac{\cos\gamma}{\sec\gamma}=1$

29. $\dfrac{(\sin\gamma+\cos\gamma)^2}{\sin\gamma\cos\gamma}=\csc\gamma\sec\gamma+2$

30. $\left(\dfrac{\sin\gamma}{\cos\gamma}+\dfrac{\cos\gamma}{\sin\gamma}\right)^2=\csc^2\gamma\sec^2\gamma$

31. $(\tan^2\alpha+1)(1+\cos^2\alpha)=\tan^2\alpha+2$

32. $(1+\cot^2\alpha)(1+\sin^2\alpha)=2+\cot^2\alpha$

33. $\sin x=(\sec x-\cos x)\cot x$

34. $\cos x=(\csc x-\sin x)\tan x$

35. $\dfrac{\tan\beta-\sin\beta}{\sin^3\beta}=\dfrac{\sec\beta}{1+\cos\beta}$

36. $\dfrac{\cot\beta-\cos\beta}{\cos^3\beta}=\dfrac{\csc\beta}{1+\sin\beta}$

37. $(\sin\alpha+\cos\alpha)^2-(\sin\beta+\cos\beta)^2$
$=2(\sin\alpha\cos\alpha-\sin\beta\cos\beta)$

38. $(\sin\alpha-\sin\beta)^2+(\cos\alpha-\cos\beta)^2$
$=2(1-\sin\alpha\sin\beta-\cos\alpha\cos\beta)$

Superset

In Exercises 39–43, express the first trigonometric function in terms of the second.

39. $\sin \alpha$ in terms of $\cot \alpha$; α in Quadrant I

40. $\cos \alpha$ in terms of $\cot \alpha$; α in Quadrant III

41. $\sec \beta$ in terms of $\sin \beta$; β in Quadrant II

42. $\tan \varphi$ in terms of $\cos \varphi$; φ in Quadrant IV

43. $\csc \gamma$ in terms of $\sec \gamma$; γ in Quadrant III

44. Express the other 5 trigonometric functions in terms of
(a) $\sin \alpha$ for α in Quadrant II;
(b) $\cos \alpha$ for α in Quadrant III.

8.2
Sum and Difference Identities

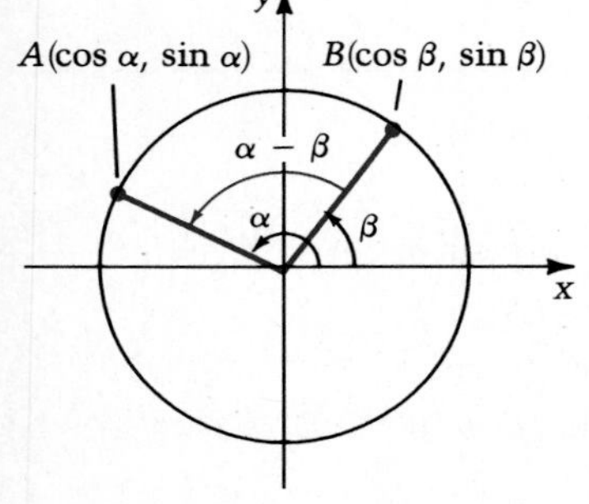

Figure 8.1

Suppose you wish to determine $\sin(\alpha + \beta)$ and you already know the trigonometric function values of α and β. Although it might be tempting to write $\sin \alpha + \sin \beta$ as your answer, it would be incorrect to do so. In this section we will derive identities for the trigonometric functions of $\alpha + \beta$ and $\alpha - \beta$, where α and β represent any real numbers or angle measurements. We begin by determining a formula for $\cos(\alpha - \beta)$, from which the other sum and difference formulas can be easily derived.

Suppose α and β are positive real numbers and $\alpha > \beta$. To visualize what we are about to do, we can think of α and β as the lengths of standard arcs, or the measures of angles in standard position on the unit circle. Interpreted this way, α and β determine points A and B on the unit circle, shown in Figure 8.1. Observe that point A has coordinates $(\cos \alpha, \sin \alpha)$, while B has coordinates $(\cos \beta, \sin \beta)$.

In Figure 8.2(a) we show $\alpha - \beta$ and a line segment connecting the points A and B. The line segment joining the endpoints of an arc is called a **chord.** We let d be the length of chord AB. In Figure 8.2(b), the angle $\alpha - \beta$ has been redrawn in standard position, and the chord of length d now has endpoints $(\cos(\alpha - \beta), \sin(\alpha - \beta))$ and $(1, 0)$, labeled A' and B' respectively.

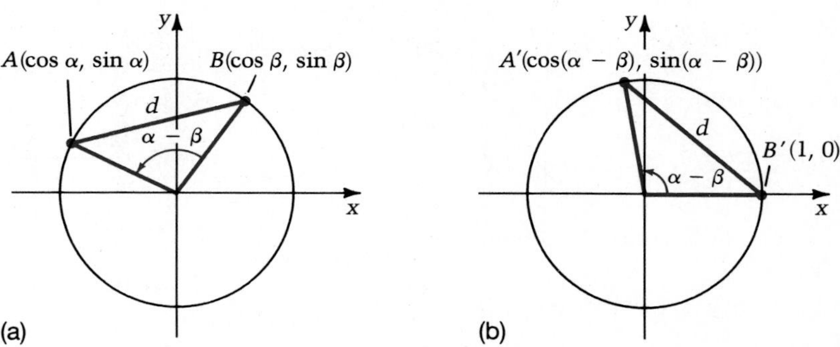

Figure 8.2

By means of the distance formula, we express d^2 in two different ways: first by using the coordinates of A and B (Figure 8.2(a)), and then by using the coordinates of A' and B' (Figure 8.2(b)). Using A and B, we have

$$\begin{aligned} d^2 &= (\text{distance between } A \text{ and } B)^2 = (\cos\alpha - \cos\beta)^2 + (\sin\alpha - \sin\beta)^2 \\ &= (\cos^2\alpha - 2\cos\alpha\cos\beta + \cos^2\beta) + (\sin^2\alpha - 2\sin\alpha\sin\beta + \sin^2\beta) \\ &= (\sin^2\alpha + \cos^2\alpha) + (\sin^2\beta + \cos^2\beta) - 2\cos\alpha\cos\beta - 2\sin\alpha\sin\beta \\ &= \quad 1 \quad + \quad 1 \quad - 2(\cos\alpha\cos\beta + \sin\alpha\sin\beta) \\ &= 2 - 2(\cos\alpha\cos\beta + \sin\alpha\sin\beta) \end{aligned}$$

Using A' and B', we have

$$\begin{aligned} d^2 &= (\text{distance between } A' \text{ and } B')^2 \\ &= (\cos(\alpha - \beta) - 1)^2 + (\sin(\alpha - \beta) - 0)^2 \\ &= \cos^2(\alpha - \beta) - 2\cos(\alpha - \beta) + 1 + \sin^2(\alpha - \beta) \\ &= \sin^2(\alpha - \beta) + \cos^2(\alpha - \beta) + 1 - 2\cos(\alpha - \beta) \\ &= \quad 1 \quad + 1 - 2\cos(\alpha - \beta) \\ &= 2 - 2\cos(\alpha - \beta) \end{aligned}$$

Equating the two expressions for d^2, we have

$$2 - 2\cos(\alpha - \beta) = 2 - 2(\cos\alpha\cos\beta + \sin\alpha\sin\beta).$$

Then, by adding (-2) to both sides and dividing by (-2), we conclude that

(12) $$\cos(\alpha - \beta) = \cos\alpha\cos\beta + \sin\alpha\sin\beta.$$

Although we have developed identity (12) for α and β that are positive with $\alpha > \beta$, a similar argument establishes (12) for all α and β.

EXAMPLE 1

Determine cos 15° without using a calculator or Table 4.

Solution

$$\begin{aligned} \cos 15^\circ &= \cos(45^\circ - 30^\circ) \\ &= \cos 45^\circ \cos 30^\circ + \sin 45^\circ \sin 30^\circ \quad \text{By identity (12)} \\ &= \left(\frac{\sqrt{2}}{2}\right)\left(\frac{\sqrt{3}}{2}\right) + \left(\frac{\sqrt{2}}{2}\right)\left(\frac{1}{2}\right) \\ &= \frac{\sqrt{6}}{4} + \frac{\sqrt{2}}{4} \end{aligned}$$

Thus, $\cos 15^\circ = \dfrac{\sqrt{6} + \sqrt{2}}{4}$.

Recall that sine is an odd function and cosine is an even function. These two facts are represented as follows: for all real numbers x,

$$\sin(-x) = -\sin x \quad \text{and} \quad \cos(-x) = \cos x.$$

We use these facts to establish another identity:

$$\begin{aligned}
\cos(\alpha + \beta) &= \cos(\alpha - (-\beta)) \\
&= \cos \alpha \cos(-\beta) + \sin \alpha(\sin(-\beta)) && \text{By (12)} \\
&= \cos \alpha \cos \beta + \sin \alpha\,(-\sin \beta) && \text{Cosine is even; sine is odd.} \\
&= \cos \alpha \cos \beta - \sin \alpha \sin \beta
\end{aligned}$$

Thus,

(13) $$\cos(\alpha + \beta) = \cos \alpha \cos \beta - \sin \alpha \sin \beta.$$

Recall that a trigonometric function value of an acute angle θ is equal to the co-trigonometric function value of the complement of θ. For example, for any acute angle θ,

$$\sin \theta = \cos\left(\frac{\pi}{2} - \theta\right).$$

By virtue of (12) we can now prove such a statement true for any θ.

$$\begin{aligned}
\cos\left(\frac{\pi}{2} - \theta\right) &= \cos \frac{\pi}{2} \cos \theta + \sin \frac{\pi}{2} \sin \theta \\
&= (0) \cdot \cos \theta + (1) \cdot \sin \theta \\
&= \sin \theta
\end{aligned}$$

In a similar manner, we can show that $\sin\left(\frac{\pi}{2} - \theta\right) = \cos \theta$.

We will use these results in deriving an identity for $\sin(\alpha + \beta)$.

$$\begin{aligned}
\sin(\alpha + \beta) &= \cos\left(\frac{\pi}{2} - (\alpha + \beta)\right) && \sin \theta = \cos\left(\frac{\pi}{2} - \theta\right). \\
&= \cos\left(\left(\frac{\pi}{2} - \alpha\right) - \beta\right) \\
&= \cos\left(\frac{\pi}{2} - \alpha\right) \cos \beta + \sin\left(\frac{\pi}{2} - \alpha\right) \sin \beta && \text{By (12)} \\
&= \sin \alpha \cos \beta + \cos \alpha \sin \beta
\end{aligned}$$

Thus,

(14) $$\sin(\alpha + \beta) = \sin \alpha \cos \beta + \cos \alpha \sin \beta.$$

Replacing β with $(-\beta)$ in (14), we have

$$\begin{aligned}\sin(\alpha - \beta) &= \sin(\alpha + (-\beta))\\ &= \sin\alpha\cos(-\beta) + \cos\alpha\sin(-\beta) && \text{By (14)}\\ &= \sin\alpha\cos\beta + \cos\alpha(-\sin\beta) && \text{Cosine is even; sine is odd.}\\ &= \sin\alpha\cos\beta - \cos\alpha\sin\beta\end{aligned}$$

Thus,

(15) $$\sin(\alpha - \beta) = \sin\alpha\cos\beta - \cos\alpha\sin\beta.$$

EXAMPLE 2

Determine $\sin\frac{5\pi}{12}$ without using a calculator or Table 4.

Solution

$$\sin\frac{5\pi}{12} = \sin\left(\frac{\pi}{4} + \frac{\pi}{6}\right) \qquad \frac{5\pi}{12} \text{ is written as a sum of special angles: } \frac{5\pi}{12} = \frac{\pi}{4} + \frac{\pi}{6}.$$

$$= \sin\frac{\pi}{4}\cos\frac{\pi}{6} + \cos\frac{\pi}{4}\sin\frac{\pi}{6}$$

$$= \frac{\sqrt{2}}{2}\cdot\frac{\sqrt{3}}{2} + \frac{\sqrt{2}}{2}\cdot\frac{1}{2}$$

Thus,

$$\sin\frac{5\pi}{12} = \frac{\sqrt{6}}{4} + \frac{\sqrt{2}}{4} = \frac{\sqrt{6} + \sqrt{2}}{4}.$$

Notice that in Example 1 we found that $\cos 15° = \frac{\sqrt{6} + \sqrt{2}}{4}$, and in Example 2 we found that $\sin\frac{5\pi}{12}$ also equals $\frac{\sqrt{6} + \sqrt{2}}{4}$. Since $\frac{5\pi}{12} = 75°$, this result is expected because of the cofunction identity:

$$\cos 15° = \sin(90° - 15°) = \sin 75°.$$

EXAMPLE 3

If $\sin\alpha = \frac{12}{13}$ for some second quadrant angle α, and $\sin\beta = \frac{3}{5}$ for some first quadrant angle β, determine (a) $\sin(\alpha - \beta)$ and (b) $\sin(\alpha + \beta)$.

Solution Begin by using Pythagorean Identity (9) to find $\cos\alpha$ and $\cos\beta$.

$$\cos\alpha = \pm\sqrt{1 - \sin^2\alpha}$$

$$\cos\alpha = -\sqrt{1 - \left(\frac{12}{13}\right)^2}$$ Since α is a second quadrant angle, $\cos\alpha$ is negative.

$$\cos\alpha = -\sqrt{1 - \frac{144}{169}}$$

$$\cos\alpha = -\sqrt{\frac{25}{169}} = -\frac{5}{13}$$

$$\cos\beta = \sqrt{1 - \left(\frac{3}{5}\right)^2}$$ Since β is a first quadrant angle, $\cos\beta$ is positive.

$$\cos\beta = \sqrt{\frac{16}{25}} = \frac{4}{5}$$

(a) $\sin(\alpha - \beta) = \sin\alpha\cos\beta - \cos\alpha\sin\beta = \left(\frac{12}{13}\right)\left(\frac{4}{5}\right) - \left(\frac{-5}{13}\right)\left(\frac{3}{5}\right) = \frac{63}{65}$

(b) $\sin(\alpha + \beta) = \sin\alpha\cos\beta + \cos\alpha\sin\beta = \left(\frac{12}{13}\right)\left(\frac{4}{5}\right) + \left(\frac{-5}{13}\right)\left(\frac{3}{5}\right) = \frac{33}{65}$

The identity for $\tan(\alpha + \beta)$ may be derived by using the Quotient Identity (7) and the identities for the sine and cosine of a sum.

$$\tan(\alpha + \beta) = \frac{\sin(\alpha + \beta)}{\cos(\alpha + \beta)}$$ By (7)

$$= \frac{\sin\alpha\cos\beta + \cos\alpha\sin\beta}{\cos\alpha\cos\beta - \sin\alpha\sin\beta}$$ By (14) and (13)

$$= \frac{\dfrac{\sin\alpha\cos\beta}{\cos\alpha\cos\beta} + \dfrac{\cos\alpha\sin\beta}{\cos\alpha\cos\beta}}{\dfrac{\cos\alpha\cos\beta}{\cos\alpha\cos\beta} - \dfrac{\sin\alpha\sin\beta}{\cos\alpha\cos\beta}}$$ Divide each term by $\cos\alpha\cos\beta$.

$$= \frac{\tan\alpha + \tan\beta}{1 - \tan\alpha\tan\beta}$$ (7) is used four times.

Therefore,

(16) $$\tan(\alpha + \beta) = \frac{\tan\alpha + \tan\beta}{1 - \tan\alpha\tan\beta}.$$

In a similar manner,

$$\tan(\alpha - \beta) = \frac{\tan\alpha - \tan\beta}{1 + \tan\alpha\tan\beta}. \tag{17}$$

The derivation of identity (17) is left as an exercise.

EXAMPLE 4

Without using a calculator or Table 4, find $\tan\frac{7\pi}{12}$.

Solution $\tan\left(\frac{7\pi}{12}\right) = \tan\left(\frac{\pi}{4} + \frac{\pi}{3}\right) = \dfrac{\tan\frac{\pi}{4} + \tan\frac{\pi}{3}}{1 - \tan\frac{\pi}{4}\tan\frac{\pi}{3}}$ By (16)

$$= \frac{1 + \sqrt{3}}{1 - \sqrt{3}}$$

When an answer has a radical in both the numerator and the denominator, it is often easier to approximate the answer if we first rationalize the denominator. In the case of Example 4, we have,

$$\frac{1 + \sqrt{3}}{1 - \sqrt{3}} = \frac{(1 + \sqrt{3})(1 + \sqrt{3})}{(1 - \sqrt{3})(1 + \sqrt{3})} = \frac{1 + 2\sqrt{3} + 3}{1 - 3} = \frac{4 + 2\sqrt{3}}{-2} = -(2 + \sqrt{3}).$$

Thus, we see that $\tan\frac{7\pi}{12}$ is approximately -3.7321. ($\sqrt{3} \approx 1.7321$.)

EXAMPLE 5

Simplify: $\dfrac{\sin(\alpha + \beta) + \sin(\alpha - \beta)}{\cos(\alpha + \beta) + \cos(\alpha - \beta)}$.

Solution $\dfrac{\sin(\alpha + \beta) + \sin(\alpha - \beta)}{\cos(\alpha + \beta) + \cos(\alpha - \beta)}$

$$= \frac{(\sin\alpha\cos\beta + \cos\alpha\sin\beta) + (\sin\alpha\cos\beta - \cos\alpha\sin\beta)}{(\cos\alpha\cos\beta - \sin\alpha\sin\beta) + (\cos\alpha\cos\beta + \sin\alpha\sin\beta)}$$

$$= \frac{2\sin\alpha\cos\beta}{2\cos\alpha\cos\beta}$$

$$= \frac{\sin\alpha}{\cos\alpha} = \tan\alpha$$

The sum and difference identities derived in this section are listed below for easy reference.

Sum and Difference Identities

(12)	$\cos(\alpha - \beta) = \cos\alpha\cos\beta + \sin\alpha\sin\beta$
(13)	$\cos(\alpha + \beta) = \cos\alpha\cos\beta - \sin\alpha\sin\beta$
(14)	$\sin(\alpha + \beta) = \sin\alpha\cos\beta + \cos\alpha\sin\beta$
(15)	$\sin(\alpha - \beta) = \sin\alpha\cos\beta - \cos\alpha\sin\beta$
(16)	$\tan(\alpha + \beta) = \dfrac{\tan\alpha + \tan\beta}{1 - \tan\alpha\tan\beta}$
(17)	$\tan(\alpha - \beta) = \dfrac{\tan\alpha - \tan\beta}{1 + \tan\alpha\tan\beta}$

EXERCISE SET 8.2

In Exercises 1–12, without using a calculator or Table 4, find the trigonometric value by using the sum and difference identities.

1. $\sin 120°$ **2.** $\cos 120°$ **3.** $\sin 15°$

4. $\cos 15°$ **5.** $\cos 75°$ **6.** $\sin 75°$

7. $\sin 195°$ **8.** $\cos 105°$ **9.** $\cos(-45°)$

10. $\sin(-45°)$ **11.** $\sin(-135°)$ **12.** $\cos(-135°)$

In Exercises 13–20, without using a calculator or Table 4, find the trigonometric value.

13. $\cos\frac{5\pi}{12}$ **14.** $\sin\frac{11\pi}{12}$ **15.** $\sin\frac{7\pi}{12}$

16. $\cos\frac{7\pi}{12}$ **17.** $\cos\left(-\frac{5\pi}{12}\right)$ **18.** $\sin\left(-\frac{5\pi}{12}\right)$

19. $\sin\left(-\frac{11\pi}{12}\right)$ **20.** $\cos\left(-\frac{11\pi}{12}\right)$

In Exercises 21–26, find the trigonometric value without using a calculator or Table 4.

21. $\tan 75°$ **22.** $\tan 15°$ **23.** $\tan\frac{19\pi}{12}$

24. $\tan\frac{\pi}{12}$ **25.** $\tan\left(-\frac{\pi}{12}\right)$ **26.** $\tan\frac{5\pi}{12}$

In Exercises 27–32, use the given information to determine: (a) $\sin(\alpha + \beta)$, (b) $\cos(\alpha + \beta)$, and (c) $\tan(\alpha + \beta)$.

27. $\sin\alpha = \frac{4}{5}$, α is a first quadrant angle, β is a second quadrant angle, and $\cos\beta = -\frac{12}{13}$.

28. $\sin\alpha = \frac{3}{5}$, α is a first quadrant angle, β is a third quadrant angle, and $\sin\beta = -\frac{5}{13}$.

29. $\cos\alpha = -\frac{4}{5}$, α is a second quadrant angle, β is a third quadrant angle, and $\tan\beta = \frac{5}{12}$.

30. $\cos\alpha = -\frac{3}{5}$, α is a second quadrant angle, β is a fourth quadrant angle, and $\tan\beta = -\frac{12}{5}$.

31. $\tan\alpha = 2$, α is a third quadrant angle, β is a fourth quadrant angle, and $\csc\beta = -\frac{3}{2}$.

32. $\tan\alpha = \frac{2}{3}$, α is a first quadrant angle, β is a fourth quadrant angle, and $\sec\beta = \frac{7}{6}$.

For Exercises 33–38, use the information from Exercises 27–32 to find the following:
(a) $\sin(\alpha - \beta)$, (b) $\cos(\alpha - \beta)$, and (c) $\tan(\alpha - \beta)$.

In Exercises 39–48, verify the identity.

39. $\sin(\alpha - \beta)\cos\beta + \cos(\alpha - \beta)\sin\beta = \sin\alpha$

40. $\sin(\alpha - \beta)\sin(\alpha + \beta) = \sin^2\alpha - \sin^2\beta$

41. $\cos(\alpha - \beta)\cos(\alpha + \beta) = \cos^2\alpha - \sin^2\beta$

42. $\cos(\alpha - \beta) + \cos(\alpha + \beta) = 2\cos\alpha\cos\beta$

43. $\tan\left(\alpha + \frac{\pi}{4}\right) = \dfrac{\cos\alpha + \sin\alpha}{\cos\alpha - \sin\alpha}$

44. $\tan\left(\frac{\pi}{4} - \alpha\right) = \frac{\cos\alpha - \sin\alpha}{\cos\alpha + \sin\alpha}$

45. $\tan\left(\frac{\pi}{4} + \alpha\right)\tan\left(\frac{\pi}{4} - \alpha\right) = \tan\frac{\pi}{4}$

46. $\frac{\tan(\alpha - \beta)}{\tan(\beta - \alpha)} = -1$

47. $\frac{\sin(x + y)}{\sin(x - y)} = \frac{\tan x + \tan y}{\tan x - \tan y}$

48. $\frac{\sin(x + y)}{\sin x \sin y} = \frac{1}{\tan x} + \frac{1}{\tan y}$

Superset

In Exercises 49–56, simplify the expression.

49. $\cos\left(\alpha + \frac{\pi}{6}\right)$ **50.** $\sin\left(\alpha + \frac{\pi}{6}\right)$

51. $\tan\left(\alpha + \frac{\pi}{4}\right)$ **52.** $\cos\left(\beta + \frac{\pi}{4}\right)$

53. $\sin\left(\frac{\pi}{2} - \varphi\right)$ **54.** $\tan\left(\frac{\pi}{2} - \beta\right)$

55. $\sin\left(\frac{\pi}{4} - \alpha\right)$ **56.** $\cos\left(\frac{\pi}{4} - \beta\right)$

In Exercises 57–62, derive an identity involving the given expression.

57. $\tan(\alpha - \beta)$ **58.** $\cot(\alpha + \beta)$

59. $\sec(\alpha + \beta)$ **60.** $\csc(\alpha + \beta)$

61. $\sin 2\alpha$ (*Hint*: $2\alpha = \alpha + \alpha$.)

62. $\cos 2\alpha$

In the examples we saw that the trigonometric values for angles of 15° and 75° can be calculated exactly using the special angles and the Sum and Difference Identities. In Exercises 63–68, compare the trigonometric function values found on the calculator, with the value found by using the special angles and identities.

63. $\sin 15°$ 64. $\cos 15°$ 65. $\cos 75°$

66. $\sin 75°$ 67. $\tan 15°$ 68. $\tan 75°$

In Exercises 69–74, suppose $\sin\alpha = -0.3472$, α a third quadrant angle, and $\cos\beta = -0.6561$, β a second quadrant angle. Compute the given expression in two ways: first determine α and β, then evaluate the trigonometric function value. Second, use the appropriate sum or difference identity.

69. $\sin(\alpha + \beta)$ 70. $\cos(\alpha + \beta)$ 71. $\cos(\alpha - \beta)$

72. $\sin(\alpha - \beta)$ 73. $\tan(\alpha + \beta)$ 74. $\tan(\alpha - \beta)$

8.3 The Double-Angle and Half-Angle Identities

In this section we will derive identities for $\sin 2\theta$, $\cos 2\theta$, $\tan 2\theta$, $\sin\frac{\theta}{2}$, $\cos\frac{\theta}{2}$, and $\tan\frac{\theta}{2}$. The first three identities are called **double-angle identities;** the other three are called **half-angle identities.**

The double-angle identities follow easily from the sum identities. For example, consider the identity $\sin(\alpha + \beta) = \sin\alpha\cos\beta + \cos\alpha\sin\beta$. If $\alpha = \beta$, we have $\sin 2\alpha = \sin(\alpha + \alpha) = \sin\alpha\cos\alpha + \cos\alpha\sin\alpha$. Thus, for any angle θ

(18) $$\sin 2\theta = 2\sin\theta\cos\theta.$$

Similarly, $\cos 2\theta = \cos(\theta + \theta) = \cos\theta\cos\theta - \sin\theta\sin\theta$, that is,

(19) $$\cos 2\theta = \cos^2\theta - \sin^2\theta.$$

Using the Pythagorean Identity, $\sin^2\theta + \cos^2\theta = 1$, we can use (19) to derive two other identities for $\cos 2\theta$:

$$\cos 2\theta = \cos^2\theta - \sin^2\theta$$
$$= (1 - \sin^2\theta) - \sin^2\theta$$

(20) $$\cos 2\theta = 1 - 2\sin^2\theta.$$

Also, $$\cos 2\theta = \cos^2\theta - \sin^2\theta$$
$$= \cos^2\theta - (1 - \cos^2\theta)$$

(21) $$\cos 2\theta = 2\cos^2\theta - 1.$$

We leave it as an exercise to show that

(22) $$\tan 2\theta = \frac{2\tan\theta}{1 - \tan^2\theta}.$$

EXAMPLE 1

If $\sin\theta = 0.60$ and $\cos\theta = 0.80$, find (a) $\sin 2\theta$ and (b) $\cos 2\theta$.

Solution

(a) $\sin 2\theta = 2\sin\theta\cos\theta = 2(0.60)(0.80) = 0.96$ By (18)

(b) $\cos 2\theta = \cos^2\theta - \sin^2\theta = (0.80)^2 - (0.60)^2 = 0.28$ By (19)

EXAMPLE 2

Use double-angle identities and the exact values of sin 30°, cos 30°, and tan 30° to determine the following: (a) sin 60°, (b) cos 60°, (c) tan 60°.

Solution

(a) $\sin 60° = \sin 2(30°) = 2\sin 30°\cos 30° = 2\left(\frac{1}{2}\right)\left(\frac{\sqrt{3}}{2}\right) = \frac{\sqrt{3}}{2}$

(b) $\cos 60° = \cos 2(30°) = \cos^2 30° - \sin^2 30° = \left(\frac{\sqrt{3}}{2}\right)^2 - \left(\frac{1}{2}\right)^2 = \frac{1}{2}$

(c) $\tan 60° = \tan 2(30°) = \dfrac{2\tan 30°}{1 - \tan^2 30°} = \dfrac{2\left(\frac{\sqrt{3}}{3}\right)}{1 - \left(\frac{\sqrt{3}}{3}\right)^2}$

$$= \frac{\frac{2\sqrt{3}}{3}}{1 - \frac{3}{9}} = \frac{2\sqrt{3}}{3}\cdot\frac{9}{6} = \sqrt{3}$$

EXAMPLE 3

Given that $\sin\frac{\pi}{9} \approx 0.3420$, find $\tan\frac{2\pi}{9}$.

Solution

$$\tan\frac{2\pi}{9} = \frac{\sin\frac{2\pi}{9}}{\cos\frac{2\pi}{9}} = \frac{\sin 2\left(\frac{\pi}{9}\right)}{\cos 2\left(\frac{\pi}{9}\right)} \qquad \text{By (7)}$$

$$= \frac{2\sin\frac{\pi}{9}\cos\frac{\pi}{9}}{1 - 2\sin^2\left(\frac{\pi}{9}\right)} \qquad \text{By (18)} \quad \text{By (20)}$$

To proceed we need to determine $\cos\frac{\pi}{9}$.

$$\sin^2\left(\frac{\pi}{9}\right) + \cos^2\left(\frac{\pi}{9}\right) = 1 \qquad \text{By (9)}$$

$$\cos\frac{\pi}{9} = \sqrt{1 - \sin^2\left(\frac{\pi}{9}\right)}$$

Since $\frac{\pi}{9}$ is a first quadrant angle, $\cos\frac{\pi}{9}$ is positive.

$$\approx \sqrt{1 - (0.3420)^2} \approx 0.9397$$

Thus, $\tan\frac{2\pi}{9} = \dfrac{2\sin\frac{\pi}{9}\cos\frac{\pi}{9}}{1 - 2\sin^2\frac{\pi}{9}} \approx \dfrac{2(0.3420)(0.9397)}{1 - 2(0.3420)^2} \approx 0.8390.$

EXAMPLE 4

Given $\sin\theta = -\frac{4}{5}$ and that θ is a third quadrant angle, determine:
(a) $\sin 2\theta$ (b) $\cos 2\theta$ (c) $\tan 2\theta$.

Solution

(a) To apply the identity $\sin 2\theta = 2\sin\theta\cos\theta$, we need to determine $\cos\theta$.

$$\cos\theta = -\sqrt{1 - \sin^2\theta}$$

By (9). Since θ is a third quadrant angle, $\cos\theta$ is negative.

$$= -\sqrt{1 - \left(-\frac{4}{5}\right)^2} = -\sqrt{1 - \frac{16}{25}} = -\frac{3}{5}$$

$$\sin 2\theta = 2\left(-\frac{4}{5}\right)\left(-\frac{3}{5}\right) = \frac{24}{25}. \qquad \text{By (18)}$$

(b) $\cos 2\theta = 1 - 2\sin^2\theta$ Identity (20)

$$= 1 - 2\left(-\frac{4}{5}\right)^2 = 1 - \frac{32}{25} = -\frac{7}{25}$$

(c) $$\tan 2\theta = \frac{\sin 2\theta}{\cos 2\theta} = \frac{\frac{24}{25}}{-\frac{7}{25}} = -\frac{24}{7}$$

In Example 4(c), we used the quotient identity (7) to determine $\tan 2\theta$, since we already knew the values of $\sin 2\theta$ and $\cos 2\theta$ from parts (a) and (b). We could also have used the double-angle identity (22),

$$\tan 2\theta = \frac{2\tan\theta}{1 - \tan^2\theta}.$$

In that case, we would first find the value of $\tan\theta$ by computing $\dfrac{\sin\theta}{\cos\theta}$.

The Half-Angle Identities

To derive the first half-angle identity, we solve the double-angle identity $\cos 2\theta = 1 - 2\sin^2\theta$ for $\sin^2\theta$:

$$\sin^2\theta = \frac{1 - \cos 2\theta}{2}.$$

If we replace θ with $\dfrac{\theta}{2}$, we have

$$\sin^2\frac{\theta}{2} = \frac{1 - \cos\theta}{2}.$$

Thus, we can state our first half-angle identity as follows:

(23) $$\sin\frac{\theta}{2} = \pm\sqrt{\frac{1 - \cos\theta}{2}}$$ The positive or negative sign is used depending on the quadrant in which $\dfrac{\theta}{2}$ is located.

EXAMPLE 5

Use identity (23) to determine sin 15°.

Solution

$$\sin 15° = \sin\frac{30°}{2} = \sqrt{\frac{1-\cos 30°}{2}}$$

We choose the positive sign since 15° is a first quadrant angle.

$$= \sqrt{\frac{1-\frac{\sqrt{3}}{2}}{2}} = \sqrt{\frac{1}{2}-\frac{\sqrt{3}}{4}} = \sqrt{\frac{2}{4}-\frac{\sqrt{3}}{4}} = \frac{\sqrt{2-\sqrt{3}}}{2}$$

To derive a half-angle identity for the cosine function, we solve the identity $\cos 2\theta = 2\cos^2\theta - 1$ for $\cos\theta$:

$$\cos\theta = \pm\sqrt{\frac{1+\cos 2\theta}{2}}.$$

Replacing θ with $\frac{\theta}{2}$, we conclude that

(24) $$\cos\frac{\theta}{2} = \pm\sqrt{\frac{1+\cos\theta}{2}}.$$

EXAMPLE 6

Use identity (24) to determine $\cos\frac{7\pi}{12}$.

Solution

$$\cos\frac{7\pi}{12} = \cos\left(\frac{1}{2}\left(\frac{7\pi}{6}\right)\right) = -\sqrt{\frac{1+\cos\frac{7\pi}{6}}{2}}$$

We choose the negative sign since $\frac{7\pi}{12}$ is a second quadrant angle and cosine is negative there.

$$= -\sqrt{\frac{1-\frac{\sqrt{3}}{2}}{2}} = -\frac{\sqrt{2-\sqrt{3}}}{2}$$

$$\cos\frac{7\pi}{6} = -\cos\frac{\pi}{6} = -\frac{\sqrt{3}}{2}.$$

Using identities (7), (23), and (24), we can establish the following:

(25) $$\tan\frac{\theta}{2} = \pm\sqrt{\frac{1-\cos\theta}{1+\cos\theta}}.$$

There are two other ways of describing $\tan\frac{\theta}{2}$ that are often useful.

$$\tan\frac{\theta}{2} = \frac{\sin\frac{\theta}{2}}{\cos\frac{\theta}{2}} = \frac{\sin\frac{\theta}{2}}{\cos\frac{\theta}{2}}\cdot\frac{2\cos\frac{\theta}{2}}{2\cos\frac{\theta}{2}} = \frac{2\sin\frac{\theta}{2}\cos\frac{\theta}{2}}{2\cos^2\left(\frac{\theta}{2}\right)}$$

$$= \frac{\sin\left(2\cdot\frac{\theta}{2}\right)}{1+\left(2\cos^2\left(\frac{\theta}{2}\right)-1\right)}$$ By (18)

Add 1 and subtract 1.

$$= \frac{\sin\theta}{1+\cos\theta}$$ By (21)

Thus,

(26) $$\tan\frac{\theta}{2} = \frac{\sin\theta}{1+\cos\theta}.$$

Multiplying the right-hand side of (26) by $\frac{1-\cos\theta}{1-\cos\theta}$ yields still another identity for $\tan\frac{\theta}{2}$.

(27) $$\tan\frac{\theta}{2} = \frac{1-\cos\theta}{\sin\theta}.$$

Identities (26) and (27) have an advantage over identity (25) in that they do not contain a radical sign.

EXAMPLE 7

Use a half-angle identity to determine $\tan(-202.5°)$.

Solution Solving this problem requires that we recognize that $-202.5°$ is $\frac{1}{2}(-405°)$ and that $-405°$ is a fourth quadrant angle which has 45° as its reference angle.

$$\tan(-202.5°) = \tan\left(\frac{-405°}{2}\right) = \frac{\sin(-405°)}{1 + \cos(-405°)} \quad \text{By (26)}$$

$$= \frac{-\frac{\sqrt{2}}{2}}{1 + \frac{\sqrt{2}}{2}} = -\frac{\sqrt{2}}{2 + \sqrt{2}} \qquad \begin{aligned} \sin(-405°) &= -\sin 45° = -\frac{\sqrt{2}}{2} \\ \cos(-405°) &= +\cos 45° = \frac{\sqrt{2}}{2} \end{aligned}$$

In Example 7 we can estimate the value of tan $(-202.5°)$ most easily if we first rationalize the denominator.

$$\frac{-\sqrt{2}}{2 + \sqrt{2}} = \frac{-\sqrt{2}}{2 + \sqrt{2}} \cdot \frac{2 - \sqrt{2}}{2 - \sqrt{2}} = \frac{-2\sqrt{2} + 2}{2}$$

$$= -\sqrt{2} + 1 \approx 1 - 1.414 = -0.414$$

That is, $\tan(-202.5°)$ is approximately -0.414.

EXAMPLE 8

Suppose that the two equal sides in an isosceles triangle have length a. What must the angles be in such a triangle, if the area is to be a maximum?

Solution

- ANALYZE We must determine the angles that yield an isosceles triangle of greatest area, where we are given that the two equal sides have length a. This problem is best organized by drawing a sketch.

- ORGANIZE

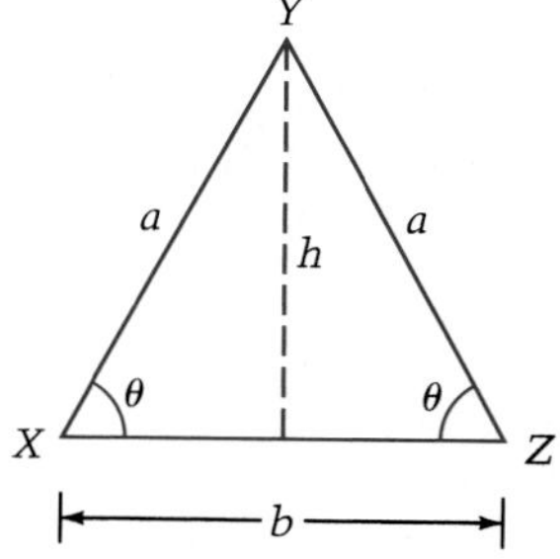

Let a = the length of each of the two equal sides.
Let θ = the measure of the equal angles.
Let b = the length of the third side.
Let h = the length of the altitude drawn to third side.

Once we find θ, the third angle is simply $180° - 2\theta$.

- MODEL Since we wish to determine the value of θ that yields the maximum area, we begin with the formula for the area of a triangle.

$$A = \frac{1}{2}hb$$

In $\Delta\, XYZ$, $\sin\theta = \frac{h}{a}$, and $\cos\theta = \frac{b/2}{a}$. Thus, $h = a\sin\theta$, and $b = a\cos\theta$

$$A = \frac{1}{2}(a\sin\theta)\,(2a\cos\theta)$$

$$A = \frac{1}{2}a^2\, 2\sin\theta\cos\theta$$

$$A = \frac{1}{2}a^2\sin 2\theta \qquad \text{Identity (18)}$$

The largest possible value for $\sin 2\theta$ is 1, and this will yield the largest value of A; $\sin 2\theta = 1$ when $2\theta = \frac{\pi}{2}$, that is, when $\theta = \frac{\pi}{4} = 45°$.

- ANSWER The triangle will have maximum area when the two equal angles (θ) are each 45°, and hence the third angle in the triangle is 90°. Thus the isosceles triangle of greatest area is a right triangle.

For easy reference, we summarize the identities presented in this section.

Double-Angle Identities	
(18)	$\sin 2\theta = 2\sin\theta\cos\theta$
(19)	$\cos 2\theta = \cos^2\theta - \sin^2\theta$
(20)	$\cos 2\theta = 1 - 2\sin^2\theta$
(21)	$\cos 2\theta = 2\cos^2\theta - 1$
(22)	$\tan 2\theta = \frac{2\tan\theta}{1 - \tan^2\theta}$

Half-Angle Identities

(23) $$\sin\frac{\theta}{2} = \pm\sqrt{\frac{1-\cos\theta}{2}}$$

(24) $$\cos\frac{\theta}{2} = \pm\sqrt{\frac{1+\cos\theta}{2}}$$

(25) $$\tan\frac{\theta}{2} = \pm\sqrt{\frac{1-\cos\theta}{1+\cos\theta}}$$

(26) $$\tan\frac{\theta}{2} = \frac{\sin\theta}{1+\cos\theta}$$

(27) $$\tan\frac{\theta}{2} = \frac{1-\cos\theta}{\sin\theta}$$

EXERCISE SET 8.3

In Exercises 1–18, given α, find $\sin 2\alpha$, $\cos 2\alpha$, and $\tan 2\alpha$ using the double-angle identities.

1. 60° **2.** 45° **3.** 120° **4.** 180°

5. 135° **6.** 150° **7.** −300° **8.** −225°

9. 585° **10.** 765° **11.** $\frac{5\pi}{6}$ **12.** $\frac{3\pi}{4}$

13. $-\frac{2\pi}{3}$ **14.** $-\frac{7\pi}{6}$ **15.** $-\frac{3\pi}{4}$ **16.** -3π

17. $\frac{22\pi}{8}$ **18.** $\frac{26\pi}{12}$

In Exercises 19–26, from the given information, determine (a) $\sin 2\alpha$, (b) $\cos 2\alpha$, (c) $\tan 2\alpha$.

19. $\cos\alpha = -\frac{4}{5}$, α in Quadrant II

20. $\cos\alpha = -\frac{4}{5}$, α in Quadrant IV

21. $\sin\alpha = -\frac{12}{13}$, α in Quadrant IV

22. $\tan\alpha = \frac{4}{3}$, α in Quadrant III

23. $\tan\alpha = \frac{9}{40}$, α in Quadrant I

24. $\sin\alpha = \frac{5}{13}$, α in Quadrant I

25. $\cot\alpha = 2$, α in Quadrant III

26. $\sec\alpha = -\frac{25}{24}$, α in Quadrant II

In Exercises 27–30, suppose that θ is an acute angle and $\sin\theta = 0.3846$. Compute the given trigonometric function value.

27. $\sin 2\theta$ **28.** $\cos 2\theta$ **29.** $\cot 2\theta$ **30.** $\tan 2\theta$

In Exercises 31–34, suppose that θ is a second quadrant angle and $\cos\theta = -0.6154$. Compute the given trigonometric function value.

31. $\cos\frac{\theta}{2}$ **32.** $\sin\frac{\theta}{2}$ **33.** $\tan\frac{\theta}{2}$ **34.** $\cot\frac{\theta}{2}$

In Exercises 35–54, given α, find $\sin\frac{\alpha}{2}$, $\cos\frac{\alpha}{2}$, and $\tan\frac{\alpha}{2}$ using the half-angle identities.

35. 60° **36.** 180° **37.** 90° **38.** 120°

39. 210° **40.** 315° **41.** −225° **42.** −495°

43. 675° **44.** 855° **45.** 4π **46.** $\frac{2\pi}{3}$

47. $\frac{4\pi}{3}$ **48.** $\frac{5\pi}{2}$ **49.** $\frac{\pi}{6}$ **50.** $\frac{5\pi}{4}$

51. $-\frac{11\pi}{6}$ **52.** $-\frac{3\pi}{4}$ **53.** $\frac{19\pi}{6}$ **54.** $\frac{16\pi}{3}$

For Exercises 55–62, use the information given in Exercises 19–26 to determine (a) $\sin \frac{\alpha}{2}$, (b) $\cos \frac{\alpha}{2}$, and (c) $\tan \frac{\alpha}{2}$.

Superset

63. Derive identity (22) for $\tan 2\alpha$.

64. Derive identity (25) for $\tan \frac{\alpha}{2}$.

65. Work out the details to show identity (27).

66. Express $\sin 3\alpha$ in terms of $\sin \alpha$ only.

67. Express $\cos 3\alpha$ in terms of $\cos \alpha$ only.

68. Express $\cos 4\alpha$ in terms of $\cos \alpha$ only.

69. Express $\sin 4\alpha$ in terms of $\sin \alpha$ and $\cos \alpha$.

70. Express $\tan 4\alpha$ in terms of $\tan \alpha$ only.

In Exercises 71–80, express the first trigonometric function as specified.

71. $\cos 12\alpha$ in terms of $\cos 6\alpha$

72. $\cos 12\alpha$ in terms of $\sin 6\alpha$

73. $\sin 8\alpha$ in terms of $\sin 4\alpha$ and $\cos 4\alpha$

74. $\cos 8\alpha$ in terms of $\sin 4\alpha$ and $\cos 4\alpha$

75. $\tan 6\alpha$ in terms of $\tan 2\alpha$

76. $\cot 6\alpha$ in terms of $\tan 2\alpha$

77. $\cot 6\alpha$ in terms of $\cot 12\alpha$ and $\csc 12\alpha$

78. $\tan 6\alpha$ in terms of $\tan 12\alpha$ and $\sec 12\alpha$

79. $\sin 8\alpha$ in terms of $\sin 4\alpha$

80. $\sin 2\alpha$ in terms of $\cot 4\alpha$ and $\csc 4\alpha$

In Exercises 81–92, simplify the expression.

81. $\dfrac{\sin 2\alpha}{2 \sin \alpha}$

82. $\dfrac{\sin 2\alpha}{2 \cos \alpha}$

83. $2 \sin \frac{\alpha}{2} \cos \frac{\alpha}{2}$

84. $2 \sin 2\alpha \cos 2\alpha$

85. $1 - 2 \sin^2 \frac{\alpha}{2}$

86. $2 \cos^2 \frac{\alpha}{2} - 1$

87. $(\sin \alpha + \cos \alpha)^2 - \sin 2\alpha$

88. $(\sin \alpha - \cos \alpha)^2 + \sin 2\alpha$

89. $\dfrac{\cos^2 \alpha}{1 + \sin \alpha}$

90. $\dfrac{\cos^3 \alpha - \sin^3 \alpha}{\cos \alpha - \sin \alpha}$

91. $\cos^4 \alpha - \sin^4 \alpha$

92. $\dfrac{\cos^4 \alpha - \sin^4 \alpha}{\sin^2 \alpha - \cos^2 \alpha}$

8.4 Identities Revisited

As we have already seen, the primary strategy used to verify a trigonometric identity is to transform the expression on one side of the identity so that it is identical to the expression on the other side. Frequently this means that you must recognize that a form present in the problem is a part of a known identity.

EXAMPLE 1

Simplify the expression $\cos 2x \cos x + \sin 2x \sin x$.

Solution Observe that $\cos 2x \cos x + \sin 2x \sin x$ is simply the right-hand side of identity (12): $\cos(\alpha - \beta) = \cos \alpha \cos \beta + \sin \alpha \sin \beta$, with α replaced by $2x$ and β replaced by x. Thus,

$$\cos 2x \cos x + \sin 2x \sin x = \cos(2x - x) = \cos x.$$

EXAMPLE 2

Simplify the expression $\dfrac{\sin 2x}{1 + \cos 2x}$.

Solution The given expression is the right-hand side of identity (26): $\tan \dfrac{\theta}{2} = \dfrac{\sin \theta}{1 + \cos \theta}$, with θ replaced by $2x$. Thus,

$$\frac{\sin 2x}{1 + \cos 2x} = \tan \frac{2x}{2} = \tan x.$$

EXAMPLE 3

Simplify the expression $1 - 2 \sin^2 3A$.

Solution This expression is simply the right-hand side of identity (20): $\cos 2\theta = 1 - 2 \sin^2 \theta$, with θ replaced by $3A$. Thus,

$$1 - 2 \sin^2 3A = \cos 2(3A) = \cos 6A.$$

In Section 1 of this chapter, we introduced techniques for verifying identities, but had only the Basic Identities (1)–(11) at our disposal. We now have, in addition, identities (12)–(27) to assist us in verifying new identities. We now consider several examples that draw upon the knowledge of all the identities established thus far.

EXAMPLE 4

Verify the identity: $\dfrac{\sin 2\theta}{1 + \cos 2\theta} = \tan \theta$.

Solution

$$\frac{\sin 2\theta}{1 + \cos 2\theta} = \frac{2 \sin \theta \cos \theta}{1 + (2 \cos^2 \theta - 1)}$$

(18) is used in the numerator. (21) is used in the denominator to produce an expression that will have only one term.

$$= \frac{2 \sin \theta \cos \theta}{2 \cos^2 \theta}$$

Denominator is simplified.

$$= \frac{\sin \theta}{\cos \theta} = \tan \theta$$

Common factors are canceled.

EXAMPLE 5

Verify the identity: $\dfrac{\tan\theta + \sin\theta}{2\tan\theta} = \cos^2\dfrac{\theta}{2}$.

Solution

$$\frac{\tan\theta + \sin\theta}{2\tan\theta} = \frac{\tan\theta}{2\tan\theta} + \frac{\sin\theta}{2\tan\theta}$$

We work with the left-hand side of the identity since it is more complicated and offers the best chance for simplification.

$$= \frac{\tan\theta}{2\tan\theta} + \frac{\sin\theta}{2}\cdot\frac{\cos\theta}{\sin\theta}$$

Since $\tan\theta = \dfrac{\sin\theta}{\cos\theta}$, we divide by $\tan\theta$ by multiplying by its reciprocal $\dfrac{\cos\theta}{\sin\theta}$.

$$= \frac{1}{2} + \frac{\cos\theta}{2} = \frac{1 + \cos\theta}{2}$$

$$= \cos^2\frac{\theta}{2}$$

By (24)

EXAMPLE 6

Verify the identity: $\sin(\alpha + \beta)\sin(\alpha - \beta) = \sin^2\alpha - \sin^2\beta$.

Solution

$\sin(\alpha + \beta)\sin(\alpha - \beta)$

$= (\sin\alpha\cos\beta + \cos\alpha\sin\beta)(\sin\alpha\cos\beta - \cos\alpha\sin\beta)$ By (14) and (15)

$= (\sin\alpha\cos\beta)^2 - (\cos\alpha\sin\beta)^2$ $(A + B)(A - B) = A^2 - B^2$.

$= \sin^2\alpha\cos^2\beta - \cos^2\alpha\sin^2\beta$

$= \sin^2\alpha(1 - \sin^2\beta) - (1 - \sin^2\alpha)(\sin^2\beta)$

$= \sin^2\alpha - \sin^2\alpha\sin^2\beta - \sin^2\beta + \sin^2\alpha\sin^2\beta$

$= \sin^2\alpha - \sin^2\beta$

We want an expression that involves only $\sin\alpha$ and $\sin\beta$.

EXAMPLE 7

Verify the identity: $\dfrac{\sec\beta}{1 + \cos\beta} = \dfrac{\tan\beta - \sin\beta}{\sin^3\beta}$.

Solution

$$\frac{\sec\beta}{1 + \cos\beta} = \frac{\sec\beta}{1 + \cos\beta}\cdot\frac{1 - \cos\beta}{1 - \cos\beta}$$

We will work with the left-hand side of the identity.

$$= \frac{\sec\beta(1 - \cos\beta)}{1 - \cos^2\beta}$$

$$= \frac{\sec \beta - \sec \beta(\cos \beta)}{\sin^2 \beta}$$ By (9)

$$= \frac{\sec \beta - 1}{\sin^2 \beta}$$ $\sec \beta(\cos \beta) = \sec \beta\left(\frac{1}{\sec \beta}\right) = 1.$

$$= \frac{\sec \beta - 1}{\sin^2 \beta} \cdot \frac{\sin \beta}{\sin \beta}$$ We know that we want $\sin^3 \beta$ in the denominator, so we multiply by $\frac{\sin \beta}{\sin \beta}$.

$$= \frac{\sec \beta \sin \beta - \sin \beta}{\sin^3 \beta}$$

$$= \frac{\tan \beta - \sin \beta}{\sin^3 \beta}$$ $\sec \beta \sin \beta = \frac{1}{\cos \beta} \sin \beta = \tan \beta.$

We now consider a group of identities that allow us to restate a product of two trigonometric function values as a sum, and a sum of two trigonometric function values as a product. These new identities are derived from the sum and difference identities for the sine and cosine (identities (12)–(15)). For example, based on identities (14) and (15), we have

$$\begin{aligned}\sin(\alpha + \beta) + \sin(\alpha - \beta) &= (\sin \alpha \cos \beta + \cos \alpha \sin \beta) \\ &\quad + (\sin \alpha \cos \beta - \cos \alpha \sin \beta) \\ &= 2 \sin \alpha \cos \beta.\end{aligned}$$

Thus, $\sin(\alpha + \beta) + \sin(\alpha - \beta) = 2 \sin \alpha \cos \beta$ so that

(28) $$\sin \alpha \cos \beta = \frac{1}{2}[\sin(\alpha + \beta) + \sin(\alpha - \beta)].$$

EXAMPLE 8

Determine $\sin 15° \cos 75°$ by using identity (28).

Solution

$$\begin{aligned}\sin 15° \cos 75° &= \frac{1}{2}[\sin(15° + 75°) + \sin(15° - 75°)] \\ &= \frac{1}{2}[\sin 90° + \sin(-60°)] \\ &= \frac{1}{2}[\sin 90° - \sin 60°] \qquad \text{Recall } \sin(-\theta) = -\sin \theta. \\ &= \frac{1}{2}\left(1 - \frac{\sqrt{3}}{2}\right) = \frac{2 - \sqrt{3}}{4}\end{aligned}$$

In identity (28), let us replace $\alpha + \beta$ with x and $\alpha - \beta$ with y. Since

$$x + y = (\alpha + \beta) + (\alpha - \beta) = 2\alpha, \quad \text{and}$$
$$x - y = (\alpha + \beta) - (\alpha - \beta) = 2\beta,$$

we have $\alpha = \dfrac{x + y}{2}$ and $\beta = \dfrac{x - y}{2}$. Thus identity (28) can be rewritten as

$$\sin\left(\frac{x + y}{2}\right) \cos\left(\frac{x - y}{2}\right) = \frac{1}{2}(\sin x + \sin y),$$

or equivalently as

(31) $$\sin x + \sin y = 2\left[\sin\left(\frac{x + y}{2}\right) \cos\left(\frac{x - y}{2}\right)\right].$$

Notice that identity (31) expresses the sum of two trigonometric function values as a product of trigonometric function values.

Employing arguments similar to those used to derive identities (28) and (31), we can develop five other identities. These identities are called **conversion identities** because they allow us to convert products of trigonometric function values into sums, and vice versa.

Conversion Identities

(28) $$\sin \alpha \cos \beta = \frac{1}{2}[\sin(\alpha + \beta) + \sin(\alpha - \beta)]$$

(29) $$\sin \alpha \sin \beta = \frac{1}{2}[\cos(\alpha - \beta) - \cos(\alpha + \beta)]$$

(30) $$\cos \alpha \cos \beta = \frac{1}{2}[\cos(\alpha + \beta) + \cos(\alpha - \beta)]$$

(31) $$\sin x + \sin y = 2 \sin\left(\frac{x + y}{2}\right) \cos\left(\frac{x - y}{2}\right)$$

(32) $$\sin x - \sin y = 2 \cos\left(\frac{x + y}{2}\right) \sin\left(\frac{x - y}{2}\right)$$

(33) $$\cos x + \cos y = 2 \cos\left(\frac{x + y}{2}\right) \cos\left(\frac{x - y}{2}\right)$$

(34) $$\cos x - \cos y = -2 \sin\left(\frac{x + y}{2}\right) \sin\left(\frac{x - y}{2}\right)$$

EXAMPLE 9

Express (a) $\cos 8\alpha + \cos 6\alpha$ as a product (b) $\cos 2\theta \sin \theta$ as a sum or difference.

Solution

(a) $$\cos 8\alpha + \cos 6\alpha = 2\cos\left(\frac{8\alpha + 6\alpha}{2}\right)\cos\left(\frac{8\alpha - 6\alpha}{2}\right)$$ Identity (33) is used with $x = 8\alpha$ and $y = 6\alpha$.

$$= 2\cos 7\alpha \cos \alpha$$

(b) $$\cos 2\theta \sin \theta = \frac{1}{2}[\sin(3\theta) + \sin(-\theta)]$$ Identity (28) is used with $\alpha = \theta$ and $\beta = 2\theta$.

$$= \frac{1}{2}[\sin 3\theta - \sin \theta]$$

EXAMPLE 10

Verify the identity: $\dfrac{\sin 4x - \sin 3x}{\cos 4x + \cos 3x} = \dfrac{1 - \cos x}{\sin x}$.

Solution

$$\frac{\sin 4x - \sin 3x}{\cos 4x + \cos 3x} = \frac{2\cos\left(\frac{4x + 3x}{2}\right)\sin\left(\frac{4x - 3x}{2}\right)}{2\cos\left(\frac{4x + 3x}{2}\right)\cos\left(\frac{4x - 3x}{2}\right)}$$ By (32) / By (33)

$$= \frac{\sin\frac{x}{2}}{\cos\frac{x}{2}} = \tan\frac{x}{2} = \frac{1 - \cos x}{\sin x}$$ By (27)

EXERCISE SET 8.4

In Exercises 1–8, simplify the expression.

1. $\cos 4\alpha \cos \alpha + \sin \alpha \sin 4\alpha$

2. $\cos 4\alpha \cos \alpha - \sin \alpha \sin 4\alpha$

3. $\sin 4\alpha \cos \alpha - \sin \alpha \cos 4\alpha$

4. $\sin 4\alpha \cos \alpha + \sin \alpha \cos 4\alpha$

5. $\dfrac{2\tan 2\varphi}{1 - \tan^2 2\varphi}$

6. $\dfrac{1 - \cos 2\alpha}{\sin 2\alpha}$

7. $\dfrac{\sec \alpha - 1}{\sin \alpha \sec \alpha}$

8. $\dfrac{\tan \beta}{\sec^2 \beta - 2}$

In Exercises 9–20, verify the identity.

9. $\tan x = \dfrac{2\tan\frac{1}{2}x}{1 - \tan^2\frac{1}{2}x}$

10. $\cot x = \dfrac{\cot^2\frac{1}{2}x - 1}{2\cot\frac{1}{2}x}$

11. $\dfrac{\cos(x + y)}{\cos x \sin y} = \cot y - \tan x$

12. $\dfrac{\sin(x + y)}{\sin(x - y)} = \dfrac{\tan x + \tan y}{\tan x - \tan y}$

13. $\dfrac{1 - \cos 2\alpha}{1 + \cos 2\alpha} = \tan^2 \alpha$

14. $\dfrac{\sin(\alpha + \beta)}{\sin \alpha \sin \beta} = \cot \alpha + \cot \beta$

15. $\dfrac{\sin \alpha + \sin \beta}{\cos \alpha - \cos \beta} = -\cot \dfrac{1}{2}(\alpha - \beta)$

16. $\dfrac{\cos(\alpha + \beta)}{\cos(\alpha - \beta)} = -\dfrac{\tan \beta - \cot \alpha}{\tan \beta + \cot \alpha}$

17. $\csc x - \sin x = \sin x \cot^2 x$

18. $\cos x \csc x \tan x = 1$

19. $\dfrac{1 + \cos \varphi}{\sin \varphi} + \dfrac{\sin \varphi}{1 + \cos \varphi} = 2 \csc \varphi$

20. $\dfrac{\sin \varphi}{1 - \sin \varphi} + \dfrac{1 + \sin \varphi}{\sin \varphi} = \dfrac{\csc \varphi}{1 - \sin \varphi}$

In Exercises 21–26, determine whether the given equation is an identity.

21. $\tan s = \dfrac{\tan s + 1}{\cot s + 1}$

22. $\dfrac{\sin s}{1 + \cos s} = \dfrac{1 - \cos s}{\sin s}$

23. $\tan 3\alpha = \dfrac{1 + \cos 2\alpha}{\sin 2\alpha}$

24. $\dfrac{\sin 3\alpha}{\sin \alpha} + \dfrac{\cos 3\alpha}{\cos \alpha} = 1$

25. $\dfrac{1 + \sin \theta - \cos \theta}{1 + \sin \theta + \cos \theta} = \tan \dfrac{1}{2}\theta$

26. $\dfrac{1 + \sin \theta}{1 - \sin \theta} - \dfrac{1 - \sin \theta}{1 + \sin \theta} = \dfrac{4 \tan \theta}{\cos \theta}$

In Exercises 27–34, evaluate by using the conversion identities.

27. $\sin 75° + \sin 15°$

28. $\cos 75° - \cos 15°$

29. $\cos 105° + \cos 15°$

30. $\sin 105° - \sin 15°$

31. $\sin \dfrac{\pi}{12} - \sin \dfrac{5\pi}{12}$

32. $\cos \dfrac{11\pi}{12} + \cos \dfrac{5\pi}{12}$

33. $\cos \dfrac{7\pi}{12} - \cos \dfrac{\pi}{12}$

34. $\sin \dfrac{13\pi}{12} + \sin \dfrac{7\pi}{12}$

For Exercises 35–38, derive the identities (29), (30), (32), and (33), respectively.

In Exercises 39–44, rewrite the trigonometric expression as a product.

39. $\sin 3x + \sin x$

40. $\sin 3x - \sin x$

41. $\sin 6x - \sin 2x$

42. $\sin 6x + \sin 2x$

43. $\cos 6x - \cos 2x$

44. $\cos 6x + \cos 2x$

In Exercises 45–48, rewrite the expression as a sum or difference involving sines and cosines.

45. $\sin 2x \cos 3x$

46. $\sin 3x \cos 2x$

47. $\sin 3x \sin 2x$

48. $\cos 3x \cos 2x$

Superset

49. Find an expression for $\tan 2x + \sec 2x$ in terms only of $\sin x$ and $\cos x$.

In Exercises 50–52, establish the result given that $0 < \alpha < \dfrac{\pi}{2}$ and $\beta = \dfrac{\pi}{2} - \alpha$.

50. $\sin \alpha \cos \beta + \cos \alpha \sin \beta = 1$

51. $\cos \alpha \cos \beta - \sin \alpha \sin \beta = 0$

52. $\sin \alpha + \sin \beta = \cos \alpha + \cos \beta$

In Exercises 53–55, verify the statement.

53. $2 \arctan\left(\dfrac{1}{3}\right) + \arctan\left(\dfrac{1}{7}\right) = \dfrac{\pi}{4}$

54. $\arctan\left(\dfrac{1}{2}\right) + \arctan\left(\dfrac{1}{3}\right) = \dfrac{\pi}{4}$

55. $\tan(\arctan(a) - \arctan(b)) = \dfrac{a - b}{1 + ab}$

In Exercises 56–60, determine nonzero numbers M and N so that the statement is an identity.

56. $M + N \cos^2 x = \sin^4 x - \cos^4 x$

57. $\cos^4 x - \sin^4 x = M - N \sin^2 x$

58. $M + N \cos x = \dfrac{\sin^2 x}{1 - \cos x}$

59. $\tan x \csc x \cos x + \cot x \sec x \sin x = M$

60. $(\tan x + \cot x)^2 \sin^2 x = M + N \tan^2 x$

8.5 Trigonometric Equations

Much of algebra is concerned with techniques for solving equations like

$$2x + 1 = 0 \quad \text{or} \quad 2x^2 - x = 1,$$

where x represents a real number. We now wish to consider equations that involve trigonometric functions, such as

$$2 \cos \alpha + 1 = 0 \quad \text{or} \quad 2 \cos^2 \alpha - \cos \alpha = 1.$$

To solve such equations, we first solve for $\cos \alpha$ using algebraic techniques, and then we use our knowledge of trigonometry to solve for α. For example, to solve the trigonometric equation

$$2 \cos \alpha + 1 = 0,$$

we begin by solving for $\cos \alpha$. This requires the same algebraic steps as solving the equation $2x + 1 = 0$ for x.

EXAMPLE 1

Solve $2 \cos \alpha + 1 = 0$ for α. Express the solution in radians.

Solution

$$2 \cos \alpha + 1 = 0 \qquad \text{Begin by solving for } \cos \alpha.$$

$$2 \cos \alpha = -1$$

$$\cos \alpha = -\frac{1}{2} \qquad \text{Having solved for } \cos \alpha \text{, we now solve for } \alpha.$$

Recall that $\cos \frac{\pi}{3} = \frac{1}{2}$, and that cosine is negative for second and third quadrant angles. Thus, α must be a second or third quadrant angle whose reference angle is $\frac{\pi}{3}$, namely $\frac{2\pi}{3}$ or $\frac{4\pi}{3}$. Since adding any multiple of 2π to either of these values produces an angle whose cosine is also $-\frac{1}{2}$, we conclude that

$$\alpha = \frac{2\pi}{3} + 2n\pi \quad \text{or} \quad \alpha = \frac{4\pi}{3} + 2n\pi \quad \text{for any integer } n.$$

When solving trigonometric equations, we often restrict the solutions to values in the interval $[0, 2\pi)$ or angles in the interval $[0°, 360°)$.

EXAMPLE 2

Determine all solutions of $2\sin^2\alpha - \sin\alpha = 1$ in the interval $[0, 2\pi)$.

Solution

$$2\sin^2\alpha - \sin\alpha = 1$$ Begin by solving for $\sin\alpha$.

$$2\sin^2\alpha - \sin\alpha - 1 = 0$$ Now factor the left side; treat $\sin\alpha$ as the variable.

$$(2\sin\alpha + 1)(\sin\alpha - 1) = 0$$

$2\sin\alpha + 1 = 0$	$\sin\alpha - 1 = 0$
$\sin\alpha = -\frac{1}{2}$	$\sin\alpha = 1$

Use the Principle of Zero Products.

To solve $\sin\alpha = -\frac{1}{2}$, recall that $\sin\frac{\pi}{6} = \frac{1}{2}$ and sine is negative in the third and fourth quadrants. Thus, α is a third or fourth quadrant angle whose reference angle is $\frac{\pi}{6}$. So $\alpha = \frac{7\pi}{6}$ or $\frac{11\pi}{6}$. To solve $\sin\alpha = 1$, recall that the only value in the interval $[0, 2\pi)$ whose sine is 1 is $\alpha = \frac{\pi}{2}$.

The solutions of the given equation in the interval $[0, 2\pi)$ are $\frac{\pi}{2}$, $\frac{7\pi}{6}$, and $\frac{11\pi}{6}$.

If the domain in the Example 2 had not been restricted to the interval $[0, 2\pi)$, the solutions would have been $\frac{7\pi}{6} + 2n\pi$, $\frac{11\pi}{6} + 2n\pi$, and $\frac{\pi}{2} + 2n\pi$, for any integer n.

When solving trigonometric equations, it is important to remember the domains and ranges of the trigonometric functions. Example 3 illustrates how knowledge of the range can be crucial.

EXAMPLE 3

Determine the solutions of $\cos^2\varphi = 2 - \cos\varphi$ in the interval $[0°, 360°)$.

Solution

$$\cos^2\varphi = 2 - \cos\varphi$$ First solve for $\cos\varphi$.

$$\cos^2\varphi + \cos\varphi - 2 = 0$$

$$(\cos\varphi + 2)(\cos\varphi - 1) = 0$$ Now use the Principle of Zero Products.

$\cos\varphi + 2 = 0$	$\cos\varphi - 1 = 0$
$\cos\varphi = -2$	$\cos\varphi = 1$

There is no solution to the equation $\cos\varphi = -2$ because the range of the cosine function is the interval $[-1, 1]$, and therefore there is no angle having a cosine of -2. To solve $\cos\varphi = 1$, recall that the only angle in $[0°, 360°)$ whose cosine is 1 is $\varphi = 0°$. The only solution of the given equation in the interval $[0°, 360°)$ is $0°$.

In the next two examples, we solve equations involving two different trigonometric functions. Given such an equation, you should try to use algebraic techniques and trigonometric identities to rewrite the equation in a form involving only one trigonometric function.

EXAMPLE 4

Determine all solutions of the equation $\sin x + 1 = \cos x$ in the interval $[0, 2\pi)$.

Solution

$$\sin x + 1 = \cos x$$

$$(\sin x + 1)^2 = (\cos x)^2$$

$$\sin^2 x + 2\sin x + 1 = \cos^2 x$$

Square both sides of the equation to produce $\cos^2 x$ on the right, which can be rewritten in terms of $\sin x$.

$$\sin^2 x + 2\sin x + 1 = 1 - \sin^2 x$$

$\cos^2 x$ is replaced with $1 - \sin^2 x$.

$$\sin^2 x + 2\sin x = -\sin^2 x$$

$$2\sin^2 x + 2\sin x = 0$$

$$2\sin x(\sin x + 1) = 0$$

$2\sin x = 0$	$\sin x + 1 = 0$
$\sin x = 0$	$\sin x = -1$
$x = 0$ or π	$x = \dfrac{3\pi}{2}$

Because we squared both sides of the original equation, we must check for extraneous solutions.

Must check:

$x = 0$	$x = \pi$	$x = \dfrac{3\pi}{2}$
$\sin 0 + 1 = \cos 0$	$\sin\pi + 1 = \cos\pi$	$\sin\dfrac{3\pi}{2} + 1 = \cos\dfrac{3\pi}{2}$
$0 + 1 = 1$ True	$0 + 1 = -1$ False	$-1 + 1 = 0$ True

The solutions of the given equation in the interval $[0, 2\pi)$ are 0 and $\dfrac{3\pi}{2}$.

EXAMPLE 5

Solve the equation $\cos 2x + \sin x = 0$ for x. Express solutions in radians.

Solution

$$\begin{aligned} \cos 2x + \sin x &= 0 && \text{Use identity (20) to rewrite } \cos 2x \text{ in terms of } \sin x. \\ 1 - 2\sin^2 x + \sin x &= 0 \\ 2\sin^2 x - \sin x - 1 &= 0 && \text{The equation was multiplied by } -1. \end{aligned}$$

This equation was solved over the interval $[0, 2\pi)$ in Example 2. The solutions were found to be $\frac{\pi}{2}$, $\frac{7\pi}{6}$, and $\frac{11\pi}{6}$. Thus the solutions of the given equation are

$$\frac{\pi}{2} + 2n\pi, \quad \frac{7\pi}{6} + 2n\pi, \quad \text{and} \quad \frac{11\pi}{6} + 2n\pi \quad \text{for any integer } n.$$

EXAMPLE 6

Solve the equation $\sin\theta + 2\sin\frac{\theta}{2} = \cos\frac{\theta}{2} + 1$ over the interval $[0°, 360°)$.

Solution We begin by making the substitution $\alpha = \frac{\theta}{2}$ in order to simplify the equation.

$$\begin{aligned} \sin 2\alpha + 2\sin\alpha &= \cos\alpha + 1 && \text{Since } \alpha = \tfrac{\theta}{2}, \theta = 2\alpha. \\ 2\sin\alpha\cos\alpha + 2\sin\alpha &= \cos\alpha + 1 && \text{By identity (18)} \\ 2\sin\alpha\cos\alpha + 2\sin\alpha - \cos\alpha - 1 &= 0 \\ 2\sin\alpha(\cos\alpha + 1) - 1(\cos\alpha + 1) &= 0 && \text{Factor.} \\ (2\sin\alpha - 1)(\cos\alpha + 1) &= 0 && \text{Factor again, and use the Principle of Zero Products.} \end{aligned}$$

$2\sin\alpha - 1 = 0$	$\cos\alpha + 1 = 0$
$\sin\alpha = \frac{1}{2}$	$\cos\alpha = -1$
$\alpha = 30°$ or $150°$	$\alpha = 180°$

By virtue of our substitution $\theta = 2\alpha$, and so the values of θ corresponding to $\alpha = 30°$, $\alpha = 150°$, and $\alpha = 180°$, are $\theta = 60°$, $\theta = 300°$, and $\theta = 360°$. Since the last value is not in the interval $[0°, 360°)$, we conclude that the solutions of the original equation in the given interval are

$$\theta = 60° \quad \text{and} \quad \theta = 300°.$$

EXAMPLE 7

Solve the equation $\sin 3x + \sin x + \cos x = 0$ for x. Express solutions in degrees.

Solution

$$\sin 3x + \sin x + \cos x = 0$$

$$\sin(2x + x) + \sin(2x - x) + \cos x = 0$$

$3x$ and x are rewritten to allow the use of identity (28).

$$2 \sin 2x \cos x + \cos x = 0$$

By (28) with $\alpha = 2x$ and $\beta = x$: $2 \sin \alpha \cos \beta = \sin(\alpha + \beta) + \sin(\alpha - \beta)$.

$$\cos x\,(2 \sin 2x + 1) = 0$$

$\cos x = 0$	$2 \sin 2x + 1 = 0$
$x = 90^\circ$ or 270°	$\sin 2x = -\frac{1}{2}$
	$2x = 210^\circ$ or 330°
	$x = 105^\circ$ or 165°

The solutions of the original equation are all angles of the form

$$90^\circ + n \cdot 360^\circ, \quad 105^\circ + n \cdot 360^\circ, \quad 165^\circ + n \cdot 360^\circ, \quad \text{and} \quad 270^\circ + n \cdot 360^\circ$$

for any integer n.

An equation such as $2 - 7 \cos x = 0$ has a solution which is not a special angle $\left(0, \frac{\pi}{6}, \frac{\pi}{4}, \text{etc}\right)$. We must use a calculator or a table of values to determine such a solution.

EXAMPLE 8

Solve the equation $2 - 7 \cos x = 0$ over the interval $[0^\circ, 360^\circ)$.

Solution

$$2 - 7 \cos x = 0$$

$$\cos x = \frac{2}{7} \approx 0.2857$$

Using a calculator or Table 4 and rounding to the nearest tenth of a degree, we find that $x \approx 73.4^\circ$. Since we are given that $\cos x$ is positive, x could be a first or fourth quadrant angle. The fourth quadrant angle having 73.4° as a reference angle is 286.6°.

The solutions rounded to the nearest tenth are 73.4° and 286.6°.

EXAMPLE 9

Solve the equation $\sin^2 \theta + 2 \sin \theta - 1 = 0$ over the interval $[0°, 360°)$.

Solution

$$\sin^2 \theta + 2 \sin \theta - 1 = 0$$

$$x^2 + 2x - 1 = 0$$

Replace $\sin \theta$ with x, and solve the quadratic equation for x.

$$x = \frac{-2 \pm \sqrt{2^2 - 4(1)(-1)}}{2(1)}$$

Use the quadratic formula with $a = 1, b = 2$, and $c = -1$.

$$x = \frac{-2 \pm \sqrt{8}}{2} = \frac{-2 \pm 2\sqrt{2}}{2} = -1 \pm \sqrt{2}$$

Using $\sqrt{2} \approx 1.41$, and recalling that x represents $\sin \theta$, we have

$$\sin \theta \approx -1 + 1.41 = 0.41 \quad \text{or} \quad \sin \theta \approx -1 - 1.41 = -2.41.$$

Since the range of the sine function is the interval $[-1, 1]$, there is no θ such that $\sin \theta = -2.41$. Thus, our only solutions result from the equation $\sin \theta = 0.41$. Using a calculator or Table 4 and rounding to the nearest tenth, we obtain $\theta \approx 24.2°$. Since 0.41 is positive, θ could also be the second quadrant angle with 24.2° as its reference angle, namely 155.8°. Thus, the solutions to the nearest tenth are

$$\theta = 24.2° \quad \text{and} \quad \theta = 155.8°.$$

The following examples illustrate techniques for solving equations that involve inverse trigonometric functions.

EXAMPLE 10

Solve $\sin^{-1}\left(\frac{5}{13}\right) + \cos^{-1}\left(\frac{3}{5}\right) = \sin^{-1} x$ for x.

Solution By the definition of $\sin^{-1}x$, we know that $\sin(\sin^{-1}x) = x$. Taking the sine of both sides of the given equation we get

$$\sin\left(\sin^{-1}\left(\frac{5}{13}\right) + \cos^{-1}\left(\frac{3}{5}\right)\right) = x.$$

Let $\alpha = \sin^{-1}\left(\frac{5}{13}\right)$, and $\beta = \cos^{-1}\left(\frac{3}{5}\right)$. Draw α and β as acute angles in two different right triangles.

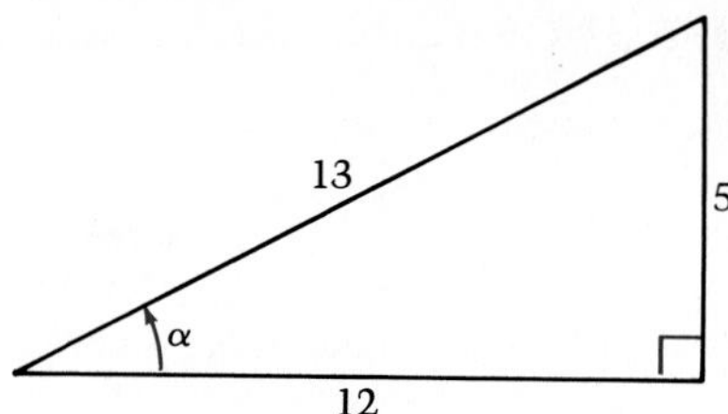

Since $\alpha = \sin^{-1}\left(\frac{5}{13}\right)$, $\sin\alpha = \sin\left(\sin^{-1}\left(\frac{5}{13}\right)\right) = \frac{5}{13}$. We label the sides of the triangle accordingly.

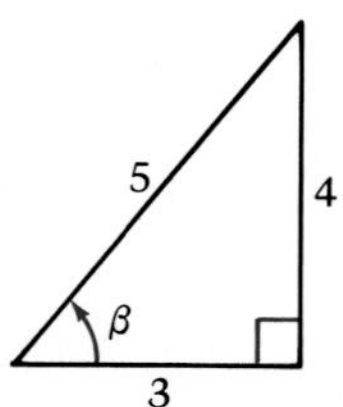

Since $\beta = \cos^{-1}\left(\frac{3}{5}\right)$, $\cos\beta = \cos\left(\cos^{-1}\left(\frac{3}{5}\right)\right) = \frac{3}{5}$. We label the triangle accordingly.

Figure 8.3

$$x = \sin\left(\sin^{-1}\left(\frac{5}{13}\right) + \cos^{-1}\left(\frac{3}{5}\right)\right)$$

$$= \sin(\alpha + \beta) = \sin\alpha\cos\beta + \cos\alpha\sin\beta \qquad \text{By identity (14).}$$

$$= \frac{5}{13}\cdot\frac{3}{5} + \frac{12}{13}\cdot\frac{4}{5} = \frac{15}{65} + \frac{48}{65} = \frac{63}{65} \qquad \text{From the two triangles.}$$

EXAMPLE 11

Find the angle x, in degrees, such that

$$\tan^{-1}\left(\frac{1}{2}\right) + \tan^{-1}\left(\frac{1}{3}\right) = x.$$

Solution Let $\alpha = \tan^{-1}\left(\frac{1}{2}\right)$ and $\beta = \tan^{-1}\left(\frac{1}{3}\right)$; the equation becomes $\alpha + \beta = x$. Then, taking the tangent of both sides of this equation, we have

$$\tan x = \tan(\alpha + \beta) = \frac{\tan\alpha + \tan\beta}{1 - \tan\alpha\tan\beta} \qquad \text{Identity (16)}$$

$$\text{Therefore, } \tan x = \frac{\frac{1}{2} + \frac{1}{3}}{1 - \left(\frac{1}{2}\right)\left(\frac{1}{3}\right)} = \frac{\frac{5}{6}}{\frac{5}{6}} = 1 \qquad \begin{array}{l}\tan\alpha = \tan\left(\tan^{-1}\left(\frac{1}{2}\right)\right) = \frac{1}{2}.\\ \tan\beta = \tan\left(\tan^{-1}\left(\frac{1}{3}\right)\right) = \frac{1}{3}.\end{array}$$

The smallest positive angle x having a tangent function value of 1 is 45°.

EXERCISE SET 8.5

In Exercises 1–10, solve the equation for x. Express your answer in radians.

1. $2 \sin x + \sqrt{3} = 0$
2. $3 \cot x - \sqrt{3} = 0$
3. $2 \sin^2 x - \sin x = 0$
4. $2 \cos^2 x - \cos x - 1 = 0$
5. $4 \cos^2 x - 3 = 0$
6. $\tan^2 x - 1 = 0$
7. $2 \cos^2 x + 5 \cos x - 3 = 0$
8. $\sin x \cos x - \cos x = 0$
9. $\sin x \cos x - \sin x + \cos x - 1 = 0$
10. $2 \sin^2 x - \sqrt{2} \sin x - 4 \sin x + 2\sqrt{2} = 0$

In Exercises 11–18, determine all the solutions of the equation in the interval $[0, 2\pi)$.

11. $2 \sin^2 x + \sin x - 1 = 0$
12. $2 \cos^2 x - \cos x - 1 = 0$
13. $2 \sin^2 x + 3 \sin x - 2 = 0$
14. $\sqrt{2} \cos^2 x + (1 - \sqrt{2})\cos x - 1 = 0$
15. $2 \cos^2 x + 3 \sin x - 2 = 0$
16. $3 \cot^2 x - 1 = 0$
17. $2 \sin^2 3\theta - 4 \sin 3\theta - 6 = 0$
18. $2 \sin^2 \frac{1}{2}\theta + 3 \sin \frac{1}{2}\theta - 2 = 0$

In Exercises 19–32, solve the equation for x in the indicated domain.

19. $\sin 2x = \cos 2x$; x in $[0, 2\pi)$
20. $\cos x \tan x = 0$; x in $[0, 360°)$
21. $3 \cos^2 x = \sin^2 x$; x in $[-90°, 90°]$
22. $\sin x \cos x = \cos x$; x in $[0, 180°]$
23. $\dfrac{\sqrt{2} \sin^2 x}{\cos x} - \tan x = 0$; x in $[0, 2\pi)$
24. $\cos^2 x - \sin x = 1$; x in $[0, \pi]$
25. $2 \sin x - \tan x = 0$; x in $[0°, 180°]$
26. $3 \sin^2 x - \cos x = -1$; x in $[0, \pi]$
27. $\cos 5x + \cos x = 0$; x in $[0°, 360°)$
28. $\cos \frac{7}{2}x + \cos \frac{1}{2}x = 0$; x in $[0, \pi)$
29. $\sin 2x + \cos x = 0$; x in $[0, 2\pi)$
30. $\cos 2x + \cos x + 1 = 0$; x in $[0, 2\pi)$
31. $\cos 2x + \sin x = 1$; x in $[0, 2\pi)$
32. $\cos 2x - 1 - \tan x = 0$; x in $[0, 2\pi)$

In Exercises 33–54, solve the equation for x in $[0°, 360°)$ using Table 4 or a calculator. Express your answers to the nearest tenth of a degree.

33. $4 \sin^2 x - 3 \sin x = 0$
34. $5 \cos^2 x - 2 \cos x = 0$
35. $3 \sin^2 x + 2 \sin x - 1 = 0$
36. $3 \cos^2 x - \cos x - 2 = 0$
37. $6 \sin^2 x - \sin x - 2 = 0$
38. $2 \sin^2 x + \frac{1}{2} \sin x - 2 = 0$
39. $(4 \sin x + 1)^2 - 3 = 0$
40. $(3 \cos x - 5)^2 - 8 = 0$
41. $2 \tan^2 x + 2 \tan x + 3 = 0$
42. $3 \tan^2 x + 4 \tan x - 4 = 0$
43. $3 \sin^2 2x - 2 = 0$
44. $(2 \cos^2 2x + 3)^2 = 1$
45. $12 \sin^2 x - \sin x - 1 = 0$
46. $15 \cos^2 x + \cos x - 6 = 0$
47. $\sin x + 3 \cos x = 0$
48. $4 \cos^2 x + 2 \sin x = 3$
49. $2 \cos 2x - 2 \sin^2 x = 1$
50. $\tan^2 2x + 2 \tan 2x = 1$
51. $\sec^2 x + 2 \tan x - 2 = 0$
52. $3 \tan^2 x + 2 \sec x = 5$
53. $\cos^2 x - 0.97 \cos x + 0.18 = 0$
54. $\sin^2 x + 0.24 \sin x - 0.1081 = 0$

In Exercises 55–66, solve the equation for x.

55. $\arcsin\left(-\frac{1}{2}\right) = x$
56. $\arccos\left(-\frac{1}{2}\right) = x$
57. $\tan^{-1}\sqrt{3} = x$
58. $\frac{1}{2}\pi - \tan^{-1}(-1) = x$
59. $\sin\left(2 \arcsin \frac{3}{5}\right) = x$
60. $\sin\left(2 \arccos \frac{3}{5}\right) = x$
61. $\cos^{-1}\frac{3}{5} + \sin^{-1}\frac{4}{5} = \cos^{-1} x$
62. $\cos^{-1}\frac{3}{5} - \sin^{-1}\frac{4}{5} = \cos^{-1} x$

63. $\arcsin\frac{3}{5} - \arccos\frac{5}{13} = \arcsin x$

64. $\arcsin\frac{3}{5} + \arccos\frac{12}{13} = \arccos x$

65. $\tan^{-1}\frac{3}{4} + \tan^{-1}\frac{1}{7} = x$

66. $\arctan\left(-\frac{1}{3}\right) - \arctan\left(\frac{2}{3}\right) = \arctan x$

Superset

67. Solve for x: $\arctan x + \arctan\frac{1}{3} = \frac{\pi}{4}$.

In Exercises 68–69, solve for x in $[0, \pi)$. Express your answer in radians, rounded to the nearest hundredth.

68. $\tan x + \sqrt{\tan x + 5} = 7$

69. $\sqrt{2 \cot x - 1} + \sqrt{\cot x - 4} = 4$

In Exercises 70–74, solve for x in $[0, 2\pi)$.

70. $\sin 4x = 4 \sin x$

71. $\sin x + \cos x = 1$

72. $\sqrt{3} \sin x - \cos x = 1$

73. $4 = \sqrt{3} (\tan x + \cot x)$

74. $2 + \csc x = 2 \cot x + \sec x$

75. Solve the following system of equations:

$$\begin{cases} \sin x + \cos x = 1 \\ 2 \sin x = \sin 2x \end{cases}$$

76. Solve simultaneously:

$$\begin{cases} r = 1 - \cos\theta \\ r = \cos\theta \end{cases}$$

where $r > 0$ and $0 \leq \theta < 2\pi$

8.6 Polar Coordinates

Up to this point, we have located points in a plane by specifying rectangular coordinates, such as $P(3, 4)$ or $Q(-1, 5)$. We now consider an alternate method for identifying points in a plane. The method depends on describing each point in terms of two numbers r and θ, known as polar coordinates. To simplify our work, we make the following definition:

Definition

A **θ-ray** is a ray which has its initial point at (0, 0) and which makes an angle of θ with the positive x-axis. The ray in the direction opposite to that of a θ-ray is called the **opposite of the θ-ray.** Note that the opposite of the θ-ray is the $(\theta + \pi)$-ray, or the $(\theta + 180°)$-ray.

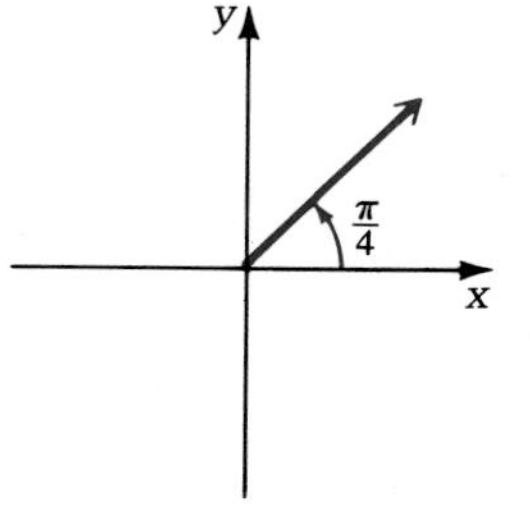

(a) The $\frac{\pi}{4}$-ray

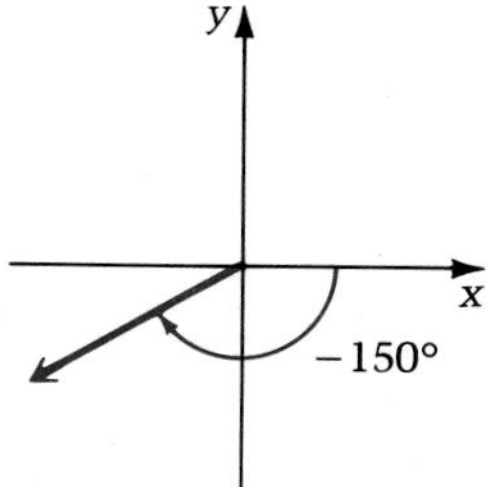

(b) The (−150°)-ray

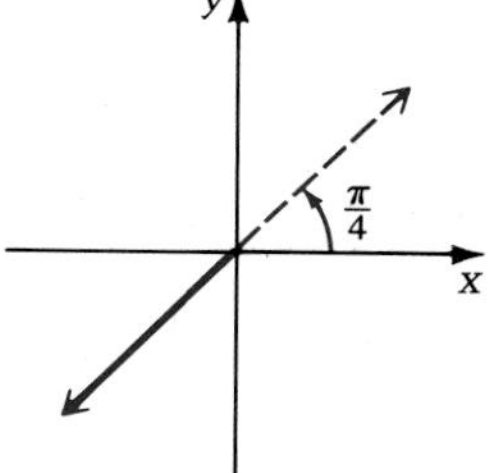

(c) The opposite of the $\frac{\pi}{4}$-ray

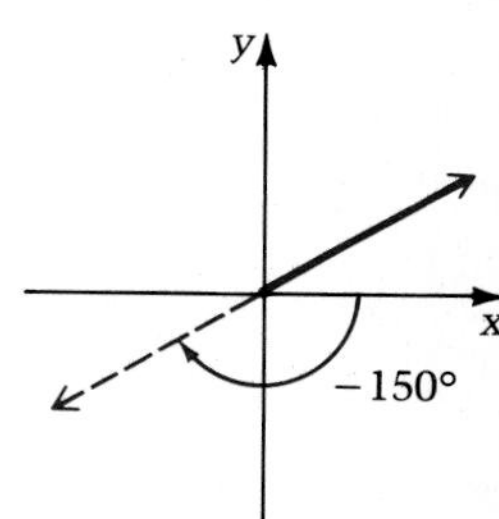

(d) The opposite of the (−150°)-ray

Figure 8.4

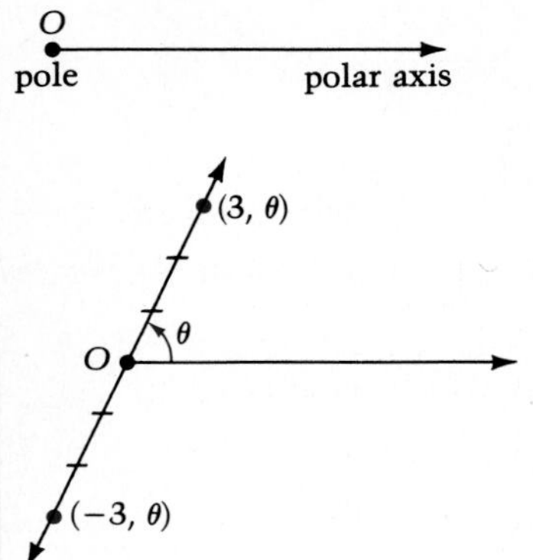

Figure 8.5

We construct what is called the *polar coordinate system* in the following way. We designate one point in the plane as the **origin** or **pole** (labeled O), and one ray emanating from O as the **polar axis.** Points in the plane can then be described in polar form as follows.

Definition

The **polar coordinates** (r, θ) specify a point that lies at a distance $|r|$ from O. If $r > 0$, the point lies on the θ-ray. If $r < 0$, the point lies on the opposite of the θ-ray. If $r = 0$, the point lies at O, regardless of the value of θ.

When plotting points given by polar coordinates, it is a common practice to use graph paper which displays concentric circles (centered at O) and rays emanating from O.

EXAMPLE 1

Graph the following points in the polar plane.

(a) $A(3, 135°)$ (b) $B(-4, 60°)$ (c) $C(-3, 315°)$ (d) $D\left(2, -\frac{3\pi}{2}\right)$

(e) $E(-5, \pi)$ (f) $F\left(0, \frac{\pi}{12}\right)$

Solution

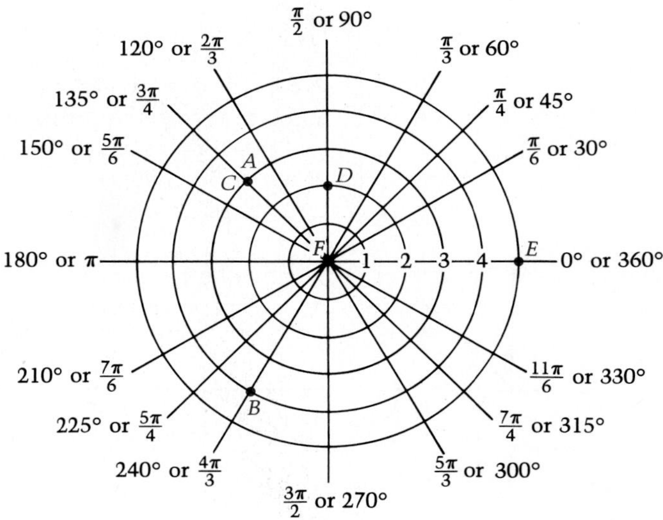

Figure 8.6

(a) Find the 135°-ray. Since 3 is positive, move 3 units away from the origin along that ray.
(b) Find the 60°-ray. Since -4 is negative, move 4 units away from the origin along the opposite of the 60°-ray.
(c) Find the 315°-ray. Since -3 is negative, move 3 units away from the origin along the opposite of the 315°-ray. Notice that this point coincides with (3, 135°).
(d) Find the $\left(-\frac{3\pi}{2}\right)$-ray (it is the same as the $\frac{\pi}{2}$-ray). Since 2 is positive, move 2 units away from the origin on that ray.
(e) Find the π-ray. Since -5 is negative, move 5 units away from the origin on the opposite of the π-ray.
(f) Since $r = 0$, this point is at the origin, regardless of the value of θ.

The description of a point in polar coordinates is not unique. See, for example, points A and C in Example 1. The coordinates (3, 135°) and (-3, 315°) describe the same point. In addition, point $A(3, 135°)$ could alternately be described by adding any multiple of 360° to 135°, and keeping $r = 3$. In fact, for all integers n, any point (r, θ) can be described as

$$(r, \theta + n \cdot 360°) \quad \text{or} \quad (-r, (\theta + 180°) + n \cdot 360°).$$

When we superimpose the polar plane on the xy-plane, we discover some relationships that are useful in translating from one system to the other.

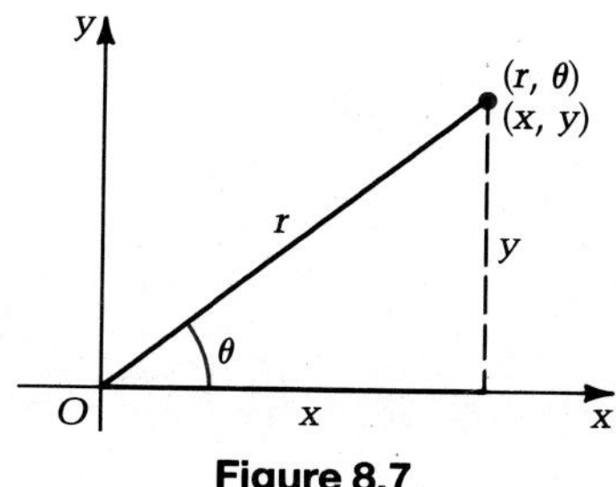

Figure 8.7

Since $\sin\theta = \frac{y}{r}$, and $\cos\theta = \frac{x}{r}$, we have

$$y = r\sin\theta,$$
$$x = r\cos\theta.$$

The following four relationships are used to solve translation problems.

Rectangular ↔ Polar Relationships

If point P has rectangular coordinates (x, y) and polar coordinates (r, θ), then

$$x = r\cos\theta, \qquad \tan\theta = \frac{y}{x}, \text{ provided } x \neq 0,$$

$$y = r\sin\theta, \qquad x^2 + y^2 = r^2.$$

EXAMPLE 2

Find the xy-coordinates of the point with polar coordinates $\left(-1, \frac{\pi}{3}\right)$.

Solution

$$x = r\cos\theta = (-1)\left(\cos\frac{\pi}{3}\right) = -\frac{1}{2}$$

$$y = r\sin\theta = (-1)\left(\sin\frac{\pi}{3}\right) = -\frac{\sqrt{3}}{2}$$

The rectangular coordinates of the polar point $\left(-1, \frac{\pi}{3}\right)$ are $\left(-\frac{1}{2}, -\frac{\sqrt{3}}{2}\right)$.

EXAMPLE 3

Find all polar coordinates of the point with rectangular coordinates $(-1, 2)$.

Solution

$$r^2 = x^2 + y^2 = (-1)^2 + (2)^2 = 5 \quad \text{First we find } r.$$

$$r = \pm\sqrt{5} \quad \text{(two cases)}$$

Case 1. Suppose $r = \sqrt{5}$:

$$x = r\cos\theta \qquad y = r\sin\theta$$

$$-1 = \sqrt{5}\cos\theta \qquad 2 = \sqrt{5}\sin\theta$$

$$\cos\theta = -\frac{1}{\sqrt{5}} \approx -0.4472 \qquad \sin\theta = \frac{2}{\sqrt{5}} \approx 0.8944.$$

The reference angle of θ is approximately 63.4°. Since $\cos\theta$ is negative and $\sin\theta$ is positive, θ is a second quadrant angle. Thus, $\theta \approx 180° - 63.4° = 116.6°$.

Case 2. If $r = -\sqrt{5}$, then $\theta \approx 116.6° + 180° = 296.6°$.

Thus, polar coordinates for the given point are of the form

$$(\sqrt{5}, 116.6° + n \cdot 360°) \quad \text{or} \quad (-\sqrt{5}, 296.6° + n \cdot 360°), \quad \text{for all integers } n.$$

Certain graphs are represented by very simple equations in polar form. For example, the polar equation $r = c$ (where c is a constant) represents a circle of radius $|c|$, centered at the origin. Also, the polar equation $\theta = \alpha$ (where α is a constant) is the line formed by joining the α-ray and the opposite of the α-ray. The slope of such a line is $\tan\alpha$.

EXAMPLE 4

Convert the following equations to rectangular form.

(a) $r = 5$ (b) $r = 6 \sin \theta$

Solution

(a)

$$\begin{aligned} r &= 5 \\ r^2 &= 25 && \text{Both sides are squared.} \\ x^2 + y^2 &= 25 && r^2 \text{ is replaced by } x^2 + y^2. \end{aligned}$$

Thus $r = 5$ is a circle with center at (0, 0) having radius 5.

(b)

$$\begin{aligned} r &= 6 \sin \theta \\ r^2 &= 6(r \sin \theta) && \text{Both sides are multiplied by } r. \\ x^2 + y^2 &= 6y && r \sin \theta \text{ is replaced by } y;\ r^2 \text{ is replaced by } x^2 + y^2. \\ x^2 + (y^2 - 6y + \square) &= \square && \text{We prepare to complete the square in } y. \\ x^2 + (y^2 - 6y + 9) &= 9 \\ x^2 + (y - 3)^2 &= 9 \end{aligned}$$

Thus $r = 6 \sin \theta$ is the polar equation of a circle of radius 3 with center at (0, 3).

Note that, in each problem in Example 4, we began by transforming the given equation so that one or more of the Rectangular ↔ Polar Relationships could be used. This is the key to handling such translation problems.

EXAMPLE 5

Convert the following rectangular equations to polar form.

(a) $y = 5x$ (b) $x^2 - y^2 = 3$

Solution

(a)

$$\begin{aligned} y &= 5x \\ \frac{y}{x} &= 5 \\ \tan \theta &= 5 \end{aligned}$$

(b)

$$\begin{aligned} x^2 - y^2 &= 3 \\ (r \cos \theta)^2 - (r \sin \theta)^2 &= 3 \\ r^2 \cos^2 \theta - r^2 \sin^2 \theta &= 3 \\ r^2(\cos^2 \theta - \sin^2 \theta) &= 3 \\ r^2 \cos 2\theta &= 3 && \cos^2 \theta - \sin^2 \theta = \cos 2\theta. \end{aligned}$$

Graphing Polar Equations

We end this section with the problem of graphing polar equations. In Example 6, we graph a polar equation by plotting a sufficient number of points to get a reasonable idea about the shape of the curve.

EXAMPLE 6

Sketch the graph of the polar equation $r = 1 + 2\cos\theta$.

Solution We will provide 4 charts showing values of r for selected values of θ, and will use these charts to sketch the graph in stages. Values of $\cos\theta$ and r are approximations.

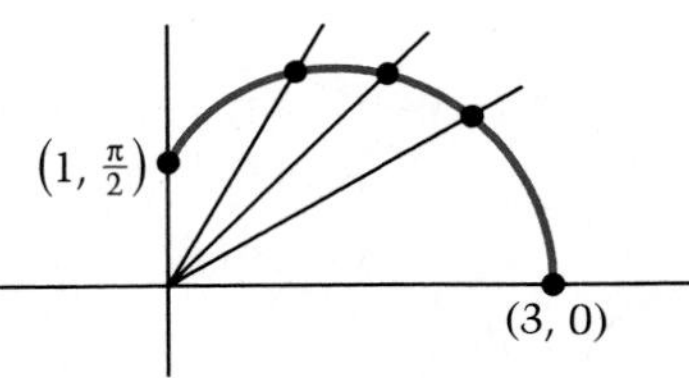

θ	0	$\frac{\pi}{6}$	$\frac{\pi}{4}$	$\frac{\pi}{3}$	$\frac{\pi}{2}$
r	3	2.7	2.4	2	1

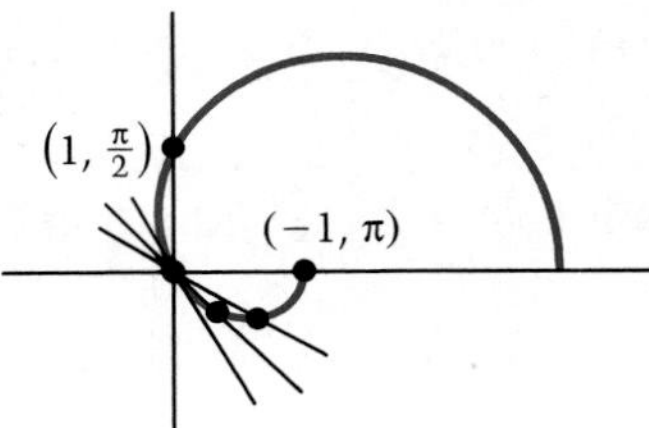

θ	$\frac{\pi}{2}$	$\frac{2\pi}{3}$	$\frac{3\pi}{4}$	$\frac{5\pi}{6}$	π
r	1	0	−0.4	−0.7	−1

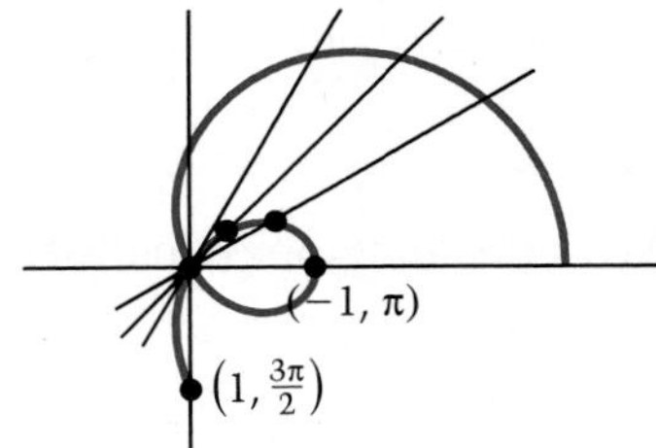

θ	π	$\frac{7\pi}{6}$	$\frac{5\pi}{4}$	$\frac{4\pi}{3}$	$\frac{3\pi}{2}$
r	−1	−0.7	−0.4	0	1

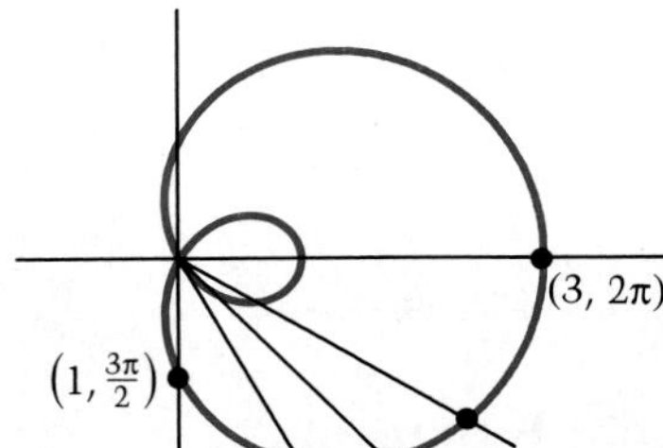

θ	$\frac{3\pi}{2}$	$\frac{5\pi}{3}$	$\frac{7\pi}{4}$	$\frac{11\pi}{6}$	2π
r	1	2	2.4	2.7	3

Figure 8.8

Here is the complete graph:

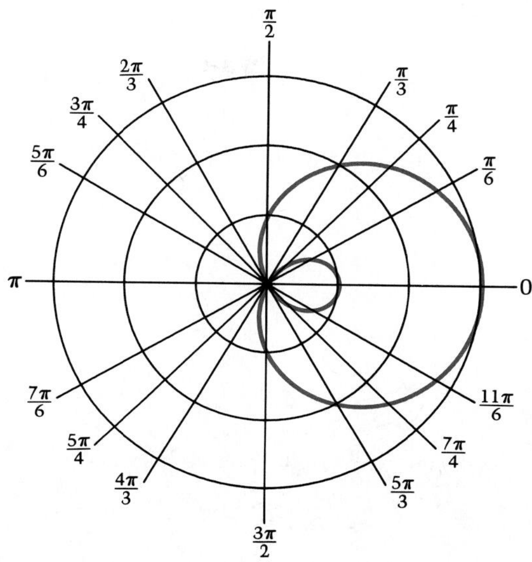

Figure 8.9

EXERCISE SET 8.6

In Exercises 1–16, graph the given point in the polar plane.

1. (2, 120°) **2.** (5, 30°) **3.** (3, 135°)
4. (4, 210°) **5.** (2, −120°) **6.** (−3, 135°)
7. (−1, −270°) **8.** (−5, −45°) **9.** (−2, −420°)
10. (4, −570°) **11.** $(3, \pi)$ **12.** $\left(4, \frac{\pi}{4}\right)$
13. $\left(-2, \frac{5\pi}{6}\right)$ **14.** $\left(-5, -\frac{3\pi}{2}\right)$ **15.** $\left(-4, -\frac{\pi}{2}\right)$
16. $\left(-2, \frac{3\pi}{4}\right)$

For Exercises 17–32, determine the x- and y-coordinates of the polar points in Exercises 1–16.

In Exercises 33–44, the rectangular coordinates of a point are given. Determine two polar representations (r, θ) for the point, with $0 \le \theta < 2\pi$.

33. (0, 4) **34.** (−2, 0) **35.** (−1, −1)
36. (1, −1) **37.** $(\sqrt{3}, -1)$ **38.** $(-1, -\sqrt{3})$
39. $(-\sqrt{2}, \sqrt{2})$ **40.** $(\sqrt{2}, -\sqrt{2})$ **41.** (−3, 4)
42. (−5, −12) **43.** (4, 6) **44.** (5, −10)

In Exercises 45–48, the point is given in polar form. Determine the rectangular coordinates of the point.

45. (−5.38, 231.7°) **46.** (4.32, 325.9°)
47. (2.19, 2.59) **48.** (−3.07, 0.93)

In Exercises 49–52, the point is written with rectangular coordinates. Determine the polar coordinates of the given point for θ in the interval $[0, \pi)$.

49. (−3.83, 5.25) **50.** (4.14, −3.86)
51. (−5.20, −3.12) **52.** (7.46, 2.39)

In Exercises 53–62, convert the given polar equation to rectangular form.

53. $r = 3$ **54.** $2r = 5$

55. $\theta = \dfrac{3\pi}{4}$

56. $\theta = \dfrac{\pi}{3}$

57. $r(\cos\theta - 2\sin\theta) = 5$

58. $r(3\sin\theta - 4\cos\theta) = 5$

59. $r(1 + \cos\theta) = 1$

60. $r(1 + \sin\theta) = 2$

61. $r(1 - 2\sin\theta) = 2$

62. $r(2 - \cos\theta) = 2$

In Exercises 63–74, convert the given rectangular equation to polar form.

63. $x = 0$

64. $y = -5$

65. $x + y = 1$

66. $x + 4 = 4y$

67. $x^2 - 2x + y^2 = 0$

68. $x^2 - 6y - y^2 = 0$

69. $x^2 = 6y + 9$

70. $y^2 = 8x + 16$

71. $2xy = 1$

72. $xy = 1$

73. $x^2 + y^2 = 16y$

74. $x^2 + y^2 = 4x$

In Exercises 75–94, sketch the graph of the polar equation.

75. $r = -3$

76. $r^2 = 4$

77. $\theta^2 = \dfrac{\pi^2}{16}$

78. $\theta = \dfrac{5\pi}{4}$

79. $r = 3\cos\theta$

80. $r = 4\sin\theta$

81. $r = 2 + \sin\theta$

82. $r = 1 - \cos\theta$

83. $r = -2\sin\theta$

84. $r\sin\theta = 2$

85. $r\cos\theta = -3$

86. $r = -2 + 5\cos\theta$

87. $r = 2 - \sin\theta$

88. $r = 1 + \cos\theta$

89. $r = -3 + 4\cos\theta$

90. $r = -3\cos\theta$

91. $r^2 = 9$

92. $r = -2$

93. $r\cos\theta = -1$

94. $r\sin\theta = -2$

Superset

95. Determine the polar equation of the line through the points $(-2, 0)$ and $(1, 2)$.

96. Determine the polar equation of the line through the points that have polar coordinates $\left(1, \dfrac{\pi}{2}\right)$ and $(-2, \pi)$.

97. (a) Find the distance between the points $\left(-2, \dfrac{\pi}{3}\right)$ and $\left(4, \dfrac{3\pi}{4}\right)$. (*Hint:* First convert the points to rectangular coordinates.)
(b) Sketch the points on the polar plane. Use the Law of Cosines to determine the distance between the two points.

98. Determine the polar equation of the circle of radius 2 centered at the point $(3, 1)$.

In Exercises 99–108, sketch the graph of the polar equation. (*Hint:* In some instances it may help to convert the equation to rectangular coordinates.)

99. $r(1 - \cos\theta) = 2$

100. $r(1 + \sin\theta) = 2$

101. $\tan\theta = 2$

102. $\cot\theta = 2$

103. $r^2\sin 2\theta = 2$

104. $r^2\cos 2\theta = 2$

105. $r = 4\sin 2\theta$

106. $r = 6\cos 3\theta$

107. $r = 8\cos\theta + 6\sin\theta$

108. $\theta = 1$

In Exercises 109–118, sketch the graph of the polar equation for the indicated values of θ.

109. $r = \theta, 0 \le \theta \le 2\pi$

110. $r = \dfrac{\theta}{2}, 0 \le \theta \le 2\pi$

111. $r^2 = 3r, 0 \le \theta \le \dfrac{\pi}{2}$

112. $r = -2, 0 \le \theta \le \dfrac{\pi}{2}$

113. $r = 3\cos 2\theta, 0 \le \theta \le \pi$

114. $r = 5\sin 2\theta, 0 \le \theta \le \pi$

115. $r\sin\theta = 2, 0 < \theta < \dfrac{\pi}{2}$

116. $r\cos\theta = 2, 0 < \theta < \dfrac{\pi}{2}$

117. $r = 4\sin\theta, \dfrac{\pi}{2} \le \theta \le \pi$

118. $r = 6\cos\theta, \dfrac{\pi}{2} \le \theta \le \pi$

8.7 Trigonometric Form of Complex Numbers

Recall that in order to visualize the real numbers, we associated each number with a point on a line, called the real line. Complex numbers can be visualized as points in a plane called the **complex plane,** shown in Figure 8.10.

Every complex number $a + bi$ can be associated with an ordered pair (a, b) in the complex plane, where a is the real part of the complex number

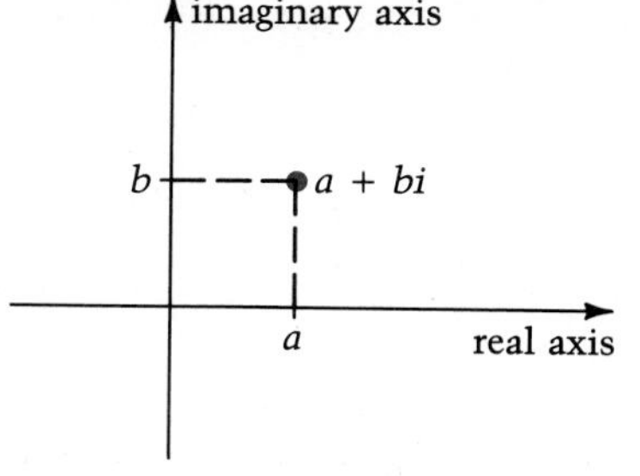

Figure 8.10 The Complex Plane.

and b is the imaginary part. The horizontal axis in the complex plane is referred to as the real axis, since real numbers are graphed there. The vertical axis is called the imaginary axis. Pure imaginary numbers (numbers of the form bi) are graphed on the imaginary axis (Figure 8.11).

We will define the absolute value of a complex number $a + bi$ as the distance between the point corresponding to $a + bi$ and the origin (Figure 8.12). For example, the absolute value of $4 + 3i$ is the distance between the point $4 + 3i$ and the origin, and can be found by the Pythagorean Theorem:

$$\sqrt{4^2 + 3^2} = \sqrt{25} = 5.$$

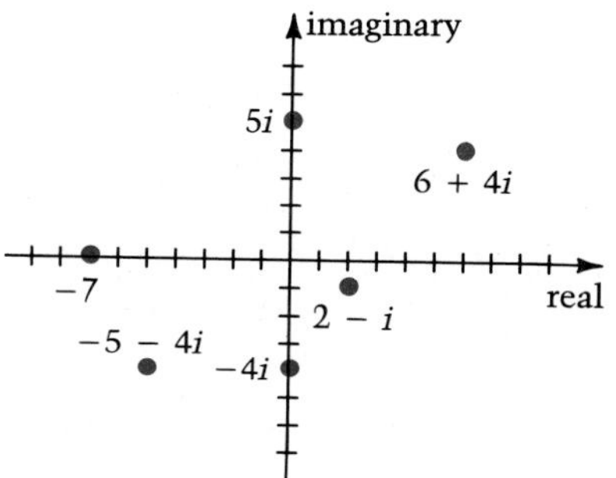

Figure 8.11

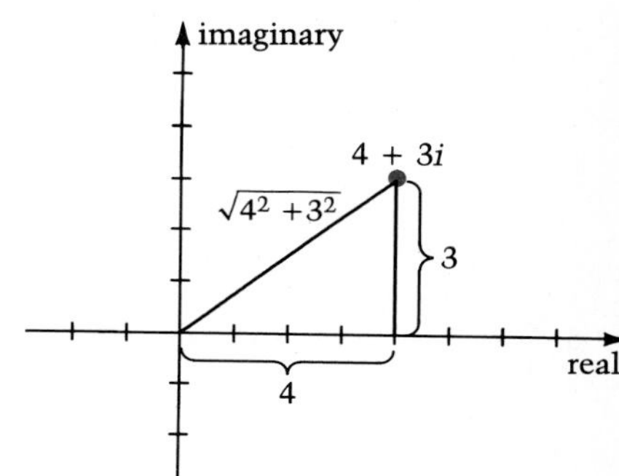

Figure 8.12

EXAMPLE 1

Represent each of the following complex numbers as a point in the complex plane.

(a) $4 - 3i$ (b) $-3 + 4i$ (c) $-3 - 4i$ (d) $-2 + 3i$

Solution

(a)

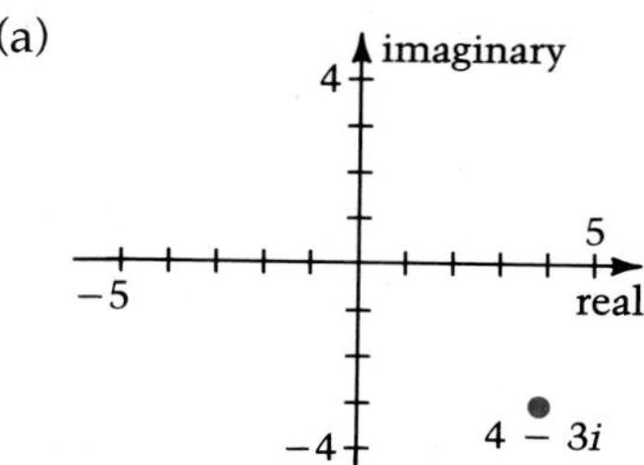

(b)

imaginary
−3 + 4i
4
5
−5
real
−4

Figure 8.13

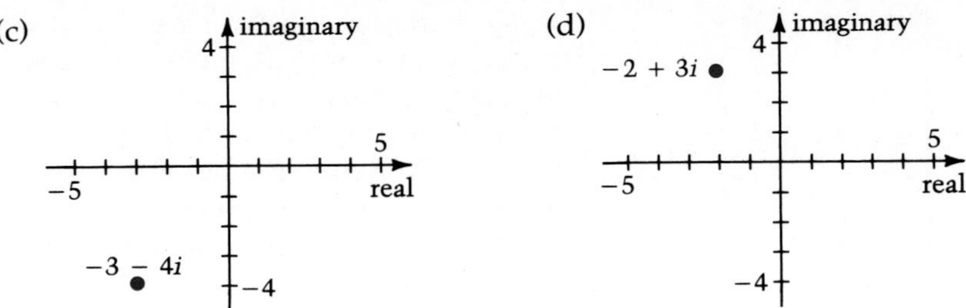

Figure 8.13 (continued)

Definition of the Absolute Value of a Complex Number

The **absolute value** of a complex number $a + bi$, denoted $|a + bi|$, is the distance between the origin and the point associated with $a + bi$. That is, $|a + bi| = \sqrt{a^2 + b^2}$.

EXAMPLE 2

Determine the absolute value of each of the following complex numbers.

(a) $2 - 3i$ (b) $-5i$ (c) $1 - i$

Solution

(a) $|2 - 3i| = \sqrt{2^2 + (-3)^2} = \sqrt{4 + 9} = \sqrt{13}$
(b) $|-5i| = \sqrt{0^2 + (-5)^2} = \sqrt{25} = 5$
(c) $|1 - i| = \sqrt{1^2 + (-1)^2} = \sqrt{1 + 1} = \sqrt{2}$

In the last section we described how each point $P(a, b)$ can be represented by polar coordinates (r, θ), where r is the distance between P and the origin,

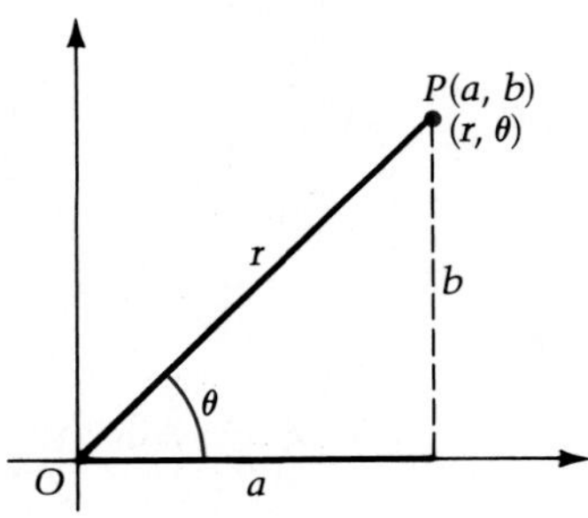

Figure 8.14

and θ is the angle that ray OP makes with the positive x-axis (see Figure 8.14). Any complex number $a + bi$ can be written in *polar* or *trigonometric form*, as illustrated below.

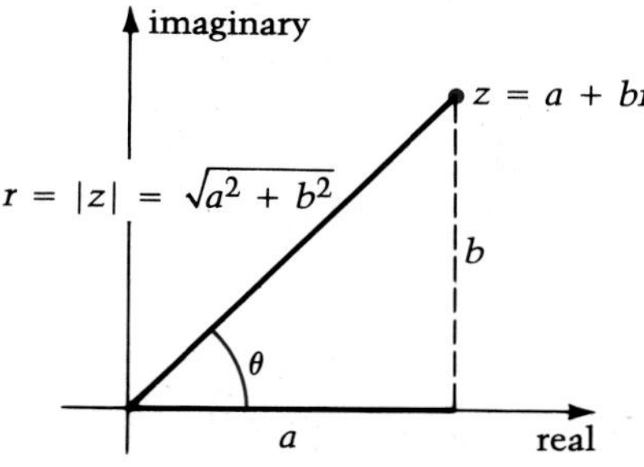

Figure 8.15

$a = r\cos\theta$ since $\cos\theta = \dfrac{a}{r}$.

$b = r\sin\theta$ since $\sin\theta = \dfrac{b}{r}$.

Therefore, $z = (r\cos\theta) + (r\sin\theta)i$.

Definition of the Trigonometric Form of a Complex Number

The complex number $z = a + bi$ has **trigonometric form**

$$z = r(\cos\theta + i\sin\theta)$$

where $r = |z| = \sqrt{a^2 + b^2}$ and $\tan\theta = \dfrac{b}{a}$. The number r is called the **modulus** of z, and θ is called the **argument** of z.

EXAMPLE 3

Express each of the following complex numbers in trigonometric form. If rounding is necessary, express θ to the nearest tenth of a degree.

(a) $2 + 2i$ (b) $-7 - 3i$

Solution

(a)

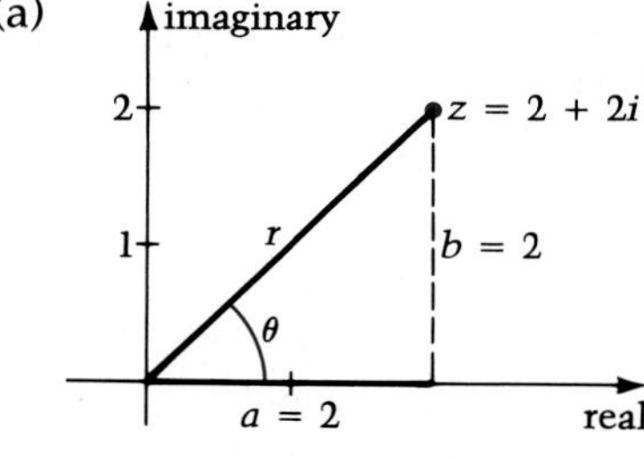

Figure 8.16

Find r and θ, then apply the above definition.

$r = |z| = \sqrt{2^2 + 2^2} = \sqrt{8} = 2\sqrt{2}$.

Since $\tan\theta = \dfrac{2}{2} = 1$, and θ is a first quadrant angle, $\theta = \dfrac{\pi}{4}$.

Thus, $z = 2\sqrt{2}\left(\cos\dfrac{\pi}{4} + i\sin\dfrac{\pi}{4}\right)$.

(b)

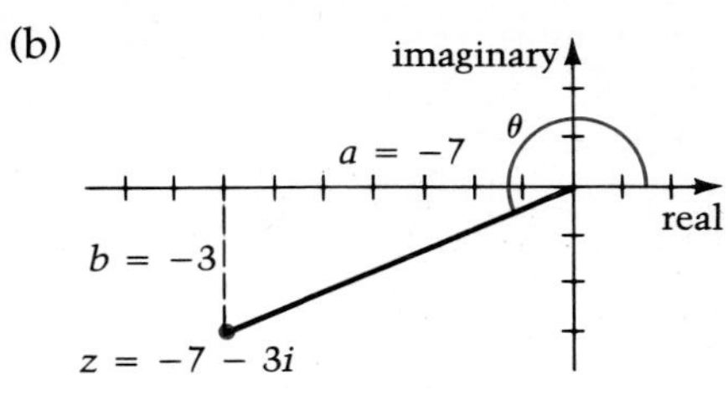

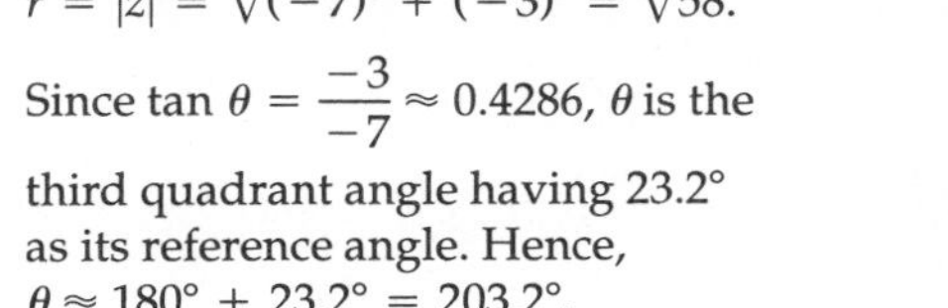

$r = |z| = \sqrt{(-7)^2 + (-3)^2} = \sqrt{58}$.

Since $\tan \theta = \frac{-3}{-7} \approx 0.4286$, θ is the third quadrant angle having $23.2°$ as its reference angle. Hence, $\theta \approx 180° + 23.2° = 203.2°$.

Thus, $z = \sqrt{58}\,(\cos 203.2° + i \sin 203.2°)$

Figure 8.16 (continued)

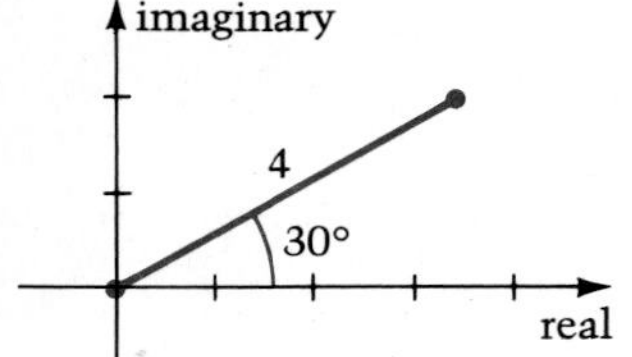

Figure 8.17

It is important to remember that the trigonometric form of a complex number is not unique. For example, all expressions of the form

$$z = 4[\cos(30° + k \cdot 360°) + i \sin(30° + k \cdot 360°)]$$

for every integer k, represent the same complex number, $2\sqrt{3} + 2i$.

Products and Quotients of Complex Numbers

Suppose $z_1 = r_1(\cos \theta_1 + i \sin \theta_1)$ and $z_2 = r_2(\cos \theta_2 + i \sin \theta_2)$ are two complex numbers. Their product can be expressed as

$$\begin{aligned} z_1 z_2 &= r_1(\cos \theta_1 + i \sin \theta_1) \cdot r_2(\cos \theta_2 + i \sin \theta_2) \\ &= r_1 r_2(\cos \theta_1 \cos \theta_2 + i \cos \theta_1 \sin \theta_2 + i \sin \theta_1 \cos \theta_2 + i^2 \sin \theta_1 \sin \theta_2) \\ &= r_1 r_2[\cos \theta_1 \cos \theta_2 - \sin \theta_1 \sin \theta_2 + i(\sin \theta_1 \cos \theta_2 + \cos \theta_1 \sin \theta_2)] \\ &= r_1 r_2[\cos(\theta_1 + \theta_2) + i \sin(\theta_1 + \theta_2)] \end{aligned}$$

Thus we have proven the following theorem.

Theorem

If $z_1 = r_1(\cos \theta_1 + i \sin \theta_1)$ and $z_2 = r_2(\cos \theta_2 + i \sin \theta_2)$ are two complex numbers, their product is given by

$$z_1 z_2 = r_1 r_2[\cos(\theta_1 + \theta_2) + i \sin(\theta_1 + \theta_2)].$$

The preceding theorem tells us that the product of two complex numbers has modulus equal to the product of the two moduli r_1 and r_2, and has argument equal to the sum of the two arguments θ_1 and θ_2.

EXAMPLE 4

Find $z_1 z_2$ by using trigonometric forms. Express the answer in the form $a + bi$.

(a) $z_1 = 2(\cos 18° + i \sin 18°)$, $z_2 = 5(\cos 162° + i \sin 162°)$

(b) $z_1 = -3 + 3i\sqrt{3}$, $z_2 = 1 - i\sqrt{3}$

Solution

(a) $z_1z_2 = 2 \cdot 5[\cos(18° + 162°) + i\sin(18° + 162°)]$
$= 10[\cos(180°) + i\sin(180°)] = 10[-1 + i(0)] = -10$

(b) First express each complex number in trigonometric form.

$$z_1 = 6\left(\cos\frac{2\pi}{3} + i\sin\frac{2\pi}{3}\right), \qquad z_2 = 2\left(\cos\frac{5\pi}{3} + i\sin\frac{5\pi}{3}\right)$$

$$\begin{aligned} z_1z_2 &= 6 \cdot 2\left[\cos\left(\frac{2\pi}{3} + \frac{5\pi}{3}\right) + i\sin\left(\frac{2\pi}{3} + \frac{5\pi}{3}\right)\right] \\ &= 12\left[\cos\left(\frac{7\pi}{3}\right) + i\sin\left(\frac{7\pi}{3}\right)\right] \\ &= 12\left(\cos\frac{\pi}{3} + i\sin\frac{\pi}{3}\right) = 12\left(\frac{1}{2} + i\frac{\sqrt{3}}{2}\right) = 6 + 6i\sqrt{3} \end{aligned}$$

By using the FOIL method, we can check the answer to part (b) of the above example.

$$\begin{aligned} (-3 + 3i\sqrt{3})(1 - i\sqrt{3}) &= (-3)(1) + (-3)(-i\sqrt{3}) \\ &\quad + (3i\sqrt{3})(1) + (3i\sqrt{3})(-i\sqrt{3}) \\ &= -3 + 3i\sqrt{3} + 3i\sqrt{3} - 9i^2 \\ &= (-3 + 9) + i(3\sqrt{3} + 3\sqrt{3}) = 6 + 6i\sqrt{3} \end{aligned}$$

In the exercises, we suggest a way of proving the following theorem, which is useful in determining the quotient of two complex numbers.

Theorem

If $z_1 = r_1(\cos\theta_1 + i\sin\theta_1)$ and $z_2 = r_2(\cos\theta_2 + i\sin\theta_2)$, then

$$\frac{z_1}{z_2} = \frac{r_1}{r_2}[\cos(\theta_1 - \theta_2) + i\sin(\theta_1 - \theta_2)], \quad \text{for } z_2 \neq 0.$$

EXAMPLE 5

Find $\frac{z_1}{z_2}$ where $z_1 = 3 - i\sqrt{3}$ and $z_2 = 4 + 4i$ by using trigonometric forms. Express the answer in trigonometric form.

Solution

Begin by expressing each complex number in trigonometric form.

$$z_1 = 2\sqrt{3}\left(\cos\frac{11\pi}{6} + i\sin\frac{11\pi}{6}\right), \qquad z_2 = 4\sqrt{2}\left(\cos\frac{\pi}{4} + i\sin\frac{\pi}{4}\right)$$

Thus,

$$\begin{aligned}\frac{z_1}{z_2} &= \frac{2\sqrt{3}}{4\sqrt{2}}\left[\cos\left(\frac{11\pi}{6} - \frac{\pi}{4}\right) + i\sin\left(\frac{11\pi}{6} - \frac{\pi}{4}\right)\right] \\ &= \frac{\sqrt{3}}{2\sqrt{2}}\cdot\frac{\sqrt{2}}{\sqrt{2}}\left[\cos\left(\frac{19\pi}{12}\right) + i\sin\left(\frac{19\pi}{12}\right)\right] = \frac{\sqrt{6}}{4}\left(\cos\frac{19\pi}{12} + i\sin\frac{19\pi}{12}\right).\end{aligned}$$

At this point you should verify that the answer in Example 5 is the same complex number you would find if you used the complex conjugate of $4 + 4i$ to simplify the quotient $\left(\frac{3 - i\sqrt{3}}{4 + 4i}\right)$. You can determine $\cos\frac{19\pi}{12}$ and $\sin\frac{19\pi}{12}$ exactly by using the appropriate sum and difference identities.

Powers and Roots of Complex Numbers

By successively multiplying a complex number $z = r(\cos\theta + i\sin\theta)$ by itself, we observe a simple pattern:

$$\begin{aligned}z^2 &= r(\cos\theta + i\sin\theta)\cdot r(\cos\theta + i\sin\theta) \\ &= r^2[\cos(\theta + \theta) + i\sin(\theta + \theta)] \\ &= r^2(\cos 2\theta + i\sin 2\theta)\end{aligned}$$

$$\begin{aligned}z^3 = z^2\cdot z &= r^2(\cos 2\theta + i\sin 2\theta)\cdot r(\cos\theta + i\sin\theta) \\ &= r^3(\cos 3\theta + i\sin 3\theta)\end{aligned}$$

This pattern holds for any positive integral power of z. This result is called *De Moivre's Theorem.*

De Moivre's Theorem

For any positive integer n,

$$[r(\cos\theta + i\sin\theta)]^n = r^n(\cos n\theta + i\sin n\theta).$$

EXAMPLE 6

Use De Moivre's Theorem to express in the form $a + bi$:

(a) $(-1 + i\sqrt{3})^5$ (b) $(2 + 2i)^6$

Solution

(a) $$-1 + i\sqrt{3} = 2\left(\cos\frac{2\pi}{3} + i\sin\frac{2\pi}{3}\right) \qquad -1 + i\sqrt{3} \text{ is expressed in trigonometric form.}$$

$$\begin{aligned}(-1 + i\sqrt{3})^5 &= \left[2\left(\cos\frac{2\pi}{3} + i\sin\frac{2\pi}{3}\right)\right]^5\\ &= 2^5\left[\cos\left(5\cdot\frac{2\pi}{3}\right) + i\sin\left(5\cdot\frac{2\pi}{3}\right)\right]\\ &= 32\left(\cos\frac{10\pi}{3} + i\sin\frac{10\pi}{3}\right) = 32\left[-\frac{1}{2} + i\left(-\frac{\sqrt{3}}{2}\right)\right]\\ &= -16 - 16i\sqrt{3}\end{aligned}$$

(b) $$2 + 2i = 2\sqrt{2}\left(\cos\frac{\pi}{4} + i\sin\frac{\pi}{4}\right) \qquad 2 + 2i \text{ is expressed in trigonometric form.}$$

$$\begin{aligned}(2 + 2i)^6 &= \left[2\sqrt{2}\left(\cos\frac{\pi}{4} + i\sin\frac{\pi}{4}\right)\right]^6\\ &= (2\sqrt{2})^6\left[\cos\left(6\cdot\frac{\pi}{4}\right) + i\sin\left(6\cdot\frac{\pi}{4}\right)\right]\\ &= 512\left(\cos\frac{3\pi}{2} + i\sin\frac{3\pi}{2}\right) = 512[0 + i(-1)] = -512i\end{aligned}$$

In Example 6, we found that $(2 + 2i)^6 = -512i$; that is, $2 + 2i$ is a sixth root of $-512i$. In general, if w and z are complex numbers such that $w^n = z$, then w is called an ***n*th root** of z.

We now consider how to determine such roots. For example, to find a cube root of $z = 1 - i\sqrt{3}$, we need $w = r(\cos\theta + i\sin\theta)$ such that

(1) $$[r(\cos\theta + i\sin\theta)]^3 = 1 - i\sqrt{3}.$$

By De Moivre's Theorem, $[r(\cos\theta + i\sin\theta)]^3 = r^3(\cos 3\theta + i\sin 3\theta)$. Also, we can rewrite $1 - i\sqrt{3}$ in general trigonometric form:

$$1 - i\sqrt{3} = 2[\cos(300° + k\cdot 360°) + i\sin(300° + k\cdot 360°)].$$

Thus, equation (1) can be rewritten as follows:

$$r^3(\cos 3\theta + i\sin 3\theta) = 2[\cos(300° + k\cdot 360°) + i\sin(300° + k\cdot 360°)].$$

We now determine r and θ so that we can write w in trigonometric form. It seems reasonable to set $r^3 = 2$ and $3\theta = 300° + k\cdot 360°$. Thus,

$$r = \sqrt[3]{2} \text{ and } \theta = 100° + k\cdot 120°.$$

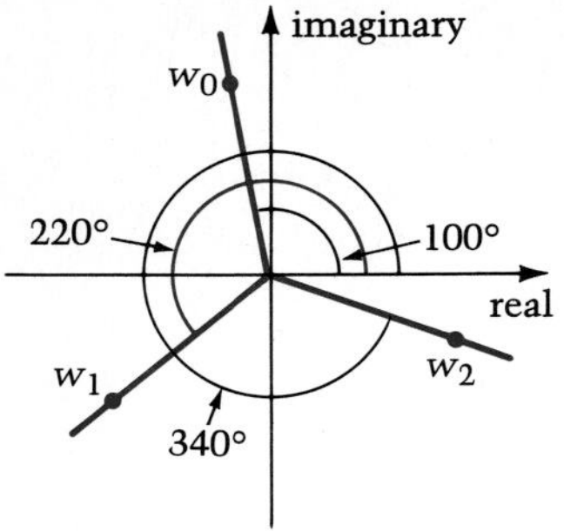

Figure 8.18 The three cube roots of $1-i\sqrt{3}$. Note that they are equally spaced (in 120° increments) and lie at the same distance $(\sqrt[3]{2})$ from the origin.

If

$$\begin{aligned} k &= 0, \quad w_0 = \sqrt[3]{2}(\cos 100^\circ + i \sin 100^\circ), \\ k &= 1, \quad w_1 = \sqrt[3]{2}(\cos 220^\circ + i \sin 220^\circ), \\ k &= 2, \quad w_2 = \sqrt[3]{2}(\cos 340^\circ + i \sin 340^\circ). \end{aligned}$$

For $k > 2$, the three complex numbers we found for $k = 0, 1, 2$, are repeated. Thus, there are exactly 3 distinct cube roots of $1 - i\sqrt{3}$ (see Figure 8.18). In general, any nonzero complex number has exactly n distinct nth roots.

The nth Root Theorem

The nth roots of the nonzero complex number $r(\cos \theta + i \sin \theta)$ are given by

$$\sqrt[n]{r}\left[\cos\left(\frac{\theta + k \cdot 360^\circ}{n}\right) + i \sin\left(\frac{\theta + k \cdot 360^\circ}{n}\right)\right]$$

where $k = 0, 1, 2, \ldots, n - 1$. We can replace 360° with 2π in the above statement to accommodate radian measure.

EXAMPLE 7

Find and plot the 4 fourth roots of $16(\cos 120^\circ + i \sin 120^\circ)$.

Solution By the nth Root Theorem, there are exactly 4 fourth roots of this complex number:

$$\begin{aligned} w_0 &= \sqrt[4]{16}\left(\cos \frac{120^\circ + 0 \cdot 360^\circ}{4} + i \sin \frac{120^\circ + 0 \cdot 360^\circ}{4}\right) \\ &= 2(\cos 30^\circ + i \sin 30^\circ) \\ w_1 &= \sqrt[4]{16}\left(\cos \frac{120^\circ + 1 \cdot 360^\circ}{4} + i \sin \frac{120^\circ + 1 \cdot 360^\circ}{4}\right) \\ &= 2(\cos 120^\circ + i \sin 120^\circ) \\ w_2 &= \sqrt[4]{16}\left(\cos \frac{120^\circ + 2 \cdot 360^\circ}{4} + i \sin \frac{120^\circ + 2 \cdot 360^\circ}{4}\right) \\ &= 2(\cos 210^\circ + i \sin 210^\circ) \\ w_3 &= \sqrt[4]{16}\left(\cos \frac{120^\circ + 3 \cdot 360^\circ}{4} + i \sin \frac{120^\circ + 3 \cdot 360^\circ}{4}\right) \\ &= 2(\cos 300^\circ + i \sin 300^\circ) \end{aligned}$$

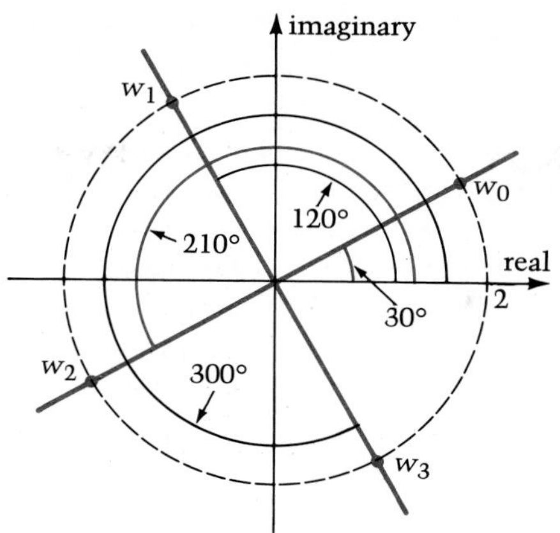

Note that the four fourth roots of $16(\cos 120° + i \sin 120°)$ are equally spaced $\left(\text{at } \frac{360°}{4} = 90° \text{ intervals}\right)$ on a circle centered at the origin, having radius 2.

Figure 8.19

If $z = 1$, then we refer to the n nth roots of z as the ***n*th roots of unity.**

EXAMPLE 8

Determine the sixth roots of unity.

Solution First write $z = 1$ in trigonometric form. Since $z = 1 + 0i$, then $r = \sqrt{1^2 + 0^2} = 1$, and $\theta = 0°$. Thus $z = 1(\cos 0° + i \sin 0°)$. Each root has modulus $\sqrt[6]{1} = 1$, and argument determined by $\dfrac{0° + k \cdot 360°}{6}$ for $k = 0, 1, 2,$ 3, 4, and 5.

$$w_0 = 1(\cos 0° + i \sin 0°) = 1$$
$$w_1 = 1(\cos 60° + i \sin 60°) = \frac{1}{2} + \frac{\sqrt{3}}{2}i$$
$$w_2 = 1(\cos 120° + i \sin 120°) = -\frac{1}{2} + \frac{\sqrt{3}}{2}i$$
$$w_3 = 1(\cos 180° + i \sin 180°) = -1$$
$$w_4 = 1(\cos 240° + i \sin 240°) = -\frac{1}{2} - \frac{\sqrt{3}}{2}i$$
$$w_5 = 1(\cos 300° + i \sin 300°) = \frac{1}{2} - \frac{\sqrt{3}}{2}i$$

EXERCISE SET 8.7

In Exercises 1–18, represent the complex number as a point in the complex plane.

1. $2 + 5i$ **2.** $3 + 2i$ **3.** $2 - 5i$ **4.** $3 - 2i$ **5.** $-2 + 5i$ **6.** $-3 + 2i$ **7.** -6 **8.** $-i$ **9.** $-2i$ **10.** -3 **11.** $\overline{2 + 5i}$ **12.** $\overline{3 + 2i}$ **13.** $\overline{5 - 4i}$ **14.** $\overline{2 - 3i}$ **15.** $\overline{-5 - 4i}$ **16.** $\overline{-5 - 2i}$ **17.** $\overline{-4 + 3i}$ **18.** $\overline{-3 + 2i}$

In Exercises 19–30, determine the absolute value of the complex number.

19. $7 + 2i$ **20.** $5 + 3i$ **21.** $7 - 2i$

22. $5 - 3i$ **23.** $2i$ **24.** $-7i$

25. $-i$ **26.** i **27.** $3 - i\sqrt{7}$

28. $\sqrt{5} - 2i$ **29.** $\sqrt{14} + i$ **30.** $2 + i\sqrt{10}$

In Exercises 31–50, express the complex number in trigonometric form.

31. $3i$ **32.** $-2i$ **33.** -6

34. 8 **35.** $-1 + i\sqrt{3}$ **36.** $-\sqrt{3} + i$

37. $-2 - 2i$ **38.** $2 - 2i$ **39.** $-1 - i$

40. $1 + i$ **41.** $-3 + i\sqrt{3}$ **42.** $-3 - i\sqrt{3}$

43. $2\sqrt{3} - 2i$ **44.** $-2 + 2i\sqrt{3}$

45. $-3\sqrt{2} - 3i\sqrt{2}$ **46.** $3\sqrt{2} - 3i\sqrt{2}$

47. $5 + 12i$ **48.** $-8 + 15i$

49. $-24 - 7i$ **50.** $-7 - 4i$

In Exercises 51–58, determine the product by multiplying the trigonometric forms. Express the answer in the form $a + bi$.

51. $(-2 + 2i)(-2 - 2i)$ **52.** $(-\sqrt{3} + i)(\sqrt{3} - i)$

53. $(3 + i\sqrt{3})(1 - i\sqrt{3})$ **54.** $(1 - i\sqrt{3})(-3 - i\sqrt{3})$

55. $(\sqrt{3} + i\sqrt{3})(-2 + 2i)$ **56.** $(3\sqrt{3} + 3i)(3 + 3i\sqrt{3})$

57. $(3 - 4i)(3i)$ **58.** $(5 + 12i)(-2i)$

In Exercises 59–66, find $\frac{z_1}{z_2}$ by using the theorem for dividing trigonometric forms.

59. $z_1 = 6(\cos 300° + i \sin 300°)$
$z_2 = 3(\cos 135° + i \sin 135°)$

60. $z_1 = 8\left(\cos \frac{7\pi}{6} + i \sin \frac{7\pi}{6}\right)$
$z_2 = 4\left(\cos \frac{2\pi}{3} + i \sin \frac{2\pi}{3}\right)$

61. $z_1 = 1 + 3i, \quad z_2 = 1 - i$

62. $z_1 = 1 - i\sqrt{3}, \quad z_2 = 1 + i$

63. $z_1 = \sqrt{3} - 3i, \quad z_2 = \sqrt{3} + i$

64. $z_1 = -3 - i\sqrt{3}, \quad z_2 = 1 - i$

65. $z_1 = 2 + 2i, \quad z_2 = 3 - i\sqrt{3}$

66. $z_1 = -2 - 2i, \quad z_2 = -3 + 3i$

In Exercises 67–72, use De Moivre's Theorem to express in the form $a + bi$.

67. $(-2 + 2i)^4$ **68.** $(\sqrt{3} - i)^5$

69. $(\sqrt{3} + i)^3$ **70.** $(-1 - i)^6$

71. $(2\sqrt{3} - 2i)^{12}$ **72.** $(-3 - i\sqrt{3})^{10}$

In Exercises 73–78, use De Moivre's Theorem to express the quantity in the form $a + bi$.

73. $(0.91 + 0.84i)^5$ **74.** $(8.17 - 1.97i)^3$

75. $(-2.12 - 0.63i)^6$ **76.** $(12.06 + 4.78i)^4$

77. $(13.41 - 5.07i)^4$ **78.** $(-0.32 - 0.68i)^5$

In Exercises 79–80, find the three cube roots of z.

79. $z = 3i$ **80.** $z = -2i$

In Exercises 81–82, find the four fourth roots of z.

81. $z = -2 + 2i$ **82.** $z = -\sqrt{3} + i$

In Exercises 83–84, find the five fifth roots of z.

83. $z = 1 + i$ **84.** $z = 1 + i\sqrt{3}$

85. Find (a) the fifth roots of unity, (b) the fourth roots of unity, (c) the eighth roots of unity.

86. Find (a) the eighth roots of -1, (b) the fifth roots of 32, (c) the fourth roots of 16.

Superset

In Exercises 87–92, solve the equation over the set of complex numbers.

87. $x^2 + 2x + 3 = 0$ **88.** $x^2 + 4x + 5 = 0$

89. $x^3 - 2i = 0$ **90.** $x^3 + 1 = 0$

91. $x^5 + 32 = 0$ **92.** $x^6 - 64 = 0$

93. Prove that if $z = r(\cos \theta + i \sin \theta)$, then $\bar{z} = r(\cos(-\theta) + i \sin(-\theta))$.

94. Prove that if $z = r(\cos \theta + i \sin \theta)$, then $\frac{1}{z} = \frac{1}{r}(\cos \theta - i \sin \theta)$ for $z \neq 0 + 0i$. Use this result to prove the theorem for dividing two complex numbers.

In Exercises 95–98, simplify the expression.

95. $\left|\frac{1 - 2i}{3 + i}\right|$ **96.** $\left|\frac{1 + 2i}{3 - i}\right|$

97. $\frac{|1 - 2i|}{|3 + i|}$ **98.** $\frac{|1 + 2i|}{|3 - i|}$

99. Let $z = a + bi$. Show that $|\bar{z}| = |z|$.

100. Let $z = a + bi$. Show that $|z|^2 = z \cdot \bar{z}$.

In Exercises 101–104, determine whether the statement is true or false. If false, illustrate this with an example.

101. If z and w represent any complex numbers, then $|z - w| = |z| - |w|$.

102. If z and w represent any complex numbers, then $|z + w| = |z| + |w|$.

103. If z and w represent any complex numbers, then $|zw| = |z| \cdot |w|$.

104. If z and w represent any complex numbers, then $|z - w| = |w - z|$.

105. Let $z = 2 + 5i$ and $w = 4 - 7i$. Represent z, w, and $z + w$ as points in the complex plane.

106. Repeat Exercise 105 with the complex numbers $z = -2 + 7i$ and $w = 8 - 4i$.

107. Let $z = 2 + i$ and $w = 3 - 2i$. Represent z, w, and zw as points in the complex plane.

108. Repeat Exercise 107 with the complex numbers $z = -2 + 4i$ and $w = 1 + i$.

8.8 Simple Harmonic Motion

We now use our knowledge of trigonometric functions to describe motion that repeats itself periodically, such as the up-and-down bobbing motion of a buoy in the ocean, or the back-and-forth motion of a simple pendulum.

As a model for this type of motion, we consider a mass m attached to a spring suspended from the ceiling (Figure 8.20). Initially the mass is at rest (Figure 8.20(a)). At that time, the mass is said to be in the **equilibrium position.** Next, the mass is pulled downward to a position A units below equilibrium, and then released (Figure 8.20(b)). The spring then causes the mass to move upward, through the equilibrium position until it reaches a point A units above equilibrium (Figures 8.20(c) and (d)). Then, the mass begins to move back downward towards the point from which it was released.

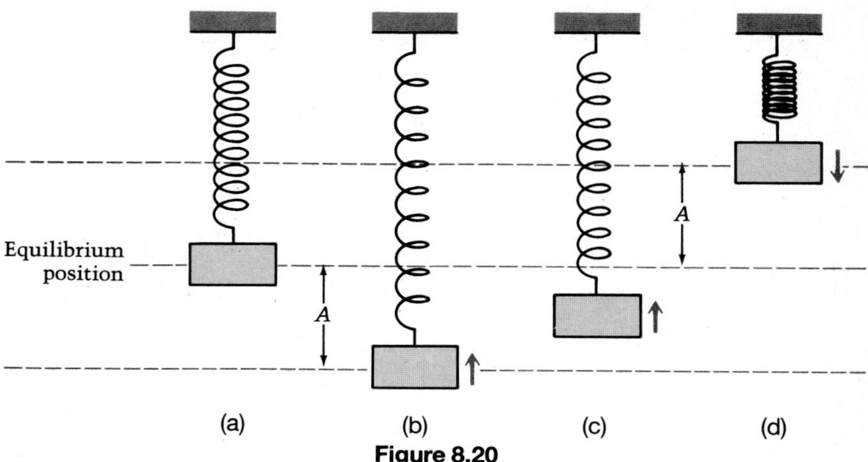

Figure 8.20

If the effects of friction in the spring and air resistance are ignored, the up-and-down motion just described would continue forever. This "idealized" motion of the mass m is called *simple harmonic motion*. It is the basic model for a variety of physical phenomena, including the vibration of a guitar string, the oscillation of atoms in a molecule, the propagation in air of sound waves, as well as various types of electromagnetic waves, such as radio waves and television signals.

The mathematical model for simple harmonic motion describes an object's displacement x, measured from the equilibrium position, as a function of time t. (The equilibrium position corresponds to the point where $x = 0$.) The sine and cosine functions are the building blocks of this model.

Simple Harmonic Motion

If an object exhibits simple harmonic motion, its displacement x from the equilibrium position is a function of time t. This function can be described by an equation of the form

$$x = A\sin(\omega t - D),$$

or alternatively,

$$x = A\cos(\omega t - D).$$

Suppose that we are considering simple harmonic motion described by the equation $x = A\sin(\omega t - D)$ where ω and D are positive numbers. Note that the equation can be rewritten as

$$x = A\sin\omega\left(t - \frac{D}{\omega}\right).$$

Recall that this form was extremely useful when we graphed transformations of the trigonometric functions. In particular, it follows from our earlier definitions that

$$\text{the amplitude is } |A|, \quad \text{and} \quad \text{the period is } \left|\frac{2\pi}{\omega}\right| = \frac{2\pi}{\omega}.$$

The **frequency** of the motion, denoted f, is the number of periods that are completed in one unit of time, and is described as the reciprocal of the period

$$f = \frac{1}{\text{period}} = \frac{1}{\frac{2\pi}{\omega}} = \frac{\omega}{2\pi}.$$

The graph of $x = A\sin(\omega t - D)$ is given in Figure 8.21, where A is assumed to be positive.

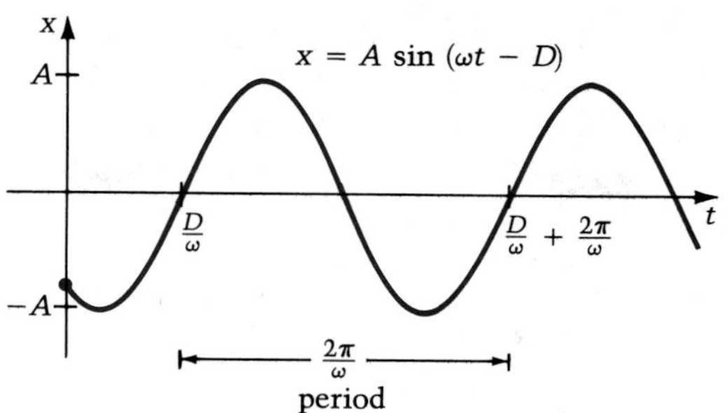

Note that this graph is a translation $\frac{D}{\omega}$ units to the right of the graph of $x = A \sin \omega t$.

Figure 8.21

EXAMPLE 1

An object suspended on a spring is pulled down 4 units from the equilibrium point. When released, the object is in simple harmonic motion, and makes one complete up-and-down oscillation every 2 sec.

(a) Determine the frequency of the motion;

(b) Write an equation for the object's displacement (x) as a function of time (t). Take displacements below the equilibrium point as negative, and those above as positive.

Solution

(a) $f = \dfrac{1}{\text{period}} = \dfrac{1}{2}$ Frequency is the reciprocal of the period.

Note that this means the object makes $\frac{1}{2}$ of an oscillation each second.

(b) We need to determine values for A, ω, and D in the general formula

$$x = A \sin(\omega t - D).$$

$A = -4$ Initial displacement is 4 units below equilibrium point, so A is negative: -4. Remember that $|A|$ is the amplitude.

$f = \dfrac{\omega}{2\pi} = \dfrac{1}{2}$ To find ω, we use the definition of f and the result from (a) that $f = \dfrac{1}{2}$.

$2\omega = 2\pi$

$\omega = \pi$

Now that we have found A and ω, we can determine D.

$x = -4 \sin(\pi t - D)$

$-4 = -4 \sin(0 - D)$ At release, $t = 0$ and $x = -4$. These values are substituted.

$1 = \sin(-D)$

$1 = -\sin D$ Sine is odd: $\sin(-x) = -\sin x$.

$-1 = \sin D$

$D = \dfrac{3\pi}{2}$ We have selected the smallest positive value of D for which $\sin D = -1$. (Recall, $D > 0$ by definition.)

An equation for the object's displacement is $x = -4\sin\left(\pi t - \dfrac{3\pi}{2}\right)$.

EXAMPLE 2

Show that the equation $x = 3\sin 2t - 4\cos 2t$, where x is measured in inches and t in seconds, describes simple harmonic motion. Determine the amplitude, period, and frequency.

Solution We wish to rewrite the equation in the form

$$x = A\sin(\omega t - D).$$

Recall: $\sin(\omega t - D) = (\sin\omega t)(\cos D) - (\cos\omega t)(\sin D)$.

Thus, $A\sin(\omega t - D) = A[(\sin\omega t)(\cos D) - (\cos\omega t)(\sin D)]$

$= (A\cos D)(\sin\omega t) - (A\sin D)(\cos\omega t)$

Compare: $x = (\quad 3 \quad)(\sin 2t) - (\quad 4 \quad)(\cos 2t)$

Therefore, $A\cos D = 3 \qquad A\sin D = 4 \qquad \text{and} \qquad \omega = 2$

$$\cos D = \frac{3}{A} \qquad \sin D = \frac{4}{A}$$

Thus, since $1 = \cos^2 D + \sin^2 D = \dfrac{3^2}{A^2} + \dfrac{4^2}{A^2} = \dfrac{25}{A^2}$,

we have $A = \pm 5$.

Let $A = 5$; then, since $\omega = 2$, we have $x = 5\sin(2t - D)$.

To determine D, notice that

$$\tan D = \frac{\sin D}{\cos D} = \frac{4A}{3A} = \frac{4}{3}, \text{ and so } D \approx 0.93 \text{ radians}$$

Thus, $x = 5\sin(2t - 0.93)$, which is an equation for simple harmonic motion.

The amplitude is 5, the period $= \dfrac{2\pi}{2} = \pi$, and the frequency $= \dfrac{1}{\pi}$.

A pendulum in its simplest form consists of a point mass suspended by a "weightless" string of length l, such that the mass moves back and forth along a small arc. For small arcs the motion of the mass can be considered approximately straight-line motion. In this case the displacement x from the

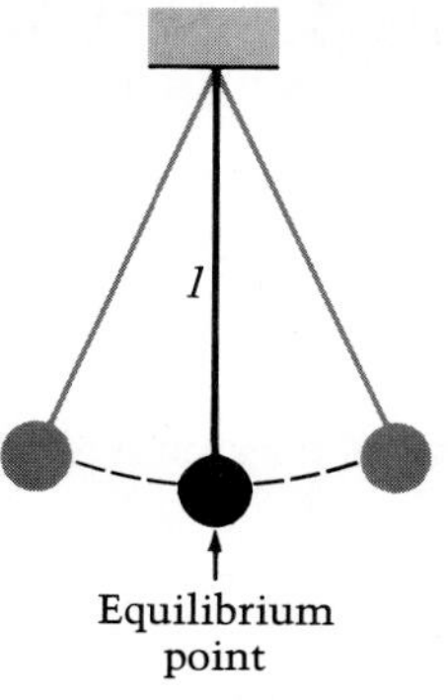

Figure 8.22

equilibrium point is given by the equation

$$x = A \cos \omega t,$$

where A is the initial displacement, and $\omega = \sqrt{\frac{g}{l}}$, with g the acceleration due to gravity: 32 ft/sec^2, or 9.8 m/sec^2.

EXAMPLE 3

A pendulum in a grandfather clock is 2 ft long. It is released on an arc with initial displacement of 0.25 ft. (a) What is the period of the pendulum? (Assume that the acceleration due to gravity is 32 ft/sec^2.) (b) Write an equation for the motion of the pendulum and sketch the graph.

Solution

(a) $\text{period} = \dfrac{2\pi}{\omega}$

$\text{period} = \dfrac{2\pi}{4} = \dfrac{\pi}{2}$ Since $\omega = \sqrt{\frac{g}{l}}$, we have $\omega = \sqrt{\frac{32}{2}} = \sqrt{16} = 4$.

This means that one back-and-forth oscillation takes approximately $\dfrac{\pi}{2} \approx 1.6$ sec.

(b) $x = A \cos \omega t$

$x = 0.25 \cos 4t$ 0.25 substituted for A; 4 for ω.

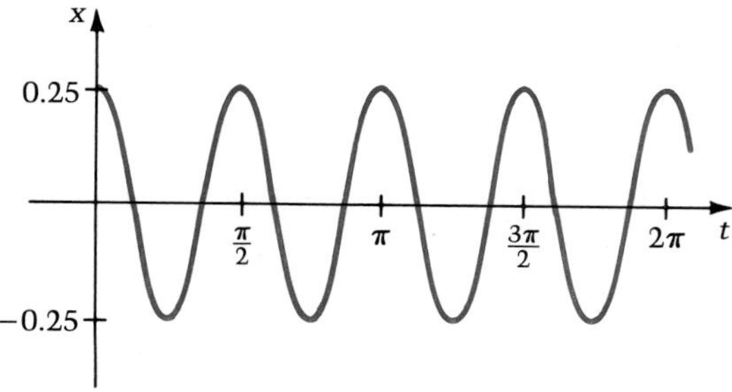

Figure 8.23

EXERCISE SET 8.8

In Exercises 1–6, a mass suspended from a spring is in simple harmonic motion, described by the given equation, where x is measured in inches and t in seconds. Use the graph of the motion to determine at each of the following times where the object is and whether it is moving up or down. (Note: take displacement below the equilibrium point as negative and above as positive.)
(a) $t = 0$, (b) $t = \frac{1}{2}$, (c) $t = 1$, (d) $t = 2$, (e) $t = 3$, (f) $t = 4$, (g) $t = 6$.

1. $x = \frac{1}{2}\sin\left(\frac{1}{2}\pi t\right)$ **2.** $x = \frac{3}{2}\cos\left(\frac{1}{2}\pi t\right)$

3. $x = 3\cos 2\pi\left(t + \frac{1}{2}\right)$ **4.** $x = 3\sin 2\pi\left(t - \frac{1}{2}\right)$

5. $x = 0.20\sin 4\pi t$ **6.** $x = 0.33\cos 6\pi t$

In Exercises 7–10, the equation represents simple harmonic motion. State the amplitude, period, and frequency. Sketch the graph.

7. $x = 6\sin 2t$ **8.** $x = 4\cos\frac{t}{2}$

9. $x = 3\cos\left(\frac{t}{2} + 2\pi\right)$ **10.** $x = 5\sin(2t - \pi)$

In Exercises 11–14, a mass suspended from a spring is in simple harmonic motion described by the given equation, where x is measured in cm and t is measured in seconds. Determine where the object is and whether it is moving up or down at $t = 0$, $t = 1.35$, $t = 2.57$, and $t = 12.39$. (Note: Take displacement below the equilibrium point as negative and above as positive.)

11. $x = 0.56\sin 4.2t$ **12.** $x = 9.32\cos(0.37t)$

13. $x = 0.24\cos\left(\frac{t}{0.25} + 1.05\right)$

14. $x = 2.36\sin(1.25t - 0.79)$

In Exercises 15–20, show that the equation, where x is measured in centimeters and t in seconds, describes simple harmonic motion. Determine the amplitude, the period, and the frequency.

15. $x = \frac{\sqrt{3}}{2}\sin t + \frac{1}{2}\cos t$ **16.** $x = \sqrt{3}\sin t - \cos t$

17. $x = \sin t - \cos t$ **18.** $x = \sin t + \cos t$

19. $x = \frac{5}{2}\sin 2t - 6\cos 2t$ **20.** $x = 4\sin 2t + \frac{15}{2}\cos 2t$

In Exercises 21–28, a situation describing simple harmonic motion is given. (a) Write an equation for the object's displacement (x) as a function of time (t). (b) Determine the frequency of the motion.

21. An object suspended on a spring is pulled down 6 in. below the equilibrium point. When released, the object is in simple harmonic motion and makes one complete oscillation every 3 sec.

22. An object attached to a spring is released from a compressed position 4 cm above its position of equilibrium. When released, the object is in simple harmonic motion and makes one complete oscillation every second.

23. A weight of 50 g is attached to a spring. The weight is pulled down 6 cm below the equilibrium point and then released. The object is then in simple harmonic motion and has a period of 1.6 sec.

24. The weight in Exercise 23 is pulled down 3 cm and released. It makes one complete oscillation every 0.8 sec.

25. A pendulum is 8 ft long and is released with an initial displacement of 0.60 ft.

26. A pendulum is 6 ft long and is released with an initial displacement of 0.62 ft.

27. A pendulum 1 m long is released with an initial displacement of 2 cm.

28. A pendulum 1.6 m long is released with an initial displacement of 6 cm.

Superset

In Exercises 29–32, write an equation for simple harmonic motion, given the amplitude and period.

29. amplitude $= 3$, period $= \frac{4\pi}{3}$

30. amplitude $= \frac{3}{5}$, period $= \pi$

31. amplitude $= \frac{3}{2}$, period $= 3\pi$

32. amplitude $= 4$, period $= \frac{3\pi}{2}$

33. If the length of a pendulum is doubled, how does the period change?

34. How does the period of a pendulum compare with that of another pendulum which is one-fourth as long?

35. A guitar string is plucked so that a point on the string makes one complete oscillation every $\frac{1}{200}$ sec. If the string is plucked by lifting the point 0.01 cm and then releasing it, write an equation for the simple harmonic motion of the point.

36. Show that the motion of a particle which moves on a line according to the equation $x = 4\sin 3t\cos 2t$ is the "sum" of two simple harmonic motions.

37. Is the motion represented by the equation $x = \sin^2 t$ simple harmonic motion?

38. A simple pendulum about 9.8 in. long has a period of 1 sec at sea level. A pendulum 4 times as long has a period of 2 sec. One that is 9 times as long has a period of 3 sec, and so on. Verify each of these statements.

Analyzing Data

The Case of the Cardboard Circles

Our definition of the trigonometric functions relied heavily on points and standard arcs on the unit circle. We found the number π to be rather prominent, not only when working with special angles in radian measure, but also when specifying the periods of the trigonometric functions. Prior to your study of trigonometry, you encountered π in the formulas for the circumference and area of a circle. Let us now attempt to use data analytic techniques to discover a formula for the area of a circle. In so doing, we will have experimentally estimated a value for π. This experience should serve to remind us that the exploration of mathematical concepts is one of the powerful uses of data analysis.

The students in Mr. Simmons' third grade class can get us started. They were asked to figure out a way of estimating the areas of various circles. (They were *not* given a formula that they could simply "plug into".) After much discussion they decided on the following indirect method.

1. Outline various circles (using drinking glasses, bowls, soda cans, etc.) on a special heavy cardboard, then cut out each circle and determine its weight W.
2. Using the fact that a 7 in. $\times$ 7 in. square (49 in.^{2}) of this cardboard weighs 4 ounces, estimate the area A of each circle by using the proportion

$$\frac{49}{4} = \frac{A}{W} \quad \frac{\text{area}}{\text{weight}}$$

The following table contains the area (estimated by the method above) and the measured radius for each of ten cardboard circles.

Area (A) (in in.^{2})	**radius (r) (in inches)**	**Area (A) (in in.^{2})**	**radius (r) (in inches)**
4.0	1.1	15	2.2
4.5	1.2	20	2.5
6.5	1.5	30	3.0
8.0	1.6	30	3.2
9.0	1.7	48	3.8

(At this point we thank the third graders for their data, but must now make use of the more advanced skills we have developed in this course.)

In Figure 8.24 we have graphed the data from the table. It seems that the relationship between area and radius is not a straight line. Let us assume that we do not know what the formula is, but that area A is proportional to some power of the radius r. Thus we wish to fit the model $A = kr^P$.

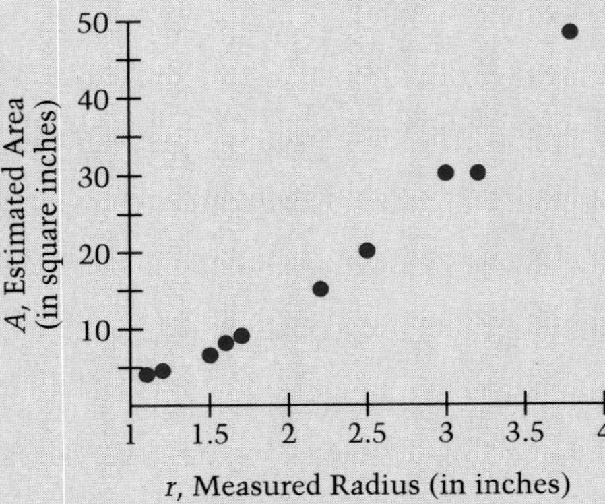

Figure 8.24

Taking the natural logarithm of each side of this equation will produce a form that looks like a linear equation.

$$\ln A = \ln kr^p$$

$$\ln A = \ln k + \ln r^p$$

$$(1) \qquad \underbrace{\ln A}_{\hat{y}} = \underbrace{\ln k}_{a} + \underbrace{p}_{b} \cdot \underbrace{\ln r}_{x}$$

Comparing equation (1) with the equation of a line, $\hat{y} = a + bx$, we see that $\ln A$ is a linear function of $\ln r$. Thus, in Figure 8.25, we will plot data points of the form $(\ln A, \ln r)$, and determine the line of best fit.

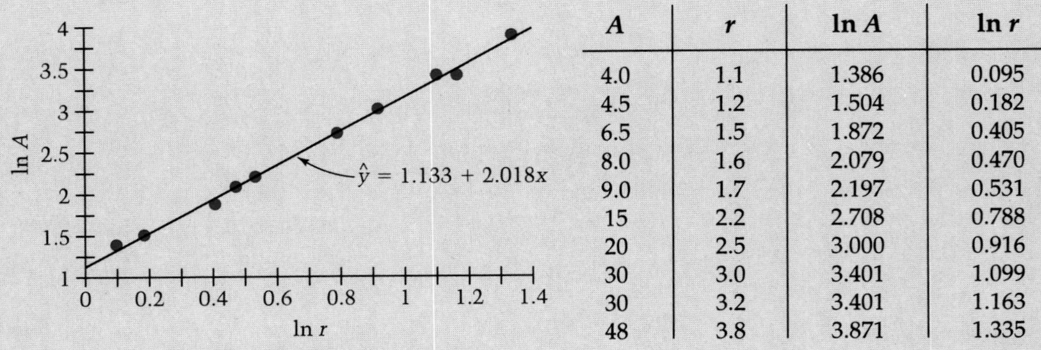

A	r	ln A	ln r
4.0	1.1	1.386	0.095
4.5	1.2	1.504	0.182
6.5	1.5	1.872	0.405
8.0	1.6	2.079	0.470
9.0	1.7	2.197	0.531
15	2.2	2.708	0.788
20	2.5	3.000	0.916
30	3.0	3.401	1.099
30	3.2	3.401	1.163
48	3.8	3.871	1.335

Figure 8.25

We now compare the line of the best fit, $\hat{y} = 1.133 + 2.018\,x$, with equation (1) above.

$$\hat{y} = 1.133 + 2.018x \qquad \ln A = \ln k + p \ln r$$

We conclude that $p = 2.018$, and since $\ln k = 1.133$, $k = e^{1.133} \approx 3.105$. Therefore our model for the area of the circle, $A = kr^p$, becomes $A = 3.105\,r^{2.018}$, not a bad estimate for the true formula $A = \pi r^2$, considering we developed it by using data gathered by a group of third graders.

Chapter Review

8.1 Basic Trigonometric Identities (pp. 461–466)

Reciprocal Identities

(1) $\sin\theta = \dfrac{1}{\csc\theta}$

(2) $\cos\theta = \dfrac{1}{\sec\theta}$

(3) $\tan\theta = \dfrac{1}{\cot\theta}$

(4) $\cot\theta = \dfrac{1}{\tan\theta}$

(5) $\sec\theta = \dfrac{1}{\cos\theta}$

(6) $\csc\theta = \dfrac{1}{\sin\theta}$

Quotient Identities

(7) $\tan\theta = \dfrac{\sin\theta}{\cos\theta}$

(8) $\cot\theta = \dfrac{\cos\theta}{\sin\theta}$

Pythagorean Identities

(9) $\sin^2\theta + \cos^2\theta = 1$

(10) $\tan^2\theta + 1 = \sec^2\theta$

(11) $1 + \cot^2\theta = \csc^2\theta$

8.2 Sum and Difference Identities (pp. 466–473)

(12) $\cos(\alpha - \beta) = \cos\alpha\cos\beta + \sin\alpha\sin\beta$

(13) $\cos(\alpha + \beta) = \cos\alpha\cos\beta - \sin\alpha\sin\beta$

(14) $\sin(\alpha + \beta) = \sin\alpha\cos\beta + \cos\alpha\sin\beta$

(15) $\sin(\alpha - \beta) = \sin\alpha\cos\beta - \cos\alpha\sin\beta$

(16) $\tan(\alpha + \beta) = \dfrac{\tan\alpha + \tan\beta}{1 - \tan\alpha\tan\beta}$

(17) $\tan(\alpha - \beta) = \dfrac{\tan\alpha - \tan\beta}{1 + \tan\alpha\tan\beta}$

8.3 The Double-Angle and Half-Angle Identities (pp. 473–482)

(18) $\sin 2\theta = 2\sin\theta\cos\theta$

(19) $\cos 2\theta = \cos^2\theta - \sin^2\theta$

(20) $\cos 2\theta = 1 - 2\sin^2\theta$

(21) $\cos 2\theta = 2\cos^2\theta - 1$

(22) $\tan 2\theta = \dfrac{2\tan\theta}{1 - \tan^2\theta}$

(23) $\sin\dfrac{\theta}{2} = \pm\sqrt{\dfrac{1 - \cos\theta}{2}}$

(24) $$\cos\frac{\theta}{2} = \pm\sqrt{\frac{1+\cos\theta}{2}}$$

(25) $$\tan\frac{\theta}{2} = \pm\sqrt{\frac{1-\cos\theta}{1+\cos\theta}}$$

(26) $$\tan\frac{\theta}{2} = \frac{\sin\theta}{1+\cos\theta}$$

(27) $$\tan\frac{\theta}{2} = \frac{1-\cos\theta}{\sin\theta}$$

8.4 Identities Revisited (pp. 482–488)

(28) $$\sin\alpha\cos\beta = \frac{1}{2}[\sin(\alpha+\beta)+\sin(\alpha-\beta)]$$

(29) $$\sin\alpha\sin\beta = \frac{1}{2}[\cos(\alpha-\beta)-\cos(\alpha+\beta)]$$

(30) $$\cos\alpha\cos\beta = \frac{1}{2}[\cos(\alpha+\beta)+\cos(\alpha-\beta)]$$

(31) $$\sin x+\sin y = 2\sin\left(\frac{x+y}{2}\right)\cos\left(\frac{x-y}{2}\right)$$

(32) $$\sin x-\sin y = 2\cos\left(\frac{x+y}{2}\right)\sin\left(\frac{x-y}{2}\right)$$

(33) $$\cos x+\cos y = 2\cos\left(\frac{x+y}{2}\right)\cos\left(\frac{x-y}{2}\right)$$

(34) $$\cos x-\cos y = -2\sin\left(\frac{x+y}{2}\right)\sin\left(\frac{x-y}{2}\right)$$

8.5 Trigonometric Equations (pp. 489–497)

A trigonometric equation is an equation that involves trigonometric functions. For example, $2\cos^2 x - \cos x = 1$ is a trigonometric equation. To solve such equations usually involves a two-step process. For example, for this equation, first solve for $\cos x$ and then, using trigonometry, solve for x.

8.6 Polar Coordinates (pp. 497–504)

The *polar coordinates* (r, θ) specify a point that lies at a distance $|r|$ from the origin. If $r > 0$, the point lies on the θ-ray. If $r < 0$, the point lies on the opposite of the θ-ray. If $r = 0$, the point lies at the origin, regardless of the value of θ. (p. 498) If point P has rectangular coordinates (x, y) and polar coordinates (r, θ), then $x = r\cos\theta$, $y = r\sin\theta$, $\tan\theta = \frac{y}{x}$ (provided $x \neq 0$) and $x^2 + y^2 = r^2$. (p. 499)

8.7 Trigonometric Form of Complex Numbers (pp. 504–515)

The complex number $z = a + bi$ has trigonometric form

$$z = r(\cos\theta + i\sin\theta),$$

where $r = |z| = \sqrt{a^2 + b^2}$ and $\tan\theta = \dfrac{b}{a}$. The number r is called the *modulus* of z and θ is called the *argument* of z. (p. 507)

De Moivre's Theorem: For any positive integer n,

$$[r(\cos\theta + i\sin\theta)]^n = r^n(\cos n\theta + i\sin n\theta). \qquad \text{(p. 510)}$$

The nth Root Theorem: The n nth roots of the nonzero complex number $r(\cos\theta + i\sin\theta)$ are given by

$$\sqrt[n]{r}\left[\cos\left(\frac{\theta + k\cdot 360^\circ}{n}\right) + i\sin\left(\frac{\theta + k\cdot 360^\circ}{n}\right)\right]$$

where $k = 0, 1, 2, \ldots, n - 1$. We can replace 360° with 2π in the above statement to accommodate radian measure. (p. 512)

8.8 Simple Harmonic Motion (pp. 515–520)

If an object exhibits *simple harmonic motion,* its displacement x from the equilibrium position is a function of t. This function can be described by either of the equations $x = A\sin(\omega t - D)$, or $x = A\cos(\omega t - D)$.

Review Exercises

In Exercises 1–4, rewrite the expression in terms of sin x only.

1. $\dfrac{\sec x}{\sec x + \tan x}$
2. $\tan^2 x$
3. $\dfrac{\cos x + \cot x}{\sec x}$
4. $\dfrac{1}{1 + \cot^2 x}$

In Exercises 5–14, verify the identity.

5. $\sin x(\cot x + \csc x) = 1 + \cos x$
6. $(1 + \sec^2 x) - \tan^2 x = 2$
7. $\dfrac{\csc x}{\cot x} - \dfrac{\cot x}{\csc x} = \sin x \tan x$
8. $\dfrac{\cos x}{1 + \sin x} + \dfrac{\sin x - 1}{\cos x} = 0$
9. $(\sin x + \cos x)^2 - 1 = \sin 2x$
10. $(\cot x + \tan x)^2 = 4\csc^2 2x$
11. $\dfrac{\sin(x + y)}{\cos x \cos y)} = \tan x + \tan y$
12. $\sin(x + y)\cos y - \cos(x + y)\sin y = \sin x$
13. $\dfrac{\sin 4x}{4\sin x} = \cos x \cos 2x$
14. $\dfrac{\cos 4x - 1}{4\cos^4 x} = -2\tan^2 x$

In Exercises 15–18, use the sum and difference identities to evaluate.

15. $\sin 105^\circ$
16. $\cos 165^\circ$
17. $\cos\left(-\dfrac{7\pi}{12}\right)$
18. $\sin\dfrac{\pi}{12}$

In Exercises 19–20, use the given information to determine (a) $\sin(\alpha + \beta)$, (b) $\cos(\alpha + \beta)$, and (c) $\tan(\alpha + \beta)$.

19. $\sin \alpha = \frac{3}{5}$, α is a second quadrant angle, β is a third quadrant angle, and $\cos \beta = -\frac{5}{13}$.

20. $\cos \alpha = \frac{4}{5}$, α is a fourth quadrant angle, β is a second quadrant angle, and $\sin \beta = \frac{12}{13}$.

In Exercises 21–22, from the given information determine (a) $\sin 2\alpha$, (b) $\cos 2\alpha$, and (c) $\tan 2\alpha$.

21. $\cos \alpha = \frac{3}{5}$, α in Quadrant IV

22. $\tan \alpha = -2$, α in Quadrant II

In Exercises 23–32, determine all the solutions of the equation in the interval $[0, 2\pi)$.

23. $\sin 5x = \frac{1}{2}$

24. $\tan 4x = -1$

25. $3 \tan^2 2x - 1 = 0$

26. $\csc^2 3x - 4 = 0$

27. $\sin 2x = \tan x$

28. $\cos 2x - \cos x + 1 = 0$

29. $2 \cos^2 x + 3 \cos x - 2 = 0$

30. $\sin^2 x - 2 \sin x - 3 = 0$

31. $6 \sin^2 x + \sin x - 12 = 0$

32. $10 \cos^2 x - \cos x + 2 = 0$

In Exercises 33–36, determine all the solutions of the equation in the interval $[0, 360°)$ to the nearest degree.

33. $3 \tan^2 x - \tan x - 2 = 0$

34. $5 \cot^2 x + \cot x - 4 = 0$

35. $3 \sin 2x = \tan x$

36. $5 \sin 2x = \cot x$

In Exercises 37–40, solve for x.

37. $\sin\left(2 \arcsin \frac{4}{5}\right) = x$

38. $\cos\left(\arctan \frac{2}{3}\right) = x$

39. $\sin^{-1}\left(\frac{4}{5}\right) + \cos^{-1}\left(\frac{5}{13}\right) = \cos^{-1} x$

40. $\sin^{-1}\left(\frac{4}{5}\right) - \cos^{-1}\left(\frac{12}{13}\right) = \sin^{-1} x$

In Exercises 41–42, the rectangular coordinates of a point are given. Determine two polar representations (r, θ) for the point with $0 \le \theta < 2\pi$.

41. $(-1, 1)$

42. $(-\sqrt{3}, -1)$

In Exercises 43–50, convert the given polar equation to rectangular form.

43. $r = 2$

44. $\theta = \frac{\pi}{6}$

45. $r(2 \sin \theta - 3 \cos \theta) = 1$

46. $r = 2 \cos \theta$

47. $\tan \theta = 2$

48. $r^2 = \cos 2\theta$

49. $r^2 + \sin 2\theta = 0$

50. $r \csc \theta = 3$

In Exercises 51–54, convert the given rectangular equation to polar form.

51. $x^2 = 3y + 6$

52. $6xy = 1$

53. $x^2 - y^2 = 8xy$

54. $x = 3$

In Exercises 55–58, sketch the graph of the polar equation.

55. $r = 2 + \sin \theta$

56. $r = 3 - \cos \theta$

57. $r = 4 \cos \theta$

58. $r(\sin \theta + \cos \theta) = 2$

In Exercises 59–60, determine the product by multiplying the trigonometric forms of the numbers. Express the answer in the form $a + bi$.

59. $(\sqrt{3} + i)(-1 + i\sqrt{3})$

60. $(2 - 2i)(-2 - 2i)$

In Exercises 61–62, find z_1/z_2 by using the theorem on dividing trigonometric forms.

61. $z_1 = 1 - i, \quad z_2 = -2 - 2i\sqrt{3}$

62. $z_1 = \sqrt{3} + i, \quad z_2 = 2 + 2i$

In Exercises 63–64, use DeMoivre's Theorem to express the quantity in the form $a + bi$.

63. $(2 - 2i)^3$

64. $(-3 + 3i\sqrt{3})^4$

In Exercises 65–66, find all values of z that satisfy the equation.

65. $z^3 = 8i$

66. $z^4 = -1 + i\sqrt{3}$

In Exercises 67–68, the equation represents simple harmonic motion. Determine the amplitude, period, and frequency. Sketch the graph.

67. $x = 3 \cos\left(\frac{t}{2} - \pi\right)$

68. $x = 5 \sin(3t + \pi)$

69. Express $\tan 6x$ in terms of $\tan 2x$.

70. Find the distance between the two polar points $\left(8, \frac{\pi}{6}\right)$ and $\left(-2, \frac{3\pi}{2}\right)$. (*Hint:* Use the Law of Cosines.)

Graphing Calculator Exercises

[Round all estimated values to the nearest hundredth.]

1. (a) Graph the functions $y = x^2 + 1$ and $y = \sin x$ on the same set of axes. Based on these graphs, why can you conclude that the equation $x^2 + 1 = \sin x$ has no solutions?
   (b) Since $x^2 + 1 = \sin x$ has no solutions, what can you conclude about the function $f(x) = x^2 + 1 - \sin x$?
2. (a) Graph the functions $y = x^2$ and $y = \sin x$ on the same set of axes, then use the graphs to determine the interval(s) where the graph of $y = x^2$ lies above that of $y = \sin x$.
   (b) Based on the graphs in part a, on what interval(s) is the inequality $\sin x < x^2$ true?
   (c) On what interval(s) is the inequality $x^2 < \sin x$ true?

In Exercises 3–8, solve the equation over the interval $[0, 2\pi)$. (*Hint:* Recall that to solve the equation $h(x) = g(x)$, sketch the graph of function $f(x) = h(x) - g(x)$ and approximate the x-intercepts.)

3. $\cos x = x$
4. $\sin 2x - \sin x = x^3 - 2x^2$
5. $\cos 2x = 4x^2 - 24x + 29$
6. $2x = \sin x$
7. $\sin^2 x = x + \ln x$
8. $\cos^2 x = \dfrac{1}{x}$

In Exercises 9–12, use a graphing technique to determine whether the equation is an identity. (*Hint:* One way of checking that an equation $h(x) = g(x)$ is an identity is to graph $y = h(x)$ and $y = g(x)$ on the same set of axes, and observe that the graphs coincide.)

9. $\cos\left(\dfrac{\pi}{2} + \theta\right) = -\sin\theta$
10. $\sin\left(\dfrac{3\pi}{2} - \theta\right) = \sin\left(\dfrac{\pi}{2} + \theta\right)$
11. $\cos\left(\dfrac{\pi}{3} + \theta\right) + \cos\left(\dfrac{\pi}{3} - \theta\right) = \cos\theta$
12. $\sin 3\theta + \sin\theta = 2\sin 2\theta\cos\theta$

Chapter Test

In Problems 1–2, verify the identity.

1. $\dfrac{1 - \tan x}{\sec x} + \sin x = \cos x$
2. $\dfrac{\tan^2 x + 1}{\tan^2 x + 2} = \dfrac{1}{1 + \cos^2 x}$
3. Evaluate $\cot\dfrac{5\pi}{12}$ without using a calculator or Table 4.

In Problems 4–5, verify the identity.

4. $\dfrac{\sin(a + b)}{\cos a\cos b} = \tan a + \tan b$
5. $\dfrac{\tan a - \tan b}{\tan a + \tan b} = \dfrac{\sin(a - b)}{\sin(a + b)}$
6. Given that $\alpha = 315°$, find $\sin 2\alpha$, $\cos 2\alpha$, and $\tan 2\alpha$ using the double-angle identities.

In Problems 7–8, from the given information, determine $\cos\dfrac{\alpha}{2}$.

7. $\sin\alpha = -\dfrac{2}{3}$, α in Quadrant III
8. $\tan\alpha = -\dfrac{3}{2}$, α in Quadrant II

In Problems 9–10, simplify the expression.

9. $\dfrac{\cot\varphi - \tan\varphi}{\cot\varphi + \tan\varphi}$
10. $\dfrac{\sin 2\alpha}{1 - \cos 2\alpha}$

In Problems 11–12, verify the identity.

11. $\dfrac{1}{\sin x} - \sin x = \dfrac{\sin x}{\tan^2 x}$
12. $\sin\left(\dfrac{\pi}{4} - \alpha\right) = \cos\left(\alpha + \dfrac{\pi}{4}\right)$
13. Evaluate $\cos\dfrac{11\pi}{12} - \cos\dfrac{5\pi}{12}$ by using the conversion identities.
14. Write $\sin 4x\cos 3x$ as a sum or difference.

In Problems 15–16, solve for x. Express the answer in radians.

15. $2\cos^2 x - \cos x = 1$
16. $\sin^2 x + 3\sin x = 4$

In Problems 17–18, solve the equation for x in $[0°, 360°)$.

17. $\tan^2 x - 2 \tan x - 3 = 0$

18. $12 \sin^2 x + 12 \sin x - 6 = 0$

In Problems 19–20, solve the equation for x.

19. $\arcsin\left(\frac{3}{5}\right) - \arccos\left(\frac{4}{5}\right) = \sin^{-1} x$

20. $\tan^{-1} 1 + \tan^{-1}\left(\frac{12}{5}\right) = x$

In Problems 21–22, graph the polar point and determine the x- and y-coordinates.

21. $(4, -135°)$

22. $(-3, 210°)$

23. Convert the polar equation $r = 3 \sin \theta$ to rectangular form.

24. Sketch the graph of the polar equation $r = 3 - 2 \cos \theta$.

25. Represent each complex number as a point in the complex plane (a) $5 - 12i$, (b) -9, (c) $-7i$.

26. Determine the absolute value of each complex number (a) $4 + 5i$, (b) $-9i$, (c) $-1 + i\sqrt{3}$.

In Problems 27–28, express the complex number in trigonometric form.

27. $3 - i\sqrt{3}$

28. $-4 - 4i$

29. Find $z_1 z_2$ and $\frac{z_1}{z_2}$ where $z_1 = 6\sqrt{3} + 6i$ and $z_2 = 1 - i\sqrt{3}$ by using trigonometric forms. Express the answers in the form $a + bi$.

30. Use De Moivre's Theorem to express $\left(-\frac{\sqrt{3}}{2} + \frac{1}{2}i\right)^5$ in the form $a + bi$.

31. The equation $x = 4 \sin\left(\frac{t}{2} - 2\pi\right)$ represents simple harmonic motion. Determine the amplitude, period, and frequency. Sketch the graph.

9 Systems of Equations and Inequalities

In this chapter we study the equations already considered in previous chapters, but in an entirely new context: two or three at a time. These systems of equations *provide realistic models for problems involving several variables, when several pieces of information about these variables are known. The method of* Linear Programming *is one such technique we present for solving such problems. At the end of the chapter we will discuss situations as diverse as international economic competition, traffic safety, and the interaction of personal fitness with everyday stress. We will see that considering more than one function at a time is a very natural way to gain an understanding of real problems.*

9.1 Systems of Equations in Two Variables

In Chapter 3 we saw that the graph of an equation of the form

$$Ax + By = C$$

(where A and B are not both zero) is a straight line. Now, let us consider a pair of such equations.

$$\begin{cases} 4x + 3y = 11 \\ -3x + 2y = -4. \end{cases}$$

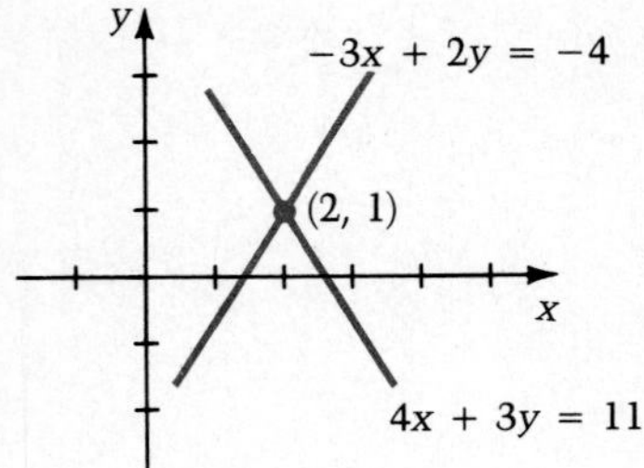

Figure 9.1

These two equations are graphed in Figure 9.1. Each point on these lines is a solution of one equation or the other. Only the point (2, 1) lies on both lines, and thus, only the ordered pair (2, 1) is a solution of both equations.

A collection of two or more equations is called a **system of equations.** The example above is a system of two equations in two variables. Those ordered pairs that are solutions of *every* equation in the system are called **solutions of the system.** We refer to the process of finding solutions of the system as **solving the system** (or solving the system simultaneously).

EXAMPLE 1

Determine whether the following ordered pairs are solutions of the system

$$\begin{cases} x + 2y + 1 = 0 \\ 3x + 4y - 3 = 0. \end{cases}$$

(a) $(3, -2)$ (b) $(5, -3)$

Solution

(a) $x + 2y + 1 = 0$

$3 + 2(-2) + 1 = 0$ $(3, -2)$ is a solution of the first equation.

$3x + 4y - 3 = 0$

$3(3) + 4(-2) - 3 \neq 0$ $(3, -2)$ is not a solution of the second equation.

Thus, $(3, -2)$ is not a solution of the system.

(b) $x + 2y + 1 = 0$

$5 + 2(-3) + 1 = 0$ $(5, -3)$ is a solution of the first equation.

$3x + 4y - 3 = 0$

$3(5) + 4(-3) - 3 = 0$ $(5, -3)$ is a solution of the second equation.

Thus, $(5, -3)$ is a solution of the system.

Recall the strategy for solving a linear equation in one variable. Given an equation such as $3(2 - x) + 5x = 18 - 4x$, we rewrite the equation as a succession of equivalent equations,

$$\begin{aligned} 6 - 3x + 5x &= 18 - 4x \\ 2x + 6 &= 18 - 4x \\ 6x &= 12 \\ x &= 2, \end{aligned}$$

until we obtain an equivalent equation (such as $x = 2$) whose solution is obvious.

Our approach to solving a system of equations is similar. We will develop techniques whereby a system such as

$$\begin{cases} 4x + 3y = 11 \\ -3x + 2y = -4 \end{cases}$$

can be written as a simpler, equivalent system, such as

$$\begin{cases} x = 2 \\ y = 1. \end{cases}$$

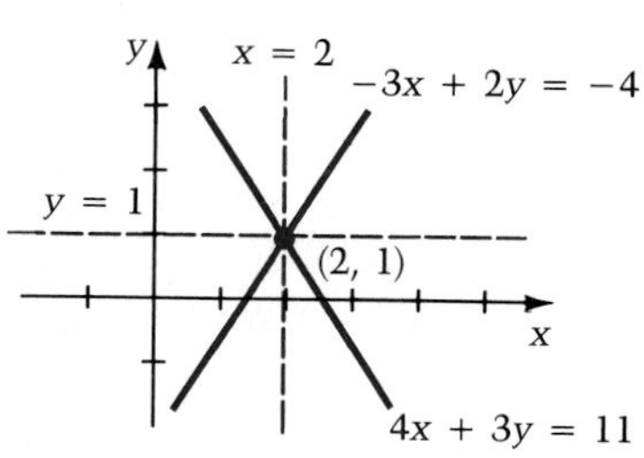

Figure 9.2

Notice that the lines $x = 2$ and $y = 1$ intersect at the point (2, 1). Thus, the lines in the original system also intersect at this point, and the ordered pair (2, 1) is a solution of the system.

The theorem below can be used to justify the procedures for writing equivalent systems. (In the statement of the theorem, we use the phrase "multiple of the equation" to mean the new equation found by multiplying both sides of the original equation by the same number.)

Theorem

Suppose lines $\mathscr{L}$ and $\mathscr{M}$ intersect at point P. Adding any nonzero multiple of the equation of line $\mathscr{L}$ to any nonzero multiple of the equation of line $\mathscr{M}$ produces an equation of a line. This line also passes through point P.

For example, for the system at the top of the page, we have the following:

$$\begin{array}{rcll} 8x + 6y &=& 22 & 2 \text{ "times" } 4x + 3y = 11 \\ + \ -15x + 10y &=& -20 & 5 \text{ "times" } -3x + 2y = -4 \\ \hline -7x + 16y &=& 2 & \text{The "sum" of the two equations.} \end{array}$$

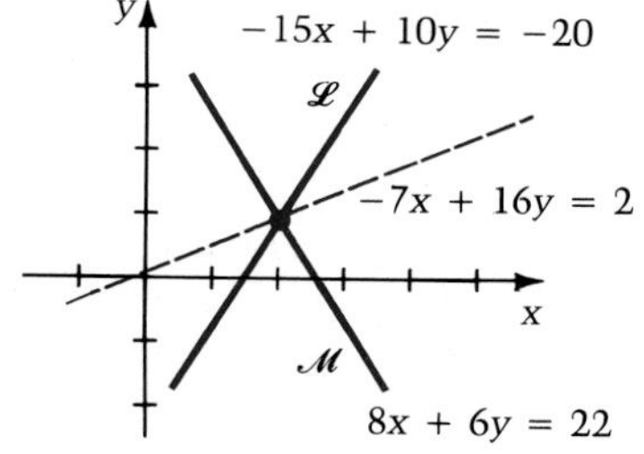

Figure 9.3

As the theorem promises, the point corresponding to the solution (2, 1) also lies on the line $-7x + 16y = 2$ (see Figure 9.3).

The first technique we shall use to solve a system of equations is called the **addition method.** This method involves two steps.

Step 1. Form multiples of the equations so that their sum is an equation *in one variable* and solve this resulting equation.

Step 2. Substitute the value found in Step 1 back into either of the original equations to find the value of the other variable.

EXAMPLE 2

Use the addition method to solve the system $\begin{cases} 4x + 3y = 11 \\ -3x + 2y = -4. \end{cases}$

Solution

$$\begin{cases} 4x + 3y = 11 \\ -3x + 2y = -4 \end{cases}$$

$$3(4x + 3y) = 3(11)$$
$$4(-3x + 2y) = 4(-4)$$

Step 1. To eliminate x, multiply the first equation by 3 and the second equation by 4.

$$\begin{array}{rr} & 12x + 9y = 33 \\ + & -12x + 8y = -16 \\ \hline & 17y = 17 \end{array}$$

Add the equations.

Solve the equation for y.

$$y = 1$$

Replace one of the equations in the original system with $y = 1$.

$$\begin{cases} 4x + 3y = 11 \\ y = 1 \end{cases}$$

$$4x + 3(1) = 11$$

Step 2. Substitute the value of y into the other equation and solve for x.

$$4x = 8$$
$$x = 2$$

Use the equation $x = 2$ to state a simpler, equivalent system.

$$\begin{cases} x = 2 \\ y = 1 \end{cases}$$

The solution is (2, 1).

We chose the multipliers 3 and 4 in Example 2 so that the sum of the resulting equations would have no x-term. Instead, we could have eliminated the y-term by using the multipliers 2 and -3:

$$\begin{array}{rcr} 2(4x + 3y) = 2(11) & \longrightarrow & 8x + 6y = 22 \\ -3(-3x + 2y) = -3(-4) & \longrightarrow & + \; 9x - 6y = 12 \\ \hline & & 17x = 34 \end{array}$$

There is always more than one way to select the multipliers.

There are three types of answers you can get when you solve a system of two linear equations in two variables. These three types are illustrated in Figure 9.4.

In the Figure 9.4(a), lines $\mathcal{L}$ and $\mathcal{M}$ are nonparallel. The corresponding system of equations has exactly one solution. In the Figure 9.4(b), lines $\mathcal{L}$ and $\mathcal{M}$ are different parallel lines. In this case, the system has no solutions. In the Figure 9.4(c), lines $\mathcal{L}$ and $\mathcal{M}$ are the same. Such a system has infinitely many solutions. We looked at the first type of system in Example 2. We will consider the second and third types in Examples 3 and 5.

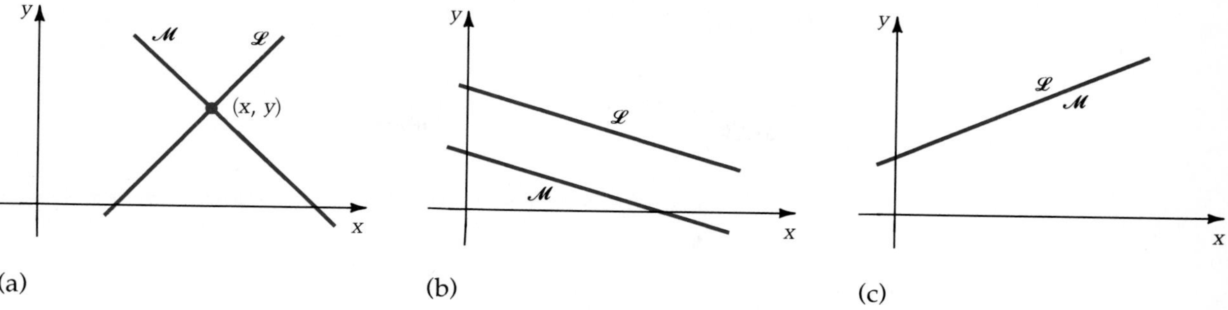

Figure 9.4 (a) The system has a unique solution. (b) The system has no solution. (c) The system has infinitely many solutions.

EXAMPLE 3

Use the addition method to solve the system $\begin{cases} 3x + 4y = 7 \\ 6x + 8y = 5. \end{cases}$

Solution

$$\begin{cases} 3x + 4y = 7 \\ 6x + 8y = 5 \end{cases}$$

$$\begin{aligned} 2(3x + 4y) &= 2(7) \\ -1(6x + 8y) &= -1(5) \end{aligned}$$

Step 1. To eliminate x, multiply the first equation by 2 and the second equation by -1.

$$\begin{array}{rr} & 6x + 8y = 14 \\ + & -6x - 8y = -5 \\ \hline & 0 = 9 \end{array}$$

Add the equations.

Note that both variables have been eliminated. Replace one of the equations in the original system with $0 = 9$.

$$\begin{cases} 3x + 4y = 7 \\ 0 = 9 \end{cases}$$

The statement $0 = 9$ is never true. A system that includes such an equation has no solutions: no values of x and y will ever make this equation true. Since the original system is equivalent to this system, it also has no solutions.

To verify our work in Example 3, we rewrite the equations in slope-intercept form.

$$y = -\frac{3}{4}x + \frac{7}{4} \quad \text{and} \quad y = -\frac{3}{4}x + \frac{5}{8}.$$

These are different parallel lines. (Slopes are the same, but y-intercepts are different.) Thus, the lines do not intersect.

The Substitution Method

The second technique we shall use to solve a system of equations is called the **substitution method.** This method involves three steps, as shown in Example 4.

EXAMPLE 4

Use the substitution method to solve the system $\begin{cases} 3y + 4x = 11 \\ 2y - 3x = -4. \end{cases}$

Solution

$$\begin{cases} 3y + 4x = 11 \\ 2y - 3x = -4 \end{cases}$$

$$\begin{aligned} 3y + 4x &= 11 \\ 4x &= -3y + 11 \\ x &= -\frac{3}{4}y + \frac{11}{4} \end{aligned}$$

Step 1. Solve one of the equations for one of the variables. We have solved for x in the first equation.

$$\begin{cases} x = -\frac{3}{4}y + \frac{11}{4} \\ 2y - 3x = -4 \end{cases}$$

$$\begin{aligned} 2y - 3\left(-\frac{3}{4}y + \frac{11}{4}\right) &= -4 \\ 2y + \frac{9}{4}y - \frac{33}{4} &= -4 \\ \frac{17}{4}y &= -4 + \frac{33}{4} \\ 17y &= 17 \\ y &= 1 \end{aligned}$$

Step 2. Substitute the expression for x from Step 1 into the second equation, and then solve for y.

$$\begin{cases} x = -\frac{3}{4}y + \frac{11}{4} \\ y = 1 \end{cases}$$

$$\begin{aligned} x &= -\frac{3}{4}(1) + \frac{11}{4} \\ x &= 2 \end{aligned}$$

Step 3. Substitute the y-value found in Step 2 into the other equation to obtain a value for x.

$$\begin{cases} x = 2 \\ y = 1 \end{cases}$$

The solution of the system is (2, 1). (You should check that (2, 1) satisfies both equations in the given system.)

Careful! In the previous examples, notice that we have restated the system as an equivalent system at various stages. You should follow this practice in order to keep track of the pair of equations you must work on next.

Let us summarize the three steps of the substitution method.

Step 1. Solve one of the equations for one of the variables.

Step 2. Substitute the result from Step 1 into the other equation of the system, and solve for the second variable.

Step 3. Substitute the value found in Step 2 into either one of the original equations and solve for the value of the first variable.

The next example illustrates a system in which the two equations describe the same line. Notice that we never reach the third step of the substitution method.

EXAMPLE 5

Use the substitution method to solve the system $\begin{cases} 3x + 4y = 7 \\ 6x + 8y = 14. \end{cases}$

Solution

$$\begin{cases} 3x + 4y = 7 \\ 6x + 8y = 14 \end{cases}$$

$$\begin{aligned} 6x + 8y &= 14 \\ 6x &= -8y + 14 \\ x &= -\frac{4}{3}y + \frac{7}{3} \end{aligned}$$

Step 1. Solve for x in the second equation.

$$\begin{cases} 3x + 4y = 7 \\ x = -\frac{4}{3}y + \frac{7}{3} \end{cases}$$

$$\begin{aligned} 3\left(-\frac{4}{3}y + \frac{7}{3}\right) + 4y &= 7 \\ -4y + 7 + 4y &= 7 \\ 7 &= 7 \end{aligned}$$

Step 2. Substitute the expression for x and (try to) solve for y.

$$\begin{cases} 7 = 7 \\ x = -\frac{4}{3}y + \frac{7}{3} \end{cases}$$

The equation $7 = 7$ is always true. The solutions of the equation $x = -\frac{4}{3}y + \frac{7}{3}$ are all the points on that line. Therefore, the solution set of the system is the set of all points on the line $x = -\frac{4}{3}y + \frac{7}{3}$.

Nonlinear Systems

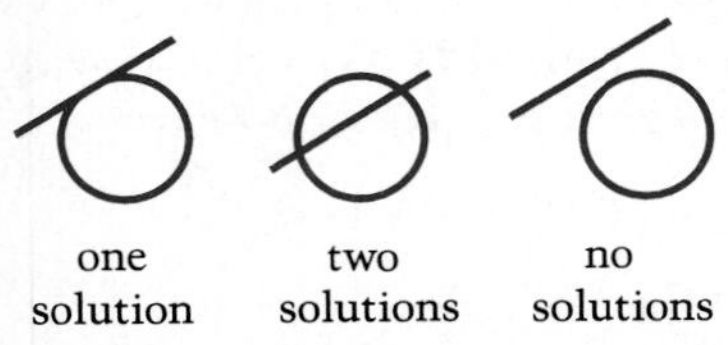

Figure 9.5

When solving a system that contains a nonlinear equation we follow this general strategy: rewrite the system as a succession of equivalent systems until a solution becomes obvious.

The equations in the next example represent a line and a circle. There are only three types of solutions that may be obtained; these are shown in Figure 9.5. Notice that there cannot be three or more points of intersection.

EXAMPLE 6

Solve the system $\begin{cases} y = x - 2 \\ (x-4)^2 + y^2 = 4. \end{cases}$

Solution

$$\begin{cases} y = x - 2 \\ (x-4)^2 + y^2 = 4 \end{cases}$$

$$(x-4)^2 + y^2 = 4$$
$$(x-4)^2 + (x-2)^2 = 4$$

Substitute the expression for y from the first equation into the second equation to obtain an equivalent system.

$$\begin{cases} y = x - 2 \\ (x-4)^2 + (x-2)^2 = 4 \end{cases}$$

$$(x-4)^2 + (x-2)^2 = 4$$

Solve the second equation for x.

$$x^2 - 8x + 16 + x^2 - 4x + 4 = 4$$
$$2x^2 - 12x + 16 = 0$$
$$x^2 - 6x + 8 = 0$$
$$(x-2)(x-4) = 0$$
$$x = 2 \quad \text{or} \quad x = 4$$

Each of these two equations produces one equivalent system.

$$\begin{cases} y = x - 2 \\ x = 2 \end{cases} \qquad \begin{cases} y = x - 2 \\ x = 4 \end{cases}$$

$$y = (2) - 2 \qquad y = (4) - 2$$
$$y = 0 \qquad y = 2$$

Substitute the x-values into the first equation.

$$\begin{cases} y = 0 \\ x = 2 \end{cases} \qquad \begin{cases} y = 2 \\ x = 4 \end{cases}$$

The given system has two solutions: (2, 0) and (4, 2). (You should check that these are, in fact, solutions of each equation in the original system.)

In the next example, the equations represent two different circles. Such a system may have no solutions, one solution, or two solutions, as illustrated in Figure 9.6.

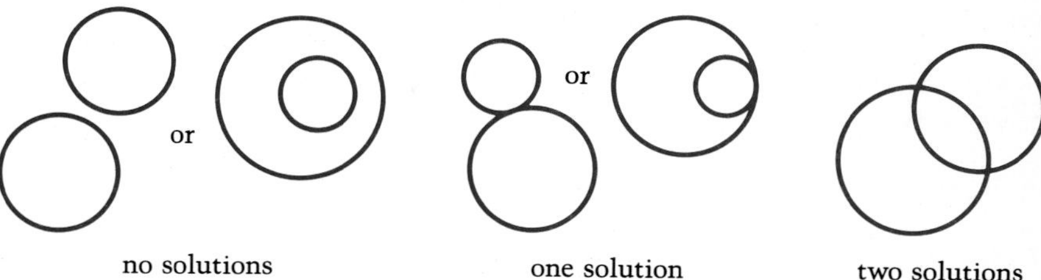

Figure 9.6

EXAMPLE 7

Solve the system $\begin{cases} (x-2)^2 + (y-2)^2 = 4 \\ (x-4)^2 + y^2 = 4. \end{cases}$

Solution

$$\begin{cases} (x-2)^2 + (y-2)^2 = 4 \\ (x-4)^2 + y^2 = 4 \end{cases}$$

$$\begin{aligned} (x^2 - 4x + 4) + (y^2 - 4y + 4) &= 4 \\ (x^2 - 8x + 16) + y^2 &= 4 \end{aligned}$$

Each expression is expanded.

$$\begin{array}{rl} x^2 + y^2 - 4x - 4y + 4 &= 0 \\ -\quad (x^2 + y^2 - 8x + 12) &= -(0) \\ \hline 4x - 4y - 8 &= 0 \end{array}$$

We subtract the second equation from the first.

$$\begin{aligned} 4y &= 4x - 8 \\ y &= x - 2 \end{aligned}$$

We simplify and then state an equivalent system.

$$\begin{cases} y = x - 2 \\ (x-4)^2 + y^2 = 4 \end{cases}$$

This is precisely the system we solved in Example 6. Our original system is equivalent to a system for which we already know the solutions: (2, 0) and (4, 2).

EXERCISE SET 9.1

In Exercises 1–4, determine whether each of the ordered pairs is a solution of the system.

1. (a) (3, 2) (b) (3, −2) $\begin{cases} 5x + 7y = 1 \\ x - 2y = 1 \end{cases}$

2. (a) (4, 1) (b) (1, 4) $\begin{cases} 3x - y = -1 \\ x + 2y = 9 \end{cases}$

3. (a) (5, 1) (b) (−4, −2) $\begin{cases} 2x - 6y = 4 \\ 3x - 9y = 6 \end{cases}$

4. (a) (5, 1) (b) (3, 0) $\begin{cases} 2x - 4y = 6 \\ -3x + 6y = 9 \end{cases}$

In Exercises 5–12, use the addition method to solve the system of equations.

5. $\begin{cases} 3x + 5y = 1 \\ 4x + 3y = 5 \end{cases}$

6. $\begin{cases} -2x + 7y = 8 \\ 3x - y = 7 \end{cases}$

7. $\begin{cases} 5x - 2y = 2 \\ 7x + 6y = 5 \end{cases}$

8. $\begin{cases} -9x + 6y = 4 \\ 7x - 2y = 2 \end{cases}$

9. $\begin{cases} 12x + 9y = 6 \\ 8x + 6y = 4 \end{cases}$

10. $\begin{cases} 2x + 4y = 5 \\ 3x + 6y = 1 \end{cases}$

11. $\begin{cases} \sqrt{14}x - \sqrt{10}y = 2 \\ \sqrt{7}x - \sqrt{5}y = -3 \end{cases}$

12. $\begin{cases} \sqrt{8}x + 2y = 8 \\ \sqrt{2}x + y = 4 \end{cases}$

In Exercises 13–20, use the substitution method to solve the system of equations.

13. $\begin{cases} 3x + 6y = 4 \\ x - 3y = -1 \end{cases}$

14. $\begin{cases} 2x - 6y = 5 \\ 6x + 7y = 0 \end{cases}$

15. $\begin{cases} 6x + 2y = 8 \\ 3x + y = 4 \end{cases}$

16. $\begin{cases} 6x - 10y = -4 \\ 3x - 5y = 2 \end{cases}$

17. $\begin{cases} 10x + 5y = 7 \\ 6x + 3y = 4 \end{cases}$

18. $\begin{cases} -15x + 9y = -3 \\ 5x - 3y = 1 \end{cases}$

19. $\begin{cases} \sqrt{3}x - y = 5 \\ x + \sqrt{3}y = -\sqrt{3} \end{cases}$

20. $\begin{cases} 2x + \sqrt{5}y = 7 \\ \sqrt{5}x - 3y = -2\sqrt{5} \end{cases}$

In Exercises 21–30, solve the system.

21. $\begin{cases} x^2 - 2x + y^2 = 3 \\ x - y = 3 \end{cases}$

22. $\begin{cases} x^2 - 10y + y^2 = 0 \\ 4y - 3x = 20 \end{cases}$

23. $\begin{cases} x^2 - 6x + y^2 = -4 \\ 2x - y = 1 \end{cases}$

24. $\begin{cases} x^2 + 2y + y^2 = 12 \\ 3x - 2y = 7 \end{cases}$

25. $\begin{cases} x^2 - 4x + y^2 + 8y = -19 \\ 4y + 5x = 20 \end{cases}$

26. $\begin{cases} x^2 - 2x + y^2 - 6y = -6 \\ 4y - x = -8 \end{cases}$

27. $\begin{cases} x^2 + 8x + y^2 - 18y = 72 \\ (x - 2)^2 + (y - 9)^2 = 1 \end{cases}$

28. $\begin{cases} (x - 3)^2 + (y + 2)^2 = 9 \\ (x + 1)^2 + y^2 = 9 \end{cases}$

29. $\begin{cases} (x + 3)^2 + (y - 1)^2 = 24 \\ (x - 1)^2 + (y + 1)^2 = 4 \end{cases}$

30. $\begin{cases} x^2 - 10x + y^2 + 6y = -30 \\ (x - 5)^2 + (y + 3)^2 = 4 \end{cases}$

In Exercises 31–34, solve the given system of equations. (Depending on the method of solution, answers may vary slightly.)

31. $\begin{cases} 5.7x - 8y = 12.2 \\ 6x + 2.7y = 20 \end{cases}$
Round answers to the nearest hundredth.

32. $\begin{cases} 10x + 22.5y = -18.6 \\ 14.3x - 0.25y = 40 \end{cases}$
Round answers to the nearest tenth.

33. $\begin{cases} 1607x - 5925y = -856 \\ 47x + 219y = 360 \end{cases}$
Round answers to the nearest thousandth.

34. $\begin{cases} 0.003x - 1.02y = 0.05 \\ -0.008x + 1.83y = 0.008 \end{cases}$
Round answers to the nearest hundredth.

In Exercises 35–40, determine the coordinates of the points of intersection of the given line and the unit circle $x^2 + y^2 = 1$. Round answers to the nearest hundredth.

35. $y = x$

36. $y = -\frac{\sqrt{3}}{3}x$

37. $y = \sqrt{3}x$

38. $y = 0.27x$

39. $y = \sqrt{2} - x$

40. $2y = \sqrt{3} + 1 - 2x$

Superset

In Exercises 41–44, solve the system.

41. $\begin{cases} (y-1)^2 = x + 1 \\ x = 4y - 9 \end{cases}$

42. $\begin{cases} y + 3 = (x-2)^2 \\ y = x - 3 \end{cases}$

43. $\begin{cases} 8(x-1) = (y+2)^2 \\ y + 2 = (x-1)^2 \end{cases}$

44. $\begin{cases} x^2 + y^2 = 4 \\ x = y^2 + 2 \end{cases}$

45. A cash register contains dimes and quarters. There are 9 more quarters than dimes, and the total value of the coins is \$5.05. How many quarters are there? How many dimes?

46. A 450 mi plane trip takes 3 hrs with the wind and 5 hrs against the wind. Find the rate of the plane in still air and the rate of the wind.

47. You have invested \$1500 in stocks and bonds, offering a return of 15% and 9%, respectively. If you receive \$195 in interest, how much is invested in bonds?

48. Find all two-digit numbers such that the sum of the digits is 13, and the number is increased by 45 if the order of the digits is reversed.

49. The medians of a triangle intersect in a point. Determine the coordinates of this point for the triangle whose vertices are $A(0, 2)$, $B(5, 3)$, and $C(8, 6)$.

50. The altitudes of a triangle intersect in a point. Determine the coordinates of this point for the triangle in Exercise 49.

51. You are positioned 5 mi due east of a lighthouse on a straight beach that runs east-west. You pick up a distress signal from a small boat which you sight as 4 mi from your position at the same time that the lighthouse keeper notes the boat as 3 mi from his position. How far offshore is the boat?

52. A rectangular plot of land is 15 ft shy of being twice as long as it is wide. What is the perimeter of the plot, if its area is 1350 ft²?

53. Find an equation for the circle which passes through the points $A(3, 5)$, $B(5, 1)$, and $C(-1, 7)$. (*Hint:* Use the fact that the perpendicular bisector of a chord in a circle passes through the center of the circle.)

54. Two small planes are flying circular holding patterns centered over the opposite ends of a 10,000 ft long runway which runs north-south. Plane A is flying a tight circle of radius 3000 ft about the northern end of the runway, while plane B is flying a more relaxed circle of radius 8000 ft about the southern end of the runway. Unfortunately, the planes are flying at the same altitude and a collision occurs just east of the runway. How far east of the runway did this mid-air collision occur?

55. If one side of a rectangle is increased by 2 in., while an adjacent side is increased by 3 in., the resulting rectangle is a square whose area is three times that of the original rectangle. What was the perimeter of the original rectangle?

56. Determine the radius of the circle which is centered at the point (1, 1) and is tangent to the line which passes through the points (6, 2) and (4, 8).

9.2
Systems of Inequalities in Two Variables

To solve an inequality in two variables, we begin by looking at the associated equality. For example, in order to solve

$$y - 2x - 3 > 0$$

we first look at

$$y - 2x - 3 = 0.$$

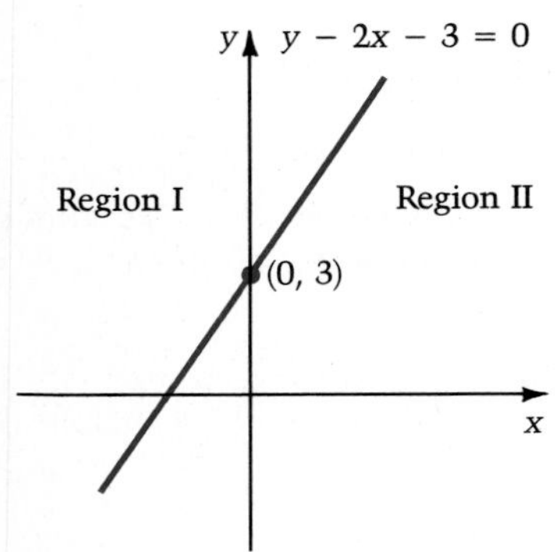

Figure 9.7

The graph of this equation is the line shown at the left. This line divides the plane into two regions; call them Region I and Region II. Since any point on the line satisfies the equation $y - 2x - 3 = 0$, it seems reasonable that any point not on the line satisfies one of the inequalities

$$y - 2x - 3 < 0 \qquad \text{or} \qquad y - 2x - 3 > 0.$$

It turns out that all the points in the same region satisfy the same inequality.

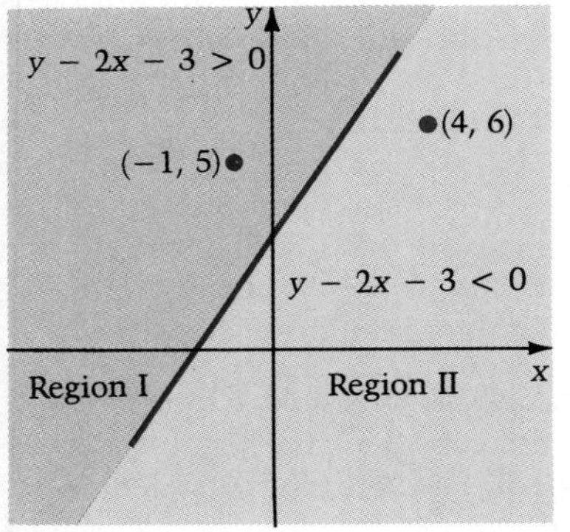

Figure 9.8

Select a point in Region I, say $(-1, 5)$. Substitution yields

$$5 - 2(-1) - 3 = 4.$$

Thus, $(-1, 5)$ satisfies $y - 2x - 3 > 0$. We can conclude that all other points in Region I satisfy this inequality as well.

Now consider the point $(4, 6)$, which lies in Region II. Substitution yields

$$6 - 2(4) - 3 = -5.$$

Thus, $(4, 6)$ satisfies $y - 2x - 3 < 0$. We conclude that all the other points in Region II satisfy this inequality. The reasoning behind these claims is based on the following theorem.

Theorem

Suppose the graph of an equation divides the xy-plane into two regions. Then all the points in any one region must satisfy the same inequality formed from the associated equality. Thus, to determine which inequality holds for a region, you need only test one point from that region.

EXAMPLE 1

Sketch the graph of each inequality.

(a) $x^2 + y^2 < 9$ (b) $y \geq 2$

Solution

(a)

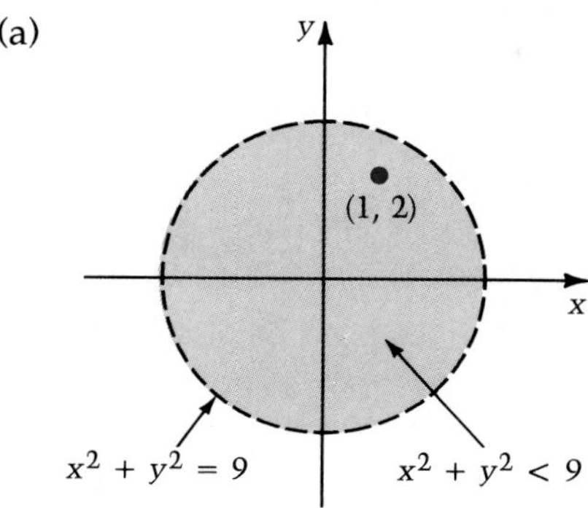

Begin by graphing the equation $x^2 + y^2 = 9$ (a circle) with dashes to indicate that the points on the circle are not in the solution set of the inequality.

Select a point outside the circle, say (3, 4). Since $3^2 + 4^2 > 9$, all points outside the circle satisfy the inequality $x^2 + y^2 > 9$.

Select a point inside the circle, say (1, 2). Since $1^2 + 2^2 < 9$, all points inside the circle satisfy the inequality $x^2 + y^2 < 9$.

We shade the interior of the circle to indicate the solution set of the inequality.

(b)

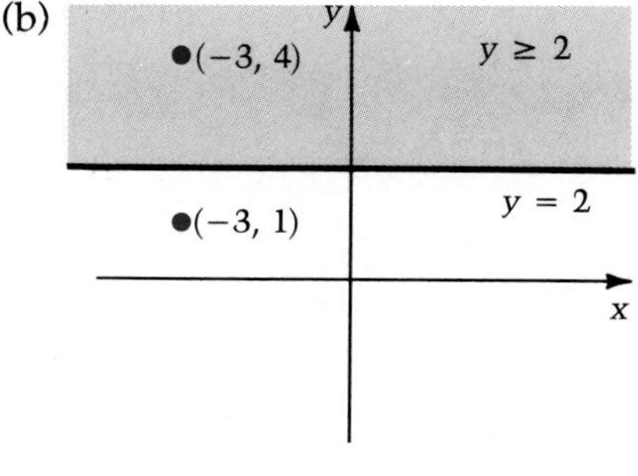

Figure 9.9

Begin by graphing the equation $y = 2$ as a solid line to indicate that the points on the line are included in the solution set of the inequality.

Select a point above the line, say $(-3, 4)$. Since its y-coordinate, 4, is greater than 2, all points above the line satisfy the inequality $y > 2$.

Select a point below the line, say $(-3, 1)$. Since its y-coordinate, 1, is less than 2, all points below the line satisfy the inequality $y < 2$.

We graph the solution set of $y \geq 2$ by shading the region above the line.

Careful! Remember that if the inequality symbol is either $\geq$ or $\leq$, we use a solid curve to indicate that the points on the boundary of the region (i.e., the solutions of the associated equation) are included in the solution set of the inequality. If the symbol is $>$ or $<$, we indicate that these points are excluded by using a dashed curve.

The following is a system of inequalities in two variables:

$$\begin{cases} y < 3x - 1 \\ y > 3 - x. \end{cases}$$

The graph of such a system lies in the xy-plane and consists of all points satisfying both inequalities. To graph this system, we graph the region of each inequality and then determine the points common to both regions.

EXAMPLE 2

Graph the solution set of the following system: $\begin{cases} y < 3x - 1 \\ y > 3 - x. \end{cases}$

Solution

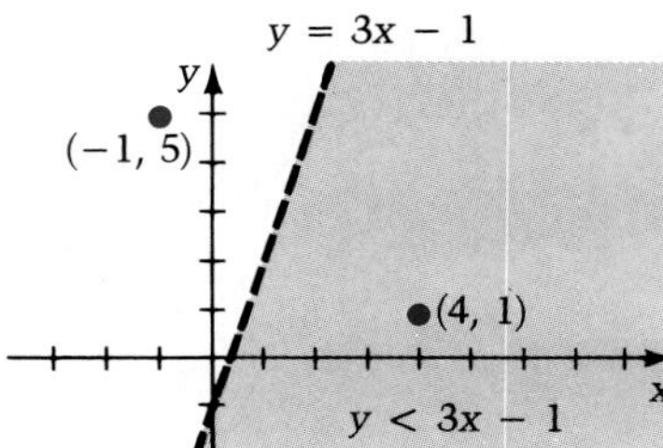

$y < 3x - 1$

Begin by graphing $y = 3x - 1$ as a dashed line.
Test a point above the line, say $(-1, 5)$. Since $(-1, 5)$ satisfies $y > 3x - 1$, all points above the line satisfy $y > 3x - 1$.

Test a point below the line, say $(4, 1)$. Since $(4, 1)$ satisfies $y < 3x - 1$, all points below the line satisfy $y < 3x - 1$.

Thus, we shade the region below the line.

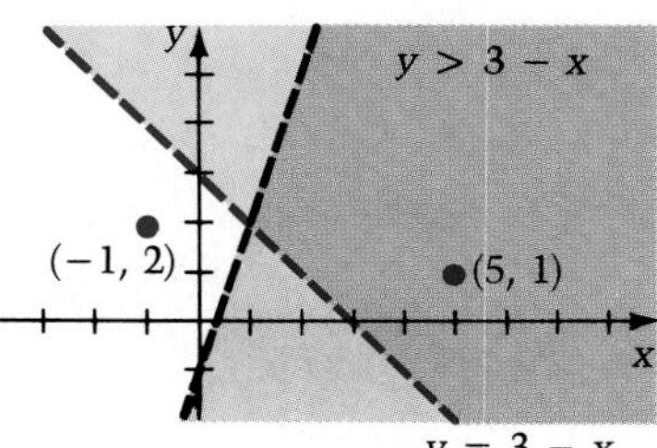

$y > 3 - x$

Begin by graphing $y = 3 - x$ as a dashed line.
Test a point above the line, say $(5, 1)$. Since $(5, 1)$ satisfies $y > 3 - x$, all points above the line satisfy $y > 3 - x$.

Test a point below the line, say $(-1, 2)$. Since $(-1, 2)$ satisfies $y < 3 - x$, all points below the line satisfy $y < 3 - x$.

Thus, we shade the region above the line.

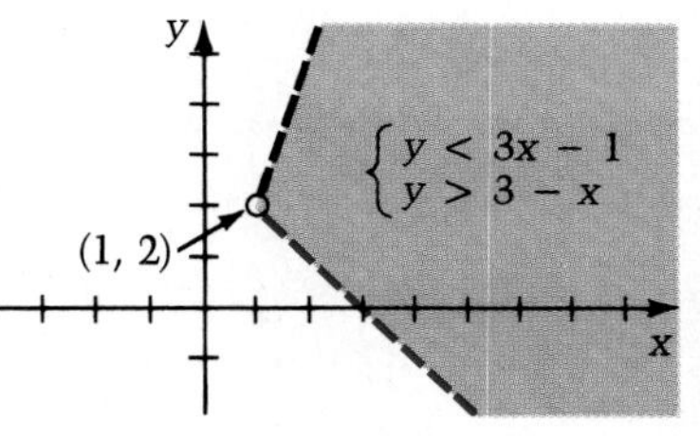

Figure 9.10

Finally, the solution set of the system consists of those points common to both regions. These points are represented by the overlap of the two regions. A good point of reference is the "corner point" $(1, 2)$, which is found by solving the system

$$\begin{cases} y = 3x - 1 \\ y = 3 - x. \end{cases}$$

To graph the solution set of a system of three or more inequalities, graph the solution set of each inequality, and identify the common points.

EXAMPLE 3

Graph the solution set of the following system: $\begin{cases} y \geq 0 \\ x \leq 5 \\ y \leq x - 1. \end{cases}$

Solution

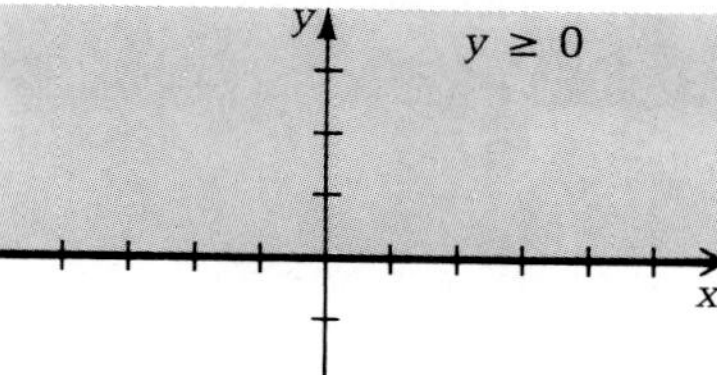

Begin by graphing the solution set of each of the inequalities on the same pair of coordinate axes. We first graph the solution set of $y \geq 0$.

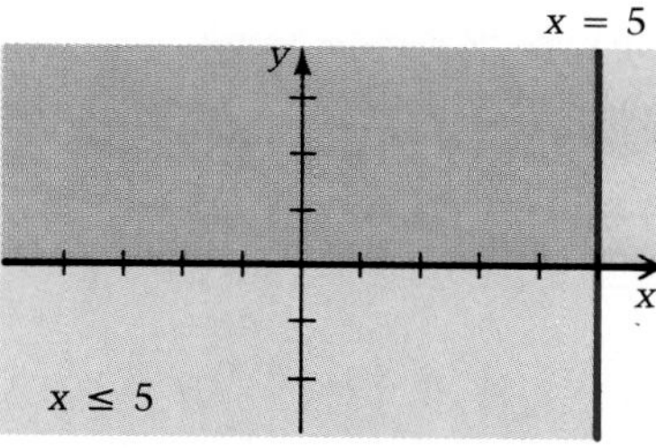

Next, we graph the solution set of $x \leq 5$.

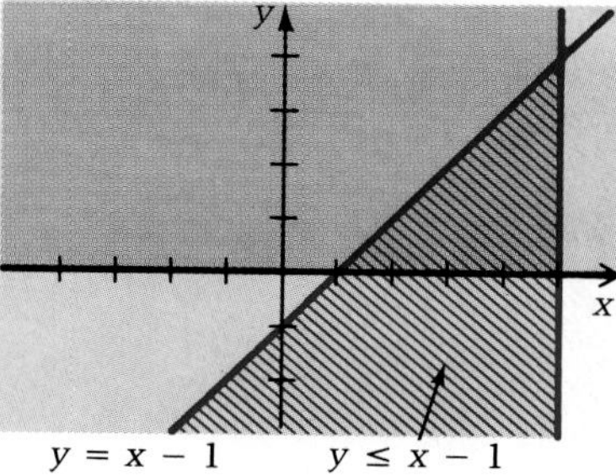

Now, sketch the solution set of $y \leq x - 1$. The point (3, 1) lies below the line. Since $1 < 3 - 1 = 2$, every point below the line satisfies the inequality $y < x - 1$.

The point (2, 5) lies above the line. Since $5 > 2 - 1 = 1$, every point above the line satisfies the inequality $y > x - 1$.

Thus, we shade the region below the line.

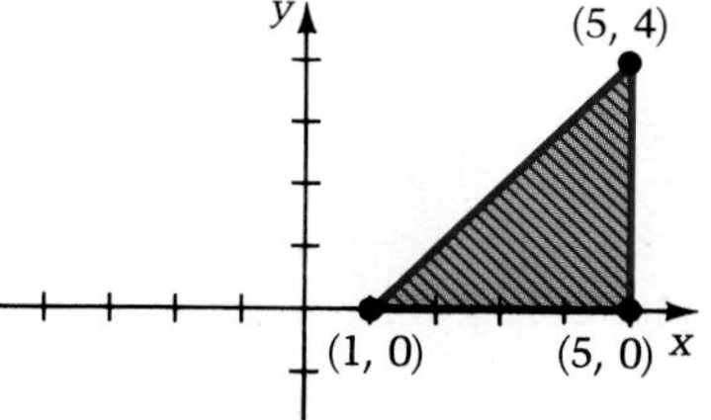

Figure 9.11

The region where all three graphs overlap is the graph of the solution set of the system. We find the coordinates of the "corners" by solving the appropriate two-equation systems. For example, we obtain the corner point (5, 4) by solving the system

$$\begin{cases} x = 5 \\ y = x - 1. \end{cases}$$

EXAMPLE 4

Graph the solution set of the following system: $\begin{cases} y \geq 2 \\ x \geq 3 \\ x > y \\ x + y < 10. \end{cases}$

Solution

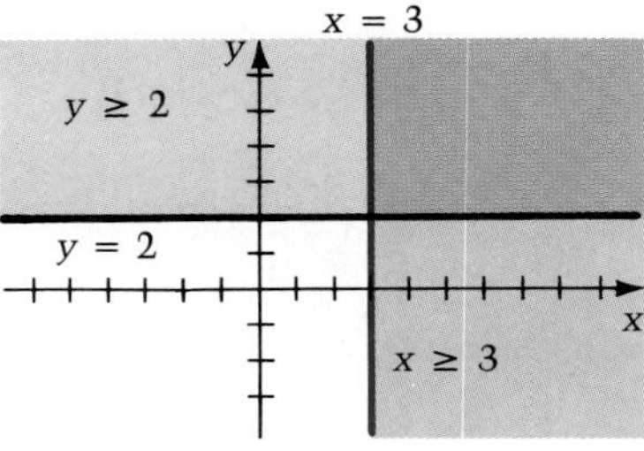

Begin by graphing the solution set of each of the inequalities on the same pair of coordinate axes. We first graph the solution set of $y \geq 2$, and the solution set of $x \geq 3$.

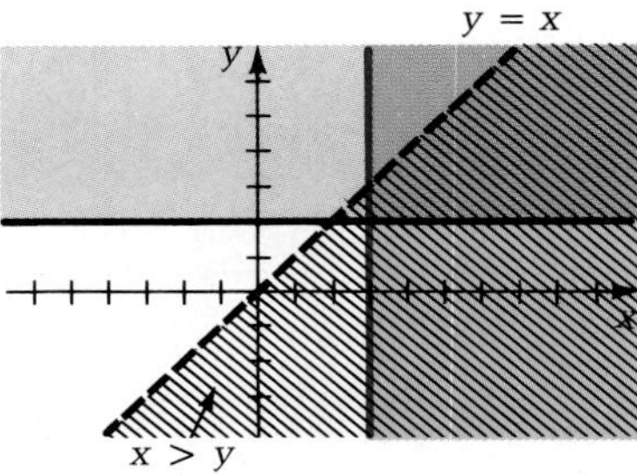

Now, we graph the solution set of $x > y$.

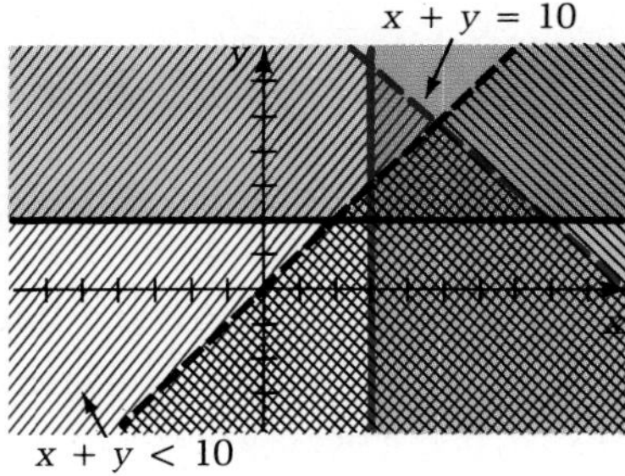

Now, using test points, we find the solution set of $x + y < 10$.

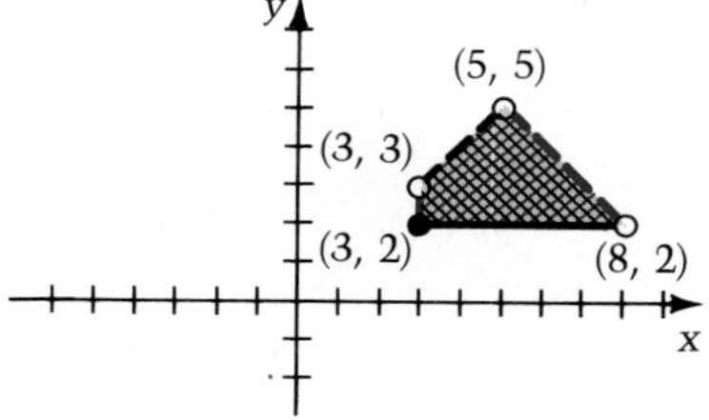

Finally, we determine the points common to the previous four solution sets. The region where all four graphs overlap is the graph of the solution set of the system.

Figure 9.12

EXERCISE SET 9.2

In Exercises 1–24, sketch the graph of the inequality.

1. $x \le 0$
2. $y > 0$
3. $x + y > 1$
4. $y < 2x + 3$
5. $y \le 3x - 1$
6. $y \ge 5 - x$
7. $x^2 + y^2 \le 9$
8. $x^2 + y^2 \ge 4$
9. $(x - 1)^2 + y^2 \ge 1$
10. $x^2 + (y + 2)^2 < 9$
11. $(x + 4)^2 + (y - 7)^2 < 4$
12. $(x + 3)^2 + (y + 5)^2 > 25$
13. $y \ge x^2$
14. $x < y^2$
15. $x - 2 \le (y + 3)^2$
16. $y - 1 < (x + 2)^2$
17. $y > |x|$
18. $x \ge -|y|$
19. $x < |6 - y|$
20. $y \le |x + 3|$
21. $xy \ge 1$
22. $xy < -9$
23. $y(x - 1) < -3$
24. $y(x + 2) \le 4$

In Exercises 25–40, graph the solution set of the system.

25. $\begin{cases} x \ge -2 \\ y < 1 \end{cases}$
26. $\begin{cases} x < 3 \\ y \ge -4 \end{cases}$
27. $\begin{cases} x + y > 1 \\ y < 1 \end{cases}$
28. $\begin{cases} x + y < 3 \\ y > 2x - 5 \end{cases}$
29. $\begin{cases} 5y + 2x \ge 10 \\ 3y - x > 3 \end{cases}$
30. $\begin{cases} 7y + 8x \le 56 \\ 2y - x < -2 \end{cases}$
31. $\begin{cases} y \ge x \\ y \le 2x \\ y \le 6 \end{cases}$
32. $\begin{cases} y \le 3x \\ y \ge -x \\ x \le 2 \end{cases}$
33. $\begin{cases} y > x - 2 \\ y < 6 - x \\ y < 2x - 3 \end{cases}$
34. $\begin{cases} y < x + 4 \\ y > -x \\ y < 4 - 2x \end{cases}$
35. $\begin{cases} 3y - x - 10 < 0 \\ y + 3x - 10 \le 0 \\ x + 2y \ge 0 \end{cases}$
36. $\begin{cases} 3y - x - 10 \le 0 \\ y - 4x + 15 < 0 \\ 2y + 3x - 3 \ge 0 \end{cases}$
37. $\begin{cases} x \ge 1 \\ y \ge -1 \\ y \le 2x - 1 \\ y \le 8 - x \end{cases}$
38. $\begin{cases} x \ge -2 \\ y \ge x \\ y \le x + 4 \\ y \le 8 - x \end{cases}$
39. $\begin{cases} 2y < 6 - x \\ 2y > x - 6 \\ y < x - 3 \\ y > -x - 3 \end{cases}$
40. $\begin{cases} 4y < 24 - x \\ 4y < 24 + x \\ y > -2x - 3 \\ y > 2x - 3 \end{cases}$

Superset

In Exercises 41–48, graph the solution set of the system.

41. $\begin{cases} y \ge x^2 - 4x + 1 \\ y \le x - 3 \end{cases}$
42. $\begin{cases} x \le y^2 - 2y \\ x \le 2 - y \end{cases}$
43. $\begin{cases} (x - 2)^2 + y^2 \le 9 \\ (x + 2)^2 + y^2 \le 9 \end{cases}$
44. $\begin{cases} x^2 + (y - 1)^2 \ge 4 \\ x^2 + (y + 1)^2 \ge 4 \end{cases}$
45. $\begin{cases} |x + y| \le 2 \\ |x| \le 3 \end{cases}$
46. $\begin{cases} |x + y| \le 3 \\ |y| \le 2 \end{cases}$
47. $\begin{cases} y \ge |x| \\ y \le |2x| - 4 \end{cases}$
48. $\begin{cases} |y - 2x| \le 4 \\ |y + x| \le 2 \end{cases}$

In Exercises 49–56, describe the graphed region by a system of inequalities.

49.

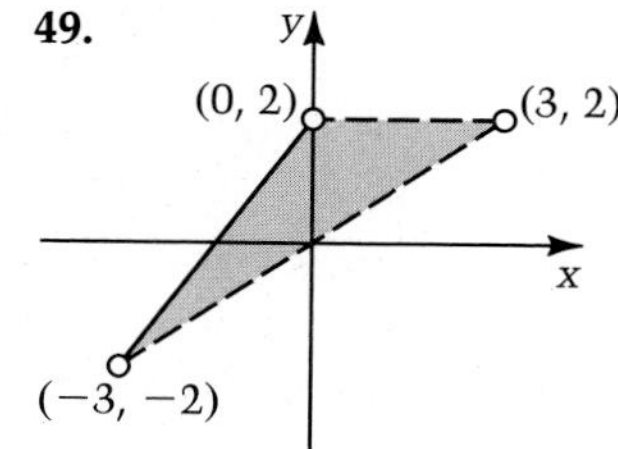

50.

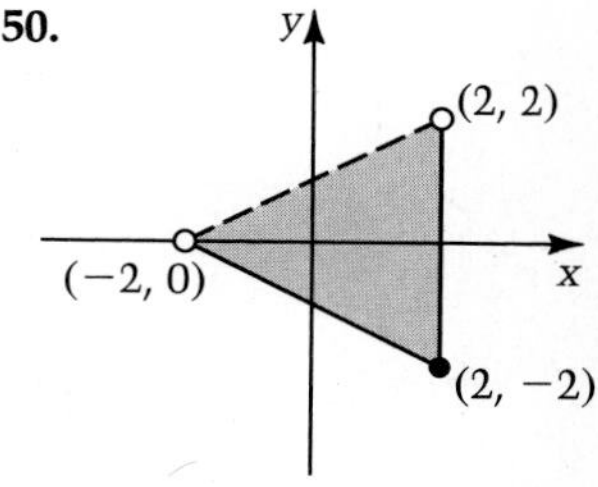

51.

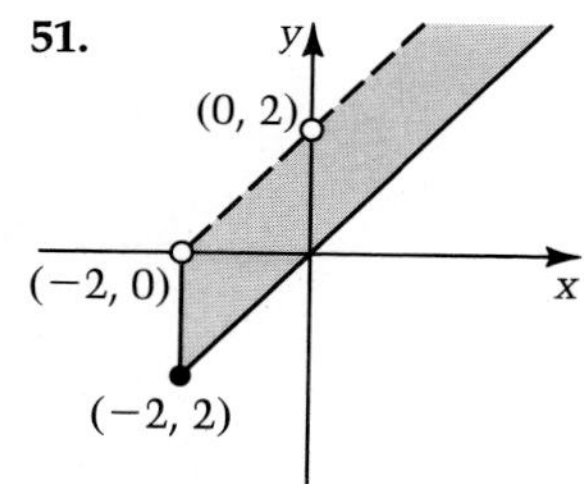

52.

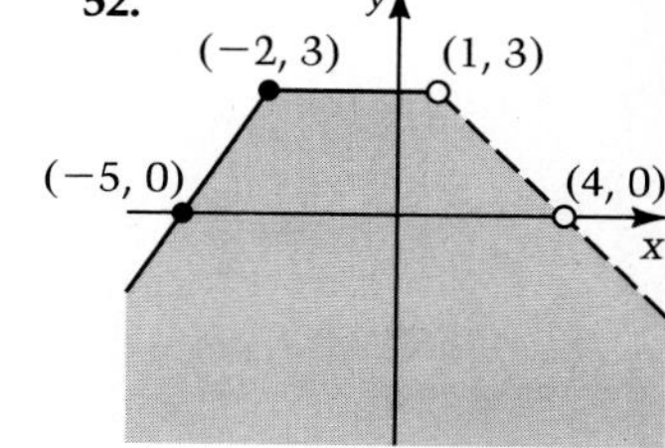

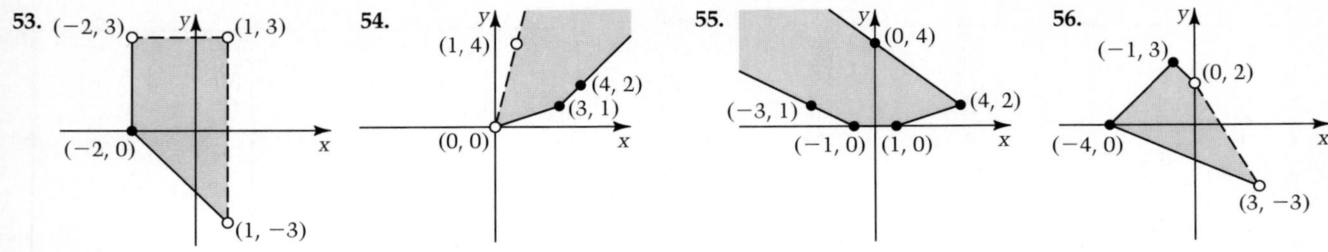

9.3 Systems of Linear Equations in Three Variables

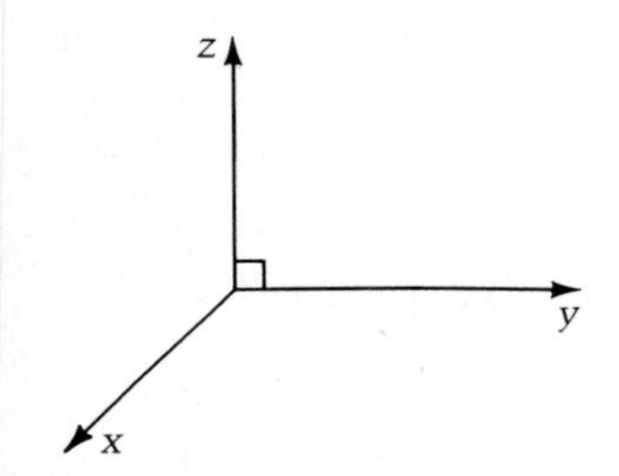

Figure 9.13 The *xyz*-coordinate system.

Up to this point all graphing has been done in the *xy*-plane. By adding a third coordinate axis, perpendicular to the *xy*-plane, we produce a coordinate system for describing the position of a point in 3-dimensional space. We usually call this third axis the **z-axis,** and consequently the entire system is called the ***xyz*-coordinate system.**

In the *xyz*-coordinate system, we graph **ordered triples** of the form

$$(x, y, z).$$

Plotting a point requires three moves: a move in the x-direction, followed by one in the y-direction, followed by another in the z-direction.

For example, suppose we must plot the point with coordinates (1, 4, 2). (See Figure 9.14.) Starting at the origin (0, 0, 0), we move 1 unit in the positive x-direction, then 4 units in the positive y-direction, and finally 2 units in the positive z-direction.

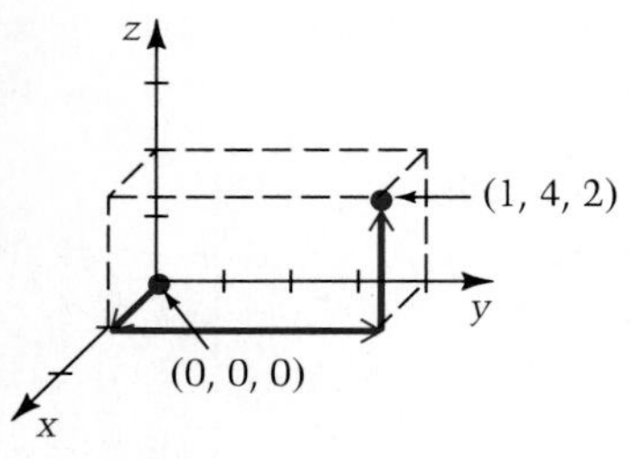

Figure 9.14 Plotting (1, 4, 2).

To help us understand point-plotting in the *xyz*-coordinate system, let us examine the figures below. Figure 9.15(a) represents a box 3 units long, 2 units wide and 1 unit high. In Figure 9.15(b), we have placed the box in the three-dimensional *xyz*-coordinate system. The corners of the box are labeled with ordered triples of the form (x, y, z).

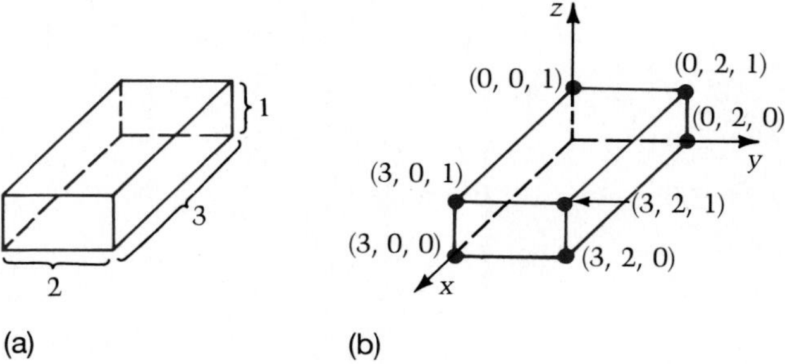

Figure 9.15

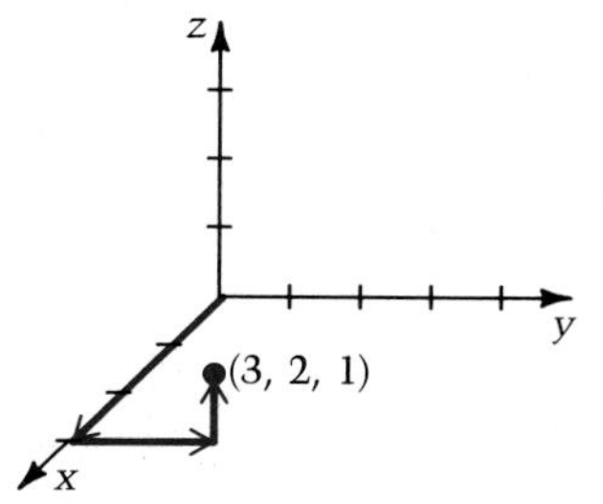

Figure 9.16 Plotting (3, 2, 1).

Notice that each corner of the bottom of the box lies in the xy-plane and can be described as an ordered triple of the form $(x, y, 0)$. Each corner of the top of the box is 1 unit above the xy-plane and has coordinates of the form $(x, y, 1)$. Starting at the origin, we find the corner $(3, 2, 1)$ by moving 3 units in the positive x-direction, 2 units in the positive y-direction, and then 1 unit in the positive z-direction.

EXAMPLE 1

Plot the following points in the xyz-coordinate system.

(a) $A(1, 0, 0)$ (b) $B(0, 0, -1)$ (c) $C(0, -2, 0)$ (d) $D(-2, 1, 3)$

Solution

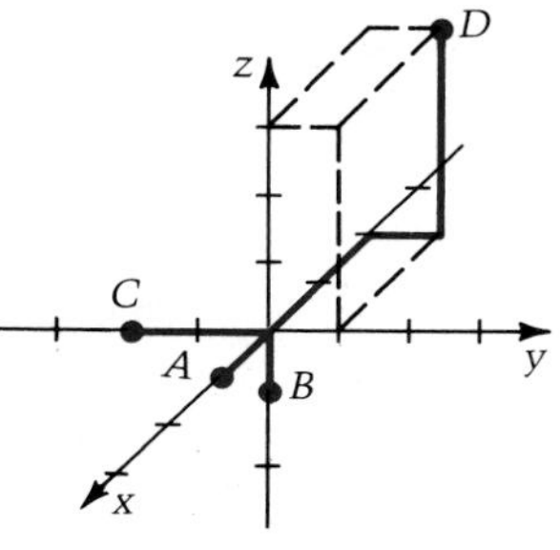

Figure 9.17

To plot point A, start at $(0, 0, 0)$ and move 1 unit in the positive x-direction.

To plot point B, start at $(0, 0, 0)$ and move 1 unit in the negative z-direction.

To plot point C, start at $(0, 0, 0)$ and move 2 units in the negative y-direction.

To plot point D, start at $(0, 0, 0)$ and move 2 units in the negative x-direction, 1 unit in the positive y-direction, and 3 units in the positive z-direction.

Our study of ordered triples will now help us understand linear equations in three variables. The graph of a linear equation in three variables is a plane in the xyz-coordinate system.

General Form of the Equation of a Plane

The **general form** of the equation of a plane in the xyz-coordinate system is

$$Ax + By + Cz + D = 0,$$

where A, B, C, and D are real numbers with A, B, and C not all zero.

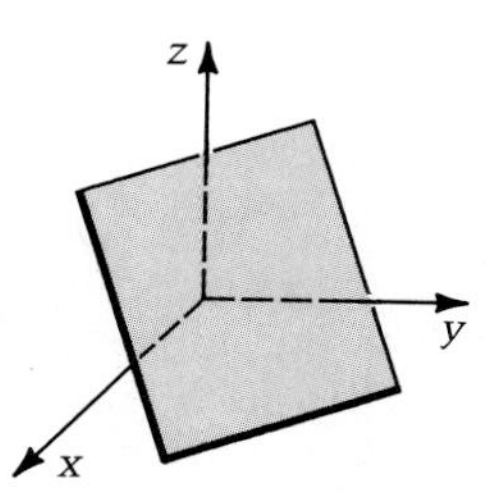

Figure 9.18

Some very special planes and their equations deserve mention. As we suggested earlier, every point in the xy-plane has coordinates of the form $(x, y, 0)$. Since every point in the xy-plane has z-coordinate 0,

$z = 0$ is the equation of the xy-plane.

Likewise,

$$x = 0 \text{ is the equation of the } yz\text{-plane, and}$$
$$y = 0 \text{ is the equation of the } xz\text{-plane.}$$

More generally, an equation of the form

$$z = c, \text{ with } c \text{ a constant,}$$

is the equation of a plane that is parallel to the xy-plane and c units above (for $c > 0$) or below (for $c < 0$) the xy-plane. Similarly, the plane $x = c$ is parallel to the yz-plane, and the plane $y = c$ is parallel to the xz-plane.

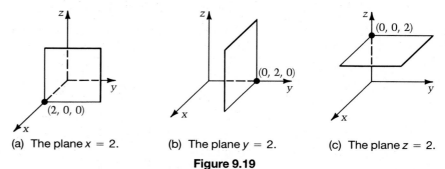

(a) The plane $x = 2$.

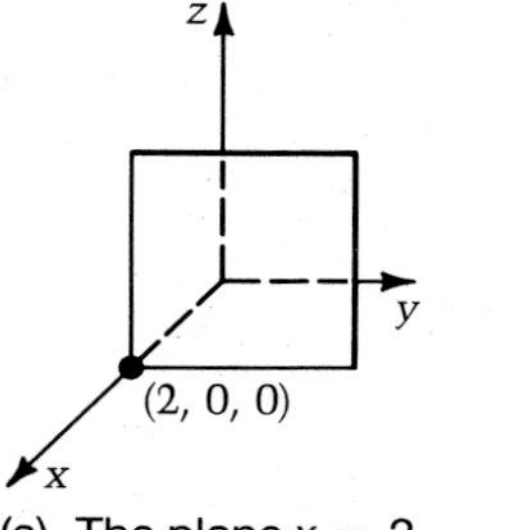

(b) The plane $y = 2$.

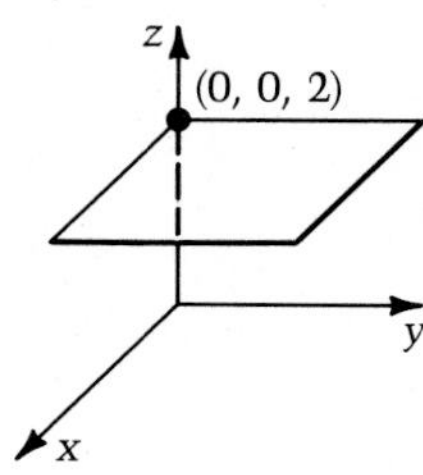

(c) The plane $z = 2$.

Figure 9.19

EXAMPLE 2

Graph the following planes on the same set of coordinate axes and then determine the point where the three planes intersect: $x = 2, y = 3$, and $z = 4$.

Solution

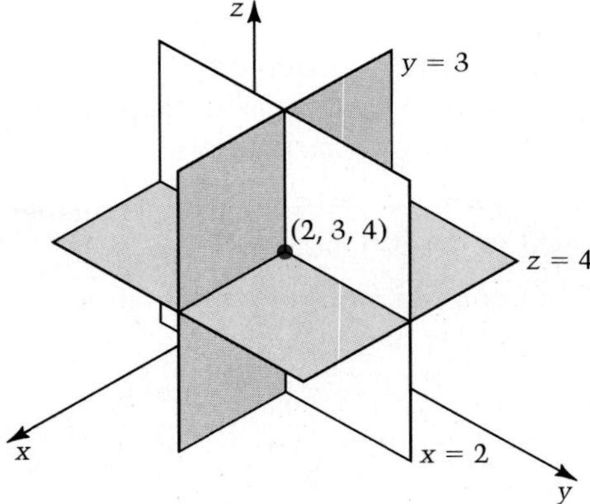

Figure 9.20

The plane $x = 2$ is parallel to and 2 units in front of the yz-plane. All points on this plane are of the form $(2, y, z)$, since x must be 2, but y and z can represent any numbers.

The plane $y = 3$ is parallel to and 3 units to the right of the xz-plane. All points on this plane are of the form $(x, 3, z)$, since y must be 3, but x and z can represent any numbers.

The plane $z = 4$ is parallel to and 4 units above the xy-plane. All points on this plane are of the form $(x, y, 4)$, since z must be 4, but x and y can represent any numbers.

The three planes intersect at the point (2, 3, 4).

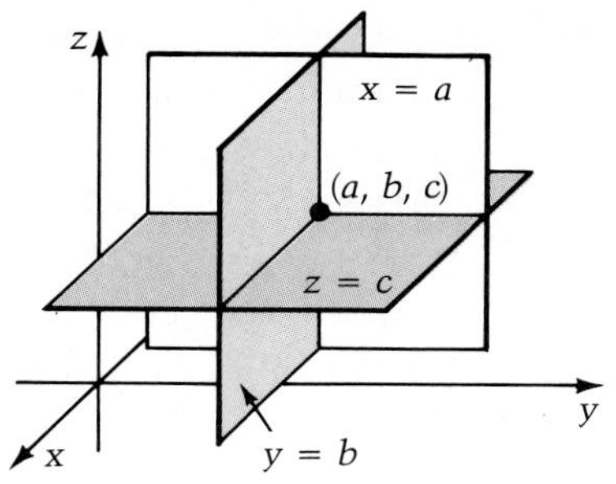

Figure 9.21

When we are solving a system of three linear equations in three variables, our objective is to produce a succession of equivalent systems until we obtain, if possible, a system of the form

$$\begin{cases} x = a \\ y = b \\ z = c \end{cases}$$

where a, b, and c are constants. We then conclude that the ordered triple (a, b, c) is the solution of the original system.

You should picture a system of three linear equations in three variables as three planes in space. Then, the solutions of the system are the ordered triples describing the points of intersection of the three planes (Figure 9.22). This is summarized in the table below.

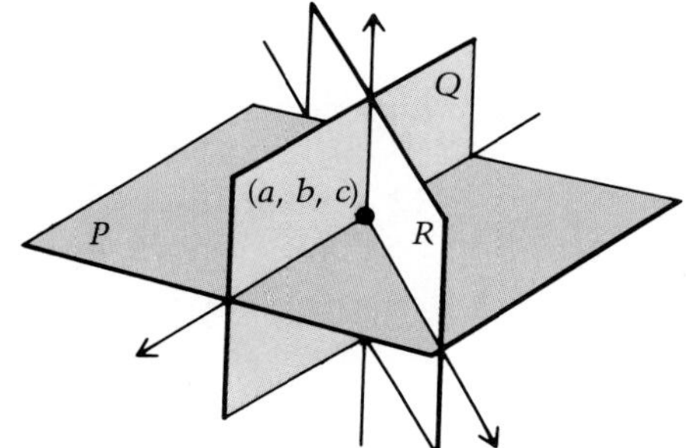

(a) One solution: (a, b, c).

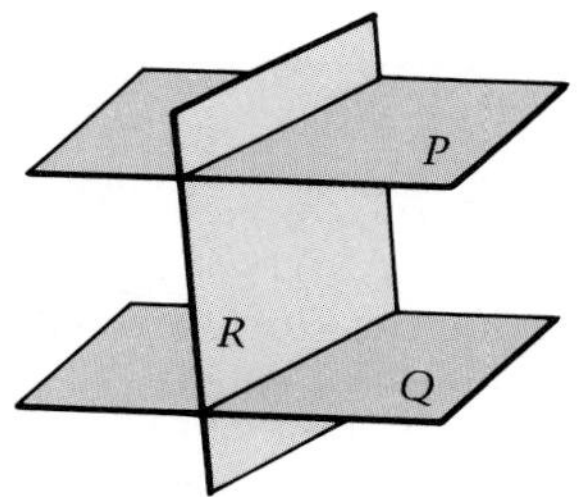

(b) No solutions.

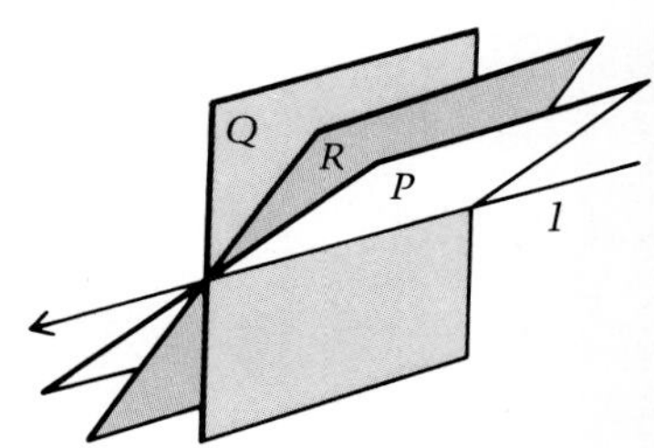

(c) Infinitely many solutions on line l.

Figure 9.22

Solutions of Linear Systems in Three Variables

Form of an Equivalent System	Number of Solutions
$x =$ a number $y =$ a number $z =$ a number	one
Includes an equation like $7 = 7$ or $0 = 0$.	infinitely many
Includes an equation like $0 = 9$ or $2 = 3$.	none

We can use either the substitution method or the addition method to solve such systems.

EXAMPLE 3

Use the substitution method to solve the system $\begin{cases} 5x + y - z = 3 \\ x + 2y + z = 0 \\ 3x - 4y - 5z = -2. \end{cases}$

Solution

$$\begin{cases} 5x + y - z = 3 \\ x + 2y + z = 0 \\ 3x - 4y - 5z = -2 \end{cases}$$

We begin by solving the first equation for y.

$$y = 3 + z - 5x \quad ①$$

Next we substitute this expression for y into the other two equations, then simplify.

$x + 2y + z = 0$	$3x - 4y - 5z = -2$
$x + 2(3 + z - 5x) + z = 0$	$3x - 4(3 + z - 5x) - 5z = -2$
$x + 6 + 2z - 10x + z = 0$	$3x - 12 - 4z + 20x - 5z = -2$
② $3x - z = 2$	③ $23x - 9z = 10$

Using the simplified equations ①, ②, and ③, the original system is restated as an equivalent system.

$$\begin{cases} y = 3 + z - 5x \\ 3x - z = 2 \\ 23x - 9z = 10 \end{cases}$$

$z = 3x - 2$ The second equation is solved for z.

$23x - 9(3x - 2) = 10$ Substitute the expression for z into the third equation and solve for x.

$-4x = -8$

$x = 2$

$$\begin{cases} y = 3 + z - 5x \\ z = 3x - 2 \\ x = 2 \end{cases}$$

$z = 3(2) - 2 = 4$ The value of x is substituted into the second equation.

$y = 3 + (4) - 5(2) = -3$ The values of x and z are substituted into the first equation.

$$\begin{cases} y = -3 \\ z = 4 \\ x = 2 \end{cases}$$

The solution of the system is $(2, -3, 4)$.

EXERCISE SET 9.3

In Exercises 1–6, plot the point in the xyz-coordinate system.

1. $A(2, 3, 0)$
2. $B(1, -2, 3)$
3. $C(-2, 1, 3)$
4. $D(3, -1, 2)$
5. $E(-1, 3, 2)$
6. $F(3, 0, 2)$

In Exercises 7–12, graph the plane.

7. $x = 3$
8. $y = 5$
9. $z = 1$
10. $x = -6$
11. $y = -4$
12. $z = -8$

In Exercises 13–18, graph the set of planes on the same set of coordinate axes and then determine the point where the three planes intersect.

13. $x = 4$, $y = 2$, $z = 6$
14. $x = -1$, $y = 5$, $z = 4$
15. $y = 7$, $x = -3$, $z = -1$
16. $y = -3$, $z = 5$, $x = -6$
17. $z = 0$, $x = -2$, $y = -10$
18. $z = -9$, $y = 0$, $x = 12$

In Exercises 19–36, solve the system.

19. $\begin{cases} 4x + 2y + 5z = 4 \\ x - y + z = 0 \\ y = 1 \end{cases}$

20. $\begin{cases} 5x + 4y + z = 2 \\ 3x + 2y - z = 0 \\ z = 2 \end{cases}$

21. $\begin{cases} 4x + 2y + z = 8 \\ 2x - y - 3z = 3 \\ x + y - z = 1 \end{cases}$

22. $\begin{cases} -x + 2y - 3z = 5 \\ 4x + 3y + z = 2 \\ x - y - z = 3 \end{cases}$

23. $\begin{cases} 5x - y + z = -1 \\ x + 3y = 7 \\ y + 2z = -6 \end{cases}$

24. $\begin{cases} 4x - 5y + 2z = -1 \\ x + 2y + z = 4 \\ 2x + 3z = 1 \end{cases}$

25. $\begin{cases} 5x - 4y - 3z = -1 \\ x + 2y + 5z = -3 \\ 3x - y - z = 2 \end{cases}$

26. $\begin{cases} 3x + 2y + z = -1 \\ 4x + y + z = 1 \\ x + 4y + z = 2 \end{cases}$

27. $\begin{cases} -5x + 3y + 2z = 2 \\ 3x + y - 4z = 3 \\ x - 2y + z = 0 \end{cases}$

28. $\begin{cases} 3x + 2y + 4z = -6 \\ -x + y + 3z = 5 \\ 2x + y - z = 1 \end{cases}$

29. $\begin{cases} -5x + 3y + 2z = 2 \\ 3x + y - 4z = 8 \\ -x + 2y - z = 5 \end{cases}$

30. $\begin{cases} 3x + y + 2z = 1 \\ 4x - y + 4z = 4 \\ x + y - 2z = -5 \end{cases}$

31. $\begin{cases} 2x + 5y + 4z = 1 \\ 5x + 3y - 3z = -2 \\ 4x + y + 2z = 3 \end{cases}$

32. $\begin{cases} 3x - 6y + 4z = -1 \\ 3x + 2y - 2z = 5 \\ 3x - 2y + z = 2 \end{cases}$

33. $\begin{cases} 3x + 4y + 2z = -8 \\ 3x + 8y + z = -6 \\ 6x - 4y - z = 8 \end{cases}$

34. $\begin{cases} -x + 2y - 7z = 9 \\ 4x + 3y + z = 8 \\ 3x - y - 3z = -7 \end{cases}$

35. $\begin{cases} x + 10y - 11z = 1 \\ 2x - y + 5z = 5 \\ x + 3y - 2z = 2 \end{cases}$

36. $\begin{cases} -3x - 7y + 7z = 0 \\ 3x + 2y - z = 1 \\ 6x - y + 4z = 3 \end{cases}$

Superset

The distance d between the point $P(x_0, y_0, z_0)$ and the plane described by the equation $Ax + By + Cz + D = 0$ is given by the formula

$$d = \frac{|Ax_0 + By_0 + Cz_0 + D|}{\sqrt{A^2 + B^2 + C^2}}.$$

In Exercises 37–40, determine the distance between the given point and the given plane. Round answers to the nearest hundredth.

37. $2x + y + z + 5 = 0$; $P(3, 5, 8)$
38. $x - 3y + 7z - 15 = 0$; $P(13, 15, -27)$
39. $2.5x - 18.6z = 7$; $P(-6.1, 22.7, 31.5)$
40. $12.2y + 15.8z = -4.8$; $P(14.7, 19.5, -3.4)$

In Exercises 41–44, determine the points of intersection of the given plane with (a) the x-axis, (b) the y-axis, and (c) the z-axis. Round the coordinates to the nearest tenth.

41. $6x + 8y - z + 17 = 0$
42. $3.2x - 0.2y + 8z - 15.7 = 0$
43. $5x + 3.7y - 8.2z = 1$
44. $\sqrt{2}x + \sqrt{3}y - \sqrt{5}z + 1 = 0$

45. Frank is 3 times as old as Ann and 15 years older than Mary. Ann is one year younger than Mary. How old is Frank?

46. A child's bank contains twice as many dimes as nickels and four times as many pennies as nickels. There are 42 coins in the bank. What is the coins' total value?

47. Find all three digit numbers such that the sum of the digits is 7, the number is increased by 99 if the order of the digits is reversed, and the hundreds digit is 3 less than the sum of the other two digits.

48. Find all three digit numbers such that the sum of the digits is 6, the hundreds digit is equal to the sum of the other two digits, and the number is decreased by 297 if the order of the digits is reversed.

49. Bob is ten years older than Ted. Last year Ted was twice as old as Carol was. Three years ago Bob was five times as old as Carol was. How old is Carol today?

50. A child's bank contains twice as many nickels as pennies and two-thirds as many dimes as nickels. The total value of the coins is at least \$3.70. Determine the smallest number of coins that could be in the bank.

51. A total of \$1000 is invested in three ventures. The first investment returns a profit of 4%, the second investment returns a profit of 7%, but a loss of 2% is suffered on the balance of the \$1000 invested in a third venture. The net gain from all three investments is \$28. If each of the gains had instead been 6% and had the loss instead been 4%, the net gain from all three investments would have been \$17.50. How much was invested in the venture that earned a 7% profit?

52. At a local fast food eatery an order of 5 burgers, 2 fries, and 3 sodas costs \$9.13. An order of 4 burgers, 3 fries, and 2 sodas costs \$7.70. An order of 6 burgers, 4 fries and 3 sodas costs \$11.32. What would it cost to purchase a single burger with a soda?

9.4 Matrices

A **matrix** is a rectangular array of numbers. Matrices provide a convenient shorthand for working with systems of equations. Just why this is so, is the subject of this section.

	columns 1	2	3	4
rows 1	3	−2	1	0
2	1	5	−3	2
3	1	0	−4	3

$$\begin{bmatrix} 3 & -2 & 1 & 0 \\ 1 & 5 & -3 & 2 \\ 1 & 0 & -4 & 3 \end{bmatrix}$$

A matrix with 3 rows and 4 columns

$$\begin{bmatrix} 7 & 2 & 1 \\ 4 & 1 & 6 \\ 3 & 1 & -5 \\ 0 & 0 & 0 \end{bmatrix}$$

A matrix with 4 rows and 3 columns

Each number in a matrix is called an **entry.** A matrix with m rows and n columns is said to be an $\boldsymbol{m \times n}$ (or $\boldsymbol{m}$ **by** $\boldsymbol{n}$) **matrix.** If the number of rows equals the number of columns, the matrix is called a **square matrix.**

Recall that when we use the addition method to solve a system of linear equations, we must line up the same variables and constant terms vertically to be sure we are combining x-terms with x-terms, y-terms with y-terms, and constants with constants. The linear system below is written in matrix form with the coefficients and the constants as entries. A dotted line in the matrix separates the coefficients from the constants.

$$\begin{cases} 2x + y - 3z = 2 \\ x + 3y = 2 \\ x - y + 4z = 5 \end{cases}$$

A linear system in 3 variables

$$\begin{array}{c} \begin{matrix} x & y & z & \end{matrix} \\ \left[\begin{array}{ccc|c} 2 & 1 & -3 & 2 \\ 1 & 3 & 0 & 2 \\ 1 & -1 & 4 & 5 \end{array}\right] \end{array}$$

The 3×4 matrix of the system

There is a zero in the second-row/third-column position in the matrix because the second equation has no z-term. If we remember to associate the first column with x, the second column with y, the third column with z, and the fourth column with the constants, then the matrix gives us the same information the system does.

EXAMPLE 1

For the given matrix, identify the following:

(a) the first-row/second-column entry

(b) the second-row/first-column entry

$$\left[\begin{array}{rrr:r} 3 & 0 & 1 & 5 \\ 2 & -3 & 7 & -8 \\ 5 & 4 & 6 & 10 \end{array}\right]$$

Solution

(a) The first-row/second-column entry is 0.

(b) The second-row/first-column entry is 2.

EXAMPLE 2

Represent the systems with matrices and the matrices with systems.

(a) $\begin{cases} 2x + 3y = 5 \\ x - 4y = 0 \end{cases}$

(b) $\begin{cases} 3x + y = 4 \\ 2x - y + 5z = 3 \\ 0 = 9 \end{cases}$

(c) $\left[\begin{array}{rr:r} 1 & 0 & 10 \\ 0 & 1 & -2 \end{array}\right]$

(d) $\left[\begin{array}{rrr:r} 1 & 0 & 2 & -3 \\ 8 & 0 & 10 & -5 \\ 12 & 1 & 0 & 16 \end{array}\right]$

Solution

(a) $\left[\begin{array}{rr:r} 2 & 3 & 5 \\ 1 & -4 & 0 \end{array}\right]$

(b) $\left[\begin{array}{rrr:r} 3 & 1 & 0 & 4 \\ 2 & -1 & 5 & 3 \\ 0 & 0 & 0 & 9 \end{array}\right]$

(c) $\begin{cases} x = 10 \\ y = -2 \end{cases}$

(d) $\begin{cases} x + 2z = -3 \\ 8x + 10z = -5 \\ 12x + y = 16 \end{cases}$

Using the following matrix as an example, we now define the following matrix terminology.

$$\left[\begin{array}{rr|r} 2 & 3 & 5 \\ 1 & -1 & 0 \end{array}\right] \quad \text{augmented matrix}$$

$$\begin{bmatrix} 2 & 3 \\ 1 & -1 \end{bmatrix} \quad \text{coefficient matrix}$$

$$\begin{bmatrix} 5 \\ 0 \end{bmatrix} \quad \text{constant matrix}$$

The entries to the left of the dashed line may be used to form another matrix called the **coefficient matrix.** The entries to the right of the dashed line may be used to form another matrix called the **constant matrix.** The entire matrix, including coefficients and constants, is called the **augmented matrix.**

EXAMPLE 3

Determine the augmented, coefficient, and constant matrices for the system

$$\begin{cases} x + 3y = 4 \\ 2x + 7y = 6. \end{cases}$$

Solution

$$\left[\begin{array}{rr|r} 1 & 3 & 4 \\ 2 & 7 & 6 \end{array}\right] \qquad \begin{bmatrix} 1 & 3 \\ 2 & 7 \end{bmatrix} \qquad \begin{bmatrix} 4 \\ 6 \end{bmatrix}$$

augmented matrix   coefficient matrix   constant matrix

If we wished to solve the system in Example 3 by the addition method, we would rewrite it as a succession of equivalent systems:

$$\begin{cases} x + 3y = 4 \\ 2x + 7y = 6 \end{cases} \longrightarrow \begin{cases} x + 3y = 4 \\ y = -2 \end{cases} \longrightarrow \begin{cases} x = 10 \\ y = -2. \end{cases}$$

At this point, you should convince yourself that these three equivalent systems may be written in matrix form.

$$\left[\begin{array}{rr|r} 1 & 3 & 4 \\ 2 & 7 & 6 \end{array}\right] \longrightarrow \left[\begin{array}{rr|r} 1 & 3 & 4 \\ 0 & 1 & -2 \end{array}\right] \longrightarrow \left[\begin{array}{rr|r} 1 & 0 & 10 \\ 0 & 1 & -2 \end{array}\right]$$

In the following table, we state the steps (referred to as "row operations") for changing the rows of an augmented matrix. These steps correspond to the steps for changing the equations of a system by the addition method. Remember that each row of an augmented matrix corresponds to one of the equations in a system.

Steps of the Addition Method	**Corresponding Row Operations for Matrices**
Write the equations in a different order.	Interchange the rows of the matrix.
Multiply (or divide) an equation by a nonzero number.	Multiply (or divide) every entry in a row by a nonzero number.
Add (or subtract) a multiple of one equation to (from) another.	Add (or subtract) a multiple of one row to (from) another.

This table states that whatever you do to the equations of a system, you can do to the corresponding rows of its augmented matrix. This process is called **row reduction.** When we use matrices to solve a system of two equations in two variables, our goal is to transform the augmented matrix into the form

$$\left[\begin{array}{cc|c} 1 & 0 & c \\ 0 & 1 & d \end{array}\right]$$

where c and d are real numbers. This form can be obtained only if the system has a unique solution.

EXAMPLE 4

Solve the following system by the addition method and by row reduction of a matrix.

$$\begin{cases} 2x - 4y = 2 \\ -3x + 7y = -1 \end{cases}$$

Solution

Addition Method	Row Reduction Method	
$\begin{cases} 2x - 4y = 2 \\ -3x + 7y = -1 \end{cases}$	$\left[\begin{array}{cc\|c} 2 & -4 & 2 \\ -3 & 7 & -1 \end{array}\right]$	Divide the first equation (row) by 2.
$\begin{cases} x - 2y = 1 \\ -3x + 7y = -1 \end{cases}$	$\left[\begin{array}{cc\|c} 1 & -2 & 1 \\ -3 & 7 & -1 \end{array}\right]$	Add 3 times the first equation (row) to the second equation (row).
$\begin{cases} x - 2y = 1 \\ y = 2 \end{cases}$	$\left[\begin{array}{cc\|c} 1 & -2 & 1 \\ 0 & 1 & 2 \end{array}\right]$	Add 2 times the second equation (row) to the first.
$\begin{cases} x = 5 \\ y = 2 \end{cases}$	$\left[\begin{array}{cc\|c} 1 & 0 & 5 \\ 0 & 1 & 2 \end{array}\right]$	

The solution of the system (by either method) is (5, 2).

EXAMPLE 5

Solve the system $\begin{cases} 4x - 4y = 8 \\ 5x - 4y = 2. \end{cases}$

Solution

$\left[\begin{array}{cc|c} 4 & -4 & 8 \\ 5 & -4 & 2 \end{array}\right]$ Begin by writing the augmented matrix for the system.

$\left[\begin{array}{cc|c} 1 & -1 & 2 \\ 5 & -4 & 2 \end{array}\right]$ The first row has been divided by 4 so that the first-row/first-column entry is 1.

$\left[\begin{array}{cc|c} 1 & -1 & 2 \\ 0 & 1 & -8 \end{array}\right]$ -5 times the first row has been added to the second row. The second-row/first-column entry is now 0.

$\left[\begin{array}{cc|c} 1 & 0 & -6 \\ 0 & 1 & -8 \end{array}\right]$ The second row has been added to the first row to obtain 0 in the first-row/second-column position.

The solution of the system is $(-6, -8)$.

When solving a system by matrix methods, you should always begin, as in Example 5, by transforming the given augmented matrix so that there is a "1" in the first-row/first-column position. For example, suppose you wish to solve the system

$$\begin{cases} \phantom{5x - {}} 3y = 14 \\ 5x - 7y = 8. \end{cases}$$

Since there is a zero in the first-row/first-column position of the system's augmented matrix, you should begin by interchanging the rows, and then dividing the resulting first row by 5. This produces a "1" in the first-row/first-column position.

$$\left[\begin{array}{cc|c} 0 & 3 & 14 \\ 5 & -7 & 8 \end{array}\right] \xrightarrow{\text{interchange rows}} \left[\begin{array}{cc|c} 5 & -7 & 8 \\ 0 & 3 & 14 \end{array}\right] \xrightarrow{\text{divide first row by 5}} \left[\begin{array}{cc|c} 1 & -\frac{7}{5} & \frac{8}{5} \\ 0 & 3 & 14 \end{array}\right]$$

It is not always possible to transform an augmented matrix into the form

$$\left[\begin{array}{cc|c} 1 & 0 & a \\ 0 & 1 & b \end{array}\right].$$

If the final coefficient matrix for a system of two equations in two unknowns has a row of zeros, then the system does not have a unique solution. Two such cases are illustrated in Examples 6 and 7. Example 6 contains a system of equations that has no solution; Example 7 contains a system of equations that has an infinite number of solutions.

EXAMPLE 6

Solve the system $\begin{cases} 3x + 4y = 7 \\ 6x + 8y = 5. \end{cases}$

Solution

$$\left[\begin{array}{cc|c} 3 & 4 & 7 \\ 6 & 8 & 5 \end{array}\right]$$ Begin by writing the augmented matrix for the system.

$$\left[\begin{array}{cc|c} 1 & \frac{4}{3} & \frac{7}{3} \\ 6 & 8 & 5 \end{array}\right]$$ The first row has been divided by 3 to produce a "1" in the first-row/first-column position.

$$\left[\begin{array}{cc|c} 1 & \frac{4}{3} & \frac{7}{3} \\ 0 & 0 & -9 \end{array}\right]$$ -6 times the first row has been added to the second row.

The second row corresponds to the statement $0 = -9$, which is never true. Thus, the system has no solutions.

EXAMPLE 7

Solve the system $\begin{cases} 3x + 4y = 7 \\ 6x + 8y = 14. \end{cases}$

Solution

$$\left[\begin{array}{cc|c} 3 & 4 & 7 \\ 6 & 8 & 14 \end{array}\right]$$ Begin by writing the augmented matrix for the system.

$$\left[\begin{array}{cc|c} 1 & \frac{4}{3} & \frac{7}{3} \\ 6 & 8 & 14 \end{array}\right]$$ The first row has been divided by 3 to produce a "1" in the first-row/first-column position.

$$\left[\begin{array}{cc|c} 1 & \frac{4}{3} & \frac{7}{3} \\ 0 & 0 & 0 \end{array}\right]$$ -6 times the first row has been added to the second row.

The second row corresponds to the statement $0 = 0$, which is always true. Thus, the system has infinitely many solutions. The first row corresponds to the statement $x + \frac{4}{3}y = \frac{7}{3}$, and the solutions of this equation are the solutions of the system. Thus, the solutions of the system are all the points on the line $x + \frac{4}{3}y = \frac{7}{3}$.

Another way to indicate the solution set of the system is as follows. Let $x = c$, where c is any real number. Since $y = \frac{7}{4} - \frac{3}{4}x$, we know that $y = \frac{7}{4} - \frac{3}{4}c$. Thus, we can describe the solution set of the system as the set of all ordered pairs of the form $\left(c, \frac{7}{4} - \frac{3}{4}c\right)$, where c is any real number.

The augmented matrix for the system shown below on the left is given on the right.

$$\begin{cases} 2x + 4y + 2z = 6 \\ 2x + y - z = 2 \\ x + y + 3z = -2 \end{cases} \qquad \left[\begin{array}{ccc|c} 2 & 4 & 2 & 6 \\ 2 & 1 & -1 & 2 \\ 1 & 1 & 3 & -2 \end{array}\right]$$

When we use matrix methods to solve the system, our goal is to transform, if possible, the augmented matrix shown above into one of the form

$$\left[\begin{array}{ccc|c} 1 & 0 & 0 & n \\ 0 & 1 & 0 & m \\ 0 & 0 & 1 & p \end{array}\right],$$

where n, m, and p are real numbers. The following example illustrates the matrix method for solving a system of three equations in three variables.

EXAMPLE 8

Solve the following system:

$$\begin{cases} 2x + 4y + 2z = 6 \\ 2x + y - z = 0 \\ x + y + 3z = -2. \end{cases}$$

Solution

$$\left[\begin{array}{ccc|c} 2 & 4 & 2 & 6 \\ 2 & 1 & -1 & 0 \\ 1 & 1 & 3 & -2 \end{array}\right]$$

Divide the first row by 2 to obtain a 1 in the first-row/first-column position.

$$\left[\begin{array}{ccc|c} 1 & 2 & 1 & 3 \\ 2 & 1 & -1 & 0 \\ 1 & 1 & 3 & -2 \end{array}\right]$$

Add -2 times the first row to the second row.

$$\left[\begin{array}{ccc|c} 1 & 2 & 1 & 3 \\ 0 & -3 & -3 & -6 \\ 1 & 1 & 3 & -2 \end{array}\right]$$ Add -1 times the first row to the third row.

$$\left[\begin{array}{ccc|c} 1 & 2 & 1 & 3 \\ 0 & -3 & -3 & -6 \\ 0 & -1 & 2 & -5 \end{array}\right]$$ Divide the second row by -3.

$$\left[\begin{array}{ccc|c} 1 & 2 & 1 & 3 \\ 0 & 1 & 1 & 2 \\ 0 & -1 & 2 & -5 \end{array}\right]$$ Add -2 times the second row to the first row.

$$\left[\begin{array}{ccc|c} 1 & 0 & -1 & -1 \\ 0 & 1 & 1 & 2 \\ 0 & -1 & 2 & -5 \end{array}\right]$$ Add the second row to the third row.

$$\left[\begin{array}{ccc|c} 1 & 0 & -1 & -1 \\ 0 & 1 & 1 & 2 \\ 0 & 0 & 3 & -3 \end{array}\right]$$ Divide the third row by 3.

$$\left[\begin{array}{ccc|c} 1 & 0 & -1 & -1 \\ 0 & 1 & 1 & 2 \\ 0 & 0 & 1 & -1 \end{array}\right]$$ Add the third row to the first row.

$$\left[\begin{array}{ccc|c} 1 & 0 & 0 & -2 \\ 0 & 1 & 1 & 2 \\ 0 & 0 & 1 & -1 \end{array}\right]$$ Subtract the third row from the second row.

$$\left[\begin{array}{ccc|c} 1 & 0 & 0 & -2 \\ 0 & 1 & 0 & 3 \\ 0 & 0 & 1 & -1 \end{array}\right]$$

Thus, the solution of the system is $(-2, 3, -1)$.

A system of three linear equations in three variables may have no solutions, exactly one solution, or infinitely many solutions. We can determine the nature of the solution set from the corresponding augmented matrix by using row reduction.

Theorem

Suppose we have a system of n linear equations in n variables. This system will have

- *a unique solution* if row reduction of the augmented matrix does *not* produce a row of zeros in the coefficient matrix.
- *no solutions* if row reduction of the augmented matrix produces a row of zeros in the coefficient matrix, and the corresponding entry in the constant matrix is not zero.
- *infinitely many solutions* if row reduction of the augmented matrix produces one or more rows of zeros in the coefficient matrix, and each corresponding entry in the constant matrix is also zero.

EXAMPLE 9

Suppose the first three scales in the figure below balance perfectly. How many identically shaped blocks, either all circles, all rectangles, or all triangles, will balance the square block on the fourth scale? (This problem appeared in *Games* magazine, July/August 1981.)

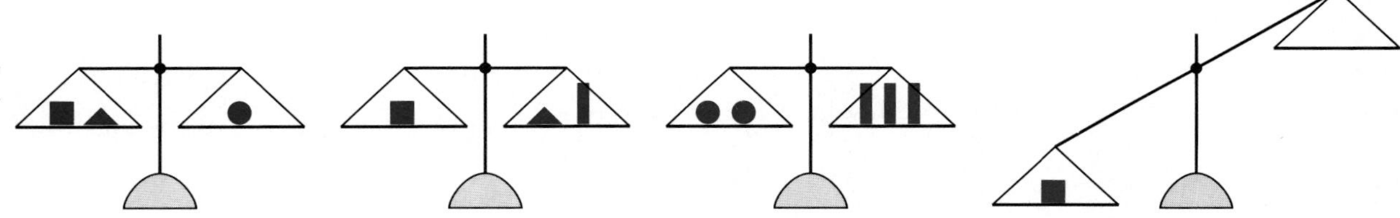

Solution

- ANALYZE The first three scales give us three pieces of information, each indicating various combinations of the blocks that yield a balance. (We can think of the blocks as being of various weights.) We wish to determine a way of balancing one square block with some integer number of identically shaped blocks.
- ORGANIZE Let t = the weight of a triangular block.
  Let c = the weight of a circular block.
  Let r = the weight of a rectangular block.
  Let s = the weight of a square block.

To construct a model for the problem we need to translate (into equations) the information in the first three scales.

■ MODEL

Translate

$$\begin{cases} s + t = c \\ s = t + r \\ 2c = 3r \end{cases} \longrightarrow \begin{cases} s + t + 0r - c = 0 \\ s - t - r + 0c = 0 \\ 0s + 0t - 3r + 2c = 0 \end{cases}$$

We rewrite the system in matrix form, then simplify by row reduction.

$$\begin{array}{ccccc} s & t & r & c & \end{array}$$

$$\begin{bmatrix} 1 & 1 & 0 & -1 & 0 \\ 1 & -1 & -1 & 0 & 0 \\ 0 & 0 & -3 & 2 & 0 \end{bmatrix}$$ Add -1 times the first row to the second row.

$$\begin{bmatrix} 1 & 1 & 0 & -1 & 0 \\ 0 & -2 & -1 & 1 & 0 \\ 0 & 0 & -3 & 2 & 0 \end{bmatrix}$$ Divide the second row by -2.

$$\begin{bmatrix} 1 & 1 & 0 & -1 & 0 \\ 0 & 1 & \frac{1}{2} & -\frac{1}{2} & 0 \\ 0 & 0 & -3 & 2 & 0 \end{bmatrix}$$ Divide the third row by -3.

$$\begin{bmatrix} 1 & 1 & 0 & -1 & 0 \\ 0 & 1 & \frac{1}{2} & -\frac{1}{2} & 0 \\ 0 & 0 & 1 & -\frac{2}{3} & 0 \end{bmatrix}$$ Add $-\frac{1}{2}$ times the third row to the second row.

$$\begin{bmatrix} 1 & 1 & 0 & -1 & 0 \\ 0 & 1 & 0 & -\frac{1}{6} & 0 \\ 0 & 0 & 1 & -\frac{2}{3} & 0 \end{bmatrix}$$ Add -1 times the second row to the first row.

$$\begin{bmatrix} 1 & 0 & 0 & -\frac{5}{6} & 0 \\ 0 & 1 & 0 & -\frac{1}{6} & 0 \\ 0 & 0 & 1 & -\frac{2}{3} & 0 \end{bmatrix}$$ Now translate this matrix back into a system of equations.

(1) $$s = \frac{5}{6}c$$

(2) $$t = \frac{1}{6}c \quad \text{or, alternatively, } c = 6t$$

(3) $$r = \frac{2}{3}c \quad \text{or, alternatively, } c = \frac{3}{2}r$$

Equation (1) says that one square is equivalent to $\frac{5}{6}$ of a circle. Therefore no integer number of circles will balance the one square.

Equations (1) and (2) taken together are useful:

Since $s = \frac{5}{6}c$ and $c = 6t$, we conclude that $s = \frac{5}{6}(6t) = 5t$. This means that one square is equivalent to 5 triangles. Thus we have found a solution to the problem.

Note that equation (1) and equation (3) taken together tell us that $s = \frac{5}{6}\left(\frac{3}{2}r\right) = \frac{5}{4}r$, that is, one square is equivalent to $\frac{5}{4}$ triangles. So there is no solution here.

■ ANSWER 5 triangles will balance the square block on the fourth scale.

EXERCISE SET 9.4

In Exercises 1–4, use the following matrix.

$$\left[\begin{array}{rrr|r} 3 & 8 & -1 & -6 \\ 4 & -5 & 2 & 7 \\ -2 & 0 & -3 & 9 \end{array}\right]$$

1. Identify the first-row/third-column entry.
2. Identify the third-row/first-column entry.
3. Identify the third-row/second-column entry.
4. Identify the second-row/fourth-column entry.

In Exercises 5–12, represent the systems with matrices and the matrices with systems.

5. $\begin{cases} 3x + 7y = 8 \\ 2x + 9y = 1 \end{cases}$

6. $\begin{cases} 3x + 4y = 11 \\ 2x - y = 0 \end{cases}$

7. $\begin{cases} 2x - y + z = 2 \\ x + 3y = 5 \\ x - 3z = 1 \end{cases}$

8. $\begin{cases} -x + 2y - 3z = 5 \\ 4x + 3y + z = 2 \\ x - y - z = 3 \end{cases}$

9. $\left[\begin{array}{rr|r} 12 & 9 & 6 \\ 8 & 6 & 4 \end{array}\right]$

10. $\left[\begin{array}{rr|r} 10 & 5 & 7 \\ -6 & -3 & -4 \end{array}\right]$

11. $\left[\begin{array}{rrr|r} -1 & 1 & -2 & 1 \\ 1 & 1 & -1 & 7 \\ 2 & 0 & 1 & 4 \end{array}\right]$

12. $\left[\begin{array}{rrr|r} -5 & 3 & 2 & 2 \\ 3 & 1 & -4 & 6 \\ -1 & 2 & -1 & 4 \end{array}\right]$

In Exercises 13–26, determine the augmented, coefficient, and constant matrices for the system.

13. $\begin{cases} -2x + 7y = 8 \\ 3x - y = 7 \end{cases}$

14. $\begin{cases} 2x + 3y = 6 \\ 5x + 7y = 13 \end{cases}$

15. $\begin{cases} 8x + 4y = 5 \\ 6x + 3y = 7 \end{cases}$

16. $\begin{cases} 2x - 3y = 5 \\ x + 2y = 3 \end{cases}$

17. $\begin{cases} 3x + 5y = 1 \\ -5 + 3y = -4x \end{cases}$

18. $\begin{cases} -4x + 10y = -2 \\ 2x - 1 = 5y \end{cases}$

19. $\begin{cases} 5x - y + z = -1 \\ x + 3y = 7 \\ y + 2z = -6 \end{cases}$

20. $\begin{cases} x + 2y + z = 4 \\ 2x + 3z = 1 \\ 5y + 2z = -1 \end{cases}$

21. $\begin{cases} 3x + 2y - z = 6 \\ -x - y + z = 1 \\ x + 2y + 2z = 5 \end{cases}$

22. $\begin{cases} 3x + y + 2z = 1 \\ 4x - y + 4z = 4 \\ x + y - 2z = -5 \end{cases}$

23. $\begin{cases} x - 10y + 14z = 2 \\ -2x - y + 2z = 5 \\ -x + 3y - 4z = 1 \end{cases}$

24. $\begin{cases} 2x - y - 3z = 3 \\ x + y - z = 1 \\ 4x + 2y + z = 8 \end{cases}$

25. $\begin{cases} 3x - 6y + 2z = -1 \\ x + 4y + z = 0 \\ 3x + 2y - 5 = 2z \end{cases}$

26. $\begin{cases} 5x - 3y - 2z = -2 \\ x - 2y + z = 0 \\ 3x - 4z - 3 = -y \end{cases}$

For Exercises 27–48, solve the systems in Exercises 5–26. (Use the matrix method.)

Superset

In Exercises 49–51, refer to the following system.

$$\begin{cases} x + 3y + z = A^2 \\ 2x + 5y + 2Az = 0 \\ x + y + A^2z = -9 \end{cases}$$

Determine all real numbers A, if any, for which the given statement is true.

49. The system has a unique solution.

50. The system has no solutions.

51. The system has infinitely many solutions.

In Exercises 52–53, refer to the following system.

$$\begin{cases} x + 2y + z = A \\ 3x + 2y + Az = -A \\ Ax + 2Ay + 3z = 9 \end{cases}$$

Determine all real numbers A, if any, for which the given statement is true.

52. The system has a unique solution.

53. The system has no solutions.

9.5
Determinants and Cramer's Rule

Associated with each $n \times n$ square matrix is a number called the **determinant.** We can use this number to determine whether a system of n linear equations in n variables has a unique solution, and to find the unique solution when it exists.

Let us begin with the case $n = 2$. For the 2×2 square matrix

$$A = \begin{bmatrix} a & b \\ c & d \end{bmatrix}$$

the value of the associated 2×2 determinant is $ad - bc$. There are several ways to denote this number:

$$\det A, \qquad |A|, \qquad \begin{vmatrix} a & b \\ c & d \end{vmatrix}.$$

Careful! We use square brackets to indicate a matrix and vertical bars to indicate the associated determinant. Thus,

$$\det A = \det \begin{bmatrix} a & b \\ c & d \end{bmatrix} = \begin{vmatrix} a & b \\ c & d \end{vmatrix} = ad - bc.$$

The determinant of a 2×2 matrix is the difference of the products of the diagonal elements of the matrix:

$$\begin{vmatrix} a & b \\ c & d \end{vmatrix} = ad - bc$$

EXAMPLE 1

Compute the determinant of each of the following matrices:

(a) $A = \begin{bmatrix} 1 & 2 \\ 3 & 4 \end{bmatrix}$ (b) $B = \begin{bmatrix} 2 & 1 \\ 3 & 4 \end{bmatrix}$ (c) $C = \begin{bmatrix} 2 & 3 \\ 4 & 6 \end{bmatrix}$

Solution

(a) $\det A = \begin{vmatrix} 1 & 2 \\ 3 & 4 \end{vmatrix} = 1 \cdot 4 - 3 \cdot 2 = 4 - 6 = -2$

(b) $\det B = \begin{vmatrix} 2 & 1 \\ 3 & 4 \end{vmatrix} = 2 \cdot 4 - 3 \cdot 1 = 8 - 3 = 5$

(c) $\det C = \begin{vmatrix} 2 & 3 \\ 4 & 6 \end{vmatrix} = 2 \cdot 6 - 4 \cdot 3 = 12 - 12 = 0$

Be careful to form the difference of the products of the diagonal entries in proper order. For

$$\begin{vmatrix} a & b \\ c & d \end{vmatrix}$$

we want $ad - bc$, not $bc - ad$.

To see how determinants can help us solve systems of linear equations, we consider the general case of two linear equations in two variables x and y:

$$\begin{cases} ax + by = e \\ cx + dy = f. \end{cases}$$

To simplify the work, let us restrict our attention to the case where none of the coefficients a, b, c, or d is zero. As it turns out, we get the same results in those cases when one or more of the coefficients are zero.

We use the addition method to solve the system. Multiplying the first equation in the system by $-c$ and the second equation by a, we get

$$\begin{cases} -acx - bcy = -ce \\ acx + ady = af. \end{cases}$$

Adding these two equations, we have

$$(ad - bc)y = af - ce.$$

Replacing the second equation in the original system with this last equation yields the equivalent system

$$\begin{cases} ax + by = e \\ (ad - bc)y = af - ce. \end{cases}$$

Solve the second equation in this system for y by multiplying each side by the reciprocal of $ad - bc$ (assuming that $ad - bc \neq 0$). Clearly this system has a unique solution if and only if

$$ad - bc \neq 0.$$

Let us suppose then that $ad - bc \neq 0$ and continue to look for the unique solution. We substitute the value for y, that is,

$$y = \frac{af - ce}{ad - bc},$$

in the other equation

$$ax + by = e$$

which can be written as

$$x = \frac{1}{a}(e - by).$$

Solving for x, we get:

$$x = \frac{1}{a}\left[e - b\left(\frac{af - ce}{ad - bc}\right)\right] = \frac{1}{a}\left[\frac{e(ad - bc) - b(af - ce)}{ad - bc}\right]$$

$$= \frac{1}{a}\left[\frac{aed - afb}{ad - bc}\right] = \frac{ed - fb}{ad - bc}.$$

By following a similar argument in those cases where one or more of the coefficients a, b, c, or d is zero, we have a proof of the following theorem.

Theorem

A system of 2 linear equations in 2 variables

$$\begin{cases} ax + by = e \\ cx + dy = f \end{cases}$$

has a unique solution if and only if $ad - bc \neq 0$. This unique solution, if it exists, is given by

$$x = \frac{ed - fb}{ad - bc}, \qquad y = \frac{af - ce}{ad - bc}.$$

Notice that the number $ad - bc$, which plays such an important role in establishing this theorem, is the determinant of the coefficient matrix of the system:

$$A = \begin{bmatrix} a & b \\ c & d \end{bmatrix}, \qquad \det A = ad - bc.$$

Furthermore, since

$$ed - fb = \begin{vmatrix} e & b \\ f & d \end{vmatrix} \quad \text{and} \quad af - ce = \begin{vmatrix} a & e \\ c & f \end{vmatrix},$$

we notice that when there is a unique solution to the system (i.e., $\det A \neq 0$), then the values of the variables can be written as quotients of determinants:

$$x = \frac{\begin{vmatrix} e & b \\ f & d \end{vmatrix}}{\begin{vmatrix} a & b \\ c & d \end{vmatrix}}$$

The x-coefficients have been replaced with the constants of the system.

The denominator is $\det A$.

$$y = \frac{\begin{vmatrix} a & e \\ c & f \end{vmatrix}}{\begin{vmatrix} a & b \\ c & d \end{vmatrix}}$$

The y-coefficients have been replaced with the constants of the system.

The denominator is $\det A$.

EXAMPLE 2

Determine whether the following system has a unique solution, and find the solution if it exists.

$$\begin{cases} 2x - 3y = -1 \\ 4x - 5y = 7 \end{cases}$$

Solution We begin by computing the determinant of the coefficient matrix

$$A = \begin{bmatrix} 2 & -3 \\ 4 & -5 \end{bmatrix}.$$

$$\det A = \begin{vmatrix} 2 & -3 \\ 4 & -5 \end{vmatrix} = (2)(-5) - (4)(-3) = -10 + 12 = 2.$$

Since $\det A \neq 0$, the system has a unique solution. It is given by

$$x = \frac{\begin{vmatrix} -1 & -3 \\ 7 & -5 \end{vmatrix}}{\det A} = \frac{(-1)(-5) - (7)(-3)}{2} = 13$$

The matrix A is modified by replacing the column of x-coefficients with the constants of the system.

$$y = \frac{\begin{vmatrix} 2 & -1 \\ 4 & 7 \end{vmatrix}}{\det A} = \frac{(2)(7) - (4)(-1)}{2} = 9$$

The matrix A is modified by replacing the column of y-coefficients with the constants of the system.

The unique solution of the system is (13, 9).

EXAMPLE 3

Determine those real numbers b, if any, for which the following system does *not* have a unique solution.

$$\begin{cases} 2x + (1 - b)y = 5 \\ \qquad bx - 3y = 4 \end{cases}$$

Solution

$$\begin{vmatrix} 2 & 1-b \\ b & -3 \end{vmatrix} = (2)(-3) - (b)(1-b) = b^2 - b - 6$$

The determinant of the coefficient matrix is computed.

$$b^2 - b - 6 = 0$$
$$(b-3)(b+2) = 0$$

The system does not have a unique solution when det $A = 0$.

$$b - 3 = 0 \qquad b + 2 = 0$$
$$b = 3 \qquad b = -2$$

When $b = 3$ or $b = -2$, the system does not have a unique solution.

Determinants of $n \times n$ Matrices

In order to compute the determinant of a 3×3 matrix or, in general, any $n \times n$ matrix with $n > 2$, we must first discuss the idea of a *signed minor* also called a *cofactor.*

Associated with each entry in an $n \times n$ matrix is a "subdeterminant" called a *minor.* The **minor** of an entry is the determinant of the matrix obtained by deleting the row and column containing the entry. For the matrix

$$\begin{bmatrix} 3 & 2 & 1 \\ 4 & 5 & 6 \\ 9 & 7 & 8 \end{bmatrix}$$

the minor for the entry in the first-row/second-column is found by deleting the first row and the second column of the matrix,

$$\begin{bmatrix} 3 & 2 & 1 \\ 4 & 5 & 6 \\ 9 & 7 & 8 \end{bmatrix},$$

and forming the determinant of the 2×2 matrix of remaining entries:

$$\begin{vmatrix} 4 & 6 \\ 9 & 8 \end{vmatrix}.$$

Similarly, the minor for the second-row/first-column entry 4 in the original 3×3 matrix is

$$\begin{vmatrix} 2 & 1 \\ 7 & 8 \end{vmatrix}.$$

Clearly, in any 3×3 matrix, since there are 9 entries, there are 9 minors. For a 4×4 matrix there are 16 minors, one for each entry.

We next consider the notion of a **signed minor,** or **cofactor.** To compute a signed minor we multiply a minor by $+1$ or -1 depending upon the position of the entry in the matrix from which the minor was formed. The choice of $+1$ or -1 is determined by a checkerboard pattern that begins with a $+$ for the entry in the first-row/first-column, as shown in Figure 9.23.

$$\begin{bmatrix} + & - & + & - & \cdots \\ - & + & - & + & \cdots \\ + & - & + & - & \cdots \\ - & + & - & + & \cdots \\ \vdots & \vdots & \vdots & \vdots & \vdots \end{bmatrix}$$

Figure 9.23

$$\begin{bmatrix} 1 & 2 & 3 \\ 4 & 5 & 6 \\ 7 & 8 & 9 \end{bmatrix}$$

For example, for the 3×3 matrix at the left, the signed minors for the entries in the first row of the matrix are

$$+\begin{vmatrix} 5 & 6 \\ 8 & 9 \end{vmatrix}, \quad -\begin{vmatrix} 4 & 6 \\ 7 & 9 \end{vmatrix}, \quad +\begin{vmatrix} 4 & 5 \\ 7 & 8 \end{vmatrix}.$$

We now can state a procedure that enables us to evaluate the determinant of any $n \times n$ matrix.

Computing a Determinant

To compute the determinant of any $n \times n$ matrix, select any row (or column) of the matrix and form the sum of the products of the entries from that row (or column) with their corresponding signed minors.

When we use this fact to evaluate a determinant, we say that we are **expanding along a row (or column).**

EXAMPLE 4

Compute det A for the matrix $A = \begin{bmatrix} 2 & 3 & 4 \\ 1 & 0 & -2 \\ 5 & -1 & -3 \end{bmatrix}$ by expanding along

(a) the first row, (b) the second row, (c) the third column.

Solution

(a) $\det A = 2\left(+\begin{vmatrix} 0 & -2 \\ -1 & -3 \end{vmatrix}\right) + 3\left(-\begin{vmatrix} 1 & -2 \\ 5 & -3 \end{vmatrix}\right) + 4\left(+\begin{vmatrix} 1 & 0 \\ 5 & -1 \end{vmatrix}\right)$

$= 2(-2) + 3(-7) + 4(-1)$

$= -29$

(b) $$\det A = 1\left(-\begin{vmatrix} 3 & 4 \\ -1 & -3 \end{vmatrix}\right) + 0\left(+\begin{vmatrix} 2 & 4 \\ 5 & -3 \end{vmatrix}\right) + (-2)\left(-\begin{vmatrix} 2 & 3 \\ 5 & -1 \end{vmatrix}\right)$$
$$= 1(5) + 0 - 2(17)$$
$$= -29$$

(c) $$\det A = 4\left(+\begin{vmatrix} 1 & 0 \\ 5 & -1 \end{vmatrix}\right) + (-2)\left(-\begin{vmatrix} 2 & 3 \\ 5 & -1 \end{vmatrix}\right) + (-3)\left(+\begin{vmatrix} 2 & 3 \\ 1 & 0 \end{vmatrix}\right)$$
$$= 4(-1) - 2(17) - 3(-3)$$
$$= -29$$

There is an alternate method for computing a 3×3 determinant. There is a "diagonal method" similar to the scheme used for 2×2 determinants. Begin by writing the determinant with the first two columns written over again to the right of the third column. Then, the determinant is the sum of products of the entries on each "falling" diagonal minus the products of the entries on each "rising" diagonal:

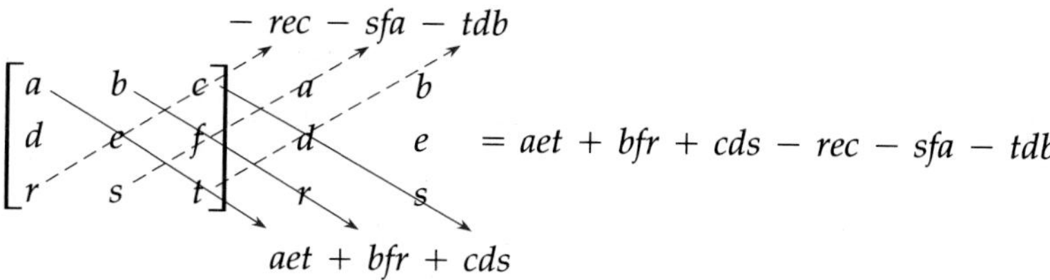

Careful! This diagonal method does *not* generalize to 4×4 determinants or larger $n \times n$ determinants.

EXAMPLE 5

Use the diagonal method to compute the determinant of the 3×3 matrix

$$\begin{vmatrix} 2 & 3 & -1 \\ 0 & -4 & 2 \\ 1 & 5 & 1 \end{vmatrix}.$$

Solution

$$\begin{vmatrix} 2 & 3 & -1 \\ 0 & -4 & 2 \\ 1 & 5 & 1 \end{vmatrix}\begin{matrix} 2 & 3 \\ 0 & -4 \\ 1 & 5 \end{matrix} = (2)(-4)(1) + (3)(2)(1) + (-1)(0)(5)$$
$$-(1)(-4)(-1) - (5)(2)(2) - (1)(0)(3)$$
$$= -8 + 6 + 0 - 4 - 20 - 0 = -26$$

Cramer's Rule

The results we found earlier for systems of two equations in two variables can be generalized to systems of n equations in n variables.

Cramer's Rule

A system of n linear equations in n variables has a unique solution if and only if the determinant of the coefficient matrix A is not zero. If a unique solution exists, the values of the variables can be written as quotients of determinants; the denominators are det A and the numerators are the determinant of A modified by replacing the column whose entries are the coefficients of the variable with the constants of the system.

In Example 6, we use Cramer's rule to determine whether a system of 3 linear equations in 3 variables has a unique solution.

EXAMPLE 6

Determine whether the following system has a unique solution, and find the solution if it exists.

$$\begin{cases} 2x + y = 3 \\ x + y + 2z = -1 \\ x + 3z = 1 \end{cases}$$

Solution We begin by computing the determinant of the coefficient matrix A.

$$\det A = \begin{vmatrix} 2 & 1 & 0 \\ 1 & 1 & 2 \\ 1 & 0 & 3 \end{vmatrix} = 2 \cdot 1 \cdot 3 + 1 \cdot 2 \cdot 1 + 0 \cdot 1 \cdot 0 - 1 \cdot 1 \cdot 0 - 0 \cdot 2 \cdot 2 - 3 \cdot 1 \cdot 1 = 5$$

Since det $A \neq 0$, the system has a unique solution. It is given by

$$x = \frac{\begin{vmatrix} 3 & 1 & 0 \\ -1 & 1 & 2 \\ 1 & 0 & 3 \end{vmatrix}}{\det A} = \frac{(9 + 2 + 0) - (0 + 0 - 3)}{5} = \frac{14}{5}$$

Replace first column of A with the column of constants.

$$y = \frac{\begin{vmatrix} 2 & 3 & 0 \\ 1 & -1 & 2 \\ 1 & 1 & 3 \end{vmatrix}}{\det A} = \frac{(-6 + 6 + 0) - (0 + 4 + 9)}{5} = -\frac{13}{5}$$

Replace second column of A with the column of constants.

$$z = \frac{\begin{vmatrix} 2 & 1 & 3 \\ 1 & 1 & -1 \\ 1 & 0 & 1 \end{vmatrix}}{\det A} = \frac{(2 - 1 + 0) - (3 + 0 + 1)}{5} = -\frac{3}{5}$$

Replace third column of A with the column of constants.

The unique solution of the system is $\left(\frac{14}{5}, -\frac{13}{5}, -\frac{3}{5}\right)$.

Notice that in Example 6, we used the diagonal method to calculate the determinants. In Example 7, we use signed minors.

EXAMPLE 7

Determine those real numbers b, if any, for which the following system does *not* have a unique solution.

$$\begin{cases} x + 2y + bz = 3 \\ \quad bx - y = 4 \\ 3x + y + 2z = 0 \end{cases}$$

Solution

$$\begin{vmatrix} 1 & 2 & b \\ b & -1 & 0 \\ 3 & 1 & 2 \end{vmatrix} = b\left(-\begin{vmatrix} 2 & b \\ 1 & 2 \end{vmatrix}\right) + (-1)\left(+\begin{vmatrix} 1 & b \\ 3 & 2 \end{vmatrix}\right)$$
$$= b[-(4 - b)] - (2 - 3b)$$
$$= b^2 - b - 2$$
$$= (b - 2)(b + 1)$$

The determinant of the coefficient matrix A is computed by expanding along the second row.

Recall that the system does not have a unique solution when $\det A = 0$.

When $b = 2$ or $b = -1$, the determinant is 0 and thus the system does not have a unique solution.

EXERCISE SET 9.5

In Exercises 1–8, compute the determinant of the matrix.

1. $\begin{bmatrix} 2 & 5 \\ 1 & 4 \end{bmatrix}$ **2.** $\begin{bmatrix} 3 & 1 \\ 5 & 3 \end{bmatrix}$ **3.** $\begin{bmatrix} 2 & 4 \\ 1 & 5 \end{bmatrix}$

4. $\begin{bmatrix} 3 & 3 \\ 5 & 1 \end{bmatrix}$ **5.** $\begin{bmatrix} 2 & -4 \\ 3 & -6 \end{bmatrix}$ **6.** $\begin{bmatrix} 2 & -1 \\ 8 & -4 \end{bmatrix}$

7. $\begin{bmatrix} 2 & -4 \\ -3 & 7 \end{bmatrix}$ **8.** $\begin{bmatrix} 5 & -4 \\ -2 & 3 \end{bmatrix}$

In Exercises 9–20, determine whether the system has a unique solution, and find the solution if it exists.

9. $\begin{cases} 2x + 5y = 3 \\ x + 4y = 2 \end{cases}$ **10.** $\begin{cases} 3x + y = 0 \\ 5x + 3y = -1 \end{cases}$

11. $\begin{cases} 2x + 6y = 4 \\ x + 3y = 2 \end{cases}$ **12.** $\begin{cases} 2x + 3y = 5 \\ 3x + 4y = 8 \end{cases}$

13. $\begin{cases} 2x + y = 1 \\ x - 4y = 0 \end{cases}$ **14.** $\begin{cases} 3x - y = 0 \\ -6x + 2y = 0 \end{cases}$

15. $\begin{cases} 2x - 3y = 4 \\ 4x - 6y = -2 \end{cases}$

16. $\begin{cases} 3x - y = 2 \\ -9x + 3y = 6 \end{cases}$

17. $\begin{cases} -2x + y = 3 \\ 3x - 4y = -5 \end{cases}$

18. $\begin{cases} 3x - y = 1 \\ -9x + 4y = -1 \end{cases}$

19. $\begin{cases} 4x + y = 2 \\ 2x - y = 4 \end{cases}$

20. $\begin{cases} 2x + y = 1 \\ x + 4y = 2 \end{cases}$

In Exercises 21–26, determine those values of b, if any, for which the system does not have a unique solution.

21. $\begin{cases} 2x + 3y = 1 \\ 3x + by = 5 \end{cases}$

22. $\begin{cases} 3x - by = 3 \\ 4x + y = -2 \end{cases}$

23. $\begin{cases} x - by = 3 \\ bx + y = 2 \end{cases}$

24. $\begin{cases} x + b^2y = 2 \\ x + by = 3 \end{cases}$

25. $\begin{cases} bx + 5y = 2 \\ 3x + (b - 2)y = 1 \end{cases}$

26. $\begin{cases} bx + 3y = 2 \\ x + by = 5 \end{cases}$

In Exercises 27–32, compute the determinant by expanding along (a) the first row, (b) the second column.

27. $\begin{vmatrix} 2 & 1 & 3 \\ 1 & 0 & 2 \\ 1 & 1 & 1 \end{vmatrix}$

28. $\begin{vmatrix} 1 & 2 & 1 \\ 0 & 1 & 3 \\ 2 & 3 & 4 \end{vmatrix}$

29. $\begin{vmatrix} 2 & 1 & 0 \\ 1 & 3 & 4 \\ 2 & 0 & 1 \end{vmatrix}$

30. $\begin{vmatrix} 1 & 2 & -1 \\ 3 & 1 & 1 \\ -1 & 3 & -3 \end{vmatrix}$

31. $\begin{vmatrix} 1 & -1 & -2 \\ 3 & -2 & 0 \\ -1 & 4 & 2 \end{vmatrix}$

32. $\begin{vmatrix} -1 & 2 & 1 \\ 0 & -1 & 2 \\ 1 & -3 & -2 \end{vmatrix}$

For Exercises 33–38, use the diagonal method to compute the determinants in Exercises 27–32.

For Exercises 39–56, using Cramer's Rule, determine whether the system of linear equations in Exercises 19–36 in Exercise Set 9.3 has a unique solution. Find the solution if it exists.

Superset

In Exercises 57–60, compute the determinant using signed minors.

57. $\begin{vmatrix} 1 & 2 & 0 & 3 \\ -1 & 1 & 2 & 0 \\ 1 & 0 & -3 & 1 \\ 4 & 1 & -3 & 5 \end{vmatrix}$

58. $\begin{vmatrix} 1 & 2 & 0 & 1 \\ 0 & 2 & 1 & -3 \\ 0 & 0 & 3 & 4 \\ 0 & 0 & 0 & -1 \end{vmatrix}$

59. $\begin{vmatrix} 0 & 0 & 0 & 2 \\ 0 & 0 & -1 & 4 \\ 0 & 1 & 1 & 0 \\ 2 & 0 & -3 & 1 \end{vmatrix}$

60. $\begin{vmatrix} 2 & 0 & 1 & 3 \\ 1 & 2 & 0 & -1 \\ 3 & -1 & 0 & 2 \\ 2 & 3 & 2 & 3 \end{vmatrix}$

The following statements are true for every $n \times n$ matrix A.

I. If A contains a row of zeros, then $\det A = 0$.

II. If two rows of A are identical, then $\det A = 0$.

III. If a matrix B is obtained from A by interchanging two rows, then $\det B = -\det A$.

IV. If a matrix B is obtained from A by multiplying each entry in some row of A by a real number k, then $\det B = k \det A$.

V. If a matrix B is obtained from A by adding a multiple of one row of A to another row of A, then $\det B = \det A$.

Statements I–V are also true if "row" is replaced with "column."

61. Using a 3×3 matrix of your choice, verify Statement I.

62. Using a 3×3 matrix of your choice, verify Statement II.

In Exercises 63–66, verify the specified statement using the 3×3 matrix

$$A = \begin{bmatrix} -1 & 0 & 4 \\ 3 & 1 & -2 \\ 0 & 1 & 1 \end{bmatrix}$$

63. Statement III, using the first and second rows.

64. Statement IV using 2 times the third row.

65. Statement V, using 2 times the first row added to the third row.

66. Statement V, using -3 times the second row added to the first row.

For Exercises 67–72, redo Exercises 61–66 with "row" replaced with "column."

73. Verify that the following equation describes a straight line through the points (x_0, y_0) and (x_1, y_1).

$$\begin{vmatrix} 1 & x & y \\ 1 & x_0 & y_0 \\ 1 & x_1 & y_1 \end{vmatrix} = 0$$

74. Use Exercise 73 to find an equation that describes the line through the points (2, 5) and (4, 1).

9.6
The Inverse of a Square Matrix

Thus far, we have used matrices to represent systems of n linear equations in n variables. In certain applications, we often need to solve several such systems, which differ from one another only in the constant terms. For example, consider the following systems.

$$\begin{cases} 2x + 3y = 4 \\ 5x + 6y = 1 \end{cases} \qquad \begin{cases} 2x + 3y = 1 \\ 5x + 6y = -2 \end{cases}$$

The augmented matrices for these systems are shown below. Notice that these augmented matrices are identical except in the third columns.

$$\left[\begin{array}{cc|c} 2 & 3 & 4 \\ 5 & 6 & 1 \end{array}\right] \qquad \left[\begin{array}{cc|c} 2 & 3 & 1 \\ 5 & 6 & -2 \end{array}\right]$$

We will see that the steps one follows to row reduce one of these matrices are precisely the steps needed to row reduce the other. In this section we develop a technique for solving such systems without any duplication of effort. This technique is based upon the idea of *matrix multiplication.*

Definition of Matrix Multiplication

Suppose A is an $m \times k$ matrix and B is a $k \times n$ matrix. The **product matrix** $C = AB$ is an $m \times n$ matrix, whose ith-row/jth-column entry (denoted c_{ij}) is the sum of the products of corresponding entries from the ith-row of A and the jth-column of B:

$$c_{ij} = a_{i1}b_{1j} + a_{i2}b_{2j} + a_{i3}b_{3j} + \cdots + a_{ik}b_{kj}.$$

Note that for the product AB to exist, the number of columns of A must equal the number of rows of B.

EXAMPLE 1

Compute the matrix product AB for the matrices $A = \begin{bmatrix} 2 & 3 & -4 \\ 5 & 1 & 2 \end{bmatrix}$ and $B = \begin{bmatrix} 4 & -2 \\ -1 & 6 \\ 3 & 0 \end{bmatrix}$.

Solution Since A is a 2×3 matrix and B is a 3×2 matrix, the product matrix $C = AB$ is a 2×2 matrix:

$$C = \begin{bmatrix} c_{11} & c_{12} \\ c_{21} & c_{22} \end{bmatrix}$$

The first digit in each subscript denotes the row; the second digit denotes the column.

We begin by computing the first-row/first-column entry c_{11}.

$$\begin{bmatrix} 2 & 3 & -4 \\ 5 & 1 & 2 \end{bmatrix} \begin{bmatrix} 4 & -2 \\ -1 & 6 \\ 3 & 0 \end{bmatrix}$$

To compute c_{11}, we form the sum of the products of the corresponding entries of the first row of A and the first column of B.

$$\begin{aligned} c_{11} &= (2)(4) + (3)(-1) + (-4)(3) \\ &= 8 - 3 - 12 = -7 \end{aligned}$$

Next, we compute the second-row/first-column entry c_{21}.

$$\begin{bmatrix} 2 & 3 & -4 \\ 5 & 1 & 2 \end{bmatrix} \begin{bmatrix} 4 & -2 \\ -1 & 6 \\ 3 & 0 \end{bmatrix}$$

To compute c_{21}, we form the sum of the products of corresponding entries of the second row of A and first column of B.

$$\begin{aligned} c_{21} &= (5)(4) + (1)(-1) + (2)(3) \\ &= 20 - 1 + 6 = 25 \end{aligned}$$

Thus far we have found that

$$\begin{matrix} A & \times & B & = & C \\ \begin{bmatrix} 2 & 3 & -4 \\ 5 & 1 & 2 \end{bmatrix} & & \begin{bmatrix} 4 & -2 \\ -1 & 6 \\ 3 & 0 \end{bmatrix} & = & \begin{bmatrix} -7 & ? \\ 25 & ? \end{bmatrix} \end{matrix}$$

The other two entries in the product C are found in a similar fashion.

$$\begin{aligned} c_{12} &= (2)(-2) + (3)(6) + (-4)(0) \\ &= -4 + 18 + 0 = 14 \end{aligned}$$

The first row of A and the second column of B are used.

$$\begin{aligned} c_{22} &= (5)(-2) + (1)(6) + (2)(0) \\ &= -10 + 6 + 0 = -4 \end{aligned}$$

The second row of A and the second column of B are used.

Thus,

$$\begin{bmatrix} 2 & 3 & -4 \\ 5 & 1 & 2 \end{bmatrix} \begin{bmatrix} 4 & -2 \\ -1 & 6 \\ 3 & 0 \end{bmatrix} = \begin{bmatrix} -7 & 14 \\ 25 & -4 \end{bmatrix}$$

Careful! In order to multiply two matrices, the number of columns in the first must equal the number of rows in the second.

$$\underset{m \times k}{A} \times \underset{k \times n}{B} = \underset{m \times n}{C}$$

$\uparrow$ $\uparrow$ must be equal for product to be defined

EXAMPLE 2

Compute the matrix product BA for the matrices A and B from Example 1.

Solution Since B is a 3×2 matrix, and A is a 2×3 matrix, the product matrix BA is a 3×3 matrix whose 9 entries we must find.

$$\begin{bmatrix} 4 & -2 \\ -1 & 6 \\ 3 & 0 \end{bmatrix} \begin{bmatrix} 2 & 3 & -4 \\ 5 & 1 & 2 \end{bmatrix} = \begin{bmatrix} c_{11} & c_{12} & c_{13} \\ c_{21} & c_{22} & c_{23} \\ c_{31} & c_{32} & c_{33} \end{bmatrix}$$

The first digit in each subscript denotes the row; the second subscript denotes the column.

$$c_{11} = 4(2) + (-2)(5) = -2 \qquad c_{12} = 4(3) + (-2)(1) = 10$$
$$c_{21} = (-1)(2) + (6)(5) = 28 \qquad c_{22} = (-1)(3) + (6)(1) = 3$$
$$c_{31} = 3(2) + (0)(5) = 6 \qquad c_{32} = 3(3) + (0)(1) = 9$$

$$c_{13} = 4(-4) + (-2)(2) = -20$$
$$c_{23} = (-1)(-4) + (6)(2) = 16$$
$$c_{33} = 3(-4) + (0)(2) = -12$$

Thus, $BA = \begin{bmatrix} -2 & 10 & -20 \\ 28 & 3 & 16 \\ 6 & 9 & -12 \end{bmatrix}.$

Two matrices are *equal* if and only if they are the same size (have the same number of rows and have the same number of columns) and corresponding entries of the two matrices are equal. Thus, the matrix products found in Examples 1 and 2 are not equal: $AB \neq BA$, since AB is a 2×2 matrix and BA is a 3×3 matrix.

An $n \times 1$ matrix (that is, a matrix consisting of a single column) is called a **column matrix.** We can use column matrices and matrix multiplication to write a system of linear equations as a single *matrix equation.*

$$\begin{cases} 2x + 3y - z = 5 \\ x - 4y = 8 \\ 3x + 2z = 9 \end{cases}$$

A system of 3 linear equations in 3 variables.

$$\begin{bmatrix} 2 & 3 & -1 \\ 1 & -4 & 0 \\ 3 & 0 & 2 \end{bmatrix} \begin{bmatrix} x \\ y \\ z \end{bmatrix} = \begin{bmatrix} 5 \\ 8 \\ 9 \end{bmatrix}$$

The system written as a matrix equation.

In general, a system of n linear equations in the n variables $x_1, x_2, \ldots x_n$ can be written as $AX = C$, where A is the coefficient matrix, C is the constant matrix, and X is the column matrix,

$$X = \begin{bmatrix} x_1 \\ x_2 \\ x_3 \\ \cdot \\ \cdot \\ \cdot \\ x_n \end{bmatrix}$$

When a system is written as a single matrix equation, we are reminded of a simple linear equation with one variable, and in particular, how we go about solving such an equation. Recall, to solve the equation $3x = 5$, we multiply each side of the equation by the multiplicative inverse of 3 $\left(\text{namely, } 3^{-1} = \frac{1}{3}\right)$ to obtain $x = 3^{-1} \cdot 5 = \frac{5}{3}$. In general, for $a \neq 0$ we multiply each side of the equation

$$ax = c$$

by the multiplicative inverse of the coefficient a (namely, a^{-1}) to obtain $x = a^{-1}c$. The natural question then is whether we can do the same sort of thing with the matrix equation

$$AX = C.$$

That is, does the matrix A have a "multiplicative inverse" A^{-1}, which would permit us to solve for X and write $X = A^{-1}C$? Under certain conditions, the answer is yes, but first we must consider the notion of a *multiplicative identity* for matrices.

Definition of Multiplicative Identities for Matrices

The $n \times n$ matrix

$$I_n = \begin{bmatrix} 1 & 0 & 0 & \cdots & 0 \\ 0 & 1 & 0 & \cdots & 0 \\ 0 & 0 & 1 & \cdots & 0 \\ \vdots & \vdots & \vdots & & \vdots \\ 0 & 0 & 0 & \cdots & 1 \end{bmatrix}$$

whose ith-row/ith-column entries are 1's (we say that it has 1's along the main diagonal), and whose other entries are 0's is called the $\boldsymbol{n \times n}$ **identity matrix.** For any $n \times n$ matrix A, we have

$$A \cdot I_n = I_n \cdot A = A.$$

When $n = 2$ and $n = 3$, we have the identity matrices

$$I_2 = \begin{bmatrix} 1 & 0 \\ 0 & 1 \end{bmatrix} \quad \text{and} \quad I_3 = \begin{bmatrix} 1 & 0 & 0 \\ 0 & 1 & 0 \\ 0 & 0 & 1 \end{bmatrix} \text{ respectively.}$$

Observe the effect of multiplying a 3×3 matrix by the 3×3 identity matrix:

$$\begin{bmatrix} 2 & 3 & 1 \\ 4 & 7 & 2 \\ 8 & 9 & 0 \end{bmatrix}\begin{bmatrix} 1 & 0 & 0 \\ 0 & 1 & 0 \\ 0 & 0 & 1 \end{bmatrix} = \begin{bmatrix} 2 & 3 & 1 \\ 4 & 7 & 2 \\ 8 & 9 & 0 \end{bmatrix}$$

If A is an $n \times n$ matrix, our objective is to find an $n \times n$ matrix B for which $BA = AB = I_n$. If such a matrix B exists, it will act as an inverse for A under matrix multiplication, since multiplying A by B produces the identity matrix I_n. We will denote this inverse by A^{-1}. We now state without proof a condition for a square matrix A to have an inverse A^{-1}.

Existence of Multiplicative Inverses for Matrices

The $n \times n$ matrix A has a multiplicative inverse, denoted by A^{-1}, if and only if

$$\det A = |A| \neq 0.$$

EXAMPLE 3

Verify that each of the following matrices has an inverse:

(a) $A = \begin{bmatrix} 2 & 3 \\ 1 & 5 \end{bmatrix}$ (b) $B = \begin{bmatrix} 1 & 4 & 2 \\ -2 & 3 & 1 \\ 2 & -1 & 4 \end{bmatrix}$

Solution

(a) $\det A = \begin{vmatrix} 2 & 3 \\ 1 & 5 \end{vmatrix} = 2(5) - 1(3) = 7 \neq 0$, so A^{-1} exists.

(b) To compute det B, we expand along the second row using signed minors to obtain

$$\begin{aligned} \det B &= \begin{vmatrix} 1 & 4 & 2 \\ -2 & 3 & 1 \\ 2 & -1 & 4 \end{vmatrix} \\ &= -2\left(-\begin{vmatrix} 4 & 2 \\ -1 & 4 \end{vmatrix}\right) + 3\left(\begin{vmatrix} 1 & 2 \\ 2 & 4 \end{vmatrix}\right) + 1\left(-\begin{vmatrix} 1 & 4 \\ 2 & -1 \end{vmatrix}\right) \\ &= -2(-18) + 3(0) + 1(9) = 45. \end{aligned}$$

Since $\det B \neq 0$, the inverse B^{-1} exists.

Before we can actually describe a method for computing A^{-1}, we need to define two simple operations on matrices.

Definition of Scalar Multiplication for Matrices

The product kA of a real number k, called a **scalar,** and any matrix A is computed by multiplying each entry of A by k.

For example,

$$2\begin{bmatrix} 1 & 4 \\ 3 & -2 \end{bmatrix} = \begin{bmatrix} 2 & 8 \\ 6 & -4 \end{bmatrix} \quad \text{and} \quad \frac{1}{3}\begin{bmatrix} 6 & -3 \\ 2 & 0 \end{bmatrix} = \begin{bmatrix} 2 & -1 \\ \frac{2}{3} & 0 \end{bmatrix}.$$

Definition of Matrix Transpose

The **transpose** A^t for any matrix A is formed by interchanging the rows with the corresponding columns of A.

For example, if

$$A = \begin{bmatrix} 2 & 7 & 5 \\ 0 & 3 & 0 \\ 0 & 0 & 4 \end{bmatrix}, \quad \text{then } A^t = \begin{bmatrix} 2 & 0 & 0 \\ 7 & 3 & 0 \\ 5 & 0 & 4 \end{bmatrix}$$

First row of A becomes the first column of A^t; second row of A becomes second column of A^t; third row of A becomes third column of A^t.

We are finally in a position to outline a method for computing A^{-1}, the inverse of matrix A.

Computation of the Multiplicative Inverse of a Matrix

Suppose that A is an $n \times n$ matrix for which $\det A \neq 0$. Then A^{-1} exists, and we can determine it in three steps:

Step 1. Replace each entry of A by its signed minor. The result is called the **cofactor matrix of *A*.**

Step 2. Form the transpose of the matrix found in Step 1. The result is called the **adjoint of *A*,** denoted by adj(A).

Step 3. $A^{-1} = \dfrac{1}{\det A}\,\text{adj}(A)$.

EXAMPLE 4

Determine A^{-1} for the matrix $A = \begin{bmatrix} 2 & 0 & 3 \\ 4 & 1 & 5 \\ 1 & 2 & 1 \end{bmatrix}$.

Solution

Step 1: Each entry of A is replaced by its signed minor.

$$\text{Cofactor matrix of } A = \begin{bmatrix} +\begin{vmatrix} 1 & 5 \\ 2 & 1 \end{vmatrix} & -\begin{vmatrix} 4 & 5 \\ 1 & 1 \end{vmatrix} & +\begin{vmatrix} 4 & 1 \\ 1 & 2 \end{vmatrix} \\ -\begin{vmatrix} 0 & 3 \\ 2 & 1 \end{vmatrix} & +\begin{vmatrix} 2 & 3 \\ 1 & 1 \end{vmatrix} & -\begin{vmatrix} 2 & 0 \\ 1 & 2 \end{vmatrix} \\ +\begin{vmatrix} 0 & 3 \\ 1 & 5 \end{vmatrix} & -\begin{vmatrix} 2 & 3 \\ 4 & 5 \end{vmatrix} & +\begin{vmatrix} 2 & 0 \\ 4 & 1 \end{vmatrix} \end{bmatrix} = \begin{bmatrix} -9 & 1 & 7 \\ 6 & -1 & -4 \\ -3 & 2 & 2 \end{bmatrix}$$

Step 2: We form the transpose of the cofactor matrix found in Step 1.

$$\text{adj}(A) = \begin{bmatrix} -9 & 1 & 7 \\ 6 & -1 & -4 \\ -3 & 2 & 2 \end{bmatrix}^t = \begin{bmatrix} -9 & 6 & -3 \\ 1 & -1 & 2 \\ 7 & -4 & 2 \end{bmatrix}$$

Step 3: We begin by computing the determinant of A.

$$\det A = 0\left(-\begin{vmatrix} 4 & 5 \\ 1 & 1 \end{vmatrix}\right) + 1\left(+\begin{vmatrix} 2 & 3 \\ 1 & 1 \end{vmatrix}\right) + 2\left(-\begin{vmatrix} 2 & 3 \\ 4 & 5 \end{vmatrix}\right)$$

We expanded along the second column of A.

$$= 0(1) + 1(-1) + 2(2) = 0 - 1 + 4 = 3$$

Thus,

$$A^{-1} = \frac{1}{\det A}\text{adj}(A) = \frac{1}{3}\begin{bmatrix} -9 & 6 & -3 \\ 1 & -1 & 2 \\ 7 & -4 & 2 \end{bmatrix} = \begin{bmatrix} -3 & 2 & -1 \\ \frac{1}{3} & -\frac{1}{3} & \frac{2}{3} \\ \frac{7}{3} & -\frac{4}{3} & \frac{2}{3} \end{bmatrix}$$

Check:

$$A^{-1}A = \begin{bmatrix} -3 & 2 & -1 \\ \frac{1}{3} & -\frac{1}{3} & \frac{2}{3} \\ \frac{7}{3} & -\frac{4}{3} & \frac{2}{3} \end{bmatrix}\begin{bmatrix} 2 & 0 & 3 \\ 4 & 1 & 5 \\ 1 & 2 & 1 \end{bmatrix} = \begin{bmatrix} 1 & 0 & 0 \\ 0 & 1 & 0 \\ 0 & 0 & 1 \end{bmatrix} = I_3$$

You should verify that $AA^{-1} = I_3$ as well.

For 2×2 matrices that have an inverse, the formula for A^{-1} is rather simple, and follows directly from the technique outlined above.

For $A = \begin{bmatrix} a & b \\ c & d \end{bmatrix}$ with $\det A \neq 0$,

$$A^{-1} = \frac{1}{ad - bc}\begin{bmatrix} d & -b \\ -c & a \end{bmatrix}.$$

For example, if $A = \begin{bmatrix} 2 & 3 \\ 4 & 5 \end{bmatrix}$, then

$$A^{-1} = \frac{1}{(2)(5) - (4)(3)} \begin{bmatrix} 5 & -3 \\ -4 & 2 \end{bmatrix} = \frac{1}{-2}\begin{bmatrix} 5 & -3 \\ -4 & 2 \end{bmatrix} = \begin{bmatrix} -\frac{5}{2} & \frac{3}{2} \\ 2 & -1 \end{bmatrix}$$

Now we are in a position to solve systems by solving matrix equations. Suppose we have a system of n equations in n unknowns. Let A be the coefficient matrix, let X be the column matrix of unknowns, and let C be the constant matrix. For example, for the system

$$\begin{cases} ax + by + cz = c_1 \\ dx + ey + fz = c_2 \\ gx + hy + iz = c_3 \end{cases}$$

we have the matrices

$$A = \begin{bmatrix} a & b & c \\ d & e & f \\ g & h & i \end{bmatrix}, \quad X = \begin{bmatrix} x \\ y \\ z \end{bmatrix}, \quad C = \begin{bmatrix} c_1 \\ c_2 \\ c_3 \end{bmatrix}.$$

We can represent the system by the simple matrix equation $AX = C$. Now if $\det A \neq 0$, A^{-1} exists, and we can solve the equation for X as follows.

$$\begin{aligned} AX &= C \\ A^{-1}AX &= A^{-1}C && \text{Both sides are multiplied on the left by } A^{-1}. \\ I_nX &= A^{-1}C && \text{Recall: } A^{-1}A = I_n. \\ X &= A^{-1}C && \text{Recall: } I_nX = X. \end{aligned}$$

Solving a System Using Inverses

If the matrix equation $AX = C$ represents a system of n equations in n unknowns, and if A^{-1} exists, then the solution of the system is $X = A^{-1}C$.

EXAMPLE 5

Solve the system $\begin{cases} 2x + 3z = 2 \\ 4x + y + 5z = -1. \\ x + 2y + z = 0 \end{cases}$

Solution With

$$A = \begin{bmatrix} 2 & 0 & 3 \\ 4 & 1 & 5 \\ 1 & 2 & 1 \end{bmatrix}, \quad X = \begin{bmatrix} x \\ y \\ z \end{bmatrix} \quad \text{and} \quad C = \begin{bmatrix} 2 \\ -1 \\ 0 \end{bmatrix},$$

the matrix equation for this system is $AX = C$. In Example 4 we found that A^{-1} exists and is given by

$$A^{-1} = \begin{bmatrix} -3 & 2 & -1 \\ \frac{1}{3} & -\frac{1}{3} & \frac{2}{3} \\ \frac{7}{3} & -\frac{4}{3} & \frac{2}{3} \end{bmatrix}$$

Thus,

$$X = A^{-1}C$$

$$\begin{bmatrix} x \\ y \\ z \end{bmatrix} = \begin{bmatrix} -3 & 2 & -1 \\ \frac{1}{3} & -\frac{1}{3} & \frac{2}{3} \\ \frac{7}{3} & -\frac{4}{3} & \frac{2}{3} \end{bmatrix}\begin{bmatrix} 2 \\ -1 \\ 0 \end{bmatrix} = \begin{bmatrix} -8 \\ 1 \\ 6 \end{bmatrix}$$

The solution of the system is $(x, y, z) = (-8, 1, 6)$.

In the final example we use a matrix inverse to solve two systems each with the same coefficient matrix.

EXAMPLE 6

Solve the following systems.

(a) $\begin{cases} 2x + 3y = 4 \\ 5x + 6y = 1 \end{cases}$ (b) $\begin{cases} 2x + 3y = 1 \\ 5x + 6y = -2 \end{cases}$

Solution With

$$A = \begin{bmatrix} 2 & 3 \\ 5 & 6 \end{bmatrix}, \quad X = \begin{bmatrix} x \\ y \end{bmatrix}, \quad C_1 = \begin{bmatrix} 4 \\ 1 \end{bmatrix} \quad \text{and} \quad C_2 = \begin{bmatrix} 1 \\ -2 \end{bmatrix},$$

we can write these systems as matrix equations:

(a) $AX = C_1$ (b) $AX = C_2$

Since $\det A = \begin{vmatrix} 2 & 3 \\ 5 & 6 \end{vmatrix} = 2(6) - 5(3) = -3 \neq 0$, A^{-1} exists.

We compute A^{-1} using the formula for the determinant of a 2×2 matrix.

$$A^{-1} = \frac{1}{ad - bc}\begin{bmatrix} d & -b \\ -c & a \end{bmatrix} = \frac{1}{-3}\begin{bmatrix} 6 & -3 \\ -5 & 2 \end{bmatrix} = \begin{bmatrix} -2 & 1 \\ \frac{5}{3} & -\frac{2}{3} \end{bmatrix}$$

(a) $X = \begin{bmatrix} x \\ y \end{bmatrix} = A^{-1}C_1 = \begin{bmatrix} -2 & 1 \\ \frac{5}{3} & -\frac{2}{3} \end{bmatrix}\begin{bmatrix} 4 \\ 1 \end{bmatrix} = \begin{bmatrix} -7 \\ 6 \end{bmatrix}$, so $x = -7$ and $y = 6$.

Check: $2(-7) + 3(6) = 4$
$5(-7) + 6(6) = 1$

$x = -7$ and $y = 6$ are substituted in the original system. The values check.

(b) $X = \begin{bmatrix} x \\ y \end{bmatrix} = A^{-1}C_2 = \begin{bmatrix} -2 & 1 \\ \frac{5}{3} & -\frac{2}{3} \end{bmatrix}\begin{bmatrix} 1 \\ -2 \end{bmatrix} = \begin{bmatrix} -4 \\ 3 \end{bmatrix}$, so $x = -4$ and $y = 3$.

Check: $2(-4) + 3(3) = 1$, $5(-4) + 6(3) = -2$ — $x = -4$ and $y = 3$ are substituted in the original system. The values check.

EXERCISE SET 9.6

In Exercises 1–10, compute the matrix products AB and BA, where defined.

1. $A = \begin{bmatrix} 5 & -2 \\ 3 & 1 \end{bmatrix}, B = \begin{bmatrix} 1 & 2 \\ 0 & 0 \end{bmatrix}$

2. $A = \begin{bmatrix} 2 & -3 \\ 4 & 1 \end{bmatrix}, B = \begin{bmatrix} 3 & 0 \\ 2 & 0 \end{bmatrix}$

3. $A = \begin{bmatrix} 2 \\ -3 \end{bmatrix}, B = [4 \quad 1]$

4. $A = \begin{bmatrix} -1 \\ 2 \end{bmatrix}, B = [3 \quad 1]$

5. $A = \begin{bmatrix} 2 & 0 & 1 \\ 0 & 3 & -4 \end{bmatrix}, B = \begin{bmatrix} 0 & 1 \\ 2 & 1 \\ 0 & 1 \end{bmatrix}$

6. $A = \begin{bmatrix} -1 & 2 & 1 \\ 0 & 0 & -3 \end{bmatrix}, B = \begin{bmatrix} 2 & 1 \\ 1 & 1 \\ 0 & 1 \end{bmatrix}$

7. $A = \begin{bmatrix} 1 & 0 & -1 \\ 0 & 1 & 2 \end{bmatrix}, B = \begin{bmatrix} 3 & 0 & 0 & 2 \\ 1 & 1 & 0 & 2 \\ 0 & 1 & 0 & -1 \end{bmatrix}$

8. $A = \begin{bmatrix} -3 & 1 & 4 \\ 2 & -2 & -1 \end{bmatrix}, B = \begin{bmatrix} 2 & 1 & 0 & 0 \\ 3 & -1 & 1 & 0 \\ 1 & 1 & 0 & 0 \end{bmatrix}$

9. $A = \begin{bmatrix} -2 & 3 & 1 \\ 1 & 1 & 1 \\ 0 & 2 & 0 \end{bmatrix}, B = \begin{bmatrix} 1 & 3 & 0 \\ 0 & 2 & 1 \\ 0 & -1 & 0 \end{bmatrix}$

10. $A = \begin{bmatrix} -2 & 1 & 0 \\ 1 & 0 & 2 \\ 3 & 0 & 0 \end{bmatrix}, B = \begin{bmatrix} 0 & 1 & -1 \\ 3 & -2 & -1 \\ 0 & 4 & -1 \end{bmatrix}$

In Exercises 11–20, for the given matrix A determine the inverse matrix A^{-1}.

11. $A = \begin{bmatrix} 2 & 3 \\ 3 & 4 \end{bmatrix}$

12. $A = \begin{bmatrix} 3 & 4 \\ 1 & 2 \end{bmatrix}$

13. $A = \begin{bmatrix} 4 & 5 \\ 1 & 2 \end{bmatrix}$

14. $A = \begin{bmatrix} 2 & 3 \\ 5 & 6 \end{bmatrix}$

15. $A = \begin{bmatrix} 1 & 2 & 1 \\ 2 & 0 & 3 \\ 1 & 4 & 1 \end{bmatrix}$

16. $A = \begin{bmatrix} 2 & 1 & 1 \\ 1 & 1 & 4 \\ 3 & 2 & 0 \end{bmatrix}$

17. $A = \begin{bmatrix} 0 & 2 & -1 \\ 3 & -3 & 4 \\ 1 & 2 & -1 \end{bmatrix}$

18. $A = \begin{bmatrix} 5 & 3 & 1 \\ 2 & -3 & -1 \\ 0 & 4 & 1 \end{bmatrix}$

19. $A = \begin{bmatrix} 1 & 0 & 1 & 0 \\ 2 & 0 & 2 & -1 \\ 5 & 3 & -3 & 4 \\ -2 & 1 & 2 & -1 \end{bmatrix}$

20. $A = \begin{bmatrix} 1 & -2 & -1 & 1 \\ 0 & 5 & 3 & 1 \\ -1 & 2 & -3 & -1 \\ 0 & 0 & 4 & 1 \end{bmatrix}$

For Exercises 21–24, solve the matrix equation

$$A\begin{bmatrix} x \\ y \end{bmatrix} = \begin{bmatrix} 3 \\ -1 \end{bmatrix}$$

for x and y using the matrix A given in Exercises 11–14.

For Exercises 25–28, solve the matrix equation

$$A\begin{bmatrix} x \\ y \\ z \end{bmatrix} = \begin{bmatrix} 2 \\ -1 \\ 3 \end{bmatrix}$$

for x, y, and z using the matrix A given in Exercises 15–18.

Superset

In Exercises 29–32, solve the system for each of the following values of t: (a) $t = 1$, (b) $t = 2$, (c) $t = -1$. (*Hint:* Write the system as a matrix equation and use the inverse of the coefficient matrix to find x and y in terms of t.)

29. $\begin{cases} 7x + 3y = t^2 + 1 \\ 4x + 2y = 3 + t \end{cases}$ **30.** $\begin{cases} 2x + 5y = t - 1 \\ x + 4y = 2t \end{cases}$

31. $\begin{cases} 4x - 3y = 2 - t \\ x - 2y = t - 2 \end{cases}$ **32.** $\begin{cases} 8x - 7y = t^2 - 4 \\ -2x + y = 2 - t \end{cases}$

In Exercises 33–36, determine the inverse A^{-1} for the given matrix A.

33. $A = \begin{bmatrix} 2 & 0 & 0 & 1 \\ 0 & 1 & 0 & 0 \\ 0 & 0 & -1 & 0 \\ 0 & 0 & 0 & 2 \end{bmatrix}$ **34.** $A = \begin{bmatrix} 1 & 0 & 0 & 0 \\ 0 & -1 & 0 & 0 \\ 0 & 0 & 3 & 0 \\ 1 & 0 & 0 & 1 \end{bmatrix}$

35. $A = \begin{bmatrix} 1 & 1 & 1 & 1 \\ 0 & 1 & 1 & 1 \\ 0 & 0 & 1 & 1 \\ 0 & 0 & 0 & 1 \end{bmatrix}$ **36.** $A = \begin{bmatrix} 1 & 0 & 0 & 0 \\ 1 & 1 & 0 & 0 \\ 1 & 1 & 1 & 0 \\ 1 & 1 & 1 & 1 \end{bmatrix}$

In Exercises 37–40, determine the real values of t (if any) for which the inverse A^{-1} of the given matrix A exists, and find A^{-1} when it exists.

37. $A = \begin{bmatrix} 6 & t \\ 2 & t \end{bmatrix}$ **38.** $A = \begin{bmatrix} 3 & t+5 \\ t & -2 \end{bmatrix}$

39. $A = \begin{bmatrix} 1 & t+1 \\ t-2 & 4 \end{bmatrix}$ **40.** $A = \begin{bmatrix} 2t & 3 \\ 6 & t \end{bmatrix}$

41. Given that $A = \begin{bmatrix} a & b \\ 0 & c \end{bmatrix}$, determine those real numbers a, b, and c for which A^{-1} exists.

42. Given that precisely six of the entries in a 3×3 matrix A are 0's and that A^{-1} exists, what can you say about the position of the nonzero entries of A?

43. An $n \times n$ diagonal matrix is obtained by replacing each entry of 1 in I_n by any real number (possibly leaving the 1). Suppose A is a diagonal matrix that has an inverse A^{-1}. Describe a procedure to specify A^{-1}.

44. Suppose that A is a 3×3 diagonal matrix for which $A = A^{-1}$. How many such matrices A are there?

9.7 Linear Programming

When trying to solve a real-world problem, one must bear in mind that the solution is very much influenced by certain physical limitations. Resources such as money, labor, raw materials, time, or fuel may be needed and are always limited. A good problem solver tries to allocate resources in such a way that the problem is solved in the most efficient or least wasteful manner.

Often limitations can be translated into a system of linear inequalities, such as

$$\begin{cases} y \geq 0 \\ x \leq 5 \\ y \leq x - 1, \end{cases}$$

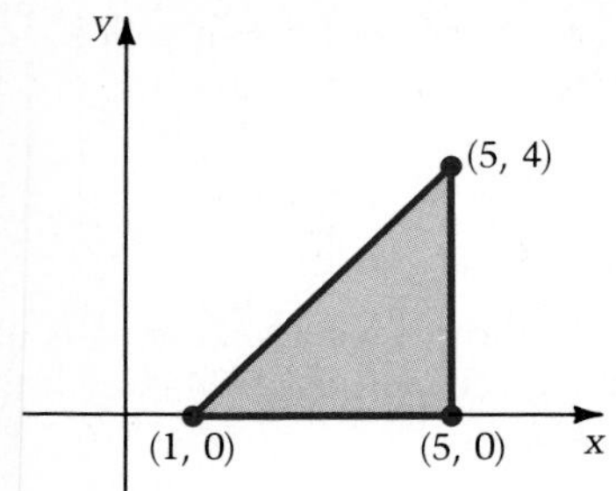

Figure 9.24 The solution set, called the **feasible region,** of the system

$$\begin{cases} y \geq 0 \\ x \leq 5 \\ y \leq x - 1. \end{cases}$$

and the function to be maximized (or minimized) can be written as a linear equation like

$$P = 17x + 12y.$$

A problem that can be described in this way is called a **linear programming problem.**

In linear programming problems, the inequalities are referred to as **constraints,** the solution set of the system of inequalities is called the **feasible region,** and the ordered pairs in this region are called **feasible solutions.** The **corner points** of the feasible region are especially important. For the feasible region in Figure 9.24, the corner points are

$$(1, 0), \quad (5, 0), \quad \text{and } (5, 4).$$

Finally, the function to be maximized (or minimized) is called the **objective function.**

The solution of a linear programming problem is that ordered pair in the feasible region which produces the greatest (or least) value of the objective function. In the table below, some ordered pairs have been selected from the feasible region sketched in Figure 9.24. These pairs have been used to determine corresponding values of the objective function. Of those ordered pairs chosen, the corner point (5, 4) produces the greatest value of P, namely 133. You will soon discover that the solution of a linear programming problem always occurs at one of the corner points.

Sample Feasible Solutions (x, y)	**Objective Function $P = 17x + 12y$**
(1, 0)	$P = 17$
(2, 1)	$P = 46$
(4, 2)	$P = 92$
(5, 0)	$P = 85$
(5, 4)	$P = 133$

Before we can state the fundamental theorem about objective functions, we must first explore the different types of regions which can exist.

A region is **convex** if, for any two points in the region, the line segment joining the two points is also part of the region.

(a) A convex region. (b) *Not* a convex region.

Figure 9.25

When we refer to feasible regions, we assume them to be convex.

A feasible region can be **bounded** or **unbounded,** as shown in Figure 9.26. We now state the theorem that is very useful in solving linear programming problems.

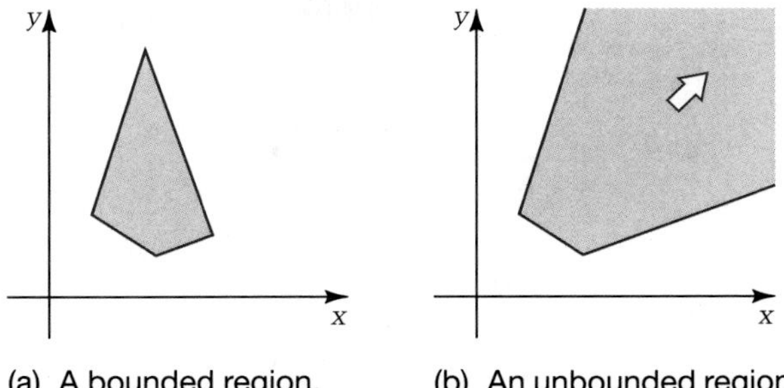

(a) A bounded region. (b) An unbounded region.

Figure 9.26

Basic Theorem of Linear Programming

If the objective function in a linear programming problem attains a maximum or minimum value, then it does so at one of the corner points of the feasible region.

EXAMPLE 1

A small furniture company manufactures two styles of waterbeds: modern and traditional. (All beds are queen-size.) Two machines are used to produce the beds: machine A cuts and forms the wood, and machine B finishes the wood. A modern bed requires two hours of machine A's time and $\frac{1}{2}$ hour of machine B's time. A traditional bed requires one hour on machine A and one hour on machine B. Limitations on the costly finishing materials for the modern bed restrict production to a maximum of 7 modern beds per day. Machines A and B can be operated a maximum of 24 and 18 hours per day, respectively. If the profit is \$90 per modern bed and \$60 per traditional bed, how many of each type should be produced each day to achieve the maximum possible profit?

Solution Begin by organizing the data in a table.

	Modern	**Traditional**	**Hours Available**
Machine A	2 hours	1 hr	24 hrs
Machine B	$\frac{1}{2}$ hr	1 hr	18 hrs

Profit per bed	\$90	\$60	Data for the objective function.

Assign Variables

Let x = the number of modern beds produced each day.
Let y = the number of traditional beds produced each day.
Let P = the daily profit on the sale of the two kinds of beds.

Constraints

$$\begin{cases} 2x + y \le 24 & \text{Line 1 of the table of values.} \\ \frac{1}{2}x + y \le 18 & \text{Line 2 of the table of values.} \\ x \le 7 & \text{"a maximum of 7 modern beds per day"} \\ x \ge 0 & \text{Nonnegativity constraints: the number of beds produced cannot be} \\ y \ge 0 & \text{negative.} \end{cases}$$

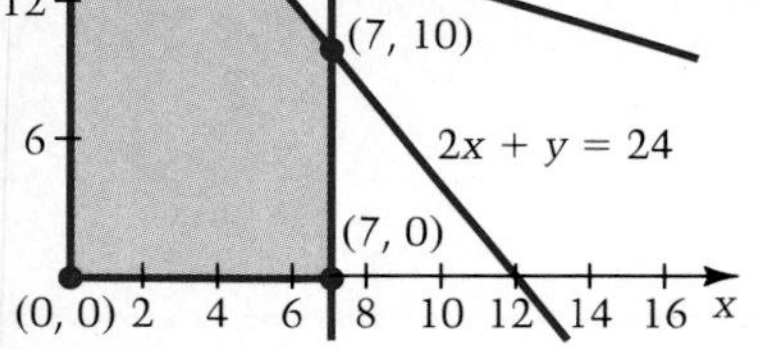

Figure 9.27 Feasible Region

Objective Function $P = 90x + 60y$ The function to be maximized.

Feasible Region Determine the feasible region by graphing the system of constraints. Corner points are found by solving systems of two of the five equations. For example, (4, 16) is the solution to the system

$$\begin{cases} 2x + y = 24 \\ \frac{1}{2}x + y = 18. \end{cases}$$

Evaluate P at Each Corner Point

Corner Point	$P = 90x + 60y$	
(0, 0)	$P = 90(0) + 60(0) = 0$	
(0, 18)	$P = 90(0) + 60(18) = 1080$	
(4, 16)	$P = 90(4) + 60(16) = 1320$	Maximum value of P.
(7, 10)	$P = 90(7) + 60(10) = 1230$	
(7, 0)	$P = 90(7) + 60(0) = 630$	

The maximum value of the objective function is \$1320. The maximum value occurs when $x = 4$ and $y = 16$. Thus, for maximum profit (\$1320 per day), the company should manufacture 4 modern and 16 traditional beds daily.

EXAMPLE 2

A health club promises: "our high energy lunch contains at least 36 grams of protein and at least 42 grams of carbohydrates." Suppose one ounce of the club's vegetable rice soup provides 2 grams of protein, 6 grams of carbohydrates, and 1 gram of fat, and one ounce of the club's tuna salad provides 9 grams of protein, 6 grams of carbohydrates, and 4 grams of fat. How many ounces each of vegetable rice soup and tuna salad constitute a lunch that fulfills the club's promise, but does so with minimum fat content?

Solution

	Vegetable rice soup	Tuna salad	Minimum
Protein	2 gm (per oz)	9 gm (per oz)	36 gm
Carbohydrates	6 gm (per oz)	6 gm (per oz)	42 gm
Fat	1 gm (per oz)	4 gm (per oz)	

Assign Variables Let x = the number of ounces of soup.
Let y = the number of ounces of tuna salad.
Let F = the number of grams of fat in the entire lunch.

Constraints

$$\begin{cases} 2x + 9y \geq 36 \\ 6x + 6y \geq 42 \\ x \geq 0 \\ y \geq 0 \end{cases}$$

Lunch must contain at least 36 gm of protein and 42 gm of carbohydrates.

Amount of protein and carbohydrates cannot be negative.

Objective Function $F = x + 4y$ The function to be minimized.

Evaluate F at Each Corner Point

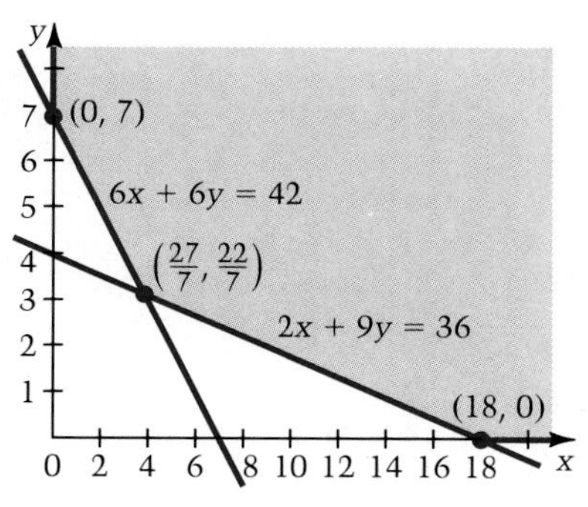

Figure 9.28 Feasible Region

Corner Point	$F = x + 4y$
$(0, 7)$	$F = (0) + 4(7) = 28$
$\left(\frac{27}{7}, \frac{22}{7}\right)$	$F = \left(\frac{27}{7}\right) + 4\left(\frac{22}{7}\right) = 16\frac{3}{7}$
$(18, 0)$	$F = (18) + 4(0) = 18$

The promise is fulfilled with minimum fat content when the lunch is composed of $\frac{27}{7}$ oz of soup and $\frac{22}{7}$ oz of tuna salad (roughly 4 oz of soup and 3 oz of tuna salad).

Procedure for Solving a Linear Programming Problem

Step 1. Assign variables; then write the constraints and objective function.

Step 2. Determine the feasible region and corner points.

Step 3. Evaluate the objective function at each corner point, and determine which corner point produces the maximum or minimum.

Step 4. Answer the question stated in the problem.

EXAMPLE 3

Suppose you have \$10,000 and wish to invest all or part of this money in stocks and bonds. You would like to invest at least \$3000 in bonds, offering a return of 8%, and at least \$2000 in stocks, offering a return of 12%. As a precaution, you decide that the investment in bonds should be as much as or more than your investment in stocks. How should the money be invested to maximize your earnings?

Solution

Assign Variables Let s = the amount invested in stocks (in dollars).
Let b = the amount invested in bonds (in dollars).
Let E = total earnings on the investments.

Constraints

$$\begin{cases} s + b \leq 10{,}000 \\ b \geq 3{,}000 \\ s \geq 2{,}000 \\ b \geq s \end{cases}$$

- $s + b \leq 10{,}000$ — You can invest up to \$10,000.
- $b \geq 3{,}000$ — You will invest at least \$3,000 in bonds.
- $s \geq 2{,}000$ — You will invest at least \$2,000 in stocks.
- $b \geq s$ — Bond investments are as much as or more than stock investments.

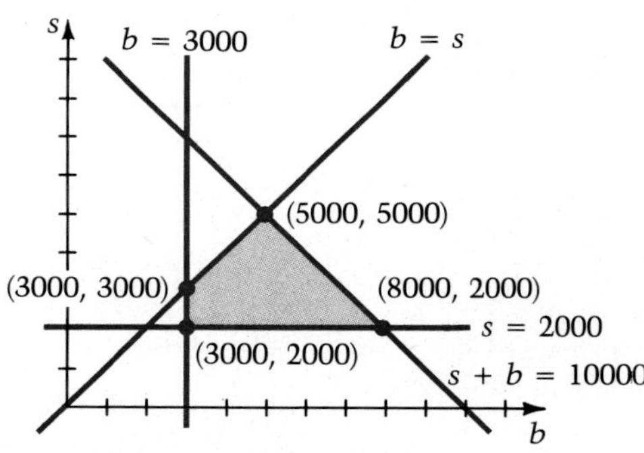

Figure 9.29 Feasible Region

Objective Function $E = 0.08b + 0.12s$

Evaluate E at Each Corner Point

Corner Point	$E = 0.08b + 0.12s$
(3000, 2000)	$E = 0.08(3000) + 0.12(2000) = 480$
(3000, 3000)	$E = 0.08(3000) + 0.12(3000) = 600$
(5000, 5000)	$E = 0.08(5000) + 0.12(5000) = 1000$
(8000, 2000)	$E = 0.08(8000) + 0.12(2000) = 880$

Since the maximum earnings occur when $b = 5{,}000$ and $s = 5{,}000$, you should invest \$5,000 in bonds and \$5,000 in stocks.

EXERCISE SET 9.7

1. A baker decides to operate a small outlet at the airport and sell jelly donuts and brownies. Jelly donuts will be sold at a profit of 15 cents and brownies at a profit of 12 cents. The shelves can hold up to 300 jelly donuts and 500 brownies. However, the delivery van can bring, at most, 600 items (jelly donuts and brownies) each morning. How should the baker stock the outlet to maximize the profit? What is the maximum profit?
2. A manufacturer of hubcaps makes a profit of \$17 on each deluxe hubcap and \$11 on each standard hubcap. The manufacturer can produce 700 hubcaps each day, but not more than 400 deluxe or more than 600 standard. How many of each type should the manufacturer produce in order to maximize profit? What is the maximum profit?
3. An "oldies" record store makes a profit of \$.37 on each 45 rpm and \$1.12 on each LP record it sells. Consumer demand suggests that the store should maintain an inventory of 5000–9000 45's and 2000–3000 LP's. Storage space in the store limits the inventory to, at most, 11,000 records. How many of each type of record should the store have to maximize the profit potential of its inventory? What is the profit?
4. The pro shop at the local country club makes \$9 profit for every golfer who uses a caddy and \$5 profit for every golfer who rents an electric cart. A golfer who wants to use a caddy will rent a cart if no caddy is available; and a golfer who wants to use a cart will use a caddy if no cart is available. The manager of the shop must decide how many carts and caddies to have available to maximize the profit. The manager can purchase 25–40 carts and contract to have 20–30 caddies on hand each day. The manager also knows that, on any given day, 65 golfers, at most, will want a cart or a caddy. How many carts and caddies should the manager have on hand? What is the profit?
5. A history exam consists of 20 multiple choice questions worth 3 points each, and 25 true/false questions worth 2 points each. You are required to answer at least 5 questions of each type. If you can answer only 30 questions in the time period, how many of each type should you answer to maximize your test score?
6. In Example 2, what would the solution have been if the fat content and the protein content of the soup were each 3 gm per ounce?
7. A baseball team in the local league is classified as semipro because some of the players are amateurs and some are professionals. The facilities used for spring training will accommodate between 35 and 60 players. League rules require that at the start of camp, each team must have at least 10 amateurs and at least 20 professionals on its roster. The daily camp costs are \$35 per amateur and \$55 per professional. How many amateurs and professionals should the team have to minimize its daily costs? How many should it have to maximize its daily costs?
8. A local trucking firm transports the products of two manufacturers, A and B. Each carton of products from manufacturer A weighs 36 pounds and is 8 cubic feet in volume. Each carton of products from manufacturer B weighs 24 pounds and is 10 cubic feet in volume. The trucking firm charges Company A \$1.70 and Company B \$2.15 for each carton shipped. Each of the trucks has a maximum capacity of 4800 cubic feet and cannot carry more than 20,000 pounds. To keep labor costs down for loading and unloading the trucks, the firm limits each truck to a maximum of 560 cartons. How many cartons from each manufacturer should be loaded on a truck to maximize the charges? What is this total maximum charge?
9. A nursery uses two brands of fertilizer for rose bushes. Fertilizer A costs \$4 per pound and provides 250 units of nutrients per pound. Fertilizer B costs \$5 per pound and provides 300 units of nutrients per pound. The nursery spends \$100 or less for fertilizer and wants to provide between 5200 and 6100 units of nutrients. Since brand A contains a special nutrient that brand B doesn't, the nursery uses at least 5 lbs of brand A. How many pounds of each brand should the nursery use to minimize cost?
10. A tailor has 100 square yards of cotton material and 120 square yards of woolen material. It takes 2 square yards of each type of material to make a sports coat. It takes 1 square yard of cotton and 3 square yards of wool to make a pleated skirt. The tailor already has orders for at least 15 sports coats and 5 skirts. If the tailor sells each coat for \$40 and each skirt for \$45, how many of each garmet should be made in order to fill the orders and maximize income?
11. A large restaurant specializes in chicken dishes and advertises chicken that is "fresh daily." Each day the restaurant uses at least 350 lb of chicken breasts, at least 290 lb of chicken legs, and at least 390 lb of chicken wings. A local poultry supplier offers two chicken-part packages. Package I contains 25 lb of breasts, 15 lb of legs, and 15 lb of wings for \$32. Package II contains

20 lb of breasts, 20 lb of legs and 45 lb of wings for \$35. How many of each package should the restaurant order to meet its chicken requirement as economically as possible?

12. In Exercise 11, what would the solution be if Package I cost \$30 and Package II cost \$40?

13. A farmer has 128 acres available for planting beans and alfalfa. Seed and fertilizer costs per acre are \$50 for alfalfa and \$70 for beans. Total labor costs per acre are \$60 for alfalfa and \$120 for beans. The farmer has \$7000 to spend for seed and fertilizer and \$11,280 to spend for labor. The farmer can realize an income (net profit) of \$200 for each acre of alfalfa harvested and \$260 for each acre of beans harvested. How many acres of each crop should the farmer plant to maximize profit? What is the profit?

14. The students in a gym class are mastering various gymnastic exercises. The students' grades will be based upon those exercises they are able to demonstrate successfully during a 48 minute gym period. The various exercises have been divided into two groups and assigned point values of 3 and 5 points. Each student must stand in line two minutes before demonstrating mastery of a 3-point exercise and three minutes before demonstrating mastery of a 5-point exercise. Each student is required to demonstrate at least three 3-point exercises. No student may demonstrate more than 20 exercises. How many exercises of each type should be demonstrated to maximize a student's point total? (Assume that the time needed to demonstrate each exercise can be ignored.)

15. A local dairy farmer has two farms, F_1 and F_2, from which milk is shipped to two retail outlet stores, S_1 and S_2. Farm F_1 produces 40 gallons of milk each day; farm F_2 produces 30 gallons each day. Store S_1 orders 25 gallons each day; store S_2 orders 20 gallons each day. The milk that the farmer does not ship to either store is sold to a consortium. The cost of shipping a gallon of milk from a farm to a store is given in the table below.

Farm	Store	Shipping Cost per Gallon
F_1	S_1	7¢
F_1	S_2	9¢
F_2	S_1	10¢
F_2	S_2	13¢

How should the farmer fulfill the stores' orders each day to minimize the shipping costs?

16. A candy manufacturer produces three kinds of fudge. The number of ounces of chocolate and nuts in a pound of each type is shown in the table below.

Type	Ounces of Chocolate	Ounces of Nuts
Smooth	13	2
Tasty	8	5
Crunchy	4	9

A pound of chocolate costs \$1.20 and a pound of nuts costs \$1.92. The manufacturer intends to market 20-ounce packages containing at least 2 ounces of each type of fudge, and advertises "contains at least as much chocolate as nuts." What is the manufacturer's minimum costs per package of chocolate and nuts?

Superset

17. Refer to Example 3. On the given feasible region, sketch the graph of the line

$$800 = 0.08b + 0.12s.$$

What can you say about all the points in the feasible region that lie on this line?

Next, sketch the graph of the line

$$E = 0.08b + 0.12s,$$

with E first as 900 and then as 1000. What do you notice? What happens if $E > 1000$? What happens if $E < 480$?

18. Refer to Example 3. Is there an objective function for which the maximum (minimum) value could occur at points other than corner points?

19. Explain why the maximum and minimum values of $E = x^2 + y^2$ for a bounded convex region in the first quadrant must occur on the boundary of the region.

20. For a bounded convex region in the first quadrant, where would you look for the maximum and minimum values of $E = x^3$? Where would you look for $E = y^3$?

21. For a bounded convex region in the first quadrant where would you look for the minimum value of $E = |x - y|$?

22. For a bounded convex region in the first quadrant where would you look for the minimum value of $E = |2x - 3y|$?

Exercises 23–24 are problems that require integer solutions. We can use linear programming methods to help us determine these solutions even though they do not occur at corner points.

23. Suppose the only resources a furniture manufacturer needs to produce tables and chairs are wood and labor. A table requires 7 square feet of wood and $3\frac{1}{2}$ hrs of work, and a chair requires 3 square feet of wood and 4 hrs of work. The manufacturer has 40 square feet of wood and can provide 28 hrs of work. If there is a \$5 profit on each table produced and \$3 profit on each chair, how many chairs and tables should the manufacturer produce to maximize profits?

24. In Exercise 23, what is the solution if the profit is \$6 per table and \$2 per chair?

Analyzing Data

The Case of the Colliding Curves

In this chapter, our concern has been with those types of problems where more than one equation is needed to model a given situation. We have seen that the geometric solution of such a system of equations is found by graphing each equation and then determining the point(s) of intersection. In published reports it is common to see graphs containing more than one curve, since research is frequently concerned with comparisons of two or more cases of the same process. We present three examples.

First, consider the graph in Figure 9.30 which presents a comparison of the (logarithm of) production values (measured in millions of U.S. dollars) of Korea and Taiwan over a period of time [Mody, 1990]. It is clear that Korea was outperforming Taiwan by 1985 (the intersection of the two graphs occurs between 1984 and 1985). Some analysts speculate that Korea's somewhat greater success in the production of high-growth products such as VCRs and microwave ovens has been responsible for its outperforming Taiwan since 1985.

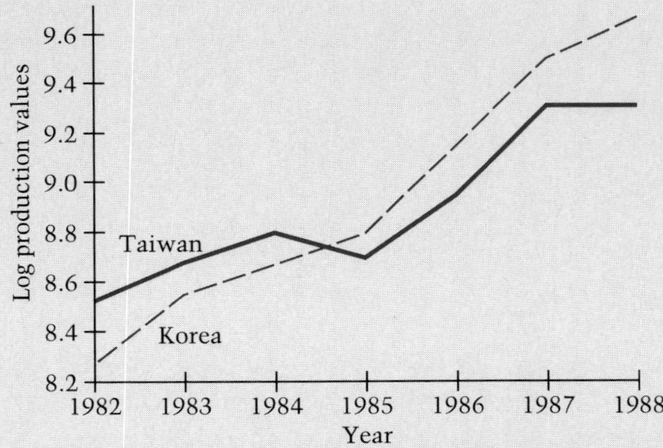

Figure 9.30 Production of electronics. Vertical axis represents log of production values measured in US $ millions. Data for 1988 are projections. Source: *Yearbook of World Electronics Data*, Elsevier Advanced Technology, Oxford, 0X27DH (formerly known as Mackintosh Yearbook of Electronics Data). Reprinted by permission.

As a second example, we present a somewhat more complicated picture in Figure 9.31 [Wilde, 1991]. This figure suggests that, depending on what data we consider, our conclusions on a certain issue may be quite different. If we look at the top graph (deaths per 100,000 inhabitants), we might conclude that despite some peaks and valleys over the years 1923–1983, the number of traffic deaths per 100,000 residents is about the same—not a very encouraging statistic.

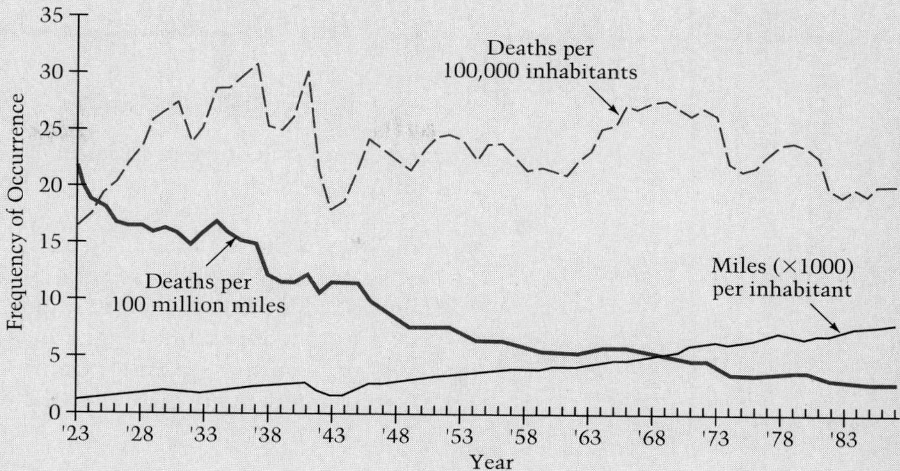

Figure 9.31 The traffic death rate per distance traveled, the traffic death rate per capita, and road distance traveled per capita in the United States, 1923–1987 (from Wilde).

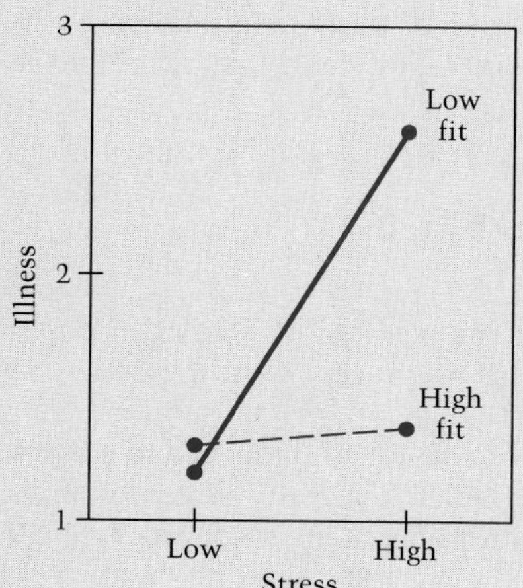

Figure 9.32 Schematic representation of the Stress × Fitness interaction in the prediction of illness (health center visits).

However, if we turn our attention to the two lower graphs in the figure, we might get a more positive impression: observe that the number of "deaths per 100 million miles" has decreased steadily. Although it might not seem so at first, this fact is consistent with the top graph (deaths per 100,000 have remained roughly the same), since the "miles per inhabitant" graph has steadily increased. There are less deaths per mile driven, but people are driving more miles; these two facts are working together to keep the death rate per 100,000 inhabitants roughly constant. As Gerald Wilde, the author of the report points out, we need to ask ourselves the question, "Do we wish to put additional miles into people's years, or do we wish to put more years into people's lives?"

Finally, in Figure 9.32 we present a rather typical "interaction graph," reported as part of a study which investigated the effects of stress and level of physical fitness on illness [Brown, 1991]. The regression line for the "Low fit" individuals indicates that as stress increases the occurrence of illness also increases. Whereas for the "High fit" group, an increase in stress is not accompanied by such a precipitous increase in illness. Once again, the positive effects of physical fitness have been shown to act as a buffer for the harmful effects of stress.

Sources: Mody, A. Institutions and dynamic comparative advantage: the electronics industry in South Korea and Taiwan, *Cambridge J. of Econ.*, **14,** 291–314 (1990). Reprinted by permission.

Wilde, G. Economics and accidents: a commentary, *J. of App. Behavior Analysis*, **24,** 81–84 (1991). © 1991 by the Society for the Experimental Analysis of Behavior, Inc.

Brown, J. Staying fit and staying well: physical fitness as a moderator of life stress, *J. of Personality and Soc. Psych*, **60,** 555–561 (1991). Copyright © 1991 by the American Psychological Association. Reprinted by permission.

Chapter Review

9.1 Systems of Equations in Two Variables (pp. 529–539)

A collection of two or more equations is called a *system of equations*. The *solutions* of a system of two equations in two variables are those ordered pairs that are solutions of *every* equation of the system. (p. 530) To solve such a system, we rewrite it as a succession of equivalent systems until we obtain an equivalent system whose solution is obvious. To do this, we use either the *addition method* (p. 531) or the *substitution method* (p. 534).

For a system of two linear equations in two variables there are three possibilities: no solutions (parallel lines), precisely one solution (nonparallel lines), or infinitely many solutions (the same line). (p. 532)

9.2 Systems of Inequalities in Two Variables (pp. 540–546)

To solve an inequality in two variables, we begin by looking at the associated equality. The graph of this equation divides the xy-plane into regions. All the points in any one region must satisfy the same inequality. To determine which inequality holds for a region, you need only test one point from that region. (p. 540)

To graph the solution set of a system of two or more inequalities in two variables, we graph the solution set of each inequality and then identify the points common to all regions. (p. 542)

9.3 Systems of Linear Equations in Three Variables (pp. 546–552)

We use the *xyz-coordinate system* to describe the position of a point in three-dimensional space. In this coordinate system, we graph *ordered triples* of the form (x, y, z) (p. 546)

We can use either the addition method or substitution method to solve a system of three linear equations in three variables. There are three possibilities: no solutions, precisely one solution, or infinitely many solutions. (p. 549)

9.4 Matrices (pp. 552–563)

A *matrix* is a rectangular array of numbers used to represent a linear system. A matrix with m rows and n columns is called an $m \times n$ *matrix*. (p. 552) *Row reduction* of matrices is a method used to solve linear systems (p. 555)

$$\left[\begin{array}{cc|c} 1 & 2 & 3 \\ 4 & 5 & 6 \end{array}\right] \qquad \begin{bmatrix} 1 & 2 \\ 4 & 5 \end{bmatrix} \qquad \begin{bmatrix} 3 \\ 6 \end{bmatrix}$$

an *augmented matrix*   its *coefficient matrix*   its *constant matrix*

9.5 Determinants and Cramer's Rule (pp. 563–572)

The *determinant* is a number associated with an $n \times n$ square matrix. The determinant of a 2×2 matrix is the difference of the products of the diagonal entries. (p. 563) *Signed minors* or *cofactors* are used to compute determinants of $n \times n$ matrices for $n > 2$. (p. 567)

Cramer's Rule is used to determine whether a system has a unique solution, and to find that solution if it exists. A system has a unique solution if and only if the determinant of the coefficient matrix is *not* zero. (p. 570)

9.6 The Inverse of a Square Matrix (pp. 573–583)

An $n \times n$ matrix A has a multiplicative inverse, A^{-1}, if and only if $\det A \neq 0$. (p. 577) We can find A^{-1} in three steps (p. 578):

Step 1. Replace each entry of A by its signed minor to produce the *cofactor matrix of A*.

Step 2. Form the transpose of the cofactor matrix to obtain the *adjoint of A*, denoted adj(A).

Step 3. $A^{-1} = \dfrac{1}{\det A}\operatorname{adj}(A)$.

9.7 Linear Programming (pp. 583–591)

To solve a linear programming problem, we

- assign variables; then write the constraints and objective function;
- determine the feasible region and the corner points;
- evaluate the objective function at each corner point, and determine which corner point produces the desired extreme value;
- answer the question stated in the problem. (p. 587)

Review Exercises

In Exercises 1–4, use the addition method to solve the system of equations.

1. $\begin{cases} 2x + 3y = 4 \\ 5x - 2y = -1 \end{cases}$

2. $\begin{cases} 3x - 2y = -1 \\ 4x + 3y = -6 \end{cases}$

3. $\begin{cases} 3x + 2y = 3 \\ 5x - y = 2 \end{cases}$

4. $\begin{cases} 4x + 3y = 0 \\ 7x + 5y = 6 \end{cases}$

For Exercises 5–8, use the substitution method to solve the system given in Exercises 1–4.

In Exercises 9–12, solve the system.

9. $\begin{cases} (x-2)^2 + (y-1)^2 = 4 \\ (x-6)^2 + (y-1)^2 = 9 \end{cases}$

10. $\begin{cases} x^2 + y^2 - 6x + 4y + 3 = 0 \\ 3x - y = 5 \end{cases}$

11. $\begin{cases} x^2 + y^2 + 6x - 4y = 12 \\ 3x + 4y = -1 \end{cases}$

12. $\begin{cases} x^2 + (y-4)^2 = 20 \\ (x-2)^2 + y^2 = 8 \end{cases}$

In Exercises 13–20, graph the solution set of the system.

13. $\begin{cases} x \geq y \\ y > 3 \end{cases}$

14. $\begin{cases} x + y \leq 5 \\ x < 3 \end{cases}$

15. $\begin{cases} x + y \geq 2 \\ y \leq 2 \\ x \leq 2 \end{cases}$

16. $\begin{cases} x \geq 0 \\ y \leq 4 \\ y \geq 2x \end{cases}$

17. $\begin{cases} y - x > 0 \\ y - 2x < 0 \\ y \leq 10 \end{cases}$

18. $\begin{cases} x + 2y > 8 \\ 2x - y < 1 \\ y \leq 5 \end{cases}$

19. $\begin{cases} x + 3y \geq 10 \\ 3x - y \leq 0 \\ y - x < 5 \\ x < 2 \end{cases}$

20. $\begin{cases} x + y < 12 \\ y + 5x \leq 28 \\ 2y - 7x < 6 \\ y \geq 3 \end{cases}$

In Exercises 21–24, solve the system.

21. $\begin{cases} 2x + y - z = 0 \\ x - 2y - 5z = 1 \\ 3x + 3y + z = -1 \end{cases}$

22. $\begin{cases} x + 3y + z = 0 \\ 3x - y + 2z = -5 \\ 2x + 4y + z = 3 \end{cases}$

23. $\begin{cases} 3x - 2y = 5 \\ 4y + 3z = 1 \\ 2x - 5z = 7 \end{cases}$

24. $\begin{cases} 4x - 5y = 9 \\ 3y + 4z = -1 \\ 2x - 3z = 5 \end{cases}$

In Exercises 25–32, compute the determinant of the matrix.

25. $\begin{bmatrix} 2 & 5 \\ 1 & 3 \end{bmatrix}$

26. $\begin{bmatrix} -3 & 1 \\ 4 & 2 \end{bmatrix}$

27. $\begin{bmatrix} 3 & 0 \\ 4 & -2 \end{bmatrix}$

28. $\begin{bmatrix} 0 & -2 \\ 0 & 4 \end{bmatrix}$

29. $\begin{bmatrix} 2 & 1 & 3 \\ 0 & -1 & 2 \\ 1 & 4 & 0 \end{bmatrix}$

30. $\begin{bmatrix} 1 & 3 & 5 \\ -2 & 4 & 0 \\ 1 & -1 & 1 \end{bmatrix}$

31. $\begin{bmatrix} 2 & -1 & 1 \\ 1 & -2 & 5 \\ -1 & 0 & 1 \end{bmatrix}$

32. $\begin{bmatrix} 0 & 4 & 1 \\ 2 & 0 & -1 \\ -1 & 3 & 0 \end{bmatrix}$

In Exercises 33–38, use Cramer's Rule to determine whether the system has a unique solution, then find the solution if it exists.

33. $\begin{cases} 3x + 2y = 1 \\ 5x - 3y = -4 \end{cases}$

34. $\begin{cases} 5x - 3y = 5 \\ 4x + 2y = -3 \end{cases}$

35. $\begin{cases} 2x - 4y = 4 \\ x - 6y + 2z = 1 \\ x - z = 3 \end{cases}$

36. $\begin{cases} 2x - 3z = 5 \\ 3x + 4y = -1 \\ y + 2z = 2 \end{cases}$

37. $\begin{cases} 2x - 4y = -1 \\ x - 6y + 2z = 0 \\ 3x - z = 2 \end{cases}$

38. $\begin{cases} 3x + 5y - z = 2 \\ 7x - 4z = -3 \\ x + 4y = 1 \end{cases}$

In Exercises 39–44, compute the matrix products AB and BA, where defined.

39. $A = \begin{bmatrix} 1 & 2 \\ -1 & 0 \end{bmatrix}$, $B = \begin{bmatrix} 3 & 1 \\ 2 & 4 \end{bmatrix}$

40. $A = \begin{bmatrix} -1 & 3 \\ 2 & 1 \end{bmatrix}$, $B = \begin{bmatrix} 1 & 0 & -2 \\ 3 & -1 & 1 \end{bmatrix}$

41. $A = \begin{bmatrix} 0 & 2 & -1 \\ 3 & 1 & 2 \end{bmatrix}$, $B = \begin{bmatrix} 1 & 0 & -1 \\ 1 & 2 & -1 \\ 0 & 1 & 2 \end{bmatrix}$

42. $A = \begin{bmatrix} 1 & 0 & 2 \\ 3 & 0 & 1 \\ 2 & 1 & -1 \end{bmatrix}$, $B = \begin{bmatrix} 2 & -1 & 0 \\ -1 & 3 & 2 \\ 0 & 2 & 0 \end{bmatrix}$

43. $A = \begin{bmatrix} 2 & -1 & 0 \\ 1 & 3 & -1 \\ -1 & 0 & 1 \end{bmatrix}$, $B = \begin{bmatrix} 0 & -2 & 3 \\ 3 & 0 & 1 \\ 4 & -1 & 2 \end{bmatrix}$

44. $A = \begin{bmatrix} 1 & 0 & -1 \\ 3 & -1 & 2 \end{bmatrix}$, $B = \begin{bmatrix} 0 & -1 \\ 2 & 5 \\ 3 & -2 \end{bmatrix}$

In Exercises 45–52, for the given matrix A determine the inverse matrix A^{-1}.

45. $A = \begin{bmatrix} 4 & 3 \\ 3 & 2 \end{bmatrix}$

46. $A = \begin{bmatrix} 6 & 5 \\ 3 & 2 \end{bmatrix}$

47. $A = \begin{bmatrix} 5 & -6 \\ -3 & 4 \end{bmatrix}$

48. $A = \begin{bmatrix} 5 & 2 \\ 8 & 3 \end{bmatrix}$

49. $A = \begin{bmatrix} 1 & 1 & -2 \\ 1 & -2 & 0 \\ 3 & 1 & -5 \end{bmatrix}$

50. $A = \begin{bmatrix} 0 & -1 & 3 \\ 2 & 0 & 1 \\ -1 & -2 & 5 \end{bmatrix}$

51. $A = \begin{bmatrix} -2 & 1 & 0 \\ 3 & 0 & -1 \\ 3 & -2 & 1 \end{bmatrix}$

52. $A = \begin{bmatrix} 2 & 3 & -1 \\ 0 & -1 & 2 \\ 2 & 0 & 4 \end{bmatrix}$

For Exercises 53–56, solve the matrix equations

(a) $A\begin{bmatrix} x \\ y \end{bmatrix} = \begin{bmatrix} 1 \\ 2 \end{bmatrix}$ and (b) $A\begin{bmatrix} x \\ y \end{bmatrix} = \begin{bmatrix} 3 \\ -1 \end{bmatrix}$

using the matrix A given in Exercises 45–48.

For Exercises 57–60, solve the matrix equations

(a) $A\begin{bmatrix} x \\ y \\ z \end{bmatrix} = \begin{bmatrix} -2 \\ 0 \\ 3 \end{bmatrix}$ and (b) $A\begin{bmatrix} x \\ y \\ z \end{bmatrix} = \begin{bmatrix} 2 \\ -1 \\ 0 \end{bmatrix}$

using the matrix A given in Exercises 49–52.

In Exercises 61–62, determine the maximum and minimum values of the objective function P on the feasible region defined by the given inequalities.

61. $P = 4x + 7y + 3;\ y + 3x \leq 36,\ y + x \leq 14,\ y - x \leq 2,\ y \geq 0$

62. $P = 6y - 5x + 12;\ 2y + x \geq 5,\ 4x - 3y \leq 9,\ 2y - x \leq 14,\ x \geq 0$

63. A firm produces two items, each of which requires three manufacturing departments: assembly, inspection, and packing. Item A requires 0.30 hrs of assembly time, 0.20 hrs of inspection time and 0.10 hrs of packing time. Item B requires 0.20 hrs of assembly time, 0.05 hrs of inspection time, and 0.05 hrs of packing time. Daily capacities are 40 hrs of assembly time, 25 hrs of inspection time, and 15 hrs of packing time. The profit per item sold is \$5 for A and \$4 for B. Any items produced are sold. How many of each item should the manufacturer produce to maximize profit?

64. A newlywed couple has received \$7000 from their parents. They want to invest all or part of this money in a savings account and in stocks. The savings account offers a return of 9.5%, and the stocks will earn 11%. They decide to invest at least as much in the savings account as they do in stocks, and they want to put at least \$2000 into their savings and at least \$1500 into stocks. To get the greatest return on their investments, how much money should be placed into each investment?

65. Determine the real number values of A, if any, such that the system below has (a) exactly one solution, (b) no solutions, and (c) infinitely many solutions.

$$\begin{cases} x - 5y = A \\ Ax + 10y = -4 \end{cases}$$

In Exercises 66–68, graph the solution set of the system.

66. $\begin{cases} y \geq x^2 - 3x - 10 \\ y < x \end{cases}$

67. $\begin{cases} x^2 + (y - 2)^2 \leq 9 \\ x^2 + (y + 2)^2 < 9 \end{cases}$

68. $\begin{cases} |x + y| \leq 7 \\ |y| \leq 5 \end{cases}$

In Exercises 69–70, describe the graphed region by a system of inequalities.

69.

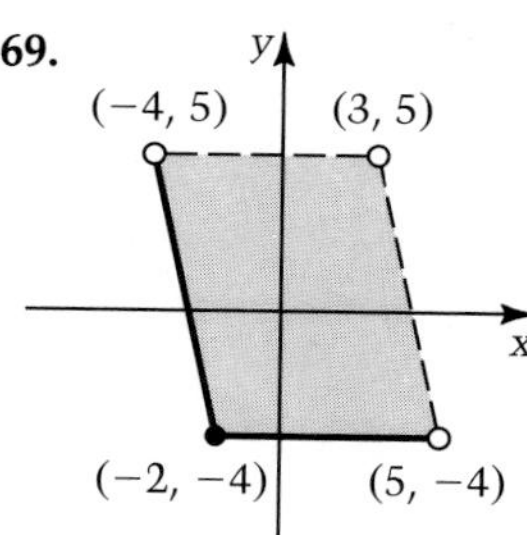

70.

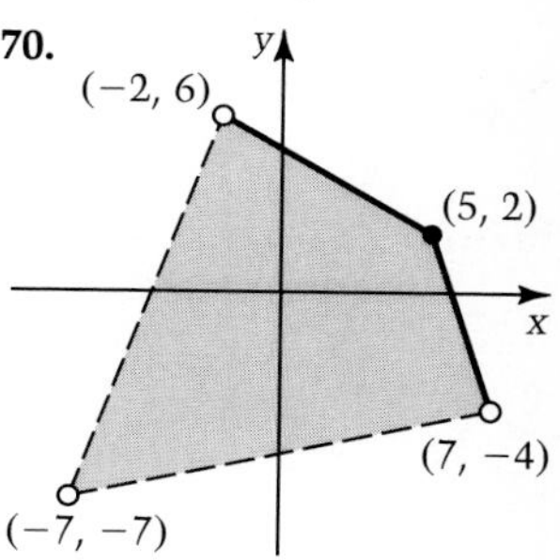

Graphing Calculator Exercises

[Round all estimated values to the nearest hundredth.]

In Exercises 1–12, use your function grapher to solve the given system of equations. (For Exercises 7 and 8, you may wish to review the statement prior to Exercise 19 on page 294.)

1. $\begin{cases} 3x - 2y = 8 \\ 5x + y = 17 \end{cases}$

2. $\begin{cases} 10x + 3y = 1 \\ 3x - 7y = 15 \end{cases}$

3. $\begin{cases} 2x - 7y = 31 \\ y + 3x = 10 \end{cases}$

4. $\begin{cases} 5x - 3y = 7 \\ 2y + 4x = 5 \end{cases}$

5. $\begin{cases} y - 2x^2 = -6 \\ y - 4x = -8 \end{cases}$

6. $\begin{cases} 6x^2 + y = 1 \\ 12x + y = 7 \end{cases}$

7. $\begin{cases} 2x^2 + 3y^2 = 1 \\ x - 2y = 0 \end{cases}$

8. $\begin{cases} x^2 - 3y^2 = 1 \\ 5x + 2y - 9 = 0 \end{cases}$

9. $\begin{cases} 3x^2 - 7x + 2y = 10 \\ y - e^x = 2 \end{cases}$

10. $\begin{cases} y = e^{-x^2/2} \\ y = \frac{1}{x} \end{cases}$

11. $\begin{cases} 0.2^{x^2} - y = 0 \\ \ln x - y = 0 \end{cases}$

12. $\begin{cases} x + y = \sin x - 1 \\ x - y = -1 \end{cases}$

Chapter Test

In Problems 1–2, determine whether the point is a solution of the system: $\begin{cases} y + 5x - 3 = 0 \\ 3y - 2x + 6 = 0 \end{cases}$

1. $(3, 0)$ **2.** $(-2, 1)$

In Problems 3–4, use the addition method to solve the system.

3. $\begin{cases} 6x + 2y = 12 \\ 3x - 4y = 6 \end{cases}$ **4.** $\begin{cases} -25x + 10y = 5 \\ 5x - 2y = -1 \end{cases}$

In Problems 5–6, use the substitution method to solve the system.

5. $\begin{cases} 4x + 2y = -5 \\ 6x + 3y = -7 \end{cases}$ **6.** $\begin{cases} 2x + 5y = -7 \\ 4x - 3y = 10 \end{cases}$

7. Solve the system: $\begin{cases} x^2 - 8x + y^2 = 0 \\ x^2 + y^2 = 9 \end{cases}$

In Problems 8–9, sketch the graph of the inequality.

8. $x^2 + y^2 > 4$ **9.** $y \le 2x + 3$

In Problems 10–11, graph the solution set of the system.

10. $\begin{cases} y \ge 5 - 3x \\ y < x - 2 \end{cases}$ **11.** $\begin{cases} y < -2 \\ x > 1 \\ -y \ge x \\ y \ge x - 10 \end{cases}$

In Problems 12–14, graph the plane.

12. $x = 1$ **13.** $y = 6$ **14.** $z = -1$

15. Solve the system: $\begin{cases} x + y + z = 5 \\ 2x - z = 0 \\ 5y + z = -1 \end{cases}$.

In Problems 16–17, determine the augmented, coefficient, and constant matrices for the system.

16. $\begin{cases} 4x - 3y = 7 \\ x + 2y = -3 \end{cases}$ **17.** $\begin{cases} 2x + y - 3z = -7 \\ x - 2y = 5 \\ 4y + 2z = 0 \end{cases}$

In Problems 18–19, use matrix methods to solve the system.

18. $\begin{cases} x + 2y = 3 \\ 3x - y = 1 \end{cases}$ **19.** $\begin{cases} 2x - y + 3z = -8 \\ 5x - 2y - z = -1 \\ x + y + z = 3 \end{cases}$

20. Compute the determinant of the matrix.

$$\begin{bmatrix} -1 & 1 \\ 3 & 3 \end{bmatrix}$$

21. Compute the determinant by expanding along (a) the first row, (b) the second column.

$$\begin{vmatrix} -1 & -1 & 0 \\ 2 & 3 & 1 \\ 3 & 1 & 1 \end{vmatrix}$$

22. Use Cramer's Rule to determine whether the following system has a unique solution. If it does, determine the solution.

$$\begin{cases} 10x - 13y - 2z = 1 \\ 2x - 7y - 4z = -1 \\ 3x + 4y + 5z = 2 \end{cases}$$

In Problems 23–24, compute the matrix products AB and BA.

23. $A = \begin{bmatrix} 1 \\ 2 \end{bmatrix}, B = \begin{bmatrix} 3 & -1 \end{bmatrix}$

24. $A = \begin{bmatrix} 2 & 0 & 3 \\ 1 & -1 & 1 \end{bmatrix}, B = \begin{bmatrix} 1 & -1 \\ 1 & 4 \\ 0 & 3 \end{bmatrix}$

In Problems 25–26, determine the multiplicative inverse of the given matrix.

25. $\begin{bmatrix} 4 & 1 \\ 5 & 2 \end{bmatrix}$ **26.** $\begin{bmatrix} 2 & 1 & -2 \\ 1 & 1 & 0 \\ 2 & -1 & -4 \end{bmatrix}$

27. A vending firm at the football stadium makes a \$0.27 profit on each hamburger and a \$0.22 profit on each hotdog sold. The firm sells between 4000 and 7000 hamburgers and between 6000 and 11,000 hotdogs per game. Refrigerator space limits the vendor's inventory to 16,000 items (hamburgers and hotdogs). How many of each type should the vendor have to maximize the potential profit of the inventory?

10

Special Topics in Algebra

In this chapter we consider several topics that will complete a core of precalculus topics. Sequences, series, and limits raise questions for which one must reconcile the concepts of "finite" and "infinite." Mathematical induction provides a technique for proving some of the most profound results in mathematics. Probability has become an essential tool in the decision-making of physicists, space scientists, engineers, economists, actuaries, medical practitioners, and market analysts. To that listing of professionals who use probability we add: lottery officials. This brings us to the question: "How can you beat the lottery?" Some have tried and claim to have succeeded. Others have tried and failed. It is the plight of a few in this latter category that we consider at the end of this chapter.

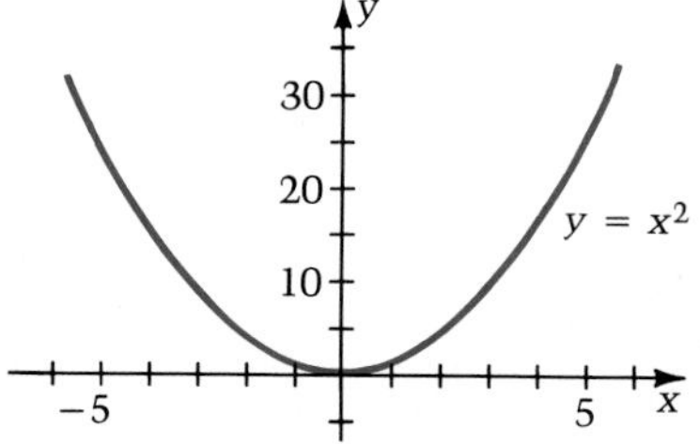

Figure 10.1

10.1
Arithmetic Sequences and Sums

Up to this point we have been concerned mainly with functions that have the set of real numbers (or intervals of real numbers) as their domains. One such function, $f(x) = x^2$, is graphed in Figure 10.1. What would happen if we changed the domain from the set of all real numbers to the set of positive integers? The resulting graph would then consist of the set of discrete points

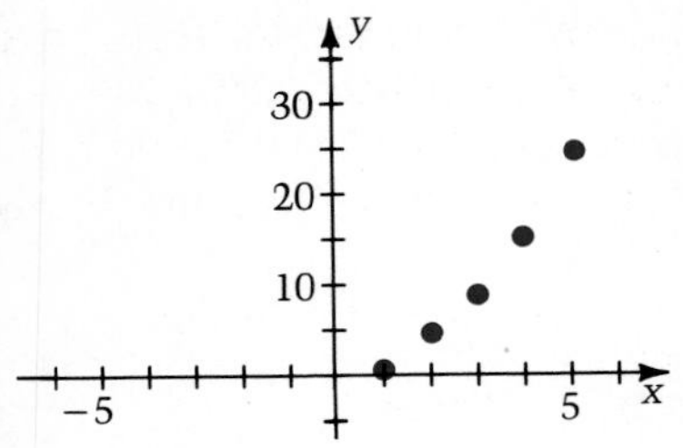

Figure 10.2

n	$a(n)$
1	1
2	4
3	9
4	16
5	25
⋮	⋮
k	k^2
⋮	⋮

(unconnected dots) shown in Figure 10.2. Such a function is an example of a *sequence*.

In general, a **sequence** is a function whose domain is the set of positive integers. To remind ourselves of this special domain, we usually use "n" instead of "x" to represent domain values.

Some values of the sequence described by the equation

$$a(n) = n^2$$

are given in the table at the left. We will adopt the common practice of referring to the listing 1, 4, 9, 16, 25, . . . as the sequence, and to the individual values 1, 4, 9, etc., as the **terms of the sequence.** A note regarding notation is essential:

in place of the symbol $a(n)$, we shall use a_n.

Thus, the subscript n represents the domain value. Applying this new notation to the above sequence, we write:

$$a_1 = 1, \qquad a_2 = 4, \qquad a_3 = 9, \qquad a_{10} = 100, \qquad a_n = n^2.$$

Notice that a_1 is the first term of the sequence, a_2 the second term, a_3 the third term, and so on. We refer to a_n as the nth term, or the **general term of the sequence.** It can be used to state a rule for generating the specific terms of the sequence. Substituting the values 1, 2, 3, 4, 5, etc., for n in the rule produces the terms of the sequence. Several examples are displayed below.

Rule	Sequence
$a_n = 2n - 1$	$1, 3, 5, 7, 9, \ldots$
$a_n = 2^n$	$2, 4, 8, 16, 32, \ldots$
$a_n = 3 - 2^n$	$1, -1, -5, -13, -29, \ldots$
$a_n = (-1)^n$	$-1, 1, -1, 1, -1, \ldots$

EXAMPLE 1

Find the first three terms and the twentieth term of the sequence whose general term is given by the rule $a_n = n^2 - n$.

Solution

$a_n = n^2 - n$ — We evaluate a_n for $n = 1, 2, 3$, and 20.

$a_1 = 1^2 - 1 = 0$

$a_2 = 2^2 - 2 = 2$

$a_3 = 3^2 - 3 = 6$

$a_{20} = 20^2 - 20 = 380$

For the sequences discussed thus far, each term is found by evaluating an expression containing the term's subscript. A sequence can be defined another way: specify the first term of the sequence, and a rule for finding any other term by using the preceding term(s). A sequence described this way is said to be defined **recursively.**

$a_1 = 1$

$a_2 = 2(1) + 1 = 3$

$a_3 = 2(3) + 1 = 7$

$a_4 = 2(7) + 1 = 15$

$a_5 = 2(15) + 1 = 31$

Consider the sequence defined recursively as follows:

$$a_1 = 1, \qquad a_n = 2(a_{n-1}) + 1 \quad \text{for } n > 1.$$

To find any term a_n after the first, multiply the preceding term, a_{n-1}, by 2 and then add 1. The first five terms of this sequence are 1, 3, 7, 15 and 31 (see the computations in the margin). Note that every term after the first is found by using the preceding term.

The following statement specifies the first two terms of a sequence, and then the recursive rule:

$$a_1 = 1, \qquad a_2 = 1, \qquad a_n = a_{n-1} + a_{n-2} \quad \text{for } n > 2.$$

According to this rule, each term after the second is the sum of the two preceding terms. Thus, we obtain the sequence

$$1, 1, 2, 3, 5, 8, 13, 21, 34, 55, 89, 144, \ldots.$$

This sequence is called the **Fibonacci sequence,** after the 13th century Italian mathematician Fibonacci. This famous sequence has extensive applications even today.

EXAMPLE 2

Find the first four terms of each sequence.

(a) $a_1 = 7, a_n = a_{n-1} + 3 \quad \text{for } n > 1$

(b) $a_1 = 1, a_n = a_{n-1} + 2n - 1 \quad \text{for } n > 1$

Solution

(a) $a_1 = 7$

$a_2 = 7 + 3 = 10$

$a_3 = 10 + 3 = 13$

$a_4 = 13 + 3 = 16$

(b) $a_1 = 1$

$a_2 = 1 + 2(2) - 1 = 4$

$a_3 = 4 + 2(3) - 1 = 9$

$a_4 = 9 + 2(4) - 1 = 16$

Notice that for the sequence 7, 10, 13, 16, . . . , the difference between successive terms is constant—it is always 3. Such a sequence is called an **arithmetic sequence,** and the constant difference, 3, is called the **common difference.** Every term of an arithmetic sequence after the first is the sum of the preceding term and the common difference.

EXAMPLE 3

Find the first five terms of the arithmetic sequence

(a) whose first term is 20, and whose common difference is 4;

(b) whose first term is 10, and whose common difference is -3.

Solution

(a) $a_1 = 20$
$a_2 = 20 + 4 = 24$
$a_3 = 24 + 4 = 28$
$a_4 = 28 + 4 = 32$
$a_5 = 32 + 4 = 36$

(b) $a_1 = 10$
$a_2 = 10 + (-3) = 7$
$a_3 = 7 + (-3) = 4$
$a_4 = 4 + (-3) = 1$
$a_5 = 1 + (-3) = -2$

EXAMPLE 4

Assuming the obvious pattern continues, determine whether each of the following sequences is arithmetic. If it is, state the common difference, d.

(a) $-1, 1, 3, 5, \ldots$

(b) $-1, 1, -1, 1, \ldots$

(c) $\frac{11}{2}, 4, \frac{5}{2}, 1, \ldots$

(d) $2, 4, 8, 16, \ldots$

Solution

(a) $-1, 1, 3, 5, \ldots$ (differences $+2$, $+2$, $+2$) The sequence is arithmetic, $d = 2$.

(b) $-1, 1, -1, 1, \ldots$ (differences $+2$, -2, $+2$) The sequence is not arithmetic. The difference between the first and second terms is not the same as the difference between the second and third terms.

(c) $\frac{11}{2}, 4, \frac{5}{2}, 1, \ldots$ (differences $-\frac{3}{2}$, $-\frac{3}{2}$, $-\frac{3}{2}$) The sequence is arithmetic, $d = -\frac{3}{2}$.

(d) $2, 4, 8, 16, \ldots$ (differences $+2$, $+4$, $+8$) The sequence is not arithmetic. There is no common difference.

Suppose an arithmetic sequence has first term a_1 and common difference d. Then each term in the sequence can be expressed in terms of a_1 and d, as shown below.

$$\begin{aligned} a_1 &= a_1 + 0d \\ a_2 &= a_1 + 1d \\ a_3 &= a_1 + 2d \\ a_4 &= a_1 + 3d \\ &\vdots \end{aligned}$$

(For example, we know that $a_3 = a_2 + d$; thus, $a_3 = (a_1 + d) + d = a_1 + 2d$. We find $a_4, a_5, a_6, \ldots$ similarly.) Notice that the coefficient of d is one less than the subscript of the term. This suggests the following result.

The *n*th Term of an Arithmetic Sequence

In an arithmetic sequence with first term a_1 and common difference d, the nth term can be described as follows:

$$a_n = a_1 + (n - 1)d \qquad (n \geq 1).$$

EXAMPLE 5

Find the twelfth and thirtieth terms of the arithmetic sequence

$$-7, -2, 3, 8, \ldots.$$

Solution

$a_n = a_1 + (n - 1)d$ — Substitute the first term, -7, for a_1 and substitute the common difference, 5, for d.

$$a_{12} = -7 + (12 - 1)(5) = 48$$

$$a_{30} = -7 + (30 - 1)(5) = 138$$

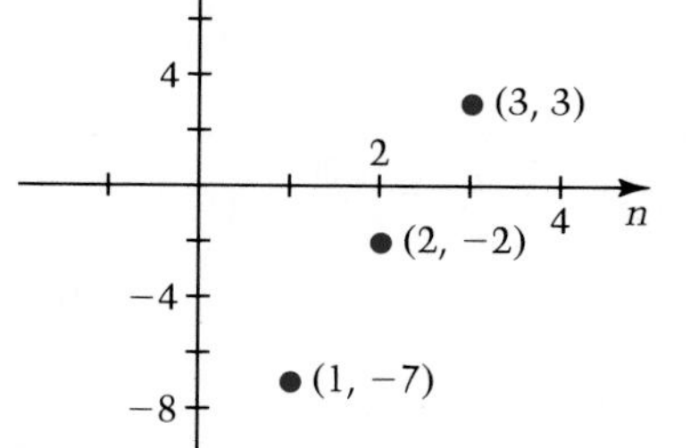

Figure 10.3

If we were to graph an arithmetic sequence, the discrete points would lie along a straight line. The graph of the sequence given in Example 5 is shown in Figure 10.3. Notice that for this sequence $a_1 = -7$ and $d = 5$. Thus, the general term is given by the rule

$$a_n = -7 + (n - 1)5, \qquad \text{or simply} \qquad a_n = -12 + 5n.$$

Therefore, in the xy-plane, the dots representing this sequence lie along the line $y = -12 + 5x$, a line with slope 5.

Adding Terms of an Arithmetic Sequence

Suppose you are asked to determine the sum of the integers from 1 to 100. An easy way to find this sum is as follows. Begin by writing the sum; then write the sum in reverse order:

$$\begin{array}{r} 1 + \quad 2 + \quad 3 + \cdots + \ 98 + \ 99 + 100 \\ \underline{100 + \ 99 + \ 98 + \cdots + \quad 3 + \quad 2 + \quad 1} \\ 101 + 101 + 101 + \cdots + 101 + 101 + 101 \end{array}$$

When we add the terms of the two sums, we obtain 100 addends, each equal to 101. Thus, twice the sum of the numbers from 1 to 100 is equal to 100(101).

Therefore

$$1 + 2 + 3 + \cdots + 100 = \frac{1}{2}(100)(101) = 5050.$$

We can apply a similar method to find the sum of the first n terms of any arithmetic sequence.

$$\begin{array}{ccccccc} a_1 & + & (a_1 + d) & + \cdots + & [a_1 + (n-1)d] \\ [a_1 + (n-1)d] & + & [a_1 + (n-2)d] & + \cdots + & a_1 \\ \hline [2a_1 + (n-1)d] & + & [2a_1 + (n-1)d] & + \cdots + & [2a_1 + (n-1)d] \end{array}$$

The result is n addends, each equal to

$$[2a_1 + (n-1)d].$$

Thus, twice the sum of the first n terms is equal to

$$n[2a_1 + (n-1)d].$$

We conclude that the sum S_n of the first n terms is

$$S_n = \frac{n}{2}[2a_1 + (n-1)d].$$

Since $2a_1 + (n-1)d = a_1 + [a_1 + (n-1)d] = a_1 + a_n$, we have

$$S_n = \frac{n}{2}(a_1 + a_n).$$

Thus, we have proven the following result.

The Sum of the First n Terms of an Arithmetic Sequence

The sum S_n of the first n terms of an arithmetic sequence is

$$S_n = \frac{n}{2}[2a_1 + (n-1)d] \quad \text{or} \quad S_n = \frac{n}{2}(a_1 + a_n),$$

where a_1 is the first term, a_n is the nth term, and d is the common difference.

EXAMPLE 6

Find the sum of the first 40 terms of the arithmetic sequence

$$11, 15, 19, 23, \ldots.$$

Solution

$$S_n = \frac{n}{2}[2a_1 + (n - 1)d] \qquad \text{Substitute 11 for } a_1\text{, 4 for } d\text{, and 40 for } n.$$

$$S_{40} = \frac{40}{2}[2(11) + (40 - 1)4]$$

$$S_{40} = 3560$$

Alternative solution: to use the alternative formula for S_n, you must first find the fortieth term, a_{40}.

$$a_n = a_1 + (n - 1)d$$

$$a_{40} = 11 + (40 - 1)4$$

$$a_{40} = 167$$

$$S_n = \frac{n}{2}(a_1 + a_n) \qquad \text{Now, substitute 11 for } a_1\text{, 167 for } a_n\text{, and 40 for } n.$$

$$S_{40} = \frac{40}{2}(11 + 167)$$

$$S_{40} = 3560$$

EXERCISE SET 10.1

In Exercises 1–16, find the first four terms, the ninth term, and the twentieth term of the sequence whose general term is given.

1. $a_n = 3n - 5$

2. $a_n = 4 - n$

3. $a_n = \dfrac{n}{2n + 1}$

4. $a_n = \dfrac{3n}{1 + n}$

5. $a_n = 1 - \dfrac{1}{n}$

6. $a_n = (-1)^n\dfrac{1}{n}$

7. $a_n = n^2 + 3$

8. $a_n = 7 - n^2$

9. $a_n = (n - 3)(n - 4)$

10. $a_n = (n + 2)(n + 3)$

11. $a_n = n(n - 1)(n - 2)$

12. $a_n = n(n + 6)(n + 1)$

13. $a_n = \dfrac{1}{n} + \dfrac{1}{n + 1}$

14. $a_n = \dfrac{1}{n} - \dfrac{1}{n + 1}$

15. $a_n = 2^{-n}$

16. $a_n = (-2)^{-n}$

In Exercises 17–26, find the first six terms of the sequence.

17. $a_1 = 3, \quad a_n = a_{n-1} + 1$ for $n \geq 2$

18. $a_1 = 4, \quad a_n = a_{n-1} - 1$ for $n \geq 2$

19. $a_1 = 1, \quad a_n = 2a_{n-1} - 3$ for $n \geq 2$

20. $a_1 = 5, \quad a_n = 2a_{n-1} - 3$ for $n \geq 2$

21. $a_1 = 3, \quad a_n = (-1)^n a_{n-1} - 7$ for $n \geq 2$

22. $a_1 = 1, \quad a_n = (-1)^n a_{n-1} + 2$ for $n \geq 2$

23. $a_1 = 1, \quad a_2 = 3, \quad a_n = a_{n-1} - a_{n-2}$ for $n \geq 3$

24. $a_1 = 3, \quad a_2 = 1, \quad a_n = a_{n-2} + a_{n-1}$ for $n \geq 3$

25. $a_1 = 1, \quad a_2 = 3, \quad a_n = a_{n-2}$ for $n \geq 3$

26. $a_1 = 1, \quad a_2 = 4, \quad a_n = a_{n-1} - a_{n-2}$ for $n \geq 3$

In Exercises 27–36, the first term a_1 and the common difference d of an arithmetic sequence are given. Find the first five terms of the sequence.

27. $a_1 = 4, d = 3$

28. $a_1 = 3, d = 6$

29. $a_1 = 5, d = -2$

30. $a_1 = -5, d = 2$

31. $a_1 = -2, d = 5$

32. $a_1 = 2, d = -5$

33. $a_1 = \dfrac{1}{2}, d = \dfrac{1}{4}$

34. $a_1 = \dfrac{1}{3}, d = \dfrac{1}{6}$

35. $a_1 = \dfrac{1}{4}, d = \dfrac{1}{8}$

36. $a_1 = \dfrac{1}{6}, d = \dfrac{1}{3}$

In Exercises 37–48, assuming the obvious pattern continues, determine whether the given sequence is arithmetic. If it is, state the common difference d.

37. $-3, 0, 3, 6, \ldots$

38. $2, 4, 8, 16, \ldots$

39. $1, 3, 9, 27, \ldots$

40. $4, -1, -6, -11, \ldots$

41. $10, 2, -6, -14, \ldots$

42. $1, 0, 1, 0, \ldots$

43. $2, 2, 2, 2, \ldots$

44. $\sqrt{3}, \quad 1 + \sqrt{3}, \quad 2 + \sqrt{3}, \quad 3 + \sqrt{3}, \ldots$

45. $0.1, 0.01, 0.001, 0.0001, \ldots$

46. $\frac{3}{4}, 2, \frac{13}{4}, 4, \ldots$

47. $\sqrt{3}, \sqrt{12}, \sqrt{27}, \sqrt{48}, \ldots$

48. $0.1, 0.11, 0.111, 0.1111, \ldots$

In Exercises 49–58, find the fifteenth and thirtieth terms of the arithmetic sequence.

49. $3, 7, 11, 15, \ldots$

50. $8, 6, 4, 2, \ldots$

51. $9, 7, 5, 3, \ldots$

52. $1, 6, 11, 16, \ldots$

53. $-7, -1, 5, 11, \ldots$

54. $-3, -1, 1, 3, \ldots$

55. $-2, -6, -10, -14, \ldots$

56. $-9, -6, -3, 0, \ldots$

57. $\frac{1}{2}, \frac{5}{6}, \frac{7}{6}, \frac{3}{2}, \ldots$

58. $-\frac{1}{4}, \frac{1}{8}, \frac{1}{2}, \frac{7}{8}, \ldots$

For Exercises 59–68, find the sum of the first six terms and the sum of the first twenty terms of the arithmetic sequences specified in Exercises 49–58.

In Exercises 69–72, determine the fiftieth and one hundredth terms in the given arithmetic sequence; then compute the sum of the first 100 terms.

69. $-8.12, -4.04, 0.04, \ldots$

70. $9.76, 9.09, 8.42, \ldots$

71. $6595, 6153, 5711, \ldots$

72. $18, 1800, 3582, \ldots$

Superset

In Exercises 73–78, find a formula for the general term of the recursively defined sequence.

73. $a_1 = 1, \quad a_n = a_{n-1} + 2n - 1$ for $n \geq 2$

74. $a_1 = 1, \quad a_n = 2a_{n-1} + 1$ for $n \geq 2$

75. $a_1 = 1, \quad a_n = 2a_{n-1}$ for $n \geq 2$

76. $a_1 = 1, \quad a_n = -a_{n-1}$ for $n \geq 2$

77. $a_1 = 1, \quad a_n = a_1 + a_2 + \cdots + a_{n-1}$ for $n \geq 2$

78. $a_1 = 1, \quad a_n = \frac{1}{2}a_{n-1} + 1$ for $n \geq 2$

In Exercises 79–80, use a calculator to solve the problem.

79. Determine x so that the following is an arithmetic sequence: $8.65, x, -4.11, \ldots$

80. Determine x so that the following is an arithmetic sequence: $10^{-2}, x, 10^{-4}, \ldots$

In Exercises 81–84, the sum S_n of the first n terms of an arithmetic sequence is given. Find the first, third, and tenth terms of the sequence.

81. $S_n = n(n + 2)$

82. $S_n = n(n - 1)$

83. $S_n = n(7 - n)$

84. $S_n = 8n$

85. You are playing blackjack in Las Vegas. You bet \$10 on the first hand and increase your bet by \$5 every time you win. Fortunately, you win every hand you play. If you began with \$50 and won ten hands, how much money do you have?

86. (Refer to Exercise 85.) How many hands in a row must you win before you have at least \$800?

87. A new drive-in movie theatre has 20 spaces for cars in the first row, 22 spaces in the second row, 24 spaces in the third row, and so on, for a total of 18 rows. What is the capacity (number of cars) of the theatre?

88. If $b_1, b_2, b_3, \ldots, b_n$ are numbers such that $a, b_1, \ldots, b_n, c$ are successive terms of an arithmetic sequence, then $b_1, \ldots, b_n$ are the n **arithmetic means** between a and c. Finding such numbers when given a, c, and n, is known as *inserting arithmetic means*. Insert 3 arithmetic means between 4 and 11.

In Exercises 89–90, use a calculator to solve the problem.

89. Determine the sum of all the odd integers between 10,000 and 11,500.

90. Determine the sum of all the even integers between 10,001 and 11,499.

10.2
Geometric Sequences and Sums

$a_{n+1} = a_n + d$
An arithmetic sequence

$a_{n+1} = ra_n$
A geometric sequence

We have already seen that in an arithmetic sequence, each term is d units more than the preceding term, where d is some constant. There is another type of sequence that can be characterized just as easily. In this second type of sequence, each term is r *times* the preceding term, where r is some nonzero constant. Such a sequence is called a **geometric sequence** and r is called the **common ratio** since the ratio of any term to the preceding one is always r. Thus, for a geometric sequence,

$$a_{n+1} = ra_n$$

$1, 2, 4, 8, 16, \ldots$

$3, \frac{9}{2}, \frac{27}{4}, \frac{81}{8}, \ldots$

$5, -10, 20, -40, \ldots$

Examples of geometric sequences are shown at the left. In the first sequence, each term (after the first) is 2 times the preceding term; thus, the common ratio is 2. In the second sequence, the common ratio is $\frac{3}{2}$, and in the third, it is -2.

EXAMPLE 1

Assuming the obvious pattern continues, determine whether each sequence is geometric.

(a) $2, -6, 18, -54, \ldots$

(b) $2, 4, 8, 24, 40, \ldots$

Solution

(a) $\dfrac{-6}{2} = \dfrac{18}{-6} = \dfrac{-54}{18} = -3$

Since the ratio of each term to the preceding term is constant, the sequence is geometric.

(b) $\dfrac{4}{2} = 2, \dfrac{8}{4} = 2$, but $\dfrac{24}{8} = 3$

Since there is no common ratio, the sequence is not geometric.

Suppose a geometric sequence has first term a_1 and common ratio r. Then each term of the sequence can be expressed in terms of a_1 and r, as shown below.

$$\begin{aligned} a_1 &= a_1r^0 \\ a_2 &= a_1r^1 \\ a_3 &= a_1r^2 \\ a_4 &= a_1r^3 \\ &\vdots \end{aligned}$$

(For example, we know that $a_3 = a_2r$; thus, $a_3 = (a_1r)r = a_1r^2$. We can find a_4, a_5, $a_6 \ldots$ similarly.) Notice that the exponent of r is always one less than the subscript of the term. This suggests the following result.

The nth Term of a Geometric Sequence

In a geometric sequence with first term a_1 and common ratio r, the nth term can be described as follows:

$$a_n = a_1 r^{n-1} \qquad (n \geq 1).$$

EXAMPLE 2

Find the seventh term and the general term of each geometric sequence.

(a) 3, 6, 12, 24, . . . (b) 2, -10, 50, -250, . . .

Solution

(a) $r = \frac{6}{3} = 2$ Since we know the series is geometric, we can find r by forming the ratio of any term to its preceding term.

$a_n = a_1 r^{n-1}$ To find the seventh term, substitute 3 for a_1, 2 for r, and 7 for n.

$a_7 = (3)(2)^{7-1}$

$a_7 = 192$

$a_n = 3(2)^{n-1}$ To find an expression for the nth term, we substituted 3 for a_1, and 2 for r, and let n remain a variable.

(b) $r = \frac{-10}{2} = -5$

$a_n = a_1 r^{n-1}$ To find the seventh term, substitute 2 for a_1, -5 for r, and 7 for n.

$a_7 = (2)(-5)^{7-1}$

$a_7 = 31{,}250$

$a_n = 2(-5)^{n-1}$ To find an expression for the nth term, we substituted 2 for a_1, -5 for r, and let n remain a variable.

Let S_n represent the sum of the first n terms of a geometric sequence. We represent this fact algebraically as follows:

$$S_n = a_1 + a_1 r + a_1 r^2 + \cdots + a_1 r^{n-1}.$$

Let us multiply both sides of this equation by r, and then subtract the result from the original equation.

$$\begin{array}{rcl} S_n & = & a_1 + a_1 r + a_1 r^2 + \cdots + a_1 r^{n-1} \\ -(S_n \cdot r) & = & \quad -(a_1 r + a_1 r^2 + \cdots + a_1 r^{n-1} + a_1 r^n) \\ \hline S_n - S_n r & = & a_1 + 0 + 0 + \cdots + 0 - a_1 r^n \end{array}$$

Notice that the terms on the right side of the second equation have been moved one place to the right to make the difference easier to compute. We conclude that

$$S_n - S_n r = a_1 - a_1 r^n$$
$$S_n(1 - r) = a_1(1 - r^n)$$
$$S_n = \frac{a_1(1 - r^n)}{1 - r} \qquad (r \neq 1).$$

Of course, if $r = 1$, then the denominator in the previous formula is zero, and this equation is undefined. In this case every term of the sequence is equal to a_1, and the sum of the first n terms is simply na_1. Thus, we have proven the following result.

The Sum of the First n Terms of a Geometric Sequence

The sum S_n of the first n terms of a geometric sequence is

$$S_n = \frac{a_1(1 - r^n)}{1 - r}$$

where a_1 is the first term and $r \neq 1$ is the common ratio.

EXAMPLE 3

Determine the sum of the first n terms and the sum of the first 10 terms of the geometric sequence 5, 15, 45, 135, ...

Solution The common ratio is 3. Thus, we can use the formula above with $a_1 = 5$ and $r = 3$.

$$S_n = \frac{5(1 - 3^n)}{1 - 3} = \frac{5(1 - 3^n)}{-2}$$

Now, we substitute $n = 10$.

$$S_{10} = \frac{5(1 - 3^{10})}{-2} = 147{,}620$$

EXERCISE SET 10.2

In Exercises 1–6, assuming the obvious pattern continues, determine whether the sequence is geometric. If it is, state the common ratio.

1. 2, 4, 6, 8, . . .

2. $-4, -2, -1, \frac{1}{2}, \ldots$

3. $12, 18, 27, \frac{81}{2}, \ldots$

4. $2, -3, 8, -12, \ldots$

5. $-1, -2, -0.04, 0.008, \ldots$

6. $-16, 24, -36, 54, \ldots$

In Exercises 7–22, find the sixth, ninth, and general terms of the geometric sequence.

7. $1, 2, 4, 8, \ldots$

8. $1, 3, 9, 27, \ldots$

9. $2, 4, 8, 16, \ldots$

10. $8, -4, 2, -1, \ldots$

11. $\frac{1}{8}, \frac{1}{2}, 2, 8, \ldots$

12. $27, -18, 12, -8, \ldots$

13. $-81, 27, -9, 3, \ldots$

14. $\frac{1}{8}, -\frac{1}{4}, \frac{1}{2}, -1, \ldots$

15. $\frac{1}{3}, 1, 3, 9, \ldots$

16. $\sqrt{2}, 2, \sqrt{8}, 4, \ldots$

17. $\sqrt{2}, \sqrt{6}, 3\sqrt{2}, 3\sqrt{6}, \ldots$

18. $-16, 24, -36, 54, \ldots$

19. $0.1, -0.01, 0.001, -0.0001, \ldots$

20. $1, \frac{1}{3}, \frac{1}{9}, \frac{1}{27}, \ldots$

21. $\frac{25}{8}, \frac{5}{4}, \frac{1}{2}, \frac{1}{5}, \ldots$

22. $0.1, 0.02, 0.004, 0.0008, \ldots$

For Exercises 23–38, determine the sum of the first n terms and the sum of the first seven terms of the geometric sequences in Exercises 7–22.

Superset

39. Suppose that $a_1, a_2, a_3, a_4, \ldots$ is a geometric sequence. Show that $a_1, a_3, a_5, \ldots$ and $a_2, a_4, a_6, \ldots$ are also geometric sequences.

40. Show that if $2^{a_1}, 2^{a_2}, 2^{a_3}, 2^{a_4}, \ldots$ is a geometric sequence, then $a_1, a_2, a_3, a_4, \ldots$ is an arithmetic sequence.

41. A rubber ball is dropped from a height of 27 feet and always bounces to a height that is $\frac{1}{3}$ the height of the previous fall. What distance does the ball fall the second time? the third time? the fifth time?

42. Every time a child jumps into a wading pool, one-tenth of the water splashes out of the pool. After the child has jumped in the pool 4 times, what percentage of the original amount of water remains in the pool?

43. (Refer to exercise 41) What is the total distance (up and down) the ball has traveled after the second fall? the third fall? the fifth fall?

44. The annual depreciation of an office computer facility is 20% of its value at the start of the year. Determine the value of the computer after 5 years, if its original cost was $60,000.

45. You have deposited $1000 in an account that earns 8% annually. You leave the interest in the account so that in subsequent years, you earn interest on your interest. How much interest do you earn the first year? the second year? the first five years?

46. Determine a value of r, not equal to 1, so that the third term of a geometric sequence is the arithmetic mean of the first and fifth terms of the sequence.

47. The **geometric mean** of two positive numbers, x and y, is defined as $\sqrt{xy}$. Suppose a and b are positive numbers with $a < b$. Show that the geometric mean of a and b is less than the arithmetic mean of a and b.

48. Let BD be the altitude from the vertex B to the hypotenuse AC of the right triangle ABC. Show that BD is the geometric mean of AD and DC. (*Hint*: Refer to Exercise 47.)

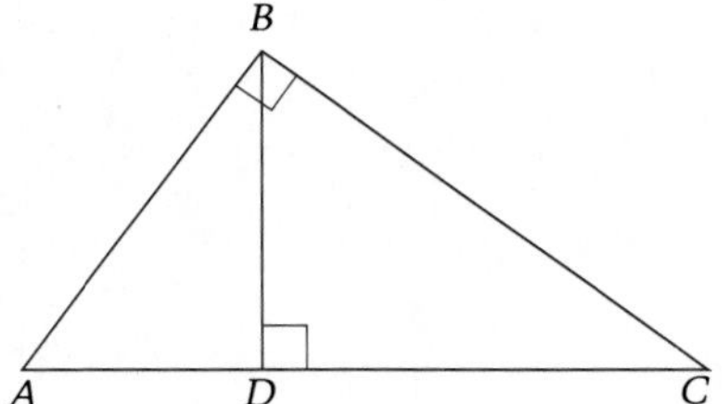

10.3 Series

We shall now consider an alternative way of writing the sum of the first n terms of a sequence, $a_1 + a_2 + \cdots + a_n$. The Greek letter *sigma*, written Σ, is used to express this sum in shorthand form:

$$\sum_{k=1}^{n} a_k$$

In the sigma notation above, the letter k is called an **index.** The statement "$k = 1$" below the sigma indicates that the sum starts with the first term of the sequence; the n above the sigma indicates that the sum ends with the nth term of the sequence. The index k increases by 1 from term to term and can be used as a subscript (as in the first sum below) or can actually be used to compute the terms of the sum (as in the second sum below).

$$\sum_{k=1}^{3} a_k = a_1 + a_2 + a_3$$

$$\sum_{k=3}^{6} k^3 = 3^3 + 4^3 + 5^3 + 6^3$$

EXAMPLE 1

Rewrite each sum without using sigma notation; then compute each sum.

(a) $\sum_{k=1}^{4} 3k$ (b) $\sum_{k=3}^{5} (k^2 - 1)$

Solution

(a) $\sum_{k=1}^{4} 3k = 3(1) + 3(2) + 3(3) + 3(4) = 3 + 6 + 9 + 12 = 30$

(b) $\sum_{k=3}^{5} (k^2 - 1) = (3^2 - 1) + (4^2 - 1) + (5^2 - 1) = 8 + 15 + 24 = 47$

EXAMPLE 2

Use sigma notation to rewrite each sum with 1 as the first value of the index.

(a) $1 + 4 + 9 + 16 + 25$ (b) $7 + 9 + 11 + 13$

Solution

(a) $1 + 4 + 9 + 16 + 25 = \sum_{k=1}^{5} k^2$

(b) $7 + 9 + 11 + 13 = \sum_{k=1}^{4} (2k + 5)$

The expression

$$a_1 + a_2 + a_3 + \cdots + a_n + \cdots$$

is called an **infinite series.** In sigma notation this series is denoted by

$$\sum_{k=1}^{\infty} a_k$$

The symbol "∞" above the sigma indicates that the index k takes on *all* integer values beginning with $k = 1$.

If a sequence is arithmetic, then its corresponding series is also called arithmetic; if a sequence is geometric, then its corresponding series is called geometric.

Partial Sums

When does it make sense to talk about the "sum" of an infinite series? To help us answer this question, our strategy will be to look at the sum of the first term, then the sum of the first two terms, then the sum of the first three terms, etc., and determine whether there is a trend.

For the series

$$a_1 + a_2 + a_3 + \cdots + a_n + \cdots,$$

we find **partial sums** as follows:

first partial sum: $S_1 = a_1$
second partial sum: $S_2 = a_1 + a_2$
third partial sum: $S_3 = a_1 + a_2 + a_3$
$\vdots$
nth partial sum: $S_n = a_1 + a_2 + a_3 + \cdots + a_n$

(Note that the subscript on S denotes the number of terms in the partial sum.)

For example, 1, 4, 9, 16, and 25 are the first five partial sums of the arithmetic series of odd integers, $1 + 3 + 5 + \cdots$:

$$S_1 = 1$$
$$S_2 = 1 + 3 = 4$$
$$S_3 = 1 + 3 + 5 = 9$$
$$S_4 = 1 + 3 + 5 + 7 = 16$$
$$S_5 = 1 + 3 + 5 + 7 + 9 = 25$$

As another example consider the geometric series

$$\frac{1}{2} + \frac{1}{4} + \frac{1}{8} + \frac{1}{16} + \frac{1}{32} + \frac{1}{64} + \cdots.$$

Its first five partial sums are $\frac{1}{2}, \frac{3}{4}, \frac{7}{8}, \frac{15}{16}$, and $\frac{31}{32}$.

EXAMPLE 3

Compute the third and sixth partial sums of each series.

(a) $1 + 3 + 9 + 27 + \cdots + 3^{n-1} + \cdots$ (b) $\sum_{n=1}^{\infty}\left(-\frac{1}{2}\right)^{n-1}$

Solution

(a) $S_3 = 1 + 3 + 9 = 13$

$$S_6 = \frac{1(1 - 3^6)}{1 - 3} = 364$$

Since the series is geometric, $S_n = \dfrac{a_1(1 - r^n)}{1 - r}$.

(b) $$\sum_{n=1}^{\infty}\left(-\frac{1}{2}\right)^{n-1} = 1 - \frac{1}{2} + \frac{1}{4} - \frac{1}{8} + \cdots + \left(-\frac{1}{2}\right)^{n-1} + \cdots$$

$$S_3 = 1 - \frac{1}{2} + \frac{1}{4} = \frac{3}{4}$$

$$S_6 = \frac{1\left(1 - \left(-\frac{1}{2}\right)^6\right)}{1 + \frac{1}{2}} = \frac{21}{32}$$

Since the series is geometric, $S_n = \dfrac{a_1(1 - r^n)}{1 - r}$.

For the arithmetic series

$$1 + 3 + 5 + 7 + 9 + \cdots,$$

we know that the first five partial sums are 1, 4, 9, 16, and 25. We can consider these values to be the first five terms of a **sequence of partial sums,** as shown at the left. For the geometric series $\frac{1}{2} + \frac{1}{4} + \frac{1}{8} + \frac{1}{16} + \frac{1}{32} + \frac{1}{64} + \cdots$, we found the first five partial sums to be $\frac{1}{2}, \frac{3}{4}, \frac{7}{8}, \frac{15}{16}$, and $\frac{31}{32}$. The sequence of partial sums for this series is also shown at the left.

1, 4, 9, 16, 25, . . .

$$\frac{1}{2}, \frac{3}{4}, \frac{7}{8}, \frac{15}{16}, \frac{31}{32}, \ldots$$

Notice that for the arithmetic series, the terms in the sequence of partial sums, 1, 4, 9, 16, 25, . . . , are getting *arbitrarily large* (that is, eventually larger than any number you can imagine). Thus, we say that this series "goes to infinity or has no sum." For the geometric series, the terms in the sequence of partial sums get *arbitrarily close* to 1 (that is, eventually as close to 1 as you wish). In this case, we say that the sum of this series is 1, and write:

$$\frac{1}{2} + \frac{1}{4} + \frac{1}{8} + \frac{1}{16} + \cdots = 1 \quad \text{or} \quad \sum_{k=1}^{\infty}\frac{1}{2^k} = 1.$$

The sum of a series is determined by the behavior of the sequence of its partial sums.

The Sum of a Series

Definition of the Sum of a Series

The infinite series

$$a_1 + a_2 + a_3 + \cdots + a_n + \cdots$$

is said to have **sum** S, if the terms in its sequence of partial sums

$$S_1, S_2, S_3, \ldots, S_n, \ldots,$$

get arbitrarily close to S as the value of n increases.

We now consider geometric series. Whether or not a geometric series has a sum depends on the value of the common ratio r.

Sum of a Geometric Series

The sum of the geometric series $a_1 + a_1r + a_1r^2 + \cdots$ is

$$S = \frac{a_1}{1 - r}$$

provided $|r| < 1$. If $|r| \geq 1$, the series has no sum.

EXAMPLE 4

Find the sum, if it exists, of each geometric series.

(a) $\dfrac{16}{125} + \dfrac{4}{25} + \dfrac{1}{5} + \dfrac{1}{4} + \ldots$ (b) $\displaystyle\sum_{k=2}^{\infty} 3(2)^{-k}$

Solution

(a) The series has no sum since $r = \frac{5}{4}$, and $\frac{5}{4} \geq 1$.

(b) $\displaystyle\sum_{k=2}^{\infty} 3(2)^{-k} = \frac{3}{4} + \frac{3}{8} + \frac{3}{16} + \frac{3}{32} + \ldots$

Begin by writing out the first few terms. Notice that $a_1 = \frac{3}{4}$ and $r = \frac{1}{2}$. Since $|r| < 1$, the series has a sum.

$$S = \frac{a_1}{1 - r} = \frac{\frac{3}{4}}{1 - \frac{1}{2}} = \frac{3}{2}$$

A repeating decimal represents a rational number. We can determine the fractional form of this rational number by viewing the decimal expansion as a geometric series. For example, the repeating decimal 0.333 . . . can be viewed as a geometric series with $a_1 = \frac{3}{10}$ and $r = \frac{1}{10}$:

$$0.333\ldots = \frac{3}{10} + \frac{3}{10^2} + \frac{3}{10^3} + \cdots$$

Computing the sum S of this infinite series, we obtain

$$S = \frac{a_1}{1 - r} = \frac{\frac{3}{10}}{1 - \frac{1}{10}} = \frac{\frac{3}{10}}{\frac{9}{10}} = \frac{1}{3}.$$

Thus, $\frac{1}{3}$ is the fractional equivalent of 0.333. . . .

EXAMPLE 5

Find the fraction represented by the decimal expansion for $x = 0.7\overline{21}$.

Solution

$$x = 0.7\overline{21} = \frac{7}{10} + 0.0\overline{21}$$

Write x as the sum of a nonrepeating part and a repeating part.

$$0.0\overline{21} = 0.021 + 0.00021 + 0.0000021 + \cdots$$

Write the repeating part as a geometric series.

$$= \frac{21}{10^3} + \frac{21}{10^5} + \frac{21}{10^7} + \cdots$$

$$a_1 = \frac{21}{10^3}, r = \frac{1}{10^2}$$

$$x = \frac{7}{10} + \frac{a_1}{1 - r} = \frac{7}{10} + \frac{\frac{21}{10^3}}{1 - \frac{1}{10^2}} = \frac{7}{10} + \frac{7}{330} = \frac{119}{165}$$

Another type of series whose sum we can compute is called the **telescoping series.** Consider the following:

$$\sum_{n=1}^{\infty} \left(\frac{1}{n} - \frac{1}{n + 1}\right) = \left(1 - \frac{1}{2}\right) + \left(\frac{1}{2} - \frac{1}{3}\right) + \cdots + \left(\frac{1}{n} - \frac{1}{n + 1}\right) + \cdots.$$

The fourth partial sum, S_4, for this series is computed as follows:

$$\begin{array}{c} 1 - \cancel{\frac{1}{2}} \\ \cancel{\frac{1}{2}} - \cancel{\frac{1}{3}} \\ \cancel{\frac{1}{3}} - \cancel{\frac{1}{4}} \\ \cancel{\frac{1}{4}} - \frac{1}{5} \\ \hline 1 - \frac{1}{5} = S_4 \end{array}$$

Notice that almost every term in the sum cancels with some other term. Viewed this way, the partial sum S_n collapses, like a telescope, from

$$S_n = \left(1 - \frac{1}{2}\right) + \left(\frac{1}{2} - \frac{1}{3}\right) + \cdots + \left(\frac{1}{n-1} - \frac{1}{n}\right) + \left(\frac{1}{n} - \frac{1}{n+1}\right)$$

to

$$S_n = 1 - \frac{1}{n+1}.$$

As n increases, the value of $\frac{1}{n+1}$ gets closer and closer to 0, and thus S_n gets closer and closer to 1. We conclude that the sum of the original series is 1.

EXAMPLE 6

Find the sum of the telescoping series

$$\left(\frac{1}{3} - \frac{1}{5}\right) + \left(\frac{1}{4} - \frac{1}{6}\right) + \cdots + \left(\frac{1}{n+2} - \frac{1}{n+4}\right) + \cdots.$$

Solution

$$S_n = \left(\frac{1}{3} - \frac{1}{5}\right) + \left(\frac{1}{4} - \frac{1}{6}\right) + \left(\frac{1}{5} - \frac{1}{7}\right) + \left(\frac{1}{6} - \frac{1}{8}\right) + \left(\frac{1}{7} - \frac{1}{9}\right) + \cdots$$
$$+ \left(\frac{1}{n} - \frac{1}{n+2}\right) + \left(\frac{1}{n+1} - \frac{1}{n+3}\right) + \left(\frac{1}{n+2} - \frac{1}{n+4}\right)$$

$$S_n = \frac{1}{3} + \frac{1}{4} - \frac{1}{n+3} - \frac{1}{n+4}$$ Note $\frac{1}{n+3}$ and $\frac{1}{n+4}$ approach zero as n increases.

Thus, $S = \frac{1}{3} + \frac{1}{4} = \frac{7}{12}$.

$$1 - \frac{1}{3} + \frac{1}{5} - \frac{1}{7} + \frac{1}{9} - \cdots = \frac{\pi}{4}$$

$$1 + \frac{1}{2} + \frac{1}{3} + \frac{1}{4} + \frac{1}{5} + \frac{1}{6} + \cdots$$

There are many other types of series, but their sums (if they exist) are not always easy to find. Some of the results are rather surprising. For example, the sum of the first series at the left is equal to $\frac{\pi}{4}$, whereas the second series (the *harmonic* series) has no sum.

EXAMPLE 7

A snowflake curve is constructed as follows.

Start with an equilateral triangle, each of whose sides has length 1 (Stage 1).

Let the middle third of each of the three sides form the base of a new equilateral triangle. Thus the new boundary curve is a polygon with twelve edges (Stage 2).

Let the middle third of each of the twelve edges form the base for a new equilateral triangle. Thus the new boundary curve is a polygon with 48 edges (Stage 3).

Continue this process indefinitely. What is the area of the resulting "infinite snowflake" (Stage n)?

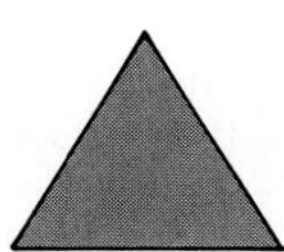
Stage 1

Stage 2

Stage 3

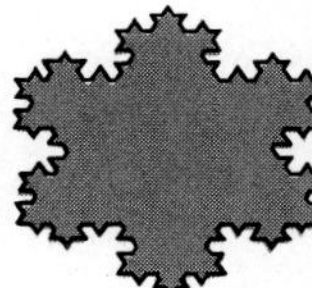
. . .Stage n

Solution

- ANALYZE We are trying to calculate the area of a region whose boundary is determined through an infinite number of stages. This process reminds us of what happens when an infinite number of terms in a geometric series "add up" to a finite sum. Although the triangles being added on at each stage get smaller and smaller, we must determine whether the sum of the areas is in fact finite, and, if so, what is the value of the sum. To organize our work, it is best to make a chart.
- ORGANIZE Before we construct the chart, let us make a number of observations.

  First, the area of an equilateral triangle of sides s is $s^2 \cdot \frac{\sqrt{3}}{4}$.

  Second, the number of edges at a given stage is 4 times the number of edges at the previous stage. (Why?)

Third, the number of edges at a given stage tells us the number of new triangles there will be at the next stage. (Why?) This number is important in calculating the total additional area accounted for by the new triangles at each stage.

Make a chart

Stage	Number of Edges	Number of New Triangles	Length of Side of New Triangles	Total Area of New Triangles
1	3	1	1	$(1)(1)^2\frac{\sqrt{3}}{4} = \frac{\sqrt{3}}{4}$
2	12	3	$\frac{1}{3}$	$(3)\left(\frac{1}{3}\right)^2\frac{\sqrt{3}}{4} = \frac{1}{3}\cdot\frac{\sqrt{3}}{4}$
3	48	12	$\frac{1}{9}$	$(12)\left(\frac{1}{9}\right)^2\frac{\sqrt{3}}{4} = \frac{4}{3^3}\cdot\frac{\sqrt{3}}{4}$
4	192	48	$\frac{1}{27}$	$(48)\left(\frac{1}{27}\right)^2\frac{\sqrt{3}}{4} = \frac{4^2}{3^5}\cdot\frac{\sqrt{3}}{4}$
5	768	192	$\frac{1}{81}$	$(192)\left(\frac{1}{81}\right)^2\frac{\sqrt{3}}{4} = \frac{4^3}{3^7}\cdot\frac{\sqrt{3}}{4}$

■ MODEL At this point we must observe the pattern and take an informed leap. Consider the area of the snowflake at the end of Stage 5:

$$A_5 = \frac{\sqrt{3}}{4} + \frac{1}{3}\cdot\frac{\sqrt{3}}{4} + \frac{4}{3^3}\cdot\frac{\sqrt{3}}{4} + \frac{4^2}{3^5}\cdot\frac{\sqrt{3}}{4} + \frac{4^3}{3^7}\cdot\frac{\sqrt{3}}{4}.$$

Rewriting, we get

$$A_5 = \frac{\sqrt{3}}{4} + \frac{\sqrt{3}}{4}\left(\frac{1}{3} + \frac{4}{3^3} + \frac{4^2}{3^5} + \frac{4^3}{3^7}\right).$$

If we were to continue adding little triangles indefinitely to the boundary of our snowflake, the area of the snowflake could be represented as

$$A = \frac{\sqrt{3}}{4} + \frac{\sqrt{3}}{4}\left(\frac{1}{3} + \frac{4}{3^3} + \frac{4^2}{3^5} + \frac{4^3}{3^7} + \cdots\right),$$

where the expression in parentheses is an infinite geometric series with first term $\frac{1}{3}$, and common ratio $\frac{4}{3^2} = \frac{4}{9}$. So the sum is $\frac{1/3}{1 - 4/9} = \frac{3}{5}$. (Recall the formula for the sum of a geometric series $S = \frac{a}{1 - r}$.) Thus,

$$A = \frac{\sqrt{3}}{4} + \frac{\sqrt{3}}{4}\left(\frac{3}{5}\right) = \frac{8\sqrt{3}}{20} = \frac{2\sqrt{3}}{5}.$$

(Note that the Model in this solution is the expression for the area, which we constructed by observing a pattern, and by using our knowledge of infinite geometric series.)

■ ANSWER The area of this infinite snowflake is $\frac{2\sqrt{3}}{5}$.

EXERCISE SET 10.3

In Exercises 1–8, rewrite the sum without using sigma notation; then compute the sum.

1. $\sum_{n=1}^{5} (3n - 2)$ **2.** $\sum_{n=1}^{6} (9 - 2n)$

3. $\sum_{n=2}^{4} (17 - n^2)$ **4.** $\sum_{n=3}^{5} (7n - n^2)$

5. $\sum_{n=1}^{4} 3$ **6.** $\sum_{n=2}^{5} (6 - n)$

7. $\sum_{n=1}^{5} (-1)^n(n - 1)$ **8.** $\sum_{n=3}^{6} 2^n n$

In Exercises 9–16, use sigma notation to rewrite the sum.

9. $3 + 4 + 5 + 6$ **10.** $7 + 8 + 9 + 10$

11. $1 + 4 + 9 + 16 + 25$ **12.** $1 + 8 + 27 + 64$

13. $2 + 4 + 6 + 8 + 10 + 12$

14. $4 + 8 + 12 + 16 + 20$

15. $-1 + 1 + (-1) + 1$ **16.** $5 + 5 + 5 + 5$

In Exercises 17–26, compute the third and sixth partial sums of the series.

17. $1 + 4 + 9 + 16 + \cdots$ **18.** $1 - 8 + 27 - 64 + \cdots$

19. $1 + 4 + 16 + 64 + \cdots$ **20.** $1 + 6 + 36 + 216 + \cdots$

21. $1 - \frac{1}{3} + \frac{1}{9} - \frac{1}{27} + \cdots$ **22.** $1 + \frac{2}{5} + \frac{4}{25} + \frac{8}{125} + \cdots$

23. $\sum_{n=1}^{\infty} 5\left(\frac{2}{3}\right)^{n-1}$ **24.** $\sum_{n=1}^{\infty} 7\left(\frac{3}{5}\right)^{n-1}$

25. $\sum_{n=1}^{\infty} (2^n - n)$ **26.** $\sum_{n=1}^{\infty} (n^2 - 8)$

In Exercises 27–38, find the sum, if it exists, of the series.

27. $25 - 10 + 4 - \frac{8}{5} + \cdots$

28. $64 + 48 + 36 + 27 + \cdots$

29. $27 + 36 + 48 + 64 + \cdots$

30. $1 - \frac{3}{2} + \frac{9}{4} - \frac{27}{8} + \cdots$

31. $\sum_{n=1}^{\infty} 4\left(-\frac{2}{5}\right)^{n-1}$ **32.** $\sum_{n=1}^{\infty} 5\left(-\frac{6}{5}\right)^{n-1}$

33. $\sum_{n=1}^{\infty} 6\left(\frac{4}{3}\right)^{n-1}$ **34.** $\sum_{n=1}^{\infty} 7\left(\frac{1}{3}\right)^{n-1}$

35. $\left(\frac{1}{2} - \frac{1}{5}\right) + \left(\frac{1}{3} - \frac{1}{6}\right) + \left(\frac{1}{4} - \frac{1}{7}\right) + \left(\frac{1}{5} - \frac{1}{8}\right) + \cdots$

36. $\left(\frac{1}{4} - \frac{1}{8}\right) + \left(\frac{1}{5} - \frac{1}{9}\right) + \left(\frac{1}{6} - \frac{1}{10}\right) + \left(\frac{1}{7} - \frac{1}{11}\right) + \cdots$

37. $\sum_{n=1}^{\infty} \left(\frac{1}{n + 2} - \frac{1}{n}\right)$ **38.** $\sum_{n=1}^{\infty} \left(\frac{1}{n + 3} - \frac{1}{n}\right)$

In Exercises 39–42, determine the sum of the given geometric series.

39. $3.2 + 1.6 + 0.8 + 0.4 + \cdots$

40. $2 - 0.5 + 0.125 - 0.03125 + \cdots$

41. $\sum_{k=3}^{\infty} (1.63)(0.3)^k$ (approximate the sum to the nearest thousandth)

42. $\sum_{k=2}^{\infty} (2.8)(0.25)^k$ (approximate the sum to the nearest thousandth)

In Exercises 43–50, find the fraction represented by the decimal expansion.

43. $0.215\overline{1}$ **44.** $0.568\overline{8}$ **45.** $0.437\overline{171}$

46. $0.236\overline{262}$ **47.** $0.\overline{132}132$ **48.** $0.\overline{395}395$

49. $0.003\overline{232}$ **50.** $0.0\overline{909}$

Superset

51. Find a geometric series whose first term is 4 and whose sum is 3.

52. Find a geometric series whose second term is 2 and whose sum is 9.

In Exercises 53–58 find the first four terms and the general term of the series whose nth partial sum is given.

53. $S_n = 2^n - 1$

54. $S_n = n^2 + 2n$

55. $S_n = 2n^2 + n$

56. $S_n = \frac{1}{2}(3^n - 1)$

57. $S_n = n^2 - 4n$

58. $S_n = 2n^2 - n$

In Exercises 59–64, use sigma notation to rewrite the series.

59. $4 + 7 + 10 + 13 + 16 + \cdots$

60. $3 + 5 + 7 + 9 + 11 + \cdots$

61. $2 + 5 + 10 + 17 + 26 + \cdots$

62. $3 + 5 + 9 + 17 + 33 + \cdots$

63. $2 + 8 + 26 + 80 + \cdots$

64. $3 + 10 + 29 + 66 + 127 + \cdots$

In Exercises 65–66, find the sum of the series. (*Hint*: write each term of the series as the difference of two fractions.)

65. $\frac{1}{2\cdot 3} + \frac{1}{3\cdot 4} + \frac{1}{4\cdot 5} + \frac{1}{5\cdot 6} + \frac{1}{6\cdot 7} + \cdots$

66. $\frac{1}{3\cdot 5} + \frac{1}{4\cdot 6} + \frac{1}{5\cdot 7} + \frac{1}{6\cdot 8} + \frac{1}{7\cdot 9} + \cdots$

67. See Exercise 41, Section 10.2. What is the total distance (up and down) the ball travels?

68. See Exercise 42, Section 10.2. Use a geometric series to determine whether the pool is "eventually empty."

In Exercises 69–72, use a calculator to solve the problem.

69. Suppose a wealthy friend agrees to pay you a penny on January 1, two cents on January 2, four cents on January 3, and so forth, doubling on any given day the amount given on the previous day. How much money will you receive during the month of January?

70. (Refer to Exercise 69) How much will you receive on January 21?

71. (Refer to Exercise 69) How much will you receive during the third week of January (the 15th through the 21st)?

72. (Refer to Exercise 69) Determine the first day on which the daily gift exceeds $1000 (use logarithms).

10.4 Mathematical Induction

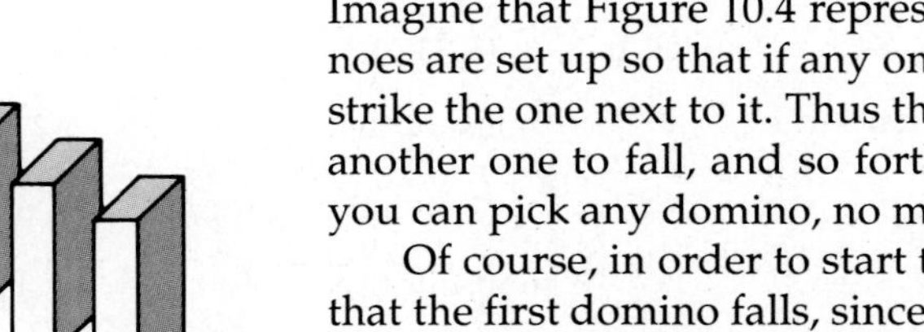

Figure 10.4

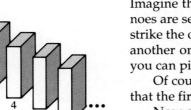

Imagine that Figure 10.4 represents an infinite row of dominoes. The dominoes are set up so that if any one of them falls to the right, it will necessarily strike the one next to it. Thus the next domino falls, which in turn causes yet another one to fall, and so forth, forever. Once the process is set in motion you can pick any domino, no matter how far out, and eventually it will fall.

Of course, in order to start the row of dominoes falling, we must specify that the first domino falls, since it has no predecessor to strike it.

Now consider the following two statements:

1. We know that the first domino falls.
2. If we know that any domino falls, then we are sure that the one following it also falls.

Taken together, these two statements assure us that *all* the dominoes will fall; i.e., any domino you pick, no matter how far out in the row, will eventually fall.

These two statements illustrate a method known as *mathematical induction*. The method is very useful for proving statements that we claim are true for every positive integer. For example, using the method of mathematical induction, we can prove that the following statement is true for all positive integers n:

$$1 + 2 + 3 + \cdots + n = \frac{n(n + 1)}{2}.$$

Principle of Mathematical Induction

A statement is true for all positive integers n if the following two requirements are met.

1. The statement is true for

$$n = 1$$

(the first domino falls);

2. If you assume the statement is true for $n = k$, then you can prove that the statement is true for

$$n = k + 1$$

(if any domino falls, the one following it also falls).

EXAMPLE 1

Prove the following by mathematical induction: for all positive integers n,

$$1 + 2 + 3 + \cdots + n = \frac{n(n + 1)}{2}.$$

Solution

Requirement 1
When $n = 1$, the given statement becomes

$$1 = \frac{1(1 + 1)}{2},$$

which is true. Thus, Requirement 1 is met.

Requirement 2
Assumption: the given statement is true for $n = k$; that is, we assume

$$1 + 2 + 3 + \cdots + k = \frac{k(k + 1)}{2}.$$

To be proved: the given statement is true for $n = k + 1$; that is, we must prove that

$$1 + 2 + 3 + \cdots + k + (k + 1) = \frac{(k + 1)(k + 2)}{2}.$$

Proof:

$$1 + 2 + 3 + \cdots + k = \frac{k(k + 1)}{2}$$

Begin by stating the assumption.

$$1 + 2 + 3 + \cdots + k + (k + 1) = \frac{k(k + 1)}{2} + (k + 1)$$

Add the $(k + 1)$st term of the sum to both sides of the equation and simplify.

$$= \frac{k^2 + k}{2} + \frac{2(k + 1)}{2}$$

$$= \frac{k^2 + 3k + 2}{2}$$

$$1 + 2 + 3 + \cdots + k + (k + 1) = \frac{(k + 1)(k + 2)}{2}$$

Thus, Requirement 2 is met. Therefore, by the Principle of Mathematical Induction, the given statement is true for all positive integers. ∎

Note that Requirement 2 does not say that the statement is true for $n = k$. Rather, it says that *if* the statement is true for $n = k$, then we can conclude that the statement is also true for $n = k + 1$. Thus, Requirement 2 does not say that the kth domino has fallen. It does say that if we know that the kth domino has fallen, then we can conclude that the $(k + 1)$st domino has also fallen.

EXAMPLE 2

Prove the following statement: for all positive integers n,

$$\frac{1}{1 \cdot 2} + \frac{1}{2 \cdot 3} + \cdots + \frac{1}{n(n + 1)} = \frac{n}{n + 1}.$$

Solution

Requirement 1
If $n = 1$, the given statement becomes

$$\frac{1}{1 \cdot 2} = \frac{1}{1 + 1},$$

which is true. Thus, Requirement 1 is met.

Requirement 2

Assumption ($n = k$): $\quad \frac{1}{1 \cdot 2} + \frac{1}{2 \cdot 3} + \cdots + \frac{1}{k(k + 1)} = \frac{k}{k + 1}.$

To be proved ($n = k + 1$): $\quad \frac{1}{1 \cdot 2} + \frac{1}{2 \cdot 3} + \cdots + \frac{1}{(k + 1)(k + 2)} = \frac{k + 1}{k + 2}$

Proof:

$$\frac{1}{1\cdot 2}+\frac{1}{2\cdot 3}+\cdots+\frac{1}{k(k+1)}=\frac{k}{k+1}$$

Begin with the assumption. Then add the $(k+1)$st term to both sides.

$$\frac{1}{1\cdot 2}+\cdots+\frac{1}{k(k+1)}+\frac{1}{(k+1)(k+2)}=\frac{k}{k+1}+\frac{1}{(k+1)(k+2)}$$

$$=\frac{k}{k+1}\cdot\frac{k+2}{k+2}+\frac{1}{(k+1)(k+2)}$$

$$=\frac{k^2+2k+1}{(k+1)(k+2)}$$

$$=\frac{\cancel{(k+1)}(k+1)}{\cancel{(k+1)}(k+2)}$$

$$=\frac{k+1}{k+2}$$

Thus, Requirement 2 is met. Therefore, by the Principle of Mathematical Induction, the given statement is true for all positive integers. ■

EXERCISE SET 10.4

In Exercises 1–16, prove the statement by mathematical induction. (Assume $n \geq 1$, unless stated otherwise.)

1. $1+3+5+7+\cdots+(2n-1)=n^2$

2. $2+4+6+8+\cdots+2n=n(n+1)$

3. $2+5+8+11+\cdots+(3n-1)=\dfrac{n(3n+1)}{2}$

4. $1+4+7+10+\cdots+(3n-2)=\dfrac{n(3n-1)}{2}$

5. $1+4+9+\cdots+n^2=\dfrac{n(n+1)(2n+1)}{6}$

6. $1+8+27+64+\cdots+n^3=\dfrac{n^2(n+1)^2}{4}$

7. $1+5+9+\cdots+(4n-3)=n(2n-1)$

8. $7+10+13+\cdots+(4+3n)=\dfrac{n}{2}(11+3n)$

9. $\dfrac{1}{2}+\dfrac{1}{4}+\dfrac{1}{8}+\cdots+\dfrac{1}{2^n}=1-\dfrac{1}{2^n}$

10. $\dfrac{1}{2}-\dfrac{1}{4}-\dfrac{1}{8}-\cdots-\dfrac{1}{2^n}=\dfrac{1}{2^n},\ n\geq 2$

11. $\dfrac{1}{3}+\dfrac{1}{9}+\dfrac{1}{27}+\cdots+\dfrac{1}{3^n}=\dfrac{1}{2}\left(1-\dfrac{1}{3^n}\right)$

12. $\dfrac{1}{2}-\dfrac{1}{4}+\dfrac{1}{8}-\dfrac{1}{16}+\cdots+(-1)^{n-1}\dfrac{1}{2^n}=\dfrac{1}{3}\left(1+(-1)^{n-1}\dfrac{1}{2^n}\right)$

13. $\dfrac{1}{1\cdot 3}+\dfrac{1}{3\cdot 5}+\dfrac{1}{5\cdot 7}+\cdots+\dfrac{1}{(2n-1)(2n+1)}=\dfrac{n}{2n+1}$

14. $\dfrac{1}{1\cdot 4}+\dfrac{1}{4\cdot 7}+\cdots+\dfrac{1}{(3n-2)(3n+1)}=\dfrac{n}{3n+1}$

15. $\dfrac{1}{3}+\dfrac{1}{15}+\dfrac{1}{35}+\cdots+\dfrac{1}{4n^2-1}=\dfrac{n}{2n+1}$

16. $\dfrac{1}{1\cdot 2\cdot 3}+\dfrac{1}{2\cdot 3\cdot 4}+\cdots+\dfrac{1}{n(n+1)(n+2)}=\dfrac{n(n+3)}{4(n+1)(n+2)}$

Superset

In Exercises 17–30, prove the statement by mathematical induction. (Assume $m, n \geq 1$ unless stated otherwise.)

17. $x^m x^n = x^{m+n}$

18. $(ab)^n = a^n b^n$

19. $2n \leq 2^n$

20. $1+2n \leq 3^n$

21. $\frac{1}{\sqrt{1}} + \frac{1}{\sqrt{2}} + \frac{1}{\sqrt{3}} + \cdots + \frac{1}{\sqrt{n}} > \sqrt{n}$

22. $a(b_1 + \cdots + b_n) = ab_1 + \cdots + ab_n$

23. $a + (a + d) + (a + 2d) + \cdots + [a + (n - 1)d] = \frac{n}{2}[2a + (n - 1)d]$

24. $a + ar + ar^2 + \cdots + ar^{n-1} = \frac{a(1 - r^n)}{1 - r}$

25. If $x > 1$, then $x^n > 1$.

26. If $0 < x < 1$, then $0 < x^n < 1$.

27. $n^2 + n$ is a multiple of 2.

28. $n^3 + 2n$ is a multiple of 3.

29. 4 is a factor of $5^n - 1$.

30. 5 is a factor of $6^n - 1$.

31. For $n \geq 3$, the sum of the angles in an n-sided convex polygon is $(n - 2)180°$.

32. For $n \geq 4$, the number of diagonals in an n-sided convex polygon is $\frac{1}{2}n(n - 3)$.

10.5 The Binomial Theorem

Suppose you wished to expand

$$(9x^2 + 6y)^{17},$$

that is, to write it as a sum of terms in x and y. It is the purpose of this section to provide you with a streamlined "formula" for determining such expansions. This formula is contained in a theorem, called the Binomial Theorem, which we will prove at the end of this section.

To motivate the Binomial Theorem, let us write the expansions of the expression $(a + b)^n$ for values of $n = 1, 2, 3, 4$, and 5.

$$\begin{aligned}
n &= 1 & (a + b)^1 &= a + b \\
n &= 2 & (a + b)^2 &= a^2 + 2ab + b^2 \\
n &= 3 & (a + b)^3 &= a^3 + 3a^2b + 3ab^2 + b^3 \\
n &= 4 & (a + b)^4 &= a^4 + 4a^3b + 6a^2b^2 + 4ab^3 + b^4 \\
n &= 5 & (a + b)^5 &= a^5 + 5a^4b + 10a^3b^2 + 10a^2b^3 + 5ab^4 + b^5
\end{aligned}$$

For these five values of n, we observe some patterns in the expansions:

- There are a total of $(n + 1)$ terms in the expansion of $(a + b)^n$.
- The first term is always a^n and the last term is always b^n.
- As you read each expansion from left to right, the exponent of a starts with n and decreases by one in each succeeding term; the exponent of b starts with 0 and increases by one in each succeeding term.
- In each term, the sum of the exponents of a and b is always n.
- In each expansion, the coefficient of the first term, a^n, is 1. Thereafter, the coefficient of any term is: the coefficient of the preceding term times the exponent of a in the preceding term, divided by the number of the preceding term.

Let us illustrate this last observation. In the expansion of $(a + b)^4$, the coefficient of the third term $(6a^2b^2)$ is 6. It is computed as follows:

$$\frac{(\text{coefficient of the second term}) \times (\text{exponent of } a \text{ in the second term})}{\text{number of the preceding term}}$$

$$= \frac{4 \times 3}{2} = 6$$

EXAMPLE 1

Use the observations just listed to guess the expansion for $(a + b)^7$.

Solution

Term:

first	second	third	fourth	fifth	sixth	seventh	eighth
$1a^7$	$7a^6b$	$21a^5b^2$	$35a^4b^3$	$35a^3b^4$	$21a^2b^5$	$7ab^6$	$1b^7$
	▲ $\frac{1 \cdot 7}{1}$	▲ $\frac{7 \cdot 6}{2}$	▲ $\frac{21 \cdot 5}{3}$	▲ $\frac{35 \cdot 4}{4}$	▲ $\frac{35 \cdot 3}{5}$	▲ $\frac{21 \cdot 2}{6}$	▲ $\frac{7 \cdot 1}{7}$

Note that in this case $n = 7$, thus the first term is a^7 and the last term is b^7. For terms after the first, the exponent on a drops by one from term to term, and the exponent on b increases by 1; the sum of the two exponents is always 7. The coefficient of each term after the first is the product of the coefficient and the power of a in the preceding term, divided by the number of the preceding term. Thus we guess that

$$(a + b)^7 = a^7 + 7a^6b + 21a^5b^2 + 35a^4b^3 + 35a^3b^4 + 21a^2b^5 + 7ab^6 + b^7.$$

You should verify by (tedious) multiplication that this guess is, in fact, true.

There is a more common way of describing the coefficients in expansions of $(a + b)^n$. This description involves *factorials*. For any positive integer n, we define ***n* factorial,** written $n!$, as the product of the first n positive integers:

$$n! = n \cdot (n - 1) \cdot (n - 2) \cdot \cdots \cdot 3 \cdot 2 \cdot 1.$$

For example, $6! = 6 \cdot 5 \cdot 4 \cdot 3 \cdot 2 \cdot 1 = 720$, $(7 - 4)! = 3! = 3 \cdot 2 \cdot 1 = 6$, and $1! = 1$. In addition, we define **zero factorial** to be equal to 1, that is, $0! = 1$. We can now define binomial coefficients as follows.

Definition of a Binomial Coefficient

If n is any positive integer, and if k is any integer such that $0 \le k \le n$, the **binomial coefficient** $\binom{n}{k}$ is defined as follows:

$$\binom{n}{k} = \frac{n!}{(n - k)!\, k!}.$$

EXAMPLE 2

Evaluate the following: $\binom{3}{0}$, $\binom{3}{1}$, $\binom{3}{2}$, and $\binom{3}{3}$.

Solution

$$\binom{3}{0} = \frac{3!}{(3-0)!0!} = \frac{3!}{3!0!} = \frac{1}{1} = 1$$
$$\binom{3}{1} = \frac{3!}{(3-1)!1!} = \frac{3!}{2!1!} = \frac{3\cdot 2\cdot 1}{(2\cdot 1)1} = 3$$
$$\binom{3}{2} = \frac{3!}{(3-2)!2!} = \frac{3!}{1!2!} = \frac{3\cdot 2\cdot 1}{(1)\cdot 2\cdot 1} = 3$$
$$\binom{3}{3} = \frac{3!}{(3-3)!3!} = \frac{3!}{0!3!} = \frac{1}{1} = 1$$

Notice that in Example 2, what we have determined are the binomial coefficients in the expansion of $(a + b)^3$: $1a^3 + 3a^2b + 3ab^2 + 1b^3$. You should verify that the coefficients for the expansion of $(a + b)^7$, which we determined in Example 1, can be expressed in terms of our definition of binomial coefficients as follows:

$$(a+b)^7 = \binom{7}{0}a^7 + \binom{7}{1}a^6b + \binom{7}{2}a^5b^2 + \binom{7}{3}a^4b^3 + \binom{7}{4}a^3b^4 + \binom{7}{5}a^2b^5 + \binom{7}{6}ab^6 + \binom{7}{7}b^7$$

Our earlier guesswork, together with the definition of binomial coefficients, leads us to the following theorem. A proof of this theorem, using mathematical induction, appears at the end of this section.

Binomial Theorem

For all numbers a and b, and for all positive integers n,

$$(a+b)^n = \binom{n}{0}a^nb^0 + \binom{n}{1}a^{n-1}b^1 + \binom{n}{2}a^{n-2}b^2 + \cdots + \binom{n}{n-1}a^1b^{n-1} + \binom{n}{n}a^0b^n.$$

Using sigma notation, we write $(a+b)^n = \sum_{k=0}^{n}\binom{n}{k}a^{n-k}b^k$.

EXAMPLE 3

Use the Binomial Theorem to expand

(a) $(2 + t)^5$ (b) $(2x - 4y^2)^3$.

Solution

(a) In the Binomial Theorem take $a = 2$, $b = t$, and $n = 5$.

$$\begin{aligned}(2 + t)^5 &= \binom{5}{0}2^5t^0 + \binom{5}{1}2^4t^1 + \binom{5}{2}2^3t^2 + \binom{5}{3}2^2t^3 + \binom{5}{4}2^1t^4 + \binom{5}{5}2^0t^5\\ &= (1)(32)\cdot 1 + (5)(16)t + (10)(8)t^2 + (10)(4)t^3 + (5)(2)t^4 + (1)(1)t^5\\ &= 32 + 80t + 80t^2 + 40t^3 + 10t^4 + t^5\end{aligned}$$

(b) In the Binomial Theorem take $a = 2x$, $b = -4y^2$, and $n = 3$.

$$\begin{aligned}(2x - 4y^2)^3 &= \binom{3}{0}(2x)^3(-4y^2)^0 + \binom{3}{1}(2x)^2(-4y^2)^1 + \binom{3}{2}(2x)^1(-4y^2)^2 + \binom{3}{3}(2x)^0(-4y^2)^3\\ &= (1)(8x^3)\cdot 1 + (3)(4x^2)(-4y^2) + (3)(2x)(16y^4) + (1)(1)(-64y^6)\\ &= 8x^3 - 48x^2y^2 + 96xy^4 - 64y^6\end{aligned}$$

Note that the mth term in the expansion of $(a + b)^n$ is given by

$$\binom{n}{m - 1}a^{n-(m-1)}b^{m-1}.$$

For example, the fifth term in the expansion of $(a + b)^7$ is $\binom{7}{4}a^3b^4$. As for any term in a binomial expansion, the bottom number in the binomial coefficient (here 4), is the power on b.

EXAMPLE 4

Determine the fourth term in the expansion of $\left(\sqrt{x} - \frac{y}{2}\right)^8$. Assume $x > 0$.

Solution The fourth term in the expansion of $(a + b)^8$ is $\binom{8}{3}a^5b^3$. Here $a = \sqrt{x}$ and $b = -\frac{y}{2}$. Since

$$\binom{8}{3}(\sqrt{x})^5\left(-\frac{y}{2}\right)^3 = \frac{8!}{5!3!}(x^{5/2})\left(-\frac{y}{2}\right)^3 = 56(x^{5/2})\left(-\frac{1}{8}y^3\right) = -7x^{5/2}y^3,$$

the desired term is $-7x^{5/2}y^3$.

Sometimes the binomial coefficients are displayed in a triangular array, known as **Pascal's Triangle.**

$$\begin{array}{ccccccccccc} & & & & 1 & & 1 & & & & \\ & & & 1 & & 2 & & 1 & & & \\ & & 1 & & 3 & & 3 & & 1 & & \\ & 1 & & 4 & & 6 & & 4 & & 1 & \\ 1 & & 5 & & 10 & & 10 & & 5 & & 1 \end{array}$$

$$\begin{array}{ccccccccccc} & & & & \binom{1}{0} & & \binom{1}{1} & & & & \\ & & & \binom{2}{0} & & \binom{2}{1} & & \binom{2}{2} & & & \\ & & \binom{3}{0} & & \binom{3}{1} & & \binom{3}{2} & & \binom{3}{3} & & \\ & \binom{4}{0} & & \binom{4}{1} & & \binom{4}{2} & & \binom{4}{3} & & \binom{4}{4} & \\ \binom{5}{0} & & \binom{5}{1} & & \binom{5}{2} & & \binom{5}{3} & & \binom{5}{4} & & \binom{5}{5} \end{array}$$

Notice that each entry other than the 1's is the sum of the two entries to the right and left of it in the row above. For instance $\binom{4}{1} + \binom{4}{2} = \binom{5}{2}$. In general, $\binom{n}{k-1} + \binom{n}{k} = \binom{n+1}{k}$, a result that we prove next.

EXAMPLE 5

For $1 \leq k \leq n$, verify that $\binom{n}{k-1} + \binom{n}{k} = \binom{n+1}{k}$.

Solution

$$\begin{aligned} \binom{n}{k-1} + \binom{n}{k} &= \frac{n!}{(n-k+1)!(k-1)!} + \frac{n!}{(n-k)!k!} \\ &= \frac{k \cdot n!}{(n-k+1)!k!} + \frac{(n-k+1)n!}{(n-k+1)!k!} \\ &= \frac{n!(k+n-k+1)}{(n-k+1)!k!} = \frac{n!(n+1)}{(n+1-k)!k!} \\ &= \frac{(n+1)!}{(n+1-k)!k!} = \binom{n+1}{k} \end{aligned}$$

We are now in a position to prove the Binomial Theorem:

$$(a+b)^n = \binom{n}{0}a^nb^0 + \binom{n}{1}a^{n-1}b^1 + \cdots + \binom{n}{n-1}a^1b^{n-1} + \binom{n}{n}a^0b^n.$$

Proof: We use mathematical induction, and thus must show that two requirements are met.

Requirement 1: Show that the statement in the theorem is true for $n = 1$. If $n = 1$, the statement becomes $(a+b)^1 = \binom{1}{0}a^1b^0 + \binom{1}{1}a^0b^1$, which is true since the expression on the left simplifies to yield $a + b$.

Requirement 2: Show that if we assume the statement to be true when $n = N$, then we can conclude it is also true when $n = N + 1$. We use our assumption to write

$$\begin{aligned}(a + b)^{N+1} &= (a + b)(a + b)^N \\ &= (a + b)\left\{\binom{N}{0}a^N b^0 + \binom{N}{1}a^{N-1}b^1\right. \\ &\quad \left. + \cdots + \binom{N}{k-1}a^{N-k+1}b^{k-1} + \binom{N}{k}a^{N-k}b^k + \cdots + \binom{N}{N}a^0 b^N.\right\}\end{aligned}$$

Next we carry out the multiplication and collect like terms to write

$$\begin{aligned}(a + b)^{N+1} &= \binom{N}{0}a^{N+1}b^0 + \binom{N}{0}a^N b^1 + \binom{N}{1}a^N b^1 + \binom{N}{1}a^{N-1}b^2 \\ &\quad + \cdots + \binom{N}{k-1}a^{N-k+1}b^k + \binom{N}{k}a^{N-k+1}b^k + \cdots + \binom{N}{N}a^0 b^{N+1} \\ &= \binom{N}{0}a^{N+1}b^0 + \cdots + \left[\binom{N}{k-1} + \binom{N}{k}\right]a^{N-k+1}b^k + \cdots + \binom{N}{N}a^0 b^{N+1}.\end{aligned}$$

By virtue of our work in Example 5, we can rewrite this last statement as

$$(a + b)^{N+1} = \binom{N}{0}a^{N+1}b^0 + \cdots + \binom{N+1}{k}a^{N+1-k}b^k + \cdots + \binom{N}{N}a^0 b^{N+1}.$$

We can rewrite the first and last coefficients since

$$\binom{N}{0} = \binom{N+1}{0} \text{ and } \binom{N}{N} = \binom{N+1}{N+1}.$$

Thus,

$$\begin{aligned}(a + b)^{N+1} &= \binom{N+1}{0}a^{N+1}b^0 + \cdots \\ &\quad + \binom{N+1}{k}a^{N+1-k}b^k + \cdots + \binom{N+1}{N+1}a^0 b^{N+1},\end{aligned}$$

and Requirement 2 is met, thus completing the proof. ■

EXERCISE SET 10.5

In Exercises 1–14, evaluate the given expression.

1. $3!$ **2.** $(6 - 2)!$ **3.** $(5 - 3)!$ **4.** $8!$

5. $0!$ **6.** $(6 - 4)!\,4!$ **7.** $(5 - 2)!\,3!$ **8.** $(2 - 2)!$

9. $\binom{4}{2}$ **10.** $\binom{5}{4}$ **11.** $\binom{7}{6}$

12. $\binom{6}{3}$ **13.** $\binom{8}{2}$ **14.** $\binom{7}{3}$

In Exercises 15–30, use the Binomial Theorem to expand the given expression.

15. $(x + y)^6$ **16.** $(x + y)^8$ **17.** $(a - b)^7$

18. $(a - b)^6$ **19.** $(3 + 2p)^4$ **20.** $(2 + 5s)^3$

21. $(2 + 3p)^4$ **22.** $(5 + 2s)^3$ **23.** $\left(\frac{t}{3} - 1\right)^3$

24. $\left(\frac{a}{2} - 1\right)^4$ **25.** $\left(z + \frac{1}{z}\right)^5$ **26.** $\left(p - \frac{1}{p}\right)^4$

27. $(1 - \sqrt{x})^6$ **28.** $(1 + \sqrt{t})^5$ **29.** $(s^{1/3} + s^{2/3})^3$

30. $(p^{1/4} + p^{3/4})^4$

In Exercises 31–36, use the Binomial Theorem to write the first four terms in the expansion of the given expression.

31. $(1 + x)^{12}$ **32.** $(1 - y)^{10}$ **33.** $(2 - s)^8$

34. $(3 + p)^9$ **35.** $(1 + 2p)^{11}$ **36.** $(1 + 3x)^{12}$

In Exercises 37–44, find the indicated term in the expansion of the given expression.

37. fourth term of $(1 + t)^{20}$

38. fifth term of $(1 - s)^{18}$

39. fifth term of $\left(x - \frac{1}{x}\right)^8$

40. third term of $(2 + y)^{10}$

41. tenth term of $(y + \sqrt{y})^{11}$

42. seventh term of $\left(2t - \frac{1}{2t}\right)^{12}$

43. ninth term of $\left(\frac{p}{3} + 3p\right)^{16}$

44. fourteenth term of $(2y + 1)^{14}$

Superset

45. If P dollars are invested at an interest rate i (expressed as a decimal), and the interest is compounded n times per year, then the amount, A, of money in the account after t years is

$$A = P\left(1 + \frac{i}{n}\right)^{nt}.$$

The *effective rate of interest* (expressed as a decimal) is

$$\left(1 + \frac{i}{n}\right)^n - 1$$

Use the first four terms in the binomial expansion of $\left(1 + \frac{i}{n}\right)^n$ to obtain an approximate expression for the effective rate of interest.

46. Use the first five terms of a binomial expansion to estimate the value after one year of \$1000, invested in a savings account that earns interest at the rate of 6% compounded daily. (*Hint:* In Exercise 45, take $n = 360$.)

47. Show that $\binom{n}{0} + \binom{n}{1} + \binom{n}{2} + \cdots + \binom{n}{n+1} + \binom{n}{n} = 2^n$ for $n \geq 1$. (*Hint:* Use the Binomial Theorem to expand $(1 + x)^n$ and evaluate the two expressions when $x = 1$.)

48. Show that $\binom{n}{0} - \binom{n}{1} + \binom{n}{2} - \binom{n}{3} + \cdots + (-1)^n\binom{n}{n} = 0$ for $n \geq 1$. (*Hint:* See Exercise 47.)

49. Evaluate the quantity $\binom{n}{0} + 2\binom{n}{1} + 4\binom{n}{2} + \cdots + 2^n\binom{n}{n}$ for $n \geq 1$. (*Hint:* See Exercise 47.)

50. Express the sum $\binom{n}{n} + \binom{n+1}{n} + \binom{n+2}{n} + \cdots + \binom{n+k}{n}$ as a single binomial coefficient. (*Hint:* Observe the position of the addends in Pascal's Triangle.)

51. Suppose p is a prime. Show that p divides each of the binomial coefficients $\binom{p}{1}, \binom{p}{2}, \binom{p}{3}, \binom{p}{4}, \ldots, \binom{p}{p-1}$. Show that a comparable statement is not true if p is not prime.

For specific values of x and y, if the value of y is relatively small compared to the value of x, the quantity $(x + y)^n$ can be approximated by adding the first few terms in the expansion. In Exercises 52–57, use the first four terms of the expansion to approximate the given quantity. Then, approximate the quantity by using the y^x key on a calculator, and compare the two estimates.

52. Expand $(1 + 0.01)^8$ to approximate $(1.01)^8$.

53. Expand $(1 + 0.02)^{10}$ to approximate $(1.02)^{10}$.

54. Expand $(1 - 0.02)^{10}$ to approximate $(0.98)^{10}$.

55. Expand $(1 - 0.01)^8$ to approximate $(0.99)^8$.

56. Expand $(2 + 0.001)^7$ to approximate $(2.001)^7$.

57. Expand $(3 + 0.001)^9$ to approximate $(3.001)^9$.

58. There is a rule stating that as long as nx is close to zero, $(1 + x)^n \approx 1 + nx$. Use a calculator to approximate $(1.002)^{15}$, and then use the rule above to approximate the same expression.

59. Repeat Exercise 58 with the expression $(1.007)^{10}$.

60. Factorials are important in certain series approximations of function values. For example, the infinite series

$$\sum_{n=0}^{\infty} \frac{x^n}{n!} = 1 + x + \frac{x^2}{2!} + \frac{x^3}{3!} + \frac{x^4}{4!} + \cdots$$

has e^x as its sum for all real numbers x. Use the first five terms of the series to approximate $e^{0.2}$, $e^{0.5}$, $e^{0.6}$, $e^{0.75}$, and $e^{0.8}$; then use a calculator to approximate these same expressions.

10.6 Limits

When we plot a point, we focus on the value of y for a particular value of x. When we find the *limit* of a function, however, we look at the larger picture; i.e., we determine whether the y-values exhibit any trend as the x-coordinates increase or decrease through an entire set of values.

For example, in the four statements below we characterize the behavior of the y-coordinates of the graph of the rational function

$$y = \frac{x - 1}{x - 2}$$

(shown at the left) as the x-coordinates get arbitrarily large, arbitrarily small, and arbitrarily close to some fixed value.

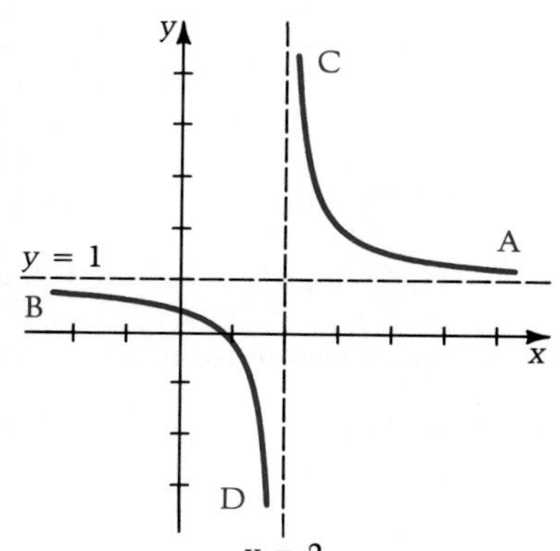

Figure 10.5

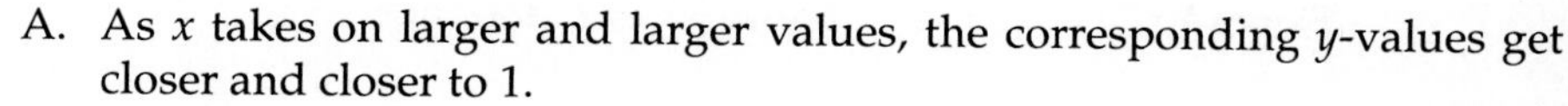

A. As x takes on larger and larger values, the corresponding y-values get closer and closer to 1.

B. As x takes on smaller and smaller (negative) values, the corresponding y-values get closer and closer to 1.

C. As the x-values get closer and closer to 2 from the right, the corresponding y-values increase without bound.

D. As the x-values get closer and closer to 2 from the left, the corresponding y-values decrease without bound.

Now, using limit notation, we restate these four statements. Remember that since the function being discussed is

$$y = \frac{x - 1}{x - 2},$$

the symbols y and $\frac{x - 1}{x - 2}$ are interchangeable.

A. $\lim\limits_{x \to \infty} \frac{x - 1}{x - 2} = 1$ — The limit of $x - 1$ divided by $x - 2$, as x "goes to" positive infinity.

B. $\lim\limits_{x \to -\infty} \frac{x - 1}{x - 2} = 1$ — The limit of $x - 1$ divided by $x - 2$, as x "goes to" negative infinity.

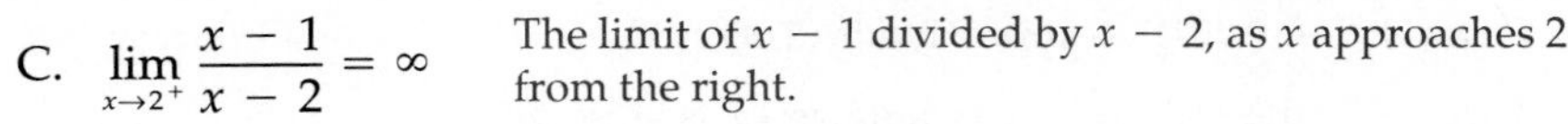

C. $\lim_{x \to 2^+} \dfrac{x-1}{x-2} = \infty$ The limit of $x - 1$ divided by $x - 2$, as x approaches 2 from the right.

D. $\lim_{x \to 2^-} \dfrac{x-1}{x-2} = -\infty$ The limit of $x - 1$ divided by $x - 2$, as x approaches 2 from the left.

The "+" in $x \to 2^+$ and the "−" in $x \to 2^-$ indicate whether the x-values approach from the right or the left. We say that the first two limits are equal to 1 and the last two do not exist.

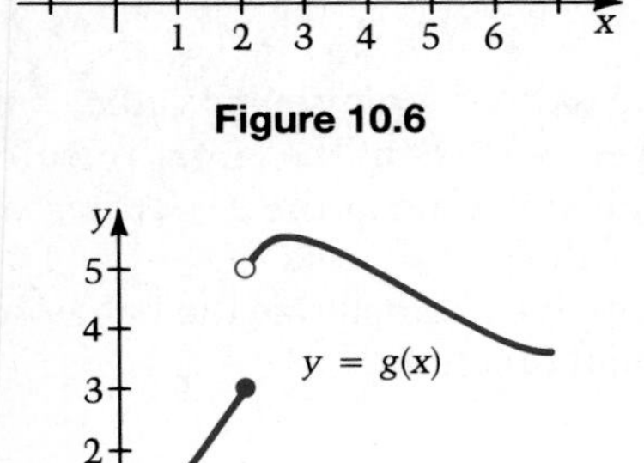

Figure 10.6

Figure 10.7

In Figure 10.6, $f(2) = 3$. But as the x-values approach 2 from both the right and left, the corresponding y-values of the graph approach 5. When determining limits, we are concerned with what happens to $f(x)$ as x *approaches* a certain value, and are not concerned with what happens *at* the x-value. We write:

$$\lim_{x \to 2^+} f(x) = 5 \quad \text{and} \quad \lim_{x \to 2^-} f(x) = 5.$$

Since the limit from the right and limit from the left are equal, we can drop the "+" and "−" and simply write

$$\lim_{x \to 2} f(x) = 5.$$

In Figure 10.7, the graph of $y = g(x)$ jumps at $x = 2$. Here, the limit from the right and the limit from the left are not equal. We write:

$$\lim_{x \to 2^+} g(x) = 5 \quad \text{and} \quad \lim_{x \to 2^-} g(x) = 3.$$

Since these two limits are not equal, we say that the limit of $g(x)$, as x approaches 2, does not exist.

EXAMPLE 1

For the function sketched below, evaluate each limit.

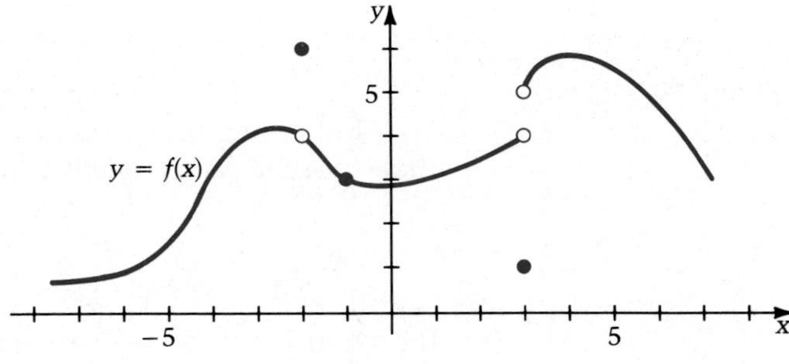

Figure 10.8

(a) $\lim_{x \to -2} f(x)$ (b) $\lim_{x \to -1^+} f(x)$ (c) $\lim_{x \to 3^+} f(x)$ (d) $\lim_{x \to 3^-} f(x)$ (e) $\lim_{x \to 3} f(x)$

Solution

(a) Since $\lim_{x\to -2^+} f(x) = \lim_{x\to -2^-} f(x) = 4$, we write $\lim_{x\to -2} f(x) = 4$.

(b) 3

(c) 5

(d) 4

(e) Since $\lim_{x\to 3^+} f(x) \neq \lim_{x\to 3^-} f(x)$, we say $\lim_{x\to 3} f(x)$ does not exist.

EXAMPLE 2

Evaluate $\lim_{x\to\infty} f(x)$ and $\lim_{x\to -\infty} f(x)$ for the functions graphed below.

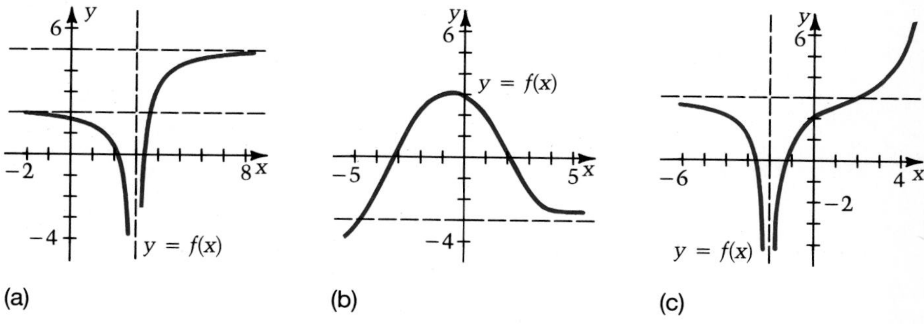

Figure 10.9

Solution

(a) $\lim_{x\to\infty} f(x) = 5$; $\lim_{x\to -\infty} f(x) = 2$.

(b) $\lim_{x\to\infty} f(x) = -3$; $\lim_{x\to -\infty} f(x) = -\infty$; thus, the limit does not exist.

(c) $\lim_{x\to\infty} f(x) = \infty$; thus, the limit does not exist; $\lim_{x\to -\infty} f(x) = 3$.

Careful! Remember that ∞ and $-\infty$ are not numbers. When we say that a limit equals ∞ or $-\infty$, we are simply specifying the reason the limit does not exist.

Since a sequence is a function, we can discuss its limit. Figures 10.10 and 10.11 on the next page suggest that sometimes the limit of a sequence can be found by determining the limit of the corresponding function of x as $x \to \infty$.

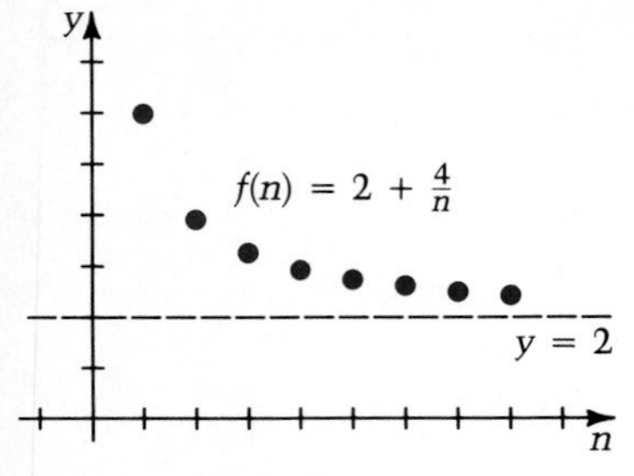

Figure 10.10

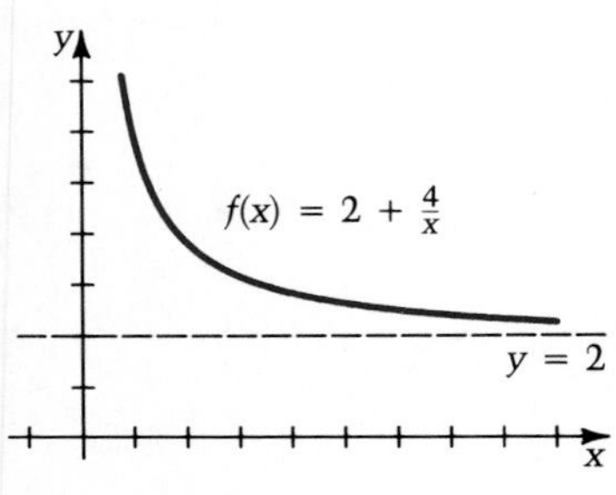

Figure 10.11

The set of discrete dots in Figure 10.10 follows the same pattern as the curve in Figure 10.11. Since

$$\lim_{x \to \infty} \left(2 + \frac{4}{x}\right) = 2,$$

we conclude that the limit of the sequence is 2, and we write

$$\lim_{n \to \infty} \left(2 + \frac{4}{n}\right) = 2.$$

Thus, the sequence whose general term is $2 + \frac{4}{n}$, i.e.,

$$6, 4, 3\frac{1}{3}, 3, \ldots, 2 + \frac{4}{n}, \ldots,$$

has a limit of 2.

Suppose we want to find the limit of a sequence whose general term is a rational expression. In such a case, we need not sketch the graph of the associated function. Instead, we can apply the Second Property of Rational Functions (Section 4.7) to find the limit.

$$a_n = \frac{3n^2 + 5}{2n^2 + 5n + 2}$$

Suppose we are given the general term at the left. By the Second Property of Rational Functions, we know that since the degrees of the numerator and denominator are equal, the line $y = \frac{3}{2}$ is a horizontal asymptote. Thus,

$$\lim_{x \to \infty} \frac{3x^2 + 5}{2x^2 + 5x + 2} = \frac{3}{2}, \quad \text{and so} \quad \lim_{n \to \infty} \frac{3n^2 + 5}{2n^2 + 5n + 2} = \frac{3}{2}.$$

EXAMPLE 3

Evaluate the following limits

(a) $\lim_{n \to \infty} \left(\frac{2n - 1}{3n + 4}\right)^2$ (b) $\lim_{n \to \infty} \frac{2n^2}{4 - n^3}$ (c) $\lim_{n \to \infty} \frac{7n - n^2 + 2}{4n + 3}$

Solution

(a) $\lim_{n \to \infty} \left(\frac{2n - 1}{3n + 4}\right)^2 = \lim_{n \to \infty} \frac{4n^2 - 4n + 1}{9n^2 + 24n + 16} = \frac{4}{9}$ Second Property of Rational Functions (ii) is used.

(b) $\lim_{n \to \infty} \frac{2n^2}{4 - n^3} = 0$ Second Property of Rational Functions (i) is used.

(c) $\lim_{n \to \infty} \frac{7n - n^2 + 2}{4n + 3}$ does not exist Second Property of Rational Functions (iii) is used.

EXERCISE SET 10.6

In Exercises 1–4, evaluate the following limits.

(a) $\lim_{x\to -1} f(x)$ (b) $\lim_{x\to 2} f(x)$ (c) $\lim_{x\to 3} f(x)$

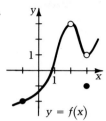

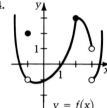

1.

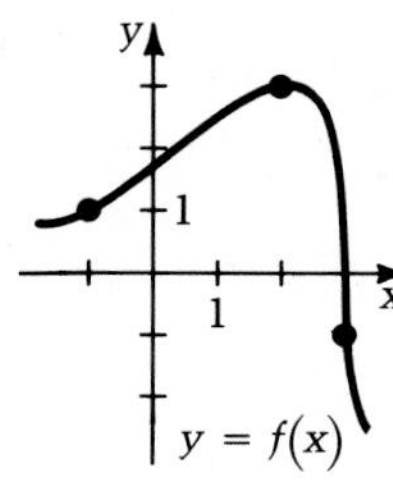

2.

3.

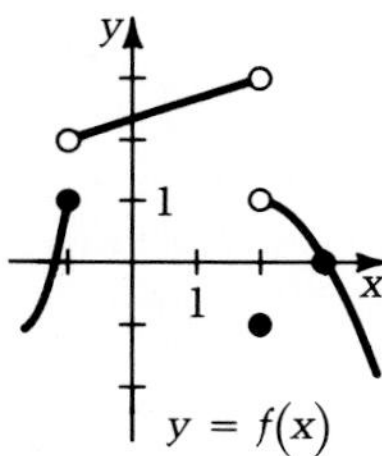

4.

In Exercises 5–8, evaluate the following:

(a) $\lim_{x\to -\infty} f(x)$ (b) $\lim_{x\to \infty} f(x)$

5.

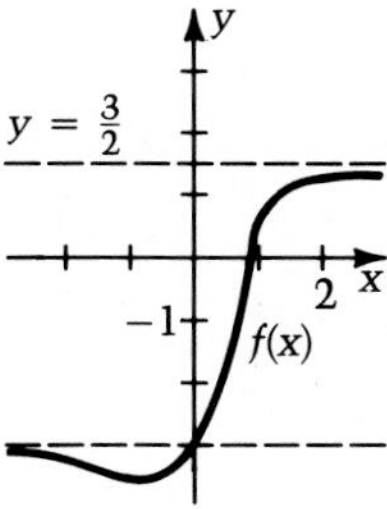

6.

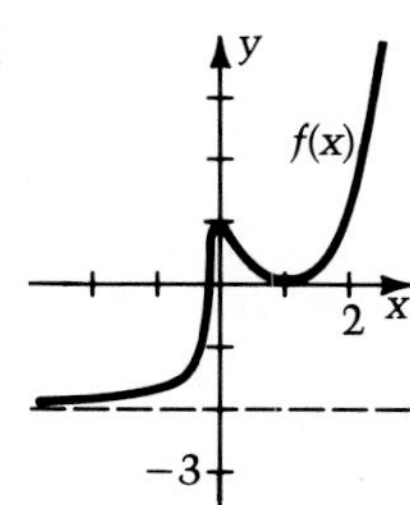

7.

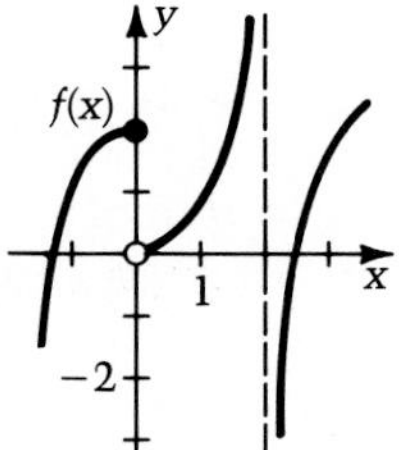

8.

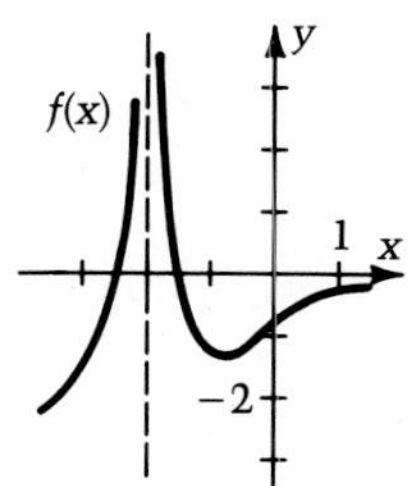

In Exercises 9–16, evaluate the limit.

9. $\lim_{n\to\infty} \dfrac{3n}{2n-1}$

10. $\lim_{n\to\infty} 3^{-n}$

11. $\lim_{n\to\infty} 2^{-n}$

12. $\lim_{n\to\infty} \dfrac{n^3}{7n+3}$

13. $\lim_{n\to\infty} \dfrac{7n^2}{4+5n}$

14. $\lim_{n\to\infty} \dfrac{7n^2}{n^2+4}$

15. $\lim_{n\to\infty} \dfrac{8n^2}{n^4+1}$

16. $\lim_{n\to\infty} \dfrac{5n^2}{2n^5+3}$

Superset

In Exercises 17–24, express the sum of the series in the form

$$\lim_{n\to\infty} S_n$$

where S_n is the nth partial sum of the series.

17. $1 - \frac{1}{2} + \frac{1}{4} - \frac{1}{8} + \cdots$

18. $1 + \frac{1}{3} + \frac{1}{9} + \frac{1}{27} + \cdots$

19. $\sum_{n=1}^{\infty} 6\left(\frac{3}{4}\right)^{n-1}$

20. $\sum_{n=1}^{\infty} 7\left(-\frac{2}{5}\right)^{n-1}$

21. $\left(\frac{1}{2} - \frac{1}{3}\right) + \left(\frac{1}{3} - \frac{1}{4}\right) + \left(\frac{1}{4} - \frac{1}{5}\right) + \cdots$

22. $\left(1 - \frac{1}{3}\right) + \left(\frac{1}{2} - \frac{1}{4}\right) + \left(\frac{1}{3} - \frac{1}{5}\right) + \cdots$

23. $\sum_{n=1}^{\infty} \left(\frac{1}{n+4} - \frac{1}{n+1}\right)$

24. $\sum_{n=1}^{\infty} \left(\frac{1}{n+6} - \frac{1}{n+5}\right)$

25. Let $M(x)$ be the slope of the line that passes through the points $A(0, 0)$ and $B(x, x^2)$ on the curve $y = x^2$.

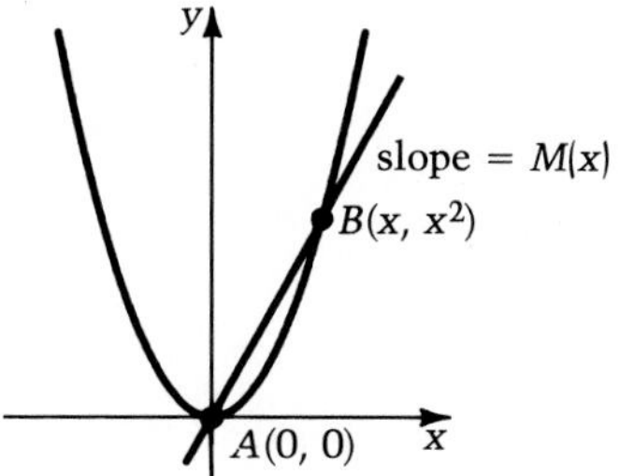

Notice that as $x \to 0$, point B slides along the curve towards point A. Compute $\lim_{x\to 0} M(x)$.

26. Repeat Exercise 25 with the point $A(1, 1)$. (Note: $x \to 1$.)

27. Repeat Exercise 25 with the point $A(3, 9)$. (Note: $x \to 3$.)

28. Repeat Exercise 25 with the point $A(a, a^2)$, for any number a. (Note: $x \to a$.)

29. Repeat Exercise 25 with the curve $y = x^3$. (Note: $x \to a$.)

30. Repeat Exercise 29 with the point $A(a, a^3)$, for any number a. (Note: $x \to a$.)

In Exercises 31–34, for the given function f, evaluate the expression

$$\frac{f(2 + h) - f(2)}{h}$$

for the h-values 0.01, 0.001, and 0.0001. As the h-values get smaller, what limiting value does the expression seem to be approaching?

31. $f(x) = x^2$ 32. $f(x) = x^3$ 33. $f(x) = x^4$

34. $f(x) = x^5$

35. For values of x such that $|x| < 1$, the infinite series $1 + x + x^2 + x^3 + \cdots$ has sum equal to $\dfrac{1}{1 - x}$.

(a) Use the first five terms of the series with $x = 0.2$ to approximate $\dfrac{1}{1 - 0.2}$. What is the difference between the approximation and the exact value of $\dfrac{1}{1 - 0.2}$?

(b) Use the first five terms of the series with $x = 0.5$ to approximate $\dfrac{1}{1 - 0.5}$. What is the difference between the approximation and the exact value of $\dfrac{1}{1 - 0.5}$?

(c) Repeat the process outlined in parts (a) and (b) for the x-values 0.6, 0.75, and 0.8.

(d) Look at the differences between the approximate values and the actual values for the x-values 0.2, 0.5, 0.6, 0.75, and 0.8. What do you conclude?

36. For values of x such that $-1 < x \leq 1$, the infinite series

$$x - \frac{x^2}{2} + \frac{x^3}{3} - \frac{x^4}{4} + \frac{x^5}{5} - \cdots$$

has sum equal to $\ln(1 + x)$. Repeat the process outlined in Exercise 35. That is, for each of the x-values 0.2, 0.5, 0.6, 0.75, and 0.8, use the first five terms of the series above to approximate $\ln(1 + x)$. Then use a calculator to evaluate $\ln(1 + x)$. (Recall that the calculator value is also an approximation.)

37. Evaluate the first ten partial sums of the series

$$\sum_{n=1}^{\infty} \left(\frac{1}{n + 1} - \frac{1}{n + 2} \right).$$

(The sum of the series is 0.5.)

38. Evaluate the first ten partial sums of the series

$$\sum_{n=2}^{\infty} \frac{1}{n^2 - 1}.$$

(The sum of the series is 0.75.)

10.7 Simple Counting, Permutations, and Combinations

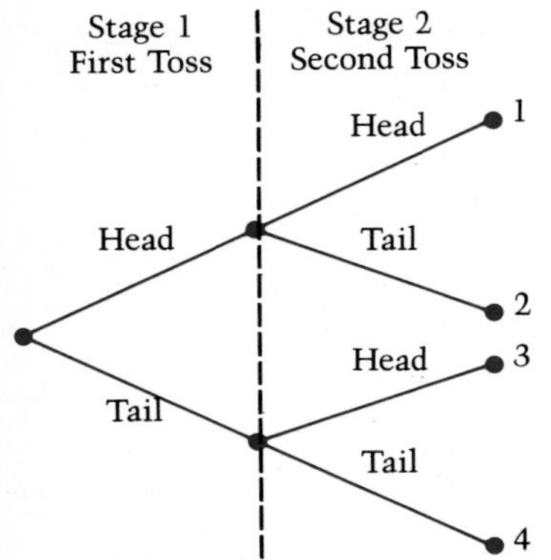

Figure 10.12 Tree diagram of outcomes of tossing a coin twice.

Many real-life problems involve counting. For example, if you are planning a vacation and wish to find the most economical way of traveling from one place to another, you might begin by determining how many different travel routes are available. Or, a college administrator may want to know the number of different ways that students can sign up for one of three lecture sections and one of four laboratory sections of biology. A *tree diagram* can provide a systematic way of listing all possible outcomes in a given process. The tree diagram in Figure 10.12 shows the four possible outcomes of the two-stage process of tossing a coin twice.

Notice that the first toss has two possible results (branches); one is heads and the other is tails. From each of these outcomes, there are two branches or outcomes associated with the second toss of the coin. Altogether there are four different paths along the branches: head-head, head-tail, tail-head, tail-tail. These are the four possible outcomes of tossing a coin twice.

EXAMPLE 1

Suppose you can travel from Durham to Boston by limousine, private car, or bus, and from Boston to New York by bus, car, train, or plane. Using a tree diagram, determine how many ways you can travel from Durham to New York via Boston.

Solution The process involves two stages: traveling from Durham to Boston, and then from Boston to New York.

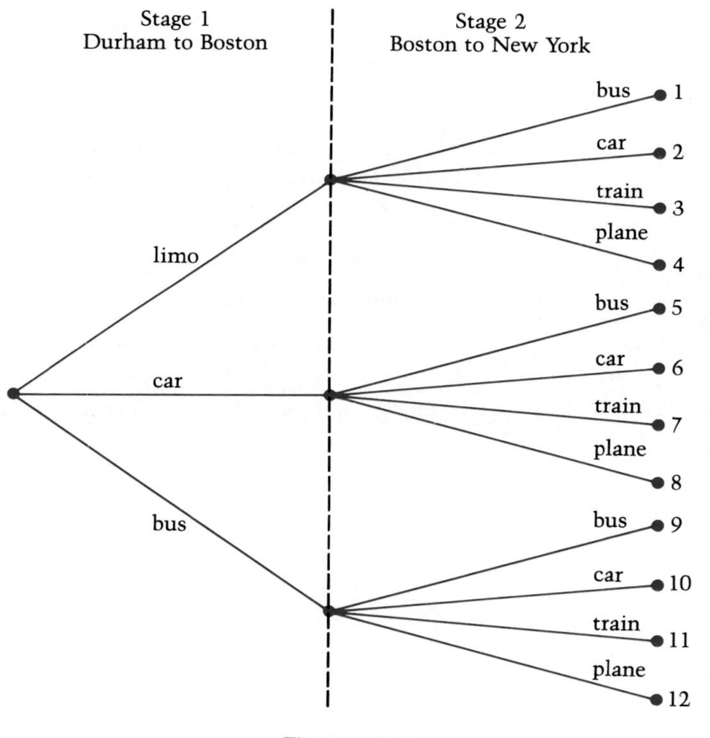

Note that there are 12 different outcomes.

Figure 10.13

Thus, there are 12 different ways to travel from Durham to New York via Boston.

In Example 1, the tree diagram has three branches representing the three possible results in Stage 1, and then, from each of these branches, there are four more branches representing the possible results in Stage 2. Thus, altogether there are $3 \cdot 4$, or 12 paths along the branches of the tree diagram, i.e., there are 12 possible ways. This suggests the following principle.

Basic Principle of Counting

If a process consists of two stages such that the first can be done in m possible ways and the second in n possible ways, then the entire process can be done in $m \cdot n$ ways.

EXAMPLE 2

The owner of an auto dealership wishes to appoint a two-person committee to consider health insurance benefits. The committee will consist of one individual selected from among 22 mechanics and the other from among 13 salespersons. How many possible committees can be formed?

Solution The owner must choose one individual from among 22 mechanics (Stage 1) and another from among 13 salespersons (Stage 2). The number of possible outcomes is

$$22 \cdot 13 = 286.$$

Thus, there are 286 different two-person committees possible.

The Basic Principle of Counting can be generalized to situations in which there are more than two stages.

EXAMPLE 3

A photographer is arranging four people around a sports car for an advertising photo. There will be one person leaning on the front fender, one sitting on the roof, one sitting in the driver's seat, and one sitting on the rear bumper. How many different arrangements are possible?

Solution Think of the four positions as stages in a process:

Stage 1 front fender		Stage 2 roof		Stage 3 driver's seat		Stage 4 rear bumper
$\square$	$\cdot$	$\square$	$\cdot$	$\square$	$\cdot$	$\square$

First, the photographer must position one of the four people at the front fender (Stage 1). Next, the photographer must pick one of the three remaining people to sit on the roof (Stage 2). Then, one of the two remaining people will be chosen to sit in the driver's seat (Stage 3). And, finally, the last person will sit on the rear bumper (Stage 4). Thus, by the Basic Principle of Counting, we know that since

$$\boxed{4} \cdot \boxed{3} \cdot \boxed{2} \cdot \boxed{1} = 24,$$

there are 24 different arrangements of the four people.

In Example 3, we saw that there were 24 different ways of arranging four people for the photo. Obviously, each person could appear only once in each arrangement. The following problem is modeled in precisely the same way as the photograph example:

> How many four-letter codes can be formed using A, B, C, and D, if each letter is used only once in any code?

The answer is 24 since one of the four letters can be in the first position, leaving three possible letters for the second position, then two possible letters for the third position, and finally one letter for the fourth position. Thus, by the Basic Principle of Counting there are $4 \cdot 3 \cdot 2 \cdot 1$, or 24 possible four-letter codes.

When objects cannot be used more than once in the stages of a process, we say that we are counting **without replacement.**

The answer to the code problem above is quite different if we permit the same letter to appear more than once in the four-letter codes (e.g., AABC or BBCB). At each stage we can choose any one of the four letters A, B, C, and D; that is, at each of the four stages, we have four choices. Thus, the number of different four-letter codes in this case is $4 \cdot 4 \cdot 4 \cdot 4$, or 256. In cases like this, we say that we are counting **with replacement.**

EXAMPLE 4

Fishing licenses in a certain state are identified by a code consisting of two letters followed by three digits, in which repeated letters and digits are permitted. How many licenses can be issued if each license has a distinct code?

Solution There are five stages:

Stage 1	Stage 2	Stage 3	Stage 4	Stage 5
Letter	Letter	Digit	Digit	Digit

At Stage 1 we can choose one of 26 letters. At Stage 2 we can still choose one of 26 letters since we are counting with replacement. At stages 3, 4, and 5 we can choose one of the ten digits 0–9. Since

$$26 \cdot 26 \cdot 10 \cdot 10 \cdot 10 = 676{,}000,$$

676,000 licenses, each with a different code, can be issued.

Permutations

Let us refer back to Example 3 once again. We concluded that there were $4 \cdot 3 \cdot 2 \cdot 1 = 24$ distinct ways to arrange the four people in the picture. Recall that the product $4 \cdot 3 \cdot 2 \cdot 1$ can be represented with the special symbol 4!, referred to as *4 factorial*. In general, for any positive integer n, we have defined ***n* factorial** as follows:

$$n! = n \cdot (n - 1) \cdot (n - 2) \cdot \cdots \cdot 3 \cdot 2 \cdot 1$$

Remember also that by definition, $0! = 1$.

It is quite useful to think of 4! as the number of ways of arranging four distinct objects in a row. Each of the 24 different arrangements of the four objects is called a *permutation* of the objects. In general, each distinct arrangement or ordering of a set of objects is called a **permutation.** Thus, ACBD and BDAC are two of the 24 different permutations of the four letters A, B, C, and D. This suggests the following theorem.

Theorem

There are $n!$ permutations of n distinct objects. That is, there are $n!$ different ways of ordering n distinct objects.

For example, suppose you have six diving trophies, and wish to arrange them in a row on the mantel. There are $6! = 720$ different ways to line up the six for display.

Now suppose there were room on the mantel for only four of your six trophies. Let's think of the trophies as A, B, C, D, E, and F. Thus, one display might be the ordering *DBCF*, while another might be *ACFB*. How many different four-trophy lineups are possible, if we have six trophies from which to choose? Remember, order makes a difference here, so *ABCD* and *ACBD* are to count as two different lineups.

Since we wish to determine the number of ways of filling four distinguishable slots without replacement, when we have six objects to choose from, we can apply the Basic Principle of Counting. As shown on the next page, there are 360 different orderings of the six trophies, if we take them four at a time.The number 360 is referred to as *the number of permutations of 6 objects taken 4 at a time,* and is represented by the symbol ${}_6P_4$.

$$ {}_6P_4 = 6 \cdot 5 \cdot 4 \cdot 3 = 360 $$

Note that ${}_6P_4$ is the product of the first 4 factors of $6!$.

Definition of Permutation; ${}_nP_r$

The **number of permutations of n objects taken r at a time,** where $0 \le r \le n$, is denoted ${}_nP_r$ and is given by the following formula:

$$ {}_nP_r = n \cdot (n-1) \cdot (n-2) \cdot \cdots \cdot [n-(r-1)]. $$

Note that the product on the right consists of the first r factors of $n!$.

As an example, let us calculate ${}_7P_3$:

$$ {}_7P_3 = 7 \cdot 6 \cdot 5 = \frac{7 \cdot 6 \cdot 5}{1}\left(\frac{4!}{4!}\right) = \frac{7!}{4!} = \frac{7!}{(7-3)!} = 210. $$

This suggests an alternative way of counting permutations. In general, this alternate definition is written as follows:

$$ {}_nP_r = \frac{n!}{(n-r)!} $$

By the previous definition, ${}_nP_n$, the number of permutations of n objects taken n at a time is

$$ {}_nP_n = \frac{n!}{(n-n)!} = \frac{n!}{0!} = n!. $$

EXAMPLE 5

Evaluate the following: (a) ${}_7P_5$ (b) ${}_7P_2$ (c) ${}_8P_0$ (d) ${}_5P_5$

Solution

(a) ${}_7P_5 = 7 \cdot 6 \cdot 5 \cdot 4 \cdot 3 = 2520$ Note that there are 5 factors.

(b) ${}_7P_2 = 7 \cdot 6 = 42$ Note that there are 2 factors.

(c) ${}_8P_0 = \dfrac{8!}{(8-0)!} = \dfrac{8!}{8!} = 1$

(d) ${}_5P_5 = \dfrac{5!}{(5-5)!} = \dfrac{5!}{0!} = 5! = 120$

EXAMPLE 6

A club has 20 members. How many different ways are there of selecting a president, vice-president, and secretary from the membership?

Solution The problem asks for the number of ways of filling 3 distinguishable slots if we can choose from a pool of 20 people without replacement. Thus, the permutation model is called for here.

$$_{20}P_3 = 20 \cdot 19 \cdot 18 = 6840.$$

There are 6840 ways of selecting the three officers.

Combinations

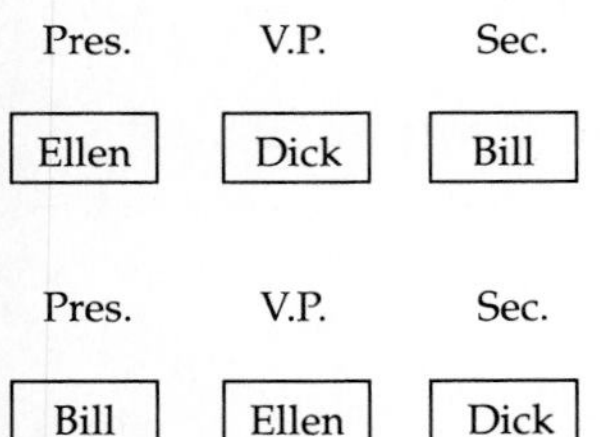

We have seen that $_5P_3$ represents the number of different permutations, or orderings, of 5 objects taken 3 at a time. For example, suppose Joan, Dick, Ellen, Mike, and Bill are five candidates for three offices in a club. We count the two groupings at the left as two distinct permutations, even though the same three people are used to fill the slots. We use the permutation model when *distinguishable slots* are filled without replacement.

Suppose instead that we simply wish to count the number of ways of selecting a committee of 3 persons from among the 5 candidates. Note that within the committee there are no distinctions among the slots. The two groupings below are the same since, in each case, the group forming the committee is the same; we should not count them twice. Thus, we are interested

in the number of different three-object groupings or subsets that can be formed from a set of 5 objects without counting the various reorderings within each grouping. We say that we are counting *the number of combinations of 5 elements taken 3 at a time,* and represent this number with the symbol $_5C_3$, or more commonly, with the symbol $\binom{5}{3}$, represented in spoken English as "5 choose 3." (For those who studied the Binomial Theorem in Section 5 of this chapter, you will recognize that the latter symbol, $\binom{5}{3}$, is also referred to as a *binomial coefficient*.)

We now consider a way of counting combinations by using what we already know about counting permutations. Below we have listed all permutations of the five letters A, B, C, D, and E, taken three at a time. (We know there are $_5P_3$, or 60 such permutations.)

ABC	ABD	ABE	ACD	ACE	ADE	BCD	BCE	BDE	CDE
ACB	ADB	AEB	ADC	AEC	AED	BDC	BEC	BED	CED
BAC	BAD	BAE	CAD	CAE	DAE	CBD	CBE	DBE	DCE
BCA	BDA	BEA	CDA	CEA	DEA	CDB	CEB	DEB	DEC
CAB	DAB	EAB	DAC	EAC	EAD	DBC	EBC	EBD	ECD
CBA	DBA	EBA	DCA	ECA	EDA	DCB	ECB	EDB	EDC

Across the top row we have listed the ten different subsets of the five letters taken three at a time. These are the distinct *combinations* of the five letters taken three at a time: $\binom{5}{3}$. Then down each column we have the $3! = 6$ permutations of the three letters listed at the top of the column: ${}_3P_3$. Multiplying the number of columns, $\binom{5}{3}$, by the number of rows, ${}_3P_3$, we obtain the number of permutations of 5 letters taken 3 at a time, ${}_5P_3$. That is,

number of combinations	$\cdot$	number of permutations of each combination	$=$	total number of permutations
$\binom{5}{3}$	$\cdot$	$3!$	$=$	${}_5P_3$.

In general, we say that the ${}_nP_r$ permutations of n objects taken r at a time are formed by multiplying the number of combinations by $r!$. That is,

$$\binom{n}{r} \cdot r! = {}_nP_r \qquad \text{or} \qquad \binom{n}{r} = \frac{{}_nP_r}{r!}.$$

Definition of Combination; $\binom{n}{r}$

The **number of combinations of n objects taken r at a time,** where $0 \le r \le n$, is denoted by $\binom{n}{r}$ and is given by the following formula:

$$\binom{n}{r} = \frac{{}_nP_r}{r!} = \frac{n!}{(n-r)!\,r!}.$$

The symbol $\binom{n}{r}$ is read, "n choose r."

It is important to remember that we use combinations when the different orderings of the r chosen objects are not to be counted.

EXAMPLE 7

Evaluate each of the following: (a) $\binom{3}{2}$ (b) $\binom{5}{2}$ (c) $\binom{7}{5}$ (d) $\binom{10}{0}$

Solution

(a) $\binom{3}{2} = \frac{3!}{(3-2)!2!} = \frac{3!}{1!2!} = \frac{3 \cdot \cancel{2!}}{1!\cancel{2!}} = 3$ Use: $\binom{n}{r} = \frac{n!}{(n-r)!r!}$.

(b) $\binom{5}{2} = \frac{5!}{(5-2)!2!} = \frac{5!}{3!2!} = \frac{5 \cdot 4 \cdot \cancel{3!}}{\cancel{3!} \cdot 2 \cdot 1} = \frac{20}{2} = 10$

(c) $\binom{7}{5} = \dfrac{7!}{(7-5)!5!} = \dfrac{7!}{2!5!} = \dfrac{7 \cdot 6 \cdot \cancel{5!}}{2 \cdot 1 \cdot \cancel{5!}} = \dfrac{42}{2} = 21$

(d) $\binom{10}{0} = \dfrac{10!}{(10-0)!0!} = \dfrac{\cancel{10!}}{\cancel{10!}0!} = \dfrac{1}{1} = 1$

EXAMPLE 8

A record club allows you to select any four bonus records from a list of twelve. How many different ways are there to choose the four records?

Solution We want to find the number of different four-record groupings and are not interested in the order within each grouping. Thus, we should use the combination model.

$$\binom{12}{4} = \frac{12!}{(12-4)!4!} = \frac{12!}{8!4!} = \frac{12 \cdot 11 \cdot 10 \cdot 9 \cdot \cancel{8!}}{\cancel{8!} \cdot 4 \cdot 3 \cdot 2 \cdot 1} = 495$$

There are 495 different ways to choose the four records.

We now consider some examples where choosing the appropriate model is the key to solving the problem.

EXAMPLE 9

How many different four-digit lottery numbers are there

(a) if no digit can be used more than once?

(b) if digits may be used more than once?

Solution

(a) This problem requires that 4 distinguishable slots be filled without replacement (e.g., 5427 is different from 7254). Thus we should use the permutation model.

$$_{10}P_4 = \frac{10!}{(10-4)!} = \frac{10 \cdot 9 \cdot 8 \cdot 7 \cdot \cancel{6!}}{\cancel{6!}} = 5040$$

There are 5040 different lottery numbers.

(b) Since digits may be repeated, we are counting with replacement, and the permutation model is not correct. By the Basic Principle of Counting, we have

$$10 \cdot 10 \cdot 10 \cdot 10 = 10{,}000.$$

Thus, there are 10,000 different lottery numbers.

EXAMPLE 10

A hospital has decided to choose a committee of 3 physicians and 4 nonphysicians from the hospital's board of directors staff of 7 physicians and 9 nonphysicians. How many such committees are possible?

Solution There are $\binom{7}{3}$ ways of selecting the physicians, and $\binom{9}{4}$ ways of selecting the nonphysicians. Constructing the committee is then a two-stage process.

Stage 1	Stage 2
Physicians	Nonphysicians

$$\binom{7}{3} \cdot \binom{9}{4} = \frac{7!}{(7-3)!3!} \cdot \frac{9!}{(9-4)!4!} = 35 \cdot 126 = 4410$$

There are 4410 possible committees.

EXERCISE SET 10.7

1. A coin is tossed three times. Use a tree diagram to show the set of all possible outcomes.
2. A quiz consists of four true/false questions. Use a tree diagram to illustrate all possible outcomes if a student decides to answer the questions by randomly selecting T or F for each question.
3. Use a tree diagram to show all the two-digit numbers that can be formed from the numbers 1, 2, 3, and 4 (a) with replacement. (b) without replacement.
4. Use a tree diagram to show all the three-letter codes that can be formed from the letters in the word MATH without replacement.
5. You have decided to buy a new car. The model you want to buy is available in four colors (red, blue, silver, and beige) and two transmission styles (standard and automatic), and can be purchased with no radio, an AM radio, or an AM/FM radio. Use a tree diagram to show the number of different choices of automobiles available to you.
6. A coin is tossed five times, and the result is recorded (e.g., HHTHT). How many different outcomes are possible?
7. An instructor makes up a five-question multiple-choice quiz, with choices a, b, c, d, and e. If each question is answered, how many different ways can the quiz be answered?
8. Fifty students are competing in an essay contest. How many ways can a first prize and a second prize be awarded?
9. Each of the eight candidates for the nomination for president is to make a five minute presentation on television. How many ways can the order of presentation be arranged?
10. License plates are issued with three letters followed by three digits. How many can be issued if
    (a) letters and digits cannot be repeated?
    (b) letters and digits can be repeated?
    (c) letters and digits can be repeated and 0 is not used as the first digit?
11. How many (a) seven-digit codes are possible? (b) ten-digit codes are possible if the first digit is neither 0 nor 1?
12. An executive is asked to select one male and one female employee to attend a seminar. If the executive selects randomly from 8 male and 12 female employees, how many different 2-person groups are there?
13. Parking stickers for students and faculty contain 5-digit codes formed by using the digits 1, 2, 3, 4, and 5 without replacement. Student codes must be odd numbers, and faculty codes must be even numbers.
    (a) How many possible student codes are there?
    (b) How many possible faculty codes are there?

14. A summit meeting is being planned for the leaders of five countries. (a) How many days must the meeting last to insure that the flags are flown in all possible arrangements? (b) If the flag for the host country is always flown from the middle of the five flagpoles, how many different arrangements are there?

15. Some people predict that the population of the United States will be 275 million by the year 1999. (a) Will there be enough social security numbers to issue a different one to each person? (Social security numbers consist of nine digits.) (b) How many different social security numbers beginning with the digits 99 can be issued?

16. There are twenty citizens on a committee to study housing for the elderly. Fourteen of them are women and six are men. (a) How many ways can they pick a chairperson and a recorder? (b) If they agree that the chairperson will be a woman, how many ways can a chairperson and a recorder be selected? (c) How many ways can they pick a chairperson and a recorder if both positions must be held by men?

In Exercises 17–22, evaluate the given expression.

17. ${}_7P_4$ **18.** ${}_{12}P_3$ **19.** ${}_{15}P_4$

20. ${}_{11}P_5$ **21.** ${}_4P_0$ **22.** ${}_4P_4$

23. How many different four-letter codes can be formed with the letters W, X, Y, and Z, if each letter is to be used exactly once?

24. You have put eight letters into envelopes and sealed them. You have eight address labels, but have forgotten which label goes on which envelope. How many ways are there of randomly applying the address labels on the envelopes?

25. How many ways can a college student select a major, a first minor, and a second minor, if there are fifteen subject areas from which to choose?

26. How many ways can a president and vice-president be chosen from a class of 324 students?

27. A raffle at a local health club advertises a first, second, third, and fourth prize, and each of the 800 club members receives one ticket. How many different ways can the prizes be awarded?

28. How many ways can a personnel officer fill the posts of office manager, administrative assistant, and executive secretary from a pool of fifty equally qualified candidates?

In Exercises 29–34 evaluate the given expression.

29. $\binom{8}{2}$ **30.** $\binom{9}{3}$ **31.** $\binom{13}{4}$

32. $\binom{10}{1}$ **33.** $\binom{3}{0}$ **34.** $\binom{7}{7}$

35. Three lawyers are to be selected from among the twelve lawyers in a local firm to represent the firm at a conference. How many different groups of three lawyers can be selected?

36. An educational researcher wishes to select a sample of six students from a class of twenty to participate in a study. How many different samples can be selected?

37. How many ways can a committee of five be chosen from a legislative body containing fifteen members?

38. A standard deck of cards contains 52 cards. How many different 5-card poker hands are possible?

39. A decorative lighting fixture contains ten 25-watt lightbulbs in a row. How many ways can exactly three of the bulbs be burned out?

40. A true/false test consists of ten items. If a student answers all ten items, how many ways can the student get exactly seven correct?

41. The letters V, W, X, Y, and Z are used to construct three-letter codes. How many different codes are possible if
(a) no letter may be used more than once?
(b) the letters may be used more than once?
(c) no letter may be used more than once, and each code must begin with a Z?

42. The digits 1, 2, 3, 4, 5, and 6 are used to form four-digit identification numbers. How many different numbers can be formed if
(a) no digit may be repeated?
(b) digits may be repeated?
(c) no digit may be repeated and the resulting number is even?

43. A committee of five students is to be formed from a group consisting of seven undergraduate and four graduate students. How many ways can the committee be formed if
(a) there must be three undergraduates and two graduates on the committee?
(b) there must be four undergraduates and one graduate on the committee?
(c) there must be no graduate students on the committee?

44. A quality control supervisor selects a sample of three digital watches from a lot of eighteen. If the lot contains four defectives, how many ways can the supervisor select a sample with (a) no defectives? (b) exactly one defective? (c) three defectives?

Superset

45. A box contains one black ball and three white balls. Use a tree diagram to illustrate the process in which balls are drawn out without replacement until the black ball is selected.

46. (a) Using the digits in 1986 without replacement, determine how many different numbers can be formed. (Be careful! How many one-digit, two-digit, three-digit, and four-digit numbers are there?) (b) How many are greater than 7000? (c) How many are less than 6000? (d) How many are even numbers?

47. Use a tree diagram to illustrate all distinguishable two-letter codes that can be formed from the letters in the word EGG.

48. Three different numbers are selected at random from among the numbers 1, 2, 3, 4, and 5. How many of the sets of three numbers have (a) a sum greater than nine? (b) a sum less than eight?

49. (a) Use a tree diagram to illustrate the process in which you continue drawing letters from the set A, B, C without replacement until the letter A is selected. (b) How does the process change if you permit replacement?

50. A box contains one black ball, one red ball, and two white balls. Use a tree diagram to illustrate the process in which balls are drawn without replacement until the black ball and the red ball (in either order) have been drawn.

51. The owner of a chain of ten drug stores wishes to visit four of the stores, but has not decided which four to visit. If the order of the visits is important, how many possible ways of making the visits are there?

52. A child has 11 identical white socks and 8 identical blue socks mixed together in a drawer. If the child selects two socks without looking, how many matching pairs are there among the 17 remaining socks?

53. Using the definition on page 625, show that

$$\binom{n}{n} = \binom{n}{0} = 1.$$

54. Using the definition on page 625, show that

$$\binom{n}{r} = \binom{n}{n-r}.$$

55. Suppose a set consists of p objects. If m_1 of them are identical of one type, m_2 of them are identical of a second type, . . . , m_n of them are identical of an nth type, and the rest are distinct, then the number of distinct permutations of the p objects is

$$\frac{p!}{m_1!\, m_2! \cdots m_n!}$$

How many distinct permutations are there of the letters in the word "assess"?

56. (Refer to Exercise 55.) How many distinct permutations are there of the letters in the word "Mississippi"?

57. A baby has five blocks, each bearing a single letter: A, B, D, E, or F. The child randomly selects two of the blocks. How many such pairs contain at least one vowel? (Do not count different orderings of the same pair.)

58. A set contains five elements. How many distinct subsets are there that contain
(a) one element? (b) two elements?
(c) three elements? (d) four elements?

59. A set contains five elements. How many distinct subsets does it have?

60. A set contains n elements. How many distinct subsets does it have?

10.8 Introduction to Probability

Consider the process of tossing a coin three times. Since this is a three-stage process, with two possible results at each stage, the total number of possible outcomes is $2 \cdot 2 \cdot 2 = 8$. These eight possible outcomes are listed below (H = head; T = tail).

$HHH \quad HHT \quad HTH \quad THH \quad TTT \quad TTH \quad THT \quad HTT$

If the coin is fair, then each of these eight possible outcomes is equally likely, and the probability of any one of them, say *HHT*, is $\frac{1}{8}$, representing a "one out of 8" chance of occurring.

We will refer to a process such as tossing a coin three times as an **experiment,** and the set of all possible outcomes of the experiment as the **sample space.** An **event** is then defined as any subset of the sample space. For example, for the coin toss experiment, the subset

$$\{HHT, HTH, THH\}$$

is the event consisting of all outcomes in which there are exactly two heads. The probability of this three-outcome event is $\frac{3}{8}$, according to the following definition.

Definition of the Probability of an Event

Suppose the sample space for an experiment contains n equally likely outcomes, and suppose event E consists of m of these outcomes. Then the **probability of event E,** denoted $P(E)$, is defined as $\dfrac{m}{n}$.

EXAMPLE 1

Three fair coins are tossed. Describe each of the following events in set notation, and determine its probability.

(a) E_1: those outcomes in which all three tosses are the same.

(b) E_2: those outcomes having exactly one tail.

(c) E_3: those outcomes having at least two tails.

(d) E_4: those outcomes having at most two tails.

Solution

(a) $E_1 = \{HHH, TTT\}$;

$$P(E_1) = \frac{\text{number of outcomes in } E_1}{\text{number of outcomes in sample space}} = \frac{2}{8} = \frac{1}{4}$$

(b) $E_2 = \{HHT, HTH, THH\}; \quad P(E_2) = \dfrac{3}{8}$

(c) $E_3 = \{TTH, THT, HTT, TTT\}; \quad P(E_3) = \dfrac{4}{8} = \dfrac{1}{2}$

(d) $E_4 = \{HHH, HHT, HTH, THH, TTH, THT, HTT\}; \quad P(E_4) = \dfrac{7}{8}$

Sometimes it is useful to illustrate a sample space with a diagram. Figure 10.14 depicts all possible outcomes of the experiment in Example 1. The entire sample space is called $\mathscr{S}$, and a point corresponding to each outcome is labeled. Events E_2 and E_4 from Example 1 are circled. Note that E_4 contains E_2.

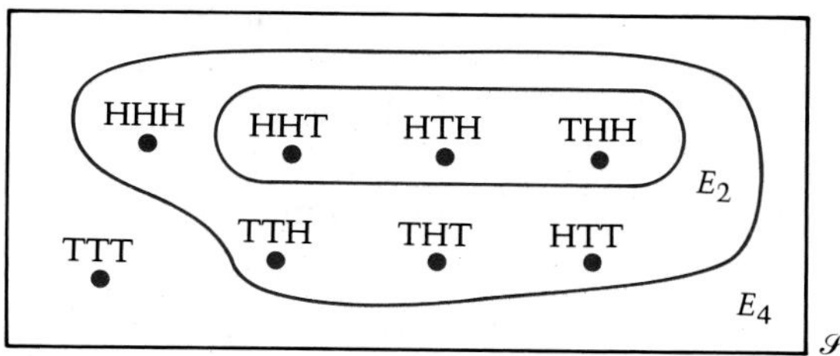

Figure 10.14

It is clear from the diagram that event E_4 contains all the points of the sample space except TTT. For this reason, the set $\{TTT\}$ is called the **complement** of E_4 and is denoted E_4'. Similarly,

$$E_2' = \{HHH, TTT, TTH, THT, HTT\}$$

is the complement of E_2.

We now summarize some of the basic principles of probability.

Basic Principles of Probability

1. The probability of any event E is between 0 and 1 inclusive: $0 \leq P(E) \leq 1$.
2. The probability of the entire sample space $\mathscr{S}$ is 1: $P(\mathscr{S}) = 1$.
3. The probability of the empty event $\emptyset$, i.e., the event containing no outcomes, is 0: $P(\emptyset) = 0$.
4. The probability of E', the complement of the event E, is 1 minus the probability of E: $P(E') = 1 - P(E)$.

Let us now turn our attention to two examples in which we determine the probability of an event by first determining the number of outcomes in the sample space and the number of outcomes in the event.

EXAMPLE 2

An executive claims to have randomly selected a four-person committee from among 8 female and 6 male employees. The resulting committee is composed of 3 men and 1 woman. What is the probability of randomly selecting such a committee?

Solution If the selection process were truly random, all four-person committees would have been equally likely. We must find the probability of randomly selecting a committee having 3 men and 1 woman.

The sample space $\mathcal{S}$ for the process is the collection of all four-person committees. Since there are 14 individuals from which to choose, there are a total of $\binom{14}{4}$ different four-person committees.

The event of interest, E, is the collection of all those four-person committees containing 3 men and 1 woman. There are $\binom{6}{3}$ ways of selecting the three men and $\binom{8}{1}$ ways of selecting the one woman. By the Basic Principle of Counting there are $\binom{6}{3} \cdot \binom{8}{1}$ ways of selecting 3 men and one woman.

$$P(E) = \frac{\binom{6}{3} \cdot \binom{8}{1}}{\binom{14}{4}} = \frac{\frac{6!}{3!3!} \cdot \frac{8!}{7!1!}}{\frac{14!}{10!4!}} = \frac{20 \cdot 8}{1001} \approx 0.16$$

Thus, if the selection process were truly random, the probability of selecting such a committee would be approximately 0.16.

EXAMPLE 3

If a student randomly guesses the answers on a 10-item true/false test, what is the probability of the student scoring 70 or better?

Solution Picture the sample space as the collection of all strings of T's and F's that are 10 letters long. For example, *TTFFFTFFTT* is one possible outcome in the sample space.

Since the process of randomly guessing answers to the 10 questions is made up of 10 stages, with two options at each stage, there are 2^{10}, or 1024, possible outcomes in the sample space. Let E be the event of interest (a score of 70 or better).

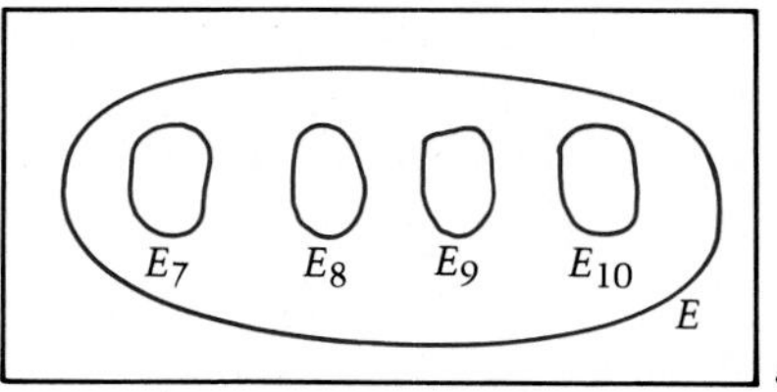

E_7: exactly 7 correct guesses
E_8: exactly 8 correct guesses
E_9: exactly 9 correct guesses
E_{10}: exactly 10 correct guesses.

Figure 10.15

Note that E is composed of the points in E_7, E_8, E_9, and E_{10}. The event E_7 contains those strings of 10 T's and F's in which exactly 7 are correct. Since there are $\binom{10}{7}$ ways of selecting 7 of the 10 to be correct, E_7 contains $\binom{10}{7}$ outcomes. Similarly, E_8 contains $\binom{10}{8}$ outcomes, E_9 contains $\binom{10}{9}$ outcomes, and E_{10} contains $\binom{10}{10}$, or 1 outcome, corresponding to a perfect paper. Since

$$\binom{10}{7} + \binom{10}{8} + \binom{10}{9} + \binom{10}{10} = 120 + 45 + 10 + 1 = 176,$$

E contains 176 outcomes, and

$$P(E) = \frac{176}{1024} \approx 0.17.$$

Thus, the probability of the student scoring 70 or better is approximately 0.17.

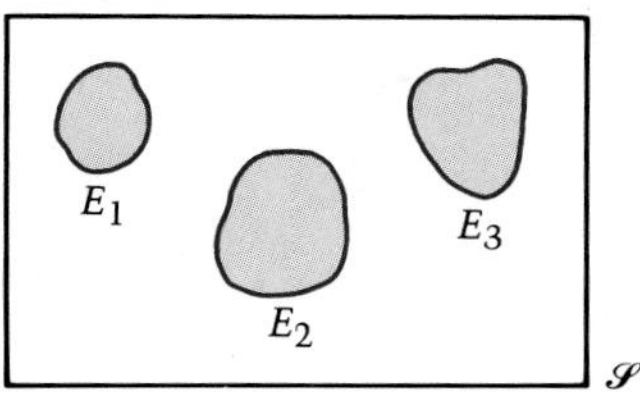

Figure 10.16
$P(E_1 \cup E_2 \cup E_3)$
$= P(E_1) + P(E_2) + P(E_3)$

The events E_7, E_8, E_9, and E_{10} in Example 3 have no outcomes in common. For example E_7 contains only those outcomes with exactly 7 correct guesses, whereas E_8 contains only those outcomes with exactly 8 correct guesses. Such events are said to be **mutually exclusive.** It seems reasonable to assume that if we know the individual probabilities of the mutually exclusive events that compose a larger event E, then we can add these probabilities to obtain the probability of the event E. This is illustrated in Figure 10.16 and suggests the following theorem. (Note that the symbol $\cup$, representing union, is used to indicate that events E_1, E_2, and E_3 are combined into one compound event.)

Theorem

Suppose an event E is composed of n mutually exclusive events $E_1, \ldots, E_n$ with probabilities $P(E_1), \ldots, P(E_n)$, respectively. Then

$$P(E) = P(E_1 \cup \cdots \cup E_n) = P(E_1) + \cdots + P(E_n).$$

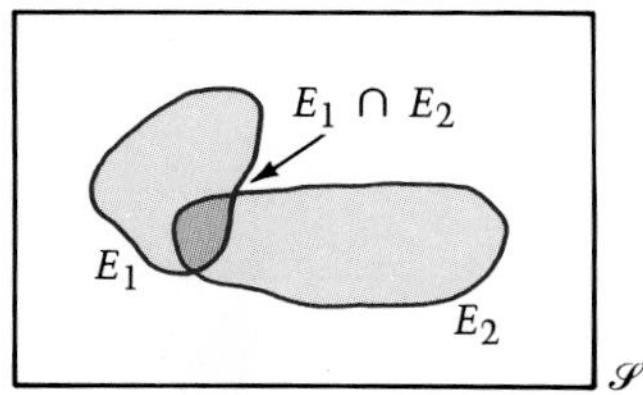

Figure 10.17
$P(E_1 \cup E_2)$
$= P(E_1) + P(E_2) - P(E_1 \cap E_2)$

If two events E_1 and E_2 are not mutually exclusive, then there will be some region of overlap of the two events as illustrated in Figure 10.17. The region of overlap, or intersection, is denoted $E_1 \cap E_2$.

To find the probability of $E_1 \cup E_2$ when E_1 and E_2 are not mutually exclusive, you add the probabilities of the two events, and then subtract the probability of $E_1 \cap E_2$, since it has been counted twice, once as part of E_1 and once as part of E_2.

EXAMPLE 4

An experiment consists of rolling two dice.

(a) Use ordered pairs to depict the sample space.

(b) Indicate the following events in your diagram:

E_1: the sum of the dots on the dice is less than 6,
E_2: the sum of the dots on the dice is 7,
E_3: the two dice show the same number of dots.

(c) Describe the event $E_1 \cup E_2$ in words and determine $P(E_1 \cup E_2)$.

(d) Describe the event $E_1 \cup E_3$ in words and determine $P(E_1 \cup E_3)$.

Solution

(a) and (b) The sample space contains 36 possible outcomes.

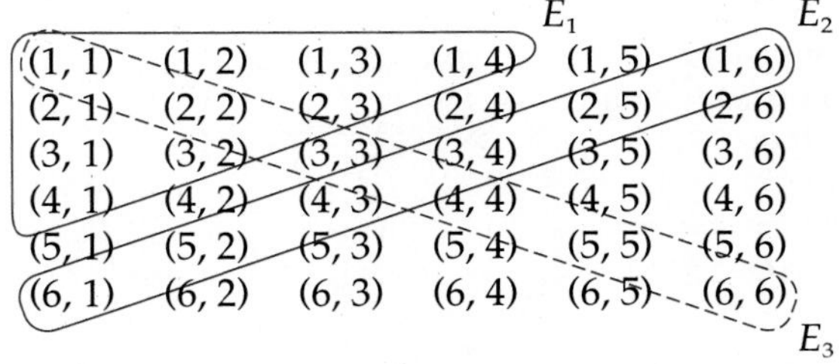

(c) $E_1 \cup E_2$: the sum of the dots is less than 6 or equal to 7. Since E_1 and E_2 are mutually exclusive, we have the following:

$$P(E_1 \cup E_2) = P(E_1) + P(E_2) = \frac{10}{36} + \frac{6}{36} = \frac{16}{36} = \frac{4}{9}$$

(d) $E_1 \cup E_3$: the sum of the dots is less than 6 or the two dice match. Notice that the two events E_1 and E_3 are not mutually exclusive since $E_1 \cap E_3$ contains the ordered pairs (1, 1) and (2, 2).

$$P(E_1 \cup E_3) = P(E_1) + P(E_3) - P(E_1 \cap E_3) = \frac{10}{36} + \frac{6}{36} - \frac{2}{36} = \frac{14}{36} = \frac{7}{18}$$

EXERCISE SET 10.8

Exercises 1–10 refer to the experiment of rolling two fair dice. Describe each of the following events with a set of ordered pairs, and determine the probability of each event.

1. E_1: the sum of the dice is less than five.
2. E_2: the sum of the dice is greater than or equal to eight.
3. E_3: one of the dice is six.
4. E_4: one of the dice is less than three.
5. E_5: both dice are even.
6. E_6: neither die is less than five.
7. E_7: the sum of the dice is one.
8. E_8: the sum of the dice is greater than one.
9. E_9: the sum of the dice is less than four or greater than eleven.
10. E_{10}: the sum of the dice is six, and both dice are the same.

In Exercises 11–16, determine each of the following probabilities if the sample space $\mathcal{S} = \{x_1, x_2, x_3, x_4, x_5\}$, $E_1 = \{x_1, x_2\}$, $E_2 = \{x_1, x_3, x_5\}$, and $E_3 = \{x_2, x_3, x_4\}$. (Assume that all outcomes in $\mathcal{S}$ are equally likely.)

11. $P(E_1)$ **12.** $P(E_2)$ **13.** $P(E_1')$

14. $P(E_3')$ **15.** $P(\mathcal{S})$ **16.** $P(\mathcal{S}')$

17. A bag contains three red cubes and five blue cubes. A pair of cubes is drawn from the bag. What is the probability of drawing
(a) one red cube and one blue cube?
(b) two red cubes?
(c) two blue cubes?

18. A container of twelve digital watches contains three defectives. If a random sample of four watches is drawn from the container, what is the probability of drawing
(a) no defectives?
(b) exactly one defective?
(c) all three defectives?
(d) exactly two defectives?

19. A fair coin is tossed four times. What is the probability of tossing
(a) all heads?
(b) all tails?
(c) one head and three tails?
(d) two heads and two tails?
(e) one tail and three heads?

20. Three cards are drawn from a well-shuffled bridge deck (52 cards). What is the probability of drawing (a) three hearts? (b) three aces? (c) a king and two queens?

21. Suppose a student randomly guesses the answers on a ten-question multiple-choice quiz with choices, a, b, and c for each question. What is the probability of the student scoring 60 or less?

22. (Refer to Exercise 21) What is the probability of the student scoring 80 or better?

In Exercises 23–28, determine each of the following probabilities if the sample space $\mathcal{S} = \{x_1, x_2, x_3, x_4, x_5\}$, $E_1 = \{x_1, x_2\}$, $E_2 = \{x_1, x_3, x_5\}$, and $E_3 = \{x_2, x_3, x_4\}$. (Assume that all outcomes in $\mathcal{S}$ are equally likely.)

23. $P(E_1 \cup E_2)$ **24.** $P(E_1 \cup E_3)$

25. $P(E_2 \cap E_3)$ **26.** $P(E_1 \cap E_2 \cap E_3)$

27. $P(E_2' \cap E_3)$ **28.** $P(E_3' \cap E_2')$

29. An experiment consists of tossing five fair coins.
(a) Use set notation to depict the sample space.
(b) Use set notation to depict each of the following events:
E_1: those outcomes having at least three heads
E_2: those outcomes having at most two heads
E_3: those outcomes in which exactly four of the tosses are the same.
(c) Describe $E_1 \cup E_2$ in words and determine $P(E_1 \cup E_2)$.
(d) Describe $E_1 \cup E_3$ in words and determine $P(E_1 \cup E_3)$.

30. Three-letter codes are formed from the letters A, B, and C with replacement.
(a) Use set notation to depict the sample space.
(b) Use set notation to depict each of the following events:
E_1: those outcomes having exactly two C's
E_2: those outcomes having at least one A
E_3: those outcomes having no C's.
(c) Describe $E_1 \cup E_3$ in words and determine $P(E_1 \cup E_3)$.
(d) Describe $E_2 \cup E_3$ in words and determine $P(E_2 \cup E_3)$.

Superset

In Exercises 31–36, determine what is wrong with the statement.

31. Mike determined that his coin was unbalanced: the probability of tossing a head was 0.75, and the probability of tossing a tail was 0.35.

32. The probability that the instructor's car will break down in the next two weeks is 0.55, and the probability that it will not break down is also 0.55.

33. The probability that the construction company will bid on the new city development project is 1.03.

34. The probability that you will pass or fail the next exam you take is 0.95.

35. The probability that Tina will get a part in the play is 0.3; the probability that Bob will get a part is 0.25. The probability that either Tina or Bob will get a part is 0.6.

36. The probability that the race horse will come in first is 0.15. The probability that the horse will come in first or second is 0.08.

37. Suppose E_1 and E_2 are two mutually exclusive events with $P(E_1) = 0.25$ and $P(E_2) = 0.42$. By shading the appropriate portion of the diagram below, represent each of the following probabilities, and then calculate the probability. (Remember, $P(\mathscr{S}) = 1$.)

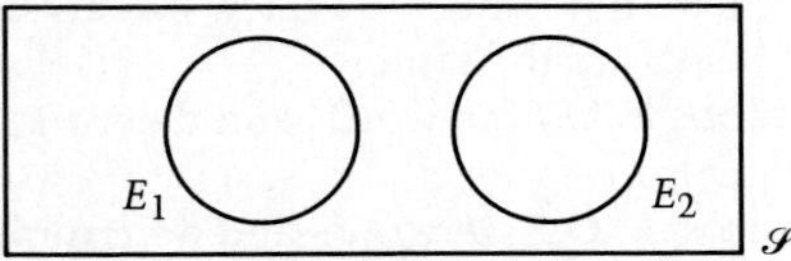

(a) $P(E_1')$ (b) $P(E_2')$
(c) $P(E_1 \cup E_2)$ (d) $P(E_1' \cup E_2')$
(e) $P((E_1 \cup E_2)')$ (f) $P(E_1 \cap E_2)$
(g) $P(E_1' \cap E_2')$ (h) $P((E_1 \cap E_2)')$

38. Suppose E_1 and E_2 are two events with $P(E_1) = 0.25$, $P(E_2) = 0.42$, and $P(E_1 \cap E_2) = 0.10$. By shading the appropriate portion of the diagram below, represent each of the following probabilities. Then calculate the probability. (Remember, $P(\mathscr{S}) = 1$.)

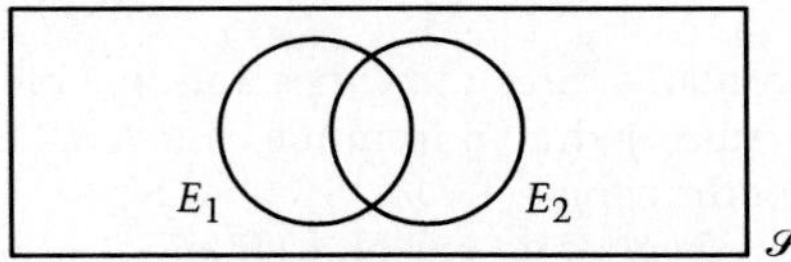

(a) $P(E_1')$ (b) $P(E_2')$
(c) $P(E_1 \cup E_2)$ (d) $P(E_1' \cup E_2')$
(e) $P((E_1 \cup E_2)')$ (f) $P(E_1' \cap E_2')$
(g) $P((E_1 \cap E_2)')$

Analyzing Data

The Case of the Nonrandom Lots

The Pennsylvania state lottery has made the news more than once. The lottery has both a 4-digit and 3-digit daily number. On two occasions, lottery administrators found it necessary to suspend play (for the remainder of the day) on 4-digit numbers that were drawing an unusually high tally of bets.

- On June 8, 1983, it was the number 3526, chosen by bettors because it represented the record number of strike-outs achieved by Phillies pitcher Steve Carlton just a few days earlier.
- On March 2, 1982, it was the number 4077, associated with the TV series M*A*S*H, which broadcast its final episode that day.

In both cases, lottery administrators felt that by allowing exorbitantly large numbers of bets, the state could face a serious problem if the number in question were to be selected. Of course the probability in either case was very small, but the potential cost was very great.

$$\text{Prob (picking 3526)} = \frac{\text{\# of ways of picking 3526}}{\text{\# of ways of picking any 4-digit number with replacement}}$$

$$= \frac{1}{10 \times 10 \times 10 \times 10} = 0.0001$$

Another incident involved a conspiracy on the part of the TV announcer who conducted the drawing of the daily numbers, together with a lottery official and four other individuals. The conspirators rigged the selection of the 3-digit daily number. The following is an excerpt from an article in the *New York Times* on September 20, 1980.

> Each of the three digits in the daily number is chosen by a machine that uses forced air to select, at random, one of the 10 numbered table tennis balls and blow it up a tube, where the number can be read. The grand jury said the rigging was achieved by injecting liquid into eight of the 10 balls, weighting them down so that they would not be blown up the tube. Because only the two unweighted balls would rise, there were only eight possible combinations of three digits, the grand jury said.

The ping-pong balls that were not weighted down bore the numbers 4 and 6. Assured that the only numbers possible in that drawing were

$$\{444, 446, 464, 644, 664, 646, 466, 666\},$$

the conspirators bought tickets, in large quantities, bearing these eight numbers. The winning number was 666, and would have paid the conspirators \$1.18 million, had they not been caught.

When the selection is fair, the probability of selecting any 3-digit number (with replacement) is simply $\frac{1}{10 \times 10 \times 10} = \frac{1}{1000}$ and thus the probability of selecting any of the 8 numbers composed of 4s and 6s would be $8\left(\frac{1}{1000}\right) = \frac{1}{125}$. One would expect the proportion of bets to be roughly consistent with this probability. But the betting on these numbers was so inconsistently heavy that suspicions were raised, and an investigation was initiated.

Source: New York Times, September 20, 1980, p. 1.

Chapter Review

10.1 Arithmetic Sequences and Sums (pp. 599–606)

A *sequence* is a function whose domain is the set of positive integers. (p. 600) The function values are called the *terms of the sequence.* We refer to a_n as the nth term, or the general term of the sequence. (p. 600)

A sequence is defined *recursively* if we specify its first term, along with a rule for finding any other term by using the preceding term(s). (p. 601)

Successive terms of an *arithmetic sequence* differ by a constant called the *common difference.* (p. 601) In an arithmetic sequence with first term a_1 and common difference d, the general term is given by $a_n = a_1 + (n - 1)d$. (p. 603) The sum S_n of the first n terms of such a sequence is given by the equation (p. 604)

$$S_n = \frac{n}{2}[2a_1 + (n - 1)d] = \frac{n}{2}(a_1 + a_n).$$

10.2 Geometric Sequences and Sums (pp. 607–610)

In a *geometric sequence,* each term is r times the preceding term, where r is some nonzero constant called the *common ratio.* In a geometric sequence with first term a_1 and common ratio r, the general term is given by the equation $a_n = a_1 r^{n-1}$. (p. 608) The sum S_n of the first n terms of such a sequence is given by the equation (p. 609)

$$S_n = \frac{a_1(1 - r^n)}{1 - r}.$$

10.3 Series (pp. 610–620)

The expression $a_1 + a_2 + \cdots + a_n + \cdots$ is called an *infinite series.* The *nth partial sum* of a series is the sum of the first n terms of the series. (p. 612) The series $a_1 + a_2 + \cdots + a_n + \cdots$ has sum S, if the terms in the sequence of partial sums, $S_1, S_2, \ldots, S_n, \ldots$, get arbitrarily close to S as the value of n increases. (p. 614)

The sum of the geometric series $a_1 + a_1 r + a_1 r^2 + \ldots$ is

$$S = \frac{a_1}{1 - r},$$

provided $|r| < 1$. If $|r| \geq 1$, the series has no sum. (p. 614)

10.4 Mathematical Induction (pp. 620–624)

Principle of Mathematical Induction:
A statement is true for all positive integers n if the following two requirements are met:

1. the statement is true for $n = 1$;

2. if you assume the statement is true for $n = k$, then you can prove that the statement is true for $n = k + 1$. (p. 621)

10.5 The Binomial Theorem (pp. 624–631)

For any positive integer n, we define *n factorial*, written $n!$, as the product of the first n positive integers:

$$n! = n \cdot (n - 1) \cdot (n - 2) \cdot \cdots \cdot 3 \cdot 2 \cdot 1.$$

Zero factorial is equal to 1: $0! = 1$. (p. 625)

If n is any positive integer, and if k is any positive integer such that $0 \le k \le n$, the *binomial coefficient* $\binom{n}{k}$ is

$$\binom{n}{k} = \frac{n!}{(n - k)!k!}$$

Binomial Theorem: For all numbers a and b, and for all positive integers n,

$$(a + b)^n = \binom{n}{0}a^n b^0 + \binom{n}{1}a^{n-1}b^1 + \binom{n}{2}a^{n-2}b^2 + \cdots + \binom{n}{n-1}a^1 b^{n-1} + \binom{n}{n}a^0 b^n.$$

10.6 Limits (pp. 631–636)

We read the statement $\lim_{x \to c} f(x) = L$ as, "The limit of $f(x)$, as x approaches c, is L." This means that, as the x-values get closer and closer to the value c, the values of $f(x)$ get closer and closer to the value L. If the values of $f(x)$ do not approach a fixed value, then the limit does not exist. (p. 632)

To find the limit of a sequence whose general term is a rational expression, we use the Second Property of Rational Functions to find the horizontal asymptote of the associated function (p. 634)

10.7 Simple Counting, Permutations, and Combinations (pp. 636–647)

The *Basic Principle of Counting* states that if a process consists of two stages such that one of them can be done in m possible ways and the other in n possible ways, then the whole process can be done in $m \cdot n$ ways. (p. 638)

When objects cannot be used more than once in the stages of a process, we say that we are counting *without replacement*. When they can be used more than once, we are counting *with replacement*. (p. 639)

There are $n!$ *permutations* (different orderings) of n distinct objects (p. 640) The *number of permutations of n objects taken r at a time*, where $0 \le r \le n$, is denoted ${}_nP_r$ and is given by the following formula (p. 641):

$${}_nP_r = n \cdot (n - 1) \cdot (n - 2) \cdot \cdots \cdot [n - (r - 1)] = \frac{n!}{(n - r)!}.$$

The *number of combinations of n objects taken r at a time,* where $0 \le r \le n$ is denoted by ${}_nC_r$ or $\binom{n}{r}$ and is given by the following formula (p. 643):

$$\binom{n}{r} = \frac{n!}{(n-r)!r!}.$$

10.8 Introduction to Probability (pp. 647–654)

If the sample space for an experiment contains n equally likely outcomes, and if event E contains m of these outcomes, then $P(E) = \frac{m}{n}$. (p. 648)

Let the event E be a subset of the sample space $\mathcal{S}$, E' be the complement of E, and $\emptyset$ be the empty event. Then (p. 649)

$$0 \le P(E) \le 1, \qquad P(\mathcal{S}) = 1, \qquad P(\emptyset) = 0, \qquad \text{and} \qquad P(E') = 1 - P(E).$$

If an event E is composed of n mutually exclusive events $E_1, E_2, \ldots, E_n$, then $P(E) = P(E_1) + P(E_2) + \cdots + P(E_n)$. If events E_1 and E_2 are not mutually exclusive, then $P(E_1 \cup E_2) = P(E_1) + P(E_2) - P(E_1 \cap E_2)$. (p. 651)

Review Exercises

In Exercises 1–4, find the first four terms, the ninth term and the twentieth term of the sequence whose general term is given.

1. $a_n = n + (-1)^n n$

2. $a_n = (n-3)(n+1)$

3. $a_n = 9 - n^2$

4. $a_n = 14 + (-1)^n n^2$

In Exercises 5–8, find the first six terms of the sequence.

5. $a_1 = 2, \quad a_n = 1 - 2a_{n-1}$ for $n \ge 2$

6. $a_1 = 1, \quad a_n = a_{n-1} + n$ for $n \ge 2$

7. $a_1 = 1, \quad a_2 = 2, \quad a_n = a_{n-1} + a_{n-2} - 1$ for $n \ge 3$

8. $a_1 = 1, \quad a_2 = 3, \quad a_n = a_{n-1} - a_{n-2} + n$ for $n \ge 3$

In Exercises 9–12, the first term a_1 and the common difference d of an arithmetic sequence are given. Find the first five terms of the sequence, the tenth term of the sequence, and the sum S_{20} of the first twenty terms of the sequence.

9. $a_1 = 2, d = 3$

10. $a_1 = -2, d = 4$

11. $a_1 = 3, d = 2$

12. $a_1 = 4, d = -2$

In Exercises 13–16, find the fifth term and the general term of the given geometric sequence and then find the sum S_6 of the first six terms of the sequence.

13. $3, 9, 27, 81, \ldots$

14. $-128, 96, -72, 54, \ldots$

15. $\frac{1}{8}, -\frac{1}{4}, \frac{1}{2}, -1, \ldots$

16. $3, 0.6, 0.12, 0.024, \ldots$

In Exercises 17–20, use sigma notation to rewrite the sum with 1 as the first value of the index.

17. $2 + 7 + 12 + 17 + 22$

18. $-2 + 5 - 8 + 11 - 14 + 17$

19. $3 - 6 + 12 - 24 + 48 - 96$

20. $20 + 2 + 0.2 + 0.02 + 0.002$

In Exercises 21–28, find the sum, if it exists, of the series.

21. $\sum_{n=1}^{\infty} 3\left(\frac{1}{2}\right)^n$

22. $\sum_{n=1}^{\infty} 2\left(\frac{1}{3}\right)^{n-1}$

23. $\sum_{n=1}^{\infty} 2\left(-\frac{1}{3}\right)^{n+1}$

24. $\sum_{n=1}^{\infty} 4\left(\frac{3}{2}\right)^n$

25. $\sum_{n=1}^{\infty} 6\left(-\frac{3}{2}\right)^n$

26. $\sum_{n=1}^{\infty} 5(-4)^{-n}$

27. $\sum_{n=1}^{\infty} \left(\frac{1}{n+3} - \frac{1}{n+1}\right)$

28. $\sum_{n=1}^{\infty} \left(\frac{1}{n+4} - \frac{1}{n}\right)$

In Exercises 29–32, find the fraction represented by the decimal expansion.

29. $0.31\overline{24}$

30. $0.451\overline{7}$

31. $0.312\overline{4}$

32. $0.4\overline{517}$

In Exercises 33–34, use mathematical induction to prove the statement.

33. For $n \ge 1$, $1 - \frac{1}{3} - \frac{1}{9} - \frac{1}{27} - \cdots - \frac{1}{3^n} = \frac{1}{2}\left(1 + \frac{1}{3^n}\right)$.

34. For $n \geq 1$, $2 + 6 + 10 + \cdots + (4n - 2) = 2n^2$.

In Exercises 35–36, use the Binomial Theorem to expand the expression.

35. $(3x - 2y)^4$ **36.** $(2x - y)^5$

In Exercises 37–40, find the indicated term in the expansion of the given expression.

37. third term of $\left(x + \frac{1}{x}\right)^{20}$

38. fifth term of $\left(y - \frac{1}{y}\right)^{16}$

39. fourth term of $\left(4p - \frac{1}{2p}\right)^9$

40. second term of $(3t - \sqrt{t})^{10}$

In Exercises 41–42, evaluate the following: (a) $\lim_{x \to -\infty} f(x)$, (b) $\lim_{x \to -1} f(x)$, (c) $\lim_{x \to 1^-} f(x)$, (d) $\lim_{x \to 3^+} f(x)$, (e) $\lim_{x \to \infty} f(x)$.

41.

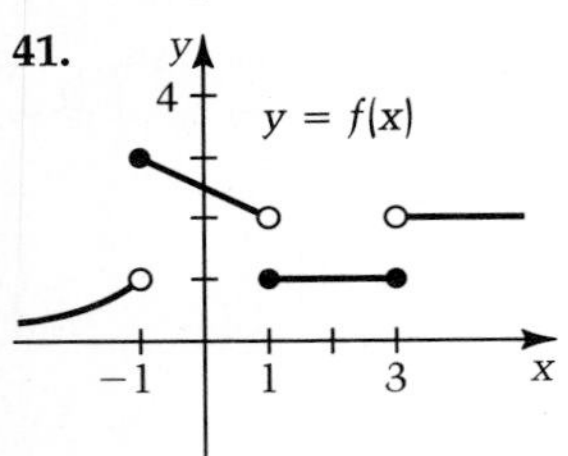

42.

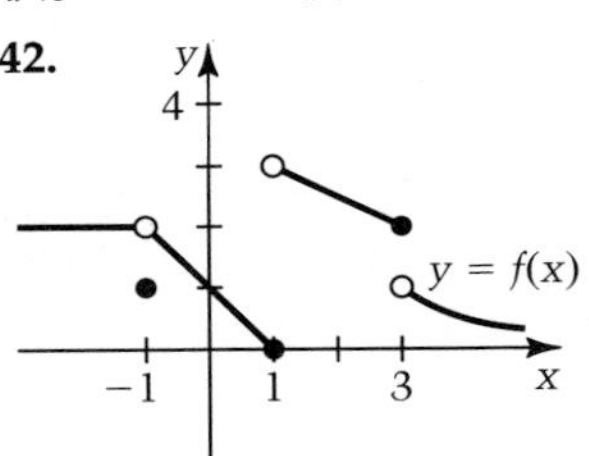

In Exercises 43–46, evaluate the limit.

43. $\lim_{n \to \infty} \frac{2n}{5n + 100}$ **44.** $\lim_{n \to \infty} \frac{8n^2}{7 + 2n}$

45. $\lim_{n \to \infty} \frac{7n^2}{9 - 5n^3}$ **46.** $\lim_{n \to \infty} \frac{7n^2}{3n^2 + 5}$

In Exercises 47–48, use a tree diagram to show the set of all possible outcomes.

47. Each of two multiple-choice questions in one part of a quiz can be answered A(always), S(Sometimes) or N(never).

48. A school lunch program gives a choice of beverage: M(milk) or S(soda), and a choice of sandwich: C(cheese), H(ham) or P(peanut butter and jelly).

49. A student selecting three courses for summer school will chose one of five literature courses, one of three science courses, and one of eight social science courses. How many different course schedules are there?

50. A three-symbol code is composed of a vowel, followed by a consonant and then a digit. How many possible codes are there?

51. A die is rolled three times. How many results are possible if the first roll is a number greater than two, the second roll is an odd number, and the third roll is different from the first roll?

52. There are nine members on the local school board. They will choose a chairperson, a vice-chairperson and a secretary. If none of the nine members may hold more than one of the three offices, in how many ways can the officers be chosen?

In Exercises 53–60, evaluate the expression.

53. ${}_6P_2$ **54.** ${}_7P_0$ **55.** ${}_{10}P_{10}$ **56.** ${}_8P_5$

57. $\binom{8}{0}$ **58.** $\binom{7}{5}$ **59.** $\binom{9}{4}$ **60.** $\binom{6}{6}$

61. Your mail for the day consists of seven letters, three of which are bills. In how many ways can you open your mail, one letter at a time?

62. In Exercise 61, in how many ways can you open your mail, one letter at a time, if you open the three bills first?

63. How many different 5-letter codes can be formed using the four letters *A*, *B*, *C* and *D*, if precisely one of the letters is used twice and appears last in the code?

64. In Exercise 63, how many different codes are possible if the repeated letter does not have to appear last?

65. In how many ways can a family of two adults and five children line up side-by-side for a photograph, if the adults stand one at each end of the lineup?

66. In how many ways can the family of Exercise 65 line up, if the adults stand side-by-side at one end of the lineup?

67. A container of nine batteries contains four defectives. If a random sample of five batteries is drawn from the container, what is the probability of drawing (a) no defectives? (b) exactly three defectives? (c) all four defectives?

68. Five cards are drawn from a well-shuffled bridge deck (52 cards). What is the probability of drawing (a) five spades? (b) four jacks? (c) exactly two aces and exactly two queens?

In Exercises 69–70, determine the probability if the sample space $\mathcal{S} = \{a, b, c, d, e, f, g\}$, and $E_1 = \{b, f, g\}$, $E_2 = \{a, c, e, f, g\}$, and $E_3 = \{a, c, e, d\}$. (Assume that all outcomes in $\mathcal{S}$ are equally likely.)

69. $P(E_1 \cup E_2)$ **70.** $P(E_1' \cap E_3')$

Graphing Calculator Exercises

[Round all estimated values to the nearest hundredth.]

In Exercises 1–14 use your function grapher to determine the given limit. (*Hint:* For example, to evaluate $\lim_{x\to 1} \frac{x-1}{(x+2)(x-1)}$, graph the function $y = \frac{x-1}{(x+2)(x-1)}$, and observe what happens to the y-values for x-values *very close* to 1.)

1. $\lim_{x\to 1} \frac{x-1}{(x+2)(x-1)}$
2. $\lim_{x\to -2^+} \frac{x-1}{(x+2)(x-1)}$
3. $\lim_{x\to -2^-} \frac{x-1}{(x+2)(x-1)}$
4. $\lim_{x\to -1} \frac{x^2-1}{x+1}$
5. $\lim_{x\to 0} \frac{x^2}{x^2+3x}$
6. $\lim_{x\to \infty} \frac{x^2}{x^2+3x}$
7. $\lim_{x\to -\infty} \frac{x^2}{x^2+3x}$
8. $\lim_{x\to 0} x \sin\left(\frac{1}{x}\right)$
9. $\lim_{x\to 0^+} \frac{e^x}{x}$
10. $\lim_{x\to 0} \frac{e^x-1}{\sin x}$
11. $\lim_{x\to 0} \frac{\sin 3x}{\sin 4x}$
12. $\lim_{x\to \infty} \frac{\ln x}{x}$
13. $\lim_{x\to \infty} \left(1 + \frac{1}{x}\right)^x$
14. $\lim_{x\to 0^+} x^x$
15. Graph each of the following pairs of functions on the same set of axes, (a) for values of x in the interval $[-1, 1)$, and (b) for values of x in the interval $[-10, 10]$.

$$y = \frac{1}{1-x} \quad \text{and} \quad y = 1 + x + x^2 + x^3$$

$$y = \frac{1}{1-x} \quad \text{and} \quad y = 1 + x + x^2 + x^3 + x^4 + x^5$$

$$y = \frac{1}{1-x} \quad \text{and} \quad y = 1 + x + x^2 + x^3 + x^4 + x^5 + x^6 + x^7$$

What do you observe?

16. For any real number x, the expression

$$1 - x + x^2 - x^3 + x^4 - \cdots$$

is a geometric series with first term 1, and common ratio $-x$.
(a) For what values of x will the resulting series have a sum?
(b) If, for some x, the series $1 - x + x^2 - x^3 + x^4 - \cdots$ *does* have a sum, what will the value of this sum be? (Answer in terms of x.)
(c) Graph $y = \frac{1}{1+x}$ and

$$y = 1 - x + x^2 - x^3 + x^4 - x^5 + x^6 - x^7$$

on the same set of axes for values of x in $[-10, 10]$. How does this graph help to explain your answers to parts (a) and (b)?

Chapter Test

1. Find the first five terms and the ninth term of the sequence whose general term is given by the rule $a_n = 2^n - n$.
2. Find the first six terms of each sequence.
(a) $a_1 = 2, a_n = 5 + (-1)^n a_{n-1}$ for $n \geq 2$.
(b) $a_1 = 1, a_2 = 3, a_3 = 2, a_n = a_{n-1} + a_{n-2} + a_{n-3}$ for $n \geq 4$.
3. Find the first five terms of the arithmetic sequence with first term $a_1 = 7$ and common difference $d = -5$.
4. Find the eleventh and twenty-first terms of the arithmetic sequence 3, 9, 15, 21, . . .
5. Find the sum of the first twelve terms of the arithmetic sequence 2, 5, 8, 11, 14, . . .
6. Determine whether each sequence is geometric.
(a) $16, 12, 9, 6, \frac{9}{2}, \ldots$
(b) $3, -6, 12, -24, 48, \ldots$
7. Find the sixth, ninth, and nth terms of the geometric sequence $\frac{27}{10}, -\frac{9}{5}, \frac{6}{5}, -\frac{4}{5}, \ldots$
8. Find the sum of the first n terms and the sum of the first five terms of the geometric sequence $2, \frac{4}{3}, \frac{8}{9}, \frac{16}{27}, \ldots$
9. Rewrite $\sum_{n=2}^{6} (n^2 - 3n)$ without using sigma notation; then compute the sum.

10. Use sigma notation to rewrite the sum

$$4 + 9 + 16 + 25 + 36.$$

11. Compute the third and sixth partial sums of the series $\sum_{n=1}^{\infty} 3\left(\frac{1}{2}\right)^{n-1}$.

12. Compute the sum, if it exists, of each series.

(a) $\sum_{n=1}^{\infty} 7\left(\frac{3}{4}\right)^{n-1}$

(b) $\sum_{n=1}^{\infty} \left(\frac{1}{2n + 3} - \frac{1}{2n + 1}\right)$.

13. Find the fraction represented by the decimal $0.37\overline{272}$.

14. Prove the following statement by mathematical induction. (Assume $n \geq 1$.)

$$3 + 6 + 9 + 12 + \cdots + 3n = \frac{3n(n + 1)}{2}.$$

15. Use the Binomial Theorem to expand $(3 + 5t)^4$.

16. Evaluate the following limits for the function whose graph is shown.

(a) $\lim_{x\to -2} f(x)$ (b) $\lim_{x\to 1} f(x)$

(c) $\lim_{x\to 3} f(x)$ (d) $\lim_{x\to 6} f(x)$

(e) $\lim_{x\to -\infty} f(x)$ (f) $\lim_{x\to \infty} f(x)$

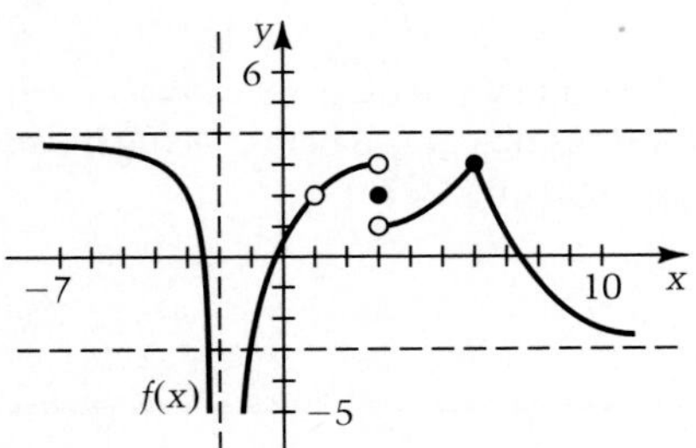

In Problems 17–19, evaluate the limit.

17. $\lim_{n\to\infty} \frac{7n - n^2}{5n^2 + 1}$

18. $\lim_{n\to\infty} \frac{8n}{n^3 + 1}$

19. $\lim_{n\to\infty} \frac{3n^2}{7n + 2}$

20. Draw a tree diagram to illustrate the following: The different possible girl/boy arrangements in a family containing three children.

21. A pizza shop offers two sizes of pizza, small and large. You can order either size pizza with regular or extra cheese, and have it plain, or topped with any or all of six toppings. How many different pizzas can be ordered?

22. Evaluate each of the following expressions.

(a) ${}_8P_4$ (b) ${}_{11}P_{11}$ (c) $\binom{9}{6}$ (d) $\binom{8}{0}$

23. Five senators will be chosen out of 100 senators to conduct a committee investigation. How many different groups of five senators can be chosen?

24. The digits 2, 4, 5, 6, and 8 are used to form four-digit numbers. How many different numbers can be formed if (a) no digit may be repeated? (b) digits may be repeated? (c) no digits may be repeated and the resulting number is odd?

25. A microcomputer user will test five software programs from a group of seven word-processing programs and four database management programs. How many ways can the user select the five programs if (a) two word-processing programs and three database management programs must be tested? (b) all four database management programs must be tested? (c) only word-processing programs must be tested?

26. A fair coin is tossed three times. What is the probability of tossing

(a) all heads?

(b) exactly one head?

(c) at least one head?

27. Suppose that for sample space $\mathcal{S} = \{x_1, x_2, x_3, x_4\}$, we have events $E_1 = \{x_1, x_3\}$, $E_2 = \{x_2, x_3, x_4\}$, and $E_3 = \{x_4\}$. Assuming that all outcomes in $\mathcal{S}$ are equally likely, determine the following probabilities.

(a) $P(E_1 \cup E_2)$

(b) $P(E_1 \cap E_2)$

(c) $P(E_1' \cup E_3)$

Linear Interpolation

x	$f(x)$
−1	1.9
1	3.4
3	4.8

Figure A.1

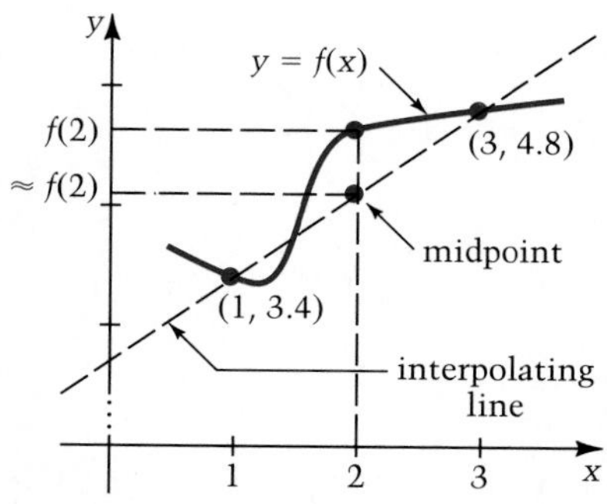

Figure A.2

Prior to the availability of the hand-held calculator, tables such as those on page A7 or page A11 were indispensable tools for mathematical applications requiring approximate logarithmic or trigonometric function values. However, a table was frequently not complete enough to include a specific value that was needed. For example, suppose we have the "mini-table" of values in Figure A.1, and we need a value for $f(2)$. Although the table provides no function value for $x = 2$, we can use the given information to estimate $f(2)$. Having no other information about the function f, we can reasonably assume that, since 2 is midway between 1 and 3, the value $f(2)$ is approximately midway between $f(1)$ and $f(3)$.

As shown in Figure A.2, the geometric meaning of our assumption is that the point $(2, f(2))$ is near the midpoint of the line segment that joins the points $(1, f(1))$ and $(3, f(3))$. As you can see, the midpoint of the line segment and the point $(2, f(2))$ do not coincide, but the y-coordinate of the midpoint is a good approximation of $f(2)$.

The use of a line to approximate function values is called **linear interpolation.** For the example above, the equation of the line is found by using the points (1, 3.4) and 3, 4.8). We find the approximation by substituting $x = 2$ into the equation of this line.

EXAMPLE 1

Use linear interpolation and the function values given in the table at the right to approximate $f(6.2)$ to the nearest tenth.

x	1	3	5	7
$f(x)$	3.4	8.7	15.2	23.5

Solution Since 6.2 is an x-value lying between 5 and 7 in the table, we use the points (5, 15.2) and (7, 23.5) to determine the equation of the interpolating line.

$$\text{slope} = \frac{23.5 - 15.2}{7 - 5} = \frac{8.3}{2} = 4.15$$

The slope of the interpolating line is computed. Now use the point-slope form to write an equation of the interpolating line.

$$y - 15.2 = 4.15(x - 5)$$

$$y - 15.2 = 4.15(6.2 - 5)$$

$x = 6.2$ is substituted into the equation.

$$y = 4.15(1.2) + 15.2$$

$$y = 20.18$$

The approximate value of $f(6.2)$ to the nearest tenth is 20.2.

In Example 1, we could have determined $f(6.2)$ without actually finding the equation for the interpolating line. The method involves setting up a proportion, using the four differences shown below.

	x	$f(x)$	
	5	15.2	
d_2	d_1 6.2	? d_3	
	7	23.5	d_4

15.2 and 23.5 are called **bracketing values** for the unknown value $f(6.2)$.

Because the relationship between x-values and function values is assumed to be linear, we can write the following proportion.

$$\frac{d_1}{d_2} = \frac{d_3}{d_4}$$

Note that d_3 is the only unknown difference.

$$\frac{6.2 - 5}{7 - 5} = \frac{d_3}{23.5 - 15.2}$$

$$\frac{1.2}{2} = \frac{d_3}{8.3}$$

$$4.98 = d_3$$

Since $d_3 = 4.98$, we can conclude that the unknown function value, $f(6.2)$, is 4.98 more than 15.2. That is,

$$f(6.2) = f(5) + d_3 = 15.2 + 4.98 = 20.18$$

As before, the approximate value of $f(6.2)$ is 20.2 (rounded to nearest tenth).

We now apply linear interpolation to approximate a logarithmic function value. You will need to refer to Table 1 on page A7 to determine the bracketing values in this example. Remember than when we write $\log x$, without stating the base explicitly, we mean $\log_{10} x$, the common logarithm of x.

EXAMPLE 2

Determine log 85.37.

Solution

$$\begin{aligned}\log 85.37 &= \log(8.537 \times 10^1)\\ &= \log 8.537 + \log 10^1\\ &= \log 8.537 + 1\end{aligned}$$

Interpolation

	x	**log x**	
d_2	8.53	0.9309	d_4
d_1	8.537	? d_3	
	8.54	0.9315	

$$\frac{d_1}{d_2} = \frac{d_3}{d_4}$$

$$\frac{0.007}{0.01} = \frac{d_3}{0.0006}$$

$$0.00042 = d_3$$

$$\begin{aligned}\log 8.537 &= 0.9309 + 0.00042\\ &\approx 0.9313\end{aligned}$$

$$\begin{aligned}\log 85.37 &\approx 0.9313 + 1\\ &= 1.9313\end{aligned}$$

Thus, log 85.37 ≈ 1.9313 (rounded to four decimal place accuracy).

In the next example we use linear interpolation to determine the value of A, given the common logarithm of A.

EXAMPLE 3

Solve for A: $\log A = -1.4132$

Solution Remember that we must first write −1.4132 as the sum of an integer and a number between 0 and 1.

$$\underbrace{-1.4132}_{\log A} = \underbrace{-2}_{\log 10^{-2}} + \underbrace{0.5868}_{\log x}$$

We must find x, given that $\log x = 0.5868$.

Interpolation

	x	$\log x$	
d_2 [d_1 [	3.86	0.5866	] d_3] d_4
	x	0.5868	
	3.87	0.5877	

$$\frac{d_1}{d_2} = \frac{d_3}{d_4}$$

$$\frac{d_1}{0.01} = \frac{0.0002}{0.0011}$$

$$d_1 \approx 0.002$$

$$\text{So, } x \approx 3.86 + 0.002 = 3.862$$

$$\log A \approx \log 10^{-2} + \log 3.862$$
$$\log A \approx \log (10^{-2} \cdot 3.862)$$
$$A \approx 10^{-2} \cdot 3.862$$

Thus, $A \approx 0.03862$.

The technique of linear interpolation can be applied to tables of trigonometric function values also. Table 4 in the appendix contains function values for angles in multiples of 10′. This means that we can determine sin 36°20′ and sin 36°30′ directly from the table. But if we wish to use the table to determine 36°27′, then we must interpolate.

EXAMPLE 4

Find: (a) sin 36°27′ (b) cos 73°48′.

Solution

(a)

	θ	$\sin \theta$	
10 [7 [	36°20′	0.5925	] d] 0.0023
	36°27′	?	
	36°30′	0.5948	

Set up a chart involving the bracketing values. Determine the three differences, and let d represent the unknown difference.

$$\frac{7}{10} = \frac{d}{0.0023}$$

$$d \approx 0.0016$$

Set up the proportion and solve. Since the values are given to four decimal places, we approximate the value of d to 4 decimal places.

$$\sin 36°27' = 0.5925 + 0.0016 = 0.5941$$

d is added to the upper function value.

(b)

	θ	$\cos\theta$	
10 [8]	73°40′	0.2812	
	73°48′	? [d]	−0.0028
	73°50′	0.2784	

Set up the proportion:

$$\frac{8}{10} = \frac{d}{-0.0028}$$

Solve to get $d = -0.0022$.

$$\cos 73°48' = 0.2812 + (-0.0022) = 0.2790$$

The method of linear interpolation can also be used to determine an approximate value for θ, given a trigonometric function value of θ.

EXAMPLE 5

If $\tan\theta = 0.8915$ and θ is an acute angle, determine θ.

Solution

	θ	$\tan\theta$	
10′ [d]	41°40′	0.8899	0.0016
	?	0.8915	
	41°50′	0.8952	0.0053

The value 0.8915 is between 0.8899 and 0.8952.

$$\frac{d}{10'} = \frac{0.0016}{0.0053}$$

Set up the proportion and solve.

$$d = 3.019'$$

Round to the nearest minute.

$$d \approx 3'$$

$$\theta = 41°40' + 3' = 41°43'$$

Add the value of d to the upper angle value.

EXERCISE SET

In Exercises 1–4, use linear interpolation and the function values given in the table to approximate A to the nearest tenth.

1.

x	1	3	7	13
$f(x)$	2.8	14.3	38.5	64.2

(a) $A = f(5.1)$ (b) $f(A) = 14.3$
(c) $A = f(2.8)$ (d) $f(A) = 53.1$

2.

x	2	5	7	10
$f(x)$	6.1	18.3	27.5	31.2

(a) $A = f(2.7)$ (b) $f(A) = 19$
(c) $A = f(9.9)$ (d) $f(A) = 28.3$

3.

x	2.1	3.8	5.6	8.1
$f(x)$	-7.1	-0.4	6.9	15.2

(a) $A = f(3)$ (b) $f(A) = 8$
(c) $A = f(4.1)$ (d) $f(A) = -0.1$

4.

x	-7.3	-2.4	5.9	12.1
$f(x)$	-17.0	3.8	9.6	16.3

(a) $A = f(-5.1)$ (b) $f(A) = 0$
(c) $A = f(1.7)$ (d) $f(A) = 14.6$

In Exercises 5–22, use Table 1 in the Appendix and linear interpolation to solve the given equation for A.

5. $\log 7.513 = A$
6. $\log 1.395 = A$
7. $\log 0.3045 = A$
8. $\log 52.36 = A$
9. $\log 478.2 = A$
10. $\log 0.02948 = A$
11. $\log 0.05173 = A$
12. $\log 332.2 = A$
13. $\log 0.91375 = A$
14. $\log 0.01112 = A$
15. $\log A = 0.9501$
16. $\log A = 0.8778$
17. $\log A = 3.9844$
18. $\log A = 4.3299$
19. $\log A = -0.6573$
20. $\log A = -0.7313$
21. $\log A = -1.5797$
22. $\log A = -9.2412$

In Exercises 23–32, evaluate the given expression.

23. sin 39°33′
24. tan 12°15′
25. tan 31°23′
26. cos 19°3′
27. cos 71°27′
28. sin 63°11′
29. cot 82°18′
30. cot 75°52′
31. sec 87°47′
32. csc 45°36′

In Exercises 33–38, determine the acute angle α.

33. $\sin \alpha = 0.0600$
34. $\cos \alpha = 0.9727$
35. $\cos \alpha = 0.3490$
36. $\tan \alpha = 1.7030$
37. $\tan \alpha = 0.5000$
38. $\sin \alpha = 0.8500$

Tables

Table 1 ■ Common Logarithms

x	0	1	2	3	4	5	6	7	8	9
1.0	.0000	.0043	.0086	.0128	.0170	.0212	.0253	.0294	.0334	.0374
1.1	.0414	.0453	.0492	.0531	.0569	.0607	.0645	.0682	.0719	.0755
1.2	.0792	.0828	.0864	.0899	.0934	.0969	.1004	.1038	.1072	.1106
1.3	.1139	.1173	.1206	.1239	.1271	.1303	.1335	.1367	.1399	.1430
1.4	.1461	.1492	.1523	.1553	.1584	.1614	.1644	.1673	.1703	.1732
1.5	.1761	.1790	.1818	.1847	.1875	.1903	.1931	.1959	.1987	.2014
1.6	.2041	.2068	.2095	.2122	.2148	.2175	.2201	.2227	.2253	.2279
1.7	.2304	.2330	.2355	.2380	.2405	.2430	.2455	.2480	.2504	.2529
1.8	.2553	.2577	.2601	.2625	.2648	.2672	.2695	.2718	.2742	.2765
1.9	.2788	.2810	.2833	.2856	.2878	.2900	.2923	.2945	.2967	.2989
2.0	.3010	.3032	.3054	.3075	.3096	.3118	.3139	.3160	.3181	.3201
2.1	.3222	.3243	.3263	.3284	.3304	.3324	.3345	.3365	.3385	.3404
2.2	.3424	.3444	.3464	.3483	.3502	.3522	.3541	.3560	.3579	.3598
2.3	.3617	.3636	.3655	.3674	.3692	.3711	.3729	.3747	.3766	.3784
2.4	.3802	.3820	.3838	.3856	.3874	.3892	.3909	.3927	.3945	.3962
2.5	.3979	.3997	.4014	.4031	.4048	.4065	.4082	.4099	.4116	.4133
2.6	.4150	.4166	.4183	.4200	.4216	.4232	.4249	.4265	.4281	.4298
2.7	.4314	.4330	.4346	.4362	.4378	.4393	.4409	.4425	.4440	.4456
2.8	.4472	.4487	.4502	.4518	.4533	.4548	.4564	.4579	.4594	.4609
2.9	.4624	.4639	.4654	.4669	.4683	.4698	.4713	.4728	.4742	.4757
3.0	.4771	.4786	.4800	.4814	.4829	.4843	.4857	.4871	.4886	.4900
3.1	.4914	.4928	.4942	.4955	.4969	.4983	.4997	.5011	.5024	.5038
3.2	.5051	.5065	.5079	.5092	.5105	.5119	.5132	.5145	.5159	.5172
3.3	.5185	.5198	.5211	.5224	.5237	.5250	.5263	.5276	.5289	.5302
3.4	.5315	.5328	.5340	.5353	.5366	.5378	.5391	.5403	.5416	.5428
3.5	.5441	.5453	.5465	.5478	.5490	.5502	.5514	.5527	.5539	.5551
3.6	.5563	.5575	.5587	.5599	.5611	.5623	.5635	.5647	.5658	.5670
3.7	.5682	.5694	.5705	.5717	.5729	.5740	.5752	.5763	.5775	.5786
3.8	.5798	.5809	.5821	.5832	.5843	.5855	.5866	.5877	.5888	.5899
3.9	.5911	.5922	.5933	.5944	.5955	.5966	.5977	.5988	.5999	.6010
4.0	.6021	.6031	.6042	.6053	.6064	.6075	.6085	.6096	.6107	.6117
4.1	.6128	.6138	.6149	.6160	.6170	.6180	.6191	.6201	.6212	.6222
4.2	.6232	.6243	.6253	.6263	.6274	.6284	.6294	.6304	.6314	.6325
4.3	.6335	.6345	.6355	.6365	.6375	.6385	.6395	.6405	.6415	.6425
4.4	.6435	.6444	.6454	.6464	.6474	.6484	.6493	.6503	.6513	.6522
4.5	.6532	.6542	.6551	.6561	.6571	.6580	.6590	.6599	.6609	.6618
4.6	.6628	.6637	.6646	.6656	.6665	.6675	.6684	.6693	.6702	.6712
4.7	.6721	.6730	.6739	.6749	.6758	.6767	.6776	.6785	.6794	.6803
4.8	.6812	.6821	.6830	.6839	.6848	.6857	.6866	.6875	.6884	.6893
4.9	.6902	.6911	.6920	.6928	.6937	.6946	.6955	.6964	.6972	.6981
5.0	.6990	.6998	.7007	.7016	.7024	.7033	.7042	.7050	.7059	.7067
5.1	.7076	.7084	.7093	.7101	.7110	.7118	.7126	.7135	.7143	.7152
5.2	.7160	.7168	.7177	.7185	.7193	.7202	.7210	.7218	.7226	.7235
5.3	.7243	.7251	.7259	.7267	.7275	.7284	.7292	.7300	.7308	.7316
5.4	.7324	.7332	.7340	.7348	.7356	.7364	.7372	.7380	.7388	.7396

Table 1 ▪ Common Logarithms (*continued*)

x	0	1	2	3	4	5	6	7	8	9
5.5	.7404	.7412	.7419	.7427	.7435	.7443	.7451	.7459	.7466	.7474
5.6	.7482	.7490	.7497	.7505	.7513	.7520	.7528	.7536	.7543	.7551
5.7	.7559	.7566	.7574	.7582	.7589	.7597	.7604	.7612	.7619	.7627
5.8	.7634	.7642	.7649	.7657	.7664	.7672	.7679	.7686	.7694	.7701
5.9	.7709	.7716	.7723	.7731	.7738	.7745	.7752	.7760	.7767	.7774
6.0	.7782	.7789	.7796	.7803	.7810	.7818	.7825	.7832	.7839	.7846
6.1	.7853	.7860	.7868	.7875	.7882	.7889	.7896	.7903	.7910	.7917
6.2	.7924	.7931	.7938	.7945	.7952	.7959	.7966	.7973	.7980	.7987
6.3	.7993	.8000	.8007	.8014	.8021	.8028	.8035	.8041	.8048	.8055
6.4	.8062	.8069	.8075	.8082	.8089	.8096	.8102	.8109	.8116	.8122
6.5	.8129	.8136	.8142	.8149	.8156	.8162	.8169	.8176	.8182	.8189
6.6	.8195	.8202	.8209	.8215	.8222	.8228	.8235	.8241	.8248	.8254
6.7	.8261	.8267	.8274	.8280	.8287	.8293	.8299	.8306	.8312	.8319
6.8	.8325	.8331	.8338	.8344	.8351	.8357	.8363	.8370	.8376	.8382
6.9	.8388	.8395	.8401	.8407	.8414	.8420	.8426	.8432	.8439	.8445
7.0	.8451	.8457	.8463	.8470	.8476	.8482	.8488	.8494	.8500	.8506
7.1	.8513	.8519	.8525	.8531	.8537	.8543	.8549	.8555	.8561	.8567
7.2	.8573	.8579	.8585	.8591	.8597	.8603	.8609	.8615	.8621	.8627
7.3	.8633	.8639	.8645	.8651	.8657	.8663	.8669	.8675	.8681	.8686
7.4	.8692	.8698	.8704	.8710	.8716	.8722	.8727	.8733	.8739	.8745
7.5	.8751	.8756	.8762	.8768	.8774	.8779	.8785	.8791	.8797	.8802
7.6	.8808	.8814	.8820	.8825	.8831	.8837	.8842	.8848	.8854	.8859
7.7	.8865	.8871	.8876	.8882	.8887	.8893	.8899	.8904	.8910	.8915
7.8	.8921	.8927	.8932	.8938	.8943	.8949	.8954	.8960	.8965	.8971
7.9	.8976	.8982	.8987	.8993	.8998	.9004	.9009	.9015	.9020	.9025
8.0	.9031	.9036	.9042	.9047	.9053	.9058	.9063	.9069	.9074	.9079
8.1	.9085	.9090	.9096	.9101	.9106	.9112	.9117	.9122	.9128	.9133
8.2	.9138	.9143	.9149	.9154	.9159	.9165	.9170	.9175	.9180	.9186
8.3	.9191	.9196	.9201	.9206	.9212	.9217	.9222	.9227	.9232	.9238
8.4	.9243	.9248	.9253	.9258	.9263	.9269	.9274	.9279	.9284	.9289
8.5	.9294	.9299	.9304	.9309	.9315	.9320	.9325	.9330	.9335	.9340
8.6	.9345	.9350	.9355	.9360	.9365	.9370	.9375	.9380	.9385	.9390
8.7	.9395	.9400	.9405	.9410	.9415	.9420	.9425	.9430	.9435	.9440
8.8	.9445	.9450	.9455	.9460	.9465	.9469	.9474	.9470	.9484	.9489
8.9	.9494	.9499	.9504	.9509	.9513	.9518	.9523	.9528	.9533	.9538
9.0	.9542	.9547	.9552	.9557	.9562	.9566	.9571	.9576	.9581	.9586
9.1	.9590	.9595	.9600	.9605	.9609	.9614	.9619	.9624	.9628	.9633
9.2	.9638	.9643	.9647	.9652	.9657	.9661	.9666	.9671	.9675	.9680
9.3	.9685	.9689	.9694	.9699	.9703	.9708	.9713	.9717	.9722	.9727
9.4	.9731	.9736	.9741	.9745	.9750	.9754	.9759	.9763	.9768	.9773
9.5	.9777	.9782	.9786	.9791	.9795	.9800	.9805	.9809	.9814	.9818
9.6	.9823	.9827	.9832	.9836	.9841	.9845	.9850	.9854	.9859	.9863
9.7	.9868	.9872	.9877	.9881	.9886	.9890	.9894	.9899	.9903	.9908
9.8	.9912	.9917	.9921	.9926	.9930	.9934	.9939	.9943	.9948	.9952
9.9	.9956	.9961	.9965	.9969	.9974	.9978	.9983	.9987	.9991	.9996

Table 2 ▪ Values of e^x and e^{-x}

x	e^x	e^{-x}	x	e^x	e^{-x}	x	e^x	e^{-x}
.00	1.00000	1.00000	.40	1.49182	.67032	.80	2.22554	.44032
.01	1.01005	.99005	.41	1.50682	.66365	.85	2.33965	.42741
.02	1.02020	.98020	.42	1.52196	.65705	.90	2.45960	.40657
.03	1.03045	.97045	.43	1.53726	.65051	.95	2.58571	.38674
.04	1.04081	.96079	.44	1.55271	.64404	1.00	2.71828	.36788
.05	1.05127	.95123	.45	1.56831	.63763	1.10	3.00416	.33287
.06	1.06184	.94176	.46	1.58407	.63128	1.20	3.32011	.30119
.07	1.07251	.93239	.47	1.59999	.62500	1.30	3.66929	.27253
.08	1.08329	.92312	.48	1.61607	.61878	1.40	4.05519	.24659
.09	1.09417	.91393	.49	1.63232	.61263	1.50	4.48168	.22313
.10	1.10517	.90484	.50	1.64872	.60653	1.60	4.95302	.20189
.11	1.11628	.89583	.51	1.66529	.60050	1.70	5.47394	.18268
.12	1.12750	.88692	.52	1.68203	.59452	1.80	6.04964	.16529
.13	1.13883	.87810	.53	1.69893	.58860	1.90	6.68589	.14956
.14	1.15027	.86936	.54	1.71601	.58275	2.00	7.38905	.13533
.15	1.16183	.86071	.55	1.73325	.57695	2.10	8.16616	.12245
.16	1.17351	.85214	.56	1.75067	.57121	2.20	9.02500	.11080
.17	1.18530	.84366	.57	1.76827	.56553	2.30	9.97417	.10025
.18	1.19722	.83527	.58	1.78604	.55990	2.40	11.02316	.09071
.19	1.20925	.82696	.59	1.80399	.55433	2.50	12.18248	.08208
.20	1.22140	.81873	.60	1.82212	.54881	3.00	20.08551	.04978
.21	1.23368	.81058	.61	1.84043	.54335	3.50	33.11545	.03020
.22	1.24608	.80252	.62	1.85893	.53794	4.00	54.59815	.01832
.23	1.25860	.79453	.63	1.87761	.53259	4.50	90.01713	.01111
.24	1.27125	.78663	.64	1.89648	.52729	5.00	148.41316	.00674
.25	1.28403	.77880	.65	1.91554	.52205	5.50	224.69193	.00409
.26	1.29693	.77105	.66	1.93479	.51685	6.00	403.42879	.00248
.27	1.30996	.76338	.67	1.95424	.51171	6.50	665.14163	.00150
.28	1.32313	.75578	.68	1.97388	.50662	7.00	1096.63316	.00091
.29	1.33643	.74826	.69	1.99372	.50158	7.50	1808.04241	.00055
.30	1.34986	.74082	.70	2.01375	.49659	8.00	2980.95799	.00034
.31	1.36343	.73345	.71	2.03399	.49164	8.50	4914.76884	.00020
.32	1.37713	.72615	.72	2.05443	.48675	9.00	8130.08392	.00012
.33	1.39097	.71892	.73	2.07508	.48191	9.50	13359.72683	.00007
.34	1.40495	.71177	.74	2.09594	.47711	10.00	22026.46579	.00005
.35	1.41907	.70469	.75	2.11700	.47237			
.36	1.43333	.69768	.76	2.13828	.46767			
.37	1.44773	.69073	.77	2.15977	.46301			
.38	1.46228	.68386	.78	2.18147	.45841			
.39	1.47698	.67706	.79	2.20340	.45384			

Table 3 ■ Natural Logarithms

x	$\ln x$	x	$\ln x$	x	$\ln x$
		4.5	1.5041	9.0	2.1972
0.1	−2.3026	4.6	1.5261	9.1	2.2083
0.2	−1.6094	4.7	1.5476	9.2	2.2192
0.3	−1.2040	4.8	1.5686	9.3	2.2300
0.4	−0.9163	4.9	1.5892	9.4	2.2407
0.5	−0.6931	5.0	1.6094	9.5	2.2513
0.6	−0.5108	5.1	1.6292	9.6	2.2618
0.7	−0.3567	5.2	1.6487	9.7	2.2721
0.8	−0.2231	5.3	1.6677	9.8	2.2824
0.9	−0.1054	5.4	1.6864	9.9	2.2925
1.0	0.0000	5.5	1.7047	10	2.3026
1.1	0.0953	5.6	1.7228	11	2.3979
1.2	0.1823	5.7	1.7405	12	2.4849
1.3	0.2624	5.8	1.7579	13	2.5649
1.4	0.3365	5.9	1.7750	14	2.6391
1.5	0.4055	6.0	1.7918	15	2.7081
1.6	0.4700	6.1	1.8083	16	2.7726
1.7	0.5306	6.2	1.8245	17	2.8332
1.8	0.5878	6.3	1.8405	18	2.8904
1.9	0.6419	6.4	1.8563	19	2.9444
2.0	0.6931	6.5	1.8718	20	2.9957
2.1	0.7419	6.6	1.8871	25	3.2189
2.2	0.7885	6.7	1.9021	30	3.4012
2.3	0.8329	6.8	1.9169	35	3.5553
2.4	0.8755	6.9	1.9315	40	3.6889
2.5	0.9163	7.0	1.9459	45	3.8067
2.6	0.9555	7.1	1.9601	50	3.9120
2.7	0.9933	7.2	1.9741	55	4.0073
2.8	1.0296	7.3	1.9879	60	4.0943
2.9	1.0647	7.4	2.0015	65	4.1744
3.0	1.0986	7.5	2.0149	70	4.2485
3.1	1.1314	7.6	2.0281	75	4.3175
3.2	1.1632	7.7	2.0412	80	4.3820
3.3	1.1939	7.8	2.0541	85	4.4427
3.4	1.2238	7.9	2.0669	90	4.4998
3.5	1.2528	8.0	2.0794	100	4.6052
3.6	1.2809	8.1	2.0919	110	4.7005
3.7	1.3083	8.2	2.1041	120	4.7875
3.8	1.3350	8.3	2.1163	130	4.8676
3.9	1.3610	8.4	2.1282	140	4.9416
4.0	1.3863	8.5	2.1401	150	5.0106
4.1	1.4110	8.6	2.1518	160	5.0752
4.2	1.4351	8.7	2.1633	170	5.1358
4.3	1.4586	8.8	2.1748	180	5.1930
4.4	1.4816	8.9	2.1861	190	5.2470

Table 4 ■ Values of Trigonometric Functions

α (degrees)	α (radians)	sin α	cos α	tan α	cot α	sec α	csc α		
0°00′	.0000	.0000	1.0000	.0000	—	1.000	—	1.5708	**90°00′**
10	.0029	.0029	1.0000	.0029	343.8	1.000	343.8	1.5679	50
20	.0058	.0058	1.0000	.0058	171.9	1.000	171.9	1.5650	40
30	.0087	.0087	1.0000	.0087	114.6	1.000	114.6	1.5621	30
40	.0116	.0116	.9999	.0116	85.94	1.000	85.95	1.5592	20
50	.0145	.0145	.9999	.0145	68.75	1.000	68.76	1.5563	10
1°00′	.0175	.0175	.9998	.0175	57.29	1.000	57.30	1.5533	**89°00′**
10	.0204	.0204	.9998	.0204	49.10	1.000	49.11	1.5504	50
20	.0233	.0233	.9997	.0233	42.96	1.000	42.98	1.5475	40
30	.0262	.0262	.9997	.0262	38.19	1.000	38.20	1.5446	30
40	.0291	.0291	.9996	.0291	34.37	1.000	34.38	1.5417	20
50	.0320	.0320	.9995	.0320	31.24	1.001	31.26	1.5388	10
2°00′	.0349	.0349	.9994	.0349	28.64	1.001	28.65	1.5359	**88°00′**
10	.0378	.0378	.9993	.0378	26.43	1.001	26.45	1.5330	50
20	.0407	.0407	.9992	.0407	24.54	1.001	24.56	1.5301	40
30	.0436	.0436	.9990	.0437	22.90	1.001	22.93	1.5272	30
40	.0465	.0465	.9989	.0466	21.47	1.001	21.49	1.5243	20
50	.0495	.0494	.9988	.0495	20.21	1.001	20.23	1.5213	10
3°00′	.0524	.0523	.9986	.0524	19.08	1.001	19.11	1.5184	**87°00′**
10	.0553	.0552	.9985	.0553	18.07	1.002	18.10	1.5155	50
20	.0582	.0581	.9983	.0582	17.17	1.002	17.20	1.5126	40
30	.0611	.0610	.9981	.0612	16.35	1.002	16.38	1.5097	30
40	.0640	.0640	.9980	.0641	15.60	1.002	15.64	1.5068	20
50	.0669	.0669	.9978	.0670	14.92	1.002	14.96	1.5039	10
4°00′	.0698	.0698	.9976	.0699	14.30	1.002	14.34	1.5010	**86°00′**
10	.0727	.0727	.9974	.0729	13.73	1.003	13.76	1.4981	50
20	.0756	.0756	.9971	.0758	13.20	1.003	13.23	1.4952	40
30	.0785	.0785	.9969	.0787	12.71	1.003	12.75	1.4923	30
40	.0814	.0814	.9967	.0816	12.25	1.003	12.29	1.4893	20
50	.0844	.0843	.9964	.0846	11.83	1.004	11.87	1.4864	10
5°00′	.0873	.0872	.9962	.0875	11.43	1.004	11.47	1.4835	**85°00′**
10	.0902	.0901	.9959	.0904	11.06	1.004	11.10	1.4806	50
20	.0931	.0929	.9957	.0934	10.71	1.004	10.76	1.4777	40
30	.0960	.0958	.9954	.0963	10.39	1.005	10.43	1.4748	30
40	.0989	.0987	.9951	.0992	10.08	1.005	10.14	1.4719	20
50	.1018	.1016	.9948	.1022	9.788	1.005	9.839	1.4690	10
6°00′	.1047	.1045	.9945	.1051	9.514	1.006	9.567	1.4661	**84°00′**
10	.1076	.1074	.9942	.1080	9.255	1.006	9.309	1.4632	50
20	.1105	.1103	.9939	.1110	9.010	1.006	9.065	1.4603	40
30	.1134	.1132	.9936	.1139	8.777	1.006	8.834	1.4573	30
40	.1164	.1161	.9932	.1169	8.556	1.007	8.614	1.4544	20
50	.1193	.1190	.9929	.1198	8.345	1.007	8.405	1.4515	10
		cos α	sin α	cot α	tan α	csc α	sec α	α (radians)	α (degrees)

Table 4 ▪ Values of Trigonometric Functions (*continued*)

α (degrees)	α (radians)	sin α	cos α	tan α	cot α	sec α	csc α		
7°00′	.1222	.1219	.9925	.1228	8.144	1.008	8.206	1.4486	**83°00′**
10	.1251	.1248	.9922	.1257	7.953	1.008	8.016	1.4457	50
20	.1280	.1276	.9918	.1287	7.770	1.008	7.834	1.4428	40
30	.1309	.1305	.9914	.1317	7.596	1.009	7.661	1.4399	30
40	.1338	.1334	.9911	.1346	7.429	1.009	7.496	1.4370	20
50	.1367	.1363	.9907	.1376	7.269	1.009	7.337	1.4341	10
8°00′	.1396	.1392	.9903	.1405	7.115	1.010	7.185	1.4312	**82°00′**
10	.1425	.1421	.9899	.1435	6.968	1.010	7.040	1.4283	50
20	.1454	.1449	.9894	.1465	6.827	1.011	6.900	1.4254	40
30	.1484	.1478	.9890	.1495	6.691	1.011	6.765	1.4224	30
40	.1513	.1507	.9886	.1524	6.561	1.012	6.636	1.4195	20
50	.1542	.1536	.9881	.1554	6.435	1.012	6.512	1.4166	10
9°00′	.1571	.1564	.9877	.1584	6.314	1.012	6.392	1.4137	**81°00′**
10	.1600	.1593	.9872	.1614	6.197	1.013	6.277	1.4108	50
20	.1629	.1622	.9868	.1644	6.084	1.013	6.166	1.4079	40
30	.1658	.1650	.9863	.1673	5.976	1.014	6.059	1.4050	30
40	.1687	.1679	.9858	.1703	5.871	1.014	5.955	1.4021	20
50	.1716	.1708	.9853	.1733	5.769	1.015	5.855	1.3992	10
10°00′	.1745	.1736	.9848	.1763	5.671	1.015	5.759	1.3963	**80°00′**
10	.1774	.1765	.9843	.1793	5.576	1.016	5.665	1.3934	50
20	.1804	.1794	.9838	.1823	5.485	1.016	5.575	1.3904	40
30	.1833	.1822	.9833	.1853	5.396	1.017	5.487	1.3875	30
40	.1862	.1851	.9827	.1883	5.309	1.018	5.403	1.3846	20
50	.1891	.1880	.9822	.1914	5.226	1.018	5.320	1.3817	10
11°00′	.1920	.1908	.9816	.1944	5.145	1.019	5.241	1.3788	**79°00′**
10	.1949	.1937	.9811	.1974	5.066	1.019	5.164	1.3759	50
20	.1978	.1965	.9805	.2004	4.989	1.020	5.089	1.3730	40
30	.2007	.1994	.9799	.2035	4.915	1.020	5.016	1.3701	30
40	.2036	.2022	.9793	.2065	4.843	1.021	4.945	1.3672	20
50	.2065	.2051	.9787	.2095	4.773	1.022	4.876	1.3643	10
12°00′	.2094	.2079	.9781	.2126	4.705	1.022	4.810	1.3614	**78°00′**
10	.2123	.2108	.9775	.2156	4.638	1.023	4.745	1.3584	50
20	.2153	.2136	.9769	.2186	4.574	1.024	4.682	1.3555	40
30	.2182	.2164	.9763	.2217	4.511	1.024	4.620	1.3526	30
40	.2211	.2193	.9757	.2247	4.449	1.025	4.560	1.3497	20
50	.2240	.2221	.9750	.2278	4.390	1.026	4.502	1.3468	10
13°00′	.2269	.2250	.9744	.2309	4.331	1.026	4.445	1.3439	**77°00′**
10	.2298	.2278	.9737	.2339	4.275	1.027	4.390	1.3410	50
20	.2327	.2306	.9730	.2370	4.219	1.028	4.336	1.3381	40
30	.2356	.2334	.9724	.2401	4.165	1.028	4.284	1.3352	30
40	.2385	.2363	.9717	.2432	4.113	1.029	4.232	1.3323	20
50	.2414	.2391	.9710	.2462	4.061	1.030	4.182	1.3294	10
		cos α	sin α	cot α	tan α	csc α	sec α	α (radians)	α (degrees)

Table 4 ▪ Values of Trigonometric Functions (*continued*)

α (degrees)	α (radians)	sin α	cos α	tan α	cot α	sec α	csc α		
14°00′	.2443	.2419	.9703	.2493	4.011	1.031	4.134	1.3265	**76°00′**
10	.2473	.2447	.9696	.2524	3.962	1.031	4.086	1.3235	50
20	.2502	.2476	.9689	.2555	3.914	1.032	4.039	1.3206	40
30	.2531	.2504	.9681	.2586	3.867	1.033	3.994	1.3177	30
40	.2560	.2532	.9674	.2617	3.821	1.034	3.950	1.3148	20
50	.2589	.2560	.9667	.2648	3.776	1.034	3.906	1.3119	10
15°00′	.2618	.2588	.9659	.2679	3.732	1.035	3.864	1.3090	**75°00′**
10	.2647	.2616	.9652	.2711	3.689	1.036	3.822	1.3061	50
20	.2676	.2644	.9644	.2742	3.647	1.037	3.782	1.3032	40
30	.2705	.2672	.9636	.2773	3.606	1.038	3.742	1.3003	30
40	.2734	.2700	.9628	.2805	3.566	1.039	3.703	1.2974	20
50	.2763	.2728	.9621	.2836	3.526	1.039	3.665	1.2945	10
16°00′	.2793	.2756	.9613	.2867	3.487	1.040	3.628	1.2915	**74°00′**
10	.2822	.2784	.9605	.2899	3.450	1.041	3.592	1.2886	50
20	.2851	.2812	.9596	.2931	3.412	1.042	3.556	1.2857	40
30	.2880	.2840	.9588	.2962	3.376	1.043	3.521	1.2828	30
40	.2909	.2868	.9580	.2994	3.340	1.044	3.487	1.2799	20
50	.2938	.2896	.9572	.3026	3.305	1.045	3.453	1.2770	10
17°00′	.2967	.2924	.9563	.3057	3.271	1.046	3.420	1.2741	**73°00′**
10	.2996	.2952	.9555	.3089	3.237	1.047	3.388	1.2712	50
20	.3025	.2979	.9546	.3121	3.204	1.048	3.356	1.2683	40
30	.3054	.3007	.9537	.3153	3.172	1.049	3.326	1.2654	30
40	.3083	.3035	.9528	.3185	3.140	1.049	3.295	1.2625	20
50	.3113	.3062	.9520	.3217	3.108	1.050	3.265	1.2595	10
18°00′	.3142	.3090	.9511	.3249	3.078	1.051	3.236	1.2566	**72°00′**
10	.3171	.3118	.9502	.3281	3.047	1.052	3.207	1.2537	50
20	.3200	.3145	.9492	.3314	3.018	1.053	3.179	1.2508	40
30	.3229	.3173	.9483	.3346	2.989	1.054	3.152	1.2479	30
40	.3258	.3201	.9474	.3378	2.960	1.056	3.124	1.2450	20
50	.3287	.3228	.9465	.3411	2.932	1.057	3.098	1.2421	10
19°00′	.3316	.3256	.9455	.3443	2.904	1.058	3.072	1.2392	**71°00′**
10	.3345	.3283	.9446	.3476	2.877	1.059	3.046	1.2363	50
20	.3374	.3311	.9436	.3508	2.850	1.060	3.021	1.2334	40
30	.3403	.3338	.9426	.3541	2.824	1.061	2.996	1.2305	30
40	.3432	.3365	.9417	.3574	2.798	1.062	2.971	1.2275	20
50	.3462	.3393	.9407	.3607	2.773	1.063	2.947	1.2246	10
20°00′	.3491	.3420	.9397	.3640	2.747	1.064	2.924	1.2217	**70°00′**
10	.3520	.3448	.9387	.3673	2.723	1.065	2.901	1.2188	50
20	.3549	.3475	.9377	.3706	2.699	1.066	2.878	1.2159	40
30	.3578	.3502	.9367	.3739	2.675	1.068	2.855	1.2130	30
40	.3607	.3529	.9356	.3772	2.651	1.069	2.833	1.2101	20
50	.3636	.3557	.9346	.3805	2.628	1.070	2.812	1.2072	10
		cos α	sin α	cot α	tan α	csc α	sec α	α (radians)	α (degrees)

Table 4 ▪ Values of Trigonometric Functions (*continued*)

α (degrees)	**α (radians)**	**sin α**	**cos α**	**tan α**	**cot α**	**sec α**	**csc α**		
21°00′	.3665	.3584	.9336	.3839	2.605	1.071	2.790	1.2043	**69°00′**
10	.3694	.3611	.9325	.3872	2.583	1.072	2.769	1.2014	50
20	.3723	.3638	.9315	.3906	2.560	1.074	2.749	1.1985	40
30	.3752	.3665	.9304	.3939	2.539	1.075	2.729	1.1956	30
40	.3782	.3692	.9293	.3973	2.517	1.076	2.709	1.1926	20
50	.3811	.3719	.9283	.4006	2.496	1.077	2.689	1.1897	10
22°00′	.3840	.3746	.9272	.4040	2.475	1.079	2.669	1.1868	**68°00′**
10	.3869	.3773	.9261	.4074	2.455	1.080	2.650	1.1839	50
20	.3898	.3800	.9250	.4108	2.434	1.081	2.632	1.1810	40
30	.3927	.3827	.9239	.4142	2.414	1.082	2.613	1.1781	30
40	.3956	.3854	.9228	.4176	2.394	1.084	2.595	1.1752	20
50	.3985	.3881	.9216	.4210	2.375	1.085	2.577	1.1723	10
23°00′	.4014	.3907	.9205	.4245	2.356	1.086	2.559	1.1694	**67°00′**
10	.4043	.3934	.9194	.4279	2.337	1.088	2.542	1.1665	50
20	.4072	.3961	.9182	.4314	2.318	1.089	2.525	1.1636	40
30	.4102	.3987	.9171	.4348	2.300	1.090	2.508	1.1606	30
40	.4131	.4014	.9159	.4383	2.282	1.092	2.491	1.1577	20
50	.4160	.4041	.9147	.4417	2.264	1.093	2.475	1.1548	10
24°00′	.4189	.4067	.9135	.4452	2.246	1.095	2.459	1.1519	**66°00′**
10	.4218	.4094	.9124	.4487	2.229	1.096	2.443	1.1490	50
20	.4247	.4120	.9112	.4522	2.211	1.097	2.427	1.1461	40
30	.4276	.4147	.9100	.4557	2.194	1.099	2.411	1.1432	30
40	.4305	.4173	.9088	.4592	2.177	1.100	2.396	1.1403	20
50	.4334	.4200	.9075	.4628	2.161	1.102	2.381	1.1374	10
25°00′	.4363	.4226	.9063	.4663	2.145	1.103	2.366	1.1345	**65°00′**
10	.4392	.4253	.9051	.4699	2.128	1.105	2.352	1.1316	50
20	.4422	.4279	.9038	.4734	2.112	1.106	2.337	1.1286	40
30	.4451	.4305	.9026	.4770	2.097	1.108	2.323	1.1257	30
40	.4480	.4331	.9013	.4806	2.081	1.109	2.309	1.1228	20
50	.4509	.4358	.9001	.4841	2.066	1.111	2.295	1.1199	10
26°00′	.4538	.4384	.8988	.4877	2.050	1.113	2.281	1.1170	**64°00′**
10	.4567	.4410	.8975	.4913	2.035	1.114	2.268	1.1141	50
20	.4596	.4436	.8962	.4950	2.020	1.116	2.254	1.1112	40
30	.4625	.4462	.8949	.4986	2.006	1.117	2.241	1.1083	30
40	.4654	.4488	.8936	.5022	1.991	1.119	2.228	1.1054	20
50	.4683	.4514	.8923	.5059	1.977	1.121	2.215	1.1025	10
27°00′	.4712	.4540	.8910	.5095	1.963	1.122	2.203	1.0996	**63°00′**
10	.4741	.4566	.8897	.5132	1.949	1.124	2.190	1.0966	50
20	.4771	.4592	.8884	.5169	1.935	1.126	2.178	1.0937	40
30	.4800	.4617	.8870	.5206	1.921	1.127	2.166	1.0908	30
40	.4829	.4643	.8857	.5243	1.907	1.129	2.154	1.0879	20
50	.4858	.4669	.8843	.5280	1.894	1.131	2.142	1.0850	10
		cos α	**sin α**	**cot α**	**tan α**	**csc α**	**sec α**	**α (radians)**	**α (degrees)**

Table 4 ■ Values of Trigonometric Functions (*continued*)

α (degrees)	α (radians)	sin α	cos α	tan α	cot α	sec α	csc α		
28°00′	.4887	.4695	.8829	.5317	1.881	1.133	2.130	1.0821	**62°00′**
10	.4916	.4720	.8816	.5354	1.868	1.134	2.118	1.0792	50
20	.4945	.4746	.8802	.5392	1.855	1.136	2.107	1.0763	40
30	.4974	.4772	.8788	.5430	1.842	1.138	2.096	1.0734	30
40	.5003	.4797	.8774	.5467	1.829	1.140	2.085	1.0705	20
50	.5032	.4823	.8760	.5505	1.816	1.142	2.074	1.0676	10
29°00′	.5061	.4848	.8746	.5543	1.804	1.143	2.063	1.0647	**61°00′**
10	.5091	.4874	.8732	.5581	1.792	1.145	2.052	1.0617	50
20	.5120	.4899	.8718	.5619	1.780	1.147	2.041	1.0588	40
30	.5149	.4924	.8704	.5658	1.767	1.149	2.031	1.0559	30
40	.5178	.4950	.8689	.5696	1.756	1.151	2.020	1.0530	20
50	.5207	.4975	.8675	.5735	1.744	1.153	2.010	1.0501	10
30°00′	.5236	.5000	.8660	.5774	1.732	1.155	2.000	1.0472	**60°00′**
10	.5265	.5025	.8646	.5812	1.720	1.157	1.990	1.0443	50
20	.5294	.5050	.8631	.5851	1.709	1.159	1.980	1.0414	40
30	.5323	.5075	.8616	.5890	1.698	1.161	1.970	1.0385	30
40	.5352	.5100	.8601	.5930	1.686	1.163	1.961	1.0356	20
50	.5381	.5125	.8587	.5969	1.675	1.165	1.951	1.0327	10
31°00′	.5411	.5150	.8572	.6009	1.664	1.167	1.942	1.0297	**59°00′**
10	.5440	.5175	.8557	.6048	1.653	1.169	1.932	1.0268	50
20	.5469	.5200	.8542	.6088	1.643	1.171	1.923	1.0239	40
30	.5498	.5225	.8526	.6128	1.632	1.173	1.914	1.0210	30
40	.5527	.5250	.8511	.6168	1.621	1.175	1.905	1.0181	20
50	.5556	.5275	.8496	.6208	1.611	1.177	1.896	1.0152	10
32°00′	.5585	.5299	.8480	.6249	1.600	1.179	1.887	1.0123	**58°00′**
10	.5614	.5324	.8465	.6289	1.590	1.181	1.878	1.0094	50
20	.5643	.5348	.8450	.6330	1.580	1.184	1.870	1.0065	40
30	.5672	.5373	.8434	.6371	1.570	1.186	1.861	1.0036	30
40	.5701	.5398	.8418	.6412	1.560	1.188	1.853	1.0007	20
50	.5730	.5422	.8403	.6453	1.550	1.190	1.844	.9977	10
33°00′	.5760	.5446	.8387	.6494	1.540	1.192	1.836	.9948	**57°00′**
10	.5789	.5471	.8371	.6536	1.530	1.195	1.828	.9919	50
20	.5818	.5495	.8355	.6577	1.520	1.197	1.820	.9890	40
30	.5847	.5519	.8339	.6619	1.511	1.199	1.812	.9861	30
40	.5876	.5544	.8323	.6661	1.501	1.202	1.804	.9832	20
50	.5905	.5568	.8307	.6703	1.492	1.204	1.796	.9803	10
34°00′	.5934	.5592	.8290	.6745	1.483	1.206	1.788	.9774	**56°00′**
10	.5963	.5616	.8274	.6787	1.473	1.209	1.781	.9745	50
20	.5992	.5640	.8258	.6830	1.464	1.211	1.773	.9716	40
30	.6021	.5664	.8241	.6873	1.455	1.213	1.766	.9687	30
40	.6050	.5688	.8225	.6916	1.446	1.216	1.758	.9657	20
50	.6080	.5712	.8208	.6959	1.437	1.218	1.751	.9628	10
		cos α	sin α	cot α	tan α	csc α	sec α	α (radians)	α (degrees)

Table 4 ■ Values of Trigonometric Functions (*continued*)

α (degrees)	α (radians)	sin α	cos α	tan α	cot α	sec α	csc α		
35°00′	.6109	.5736	.8192	.7002	1.428	1.221	1.743	.9599	**55°00′**
10	.6138	.5760	.8175	.7046	1.419	1.223	1.736	.9570	50
20	.6167	.5783	.8158	.7089	1.411	1.226	1.729	.9541	40
30	.6196	.5807	.8141	.7133	1.402	1.228	1.722	.9512	30
40	.6225	.5831	.8124	.7177	1.393	1.231	1.715	.9483	20
50	.6254	.5854	.8107	.7221	1.385	1.233	1.708	.9454	10
36°00′	.6283	.5878	.8090	.7265	1.376	1.236	1.701	.9425	**54°00′**
10	.6312	.5901	.8073	.7310	1.368	1.239	1.695	.9396	50
20	.6341	.5925	.8056	.7355	1.360	1.241	1.688	.9367	40
30	.6370	.5948	.8039	.7400	1.351	1.244	1.681	.9338	30
40	.6400	.5972	.8021	.7445	1.343	1.247	1.675	.9308	20
50	.6429	.5995	.8004	.7490	1.335	1.249	1.668	.9279	10
37°00′	.6458	.6018	.7986	.7536	1.327	1.252	1.662	.9250	**53°00′**
10	.6487	.6041	.7969	.7581	1.319	1.255	1.655	.9221	50
20	.6516	.6065	.7951	.7627	1.311	1.258	1.649	.9192	40
30	.6545	.6088	.7934	.7673	1.303	1.260	1.643	.9163	30
40	.6574	.6111	.7916	.7720	1.295	1.263	1.636	.9134	20
50	.6603	.6134	.7898	.7766	1.288	1.266	1.630	.9105	10
38°00′	.6632	.6157	.7880	.7813	1.280	1.269	1.624	.9076	**52°00′**
10	.6661	.6180	.7862	.7860	1.272	1.272	1.618	.9047	50
20	.6690	.6202	.7844	.7907	1.265	1.275	1.612	.9018	40
30	.6720	.6225	.7826	.7954	1.257	1.278	1.606	.8988	30
40	.6749	.6248	.7808	.8002	1.250	1.281	1.601	.8959	20
50	.6778	.6271	.7790	.8050	1.242	1.284	1.595	.8930	10
39°00′	.6807	.6293	.7771	.8098	1.235	1.287	1.589	.8901	**51°00′**
10	.6836	.6316	.7753	.8146	1.228	1.290	1.583	.8872	50
20	.6865	.6338	.7735	.8195	1.220	1.293	1.578	.8843	40
30	.6894	.6361	.7716	.8243	1.213	1.296	1.572	.8814	30
40	.6923	.6383	.7698	.8292	1.206	1.299	1.567	.8785	20
50	.6952	.6406	.7679	.8342	1.199	1.302	1.561	.8756	10
40°00′	.6981	.6428	.7660	.8391	1.192	1.305	1.556	.8727	**50°00′**
10	.7010	.6450	.7642	.8441	1.185	1.309	1.550	.8698	50
20	.7039	.6472	.7623	.8491	1.178	1.312	1.545	.8668	40
30	.7069	.6494	.7604	.8541	1.171	1.315	1.540	.8639	30
40	.7098	.6517	.7585	.8591	1.164	1.318	1.535	.8610	20
50	.7127	.6539	.7566	.8642	1.157	1.322	1.529	.8581	10
41°00′	.7156	.6561	.7547	.8693	1.150	1.325	1.524	.8552	**49°00′**
10	.7185	.6583	.7528	.8744	1.144	1.328	1.519	.8523	50
20	.7214	.6604	.7509	.8796	1.137	1.332	1.514	.8494	40
30	.7243	.6626	.7490	.8847	1.130	1.335	1.509	.8465	30
40	.7272	.6648	.7470	.8899	1.124	1.339	1.504	.8436	20
50	.7301	.6670	.7451	.8952	1.117	1.342	1.499	.8407	10
		cos α	sin α	cot α	tan α	csc α	sec α	α (radians)	α (degrees)

Table 4 ■ Values of Trigonometric Functions (*continued*)

α (degrees)	α (radians)	sin α	cos α	tan α	cot α	sec α	csc α		
42°00′	.7330	.6691	.7431	.9004	1.111	1.346	1.494	.8378	**48°00′**
10	.7359	.6713	.7412	.9057	1.104	1.349	1.490	.8348	50
20	.7389	.6734	.7392	.9110	1.098	1.353	1.485	.8319	40
30	.7418	.6756	.7373	.9163	1.091	1.356	1.480	.8290	30
40	.7447	.6777	.7353	.9217	1.085	1.360	1.476	.8261	20
50	.7476	.6799	.7333	.9271	1.079	1.364	1.471	.8232	10
43°00′	.7505	.6820	.7314	.9325	1.072	1.367	1.466	.8203	**47°00′**
10	.7534	.6841	.7294	.9380	1.066	1.371	1.462	.8174	50
20	.7563	.6862	.7274	.9435	1.060	1.375	1.457	.8145	40
30	.7592	.6884	.7254	.9490	1.054	1.379	1.453	.8116	30
40	.7621	.6905	.7234	.9545	1.048	1.382	1.448	.8087	20
50	.7650	.6926	.7214	.9601	1.042	1.386	1.444	.8058	10
44°00′	.7679	.6947	.7193	.9657	1.036	1.390	1.440	.8029	**46°00′**
10	.7709	.6967	.7173	.9713	1.030	1.394	1.435	.7999	50
20	.7738	.6988	.7153	.9770	1.024	1.398	1.431	.7970	40
30	.7767	.7009	.7133	.9827	1.018	1.402	1.427	.7941	30
40	.7796	.7030	.7112	.9884	1.012	1.406	1.423	.7912	20
50	.7825	.7050	.7092	.9942	1.006	1.410	1.418	.7883	10
45°00′	.7854	.7071	.7071	1.000	1.000	1.414	1.414	.7854	**45°00′**
		cos α	sin α	cot α	tan α	csc α	sec α	α (radians)	α (degrees)

Solutions to Chapter Tests

Chapter 1 Test, p. 64

1. $2x + 13 = 56$
For further practice, see Example 1, Section 1.1.

2. $n + (n + 1) + (n + 2) = 1.50(n + 2)$
For further practice, see Example 2, Section 1.1.

3. **(a)** integer, rational, real
(b) whole number, integer, rational, real
(c) irrational, real
For further practice, see Example 1, Section 1.2.

4. $-|-3| - |4| = -3 - 4 = -7$
For further practice, see Example 1, Section 1.3.

5. $\sqrt{0.0004} = 0.02$
For further practice, see Example 1, Section 1.7.

6. **(a)** $d(A, O) = |-6 - 0| = |-6| = 6$
(b) $d(B, O) = |9 - 0| = |9| = 9$
For further practice, see Example 5, Section 1.3.

7. **(a)** $0.0023 \times 10^{-4} = 2.3 \times 10^{-7}$
(b) $\dfrac{1}{5 \times 10^{-3}} = \dfrac{1}{5} \times 10^{3} = 2.0 \times 10^{2}$
For further practice, see Example 5, Section 1.4.

8. $(5x^2)(y^3)(3x^{-5}) = (5 \cdot 3)(x^2 \cdot x^{-5})(y^3)$
$= 15x^{-3}y^3 = \dfrac{15y^3}{x^3}$
For further practice, see Example 3, Section 1.4.

9. $(3xy^2z)^4 = (3)^4(x)^4(y^2)^4(z)^4 = 81x^4y^8z^4$
For further practice, see Example 3, Section 1.4.

10. $\left(\dfrac{2x}{5y^2}\right)^{-3} = \left(\dfrac{5y^2}{2x}\right)^{3} = \dfrac{125y^6}{8x^3}$
For further practice, see Example 3, Section 1.4.

11. $(u^2 + 3u - 7) - (u^3 - 2u^2 + u) =$
$-u^3 + 3u^2 + 2u - 7$
For further practice, see Example 3, Section 1.5.

12. $3x(x^2 - 7x + 4) = 3x^3 - 21x^2 + 12x$
For further practice, see Example 4, Section 1.5.

13. $(8m - 2)(3m + 5) =$
$(8m)(3m) + (8m)(5) - (2)(3m) - (2)(5) =$
$24m^2 + 34m - 10$
For further practice, see Example 6, Section 1.5.

14. $(9z - 2)^2 = (9z - 2)(9z - 2) =$
$(9z)^2 - 2(9z)(2) + (-2)^2 =$
$81z^2 - 36z + 4$
For further practice, see Example 7, Section 1.5.

15. $\left(\dfrac{10u + 10v}{u^3 + 8v^3}\right)\left(\dfrac{u^2 + 3uv + 2v^2}{u^2 - 5uv - 6v^2}\right) =$

$\dfrac{10(u + v)}{(u + 2v)(u^2 - 2uv + 4v^2)} \cdot \dfrac{(u + v)(u + 2v)}{(u + v)(u - 6v)} =$

$\dfrac{10(u + v)}{(u - 6v)(u^2 - 2uv + 4v^2)}$

For further practice, see Example 2, Section 1.8.

16. $12x^2yz = 2^2 \cdot 3 \cdot x^2 \cdot y \cdot z$

$8x^2z^3 = 2^3 \cdot x^2 \cdot z^3$

$2x^5yz^2 = 2 \cdot x^5 \cdot y \cdot z^2$

GCF is $2 \cdot x^2 \cdot z = 2x^2z$

For further practice, see Example 1, Section 1.6.

17. $6x^2y - 2xy^2 + 7x^2y^2 = xy(6x - 2y + 7xy)$

For further practice, see Example 3, Section 1.6.

18. $6t^2 - 5t - 4 = (2t + 1)(3t - 4)$

For further practice, see Example 7, Section 1.6.

19. $\sqrt{24x^2y^2z} = \sqrt{2^2 \cdot 6}\sqrt{x^2}\sqrt{y^2}\sqrt{z} = 2xy\sqrt{6z}$

For further practice, see Example 2, Section 1.7.

20. $\sqrt[4]{\dfrac{u^3}{32v^5}} = \sqrt[4]{\dfrac{u^3}{32v^5} \cdot \dfrac{8v^3}{8v^3}} = \dfrac{\sqrt[4]{8u^3v^3}}{\sqrt[4]{(2^2)^4(v^2)^4}} = \dfrac{\sqrt[4]{8u^3v^3}}{4v^2}$

For further practice, see Example 3, Section 1.7.

21. $25x^{1/2}(2y^{1/3})^2 = 25x^{1/2} \cdot 4y^{2/3} = 100x^{1/2}y^{2/3}$

For further practice, see Example 7, Section 1.7.

22. $\dfrac{3x^{2/3}}{8x^{1/9}y^{-1/2}} = \dfrac{3}{8}x^{2/3-1/9}y^{1/2} = \dfrac{3}{8}x^{5/9}y^{1/2}$

For further practice, see Example 7, Section 1.7.

23. $\dfrac{m^3 - 9m}{3m^2 + 8m - 3} = \dfrac{m(m + 3)(m - 3)}{(m + 3)(3m - 1)} = \dfrac{m(m - 3)}{3m - 1}$

For further practice, see Example 1, Section 1.8.

24. $\dfrac{2 + \dfrac{1}{u-5}}{\dfrac{19}{u - 5} - 12} = \dfrac{2 + \dfrac{1}{u-5}}{\dfrac{19}{u - 5} - 12} \times \dfrac{u - 5}{u - 5} =$

$\dfrac{2(u - 5) + 1}{19 - 12(u - 5)} = \dfrac{2u - 9}{79 - 12u}$

For further practice, see Example 5, Section 1.8.

Chapter 2 Test, p. 124

1. $(4x - 1)x = (2x + 3)(2x - 1)$

$4x^2 - x = 4x^2 + 4x - 3$

$-5x = -3$

$x = \dfrac{3}{5}$

For further practice, see Example 3, Section 2.1.

2. $ax - 1 = b(x - 4)$

$ax - 1 = bx - 4b$

$(a - b)x = 1 - 4b$

$x = \dfrac{1 - 4b}{a - b}$

For further practice, see Example 5, Section 2.1.

3. $\dfrac{x}{x + 2} = 2 - \dfrac{1}{x + 2}$

$\dfrac{x}{x + 2}(x + 2) = 2(x + 2) - \dfrac{1}{x + 2}(x + 2)$

$x = 2x + 4 - 1$

$x = -3$

Must Check:

$\dfrac{x}{x + 2} = 2 - \dfrac{1}{x + 2}$

$\dfrac{-3}{-3 + 2} = 2 - \dfrac{1}{-3 + 2}$

$3 = 3$

The solution is -3.

For further practice, see Example 6, Section 2.1.

4. $|6x - 1| = 8x + 5$

$$\begin{aligned} 6x - 1 &= 8x + 5 \\ -2x &= 6 \\ x &= -3 \end{aligned} \qquad \text{or} \qquad \begin{aligned} 6x - 1 &= -(8x + 5) \\ 6x - 1 &= -8x - 5 \\ 14x &= -4 \\ x &= -\frac{2}{7} \end{aligned}$$

Must Check:

$$\begin{aligned} |6x - 1| &= 8x + 5 \\ |6(-3) - 1| &= 8(-3) + 5 \\ 19 &= -19 \quad \text{False} \end{aligned} \qquad \begin{aligned} |6x - 1| &= 8x + 5 \\ \left|6\left(-\frac{2}{7}\right) - 1\right| &= 8\left(-\frac{2}{7}\right) + 5 \\ \frac{19}{7} &= \frac{19}{7} \quad \text{True} \end{aligned}$$

The only solution is $-\frac{2}{7}$.

For further practice, see Example 7, Section 2.1.

5. $(-i)^{27} = (-1)^{27}i^{27} = (-1)i^{24}i^3 = (-1)(1)(-i) = i$

For further practice, see Example 2, Section 2.2.

6. $\sqrt{-3} \cdot \sqrt{-21} \cdot \sqrt{-7} = (i\sqrt{3})(i\sqrt{21})(i\sqrt{7})$

$= (i^3)(\sqrt{3})(\sqrt{3} \cdot \sqrt{7})(\sqrt{7})$

$= -i \cdot 3 \cdot 7 = -21i$

For further practice, see Example 2, Section 2.2.

7. $(9 - 5i) + (-3 - 4i) = (9 - 3) + (-5 - 4)i =$
$6 - 9i$

For further practice, see Example 3, Section 2.2.

8. $(6 - i)(7 + 3i) = 42 + 18i - 7i - 3i^2$

$= 42 + 11i - 3(-1) = 45 + 11i$

For further practice, see Example 4, Section 2.2.

9. $\frac{9 + i}{2 - 3i} = \frac{9 + i}{2 - 3i} \cdot \frac{2 + 3i}{2 + 3i} = \frac{18 + 29i + 3i^2}{4 - 9i^2}$

$= \frac{15 + 29i}{13} = \frac{15}{13} + \frac{29}{13}i$

For further practice, see Example 6, Section 2.2.

10.

$$3v^2 + 11v - 20 = 0$$
$$(3v - 4)(v + 5) = 0$$

$3v - 4 = 0$	$v + 5 = 0$
$v = \frac{4}{3}$	$v = -5$

The solutions are $\frac{4}{3}$ and -5.

For further practice, see Example 1, Section 2.3.

11. $2x^2 - 6x + 1 = 0$

$$x^2 - 3x + \frac{1}{2} = 0$$

$$x^2 - 3x + \square = -\frac{1}{2} + \square$$

$$x^2 - 3x + \frac{9}{4} = -\frac{1}{2} + \frac{9}{4}$$

$$\left(x - \frac{3}{2}\right)^2 = \frac{7}{4}$$

$$x - \frac{3}{2} = \pm\sqrt{\frac{7}{4}}$$

$$x = \frac{3}{2} \pm \frac{1}{2}\sqrt{7}$$

The solutions are $\frac{3}{2} \pm \frac{1}{2}\sqrt{7}$.

For further practice, see Example 5, Section 2.3.

12. $x^2 - 3x + 3 = 0$

$$x = \frac{-b \pm \sqrt{b^2 - 4ac}}{2a}$$

$$x = \frac{-(-3) \pm \sqrt{(-3)^2 - 4(1)(3)}}{2(1)}$$

$$x = \frac{3 \pm \sqrt{-3}}{2}$$

The solutions are $\frac{3}{2} \pm \frac{\sqrt{3}}{2}i$.

For further practice, see Example 7, Section 2.3.

13.

$$\frac{2}{y+5} = \frac{5}{6y+3} + \frac{3}{2y+1}$$

$$\frac{2}{y+5}[(y+5)(6y+3)(2y+1)] = \frac{5}{6y+3}[(y+5)(6y+3)(2y+1)] + \frac{3}{2y+1}[(y+5)(6y+3)(2y+1)]$$

$$2(6y+3)(2y+1) = 5(y+5)(2y+1) + 3(y+5)(6y+3)$$

$$24y^2 + 24y + 6 = 28y^2 + 154y + 70$$

$$4y^2 + 130y + 64 = 0$$

$$2y^2 + 65y + 32 = 0$$

$$(y+32)(2y+1) = 0$$

$y + 32 = 0$	$2y + 1 = 0$
$y = -32$	$2y = -1$
	$y = -\frac{1}{2}$

Must Check:

$$\frac{2}{y+5} = \frac{5}{6y+3} + \frac{3}{2y+1}$$

$$\frac{2}{(-32)+5} = \frac{5}{6(-32)+3} + \frac{3}{2(-32)+1}$$

$$\frac{-14}{189} = \frac{-14}{189}$$

$$\frac{2}{y+5} = \frac{5}{6y+3} + \frac{3}{2y+1}$$

$$\frac{2}{\left(-\frac{1}{2}\right)+5} = \frac{5}{6\left(-\frac{1}{2}\right)+3} + \frac{3}{2\left(-\frac{1}{2}\right)+1}$$

$$\frac{4}{9} = \frac{5}{0} + \frac{3}{0} \quad \text{Undefined}$$

The solution is -32.

For further practice, see Example 3, Section 2.4.

14.

$$10x^{-2} + 4x^{-4} = 6$$

$$2(x^{-2})^2 + 5x^{-2} - 3 = 0$$

$$2u^2 + 5u - 3 = 0$$

$$(2u - 1)(u + 3) = 0$$

$2u - 1 = 0$	$u + 3 = 0$
$u = \frac{1}{2}$	$u = -3$
$x^{-2} = \frac{1}{2}$	$x^{-2} = -3$
$x^2 = 2$	$x^2 = -\frac{1}{3}$
$x = \pm\sqrt{2}$	$x = \pm\frac{i\sqrt{3}}{3}$

The solutions are $\pm\sqrt{2}, \pm\frac{\sqrt{3}}{3}i$.

For further practice, see Example 5, Section 2.4.

15.

$$\sqrt{2y-1} = 2 + \sqrt{y-4}$$

$$2y - 1 = 4 + 4\sqrt{y-4} + y - 4$$

$$y - 1 = 4\sqrt{y-4}$$

$$y^2 - 2y + 1 = 16(y-4)$$

$$y^2 - 18y + 65 = 0$$

$$(y-5)(y-13) = 0$$

$y - 5 = 0$	$y - 13 = 0$
$y = 5$	$y = 13$

Must Check:

$$\sqrt{2y-1} = 2 + \sqrt{y-4}$$

$$\sqrt{2(5)-1} = 2 + \sqrt{5-4}$$

$$3 = 2 + 1 \quad \text{True}$$

$$\sqrt{2y-1} = 2 + \sqrt{y-4}$$

$$\sqrt{2(13)-1} = 2 + \sqrt{13-4}$$

$$5 = 2 + 3 \quad \text{True}$$

The solutions are 5 and 13.

For further practice, see Example 7, Section 2.4.

16. 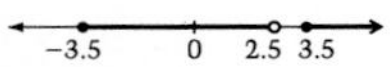 [number line: −3.5, 0, 2.5, 3.5]

For further practice, see Example 4, Section 2.5.

17. $3 + 2m > 15 + 3m$

$-12 > m$

The solution set is $(-\infty, -12)$.
For further practice, see Example 1, Section 2.5.

18. $|3 - 2x| > 8$

$3 - 2x > 8$ or $3 - 2x < -8$

$-2x > 5$ $\qquad$ $-2x < -11$

$x < -\frac{5}{2}$ $\qquad$ $x > \frac{11}{2}$

The solution set is $\left(-\infty, -\frac{5}{2}\right) \cup \left(\frac{11}{2}, \infty\right)$.
For further practice, see Example 5, Section 2.5.

19.

$$x^2 + 9x \leq -18$$
$$x^2 + 9x + 18 \leq 0$$
$$x^2 + 9x + 18 = 0$$
$$(x + 3)(x + 6) = 0$$

$x + 3 = 0$	$x + 6 = 0$
$x = -3$	$x = -6$

Interval	Test Value	Value of $x^2 + 9x + 18$	Sign
$(-\infty, -6)$	-10	$(-10)^2 + 9(-10) + 18 = 28$	+
$(-6, -3)$	-5	$(-5)^2 + 9(-5) + 18 = -2$	−
$(-3, \infty)$	0	$0^2 + 9(0) + 18 = 18$	+

The solution of the given inequality is $[-6, -3]$.
For further practice, see Example 7, Section 2.5.

20.

$$\frac{11}{x+3} \leq 2$$
$$\frac{11}{x+3} - \frac{2(x+3)}{x+3} \leq 0$$
$$\frac{5 - 2x}{x+3} \leq 0$$

The numerator is 0 when $x = \frac{5}{2}$, and the denominator is 0 when $x = -3$.

Interval	Test Value	Value of $\frac{5 - 2x}{x + 3}$	Sign
$(-\infty, -3)$	-4	$\frac{5 - 2(-4)}{-4 + 3} = -13$	−
$\left(-3, \frac{5}{2}\right)$	0	$\frac{5 - 2(0)}{0 + 3} = \frac{5}{3}$	+
$\left(\frac{5}{2}, \infty\right)$	3	$\frac{5 - 2(3)}{3 + 3} = -\frac{1}{6}$	−

The solution of the given inequality is $(-\infty, -3) \cup \left[\frac{5}{2}, \infty\right)$.
For further practice, see Example 8, Section 2.5.

21. x = distance from airport when jet catches up to small plane.

$$\text{time} = \frac{\text{distance}}{\text{rate}}$$

time equation:

$$\frac{x}{190} = 2 + \frac{x}{550}$$
$$550x = 2(190)(550) + 190x.$$
$$360x = 2(190)(550)$$
$$9x = 19(275)$$
$$x = \frac{5225}{9} = 580\frac{5}{9}$$

The jet catches the small plane $580\frac{5}{9}$ miles from the airport.
For further practice, see Example 2, Section 2.6.

Chapter 3 Test, pp. 215–216

1. $f(x) = x^2 - 3x - 8$

$$= \left(x - \frac{3}{2}\right)^2 - 8 - \frac{9}{4}$$

$$= \left(x - \frac{3}{2}\right)^2 - \frac{41}{4}$$

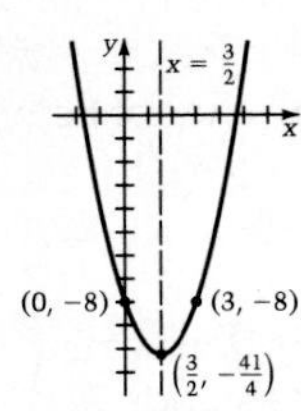

For further practice, see Example 2, Section 3.1.

2. $g(x) = 11 - 4|x|$

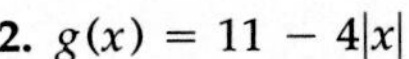

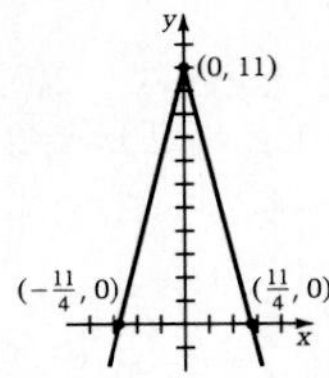

For further practice, see Example 3, Section 3.1.

3. (a)

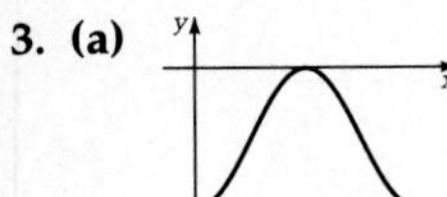

(b)

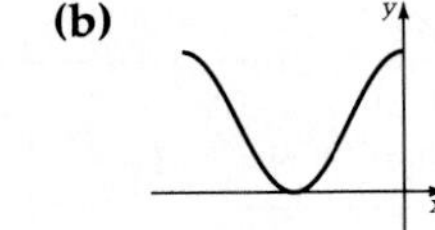

(c)

For further practice, see Example 4, Section 3.1.

4. $x^2 - y^2 + 8 = 0$

x-axis?

$$x^2 - (-y)^2 + 8 = 0$$
$$x^2 - y^2 + 8 = 0$$

yes

y-axis?

$$(-x)^2 - y^2 + 8 = 0$$
$$x^2 - y^2 + 8 = 0$$

yes

origin?

$$(-x)^2 - (-y)^2 + 8 = 0$$
$$x^2 - y^2 + 8 = 0$$

yes

For further practice, see Example 9, Section 3.1.

5. $x^3y = -5$

x-axis?

$$x^3(-y) = -5$$
$$-x^3y = -5$$

no

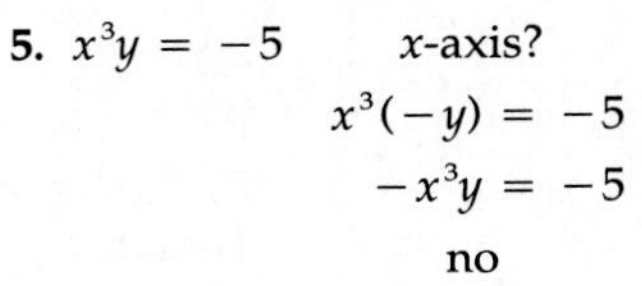

y-axis?

$$(-x)^3y = -5$$
$$-x^3y = -5$$

no

origin?

$$(-x)^3(-y) = -5$$
$$x^3y = -5$$

yes

For further practice, see Example 9, Section 3.1.

6.

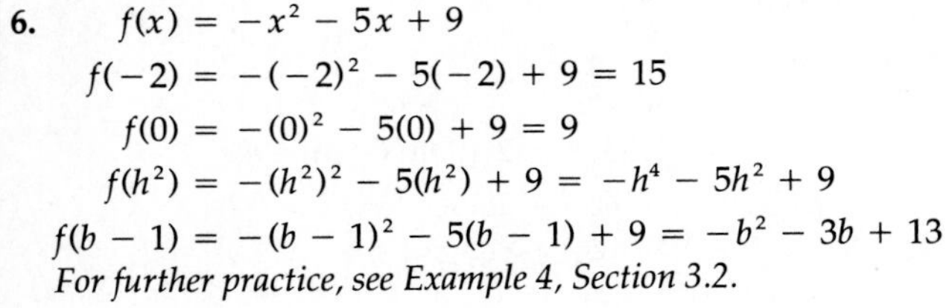

$$f(x) = -x^2 - 5x + 9$$
$$f(-2) = -(-2)^2 - 5(-2) + 9 = 15$$
$$f(0) = -(0)^2 - 5(0) + 9 = 9$$
$$f(h^2) = -(h^2)^2 - 5(h^2) + 9 = -h^4 - 5h^2 + 9$$
$$f(b - 1) = -(b - 1)^2 - 5(b - 1) + 9 = -b^2 - 3b + 13$$

For further practice, see Example 4, Section 3.2.

7.

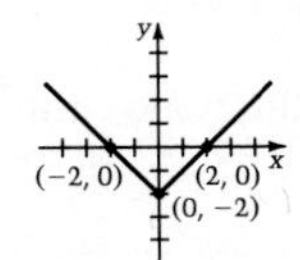

For further practice, see Example 1, Section 3.3.

8.

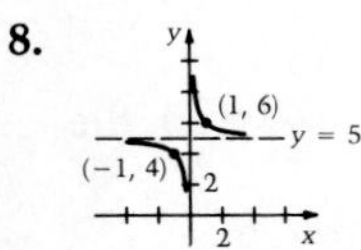

For further practice, see Example 1, Section 3.3.

9.

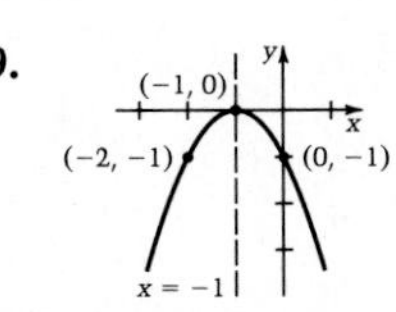

For further practice, see Example 3, Section 3.3.

10. (a)

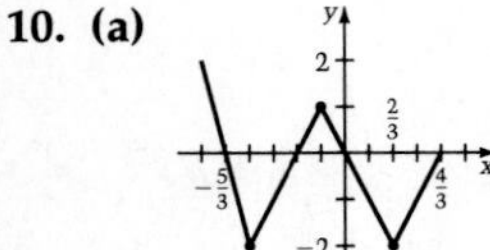

(b)

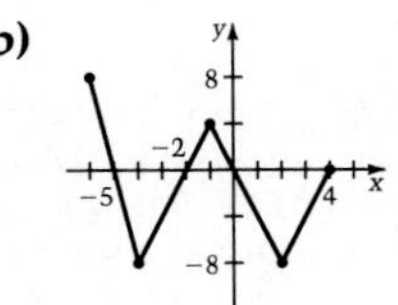

(c)

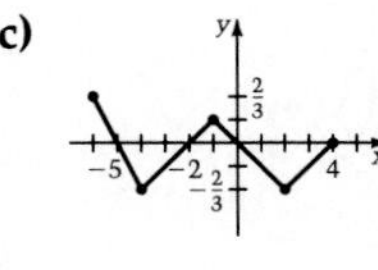

For further practice, see Example 4, Section 3.3.

11. $m = \dfrac{4 - (-1)}{5 - 3} = \dfrac{5}{2}$

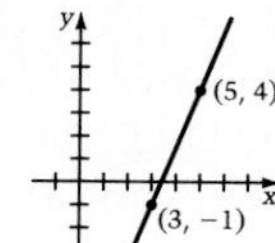

For further practice, see Example 1, Section 3.4.

12. Since the line is perpendicular to the horizontal line $y = -5$, it is a vertical line with the form $x = c$.
(a)–(b) cannot be written.
(c) $x - 6 = 0$
For further practice, see Example 5, Section 3.4.

13.
$$\begin{aligned} y &= x^2 - 4x - 21 \\ &= (x^2 - 4x + 4) - 21 - 4 \\ &= (x - 2)^2 - 25 \end{aligned}$$

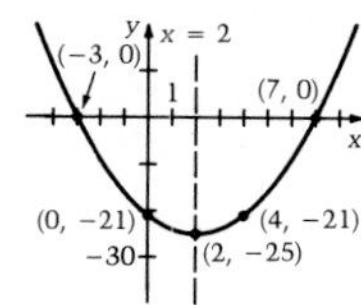

For further practice, see Example 3, Section 3.5.

14. **(a)** exhaling
(b) 12 seconds
(c) 0.6 liters
For further practice, see Example 2, Section 3.8.

15. $A = \dfrac{kb^2}{c}$
For further practice, see Example 2, Section 3.9.

16.
$$m = knp^2 \qquad m = \frac{1}{2}np^2$$
$$6 = k(3)(2)^2 \qquad m = \frac{1}{2}(4)(9)^2$$
$$k = \frac{1}{2} \qquad m = 162$$

When $n = 4$ and $p = 9$, $m = 162$.
For further practice, see Example 3, Section 3.9.

17. $(f \circ g)(-2) = f(g(-2)) = f(|-2| - 2) = f(0) = 3(0) + 3 = 3$
For further practice, see Example 2, Section 3.6.

18. $(g \circ f)(0) = g(f(0)) = g(3(0) + 3) = g(3) = |3| - 2 = 1$
For further practice, see Example 2, Section 3.6.

19. $(f \circ g)(x) = f(g(x)) = f(\sqrt{x + 6}) = \dfrac{1}{2 - \sqrt{x + 6}}$

Since the range of g must be contained in the domain of f:
$$\begin{aligned} \sqrt{x + 6} &\neq 2 \\ x + 6 &\neq 4 \\ x &\neq -2 \end{aligned}$$

The domain of $f \circ g$ consists of those values in both the domain of g, $[-6, \infty)$, and in the set found above, $(-\infty, -2) \cup (-2, \infty)$. Thus, the domain of $f \circ g$ is $[-6, -2) \cup (-2, \infty)$.
For further practice, see Example 4, Section 3.6.

20. $F(x) = (f \circ h \circ g)(x)$
For further practice, see Example 6, Section 3.6.

21.
$$\begin{aligned} f(x) &= \sqrt{x} + 3 \\ y &= \sqrt{x} + 3 \\ \sqrt{x} &= y - 3 \\ x &= (y - 3)^2 \\ y &= (x - 3)^2 \\ f^{-1}(x) &= (x - 3)^2 \end{aligned}$$

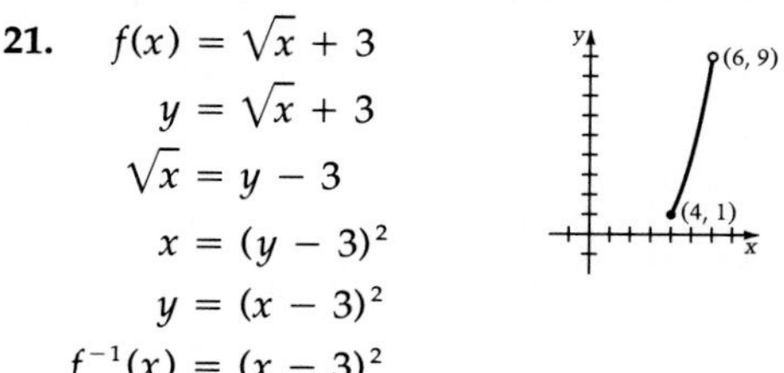

domain of $f^{-1} = [4, 6)$
range of $f^{-1} = [1, 9)$
For further practice, see Example 4, Section 3.7.

Chapter 4 Test, p. 295

1. Graph No. 4 since the leading term is x^3.
For further practice, see Example 3, Section 4.1.

2. Graph No. 1 since the leading term is $-x^3$.
For further practice, see Example 3, Section 4.1.

3. Graph No. 3 since the leading term is $-x^4$.
For further practice, see Example 3, Section 4.1.

4. Graph No. 2 since the leading term is x^4.
For further practice, see Example 3, Section 4.1.

5. By the Third Property, we know that the graph can have, at most, 2 relative maxima and minima. Thus, we can eliminate Graphs 1 and 3. The y-intercept is $f(0) = 0$, so we can eliminate Graph 4. Thus, Graph 2 is the correct choice.
For further practice, see Example 2, Section 4.2.

6. By the Third Property, we know that the graph can have, at most, 2 relative maxima and minima. Thus, we can eliminate Graphs 1 and 3. The y-intercept is $f(0) = 6$, so we can eliminate Graph 2. Thus, Graph 4 is the correct choice.
For further practice, see Example 2, Section 4.2.

7. $f(x) = x(x - 2)(x + 4)^4$
The zeros are -4, 0, and 2.

Interval	Test Value t	$f(t)$	
$(-\infty, -4)$	-5	35	$f(t) > 0$; graph lies above x-axis
$(-4, 0)$	-3	15	$f(t) > 0$; graph lies above x-axis
$(0, 2)$	1	-625	$f(t) < 0$; graph lies below x-axis
$(2, \infty)$	3	7203	$f(t) > 0$; graph lies above x-axis

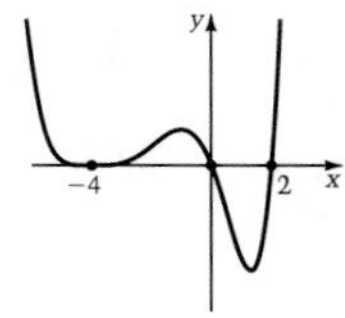

For further practice, see Example 5, Section 4.2.

8.
$$\begin{array}{r|rrrrrr} -2 & 1 & 0 & 0 & -3 & 4 & -7 \\ & & -2 & 4 & -8 & 22 & -52 \\ \hline & 1 & -2 & 4 & -11 & 26 & -59 \end{array}$$
Thus, $f(-2) = -59$.
For further practice, see Example 5, Section 4.3.

9. $f(x) = x^4 - 2x^3 - 3x + 5$
(1) (2) So, $P = 2$.
$f(-x) = x^4 + 2x^3 + 3x + 5$
So, $N = 0$.
Thus, $f(x)$ has either two or no positive real zeros, and has no negative real zeros.
For further practice, see Example 3, Section 4.4.

10. $f(x) = 4x^3 - 12x^2 + 9x - 2$
Possible rational zeros: $\pm 1, \pm 2, \pm\frac{1}{2}, \pm\frac{1}{4}$
$$\begin{array}{r|rrrr} \frac{1}{2} & 4 & -12 & 9 & -2 \\ & & 2 & -5 & 2 \\ \hline & 4 & -10 & 4 & 0 \end{array}$$
$$\begin{aligned} f(x) &= \left(x - \frac{1}{2}\right)(4x^2 - 10x + 4) \\ &= (2x - 1)(2x^2 - 5x + 2) \\ &= (2x - 1)(2x - 1)(x - 2) \\ &= (2x - 1)^2(x - 2) \end{aligned}$$
The zeros are $\frac{1}{2}$ (of multiplicity 2) and 2.
For further practice, see Example 1, Section 4.4.

11. $f(x) = 4x^4 + 4x^3 + 5x^2 + 8x - 6$
Possible rational zeros: $\pm 1, \pm 2, \pm 3, \pm 6, \pm\frac{1}{2}, \pm\frac{1}{4}, \pm\frac{3}{2}, \pm\frac{3}{4}$
$$\begin{array}{r|rrrrr} \frac{1}{2} & 4 & 4 & 5 & 8 & -6 \\ & & 2 & 3 & 4 & 6 \\ \hline & 4 & 6 & 8 & 12 & 0 \end{array}$$
$$\begin{aligned} f(x) &= \left(x - \frac{1}{2}\right)(4x^3 + 6x^2 + 8x + 12) \\ &= (2x - 1)(2x^3 + 3x^2 + 4x + 6) \\ &= (2x - 1) \cdot [x^2(2x + 3) + 2(2x + 3)] \\ &= (2x - 1)(x^2 + 2)(2x + 3) \\ &= (2x - 1)(x - i\sqrt{2})(x + i\sqrt{2})(2x + 3) \end{aligned}$$
The zeros are $\frac{1}{2}, i\sqrt{2}, -i\sqrt{2}$, and $-\frac{3}{2}$.
For further practice, see Example 3, Section 4.5.

12. $f(x) = [x - (2)] \cdot [x - (2 - 3i)] \cdot [x - (2 + 3i)]$
$= (x - 2)(x - 2 + 3i)(x - 2 - 3i)$
$= x^3 - 6x^2 + 21x - 26$
For further practice, see Example 2, Section 4.5.

14. The denominator $6x^2 - 5x + 1 = 0$ when $x = \frac{1}{2}$ or $x = \frac{1}{3}$. For $x = \frac{1}{2}$, the numerator $3(\frac{1}{2})^2 - 12 \neq 0$. For $x = \frac{1}{3}$, the numerator $3(\frac{1}{3})^2 - 12 \neq 0$. Thus, the vertical asymptotes are $x = \frac{1}{2}$ and $x = \frac{1}{3}$. By (ii) of the Second Property of Rational Functions, $y = \frac{1}{2}$ is the horizontal asymptote.
For further practice, see Examples 2 and 3, Section 4.7.

16. $f(x) = \dfrac{1}{x^2 - 4x - 5}$
Vertical asymptotes: $x = -1$ and $x = 5$; Horizontal asymptote: $y = 0$
The function has no zeros. The function is undefined at $x = -1$ and $x = 5$.

Interval	Test Value	$f(t)$	
$(-\infty, -1)$	-2	$\frac{1}{7}$	$f(t) > 0$; graph lies above x-axis
$(-1, 5)$	0	$-\frac{1}{5}$	$f(t) < 0$; graph lies below x-axis
$(5, \infty)$	6	$\frac{1}{7}$	$f(t) > 0$; graph lies above x-axis

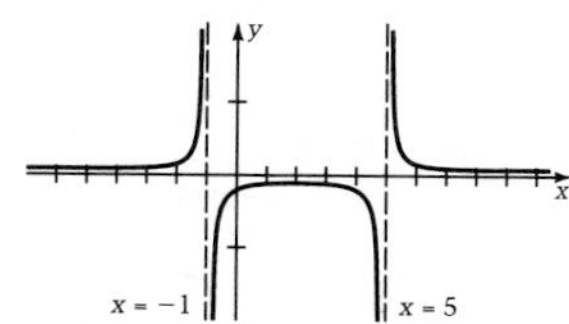

For further practice, see Example 4, Section 4.7.

18.
$$x^2 + 2x + y^2 - 6y + 7 = 0$$
$$(x^2 + 2x + 1) + (y^2 - 6y + 9) + 7 - 1 - 9 = 0$$
$$(x + 1)^2 + (y - 3)^2 = (\sqrt{3})^2$$

Center: $(-1, 3)$; Radius: $\sqrt{3}$

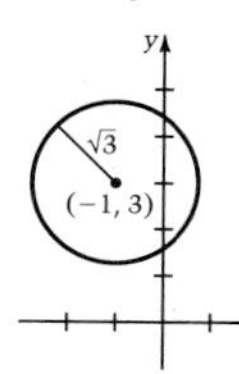

For further practice, see Example 4, Section 4.8.

13. $[3, 4]$ since $f(3) = -6 < 0$ and $f(4) = 29 > 0$.
For further practice, see Example 1, Section 4.6.

15. The denominator $2x^2 - 18 = 0$ when $x = 3$ or $x = -3$. For $x = 3$, the numerator $5(3) \neq 0$. For $x = -3$, the numerator $5(-3) \neq 0$. Thus, the vertical asymptotes are $x = 3$ and $x = -3$. By (i) of the Second Property of Rational Functions, $y = 0$ is the horizontal asymptote.
For further practice, see Examples 2 and 3, Section 4.7.

17. $y^2 - 6y - x + 5 = 0$
$$x = (y^2 - 6y + 9) + 5 - 9$$
$$x = (y - 3)^2 + (-4)$$

Vertex: $(-4, 3)$; Axis of symmetry: $y = 3$

x-intercept:
$x = (0 - 3)^2 + (-4)$
$x = 5$

y-intercepts:
$0 = (y - 3)^2 - 4$
$4 = (y - 3)^2$
$\pm 2 = y - 3$
$y = 5$ or $y = 1$

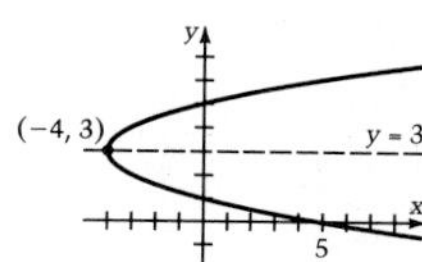

For further practice, see Example 2, Section 4.8.

19. $\dfrac{(x - 5)^2}{25} + \dfrac{y^2}{4} = 1$
$$\frac{(x - 5)^2}{5^2} + \frac{y^2}{2^2} = 1$$

Center: $(5, 0)$; Vertices: $(0, 0)$, $(10, 0)$, $(5, 2)$, $(5, -2)$

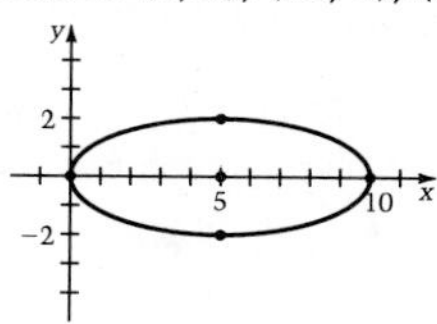

For further practice, see Example 2, Section 4.9.

20. $\dfrac{(y+2)^2}{24} - \dfrac{(x-1)^2}{4} = 1$

$\dfrac{(y+2)^2}{(2\sqrt{6})^2} - \dfrac{(x-1)^2}{2^2} = 1$

Center: $(1, -2)$; Vertices: $(1, -2 - 2\sqrt{6})$ and $(1, -2 + 2\sqrt{6})$; Asymptotes: $y + 2 = \pm\sqrt{6}(x - 1)$

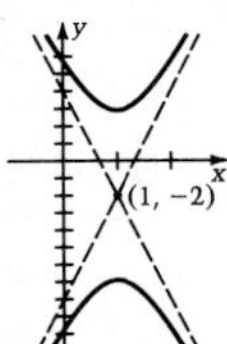

For further practice, see Example 4, Section 4.9.

Chapter 5 Test, p. 330

1. $4^{x-1} \cdot 8^{2-3x} = (2^2)^{x-1} \cdot (2^3)^{2-3x} = 2^{2x-2} \cdot 2^{6-9x} = 2^{-7x+4}$
For further practice, see Example 1, Section 5.1.

2. $\dfrac{9^{x+2}}{27^{2x}} = (3^2)^{x+2}(3^3)^{-2x} = 3^{2x+4} \cdot 3^{-6x} = 3^{-4x+4}$
For further practice, see Example 1, Section 5.1.

3.

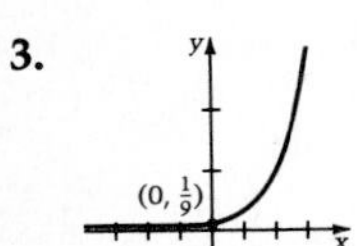

For further practice, see Example 3, Section 5.1.

4.

(−2, 0)
(0, −3)
y = −4

For further practice, see Example 3, Section 5.1.

5. $3^{x+1} = 0$

Since $3^{x+1} > 0$ for all real numbers x, there are no solutions.
For further practice, see Example 4, Section 5.1.

6.
$$\begin{aligned} 4^{-2x} &= 8^{5x-2} \\ (2^2)^{-2x} &= (2^3)^{5x-2} \\ 2^{-4x} &= 2^{15x-6} \\ -4x &= 15x - 6 \\ x &= \frac{6}{19} \end{aligned}$$
For further practice, see Example 4, Section 5.1.

7. $\log_x 125 = 3$
For further practice, see Example 1, Section 5.2.

8. $\log_4 64 = x$
For further practice, see Example 1, Section 5.2.

9. $\log_7 x = 4$
For further practice, see Example 1, Section 5.2.

10.
$$\begin{aligned} \log_3 27 &= x \\ 3^x &= 27 \\ 3^x &= 3^3 \\ x &= 3 \end{aligned}$$
For further practice, see Example 2, Section 5.2.

11.
$$\begin{aligned} \log_x 16 &= -2 \\ x^{-2} &= 16 \\ \frac{1}{x^2} &= 16 \\ x^2 &= \frac{1}{16} \\ x &= \pm\frac{1}{4} \end{aligned}$$
Since $x > 0$, the only solution is $\frac{1}{4}$.
For further practice, see Example 2, Section 5.2.

12. 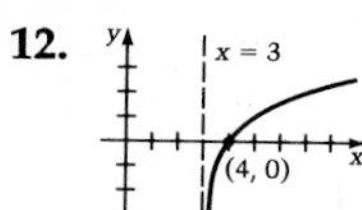

For further practice, see Example 3, Section 5.2.

13.

For further practice, see Example 3, Section 5.2.

14. $\log_2 \left(\dfrac{8x}{yz}\right) = \log_2 8x - \log_2 yz =$
$\log_2 8 + \log_2 x - (\log_2 y + \log_2 z) =$
$3 + 4.3 - [2.8 + (-3.1)] = 7.6$
For further practice, see Example 4, Section 5.2.

15. $\log_2 (xy^2z^3) = \log_2 x + \log_2 y^2 + \log_2 z^3 =$
$\log_2 x + 2 \log_2 y + 3 \log_2 z =$
$4.3 + 2(2.8) + 3(-3.1) = 0.6$
For further practice, see Example 4, Section 5.2.

16. $\log_4 2^x = -5$

$x \log_4 2 = -5$

$\dfrac{x}{2} = -5$

$x = -10$
For further practice, see Example 1, Section 5.3.

17. $\log_3 (\log_2 x) = 0$

$3^0 = \log_2 x$

$x = 2^1 = 2$
For further practice, see Example 2, Section 5.3.

18. $\log_2 x^3 = \log_2 9x$

$x^3 = 9x$

$x^3 - 9x = 0$

$x(x + 3)(x - 3) = 0$

$x = 0$	$x + 3 = 0$	$x - 3 = 0$
	$x = -3$	$x = 3$

Since the x-values 0 and -3 produce undefined expressions in the original equation, the only solution is 3.
For further practice, see Example 3, Section 5.3.

19. $\log_2 8x = \log_2 8 + \log_2 x$

$\log_2 8x = \log_2 8x$

$8x = 8x$

The solutions are all real numbers $x > 0$.
For further practice, see Example 4, Section 5.3.

20. $\ln (x - 3) + \ln (x - 4) = \ln (7 - x)$

$\ln [(x - 3)(x - 4)] = \ln (7 - x)$

$(x - 3)(x - 4) = 7 - x$

$x^2 - 6x + 5 = 0$

$(x - 1)(x - 5) = 0$

$x = 1 \text{ or } 5$

When $x = 1$, the original equation is not defined. Thus, the only solution is 5.
For further practice, see Example 4, Section 5.3.

21. $2 \ln x = \ln (3x + 4)$

$\ln x^2 = \ln (3x + 4)$

$x^2 = 3x + 4$

$x^2 - 3x - 4 = 0$

$(x - 4)(x + 1) = 0$

$x = 4 \text{ or } -1$

When $x = -1$, the original equation is not defined. Thus, the only solution is 4.
For further practice, see Example 3, Section 5.3.

22. $\ln (e^2x) = 5$

$\ln e^2 + \ln x = 5$

$2 + \ln x = 5$

$\ln x = 3$

$x = e^3$
For further practice, see Example 7, Section 5.3.

23. $\ln \left(\dfrac{e}{x}\right) = 3$

$e^3 = \dfrac{e}{x}$

$x = e^{-2}$
For further practice, see Example 7, Section 5.3.

24.

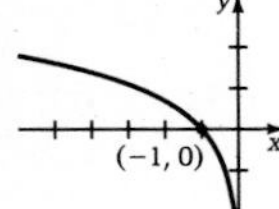

For further practice, see Example 5, Section 5.2.

25.

$x = -3$
(−2, 0)
(0, ln 3)

For further practice, see Example 5, Section 5.2.

26.

$$x^2 = 3$$
$$\ln (x^2) = \ln 3$$
$$2 \ln x = \ln 3$$
$$\ln x = \frac{1}{2} \ln 3$$
$$\ln x \approx \frac{1}{2}(1.0986)$$
$$x \approx e^{0.5493}$$
$$x \approx 1.73$$

For further practice, see Example 7, Section 5.3.

27.

$$2^x = 6$$
$$\ln (2^x) = \ln 6$$
$$x \ln 2 = \ln 6$$
$$x = \frac{\ln 6}{\ln 2}$$
$$x \approx \frac{1.7918}{0.6931}$$
$$x \approx 2.59$$

For further practice, see Example 5, Section 5.3.

28.

$$N = B_0 2^t$$
$$2{,}400{,}000 = (1{,}600{,}000)2^t$$
$$1.5 = 2^t$$
$$\ln 1.5 = \ln 2^t$$
$$\ln 1.5 = t \ln 2$$
$$t = \frac{\ln 1.5}{\ln 2}$$
$$t \approx \frac{0.4055}{0.6931} \approx 0.59$$

It will take approximately 0.6 hours.
For further practice, see Example 1, Section 5.4.

29.

$$A = P\left[1 + \frac{i}{n}\right]^{nt}$$
$$A = 500\left[1 + \frac{0.08}{4}\right]^{(4)(6)}$$
$$A = 500(1.02)^{24}$$
$$\ln A = \ln 500 + 24 \ln (1.02)$$
$$\ln A \approx 6.2146 + 24(0.0198)$$
$$A \approx e^{6.6898}$$
$$A \approx 804.16$$

The value of the account is approximately $804.
For further practice, see Example 3, Section 5.4.

30. $A = Pe^{it}$

$$A = 500e^{(0.08)(6)}$$
$$A \approx 500(1.616)$$
$$A \approx 808$$

The value of the account is approximately $808.
For further practice, see Example 4, Section 5.4.

31. $y = y_0 e^{kt}$

$$y \approx 80e^{(-0.00012)(300{,}000)}$$
$$y \approx 80(2.320 \times 10^{-16})$$
$$y \approx 1.856 \times 10^{-14}$$

Approximately 1.856×10^{-14} grams remain today.
For further practice, see Example 5, Section 5.4.

Chapter 6 Test, pp. 390–391

1. $c^2 = a^2 + b^2$

$$c^2 = 12^2 + 16^2$$
$$c^2 = 144 + 256$$
$$c^2 = 400$$
$$c = 20$$

For further practice, see Example 3, Section 6.1.

2. $c^2 = a^2 + b^2$

$$26^2 = 10^2 + b^2$$
$$676 = 100 + b^2$$
$$576 = b^2$$
$$b = 24$$

For further practice, see Example 3, Section 6.1.

3. $\Delta ABC \sim \Delta A'B'C'$

$$\frac{2}{12} = \frac{5}{b'}$$

$$2b' = 60$$

$$b' = 30$$

For further practice, see Example 4, Section 6.1.

4. $\Delta ABC \sim \Delta A'B'C'$

$$\frac{a'}{7} = \frac{20}{8}$$

$$8a' = 140$$

$$a' = \frac{35}{2}$$

For further practice, see Example 4, Section 6.1.

5. $41^2 = a^2 + 40^2$

$$81 = a^2$$

$$a = 9$$

$$\sin\alpha = \frac{\text{opp}}{\text{hyp}} = \frac{a}{c} = \frac{9}{41}$$

$$\cos\alpha = \frac{\text{adj}}{\text{hyp}} = \frac{b}{c} = \frac{40}{41}$$

$$\tan\alpha = \frac{\text{opp}}{\text{adj}} = \frac{a}{b} = \frac{9}{40}$$

$$\cot\alpha = \frac{\text{adj}}{\text{opp}} = \frac{b}{a} = \frac{40}{9}$$

$$\sec\alpha = \frac{\text{hyp}}{\text{adj}} = \frac{c}{b} = \frac{41}{40}$$

$$\csc\alpha = \frac{\text{hyp}}{\text{opp}} = \frac{c}{a} = \frac{41}{9}$$

For further practice, see Example 1, Section 6.2.

6. $c^2 = 21^2 + 20^2$

$$c^2 = 841$$

$$c = 29$$

$$\sin\alpha = \frac{\text{opp}}{\text{hyp}} = \frac{a}{c} = \frac{21}{29}$$

$$\cos\alpha = \frac{\text{adj}}{\text{hyp}} = \frac{b}{c} = \frac{20}{29}$$

$$\tan\alpha = \frac{\text{opp}}{\text{adj}} = \frac{a}{b} = \frac{21}{20}$$

$$\cot\alpha = \frac{\text{adj}}{\text{opp}} = \frac{b}{a} = \frac{20}{21}$$

$$\sec\alpha = \frac{\text{hyp}}{\text{adj}} = \frac{c}{b} = \frac{29}{20}$$

$$\csc\alpha = \frac{\text{hyp}}{\text{opp}} = \frac{c}{a} = \frac{29}{21}$$

For further practice, see Example 1, Section 6.2.

7. $\sin 30^\circ = \frac{24}{c}$

$$\frac{1}{2} = \frac{24}{c}$$

$$c = 48$$

For further practice, see Example 3, Section 6.2.

8. $\sin 45^\circ = \frac{a}{18}$

$$\frac{\sqrt{2}}{2} = \frac{a}{18}$$

$$a = 9\sqrt{2}$$

For further practice, see Example 3, Section 6.2.

9. $7^2 = b^2 + 3^2$

$$40 = b^2$$

$$b = 2\sqrt{10}$$

$$\tan\alpha = \frac{\text{side opp }\alpha}{\text{side adj }\alpha} = \frac{3}{2\sqrt{10}} = \frac{3\sqrt{10}}{20}$$

$$\tan\beta = \frac{\text{side opp }\beta}{\text{side adj }\beta} = \frac{2\sqrt{10}}{3}$$

For further practice, see Example 5, Section 6.2.

10. $7^2 = a^2 + 5^2$

$$24 = a^2$$

$$a = 2\sqrt{6}$$

$$\tan\alpha = \frac{\text{side opp }\alpha}{\text{side adj }\alpha} = \frac{2\sqrt{6}}{5}$$

$$\tan\beta = \frac{\text{side opp }\beta}{\text{side adj }\beta} = \frac{5}{2\sqrt{6}} = \frac{5\sqrt{6}}{12}$$

For further practice, see Example 5, Section 6.2.

11. $r = \sqrt{x^2 + y^2} = \sqrt{(-10)^2 + 24^2}$
$= \sqrt{676} = 26$

$\sin\theta = \frac{y}{r} = \frac{24}{26} = \frac{12}{13}$

$\tan\theta = \frac{y}{x} = \frac{24}{-10} = -\frac{12}{5}$

For further practice, see Example 3, Section 6.3.

12. $r = \sqrt{x^2 + y^2} = \sqrt{(-12)^2 + (-9)^2}$
$= \sqrt{225} = 15$

$\sin\theta = \frac{y}{r} = \frac{-9}{15} = -\frac{3}{5}$

$\tan\theta = \frac{y}{x} = \frac{-9}{-12} = \frac{3}{4}$

For further practice, see Example 3, Section 6.3.

13. For the angle $-270°$, use $(0, 1)$.

$\cos(-270°) = \frac{x}{r} = \frac{0}{1} = 0$

For further practice, see Example 4, Section 6.3.

14. For the angle $180°$, use $(-1, 0)$.

$\sin 180° = \frac{y}{r} = \frac{0}{1} = 0$

For further practice, see Example 4, Section 6.3.

15. $\sec 18° = \frac{1}{\cos 18°} = \frac{1}{\sin(90° - 18°)} = \frac{1}{\sin 72°} =$

$\frac{1}{0.95} \approx 1.05$

For further practice, see Example 4, Section 6.4.

16. $\cot 18° = \tan(90° - 18°) = \tan 72° = \frac{\sin 72°}{\cos 72°} =$

$\frac{0.95}{0.31} \approx 3.06$

For further practice, see Example 4, Section 6.4.

17. $\cot\alpha = \frac{1}{\tan\alpha} = \frac{1}{\frac{3}{2}} = \frac{2}{3}$

$\sec^2\alpha = \tan^2\alpha + 1$

$\sec^2\alpha = \left(\frac{3}{2}\right)^2 + 1$

$\sec^2\alpha = \frac{9}{4} + 1$

$\sec^2\alpha = \frac{13}{4}$

$\sec\alpha = \frac{\sqrt{13}}{2}$

$\cos\alpha = \frac{1}{\sec\alpha} = \frac{1}{\frac{\sqrt{13}}{2}} = \frac{2}{\sqrt{13}} = \frac{2\sqrt{13}}{13}$

$\csc^2\alpha = \cot^2\alpha + 1$

$\csc^2\alpha = \left(\frac{2}{3}\right)^2 + 1$

$\csc^2\alpha = \frac{4}{9} + 1$

$\csc^2\alpha = \frac{13}{9}$

$\csc\alpha = \frac{\sqrt{13}}{3}$

$\sin\alpha = \frac{1}{\csc\alpha} = \frac{1}{\frac{\sqrt{13}}{3}} = \frac{3}{\sqrt{13}} = \frac{3\sqrt{13}}{13}$

For further practice, see Example 2, Section 6.4.

18. $\csc\alpha = \frac{1}{\sin\alpha} = \frac{1}{\frac{5}{6}} = \frac{6}{5}$

$\cos^2\alpha + \sin^2\alpha = 1$

$\cos^2\alpha = 1 - \left(\frac{5}{6}\right)^2$

$\cos^2\alpha = 1 - \frac{25}{36}$

$\cos^2\alpha = \frac{11}{36}$

$\cos\alpha = \frac{\sqrt{11}}{6}$

$\sec\alpha = \frac{1}{\cos\alpha} = \frac{1}{\frac{\sqrt{11}}{6}} = \frac{6}{\sqrt{11}} = \frac{6\sqrt{11}}{11}$

$\tan^2\alpha = \sec^2\alpha - 1$

$\tan^2\alpha = \left(\frac{6\sqrt{11}}{11}\right)^2 - 1$

$\tan^2\alpha = \frac{36}{11} - 1$

$\tan^2\alpha = \frac{25}{11}$

$\tan\alpha = \frac{5}{\sqrt{11}} = \frac{5\sqrt{11}}{11}$

$\cot\alpha = \frac{1}{\tan\alpha} = \frac{1}{\frac{5\sqrt{11}}{11}} = \frac{11}{5\sqrt{11}} = \frac{\sqrt{11}}{5}$

For further practice, see Example 2, Section 6.4.

19. 0.7387
For further practice, see Example 7, Section 6.4.

20. 2.2650
For further practice, see Example 7, Section 6.4.

21. 1.0209
For further practice, see Example 7, Section 6.4.

22. 0.3371
For further practice, see Example 7, Section 6.4.

23. $A = 90° - 60° = 30°$

$$\sin 60° = \frac{12}{c} \qquad \tan 60° = \frac{12}{a}$$

$$\frac{\sqrt{3}}{2} = \frac{12}{c} \qquad \sqrt{3} = \frac{12}{a}$$

$$c = 8\sqrt{3} \qquad a = 4\sqrt{3}$$

For further practice, see Example 1, Section 6.5.

24. $B = 90° - 33°42' = 56°18'$

$$\cos 33°42' = \frac{96.8}{c}$$

$$0.8320 = \frac{96.8}{c}$$

$$c \approx 116$$

$$\tan 33°42' = \frac{a}{96.8}$$

$$0.6669 = \frac{a}{96.8}$$

$$a = 64.6$$

For further practice, see Example 1, Section 6.5.

25. $c = \sqrt{a^2 + b^2} = \sqrt{(36.3)^2 + (46.9)^2}$
$\approx \sqrt{3517.3} \approx 59.3$

$$\tan A = \frac{36.3}{46.9}$$

$$\tan A \approx 0.7740$$

$$A \approx 37°40'$$

$B \approx 90° - 37°40' = 52°20'$
For further practice, see Example 2, Section 6.5.

26.

$$\tan \theta = \frac{68}{25}$$

$$\tan \theta = 2.72$$

$$\theta \approx 70°$$

The angle of elevation is approximately 70°.
For further practice, see Example 3, Section 6.5.

27.

$$\cos \theta = \frac{6}{25}$$

$$\cos \theta = 0.24$$

$$\theta \approx 76°$$

The ladder makes an angle of approximately 76° with the ground.
For further practice, see Example 3, Section 6.5.

28. $\tan 135° = -\tan(180° - 135°) = -\tan 45° = -1$
For further practice, see Example 3, Section 6.6.

29. $-240° + 1(360°) = 120°$

$\cot(-240°) = \cot 120° = -\cot(180° - 120°) =$
$-\cot 60° = -\dfrac{\sqrt{3}}{3}$

For further practice, see Example 3, Section 6.6.

30. $C = 180° - (41° + 83°) = 56°$

$$\frac{\sin 41°}{a} = \frac{\sin 56°}{44.6}$$

$$\frac{0.6561}{a} = \frac{0.8290}{44.6}$$

$$a \approx 35$$

$$\frac{\sin 83°}{b} = \frac{\sin 56°}{44.6}$$

$$\frac{0.9925}{b} = \frac{0.8290}{44.6}$$

$$b \approx 53$$

For further practice, see Example 1, Section 6.7.

31. $\dfrac{\sin 24°18'}{6.2} = \dfrac{\sin C}{9.4}$

$\dfrac{0.4115}{6.2} = \dfrac{\sin C}{9.4}$

$\sin C \approx 0.6239$

$C \approx 38°36'$ or $141°24'$

Case 1: $C \approx 38°36'$
$B \approx 180° - (24°18' + 38°36') = 117°06'$

$\dfrac{\sin 24°18'}{6.2} = \dfrac{\sin 117°06'}{b}$

$\dfrac{0.4115}{6.2} = \dfrac{0.8902}{b}$

$b \approx 13.4$

Case 2: $C \approx 141°24'$
$B \approx 180° - (24°18' + 141°24') = 14°18'$

$\dfrac{\sin 24°18'}{6.2} = \dfrac{\sin 14°18'}{b}$

$\dfrac{0.4115}{6.2} = \dfrac{0.2470}{b}$

$b \approx 3.7$

Thus, there are two possible triangles: one with $C \approx 38°36'$, $B \approx 117°06'$, and $b \approx 13.4$, and one with $C \approx 141°24'$, $B \approx 14°18'$, and $b \approx 3.7$.
For further practice, see Example 2, Section 6.7.

32. $\dfrac{\sin A}{12.4} = \dfrac{\sin 46°54'}{10.6}$

$\dfrac{\sin A}{12.4} = \dfrac{0.7302}{10.6}$

$\sin A \approx 0.8542$

$A \approx 58°40'$ or $121°20'$

Case 1: $A \approx 58°40'$
$C \approx 180° - (46°54' + 58°40') = 74°26'$

$\dfrac{\sin 46°54'}{10.6} = \dfrac{\sin 74°26'}{c}$

$\dfrac{0.7302}{10.6} = \dfrac{0.9633}{c}$

$c \approx 14$

Case 2: $A \approx 121°20'$
$C \approx 180° - (46°54' + 121°20') = 11°46'$

$\dfrac{\sin 46°54'}{10.6} = \dfrac{\sin 11°46'}{c}$

$\dfrac{0.7302}{10.6} = \dfrac{0.2039}{c}$

$c \approx 3$

Thus, there are two possible triangles: one with $A \approx 58°40'$, $C \approx 74°26'$, and $c \approx 14$, and one with $A \approx 121°20'$, $C \approx 11°46'$, and $c \approx 3$.
For further practice, see Example 2, Section 6.7.

33.

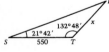

$B = 180° - (21°42' + 132°48') = 25°30'$

$\dfrac{\sin 25°30'}{550} = \dfrac{\sin 21°42'}{x}$

$\dfrac{0.4305}{550} = \dfrac{0.3697}{x}$

$x \approx 472$

The distance from the boat to Station T is approximately 472 yds.
For further practice, see Example 3, Section 6.7.

34.

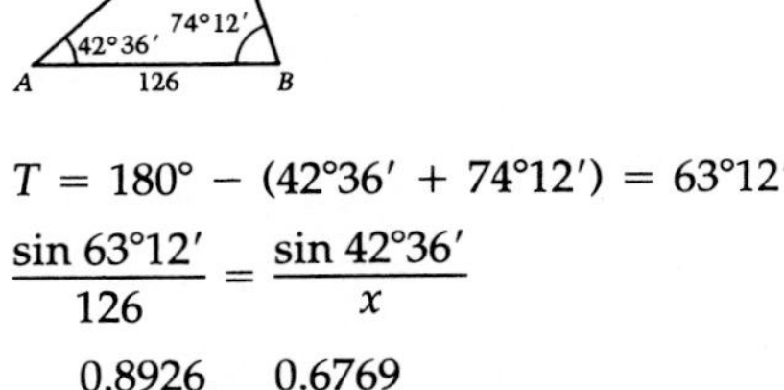

$T = 180° - (42°36' + 74°12') = 63°12'$

$\dfrac{\sin 63°12'}{126} = \dfrac{\sin 42°36'}{x}$

$\dfrac{0.8926}{126} = \dfrac{0.6769}{x}$

$x \approx 95.6$

The distance from point T to point B is approximately 95.6 ft.
For further practice, see Example 3, Section 6.7.

35. $c^2 = a^2 + b^2 - 2ab\cos C$

$c^2 = (1.52)^2 + (2.31)^2 - 2(1.52)(2.31)\cos 119°$

$c^2 = 7.6465 - 7.0224(-0.4848)$

$c^2 \approx 11.05$

$c \approx 3.32$

$a^2 = b^2 + c^2 - 2bc\cos A$

$(1.52)^2 = (2.31)^2 + (3.32)^2 - 2(2.31)(3.32)\cos A$

$\cos A \approx 0.9159$

$A \approx 23°40'$

$B \approx 180° - (23°40' + 119°) = 37°20'$

For further practice, see Example 1, Section 6.8.

36. $a^2 = b^2 + c^2 - 2bc\cos A$

$(26.7)^2 = (34.2)^2 + (41.8)^2 - 2(34.2)(41.8)\cos A$

$\cos A \approx 0.7709$

$A \approx 39°34'$

$b^2 = a^2 + c^2 - 2ac\cos B$

$(34.2)^2 = (26.7)^2 + (41.8)^2 - 2(26.7)(41.8)\cos B$

$\cos B \approx 0.5781$

$B \approx 54°41'$

$C \approx 180° - (39°34' + 54°41') = 85°45'$

For further practice, see Example 2, Section 6.8.

Chapter 7 Test, p. 460

1. $x^2 + y^2 = \left(\frac{\sqrt{2}}{6}\right)^2 + \left(\frac{4}{6}\right)^2 = \frac{2}{36} + \frac{16}{36} = \frac{18}{36} \neq 1$

Thus, $\left(\frac{\sqrt{2}}{6}, \frac{4}{6}\right)$ is not on the unit circle.

For further practice, see Example 1, Section 7.1.

2. $\alpha = \frac{d}{180°}\cdot\pi = -\frac{135°}{180°}\cdot\pi = -\frac{3\pi}{4}$

For further practice, see Example 3, Section 7.1.

3. $\alpha = \frac{d}{180°}\cdot\pi = \frac{240°}{180°}\cdot\pi = \frac{4\pi}{3}$

For further practice, see Example 3, Section 7.1.

4. $\alpha = \frac{d}{180°}\cdot\pi = -\frac{735°}{180°}\cdot\pi = -\frac{49\pi}{12}$

For further practice, see Example 3, Section 7.1.

5. $d = \frac{\alpha}{\pi}\cdot 180° = \frac{\frac{9\pi}{2}}{\pi}\cdot 180° = 810°$

For further practice, see Example 3, Section 7.1.

6. $d = \frac{\alpha}{\pi}\cdot 180° = \frac{\frac{16\pi}{3}}{\pi}\cdot 180° = 960°$

For further practice, see Example 3, Section 7.1.

7. $d = \frac{\alpha}{\pi}\cdot 180° = \frac{-\frac{17\pi}{6}}{\pi}\cdot 180° = -510°$

For further practice, see Example 3, Section 7.1.

8. $\theta = \frac{s}{r} = \frac{12\pi}{8} = \frac{3\pi}{2}$

The angle measures $\frac{3\pi}{2}$ radians or 270°.

For further practice, see Example 4, Section 7.1.

9. $\sin\frac{5\pi}{4} = -\sin\left(\frac{5\pi}{4} - \pi\right) = -\sin\frac{\pi}{4} = -\frac{\sqrt{2}}{2}$

For further practice, see Example 3, Section 7.2.

10. $\cos\left(-\frac{4\pi}{3}\right) = \cos\left(2\pi - \frac{4\pi}{3}\right) = \cos\left(\frac{2\pi}{3}\right) =$

$-\cos\left(\pi - \frac{2\pi}{3}\right) = -\cos\frac{\pi}{3} = -\frac{1}{2}$

For further practice, see Example 3, Section 7.2.

11. $\sec\left(\frac{19\pi}{2}\right) = \sec\left(5\cdot 2\pi - \frac{19}{2}\pi\right) = \sec\frac{\pi}{2}$: undefined

For further practice, see Example 3, Section 7.2.

12. $\tan\left(-\frac{17\pi}{6}\right) = \tan\left(2\cdot 2\pi - \frac{17\pi}{6}\right) = \tan\frac{7\pi}{6} =$

$\tan\left(\frac{7\pi}{6} - \pi\right) = \tan\frac{\pi}{6} = \frac{\sqrt{3}}{3}$

For further practice, see Example 3, Section 7.2.

13. $\sin 4.3 = \sin(4.3 - \pi) \approx -0.9162$

For further practice, see Example 4, Section 7.2.

14. $\cos 5.6 = \cos(2\pi - 5.6) \approx 0.7756$

For further practice, see Example 4, Section 7.2.

15. 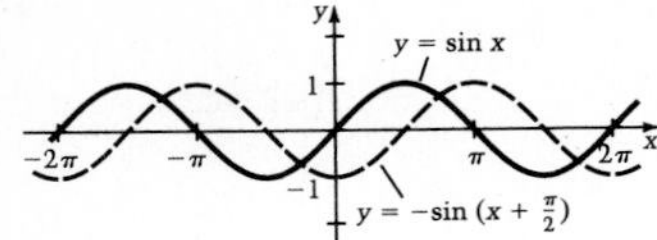

For further practice, see Example 1, Section 7.3.

16. 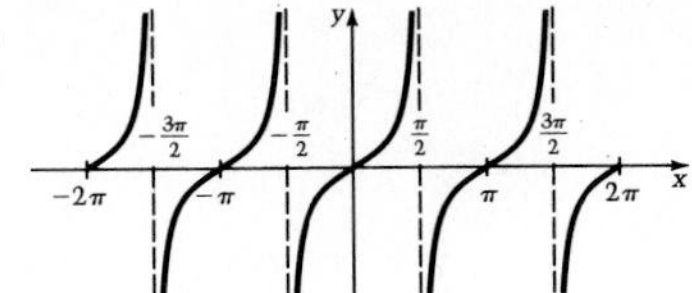

Graphs of $y = \tan x$ and $y = \tan(x - \pi)$ are the same.
For further practice, see Example 3, Section 7.3.

17. False; $\tan(-3\pi) = \tan(2 \cdot 2\pi - 3\pi)$
$= \tan \pi = 0$
For further practice, see Example 4, Section 7.3.

18. False; $\cot\left(-\frac{3\pi}{2}\right) = \cot\left(2\pi - \frac{3\pi}{2}\right) = \cot\frac{\pi}{2} = 0$
For further practice, see Example 4, Section 7.3.

19. True
For further practice, see Example 3, Section 7.3.

20. False; $\cos(-x) = \cos x$.
For further practice, see Example 3, Section 7.3.

21. amplitude 3, period 4π

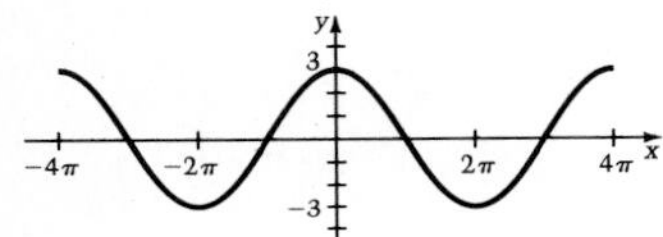

For further practice, see Example 2, Section 7.4.

22. amp: 4
period: π
phase shift: $-\frac{\pi}{2}$

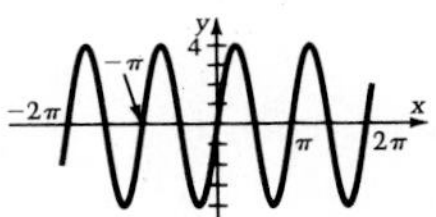

For further practice, see Example 4, Section 7.4.

23. 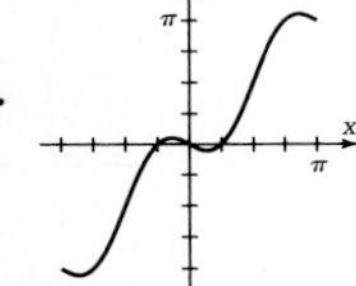

For further practice, see Example 5, Section 7.4.

24. $\arcsin\left(-\frac{1}{2}\right) = -\frac{\pi}{6}$
For further practice, see Example 1, Section 7.5.

25. $\cos^{-1}\left(\frac{\sqrt{3}}{2}\right) = \frac{\pi}{6}$
For further practice, see Example 1, Section 7.5.

26. $\sin^{-1}\left(\cot\frac{3\pi}{4}\right) = \sin^{-1}(-1)$
$= -\frac{\pi}{2}$
For further practice, see Example 2, Section 7.5.

27. 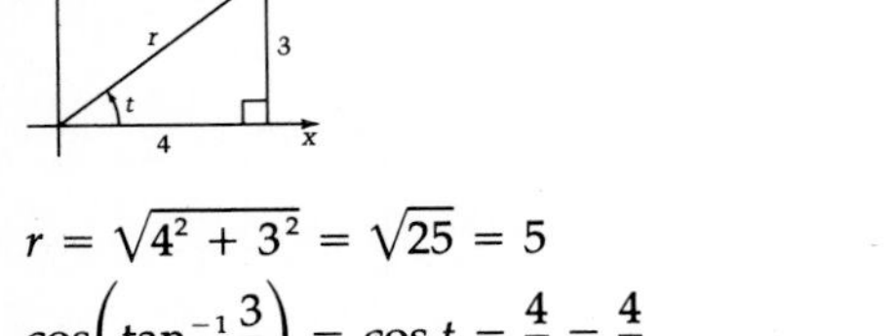

$r = \sqrt{4^2 + 3^2} = \sqrt{25} = 5$

$\cos\left(\tan^{-1}\frac{3}{4}\right) = \cos t = \frac{4}{r} = \frac{4}{5}$
For further practice, see Example 3, Section 7.5.

28. 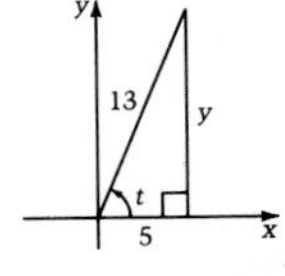

$y = \sqrt{13^2 - 5^2} = \sqrt{144} = 12$

$\sin\left(\operatorname{arcsec}\frac{13}{5}\right) = \sin t = \frac{y}{13} = \frac{12}{13}$
For further practice, see Example 3, Section 7.5.

29. (a) $d = \sqrt{(-3-0)^2 + (3-0)^2} = \sqrt{18} = 3\sqrt{2}$

$$\tan\theta = \frac{y}{x}$$

$$\tan\theta = \frac{3}{-3}$$

$$\tan\theta = -1$$

$$\theta = 135°$$

For further practice, see Examples 1 and 2, Section 7.6.

(b)

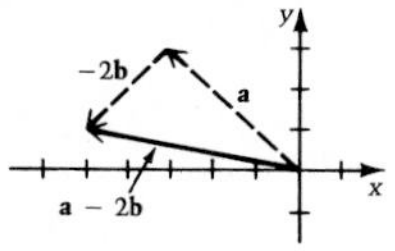

30. $\overrightarrow{AB} = \langle 2-(-3), 7-(-2)\rangle = \langle 5, 9\rangle$

For further practice, see Example 3, Section 7.6.

31. $\overrightarrow{AB} = \langle -1-6, 3-(-3)\rangle = \langle -7, 6\rangle$

For further practice, see Example 3, Section 7.6.

32. $\|\mathbf{a}\| = \sqrt{\left(-\frac{1}{2}\right)^2 + \left(\frac{7}{2}\right)^2} = \sqrt{\frac{50}{4}} = \frac{5\sqrt{2}}{2}$

$$2\mathbf{a} - \frac{1}{2}\mathbf{b} = 2\left\langle -\frac{1}{2}, \frac{7}{2}\right\rangle - \frac{1}{2}\langle 6, -1\rangle$$

$$= \langle -1, 7\rangle + \left\langle -3, \frac{1}{2}\right\rangle$$

$$= \left\langle -4, \frac{15}{2}\right\rangle$$

For further practice, see Examples 4–5, Section 7.6.

33. $\|\mathbf{b}\| = \sqrt{6^2 + (-1)^2} = \sqrt{37}$

$$\mathbf{b} - 4\mathbf{a} = \langle 6, -1\rangle - 4\left\langle -\frac{1}{2}, \frac{7}{2}\right\rangle$$

$$= \langle 6, -1\rangle + \langle 2, -14\rangle$$

$$= \langle 8, -15\rangle$$

For further practice, see Examples 4–5, Section 7.6.

34. $\mathbf{v}_x = (8.6\cos 148°)\mathbf{i}$

$$= (8.6)(-0.8480)\mathbf{i}$$

$$\approx -7.3\mathbf{i}$$

$\mathbf{v}_y = (8.6\sin 148°)\mathbf{j}$

$$= (8.6)(0.5299)\mathbf{j}$$

$$\approx 4.6\mathbf{j}$$

For further practice, see Examples 2, Section 7.7.

35. $\mathbf{v}_x = (10.8\cos 206°)\mathbf{i}$

$$= (10.8)(-0.8988)\mathbf{i}$$

$$\approx -9.7\mathbf{i}$$

$\mathbf{v}_y = (10.8\sin 206°)\mathbf{j}$

$$= (10.8)(-0.4384)\mathbf{j}$$

$$\approx -4.7\mathbf{j}$$

For further practice, see Example 2, Section 7.7.

36.

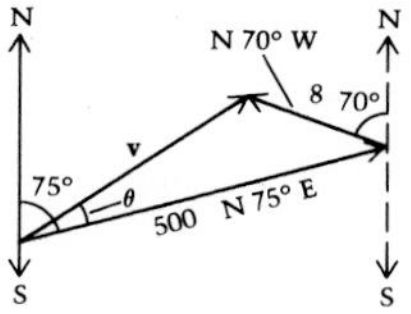

$$\|\mathbf{v}\|^2 = 500^2 + 8^2 - 2(500)(8)\cos 35°$$

$$\|\mathbf{v}\|^2 = 250{,}064 - 8000(0.8192)$$

$$\|\mathbf{v}\|^2 = 243{,}510.4$$

$$\|\mathbf{v}\| = 493$$

$$\frac{\sin\theta}{8} = \frac{\sin 35°}{493}$$

$$\frac{\sin\theta}{8} = \frac{0.5736}{493}$$

$$\sin\theta \approx 0.0093$$

$$\theta \approx 0°32'$$

$$75° - \theta \approx 75° - 0°32' = 74°28'$$

The jet's ground speed is approximately 493 mph, and its true bearing is approximately N 74°28′ E.

For further practice, see Example 1, Section 7.7.

37. $\|\mathbf{F}\| = (\sin 25°)(120) = (0.4226)(120) \approx 51$

The required force is approximately 51 lb.

For further practice, see Example 3, Section 7.7.

38. $\langle 3, -1\rangle \cdot \langle -2, -2\rangle = 3(-2) + (-1)(-2) = -4$

$$\cos\theta = \frac{\mathbf{v}\cdot\mathbf{w}}{\|\mathbf{v}\|\cdot\|\mathbf{w}\|}$$

$$\cos\theta = \frac{-4}{(\sqrt{3^2 + (-1)^2})(\sqrt{(-2)^2 + (-2)^2})}$$

$$\cos\theta \approx -0.4472$$

$$\theta \approx 117°$$

For further practice, see Example 5, Section 7.7.

39. $\langle 4, 6\rangle \cdot \langle -3, 2\rangle = 4(-3) + 6(2) = 0$

$$\cos\theta = \frac{\mathbf{v}\cdot\mathbf{w}}{\|\mathbf{v}\|\cdot\|\mathbf{w}\|}$$

$$\cos\theta = \frac{0}{(\sqrt{4^2 + 6^2})(\sqrt{(-3)^2 + 2^2})}$$

$$\cos\theta = 0$$

$$\theta = 90°$$

For further practice, see Example 5, Section 7.7.

Chapter 8 Test, pp. 527–528

1. $$\frac{1 - \tan x}{\sec x} + \sin x = \frac{1 - \dfrac{\sin x}{\cos x}}{\dfrac{1}{\cos x}} + \sin x =$$

$$\cos x - \sin x + \sin x = \cos x$$

For further practice, see Example 3, Section 8.1.

2. $$\frac{\tan^2 x + 1}{\tan^2 x + 2} = \frac{\sec^2 x}{(\tan^2 x + 1) + 1} = \frac{\sec^2 x}{\sec^2 x + 1} =$$

$$\frac{\dfrac{1}{\cos^2 x}}{\dfrac{1}{\cos^2 x} + 1} \cdot \frac{\cos^2 x}{\cos^2 x} = \frac{1}{1 + \cos^2 x}$$

For further practice, see Example 4, Section 8.1.

3. $$\cot\frac{5\pi}{12} = \cot\left(\frac{\pi}{4} + \frac{\pi}{6}\right) = \frac{1}{\tan\left(\dfrac{\pi}{4} + \dfrac{\pi}{6}\right)} =$$

$$\frac{1}{\dfrac{\tan\frac{\pi}{4} + \tan\frac{\pi}{6}}{1 - \tan\frac{\pi}{4}\tan\frac{\pi}{6}}} = \frac{1 - \tan\frac{\pi}{4}\tan\frac{\pi}{6}}{\tan\frac{\pi}{4} + \tan\frac{\pi}{6}} =$$

$$\frac{1 - (1)\left(\dfrac{\sqrt{3}}{3}\right)}{1 + \dfrac{\sqrt{3}}{3}} = \frac{3 - \sqrt{3}}{3 + \sqrt{3}}$$

For further practice, see Example 4, Section 8.2.

4. $$\frac{\sin(a + b)}{\cos a \cos b} = \frac{\sin a \cos b + \cos a \sin b}{\cos a \cos b} =$$

$$\frac{\sin a \cos b}{\cos a \cos b} + \frac{\cos a \sin b}{\cos a \cos b} =$$

$$\frac{\sin a}{\cos a} + \frac{\sin b}{\cos b} = \tan a + \tan b$$

For further practice, see Example 5, Section 8.2.

5. $$\frac{\tan a - \tan b}{\tan a + \tan b} = \frac{\dfrac{\sin a}{\cos a} - \dfrac{\sin b}{\cos b}}{\dfrac{\sin a}{\cos a} + \dfrac{\sin b}{\cos b}} \cdot \frac{\cos a \cos b}{\cos a \cos b} =$$

$$\frac{\sin a \cos b - \sin b \cos a}{\sin a \cos b + \sin b \cos a} = \frac{\sin(a - b)}{\sin(a + b)}$$

For further practice, see Example 5, Section 8.2.

6. $$\sin(2 \cdot 315°) = 2 \sin 315° \cos 315° = 2\left(-\frac{\sqrt{2}}{2}\right)\left(\frac{\sqrt{2}}{2}\right) = -1$$

$$\cos(2 \cdot 315°) = 2\cos^2 315° - 1 = 2\left(\frac{\sqrt{2}}{2}\right)^2 - 1 = 0$$

$$\tan(2 \cdot 315°) = \frac{2 \tan 315°}{1 - \tan^2 315°} = \frac{2(-1)}{1 - 1^2} = -\frac{2}{0} \quad \text{Undefined}$$

For further practice, see Example 2, Section 8.3.

7. $\sin\alpha = -\dfrac{2}{3}$

$$\frac{y}{r} = \frac{-2}{3}$$

$$x^2 = r^2 - y^2 = 3^2 - (-2)^2 = 5$$

$$x = -\sqrt{5}$$

$$\cos\alpha = \frac{x}{r} = -\frac{\sqrt{5}}{3}$$

$$\cos\frac{\alpha}{2} = -\sqrt{\frac{1+\cos\alpha}{2}} = -\sqrt{\frac{1-\frac{\sqrt{5}}{3}}{2}}$$

$$= -\sqrt{\frac{\frac{3-\sqrt{5}}{3}}{2}} = -\sqrt{\frac{3-\sqrt{5}}{6}}$$

For further practice, see Example 5, Section 8.3.

8. $\tan\alpha = -\dfrac{3}{2}$

$$\frac{y}{x} = \frac{3}{-2}$$

$$r = \sqrt{x^2+y^2} = \sqrt{(-2)^2+3^2} = \sqrt{13}$$

$$\cos\alpha = \frac{x}{r} = \frac{-2}{\sqrt{13}} = -\frac{2\sqrt{13}}{13}$$

$$\cos\frac{\alpha}{2} = \sqrt{\frac{1+\cos\alpha}{2}} = -\sqrt{\frac{1-\frac{2\sqrt{13}}{13}}{2}}$$

$$= \sqrt{\frac{\frac{13-2\sqrt{13}}{13}}{2}} = \sqrt{\frac{13-2\sqrt{13}}{26}}$$

For further practice, see Example 7, Section 8.3.

9. $$\frac{\cot\varphi - \tan\varphi}{\cot\varphi + \tan\varphi} = \frac{\frac{\cos\varphi}{\sin\varphi} - \frac{\sin\varphi}{\cos\varphi}}{\frac{\cos\varphi}{\sin\varphi} + \frac{\sin\varphi}{\cos\varphi}} \cdot \frac{\sin\varphi\cos\varphi}{\sin\varphi\cos\varphi} =$$

$$\frac{\cos^2\varphi - \sin^2\varphi}{\cos^2\varphi + \sin^2\varphi} = \frac{\cos 2\varphi}{1} = \cos 2\varphi$$

For further practice, see Example 7, Section 8.4.

10. $$\frac{\sin 2\alpha}{1-\cos 2\alpha} = \frac{2\sin\alpha\cos\alpha}{1-(1-2\sin^2\alpha)} = \frac{2\sin\alpha\cos\alpha}{2\sin^2\alpha} =$$

$$\frac{\cos\alpha}{\sin\alpha} = \cot\alpha$$

For further practice, see Example 4, Section 8.4.

11. $$\frac{1}{\sin x} - \sin x = \frac{1-\sin^2 x}{\sin x} = \frac{\cos^2 x}{\sin x} \cdot \frac{\frac{\sin x}{\cos^2 x}}{\frac{\sin x}{\cos^2 x}} = \frac{\frac{\sin x}{\cos^2 x}}{\frac{\sin^2 x}{\cos^2 x}} =$$

$$\frac{\sin x}{\tan^2 x}$$

For further practice, see Example 7, Section 8.4.

12. $$\sin\left(\frac{\pi}{4} - \alpha\right) = \sin\frac{\pi}{4}\cos\alpha - \cos\frac{\pi}{4}\sin\alpha$$

$$= \frac{\sqrt{2}}{2}\cos\alpha - \frac{\sqrt{2}}{2}\sin\alpha$$

$$= \cos\frac{\pi}{4}\cos\alpha - \sin\frac{\pi}{4}\sin\alpha$$

$$= \cos\left(\frac{\pi}{4} + \alpha\right)$$

For further practice, see Example 6, Section 8.4.

13. $$\cos\frac{11\pi}{12} - \cos\frac{5\pi}{12} =$$

$$-2\sin\left(\frac{\frac{11\pi}{12} + \frac{5\pi}{12}}{2}\right)\sin\left(\frac{\frac{11\pi}{12} - \frac{5\pi}{12}}{2}\right) =$$

$$-2\sin\frac{2\pi}{3}\sin\frac{\pi}{4} = -2\left(\frac{\sqrt{3}}{2}\right)\left(\frac{\sqrt{2}}{2}\right) = -\frac{\sqrt{6}}{2}$$

For further practice, see Example 9, Section 8.4.

14. $$\sin 4x\cos 3x = \frac{1}{2}[\sin(4x+3x) + \sin(4x-3x)] =$$

$$\frac{1}{2}\sin 7x + \frac{1}{2}\sin x$$

For further practice, see Example 9, Section 8.4.

15.
$$2\cos^2 x - \cos x = 1$$
$$2\cos^2 x - \cos x - 1 = 0$$
$$(2\cos x + 1)(\cos x - 1) = 0$$

$2\cos x + 1 = 0$	$\cos x - 1 = 0$
$\cos x = -\frac{1}{2}$	$\cos x = 1$
$x = \frac{2\pi}{3}$ or $\frac{4\pi}{3}$	$x = 0$

The solutions are $\frac{2\pi}{3} + 2n\pi$, $\frac{4\pi}{3} + 2n\pi$, and $0 + 2n\pi$.

For further practice, see Example 2, Section 8.5.

16.
$$\sin^2 x + 3\sin x = 4$$
$$\sin^2 x + 3\sin x - 4 = 0$$
$$(\sin x - 1)(\sin x + 4) = 0$$

$\sin x - 1 = 0$	$\sin x + 4 = 0$
$\sin x = 1$	$\sin x = -4$
$x = \frac{\pi}{2}$	no solutions

The solutions are $\frac{\pi}{2} + 2n\pi$.

For further practice, see Example 3, Section 8.5.

17.
$$\tan^2 x - 2\tan x - 3 = 0$$
$$(\tan x - 3)(\tan x + 1) = 0$$

$\tan x - 3 = 0$	$\tan x + 1 = 0$
$\tan x = 3$	$\tan x = -1$
$x \approx 71.6°$ or $251.6°$	$x = 135°$ or $315°$

The solutions are 71.6°, 251.6°, 135°, and 315°.
For further practice, see Example 8, Section 8.5.

18.
$$12\sin^2 x + 12\sin x - 6 = 0$$
$$2\sin^2 x + 2\sin x - 1 = 0$$
$$\sin x = \frac{-2 \pm \sqrt{2^2 - (4)(2)(-1)}}{2(2)}$$
$$\sin x = \frac{-2 \pm 2\sqrt{3}}{4}$$
$$\sin x = \frac{-1 \pm \sqrt{3}}{2}$$
$$x \approx 21.5° \quad \text{or} \quad 158.5°$$

The solutions are 21.5° and 158.5°.
For further practice, see Example 9, Section 8.5.

19.
$$\arcsin\frac{3}{5} - \arccos\frac{4}{5} = \sin^{-1}x$$
$$\sin\left(\arcsin\frac{3}{5} - \arccos\frac{4}{5}\right) = \sin(\sin^{-1}x)$$
$$\sin\left(\arcsin\frac{3}{5}\right)\cos\left(\arccos\frac{4}{5}\right) - \cos\left(\arcsin\frac{3}{5}\right)\sin\left(\arccos\frac{4}{5}\right) = x$$
$$\frac{3}{5}\cdot\frac{4}{5} - \frac{4}{5}\cdot\frac{3}{5} = x$$
$$\frac{12}{25} - \frac{12}{25} = x$$
$$x = 0$$

For further practice, see Example 10, Section 8.5.

20.
$$\tan^{-1}1 + \tan^{-1}\frac{12}{5} = x$$
$$\tan\left(\tan^{-1}1 + \tan^{-1}\frac{12}{5}\right) = \tan x$$
$$\frac{\tan(\tan^{-1}1) + \tan\left(\tan^{-1}\frac{12}{5}\right)}{1 - \tan(\tan^{-1}1)\tan\left(\tan^{-1}\frac{12}{5}\right)} = \tan x$$
$$\frac{1 + \frac{12}{5}}{1 - (1)\left(\frac{12}{5}\right)} = \tan x$$
$$-\frac{17}{7} = \tan x$$
$$x \approx 112.4^\circ \quad \text{or} \quad 292.4^\circ$$
For further practice, see Example 11, Section 8.5.

21.

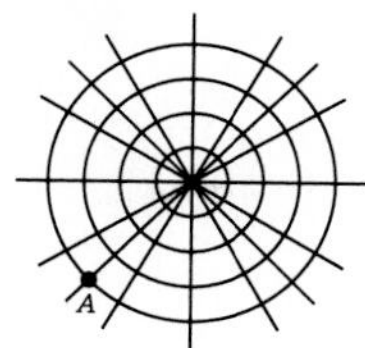

$$x = r\cos\theta = 4\cos(-135^\circ)$$
$$= 4\left(-\frac{\sqrt{2}}{2}\right) = -2\sqrt{2}$$
$$y = r\sin\theta = 4\sin(-135^\circ)$$
$$= 4\left(-\frac{\sqrt{2}}{2}\right) = -2\sqrt{2}$$
For further practice, see Example 2, Section 8.6.

22.

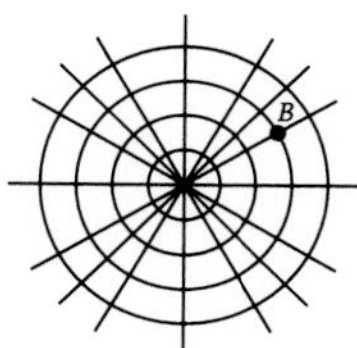

$$x = r\cos\theta = -3\cos 210^\circ$$
$$= -3\left(-\frac{\sqrt{3}}{2}\right) = \frac{3\sqrt{3}}{2}$$
$$y = r\sin\theta = -3\sin 210^\circ$$
$$= -3\left(-\frac{1}{2}\right) = \frac{3}{2}$$
For further practice, see Example 2, Section 8.6.

23.
$$r = 3\sin\theta$$
$$r^2 = 3r\sin\theta$$
$$x^2 + y^2 = 3y$$
For further practice, see Example 4, Section 8.6.

24.

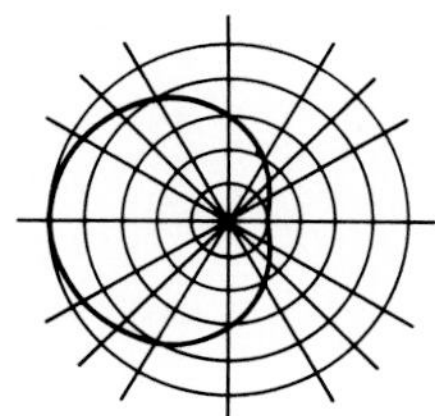

For further practice, see Example 6, Section 8.6.

25.

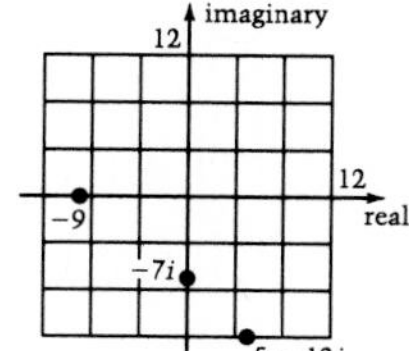

For further practice, see Example 1, Section 8.7.

26. **(a)** $|4 + 5i| = \sqrt{4^2 + 5^2} = \sqrt{41}$

(b) $|-9i| = \sqrt{0^2 + (-9)^2} = 9$

(c) $|-1 + i\sqrt{3}| = \sqrt{(-1)^2 + (\sqrt{3})^2} = 2$

For further practice, see Example 2, Section 8.7.

27. $r = \sqrt{3^2 + (-\sqrt{3})^2} = \sqrt{12} = 2\sqrt{3}$

$\tan\theta = -\dfrac{\sqrt{3}}{3}$

$\theta = \dfrac{11\pi}{6}$

$3 - i\sqrt{3} = 2\sqrt{3}\left(\cos\dfrac{11\pi}{6} + i\sin\dfrac{11\pi}{6}\right)$

For further practice, see Example 3, Section 8.7.

28. $r = \sqrt{(-4)^2 + (-4)^2} = 4\sqrt{2}$

$\tan\theta = \dfrac{-4}{-4}$

$\tan\theta = 1$

$\theta = \dfrac{5\pi}{4}$

$-4 - 4i = 4\sqrt{2}\left(\cos\dfrac{5\pi}{4} + i\sin\dfrac{5\pi}{4}\right)$

For further practice, see Example 3, Section 8.7.

29. $z_1 = 6\sqrt{3} + 6i$

$r_1 = \sqrt{(6\sqrt{3})^2 + 6^2} = \sqrt{144} = 12$

$\tan\theta_1 = \dfrac{6}{6\sqrt{3}} = \dfrac{\sqrt{3}}{3}$. So $\theta_1 = \dfrac{\pi}{6}$.

Thus, $z_1 = 12\left(\cos\dfrac{\pi}{6} + i\sin\dfrac{\pi}{6}\right)$.

$z_2 = 1 - i\sqrt{3}$

$r_2 = \sqrt{1^2 + (-\sqrt{3})^2} = \sqrt{4} = 2$

$\tan\theta_2 = -\dfrac{\sqrt{3}}{1} = -\sqrt{3}$. So, $\theta_2 = \dfrac{5\pi}{3}$.

Thus, $z_2 = 2\left(\cos\dfrac{5\pi}{3} + i\sin\dfrac{5\pi}{3}\right)$.

$$\begin{aligned} z_1z_2 &= 12\cdot 2\left[\cos\left(\frac{\pi}{6} + \frac{5\pi}{3}\right) + i\sin\left(\frac{\pi}{6} + \frac{5\pi}{3}\right)\right] \\ &= 24\left(\cos\frac{11\pi}{6} + i\sin\frac{11\pi}{6}\right) \\ &= 24\left(\frac{\sqrt{3}}{2} - \frac{1}{2}i\right) = 12\sqrt{3} - 12i \end{aligned}$$

$$\begin{aligned} \frac{z_1}{z_2} &= \frac{12}{2}\left[\cos\left(\frac{\pi}{6} - \frac{5\pi}{3}\right) + i\sin\left(\frac{\pi}{6} - \frac{5\pi}{3}\right)\right] \\ &= 6\left[\cos\left(-\frac{3\pi}{2}\right) + i\sin\left(-\frac{3\pi}{2}\right)\right] \\ &= 6\left(\cos\frac{\pi}{2} + i\sin\frac{\pi}{2}\right) = 6(0 + 1i) = 6i \end{aligned}$$

For further practice, see Examples 4–5, Section 8.7.

30. $r = \sqrt{\left(-\dfrac{\sqrt{3}}{2}\right)^2 + \left(\dfrac{1}{2}\right)^2} = \sqrt{1} = 1$

$\tan\theta = \dfrac{\frac{1}{2}}{-\frac{\sqrt{3}}{2}}$

$\tan\theta = -\dfrac{\sqrt{3}}{3}$

$\theta = \dfrac{5\pi}{6}$

$-\dfrac{\sqrt{3}}{2} + \dfrac{1}{2}i = 1\left(\cos\dfrac{5\pi}{6} + i\sin\dfrac{5\pi}{6}\right)$

$$\begin{aligned} \left(-\frac{\sqrt{3}}{2} + \frac{1}{2}i\right)^5 &= \left[1\left(\cos\frac{5\pi}{6} + i\sin\frac{5\pi}{6}\right)\right]^5 \\ &= 1^5\left[\cos\left(5\cdot\frac{5\pi}{6}\right) + i\sin\left(5\cdot\frac{5\pi}{6}\right)\right] \\ &= 1\left(\cos\frac{25\pi}{6} + i\sin\frac{25\pi}{6}\right) \\ &= 1\left(\cos\frac{\pi}{6} + i\sin\frac{\pi}{6}\right) \\ &= 1\left(\frac{\sqrt{3}}{2} + \frac{1}{2}i\right) = \frac{\sqrt{3}}{2} + \frac{1}{2}i \end{aligned}$$

For further practice, see Example 6, Section 8.7.

31. amp: 4; period: 4π, freq: $\dfrac{1}{4\pi}$

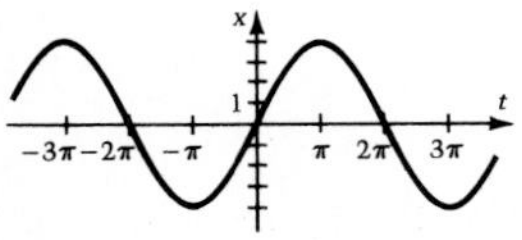

For further practice, see Example 2, Section 8.8.

Chapter 9 Test, p. 598

1.

$y + 5x - 3 = 0$	$3y - 2x + 6 = 0$
$(0) + 5(3) - 3 = 0$	$3(0) - 2(3) + 6 = 0$
$12 = 0$ False	$0 = 0$ True

Thus, (3, 0) is not a solution.

For further practice, see Example 1, Section 9.1.

2.

$$y + 5x - 3 = 0 \qquad 3y - 2x + 6 = 0$$
$$(1) + 5(-2) - 3 = 0 \qquad 3(1) - 2(-2) + 6 = 0$$
$$-12 = 0 \quad \text{False} \qquad 13 = 0 \quad \text{False}$$

Thus, $(-2, 1)$ is not a solution.
For further practice, see Example 1, Section 9.1.

3. $\begin{cases} 6x + 2y = 12 \\ 3x - 4y = 6 \end{cases}$ $\quad \begin{aligned} 6x + 2y &= 12 \\ -2(3x - 4y) &= -2(6) \end{aligned}$ $\quad \begin{aligned} 6x + 2y &= 12 \\ + \quad -6x + 8y &= -12 \\ \hline 10y &= 0 \\ y &= 0 \end{aligned}$

$\begin{cases} 6x + 2y = 12 \\ y = 0 \end{cases}$ $\quad \begin{aligned} 6x + 2(0) &= 12 \\ 6x &= 12 \\ x &= 2 \end{aligned}$

$\begin{cases} x = 2 \\ y = 0 \end{cases}$

The solution is $(2, 0)$.
For further practice, see Example 2, Section 9.1.

4. $\begin{cases} -25x + 10y = 5 \\ 5x - 2y = -1 \end{cases}$ $\quad \begin{aligned} -25x + 10y &= 5 \\ 5(5x - 2y) &= 5(-1) \end{aligned}$ $\quad \begin{aligned} -25x + 10y &= 5 \\ + \quad 25x - 10y &= -5 \\ \hline 0 &= 0 \end{aligned}$

$\begin{cases} 0 = 0 \\ 5x - 2y = -1 \end{cases}$

Since $0 = 0$ is always true, the solutions are all those points on the line $-25x + 10y = 5$ (or $5x - 2y = -1$ in simplified form).
For further practice, see Examples 2–3, Section 9.1.

5. $\begin{cases} 4x + 2y = -5 \\ 6x + 3y = -7 \end{cases}$ $\quad \begin{aligned} 4x + 2y &= -5 \\ 2y &= -5 - 4x \\ y &= -\frac{5}{2} - 2x \end{aligned}$

$\begin{cases} y = -\frac{5}{2} - 2x \\ 6x + 3y = -7 \end{cases}$ $\quad \begin{aligned} 6x + 3\left(-\frac{5}{2} - 2x\right) &= -7 \\ 0 &= \frac{1}{2} \end{aligned}$

$\begin{cases} y = -\frac{5}{2} - 2x \\ 0 = \frac{1}{2} \end{cases}$

Since $0 = \frac{1}{2}$ is never true, there are no solutions.
For further practice, see Example 5, Section 9.1.

6. $\begin{cases} 2x + 5y = -7 \\ 4x - 3y = 10 \end{cases}$ $\quad \begin{aligned} 2x + 5y &= -7 \\ 2x &= -7 - 5y \\ x &= -\frac{7}{2} - \frac{5}{2}y \end{aligned}$

$\begin{cases} x = -\frac{7}{2} - \frac{5}{2}y \\ 4x - 3y = 10 \end{cases}$ $\quad \begin{aligned} 4\left(-\frac{7}{2} - \frac{5}{2}y\right) - 3y &= 10 \\ -13y &= 24 \\ y &= -\frac{24}{13} \end{aligned}$

$\begin{cases} x = -\frac{7}{2} - \frac{5}{2}y \\ y = -\frac{24}{13} \end{cases}$ $\quad \begin{aligned} x &= -\frac{7}{2} - \frac{5}{2}\left(-\frac{24}{13}\right) \\ x &= \frac{29}{26} \end{aligned}$

$\begin{cases} x = \frac{29}{26} \\ y = -\frac{24}{13} \end{cases}$ The solution is $\left(\frac{29}{26}, -\frac{24}{13}\right)$.

For further practice, see Example 4, Section 9.1.

7. $\begin{cases} x^2 - 8x + y^2 = 0 \\ x^2 + y^2 = 9 \end{cases}$ $\qquad$ $\begin{aligned} x^2 - 8x + y^2 &= 0 \\ -1(x^2 + y^2) &= -1(9) \end{aligned}$

$$\begin{array}{rr} & x^2 - 8x + y^2 = 0 \\ + & -x^2 - y^2 = -9 \\ \hline & -8x = -9 \\ & x = \dfrac{9}{8} \end{array}$$

$\begin{cases} x = \dfrac{9}{8} \\ x^2 + y^2 = 9 \end{cases}$ $\qquad$ $\begin{aligned} \left(\frac{9}{8}\right)^2 + y^2 &= 9 \\ y^2 &= \frac{495}{64} \\ y &= \pm\frac{3\sqrt{55}}{8} \end{aligned}$

$\begin{cases} x = \dfrac{9}{8} \\ y = \pm\dfrac{3\sqrt{55}}{8} \end{cases}$

The solutions are $\left(\frac{9}{8}, \frac{3\sqrt{55}}{8}\right)$ and $\left(\frac{9}{8}, -\frac{3\sqrt{55}}{8}\right)$.

For further practice, see Example 7, Section 9.1.

8.

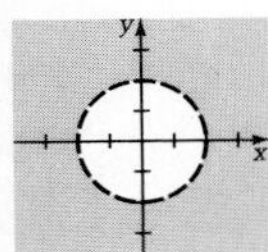

For further practice, see Example 1, Section 9.2.

9.

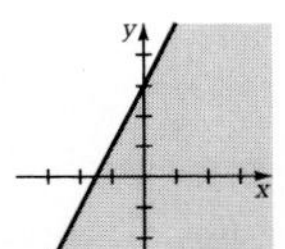

For further practice, see Example 1, Section 9.2.

10.

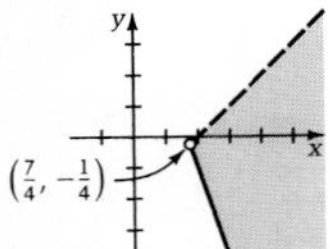

For further practice, see Example 2, Section 9.2.

11.

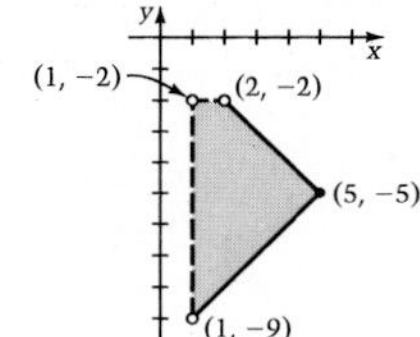

For further practice, see Example 4, Section 9.2.

12.

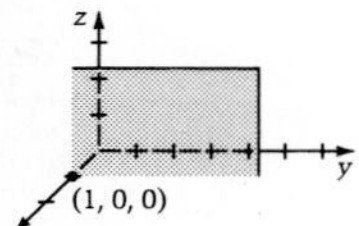

For further practice, see Example 2, Section 9.3.

13.

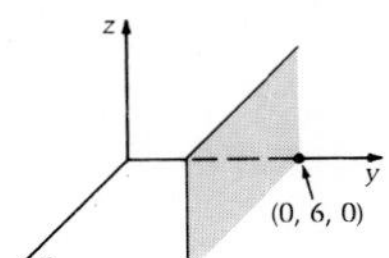

For further practice, see Example 2, Section 9.3.

14.

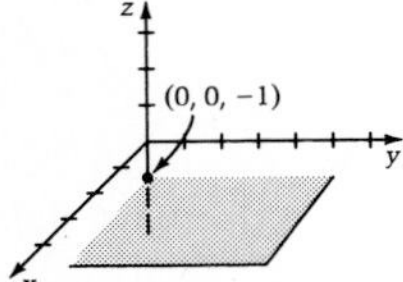

For further practice, see Example 2, Section 9.3.

15. $\begin{cases} x + y + z = 5 \\ 2x - z = 0 \\ 5y + z = -1 \end{cases}$

$2x - z = 0$
$2x = z$
$x = \frac{z}{2}$

$5y + z = -1$
$5y = -1 - z$
$y = -\frac{1}{5} - \frac{z}{5}$

$\begin{cases} x + y + z = 5 \\ x = \frac{z}{2} \\ y = -\frac{1}{5} - \frac{z}{5} \end{cases}$

$\left(\frac{z}{2}\right) + \left(-\frac{1}{5} - \frac{z}{5}\right) + z = 5$
$\frac{13z}{10} = \frac{26}{5}$
$z = 4$

$\begin{cases} z = 4 \\ x = \frac{z}{2} \\ y = -\frac{1}{5} - \frac{z}{5} \end{cases}$

$x = \frac{(4)}{2}$
$x = 2$

$y = -\frac{1}{5} - \frac{(4)}{5}$
$y = -1$

$\begin{cases} z = 4 \\ x = 2 \\ y = -1 \end{cases}$

The solution is $(2, -1, 4)$.
For further practice, see Example 3, Section 9.3.

16. $\left[\begin{array}{cc|c} 4 & -3 & 7 \\ 1 & 2 & -3 \end{array}\right]$ augmented $\begin{bmatrix} 4 & -3 \\ 1 & 2 \end{bmatrix}$ coefficient $\begin{bmatrix} 7 \\ -3 \end{bmatrix}$ constant

For further practice, see Example 3, Section 9.4.

17. $\left[\begin{array}{ccc|c} 2 & 1 & -3 & -7 \\ 1 & -2 & 0 & 5 \\ 0 & 4 & 2 & 0 \end{array}\right]$ augmented $\begin{bmatrix} 2 & 1 & -3 \\ 1 & -2 & 0 \\ 0 & 4 & 2 \end{bmatrix}$ coefficient $\begin{bmatrix} -7 \\ 5 \\ 0 \end{bmatrix}$ constant

For further practice, see Example 3, Section 9.4.

18. $\left[\begin{array}{cc|c} 1 & 2 & 3 \\ 3 & -1 & 1 \end{array}\right] \rightarrow \left[\begin{array}{cc|c} 1 & 2 & 3 \\ 0 & -7 & -8 \end{array}\right] \rightarrow \left[\begin{array}{cc|c} 1 & 2 & 3 \\ 0 & 1 & \frac{8}{7} \end{array}\right] \rightarrow$

$\left[\begin{array}{cc|c} 1 & 0 & \frac{5}{7} \\ 0 & 1 & \frac{8}{7} \end{array}\right]$

The solution is $\left(\frac{5}{7}, \frac{8}{7}\right)$.
For further practice, see Examples 4 and 5, Section 9.4.

19. $\left[\begin{array}{rrr|r} 2 & -1 & 3 & -8 \\ 5 & -2 & -1 & -1 \\ 1 & 1 & 1 & 3 \end{array}\right] \rightarrow \left[\begin{array}{rrr|r} 1 & -\frac{1}{2} & \frac{3}{2} & -4 \\ 5 & -2 & -1 & -1 \\ 1 & 1 & 1 & 3 \end{array}\right] \rightarrow$

$\left[\begin{array}{rrr|r} 1 & -\frac{1}{2} & \frac{3}{2} & -4 \\ 0 & \frac{1}{2} & -\frac{17}{2} & 19 \\ 1 & 1 & 1 & 3 \end{array}\right] \rightarrow \left[\begin{array}{rrr|r} 1 & -\frac{1}{2} & \frac{3}{2} & -4 \\ 0 & \frac{1}{2} & -\frac{17}{2} & 19 \\ 0 & \frac{3}{2} & -\frac{1}{2} & 7 \end{array}\right] \rightarrow$

$\left[\begin{array}{rrr|r} 1 & -\frac{1}{2} & \frac{3}{2} & -4 \\ 0 & 1 & -17 & 38 \\ 0 & \frac{3}{2} & -\frac{1}{2} & 7 \end{array}\right] \rightarrow \left[\begin{array}{rrr|r} 1 & 0 & -7 & 15 \\ 0 & 1 & -17 & 38 \\ 0 & \frac{3}{2} & -\frac{1}{2} & 7 \end{array}\right] \rightarrow$

$\left[\begin{array}{rrr|r} 1 & 0 & -7 & 15 \\ 0 & 1 & -17 & 38 \\ 0 & 0 & 25 & -50 \end{array}\right] \rightarrow \left[\begin{array}{rrr|r} 1 & 0 & -7 & 15 \\ 0 & 1 & -17 & 38 \\ 0 & 0 & 1 & -2 \end{array}\right] \rightarrow$

$\left[\begin{array}{rrr|r} 1 & 0 & 0 & 1 \\ 0 & 1 & -17 & 38 \\ 0 & 0 & 1 & -2 \end{array}\right] \rightarrow \left[\begin{array}{rrr|r} 1 & 0 & 0 & 1 \\ 0 & 1 & 0 & 4 \\ 0 & 0 & 1 & -2 \end{array}\right]$

The solution is $(1, 4, -2)$.
For further practice, see Example 8, Section 9.4.

20. $\det \begin{bmatrix} -1 & 1 \\ 3 & 3 \end{bmatrix} = -1(3) - 3(1) = 3 - 3 = -6$
For further practice, see Example 1, Section 9.5.

21. (a) $\begin{vmatrix} -1 & -1 & 0 \\ 2 & 3 & 1 \\ 3 & 1 & 1 \end{vmatrix} = -1\left(+\begin{vmatrix} 3 & 1 \\ 1 & 1 \end{vmatrix}\right) + (-1)\left(-\begin{vmatrix} 2 & 1 \\ 3 & 1 \end{vmatrix}\right) + 0\left(\begin{vmatrix} 2 & 3 \\ 3 & 1 \end{vmatrix}\right)$

$= -1(2) - 1(1) + 0(-7) = -3$

(b) $\begin{vmatrix} -1 & -1 & 0 \\ 2 & 3 & 1 \\ 3 & 1 & 1 \end{vmatrix} = -1\left(-\begin{vmatrix} 2 & 1 \\ 3 & 1 \end{vmatrix}\right) + 3\left(+\begin{vmatrix} -1 & 0 \\ 3 & 1 \end{vmatrix}\right) + 1\left(-\begin{vmatrix} -1 & 0 \\ 2 & 1 \end{vmatrix}\right)$

$= -1(1) + 3(-1) + 1(1) = -3$
For further practice, see Example 4, Section 9.5.

22. $A = \begin{bmatrix} 10 & -13 & -2 \\ 2 & -7 & -4 \\ 3 & 4 & 5 \end{bmatrix}$

$$\det A = \begin{vmatrix} 10 & -13 & -2 \\ 2 & -7 & -4 \\ 3 & 4 & 5 \end{vmatrix}$$

$$= (10)(-7)(5) + (-13)(-4)(3) + (-2)(2)(4) - (3)(-7)(-2) - (4)(-4)(10) - (5)(2)(-13) = 38$$

$$x = \frac{\begin{vmatrix} 1 & -13 & -2 \\ -1 & -7 & -4 \\ 2 & 4 & 5 \end{vmatrix}}{\det A}$$

$$= \frac{(1)(-7)(5) + (-13)(-4)(2) + (-2)(-1)(4) - (2)(-7)(-2) - (4)(-4)(1) - (5)(-1)(-13)}{38} = 0$$

$$y = \frac{\begin{vmatrix} 10 & 1 & -2 \\ 2 & -1 & -4 \\ 3 & 2 & 5 \end{vmatrix}}{\det A}$$

$$= \frac{(10)(-1)(5) + (1)(-4)(3) + (-2)(2)(2) - (3)(-1)(-2) - (2)(-4)(10) - (5)(2)(1)}{38} = -\frac{3}{19}$$

$$z = \frac{\begin{vmatrix} 10 & -13 & 1 \\ 2 & -7 & -1 \\ 3 & 4 & 2 \end{vmatrix}}{\det A}$$

$$= \frac{(10)(-7)(2) + (-13)(-1)(3) + (1)(2)(4) - (3)(-7)(1) - (4)(-1)(10) - (2)(2)(-13)}{38} = \frac{10}{19}$$

The solution is $\left(0, -\frac{3}{19}, \frac{10}{19}\right)$.

For further practice, see Example 6, Section 9.5.

23. $AB = \begin{bmatrix} 1 \\ 2 \end{bmatrix} [3 \quad -1] = \begin{bmatrix} 1(3) & 1(-1) \\ 2(3) & 2(-1) \end{bmatrix} = \begin{bmatrix} 3 & -1 \\ 6 & -2 \end{bmatrix}$

$BA = [3 \quad -1] \begin{bmatrix} 1 \\ 2 \end{bmatrix} = [3(1) + (-1)(2)] = [1]$

For further practice, see Examples 1–2, Section 9.6.

24. $AB = \begin{bmatrix} 2 & 0 & 3 \\ 1 & -1 & 1 \end{bmatrix} \begin{bmatrix} 1 & -2 \\ 1 & 4 \\ 0 & 3 \end{bmatrix}$

$$= \begin{bmatrix} 2(1) + 0(1) + 3(0) & 2(-2) + 0(4) + 3(3) \\ 1(1) + (-1)(1) + 1(0) & 1(-2) + (-1)(4) + 1(3) \end{bmatrix}$$

$$= \begin{bmatrix} 2 & 5 \\ 0 & -3 \end{bmatrix}$$

$$BA = \begin{bmatrix} 1 & -2 \\ 1 & 4 \\ 0 & 3 \end{bmatrix} \begin{bmatrix} 2 & 0 & 3 \\ 1 & -1 & 1 \end{bmatrix}$$

$$= \begin{bmatrix} 1(2) + (-2)(1) & 1(0) + (-2)(-1) & 1(3) + (-2)(1) \\ 1(2) + 4(1) & 1(0) + 4(-1) & 1(3) + 4(1) \\ 0(2) + 3(1) & 0(0) + 3(-1) & 0(3) + 3(1) \end{bmatrix}$$

$$= \begin{bmatrix} 0 & 2 & 1 \\ 6 & -4 & 7 \\ 3 & -3 & 3 \end{bmatrix}$$

For further practice, see Examples 1–2 , Section 9.6.

25. $A^{-1} = \dfrac{1}{4(2) - 1(5)} \begin{bmatrix} 2 & -1 \\ -5 & 4 \end{bmatrix} = \dfrac{1}{3}\begin{bmatrix} 2 & -1 \\ -5 & 4 \end{bmatrix}$

$= \begin{bmatrix} \frac{2}{3} & -\frac{1}{3} \\ -\frac{5}{3} & \frac{4}{3} \end{bmatrix}$

For further practice, see Example 6, Section 9.6.

26. *Step 1:*

$$\text{Cofactor matrix of } A = \begin{bmatrix} +\begin{vmatrix} 1 & 0 \\ -1 & -4 \end{vmatrix} & -\begin{vmatrix} 1 & 0 \\ 2 & -4 \end{vmatrix} & +\begin{vmatrix} 1 & 1 \\ 2 & -1 \end{vmatrix} \\ -\begin{vmatrix} 1 & -2 \\ -1 & -4 \end{vmatrix} & +\begin{vmatrix} 2 & -2 \\ 2 & -4 \end{vmatrix} & -\begin{vmatrix} 2 & 1 \\ 2 & -1 \end{vmatrix} \\ +\begin{vmatrix} 1 & -2 \\ 1 & 0 \end{vmatrix} & -\begin{vmatrix} 2 & -2 \\ 1 & 0 \end{vmatrix} & +\begin{vmatrix} 2 & 1 \\ 1 & 1 \end{vmatrix} \end{bmatrix} = \begin{bmatrix} -4 & 4 & -3 \\ 6 & -4 & 4 \\ 2 & -2 & 1 \end{bmatrix}$$

Step 2:

$$\operatorname{adj}(A) = \begin{bmatrix} -4 & 4 & -3 \\ 6 & -4 & 4 \\ 2 & -2 & 1 \end{bmatrix}^t = \begin{bmatrix} -4 & 6 & 2 \\ 4 & -4 & -2 \\ -3 & 4 & 1 \end{bmatrix}$$

Step 3: Expanding along the second row, $\det A =$

$$-1\begin{vmatrix} 1 & -2 \\ -1 & -4 \end{vmatrix} + 1\begin{vmatrix} 2 & -2 \\ 2 & -4 \end{vmatrix} = -1(-6) + 1(-4) = 2.$$

Thus,

$$A^{-1} = \frac{1}{\det A}\operatorname{adj}(A) = \frac{1}{2}\begin{bmatrix} -4 & 6 & 2 \\ 4 & -4 & -2 \\ -3 & 4 & 1 \end{bmatrix}$$

$$= \begin{bmatrix} -2 & 3 & 1 \\ 2 & -2 & -1 \\ -\frac{3}{2} & 2 & \frac{1}{2} \end{bmatrix}$$

For further practice, see Example 4, Section 9.6.

27. Assign Variables:

Let x = the number of hamburgers.
Let y = the number of hotdogs.
Let P = the total profit.

Constraints:

$$\begin{cases} x \geq 4000 \\ x \leq 7000 \\ y \geq 6000 \\ y \leq 11{,}000 \\ x + y \leq 16{,}000 \end{cases}$$

Objective Function: $P = 0.27x + 0.22y$

Feasible Region:

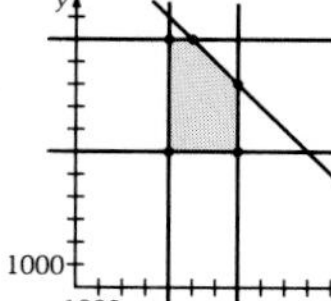

Corner Points	$P = 0.27x + 0.22y$
(4000, 6000)	$P = 0.27(4000) + 0.22(6000) = 2400$
(4000, 11,000)	$P = 0.27(4000) + 0.22(11{,}000) = 3500$
(5000, 11,000)	$P = 0.27(5000) + 0.22(11{,}000) = 3770$
(7000, 6000)	$P = 0.27(7000) + 0.22(6000) = 3210$
(7000, 9000)	$P = 0.27(7000) + 0.22(9000) = 3870$

The vendor should have 7000 hamburgers and 9000 hotdogs.
For further practice, see Example 3, Section 9.7.

Chapter 10 Test, pp. 661–662

1. $a_n = 2^n - n$

$a_1 = 2^1 - 1 = 1$

$a_2 = 2^2 - 2 = 2$

$a_3 = 2^3 - 3 = 5$

$a_4 = 2^4 - 4 = 12$

$a_5 = 2^5 - 5 = 27$

$a_9 = 2^9 - 9 = 503$

For further practice, see Example 1, Section 10.1.

2. (a) $a_n = 5 + (-1)^n a_{n-1}$

$a_1 = 2$

$a_2 = 5 + (-1)^2(2) = 7$

$a_3 = 5 + (-1)^3(7) = -2$

$a_4 = 5 + (-1)^4(-2) = 3$

$a_5 = 5 + (-1)^5(3) = 2$

$a_6 = 5 + (-1)^6(2) = 7$

(b) $a_n = a_{n-1} + a_{n-2} + a_{n-3}$

$a_1 = 1$

$a_2 = 3$

$a_3 = 2$

$a_4 = 2 + 3 + 1 = 6$

$a_5 = 6 + 2 + 3 = 11$

$a_6 = 11 + 6 + 2 = 19$

For further practice, see Example 2, Section 10.1.

3. $a_1 = 7$

$a_2 = 7 + (-5) = 2$

$a_3 = 2 + (-5) = -3$

$a_4 = -3 + (-5) = -8$

$a_5 = -8 + (-5) = -13$

For further practice, see Example 3, Section 10.1.

4. $d = 9 - 3 = 6$

$a_n = a_1 + (n - 1)d$

$a_{11} = 3 + (11 - 1)6 = 63$

$a_{21} = 3 + (21 - 1)6 = 123$

For further practice, see Example 5, Section 10.1.

5. $S_n = \frac{n}{2}[2a_1 + (n - 1)d]$

$S_{12} = \frac{12}{2}[2(2) + (12 - 1)3] = 222$

For further practice, see Example 6, Section 10.1.

6. (a) $\frac{12}{16} = \frac{3}{4},\ \frac{9}{12} = \frac{3}{4},\ \frac{6}{9} = \frac{2}{3},\ \frac{9/2}{6} = \frac{3}{4}$

The sequence is not geometric.

(b) $\frac{-6}{3} = \frac{12}{-6} = \frac{-24}{12} = \frac{48}{-24} = -2$

The sequence is geometric.

For further practice, see Example 1, Section 10.2.

7. $r = \dfrac{-9/5}{27/10} = -\dfrac{2}{3}$

$a_n = a_1 r^{n-1}$

$a_6 = \dfrac{27}{10}\left(-\dfrac{2}{3}\right)^{6-1} = -\dfrac{16}{45}$

$a_9 = \dfrac{27}{10}\left(-\dfrac{2}{3}\right)^{9-1} = \dfrac{128}{1215}$

$a_n = \dfrac{27}{10}\left(-\dfrac{2}{3}\right)^{n-1}$

For further practice, see Example 2, Section 10.2.

8. $S_n = \dfrac{a_1(1 - r^n)}{1 - r}$

$S_n = \dfrac{2[1 - (2/3)^n]}{1 - (2/3)} = 6\left[1 - \left(\dfrac{2}{3}\right)^n\right]$

$S_5 = 6\left[1 - \left(\dfrac{2}{3}\right)^5\right] = \dfrac{422}{81}$

For further practice, see Example 3, Section 10.2.

9. $\displaystyle\sum_{n=2}^{6} (n^2 - 3n) = [2^2 - 3(2)] + [3^2 - 3(3)] + [4^2 - 3(4)] + [5^2 - 3(5)] + [6^2 - 3(6)]$

$= -2 + 0 + 4 + 10 + 18 = 30$

For further practice, see Example 1, Section 10.3.

10. $4 + 9 + 16 + 25 + 36 = \displaystyle\sum_{n=2}^{6} n^2$

For further practice, see Example 2, Section 10.3.

11. $S_3 = 3 + \dfrac{3}{2} + \dfrac{3}{4} = \dfrac{21}{4}$

$S_n = \dfrac{a_1(1 - r^n)}{1 - r}$

$S_6 = \dfrac{3[1 - (1/2)^6]}{1 - (1/2)} = \dfrac{189}{32}$

For further practice, see Example 3, Section 10.3.

12. (a) $r = \dfrac{3}{4}$

$S = \dfrac{a_1}{1 - r} = \dfrac{7}{1 - \left(\dfrac{3}{4}\right)} = 28$

(b) $S_n = \left(\dfrac{1}{5} - \dfrac{1}{3}\right) + \left(\dfrac{1}{7} - \dfrac{1}{5}\right) + \cdots + \left(\dfrac{1}{2n+1} - \dfrac{1}{2n-1}\right) + \left(\dfrac{1}{2n+3} - \dfrac{1}{2n+1}\right)$

$= -\dfrac{1}{3} + \dfrac{1}{2n+3}$

Thus, $S = -\dfrac{1}{3}$.

For further practice, see Examples 4 and 6, Section 10.3.

13. $0.37\overline{272} = \dfrac{372}{1000} + 0.00\overline{072}$

$= \dfrac{372}{1000} + \dfrac{72}{10^5} + \dfrac{72}{10^7} + \dfrac{72}{10^9} + \cdots$

$= \dfrac{372}{1000} + \dfrac{72/10^5}{1 - (1/10^2)} = \dfrac{372}{1000} + \dfrac{1}{1375} = \dfrac{41}{110}$

For further practice, see Example 5, Section 10.3.

14. *Requirement 1:* When $n = 1$, the given statement becomes $3 = \dfrac{3(1)(1 + 1)}{2}$, which is true. Thus, Requirement 1 is met.

Requirement 2: We assume the given statement is true for $n = k$:

$$3 + 6 + 9 + 12 + \cdots + 3k = \frac{3k(k + 1)}{2}$$

We must prove that it is true for $n = k + 1$:

$$3 + 6 + 9 + 12 + \cdots + 3(k + 1) = \frac{3(k + 1)[(k + 1) + 1]}{2}$$

Proof:

$$3 + 6 + 9 + 12 + \cdots + 3k = \frac{3k(k + 1)}{2}$$

$$3 + 6 + 9 + 12 + \cdots + 3k + 3(k + 1) = \frac{3k(k + 1)}{2} + 3(k + 1) = \frac{3k(k + 1)}{2} + \frac{6(k + 1)}{2}$$

$$= \frac{3k^2 + 9k + 6}{2} = \frac{3(k + 1)(k + 2)}{2} = \frac{3(k + 1)[(k + 1) + 1]}{2}$$

Thus, Requirement 2 is met, and by the Principle of Mathematical Induction, the given statement is true for all positive integers.
For further practice, see Example 1, Section 10.4.

15. $(3 + 5t)^4 = \binom{4}{0} 3^4(5t)^0 + \binom{4}{1}3^3(5t)^1 + \binom{4}{2}3^2(5t)^2 + \binom{4}{3}3^1(5t)^3 + \binom{4}{4}3^0(5t)^4$

$= 1(81)(1) + 4(27)(5t) + 6(9)(25t^2) + 4(3)(125t^3) + 1(1)(625t^4)$

$= 81 + 540t + 1350t^2 + 1500t^3 + 625t^4$
For further practice, see Example 3, Section 10.5.

16. **(a)** $\lim_{x\to -2^+} f(x) = -\infty$

$\lim_{x\to -2^-} f(x) = -\infty$

$\lim_{x\to -2} f(x)$ does not exist.

(b) $\lim_{x\to 1^+} f(x) = 2$

$\lim_{x\to 1^-} f(x) = 2$

$\lim_{x\to 1} f(x) = 2$

(c) $\lim_{x\to 3^+} f(x) = 1$

$\lim_{x\to 3^-} f(x) = 3$

$\lim_{x\to 3} f(x)$ does not exist.

(d) $\lim_{x\to 6^+} f(x) = 3$

$\lim_{x\to 6^-} f(x) = 3$

$\lim_{x\to 6} f(x) = 3$

(e) $\lim_{x\to -\infty} f(x) = 4$

(f) $\lim_{x\to \infty} f(x) = -3$

For further practice, see Examples 1–2, Section 10.6.

17. $\lim_{n\to\infty} \dfrac{7n - n^2}{5n^2 + 1} = -\dfrac{1}{5}$
For further practice, see Example 3, Section 10.6.

18. $\lim_{n\to\infty} \dfrac{8n}{n^3 + 1} = 0$
For further practice, see Example 3, Section 10.6.

19. $\lim_{n\to\infty} \dfrac{3n^2}{7n + 2}$ does not exist.
For further practice, see Example 3, Section 10.6.

20. The tree has branches GGG, GGB, GBG, GBB, BGG, BGB, BBG, BBB.
For further practice, see Example 1, Section 10.7.

21. $2 \cdot 2 \cdot 2^6 = 2^8 = 256$

There are 256 different pizzas.

For further practice, see Example 4, Section 10.7.

22. (a) ${}_8P_4 = 8 \cdot 7 \cdot 6 \cdot 5 = 1680$

(b) ${}_{11}P_{11} = 11! = 39{,}916{,}800$

(c) $\binom{9}{6} = \dfrac{9!}{(9-6)!\,6!} = \dfrac{9 \cdot 8 \cdot 7 \cdot 6!}{3!\,6!} = 84$

(d) $\binom{8}{0} = \dfrac{8!}{(8-0)!\,0!} = \dfrac{8!}{8!\,0!} = 1$

For further practice, see Examples 5 and 7, Section 10.7.

23. $\binom{100}{5} = \dfrac{100!}{(100-5)!\,5!} = \dfrac{100 \cdot 99 \cdot 98 \cdot 97 \cdot 96 \cdot 95!}{95!\,5!}$

$= 75{,}287{,}520$

A total of 75,287,520 different groups can be chosen.

For further practice, see Example 8, Section 10.7.

24. (a) ${}_5P_4 = 5 \cdot 4 \cdot 3 \cdot 2 = 120$

A total of 120 different numbers can be formed.

(b) $5 \cdot 5 \cdot 5 \cdot 5 = 625$

A total of 625 different numbers can be formed.

(c) $4 \cdot 3 \cdot 2 \cdot 1 = 24$

A total of 24 different numbers can be formed.

For further practice, see Example 9, Section 10.7.

25. (a) $\binom{7}{2} \cdot \binom{4}{3} = \dfrac{7!}{(7-2)!\,2!} \cdot \dfrac{4!}{(4-3)!\,3!}$

$= \dfrac{7 \cdot 6 \cdot 5!}{5!\,2!} \cdot \dfrac{4 \cdot 3!}{1!\,3!} = 84$

The user can select the programs 84 ways.

(b) $\binom{4}{4} \cdot \binom{7}{1} = \dfrac{4!}{(4-4)!\,4!} \cdot \dfrac{7!}{(7-1)!\,1!}$

$= \dfrac{4!}{0!\,4!} \cdot \dfrac{7 \cdot 6!}{6!\,1!} = 7$

The user can select the programs 7 ways.

(c) $\binom{7}{5} = \dfrac{7!}{(7-5)!\,5!} = \dfrac{7 \cdot 6 \cdot 5!}{2!\,5!} = 21$

The user can select the programs 21 ways.

For further practice, see Example 10, Section 10.7.

26. (a) $P(E) = \dfrac{\binom{3}{3}}{2 \cdot 2 \cdot 2} = \dfrac{1}{8}$

(b) $P(E) = \dfrac{\binom{3}{1}\binom{2}{2}}{2 \cdot 2 \cdot 2} = \dfrac{3 \cdot 1}{8} = \dfrac{3}{8}$

(c) $P(E) = 1 - P(\text{no heads}) = 1 - \dfrac{1}{8} = \dfrac{7}{8}$

For further practice, see Example 2, Section 10.8.

27. (a) $P(E_1 \cup E_2) = P(\mathcal{S}) = 1$

(b) $P(E_1 \cap E_2) = P(\{x_3\}) = \dfrac{1}{4}$

(c) $P(E_1' \cup E_3) = P(\{x_2\,x_4\}) = \dfrac{1}{2}$

For further practice, see Example 4, Section 10.8.

Answers To Exercises

Chapter 1

Exercise Set 1.1, pages 5–6

1. $8x = 47$ **3.** $x + 5 = 2x$ **5.** $1.20x = x + 4$ **7.** $n + (n + 1) = 3n$ **9.** $x + y = 11$ **11.** $2x = y - 15$
2. $x - 7 = 14$ **4.** $x + 10 = 5x$ **6.** $0.75x = x - 12$ **8.** $n(n + 1) = n^2 + 7$ **10.** $xy = 53$ **12.** $5x = y + 10$

13. $x + 6 = 2y - 8$ **15.** $\frac{1}{2}(x + y) = xy + 19$ **17.** $x + 2 = 0.05y$ **19.** $x = 1.02y$ **21.** $x + 6 = 0.70y$

14. $x + 15 = 2y + 2$ **16.** $\frac{1}{2}(x + y) = xy - 11$ **18.** $x - 5 = 0.63y$ **20.** $x = 0.41y$ **22.** $x - 7 = 1.06y$

23. $n = 2b$ **25.** $b + 3 = n$ **27.** $0.10D$ **29.** $\frac{X}{12}$ **31.** $\frac{d}{12}$ **33.** $2(0.20L + 0.15W)$ **35.** $3n + 7m$

24. $b = \frac{n}{3}$ **26.** $n = b + 7$ **28.** $0.05N$ **30.** $36Y$ **32.** $(\$19.95)S$ **34.** $0.30\pi d$ **36.** $5W + 3S$

37. $0.05n + 0.09m$ ***39.** $x + 22 = 36$ ***41.** $99 + 0.20m = 198$ ***43.** $44 = 0.80P$ ***45.** $\frac{300}{N}$

38. $0.80M + 0.75N$ ***40.** $p - 18 = 32$ ***42.** $4a = 82.9$ ***44.** $0.05C = 2.45$ ***46.** $\frac{4(17) - M - P}{2}$

Exercise Set 1.2, Pages 11–13

1. rational, real **3.** whole number, integer, rational, real **5.** irrational, real **7.** rational, real **9.** True
2. integer, rational, real **4.** rational, real **6.** rational, real **8.** integer, rational, real **10.** True

11. False **13.** True **15.** False **17.** False **19.** $-5, \frac{1}{5}$ **21.** $\frac{2}{3}, -\frac{3}{2}$ **23.** 0, none **25.** $-\pi, \frac{1}{\pi}$ **27.** $x, -\frac{1}{x}$

12. True **14.** False **16.** False **18.** False **20.** $3, -\frac{1}{3}$ **22.** $1, -1$ **24.** $-\frac{5}{2}, \frac{2}{5}$ **26.** $-\sqrt{2}, \frac{1}{\sqrt{2}}$ **28.** $-x, \frac{1}{x}$

---

*denotes Superset exercises

29. $-\frac{1}{x}, x$ **31.** 21 **33.** 14 **35.** 10 **37.** 3 **39.** $\frac{304}{189}$ **41.** 29.2 **43.** -6.2 **45.** 0.3 **47.** 98.4

30. $xy, -\frac{1}{xy}$ **32.** $-\frac{1}{4}$ **34.** $32\frac{3}{4}$ **36.** 7 **38.** $\frac{15}{88}$ **40.** $\frac{36}{19}$ **42.** -14.8 **44.** -0.2 **46.** -3.4 **48.** 1.3

49. -10590 **51.** -3 **53.** 10 **55.** -2 **57.** -3 **59.** 16 **61.** $-\frac{1}{3}$ **63.** -3 **65.** 0.329 **67.** 2.935

50. -3509 **52.** -21 **54.** -2 **56.** -10 **58.** -3 **60.** 5 **62.** -80 **64.** -34 **66.** 0.385 **68.** 5.447

***69.** $\frac{1}{4}x = 5\left(\frac{1}{x}\right)$ ***71.** $3\left(x + \frac{1}{x}\right) = x + 4$ ***73.** $-(x + y) = xy - 5$ ***75.** $x + 0.54y = 0.83xy$

***70.** $x + 10 = 24\left(\frac{1}{x}\right)$ ***72.** $2\left(x + \frac{1}{x}\right) = x - 1$ ***74.** $-xy = 2(x + y) + 6$ ***76.** $\frac{1}{xy} = 0.20\left(\frac{x + y}{2}\right)$

***77.** False, $(2)(3)(2)(4) \neq (2)(3)(4)$ ***79.** True ***81.** False, $\frac{1}{2(2)} \neq \left(\frac{1}{2}\right)(2)$ ***83.** $3(7 + 2 - 5)$ ***85.** $((12 - 3) \cdot 8) \div 6$

***78.** False, $3(0 + 7) \neq 3(0) + 7$ ***80.** False, $5 - (0 + 2) \neq 5 - 0 + 2$ ***82.** True ***84.** $(3 \cdot 5) - (10 - 7)$

***86.** $(12 + (3 \cdot 8)) \div 3$

***87.** (a) Distributive (b) Commutative (add.) (c) Associative (add.) (d) Additive Inverse (e) Additive Identity

***88.** (a) Commutative (mult.) (b) Associative (mult.) (c) Multiplicative Inverse (d) Multiplicative Identity

***89.** (a) Distributive (b) Associative (mult.) (c) Commutative (mult.) (d) Associative (mult.) (e) Associative (add.) (f) Distributive (g) Additive Inverse (h) Additive Identity

***90.** (a) Distributive (b) Associative (add.) (c) Commutative (add.) (d) Associative (add.) (e) Distributive

***91.** Use hint and follow steps

***92.** Similar to Exercise 91, use hint

Section 1.3, pages 17–18

1. 4 **3.** 0 **5.** $-\frac{1}{5}$ **7.** 13 **9.** 30 **11.** $0 > -11$ **13.** $-0.5 < -0.05$ **15.** $-0.624 > -\frac{5}{8}$

2. $\frac{5}{9}$ **4.** $-\sqrt{3}$ **6.** -2 **8.** -1 **10.** $-\frac{9}{5}$ **12.** $-30 < -20$ **14.** $\frac{3}{5} > \frac{4}{7}$ **16.** $\frac{\pi}{2} < 1.577$

17. $\frac{6.8 + 5.7}{10.02} < |8.01 - 9.26|$ **19.** $\frac{(6.07)(0.889)}{1 + |1 - 6.19|} > |8.14 - 9.011|$ **21.** $x + 15 < -10$ **23.** $2x - 5 \geq 0$

18. $\frac{6.7 - |3 - 7.5|}{16} > |0.28 - 5| \div 40$ **20.** $\frac{5 - 3(9.07)}{7.9 - 2.0} < \left|5.908 - \frac{(16.1)(8.2)}{17.04}\right|$ **22.** $2x - 7 > 20$ **24.** $x + 6 > 0$

25. 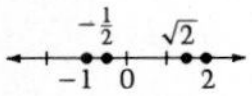

27. 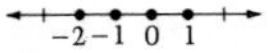

29. 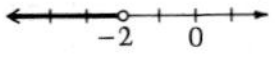

31. 12 **33.** 18 **35.** 18 **37.** -0.0475 **39.** 0.0175

26. 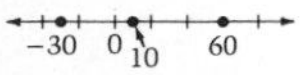

28. 

30. 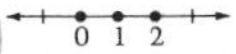

32. 8 **34.** 4 **36.** 0 **38.** -0.0025 **40.** 0.009375

41. B ***43.** $a = 2, b = -5, c = 5$ ***45.** False, $x = -1$ ***47.** True

42. E ***44.** $d(A, B) = |a - b| = |b - a| = d(B, A)$ ***46.** False, $|0| = 0$ ***48.** False, $(-1)(-1) = 1$

***49.** True ***51.** False, $|-1| \neq -1$

***50.** False, $d(1, 1) = 0$ ***52.** False, $|-1| \not> -(-1)$

Section 1.4, pages 25–27

1. 625 **3.** -625 **5.** $\frac{1}{8}$ **7.** 0 **9.** not defined **11.** $\frac{1}{9}$ **13.** $\frac{3}{2}$ **15.** 8 **17.** 9 **19.** 151,263

2. 81 **4.** -81 **6.** $\frac{1}{9}$ **8.** 1 **10.** not defined **12.** $\frac{1}{81}$ **14.** $\frac{8}{5}$ **16.** 81 **18.** 128 **20.** $\frac{36}{625}$

21. 0.000001 **23.** 100,000 **25.** $\frac{1}{6^2}$ **27.** $\left(\frac{7}{3}\right)^5$ **29.** $\frac{1}{a^2}$ **31.** x^3 **33.** -3^4 **35.** $\frac{1}{x}$ **37.** 8^{-3}

---

*denotes Superset exercises

22. 100 **24.** 100,000 **26.** $\frac{1}{(-8)^3}$ **28.** $\left(\frac{8}{5}\right)^3$ **30.** $\frac{1}{b^{10}}$ **32.** y^7 **34.** $-\frac{1}{2^3}$ **36.** $-\frac{1}{y}$ **38.** $(-10)^{-4}$
39. $\frac{1}{4^{-5}}$ **41.** x^{-5} **43.** -3^{-8} **45.** $\left(\frac{3}{2}\right)^{-5}$ **47.** y^{-1} **49.** 34 **51.** $-\frac{1}{2}$ **53.** -125 **55.** $-\frac{4}{45}$
40. $\frac{1}{3^{-9}}$ **42.** $(-y)^{-7}$ **44.** $(-3)(4^{-3})$ **46.** $\left(\frac{5}{3}\right)^{-3}$ **48.** $-2x^{-1}$ **50.** -16 **52.** 3 **54.** 1 **56.** $\frac{28}{75}$
57. 63.87 **59.** 65.01 **61.** 3^7 **63.** 3^0 **65.** 3^{30} **67.** 3^{14} **69.** $-4x^3y^5$ **71.** $12xyz^4$ **73.** $\frac{15y}{x^2}$
58. 4.09 **60.** 19.38 **62.** 3^7 **64.** 3^{-1} **66.** 3^{-18} **68.** 3^1 **70.** $14x^4y$ **72.** $10x^3yz$ **74.** $\frac{12y^3}{x}$
75. $\frac{12x}{z^5}$ **77.** $8x^3$ **79.** $36x^2y^4z^2$ **81.** $-125z^{12}$ **83.** $\frac{2x}{3yz}$ **85.** $\frac{7x^3z^2}{10y^5}$ **87.** $\frac{x^7}{yz^2}$ **89.** $\frac{27z^3}{8x^3y^3}$
76. $\frac{2y^2}{z^2}$ **78.** $9y^2$ **80.** $16x^6y^2z^2$ **82.** $1296x^{12}y^4$ **84.** $\frac{8z^4}{11y^3}$ **86.** $\frac{5x^4z^{10}}{6y^3}$ **88.** $\frac{x^4}{y^2z^7}$ **90.** $\frac{49y^2z^2}{25x^2}$
91. $\frac{243z^{15}}{x^{10}}$ **93.** 1.0×10^{-4} **95.** 3.5×10^9 **97.** 1.6×10^3 **99.** 1.18×10^{-3} **101.** 2.0×10^{-4}
92. $\frac{16x^8}{y^{12}}$ **94.** 7.19×10^4 **96.** 3.6×10^{-2} **98.** 1.81×10^5 **100.** 5.0×10^{-3} **102.** 5.0×10^1
103. 4.0×10^2 **105.** 6.0×10^{-4} **107.** 641,000 **109.** 0.000099 **111.** 300,000,000 **113.** 5,555,000,000
104. 6.0×10^5 **106.** 0.00641 **108.** 20,100 **110.** 0.000008 **112.** 1,025,000 **114.** 0.00000014
115. 0.0000081 **117.** 0.0000849 **119.** 3.52×10^8 **121.** 3.22×10^5 **123.** 9.53×10^9 ***125.** True
116. 0.0000000034 **118.** 2.84×10^5 **120.** 2.08×10^5 **122.** 5.00×10^{-1} ***124.** False ***126.** False
***127.** False ***129.** True ***131.** False ***133.** False ***135.** $\frac{3y^2}{343x^7z^{11}}$ ***137.** $\frac{9x^6y^{23}}{z^6}$ ***139.** $x^{15}y^{10}z^8$
***128.** True ***130.** False ***132.** False ***134.** $\frac{4x^3}{yz^2}$ ***136.** $\frac{16x^8y}{z^5}$ ***138.** $\frac{1}{x^9y^6z^{26}}$ ***140.** $\frac{1}{(x+y)^3}$
***141.** $\frac{1}{(y-z)^4}$ ***143.** True ***145.** True ***147.** True ***149.** True ***151.** $1 + (0.02)^{-3}$ ***153.** $1 - 0.02$
***142.** True ***144.** True ***146.** False ***148.** False ***150.** $1 - (0.004)^2$ ***152.** $(1 + 0.02)^2$ ***154.** 6.9
***155.** $\frac{3.2 \times 10^{-40}}{7.5 \times 10^{-38}}$ ***157.** $\frac{1}{5}$ ***159.** $36\pi a^6$ ft³ ***161.** $0.001y^9$ ft³ ***163.** 0.0016π ft³
***156.** 0.001 ***158.** $4\pi x^6$ cm² ***160.** $25p^{-2}$ in² ***162.** $100\pi n^4$ cm³ ***164.** $V_{\text{cylinder}} > V_{\text{sphere}}$
***165.** $A_{\text{cube}} > A_{\text{circle}}$ ***167.** 1.5×10^{15}; 1500 trillion gallons ***169.** 9.1×10^8; 910 million dollars
***166.** 7.3×10^{11}; 730 billion dollars ***168.** 1.1×10^{11}; 110 billion gallons ***170.** 1.8×10^7; 18 million TV sets
***171.** 1.7×10^{10}; 17 billion hot dogs ***173.** 8.3 minutes
***172.** 5.87×10^{12} miles/year

Section 1.5, pages 33–34

1. 1 term; degree 8 **3.** not a polynomial **5.** 3 terms; degree 4 **7.** 4 terms; degree 5 **9.** 1 term; degree 0
2. 2 terms; degree 9 **4.** 3 terms; degree 4 **6.** 3 terms; degree 4 **8.** not a polynomial **10.** 4 terms; degree 5
11. $3x^3 - 5x^2 + 10x - 4$ **13.** $4u^4 - 3u^2 + u$ **15.** $6x^4 + x^2 - 8x + 11$ **17.** $-2v^2 - 6v + 20$
12. $-x^3 - 8x^2 - 3x + 19$ **14.** $4x^4 - x^3 - 3x^2 + 2$ **16.** $21a^2 - a - 23$ **18.** $y^3 + 4y^2 - 3y + 6$
19. $-x^2 + 6y - y^2$ **21.** $10x^3 - 14x^2 + 2x$ **23.** 0 **25.** $21x^2 - 5x - 6$ **27.** $20m^2 - 17m + \frac{3}{2}$
20. $-x^2 - 4x + y^2$ **22.** $-3a^3 - 27a^2 + 3a$ **24.** 0 **26.** $10x^2 + 23x - 42$ **28.** $18n^2 - 14n + \frac{8}{3}$

---

*denotes Superset exercises

29. $90x^3 + 135x^2y + 50xy^2$ **31.** $\frac{3}{2}b^2 - b + \frac{1}{8}$ **33.** $3x^3 + \frac{47}{5}x^2 - \frac{54}{5}x - \frac{8}{5}$
30. $84u^3 + 136u^2v + 32uv^2$ **32.** $\frac{1}{2}n^2 + \frac{7}{30}n - \frac{2}{15}$ **34.** $2y^3 - \frac{31}{3}y^2 + \frac{17}{3}y - \frac{2}{3}$
35. $8m^4 - 6m^3 + 11m^2 - m - 12$ **37.** $3x^4 - 2x^3 - 6x^2 + 7x - 2$ **39.** $9x^4 - 4y^4$ **41.** $9a^2 + 42a + 49$
36. $5x^5 - 2x^4 - 10x^3 + 9x^2 - 52x + 20$ **38.** $v^4 + 2v^3 + 2v^2 + 23v - 14$ **40.** $4m^4 - 9n^4$ **42.** $16m^2 - 24m + 9$
43. $25x^2 + 30xy + 9y^2$ **45.** $x^3 + 6x^2 + 12x + 8$ **47.** $x^2 - 4xy + 4y^2 + 2x - 4y + 1$
44. $49a^2 - 56ab + 16b^2$ **46.** $x^3 - 6x^2 + 12x - 8$ **48.** $9s^2 + 12st + 4t^2 - 6s - 4t + 1$
49. $8x^6 + 60x^4y^3 + 150x^2y^6 + 125y^9$ **51.** $-8x$ **53.** $x^4 - 18x^2 + 81$ **55.** $4x^3 - 4x^2 - 39x - 36$
50. $64x^9 - 144x^6y^2 + 108x^3y^4 - 27y^6$ **52.** $6x + 21$ **54.** $x^4 - 2x^2 + 1$ **56.** $9x^3 + 33x^2 - 185x + 175$
57. $x^2 - 2(a + b)x + (a + b)^2$ ***59.** $x^2 + 4x + 4$ ***61.** $x^2 - 2x$ ***63.** $36\pi x^3 + 72\pi x^2 + 48\pi x + \frac{32\pi}{3}$
58. $x^3 - 3(a + b)x^2 + 3(a + b)^2x - (a + b)^3$ ***60.** $4x^2 + 4x + 1$ ***62.** $\pi x^2 + 6\pi x + 9\pi$ ***64.** $x^3 + 9x^2 + 27x + 27$
***65.** $2x^3 - x^2 + 8x - 4$ ***67.** $6x^2 + 36x + 54$ ***69.** $6x^3 + 21x^2 - 12x$ ***71.** 450
***66.** $36\pi x^2 + 48\pi x + 16\pi$ ***68.** $22x^2 + 32x - 8$ ***70.** -540 ***72.** -24
***73.** -28 ***75.** No; when $x = 2$, $(x - 1)^3 = 1$ while $(1 - x)^3 = -1$.
***74.** Yes; $(1 - x)^2 = [(-1)(x - 1)]^2 = (x - 1)^2$. ***76.** $(a + b + c)^2 = a^2 + b^2 + c^2 + 2ab + 2bc + 2ac$
***77.** $(a + b)^4 = a^4 + 4a^3b + 6a^2b^2 + 4ab^3 + b^4$ ***79.** $A^3 - B^3$
***78.** $(a - b)^4 = a^4 - 4a^3b + 6a^2b^2 - 4ab^3 + b^4$ ***80.** $A^3 + B^3$

Section 1.6, pages 41–42

1. $5ab$ **3.** $2x^2y^3$ **5.** 15 **7.** $m(5 - m)$ **9.** $v(4v + 3)(v - 2)$ **11.** $(x + 10)(x - 10)$
2. $6uv$ **4.** $5m^3n^2$ **6.** $6x$ **8.** $3m(m - 5)$ **10.** $v(3v + 2)(v - 3)$ **12.** $(x + 8)(x - 8)$
13. $(2t + 0.5)(2t - 0.5)$ **15.** $(p + q)(p - q)$ **17.** cannot be factored **19.** $2(2v + 3u)(2v - 3u)$
14. $(0.4t + 7)(0.4t - 7)$ **16.** $(q + 2p)(q - 2p)$ **18.** cannot be factored **20.** $2(3v + 5u)(3v - 5u)$
21. $(s - 2)(s^2 + 2s + 4)$ **23.** $(1 - 4a)(1 + 4a + 16a^2)$ **25.** $(2x + 3y)(4x^2 - 6xy + 9y^2)$
22. $(s - 3)(s^2 + 3s + 9)$ **24.** $(1 - 5a)(1 + 5a + 25a^2)$ **26.** $(4x + 3y)(16x^2 - 12xy + 9y^2)$
27. $v(0.1u + v)(0.01u^2 - 0.1uv + v^2)$ **29.** $(m + 7)^2$ **31.** $(z - 7)^2$ **33.** $(2b + 3)^2$ **35.** $3(s - 3)^2$
28. $v^2(u + 0.2v)(u^2 - 0.2uv + 0.04v^2)$ **30.** $(m + 5)^2$ **32.** $(z - 5)^2$ **34.** $(4b + 5)^2$ **36.** $2(s - 2)^2$
37. $(2x + y)(x - y)$ **39.** $(p^2 + 1)(p - 2)$ **41.** $(t + 3)(t + 4)$ **43.** $a^2(a - 0.3b)(a^2 + 0.3ab + 0.09b^2)$
38. $(3x - y)(x + y)$ **40.** $(p^2 + 2)(p - 3)$ **42.** $(t + 2)(t + 5)$ **44.** $b(b - 5a)(b^2 + 5ab + 25a^2)$
45. $(2x - 3y)(4x^2 + 6xy + 9y^2)$ **47.** $(y + 5)(y - 3)$ **49.** cannot be factored **51.** $(p + 2)(p - 3)$
46. $(4x - 3y)(16x^2 + 12xy + 9y^2)$ **48.** $(y + 7)(y - 2)$ **50.** cannot be factored **52.** $(p + 3)(p - 4)$
53. $(2d + 1)(d - 3)$ **55.** $(m + 3)(m + 12)$ **57.** $(2v - 1)(2v + 3)$ **59.** $(3y - 4)(5y + 2)$ **61.** $(3a - 4b)(2a + b)$
54. $(4d - 3)(d - 1)$ **56.** $(m + 2)(m + 14)$ **58.** $(2v - 1)(3v + 2)$ **60.** $(2y - 3)(6y + 5)$ **62.** $(2a - 7b)(2a + 5b)$
63. $(t - 3)(t + 3)(t^2 + 1)$ **65.** $n^2(n - 7)(n + 2)$ **67.** $2z(3z - 1)^2$ **69.** $(2u - v)(2u + v)(16u^4 + 4u^2v^2 + v^4)$
64. $(t - 2)(t + 2)(t^2 + 1)$ **66.** $n^3(n - 5)(n + 3)$ **68.** $3z^2(2z - 1)^2$ **70.** $(2u^2 - 5v)(4u^4 + 10u^2v + 25v^2)$
71. $(a + b - 2)(a - b + 2)$ **73.** $(4d + s^2)(4d - s^2)$ ***75.** $x^n(1 - 8x)$ ***77.** $(x^m + y^m)^2$
72. $(a + b - 3)(a - b + 3)$ **74.** $(3s^2 + d)(3s^2 - d)$ ***76.** $y^m(5 + y^m)$ ***78.** $(y^n - z^n)^2$
***79.** $(x^r + 3y^s)(x^r - 3y^s)$ ***81.** $(6z^t - y^s)^2$ ***83.** $x(x^n + y^n)(x^n - y^n)$ ***85.** $(x^s - y^t)(x^{2s} + x^sy^t + y^{2t})$
***80.** $(2x^m + y^n)(2x^m - y^n)$ ***82.** $(y^r - 5x^m)^2$ ***84.** $y(y^{n+1} + x^{n+1})(y^{n+1} - x^{n+1})$ ***86.** $(x^{2r} + y^{2t})(x^r + y^t)(x^r - y^t)$
***87.** $x(x^{2n+2} + y^{2n+2})(x^{n+1} + y^{n+1})(x^{n+1} - y^{n+1})$ ***89.** $(a + b - c)^2$ ***91.** $x + 4$ ***93.** $2c + 5$ ***95.** $4a^2 + 6a + 9$
***88.** $x(x^{m+2} - y^{m+2})(x^{2m+4} + x^{m+2}y^{m+2} + y^{2m+4})$ ***90.** $4b^2$ ***92.** $2x - 3$ ***94.** $2t + 3$ ***96.** 12
***97.** $3x + 7$
***98.** The area of a square with side x is x^2. The area of a square with side y is y^2. The area of the entire square is $(x + y)^2$.
***99.** $(100 - 99)(100 + 99) = 199$ ***101.** $(1000 - 990)(1000 + 990) = 19{,}900$ ***103.** $(13 + 2)(13 - 2) = 165$
***100.** $(1000 - 999)(1000 + 999) = 1999$ ***102.** $(100 - 90)(100 + 90) = 1900$ ***104.** $(11 + 2)(11 - 2) = 117$
***105.** $(20 + 1)(20 - 1) = 399$
***106.** $(30 + 1)(30 - 1) = 899$

---

*denotes Superset exercises

Section 1.7, pages 49–51

1. 10 **3.** 0.04 **5.** $-\frac{6}{7}$ **7.** 5 **9.** not a real number **11.** -0.3 **13.** $2\sqrt[3]{5}$ **15.** $-3\sqrt{3}$ **17.** $\frac{3\sqrt{10}}{10}$

2. 2 4. 0.2 6. $-\frac{9}{5}$ 8. -4 10. not a real number 12. -0.5 14. $2\sqrt[5]{3}$ 16. $-5\sqrt{2}$ 18. $\frac{2\sqrt{7}}{7}$

19. $\frac{-\sqrt[3]{75}}{3}$ **21.** $4\sqrt[3]{4}$ **23.** 3 **25.** $2x\sqrt[3]{x}$ **27.** $\frac{\sqrt{xy}}{|y|}$ **29.** $\frac{3w\sqrt[3]{4z^2}}{2z}$ **31.** $\sqrt{3}$ **33.** $-\frac{11}{2}\sqrt{5}$ **35.** $4\sqrt{5}$

20. $-\frac{\sqrt[3]{36}}{4}$ 22. $4\sqrt[3]{2}$ 24. $3\sqrt[4]{3}$ 26. $3y\sqrt[3]{y}$ 28. $\frac{\sqrt[4]{xy}}{|x|}$ 30. $\frac{2z\sqrt[3]{25w^2}}{5w}$ 32. $29\sqrt{3}$ 34. $\frac{5}{2}\sqrt{2}$ 36. $6\sqrt{2}$

37. $5\sqrt[3]{6}$ **39.** $\frac{\sqrt{6}}{3}$ **41.** $3\sqrt{2} - 2\sqrt{3}$ **43.** -1 **45.** $9 + 2\sqrt{14}$ **47.** $2pq^2\sqrt[3]{9p^2}$ **49.** $4a^2b^3\sqrt{5}$

38. $3\sqrt{14}$ 40. $\frac{\sqrt{2}}{2}$ 42. $5\sqrt{2} - 2\sqrt{3}$ 44. 2 46. $5 + 2\sqrt{6}$ 48. $12p^2q\sqrt[3]{3}$ 50. $4a^2b^2\sqrt{3b}$

51. $4 + 4\sqrt{x} + x$ **53.** $9x - y$ **55.** $10\sqrt[3]{10}$ **57.** $\frac{216}{125}$ **59.** $9\sqrt[3]{3}$ **61.** 4 **63.** $\sqrt[12]{5^{11}}$ **65.** $50\sqrt[3]{2}$ **67.** $10x^{5/6}$

52. $9 - 6\sqrt{x} + x$ 54. $x - 4y$ 56. 125 58. $\frac{64}{27}$ 60. $9\sqrt[5]{81}$ 62. 3 64. $\sqrt[12]{3^{11}}$ 66. $12\sqrt[3]{3}$ 68. $12x^{9/20}$

69. $-20x^{4/3}y^2$ **71.** $512x^{-3/2}y^{3/4}$ **73.** $8x^{3/8}y^{3/5}$ **75.** $\frac{1}{2}x^{1/2}y^{7/6}$ **77.** $\frac{\sqrt[3]{18}}{3}x^{-2/3}$ **79.** $x^{1/8}$ **81.** $x^{2/3}y^{1/9}$

70. $-42xy^{3/2}$ 72. $4x^{6/5}y^{16}$ 74. $2x^{1/12}y^{1/8}$ 76. $\frac{7}{9}x^{5/4}y^{1/8}$ 78. $\frac{\sqrt[4]{40}}{2}x^{-3/4}$ 80. $y^{1/6}$ 82. $y^{3/2}x^{1/6}$

83. $2x^{1/12}y^{7/12}$ ***85.** $\sqrt{1 - 0.02}$ ***87.** $\sqrt[8]{900}$ ***89.** $\sqrt{0.8}$ ***91.** False, $\sqrt{(-1)^2} \neq -1$
84. $2x^{1/3}y^{1/6}$ ***86.** $1 + 0.02$ ***88.** $\sqrt{7}$ ***90.** $\sqrt[4]{4}$ ***92.** True

***93.** False, $\sqrt{1+1} \neq \sqrt{1} + \sqrt{1}$ ***95.** False, $\frac{1}{\sqrt{1}+\sqrt{1}} \neq \frac{1}{\sqrt{1}} + \frac{1}{\sqrt{1}}$ ***97.** False, $2^{1/1} \neq \frac{1}{2^1}$ ***99.** False, $\sqrt{3^2 - 4} \neq 3 - 2$

***94.** False, $\sqrt{1^2 + 1^2} \neq 1 + 1$ ***96.** True ***98.** False, $\sqrt[3]{2} \cdot \sqrt{2} \neq \sqrt[6]{2}$ ***100.** False, $\sqrt{(-2)^2 + 2(-2) + 1} \neq -2 + 1$
***101.** The square root is multiplied by $\sqrt{2}$. ***103.** Multiply it by 8. ***105.** yes; yes; no
***102.** The square root is multiplied by $\sqrt{3}$. ***104.** Multiply it by 27. ***106.** follow the hint
***107.** No, because $\sqrt{5 - 2\sqrt{6}}$ is positive, but $\sqrt{2} - \sqrt{3}$ is negative.
***108.** Yes, because both sides of the original equation are positive, and after squaring both sides, we get an identity.
***109.** square both sides ***111.** Successive roots get closer and closer to 1.
***110.** 1 sq ft ***112.** yes
***113.** No. The successive squares are getting "arbitrarily" large: larger and larger without limit.
***114.** The successive square roots of $\frac{1}{2}$ get closer and closer to 1. The successive squares of $\frac{1}{2}$ get closer and closer to 0.

Section 1.8, pages 56–58

1. $x - 3$ **3.** $2(m - 3)$ **5.** $\frac{v(v-2)}{v-3}$ **7.** $\frac{x^2 + 3x + 9}{x + 3}$ **9.** $\frac{2(a+2)}{3a-2}$ **11.** $\frac{2 - x^2}{3(x+2)}$ **13.** $\frac{-(v-15)}{2(2v+1)}$

2. $2 - y$ 4. $5(m + 3)$ 6. $\frac{u^2(u-3)}{u+4}$ 8. $\frac{x^2 - 4x + 16}{x - 4}$ 10. $\frac{a+9}{3(a-6)}$ 12. $\frac{5}{x-1}$ 14. $\frac{-(v-2)}{2(2v-11)}$

15. $m^2(n - 4)$ **17.** $-\frac{15}{a^2b}$ **19.** $\frac{x^2-1}{x^2}$ **21.** $\frac{5x}{2y}$ **23.** $\frac{3}{4n^2}$ **25.** $\frac{6(v-2)}{5}$ **27.** $-\frac{a}{b^2}$ **29.** $\frac{(2m-5)^2}{4}$

16. $n^2(m - 5)$ 18. $-\frac{18}{a^2b}$ 20. $\frac{9 + 6y + y^2}{3y - y^2}$ 22. $\frac{14z}{3xy^2}$ 24. $\frac{10m^2}{9}$ 26. $\frac{-1}{2(v+2)}$ 28. $-\frac{a^2}{b}$ 30. $\frac{(3m-8)^2}{4}$

31. $\frac{-v - 14}{(v+2)(v-2)}$ **33.** $\frac{2x^3 + 2x^2 - 35x - 70}{(x-5)(x+3)(2x+7)}$ **35.** $2 + \frac{\sqrt{3}}{3}$ **37.** $2 - \frac{2}{3}\sqrt{6} - \sqrt{3} + \sqrt{2}$ **39.** $-\frac{1}{6}\sqrt{2} + \frac{1}{3}\sqrt{5}$

*denotes Superset exercises

32. $\frac{-3(v+7)}{(v+3)(v-3)}$ **34.** $\frac{3(5x-2)}{(2x-3)(x+4)(2x-1)}$ **36.** $8-2\sqrt{7}$ **38.** $2+\frac{1}{2}\sqrt{10}+\frac{2}{3}\sqrt{5}+\frac{5}{6}\sqrt{2}$ **40.** $\frac{2}{19}\sqrt{7}+\frac{2}{57}\sqrt{6}$

41. $\frac{x(\sqrt{x}-1)}{x-1}$ **43.** $\frac{w+5\sqrt{w}+6}{w-4}$ **45.** $\frac{x+\sqrt{xy}}{x-y}$ **47.** $\frac{6+2\sqrt{2t}-t}{2-t}$ **49.** $\frac{3(\sqrt{x+y}+2)}{x+y-4}$

42. $\frac{y(7+\sqrt{y})}{49-y}$ **44.** $\frac{w-7\sqrt{w}+12}{w-9}$ **46.** $\frac{\sqrt{xy}-y}{x-y}$ **48.** $\frac{6-\sqrt{3t}-t}{12-t}$ **50.** $\frac{2x(1+\sqrt{x+y})}{1-x-y}$

51. $\frac{x(\sqrt{x}+2\sqrt{y})}{x-4y}$ **53.** $\frac{a+2-\sqrt{b}}{a^2+4a+4-b}$ **55.** $\frac{u^3-3}{u(u^2+2)}$ **57.** $\frac{v-1}{v+1}$ **59.** $\frac{x-5}{x+3}$ **61.** $x+1$

52. $\frac{(x-2y)(2\sqrt{x}+\sqrt{y})}{4x-y}$ **54.** $\frac{a-2\sqrt{ab}+b}{(a-b)^2}$ **56.** $\frac{u^3-2}{u(u^2+5)}$ **58.** $\frac{(v+2)(v-1)}{(v+1)(v-2)}$ **60.** $\frac{x-2}{x-1}$ **62.** $\frac{x-1}{x-4}$

63. $\frac{-m^2-2m-2}{m^2+m+2}$ **65.** -0.20 **67.** 2.20 **69.** 0.07 ***71.** False, $\frac{1}{1}+\frac{1}{1}\neq\frac{1}{1+1}$

64. $\frac{1}{2n^2-1}$ **66.** -4.89 **68.** -1.03 **70.** 1.51 ***72.** False, $\frac{3}{4}+\frac{1}{1}\neq\frac{3+1}{4+1}$

***73.** False, $(1+1)^{-2}\neq 1^{-2}+1^{-2}$ ***75.** True ***77.** $\frac{4x-1}{2x^2}$ ***79.** $\frac{x+1}{x}$

***74.** False, $\left(\frac{1}{1}+\frac{2}{3}\right)^{-1}\neq\frac{1}{1}+\frac{3}{2}$ ***76.** False, $(1\div 1)\div 2\neq 1\div(1\div 2)$ ***78.** $\frac{2x-1}{2x}$ ***80.** $\frac{2x^2+2x-\pi}{2x^2}$

***81.** $\frac{1-x}{x(1-\sqrt{x})}$ ***83.** $\frac{1}{\sqrt{x+2}+\sqrt{x}}$ ***85.** 3.0 ***87.** $28,142 ***89.** no

***82.** $\frac{1-y}{\sqrt{y}(1+\sqrt{y})}$ ***84.** $\frac{1}{x+1+\sqrt{x(x+2)}}$ ***86.** 7.9% ***88.** The sum gets closer and closer to 1.

Review Exercises, pages 62–63

1. $2x-3=50$ **3.** $x=1.50y$ **5.** $c=2p$ **7.** rational, real

2. $\frac{x+y}{2}=2x$ **4.** $n+(n+1)+(n+2)=15$ **6.** $p+10=c$ **8.** irrational, real

9. rational, real **11.** -1 **13.** 1 **15.** $-\frac{5}{3}$ **17.** -4 **19.** $\frac{x+y}{2}<10$ **21.** 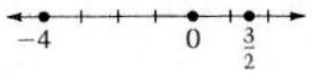

10. whole number, integer, rational, real **12.** $\frac{7}{5}$ **14.** 0 **16.** 0 **18.** $\frac{-1}{70}$ **20.** $xy>0$ **22.** 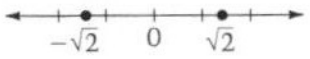

23. 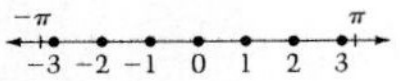

25. 5 **27.** -8 **29.** $\frac{64}{27}$ **31.** 19 **33.** $\left(\frac{c}{b}\right)^2$ **35.** $12x^4y^2$ **37.** $\frac{36y^2}{x^4}$

24. 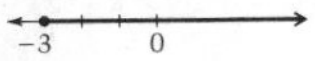

26. 7 **28.** $\frac{1}{25}$ **30.** 48 **32.** -8 **34.** c **36.** $64x^3y^6z^9$ **38.** $\frac{3x^7z^5}{2y^2}$

39. $-3x^2-x-6$ **41.** y^3-24y^2-y-2 **43.** $2x^3-12x$ **45.** $7x^2-4x-3$

40. $-2x^3+9x^2-x+1$ **42.** $-y^3+6y^2-8y+2$ **44.** $-3x^2+3x^3$ **46.** $2x^2+x-15$

47. $3x^3+5x^2-27x+4$ **49.** $25n^2-10n+1$ **51.** $2y(xy+3x^2z-2z^2)$ **53.** $(z-3)(z+6)$

48. $4x^4-5x^3-6x^2-2x+4$ **50.** $9-12n+4n^2$ **52.** $2(4m^2n^2-m^2-n^2)$ **54.** $(p-1)(p+4)$

55. $(x-4)(2x+3)$ **57.** $(3x-5y)(x+y)$ **59.** $(x+7)(x-7)$ **61.** $2(m-2)(m^2+2m+4)$

56. $(3d-1)(d+2)$ **58.** $(3t-8)(4t+1)$ **60.** cannot be factored **62.** $(1+10x^2)(1-10x^2+100x^4)$

63. $(x+y+1)(x-y-1)$ **65.** -10 **67.** $-\frac{6\sqrt{5}}{5}$ **69.** $3xy\sqrt{2xz}$ **71.** $\frac{5x\sqrt[3]{4y^2}}{2y}$ **73.** $11\sqrt{2}-15\sqrt{3}$

64. $(3m+n-1)(3m-n+1)$ **66.** 0.2 **68.** not a real number **70.** $\frac{\sqrt[3]{75x}}{5x}$ **72.** $\frac{2m\sqrt[4]{12mn}}{3n}$ **74.** $5\sqrt[3]{6}$

---

*denotes Superset exercises

75. $64 - 32\sqrt{x} + 4x$ **77.** $10x + 3\sqrt{xy} - y$ **79.** $-14x^{2/3}y$ **81.** $\frac{\sqrt[5]{1875}}{5}x^{-2/5}$ **83.** $\frac{3x - 1}{2x + 7}$

76. $x - 10\sqrt{x} + 25$ **78.** $1 - 3\sqrt{x} + 3x - x\sqrt{x}$ **80.** $\sqrt[4]{27}x^{3/8}y^{1/2}$ **82.** $2x^{3/4}y^{1/2}$ **84.** $\frac{x^2 + 2x + 4}{x + 2}$

85. $-3 - 3\sqrt{2}$ **87.** $\frac{x(2 + 3\sqrt{x})}{4 - 9x}$ **89.** $\frac{x - 1}{x + 1}$ **91.** $\frac{4}{3y - 1}$ **93.** $\frac{1}{x^2 + mx + m^2}$

86. $-\frac{1}{4}\sqrt{2} + \frac{1}{4}\sqrt{10}$ **88.** $\frac{3x(2 + \sqrt{x + y})}{4 - x - y}$ **90.** x **92.** $\frac{2x - 4}{(1 - x)(1 + x)(x - 3)}$ **94.** $y^4 + 4y^2 + 16$

95. $\frac{(n - 3)^2}{n^2(n - 4)}$ **97.** (a) False, $3 + (-1) = 2$ (b) False, $|-1| \neq -1$ (c) False, $\sqrt{(-2)^2} \neq -2$

96. $\frac{x^2}{2(x^2 + 3x + 9)}$ **98.** $5\frac{1}{2}$

99. $(4 - \pi)x^2 - (8 - 2\pi)x + (4 - \pi)$

Chapter Test, page 64

1. $2x + 13 = 56$ **3.** (a) integer, rational, real (b) whole number, integer, rational, real (c) irrational, real **5.** 0.02

2. $n + (n + 1) + n + 2 = 1.50(n + 2)$ **4.** -7 **6.** (a) 6 (b) 9

7. (a) $2.3 \times 10^{-}$ (b) 2.0×10^2 **9.** $81x^4y^8z^4$ **11.** $-u^3 + 3u^2 + 2u - 7$ **13.** $24m^2 + 34m - 10$

8. $\frac{15y^3}{x^3}$ **10.** $\frac{125y^6}{8x^3}$ **12.** $3x^3 - 21x^2 + 12x$ **14.** $81z^2 - 36z + 4$

15. $\frac{10(u + v)}{(u - 6v)(u^2 - 2uv + 4v^2)}$ **17.** $xy(6x - 2y + 7xy)$ **19.** $2xy\sqrt{6z}$ **21.** $100x^{1/2}y^{2/3}$ **23.** $\frac{m(m - 3)}{3m - 1}$

16. $2x^2z$ **18.** $(2t + 1)(3t - 4)$ **20.** $\frac{\sqrt[4]{8u^3v^3}}{4v^2}$ **22.** $\frac{3}{8}x^{5/9}y^{1/2}$ **24.** $\frac{2u - 9}{79 - 12u}$

Chapter 2

Section 2.1, pages 71–74

1. 2 is not a solution, -2 is a solution. **3.** 2 is not a solution, -2 is not a solution.

2. 2 is a solution, -2 is not a solution. **4.** 2 is not a solution, -2 is not a solution.

5. 2 is a solution, -2 is not a solution. **7.** 2 is a solution, -2 is not a solution. **9.** -4 **11.** $\frac{2}{7}$ **13.** $\frac{19}{11}$

6. 2 is not a solution, -2 is a solution. **8.** 2 is not a solution, -2 is a solution. **10.** 3 **12.** $-\frac{11}{3}$ **14.** $-\frac{1}{5}$

15. 14 **17.** no solutions **19.** 12 **21.** -4 **23.** $\frac{21}{25}$ **25.** $\frac{7 + b}{3}$ **27.** $\frac{2 - c}{ab}$ **29.** $\frac{a}{m - 5}$

16. $\frac{5}{9}$ **18.** $\frac{5}{9}$ **20.** no solutions **22.** $-\frac{1}{11}$ **24.** $\frac{9}{17}$ **26.** $\frac{a - 9}{2}$ **28.** $\frac{c - a^2}{b^2}$ **30.** $\frac{b}{m + 3}$

31. all real numbers **33.** $\frac{2}{2b - 2 - a}$ **35.** $\frac{a}{a - b - 2}$ **37.** -1 **39.** $F = \frac{9}{5}C + 32$ **41.** $W = \frac{P}{2} - L$

32. $\frac{2(b - a)}{a - 2b}$ **34.** all real numbers **36.** $\frac{-2b}{b + 2}$ **38.** -1 **40.** $z = 3A - x - y$ **42.** $\frac{a - P}{rP} = t$

43. $a = \frac{2s}{t^2}$ **45.** $h = \frac{P - 2lw}{2(w + l)}$ **47.** $v_0 = \frac{1}{t}\left(s - s_0 - \frac{1}{2}at^2\right)$ **49.** $n = \frac{t - a}{d} + 1$ **51.** 10 **53.** $-\frac{7}{9}$ **55.** $-\frac{3}{2}$

44. $h = \frac{3V}{\pi r^2}$ **46.** $R_1 = \frac{RR_2}{R_2 - R}$ **48.** $r = 1 - \frac{a}{S}$ **50.** $t = \frac{x - a^2b^2}{a^2 + b^2}$ **52.** 8 **54.** -7 **56.** $-\frac{3}{13}$

57. $\frac{5}{8}$ **59.** all real numbers except 0 **61.** $\frac{5}{3}$ **63.** -1 **65.** -4 **67.** no solutions **69.** no solutions

58. $\frac{5}{14}$ 60. $-\frac{3}{2}$ 62. all real numbers except 0 64. -5 66. $-\frac{27}{5}$ 68. no solutions 70. 2

71. $\frac{24}{11}$ **73.** $\frac{5}{2}, -\frac{11}{2}$ **75.** $\frac{1}{3}, 5$ **77.** $-1, \frac{17}{5}$ **79.** $\frac{1}{2}$ **81.** no solutions **83.** no solutions **85.** -2.9

72. no solutions 74. $5, -\frac{5}{2}$ 76. $-\frac{19}{7}, \frac{43}{7}$ 78. $-\frac{5}{2}, 3$ 80. no solutions 82. 5 84. 1 86. -0.1

87. 8.2 **89.** 4.6 **91.** 1.651, 1.649 ***93.** -6 ***95.** 5 ***97.** $\frac{1}{2}, \frac{1}{4}$ ***99.** $-1, \frac{5}{6}$ ***101.** $-\frac{3}{4}$ ***103.** $-\frac{1}{2}$

88. -1.4 90. 4.0 92. 4.50505, 4.50495 ***94.** 0 ***96.** 2 ***98.** $-1, \frac{9}{5}$ ***100.** 0, 5 ***102.** $-\frac{3}{4}, 5$ ***104.** $\frac{1}{2}$

***105.** $-4, \frac{1}{2}$ ***107.** $0, \frac{10}{3}$ ***109.** $xy - \frac{2x}{5} = 10$ ***111.** 8 meters, 10 meters ***113.** (a) $G = \frac{S + 245}{4}$ (b) 95

***106.** $\frac{3}{5}$ ***108.** $-\frac{15}{7}, \frac{5}{7}$ ***110.** $yz - \pi x^2 = 10$ ***112.** $\frac{9}{2}$ meters ***114.** (a) $P = 1.20S$ (b) \$37.50

***115.** (a) $T = 200 + 0.05V$; \$3,450 (b) \$36,000 ***117.** (a) $T = 0.70S - 40$; \$380 (b) 34 hours

***116.** (a) $I = 180 + 0.01S$; \$1,055 (b) \$32,000 ***118.** (a) $P = 0.25(X - 4500)$ (b) \$7,780

Section 2.2, pages 81–82

1. $i\sqrt{6}$ **3.** $5i$ **5.** $7i\sqrt{2}$ **7.** $-3i$ **9.** $-2i\sqrt{5}$ **11.** $-3i\sqrt{7}$ **13.** $\frac{1}{2}, 6$ **15.** 0, 6 **17.** $-\sqrt{3}, 0$

2. $i\sqrt{21}$ 4. $9i$ 6. $5i\sqrt{3}$ 8. $-6i$ 10. $-3i\sqrt{2}$ 12. $-4i\sqrt{2}$ 14. $-7, -\frac{3}{5}$ 16. 14, 0 18. 0, 1

19. $\frac{1}{4}, -\sqrt{10}$ **21.** 1 **23.** $-i$ **25.** $-8i$ **27.** -14 **29.** $2i\sqrt{3}$ **31.** $-6i\sqrt{5}$ **33.** $-2 + 8i$ **35.** $8 + 2i$

20. $-\sqrt{6}, \frac{2}{7}$ 22. -1 24. i 26. $64i$ 28. -6 30. $5i\sqrt{2}$ 32. $-15i\sqrt{2}$ 34. $13 - i$ 36. $12i$

37. 14 **39.** $-24 + 32i$ **41.** $3 + 6i$ **43.** $13 + 11i$ **45.** $-4 + 39i$ **47.** $-6 - 10i$ **49.** 41 **51.** 4

38. $-18 - 3i$ 40. $-10 + 18i$ 42. $32 - 4i$ 44. $-7 + 6i$ 46. $-15 + 15i$ 48. $12 - 24i$ 50. 13 52. 8

53. $-5 + 12i$ **55.** $24 - 10i$ **57.** $-4 - 3i$ **59.** $-27 - 54i$ **61.** $\frac{2}{13} - \frac{3}{13}i$ **63.** $\frac{5}{29} + \frac{2}{29}i$ **65.** $0 + 7i$

54. $-7 + 24i$ 56. $35 - 12i$ 58. $-40 - 42i$ 60. $-24 - 56i$ 62. $\frac{1}{17} - \frac{4}{17}i$ 64. $\frac{3}{58} + \frac{7}{58}i$ 66. $0 - \frac{3}{2}i$

67. $\frac{\sqrt{3}}{7} + \frac{2}{7}i$ **69.** $2 + 2i$ **71.** $\frac{14}{17} - \frac{5}{17}i$ **73.** $\frac{5}{29} - \frac{27}{29}i$ **75.** $5 + 3i$ **77.** $-\frac{8}{41} - \frac{10}{41}i$ **79.** $1 + \frac{3}{7}i$

68. $-\frac{\sqrt{11}}{36} - \frac{5}{36}i$ 70. $1 + 2i$ 72. $\frac{11}{13} + \frac{10}{13}i$ 74. $\frac{14}{37} - \frac{27}{37}i$ 76. $-7 - 2i$ 78. $1 + 3i$ 80. $\frac{8}{3} + 3i$

81. $-\frac{3}{20} + \frac{1}{20}i$

82. $-\frac{1}{10} + \frac{1}{5}i$

83. (a) $\overline{z + w} = 3 + i = \bar{z} + \bar{w}$ (b) $\overline{z - w} = -1 - 7i = \bar{z} - \bar{w}$ (c) $\overline{z \cdot w} = 14 - 2i = \bar{z} \cdot \bar{w}$ (d) $\overline{\left(\frac{z}{w}\right)} = -\frac{1}{2} - \frac{1}{2}i = \frac{\bar{z}}{\bar{w}}$

84. (a) $\overline{z + w} = 7 - 5i = \bar{z} + \bar{w}$ (b) $\overline{z - w} = 3 + i = \bar{z} - \bar{w}$ (c) $\overline{z \cdot w} = 4 - 19i = \bar{z} \cdot \bar{w}$ (d) $\overline{\left(\frac{z}{w}\right)} = \frac{16}{13} + \frac{11}{13}i = \frac{\bar{z}}{\bar{w}}$

*denotes Superset exercises

85. (a) $\overline{z+w} = 17 - 6i = \bar{z} + \bar{w}$ (b) $\overline{z-w} = 7 + 12i = \bar{z} - \bar{w}$ (c) $\overline{z \cdot w} = 87 - 93i = \bar{z} \cdot \bar{w}$

(d) $\overline{\left(\frac{z}{w}\right)} = \frac{33}{106} + \frac{123}{106}i = \frac{\bar{z}}{\bar{w}}$

86. (a) $\overline{z+w} = -9 + 6i = \bar{z} + \bar{w}$ (b) $\overline{z-w} = -13 - 8i = \bar{z} - \bar{w}$ (c) $\overline{z \cdot w} = -15 - 79i = \bar{z} \cdot \bar{w}$

(d) $\overline{\left(\frac{z}{w}\right)} = -\frac{29}{53} + \frac{75}{53}i = \frac{\bar{z}}{\bar{w}}$

87. (a) $\overline{z+w} = -7 + 9i = \bar{z} + \bar{w}$ (b) $\overline{z-w} = -5 + i = \bar{z} - \bar{w}$ (c) $\overline{z \cdot w} = -14 - 29i = \bar{z} \cdot \bar{w}$

(d) $\overline{\left(\frac{z}{w}\right)} = \frac{26}{17} + \frac{19}{17}i = \frac{\bar{z}}{\bar{w}}$ ***89.** $-2, 3$ ***91.** $\frac{3}{4}, -2$ ***93.** $-2, 5$ ***95.** $20, 12$ ***97.** 0 ***99.** i

88. (a) $\overline{z+w} = 1 + 6i = \bar{z} + \bar{w}$ (b) $\overline{z-w} = 17 + 12i = \bar{z} - \bar{w}$ (c) $\overline{z \cdot w} = -45 - 99i = \bar{z} \cdot \bar{w}$

(d) $\overline{\left(\frac{z}{w}\right)} = -\frac{99}{73} - \frac{45}{73}i = \frac{\bar{z}}{\bar{w}}$ ***90.** $2, -1$ ***92.** $\frac{3}{2}, -\frac{7}{5}$ ***94.** $1, 3$ ***96.** $\frac{5}{4}, \frac{7}{4}$ ***98.** $-1 + i$ ***100.** -1

***101.** True ***103.** False, $(i)(i) = -1$ ***105.** True

***102.** False, $\sqrt{-1} \cdot \sqrt{-1} \neq \sqrt{(-1)(-1)}$ ***104.** False, $i(1 + i) = -1 + i$ ***106.** False, $(1 + 0i)^2 = 1$

***107.** $-\frac{7}{25} + \frac{1}{25}i$ ***109.** 1 ***111.** i ***113.** $-\frac{23}{26} + \frac{11}{26}i$ ***115.** $\frac{1}{10} - \frac{11}{10}i$ ***117.** $-\frac{5}{2} - \frac{9}{2}i$ ***119.** $\frac{9}{13} + \frac{7}{13}i$

***108.** $-\frac{7}{25} - \frac{1}{25}i$ ***110.** -1 ***112.** $-2i$ ***114.** $\frac{12}{13} - \frac{5}{13}i$ ***116.** $\frac{2}{5} + \frac{8}{5}i$ ***118.** $\frac{23}{10} + \frac{41}{10}i$ ***120.** $-\frac{3}{5} - \frac{1}{5}i$

***121.** $\frac{4}{17} - \frac{1}{17}i$ ***123.** $\frac{15}{7} + 3i$ ***125.** $2 + 2i, -2 - 2i$

***122.** $\frac{6}{5} + \frac{2}{5}i$ ***124.** $-1 + 2i$ ***126.** $3 - 3i, -3 + 3i$

***127.** $\overline{z - w} = \overline{(a + bi) - (c + di)} = \overline{(a - c) + (b - d)i} = (a - c) - (b - d)i = (a - bi) - (c - di)$
$= \overline{(a + bi)} - \overline{(c + di)} = \bar{z} - \bar{w}$

***128.** $\overline{\left(\frac{z}{w}\right)} = \overline{\left(\frac{a + bi}{c + di}\right)} = \overline{\left(\frac{a + bi}{c + di}\right)\left(\frac{c - di}{c - di}\right)} = \overline{\left(\frac{ac - adi + bci - bdi^2}{c^2 - di^2}\right)} = \overline{\left(\frac{(ac + bd) + (bc - ad)i}{c^2 + d^2}\right)}$

$= \overline{\frac{ac + bd}{c^2 + d^2} + \frac{bc - ad}{c^2 + d^2}i} = \frac{ac + bd}{c^2 + d^2} - \frac{bc - ad}{c^2 + d^2}i = \frac{ac + adi - bci + bd}{c^2 + d^2}$

$= \frac{ac + adi - bci - bdi^2}{c^2 - d^2i^2} = \left(\frac{a - bi}{c - di}\right)\left(\frac{c + di}{c + di}\right) = \frac{a - bi}{c - di} = \frac{\overline{a + bi}}{\overline{c + di}} = \frac{\bar{z}}{\bar{w}}$

***129.** $\bar{\bar{z}} = \overline{\overline{(a + bi)}} = \overline{(a - bi)} = a + bi = z$

***130.** $z \cdot \bar{z} = (a + bi)\overline{(a + bi)} = (a + bi)(a - bi) = a^2 - b^2i^2 = a^2 - b^2(-1) = a^2 + b^2$

***131.** $\frac{1}{2}(z + \bar{z}) = \frac{1}{2}[(a + bi) + \overline{(a + bi)}] = \frac{1}{2}(a + bi + a - bi) = \frac{1}{2}(2a) = a$

***132.** $\frac{1}{2}i(\bar{z} - z) = \frac{1}{2}i[\overline{(a + bi)} - (a + bi)] = \frac{1}{2}i[a - bi - (a + bi)] = \frac{1}{2}i(-2bi) = -bi^2 = -b(-1) = b$

Section 2.3, pages 89–91

1. $-8, -1$ **3.** $5, -2$ **5.** 4 **7.** $8, -8$ **9.** $1, 5$ **11.** $\frac{2}{5}, -4$ **13.** $\frac{7}{4}i, -\frac{7}{4}i$ **15.** $0, -10$ **17.** $-\frac{1}{2}$

2. $3, 2$ **4.** $-7, 3$ **6.** -6 **8.** $10i, -10i$ **10.** $-9, 1$ **12.** $\frac{3}{5}, -\frac{3}{5}$ **14.** $\frac{1}{3}$ **16.** $0, \frac{1}{2}$ **18.** $\frac{6}{7}, 1$

---

*denotes Superset exercises

19. $-\frac{1}{5}, \frac{8}{3}$ **21.** $3i, -3i$ **23.** $-12, 2$ **25.** $4, -1$ **27.** $9, -6$ **29.** $-\frac{1}{12}, 3$ **31.** $\frac{2}{3}$ **33.** $3, -7$

20. $-\frac{5}{6}, -4$ **22.** $4, -3$ **24.** $-5, -9$ **26.** $\sqrt{2}, -\sqrt{2}$ **28.** $-15, -4$ **30.** $\frac{3}{4}, 5$ **32.** $\frac{1}{9}$ **34.** $2, -12$

35. $15, -1$ **37.** $-3 + i\sqrt{2}, -3 - i\sqrt{2}$ **39.** $2, 3$ **41.** $-\frac{3}{2}, -\frac{5}{2}$ **43.** $\frac{2}{5} + \frac{\sqrt{19}}{5}, \frac{2}{5} - \frac{\sqrt{19}}{5}$ **45.** $-4, -1$

36. $18, -2$ **38.** $4 + i\sqrt{5}, 4 - i\sqrt{5}$ **40.** $-5, -6$ **42.** $8, -\frac{4}{3}$ **44.** $-1 + \frac{2\sqrt{3}}{3}, -1 - \frac{2\sqrt{3}}{3}$ **46.** $-9, -2$

47. 4 **49.** $\frac{-1 + i\sqrt{39}}{2}, \frac{-1 - i\sqrt{39}}{2}$ **51.** $8, -8$ **53.** $4 + \sqrt{3}, 4 - \sqrt{3}$ **55.** $3, -\frac{2}{3}$ **57.** $\sqrt{11}, -\sqrt{11}$

48. $-10, 6$ **50.** $\frac{-1 + \sqrt{41}}{2}, \frac{-1 - \sqrt{41}}{2}$ **52.** $8i, -8i$ **54.** $-\frac{1}{2}$ **56.** $\frac{7 + i\sqrt{23}}{6}, \frac{7 - i\sqrt{23}}{6}$ **58.** $-11, 0$

59. $\frac{-3 + \sqrt{29}}{2}, \frac{-3 - \sqrt{29}}{2}$ **61.** $-\frac{1}{6}$ **63.** $1 - i, 1 + i$ **65.** $\frac{9}{2}, 0$ **67.** $\frac{4 + \sqrt{22}}{3}, \frac{4 - \sqrt{22}}{3}$

60. $2, \frac{1}{5}$ **62.** $i\sqrt{7}, -i\sqrt{7}$ **64.** $-1 + \sqrt{3}, -1 - \sqrt{3}$ **66.** $\frac{7 + \sqrt{14}}{5}, \frac{7 - \sqrt{14}}{5}$ **68.** $\frac{3}{4}, -\frac{3}{4}$

69. $\frac{3 + i\sqrt{7}}{2}, \frac{3 - i\sqrt{7}}{8}$ **71.** $4 + \sqrt{2}, 4 - \sqrt{2}$ **73.** $\frac{4 + i\sqrt{11}}{5}, \frac{4 - i\sqrt{11}}{5}$ **75.** $-3 + \sqrt{6}, -3 - \sqrt{6}$

70. $\frac{-8 + 5\sqrt{2}}{2}, \frac{-8 - 5\sqrt{2}}{2}$ **72.** $6, -4$ **74.** $\frac{3 + i\sqrt{6}}{5}, \frac{3 - i\sqrt{6}}{5}$ **76.** $\frac{3}{4}$

77. two different real number solutions **79.** two imaginary number solutions
78. one real number solution of multiplicity two **80.** two different real number solutions
81. two different real number solutions **83.** one real number solution of multiplicity 2
82. two different real number solutions **84.** two imaginary number solutions
85. two imaginary number solutions **87.** one real number solution of multiplicity 2
86. two different real number solutions **88.** two imaginary number solutions
89. two imaginary number solutions **91.** $-0.4, -1.4$ **93.** $1.2, -2.2$ **95.** $0.8, 0.5$
90. two different real number solutions **92.** $0.6, -1.4$ **94.** $4.8, 0.2$ **96.** $1.1, -1.0$ **98.** 9 seconds
97. After $2\frac{1}{2}$ seconds, the ball will be 620 ft above the ground and travelling up. After 3 seconds, the ball will be 620 ft above the ground and travelling down.

99. $\frac{10}{7}$ seconds ***101.** no such value exists ***103.** $2, -2$ ***105.** $-3, 1$

100. $\frac{-v_0 + \sqrt{v_0^2 + 39.2s_0}}{19.6}$ seconds ***102.** no such value exists ***104.** $0, 8$ ***106.** any real number $a \neq 0$

***107.** $r = \frac{-2 + V \pm \sqrt{V^2 - 4V + 16}}{2}$ ***109.** $x = A \pm \sqrt{A^2 + A}$ ***111.** $x = \frac{-2y - 3 \pm \sqrt{12y + 9}}{2}$

***108.** $s = \frac{A \pm \sqrt{A^2 + 20B - 20}}{10}$ ***110.** $y = \frac{-1 \pm \sqrt{1 + 2m}}{m}$ ***112.** $y = -x \pm \sqrt{-3x}$

***113.** $-4 + \sqrt{26}$ ft ***115.** 2 ft

***114.** $3 - \frac{\sqrt{26}}{2}$ ft, $3 + \frac{\sqrt{26}}{2}$ ft ***116.** 2 ft

***117.** (a) Sum of roots $= \frac{-b + \sqrt{b^2 - 4ac}}{2a} + \frac{-b - \sqrt{b^2 - 4ac}}{2a} = \frac{-b}{a}$

(b) Products of roots $= \frac{-b + \sqrt{b^2 - 4ac}}{2a} \cdot \frac{-b - \sqrt{b^2 - 4ac}}{2a} = \frac{b^2 - (b^2 - 4ac)}{4a^2} = \frac{c}{a}$ ***119.** 18

***118.** $m = -\frac{8}{3}, p = -16$ ***120.** 10

---

*denotes Superset exercises

Section 2.4, pages 97–98

1. $0, \frac{2}{3}, 4$ **3.** $0, \frac{-1 \pm i\sqrt{3}}{2}$ **5.** $0, \frac{7 \pm \sqrt{17}}{4}$ **7.** $\pm 3, \frac{1}{2}$ **9.** $\pm 4i, \frac{2}{3}$ **11.** $-3, \frac{3 \pm 3i\sqrt{3}}{2}, 1$ **13.** $0, \pm \frac{2\sqrt{5}}{5}, 4$

2. $0, \frac{6}{5}, -2$ 4. $0, \frac{1 \pm \sqrt{5}}{2}$ 6. $0, \frac{5 \pm i\sqrt{7}}{2}$ 8. $\pm 5, -3$ 10. $\pm 2, \frac{3}{8}$ 12. $2, -1 \pm i\sqrt{3}, -7$ 14. $0, \pm 2i, \frac{6}{7}$

15. $-\frac{1}{2}, 3$ **17.** $\frac{1}{2}, 1$ **19.** $0, 7$ **21.** 1 **23.** $\frac{9}{2}$ **25.** $\frac{-1 \pm i\sqrt{11}}{2}$ **27.** $\pm\sqrt{7}, \pm i\sqrt{2}$ **29.** $\pm\sqrt{3}, \pm 2$

16. $7, 5$ 18. $\frac{7}{4}, -4$ 20. $10, -3$ 22. $-\frac{5}{2}$ 24. $\frac{7 \pm i\sqrt{3}}{2}$ 26. $-\frac{1}{5}$ 28. $\pm 2\sqrt{2}, \pm\sqrt{3}$ 30. $\pm\sqrt{5}, \pm i\sqrt{3}$

31. $-2, 0, -3 \pm i\sqrt{3}$ **33.** $\pm 3, \pm 3i$ **35.** $\frac{3}{4}$ **37.** $-2, 0$ **39.** $\frac{1}{125}, 8$ **41.** 4 **43.** 9 **45.** 4 **47.** $7, 3$

32. $1, 4, \frac{-1 \pm 3i\sqrt{3}}{2}$ 34. $\pm 2, \pm 2i$ 36. $3, 2$ 38. $-1, 3$ 40. $625, 81$ 42. 171 44. 4 46. 2 48. $5, 1$

49. 1 ***51.** $0, 1$ ***53.** 3 ***55.** 25 ***57.** 7.8 ft ***59.** $4, 1$ ***61.** $\frac{1}{64}$

50. 19 ***52.** $3, 2$ ***54.** 4 ***56.** $\frac{1}{9}$ ***58.** 30.25 ft ***60.** $\frac{81}{16}, 16$ ***62.** 0

Section 2.5, pages 108–109

1. (a) $\{x \mid 3 < x < 8\}$ (b) **3.** (a) $\{x \mid -2.5 < x \le 6\}$ (b)

2. (a) $\{x \mid 3 \le x \le 10\}$ (b) 4. (a) $\{x \mid -4 \le x < 5.5\}$ (b)

5. (a) $\{x \mid x < -2\}$ (b) **7.** (a) $\{x \mid -\sqrt{3} \le x \le \sqrt{3}\}$ (b)

6. (a) $\{x \mid x > -2\}$ (b) 8. (a) $\{x \mid 1 < x < \pi\}$ (b)

9. (a) $\{x \mid x > 4\}$ (b) $(4, \infty)$ (c) **11.** (a) $\{v \mid v \le -1\}$ (b) $(-\infty, -1]$ (c)

10. (a) $\left\{m \mid m < \frac{5}{9}\right\}$ (b) $\left(-\infty, \frac{5}{9}\right)$ (c) **12.** (a) $\left\{w \mid w < \frac{15}{2}\right\}$ (b) $\left(-\infty, \frac{15}{2}\right)$ (c)

13. (a) $\left\{x \mid x > \frac{6}{5}\right\}$ (b) $\left(\frac{6}{5}, \infty\right)$ (c) **15.** (a) $\left\{x \mid x \le \frac{13}{23}\right\}$ (b) $\left(-\infty, \frac{13}{23}\right]$ (c)

14. (a) $\left\{u \mid u \ge -\frac{3}{2}\right\}$ (b) $\left[-\frac{3}{2}, \infty\right)$ (c) 16. (a) $\left\{m \mid m < \frac{105}{92}\right\}$ (b) $\left(-\infty, \frac{105}{92}\right)$ (c)

17. no solutions

18. (a) $\{w \mid w \text{ is a real number}\}$ (b) $(-\infty, \infty)$ (c)

19. (a) $\{y \mid y \text{ is a real number}\}$ (b) $(-\infty, \infty)$ (c) **21.** (a) (b) $[-7, 10]$

20. no solutions 22. (a) (b) $(3, 12)$

23. (a) (b) $\left(-\infty, -\frac{8}{5}\right] \cup \left[\frac{7}{2}, \infty\right)$ **25.** (a) (b) $(-\infty, -\sqrt{2}) \cup \left[\frac{\sqrt{3}}{2}, \infty\right)$

24. (a) (b) $(-\infty, 3) \cup (\pi, \infty)$ 26. (a) (b) $\left(-\infty, -\frac{\sqrt{3}}{2}\right) \cup [\sqrt{2}, \infty)$

27. (a) (b) $(-3, 3]$ **29.** (a) (b) $(-6, \sqrt{7})$

28. (a) (b) $[-\sqrt{5}, 2\sqrt{5}]$ 30. (a) (b) $[-5, -2)$

31. **33.** **35.**

---

*denotes Superset exercises

32. 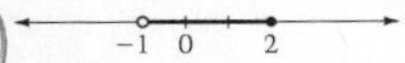 **34.** 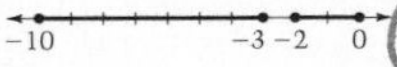 **36.** 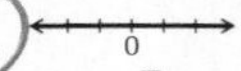

37. (a) $(-\infty, -1) \cup (1, \infty)$ (b) 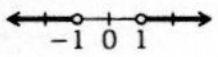 **39.** (a) $\left[-\frac{7}{2}, \frac{7}{2}\right]$ (b) 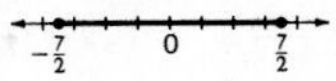 **41.** (a) $(-\infty, \infty)$ (b) 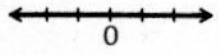

38. (a) $(-5, 5)$ (b) 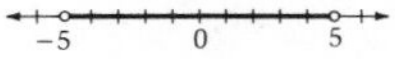 **40.** (a) $\left(-\infty, -\frac{10}{3}\right] \cup \left[\frac{10}{3}, \infty\right)$ (b) 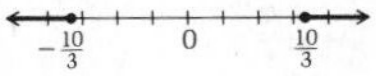 **42.** no solutions

43. (a) $\left(-\infty, -\frac{5}{3}\right) \cup (5, \infty)$ (b) 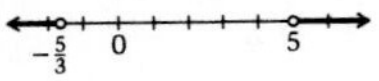 **45.** (a) $\left[-\frac{5}{3}, 7\right]$ (b) 

44. (a) $\left[-\frac{11}{5}, 5\right]$ (b)  **46.** (a) $\left(-\infty, -\frac{9}{10}\right) \cup \left(\frac{3}{2}, \infty\right)$ (b) 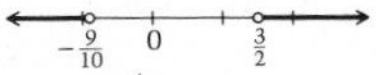

47. no solutions **49.** (a) $[7, 7]$ (b)  **51.** $(-\infty, 4.3)$ **53.** $(2.333, 2.334)$

48. (a) $(-\infty, \infty)$ (b) 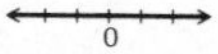 **50.** no solutions **52.** $(-\infty, 6.6]$ **54.** $(0.7749, 0.7751)$

55. $(-\infty, -0.5] \cup [0.9, \infty)$ **57.** $(-\infty, -3) \cup (0, \infty)$ **59.** $(0, 9)$ **61.** $(-\infty, -1] \cup [8, \infty)$

56. $[-14, 42]$ **58.** $(-11, 0)$ **60.** $(-\infty, 0] \cup [6, \infty)$ **62.** $(-\infty, 4) \cup (9, \infty)$

63. $\left[\frac{2-\sqrt{2}}{2}, \frac{2+\sqrt{2}}{2}\right]$ **65.** $(-\infty, -3) \cup (0, 3)$ **67.** $(-\infty, 1) \cup (3, \infty)$ **69.** $(6, 9)$ **71.** $\left(-\infty, \frac{1}{3}\right) \cup [1, \infty)$

64. $(-3-\sqrt{6}, -3+\sqrt{6})$ **66.** $[-6, 0] \cup [6, \infty)$ **68.** $(-4, -2]$ **70.** $(-\infty, 3) \cup (5, \infty)$ **72.** $\left(-\frac{3}{4}, \frac{1}{2}\right)$

73. $(30, 70]$ ***75.** False; $0 < 1$, but $(0)(-1) > (1)(-1)$ ***77.** False; $-2 < -1$, but $\frac{-2}{-1} > 1$ ***79.** $(1.725, 1.775)$

74. $(-\infty, 0] \cup \left(\frac{9}{2}, \infty\right)$ ***76.** False; $1 < 2$, but $\frac{1}{1} > \frac{1}{2}$ ***78.** False; $-2 < 1$, but $(-2)^2 > (1)^2$ ***80.** $(2.81\overline{6}, 2.85)$

***81.** $(4.6\overline{3}, 4.7)$ ***83.** $(-22°, 86°)$ ***85.** $(2, 6)$ ***87.** $[-2, -\sqrt{2}] \cup [\sqrt{2}, 2]$

***82.** $(2.78, 2.82)$ ***84.** $(-17.\overline{7}°, 31.\overline{1}°)$ ***86.** $[0, 1]$ and $[7, 8]$ ***88.** $(-\infty, -3) \cup (-1, 1) \cup (3, \infty)$

***89.** $(-1-\sqrt{2}, -1) \cup (-1, -1+\sqrt{2})$ ***91.** $(-\infty, -2) \cup (0, 2) \cup (4, \infty)$

***90.** $(-\infty, -1) \cup (3, \infty)$ ***92.** $[-4, -2-\sqrt{2}] \cup [-2+\sqrt{2}, 0]$

Section 2.6, pages 116–118

1. 5 **3.** 11, 12, 13 **5.** The sports car costs \$14,904, the station wagon costs \$11,904, and the subcompact costs \$5,952.

2. 1 **4.** −6, −4, −2 **6.** 9.5 ft

7. 30 ft × 30 ft **9.** \$7000 in the first account, \$3000 in the second account **11.** 160 mph

8. 13 in. × 13 in. **10.** 63 pairs of jeans at the 10% discount, 37 pairs at the 25% discount **12.** 2 hours, 90 miles

13. 1.5 mph **15.** $23\frac{1}{3}$ ft **17.** 71 **19.** \$61,500

14. 12 ounces of Feed A with 18 ounces of Feed B **16.** either $12\frac{1}{2}$ ft × 36 ft or 18 ft × 25 ft **18.** \$44 **20.** 65

21. 20 minutes **23.** Jim takes 9.5 minutes; Jules takes 10.5 minutes.

22. Maryann takes 2 hours; Margy takes 4 hours. **24.** 14.6 ft

25. 500 ft × 1000 ft ***27.** 27 in^2 ***29.** 8.4 mph

26. The slower driver averaged 45.2 mph; the other driver averaged 53.2 mph ***28.** 14.4 miles ***30.** 46.1 ft

***31.** between 2 hours and 2 hours 40 minutes

***32.** The current was flowing at 0.5 mph. The canoeist's still water paddling rate was faster than 0.5 mph, since he did go upstream.

***33.** 9.3 mph ***35.** $12 - 12\sqrt{1-R}$ ***37.** 10 ounces ***39.** $\frac{6500 - 45R}{600 - 6R}$ ounces

***34.** $12 - 4\sqrt{3}$ ***36.** 6 ounces ***38.** $14\frac{1}{6}$ ounces

---

*denotes Superset exercises

Review Exercises, pages 122–124

1. $\frac{13}{12}$ **3.** $\frac{5}{19}$ **5.** -4 **7.** $-\frac{3}{5}$ **9.** $-\frac{10}{3}$ **11.** $m = \frac{y - y_0}{x - x_0}$ **13.** $t = \frac{v_0 - v}{g}$ **15.** $-2, 3$ **17.** no solutions

2. -11 **4.** 10 **6.** 18 **8.** -4 **10.** $-\frac{3}{5}$ **12.** $m = \frac{2K}{v^2}$ **14.** $r = \frac{A}{\pi s} - R$ **16.** $2, -\frac{2}{7}$ **18.** $\frac{3}{2}, -5$

19. $\frac{17}{2}, \frac{13}{4}$ **21.** $-i$ **23.** -1 **25.** 81 **27.** $-3\sqrt{10}$ **29.** $-1 - 2i$ **31.** $10 + 20i$ **33.** $3 - 6i$

20. $4, \frac{20}{9}$ **22.** i **24.** i **26.** $-128i$ **28.** $60i\sqrt{2}$ **30.** $-2 - i$ **32.** $-24 - 12i$ **34.** 10

35. $11 - 10i$ **37.** 65 **39.** 6 **41.** $9 - 40i$ **43.** $\frac{2}{29} + \frac{5}{29}i$ **45.** $2 - 2i$ **47.** $-\frac{5}{17} + \frac{3}{17}i$ **49.** $\frac{1 \pm i\sqrt{7}}{2}$

36. $41 + 23i$ **38.** 85 **40.** 15 **42.** $-55 + 48i$ **44.** $\frac{3}{130} - \frac{11}{130}i$ **46.** $\frac{14}{37} + \frac{27}{37}i$ **48.** $\frac{1}{5} + \frac{2}{5}i$ **50.** $\frac{3}{2}, -2$

51. $-\frac{1}{2}, 7$ **53.** $-\frac{1}{4}, 5$ **55.** $\frac{-1 \pm i\sqrt{11}}{2}$ **57.** $\frac{5 \pm \sqrt{17}}{4}$ **59.** $\frac{-1 \pm \sqrt{22}}{7}$ **61.** $\frac{3}{2} \pm \frac{1}{2}i$ **63.** $\pm i\sqrt{5}$

52. $\frac{11 \pm \sqrt{142}}{3}$ **54.** $1 \pm i\sqrt{2}$ **56.** $\frac{13 \pm \sqrt{233}}{4}$ **58.** $\frac{-3 \pm \sqrt{5}}{2}$ **60.** $\frac{9}{10}, -1$ **62.** $-\frac{1}{5} \pm \frac{\sqrt{14}}{5}i$ **64.** $\pm\sqrt{5}$

65. two different real number solutions **67.** two imaginary number solutions **69.** $0, \frac{5}{2}, -1$ **71.** $\pm\sqrt{2}, \pm\sqrt{5}$

66. two imaginary number solutions **68.** two imaginary number solutions **70.** 0, 5 **72.** $\frac{3}{5}$

73. $512, -8$ **75.** $-\frac{2}{5}, 1$ **77.** $-\frac{1}{3}$ **79.** 3 **81.** no solutions **83.** (a) $\left(-\infty, \frac{5}{7}\right)$ (b)

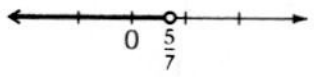

74. 0 **76.** 2 **78.** 1 **80.** $3, -1$ **82.** 13 **84.** (a) $\left(-\infty, \frac{7}{6}\right]$ (b)

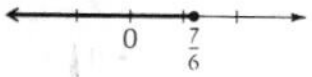

85. (a) $\left[-\frac{1}{3}, \infty\right)$ (b) **87.** (a) $[-2, 3]$ (b) 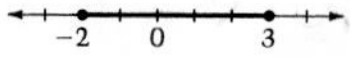 **89.** (a) $[-2, 5]$ (b)

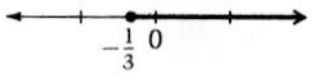

86. (a) $(-\infty, \infty)$ (b) 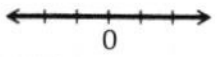 **88.** (a) $[-2, 1)$ (b) **90.** (a) $(-\infty, -3) \cup \left(\frac{7}{3}, \infty\right)$ (b)

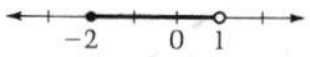

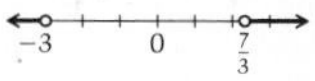

91. (a) $\left(-\infty, -\frac{1}{2}\right] \cup [4, \infty)$ (b) **93.** (a) $(-1, 7)$ (b)

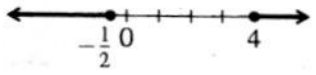

92. (a) $\left(-\frac{1}{3}, \frac{11}{3}\right)$ (b) 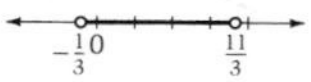 **94.** (a) $(-\infty, 4 - 2\sqrt{5}] \cup [4 + 2\sqrt{5}, \infty)$ (b)

95. (a) $(-\infty, 4 - \sqrt{7}) \cup (4 + \sqrt{7}, \infty)$ (b)  **97.** (a) $\left(-\infty, \frac{1}{2}\right) \cup (2, \infty)$ (b)

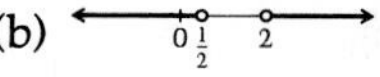

96. (a) $\left[-\frac{13}{3}, -2\right]$ (b) 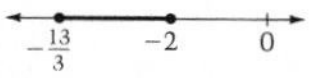 **98.** (a) $(3, 6]$ (b)

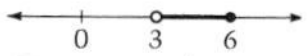

99. (a) $(-1, 1)$ (b) 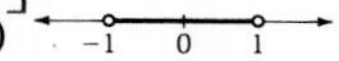 **101.** 16, 17, 18 **103.** \$10,000 in the first account; \$40,000 in the second account

100. (a) $(-\infty, -6] \cup (3, \infty)$ (b) 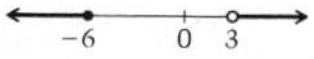 **102.** $1, -1$ **104.** squares of side $\frac{5}{2}$

105. 2:50 PM, $16\frac{2}{3}$ miles **107.** Glen takes $10\frac{2}{3}$ hours; Jerry takes 32 hours. **109.** $\frac{1}{6}$ **111.** 10 in × 4 in

106. 4 feet **108.** No, if $n + (n + 1) + (n + 2) = 52$, then $n = \frac{49}{3}$. **110.** $\frac{1}{4}$ **112.** no solutions

113. 9 **115.** 3 **117.** 1 mph, 4 miles **119.** 120 minutes

114. all real numbers less than or equal to 3 **116.** (a) True (b) False, $|-1 + 1| \neq |-1| + |1|$ (c) True **118.** 6 ml

Chapter Test, page 124

1. $\frac{3}{5}$ **3.** -3 **5.** i **7.** $6 - 9i$ **9.** $\frac{15}{13} + \frac{29}{13}i$ **11.** $\frac{3}{2} \pm \frac{1}{2}\sqrt{7}$ **13.** -32

2. $\frac{1-4b}{a-b}$ **4.** $-\frac{2}{7}$ **6.** $-21i$ **8.** $45 + 11i$ **10.** $\frac{4}{3}, -5$ **12.** $\frac{3}{2} \pm \frac{\sqrt{3}}{2}i$ **14.** $\pm\sqrt{2}, \pm\frac{\sqrt{3}}{3}i$

15. 5, 13 **17.** $(-\infty, -12)$ **19.** $[-6, -3]$ **21.** $580\frac{5}{9}$ miles

16. −3.5 0 2.5 3.5 **18.** $\left(-\infty, -\frac{5}{2}\right) \cup \left(\frac{11}{2}, \infty\right)$ **20.** $(-\infty, -3) \cup \left[\frac{5}{2}, \infty\right)$

Chapter 3

Section 3.1, pages 135–136

1.
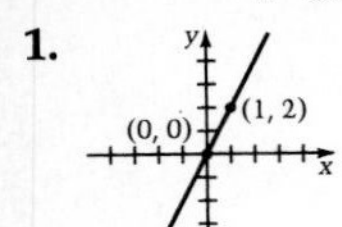

3.
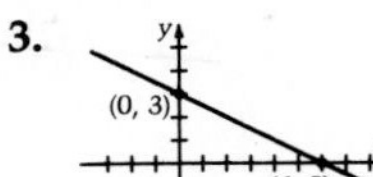

5.
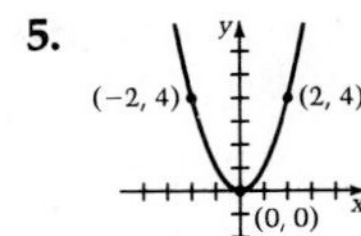

7.
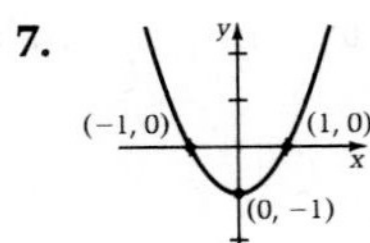

9.
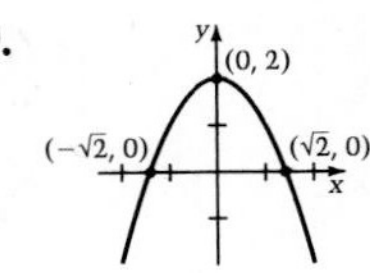

11.
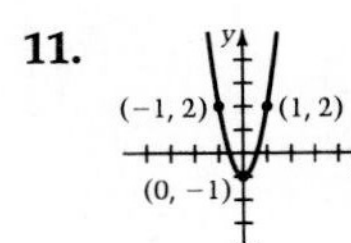

2.
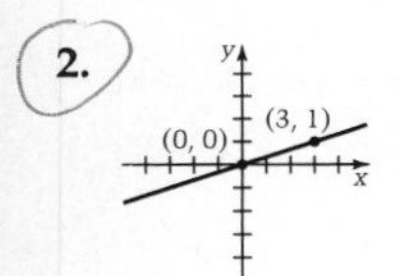

4.
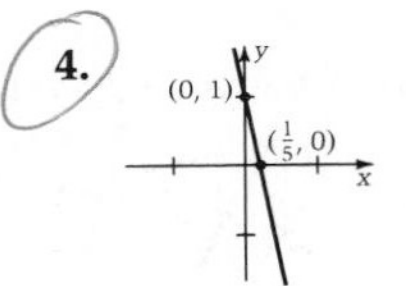

6.
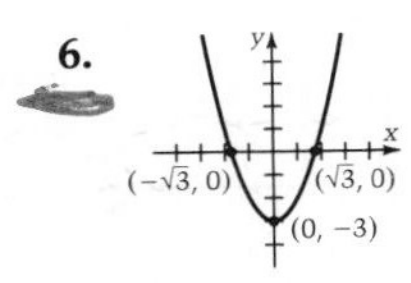

8.
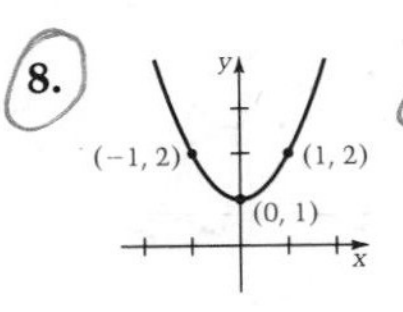

10.
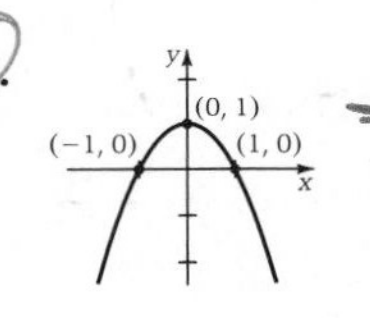

12
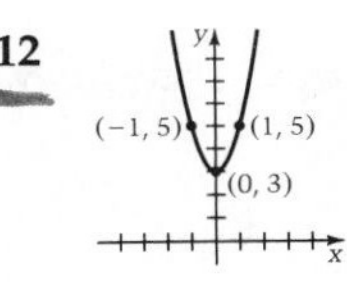

13.
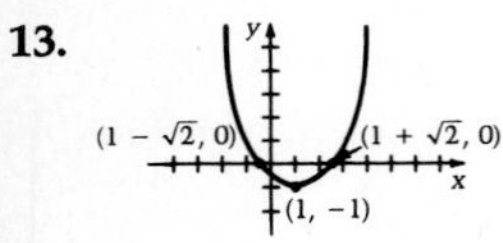

15.
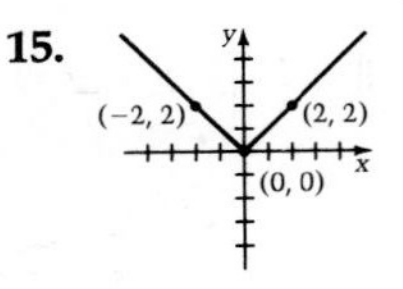

17.
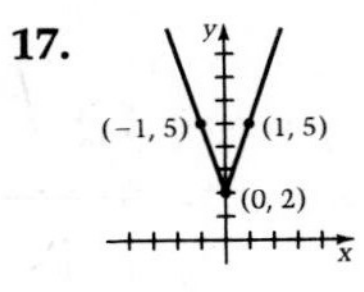

19.
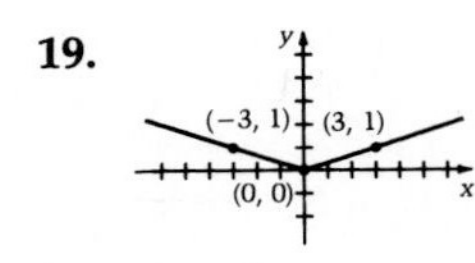

21.
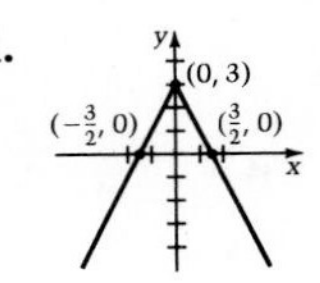

14.
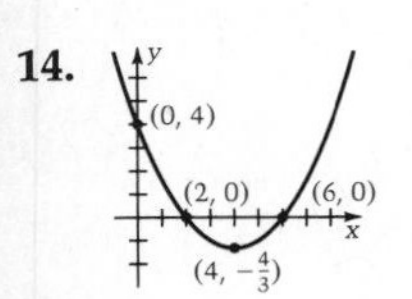

16.
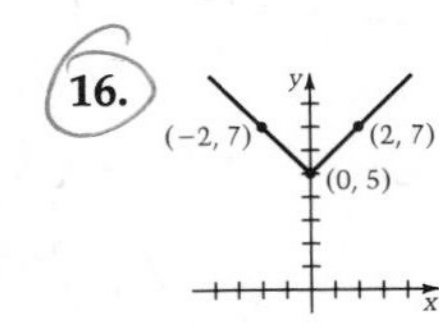

18.
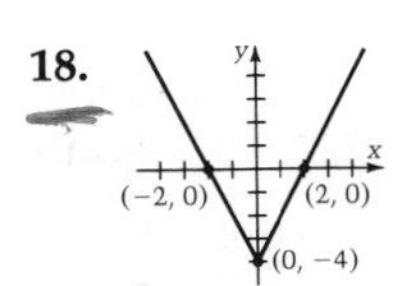

20.
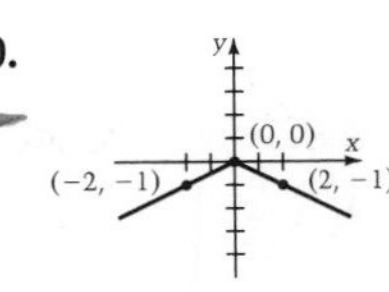

22.
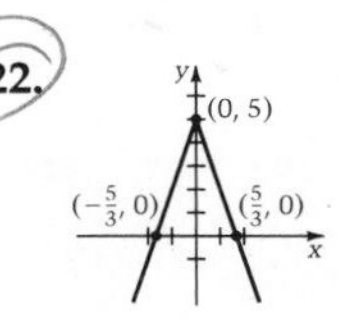

23. $5, \left(-\frac{3}{2}, 2\right)$ **25.** $\sqrt{29}, \left(\frac{3}{2}, -6\right)$ **27.** $2, (7, -3)$ **29.** $2\sqrt{13}, (1, 1)$ **31.** $\sqrt{14}, \left(\frac{3}{2}\sqrt{2}, 0\right)$

24. $5, \left(3, \frac{1}{2}\right)$ **26.** $10, (4, 3)$ **28.** $12, (7, 4)$ **30.** $\sqrt{10}, \left(\frac{9}{2}, \frac{13}{2}\right)$ **32.** $4\sqrt{3}, \left(\frac{1}{2}\sqrt{5}, \frac{3}{2}\sqrt{3}\right)$

33. (a) $(3, 2)$; (b) $(-3, -2)$; (c) $(-3, 2)$ **35.** (a) $(4, -3)$; (b) $(-4, 3)$; (c) $(-4, -3)$

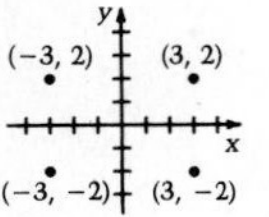

(−4, 3) (4, 3)

(−4, −3) (4, −3)

34. (a) $(-2, -1)$; (b) $(2, 1)$; (c) $(2, -1)$

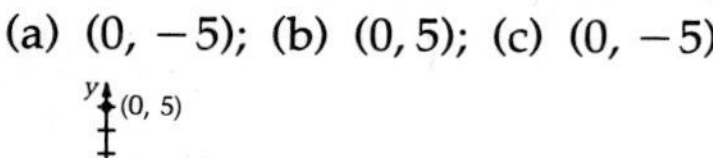

36. (a) $(-4, 8)$; (b) $(4, -8)$; (c) $(4, 8)$

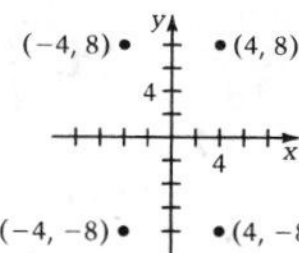

37. (a) $(0, -5)$; (b) $(0, 5)$; (c) $(0, -5)$

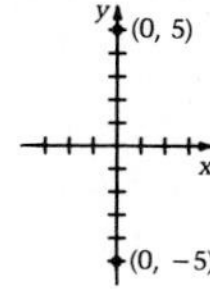

39. y-axis only **41.** all three **43.** y-axis only **45.** none **47.** all three

38. (a) $(-3, 0)$; (b) $(3, 0)$; (c) $(3, 0)$

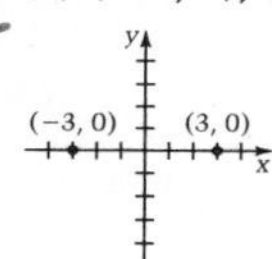

40. x-axis only **42.** all three

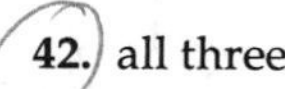

44. y-axis only

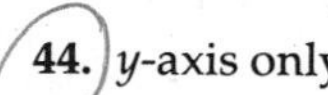

46. all three **48.** none

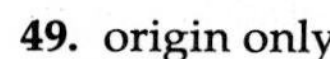

49. origin only

51.

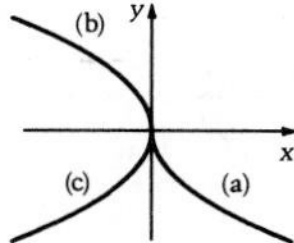

53.

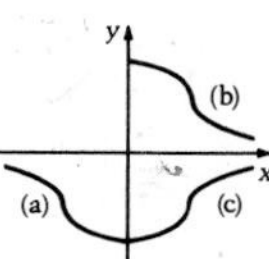

55.

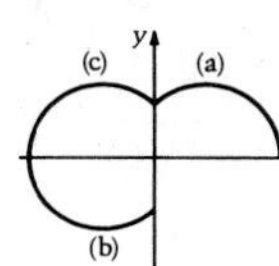

57. Only x-axis symmetry

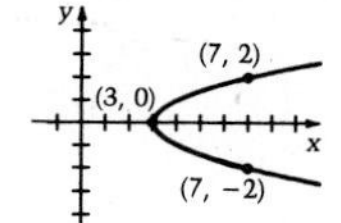

50. y-axis only

52.

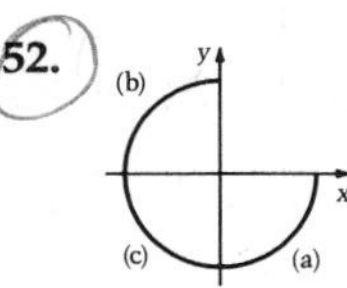

54.

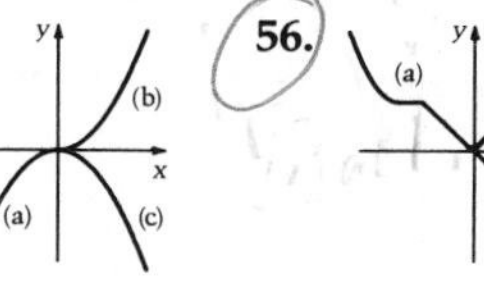

56.

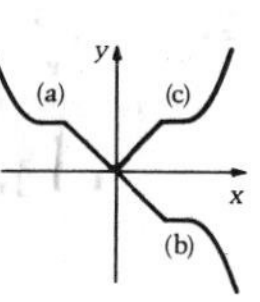

58. Only x-axis symmetry

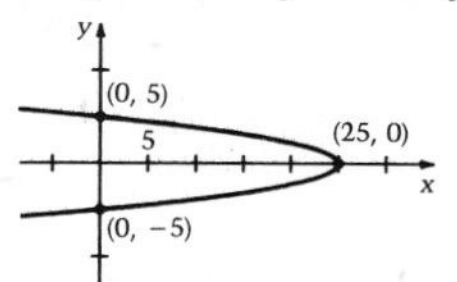

59. Only origin symmetry

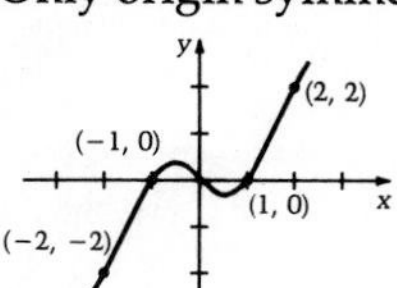

61. All 3 types of symmetry

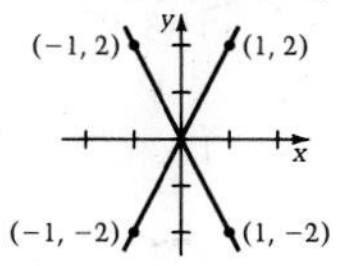

63. Only origin symmetry

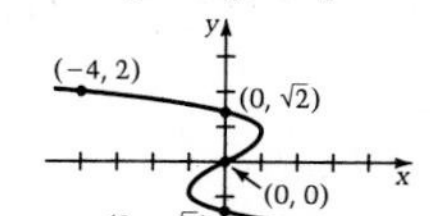

***65.** $(5, 2)$

60. Only origin symmetry

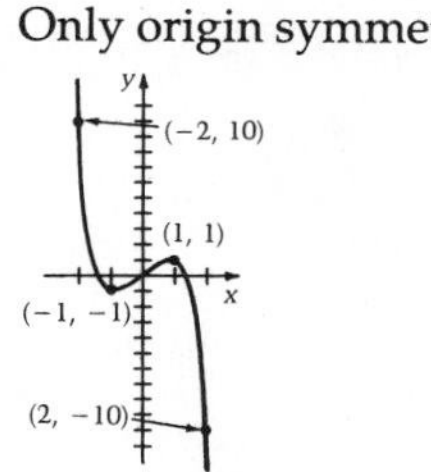

62. Only origin symmetry

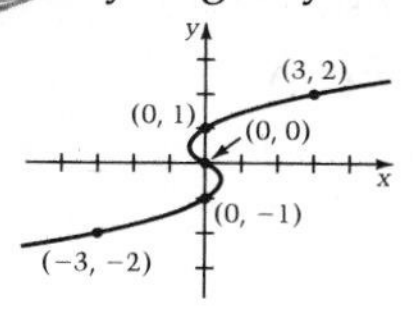

64. All 3 types of symmetry

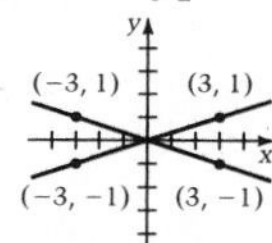

***66.** $(10, 7)$

*denotes Superset exercises

***67.** $(-6, -3)$ ***69.** $(3, 8)$ ***71.** collinear ***73.** not collinear ***75.** collinear ***77.** yes, right angle at P
***68.** $(-4, -5)$ ***70.** $(2, -15)$ ***72.** not collinear ***74.** collinear ***76.** not collinear ***78.** no
***79.** yes, right angle at Q ***81.** ***83.** ***85.** ***87.**

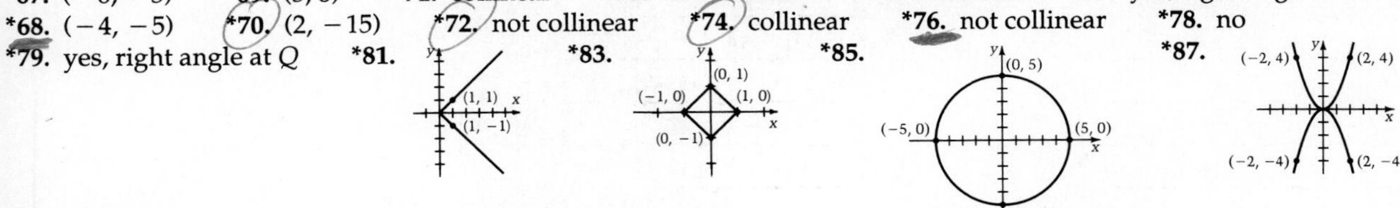

***80.** yes, right angle at Q ***82.** ***84.** ***86.** ***88.**

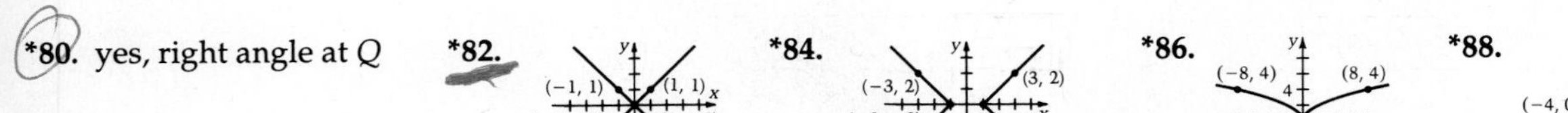

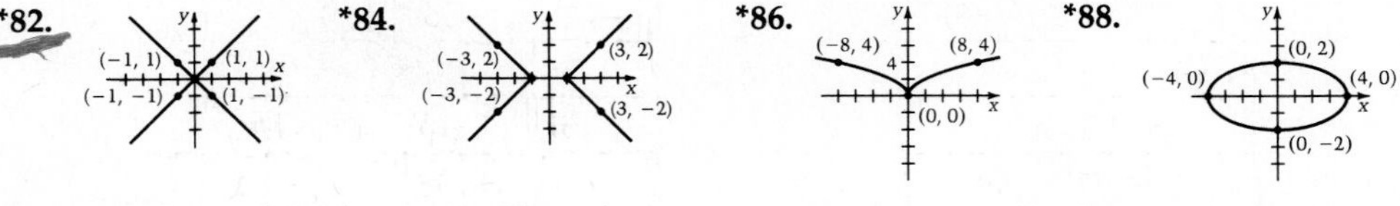

***89.** ***91.**

***90.** ***92.**

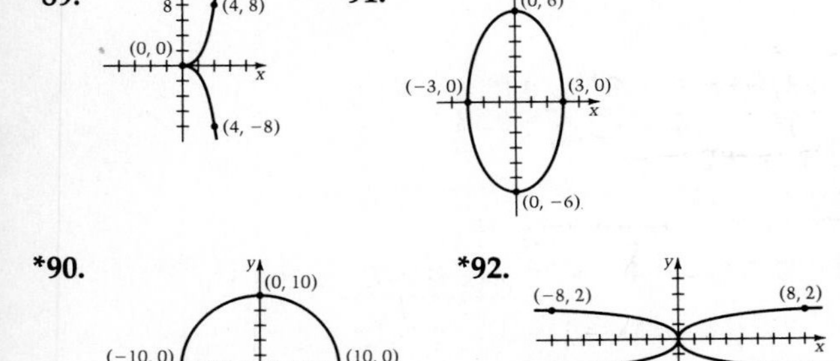

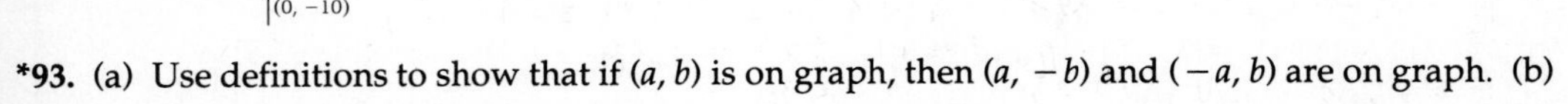

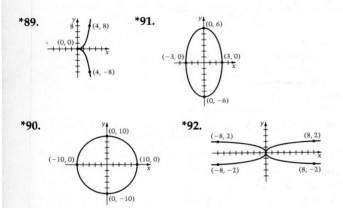

***93.** (a) Use definitions to show that if (a, b) is on graph, then $(a, -b)$ and $(-a, b)$ are on graph. (b)

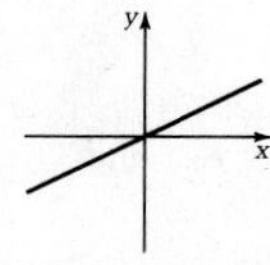

***94.** (a) There are 3 possibilities; one of them was covered in Exercise 93. The other two cases follow directly from the definitions, too.
(b) No symmetry All 3 types of symmetry x-axis symmetry only y-axis symmetry only origin symmetry only

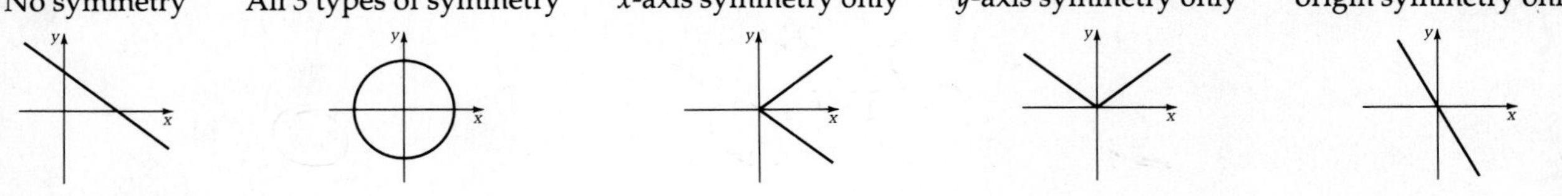

Section 3.2, pages 143–145

1. function: $\{(1, A), (2, B), (3, C), (4, A)\}$; domain: $\{1, 2, 3, 4\}$; range: $\{A, B, C\}$
2. function: $\{(1, 40), (2, 30), (3, 20), (4, 10), (5, 10)\}$; domain: $\{1, 2, 3, 4, 5\}$; range: $\{10, 20, 30, 40\}$
3. function: $\{(81, 10), (82, 20), (83, 20), (84, 35), (85, 30)\}$; domain: $\{81, 82, 83, 84, 85\}$; range: $\{10, 20, 30, 35\}$
4. function: $\{(1, 130), (2, 120), (3, 110), (4, 110), (5, 100)\}$; domain: $\{1, 2, 3, 4, 5\}$; range: $\{100, 110, 120, 130\}$

---

*denotes Superset exercises

5.

h	t	(h, t)
-1	-3	$(-1, -3)$
0	-1	$(0, -1)$
3	5	$(3, 5)$
4	7	$(4, 7)$

7.

x	y	(x, y)
-5	41	$(-5, 41)$
-2	11	$(-2, 11)$
1	-1	$(1, -1)$
4	5	$(4, 5)$

6.

t	s	(t, s)
-2	-68	$(-2, -68)$
-1	-18	$(-1, -18)$
0	0	$(0, 0)$
2	-60	$(2, -60)$

8.

x	y	(x, y)
-10	-55	$(-10, -55)$
-5	0	$(-5, 0)$
5	-40	$(5, -40)$
10	-135	$(10, -135)$

9. $f(1) = 1, f(1.2) = 2.2, f(1.4) = 3.4, f(1.6) = 4.6, f(1.8) = 5.8, f(2) = 7$

10. $f(1) = -3.74, f(1.2) = -3.736, f(1.4) = -3.732, f(1.6) = -3.728, f(1.8) = -3.724, f(2) = -3.72$

11. $f(1) = -0.34, f(1.2) = -0.9, f(1.4) = -1.38, f(1.6) = -1.78, f(1.8) = -2.1, f(2) = -2.34$

12. $f(1) = 4.45, f(1.2) = 5.98, f(1.4) = 7.79, f(1.6) = 9.88, f(1.8) = 12.25, f(2) = 14.9$

13. $f(1) \approx 0.678, f(1.2) \approx 0.748, f(1.4) \approx 0.812, f(1.6) \approx 0.872, f(1.8) \approx 0.927, f(2) \approx 0.980$

14. $f(1) \approx 14.697, f(1.2) \approx 13.971, f(1.4) \approx 13.206, f(1.6) \approx 12.394, f(1.8) \approx 11.524, f(2) \approx 10.583$

15. $f(-2) = 14, f(0) = 8, f(5) = -7, f(c^2) = 8 - 3c^2, [f(c)]^2 = (8 - 3c)^2, f(2 + h) = 2 - 3h$

16. $f(-2) = 3, f(0) = 4, f(5) = \dfrac{13}{2}, f(c^2) = \dfrac{1}{2}c^2 + 4, [f(c)]^2 = \left[\dfrac{1}{2}c + 4\right]^2, f(2 + h) = 5 + \dfrac{h}{2}$

17. $f(-2) = 6, f(0) = 2, f(5) = 8, f(c^2) = 2|c^2 - 1|, [f(c)]^2 = 4(c - 1)^2, f(2 + h) = 2|1 + h|$

18. $f(-2) = 12, f(0) = 6, f(5) = 9, f(c^2) = 3|c^2 - 2|, [f(c)]^2 = 9(c - 2)^2, f(2 + h) = 3|h|$

19. $f(-2) = 7, f(0) = -1, f(5) = 49, f(c^2) = 2c^4 - 1, [f(c)]^2 = 4c^4 - 4c^2 + 1, f(2 + h) = 2h^2 + 8h + 7$

20. $f(-2) = -2, f(0) = 6, f(5) = -44, f(c^2) = 6 - 2c^4, [f(c)]^2 = 4c^4 - 24c^2 + 36, f(2 + h) = -2 - 8h - 2h^2$

21. $f(-2) = 10, f(0) = 10, f(5) = 10, f(c^2) = 10, [f(c)]^2 = 100, f(2 + h) = 10$

22. $f(-2) = -7, f(0) = -7, f(5) = -7, f(c^2) = -7, [f(c)]^2 = 49, f(2 + h) = -7$

23. $f(-2) = 20, f(0) = 6, f(5) = 6, f(c^2) = c^4 - 5c^2 + 6, [f(c)]^2 = [c^2 - 5c + 6]^2, f(2 + h) = -h + h^2$

24. $f(-2) = -12, f(0) = -6, f(5) = 44, f(c^2) = c^4 + 5c^2 - 6, [f(c)]^2 = [c^2 + 5c - 6]^2, f(2 + h) = 8 + 9h + h^2$

25. $f(-2) = \sqrt{7}, f(0) = 1$, $f(5)$ is not a real number. $f(c^2) = \sqrt{1 - 3c^2}, [f(c)]^2 = 1 - 3c, f(2 + h) = \sqrt{-3h - 5}$

26. $f(-2)$ is not a real number, $f(0)$ is not a real number, $f(5) = \sqrt{14}, f(c^2) = \sqrt{3c^2 - 1}, [f(c)]^2 = 3c - 1, f(2 + h) = \sqrt{3h + 5}$

27. $f(-2) = \sqrt{5}, f(0) = 1, f(5) = \sqrt{26}, f(c^2) = \sqrt{1 + c^4}, [f(c)]^2 = 1 + c^2, f(2 + h) = \sqrt{h^2 + 4h + 5}$

28. $f(-2) = \sqrt{11}, f(0) = \sqrt{3}, f(5) = \sqrt{53}, f(c^2) = \sqrt{2c^4 + 3}, [f(c)]^2 = 2c^2 + 3, f(2 + h) = \sqrt{2h^2 + 8h + 11}$

29. $f(-2)$ is not defined, $f(0) = \dfrac{1}{4}, f(5) = \dfrac{1}{14}, f(c^2) = \dfrac{1}{2c^2 + 4}, [f(c)]^2 = \dfrac{1}{(2c + 4)^2}, f(2 + h) = \dfrac{1}{2h + 8}$ **31.** $\dfrac{9}{4}$

30. $f(-2)$ is not defined, $f(0) = -\dfrac{1}{4}, f(5) = \dfrac{1}{21}, f(c^2) = \dfrac{1}{c^4 - 4}, [f(c)]^2 = \dfrac{1}{(c^2 - 4)^2}, f(2 + h) = \dfrac{1}{h^2 + 4h}$ **32.** $-\sqrt{2}$

33. $-\dfrac{1}{5}$ **35.** 1 **37.** yes **39.** no **41.** yes **43.** no **45.** $(-\infty, \infty)$ **47.** $(-\infty, \infty)$ **49.** $(-\infty, \infty)$

34. $\dfrac{35}{34}\sqrt{2} + \dfrac{3}{17}$ **36.** 0 **38.** no **40.** yes **42.** yes **44.** no **46.** $(-\infty, \infty)$ **48.** $(-\infty, \infty)$ **50.** $(-\infty, \infty)$

51. $(-\infty, \infty)$ **53.** $[-3, \infty)$ **55.** $(-\infty, 4]$ **57.** $\left[\dfrac{9}{2}, \infty\right)$ **59.** $(-\infty, -4) \cup (-4, \infty)$ **61.** $(-\infty, 7) \cup (7, \infty)$

52. $(-\infty, \infty)$ **54.** $[1, \infty)$ **56.** $[-2, \infty)$ **58.** $[5, \infty)$ **60.** $(-\infty, 8) \cup (8, \infty)$ **62.** $(-\infty, -14) \cup (-14, \infty)$

63. $(-\infty, -3) \cup (-3, 3) \cup (3, \infty)$ **65.** $(-\infty, \infty)$ **67.** $(-\infty, -2] \cup [2, \infty)$ **69.** $[-1, 1]$ **71.** $\left(-\infty, \frac{3}{2}\right)$

64. $(-\infty, -3) \cup (-3, -2) \cup (-2, \infty)$ **66.** $(-\infty, \infty)$ **68.** $[-\sqrt{3}, \sqrt{3}]$ **70.** $[-3, 3]$ **72.** $(-\sqrt{2}, \sqrt{2})$

73. $f(-2) = -2, f(-1) = -1, f(0) = 2, f(1) = 1, f(2) = 4$ **75.** $f(-2) = 1, f(-1) = 2, f(0) = 1, f(1) = 2, f(2) = 5$

74. $f(-2) = 3, f(-1) = 1, f(0) = 0, f(1) = 1, f(2) = 3$ **76.** $f(-2) = 1, f(-1) = 1, f(0) = 1, f(1) = -1, f(2) = 1$

***77.** (a) 26 (b) 12 (c) $4t^2 - 8t + 5$ ***79.** 2

***78.** $\dfrac{f(3+h) - f(3)}{h} = \dfrac{\sqrt{3+h} - \sqrt{3}}{h} \cdot \dfrac{\sqrt{3+h} + \sqrt{3}}{\sqrt{3+h} + \sqrt{3}} = \dfrac{h}{h(\sqrt{3+h} + \sqrt{3})} = \dfrac{1}{f(3+h) + f(3)}$ ***80.** -4

***81.** 0 ***83.** $3 - 2x - h$ ***85.** $\dfrac{-1}{(x+h+1)(x+1)}$

***82.** $2x + h - 2$ ***84.** $3x^2 + 3xh + h^2$ ***86.** $\dfrac{-1}{(x+h-2)(x-2)}$

***87.** (a) $f(x) = 0$ for all real x (b) $f(0) = 1$ (c) $f(n) = c^n$ for all integers n

Section 3.3, pages 155–157

1. 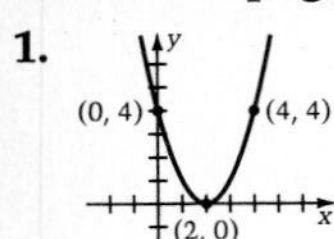

3. 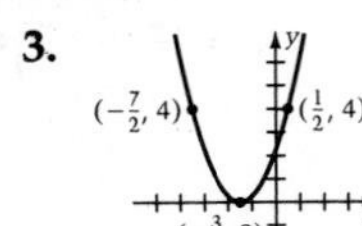

5. 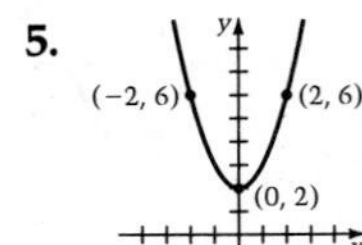

7. 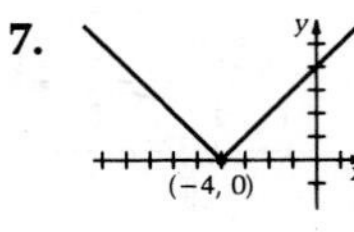

9. 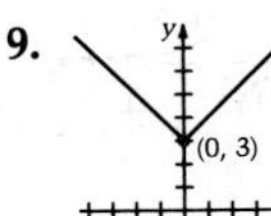

2. 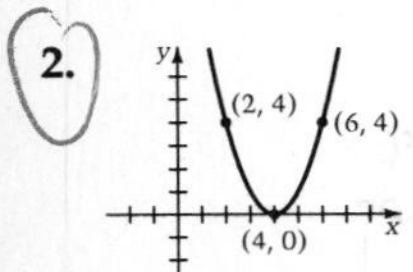

4. 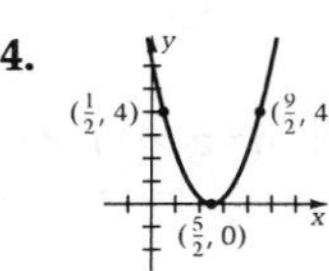

6. 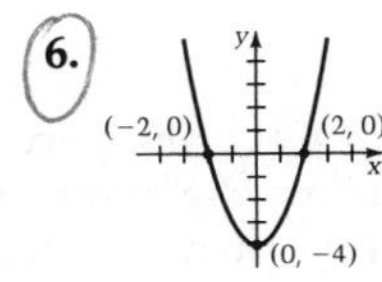

8. 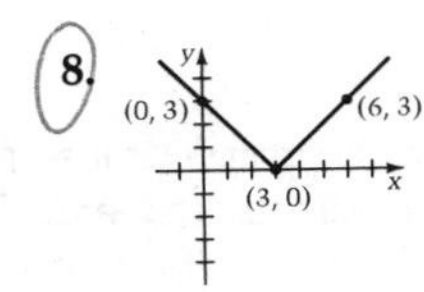

10. 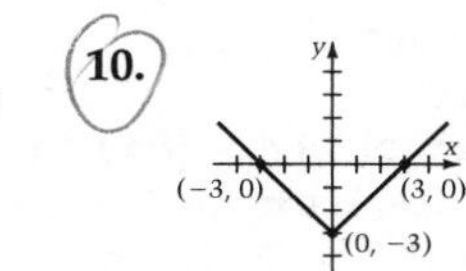

11. 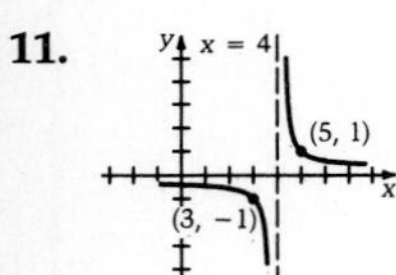

13. 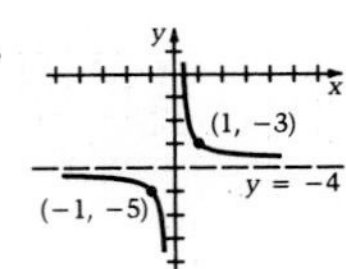

15. 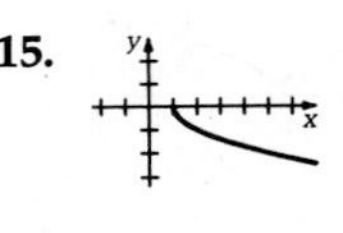

17. 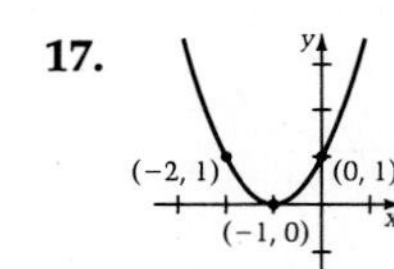

19. 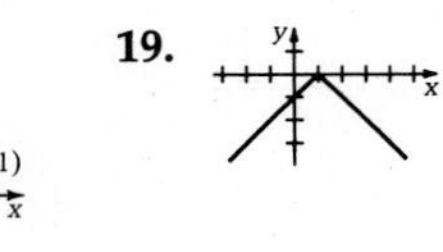

12. 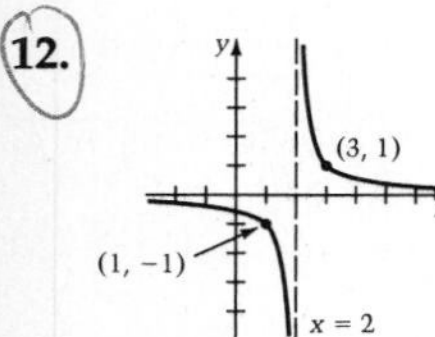

14. 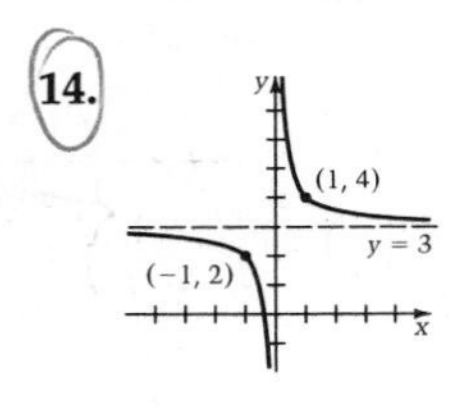

16. 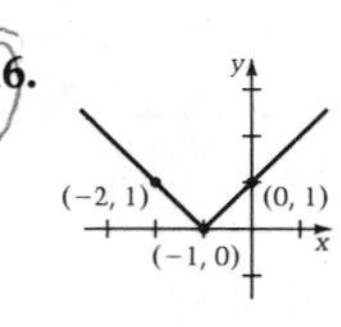

18. **20.** 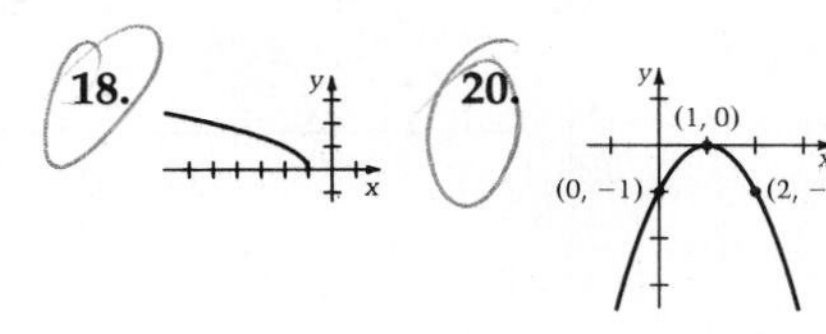

21. 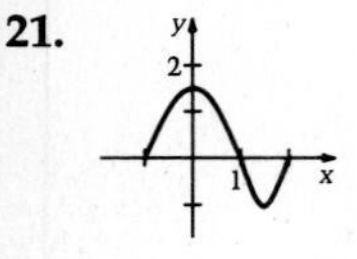

23. 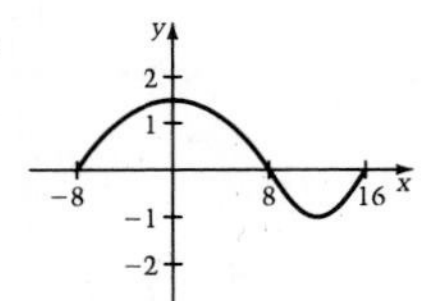

25. 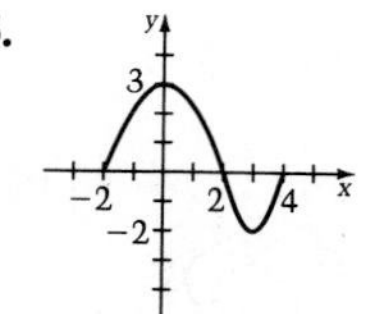

27. 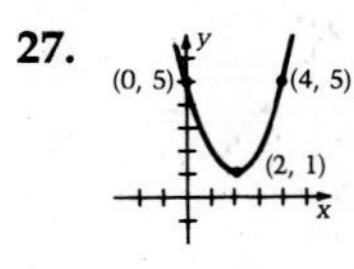

29. 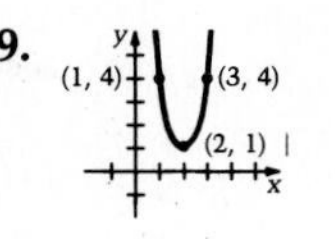

31. 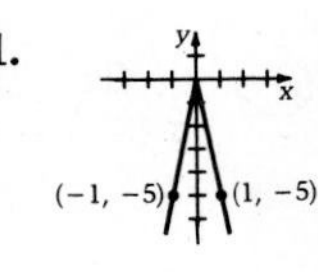

*denotes Superset exercises

22. 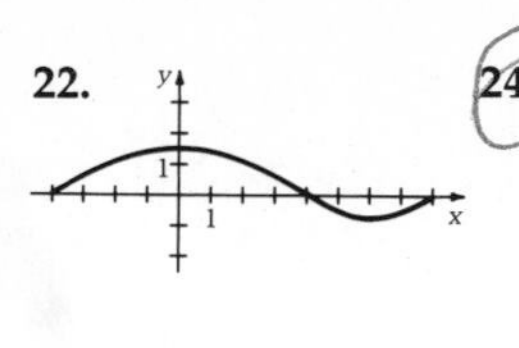

24. 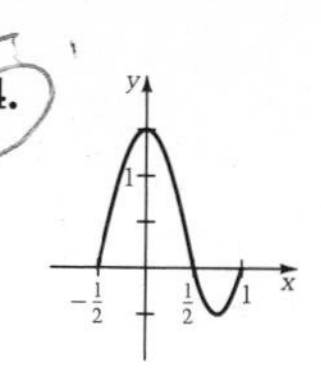

26. 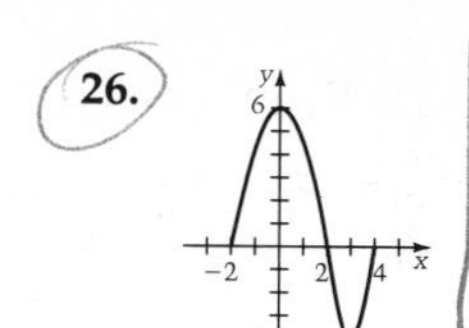

28. 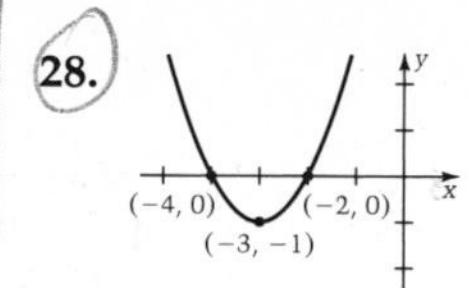

30. 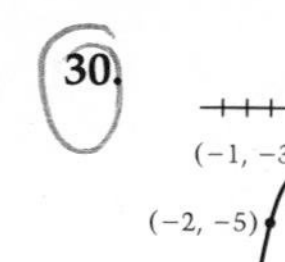

32. 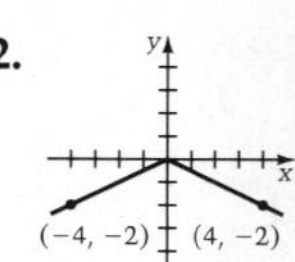

33. 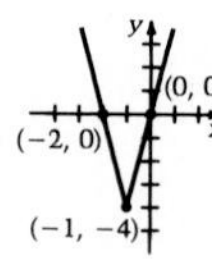

35. 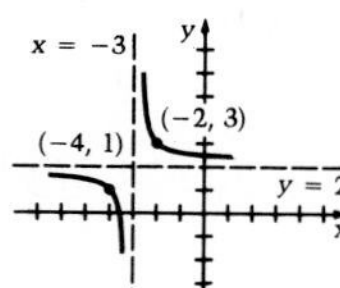

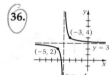

37. 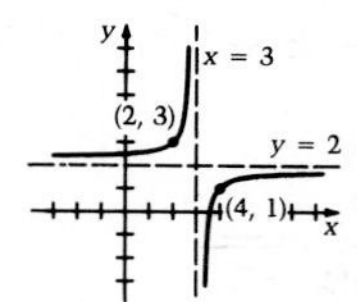

39. 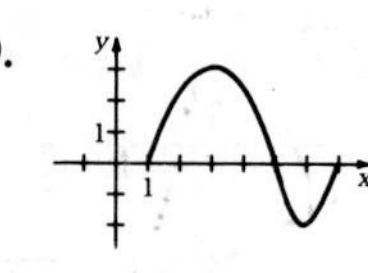

41. 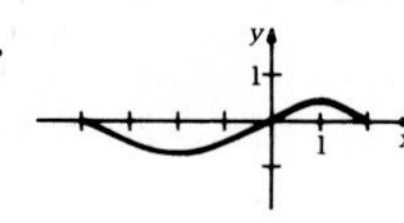

43. odd

34. 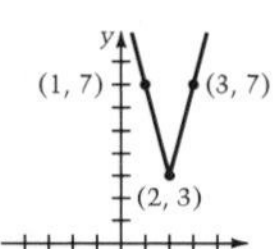

36. 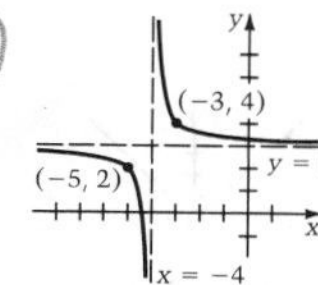

38. 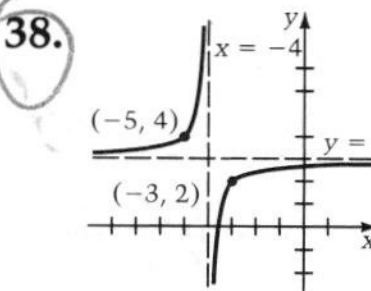

40. 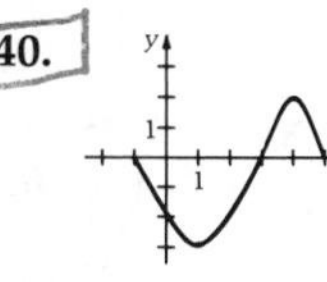

42. 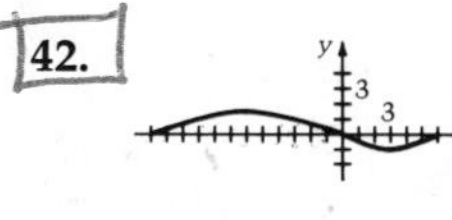

44. even

45. neither **47.** even

46. even **48.** odd

***49.** To obtain the graph of $y = |f(x)|$ from the graph of $y = f(x)$, replace any portion of the graph of $y = f(x)$ lying below the x-axis by its reflection across the x-axis. To obtain the graph of $y = f(|x|)$ from the graph of $y = f(x)$, replace the graph of $y = f(x)$ for $x < 0$ by the reflection of the graph of $y = f(x)$ for $x > 0$.

***50.** 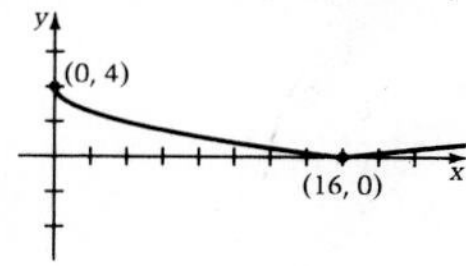

***52.** 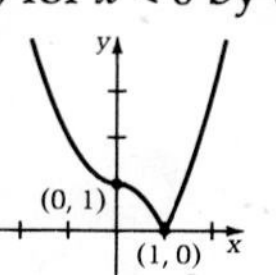

***54.** 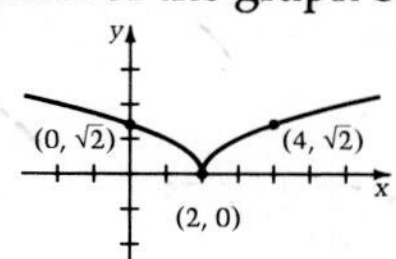

***56.** 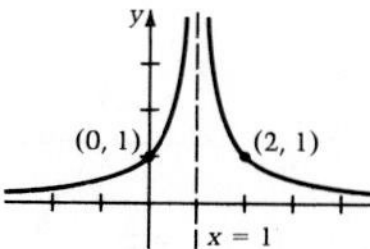

***51.** 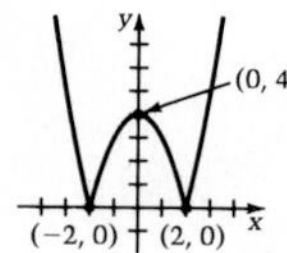

***53.** 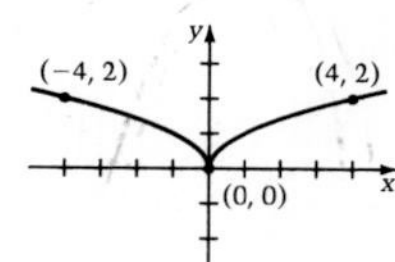

***55.** 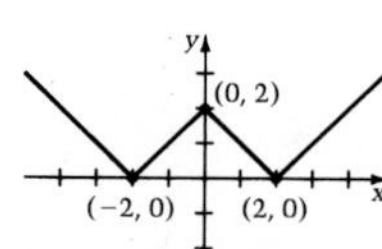

***57.** 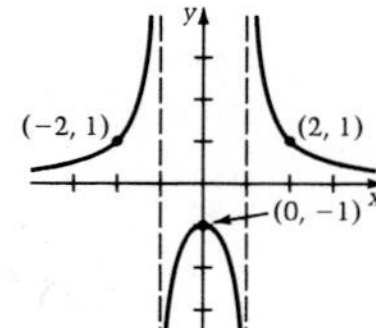

***58.** 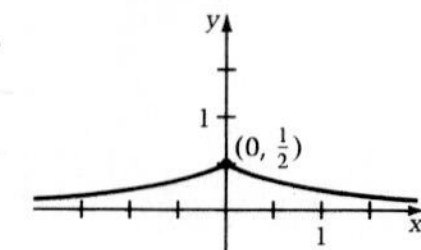

***60.** 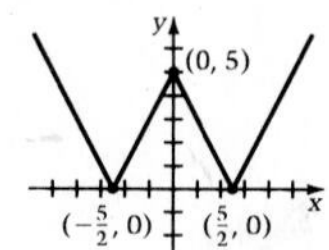

---

*denotes Superset exercises

***59.**

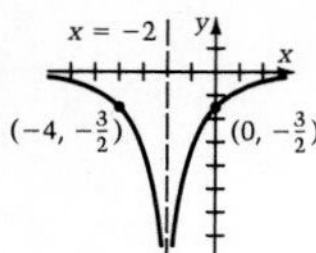

***61.** (a)

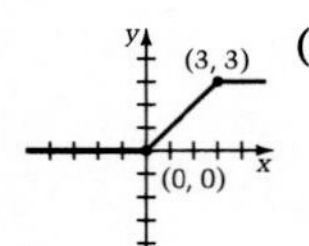

(b)

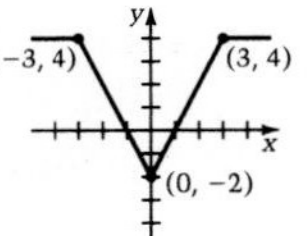

(c)

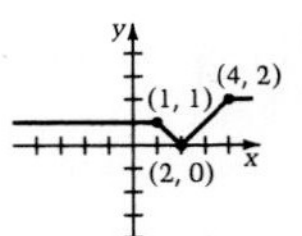

(d)

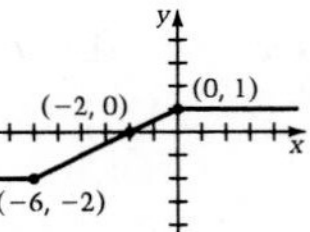

***62.** (a)

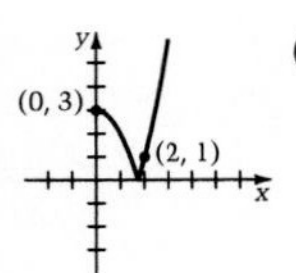

(b)

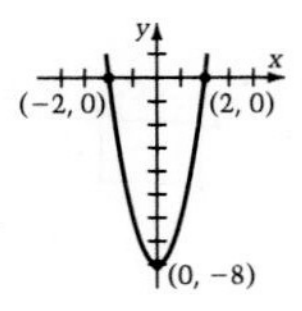

(c)

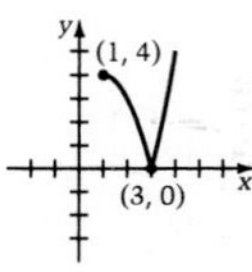

(d)

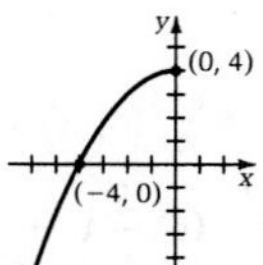

***63.** (a)

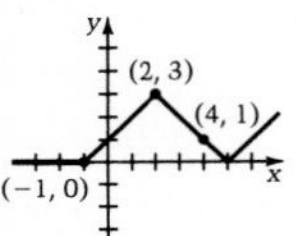

(b)

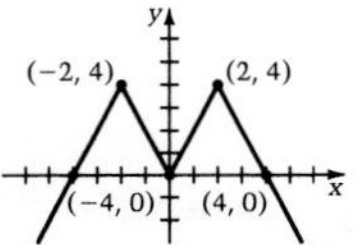

(c)

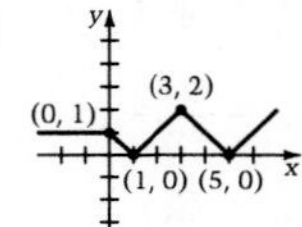

(d)

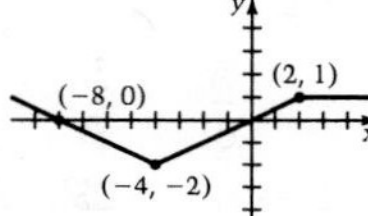

***64.** (a)

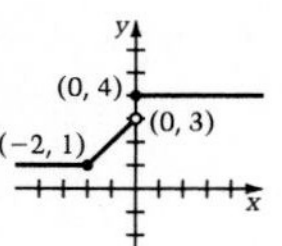

(b)

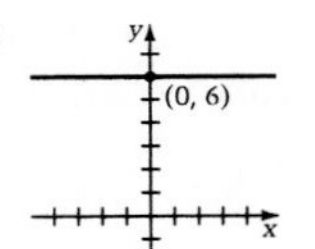

(c)

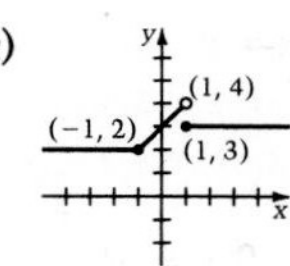

(d)

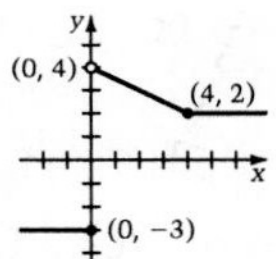

Section 3.4, pages 164–166

1. $m = 5$ **3.** $m = -\frac{1}{3}$ **5.** $m = -\frac{1}{4}$ **7.** m undefined **9.** $m = 0$ **11.** $m = \frac{1}{8}$

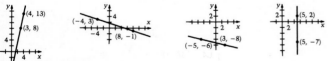

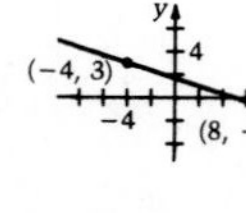

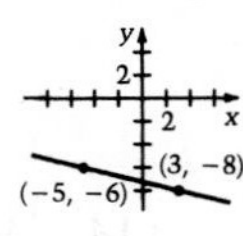

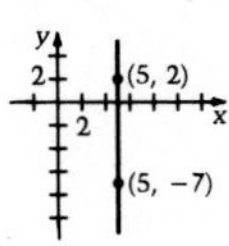

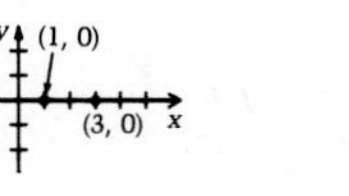

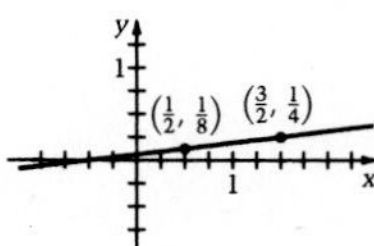

2. $m = 4$ **4.** $m = \frac{3}{5}$ **6.** $m = -\frac{5}{6}$ **8.** $m = 0$ **10.** m undefined **12.** $m = \frac{3}{8}$

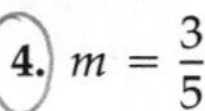

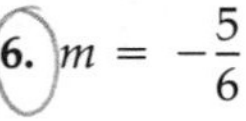

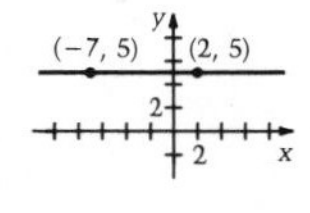

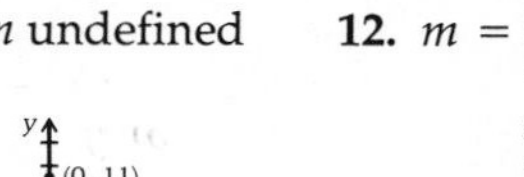

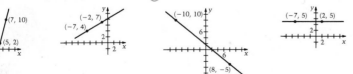

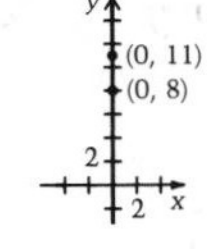

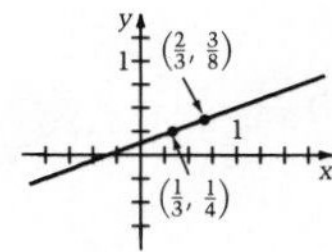

---

*denotes Superset exercises

13. $m = 1.1$

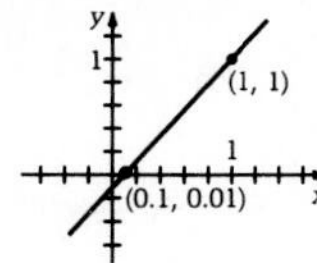

15. (a) $y - 8 = 5(x - 3)$ (b) $y = 5x - 7$ (c) $5x - y - 7 = 0$

14. $m = 0.125$

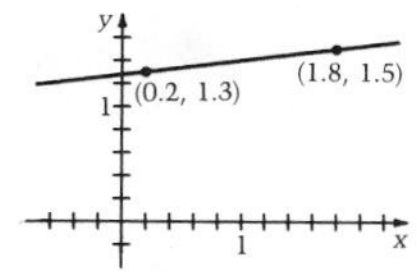

16. (a) $y - 2 = 4(x - 5)$ (b) $y = 4x - 18$ (c) $4x - y - 18 = 0$

17. (a) $y - (-1) = -\frac{1}{3}(x - 8)$ (b) $y = -\frac{1}{3}x + \frac{5}{3}$ (c) $\frac{1}{3}x + y - \frac{5}{3} = 0$

18. (a) $y - 7 = \frac{3}{5}(x - (-2))$ (b) $y = \frac{3}{5}x + \frac{41}{5}$ (c) $\frac{3}{5}x - y + \frac{41}{5} = 0$

19. (a) $y - (-6) = -\frac{1}{4}(x - (-5))$ (b) $y = -\frac{1}{4}x - \frac{29}{4}$ (c) $\frac{1}{4}x + y + \frac{29}{4} = 0$

20. (a) $y - (-5) = -\frac{5}{6}(x - 8)$ (b) $y = -\frac{5}{6}x + \frac{5}{3}$ (c) $\frac{5}{6}x + y - \frac{5}{3} = 0$

21. (a)–(b) cannot be written (c) $x - 5 = 0$

22. (a) $y - 5 = 0(x - 2)$ (b) $y = 5$ (c) $y - 5 = 0$

23. (a) $y - 0 = 0(x - 3)$ (b) $y = 0$ (c) $y = 0$

24. (a)–(b) cannot be written (c) $x = 0$

25. (a) $y - \frac{1}{8} = \frac{1}{8}\left(x - \frac{1}{2}\right)$ (b) $y = \frac{1}{8}x + \frac{1}{16}$ (c) $\frac{1}{8}x - y + \frac{1}{16} = 0$

26. $y - \frac{1}{4} = \frac{3}{8}\left(x - \frac{1}{3}\right)$ (b) $y = \frac{3}{8}x + \frac{1}{8}$ (c) $\frac{3}{8}x - y + \frac{1}{8} = 0$

27. (a) $y - 1 = 1.1(x - 1)$ (b) $y = 1.1x - 0.1$ (c) $1.1x - y - 0.1 = 0$

28. (a) $y - 1.3 = 0.125(x - 0.2)$ (b) $y = 0.125x + 1.275$ (c) $0.125x - y + 1.275 = 0$

29. (a) $y - 5 = -2(x - 3)$ (b) $y = -2x + 11$ (c) $2x + y - 11 = 0$

30. (a) $y - (-1) = 3(x - 5)$ (b) $y = 3x - 16$ (c) $3x - y - 16 = 0$

31. (a) $y - 1 = -\frac{2}{3}(x - 1)$ (b) $y = -\frac{2}{3}x + \frac{5}{3}$ (c) $\frac{2}{3}x + y - \frac{5}{3} = 0$

32. (a) $y - 1 = \frac{1}{2}(x - (-1))$ (b) $y = \frac{1}{2}x + \frac{3}{2}$ (c) $\frac{1}{2}x - y + \frac{3}{2} = 0$

33. (a) $y - (-2) = 0(x - 2)$ (b) $y = -2$ (c) $y + 2 = 0$

34. (a)–(b) cannot be written (c) $x - 2 = 0$

35. (a) $y - 3 = \frac{1}{2}(x - 0)$ (b) $y = \frac{1}{2}x + 3$ (c) $\frac{1}{2}x - y + 3 = 0$

36. (a) $y - (-2) = \frac{3}{4}(x - 0)$ (b) $y = \frac{3}{4}x - 2$ (c) $\frac{3}{4}x - y - 2 = 0$

37. (a) $y - 0 = -2(x - 0)$ (b) $y = -2x$ (c) $2x + y = 0$

38. (a) $y - (-2) = 0(x - 0)$ (b) $y = -2$ (c) $y + 2 = 0$

39. (a) $y - (-2) = \frac{3}{7}(x - 1)$ (b) $y = \frac{3}{7}x - \frac{17}{7}$ (c) $\frac{3}{7}x - y - \frac{17}{7} = 0$

40. (a) $y - 0 = -2(x - 0)$ (b) $y = -2x$ (c) $2x + y = 0$

41. (a)–(b) cannot be written (c) $x = 0$

42. (a) $y - (-8) = 0(x - (-2))$ (b) $y = -8$ (c) $y + 8 = 0$

43. (a) $y - (-2) = -\frac{1}{6}(x - 5)$ (b) $y = -\frac{1}{6}x - \frac{7}{6}$ (c) $\frac{1}{6}x + y + \frac{7}{6} = 0$

44. (a) $y - 3 = -\frac{5}{3}(x - (-1))$ (b) $y = -\frac{5}{3}x + \frac{4}{3}$ (c) $\frac{5}{3}x + y - \frac{4}{3} = 0$

45. (a) $y - (-10) = 0(x - 5)$ (b) $y = -10$ (c) $y + 10 = 0$ **47.** $y = 0.71x + 5.79$ **49.** $y = -1.25x + 4.12$

46. (a)–(b) cannot be written (c) $x - 5 = 0$ **48.** $y = 0.59x - 0.86$ **50.** $y = -0.87x + 0.82$

51. $y = -0.89x + 8.16$ **53.** $y = 5.25x - 3.50$ **55.** (c) $f(2.5) = -2$ **57.** (c) $f(2.5) = 1.9$

52. $y = 0.63x - 4.51$ **54.** $y = -0.44x - 7.60$ **56.** (c) $f(2.5) = -2.5$ **58.** (c) $f(2.5) = 10.2$

59. (c) $f(2.5) = 5.6$ **61.** (c) $f(2.5) \approx 0.394$ **63.** $P(x) = 10.5x - 28770, x \geq 0$; 2740 **65.** $P(x) = 25 - x, 0 \leq x \leq 10$

60. (c) $f(2.5) = -0.45$ **62.** (c) $f(2.5) \approx -0.394$ **64.** $P(x) = \frac{48000}{x}, x > 0$ **66.** $y = f(x) = -2720x + 37336, x > 0$

67. $P(x) = 13400 - 24x, 0 \leq x \leq 400$

68. $P(x) = 14.8x - 240500, x \geq 15000$; 16250

***69.** The slope of the segment with endpoints (6, 3) and (5, 1) is 2. The slope of the segment with endpoints (1, 3) and (5, 1) is $-1/2$. Since the slopes are negative reciprocals, the segments are perpendicular, and the triangle is a right triangle.

***70.** The slope of the line containing (1, 4) and (2, 7) is 3. The slope of the line containing $(-3, -8)$ and (2, 7) is 3. Since there is only one line having slope 3 and containing (2, 7), all three points must lie on this line.

***71.** The ant did not pass through (1, 5) because (1, 5) is not on the line $y = 3x + 4$. The point (1, 7) is on the line. The distance between the points (1, 7) and (0, 4) is $\sqrt{10}$. Since $\sqrt{10} < 4$, the ant passed through (1, 7).

***72.** (a) $y = 3x$ (b) Since the distance from the origin to (1, 3) is $\sqrt{10}$, which is greater than 3, the ant did not pass through the point (1, 3). (c) $\frac{9\sqrt{10}}{10}$ units

***73.** $y = \frac{7}{3} - \frac{x}{3}$; $[1, \infty)$ ***75.** $y = \frac{8}{7}x + \frac{100}{7}$

***74.** (a) $y = \frac{1}{2}x + \frac{11}{2}$ (b) $y = 5 - 6x$ (c) $y = -x + 4$ (d) $y = x + 7$ ***76.** $\frac{1}{2\pi}$

***77.** Use the fact that the diagonals of a parallelogram bisect one another. The fourth vertex can be positioned at $(2, -2)$, (0, 2), or (4, 10).

***78.** $(-1, -5)$, $(-3, 7)$, or (3, 3)

Section 3.5, pages 174–175

1. $y = 3(x - 1)^2 + 4$. The vertex is (1, 4), the axis of symmetry is $x = 1$, and the parabola opens upward.

2. $y = -2(x - (-3))^2 + (-4)$. The vertex is $(-3, -4)$, the axis of symmetry is $x = -3$, and the parabola opens downward.

3. $y = -2(x - (-6))^2 + 0$. The vertex is $(-6, 0)$, the axis of symmetry is $x = -6$, and the parabola opens downward.

4. $y = 5(x - 7)^2 + 0$. The vertex is (7, 0), the axis of symmetry is $x = 7$, and the parabola opens upward.

5. $y = 2(x - 3)^2 + 0$. The vertex is (3, 0), the axis of symmetry is $x = 3$, and the parabola opens upward.

6. $y = -3(x - (-7))^2 + 0$. The vertex is $(-7, 0)$, the axis of symmetry is $x = -7$, and the parabola opens downward.

7. $y = -1(x - (-5))^2 + (-1)$. The vertex is $(-5, -1)$, the axis of symmetry is $x = -5$, and the parabola opens downward.

8. $y = -1(x - 2)^2 + 3$. The vertex is (2, 3), the axis of symmetry is $x = 2$, and the parabola opens downward.

9. $y = 3(x - (-4))^2 + (-2)$. The vertex is $(-4, -2)$, the axis of symmetry is $x = -4$, and the parabola opens upward.

10. $y = 1(x - 3)^2 + 10$. The vertex is (3, 10), the axis of symmetry is $x = 3$, and the parabola opens upward.

11. $y = -\frac{1}{3}(x - (-1))^2 + \left(-\frac{5}{3}\right)$. The vertex is $(-1, -5/3)$, the axis of symmetry is $x = -1$, and the parabola opens downward.

12. $y = \frac{1}{2}(x - 4)^2 + \frac{7}{2}$. The vertex is (4, 7/2), the axis of symmetry is $x = 4$, and the parabola opens upward.

13. $y = \frac{1}{3}(x - 9)^2 + (-1)$. The vertex is $(9, -1)$, the axis of symmetry is $x = 9$, and the parabola opens upward.

*denotes Superset exercises

14. $y = \frac{1}{4}(x - (-2))^2 + 6$. The vertex is $(-2, 6)$, the axis of symmetry is $x = -2$, and the parabola opens upward.

15. $y = -\frac{2}{7}(x - (-6))^2 + 0$. The vertex is $(-6, 0)$, the axis of symmetry is $x = -6$, and the parabola opens downward.

16. $y = \frac{3}{10}(x - (-1))^2 + 0$. The vertex is $(-1, 0)$, the axis of symmetry is $x = -1$, and the parabola opens upward.

17. $y = 1(x - 0)^2 + 4$. The vertex is $(0, 4)$, the axis of symmetry is $x = 0$, and the parabola opens upward.

18. $y = \frac{3}{2}(x - 0)^2 + (-4)$. The vertex is $(0, -4)$, the axis of symmetry is $x = 0$, and the parabola opens upward.

19. None **21.** -6 **23.** 3 **25.** None **27.** $-4 \pm \frac{\sqrt{6}}{3}$ **29.** None **31.** $9 \pm \sqrt{3}$ **33.** -6 **35.** None

20. None **22.** 7 **24.** -7 **26.** $2 \pm \sqrt{3}$ **28.** None **30.** None **32.** None **34.** -1 **36.** $\pm\frac{2\sqrt{6}}{3}$

37. $x = 0$; $(-1, 2)$; $(1, 2)$; $(-\frac{\sqrt{15}}{5}, 0)$; $(\frac{\sqrt{15}}{5}, 0)$; $(0, -3)$

39. $x = 2$; $(5, 0)$; $(-1, 0)$; $(4, -5)$; $(1, -8)$; $(2, -9)$

41. $x = 2$; $(2, 0)$; $(0, -4)$; $(4, -4)$

43. $x = 3$; $(1, 4)$; $(5, 4)$; $(3, 0)$

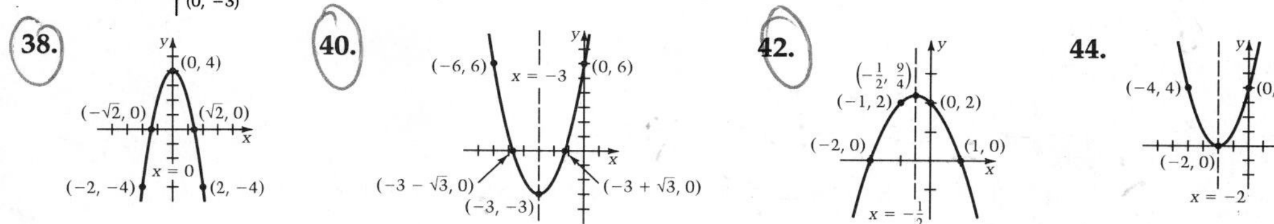

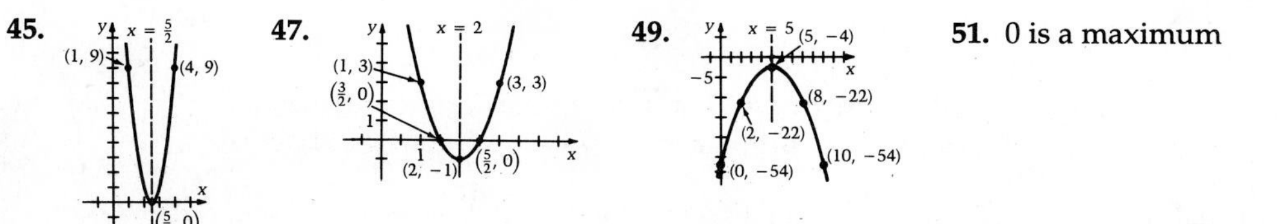

51. 0 is a maximum

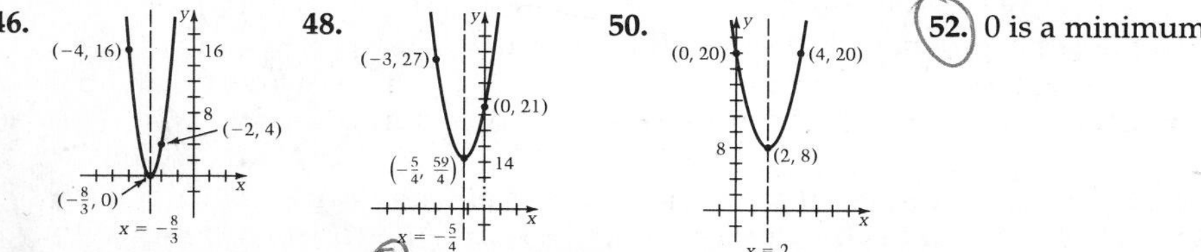

52. 0 is a minimum

53. -9 is a minimum **55.** (a) 1936 ft (b) after 11 seconds (c) after 22 seconds **57.** (c) $f(2.5) = 2.75$
54. 4 is a maximum **56.** (a) $x, 12 - x$ (b) $P(x) = 12x - x^2$, (c) 6, 6 **58.** (c) $f(2.5) = 4.875$
59. (c) $f(2.5) = 2.75$ **61.** (c) $f(2.5) = -0.25$ **63.** (c) $f(2.5) = -1.055$ **65.** 101.3 feet in $t = 2.5$ seconds
60. (c) $f(2.5) = 0.75$ **62.** (c) $f(2.5) = 12.625$ **64.** (c) $f(2.5) = -2.87$ **66.** 8.0 feet in $t = 0.7$ seconds
67. 39.4 feet in $t = 1.6$ seconds ***69.** (a) $P(T) = -20(T^2 - 58T + 100)$ (b) 29 tons
68. 608.9 feet in $t = 6.2$ seconds ***70.** (a) $R(x) = (528 - 150x)x$ (b) \$1.76
***71.** (a) $A(x) = 30x - x^2$ (b) 225 ft² when the dimensions are 15 ft by 15 ft
***72.** $A(x) = x(100 - 2x)$; the maximum possible area is 1250 ft² and occurs when the dimensions are 25 ft by 50 ft.

---

*denotes Superset exercises

*73. (a) $P(x) = -200x^2 + 15040x - 191160$ (b) \$37.60 *75. $V(x) = -12x^2 + 1008x$ (b) 42 inches each
*74. $R(x) = (11800 - 200x)x$; the price of \$29.50 will maximize revenues. This price is different from the price found in Exercise 73 to maximize profit. *76. (a) $S(x) = 2016 + 168x - 2x^2$ (b) 42 inches each
*77. $\pm 2\sqrt{6}$ *79. $\frac{125}{9}$
*78. $C = -35$; other x-intercept is $-\frac{7}{2}$

Section 3.6, pages 181–182

1. 2 3. -16 5. undefined 7. $\frac{7}{5}$ 9. 16 11. $-a$
2. 4 4. -9 6. -1 8. undefined 10. 12 12. $3a^2 - a - 8$
13. (a) $(f+g)(x) = 5x - 6$; domain is $(-\infty, \infty)$. (b) $(f-g)(x) = x + 8$; domain is $(-\infty, \infty)$.
(c) $(fg)(x) = 6x^2 - 19x - 7$; domain is $(-\infty, \infty)$. (d) $\left(\frac{f}{g}\right)(x) = \frac{3x+1}{2x-7}$; domain is $\left(-\infty, \frac{7}{2}\right) \cup \left(\frac{7}{2}, \infty\right)$.
14. (a) $(f+g)(x) = 7x + 1$; domain is $(-\infty, \infty)$. (b) $(f-g)(x) = 3x - 5$; domain is $(-\infty, \infty)$.
(c) $(fg)(x) = 10x^2 + 11x - 6$; domain is $(-\infty, \infty)$. (d) $\left(\frac{f}{g}\right)(x) = \frac{5x-2}{2x+3}$; domain is $\left(-\infty, -\frac{3}{2}\right) \cup \left(-\frac{3}{2}, \infty\right)$.
15. (a) $(f+g)(x) = 4x$; domain is $(-\infty, \infty)$. (b) $(f-g)(x) = 6x - 2$; domain is $(-\infty, \infty)$.
(c) $(fg)(x) = -5x^2 + 6x - 1$; domain is $(-\infty, \infty)$. (d) $\left(\frac{f}{g}\right)(x) = \frac{5x-1}{1-x}$; domain is $(-\infty, 1) \cup (1, \infty)$.
16. (a) $(f+g)(x) = x + 3$; domain is $(-\infty, \infty)$. (b) $(f-g)(x) = 5x + 1$; domain is $(-\infty, \infty)$.
(c) $(fg)(x) = -6x^2 - x + 2$; domain is $(-\infty, \infty)$. (d) $\left(\frac{f}{g}\right)(x) = \frac{3x+2}{1-2x}$; domain is $\left(-\infty, \frac{1}{2}\right) \cup \left(\frac{1}{2}, \infty\right)$.
17. (a) $(f+g)(x) = x^2 + x + 5$; domain is $(-\infty, \infty)$. (b) $(f-g)(x) = x^2 - x - 5$; domain is $(-\infty, \infty)$.
(c) $(fg)(x) = x^3 + 5x^2$; domain is $(-\infty, \infty)$. (d) $\left(\frac{f}{g}\right)(x) = \frac{x^2}{x+5}$; domain is $(-\infty, -5) \cup (-5, \infty)$.
18. (a) $(f+g)(x) = x^2 - x + 4$; domain is $(-\infty, \infty)$. (b) $(f-g)(x) = x^2 + x$; domain is $(-\infty, \infty)$.
(c) $(fg)(x) = -x^3 + 2x^2 - 2x + 4$; domain is $(-\infty, \infty)$. (d) $\left(\frac{f}{g}\right)(x) = \frac{x^2+2}{2-x}$; domain is $(-\infty, 2) \cup (2, \infty)$.
19. (a) $(f+g)(x) = \sqrt{x} + x - 1$; domain is $[0, \infty)$. (b) $(f-g)(x) = \sqrt{x} - x + 1$; domain is $[0, \infty)$.
(c) $(fg)(x) = x\sqrt{x} - \sqrt{x}$; domain is $[0, \infty)$. (d) $\left(\frac{f}{g}\right)(x) = \frac{\sqrt{x}}{x-1}$; domain is $[0, 1) \cup (1, \infty)$.
20. (a) $(f+g)(x) = \frac{1+x^3}{x}$; domain is $(-\infty, 0) \cup (0, \infty)$. (b) $(f-g)(x) = \frac{1-x^3}{x}$; domain is $(-\infty, 0) \cup (0, \infty)$.
(c) $(fg)(x) = x$; domain is $(-\infty, 0) \cup (0, \infty)$. (d) $\left(\frac{f}{g}\right)(x) = \frac{1}{x^3}$; domain is $(-\infty, 0) \cup (0, \infty)$.
21. (a) $(f+g)(x) = 2x^2 - 3x + 3$; domain is $(-\infty, \infty)$. (b) $(f-g)(x) = -3x + 1$; domain is $(-\infty, \infty)$.
(c) $(fg)(x) = x^4 - 3x^3 + 3x^2 - 3x + 2$; domain is $(-\infty, \infty)$. (d) $\left(\frac{f}{g}\right)(x) = \frac{x^2-3x+2}{x^2+1}$; domain is $(-\infty, \infty)$.
22. (a) $(f+g)(x) = 2x^2 - x - 3$; domain is $(-\infty, \infty)$. (b) $(f-g)(x) = 5 - x$; domain is $(-\infty, \infty)$.
(c) $(fg)(x) = x^4 - x^3 - 3x^2 + 4x - 4$; domain is $(-\infty, \infty)$.
(d) $\left(\frac{f}{g}\right)(x) = \frac{x^2-x+1}{x^2-4}$; domain is $(-\infty, -2) \cup (-2, 2) \cup (2, \infty)$.

*denotes Superset exercises

23. (a) $(f+g)(x)=\dfrac{2x-2}{x+2}$; domain is $(-\infty,-2)\cup(-2,\infty)$. (b) $(f-g)(x)=\dfrac{-2}{x+2}$; domain is $(-\infty,-2)\cup(-2,\infty)$.
(c) $(fg)(x)=\dfrac{x^2-2x}{(x+2)^2}$; domain is $(-\infty,-2)\cup(-2,\infty)$. (d) $\left(\dfrac{f}{g}\right)(x)=\dfrac{x-2}{x}$; domain is $(-\infty,-2)\cup(-2,0)\cup(0,\infty)$.

24. (a) $(f+g)(x)=\dfrac{x^3-3x^2-8x+27}{x-3}$; domain is $(-\infty,3)\cup(3,\infty)$.
(b) $(f-g)(x)=\dfrac{-x^3+3x^2+10x-27}{x-3}$; domain is $(-\infty,3)\cup(3,\infty)$.
(c) $(fg)(x)=x^2+3x$; domain is $(-\infty,3)\cup(3,\infty)$.
(d) $\left(\dfrac{f}{g}\right)(x)=\dfrac{x}{(x-3)(x^2-9)}$; domain is $(-\infty,-3)\cup(-3,3)\cup(3,\infty)$.

25. 3 **27.** 5 **29.** $-\frac{1}{3}$ **31.** 23 **33.** 23 **35.** 3 **37.** $-\frac{199}{25}$ **39.** $-\frac{31}{4}$

26. 1 **28.** -1 **30.** $\frac{1}{3}$ **32.** 3 **34.** 3 **36.** 23 **38.** $-\frac{71}{9}$ **40.** $-\frac{1}{12}$

41. not defined **43.** 0 **45.** $-\frac{2}{9}$ **47.** 8 **49.** not defined

42. $-\frac{1}{11}$ **44.** not defined **46.** 56 **48.** not defined **50.** -8

51. $(f\circ g)(x)=x^2-6x+10$; domain: all real numbers **53.** $(f\circ g)(x)=\left|\dfrac{1}{x-3}\right|$; domain: all real numbers except 3

52. $(f\circ g)(x)=2x^2+3$; domain: all real numbers **54.** $(f\circ g)(x)=\dfrac{1}{|x+1|+3}$; domain: all real numbers

55. $(f\circ g)(x)=1-\sqrt{x-5}$; domain: $[5,\infty)$ **57.** $(f\circ g)(x)=\dfrac{1}{x^2+9}$; domain: all real numbers

56. $(f\circ g)(x)=\sqrt{4-x}$; domain: $(-\infty,4]$. **58.** $(f\circ g)(x)=\dfrac{1}{x^2+4}$; domain: all real numbers

59. $(f\circ g)(x)=\dfrac{1}{\sqrt{x}}$; domain: $(0,\infty)$ **61.** $F(x)=(f\circ j)(x)$ **63.** $F(x)=(h\circ k)(x)$

60. $(f\circ g)(x)=\dfrac{1}{\sqrt{x-3}+1}$; domain: $[3,\infty)$. **62.** $F(x)=(k\circ j)(x)$ **64.** $F(x)=(h\circ f)(x)$

65. $F(x)=(k\circ f\circ j)(x)$ **67.** $F(x)=(k\circ f\circ f)(x)$ ***69.** $y=-5-15t$ ***71.** 40 seconds

66. $F(x)=(h\circ k\circ f)(x)$ **68.** $F(x)=(f\circ g\circ g)(x)$ ***70.** $y=17/6-x/6$ ***72.** $(f\circ g)(x)=3x-16$; $[4,5]$

***73.** $(f\circ g)(x)=2x+3$; $[-1,1]$ ***75.** $(f\circ g)(x)=\sqrt{x+3}$; $[-3,0]$

***74.** $(f\circ g)(x)=\sqrt{x-4}$; $[4,10]$ ***76.** $(f\circ g)(x)=-x$; $[-100,0]$

***77.** $(f\circ g)(-x)=f(g(-x))=f(-g(x))=-f(g(x))=-(f\circ g)(x)$

***78.** Suppose f is odd and g is even. Then $(f\circ g)(-x)=f(g(-x))=f(g(x))=(f\circ g)(x)$. Thus, $f\circ g$ is even. Similarly, $g\circ f$ is even.

Section 3.7, pages 188–189

1. one-to-one **3.** one-to-one **5.** not one-to-one **7.** not one-to-one **9.** one-to-one **11.** No

2. one-to-one **4.** not one-to-one **6.** one-to-one **8.** one-to-one **10.** not one-to-one **12.** No

---

*denotes Superset exercises

13. Yes

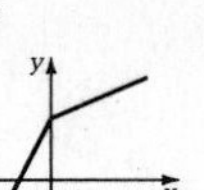

15. No **17.** Yes

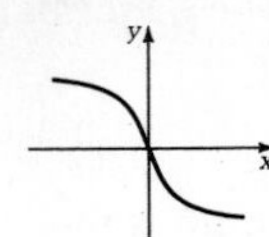

19. Yes

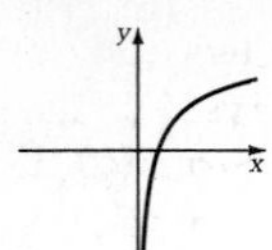

21. Yes

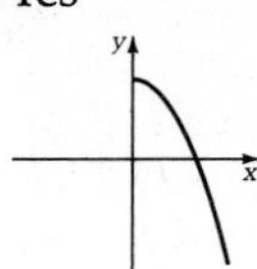

14. Yes

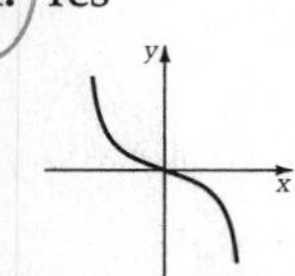

16. Yes

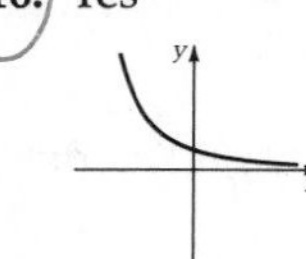

18. Yes

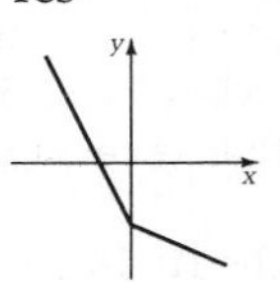

20. No **22.** Yes

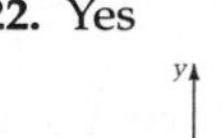

23. $f^{-1}(x) = \frac{6-x}{2}$, domain of $f^{-1} = [0, 4]$, range of $f^{-1} = [1, 3]$

24. $f^{-1}(x) = x^2 - 3$, domain of $f^{-1} = [0, \infty)$, range of $f^{-1} = [-3, \infty)$

25. $f^{-1}(x) = \sqrt{x}$, domain of $f^{-1} = [0, 9]$, range of $f^{-1} = [0, 3]$

26. $f^{-1}(x) = \frac{x-7}{4}$, domain of $f^{-1} = [-1, 11]$, range of $f^{-1} = [-2, 1]$

27. $f^{-1}(x) = 4 - x^2$, domain of $f^{-1} = [0, \infty)$, range of $f^{-1} = (-\infty, 4]$

28. $f^{-1}(x) = 2 - \sqrt{x}$, domain of $f^{-1} = [0, 9]$, range of $f^{-1} = [-1, 2]$

29. $f^{-1}(x) = 1 + \frac{1}{2}\sqrt[3]{x}$, domain of $f^{-1} = (-8, 64)$, range of $f^{-1} = (0, 3)$

30. $f^{-1}(x) = -\frac{1}{2}\sqrt[4]{x}$, domain of $f^{-1} = [0, 16]$, range of $f^{-1} = [-1, 0]$

31. $f^{-1}(x) = (4 - x)^2$, domain of $f^{-1} = [1, 3)$, range of $f^{-1} = (1, 9]$

32. $f^{-1}(x) = (x - 3)^{3/2}$, domain of $f^{-1} = (3, 4]$, range of $f^{-1} = [-1, 0)$

33. $f^{-1}(x) = (x + 1)^{3/2}$, domain of $f^{-1} = [0, 3)$, range of $f^{-1}[1, 8)$

34. $f^{-1}(x) = (x + 2)^2$, domain of $f^{-1} = (0, 1]$, range of $f^{-1} = (4, 9]$

35. $f^{-1}(x) = -\sqrt{7 - x}$, domain of $f^{-1} = [-2, 6]$, range of $f^{-1} = [-3, -1]$

36. $f^{-1}(x) = 3 - \sqrt[3]{x}$, domain of $f^{-1} = (0, 8)$, range of $f^{-1} = (1, 3)$

37. $f^{-1}(x) = 2 - x$, domain of $f^{-1} = [1, 5]$, range of $f^{-1} = [-3, 1]$ **39.** $f(1) = 1 = f(2)$

38. $f(-2) = 1 = f(0)$ **40.** $f^{-1}(x) = \frac{8-x}{3}$, domain of $f^{-1} = [2, 8]$, range of $f^{-1} = [0, 2]$

41. $f^{-1}(x) = \frac{x+2}{1-2x}$, domain of $f^{-1} = \left[-2, \frac{1}{7}\right)$, range of $f^{-1} = [0, 3)$

42. $f^{-1}(x) = \frac{x-2}{2x-3}$, domain of $f^{-1} = \left(1, \frac{10}{7}\right]$, range of $f^{-1} = (1, 4]$

43. $f(0) = 0 = f(4)$

44. $f^{-1}(x) = 3 - \sqrt{9 - x}$, domain of $f^{-1} = [0, 8)$, range of $f^{-1} = [0, 2)$

45. $f^{-1}(x) = -4 + \sqrt{x + 9}$, domain of $f^{-1} = [-5, 7]$, range of $f^{-1} = [-2, 0]$ ***47.** $A \le -8$ or $A \ge 2$

46. $f(-2) = -3 = f(0)$ ***48.** $-3 \le A \le 2$

***49.** $-3 \le A \le 2$ ***51.** $A \le -3$ or $A \ge 6$ ***53.** an interval of the form $\left(a, \frac{1}{a}\right)$ where $0 < a < 1$ or $a < -1$

***50.** $A \le -2$ or $A \ge 4$ ***52.** $1 \le A \le \frac{5}{2}$

***54.** Yes. Two examples are $f(x) = x$ on any open interval and $f(x) = -x$ on any open interval $(-a, a)$.

---

*denotes Superset exercises

Section 3.8, pages 195–201

1. (a) 2 inches (b) 3 inches (c) 8 am Tuesday (d) 4 inches (e) 3 am Tuesday
2. (a) days 35–42 (b) days 28–35 (c) 21 days (d) 28 days (e) day 55
3. (a) 3 gallons (b) 2.5 gallons (c) either 10 mph or 26 mph (d) 12.5 gallons (e) The speed of 20 mph is more economical than 50 mph; the speed of 50 mph is more economical than 30 mph.
4. (a) 750 feet (b) during (20, 60) and (90, 110) (c) 41 minutes (d) during (41, 75) and (99, 120) (e) 80 minutes
5. (a) $t = 7$ hours (b) during (8, 14) and (20, 27) (c) during (0, 11) and (17.5, 23) (d) 23 hours
6. (a) 40 calories (b) 8 minutes (c) 11.5 minutes (d) during (10, 15) (e) during (15, 20)
7. (a) 8 pounds (b) 4 months (c) months 3–6 (d) yes, months 0–3 by 62.5% (e) Since the actual gain was the same during each period, the percentage gain was greater during the 9–12 month span when the child's weight was less.
8. (a) 19 inches (b) 11 months (c) months 0–3 (d) yes, months 12–18 by only 6.5% (e) Since the actual increase was the same during each period, the percentage gain was greater during the 0–3 month span when the child's length was less.
9. (a) 18 cm (b) at $t = 2$, 10, and 14 seconds (c) during (0, 6) and (12, 18) (d) at $t = 6$ and 18 seconds
10. (a) June (b) March 11 (c) 120 days (d) 247 minutes
11. (a) 22 miles (b) 8 miles (c) 8 mph (d) during minutes 90–120 (e) $2\frac{3}{4}$ hours
12. (a) 2 feet (b) 4 feet (c) grew only 3 feet in third year (d) yes, a height of 26 feet (e) yes, 4 feet in each of seventh and eighth years
13. (a) plane B (b) plane C (c) 3 pm (d) 9:15 pm (e) from 1 pm to 2 pm, and from 9:30 pm to 11 pm
14. (a) plane A (b) plane A (c) 4:15 pm (d) 2:15 pm (e) from 2 pm to 5 pm, and from 7:15 pm to 9:30 pm
15. (a) planes A and C (b) plane B (c) 5 pm (d) plane B (e) from 1 pm to 9:15 pm
16. (a) plane C (b) plane A (c) 7 pm (d) plane C (e) from 2:15 pm to 11 pm
17. (a) A (b) C (c) A (d) C (e) A
18. (a) 105° (b) from 8 am to 11 am (c) 13 hours (d) approx. 28 points (e) from 11 am to 1 pm, and from 7 pm to midnight

*19. (a) B (b) A (c) C (d) A *20. Graph F *21. Graph G *22. Graph C *23. Graph E *24. Graph C *25. Graph B *26. Graph H *27. Graph A *28. Graph C

*29.
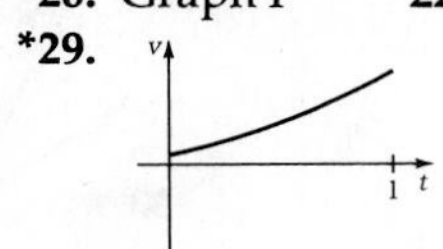

*31.
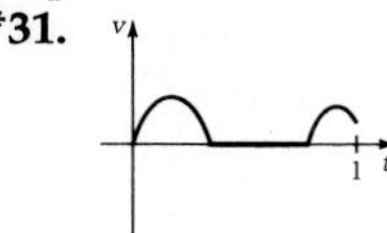

*33.
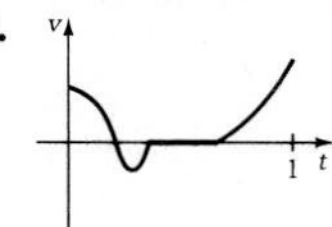

*35.
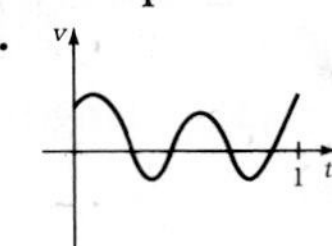

*30.
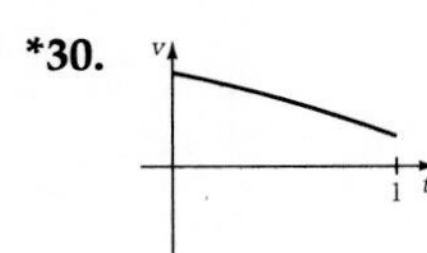

*32.
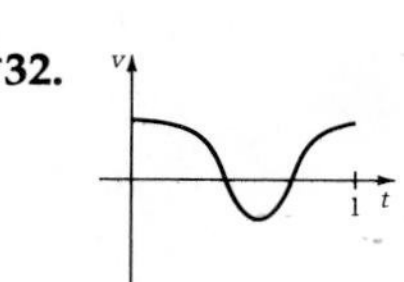

*34.
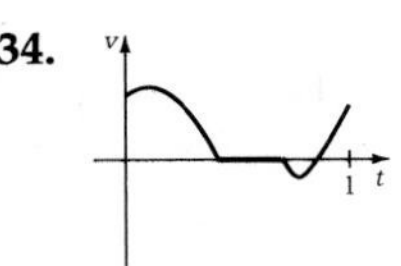

*36.
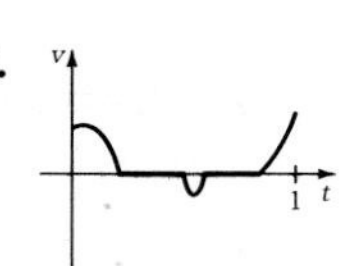

Section 3.9, pages 204–206

1. $y = kx^2$ 2. $x = \frac{k}{y^3}$ 3. $F = \frac{k}{R^{3/2}}$ 4. $D = kC^2$ 5. $m = kpq^2$ 6. $n = \frac{kp^2}{qs}$ 7. $a = \frac{kb^2}{c^3}$ 8. $z = \frac{k}{xt^{1/3}}$ 9. $y = \frac{kx^2z^3}{w}$ 10. $p = kr^2s$ 11. $k = 24, y = \frac{24}{5}$ 12. $k = \frac{9}{4}, x = \frac{81}{4}$ 13. $k = 2, m = 24$ 14. $k = 72, z = 4$

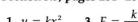

*denotes Superset exercises

15. $k = \frac{3}{2}, a = \frac{8}{3}$ **17.** $k = \frac{2}{3}, x = \frac{4}{9}$ **19.** 1600 feet **21.** 4410 centimeters **23.** 45 pounds **25.** $\frac{25}{12}$ inches

16. $k = \frac{1}{3}, m = \frac{1}{4}$ **18.** $k = \frac{2}{3}, w = \frac{16}{9}$ **20.** 1960 meters **22.** 1728 inches **24.** 15 pounds **26.** $\frac{25}{3}$ inches

***27.** $y = \frac{2}{3}x$ ***29.** $y = -2x^3$ ***31.** $y = \frac{12}{x}$ ***33.** $z = 2xy$ ***35.** $z = 3x^2y$ ***37.** $z = \frac{8x}{y}$

***28.** $y = \frac{1}{2}x^2$ ***30.** $y = -\frac{18}{x}$ ***32.** $y = 5x$ ***34.** $z = -xy$ ***36.** $z = \frac{6x}{y}$ ***38.** $z = 4xy^2$

***39.** $k \approx 1.7, D \approx 59.2$ ***41.** $k \approx 0.4, p \approx 1.9$ ***43.** $k \approx 42.4, z \approx 3.0$ ***45.** $k \approx -2.8, n \approx -82.2$ ***47.** 29.3%

***40.** $k \approx 0.7, a \approx 0.3$ ***42.** $k \approx 2286.3, F \approx 204.5$ ***44.** $k \approx -2.1, m \approx -45.5$ ***46.** $k \approx 0.3, y \approx 132.7$ ***48.** 600%

***49.** 351.3 cm. ***51.** 15.8 ft/sec.

***50.** 0.61 earth years ***52.** 9.1%

Review Exercises, pages 212–214

1. origin only **3.** x-axis only **5.** $\sqrt{85}$; midpoint $\left(\frac{5}{2}, 2\right)$

2. all three **4.** y-axis only **6.** $\sqrt{53}$; midpoint $\left(5, -\frac{3}{2}\right)$

7. $f(-3) = -\frac{1}{5}, f(0) = -\frac{1}{2}, f(2)$ is not defined, $f(a) = \frac{1}{a-2}, f(3+h) = \frac{1}{h+1}$

8. $f(-3) = 0, f(0) = 9, f(2) = 5, f(a) = 9 - a^2, f(3+h) = -6h - h^2$

9. $f(-3) = 4, f(0) = 4, f(2) = 4, f(a) = 4, f(3+h) = 4$ **11.** yes **13.** yes **15.** yes

10. $f(-3) = 2, f(0) = 1, f(2)$ is not a real number, $f(a) = \sqrt{1-a}, f(3+h) = \sqrt{-2-h}$ **12.** no **14.** yes **16.** no

17. yes **19.** $[-2, 2]$ **21.** $(-\infty, \infty)$ **23.**

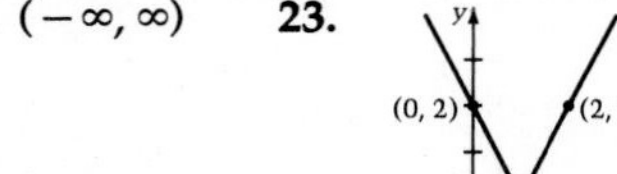

25.

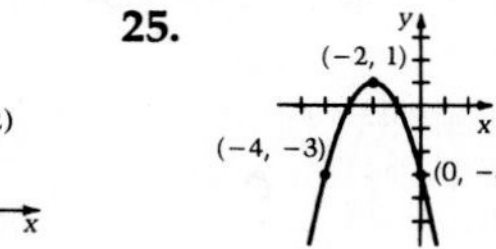

18. no **20.** $(-3, 3)$ **22.** $(-\infty, -4) \cup (-4, -1) \cup (-1, \infty)$ **24.**

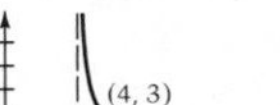

26.

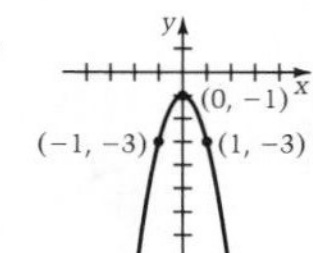

27.

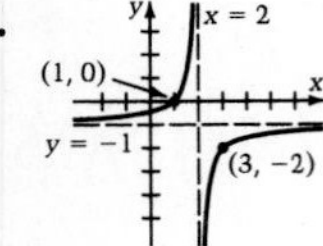

29. (a) $y - 1 = -3(x - (-2))$ (b) $y = -3x - 5$ (c) $3x + y + 5 = 0$

28. (−1, 4) (−5, 0) (3, 0)

30. (a) $y - (-4) = 0\,(x - 3)$ (b) $y = 4$ (c) $y - 4 = 0$

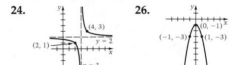

*denotes Superset exercises

31. (a)–(b) cannot be written (c) $x - 3 = 0$

32. (a) $y - (-2) = \frac{1}{2}(x - 0)$ (b) $y = \frac{1}{2}x - 2$ (c) $\frac{1}{2}x - y - 2 = 0$

33. (a) $y - (-4) = \frac{2}{3}(x - (-1))$ (b) $y = \frac{2}{3}x - \frac{10}{3}$ (c) $\frac{2}{3}x - y - \frac{10}{3} = 0$

34. (a) $y - 3 = -\frac{4}{3}(x - 2)$ (b) $y = -\frac{4}{3}x + \frac{17}{3}$ (c) $\frac{4}{3}x + y - \frac{17}{3} = 0$

35. (a)–(b) cannot be written (c) $x - 6 = 0$ **37.** $y = 2(x - 1)^2 + (-3)$ **39.** $y = -(x - 2)^2 + 11$

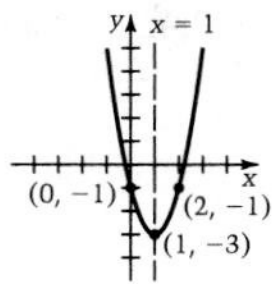

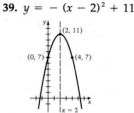

36. (a) $y - 5 = 0(x - 3)$ (b) $y = 5$ (c) $y - 5 = 0$ **38.** $y = -(x + 2)^2 + 4$ **40.** $y = (x + 2)^2 + 3$

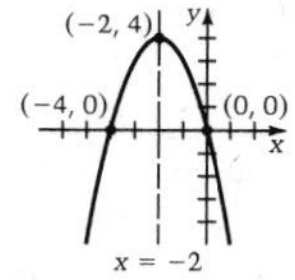

41. $y = 2(x + 1)^2 + 4$ **43.** 7 **45.** 3 **47.** 41 **49.** 19 **51.** 14 **53.** 15 **55.** one-to-one

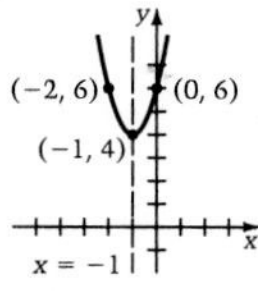

42. $y = -2(x + 2)^2 + 18$ **44.** 3 **46.** 5 **48.** 11 **50.** $\sqrt{14}$ **52.** 7 **54.** 905 **56.** not one-to-one

57. not one-to-one **59.** $f^{-1}(x) = \frac{7 - x}{3}$, domain of $f^{-1} = (-5, 4]$, range of $f^{-1} = [1, 4)$

58. one-to-one **60.** $f^{-1}(x) = (3 - x)^3$, domain of $f^{-1} = [1, 4)$, range of $f^{-1} = (-1, 8]$

61. $f^{-1}(x) = \frac{1}{4}(x - 6)^2$, domain of $f^{-1} = (8, 10]$, range of $f^{-1} = (1, 4]$

62. $f^{-1}(x) = \sqrt{x} - 3$, domain of $f^{-1} = [4, 25]$, range of $f^{-1} = [-1, 2]$

63. $f^{-1}(x) = 2 - \sqrt{x + 4}$, domain of $f^{-1} = [-3, 12]$, range of $f^{-1} = [-2, 1]$

64. $f^{-1}(x) = 4 - x$, domain of $f^{-1} = (1, 6]$, range of $f^{-1} = [-2, 3)$

65. (a) 80 beats/min (b) 15 minutes (c) 40 beats/min (d) during (5, 15) and (25, 30) **67.** $k = \frac{4}{9}, y = \frac{16}{9}$

66. (a) 3rd floor (b) 8 times (c) 8 minutes (d) 6 times **68.** $k = 21, m = \frac{21}{4}$

69. $k = 384, z = \frac{8}{9}$. **71.**

70. $k = \frac{1}{3}, p = 36$

72. $y = 2x - 7$. The ant's trip is along the ray $y = 2x - 7$ for $x \geq 2$. At time $t = 0$, the ant stops momentarily at the point $(2, -3)$, and then for $t > 0$ retraces its steps for $t < 0$.

73. Use the fact that the diagonals of a square bisect one another at right angles. The other three vertices are $(-1, 0)$, $(4, -3)$, and $(7,2)$. **75.** $y = \frac{1}{2}(x - 4)^2 + \frac{5}{2}$

74. The diagonal is shortest when the rectangle is a square 10 units on a side. **76.** 3

Chapter Test, pages 215–216

1.

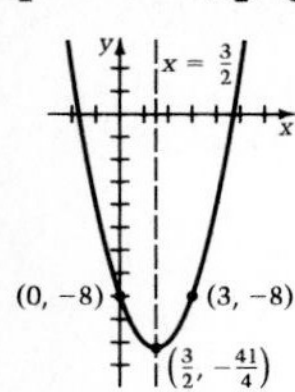

3. (a)

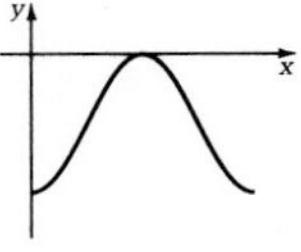

(b)

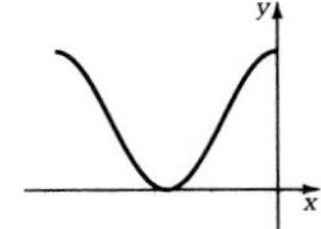

(c)

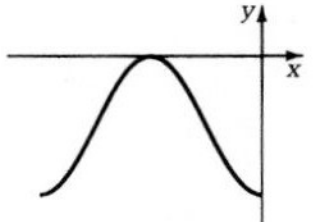

5. origin only

2.

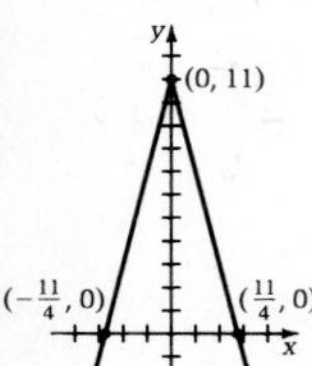

4. all three **6.** $f(-2) = 15, f(0) = 9, f(h^2) = -h^4 - 5h^2 + 9, f(b) - 1) = -b^2 - 3b + 13$

7.

9.

8.

10. (a)

(b)

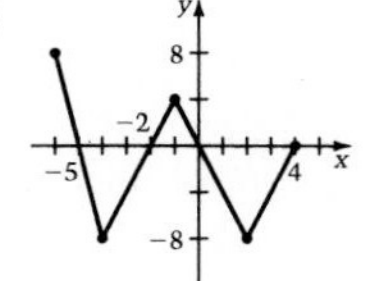

(c)

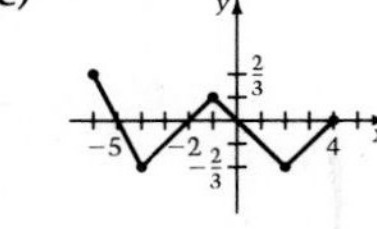

11. $m = \dfrac{5}{2}$

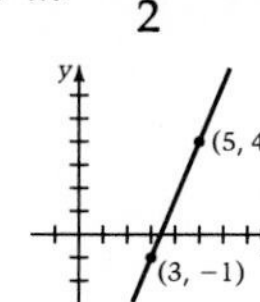

13.

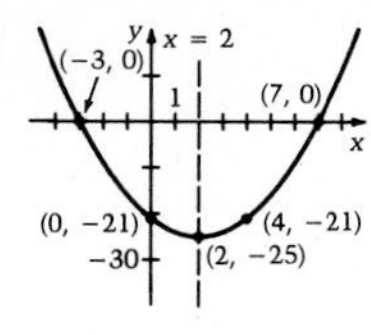

15. $A = \dfrac{kb^2}{c}$ **17.** 3

12. (a)–(b) cannot be written (c) $x - 6 = 0$ **14.** (a) exhaling (b) 12 seconds (c) 0.6 liters **16.** 162 **18.** 1

19. $(f \circ g)(x) = \dfrac{1}{2 - \sqrt{x + 6}}$; $[-6, -2) \cup (-2, \infty)$ **21.** $f^{-1}(x) = (x - 3)^2$, domain of $f^{-1} = [4, 6)$, range of $f^{-1} = [1, 9)$

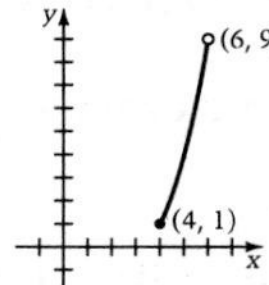

20. $F(x) = (f \circ h \circ g)(x)$

Chapter 4

Section 4.1, pages 224–225

1. yes; $f(x) = -x^3 + 7x + 2$; $-x^3$ **3.** no **5.** yes; $H(x) = -2x^5 + 3x^3 + 9x^2 + 1$; $-2x^5$
2. yes; $g(x) = -5x^4 + 2x^2 + 9x - 4$; $-5x^4$ **4.** yes; $f(x) = 3x^3 - 10x^2 + 5x$; $3x^3$ **6.** no
7. no **9.** yes; $f(x) = 9x^2 + 13x + 11$; $9x^2$ **11.** yes; $k(x) = -\sqrt{5}x^4 + 3x^3 + 7x - 3$; $-\sqrt{5}x^4$ **13.** Type III
8. yes; $F(x) = -x^4 + x^2 - x + 1$; $-x^4$ **10.** yes; $g(x) = -25x^2 - 22x + 2$; $-25x^2$ **12.** no **14.** Type II
15. Type II **17.** Type I **19.** Type II **21.** Type III **23.** Type IV **25.** Type IV **27.** Type I **29.** No. 2
16. Type IV **18.** Type III **20.** Type I **22.** Type IV **24.** Type I **26.** Type II **28.** Type III **30.** No. 3
31. No. 1 **33.** No. 1 **35.** No. 2 ***37.**

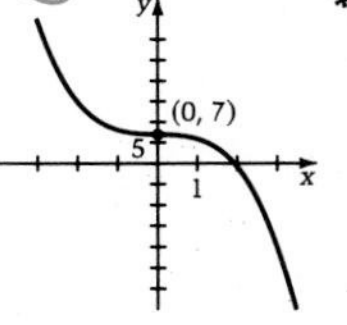

***39.**

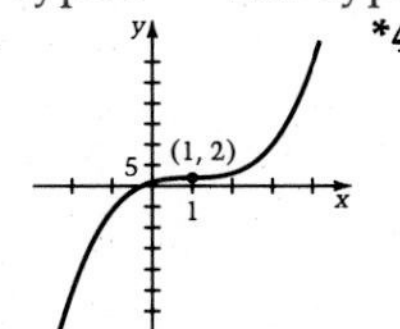

***41.**

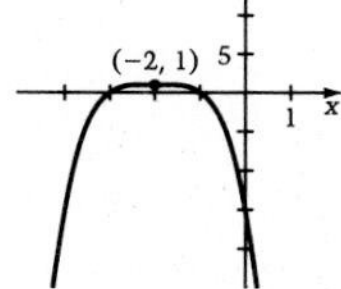

32. No. 4 **34.** No. 3

36. No. 4 ***38.** ***40.** ***42.**

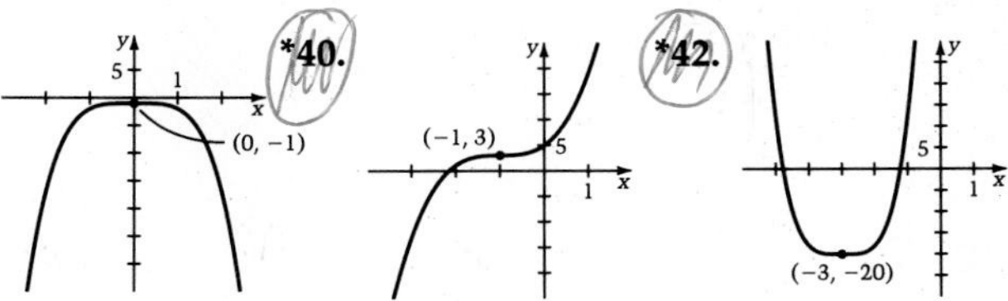

***43.**

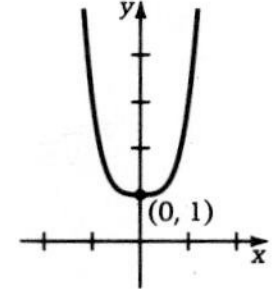

***45.**

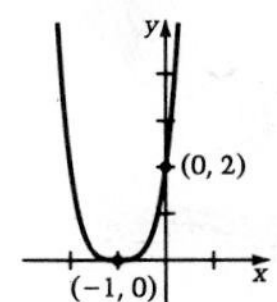

***47.** No; graph has a sharp corner ***49.** Yes

---

*denotes Superset exercises

***44.**

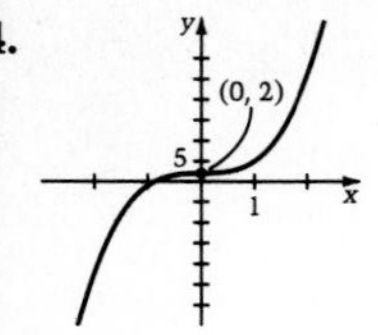

***46.**

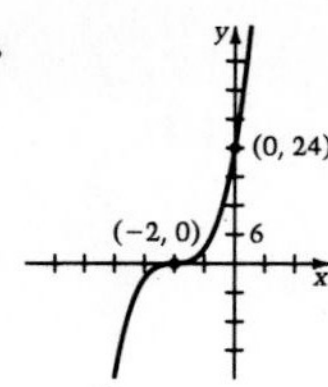

***48.** No; graph has a jump ***50.** No; graph has a sharp corner

***51.** No; y-values approach 0 as x approaches $-\infty$ ***53.** No. 2 ***55.** No. 1
***52.** Yes ***54.** No. 3 ***56.** No. 1

Section 4.2, pages 234–235

1. Graph No. 5 **3.** Graph No. 6 **5.** Graph No. 6 **7.** Graph No. 3 **9.** $-3, 5, -7$ **11.** $0, \frac{2}{3}, -2, -1$

2. Graph No. 4 **4.** Graph No. 1 **6.** Graph No. 2 **8.** Graph No. 5 **10.** $0, -2, \frac{1}{3}$ **12.** $0, 2, -3, -1$

13. $5, -\frac{5}{2}$ **15.** $-2, -\sqrt{3}, \sqrt{3}, 2$ **17.**

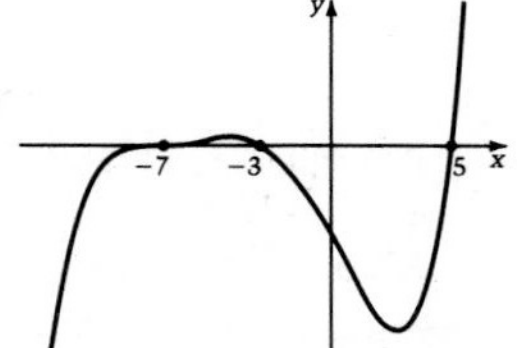

19.

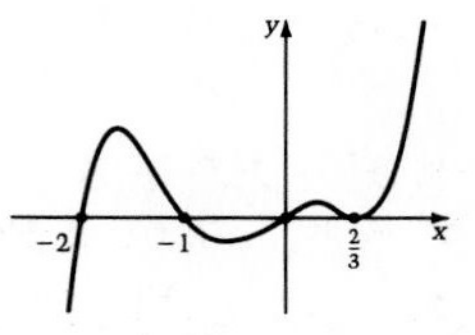

21.

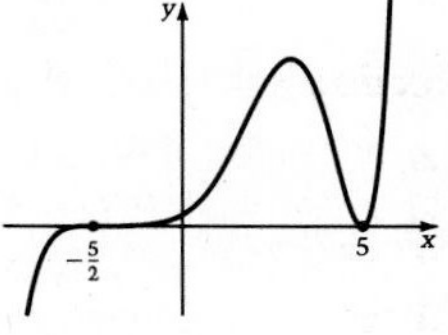

14. $0, \frac{7}{2}$ **16.** $-3, -\sqrt{2}, \sqrt{2}, 3$ **18.**

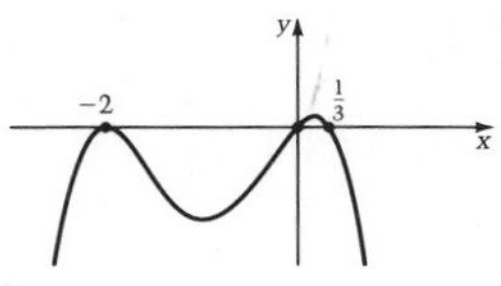

20.

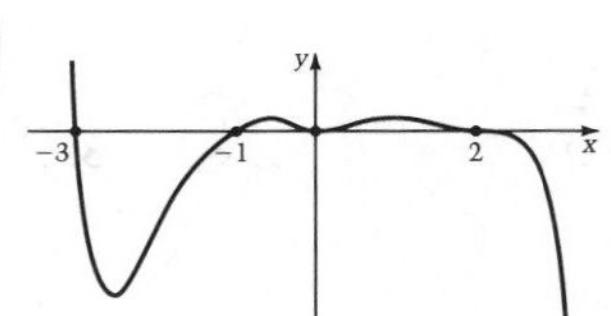

22.

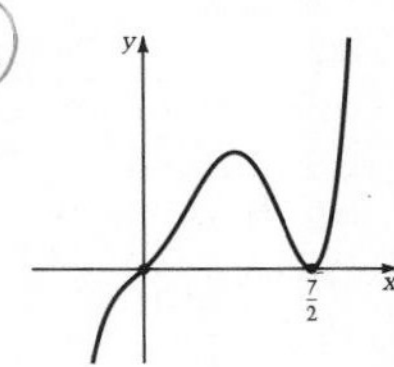

23.

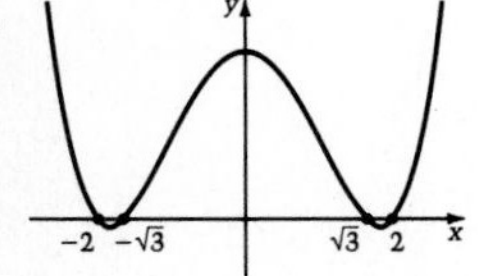

25.

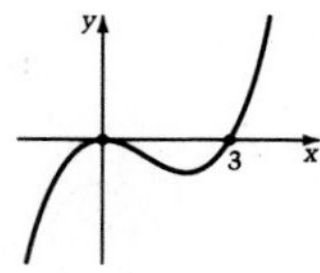

27.

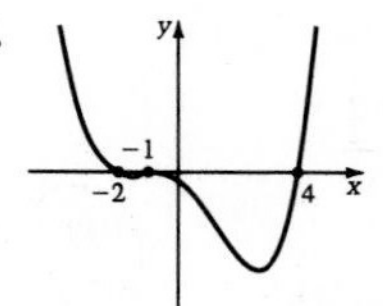

29.

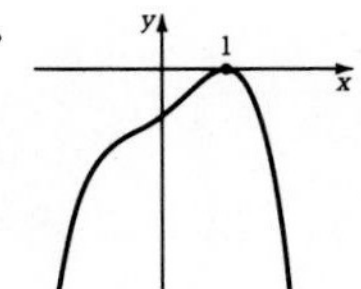

31.

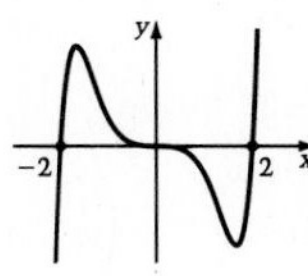

24.

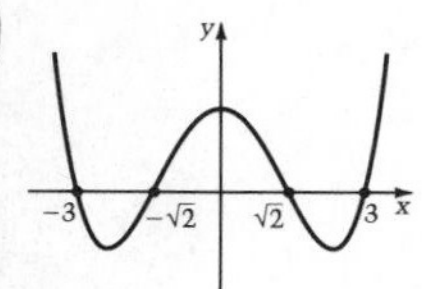

26.

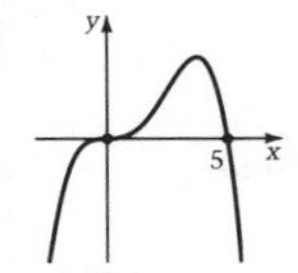

28.

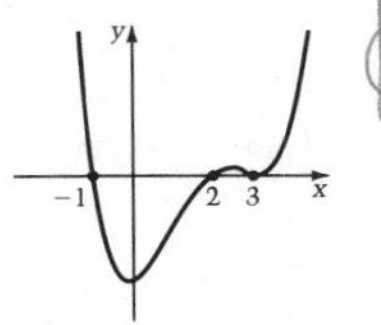

30.

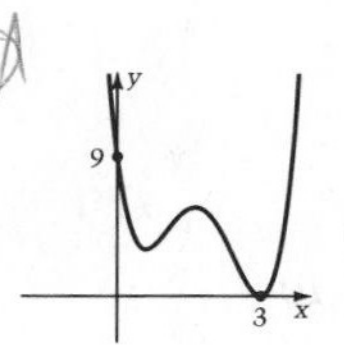

32.

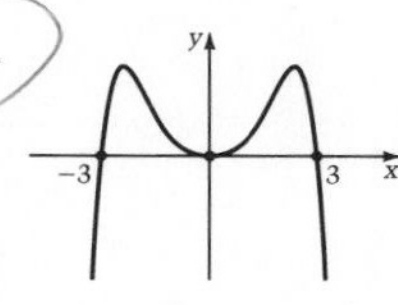

*denotes Superset exercises

33. 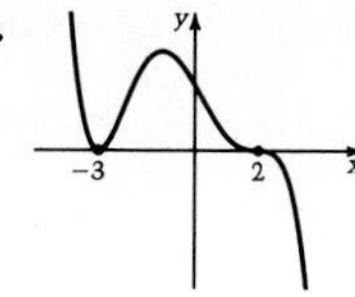

35. 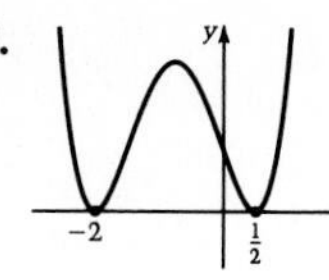

37. 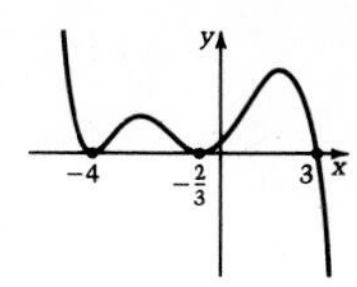

***39.** $f(x) = (x - 2)^2$ ***41.** $f(x) = -x^2(x - 2)$

34. 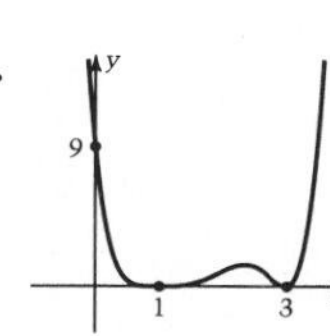

36. 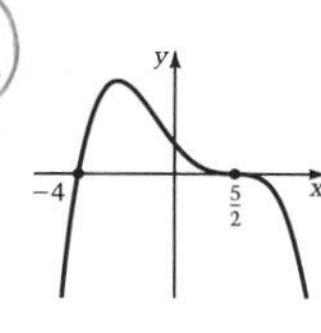

38. 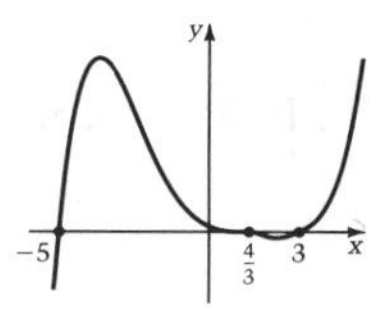

***40.** $f(x) = -x(x - 3)$ ***42.** $f(x) = -(x + 3)^2$

***43.** $f(x) = -x(x + 2)(x - 3)$ ***45.** $f(x) = x^2(x + 4)(x - 4)$ ***47.** Positive, $n = 3, m = 3$
***44.** $f(x) = x^2(x + 1)$ ***46.** $f(x) = x(x + 3)(x - 2)^2$ ***48.** Positive, $n = 6, m = 4$.
***49.** Negative, $n = 7, m = 7$
***50.** Negative, $n = 10, m = 6$.

Section 4.3, pages 242–243

1. $x^2 - 3x + 8$; Rem: -2 **3.** $x^3 + 2$; Rem: 5 **5.** $x^3 + 3x^2 + 3x - 2$; Rem: 0 **7.** $x - 3$; Rem: $2x + 6$
2. $x^2 - x + 2$; Rem: 5 **4.** $x^3 - 3x^2 + 2x - 2$; Rem: 0 **6.** $x^3 - 2x^2 - x + 5$; Rem: -3 **8.** x; Rem: $2x - 2$
9. $x^2 + 2x + 2$; Rem: 9 **11.** $x^4 - x^3 + x - 1$; Rem: 0 **13.** $x + 3$ **15.** $x + 2, x - 2$ **17.** $x - 2, x + 3$
10. $x^2 - 2x + 2$; Rem: 1 **12.** $x^3 + x^2$; Rem: 0 **14.** $x - 1$ **16.** $x - 1, x + 3$ **18.** $x - 1, x - 2$

19.

$$\begin{array}{r|rrrr} -1 & 2 & -3 & 6 & -7 \\ & & -2 & 5 & -11 \\ \hline & 2 & -5 & 11 & -18 \end{array}$$

Thus, $f(-1) = -18$.

$$\begin{array}{r|rrrr} 1 & 2 & -3 & 6 & -7 \\ & & 2 & -1 & 5 \\ \hline & 2 & -1 & 5 & -2 \end{array}$$

Thus, $f(1) = -2$.

20.

$$\begin{array}{r|rrrrr} -2 & 1 & -3 & 4 & -1 & -1 \\ & & -2 & 10 & -28 & 58 \\ \hline & 1 & -5 & 14 & -29 & 57 \end{array}$$

Thus, $g(-2) = 57$.

$$\begin{array}{r|rrrrr} 2 & 1 & -3 & 4 & -1 & -1 \\ & & 2 & -2 & 4 & 6 \\ \hline & 1 & -1 & 2 & 3 & 5 \end{array}$$

Thus, $g(2) = 5$.

21.

$$\begin{array}{r|rrrrrr} -2 & 1 & 0 & -3 & 1 & -4 & 10 \\ & & -2 & 4 & -2 & 2 & 4 \\ \hline & 1 & -2 & 1 & -1 & -2 & 14 \end{array}$$

Thus, $G(-2) = 14$.

$$\begin{array}{r|rrrrrr} 1 & 1 & 0 & -3 & 1 & -4 & 10 \\ & & 1 & 1 & -2 & -1 & -5 \\ \hline & 1 & 1 & -2 & -1 & -5 & 5 \end{array}$$

Thus, $G(1) = 5$.

---

*denotes Superset exercises

22. $$\begin{array}{r|rrrrrr} -1 & -10 & 0 & 8 & 0 & -1 & 1 \\ & & 10 & -10 & 2 & -2 & 3 \\ \hline & -10 & 10 & -2 & 2 & -3 & 4 \end{array}$$ Thus, $F(-1) = 4$.

$$\begin{array}{r|rrrrrr} 2 & -10 & 0 & 8 & 0 & -1 & 1 \\ & & -20 & -40 & -64 & -128 & -258 \\ \hline & -10 & -20 & -32 & -64 & -129 & -257 \end{array}$$ Thus, $F(2) = -257$.

23. $$\begin{array}{r|rrrrrr} -1 & -3 & 1 & 0 & -2 & -1 & 3 \\ & & 3 & -4 & 4 & -2 & 3 \\ \hline & -3 & 4 & -4 & 2 & -3 & 6 \end{array}$$ Thus, $g(-1) = 6$.

$$\begin{array}{r|rrrrrr} 1 & -3 & 1 & 0 & -2 & -1 & 3 \\ & & -3 & -2 & -2 & -4 & -5 \\ \hline & -3 & -2 & -2 & -4 & -5 & -2 \end{array}$$ Thus, $g(1) = -2$.

24. $$\begin{array}{r|rrrrr} -1 & -1 & 4 & 5 & 7 & 3 \\ & & 1 & -5 & 0 & -7 \\ \hline & -1 & 5 & 0 & 7 & -4 \end{array}$$ Thus, $f(-1) = -4$.

$$\begin{array}{r|rrrrr} 1 & -1 & 4 & 5 & 7 & 3 \\ & & -1 & 3 & 8 & 15 \\ \hline & -1 & 3 & 8 & 15 & 18 \end{array}$$ Thus, $f(1) = 18$.

25. $f(x) = (x + 5)(x + 1) - 3$ **27.** $f(x) = (x^2 + 2x + 4)(x - 2) + 16$ **29.** $f(x) = (x^2 - 4x + 3)(x - 2) + 0$
26. $f(x) = (x + 7)(x - 3) + 24$ **28.** $f(x) = (x^2 - 3x + 9)(x + 3) - 54$ **30.** $f(x) = (x^3 + 3x^2 - x - 3)(x - 3) - 4$
31. $f(x) = (x^3 - 2x^2 - x + 2)(x + 2) + 6$ **33.** $-4, 7$ **35.** $\frac{2}{3}, \frac{1}{4}$ **37.** $-2, -3 + \sqrt{7}, -3 - \sqrt{7}$ **39.** $2, -4, -5$
32. $f(x) = (x^2 + x)(x + 2) + 0$ **34.** $6, 5$ **36.** $-\frac{3}{2}, \frac{4}{5}$ **38.** $-3, -2 + \sqrt{3}, -2 - \sqrt{3}$ **40.** $2, -3, 6$
***41.** (a) any positive integer (b) no values (c) any even positive integer (d) any odd positive integer
***42.** $k = -\frac{9}{4}$ ***44.** $A = 0, B = 2$
***43.** $b = \pm \frac{1}{2}\sqrt{10}$

Section 4.4, pages 250–251

1. $\pm 1, \pm 2, \pm 3, \pm 6, \pm \frac{1}{2}, \pm \frac{3}{2}$ **3.** $\pm 1, \pm \frac{1}{2}, \pm \frac{1}{4}, \pm \frac{1}{8}$
2. $\pm 1, \pm 3, \pm \frac{1}{2}, \pm \frac{3}{2}, \pm \frac{1}{3}, \pm \frac{1}{4}, \pm \frac{3}{4}, \pm \frac{1}{6}, \pm \frac{1}{12}$ **4.** $\pm 1, \pm 2, \pm 4, \pm \frac{1}{2}, \pm \frac{1}{4}, \pm \frac{1}{8}$
5. $\pm 1, \pm 2, \pm 3, \pm 4, \pm 6, \pm 12, \pm \frac{1}{2}, \pm \frac{3}{2}, \pm \frac{1}{3}, \pm \frac{2}{3}, \pm \frac{4}{3}, \pm \frac{1}{6}, \pm \frac{1}{9}, \pm \frac{2}{9}, \pm \frac{4}{9}, \pm \frac{1}{18}$
6. $\pm 1, \pm \frac{1}{2}, \pm \frac{1}{3}, \pm \frac{1}{6}$
7. either two or no positive real zeros, and precisely one negative real zero
8. either two or no positive real zeros, and precisely one negative real zero

*denotes Superset exercises

9. precisely one positive real zero, and precisely one negative real zero
10. either two or no positive real zeros, and no negative real zeros
11. precisely one positive real zero, and either two or no negative real zeros
12. precisely one positive real zero, and precisely one negative real zero
13. either two or no positive real zeros, and no negative real zeros
14. precisely one positive real zero, and either two or no negative real zeros
15. $[-1, 5]$ 16. $[-3, 3]$ 17. $[-2, 1]$ 18. $[-4, 2]$ 19. $[-3, 1]$ 20. $[-3, 3]$

21. $-1, 2$ (multiplicity 2), 3 22. $-2, \sqrt{3}, -\sqrt{3}, -5$ 23. $3, \sqrt{5}, -\sqrt{5}, -4$ 24. $0, -6, \frac{1}{3}$ 25. -1 (of multiplicity 3) 26. $2, -\frac{1}{3}, -\frac{1}{4}$ 27. $2, -3, -\frac{5}{2} + \frac{1}{2}\sqrt{21}, -\frac{5}{2} - \frac{1}{2}\sqrt{21}$ 28. 2 (of multiplicity 3)

29. $-1, \frac{1}{2}, \frac{1}{3}$ 30. $1, -2$ (of multiplicity 2), 3 31. $-4, -\frac{1}{2}$ 32. $0, 4, -\frac{2}{5}, \frac{3}{2}$ 33. $0, -4, -\frac{5}{2}, \frac{2}{3}$ 34. $1, 4, \frac{3}{2} + \frac{1}{2}\sqrt{5}, \frac{3}{2} - \frac{1}{2}\sqrt{5}$ 35. $1, \frac{1}{3}, -\frac{1}{2}$ (of multiplicity 2) 36. $-1, \frac{1}{2}$ (of multiplicity 2), $\frac{1}{3}$

*37. Follow the suggested steps. *38. By the Factor Theorem, $f(x) = 2(x - r_1)(x - r_2) = 2x^2 - (2r + 2r_2)x + 2r_1r_2$. Thus, $b = -2(r_1 + r_2)$ and $c = 2r_1, r_2$ so that b and c are even integers. *39. no possible values of b *40. $f(x) = x^2 + \sqrt{2}$ *41. $f(x) = x^3 + 3x^2 + 3x + 1 = (x + 1)^3$ *42. A is a multiple of 42, C is a multiple of 10.

*43. $f(x) = \frac{1}{2}(x + 2)^2$ *44. $f(x) = -x(x - 3)$ *45. $f(x) = -2(x + 1)(x - 2)$ *46. $f(x) = \frac{1}{2}(x + 2)(x - 3)$ *47. $f(x) = \frac{1}{2}(x + 1)(x - 1)(x - 4)$ *48. $f(x) = -\frac{1}{2}(x - 1)^2(x - 4)$

*49. $f(x) = \frac{3}{2}(x + 2)^2(x - 1)$
*50. $f(x) = x(x + 3)(x - 2)$

Section 4.5, pages 257–258

1. $f(x) = \frac{3}{2}(x - i)(x + 2i)$ 2. $f(x) = \sqrt{2}(x - 1)(x - \sqrt{2})$ 3. $f(x) = -4i(x - 3)(x - i)$ 4. $f(x) = -\frac{1}{2}(x - i\sqrt{2})(x + 3i\sqrt{2})$ 5. $f(x) = -12\left(x - \frac{1}{2}\right)\left(x - \frac{1}{2}\right)(x + 1)$ 6. $f(x) = -10\left(x - \frac{1}{4}\right)\left(x + \frac{2}{5}\right)(x - 6)$

7. $f(x) = \sqrt{2}(x - 1)(x + 5i)(x - i\sqrt{2})$ 8. $f(x) = -9i(x + 2)(x + 2i)\left(x + \frac{1}{3}\right)$ 9. $-\frac{3}{2} + \frac{1}{2}i\sqrt{19}, -\frac{3}{2} - \frac{1}{2}i\sqrt{19}$ 10. $\frac{5}{2} + \frac{1}{2}i\sqrt{3}, \frac{5}{2} - \frac{1}{2}i\sqrt{3}$ 11. $\frac{8}{3}i, -\frac{8}{3}i$ 12. $\frac{5}{2}i, -\frac{5}{2}i$ 13. $\frac{2}{5} + \frac{1}{5}i\sqrt{6}, \frac{2}{5} - \frac{1}{5}i\sqrt{6}$ 14. $\frac{1}{6} + \frac{1}{6}i\sqrt{3}, \frac{1}{6} - \frac{1}{6}i\sqrt{3}$

15. $3, 3 + i\sqrt{2}, 3 - i\sqrt{2}$ 16. $-8, 1 + i\sqrt{6}, 1 - i\sqrt{6}$ 17. $-\frac{1}{8} + \frac{1}{8}i\sqrt{15}, -\frac{1}{8} - \frac{1}{8}i\sqrt{15}, \frac{1}{2} + \frac{1}{2}i\sqrt{3}, \frac{1}{2} - \frac{1}{2}i\sqrt{3}$ 18. $\frac{3}{2} + \frac{1}{2}i\sqrt{2}, \frac{3}{2} - \frac{1}{2}i\sqrt{2}, 2i\sqrt{3}, -2i\sqrt{3}$ 19. $-1, \frac{1}{2} + \frac{1}{2}i\sqrt{3}, \frac{1}{2} - \frac{1}{2}i\sqrt{3}$ 20. $2, -1 + i\sqrt{3}, -1 - i\sqrt{3}$

21. $f(x) = (x - 4 - 3i)(x - 4 + 3i)$ 22. $f(x) = (x - 3 + 2i)(x - 3 - 2i)$ 23. $f(x) = (x - 4)(x + 5i)(x - 5i)$ 24. $f(x) = (x - 4)(x - 2 + i\sqrt{5})(x - 2 - i\sqrt{5})$
25. $f(x) = (x + 3i)^2(x - 3i)^2$ 26. $f(x) = (x - 2 - i)(x - 2 + i)(x - 1 + i\sqrt{3})(x - 1 - i\sqrt{3})$ 27. $f(x) = (x + 2)(x - 1 - 2i)(x - 1 + 2i)(x - i\sqrt{3})(x + i\sqrt{3})$ 28. $f(x) = (x + 1)(x - 2 - i)^2(x - 2 + i)^2$

29. $3, \frac{1}{2} + \frac{1}{2}i\sqrt{7}, \frac{1}{2} - \frac{1}{2}i\sqrt{7}$ 30. $-5, -\frac{3}{2} + \frac{1}{2}i\sqrt{11}, -\frac{3}{2} - \frac{1}{2}i\sqrt{11}$ 31. $-2, -1, -1 + i\sqrt{2}, -1 - i\sqrt{2}$ 32. $1, -2, \frac{1}{2} + \frac{1}{2}i\sqrt{7}, \frac{1}{2} - \frac{1}{2}i\sqrt{7}$ 33. $2i, -2i, 3$ (of multiplicity 2) 34. $1 + 3i, 1 - 3i, 1 + i, 1 - i$

*denotes Superset exercises

*35. $f(\bar{z}) = a_n(\bar{z})^n + a_{n-1}(\bar{z})^{n-1} + \cdots + a_1\bar{z} + a_0 = a_n \cdot \overline{z^n} + a_{n-1} \cdot \overline{z^{n-1}} + \cdots + a_1 \cdot z + a_0$
$= \overline{a_n} \cdot \overline{z^n} + \overline{a_{n-1}} \cdot \overline{z^{n-1}} + \cdots + \overline{a_1} \cdot \bar{z} + \overline{a_0} = \overline{a_n z^n} + \overline{a_{n-1}z^{n-1}} + \cdots + \overline{a_1 z} + \overline{a_0}$
$= \overline{a_n z^n + a_{n-1}z^{n-1} + \cdots + a_1 z + a_0} = \overline{f(z)} = \bar{0} = 0$

*36. By the Conjugate Roots Theorem, the imaginary zeros of a polynomial equation with real coefficients "occur in pairs." That is, the total number of imaginary zeros is even. But an nth degree polynomial equation has exactly n zeros, counting multiplicities. In this case, the number of zeros is odd, so the number of real zeros is also odd, that is, at least one.

*37. $1, -2, 2i, -2i$ *39. $-1, \frac{3}{2}, -\frac{1}{2} + \frac{1}{2}i\sqrt{3}, -\frac{1}{2} - \frac{1}{2}i\sqrt{3}$ *41. $-\frac{1}{2}, i, -i, 3$

*38. $2, 3, i\sqrt{2}, -i\sqrt{2}$ *40. $-1, \frac{1}{2} + \frac{1}{2}i\sqrt{3}, \frac{1}{2} - \frac{1}{2}i\sqrt{3}, \frac{1}{3}$ *42. $-2, i\sqrt{2}, -i\sqrt{2}, \frac{1}{3}$

*43. $(x - c_1)(x - c_2)(x - c_3)(x - c_4) = x^4 - (c_1 + c_2 + c_3 + c_4)x^3 + (c_1c_2 + c_1c_3 + c_1c_4 + c_2c_3 + c_2c_3 + c_3c_4)x^2 - (c_1c_2c_3 + c_1c_2c_4 + c_1c_3c_4 + c_2c_3c_4)x + c_1c_2c_3c_4$. In general, a_0 equals $(-1)^n$ times the product of the zeros, and a_{n-1} equals the negative of the sum of the zeros. *45. $A = -4, B = 1$ *47. $-n$ *49. 0

*44. $A = -4, B = -12$ *46. $A = -8, B = 17, C = 2, D = -24$ *48. $2n$ *50. -4

Section 4.6, page 263

1. -1.488303 3. 1.789948 5. 2.408357 7. 2.332038 9. $f(x) = x^2 - 3$; 1.732051
2. 1.854865 4. 1.554782 6. -1.604730 8. 0.605830 10. $f(x) = x^2 - 5$; 2.236068
11. $f(x) = x^2 - 6x + 3$; 0.550510 13. $f(x) = x^3 - 2$; 1.259921 15. $f(x) = x^3 - 3x^2 + 3x - 4$; 2.442250
12. $f(x) = x^2 + 2x - 6$; 1.645751 14. $f(x) = x^3 - 3$; 1.442250 16. $f(x) = x^3 - 6x^2 + 12x - 10$; 3.259921

*17. $-3 - \sqrt{6} \approx -5.449490, -3 + \sqrt{6} \approx -0.550510$ *19. $-\frac{4}{3} \approx -1.333333, 1$

*18. $\frac{3 - \sqrt{13}}{2} \approx -0.302776, \frac{3 + \sqrt{13}}{2} \approx 3.302776$ *20. $-1, \frac{3}{5} = 0.6$

*21. $\frac{3.7 - \sqrt{17.53}}{2.4} \approx -0.202868, \frac{3.7 + \sqrt{17.53}}{2.4} \approx 3.286202$

*22. $\frac{-7.6 - \sqrt{34.2}}{6.2} \approx -2.169045, \frac{-7.6 + \sqrt{34.2}}{6.2} \approx -0.282568$

*23. $\frac{2}{3} \approx 0.666667, \frac{-1 - \sqrt{5}}{2} \approx -1.618034, \frac{-1 + \sqrt{5}}{2} \approx 0.618034$

*24. $\frac{5}{2} = 2.5, \frac{3 - \sqrt{41}}{4} \approx -0.850781, \frac{3 + \sqrt{41}}{4} \approx 2.350781$

*25. $\frac{2}{5} = 0.4, \frac{-5 - \sqrt{17}}{2} \approx -4.561553, \frac{-5 + \sqrt{17}}{2} \approx -0.438447$

*26. $-\frac{2}{3} \approx -0.666667, -3 - \sqrt{13} \approx -6.605551, -3 + \sqrt{13} \approx 0.605551$

*27. $-\frac{7}{2} = -3.5, \frac{-7 - \sqrt{13}}{6} \approx -1.767592, \frac{-7 + \sqrt{13}}{6} \approx -0.565741$

*28. $-\frac{3}{4} = -0.75, \frac{1 - \sqrt{5}}{2} \approx -0.618034, \frac{1 + \sqrt{5}}{2} \approx 1.618034$

Section 4.7, pages 271–272

1. Ver: none; Hor: $y = 4$ 3. Ver: $x = 0, x = 3$; Hor: $y = 0$ 5. Ver: $x = -9$, Hor: $y = 0$
2. Ver: $x = 0$; Hor: $y = 0$ 4. Ver: $x = 0$; Hor: $y = 2$ 6. Ver: $x = 3$; Hor: $y = 0$
7. Ver: $x = -2, x = -1$; Hor: $y = 0$ 9. Ver: $x = \frac{1}{2}$; Hor: $y = \frac{1}{4}$ 11. Ver: $x = -\frac{1}{5}, x = -1$; Hor: $y = 2$

*denotes Superset exercises

8. Ver: $x = -3, x = 3$; Hor: $y = 0$ **10.** Ver: $x = -2, x = 2$; Hor: $y = \frac{1}{2}$ **12.** Ver: $x = \frac{1}{3}$; Hor: $y = \frac{1}{27}$

13. Ver: $x = \frac{1}{2}, x = -1$; Hor: none **15.** **17.** **19.**

14. Ver: $x = 3, x = -5$; Hor: none

16.
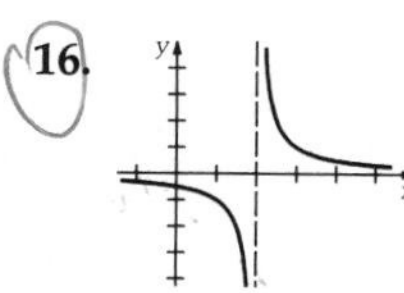

18.
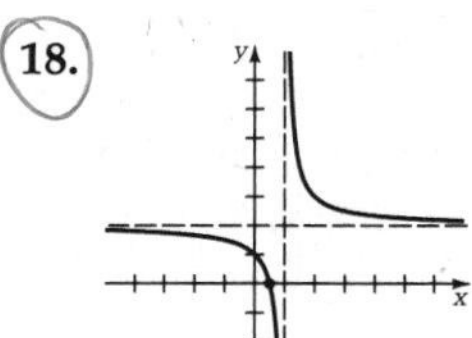

20.
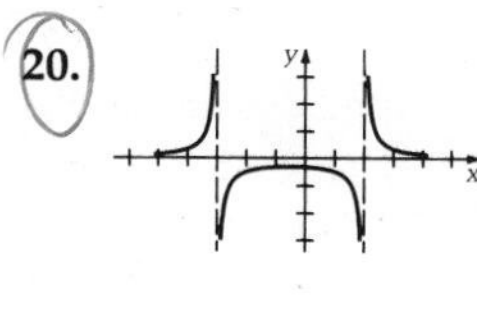

21.
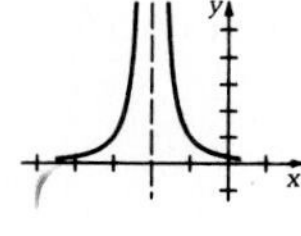

23.
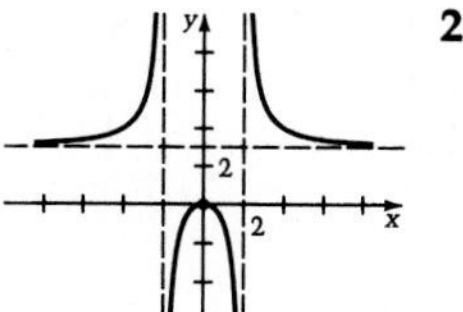

25.

27.
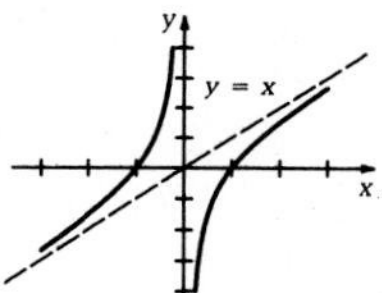

29.
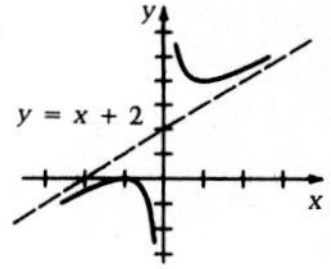

22.
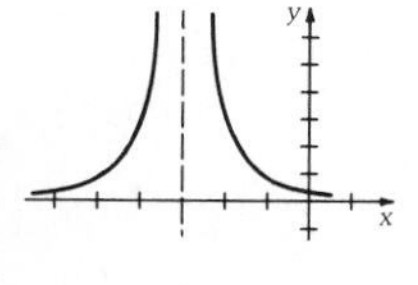

24.
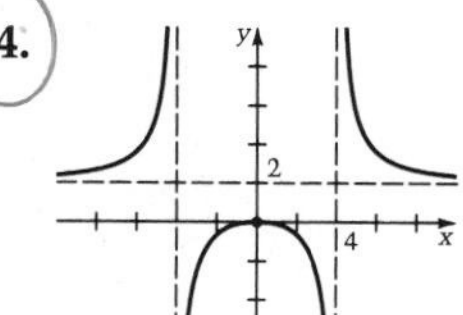

26.
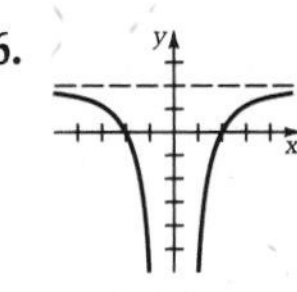

28.
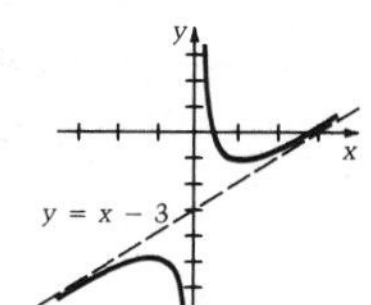

30.
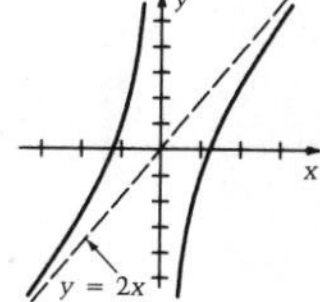

31.
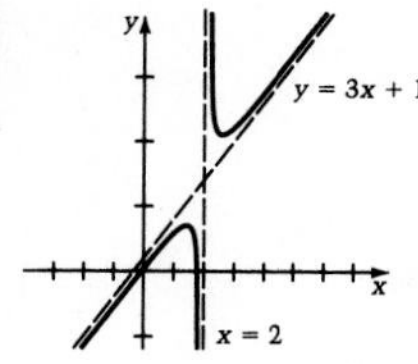

33.
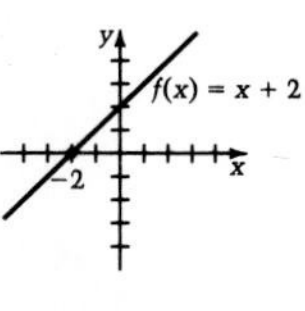

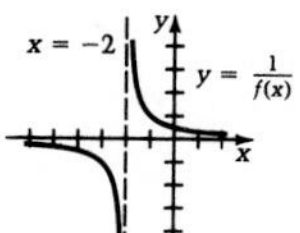

***35.**
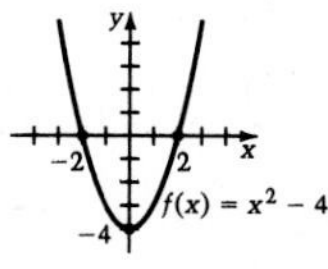

***37.**
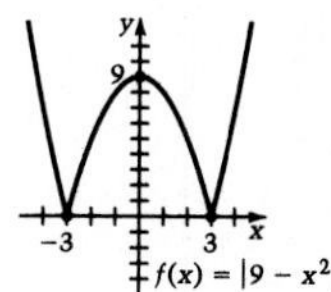

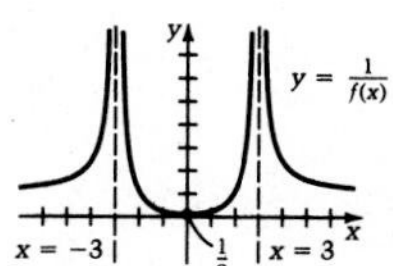

---

*denotes Superset exercises

32.

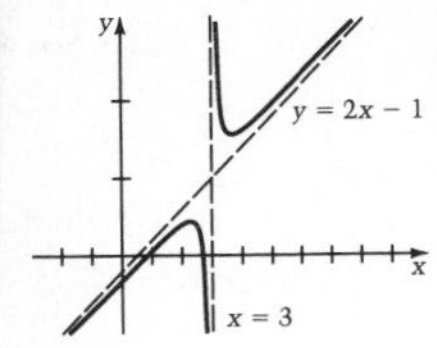

*34.

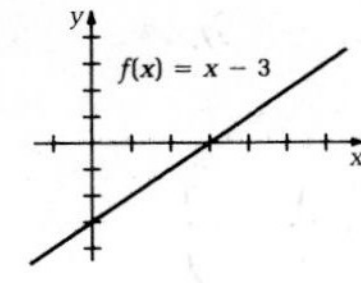

*36.

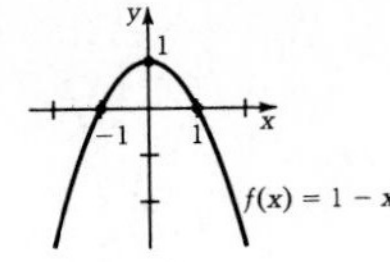

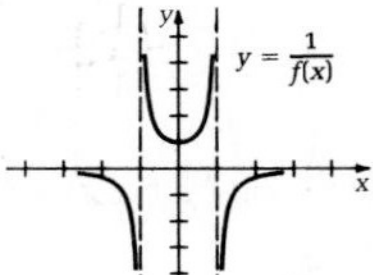

*38.

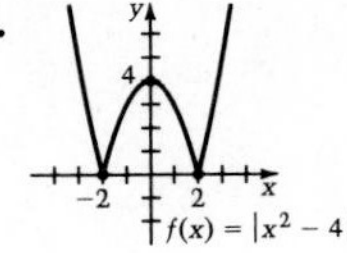

Section 4.8, pages 277–278

1. vertex: $(0, 0)$
axis of symmetry: $y = 0$
x-intercept: 0
y-intercept: 0

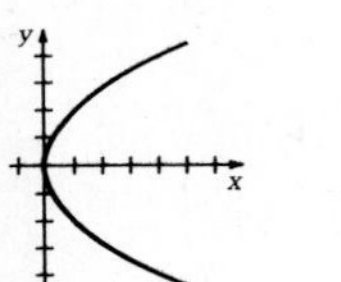

3. vertex: $(0, 0)$
axis of symmetry: $y = 0$
x-intercept: 0
y-intercept: 0

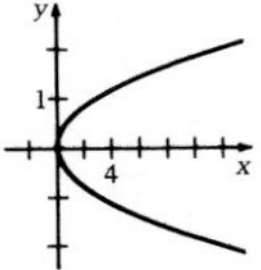

5. vertex: $(5, 3)$
axis of symmetry: $y = 3$
x-intercept: -13
y-intercepts: $3 \pm \frac{\sqrt{10}}{2}$

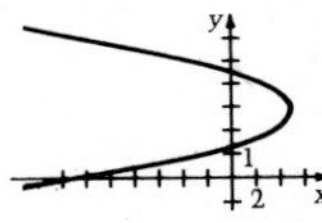

7. vertex: $\left(\frac{3}{2}, -1\right)$
axis of symmetry: $y = -1$
x-intercept: $\frac{7}{2}$
y-intercepts: none

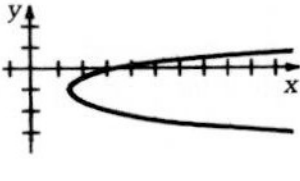

2. vertex: $(0, 0)$
axis of symmetry: $y = 0$
x-intercept: 0
y-intercept: 0

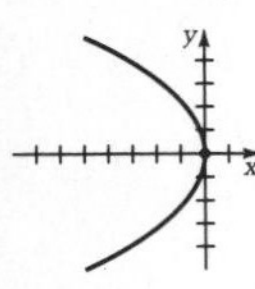

4. vertex: $(0, 0)$
axis of symmetry: $y = 0$
x-intercept: 0
y-intercept: 0

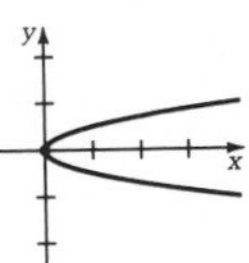

6. vertex: $(-2, -1)$
axis of symmetry: $y = -1$
x-intercept: 1
y-intercepts: $-1 \pm \frac{\sqrt{6}}{3}$

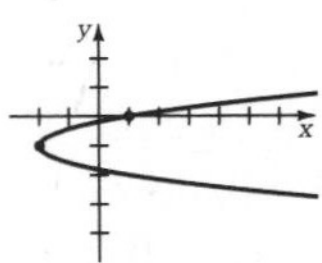

8. vertex: $(4, 2)$
axis of symmetry: $y = 2$
x-intercept: 2
y-intercepts: $2 \pm 2\sqrt{2}$

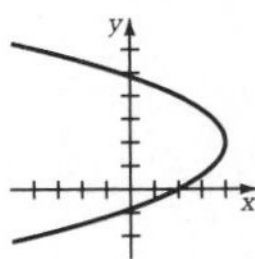

9. vertex: $(-4, 3)$
axis of symmetry: $y = 3$
x-intercept: -13
y-intercepts: none

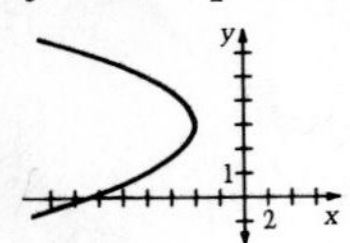

11. vertex: $(3, -1)$
axis of symmetry: $y = -1$
x-intercept: $\frac{8}{3}$
y-intercepts: $2, -4$

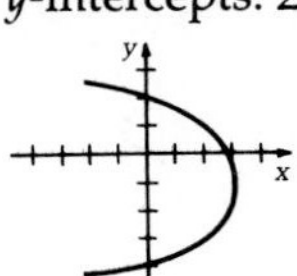

13. center $(0, 0)$; radius 4

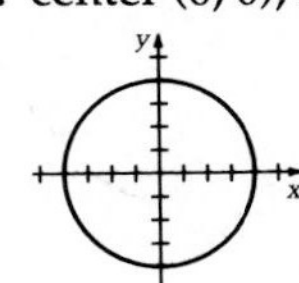

15. center $(-1, 3)$; radius 3

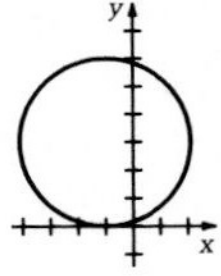

---

*denotes Superset exercises

10. vertex: $(7, -2)$
axis of symmetry: $y = -2$
x-intercept: 3
y-intercepts: $-2 \pm \sqrt{7}$

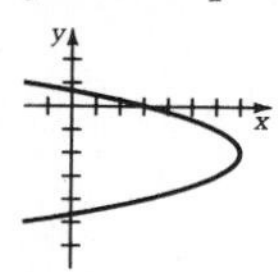

12. vertex: $(-9, 3)$
axis of symmetry: $y = 3$
x-intercept: $-\frac{9}{2}$
y-intercepts: $3 \pm 3\sqrt{2}$

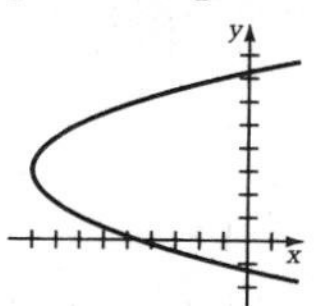

14. center $(0, 0)$; radius 5

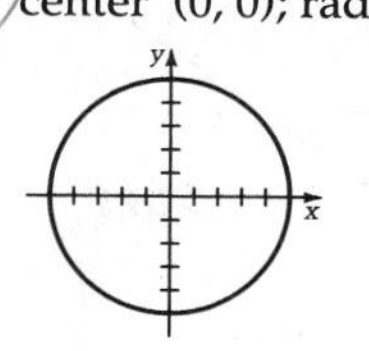

16. center $(3, -4)$; radius 1

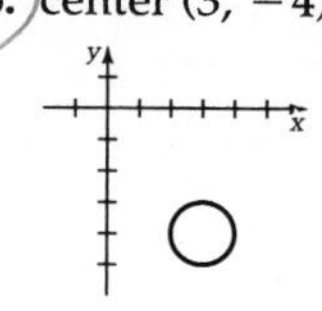

17. center $(2, 0)$; radius 2

19. center $(-2, 1)$; radius 4

21. center $\left(\frac{1}{2}, -\frac{1}{2}\right)$; radius $\sqrt{3}$

23. center $\left(\frac{1}{5}, 3\right)$; radius 3

18. center $(0, 3)$; radius 2

20. center $\left(\frac{3}{2}, \frac{5}{2}\right)$; radius 3

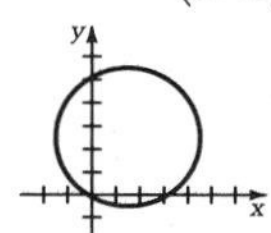

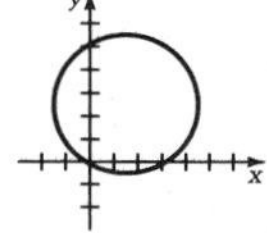

22. center $\left(\frac{1}{3}, -1\right)$; radius $\sqrt{2}$

24. center $\left(\frac{1}{6}, \frac{1}{3}\right)$; radius $\frac{1}{2}$

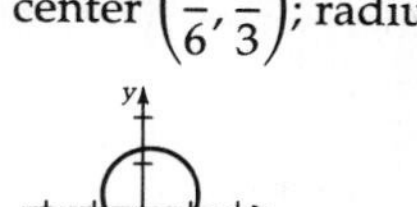

25. outside **27.** inside **29.** inside **31.** outside ***33.** $(x + 3)^2 + (y - 3)^2 = 9$

26. inside **28.** inside **30.** outside **32.** inside ***34.** $y - 5 = -\frac{4}{3}(x - 1)$

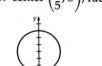

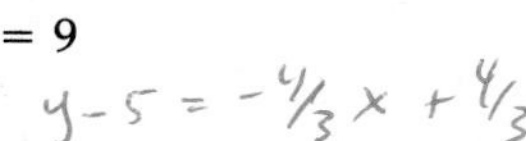

***35.** Follow the hint and compute the product of the slopes of the two sides not on the diameter.

***36.** $(x - 2)^2 + (y - 2)^2 = 4$, $(x - 10)^2 + (y - 10)^2 = 100$

***37.** Simplify $2\sqrt{(x - 0)^2 + (y - 0)^2} = \sqrt{(x - 4)^2 + (y - 0)^2}$ to obtain $\left(x + \frac{4}{3}\right)^2 + y^2 = \left(\frac{8}{3}\right)^2$.

***38.** Follow the hint to obtain $ax + by = 0$ as an equation for the line.

***39.** $y = \frac{1}{4}(x - 0)^2 + 1$
vertex: $(0, 1)$
axis of symmetry: $x = 0$

***40.** $x = \frac{1}{6}(y - 0)^2 + \frac{3}{2}$
vertex: $\left(\frac{3}{2}, 0\right)$
axis of symmetry: $y = 0$

***41.** $y = -\frac{1}{4}(x - 0)^2 + (-1)$
vertex: $(0, -1)$
axis of symmetry: $x = 0$

***42.** $x = -\frac{1}{6}(y - 0)^2 + \left(-\frac{3}{2}\right)$
vertex: $\left(-\frac{3}{2}, 0\right)$
axis of symmetry: $y = 0$

***43.** $\frac{1}{8}(y - 2)^2 + (-1) = x$
vertex: $(-1, 2)$
axis of symmetry: $y = 2$

***44.** $\frac{1}{12}(x - 3)^2 + 1 = y$
vertex: $(3, 1)$
axis of symmetry: $x = 3$

***45.** $\frac{1}{4c}(y - 0)^2 + 0 = x$
vertex: $(0, 0)$
axis of symmetry: $y = 0$

***46.** $\frac{1}{4c}(x - 0)^2 + 0 = y$
vertex: $(0, 0)$
axis of symmetry: $x = 0$

---

*denotes Superset exercises

Section 4.9, pages 284–286

1. center: (0, 0)
vertices: (3, 0), (−3, 0), (0, 4), (0, −4)

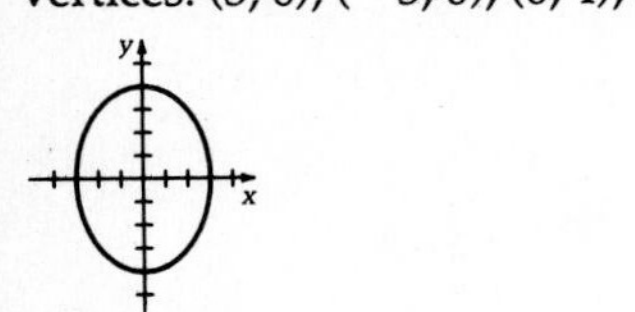

2. center: (0, 0)
vertices: (5, 0), (−5, 0), (0, 2), (0, −2)

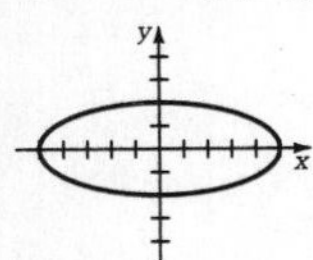

3. center: (2, −3)
vertices: (4, −3), (0, −3), (2, 7), (2, −13)

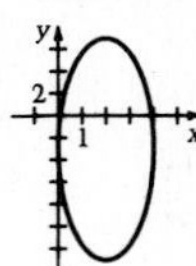

4. center: (−1, 1)
vertices: (5, 1), (−7, 1), (−1, 5), (−1, −3)

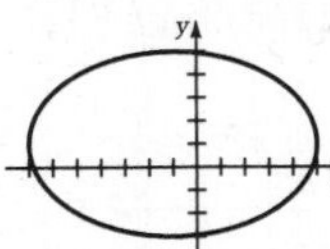

5. center: (5, 0)
vertices: $(5 + \sqrt{10}, 0)$, $(5 - \sqrt{10}, 0)$, (5, 4), (5, −4)

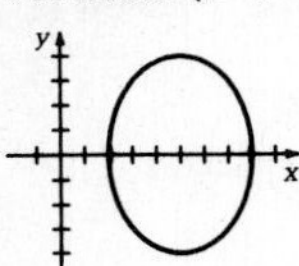

6. center: (0, 2)
vertices: $(\sqrt{3}, 2)$, $(-\sqrt{3}, 2)$, (0, 3), (0, 1)

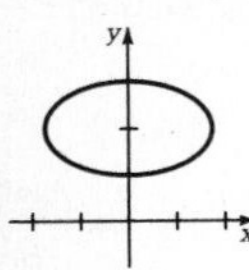

7. center: (0, 0)
vertices: $\left(\frac{11}{5}, 0\right)$, $\left(-\frac{11}{5}, 0\right)$, (0, 4), (0, −4)

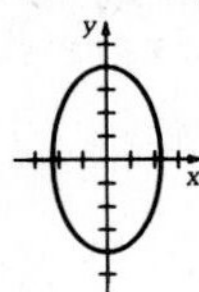

8. center: (0, 0)
vertices: (1, 0), (−1, 0), $\left(0, \frac{9}{4}\right)$, $\left(0, -\frac{9}{4}\right)$

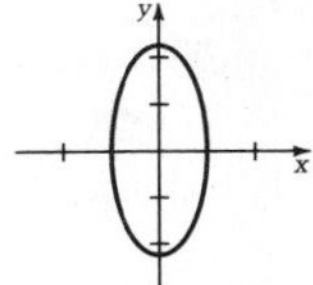

9. center: (1, 2)
vertices: (4, 2), (−2, 2), (1, 6), (1, −2)

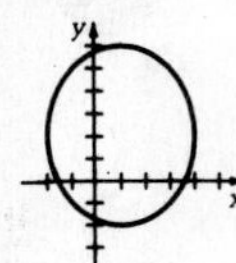

10. center: (−3, 6)
vertices: (−1, −6), (−5, −6), (−3, 4), (−3, −16)

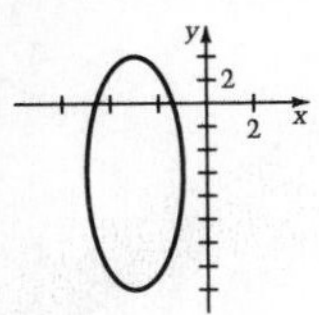

11. center: (−3, 0)
vertices: $(-3 + \sqrt{2}, 0)$, $(-3 - \sqrt{2}, 0)$, (−3, 4), (−3, 4)

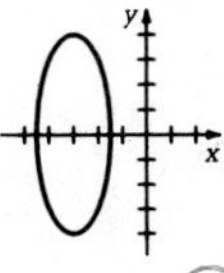

12. center: $\left(\frac{1}{2}, 0\right)$
vertices: $\left(\frac{3}{2}, 0\right)$, $\left(-\frac{1}{2}, 0\right)$, $\left(\frac{1}{2}, 6\right)$, $\left(\frac{1}{2}, -6\right)$

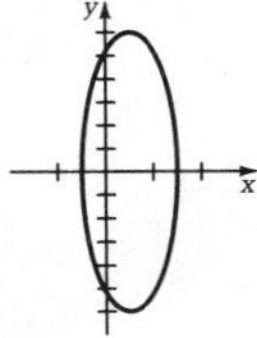

13. center: $(-1, 1)$
vertices: $(2, 1), (-4, 1), (-1, 2), (-1, 0)$

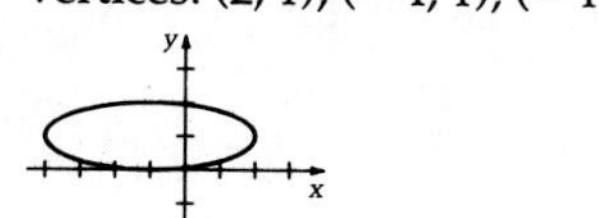

14. center: $(2, -1)$
vertices: $(4, -1), (0, -1), (2, 2), (2, -4)$

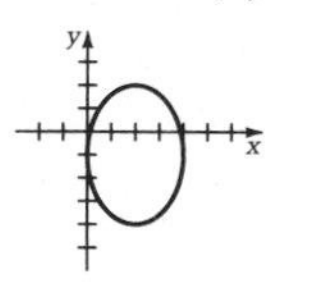

17. center: $(0, 0)$
vertices: $(\sqrt{2}, 0), (-\sqrt{2}, 0)$
asymptotes: $y = \sqrt{5}x, y = -\sqrt{5}x$

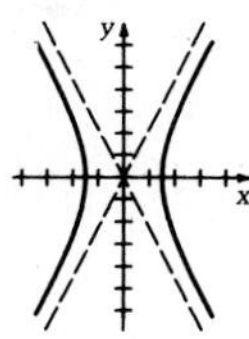

18. center: $(0, 0)$
vertices: $(0, 2\sqrt{5}), (0, -2\sqrt{5})$
asymptotes: $y = \frac{\sqrt{15}}{3}x, y = -\frac{\sqrt{15}}{3}x$

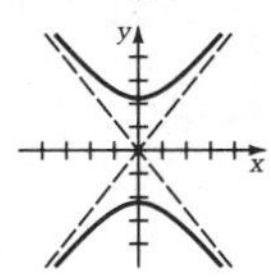

21. center: $(2, 3)$
vertices: $\left(\frac{13}{4}, 3\right), \left(\frac{3}{4}, 3\right)$
asymptotes: $y = \frac{7}{10}x + \frac{8}{5}, y = -\frac{7}{10}x + \frac{22}{5}$

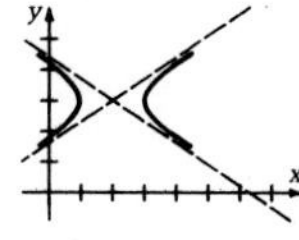

15. center: $(0, 0)$
vertices: $(4, 0), (-4, 0)$
asymptotes: $y = \frac{3}{4}x, y = -\frac{3}{4}x$

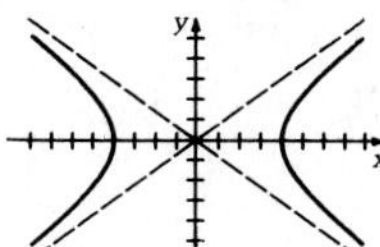

16. center: $(0, 0)$
vertices: $(0, 3), (0, -3)$
asymptotes: $y = \frac{3}{4}x, y = -\frac{3}{4}x$

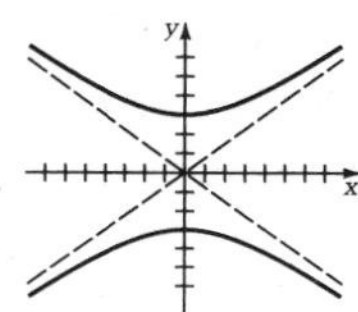

19. center: $(1, 2)$
vertices: $(1, 7), (1, -3)$
asymptotes: $y = \frac{5}{2}x - \frac{1}{2}, y = -\frac{5}{2}x + \frac{9}{2}$

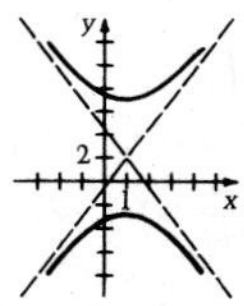

20. center: $(3, 0)$
vertices: $(9, 0), (-3, 0)$
asymptotes: $y = \frac{2}{3}x - 2, y = -\frac{2}{3}x + 2$

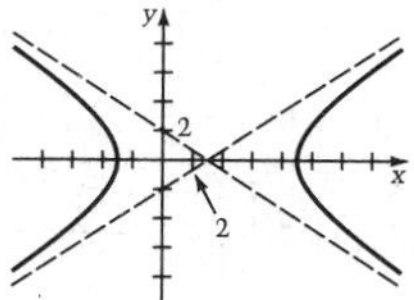

23. center: $(0, 0)$
vertices: $(5, 0), (-5, 0)$
asymptotes: $y = \frac{2}{5}x, y = -\frac{2}{5}x$

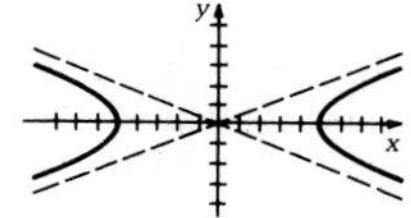

22. center: $\left(\frac{1}{2}, 0\right)$
vertices: $\left(\frac{2}{3}, 0\right), \left(\frac{1}{3}, 0\right)$
asymptotes: $y = 4x - 2, y = -4x + 2$

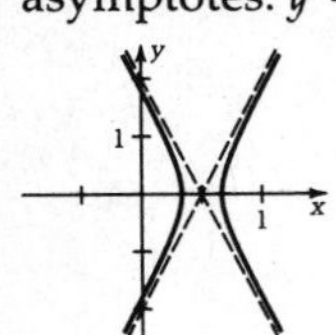

24. center: $(0, 0)$
vertices: $(0, \sqrt{10}), (0, -\sqrt{10})$
asymptotes: $y = \sqrt{10x}, y = -\sqrt{10}x$

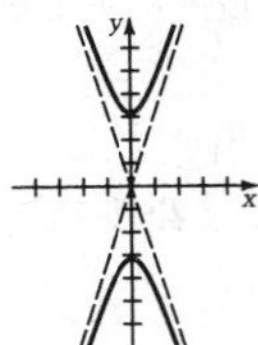

25. center: $\left(0, \frac{3}{2}\right)$
vertices: $\left(0, \frac{7}{2}\right), \left(0, -\frac{1}{2}\right)$
asymptotes: $y = \frac{1}{4}x + \frac{3}{2}, y = -\frac{1}{4}x + \frac{3}{2}$

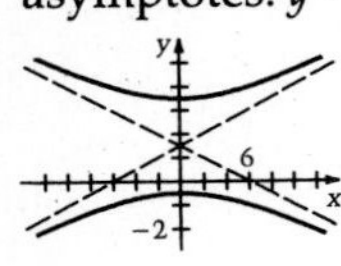

27. center: $(-2, -1)$
vertices: $(-2, 1), (-2, -3)$
asymptotes: $y = \frac{2}{3}x + \frac{1}{3}, y = -\frac{2}{3}x - \frac{7}{3}$

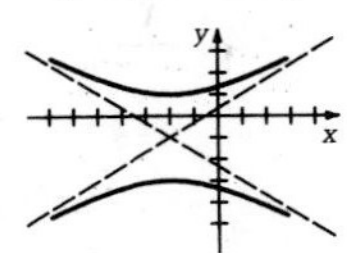

26. center: $\left(\frac{7}{4}, 0\right)$
vertices: $\left(\frac{23}{4}, 0\right), \left(-\frac{9}{4}, 0\right)$
asymptotes: $y = 4x - 7, y = -4x + 7$

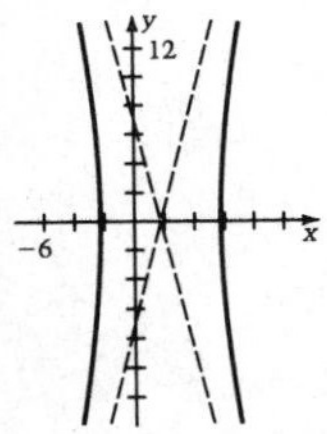

28. center: $(2, -2)$
vertices: $(2 + 2\sqrt{2}, -2), (2 - 2\sqrt{2}, -2)$
asymptotes: $y = \frac{\sqrt{2}}{4}x - \frac{4 + \sqrt{2}}{2}, y = -\frac{\sqrt{2}}{4}x - \frac{4 - \sqrt{2}}{2}$

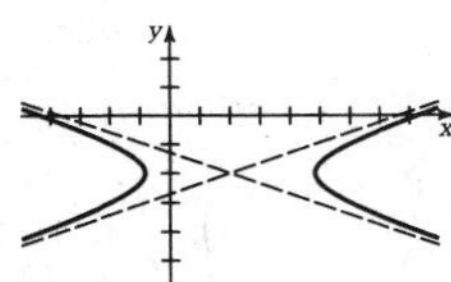

***29.** circle

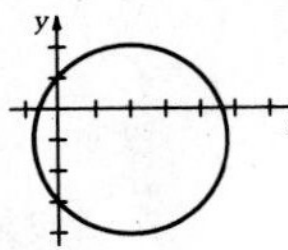

***31.** parabola

***33.** ellipse

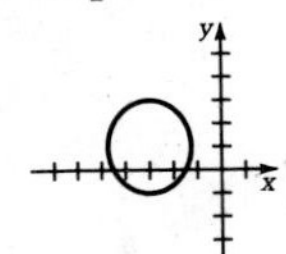

***35.** hyperbola

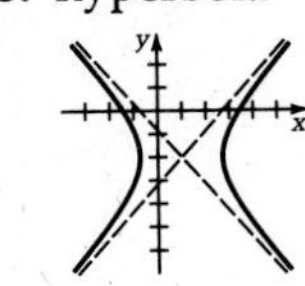

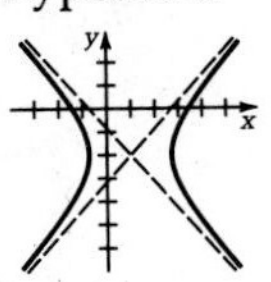

***37.** yes

***39.** no

***30.** ellipse

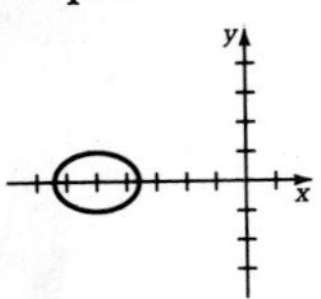

***32.** hyperbola

***34.** circle

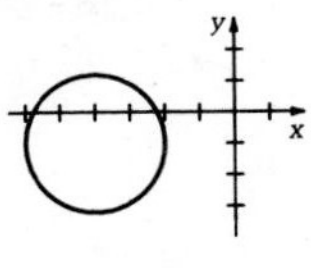

***36.** parabola

***38.** no

***40.** yes

***41.** yes
***42.** yes
***43.** no
***44.** no

---

*denotes Superset exercises

*45. $\dfrac{(x-0)^2}{3^2}+\dfrac{(y-4)^2}{5^2}=1$
center: (0, 4)
vertices: (3, 4), (−3, 4), (0, 9), (0, −1)

*46. $\dfrac{(x-3)^2}{5^2}+\dfrac{(y-0)^2}{4^2}=1$
center: (3, 0)
vertices: (8, 0), (−2, 0), (3, 4), (3, − 4)

*47. $\dfrac{(x-3)^2}{(13)^2}+\dfrac{(y-0)^2}{(12)^2}=1$
center: (3, 0)
vertices: (16, 0), (−10, 0), (3, 12), (3, −12)

*48. $\dfrac{(x-1)^2}{(12)^2}+\dfrac{(y-2)^2}{(13)^2}=1$
center: (1, 2)
vertices: (13, 2), (−11, 2), (1, 15), (1, −11)

*49. $\dfrac{(x-0)^2}{a^2}+\dfrac{(y-0)^2}{(\sqrt{a^2-c^2})^2}=1$
center: (0, 0)
vertices: $(a, 0), (-a, 0), (0, \sqrt{a^2-c^2}), (0, -\sqrt{a^2-c^2})$

*50. $\dfrac{(x-0)^2}{(\sqrt{b^2-c^2})^2}+\dfrac{(y-0)^2}{b^2}=1$
center: (0, 0)
vertices: $(\sqrt{b^2-c^2}, 0), (-\sqrt{b^2-c^2}, 0), (0, b), (0, -b)$

*51. $\dfrac{(x-5)^2}{4^2}-\dfrac{(y-0)^2}{3^2}=1$
center: (5, 0)
vertices: (1, 0), (9, 0)
asymptotes: $y=\dfrac{3}{4}(x-5), y=-\dfrac{3}{4}(x-5)$

*52. $\dfrac{(x-5)^2}{3^2}-\dfrac{(y-0)^2}{4^2}=1$
center: (5, 0)
vertices: (2, 0), (8, 0)
asymptotes: $y=\dfrac{4}{3}(x-5), y=-\dfrac{4}{3}(x-5)$

*53. $\dfrac{(y-6)^2}{5^2}-\dfrac{(x-1)^2}{(12)^2}=1$
center: (1, 6)
vertices: (1, 11), (1, 1)
asymptotes: $y-6=\dfrac{5}{12}(x-1), y-6=-\dfrac{5}{12}(x-1)$

*54. $\dfrac{(x-3)^2}{(12)^2}-\dfrac{(y-3)^2}{5^2}=1$
center: (3, 3)
vertices: (−9, 3), (15, 3)
asymptotes: $y-3=\dfrac{5}{12}(x-3), y-3=-\dfrac{5}{12}(x-3)$

*55. $\dfrac{(x-0)^2}{a^2}-\dfrac{(y-0)^2}{(\sqrt{c^2-a^2})^2}=1$
center: (0, 0)
vertices: $(a, 0), (-a, 0)$
asymptotes: $y=\dfrac{\sqrt{c^2-a^2}}{a}x, y=-\dfrac{\sqrt{c^2-a^2}}{a}x$

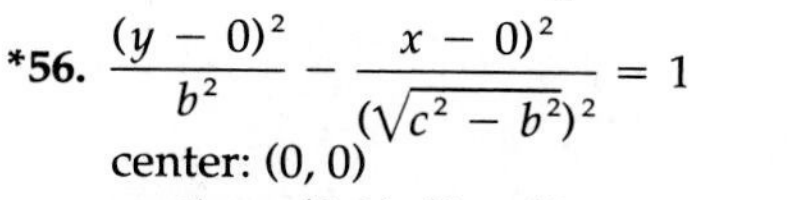

*56. $\dfrac{(y-0)^2}{b^2}-\dfrac{x-0)^2}{(\sqrt{c^2-b^2})^2}=1$
center: (0, 0)
vertices: $(0, b), (0, -b)$
asymptotes: $y=\dfrac{bx}{\sqrt{c^2-b^2}}, y=-\dfrac{bx}{\sqrt{c^2-b^2}}$

Chapter 4 Review Exercises, pages 292–294

1. Graph No. 5
2. Graph No. 2
3. Graph No. 6
4. Graph No. 6
5.
6.
7.
8.
9.

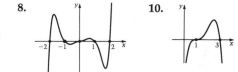

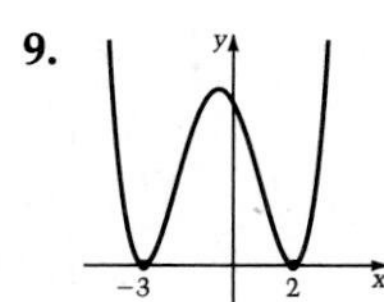

10.
11.

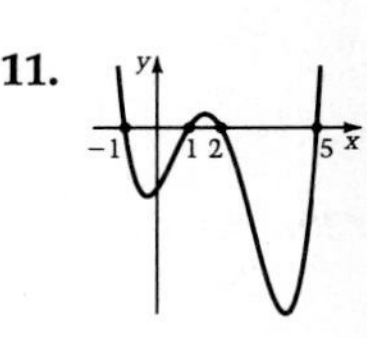

12.

*denotes Superset exercises

13. 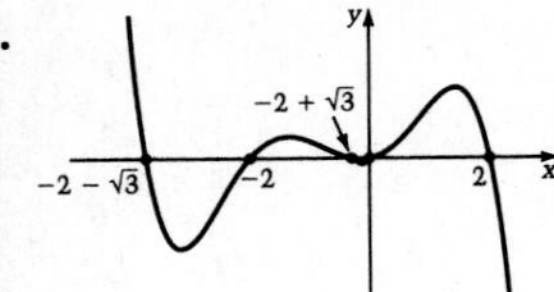

15. 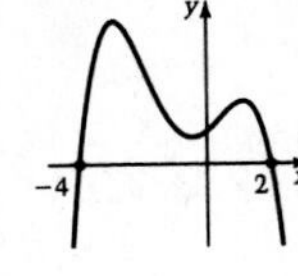

17. $x - 1, x + 2$ **19.** $x + 1, x - 1$

14. 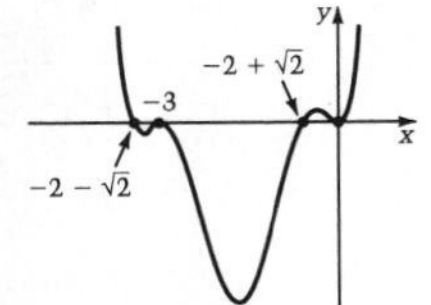

16. 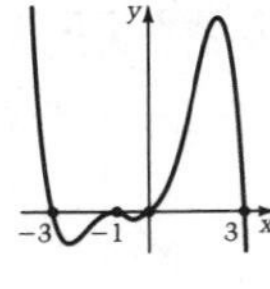

18. $x + 2, x - 2$ **20.** $x - 2$

21.
$$\begin{array}{r|rrrr} 2 & 3 & -4 & 5 & 8 \\ & & 6 & 4 & 18 \\ \hline & 3 & 2 & 9 & 26 \end{array} \quad \text{Thus, } f(2) = 26.$$

$$\begin{array}{r|rrrr} -1 & 3 & -4 & 5 & 8 \\ & & -3 & 7 & -12 \\ \hline & 3 & -7 & 12 & -4 \end{array} \quad \text{Thus, } f(-1) = -4.$$

22.
$$\begin{array}{r|rrrrr} -2 & 1 & 0 & -3 & -2 & 9 \\ & & -2 & 4 & -2 & 8 \\ \hline & 1 & -2 & 1 & -4 & 17 \end{array} \quad \text{Thus, } g(-2) = 17.$$

$$\begin{array}{r|rrrrr} 1 & 1 & 0 & -3 & -2 & 9 \\ & & 1 & 1 & -2 & -4 \\ \hline & 1 & 1 & -2 & -4 & 5 \end{array} \quad \text{Thus, } g(1) = 5.$$

23.
$$\begin{array}{r|rrrrrrr} -2 & -1 & 0 & 0 & 2 & 0 & -3 & 7 \\ & & 2 & -4 & 8 & -20 & 40 & -74 \\ \hline & -1 & 2 & -4 & 10 & -20 & 37 & -67 \end{array} \quad \text{Thus, } G(-2) = -67.$$

$$\begin{array}{r|rrrrrrr} 1 & -1 & 0 & 0 & 2 & 0 & -3 & 7 \\ & & -1 & -1 & -1 & 1 & 1 & -2 \\ \hline & -1 & -1 & -1 & 1 & 1 & -2 & 5 \end{array} \quad \text{Thus, } G(1) = 5.$$

24.
$$\begin{array}{r|rrrrrr} 2 & -1 & 0 & 5 & -1 & 0 & 4 \\ & & -2 & -4 & 2 & 2 & 4 \\ \hline & -1 & -2 & 1 & 1 & 2 & 8 \end{array} \quad \text{Thus, } F(2) = 8.$$

$$\begin{array}{r|rrrrrr} -1 & -1 & 0 & 5 & -1 & 0 & 4 \\ & & 1 & -1 & -4 & 5 & -5 \\ \hline & -1 & 1 & 4 & -5 & 5 & -1 \end{array} \quad \text{Thus, } F(-1) = -1.$$

25. $\pm 1, \pm 2, \pm 3, \pm 6, \pm \frac{1}{3}, \pm \frac{2}{3}$ **27.** $\pm 1, \pm 3, \pm \frac{1}{2}, \pm \frac{3}{2}$

26. $\pm 1, \pm 2, \pm\frac{1}{3}, \pm\frac{2}{3}$ **28.** $\pm 1, \pm 2, \pm 3, \pm 6, \pm\frac{1}{2}, \pm\frac{3}{2}, \pm\frac{1}{3}, \pm\frac{2}{3}, \pm\frac{1}{6}$

29. either two or no positive real zeros, and precisely one negative real zero.

30. either two or no positive real zeros, and no negative real zeros.

31. precisely one positive real root, and either two or no negative real roots. **33.** $f(x) = (x + 2)(x - 3 + i)(x - 3 - i)$

32. precisely one positive real root, and either two or no negative real roots. **34.** $f(x) = (x - 2 - i)^2(x - 2 + i)^2$

35. $f(x) = (x - 1 - i)(x - 1 + i)(x - 4 + i\sqrt{3})(x - 4 - i\sqrt{3})$ **37.** $-\frac{5}{2} - \frac{1}{2}i\sqrt{3}, -\frac{5}{2} + \frac{1}{2}i\sqrt{3}$

36. $f(x) = (x - 3)(x + 2 - i\sqrt{2})(x + 2 + i\sqrt{2})$ **38.** $-\frac{5}{3}i, \frac{5}{3}i$

39. $0, 1 - \frac{1}{2}i\sqrt{2}, 1 + \frac{1}{2}i\sqrt{2}$ **41.** $1, -\frac{1}{2} - \frac{1}{2}i\sqrt{3}, -\frac{1}{2} + \frac{1}{2}i\sqrt{3}$ **43.** $-\frac{3}{4}i, \frac{3}{4}i$ **45.** $-1, 3, -3, -2$

40. $0, 1 - \frac{1}{2}i\sqrt{6}, 1 + \frac{1}{2}i\sqrt{6}$ **42.** $-3 - i, -3 + i$ **44.** $3, -\frac{3}{2} - \frac{3}{2}i\sqrt{3}, -\frac{3}{2} + \frac{3}{2}i\sqrt{3}$ **46.** $1, -1, 4, 5$

47. $i, -i, -5, 2$ **49.** -2.195823 **51.** 1.732051 **53.** Ver: $x = -1, x = -4$; Hor: $y = 0$

48. $2 + i, 2 - i, 3, -3$ **50.** 2.090489 **52.** -2.236068 **54.** Ver: $x = 0, x = 6, x = -1$; Hor: $y = 27$

55. Ver: $x = 0, x = \frac{3}{2}$; Hor: $y = \frac{1}{4}$ **57.** Ver: $x = 2, x = \frac{1}{3}$; Hor: none **59.**

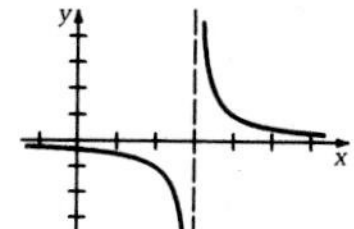

56. Ver: $x = 0, x = 3, x = -3$; Hor: None **58.** Ver: $x = 1, x = 3$; Hor: $y = 0$ **60.**

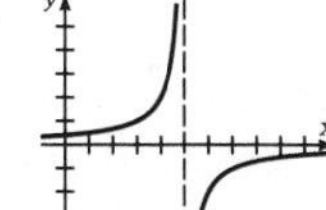

61.

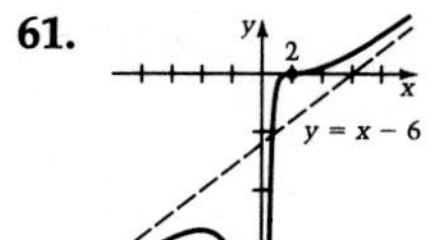

63.

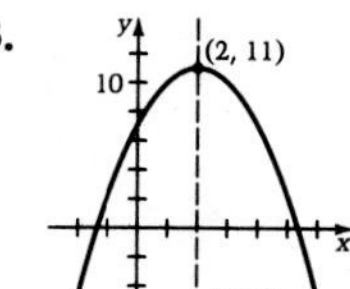

65.

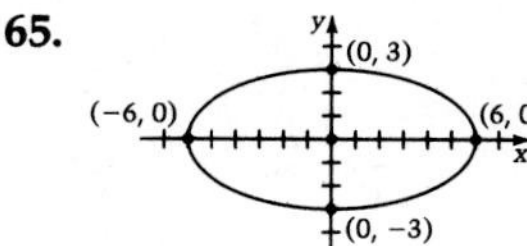

67.

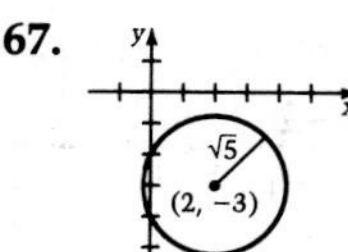

69.

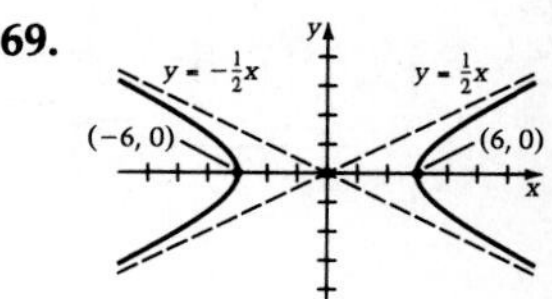

62.

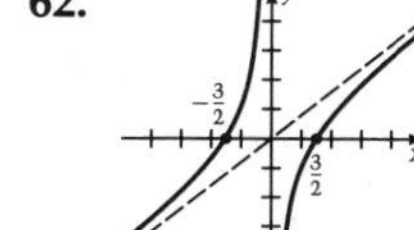

64.

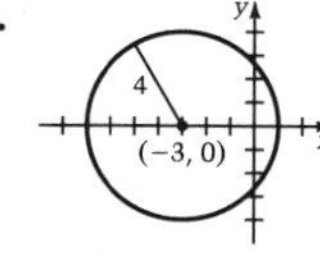

66.

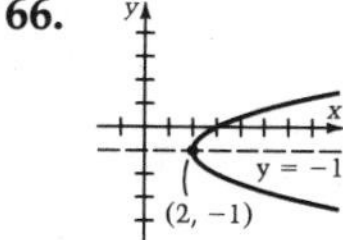

68.

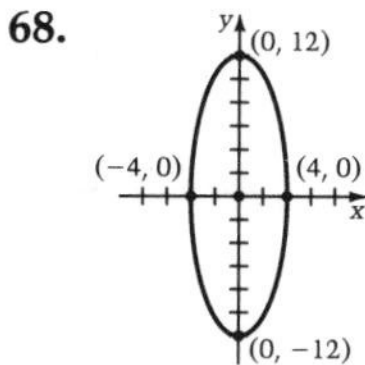

70.

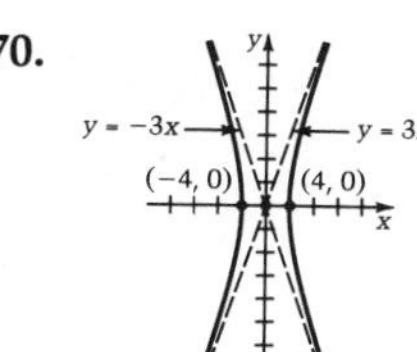

71. 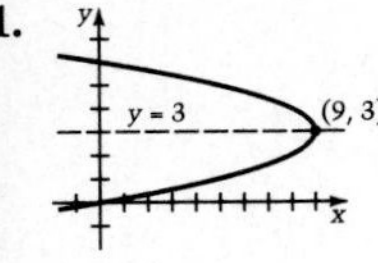

73. 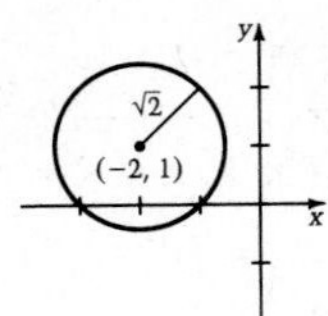

75. 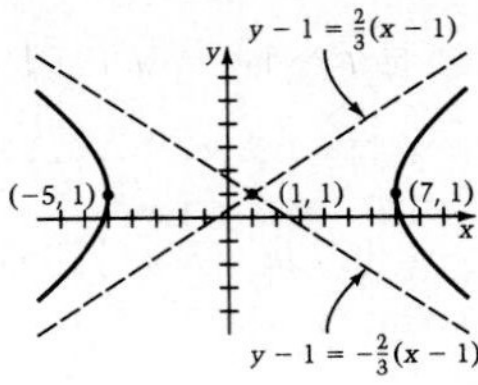

77. 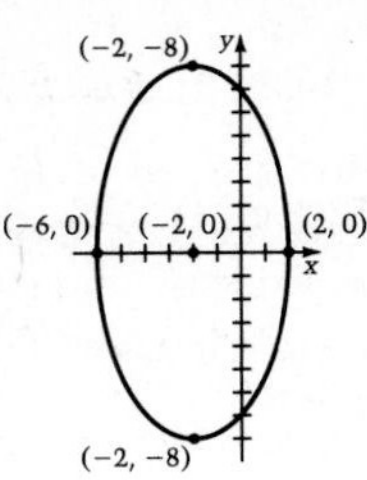

72. 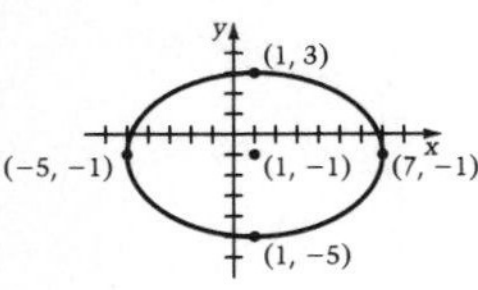

74. 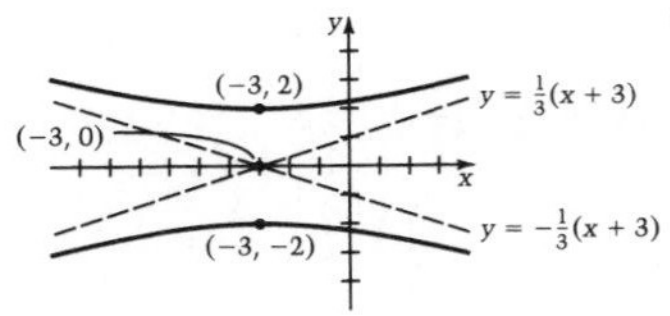

76. 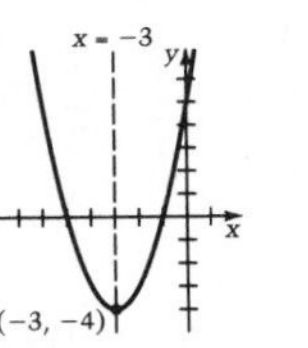

78. 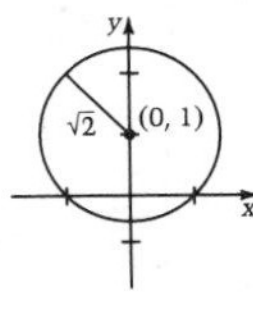

79. 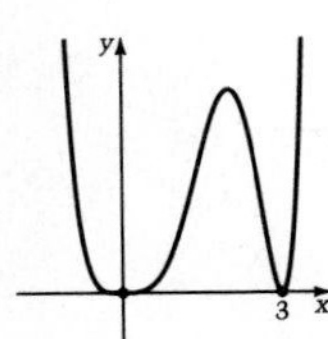

81. 5 **83.** $y = \dfrac{4x^2}{(x - 2)(x + 1)}$ **85.** $(1 + 2\sqrt{3}, 2), (1 + 2\sqrt{3}, -2), (1 - 2\sqrt{3}, 2), (1 - 2\sqrt{3}, -2)$

80. The graph of f crosses (or touches) the line $y = c$ at each of the three different x-values that led to the remainder of c. **82.** Since the coefficients of the equation are real, the pure imaginary zeros will occur in conjugate pairs. Suppose the roots are ai, $-ai$, bi, and $-bi$. By the Linear Factorization Theorem $0 = (x - ai)(x + ai)(x - bi)(x + bi) = x^4 + (a^2 + b^2)x^2 + a^2b^2$. **84.** $0 < P < 4$ **86.** The student will conclude that a value approximately equal to a is a root.

Chapter 4 Test, page 295

1. Graph No. 4 **3.** Graph No. 3 **5.** Graph No. 2 **7.** 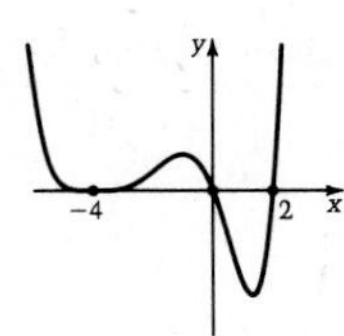

2. Graph No. 1 **4.** Graph No. 2 **6.** Graph No. 4 **8.** $f(-2) = -59$

9. either two or no positive real zeros, and no negative real zeros. **11.** $\frac{1}{2}, i\sqrt{2}, -i\sqrt{2}, -\frac{3}{2}$ **13.** [3, 4]

10. $\frac{1}{2}$ (of multiplicity 2), 2 **12.** $f(x) = (x - 2)(x - 2 + 3i)(x - 2 - 3i)$ **14.** Ver: $x = \frac{1}{2}, x = \frac{1}{3}$; Hor: $y = \frac{1}{2}$

15. Ver: $x = 3, x = -3$; Hor: $y = 0$

17. vertex: $(-4, 3)$
axis of symmetry: $y = 3$
x-intercept: 5
y-intercepts: 1, 5

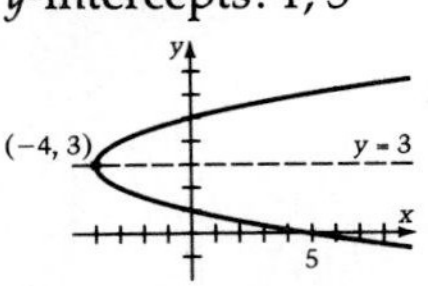

19. center: (5, 0)
vertices: (0, 0), (10, 0), (5, 2), (5, −2)

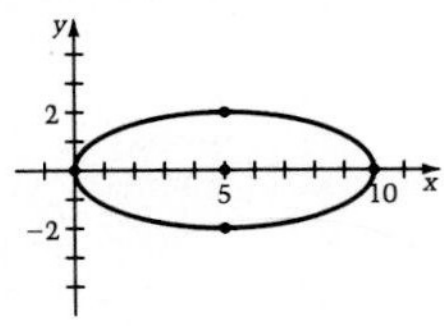

16. 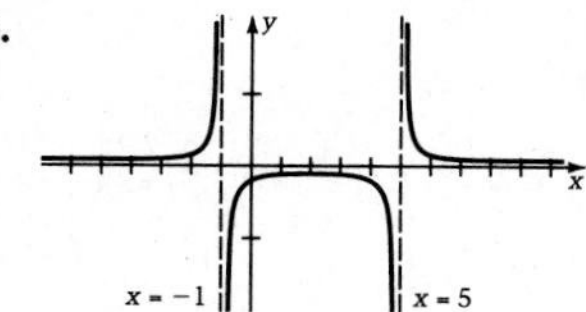

18. center: $(-1, 3)$; radius: $\sqrt{3}$

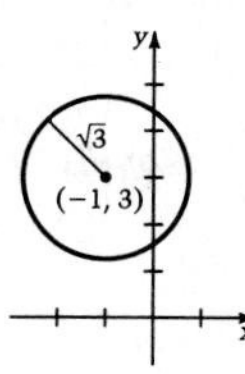

20. center: $(1, -2)$
vertices: $(1, -2 - 2\sqrt{6})$ and $(1, -2 + 2\sqrt{6})$
asymptotes: $y + 2 = \pm\sqrt{6}(x - 1)$

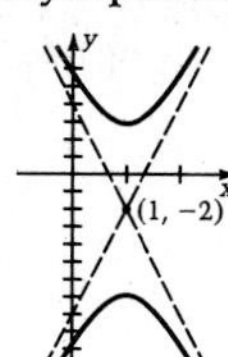

Chapter 5

Section 5.1, pages 302–303

1. 2^{-3x} 3. 5^{2x-4} 5. 2^{2-x} 7. 3^{2x-2} 9. 2^{8-x} 11. 3^{4x-7} 13. $2^x(1 + 2^x)$ 15.

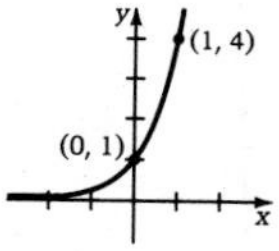

2. 3^{-2x} 4. 7^{-5} 6. 3^{3x-2} 8. 2^{-2x-4} 10. 3^{-3x-8} 12. 2^{-x-9} 14. $3^x(3^x - 1)$ 16.

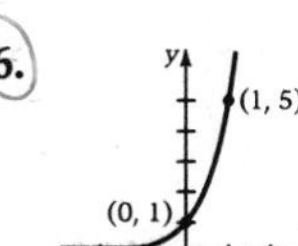

17. 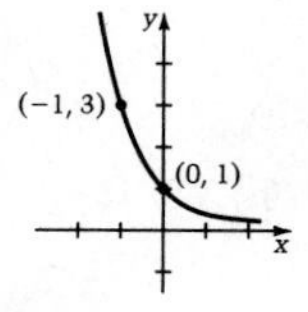

19. 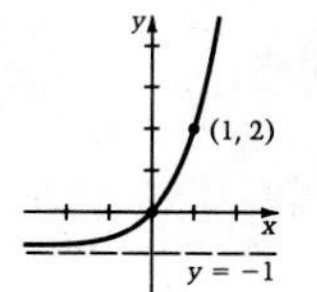

21. 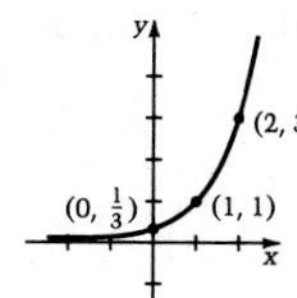

23. 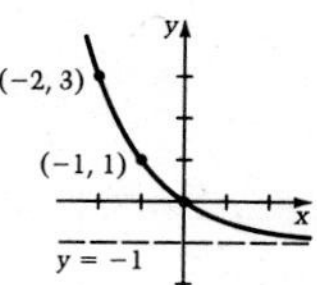

18. 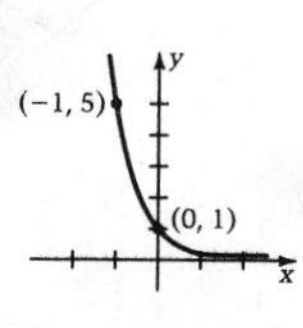

20. 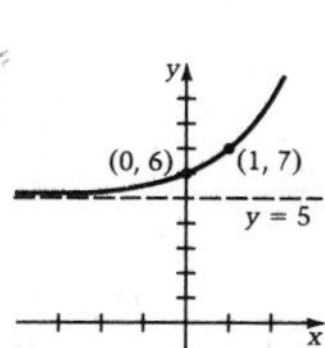

22. 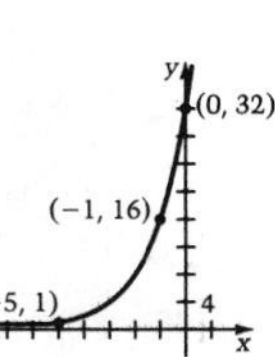

24. 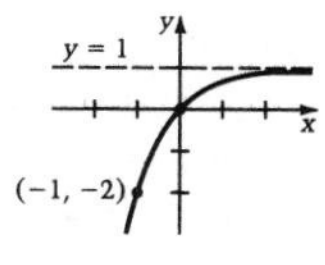

25. 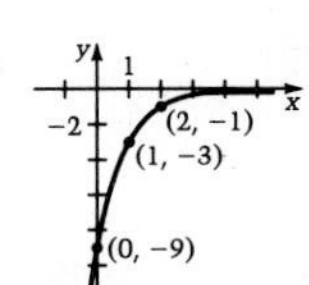

27. 3 29. $\frac{2}{3}$ 31. -8 33. $\frac{7}{2}$ 35. no solution 37. -2

26. 28. 3 30. $\frac{3}{2}$ 32. $\frac{15}{4}$ 34. -2 36. no solution 38. $-\frac{2}{3}$

*39. (a) approx. 8.866 mg; (b) approx. 3.002 mg; (c) approx. 0.024 mg *41. $\pm\sqrt{2}$ *43. 0 *45. 2, 0
*40. (a) 200; (b) approx. 1131; (c) approx. 1.68×10^9 *42. -1 *44. 2 *46. 1, 2
*47. (a) $y \approx 2.7$; (b) $y \approx 8.8$; (c) $y \approx 0.3$
*48. (a) $x \approx 1.7$; (b) $x \approx 1.9$; (c) $x \approx 0.7$
*49. If k is doubled, Q is squared. If k is tripled, Q is cubed. If k is increased tenfold, Q is raised to the tenth power.
*50. Since the graphs of $y = 3^x$ and $y = x$ do not intersect, the equation $3^x = x$ has no solutions.

*51. 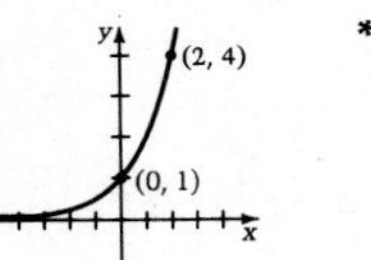

*53. 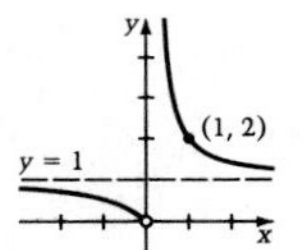

*55. 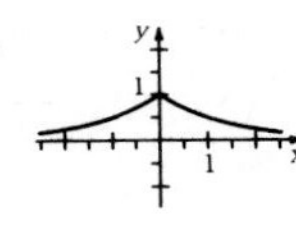

*57. 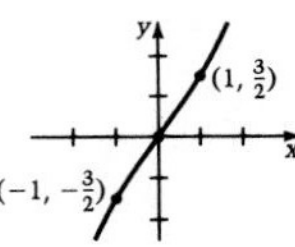

*52. *54. *56. *58.

*59. The graph of f is symmetric with respect to the y-axis.
*60. $2^{f(-x)} \neq -2^{f(x)}$ for all x, since $2^A > 0$ for all real numbers A.
*61. $\frac{b^x}{b^y} = b^x\left(\frac{1}{b^y}\right) = b^x(b^{-y}) = b^{x+(-y)} = b^{x-y}$ *63. $2^{3.14} \approx 8.815$ *65. 2.67 *67. 31.71 *69. 1.96
*62. By Property 1, $1 = b^0 = b^{x+(-x)} = b^x \cdot b^{-x}$. Thus, $b^{-x} = \frac{1}{b^x}$. *64. 3^π *66. 3.32 *68. 0.63 *70. 225.18
*71. 0.78 *73. 1.63 *75. -0.17 *77. 1.5
*72. 0.04 *74. 36.46 *76. 0.35 *78. -0.4

Section 5.2, pages 309–310

1. $\log_x 25 = 2$ 3. $\log_{16} 4 = x$ 5. $\log_5 x = 3$ 7. 8 9. $2\sqrt{2}$ 11. 2 13. $\frac{2}{3}$ 15. $\frac{\sqrt{3}}{3}$ 17. 2 19. $\frac{3}{2}$
2. $\log_x 16 = 4$ 4. $\log_{27} 3 = x$ 6. $\log_{10} x = 5$ 8. 64 10. 3 12. 2 14. $\frac{3}{2}$ 16. $\frac{1}{2}$ 18. 2 20. $\frac{2}{3}$

21. 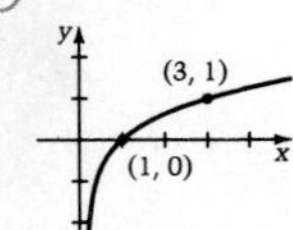

23. 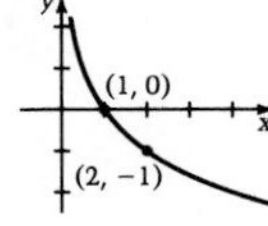

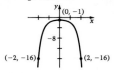

25. 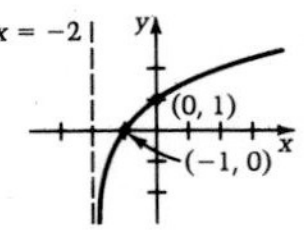

27. 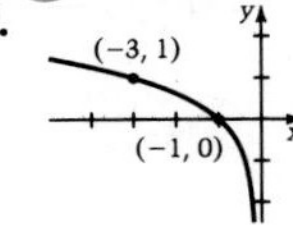

29. 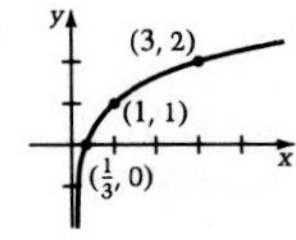

*denotes Superset exercises

22.
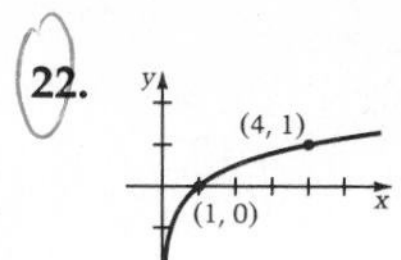

24.
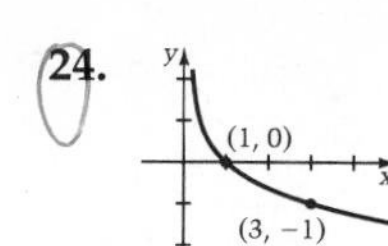

26.
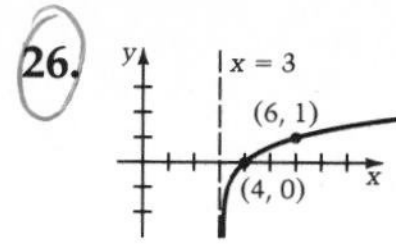

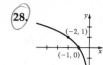

28. (−2, 1) (−1, 0)

30. (4, 1) (2, 0) (1, −1)

31.
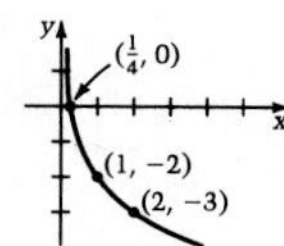

33.
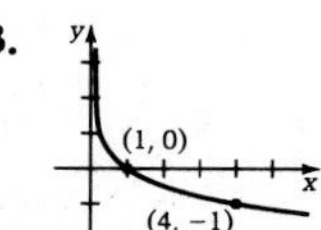

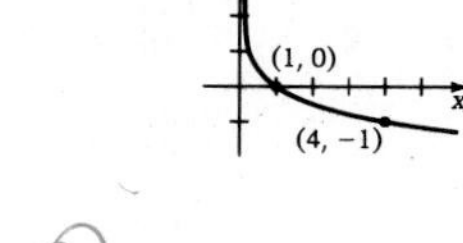

35. 2 **37.** 13.6 **39.** 4.8 **41.** 0.2 **43.** 2.6 **45.** −0.4375

32.
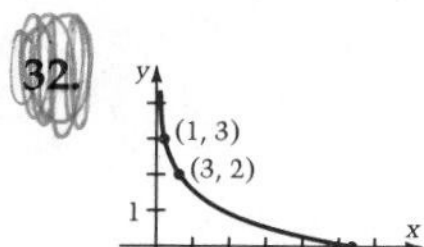

34.
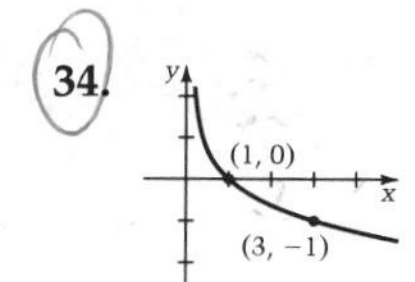

36. 0 **38.** 7.0 **40.** −7.6 **42.** 3.9 **44.** 1.4 **46.** −0.26

47. 3.41 **49.** 3 **51.** $B + C$ **53.** $4A$ **55.** $3A + B + 2C$ **57.** B **59.** $-B$ **61.** $2A - C$
48. 20.491 **50.** 2 **52.** $A + C$ **54.** $4B$ **56.** $A + 2B + 2C$ **58.** C **60.** $-A$ **62.** $C - 3A$
63. $B - A - C$ **65.** $A + B - C$ **67.** $4B - 3A - 3C$

***69.** (3, 0) (5, −1) x = 2

***71.** (1, 2) (4, 0)

64. $B - 2A - 2C$ **66.** $5B - A - C$ **68.** $B - 2C$

***70.**
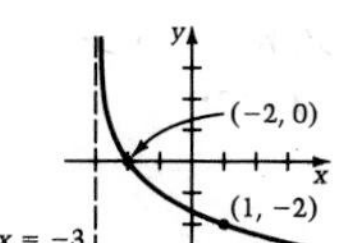

***72.** (1, 3) 1 5 (27, 0)

***73.** $y = \log_3(x^2) = 2\log_3|x|$
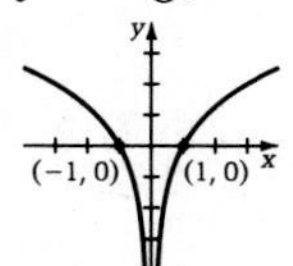

***75.**
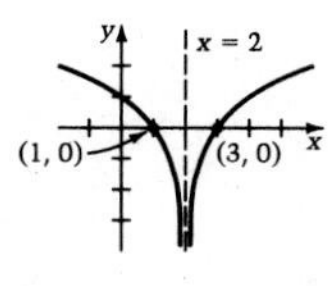

***77.**
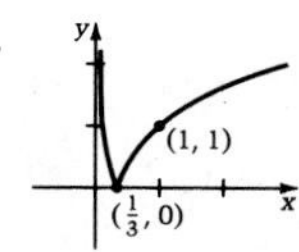

***74.** $y = \log_2(x^3) = 3\log_2 x$
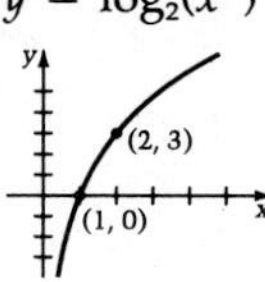

***76.**
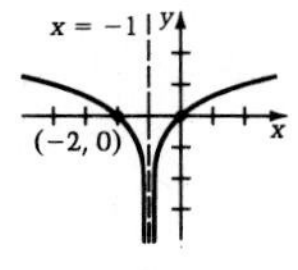

***78.**
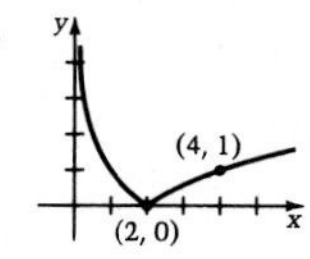

*denotes Superset exercises

*79. $y = \log_8(2^x) = x \cdot \log_8 2 = \dfrac{x}{3}$

*81. $y = 4^{\log_2 x} = 2^{(2\log_2 x)} = x^2, x > 0$

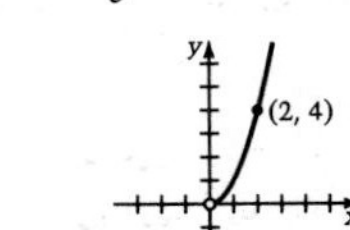

*83. $y = 2^{\log_8 x} = 8^{(1/3\log_8 x)} = x^{1/3}, x > 0$

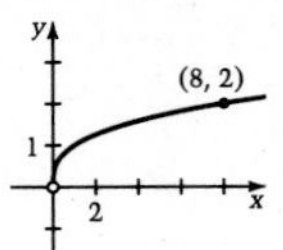

*80. $y = \log_2(8^x) = x \cdot \log_2 8 = 3x$

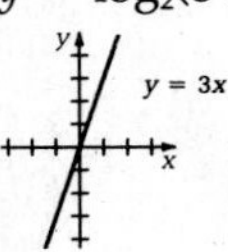

*82. $y = 2^{\log_4 x} = 4^{(1/2\log_4 x)} = \sqrt{x}, x > 0$

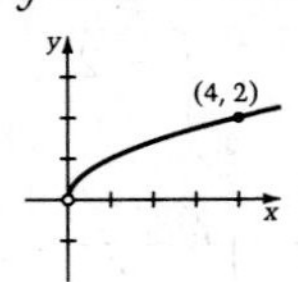

*84. $y = 8^{\log_2 x} = 2^{(3\log_2 x)} = x^3, x > 0$

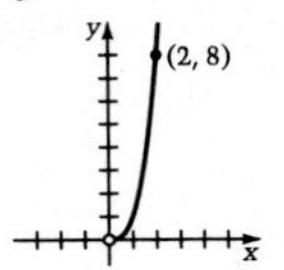

*85. From $b^1 = b$, we get $\log_b b = 1$ by the Log/Exp Principle.

*86. $\log_b\left(\dfrac{x}{y}\right) = \log_b(xy^{-1}) = \log_b x + \log_b(y^{-1}) = \log_b x - \log_b y$

*87. $\log_b\left(\dfrac{1}{x}\right) = \log_b 1 - \log_b x = -\log_b x.$

*88. Let $s = \log_b x$ and $t = \log_B b$. Then, by the Log/Exp Principle, we have $x = b^s$ and $b = B^t$ so that $x = b^s = (B^t)^s = B^{ts}$. Use the Log/Exp Principle again to get $\log_B x = ts = (\log_B b)\cdot(\log_b x)$ so that $\log_b x = \dfrac{\log_B x}{\log_B b}$.

Section 5.3, pages 316–317

1. $\sqrt{5}$ 3. 25 5. -1 7. -4 9. $\sqrt{3}$ 11. $2\sqrt{2}$ 13. -3 15. $x > 0$ 17. 8, 1 19. 1 21. 1, 2

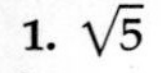

2. 4 4. $\sqrt{2}$ 6. -6 8. $-\dfrac{3}{2}$ 10. $\sqrt{5}$ 12. 5 14. $x > 0$ 16. -4 18. 32, 1 20. -1 22. $\sqrt{2}, \sqrt{3}$

23. 2 25. e^2 27. $\dfrac{1}{3e}$ 29.

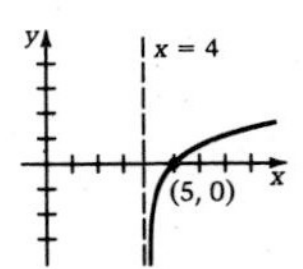

31.

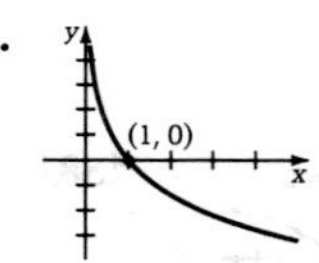

33.

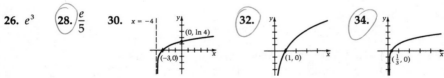

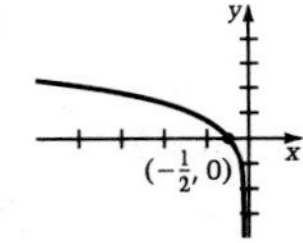

24. 3 26. e^3

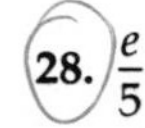

28. $\dfrac{e}{5}$ 30.

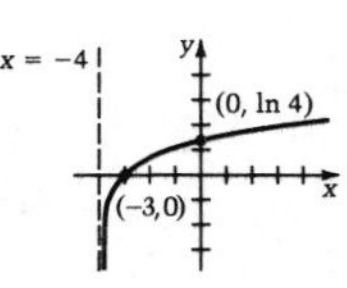

32.

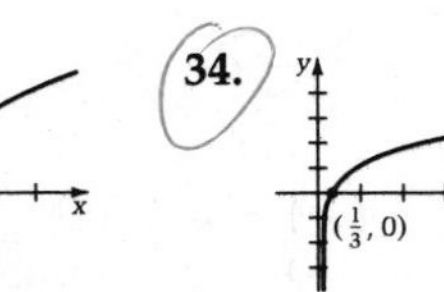

34. ($\frac{1}{3}$, 0)

35. $y = \ln(e^x) = x\ln e = x$

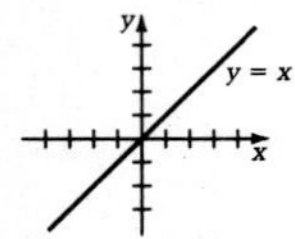

37. $y = \ln(e^2x) = 2 + \ln x$

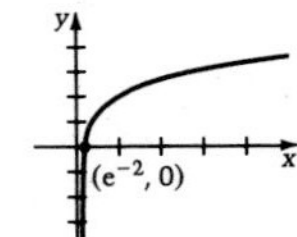

39. $y = e^{3\ln x} = x^3, x > 0$

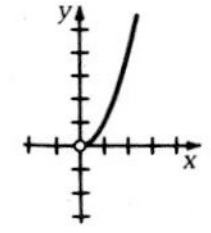

41. $y = \ln(e^{x^2}) = x^2$

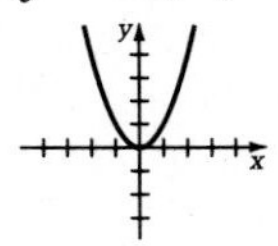

*denotes Superset exercises

36. $y = \ln(e^{-x}) = -x \ln e = -x$

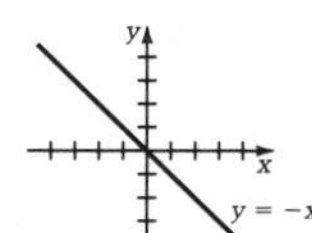

38. $y = e^{2 \ln x} = x^2, x > 0$

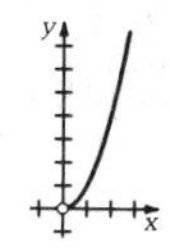

40. $y = \ln\left(\frac{x}{e}\right) = -1 + \ln x$

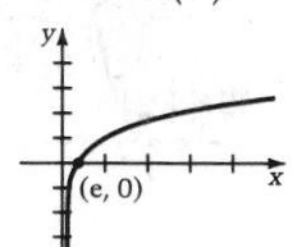

42. $y = \ln (e^{x^3}) = x^3$

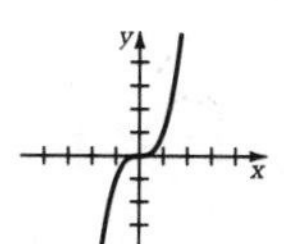

***43.** $x > 0$ and $x \neq 1$ ***45.** 2 ***47.** $\sqrt{3}$ ***49.** $e^{1/e}$ ***51.** $y = \ln \frac{1}{x-2} = -\ln (x-2)$

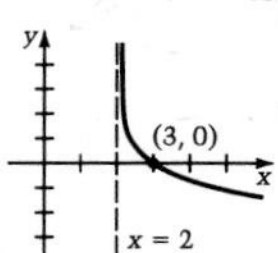

***44.** $x > 0$ ***46.** 10 ***48.** 3 ***50.** 2^e ***52.** $y = \ln \frac{1}{x+4} = -\ln (x+4)$

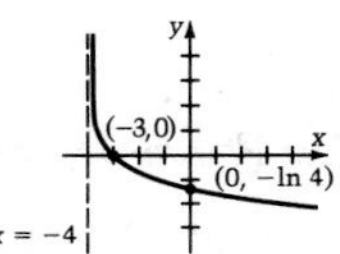

***53.** $y = \ln\sqrt{x+3} = \frac{1}{2} \ln (x+3)$

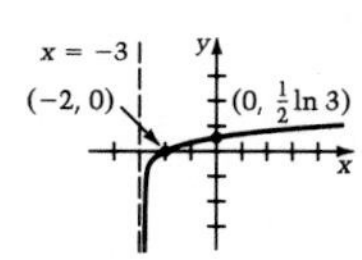

***55.**

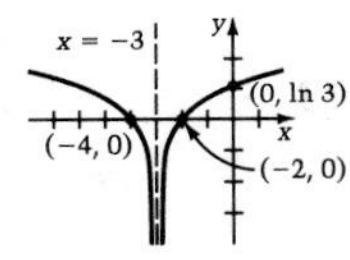

***57.**

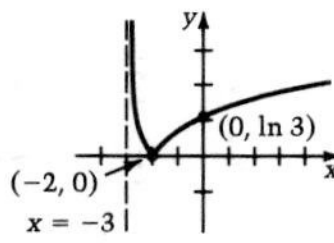

***54.** $y = \ln\sqrt{-x} = \frac{1}{2} \ln(-x)$

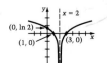

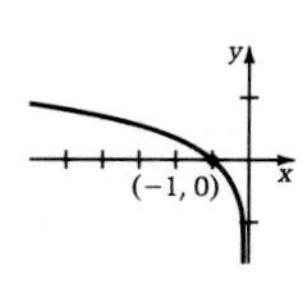

***56.**

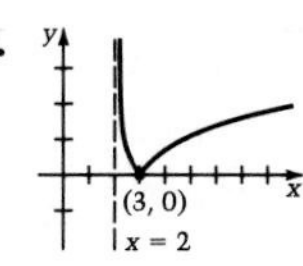

***58.**

***59.** False, $\ln 1 + \ln 1 = 0$ while $\ln (1 + 1) = \ln 2 > 0$. ***61.** False, $(\ln e)(\ln e) = 1$ while $\ln(e \cdot e) = 2$.
***60.** False, $\ln 1 - \ln 1 = 0$ while $\ln (1 - 1) = \ln 0$ is undefined. ***62.** False, $(\ln e)^2 = 1$ while $2 \ln e = 2$.
***63.** True, (provided $x > 0$). ***65.** False, $\ln(e \cdot 1) = 1$ while $e \ln 1 = 0$. ***67.** False, $\ln e^1 = 1$ while $\ln 1^e = 0$.
***64.** True, (provided $x > 0$) ***66.** True ***68.** False, $\ln(e + 1) > 1$ while $1 + \ln 1 = 1$.
***69.** (a) 7 digits (b) 3 digits
***70.** (a) 3 digits (b) 7 digits (c) 19 digits
***71.** $f(x)$ is an approximation of $\ln(1 + x)$, and this approximation becomes less accurate as x increases.
***72.** If x is doubled, Q is increased by 1. If x is tripled, Q is increased by $\log_2 3$. If x is quadrupled, Q is increased by 2.
***73.** 1.26 ***75.** 1.95 ***77.** 0.65 ***79.** −1.59 ***81.** 2.59 ***83.** 4.82 ***85.** 1.63 ***87.** 1.27 ***89.** 1.65
***74.** 1.48 ***76.** 1.89 ***78.** 0.81 ***80.** −1.26 ***82.** 0.35 ***84.** 3.51 ***86.** 1.17 ***88.** 0.58 ***90.** 1.95

*denotes Superset exercises

Section 5.4, pages 323–324

1. 172,800 **3.** 1.6 hours **5.** \$2208 **7.** \$1772 **9.** 14 **11.** \$2226 **13.** 5 years **15.** 45.2 grams
2. 928,000 **4.** 2.9 hours ago **6.** \$1017 **8.** \$621 **10.** 9 **12.** \$1027 **14.** 8 years **16.** 27.0
17. 16,451 years ago **19.** \$13.50 ***21.** At least \$16,172 ***23.** 6.14% ***25.** 7.7%
18. 8174 **20.** 303 million ***22.** At least \$16,296 ***24.** 6.18% ***26.** 2199
***27.** (a) \$27,899.83 (b) 12 years and 12 days ***29.** over 230 days ***31.** 4.8 ***33.** 181°
***28.** 6674 years ago ***30.** slightly over 26 years ***32.** 631 times as intense ***34.** almost 8 minutes
***35.** (a) 65 decibels (b) 89 decibels
***36.** 3.2×10^{-4} watt/cm²

Chapter 5 Review Exercises, pages 328–329

1. 5^{2x} **3.** 2^{-5x+3} **5.** 3^{-4x+5} **7.** $\frac{5}{2}$ **9.** 2 **11.** no solutions **13.**

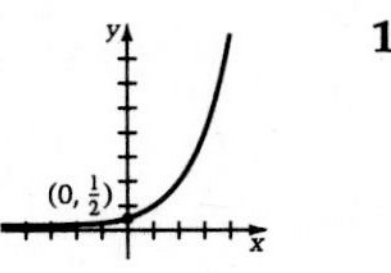

15.

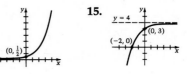

2. 3^{4x} **4.** 3^{-x-1} **6.** 2^{7x-2} **8.** $\frac{6}{5}$ **10.** $\frac{2}{3}$ **12.** no solutions **14.**

16.

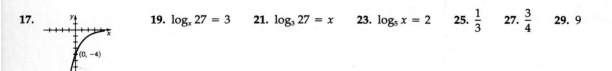

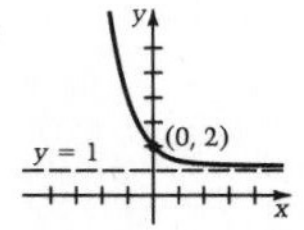

17.

19. $\log_x 27 = 3$ **21.** $\log_3 27 = x$ **23.** $\log_5 x = 2$ **25.** $\frac{1}{3}$ **27.** $\frac{3}{4}$ **29.** 9

18.

20. $\log_9 27 = x$ **22.** $\log_2 x = 5$ **24.** $\log_x 27 = 9$ **26.** $\frac{4}{3}$ **28.** $\left(\frac{1}{2}\right)^{1/3}$ **30.** $\left(\frac{1}{2}\right)^{1/3}$

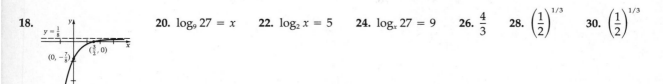

31. 8 **33.**

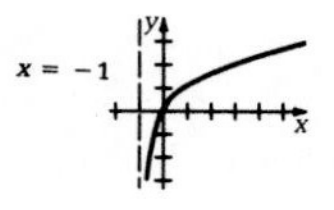

35.

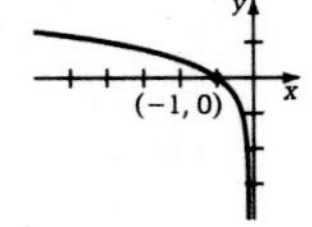

37.

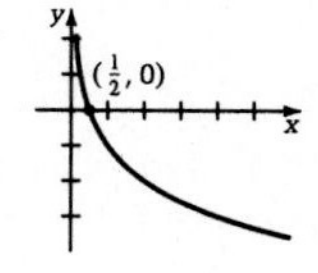

39.

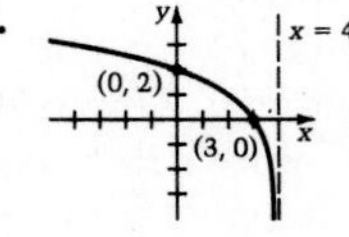

---

*denotes Superset exercises

32. 9 **34.**

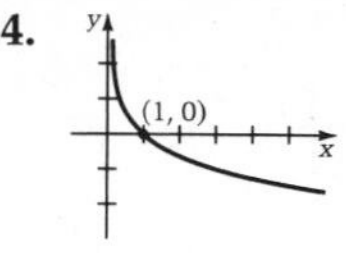

36.

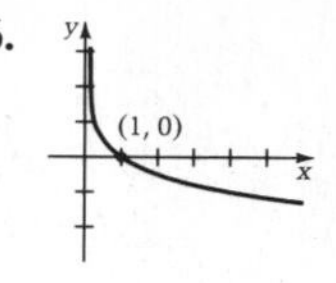

38.

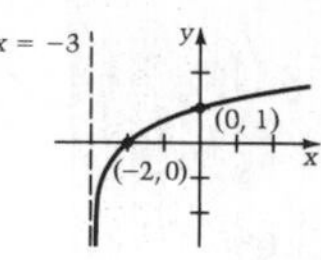

40.

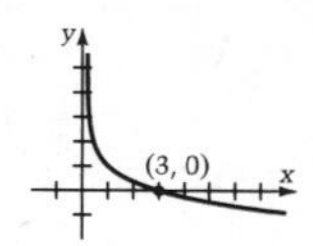

41. -5.5 **43.** 8.2 **45.** -0.5 **47.** $2B + C - A$ **49.** $3C - 2A - B$ **51.** $3A + C + 2$ **53.** 4 **55.** $2^{1/3}$
42. 3.1 **44.** 2.2 **46.** -9.8 **48.** $3B - 2A - 2C$ **50.** $4A + 2C$ **52.** $3 - B - C$ **54.** 64 **56.** 125

57. $1, \sqrt{3}$ **59.** $\frac{1}{4}e^{-3}$ **61.** $\sqrt{5}$ **63.**

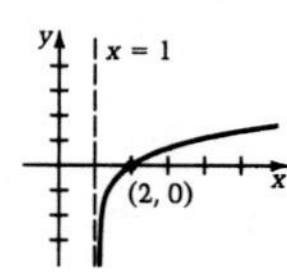

65. $y = \ln(e^{2x}) = 2x$

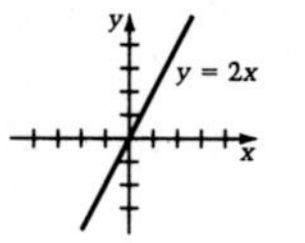

67. $y = e^{\ln 2x} = 2x, x > 0$

58. 3 **60.** $\frac{1}{2}$ **62.** e **64.**

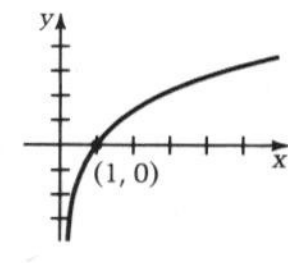

66. $y = \ln(e^3x) = 3 + \ln x$

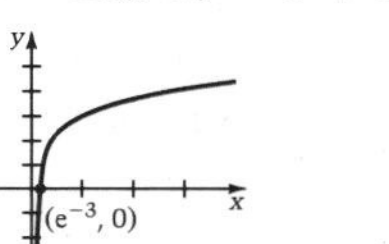

68. $y = e^{\ln(e^x)} = e^x$

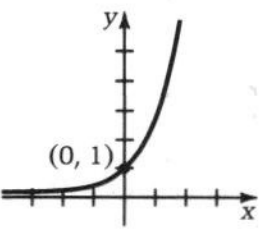

69. $y = \ln\left(\frac{e^2}{x}\right) = 2 - \ln x$ **71.** \$2044.69 **73.** 19,188 years ago. **75.** $-1, 2$ **77.** 1, 2 **79.** 2.29 **81.** 2.33

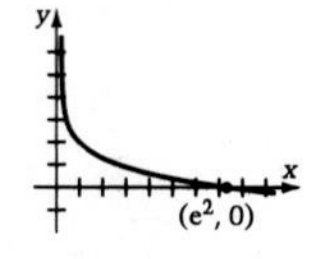

70.

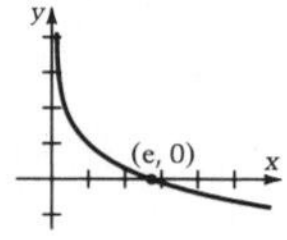

72. \$349.86 **74.** 8.9% **76.** $\pm\sqrt{3}$ **78.** 0 **80.** 2.65 **82.** -2.32

83. 0.54 **85.** 2.71
84. 3823.04 **86.** -10.32

Chapter 5 Test, page 330

1. 2^{-7x+4} **3.**

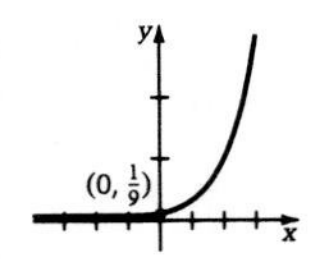

5. no solutions **7.** $\log_x 125 = 3$ **9.** $\log_7 x = 4$ **11.** $\frac{1}{4}$ **13.**

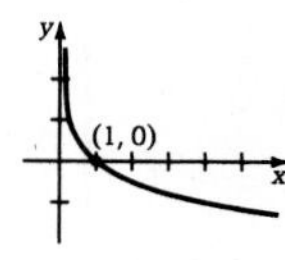

2. 3^{-4x+4} **4.** 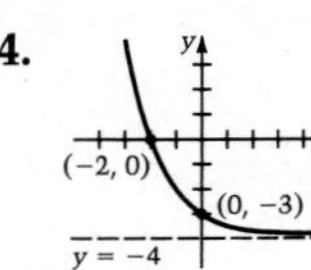

6. $\frac{6}{19}$ **8.** $\log_4 64 = x$ **10.** 3 **12.** 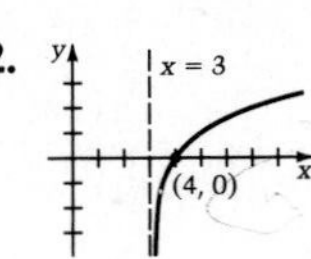

14. 7.6

15. 0.6 **17.** 2 **19.** $x > 0$ **21.** 4 **23.** e^{-2} **25.** **27.** 2.59 **29.** $804

16. -10 **18.** 3 **20.** 5 **22.** e^3 **24.** **26.** 1.73 **28.** 0.6 hours **30.** $808

31. 1.856×10^{-14} grams

Chapter 6

Section 6.1, pages 337–338

1. 67° **3.** 75° **5.** 18 **7.** 10 **9.** 36 **11.** 12 **13.** $\sqrt{13}$ **15.** $\frac{3}{4}$ **17.** 5 **19.** $\frac{32}{5}$ **21.** 10 **23.** 24

2. 106° **4.** 49° **6.** 45° **8.** 9 **10.** 8 **12.** 16 **14.** $5\sqrt{2}$ **16.** 2 **18.** 8.5 **20.** 8 **22.** $\frac{27}{2}$ **24.** 15

25. $\frac{36\sqrt{5}}{5}$ **27.** $\frac{9\sqrt{10}}{5}$ **29.** 42 ft ***31.** 17 mi ***33.** $\frac{19\sqrt{2}}{2}$ in $\times$ $\frac{19\sqrt{2}}{2}$ in, $180\frac{1}{2}$ sq. in. ***35.** 30° ***37.** $\frac{5\sqrt{3}}{2}$

26. $6\sqrt{5}$ **28.** $5\sqrt{6}$ **30.** 64 ft ***32.** 32 ft ***34.** 90° ***36.** 30° ***38.** $\frac{5}{2}$

***39.** $\frac{5}{2}$ ***41.** $\frac{25\sqrt{3}}{8}$ ***43.** 240 ***45.** $12\sqrt{2}$

***40.** $\frac{25\sqrt{3}}{4}$ ***42.** $6\sqrt{3}$ ***44.** 40 ***46.** $\frac{2\sqrt{3}}{3}h$

Section 6.2, pages 344–346

1.	**3.**	**5.**	**7.**	**9.**	**11.**
$\sin\alpha = \frac{3}{5}$	$\sin\alpha = \frac{8}{17}$	$\sin\alpha = \frac{35}{37}$	$\sin\alpha = \frac{3}{5}$	$\sin\alpha = \frac{\sqrt{2}}{2}$	$\sin\alpha = \frac{1}{3}$
$\cos\alpha = \frac{4}{5}$	$\cos\alpha = \frac{15}{17}$	$\cos\alpha = \frac{12}{37}$	$\cos\alpha = \frac{4}{5}$	$\cos\alpha = \frac{\sqrt{2}}{2}$	$\cos\alpha = \frac{2\sqrt{2}}{3}$
$\tan\alpha = \frac{3}{4}$	$\tan\alpha = \frac{8}{15}$	$\tan\alpha = \frac{35}{12}$	$\tan\alpha = \frac{3}{4}$	$\tan\alpha = 1$	$\tan\alpha = \frac{\sqrt{2}}{4}$
$\csc\alpha = \frac{5}{3}$	$\csc\alpha = \frac{17}{8}$	$\csc\alpha = \frac{37}{35}$	$\csc\alpha = \frac{5}{3}$	$\csc\alpha = \sqrt{2}$	$\csc\alpha = 3$
$\sec\alpha = \frac{5}{4}$	$\sec\alpha = \frac{17}{15}$	$\sec\alpha = \frac{37}{12}$	$\sec\alpha = \frac{5}{4}$	$\sec\alpha = \sqrt{2}$	$\sec\alpha = \frac{3\sqrt{2}}{4}$
$\cot\alpha = \frac{4}{3}$	$\cot\alpha = \frac{15}{8}$	$\cot\alpha = \frac{12}{35}$	$\cot\alpha = \frac{4}{3}$	$\cot\alpha = 1$	$\cot\alpha = 2\sqrt{2}$

*denotes Superset exercises

2. $\sin\alpha = \frac{7}{25}$, $\cos\alpha = \frac{24}{25}$, $\tan\alpha = \frac{7}{24}$, $\csc\alpha = \frac{25}{7}$, $\sec\alpha = \frac{25}{24}$, $\cot\alpha = \frac{24}{7}$

4. $\sin\alpha = \frac{12}{13}$, $\cos\alpha = \frac{5}{13}$, $\tan\alpha = \frac{12}{5}$, $\csc\alpha = \frac{13}{12}$, $\sec\alpha = \frac{13}{5}$, $\cot\alpha = \frac{5}{12}$

6. $\sin\alpha = \frac{24}{25}$, $\cos\alpha = \frac{7}{25}$, $\tan\alpha = \frac{24}{7}$, $\csc\alpha = \frac{25}{24}$, $\sec\alpha = \frac{25}{7}$, $\cot\alpha = \frac{7}{24}$

8. $\sin\alpha = \frac{40}{41}$, $\cos\alpha = \frac{9}{41}$, $\tan\alpha = \frac{40}{9}$, $\csc\alpha = \frac{41}{40}$, $\sec\alpha = \frac{41}{9}$, $\cot\alpha = \frac{9}{40}$

10. $\sin\alpha = \frac{\sqrt{2}}{2}$, $\cos\alpha = \frac{\sqrt{2}}{2}$, $\tan\alpha = 1$, $\csc\alpha = \sqrt{2}$, $\sec\alpha = \sqrt{2}$, $\cot\alpha = 1$

12. $\sin\alpha = \frac{\sqrt{3}}{2}$, $\cos\alpha = \frac{1}{2}$, $\tan\alpha = \sqrt{3}$, $\csc\alpha = \frac{2\sqrt{3}}{3}$, $\sec\alpha = 2$, $\cot\alpha = \frac{\sqrt{3}}{3}$

13. $\sin\alpha = \frac{\sqrt{3}}{2}$, $\cos\alpha = \frac{1}{2}$, $\tan\alpha = \sqrt{3}$, $\csc\alpha = \frac{2\sqrt{3}}{3}$, $\sec\alpha = 2$, $\cot\alpha = \frac{\sqrt{3}}{3}$

15. $\sin\alpha = \frac{\sqrt{3}}{3}$, $\cos\alpha = \frac{\sqrt{6}}{3}$, $\tan\alpha = \frac{\sqrt{2}}{2}$, $\csc\alpha = \sqrt{3}$, $\sec\alpha = \frac{\sqrt{6}}{2}$, $\cot\alpha = \sqrt{2}$

17. $c = 53.18$, sin A = 0.8033, cos A = 0.5957, tan A = 1.3485, csc A = 1.2449, sec A = 1.6787, cot A = 0.7416

19. $b = 486.7$, sin A = 0.5163, cos A = 0.8564, tan A = 0.6029, csc A = 1.9370, sec A = 1.1677, cot A = 1.6588

21. $c = 0.9568$, sin A = 0.7508, cos A = 0.6604, tan A = 1.1369, csc A = 1.3318, sec A = 1.5142, cot A = 0.8796

14. $\sin\alpha = \frac{\sqrt{15}}{4}$, $\cos\alpha = \frac{1}{4}$, $\tan\alpha = \sqrt{15}$, $\csc\alpha = \frac{4\sqrt{15}}{15}$, $\sec\alpha = 4$, $\cot\alpha = \frac{\sqrt{15}}{15}$

16. $\sin\alpha = \frac{\sqrt{6}}{6}$, $\cos\alpha = \frac{\sqrt{30}}{6}$, $\tan\alpha = \frac{\sqrt{5}}{5}$, $\csc\alpha = \sqrt{6}$, $\sec\alpha = \frac{\sqrt{30}}{5}$, $\cot\alpha = \sqrt{5}$

18. $c = 72.15$, sin A = 0.4761, cos A = 0.8794, tan A = 0.5414, csc A = 2.1004, sec A = 1.1371, cot A = 1.8471

20. $a = 533.9$, sin A = 0.8366, cos A = 0.5478, tan A = 1.5271, csc A = 1.1954, sec A = 1.8255, cot A = 0.6548

22. $c = 2.258$, sin A = 0.5452, cos A = 0.8384, tan A = 0.6503, csc A = 1.8343, sec A = 1.1928, cot A = 1.5378

23. $a = 3267$, sin A = 0.7986, cos A = 0.6018, tan A = 1.3270, csc A = 1.2522, sec A = 1.6617, cot A = 0.7536

25. $b = 18.931$, sin A = 0.5488, cos A = 0.8360, tan A = 0.6565, csc A = 1.8222, sec A = 1.1962, cot A = 1.5233

27. $\tan\alpha = \frac{a}{b} = \cot\beta$ **29.** $\sin\beta = \frac{b}{c} = \cos\alpha$ **31.** $3\sqrt{3}$ **33.** $4\sqrt{3}$

24. $b = 2659$, sin A = 0.8829, cos A = 0.4695, tan A = 1.8805, csc A = 1.1326, sec A = 2.1297, cot A = 0.5318

26. $a = 25.609$, sin A = 0.3071, cos A = 0.9517, tan A = 0.3226, csc A = 3.2559, sec A = 1.0508, cot A = 3.0997

28. $\csc\alpha = \frac{c}{a} = \sec\beta$ **30.** $\tan\beta = \frac{b}{a} = \cot\alpha$ **32.** $6\sqrt{3}$ **34.** $10\sqrt{3}$

35. 30 **37.** 5.498 **39.** 90.0001 **41.** $14\sqrt{3}$ **43.** $24\sqrt{2}$ **45.** 15 **47.** 45° **49.** 60° **51.** $\frac{4}{5}, \frac{3}{4}$

36. $27\sqrt{3}$ **38.** 17.620 **40.** 51.9666 **42.** $29\sqrt{3}$ **44.** $25\sqrt{2}$ **46.** $24\sqrt{3}$ **48.** 60° **50.** 60° **52.** $\frac{12}{13}, \frac{5}{13}$

53. $\frac{15}{8}, \frac{15}{17}$ **55.** $\frac{4}{5}, \frac{4}{5}$ **57.** $\frac{24}{25}, \frac{7}{24}$ **59.** $\frac{\sqrt{19}}{10}, \frac{9}{10}$ **61.** 0.9231, 0.3846 **63.** 1.8755, 0.8824 **65.** 0.3430, 0.9459

54. $\frac{21}{20}, \frac{20}{21}$ **56.** $\frac{4}{5}, \frac{5}{4}$ **58.** $\frac{12}{35}, \frac{35}{37}$ **60.** $\frac{3}{10}, \frac{3\sqrt{91}}{91}$ **62.** 0.9035, 0.4744 **64.** 1.0499, 0.9525 **66.** 0.9600, 0.2917

***67.** $\sin 60° = \frac{h}{s}$ gives $h = \frac{\sqrt{3}}{2}s$ ***69.** 6 ***71.** $30\sqrt{3}$ ***73.** $36\sqrt{3}$ ft, 36 ft

***68.** 36 ***70.** $16, 16\sqrt{3}/3, 32\sqrt{3}/3$ ***72.** 24

***74.** Wires are $40\sqrt{3}$ ft and 120 ft long and fastened at heights of $20\sqrt{3}$ ft and $60\sqrt{3}$ ft, respectively.

Section 6.3, pages 350–351

1.
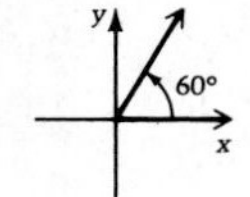

3.
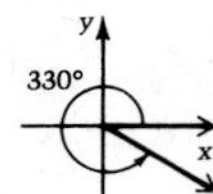

5.
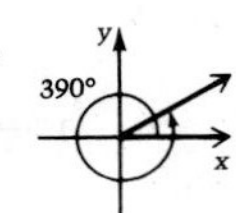

7.
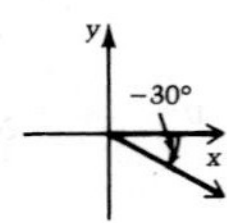

9.
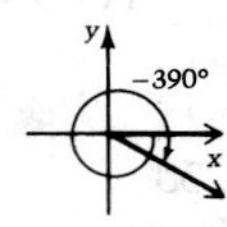

11.
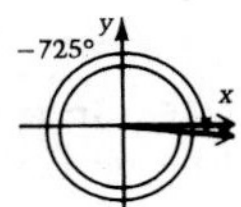

13. 45°

2. **4.**
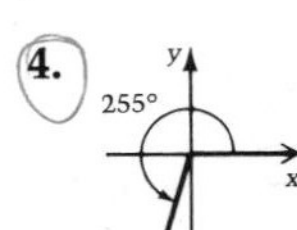

6.
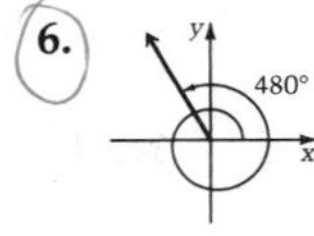

8. **10.**
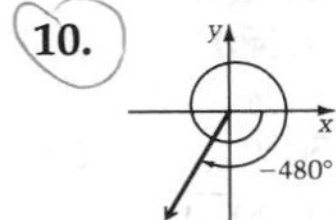

12. **14.** 125°

15. 3° **17.** 322° **19.** 60° **21.** 20°, −340° **23.** 225°, −135° **25.** 310°, −50° **27.** 89°, −271°

16. 270° **18.** 180° **20.** 111° **22.** 102°, −258° **24.** 270°, −90° **26.** 270°, −90° **28.** 220°, −140°

29. 360°, −360° **31.** 135°, −225° **33.** $\frac{1}{2}°, -359\frac{1}{2}°$

35. $\sin\alpha = \frac{3}{5}$, $\cos\alpha = -\frac{4}{5}$, $\tan\alpha = -\frac{3}{4}$, $\csc\alpha = \frac{5}{3}$, $\sec\alpha = -\frac{5}{4}$, $\cot\alpha = -\frac{4}{3}$

37. $\sin\alpha = -\frac{5}{13}$, $\cos\alpha = -\frac{12}{13}$, $\tan\alpha = \frac{5}{12}$, $\csc\alpha = -\frac{13}{5}$, $\sec\alpha = -\frac{13}{12}$, $\cot\alpha = \frac{12}{5}$

39. $\sin\alpha = \frac{3}{5}$, $\cos\alpha = -\frac{4}{5}$, $\tan\alpha = -\frac{3}{4}$, $\csc\alpha = \frac{5}{3}$, $\sec\alpha = -\frac{5}{4}$, $\cot\alpha = -\frac{4}{3}$

30. 360°, −360° **32.** 140°, −220° **34.** $359\frac{1}{2}°, -\frac{1}{2}°$

36. $\sin\alpha = -\frac{3}{5}$, $\cos\alpha = \frac{4}{5}$, $\tan\alpha = -\frac{3}{4}$, $\csc\alpha = -\frac{5}{3}$, $\sec\alpha = \frac{5}{4}$, $\cot\alpha = -\frac{4}{3}$

38. $\sin\alpha = \frac{\sqrt{3}}{2}$, $\cos\alpha = -\frac{1}{2}$, $\tan\alpha = -\sqrt{3}$, $\csc\alpha = \frac{2\sqrt{3}}{3}$, $\sec\alpha = -2$, $\cot\alpha = -\frac{\sqrt{3}}{3}$

40. $\sin\alpha = \frac{12}{13}$, $\cos\alpha = -\frac{5}{13}$, $\tan\alpha = -\frac{12}{5}$, $\csc\alpha = \frac{13}{12}$, $\sec\alpha = -\frac{13}{5}$, $\cot\alpha = -\frac{5}{12}$

41. $\sin\alpha = \frac{\sqrt{2}}{2}$, $\cos\alpha = \frac{\sqrt{2}}{2}$, $\tan\alpha = 1$, $\csc\alpha = \sqrt{2}$, $\sec\alpha = \sqrt{2}$, $\cot\alpha = 1$

43. $\sin\alpha = -\frac{\sqrt{2}}{2}$, $\cos\alpha = \frac{\sqrt{2}}{2}$, $\tan\alpha = -1$, $\csc\alpha = -\sqrt{2}$, $\sec\alpha = \sqrt{2}$, $\cot\alpha = -1$

45. $\sin\alpha = -\frac{3\sqrt{10}}{10}$, $\cos\alpha = -\frac{\sqrt{10}}{10}$, $\tan\alpha = 3$, $\csc\alpha = -\frac{\sqrt{10}}{3}$, $\sec\alpha = -\sqrt{10}$, $\cot\alpha = \frac{1}{3}$

47. $\sin\alpha = \frac{2}{3}$, $\cos\alpha = -\frac{\sqrt{5}}{3}$, $\tan\alpha = -\frac{2\sqrt{5}}{5}$, $\csc\alpha = \frac{3}{2}$, $\sec\alpha = -\frac{3\sqrt{5}}{5}$, $\cot\alpha = -\frac{\sqrt{5}}{2}$

49. $\sin\alpha = 0.3656$, $\cos\alpha = -0.9308$, $\tan\alpha = -0.3928$, $\csc\alpha = 2.7352$, $\sec\alpha = -1.0744$, $\cot\alpha = -2.5459$

---

*denotes Superset exercises

42. $\sin\alpha = \frac{3\sqrt{10}}{10}$, $\cos\alpha = \frac{\sqrt{10}}{10}$, $\tan\alpha = 3$, $\csc\alpha = \frac{\sqrt{10}}{3}$, $\sec\alpha = \sqrt{10}$, $\cot\alpha = \frac{1}{3}$

44. $\sin\alpha = -\frac{2\sqrt{5}}{5}$, $\cos\alpha = \frac{\sqrt{5}}{5}$, $\tan\alpha = -2$, $\csc\alpha = -\frac{\sqrt{5}}{2}$, $\sec\alpha = \sqrt{5}$, $\cot\alpha = -\frac{1}{2}$

46. $\sin\alpha = -\frac{1}{2}$, $\cos\alpha = -\frac{\sqrt{3}}{2}$, $\tan\alpha = \frac{\sqrt{3}}{3}$, $\csc\alpha = -2$, $\sec\alpha = -\frac{2\sqrt{3}}{3}$, $\cot\alpha = \sqrt{3}$

48. $\sin\alpha = \frac{2\sqrt{7}}{7}$, $\cos\alpha = \frac{\sqrt{21}}{7}$, $\tan\alpha = -\frac{2\sqrt{3}}{3}$, $\csc\alpha = \frac{\sqrt{7}}{2}$, $\sec\alpha = -\frac{\sqrt{21}}{3}$, $\cot\alpha = -\frac{\sqrt{3}}{2}$

50. $\sin\alpha = -0.8682$, $\cos\alpha = -0.4961$, $\tan\alpha = 1.7500$, $\csc\alpha = -1.1518$, $\sec\alpha = -2.0155$, $\cot\alpha = 0.5712$

51. $\sin\alpha = -0.9442$, $\cos\alpha = 0.3293$, $\tan\alpha = -2.8678$, $\csc\alpha = -1.0591$, $\sec\alpha = 3.0372$, $\cot\alpha = -0.3487$

53. $\sin\alpha = -0.3780$, $\cos\alpha = -0.9258$, $\tan\alpha = 0.4083$, $\csc\alpha = -2.6456$, $\sec\alpha = -1.0801$, $\cot\alpha = 2.4494$

55. 1 **57.** undefined **59.** undefined **61.** 0 **63.** 0

52. $\sin\alpha = -0.9135$, $\cos\alpha = 0.4068$, $\tan\alpha = -2.2455$, $\csc\alpha = -1.0947$, $\sec\alpha = 2.4581$, $\cot\alpha = -0.4453$

54. $\sin\alpha = 0.5009$, $\cos\alpha = -0.8655$, $\tan\alpha = -0.5788$, $\csc\alpha = 1.9962$, $\sec\alpha = -1.1554$, $\cot\alpha = -1.7277$

56. -1 **58.** undefined **60.** 1 **62.** 1 **64.** 1

65. undefined ***67.** 120° ***69.** $-450°$

***71.** $\sin\varphi = \frac{5\sqrt{26}}{26}$, $\cos\varphi = -\frac{\sqrt{26}}{26}$, $\csc\varphi = \frac{\sqrt{26}}{5}$, $\sec\varphi = -\sqrt{26}$, $\cot\varphi = -\frac{1}{5}$

***73.** $\sin\varphi = \frac{3}{5}$, $\cos\varphi = -\frac{4}{5}$, $\tan\varphi = -\frac{3}{4}$, $\sec\varphi = -\frac{5}{4}$, $\cot\varphi = -\frac{4}{3}$

***75.** $\sin\varphi = \frac{3\sqrt{34}}{34}$, $\cos\varphi = -\frac{5\sqrt{34}}{34}$, $\tan\varphi = -\frac{3}{5}$, $\csc\varphi = \frac{\sqrt{34}}{3}$, $\sec\varphi = -\frac{\sqrt{34}}{5}$

66. undefined ***68.** $-120°$ ***70.** 108°

***72.** $\cos\varphi = -\frac{3}{5}$, $\tan\varphi = -\frac{4}{3}$, $\csc\varphi = \frac{5}{4}$, $\sec\varphi = -\frac{5}{3}$, $\cot\varphi = -\frac{3}{4}$

***74.** $\sin\varphi = \frac{4}{5}$, $\cos\varphi = -\frac{3}{5}$, $\tan\varphi = -\frac{4}{3}$, $\csc\varphi = \frac{5}{4}$, $\sec\varphi = -\frac{5}{3}$

***76.** $\sin\varphi = \frac{\sqrt{2}}{2}$, $\cos\varphi = -\frac{\sqrt{2}}{2}$, $\csc\varphi = \sqrt{2}$, $\sec\varphi = -\sqrt{2}$, $\cot\varphi = -1$

***77.** $\sin\varphi = \frac{4}{5}$, $\tan\varphi = -\frac{4}{3}$, $\csc\varphi = \frac{5}{4}$, $\sec\varphi = -\frac{5}{3}$, $\cot\varphi = -\frac{3}{4}$

***79.** $\cos\varphi = -\frac{\sqrt{21}}{5}$, $\tan\varphi = \frac{2\sqrt{21}}{21}$, $\csc\varphi = -\frac{5}{2}$, $\sec\varphi = -\frac{5\sqrt{21}}{21}$, $\cot\varphi = \frac{\sqrt{21}}{2}$

***81.** $\sin\varphi = \frac{4}{5}$, $\tan\varphi = -\frac{4}{3}$, $\csc\varphi = \frac{5}{4}$, $\sec\varphi = -\frac{5}{3}$, $\cot\varphi = -\frac{3}{4}$

***83.** $\sin\varphi = -\frac{\sqrt{11}}{6}$, $\cos\varphi = \frac{5}{6}$, $\tan\varphi = -\frac{\sqrt{11}}{5}$, $\csc\varphi = -\frac{6\sqrt{11}}{11}$, $\cot\varphi = -\frac{5\sqrt{11}}{11}$

***85.** $\sin\varphi = -\frac{\sqrt{2}}{2}$, $\cos\varphi = \frac{\sqrt{2}}{2}$, $\csc\varphi = -\sqrt{2}$, $\sec\varphi = \sqrt{2}$, $\cot\varphi = -1$

---

*denotes Superset exercises

*78. $\sin\varphi = \frac{\sqrt{6}}{3}$
$\cos\varphi = -\frac{\sqrt{3}}{3}$
$\tan\varphi = -\sqrt{2}$
$\csc\varphi = \frac{\sqrt{6}}{2}$
$\cot\varphi = -\frac{\sqrt{2}}{2}$

*80. $\cos\varphi = -\frac{4}{5}$
$\tan\varphi = -\frac{3}{4}$
$\csc\varphi = \frac{5}{3}$
$\sec\varphi = -\frac{5}{4}$
$\cot\varphi = -\frac{4}{3}$

*82. $\sin\varphi = -\frac{\sqrt{51}}{10}$
$\tan\varphi = -\frac{\sqrt{51}}{7}$
$\csc\varphi = -\frac{10\sqrt{51}}{51}$
$\sec\varphi = \frac{10}{7}$
$\cot\varphi = -\frac{7\sqrt{51}}{51}$

*84. $\sin\varphi = -\frac{\sqrt{2}}{2}$
$\cos\varphi = -\frac{\sqrt{2}}{2}$
$\csc\varphi = -\sqrt{2}$
$\sec\varphi = -\sqrt{2}$
$\cot\varphi = 1$

*86. $\sin\varphi = \frac{2}{3}$
$\cos\varphi = -\frac{\sqrt{5}}{3}$
$\tan\varphi = -\frac{2\sqrt{5}}{5}$
$\sec\varphi = -\frac{3\sqrt{5}}{5}$
$\cot\varphi = -\frac{\sqrt{5}}{2}$

Section 6.4, pages 358–360

1. $\tan 32° = 0.62$
$\cot 32° = 1.60$
$\sec 32° = 1.18$
$\csc 32° = 1.89$

3. $\sin 66° = 0.91$
$\cos 66° = 0.41$
$\cot 66° = 0.44$
$\csc 66° = 1.09$

5. 36°
7. 52°
9. 45°
11. sin 63°
13. tan 52°
15. sec 87°

2. $\cos 55° = 0.57$
$\cot 55° = 0.70$
$\sec 55° = 1.74$
$\csc 55° = 1.22$

4. $\sin 23° = 0.39$
$\cot 23° = 2.38$
$\sec 23° = 1.09$
$\csc 23° = 2.59$

6. 78°
8. 9°
10. 19°
12. csc 78°
14. cot 49°
16. sin 47°

17. cos 71°
19. 0.85
21. 2.56
23. 2.38

25. $\csc\alpha = 2$
$\cos\alpha = \frac{\sqrt{3}}{2}$
$\sec\alpha = \frac{2\sqrt{3}}{3}$
$\tan\alpha = \frac{\sqrt{3}}{3}$
$\cot\alpha = \sqrt{3}$

27. $\cot\alpha = \frac{2}{3}$
$\sec\alpha = \frac{\sqrt{13}}{2}$
$\cos\alpha = \frac{2\sqrt{13}}{13}$
$\sin\alpha = \frac{3\sqrt{13}}{13}$
$\csc\alpha = \frac{\sqrt{13}}{3}$

29. $\sin\alpha = \frac{1}{3}$
$\cos\alpha = \frac{2\sqrt{2}}{3}$
$\sec\alpha = \frac{3\sqrt{2}}{4}$
$\tan\alpha = \frac{\sqrt{2}}{4}$
$\cot\alpha = 2\sqrt{2}$

18. cos 67°
20. 1.89
22. 1.60
24. 0.42

26. $\sec\alpha = \frac{10}{3}$
$\sin\alpha = \frac{\sqrt{91}}{10}$
$\csc\alpha = \frac{10\sqrt{91}}{91}$
$\tan\alpha = \frac{\sqrt{91}}{3}$
$\cot\alpha = \frac{3\sqrt{91}}{91}$

28. $\tan\alpha = 2$
$\sec\alpha = \sqrt{5}$
$\cos\alpha = \frac{\sqrt{5}}{5}$
$\csc\alpha = \frac{\sqrt{5}}{2}$
$\sin\alpha = \frac{2\sqrt{5}}{5}$

30. $\cos\alpha = \frac{1}{10}$
$\sin\alpha = \frac{3\sqrt{11}}{10}$
$\csc\alpha = \frac{10\sqrt{11}}{33}$
$\tan\alpha = 3\sqrt{11}$
$\cot\alpha = \frac{\sqrt{11}}{33}$

31. $\sec\alpha = \frac{5}{4}$
$\sin\alpha = \frac{3}{5}$
$\csc\alpha = \frac{5}{3}$
$\tan\alpha = \frac{3}{4}$
$\cot\alpha = \frac{4}{3}$

33. 18.782°
35. 71.227°
37. 24.851°
39. 43.008°
41. 45.265°
43. 69°30′36″

32. $\csc\alpha = \frac{10}{7}$
$\cos\alpha = \frac{\sqrt{51}}{10}$
$\sec\alpha = \frac{10\sqrt{51}}{51}$
$\tan\alpha = \frac{7\sqrt{51}}{51}$
$\cot\alpha = \frac{\sqrt{51}}{7}$

34. 82.436°
36. 9.014°
38. 32.971°
40. 51.632°
42. 30.760°
44. 38°47′28″

---

*denotes Superset exercises

45. 20°56′31″ **47.** 8°22′19″ **49.** 86°42′14″ **51.** 47°9′54″ **53.** 0.5175 **55.** 0.3973 **57.** 1.828 **59.** 0.8952
46. 74°8′35″ **48.** 13°27′0″ **50.** 2°37′8″ **52.** 53°16′52″ **54.** 0.7214 **56.** 1.164 **58.** 1.017 **60.** 3.703
61. 0.6111 **63.** 1.402 **65.** 6°30′ **67.** 71° **69.** 60°10′ **71.** 44°50′ **73.** 13°50′ **75.** 62°30′
62. 0.9853 **64.** 0.1974 **66.** 17° **68.** 45°40′ **70.** 23°10′ **72.** 77°10′ **74.** 83°30′ **76.** 34°

77. tan 32° = 0.6249, csc 32° = 1.8871, sec 32° = 1.1792, cot 32° = 1.6003
79. csc 55° = 1.2208, sec 55° = 1.7435, cos 55° = 0.5738, cot 55° = 0.7002
81. $\cos\alpha = 0.8587$, $\tan\alpha = 0.5967$, $\csc\alpha = 1.9516$, $\sec\alpha = 1.1646$, $\cot\alpha = 1.6759$
83. $\sin\alpha = 0.5672$, $\tan\alpha = 0.6886$, $\csc\alpha = 1.7631$, $\sec\alpha = 1.2142$, $\cot\alpha = 1.4521$
78. sin 66° = 0.9135, cos 66° = 0.4067, csc 66° = 1.0946, cot 66° = 0.4452
80. sin 23° = 0.3907, cos 23° = 0.9205, sec 23° = 1.0864, cot 23° = 2.3557
82. $\sin\alpha = 0.9523$, $\tan\alpha = 3.1213$, $\csc\alpha = 1.0501$, $\sec\alpha = 3.2776$, $\cot\alpha = 0.3204$
84. $\cos\alpha = 0.6446$, $\tan\alpha = 1.1860$, $\csc\alpha = 1.3080$, $\sec\alpha = 1.5513$, $\cot\alpha = 0.8432$

85. 0.5180 **87.** 0.3971 **89.** 0.6122 **91.** 1.2020 **93.** 0.8941 **95.** 1.4267 **97.** 20.40° **99.** 38.59°
86. 0.7218 **88.** 1.1638 **90.** 0.9862 **92.** 5.4874 **94.** 1.0388 **96.** 0.1962 **98.** 81.13° **100.** 31.92°
101. 63.84° **103.** 38.01° **105.** 40.24° **107.** 60.00° **109.** 0.6368 **111.** 0.6100 **113.** 0.3181 **115.** 0.1352
102. 65.62° **104.** 23.10° **106.** 15.64° **108.** 60.00° **110.** 0.2171 **112.** 0.9452 **114.** 0.8925 **116.** 0.2518
117. 25.8542 **119.** 2.5699 ***121.** yes ***123.** yes ***125.** yes ***127.** yes ***129.** $\frac{\sqrt{3}}{2}, \frac{1}{2}$ ***131.** $-\sqrt{3}, \frac{\sqrt{3}}{3}$
118. 1.3996 **120.** 1.2639 ***122.** no ***124.** yes ***126.** yes ***128.** no ***130.** $1, \frac{\sqrt{3}+1}{2}$ ***132.** $0, \sqrt{2}$
***133–136.** substitute function values for special angles ***137.** 0.6691 ***139.** 0.6249 ***141.** −0.8090
***133–136.** substitute function values for special angles ***138.** 1.5399 ***140.** −0.6427 ***142.** 0.5878
***143.** −0.9603 ***145.** 1.0457 ***147.** 135.60° ***149.** 275.20° ***151.** 321.60° ***153.** 196.30°
***144.** −0.2790 ***146.** −1.3902 ***148.** 229.40° ***150.** 101.50° ***152.** 351.81° ***154.** 249.16°

Section 6.5, pages 366–367

1. $A = 45°, b = 32\sqrt{2}, a = 32\sqrt{2}$ **3.** $B = 30°, c = 88, a = 44\sqrt{3}$ **5.** $c = 30, A = 30°, B = 60°$
2. $B = 60°, a = 24, b = 24\sqrt{3}$ **4.** $A = 30°, c = 8\sqrt{3}, a = 4\sqrt{3}$ **6.** $a = 40\sqrt{3}, B = 30°, A = 60°$
7. $B = 45°, c = 44\sqrt{2}, b = 44$ **9.** $b = 25\sqrt{2}, B = 45°, A = 45°$ **11.** $B = 54°, c \approx 120, b \approx 96$
8. $A = 60°, c = 16\sqrt{3}, b = 8\sqrt{3}$ **10.** $c = 25\sqrt{2}, A = 45°, B = 45°$ **12.** $A = 18°, c \approx 130, b \approx 120$
13. $A = 78°, c \approx 9.2, b \approx 1.9$ **15.** $B = 52°, a \approx 7.6, b \approx 9.7$ **17.** $B = 61°40', c \approx 0.277, a \approx 0.132$
14. $B = 37°, a \approx 16, b \approx 12$ **16.** $B = 23°, c \approx 46, b \approx 18$ **18.** $B = 38°20', c \approx 12.7, a \approx 9.93$
19. $A = 18°40', b \approx 135.0, a \approx 45.61$ **21.** $b = 36.42, c = 43.33, \angle B = 57.2°$ **23.** $a = 1422.5, c = 1922.3, \angle B = 42.27°$
20. $A = 66°20', b \approx 2.722, a \approx 6.212$ **22.** $b = 2.953, c = 3.475, \angle A = 31.8°$ **24.** $a = 0.633, c = 2.898, \angle B = 77.38°$
25. $a = 9.353, b = 17.451, \angle B = 61.81°$ **27.** $c = 3.3780, \angle A = 43.52°, \angle B = 46.48°$
26. $a = 378.97, b = 333.88, \angle A = 48.62°$ **28.** $b = 293.087, \angle A = 16.15°, \angle B = 73.85°$
29. $c = 1.0097, \angle A = 12.77°, \angle B = 77.22°$ **31.** 21 ft **33.** 5800 ft **35.** 5.2 ft **37.** 57°
30. $a = 1718.11, \angle A = 76.95°, \angle B = 13.05°$ **32.** 30° **34.** 52 ft **36.** 14 ft **38.** 510 ft
39. (a) 60° (b) 40° (c) 76° (d) 22° ***41.** 2700 feet ***43.** 23 feet ***45.** 48.4 feet
40. (a) 282 ft (b) 971 ft ***42.** 84 feet ***44.** 17° ***46.** 1105 feet

Section 6.6, page 371

1. positive **3.** positive **5.** positive **7.** positive **9.** positive **11.** positive **13.** negative
2. negative **4.** negative **6.** negative **8.** positive **10.** negative **12.** negative **14.** negative
15. positive **17.** 20° **19.** 40° **21.** 55° **23.** 70° **25.** 84° **27.** 5° **29.** 85°35′ **31.** 29.7° **33.** $-\frac{1}{2}$

---

*denotes Superset exercises

16. negative **18.** 5° **20.** 17° **22.** 24° **24.** 5° **26.** 5° **28.** 1° **30.** 84°2′ **32.** 67.5° **34.** $-\frac{2\sqrt{3}}{3}$

35. $-\sqrt{3}$ **37.** 1 **39.** $\frac{2\sqrt{3}}{3}$ **41.** $-\frac{2\sqrt{3}}{3}$ **43.** $\frac{\sqrt{3}}{3}$ **45.** $\frac{1}{2}$ **47.** −0.4040 **49.** 0.6691 **51.** −1.192

36. $\frac{\sqrt{3}}{2}$ **38.** $-\sqrt{2}$ **40.** 0 **42.** 2 **44.** $-\frac{\sqrt{3}}{2}$ **46.** $\frac{1}{2}$ **48.** −0.2588 **50.** −1.072 **52.** 0.6249

53. −0.9994 **55.** 5.759 **57.** −0.8090 **59.** −1.494 **61.** 0.9957 **63.** 0.1853 **65.** −0.7581 ***67.** negative
54. −0.5736 **56.** 1.001 **58.** 1.376 **60.** 2.366 **62.** 0.4522 **64.** −0.3393 **66.** 0.4578 ***68.** negative
***69.** negative ***71.** negative ***73.** negative ***75.** negative ***77.** positive ***79.** negative ***81.** negative
***70.** positive ***72.** negative ***74.** negative ***76.** positive ***78.** positive ***80.** negative ***82.** positive

Section 6.7, pages 377–378

1. $C = 105°, b = 20, c \approx 27$ **3.** $A = 15°, a \approx 2.9, c = 8.0$ **5.** $C = 56°, b \approx 9.9, c \approx 20$ **7.** $A = 80°, a \approx 46, b \approx 27$
2. $C = 75°, a = 24, c \approx 27$ **4.** $A = 15°, a \approx 2.9, b = 8.0$ **6.** $C = 63°, a \approx 80, c \approx 73$ **8.** $B = 54°, b \approx 80, c \approx 36$
9. $A = 41°40', b \approx 1.04, c \approx 1.01$ **11.** $a \approx 135.4, b \approx 45.69, \angle C = 59.0°$ **13.** $b \approx 4485.6, c \approx 2535.9, \angle A = 71.2°$
10. $B = 50°30', a \approx 0.391, c \approx 0.645$ **12.** $a \approx 63.16, c \approx 72.34, \angle B = 86.1°$ **14.** $a \approx 350.11, b \approx 465.54, \angle C = 37.3°$
15. $c \approx 19.66, \angle A \approx 47.1°, \angle C \approx 74.3°$ **17.** $b \approx 12860., \angle A \approx 23.5°, \angle B \approx 124.8°$ **19.** $a \approx 360.9, \angle A \approx 26.2°, \angle C \approx 41.5°$
16. $c \approx 0.2933, \angle A \approx 64.2°, \angle C \approx 37.4°$ **18.** $b \approx 3.5518, \angle B \approx 69.8°, \angle C \approx 47.7°$ **20.** $a \approx 584.9, \angle A \approx 61.8°, \angle B \approx 44.4°$
21. $B = 90°, C = 60°, c = 12\sqrt{3}$ **23.** no triangle is possible **25.** $B \approx 51°, C \approx 9°, c \approx 9.8$
22. $B \approx 35°, C \approx 85°, c \approx 41$ **24.** no triangle is possible **26.** $B \approx 35°, C \approx 10°, c \approx 11$
27. two possible triangles: $B \approx 63°50', C \approx 63°40', c \approx 36.4$, and $B \approx 116°10', C \approx 11°20'$, and $c \approx 7.97$.
28. two possible triangles: $B \approx 48°30', C \approx 93°00', c \approx 33.0$, and $B \approx 131°30', C \approx 10°00', c \approx 5.74$
29. 431 yd **31.** N 0°30′ E **33.** from first ship: 4.4 mi, from second ship: 6.2 mi **35.** 60.8 feet
30. from A to boat: 0.4820 mi, from B to boat: 0.5163 mi **32.** 1.2 mi
34. from first ship: 4.7 mi, from second ship: 6.7 mi **36.** 8.0 miles

***37.** $x = \dfrac{m \tan\theta \tan\varphi}{\tan\varphi - \tan\theta}$ ***39.** Use $\frac{a}{b} = \frac{\sin A}{\sin B}$ from Law of Sines to evaluate left side of statement.

***38.** Use $\frac{a}{b} = \frac{\sin A}{\sin B}$ from Law of Sines to evaluate left side of statement.

***40.** Use Area $= \frac{1}{2}hc$, where $h = a \sin B$ = length of altitude from C and $c = \dfrac{a \sin C}{\sin A}$ by Law of Sines

Section 6.8, pages 382–383

1. $c \approx 11, a \approx 70°, B \approx 50°$ **3.** $a \approx 31, B \approx 35°, C \approx 25°$ **5.** $b \approx 95, A \approx 72°, C \approx 54°$
2. $c \approx 6, A \approx 94°, B \approx 56°$ **4.** $a \approx 50, B \approx 20°, C \approx 25°$ **6.** $b \approx 184, A \approx 19°, C \approx 33°$
7. $a \approx 27.8, B \approx 13°50', C \approx 18°30'$ **9.** $A \approx 37°, B \approx 53°, C \approx 90°$ **11.** $A \approx 66°, B \approx 62°, C \approx 52°$
8. $b \approx 12.7, A \approx 90°30', C \approx 54°40'$ **10.** $A \approx 23°, B \approx 67°, C \approx 90°$ **12.** $A \approx 42°, B \approx 75°, C \approx 63°$
13. $A \approx 29°, B \approx 47°, C \approx 104°$ **15.** $A \approx 62°40', B \approx 36°20', C \approx 81°00'$ **17.** $c \approx 11264, \angle A \approx 29.1°, \angle B \approx 41.6°$
14. $A \approx 100°, B \approx 44°, C \approx 36°$ **16.** $A \approx 58°50', B \approx 34°50', C \approx 86°20'$ **18.** $c \approx 48.92, \angle A \approx 52.1°, \angle B \approx 47.2°$
19. $a \approx 7.558, \angle B \approx 83.3°, \angle C \approx 45.2°$ **21.** $b \approx 0.9504, \angle A \approx 63.6°, \angle C \approx 54.8°$
20. $a \approx 319.37, \angle B \approx 40.0°, \angle C \approx 41.6°$ **22.** $b \approx 8609.3, \angle A \approx 64.60°, \angle C \approx 46.69°$
23. $\angle A \approx 39.12°, \angle B \approx 77.75°, \angle C \approx 63.13°$ **25.** $\angle A \approx 45.04°, \angle B \approx 32.24°, \angle C \approx 102.72°$ **27.** 170 m
24. $\angle A \approx 81.06°, \angle B \approx 54.62°, \angle C \approx 44.32°$ **26.** $\angle A \approx 51.71°, \angle B \approx 60.12°, \angle C \approx 68.17°$ **28.** 70 ft
29. 3.9, 7.2 **31.** 47° **33.** 99.8 **35.** 52.1 **37.** 672.0 **39.** 7737 **41.** 506.0 **43.** 6.329
30. 51.6 in., 38.2 in. **32.** 110°, 70° **34.** 137.4 **36.** 32.3 **38.** 252.6 **40.** 0.019 **42.** 91,000 **44.** 20.77
***45.** $A \approx 65°30', B \approx 69°50', C = 44°40', a \approx 15.5$ in or $A \approx 114°30', B \approx 38°10', C \approx 27°20', a \approx 23.6$ in. ***47.** 47.0

*denotes Superset exercises

*46. 7.5 *48. $b = a\sqrt{2(1 - \cos\theta)}$
*49. use Law of Sines *51. 151 in. *53. 1710 sq. in
*50. use Law of Cosines *52. 220 cm *54. 318.5

Chapter 6 Review Exercises, pages 388–389

1. $4\sqrt{34}$ **3.** $\frac{27}{2}$ **5.** $\frac{64\sqrt{3}}{3}$ **7.** $\sin\alpha = \frac{2\sqrt{13}}{13}$, $\cos\alpha = \frac{3\sqrt{13}}{13}$, $\tan\alpha = \frac{2}{3}$, $\csc\alpha = \frac{\sqrt{13}}{2}$, $\sec\alpha = \frac{\sqrt{13}}{3}$, $\cot\alpha = \frac{3}{2}$ **9.** $\sin\alpha = \frac{\sqrt{3}}{2}$, $\cos\alpha = \frac{1}{2}$, $\tan\alpha = \sqrt{3}$, $\csc\alpha = \frac{2\sqrt{3}}{3}$, $\sec\alpha = 2$, $\cot\alpha = \frac{\sqrt{3}}{3}$ **11.** $\frac{10\sqrt{3}}{3}$ **13.** 60° **15.** $\frac{\sqrt{21}}{5}, \frac{\sqrt{21}}{2}$

2. $2\sqrt{3}$ **4.** $2\sqrt{13}$ **6.** $3\sqrt{3}$ **8.** $\sin\alpha = \frac{1}{2}$, $\cos\alpha = \frac{\sqrt{3}}{2}$, $\tan\alpha = \frac{\sqrt{3}}{3}$, $\csc\alpha = 2$, $\sec\alpha = \frac{2\sqrt{3}}{3}$, $\cot\alpha = \sqrt{3}$ **10.** $\sin\alpha = \frac{5\sqrt{26}}{26}$, $\cos\alpha = \frac{\sqrt{26}}{26}$, $\tan\alpha = 5$, $\csc\alpha = \frac{\sqrt{26}}{5}$, $\sec\alpha = \sqrt{26}$, $\cot\alpha = \frac{1}{5}$ **12.** $10\sqrt{2}$ **14.** 30° **16.** $\frac{\sqrt{34}}{3}, \frac{5\sqrt{34}}{34}$

17. $\frac{2\sqrt{2}}{3}, 2\sqrt{2}$ **19.** $\sin\alpha = \frac{2\sqrt{29}}{29}$, $\cos\alpha = -\frac{5\sqrt{29}}{29}$, $\tan\alpha = -\frac{2}{5}$, $\csc\alpha = \frac{\sqrt{29}}{2}$, $\sec\alpha = -\frac{\sqrt{29}}{5}$, $\cot\alpha = -\frac{5}{2}$ **21.** $\sin\alpha = -1$, $\cos\alpha = 0$, $\tan\alpha$ is undefined, $\csc\alpha = -1$, $\sec\alpha$ is undefined, $\cot\alpha = 0$ **23.** cos 58° **25.** tan 72° **27.** 23.76°

18. $\frac{\sqrt{3}}{3}, \frac{1}{2}$ **20.** $\sin\alpha = -\frac{\sqrt{17}}{17}$, $\cos\alpha = \frac{4\sqrt{17}}{17}$, $\tan\alpha = -\frac{1}{4}$, $\csc\alpha = -\sqrt{17}$, $\sec\alpha = \frac{\sqrt{17}}{4}$, $\cot\alpha = -4$ **22.** $\sin\alpha = 0$, $\cos\alpha = -1$, $\tan\alpha = 0$, $\csc\alpha$ is undefined, $\sec\alpha = -1$, $\cot\alpha$ is undefined **24.** csc 86° **26.** sec 46° **28.** 47.455°

29. 73°15′54″ **31.** 0.5760 **33.** 1.0116 **35.** 5.0658 **37.** 24° **39.** 39° **41.** 15° **43.** $\sin\alpha = 0.8275$, $\tan\alpha = 1.4741$, $\csc\alpha = 1.2084$, $\sec\alpha = 1.7813$, $\cot\alpha = 0.6784$

*denotes Superset exercises

30. $11°48'18''$ **32.** 0.6371 **34.** 0.4173 **36.** 3.3565 **38.** 56° **40.** 48° **42.** 65°

44. $\cos\alpha = 0.9385$
$\tan\alpha = 0.3678$
$\csc\alpha = 2.8969$
$\sec\alpha = 1.0655$
$\cot\alpha = 2.7188$

45. $\sin\alpha = 0.8808$
$\cos\alpha = 0.4735$
$\csc\alpha = 1.1353$
$\sec\alpha = 2.1120$
$\cot\alpha = 0.5376$

47. $\sin\alpha = 0.2447$
$\cos\alpha = 0.9696$
$\tan\alpha = 0.2523$
$\sec\alpha = 1.0313$
$\cot\alpha = 3.9629$

49. $\csc x = \frac{4}{3}$
$\cos x = \frac{\sqrt{7}}{4}$
$\sec x = \frac{4\sqrt{7}}{7}$
$\tan x = \frac{3\sqrt{7}}{7}$
$\cot x = \frac{\sqrt{7}}{3}$

51. $\cot x = \frac{3}{4}$
$\sec x = \frac{5}{3}$
$\cos x = \frac{3}{5}$
$\csc x = \frac{5}{4}$
$\sin x = \frac{4}{5}$

53. 0.9527 **55.** 2.0890

46. $\sin\alpha = 0.9238$
$\cos\alpha = 0.3830$
$\tan\alpha = 2.4122$
$\csc\alpha = 1.0825$
$\cot\alpha = 0.4146$

48. $\sin\alpha = 0.9691$
$\cos\alpha = 0.2468$
$\tan\alpha = 3.9262$
$\csc\alpha = 1.0319$
$\sec\alpha = 4.0515$

50. $\sec x = \frac{10}{9}$
$\sin x = \frac{\sqrt{19}}{10}$
$\csc x = \frac{10\sqrt{19}}{19}$
$\tan x = \frac{\sqrt{19}}{9}$
$\cot x = \frac{9\sqrt{19}}{19}$

52. $\tan x = 3$
$\sec x = \sqrt{10}$
$\cos x = \frac{\sqrt{10}}{10}$
$\csc x = \frac{\sqrt{10}}{3}$
$\sin x = \frac{3\sqrt{10}}{0}$

54. 0.7481 **56.** 5.1484

57. 29.39° **59.** 38.05° **61.** 30 ft **63.** $\frac{\sqrt{2}}{2}$ **65.** $-\frac{2\sqrt{3}}{3}$ **67.** 0 **69.** 0 **71.** $C = 69°, b \approx 22, c \approx 21$

58. 56.48° **60.** 2.76° **62.** 6.7 ft **64.** $\frac{\sqrt{3}}{2}$ **66.** $\frac{2\sqrt{3}}{3}$ **68.** 0 **70.** 1 **72.** $A = 53°, b \approx 15, c \approx 21$

73. two possible triangles: $B \approx 69°, C \approx 71°, c \approx 16$, and $B \approx 111°, C \approx 29°, c \approx 8.3$ **75.** $c \approx 4.1, A \approx 49°, c \approx 105°$

74. no triangle is possible **76.** $c \approx 18, A \approx 33°, C \approx 47°$

77. $A \approx 28°, B \approx 36°, C \approx 116°$

79. $\cos\alpha = \sqrt{1 - t^2}$
$\tan\alpha = \frac{t}{\sqrt{1 - t^2}}$
$\csc\alpha = \frac{1}{t}$
$\sec\alpha = \frac{1}{\sqrt{1 - t^2}}$
$\cot\alpha = \frac{\sqrt{1 - t^2}}{t}$

81. $\sin\alpha = -\frac{t}{\sqrt{1 + t^2}}$
$\cos\alpha = -\frac{1}{\sqrt{1 + t^2}}$
$\csc\alpha = -\frac{\sqrt{1 + t^2}}{t}$
$\sec\alpha = -\sqrt{1 + t^2}$
$\cot\alpha = \frac{1}{t}$

83. 6.18 **85.** 110.7°

78. $A \approx 47°, B \approx 89°, C \approx 44°$

80. $\sin\alpha = \frac{t}{\sqrt{1 + t^2}}$
$\cos\alpha = \frac{1}{\sqrt{1 + t^2}}$
$\csc\alpha = \frac{\sqrt{1 + t^2}}{t}$
$\sec\alpha = \sqrt{1 + t^2}$
$\cot\alpha = \frac{1}{t}$

82. $\sin\alpha = -\frac{\sqrt{t^2 - 1}}{t}$
$\cos\alpha = \frac{1}{t}$
$\tan\alpha = -\sqrt{t^2 - 1}$
$\csc\alpha = -\frac{t}{\sqrt{t^2 - 1}}$
$\cot\alpha = -\frac{1}{\sqrt{t^2 - 1}}$

84. 3 **86.** 1.6

Chapter 6 Test, pages 390–391

1. 20 **3.** 30 **5.** $\sin\alpha = \frac{9}{41}$, $\cos\alpha = \frac{40}{41}$, $\tan\alpha = \frac{9}{40}$, $\cot\alpha = \frac{40}{9}$, $\sec\alpha = \frac{41}{40}$, $\csc\alpha = \frac{41}{9}$ **7.** 48 **9.** $\frac{3\sqrt{10}}{20}, \frac{2\sqrt{10}}{3}$ **11.** $\frac{12}{13}, -\frac{12}{5}$ **13.** 0 **15.** 1.05 **17.** $\cot\alpha = \frac{2}{3}$, $\sec\alpha = \frac{\sqrt{13}}{2}$, $\cos\alpha = \frac{2\sqrt{13}}{13}$, $\csc\alpha = \frac{\sqrt{13}}{3}$, $\sin\alpha = \frac{3\sqrt{13}}{13}$

2. 24 **4.** $\frac{35}{2}$ **6.** $\sin\alpha = \frac{21}{29}$, $\cos\alpha = \frac{20}{29}$, $\tan\alpha = \frac{21}{20}$, $\cot\alpha = \frac{20}{21}$, $\sec\alpha = \frac{29}{20}$, $\csc\alpha = \frac{29}{21}$ **8.** $9\sqrt{2}$ **10.** $\frac{2\sqrt{6}}{5}, \frac{5\sqrt{6}}{12}$ **12.** $-\frac{3}{5}, \frac{3}{4}$ **14.** 0 **16.** 3.06 **18.** $\csc\alpha = \frac{6}{5}$, $\cos\alpha = \frac{\sqrt{11}}{6}$, $\sec\alpha = \frac{6\sqrt{11}}{11}$, $\tan\alpha = \frac{5\sqrt{11}}{11}$, $\cot\alpha = \frac{\sqrt{11}}{5}$

19. 0.7387 **21.** 1.0209 **23.** $A = 30°, c = 8\sqrt{3}, a = 4\sqrt{3}$ **25.** $c \approx 59.3, A \approx 37°40', B \approx 52°20'$ **27.** 76°
20. 2.2650 **22.** 0.3371 **24.** $B = 56°18', c \approx 116, a \approx 64.6$ **26.** 70° **28.** -1
29. $-\frac{\sqrt{3}}{3}$
30. $C = 56°, a \approx 35, b \approx 53$
31. two possible triangles: $C \approx 38°36', B \approx 117°06', b \approx 13.4$, and $C \approx 141°24', B \approx 14°18', b \approx 3.7$. **33.** 472 yds.
32. two possible triangles: $A \approx 58°40', C \approx 74°26', c \approx 14$, and $A \approx 121°20', C \approx 11°46', c \approx 3$ **34.** 95.6 ft
35. $c \approx 3.32, A \approx 23°40', B \approx 37°20'$
36. $A \approx 39°34', B \approx 54°41', C \approx 85°45'$

Chapter 7

Section 7.1, pages 399–400

1. yes **3.** no **5.** -0.9301 **7.** -0.8922 **9.**

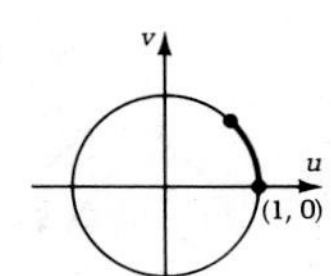

11.

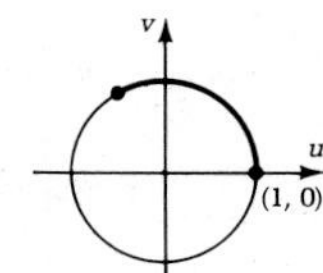

13.

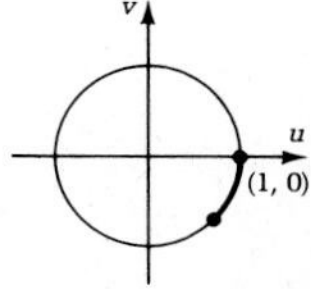

2. no **4.** no **6.** -0.6921 **8.** 0.7414 **10.**

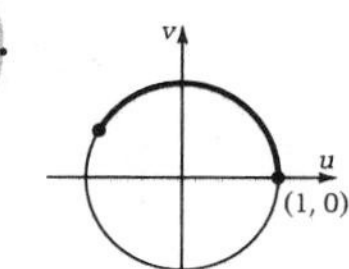

12.

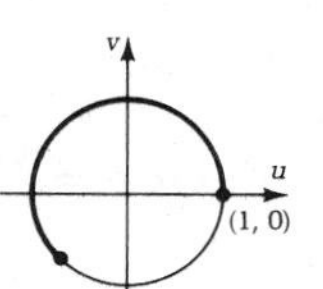

14.

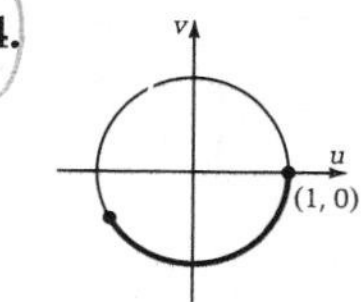

15.

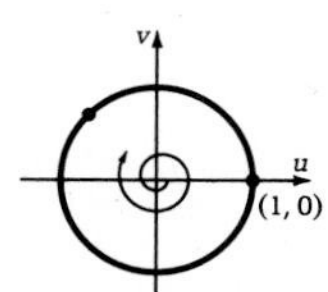

17. $\frac{\pi}{3}$ **19.** $\frac{7\pi}{4}$ **21.** $\frac{7\pi}{6}$ **23.** $-\frac{\pi}{6}$ **25.** $\frac{5\pi}{12}$ **27.** -3π **29.** 270° **31.** 150°

16. **18.** $\frac{\pi}{4}$ **20.** 2π **22.** $\frac{11\pi}{6}$ **24.** $-\frac{\pi}{12}$ **26.** $\frac{5\pi}{3}$ **28.** $-\frac{10\pi}{9}$ **30.** 60° **32.** 450°

33. −540° **35.** −750° **37.** 86° **39.** 0.625 **41.** 4.339 **43.** 720.002° **45.** −310.910° **47.** 2π in.
34. −135° **36.** −380° **38.** 286° **40.** 2.953 **42.** −5.459 **44.** 900.002° **46.** −1040.955° **48.** $\pi/3$ cm
49. 8 m **51.** $-\frac{7\pi}{5}$ ***53.** (a) Min. hand: -24π, Hr. hand: -2π (b) $-2\pi, -\frac{\pi}{6}$ (c) $-\pi, -\frac{\pi}{12}$ (d) $-\frac{\pi}{6}, -\frac{\pi}{72}$
50. 10 ft **52.** −300° ***54.** approximately 12:16 and 22 seconds
***55.** $120\pi, \frac{3\pi}{5}$ ***57.** 5280 radians ***59.** (a) $\left(\frac{\sqrt{3}}{2}, \frac{1}{2}\right)$ (b) $\left(\frac{\sqrt{2}}{2}, \frac{\sqrt{2}}{2}\right)$ (c) $\left(\frac{1}{2}, \frac{\sqrt{3}}{2}\right)$ ***61.** IV ***63.** III ***65.** IV
***56.** 20 radians ***58.** 5.5 radians ***60.** 24 times ***62.** I ***64.** III ***66.** III
***67.** II ***69.** I
***68.** II ***70.** I

Section 7.2, pages 407–408

1. Undefined **3.** 0 **5.** $\frac{\sqrt{3}}{2}$ **7.** 2 **9.** 1 **11.** $-\frac{1}{2}$ **13.** $\frac{2\sqrt{3}}{3}$ **15.** −1 **17.** $-\frac{\sqrt{2}}{2}$ **19.** $-\sqrt{3}$
2. 1 **4.** Undefined **6.** 1 **8.** $\frac{2\sqrt{3}}{3}$ **10.** $\frac{1}{2}$ **12.** $-\frac{\sqrt{3}}{3}$ **14.** $\frac{\sqrt{3}}{3}$ **16.** $-\sqrt{2}$ **18.** $-\sqrt{2}$ **20.** $\frac{\sqrt{3}}{2}$
21. $\sqrt{2}$ **23.** 0 **25.** $\frac{\sqrt{3}}{3}$ **27.** $-\sqrt{3}$ **29.** $\frac{2\sqrt{3}}{3}$ **31.** −0.53 **33.** −1.66 **35.** 1.05 **37.** −0.36 ***45.** $\frac{3\pi}{4}$
22. $\frac{2\sqrt{3}}{3}$ **24.** 0 **26.** $\sqrt{2}$ **28.** $-\sqrt{2}$ **30.** $-\frac{\sqrt{2}}{2}$ **32.** −3.38 **34.** −0.48 **36.** 0.78 **38.** 0.23 ***46.** $\frac{5\pi}{4}$
***47.** $\frac{5\pi}{6}$ ***49.** $\frac{\pi}{4}$
***48.** $\frac{2\pi}{3}$ ***50.** $\frac{\pi}{6}$

Section 7.3, pages 420–421

1.
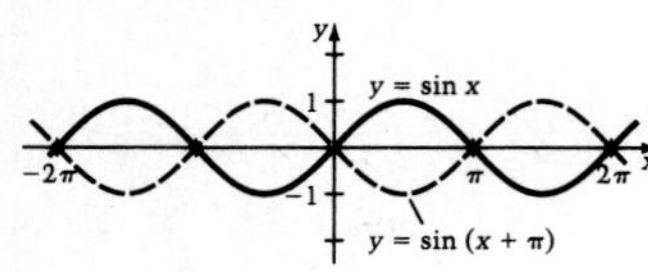

3.
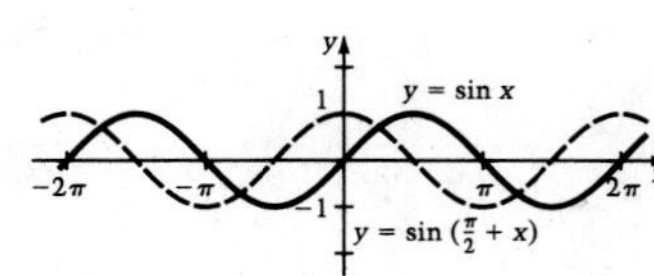

2.
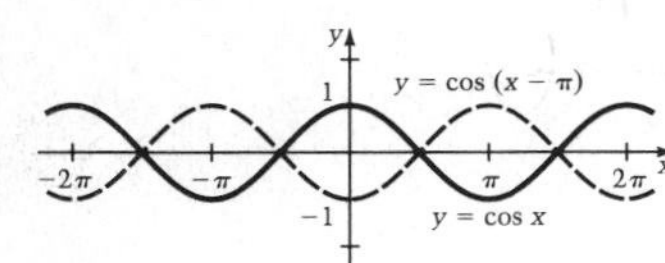

4.
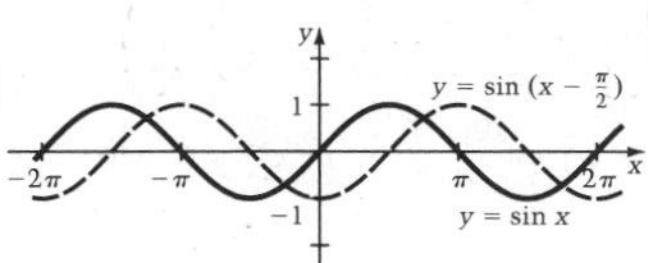

---
*denotes Superset exercises

5. Graphs of $y = \cos x$ and $y = \cos(-x)$ are the same.

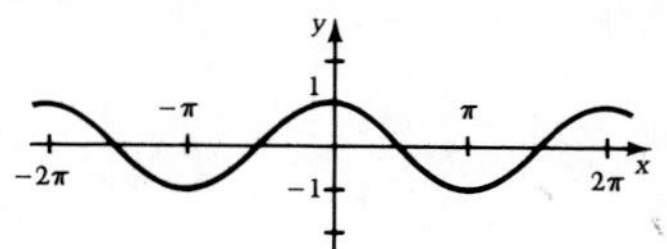

7.

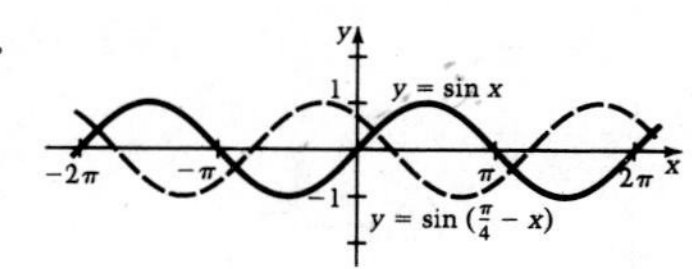

6.

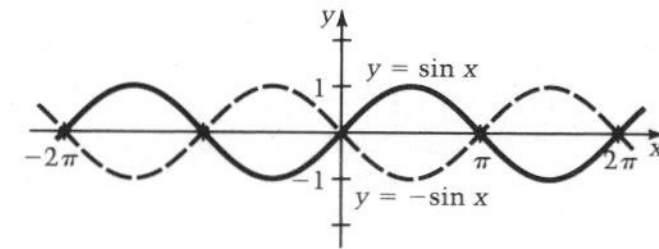

8.

$y = \cos x$

$y = \cos\left(-x - \frac{\pi}{4}\right)$

9. Graphs of $y = \cos x$ and $y = -\cos(-x - \pi)$ are the same.

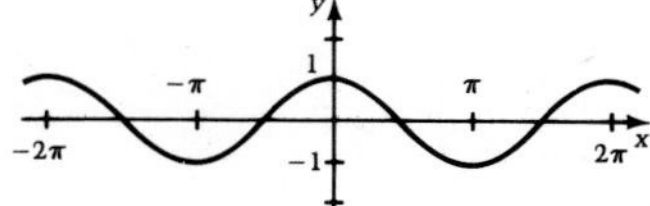

10.

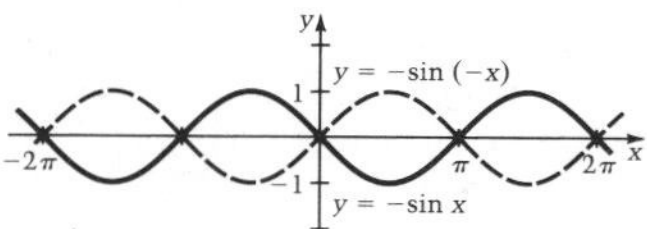

11. Graphs of $y = \sin x$ and $y = \sin(x - 2\pi)$ are the same.

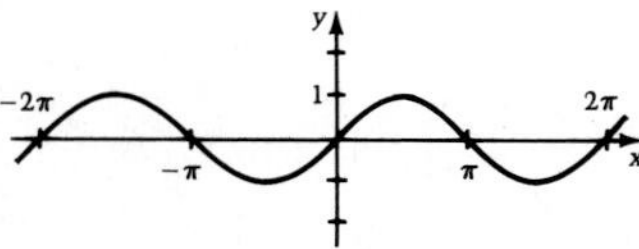

13.

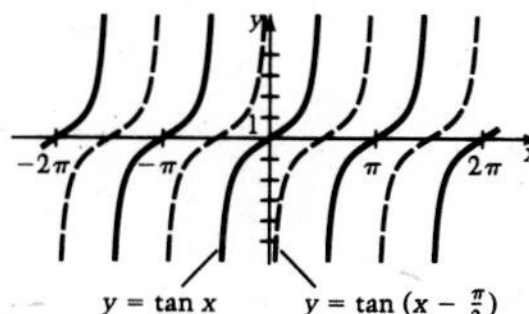

12. Graphs of $y = \cos x$ and $y = \cos(x + 4\pi)$ are the same.

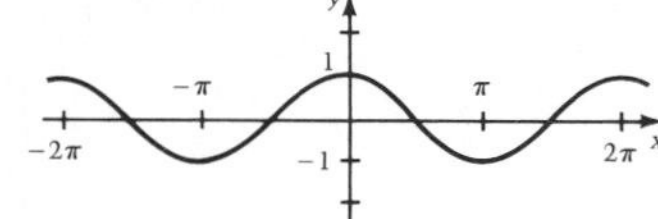

14.

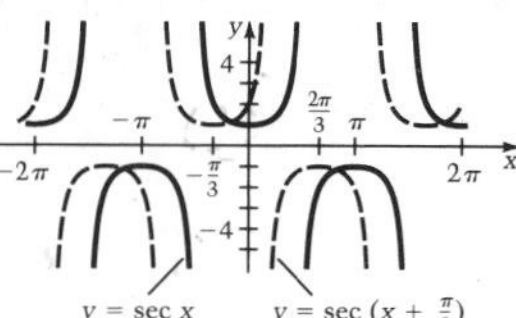

15.

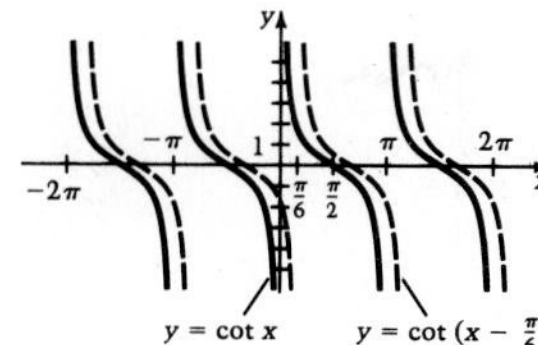

17.

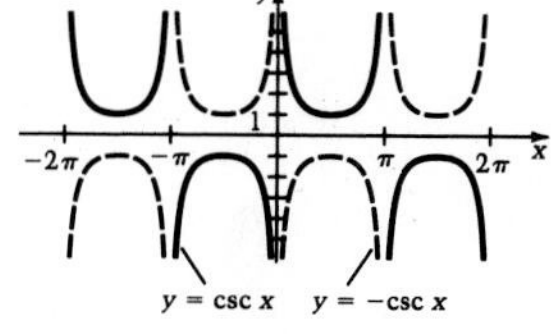

19.

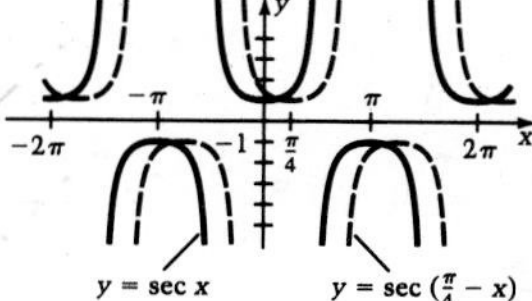

16.

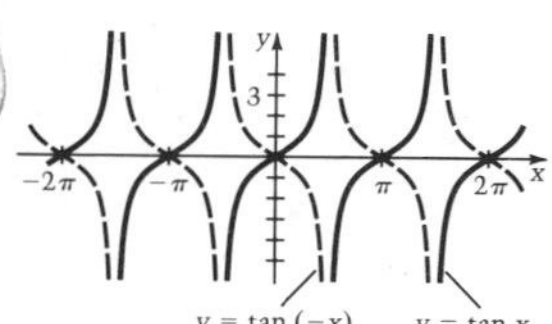

18.

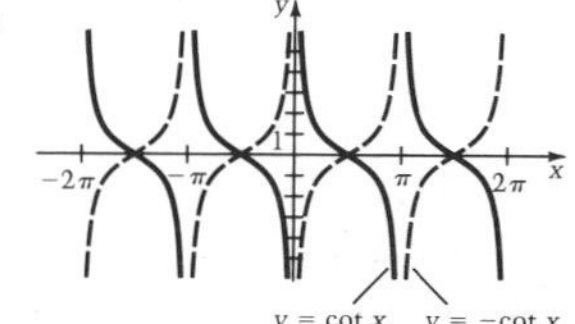

20.

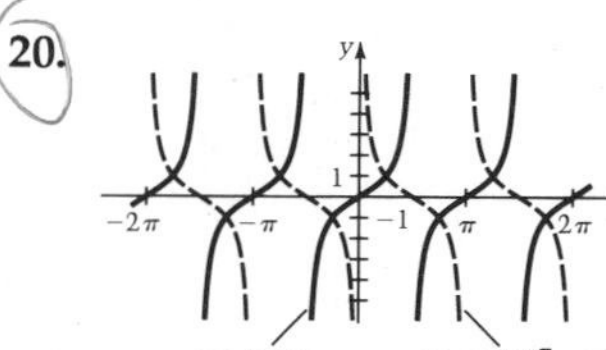

21.

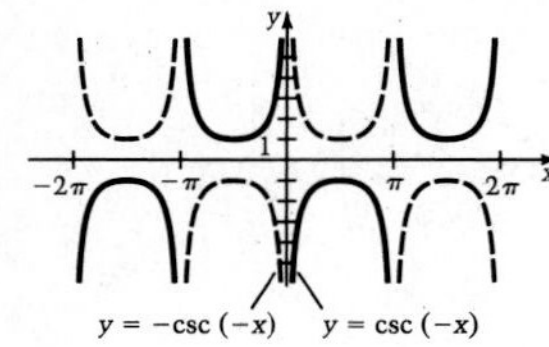

23. False **25.** True **27.** True **29.** False

***31.**

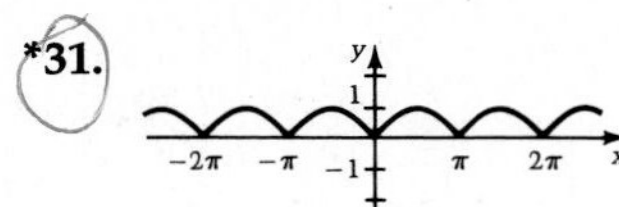

22.

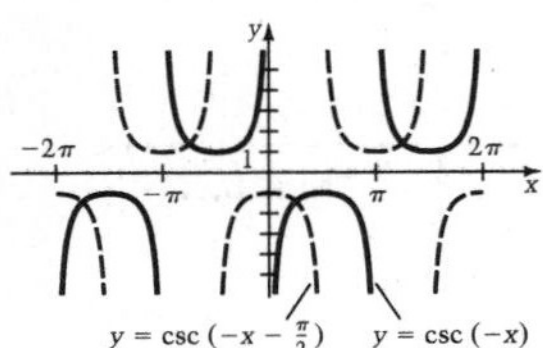

24. False **26.** True **28.** True **30.** False

***32.**

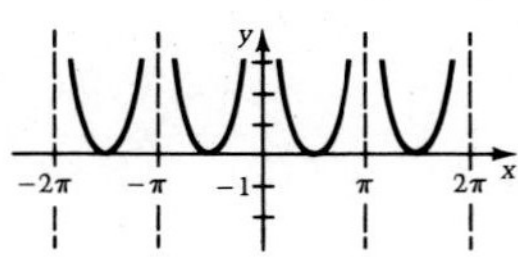

***33.**

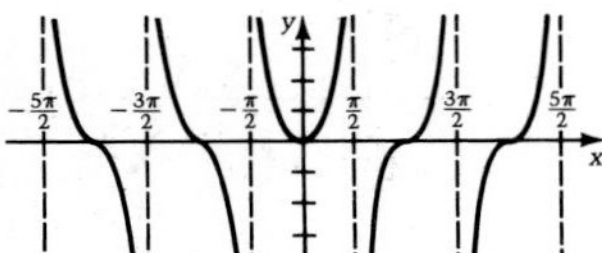

***35.**

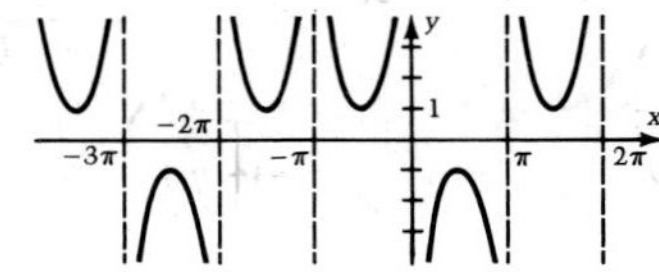

***37.**

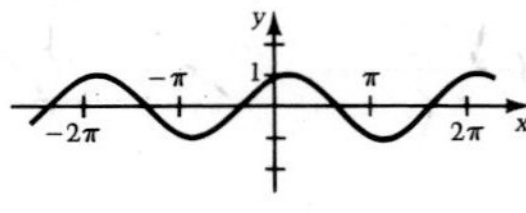

***34.**

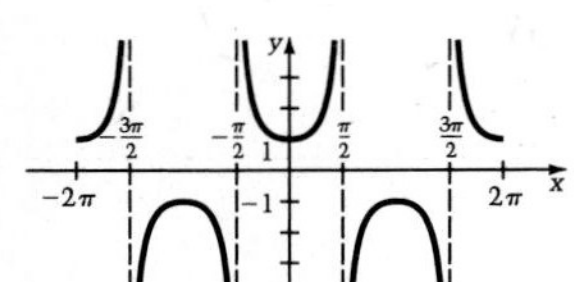

***36.**

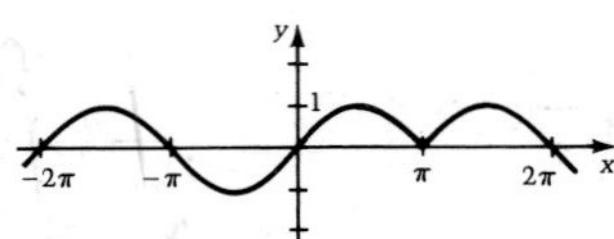

***38.**

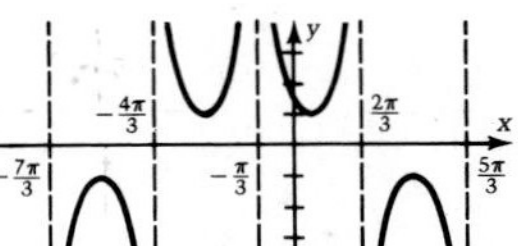

***39.**

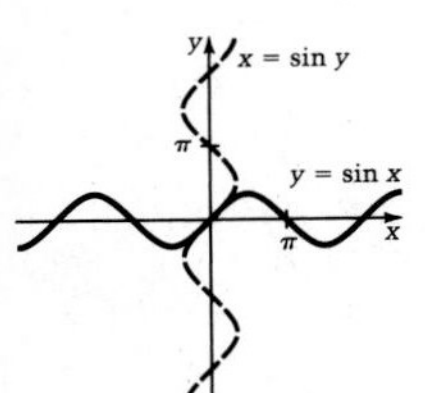

***41.**

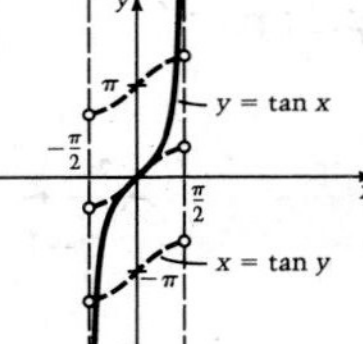

***40.**

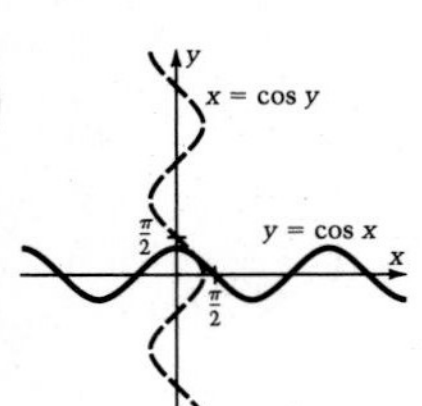

***42.**

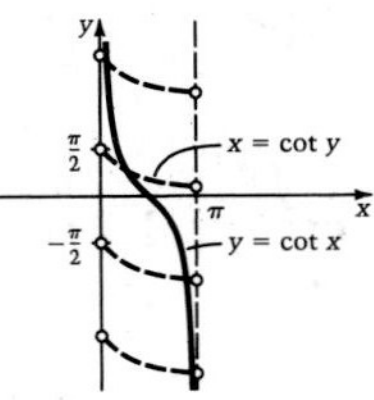

***43.** $-0.8415, 0.1411, -0.8367, -0.7568, -0.0000, -0.5806$
***44.** $-0.5403, 0.9900, 0.5477, -0.6536, 1.0000, 0.8142$
***45.** $-0.2130, -0.6002, -0.9789, -0.0730, -0.7071, -0.9863$
***46.** $0.2130, 0.6002, 0.9789, 0.0730, 0.7071, 0.9863$
***47.** $-0.6421, -7.0153, 0.6547, -0.8637, 27224.2, 1.4024$
***48.** $1.9378, -2.4892, -0.0561, 2.8743, -1.7319, -0.4559$

---

*denotes Superset exercises

Section 7.4, page 428

1.

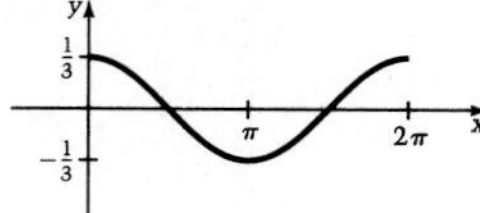

3.

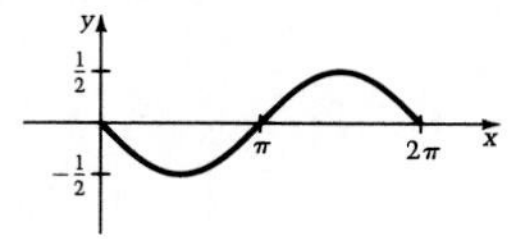

5.

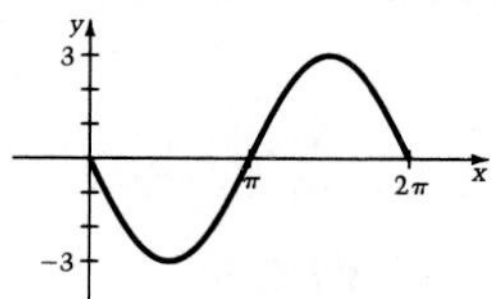

7.

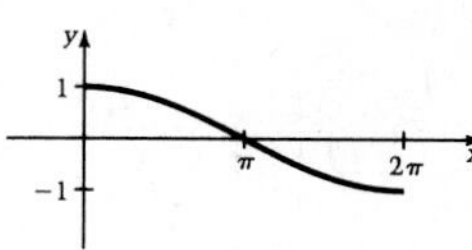

2.

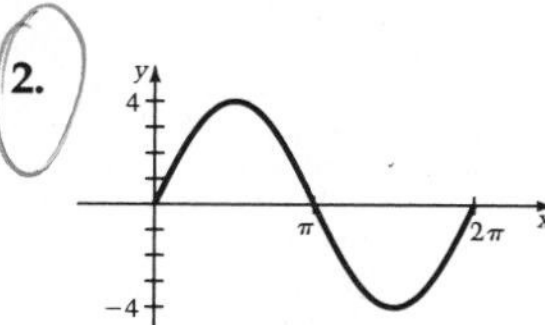

4.

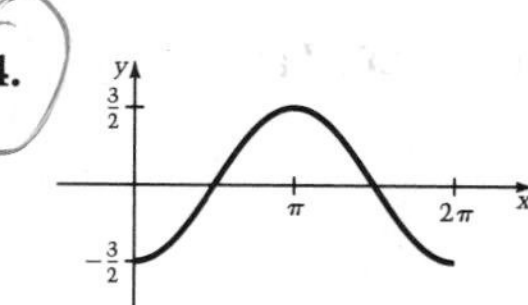

6.

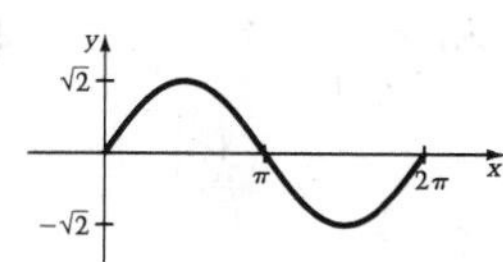

8.

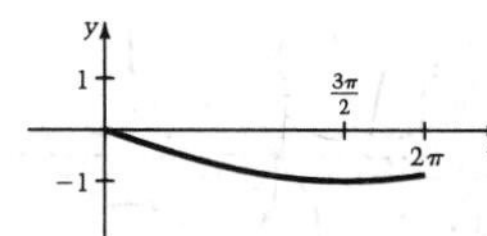

9.

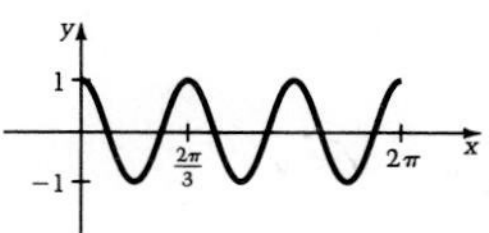

11.

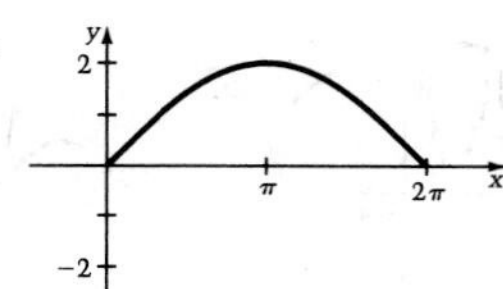

13.

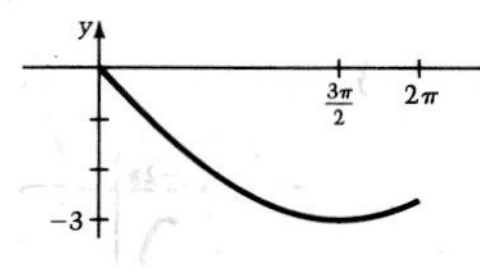

15.

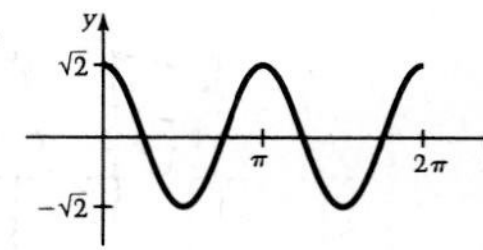

10.

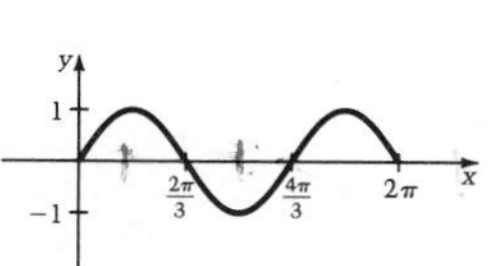

12.

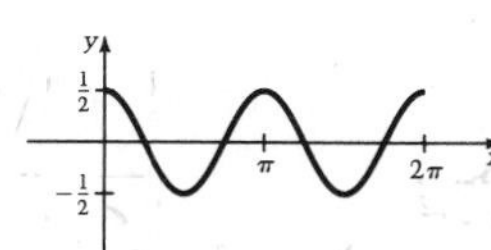

14.

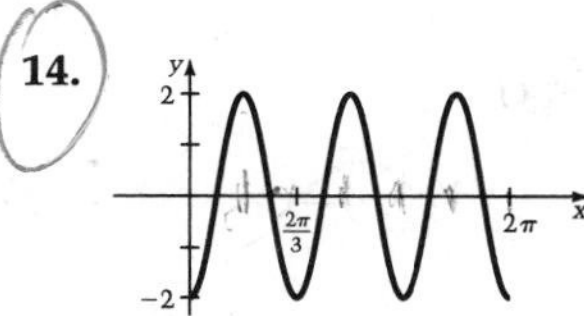

16.

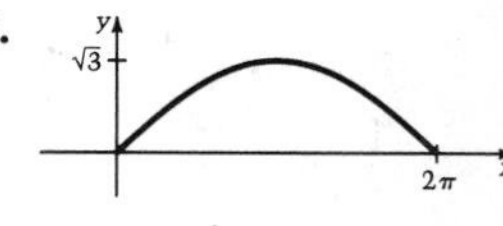

17. amp: 2; per: 2π; ps: $-\dfrac{\pi}{2}$

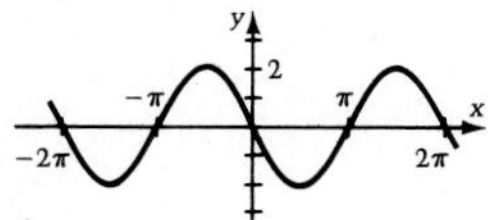

19. amp: 2; per: 4π; ps: $\dfrac{\pi}{3}$

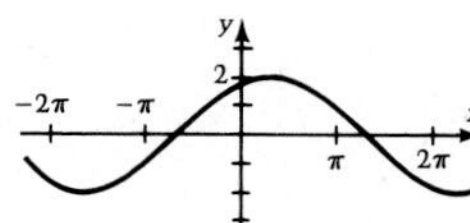

21. amp: 2; per: 4π; ps: $-\dfrac{\pi}{2}$

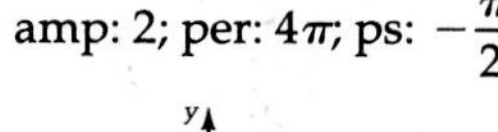

18. amp: 3; per: 2π; ps: $\dfrac{\pi}{2}$

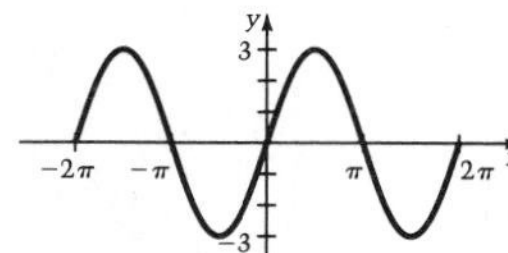

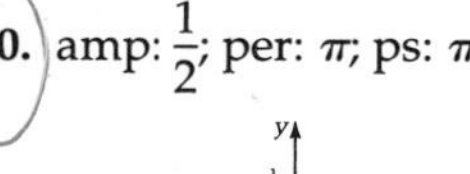

20. amp: $\frac{1}{2}$; per: π; ps: π

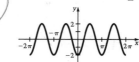

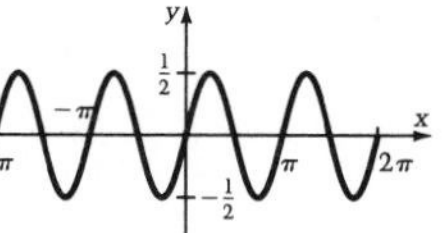

22. amp: 2; per: π; ps: $-\dfrac{\pi}{2}$

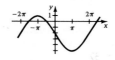

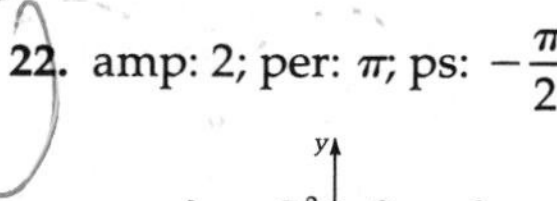

23. amp: $\sqrt{3}$; per: 4π; ps: $-\pi$

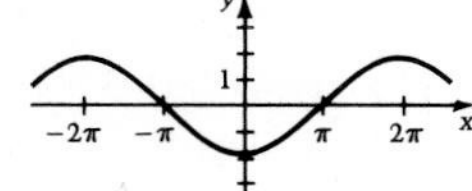

25. amp: 3; per: π; ps: $\dfrac{\pi}{2}$

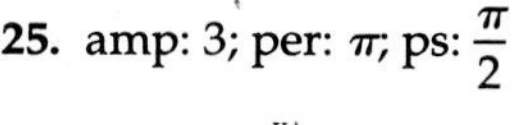

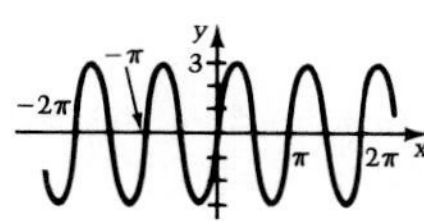

27. amp: 3; per: 4π; ps: $-\pi$

29. amp: 2; per: 2π; ps: $\dfrac{\pi}{2}$

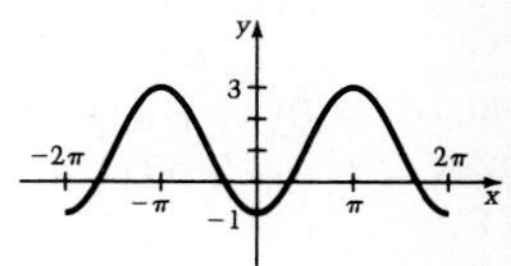

24. amp: 3; per: 2π; ps: $\frac{\pi}{2}$

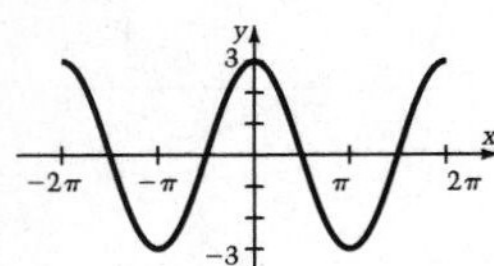

26. amp: 2; per: 4π; ps: $-\pi$

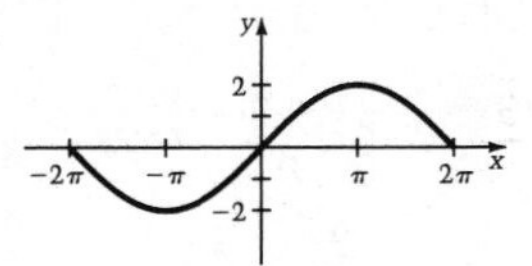

28. amp: 3; per: π; ps: $\frac{\pi}{4}$

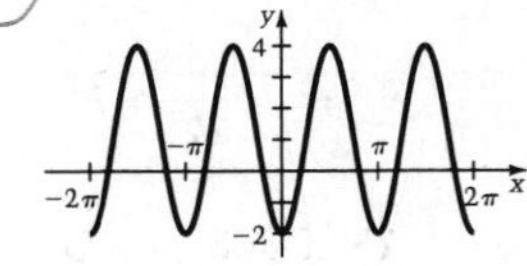

30. amp: 2; per: 4π; ps: $-\pi$

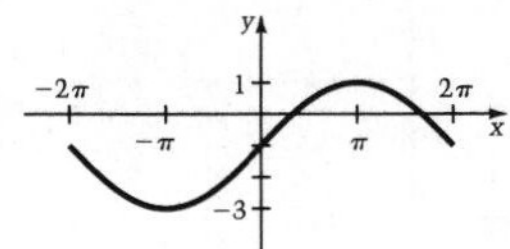

31.

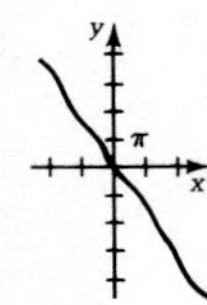

33.

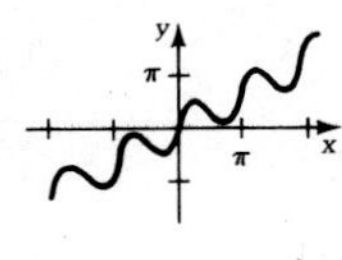

35.

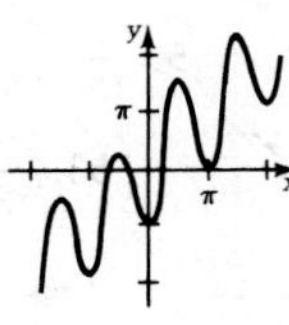

37.

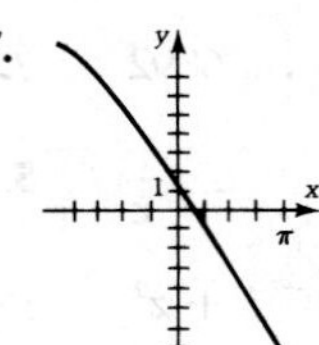

***39.** 8, -8

32.

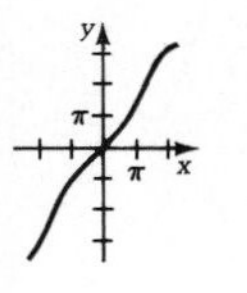

34.

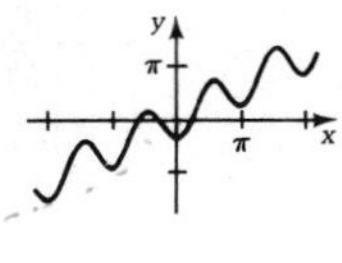

36.

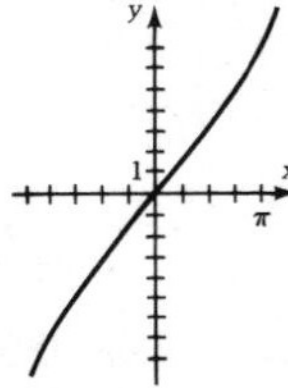

38.

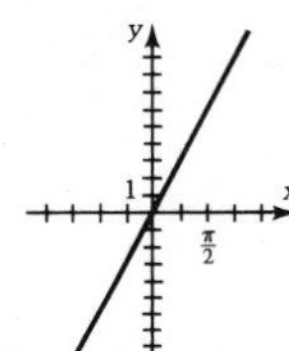

***40.** $\frac{1}{2}, -\frac{1}{2}$

***41.**

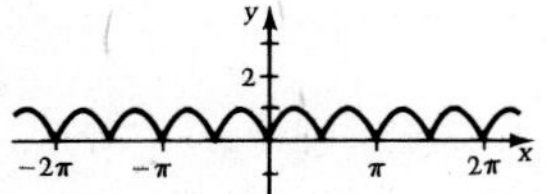

***43.**

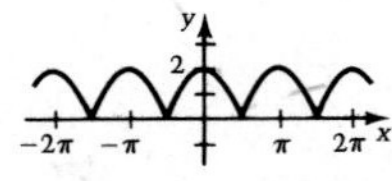

***45.**

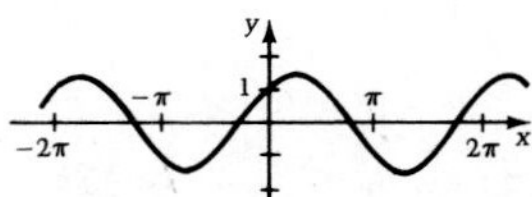

***42.**

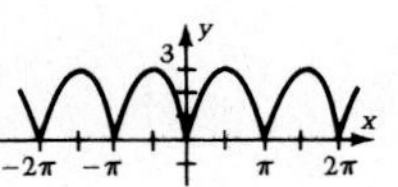

***44.**

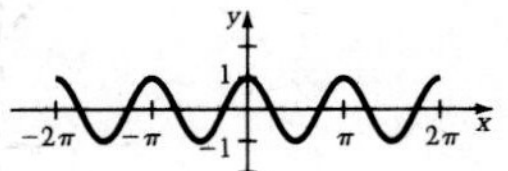

***46.**

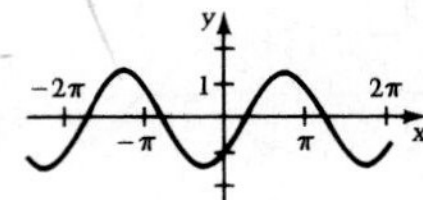

***47.**

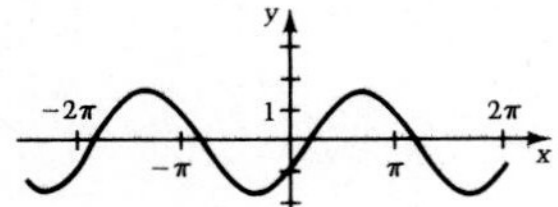

***49.** False ***51.** True

***48.**

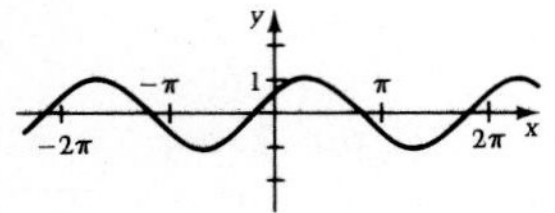

***50.** False ***52.** False

***53.** 1.9191, -0.8961, 0.2001, 1.0522, -1.3088, 0.5222 ***55.** -0.8304, 1.6436, -1.3411, 0.3972, -0.1247, 0.8625

***54.** 0.8323, -0.5673, 1.8415, -1.2029, -1.9175, 0.9836 ***56.** 0.8105, -1.2017, -0.2986, 1.3422, 1.4845, 0.7561

---

*denotes Superset exercises

Section 7.5, pages 436–437

1. $\frac{\pi}{3}$ **3.** π **5.** $-\frac{\pi}{6}$ **7.** $-\frac{\pi}{4}$ **9.** 0 **11.** $\frac{1}{2}$ **13.** $-\sqrt{3}$ **15.** 0 **17.** $-\frac{\sqrt{3}}{2}$ **19.** $-\frac{\sqrt{2}}{2}$ **21.** $-\frac{\pi}{4}$

2. $\frac{\pi}{2}$ **4.** $\frac{\pi}{3}$ **6.** $\frac{3\pi}{4}$ **8.** $-\frac{\pi}{3}$ **10.** $\frac{\pi}{2}$ **12.** $\frac{\sqrt{3}}{3}$ **14.** $\frac{\sqrt{3}}{2}$ **16.** $\frac{\sqrt{2}}{2}$ **18.** $\frac{\sqrt{2}}{2}$ **20.** $\frac{\sqrt{3}}{2}$ **22.** $\frac{\pi}{3}$

23. $\frac{\pi}{6}$ **25.** 0.1874 **27.** -1.2780 **29.** 0.9533 **31.** 0.6405 **33.** $\frac{4}{5}$ **35.** $\frac{4}{5}$ **37.** $\frac{4}{5}$ **39.** $\frac{5}{13}$ **41.** $-\frac{2\sqrt{5}}{15}$

24. $-\sqrt{3}$ **26.** 1.0799 **28.** 1.1731 **30.** 0.6892 **32.** -24 **34.** $\frac{3}{5}$ **36.** $\frac{3}{4}$ **38.** $\frac{7}{25}$ **40.** $\frac{5}{12}$ **42.** $\frac{2\sqrt{5}}{5}$

43. $\frac{z^2}{1 + z^2}$ **45.** $1 - z^2$ **47.** $-\frac{z}{\sqrt{1 - z^2}}$ **49.** $\frac{1}{z}$ **51.** $\frac{1 + 4z^2}{4z^2}$ ***53.** $\frac{2\pi}{3}$ ***55.** $-\frac{5}{13}$ ***57.** $\frac{\sqrt{5}}{3}$

44. $\frac{1}{\sqrt{1 + z^2}}$ **46.** $\frac{z^2}{1 - z^2}$ **48.** $-z$ **50.** $1 + z^2$ **52.** $\frac{\sqrt{1 - 9z^2}}{3z}$ ***54.** $-\frac{\pi}{4}$ ***56.** $-\frac{12}{13}$ ***58.** $\frac{3\sqrt{10}}{20}$

***59.** $\frac{\sqrt{5}}{2}$ ***61.** $-\frac{\sqrt{165}}{11}$ ***63.** 0.3230 ***65.** 1.4312 ***67.** 0.0814 ***69.**

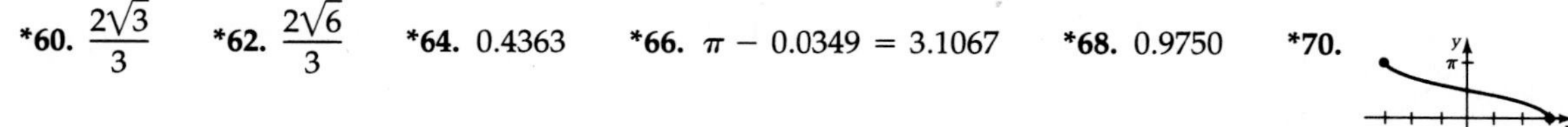

***60.** $\frac{2\sqrt{3}}{3}$ ***62.** $\frac{2\sqrt{6}}{3}$ ***64.** 0.4363 ***66.** $\pi - 0.0349 = 3.1067$ ***68.** 0.9750 ***70.**

***71.** ***73.** ***75.** ***77.**

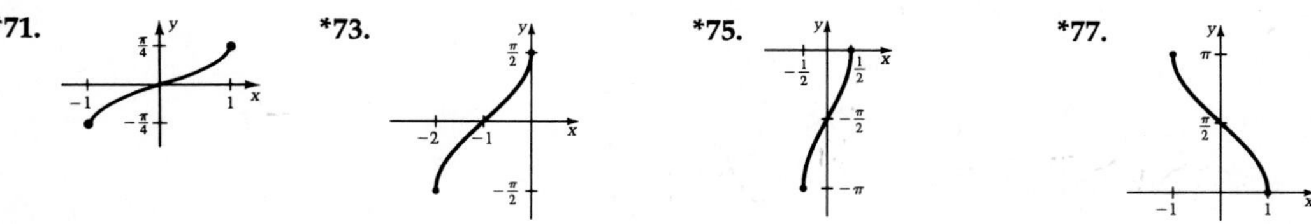

***72.** ***74.** ***76.** ***78.**

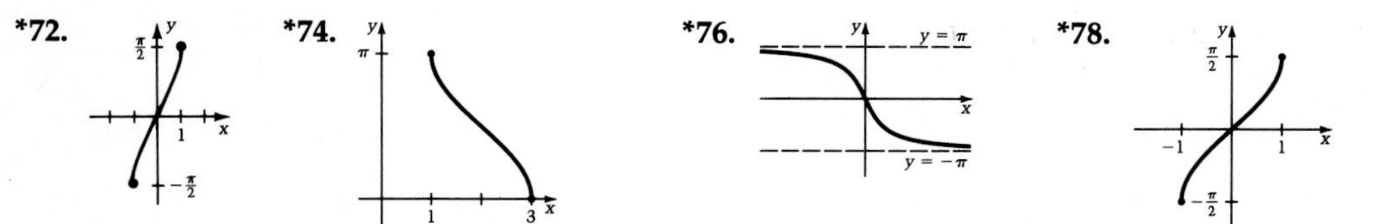

Section 7.6, pages 443–444

1. $2\sqrt{34}$, 31° **3.** $2\sqrt{13}$, 326° **5.** 13, 247° **7.** 11, 0° **9.** **11.**

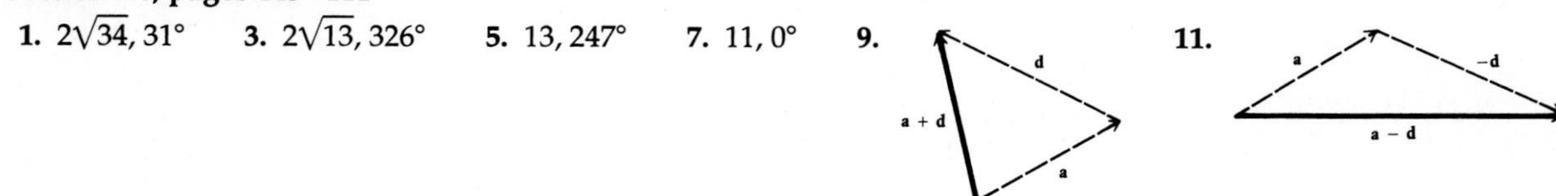

*denotes Superset exercises

2. $2\sqrt{41}, 129°$ **4.** $2\sqrt{17}, 14°$ **6.** $4\sqrt{13}, 304°$ **8.** $11, 90°$ **10.**

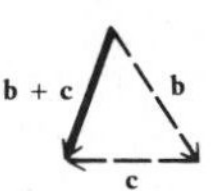

12.

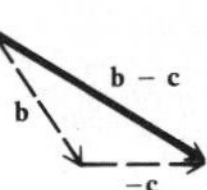

13.

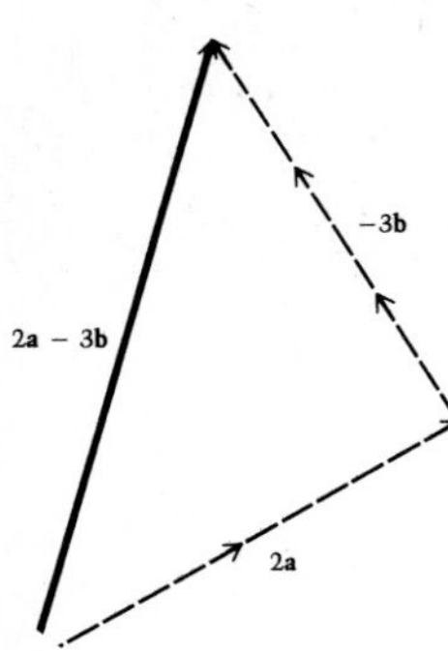

15.

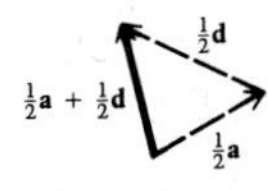

17.

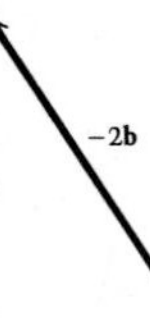

14.

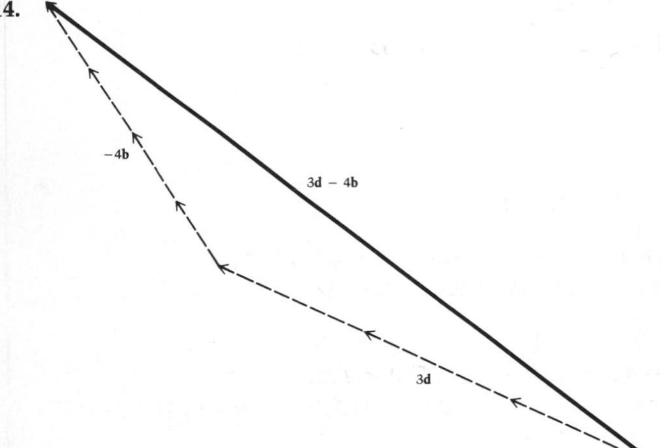

16.

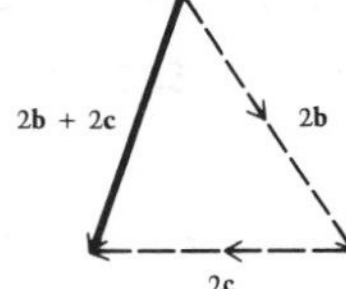

18.

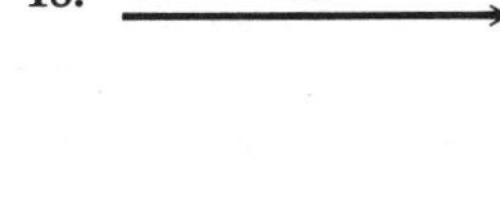

19.

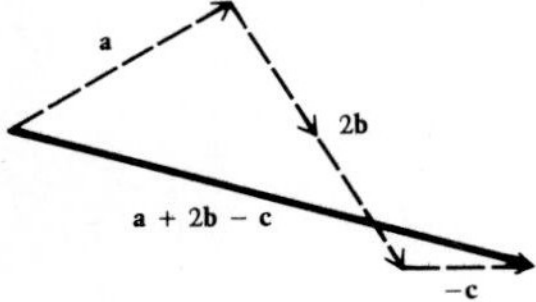

21. $\langle 5, 12\rangle$ **23.** $\langle 3, -4\rangle$ **25.** $\langle 1, -8\rangle$

20.

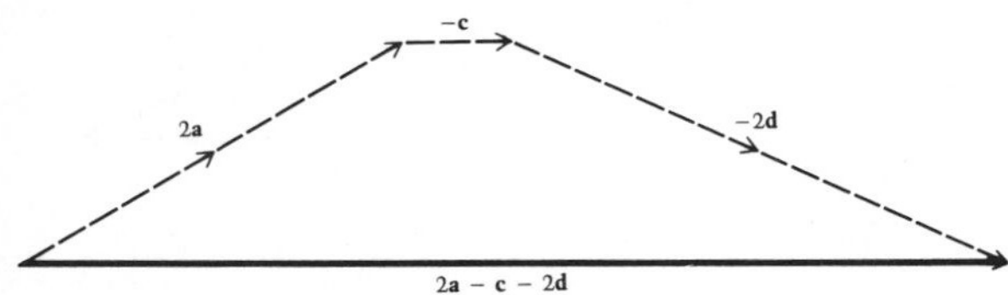

22. $\langle -6, 8\rangle$ **24.** $\langle 8, 6\rangle$ **26.** $\left\langle -3, \frac{1}{2}\right\rangle$

27. $\langle -5.12, 5.34\rangle$, magnitude 7.40, direction 133.8°
28. $\langle 6.37, 5.11\rangle$, magnitude 8.17, direction 38.7°

29. $\langle 7.53, -1.25\rangle$, magnitude 7.63, direction $-9.43°$ **31.** $(2, 2)$ **33.** $\left(-2\frac{1}{2}, -1\frac{1}{2}\right)$ **35.** $\left(\frac{5}{2}, \frac{5}{2}\right)$

30. $\langle -11.35, -4.16\rangle$, magnitude 12.1, direction $-159.9°$ **32.** $(-3, 7)$ **34.** $\left(-3\frac{1}{2}, 5\frac{1}{2}\right)$ **36.** $\left(-\frac{1}{2}, -\frac{11}{2}\right)$

37. $\langle -20, 4\rangle$ **39.** $\langle 5, 19\rangle$ **41.** $\langle 13, 17\rangle$ **43.** $\langle -21, 2\rangle$ **45.** $10, \left\langle -\frac{3}{5}, \frac{4}{5}\right\rangle, \left\langle \frac{6}{5}, -\frac{8}{5}\right\rangle$ **47.** $4, \langle 0, -1\rangle, \langle 0, 2\rangle$

38. $\langle -13, 10\rangle$ **40.** $\langle 13, 16\rangle$ **42.** $\langle -7, 6\rangle$ **44.** $\langle -8, 8\rangle$ **46.** $5, \left\langle -\frac{4}{5}, -\frac{3}{5}\right\rangle, \left\langle \frac{8}{5}, \frac{6}{5}\right\rangle$ **48.** $6, \langle -1, 0\rangle, \langle 2, 0\rangle$

49. $2\sqrt{3}, \left\langle \frac{\sqrt{6}}{3}, \frac{\sqrt{3}}{3}\right\rangle, \left\langle -\frac{2\sqrt{6}}{3}, -\frac{2\sqrt{3}}{3}\right\rangle$ **51.** $\frac{10}{3}, \left\langle \frac{4}{5}, -\frac{3}{5}\right\rangle, \left\langle -\frac{8}{5}, \frac{6}{5}\right\rangle$ **53.** $-\mathbf{i} + 8\mathbf{j}, \sqrt{2}, \frac{\sqrt{2}}{2}\mathbf{i} + \frac{\sqrt{2}}{2}\mathbf{j}$

50. $4, \left\langle \frac{3}{4}, \frac{\sqrt{7}}{4}\right\rangle, \left\langle -\frac{3}{2}, -\frac{\sqrt{7}}{2}\right\rangle$ **52.** $\frac{13}{2}, \left\langle -\frac{5}{13}, \frac{12}{13}\right\rangle, \left\langle \frac{10}{13}, -\frac{24}{13}\right\rangle$ **54.** $\mathbf{i} + 4\mathbf{j}, \sqrt{5}, \frac{2\sqrt{5}}{5}\mathbf{i} - \frac{\sqrt{5}}{5}\mathbf{j}$

55. $3\mathbf{i} - 16\mathbf{j}, \sqrt{13}, \frac{3\sqrt{13}}{13}\mathbf{i} - \frac{2\sqrt{13}}{13}\mathbf{j}$ **57.** $-\frac{19}{2}\mathbf{i} + \frac{13}{2}\mathbf{j}, \frac{\sqrt{26}}{2}, \frac{\sqrt{26}}{26}\mathbf{i} + \frac{5\sqrt{26}}{26}\mathbf{j}$ ***59.** $\langle -6, 6\sqrt{3}\rangle$

56. $7\mathbf{i} - 18\mathbf{j}, \sqrt{13}, \frac{2\sqrt{13}}{13}\mathbf{i} - \frac{3\sqrt{13}}{13}\mathbf{j}$ **58.** $-\frac{8}{3}\mathbf{i} + \frac{13}{2}\mathbf{j}, \frac{2\sqrt{37}}{3}, \frac{\sqrt{37}}{37}\mathbf{i} + \frac{6\sqrt{37}}{37}\mathbf{j}$ ***60.** $\langle -9\sqrt{3}, -9\rangle$

***61.** $\langle -15\sqrt{2}/2, -15\sqrt{2}/2\rangle$ ***63.** $\langle 2.7, -7.4\rangle$ ***65.** $r = 1, s = -2$ ***67.** $\langle -3, 2\rangle$ or $\langle 3, -2\rangle$ ***69.** $\langle 1.33, 12.5\rangle$

***62.** $\langle -5\sqrt{2}, 5\sqrt{2}\rangle$ ***64.** $\langle -3.4, 4.9\rangle$ ***66.** $100°$ ***68.** $\mathbf{i} = \frac{4}{23}\mathbf{u} + \frac{3}{23}\mathbf{v}$ ***70.** $\langle 83.1, 115\rangle$

***71.** $\langle -831.8, -80.85\rangle$
***72.** $\langle 65.72, 3.091\rangle$

Section 7.7, pages 452–453

1. 569 mph, N 82°06′ E **3.** 427 mph, N 16°19′ W **5.** 391 N **7.** 211 lbs **9.** 98° **11.** 18° **13.** 90°
2. 332 mph, S 76°24′ W **4.** 443 mph, N 28°01′ E **6.** 1558 N **8.** 9°40′ **10.** 63° **12.** 60° **14.** 90°
15. 112° ***17.** perpendicular ***19.** neither ***21.** parallel ***23.** 4827 ft-lbs ***25.** 54,067 ft-lbs
16. 75° ***18.** neither ***20.** perpendicular ***22.** parallel ***24.** 666 ft-lbs ***26.** 564 ft-lbs
***27.** show that $(\mathbf{a} + \mathbf{b}) \cdot (\mathbf{a} - \mathbf{b}) = \|\mathbf{a}\|^2 - \|\mathbf{b}\|^2$
***28.** follow the hint and show that $\mathbf{c} \cdot \mathbf{d} = \|\mathbf{b}\|^2 - \|\mathbf{a}\|^2$
***29.** routine with hint ***31.** 0.52i, 9.88j ***33.** −883.6i, − 423.4j ***35.** S 38.7°W, 6.4 mph
***30.** show that $\|\mathbf{a} + \mathbf{b}\|^2 + \|\mathbf{a} - \mathbf{b}\|^2 = 2\|\mathbf{a}\|^2 + 2\|\mathbf{b}\|^2$, where $\mathbf{a} = \langle a_1, a_2\rangle$ and $\mathbf{b} = \langle b_1, b_2\rangle$ are adjacent sides of parallelogram
***32.** 48.6i, 43.8j ***34.** −55.6i, 280.8j ***36.** 53.5 N, 45.8°
***37.** 85.13 lbs along rope to left, 45.57 lbs along rope to right

Chapter 7 Review Exercises, pages 458–459

1. $\frac{2\pi}{3}$ **3.** $\frac{5\pi}{4}$ **5.** 135° **7.** −120° **9.** $\frac{12\pi}{5}$ cm **11.** $\frac{\sqrt{2}}{2}$ **13.** $\frac{1}{2}$ **15.** $\frac{\sqrt{3}}{3}$ **17.** −1

2. $-\frac{3\pi}{4}$ **4.** $\frac{17\pi}{4}$ **6.** 990° **8.** −150° **10.** 35 in. **12.** $-\frac{\sqrt{2}}{2}$ **14.** $\sqrt{3}$ **16.** $-\frac{1}{2}$ **18.** 0

19. $-\sqrt{2}$ **21.** $\sqrt{3}$ **23.** 

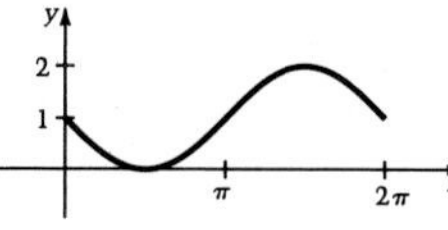

25. 

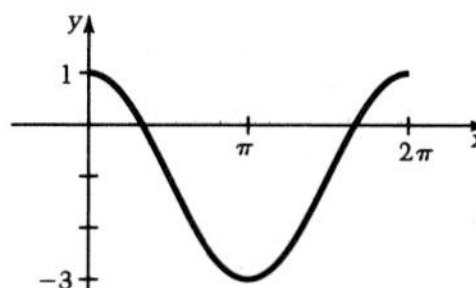

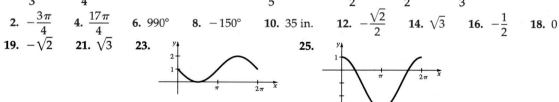

---

*denotes Superset exercises

20. $-\dfrac{2\sqrt{3}}{3}$ **22.** -2 **24.**

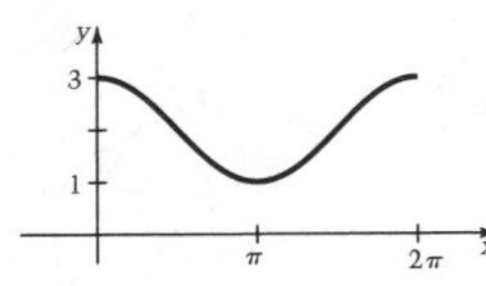

26.

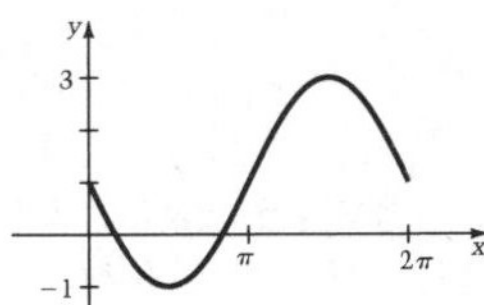

27.

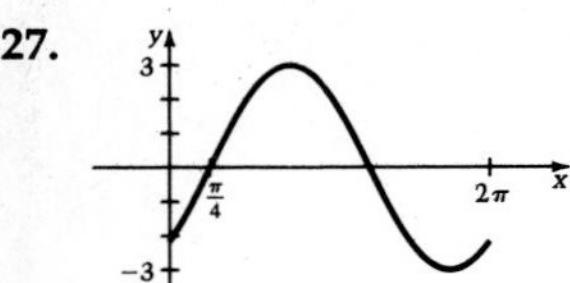

29.

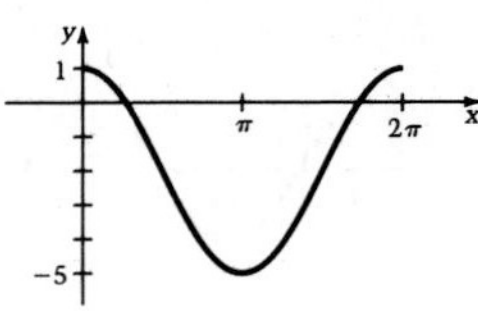

31.

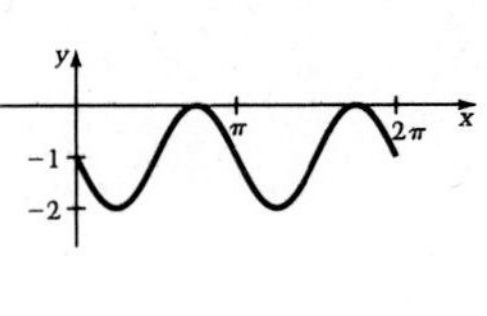

33.

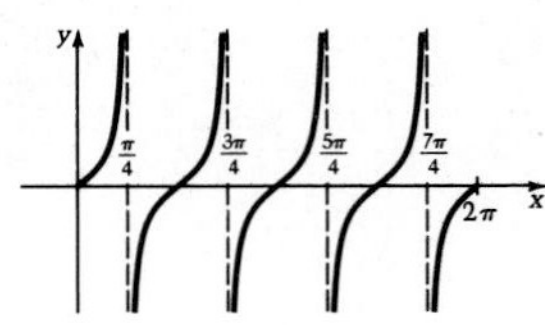

28.

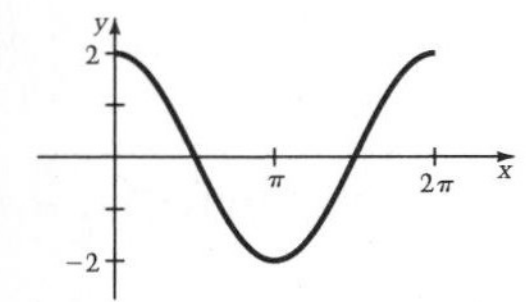

30.

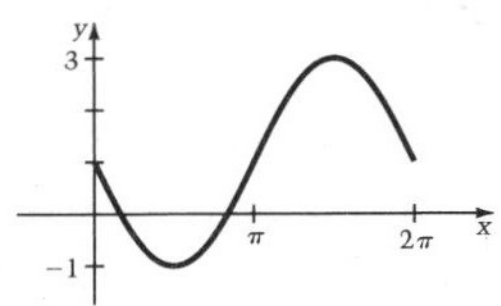

32.

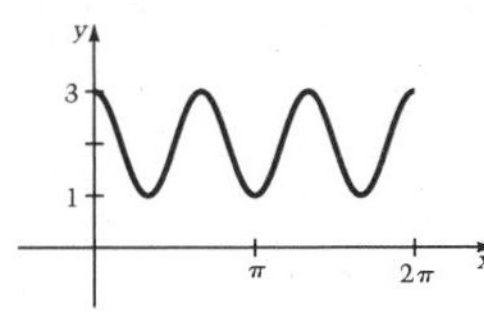

34.

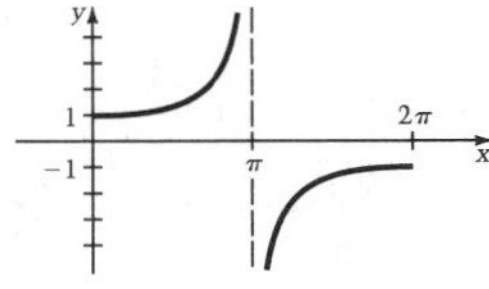

35.

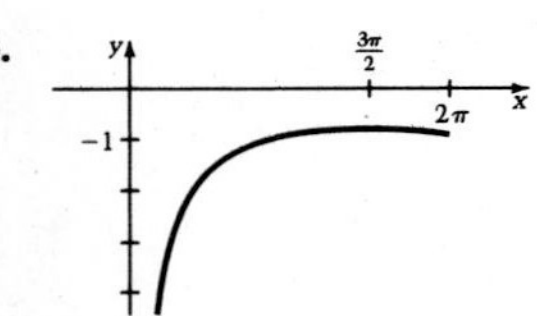

37.

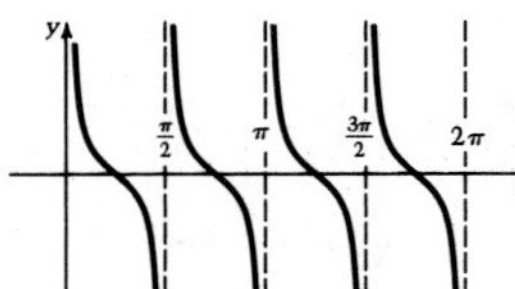

39.

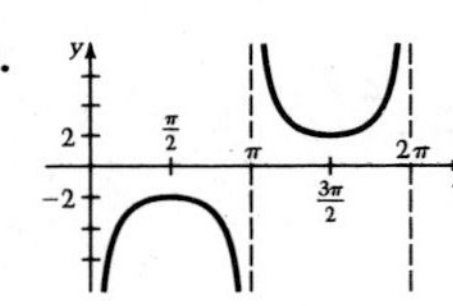

41. $-\dfrac{\pi}{6}$ **43.** 0

36.

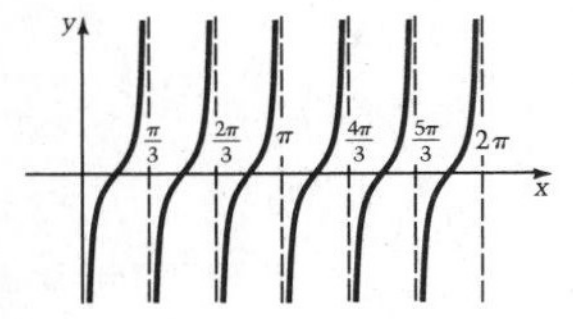

38.

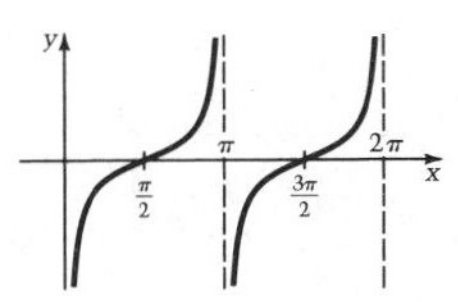

40.

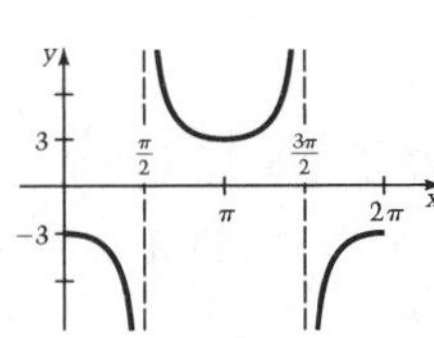

42. $\dfrac{\pi}{6}$ **44.** $-\dfrac{\pi}{4}$

45. 0 **47.** $\dfrac{\pi}{6}$ **49.** $\dfrac{\pi}{6}$ **51.** $\dfrac{\sqrt{5}}{3}$ **53.** $\dfrac{4\sqrt{15}}{15}$ **55.** $\dfrac{5\sqrt{26}}{26}$ **57.** $\dfrac{1-z^2}{z^2}$ **59.** $\langle 14, -7\rangle$ **61.** $(6, -2)$

46. $\dfrac{2\pi}{3}$ **48.** $\dfrac{\pi}{4}$ **50.** $\dfrac{\pi}{3}$ **52.** $\dfrac{\sqrt{29}}{5}$ **54.** $-\dfrac{3}{4}$ **56.** $2\sqrt{2}$ **58.** $\sqrt{1+z^2}$ **60.** $\langle -32, 6\rangle$ **62.** $(5, 2)$

63. $\langle -8, 2\rangle$ **65.** $\left\langle -\dfrac{4\sqrt{17}}{17}, \dfrac{\sqrt{17}}{17}\right\rangle$ **67.** $\sqrt{37}$ **69.** $\langle -2, 3\rangle$ **71.** $2\mathbf{i}$ **73.** -11

64. $\langle -2, 0\rangle$ **66.** $\langle 1, 0\rangle$ **68.** 1 **70.** $c = -\dfrac{3}{5}, k = -\dfrac{9}{5}$ **72.** $\mathbf{j}$ **74.** -23

75. $\dfrac{1}{2}$ **77.** 138° **79.** 235 N

76. two such vectors $\mathbf{x}$: $\left\langle \dfrac{3\sqrt{13}}{13}, \dfrac{2\sqrt{13}}{13}\right\rangle$ and $\left\langle -\dfrac{3\sqrt{13}}{13}, -\dfrac{2\sqrt{13}}{13}\right\rangle$. **78.** 525 mph, N 74°7′ E **80.** True

81. False **83.** 4, −4

82. 6, −6 **84.** 2π in.

Chapter 7 Test, page 460

1. no **3.** $\frac{4\pi}{3}$ **5.** 810° **7.** −510° **9.** $-\frac{\sqrt{2}}{2}$ **11.** Undefined **13.** −0.9162

2. $-\frac{3\pi}{4}$ **4.** $-\frac{49\pi}{12}$ **6.** 960° **8.** $3\pi/2$ radians or 270° **10.** $-\frac{1}{2}$ **12.** $\frac{\sqrt{3}}{3}$ **14.** 0.7756

15.

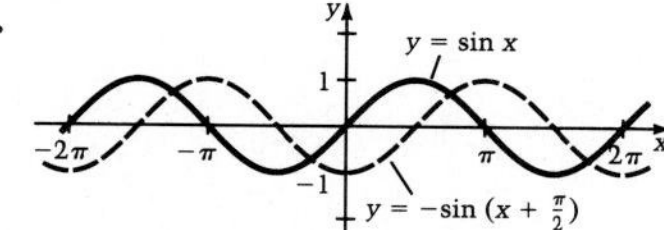

17. False **19.** True **21.** amplitude 3, period 4π

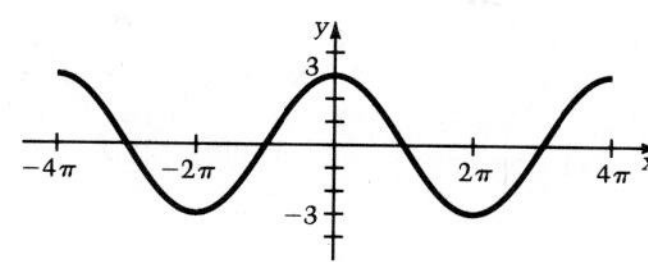

16. Graphs of $y = \tan x$ and $y = \tan(x - \pi)$ are the same **18.** False **20.** False **22.** amp: 4, per: π, ps: $-\frac{\pi}{2}$

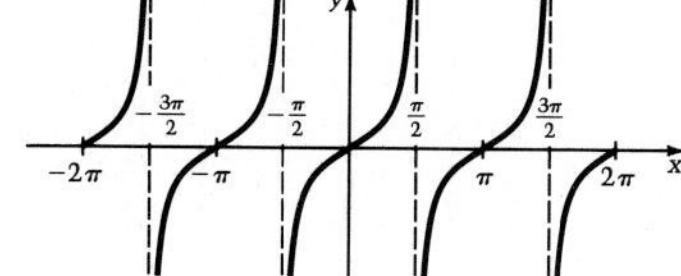

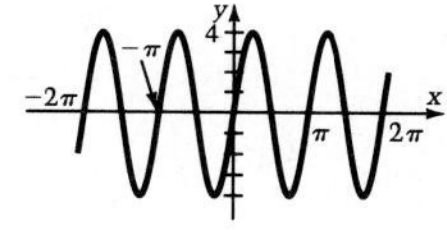

23.

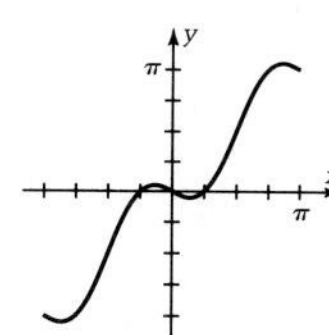

25. $\frac{\pi}{6}$ **27.** $\frac{4}{5}$ **29.** (a) $3\sqrt{2}$, 135° (b)

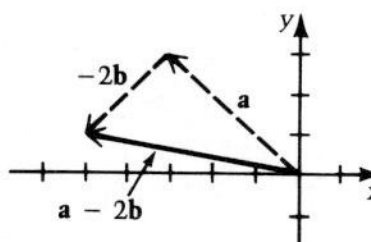

31. $\langle -7, 6 \rangle$

24. $-\frac{\pi}{6}$ **26.** $-\frac{\pi}{2}$ **28.** $\frac{12}{13}$ **30.** $\langle 5, 9 \rangle$ **32.** $\frac{5\sqrt{2}}{2}, \left\langle -4, \frac{15}{2} \right\rangle$

33. $\sqrt{37}$, $\langle 8, -15 \rangle$ **35.** −9.7**i**, −4.7**j** **37.** 51 lbs. **39.** 0, 90°

34. −7.3**i**, 4.6**j** **36.** 493 mph, N 74°28′ E **38.** −4, 117°

Chapter 8

Section 8.1, pages 465–466

9. $\sin\alpha$ **11.** $\tan\beta$ **13.** $\sin\beta$ **15.** $\tan^2\varphi$ **17.** $\tan\alpha + 1$ ***39.** $\frac{\sqrt{1 + \cot^2\alpha}}{1 + \cot^2\alpha}$

10. $\sin\alpha$ **12.** $\tan^2\alpha - 1$ **14.** $\sin\beta$ **16.** $\sin^2\varphi$ **18.** $1 + \tan\alpha$ ***40.** $-\frac{\cot\alpha\sqrt{1 + \cot^2\alpha}}{1 + \cot^2\alpha}$

***41.** $-\frac{\sqrt{1 - \sin^2\beta}}{1 - \sin^2\beta}$ ***43.** $\frac{\sec\gamma\sqrt{\sec^2\gamma - 1}}{\sec^2\gamma - 1}$

***42.** $-\frac{\sqrt{1 - \cos^2\varphi}}{\cos\varphi}$

*denotes Superset exercises

***44.** (a) $\cos\alpha = -\sqrt{1-\sin^2\alpha}$; $\tan\alpha = -\dfrac{\sin\alpha\sqrt{1-\sin^2\alpha}}{1-\sin^2\alpha}$; $\csc\alpha = \dfrac{1}{\sin\alpha}$; $\sec\alpha = -\dfrac{\sqrt{1-\sin^2\alpha}}{1-\sin^2\alpha}$; $\cot\alpha = -\dfrac{\sqrt{1-\sin^2\alpha}}{\sin\alpha}$; (b) $\sin\alpha = -\sqrt{1-\cos^2\alpha}$; $\tan\alpha = -\dfrac{\sqrt{1-\cos^2\alpha}}{\cos\alpha}$; $\csc\alpha = -\dfrac{\sqrt{1-\cos^2\alpha}}{1-\cos^2\alpha}$; $\sec\alpha = \dfrac{1}{\cos\alpha}$; $\cot\alpha = -\dfrac{\cos\alpha\sqrt{1-\cos^2\alpha}}{1-\cos^2\alpha}$

Section 8.2, pages 472–473

1. $\frac{\sqrt{3}}{2}$ **3.** $\frac{\sqrt{6}-\sqrt{2}}{4}$ **5.** $\frac{\sqrt{6}-\sqrt{2}}{4}$ **7.** $\frac{\sqrt{2}-\sqrt{6}}{4}$ **9.** $\frac{\sqrt{2}}{2}$ **11.** $-\frac{\sqrt{2}}{2}$ **13.** $\frac{\sqrt{6}-\sqrt{2}}{4}$ **15.** $\frac{\sqrt{6}+\sqrt{2}}{4}$

2. $-\frac{1}{2}$ **4.** $\frac{\sqrt{6}+\sqrt{2}}{4}$ **6.** $\frac{\sqrt{6}+\sqrt{2}}{4}$ **8.** $\frac{\sqrt{2}-\sqrt{6}}{4}$ **10.** $-\frac{\sqrt{2}}{2}$ **12.** $-\frac{\sqrt{2}}{2}$ **14.** $\frac{\sqrt{6}-\sqrt{2}}{4}$ **16.** $\frac{\sqrt{2}-\sqrt{6}}{4}$

17. $\frac{\sqrt{6}-\sqrt{2}}{4}$ **19.** $\frac{\sqrt{2}-\sqrt{6}}{4}$ **21.** $2+\sqrt{3}$ **23.** $-2-\sqrt{3}$ **25.** $\sqrt{3}-2$ **27.** (a) $-\frac{33}{65}$ (b) $-\frac{56}{65}$ (c) $\frac{33}{56}$

18. $\frac{-\sqrt{6}-\sqrt{2}}{4}$ **20.** $\frac{-\sqrt{6}-\sqrt{2}}{4}$ **22.** $2-\sqrt{3}$ **24.** $2-\sqrt{3}$ **26.** $2+\sqrt{3}$ **28.** (a) $-\frac{56}{65}$ (b) $-\frac{33}{65}$ (c) $\frac{56}{33}$

29. (a) $-\frac{16}{65}$ (b) $\frac{63}{65}$ (c) $-\frac{16}{63}$ **31.** (a) $-\frac{2}{3}+\frac{2\sqrt{5}}{15}$ (b) $-\frac{1}{3}-\frac{4\sqrt{5}}{15}$ (c) $\frac{10\sqrt{5}-18}{11}$ **33.** (a) $-\frac{63}{65}$ (b) $-\frac{16}{65}$ (c) $\frac{63}{16}$

30. (a) $\frac{56}{65}$ (b) $\frac{33}{65}$ (c) $\frac{56}{33}$ **32.** (a) $\frac{12\sqrt{13}}{91}-\frac{3}{7}$ (b) $\frac{18\sqrt{13}}{91}+\frac{2}{7}$ (c) $\frac{147-39\sqrt{13}}{136}$ **34.** (a) $-\frac{16}{65}$ (b) $-\frac{63}{65}$ (c) $\frac{16}{63}$

35. (a) $-\frac{56}{65}$ (b) $\frac{33}{65}$ (c) $-\frac{56}{33}$ **37.** (a) $-\frac{2}{3}-\frac{2\sqrt{5}}{15}$ (b) $-\frac{1}{3}+\frac{4\sqrt{15}}{15}$ (c) $-\frac{18+10\sqrt{5}}{11}$ ***49.** $\frac{\sqrt{3}}{2}\cos\alpha-\frac{1}{2}\sin\alpha$

36. (a) $-\frac{16}{65}$ (b) $-\frac{63}{65}$ (c) $\frac{16}{63}$ **38.** (a) $\frac{12\sqrt{13}}{91}+\frac{3}{7}$ (b) $\frac{18\sqrt{13}}{91}-\frac{2}{7}$ (c) $\frac{147+39\sqrt{13}}{136}$ ***50.** $\frac{\sqrt{3}}{2}\sin\alpha+\frac{1}{2}\cos\alpha$

***51.** $\frac{\tan\alpha+1}{1-\tan\alpha}$ ***53.** $\cos\phi$ ***55.** $\frac{\sqrt{2}}{2}(\cos\alpha-\sin\alpha)$ ***57.** $\tan(\alpha-\beta)=\frac{\tan\alpha-\tan\beta}{1+\tan\alpha\tan\beta}$

***52.** $\frac{\sqrt{2}}{2}(\cos\beta-\sin\beta)$ ***54.** $\cot\beta$ ***56.** $\frac{\sqrt{2}}{2}(\cos\beta+\sin\beta)$ ***58.** $\cot(\alpha+\beta)=\frac{\cot\alpha\cot\beta-1}{\cot\beta+\cot\alpha}$

***59.** $\sec(\alpha+\beta)=\frac{\sec\alpha\sec\beta}{1-\tan\alpha\tan\beta}$ ***61.** $\sin 2\alpha = 2\sin\alpha\cos\alpha$ ***63.** 0.2588 ***65.** 0.2588 ***67.** 0.2679

***60.** $\csc(\alpha+\beta)=\frac{\csc\alpha\csc\beta}{\cot\beta+\cot\alpha}$ ***62.** $\cos 2\alpha=\cos^2\alpha-\sin^2\alpha$ ***64.** 0.9659 ***66.** 0.9659 ***68.** 3.7321

***69.** −0.4799 ***71.** 0.3531 ***73.** −0.5470

***70.** 0.8773 ***72.** 0.9356 ***74.** 2.6492

Section 8.3, pages 481–482

1. $\frac{\sqrt{3}}{2}, -\frac{1}{2}, -\sqrt{3}$ **3.** $-\frac{\sqrt{3}}{2}, -\frac{1}{2}, \sqrt{3}$ **5.** −1, 0, undefined **7.** $\frac{\sqrt{3}}{2}, -\frac{1}{2}, -\sqrt{3}$ **9.** 1, 0, undefined

2. 1, 0, undefined **4.** 0, 1, 0 **6.** $-\frac{\sqrt{3}}{2}, \frac{1}{2}, -\sqrt{3}$ **8.** −1, 0, undefined **10.** 1, 0, undefined

11. $-\frac{\sqrt{3}}{2}, \frac{1}{2}, -\sqrt{3}$ **13.** $\frac{\sqrt{3}}{2}, -\frac{1}{2}, -\sqrt{3}$ **15.** 1, 0, undefined **17.** −1, 0, undefined **19.** (a) $-\frac{24}{25}$ (b) $\frac{7}{25}$ (c) $-\frac{24}{7}$

12. −1, 0, undefined **14.** $-\frac{\sqrt{3}}{2}, \frac{1}{2}, -\sqrt{3}$ **16.** 0, 1, 0 **18.** $\frac{\sqrt{3}}{2}, \frac{1}{2}, \sqrt{3}$ **20.** (a) $-\frac{24}{25}$ (b) $\frac{7}{25}$ (c) $-\frac{24}{7}$

21. (a) $-\frac{120}{169}$ (b) $-\frac{119}{169}$ (c) $\frac{120}{119}$ **23.** (a) $\frac{720}{1681}$ (b) $\frac{1519}{1681}$ (c) $\frac{720}{1519}$ **25.** (a) $\frac{4}{5}$ (b) $\frac{3}{5}$ (c) $\frac{4}{3}$ **27.** 0.7100

---

*denotes Superset exercises

22. (a) $\frac{24}{25}$ (b) $-\frac{7}{25}$ (c) $-\frac{24}{7}$ **24.** (a) $\frac{120}{169}$ (b) $\frac{119}{169}$ (c) $\frac{120}{119}$ **26.** (a) $-\frac{336}{625}$ (b) $\frac{527}{625}$ (c) $-\frac{336}{527}$ **28.** 0.7042

29. 0.9917 **31.** 0.4385 **33.** 2.0494 **35.** $\frac{1}{2}, \frac{\sqrt{3}}{2}, \frac{\sqrt{3}}{3}$ **37.** $\frac{\sqrt{2}}{2}, \frac{\sqrt{2}}{2}, 1$ **39.** $\frac{\sqrt{2+\sqrt{3}}}{2}, -\frac{\sqrt{2-\sqrt{3}}}{2}, -2-\sqrt{3}$

30. 1.0083 **32.** 0.8987 **34.** 0.4879 **36.** 1, 0, undefined **38.** $\frac{\sqrt{3}}{2}, \frac{1}{2}, \sqrt{3}$ **40.** $\frac{\sqrt{2-\sqrt{2}}}{2}, -\frac{\sqrt{2+\sqrt{2}}}{2}, 1-\sqrt{2}$

41. $-\frac{\sqrt{2+\sqrt{2}}}{2}, -\frac{\sqrt{2-\sqrt{2}}}{2}, 1+\sqrt{2}$ **43.** $-\frac{\sqrt{2-\sqrt{2}}}{2}, \frac{\sqrt{2+\sqrt{2}}}{2}, 1-\sqrt{2}$ **45.** 0, 1, 0 **47.** $\frac{\sqrt{3}}{2}, -\frac{1}{2}, -\sqrt{3}$

42. $\frac{\sqrt{2+\sqrt{2}}}{2}, -\frac{\sqrt{2-\sqrt{2}}}{2}, -1-\sqrt{2}$ **44.** $\frac{\sqrt{2+\sqrt{2}}}{2}, \frac{\sqrt{2-\sqrt{2}}}{2}, 1+\sqrt{2}$ **46.** $\frac{\sqrt{3}}{2}, \frac{1}{2}, \sqrt{3}$ **48.** $-\frac{\sqrt{2}}{2}, -\frac{\sqrt{2}}{2}, 1$

49. $\frac{\sqrt{2-\sqrt{3}}}{2}, \frac{\sqrt{2+\sqrt{3}}}{2}, 2-\sqrt{3}$ **51.** $-\frac{\sqrt{2-\sqrt{3}}}{2}, -\frac{\sqrt{2+\sqrt{3}}}{2}, 2-\sqrt{3}$ **53.** $-\frac{\sqrt{2+\sqrt{3}}}{2}, \frac{\sqrt{2-\sqrt{3}}}{2}, -2-\sqrt{3}$

50. $\frac{\sqrt{2+\sqrt{2}}}{2}, -\frac{\sqrt{2-\sqrt{2}}}{2}, -1-\sqrt{2}$ **52.** $-\frac{\sqrt{2+\sqrt{2}}}{2}, \frac{\sqrt{2-\sqrt{2}}}{2}, -1-\sqrt{2}$ **54.** $\frac{\sqrt{3}}{2}, -\frac{1}{2}, -\sqrt{3}$

55. $\frac{3\sqrt{10}}{10}, \frac{\sqrt{10}}{10}, 3$ **57.** $\frac{2\sqrt{13}}{13}, -\frac{3\sqrt{13}}{3}, -\frac{2}{3}$ **59.** $\frac{\sqrt{82}}{82}, \frac{9\sqrt{82}}{82}, \frac{1}{9}$ **61.** $\sqrt{\frac{5+2\sqrt{5}}{10}}, -\sqrt{\frac{5-2\sqrt{5}}{10}}, -\sqrt{9+4\sqrt{5}}$

56. $\frac{\sqrt{10}}{10}, -\frac{3\sqrt{10}}{10}, -\frac{1}{3}$ **58.** $\frac{2\sqrt{5}}{5}, -\frac{\sqrt{5}}{5}, -2$ **60.** $\frac{\sqrt{26}}{26}, \frac{5\sqrt{26}}{26}, \frac{1}{5}$ **62.** $\frac{7\sqrt{2}}{10}, \frac{\sqrt{2}}{10}, 7$

***66.** $3\sin\alpha - 4\sin^3\alpha$ ***68.** $8\cos^4\alpha - 8\cos^2\alpha + 1$ ***70.** $\frac{4\tan\alpha - 4\tan^3\alpha}{1 - 6\tan^2\alpha + \tan^4\alpha}$ ***72.** $1 - 2\sin^2 6\alpha$

***67.** $4\cos^3\alpha - 3\cos\alpha$ ***69.** $4\sin\alpha\cos\alpha - 8\sin^3\alpha\cos\alpha$ ***71.** $2\cos^2 6\alpha - 1$ ***73.** $2\sin 4\alpha\cos 4\alpha$

***74.** $\cos^2 4\alpha - \sin^2 4\alpha$ ***76.** $\frac{1 - 3\tan^2 2\alpha}{3\tan 2\alpha - \tan^3 2\alpha}$ ***78.** $\frac{\tan 12\alpha}{\sec 12\alpha + 1}$ ***80.** $\pm\sqrt{\frac{\csc 4\alpha - \cot 4\alpha}{2\csc 4\alpha}}$ ***82.** $\sin\alpha$

***75.** $\frac{3\tan 2\alpha - \tan^3 2\alpha}{1 - 3\tan^2 2\alpha}$ ***77.** $\csc 12\alpha + \cot 12\alpha$ ***79.** $\pm 2\sin(4\alpha)\sqrt{1 - \sin^2 4\alpha}$ ***81.** $\cos\alpha$ ***83.** $\sin\alpha$

***84.** $\sin 4\alpha$ ***86.** $\cos\alpha$ ***88.** 1 ***90.** $1 + \sin\alpha\cos\alpha$ ***92.** -1

***85.** $\cos\alpha$ ***87.** 1 ***89.** $1 - \sin\alpha$ ***91.** $\cos 2\alpha$

Section 8.4, pages 487–488

1. $\cos 3\alpha$ **3.** $\sin 3\alpha$ **5.** $\tan 4\varphi$ **7.** $\tan\frac{\alpha}{2}$ **21.** yes **23.** no **25.** yes **27.** $\frac{\sqrt{6}}{2}$ **29.** $\frac{\sqrt{2}}{2}$

2. $\cos 5\alpha$ **4.** $\sin 5\alpha$ **6.** $\tan\alpha$ **8.** $-\frac{1}{2}\tan 2\beta$ **22.** yes **24.** no **26.** yes **28.** $-\frac{\sqrt{2}}{2}$ **30.** $\frac{\sqrt{2}}{2}$

31. $-\frac{\sqrt{2}}{2}$ **33.** $-\frac{\sqrt{6}}{2}$ **39.** $2\sin 2x\cos x$ **41.** $2\cos 4x\sin 2x$ **43.** $-2\sin 4x\sin 2x$ **45.** $\frac{1}{2}\sin 5x - \frac{1}{2}\sin x$

32. $-\frac{\sqrt{2}}{2}$ **34.** $\frac{\sqrt{2}}{2}$ **40.** $2\cos 2x\sin x$ **42.** $2\sin 4x\cos 2x$ **44.** $2\cos 4x\cos 2x$ **46.** $\frac{1}{2}\sin 5x + \frac{1}{2}\sin x$

47. $\frac{1}{2}\cos x - \frac{1}{2}\cos 5x$ ***49.** $\frac{2\sin x\cos x + 1}{\cos^2 x - \sin^2 x}$ ***57.** $M = 1, N = 2$ ***59.** $M = 2$

48. $\frac{1}{2}\cos 5x + \frac{1}{2}\cos x$ ***56.** $M = 1, N = -2$ ***58.** $M = 1, N = 1$ ***60.** $M = 1, N = 1$

---

*denotes Superset exercises

Section 8.5, pages 496–497

1. $\frac{4\pi}{3} + 2n\pi, \frac{5\pi}{3} + 2n\pi$ **3.** $0 + n\pi, \frac{\pi}{6} + 2n\pi, \frac{5\pi}{6} + 2n\pi$ **5.** $\frac{\pi}{6} + n\pi, \frac{5\pi}{6} + n\pi$ **7.** $\frac{\pi}{3} + 2n\pi, \frac{5\pi}{3} + 2n\pi$

2. $\frac{\pi}{3} + n\pi$ **4.** $\frac{2\pi}{3} + 2n\pi, \frac{4\pi}{3} + 2n\pi, 0 + 2n\pi$ **6.** $\frac{\pi}{4} + n\pi, \frac{3\pi}{4} + n\pi$ **8.** $\frac{\pi}{2} + n\pi$

9. $\frac{3\pi}{2} + 2n\pi, 0 + 2n\pi$ **11.** $\frac{\pi}{6}, \frac{5\pi}{6}, \frac{3\pi}{2}$ **13.** $\frac{\pi}{6}, \frac{5\pi}{6}$ **15.** $0, \pi$ **17.** $\frac{\pi}{2}, \frac{7\pi}{6}, \frac{11\pi}{6}$ **19.** $\frac{\pi}{8}, \frac{5\pi}{8}, \frac{9\pi}{8}, \frac{13\pi}{8}$

10. $\frac{\pi}{4} + 2n\pi, \frac{3\pi}{4} + 2n\pi$ **12.** $\frac{2\pi}{3}, \frac{4\pi}{3}, 0$ **14.** $\frac{3\pi}{4}, \frac{5\pi}{4}, 0$ **16.** $\frac{\pi}{3}, \frac{2\pi}{3}, \frac{4\pi}{3}, \frac{5\pi}{3}$ **18.** $\frac{\pi}{3}, \frac{5\pi}{3}$ **20.** 0°, 180°

21. −60°, 60° **23.** $0, \pi, \frac{\pi}{4}, \frac{3\pi}{4}$ **25.** 0°, 180°, 60° **27.** 30°, 90°, 150°, 210°, 270°, 330°, 45°, 135°, 225°, 315°

22. 90° **24.** $0, \pi$ **26.** 0 **28.** $\frac{\pi}{4}, \frac{3\pi}{4}, \frac{\pi}{3}$

29. $\frac{\pi}{2}, \frac{3\pi}{2}, \frac{7\pi}{6}, \frac{11\pi}{6}$ **31.** $0, \pi, \frac{\pi}{6}, \frac{5\pi}{6}$ **33.** 0°, 180°, 48.6°, 131.4° **35.** 19.5°, 160.5°, 270°

30. $\frac{\pi}{2}, \frac{3\pi}{2}, \frac{2\pi}{3}, \frac{4\pi}{3}$ **32.** $0, \pi, \frac{3\pi}{4}, \frac{7\pi}{4}$ **34.** 90°, 270°, 66.4°, 293.6° **36.** 131.8°, 228.2°, 0°

37. 41.8°, 138.2°, 210°, 330° **39.** 10.5°, 169.5°, 223.1°, 316.9° **41.** no solutions

38. 62.0°, 118.0° **40.** 43.6°, 316.4° **42.** 33.7°, 213.7°, 116.6°, 296.6°

43. 27.4°, 62.6°, 117.4°, 152.6°, 207.4°, 242.6°, 297.4°, 332.6° **45.** 19.5°, 160.5°, 194.5°, 345.5° **47.** 108.4°, 288.4°

44. no solutions **46.** 131.8°, 228.2°, 53.1°, 306.9° **48.** 54°, 126°, 198°, 342°

49. 24.1°, 155.9° 204.1°, 335.9° **51.** 22.5°, 202.5°, 112.5°, 292.5°

50. 11.25°, 101.25°, 191.25°, 281.25°, 56.25°, 146.25°, 236.25°, 326.25° **52.** 41.4°, 318.6°, 120°, 240°

53. 43.9°, 316.1°, 75.5°, 284.5° **55.** $-\frac{\pi}{6}$ **57.** $\frac{\pi}{3}$ **59.** $\frac{24}{25}$ **61.** $-\frac{7}{25}$ **63.** $-\frac{33}{65}$ **65.** $\frac{\pi}{4}$ ***67.** $\frac{1}{2}$ ***69.** 0.20

54. 208.0°, 332.0°, 13.3°, 166.7° **56.** $\frac{2\pi}{3}$ **58.** $\frac{3\pi}{4}$ **60.** $\frac{24}{25}$ **62.** 1 **64.** $\frac{33}{65}$ **66.** $-\frac{9}{7}$ ***68.** 1.33 ***70.** $0, \pi$

***71.** $0, \frac{\pi}{2}$ ***73.** $\frac{\pi}{6}, \frac{\pi}{3}, \frac{7\pi}{6}, \frac{4\pi}{3}$ ***75.** $0 + 2n\pi$

***72.** $\frac{\pi}{3}, \pi$ ***74.** $\frac{\pi}{3}, \frac{5\pi}{3}, \frac{\pi}{4}, \frac{5\pi}{4}$ ***76.** $\theta = \frac{\pi}{3}$ or $\frac{5\pi}{3}, r = \frac{1}{2}$

Section 8.6, pages 503–504

1. $A(2, 120°)$ **3.** $C(3, 135°)$ **5.** $E(2, -120°)$ **7.** $G(-1, -270°)$ **9.** $I(-2, -420°)$ **11.** $K(3, \pi)$

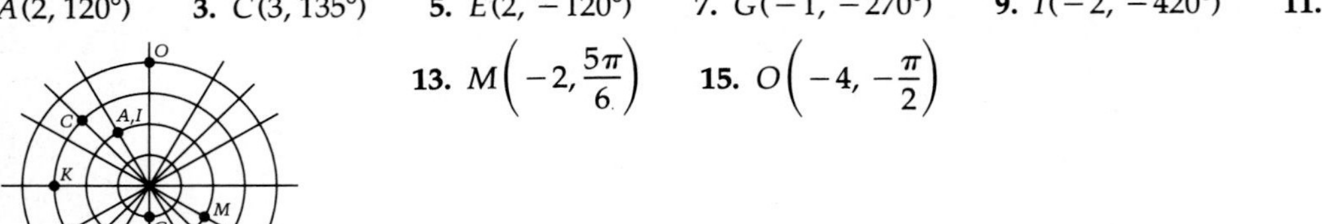

13. $M\left(-2, \frac{5\pi}{6}\right)$ **15.** $O\left(-4, -\frac{\pi}{2}\right)$

---

*denotes Superset exercises

2. $B(5, 30°)$ **4.** $D(4, 210°)$ **6.** $F(-3, 135°)$ **8.** $H(-5, -45°)$ **10.** $J(4, -570°)$ **12.** $L\left(4, \frac{\pi}{4}\right)$ **14.** $N\left(-5, -\frac{3\pi}{2}\right)$ **16.** $P\left(-2, \frac{3\pi}{4}\right)$

17. $(-1, \sqrt{3})$ **19.** $\left(-\frac{3\sqrt{2}}{2}, \frac{3\sqrt{2}}{2}\right)$ **21.** $(-1, -\sqrt{3})$ **23.** $(0, -1)$ **25.** $(-1, \sqrt{3})$ **27.** $(-3, 0)$

18. $\left(\frac{5\sqrt{3}}{2}, \frac{5}{2}\right)$ **20.** $(-2\sqrt{3}, -2)$ **22.** $\left(\frac{3\sqrt{2}}{2}, -\frac{3\sqrt{2}}{2}\right)$ **24.** $\left(-\frac{5\sqrt{2}}{2}, \frac{5\sqrt{2}}{2}\right)$ **26.** $(-2\sqrt{3}, 2)$ **28.** $(2\sqrt{2}, 2\sqrt{2})$

29. $(\sqrt{3}, -1)$ **31.** $(0, 4)$ **33.** $\left(4, \frac{\pi}{2}\right), \left(-4, \frac{3\pi}{2}\right)$ **35.** $\left(\sqrt{2}, \frac{5\pi}{4}\right), \left(-\sqrt{2}, \frac{\pi}{4}\right)$ **37.** $\left(2, \frac{11\pi}{6}\right), \left(-2, \frac{5\pi}{6}\right)$

30. $(0, -5)$ **32.** $(\sqrt{2}, -\sqrt{2})$ **34.** $(2, \pi), (-2, 0)$ **36.** $\left(\sqrt{2}, \frac{7\pi}{4}\right), \left(-\sqrt{2}, \frac{3\pi}{4}\right)$ **38.** $\left(2, \frac{4\pi}{3}\right), \left(-2, \frac{\pi}{3}\right)$

39. $\left(2, \frac{3\pi}{4}\right), \left(-2, \frac{7\pi}{4}\right)$ **41.** $(5, 2.21), (-5, 5.36)$ **43.** $(2\sqrt{13}, 0.98), (-2\sqrt{13}, 4.12)$ **45.** $(3.33, 4.22)$

40. $\left(2, \frac{7\pi}{4}\right), \left(-2, \frac{3\pi}{4}\right)$ **42.** $(13, 4.32), (-13, 1.18)$ **44.** $(5\sqrt{5}, 5.18), (-5\sqrt{5}, 2.03)$ **46.** $(3.58, -2.42)$

47. $(-1.87, 1.15)$ **49.** $(6.50, 2.20)$ **51.** $(-6.06, 0.54)$ **53.** $x^2 + y^2 = 9$ **55.** $y = -x$ **57.** $x - 2y = 5$

48. $(-1.84, -2.46)$ **50.** $(-5.66, 2.39)$ **52.** $(7.83, 0.31)$ **54.** $4x^2 + 4y^2 = 25$ **56.** $y = \sqrt{3}\,x$ **58.** $3y - 4x = 5$

59. $y^2 + 2x = 1$ **61.** $x^2 - 3y^2 - 8y - 4 = 0$ **63.** $\theta = \frac{\pi}{2}$ **65.** $r(\cos\theta + \sin\theta) = 1$

60. $x^2 + 4y = 4$ **62.** $3x^2 + 4y^2 - 4x - 4 = 0$ **64.** $r\sin\theta = -5$ **66.** $r(4\sin\theta - \cos\theta) = 4$

67. $r^2 - 2r\cos\theta = 0$ **69.** $r^2\cos^2\theta = 6r\sin\theta + 9$ **71.** $r^2\sin 2\theta = 1$ **73.** $r^2 = 16r\sin\theta$ **75.**

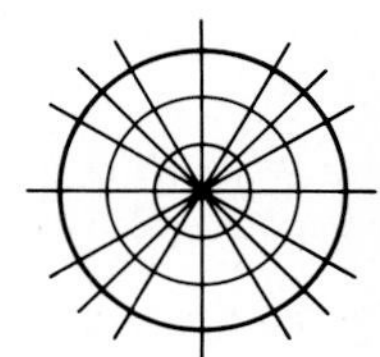

68. $r^2\cos 2\theta - 6r\sin\theta = 0$ **70.** $r^2\sin^2\theta = 8r\cos\theta + 16$ **72.** $r^2\sin 2\theta = 2$ **74.** $r^2 = 4r\cos\theta$ **76.**

77.

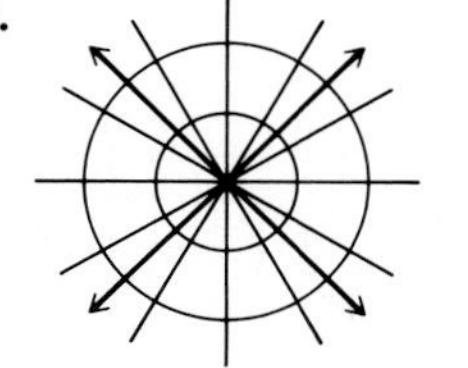

79.

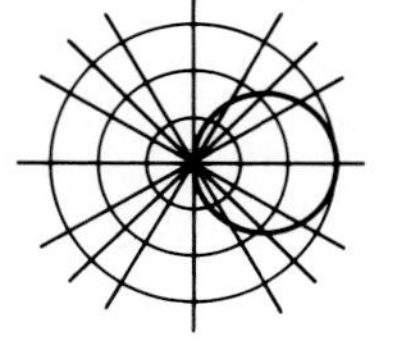

81.

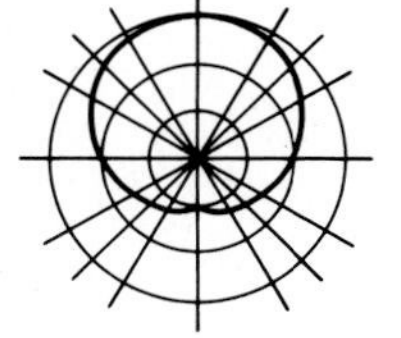

83.

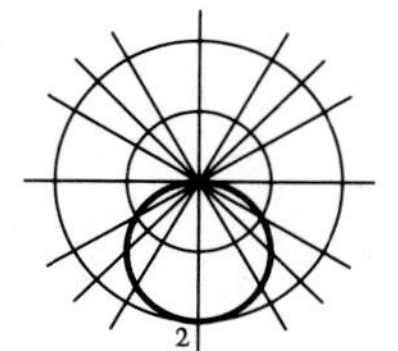

85.

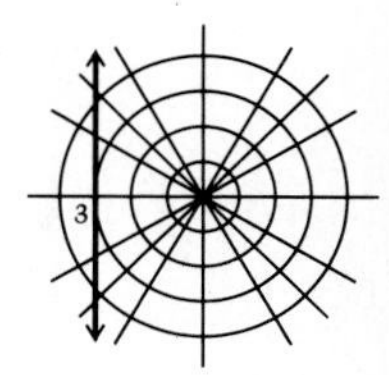

78.

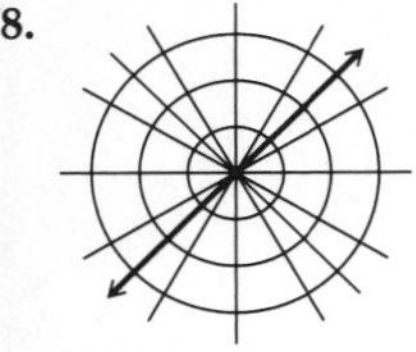

80.

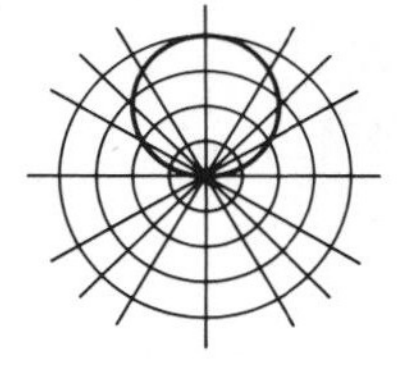

82.

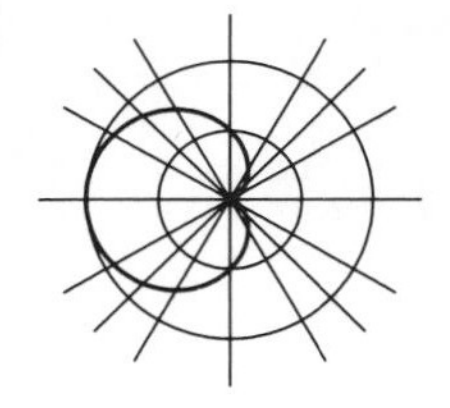

84.

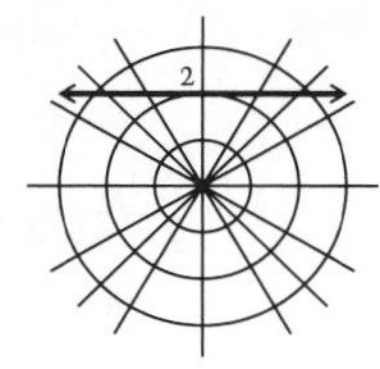

86.

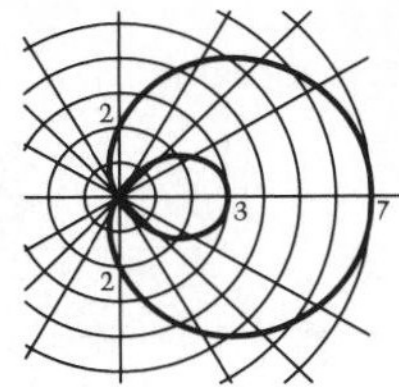

87.

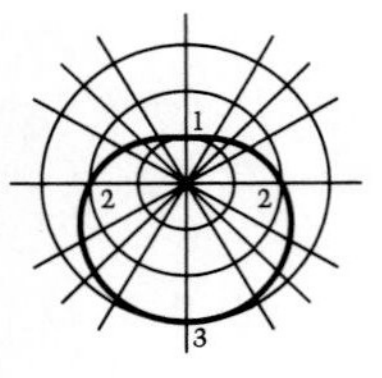

89.

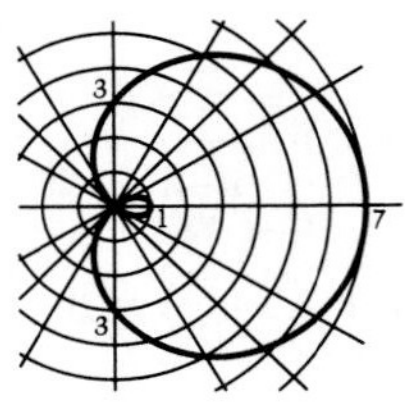

91.

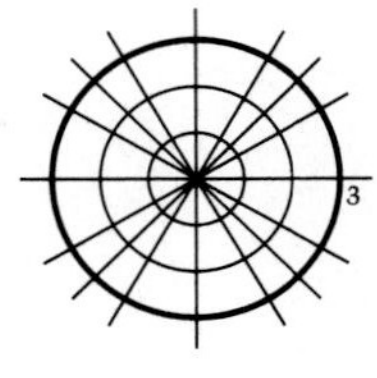

93.

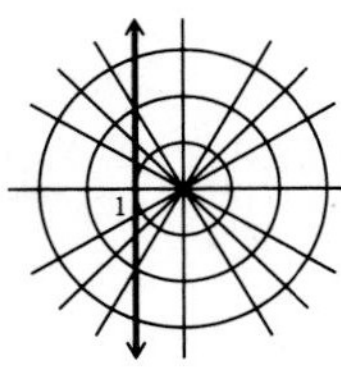

88.

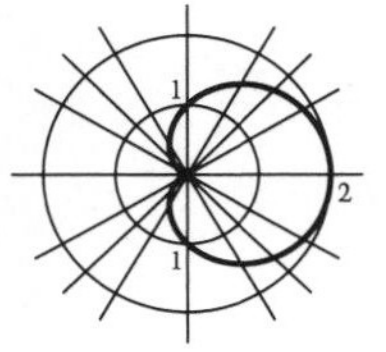

90.

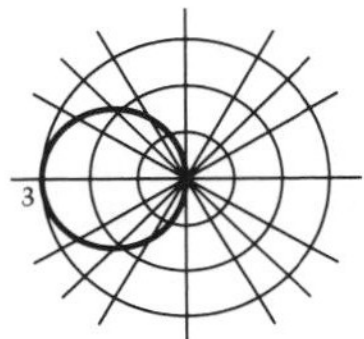

92.

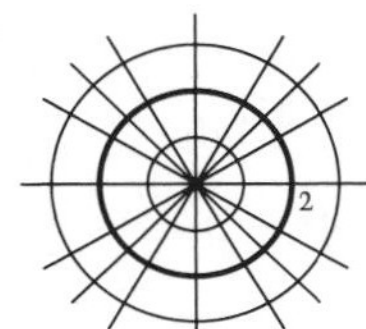

94.

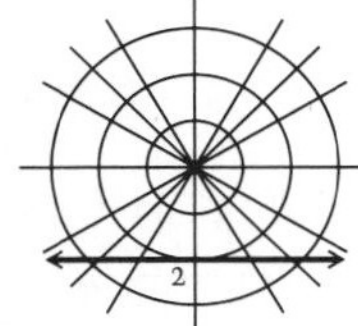

***95.** $2r\cos\theta - 3r\sin\theta = -4$ ***97.** (a) 4.91 (b) 4.91 ***99.** parabola: $y^2 = 4x + 4$

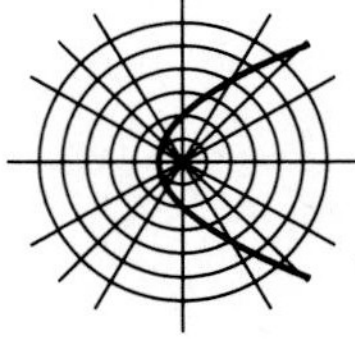

***96.** $r\cos\theta + 2r\sin\theta = 2$ ***98.** $r^2 - 2r(3\cos\theta + \sin\theta) + 6 = 0$ ***100.** parabola: $x^2 = 4 - 4y$

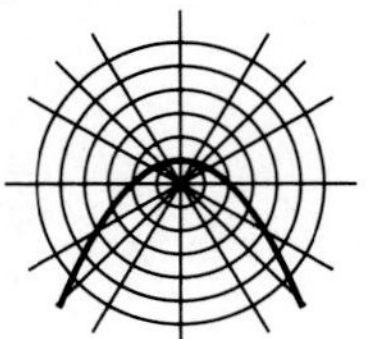

***101.** line: $y = 2x$

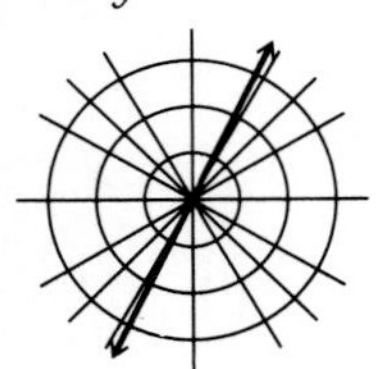

***103.** hyperbola: $xy = 1$

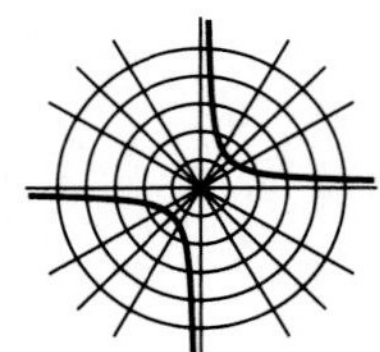

***105.**

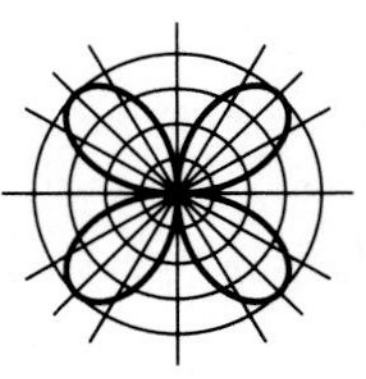

*denotes Superset exercises

***102.** line: $x = 2y$

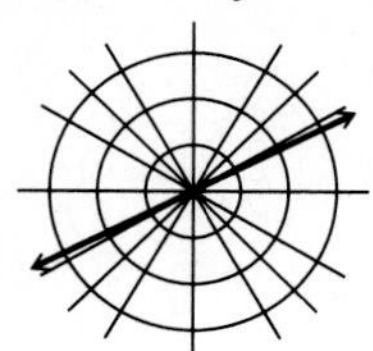

***104.** hyperbola: $x^2 - y^2 = 2$

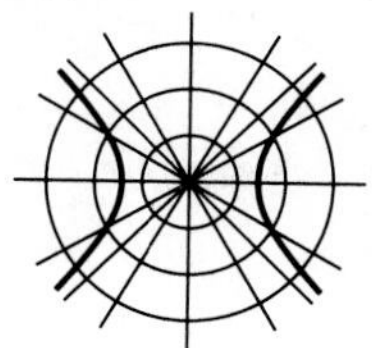

***106.**

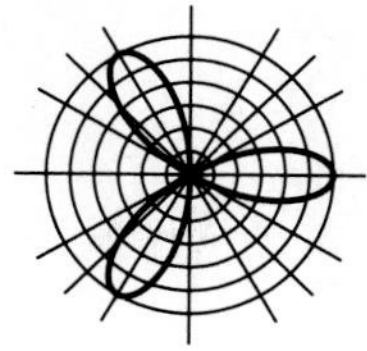

***107.** circle: $(x - 4)^2 + (y - 3)^2 = 5^2$

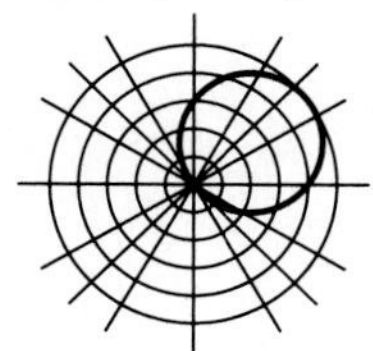

***109.**

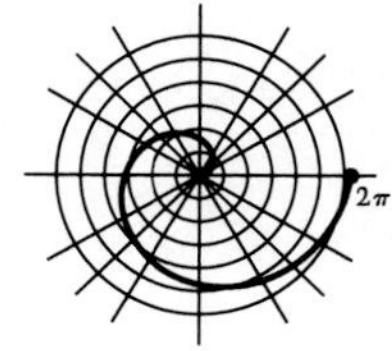

***111.**

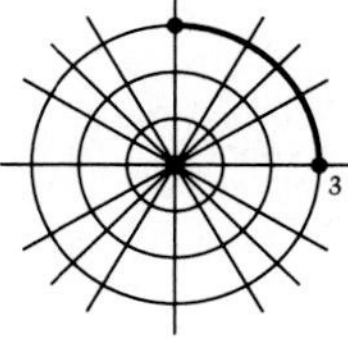

***108.** line

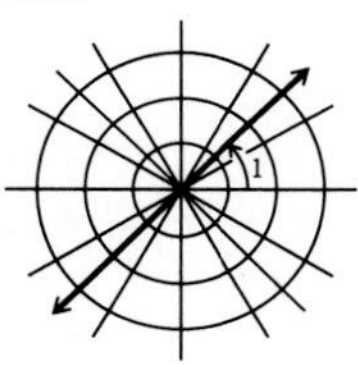

***110.**

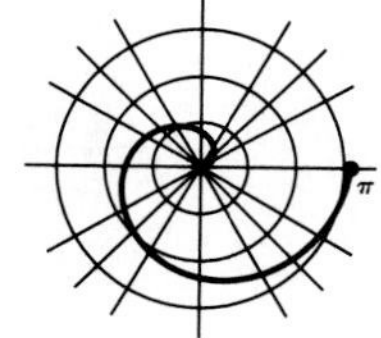

***112.**

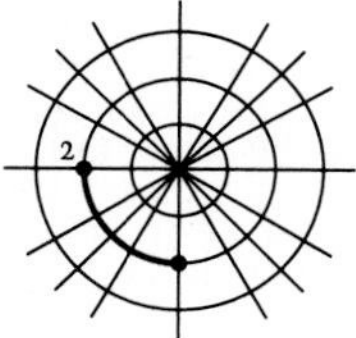

***113.**

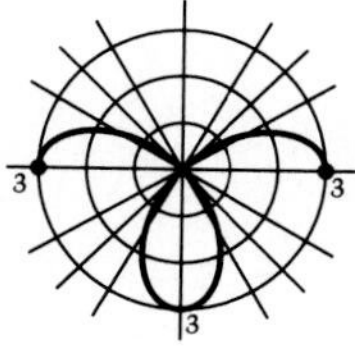

***115.**

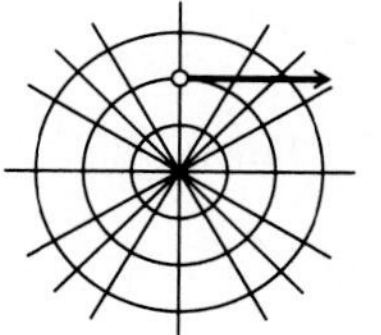

***117.**

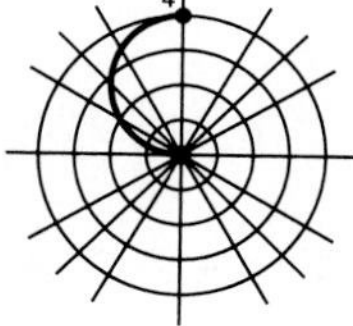

***114.**

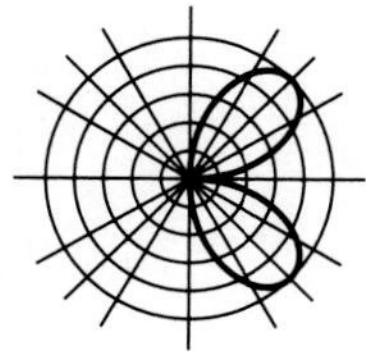

***116.**

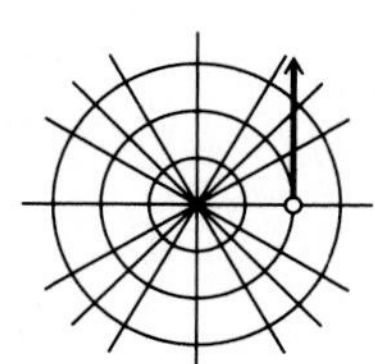

***118.**

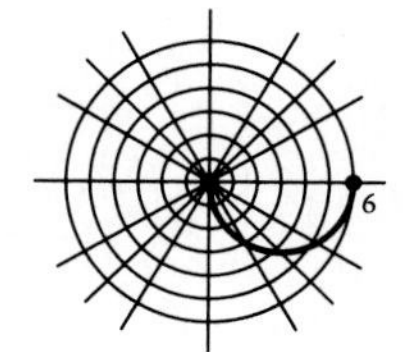

---

*denotes Superset exercises

Section 8.7, pages 513–515

1. $A = 2 + 5i$ **3.** $C = 2 - 5i$ **5.** $E = -2 + 5i$ **7.** $G = -6$ **9.** $I = -2i$ **11.** $K = \overline{2 + 5i}$
2. $B = 3 + 2i$ **4.** $D = 3 - 2i$ **6.** $F = -3 + 2i$ **8.** $H = -i$ **10.** $J = -3$ **12.** $L = \overline{3 + 2i}$
13. $M = \overline{5 - 4i}$ **15.** $O = \overline{-5 - 4i}$ **17.** $Q = \overline{-4 + 3i}$ **19.** $\sqrt{53}$ **21.** $\sqrt{53}$ **23.** 2 **25.** 1 **27.** 4

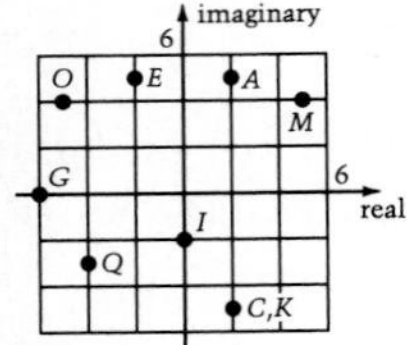

14. $N = \overline{2 - 3i}$ **16.** $P = \overline{-5 - 2i}$ **18.** $R = \overline{-3 + 2i}$ **20.** $\sqrt{34}$ **22.** $\sqrt{34}$ **24.** 7 **26.** 1 **28.** 3

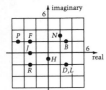

29. $\sqrt{15}$ **31.** $3\left(\cos\frac{\pi}{2} + i\sin\frac{\pi}{2}\right)$ **33.** $6(\cos\pi + i\sin\pi)$ **35.** $2\left(\cos\frac{2\pi}{3} + i\sin\frac{2\pi}{3}\right)$

30. $\sqrt{14}$ **32.** $2\left(\cos\frac{3\pi}{2} + i\sin\frac{3\pi}{2}\right)$ **34.** $8(\cos 0 + i\sin 0)$ **36.** $2\left(\cos\frac{5\pi}{6} + i\sin\frac{5\pi}{6}\right)$

37. $2\sqrt{2}\left(\cos\frac{5\pi}{4} + i\sin\frac{5\pi}{4}\right)$ **39.** $\sqrt{2}\left(\cos\frac{5\pi}{4} + i\sin\frac{5\pi}{4}\right)$ **41.** $2\sqrt{3}\left(\cos\frac{5\pi}{6} + i\sin\frac{5\pi}{6}\right)$ **43.** $4\left(\cos\frac{11\pi}{6} + i\sin\frac{11\pi}{6}\right)$

38. $2\sqrt{2}\left(\cos\frac{7\pi}{4} + i\sin\frac{7\pi}{4}\right)$ **40.** $\sqrt{2}\left(\cos\frac{\pi}{4} + i\sin\frac{\pi}{4}\right)$ **42.** $2\sqrt{3}\left(\cos\frac{7\pi}{6} + i\sin\frac{7\pi}{6}\right)$ **44.** $4\left(\cos\frac{2\pi}{3} + i\sin\frac{2\pi}{3}\right)$

45. $6\left(\cos\frac{5\pi}{4} + i\sin\frac{5\pi}{4}\right)$ **47.** $13(\cos 67.4° + i\sin 67.4°)$ **49.** $25(\cos 196.3° + i\sin 196.3°)$ **51.** 8

46. $6\left(\cos\frac{7\pi}{4} + i\sin\frac{7\pi}{4}\right)$ **48.** $17(\cos 118.1° + i\sin 118.1°)$ **50.** $\sqrt{65}(\cos 209.7° + i\sin 209.7°)$ **52.** $-2 + 2i\sqrt{3}$

53. $6 - 2i\sqrt{3}$ **55.** $-4\sqrt{3}$ **57.** $12 + 9i$ **59.** $2(\cos 165° + i\sin 165°)$ **61.** $\sqrt{5}(\cos(-243.4°) + i\sin(-243.4°))$

54. $-6 + 2i\sqrt{3}$ **56.** $36i$ **58.** $24 - 10i$ **60.** $2\left(\cos\frac{\pi}{2} + i\sin\frac{\pi}{2}\right)$ **62.** $\sqrt{2}\left(\cos\frac{17\pi}{12} + i\sin\frac{17\pi}{12}\right)$

63. $\sqrt{3}\left(\cos\frac{3\pi}{2} - i\sin\frac{3\pi}{2}\right)$ **65.** $\frac{\sqrt{6}}{3}\left(\cos\left(-\frac{19\pi}{12}\right) + i\sin\left(-\frac{19\pi}{12}\right)\right)$ **67.** -64 **69.** $8i$

64. $\sqrt{6}\left(\cos\left(-\frac{7\pi}{12}\right) + i\sin\left(-\frac{7\pi}{12}\right)\right)$ **66.** $\frac{2}{3}\left(\cos\frac{\pi}{2} + i\sin\frac{\pi}{2}\right)$ **68.** $-16\sqrt{3} - 16i$ **70.** $-8i$

71. 16,777,216 **73.** $-2.42 - 1.61i$ **75.** $-18.9 + 115.6i$ **77.** $5264.7 - 41909.5i$
72. $124{,}416 - 124{,}416i\sqrt{3}$ **74.** $450.1 - 386.9i$ **76.** $1739.0 + 28271.0i$ **78.** $-0.19 + 0.14i$
79. $\sqrt[3]{3}(\cos 30° + i\sin 30°)$, $\sqrt[3]{3}(\cos 150° + i\sin 150°)$, $\sqrt[3]{3}(\cos 270° + i\sin 270°)$
80. $\sqrt[3]{2}(\cos 90° + i\sin 90°)$, $\sqrt[3]{2}(\cos 210° + i\sin 210°)$, $\sqrt[3]{2}(\cos 330° + i\sin 330°)$
81. $\sqrt[8]{8}(\cos 33.75° + i\sin 33.75°)$, $\sqrt[8]{8}(\cos 123.75° + i\sin 123.75°)$, $\sqrt[8]{8}(\cos 213.75° + i\sin 213.75°)$, $\sqrt[8]{8}(\cos 303.75° + i\sin 303.75°)$
82. $\sqrt[4]{2}(\cos 37.5° + i\sin 37.5°)$, $\sqrt[4]{2}(\cos 127.5° + i\sin 127.5°)$, $\sqrt[4]{2}(\cos 217.5° + i\sin 217.5°)$, $\sqrt[4]{2}(\cos 307.5° + i\sin 307.5°)$
83. $\sqrt[10]{2}(\cos 9° + i\sin 9°)$, $\sqrt[10]{2}(\cos 81° + i\sin 81°)$, $\sqrt[10]{2}(\cos 153° + i\sin 153°)$, $\sqrt[10]{2}(\cos 225° + i\sin 225°)$, $\sqrt[10]{2}(\cos 297° + i\sin 297°)$

84. $\sqrt[5]{2}(\cos 12° + i \sin 12°)$, $\sqrt[5]{2}(\cos 84° + i \sin 84°)$, $\sqrt[5]{2}(\cos 156° + i \sin 156°)$, $\sqrt[5]{2}(\cos 228° + i \sin 228°)$, $\sqrt[5]{2}(\cos 300° + i \sin 300°)$

85. (a) $1(\cos 0° + i \sin 0°)$, $1(\cos 72° + i \sin 72°)$, $1(\cos 144° + i \sin 144°)$, $1(\cos 216° + i \sin 216°)$, $1(\cos 288° + i \sin 288°)$
(b) $1, i, -1, -i$ (c) $1, \frac{\sqrt{2}}{2} + \frac{\sqrt{2}}{2}i, i, -\frac{\sqrt{2}}{2} + \frac{\sqrt{2}}{2}i, -1, -\frac{\sqrt{2}}{2} - \frac{\sqrt{2}}{2}i, -i, \frac{\sqrt{2}}{2} - \frac{\sqrt{2}}{2}i$

86. (a) $1(\cos 22.5° + i \sin 22.5°)$, $1(\cos 67.5° + i \sin 67.5°)$, $1(\cos 112.5° + i \sin 112.5°)$, $1(\cos 157.5° + i \sin 157.5°)$, $1(\cos 202.5° + i \sin 202.5°)$, $1(\cos 247.5° + i \sin 247.5°)$, $1(\cos 292.5° + i \sin 292.5°)$, $1(\cos 337.5° + i \sin 337.5°)$
(b) $2(\cos 0° + i \sin 0°)$, $2(\cos 72° + i \sin 72°)$, $2(\cos 144° + i \sin 144°)$, $2(\cos 216° + i \sin 216°)$, $2(\cos 288° + i \sin 288°)$
(c) $2, 2i, -2, -2i$

***87.** $-1 \pm i\sqrt{2}$

***88.** $-2 \pm i$

***89.** $\sqrt[3]{2}(\cos 30° + i \sin 30°)$, $\sqrt[3]{2}(\cos 150° + i \sin 150°)$, $\sqrt[3]{2}(\cos 270° + i \sin 270°)$

***90.** $\frac{1}{2} + \frac{\sqrt{3}}{2}i, -1, \frac{1}{2} - \frac{\sqrt{3}}{2}i$

***91.** $2(\cos 36° + i \sin 36°)$, $2(\cos 108° + i \sin 108°)$, $2(\cos 180° + i \sin 180°)$, $2(\cos 252° + i \sin 252°)$, $2(\cos 324° + i \sin 324°)$

***92.** $2, 1 + i\sqrt{3}, -1 + i\sqrt{3}, -2, -1 - i\sqrt{3}, 1 - i\sqrt{3}$

***93.** recall, $\cos(-\theta) = \cos\theta$ and $\sin(-\theta) = -\sin\theta$

***94.** use fact that $(\cos\theta + i\sin\theta)(\cos\theta - i\sin\theta) = 1$

***95.** $\frac{\sqrt{2}}{2}$ ***96.** $\frac{\sqrt{2}}{2}$ ***97.** $\frac{\sqrt{2}}{2}$ ***98.** $\frac{\sqrt{2}}{2}$

***99.** $|\bar{z}| = |\overline{a + bi}| = |a - bi| = \sqrt{a^2 + (-b)^2} = \sqrt{a^2 + b^2} = |a + bi| = |z|$

***100.** $|z|^2 = |a + bi|^2 = (\sqrt{a^2 + b^2})^2 = a^2 + b^2 = a^2 - b^2i^2 = (a + bi)(a - bi) = (a + bi)(\overline{a + bi}) = z \cdot \bar{z}$

***101.** False, $|(-1) - 1| \neq |-1| - |1|$

***102.** False, $|1 + (-1)| \neq |1| + |-1|$

***103.** True

***104.** True

***105.** z: $P(2, 5)$; w: $P(4, -7)$; $z + w$: $P(6, -2)$

***106.** z: $P(-2, 7)$; w: $P(8, -4)$; $z + w$: $P(6, 3)$

***107.** z: $P(2, 1)$; w: $P(3, -2)$; zw: $P(8, -1)$

***108.** z: $P(-2, 4)$; w: $P(1, 1)$; zw: $P(-6, 2)$

Section 8.8, pages 519–520

1. (a) 0, up (b) $\frac{\sqrt{2}}{4}$, up (c) $\frac{1}{2}$, stationary (d) 0, up (e) $-\frac{1}{2}$, stationary (f) 0, up (g) 0, down

2. (a) $\frac{3}{2}$; stationary (b) $\frac{3\sqrt{2}}{2}$; down (c) 0; down (d) $-\frac{3}{2}$; stationary (e) 0; up (f) $\frac{3}{2}$; stationary (g) $-\frac{3}{2}$; stationary

3. (a) -3; stationary (b) 3; stationary; (c) -3; stationary (d) -3; stationary (e) -3, stationary (f) -3, stationary (g) -3, stationary

4. (a) 0, down (b) 0, up (c) 0, down (d) 0, down (e) 0, down (f) 0, down (g) 0, down

5. (a) 0; up (b) 0; up (c) 0; up (d) 0; up (e) 0; up (f) 0; up (g) 0; up

6. (a) 0.33, stationary (b) -0.33, stationary (c) 0.33, stationary (d) 0.33, stationary (e) 0.33, stationary (f) 0.33, stationary (g) 0.33, stationary

7. amp: 6; per: π; freq: $\frac{1}{\pi}$

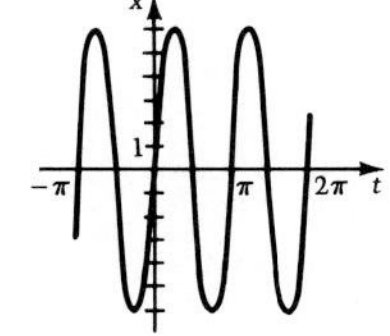

9. amp: 3; per: 4π; freq: $\frac{1}{4\pi}$

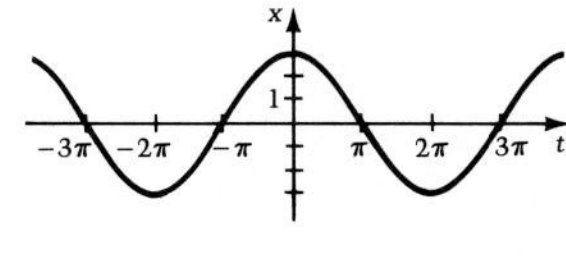

---

*denotes Superset exercises

8. amp: 4; per: 4π; freq: $\frac{1}{4\pi}$

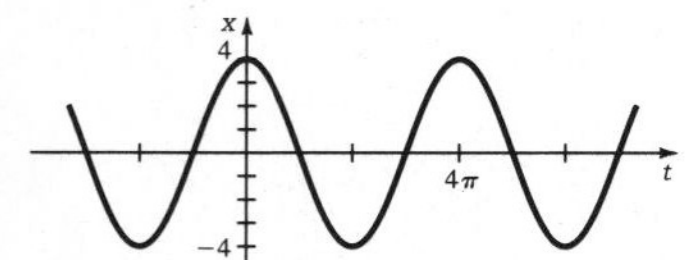

10. amp: 5; per: π; freq: $\frac{1}{\pi}$

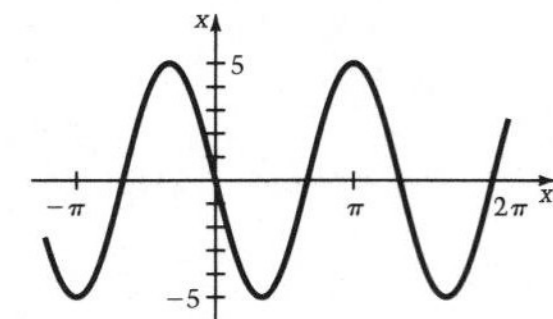

11. at equilibrium, moving up; 0.32 cm below equilibrium, moving up; 0.55 cm below equilibrium, moving down; 0.55 cm above equilibrium, moving down.
12. 9.32 cm above equilibrium, stationary; 8.18 cm above equilibrium, moving down; 541 cm above equilibrium, moving down; 1.19 cm below equilibrium, moving up.
13. 0.12 cm above equilibrium, moving down; 0.24 cm above equilibrium, moving down; 0.08 cm above equilibrium, moving up; 0.23 cm above equilibrium, moving down.
14. 1.68 cm below equilibrium, moving up; 1.84 cm above equilibrium, moving up; 1.55 cm above equilibrium, moving down; 2.00 cm above equilibrium, moving down.

15. $x = \sin\left(t - \frac{5\pi}{6}\right)$; amplitude $= 1$, period $= 2\pi$, frequency $= \frac{1}{2\pi}$

16. $x = 2\sin\left(t - \frac{\pi}{6}\right)$; amplitude $= 2$, period $= 2\pi$, frequency $= \frac{1}{2\pi}$

17. $x = \sqrt{2}\sin\left(t - \frac{\pi}{4}\right)$; amplitude $= \sqrt{2}$, period $= 2\pi$, frequency $= \frac{1}{2\pi}$

18. $x = \sqrt{2}\sin\left(t - \frac{3\pi}{4}\right)$; amplitude $= \sqrt{2}$, period $= 2\pi$, frequency $= \frac{1}{2\pi}$

19. $x = \frac{13}{2}\sin(2t - 1.18)$; amplitude $= \frac{13}{2}$, period $= \pi$, frequency $= \frac{1}{\pi}$

20. $x = \frac{17}{2}\sin(2t - 2.06)$; amplitude $= \frac{17}{2}$, period $= \pi$, frequency $= \frac{1}{\pi}$

21. (a) $x = -6\cos\frac{2\pi}{3}t$ (b) $\frac{1}{3}$

22. (a) $x = 4\cos 2\pi t$ (b) 1

23. (a) $x = -6\cos 1.25\pi t$ (b) 0.625

24. $x = -3\cos\frac{5\pi}{2}t$ (b) $\frac{5}{4}$

25. (a) $x = 0.60\cos 2t$ (b) $\frac{1}{\pi}$

26. (a) $x = 0.62\cos\frac{4\sqrt{3}}{3}t$ (b) $\frac{2\sqrt{3}}{3\pi}$

27. (a) $x = 0.02\cos\sqrt{9.8}\,t$ (b) $\frac{\sqrt{9.8}}{2\pi}$

28. (a) $x = 0.06\cos\sqrt{6.125}\,t$ (b) $\frac{\sqrt{6.125}}{2\pi}$

***29.** $x = 3\sin\frac{3}{2}t$

***30.** $x = \frac{3}{5}\sin 2t$

***31.** $x = \frac{3}{2}\sin\frac{2}{3}t$

***32.** $x = 4\sin\frac{4}{3}t$

***33.** multiplied by $\sqrt{2}$

***34.** two times as long

***35.** $x = 0.01\cos 400\pi t$

***36.** $x = 2\sin 5t + 2\sin t$

***37.** no

***38.** use: period $= \frac{2\pi\sqrt{l}}{\sqrt{g}}$

Chapter 8 Review Exercises, pages 525–527

1. $\frac{1}{1 + \sin x}$ **2.** $\frac{\sin^2 x}{1 - \sin^2 x}$ **3.** $(1 - \sin^2 x)\left(1 + \frac{1}{\sin x}\right)$ **4.** $\sin^2 x$ **15.** $\frac{\sqrt{6} + \sqrt{2}}{4}$ **16.** $-\frac{\sqrt{6} + \sqrt{2}}{4}$ **17.** $\frac{\sqrt{2} - \sqrt{6}}{4}$ **18.** $\frac{\sqrt{6} - \sqrt{2}}{4}$ **19.** (a) $\frac{33}{65}$ (b) $\frac{56}{65}$ (c) $\frac{33}{56}$ **20.** (a) $\frac{63}{65}$ (b) $\frac{16}{65}$ (c) $\frac{63}{16}$

21. (a) $-\frac{24}{25}$ (b) $-\frac{7}{25}$ (c) $\frac{24}{7}$ **23.** $\frac{\pi}{30}, \frac{\pi}{6}, \frac{13\pi}{30}, \frac{17\pi}{30}, \frac{5\pi}{6}, \frac{29\pi}{30}, \frac{37\pi}{30}, \frac{41\pi}{30}, \frac{49\pi}{30}, \frac{53\pi}{30}$

*denotes Superset exercises

22. (a) $-\frac{4}{5}$ (b) $-\frac{3}{5}$ (c) $\frac{4}{3}$ **24.** $\frac{3\pi}{16}, \frac{7\pi}{16}, \frac{11\pi}{16}, \frac{15\pi}{16}, \frac{19\pi}{16}, \frac{23\pi}{16}, \frac{27\pi}{16}, \frac{31\pi}{16}$

25. $\frac{\pi}{12}, \frac{7\pi}{12}, \frac{13\pi}{12}, \frac{19\pi}{12}, \frac{5\pi}{12}, \frac{11\pi}{12}, \frac{17\pi}{12}, \frac{23\pi}{12}$ **27.** $0, \pi, \frac{\pi}{4}, \frac{3\pi}{4}, \frac{5\pi}{4}, \frac{7\pi}{4}$ **29.** $\frac{\pi}{3}, \frac{5\pi}{3}$ **31.** no solutions

26. $\frac{\pi}{18}, \frac{5\pi}{18}, \frac{13\pi}{18}, \frac{17\pi}{18}, \frac{25\pi}{18}, \frac{29\pi}{18}, \frac{7\pi}{18}, \frac{11\pi}{18}, \frac{19\pi}{18}, \frac{23\pi}{18}, \frac{31\pi}{18}, \frac{35\pi}{18}$ **28.** $\frac{\pi}{2}, \frac{3\pi}{2}, \frac{\pi}{3}, \frac{5\pi}{3}$ **30.** $\frac{3\pi}{2}$ **32.** no solutions

33. 45°, 225°, 146.3°, 326.3° **35.** 0°, 180°, 65.9°, 294.1°, 114.1°, 245.9° **37.** $\frac{24}{25}$ **39.** $-\frac{33}{65}$ **41.** $\left(\sqrt{2}, \frac{3\pi}{4}\right), \left(-\sqrt{2}, \frac{7\pi}{4}\right)$

34. 135°, 315°, 51.3°, 231.3° **36.** 90°, 270°, 18.4°, 161.6°, 198.4°, 341.6° **38.** $\frac{3\sqrt{13}}{13}$ **40.** $\frac{33}{65}$ **42.** $\left(2, \frac{7\pi}{6}\right), \left(-2, \frac{\pi}{6}\right)$

43. $x^2 + y^2 = 4$ **45.** $2y - 3x = 1$ **47.** $y = 2x$ **49.** $(x^2 + y^2)^2 + 2xy = 0$ **51.** $r^2 \cos^2 \theta = 3r \sin \theta + 6$

44. $y = \frac{\sqrt{3}}{3}x$ **46.** $x^2 + y^2 = 2x$ **48.** $(x^2 + y^2)^2 = x^2 - y^2$ **50.** $x^2 + y^2 = 3y$ **52.** $3r^2 \sin 2\theta = 1$

53. $\tan 2\theta = \frac{1}{4}$ **55.** **57.** **59.** $-2\sqrt{3} + 2i$

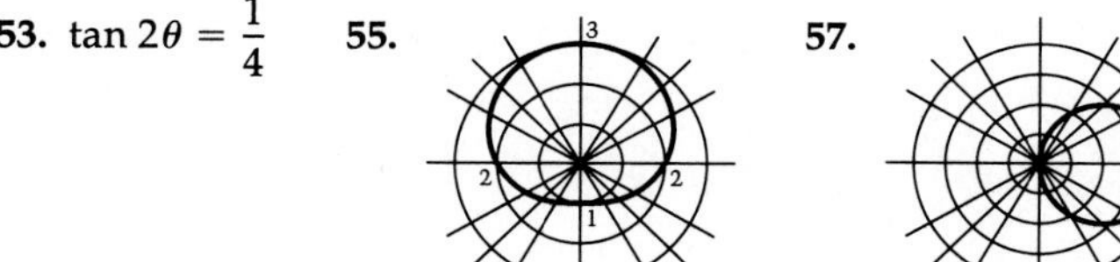

54. $r \cos \theta = 3$ **56.** **58.** **60.** -8

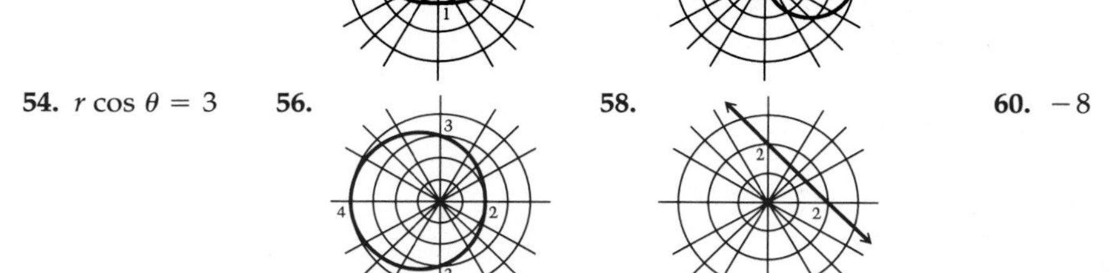

61. $\frac{\sqrt{2}}{4}\left(\cos \frac{5\pi}{12} + i \sin \frac{5\pi}{12}\right)$ **63.** $-16 - 16i$

62. $\frac{\sqrt{2}}{2}\left(\cos\left(-\frac{\pi}{12}\right) + i \sin\left(-\frac{\pi}{12}\right)\right)$ **64.** $-648 + 648i\sqrt{3}$

65. $\sqrt{3} + i, -\sqrt{3} + i, -2i$

66. $\sqrt[4]{2}\left(\frac{\sqrt{3}}{2} + \frac{1}{2}i\right); \sqrt[4]{2}\left(-\frac{1}{2} + \frac{\sqrt{3}}{2}i\right); \sqrt[4]{2}\left(-\frac{\sqrt{3}}{2} - \frac{1}{2}i\right); \sqrt[4]{2}\left(\frac{1}{2} - \frac{\sqrt{3}}{2}i\right)$

67. amp: 3; per: 4π; freq: $\frac{1}{4\pi}$ **69.** $\frac{3 \tan 2x - \tan^3 2x}{1 - 3\tan^2 2x}$

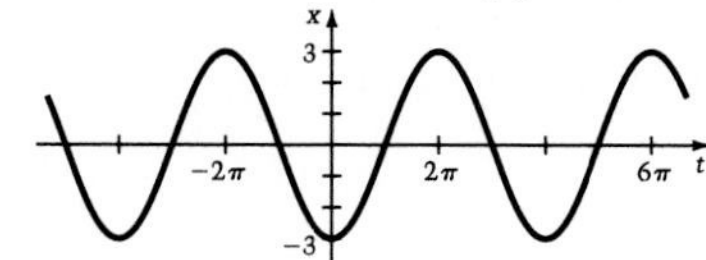

68. amp: 5; per: $\frac{2\pi}{3}$; freq: $\frac{3}{2\pi}$ **70.** $2\sqrt{13}$

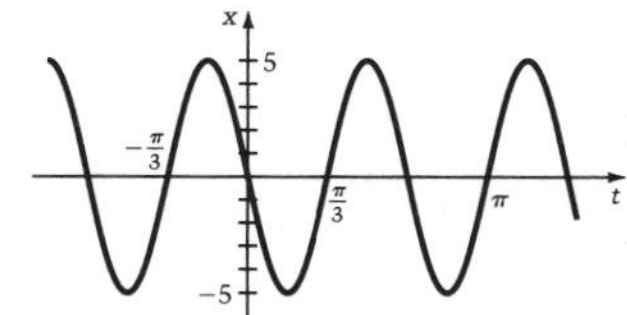

Chapter 8 Test, pages 527–528

3. $\dfrac{3-\sqrt{3}}{3+\sqrt{3}}$ **7.** $-\sqrt{\dfrac{3-\sqrt{5}}{6}}$ **9.** $\cos 2\varphi$ **13.** $-\dfrac{\sqrt{6}}{2}$ **15.** $\dfrac{2\pi}{3}+2n\pi, \dfrac{4\pi}{3}+2n\pi, 0+2n\pi$

6. $-1, 0$, undefined **8.** $\sqrt{\dfrac{13-2\sqrt{13}}{26}}$ **10.** $\cot \alpha$ **14.** $\dfrac{1}{2}\sin 7x+\dfrac{1}{2}\sin x$ **16.** $\dfrac{\pi}{2}+2n\pi$

17. 71.6°, 251.6°, 135°, 315° **19.** 0 **21.** $(-2\sqrt{2}, -2\sqrt{2})$ **23.** $x^2+y^2=3y$ **25.**

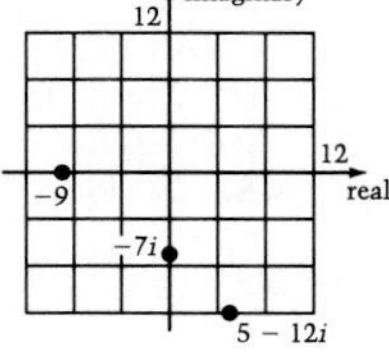

18. 21.5°, 158.5° **20.** 112.4°, 292.4° **22.** $\left(\dfrac{3\sqrt{3}}{2}, \dfrac{3}{2}\right)$ **24.** **26.** (a) $\sqrt{41}$ (b) 9 (c) 2

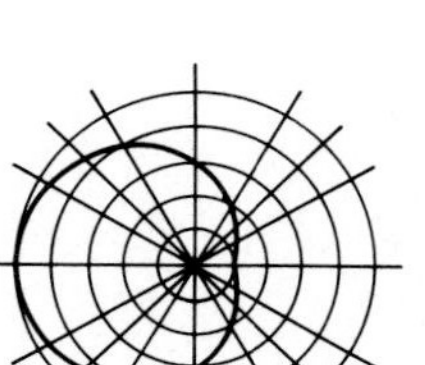

27. $2\sqrt{3}\left(\cos\dfrac{11\pi}{6}+i\sin\dfrac{11\pi}{6}\right)$ **29.** $z_1z_2=12\sqrt{3}-12i, \dfrac{z_1}{z_2}=6i$ **31.** amp: 4; per: 4π, freq: $\dfrac{1}{4\pi}$

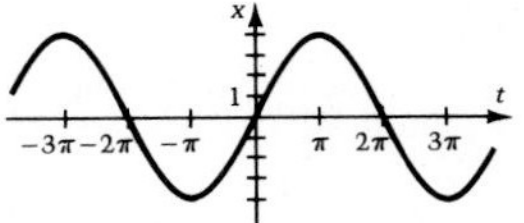

28. $4\sqrt{2}\left(\cos\dfrac{5\pi}{4}+i\sin\dfrac{5\pi}{4}\right)$ **30.** $\dfrac{\sqrt{3}}{2}+\dfrac{1}{2}i$

Chapter 9

Section 9.1, p. 529

1. (a) no (b) no **3.** (a) yes (b) yes **5.** $(2, -1)$ **7.** $\left(\dfrac{1}{2}, \dfrac{1}{4}\right)$ **9.** all the points on the line $12x+9y=6$

2. (a) no (b) yes **4.** (a) no (b) no **6.** $(3, 2)$ **8.** $\left(\dfrac{5}{6}, \dfrac{23}{12}\right)$ **10.** no solutions

11. no solutions **13.** $\left(\dfrac{2}{5}, \dfrac{7}{15}\right)$ **15.** all the points on the line $y=4-3x$

12. all the points on the line $\sqrt{8}x+2y=8$ **14.** $\left(\dfrac{7}{10}, -\dfrac{3}{5}\right)$ **16.** no solutions

17. no solutions **19.** $(\sqrt{3}, -2)$ **21.** $(3, 0), (1, -2)$ **23.** $(1, 1)$

18. all the points on the line $x=\dfrac{1}{5}+\dfrac{3}{5}y$ **20.** $(1, \sqrt{5})$ **22.** $(-4, 2), (4, 8)$ **24.** $\left(-\dfrac{9}{13}, -\dfrac{59}{13}\right), (3, 1)$

25. no real number solutions **27.** no real number solutions

26. no real number solutions **28.** $\left(1+\frac{2}{5}\sqrt{5}, -1+\frac{4}{5}\sqrt{5}\right), \left(1-\frac{2}{5}\sqrt{5}, -1-\frac{4}{5}\sqrt{5}\right)$

29. $\left(1+\frac{2}{5}\sqrt{5}, -1+\frac{4}{5}\sqrt{5}\right), \left(1-\frac{2}{5}\sqrt{5}, -1-\frac{4}{5}\sqrt{5}\right)$ **31.** (3.04, 0.64) **33.** (3.086, 0.982)

30. all the points on the circle $(x-5)^2+(y+3)^2=4$ **32.** (2.8, −2.1) **34.** (−37.33, −0.16)

35. (0.71, 0.71), (−0.71, −0.71) **37.** (0.5, 0.87), (−0.5, −0.87) **39.** (0.71, 0.71) ***41.** (3, 3)

36. (−0.87, 0.5), (0.87, −0.5) **38.** (0.97, 0.26), (−0.97, −0.26) **40.** (0.5, 0.87), (0.87, 0.5) ***42.** (4, 1), (1, −2)

***43.** (1, −2), (3, 2) ***45.** 17 quarters, 8 dimes ***47.** $500 ***49.** $\left(\frac{13}{3}, \frac{11}{3}\right)$ ***51.** 2.4 mi

***44.** (2, 0) ***46.** 120 mph, 30 mph ***48.** 49 ***50.** (11, −9) ***52.** 150 ft

***53.** $(x+2)^2+y^2=50$ ***55.** 14 m

***54.** $250\sqrt{63}$ ft ***56.** $\frac{8}{5}\sqrt{10}$

Section 9.2, p. 540

1.

3. 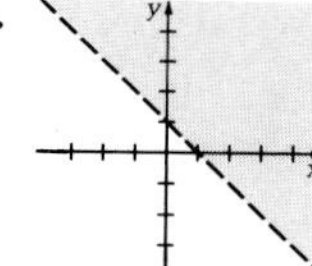

5. 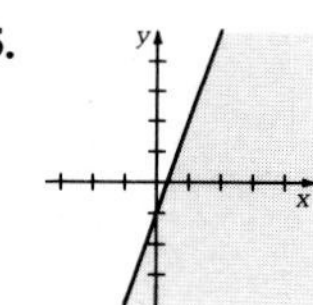

7. 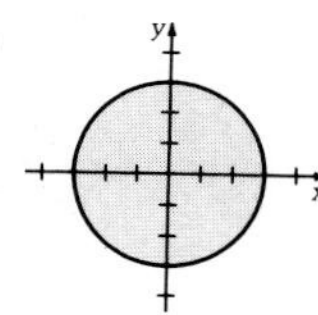

9. 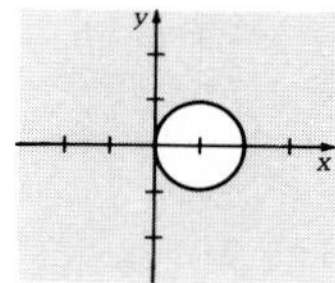

2.

4. 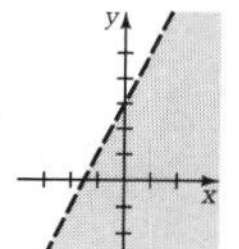

6. 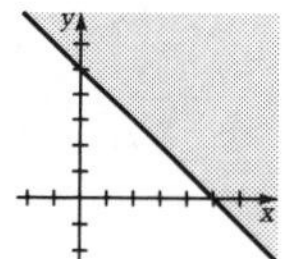

8. 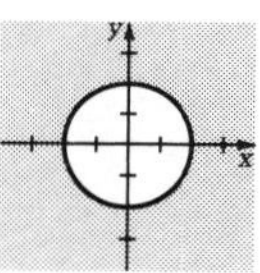

10. 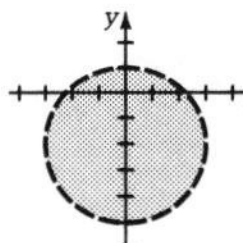

11. 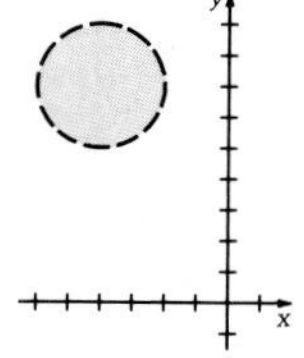

13. 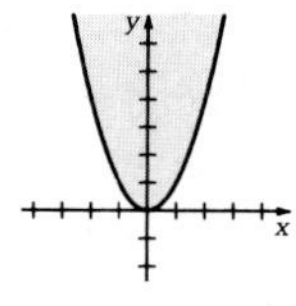

15. 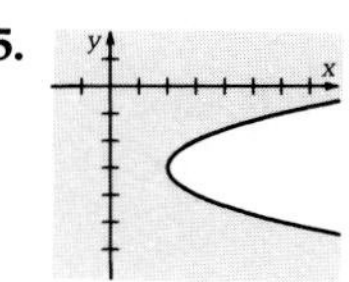

17. 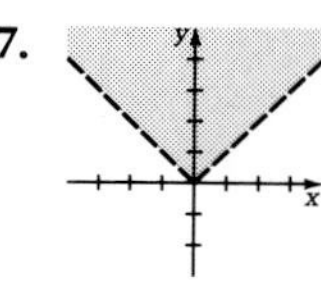

19. 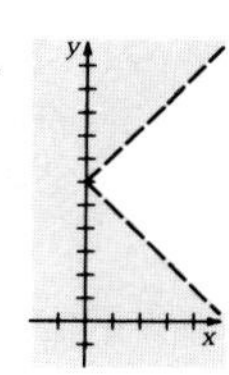

12. 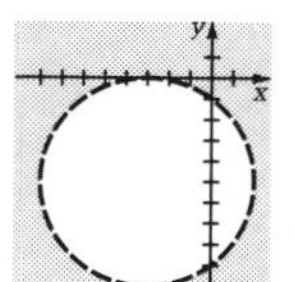

14. 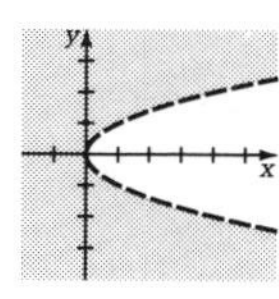

16. 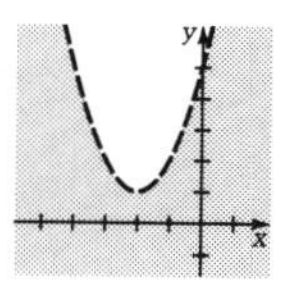

18. 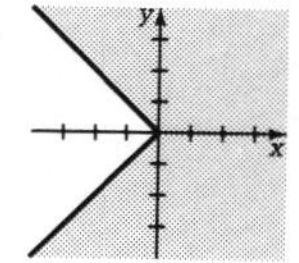

20. 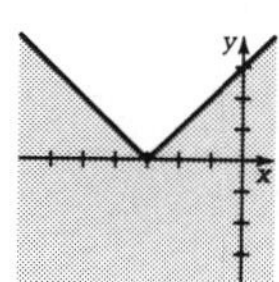

21. 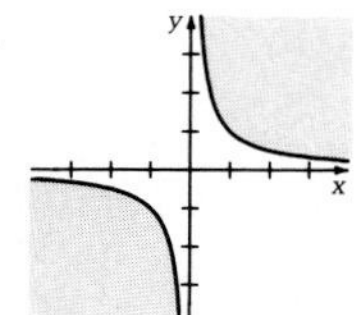

23. 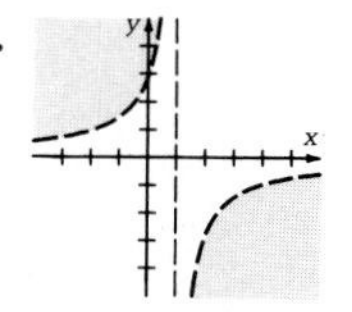

25. 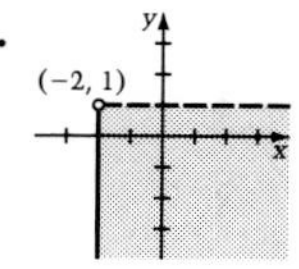

27. 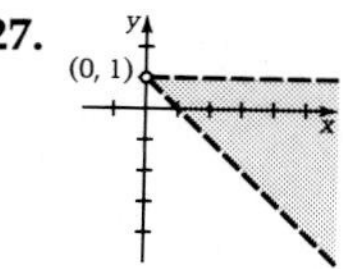

29. 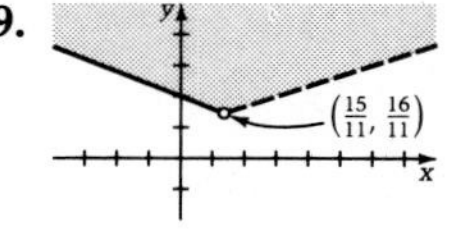

---

*denotes Superset exercises

22. 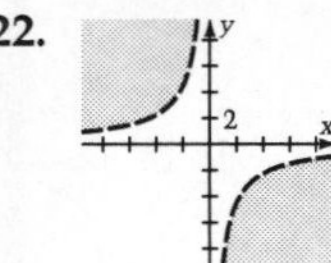

24. 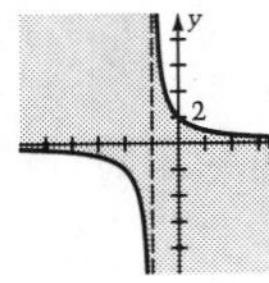

26. 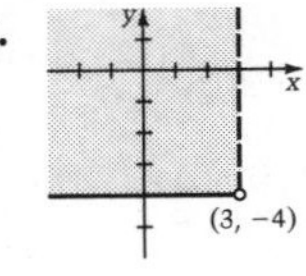

28. 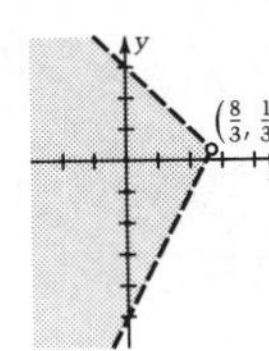

30. 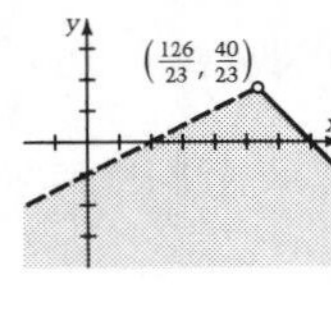

31. 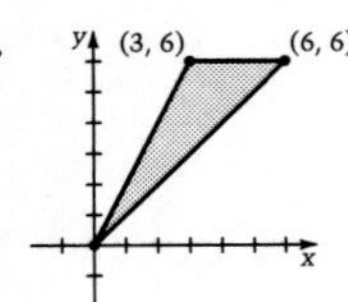

33. 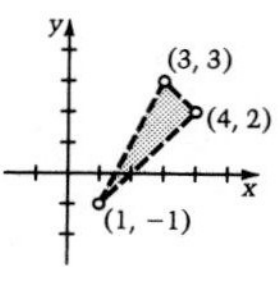

35. 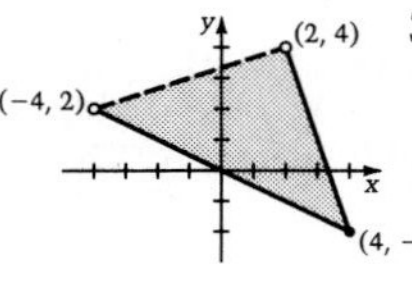

37. 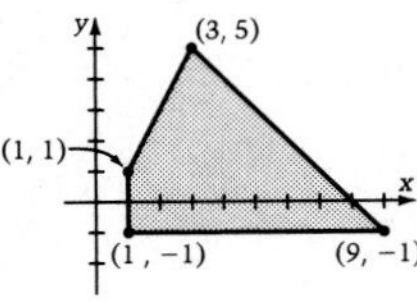

32. 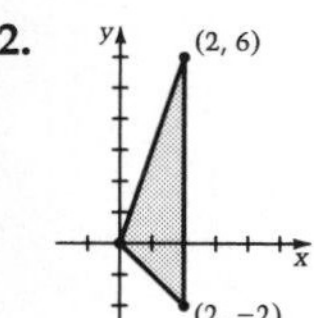

34. 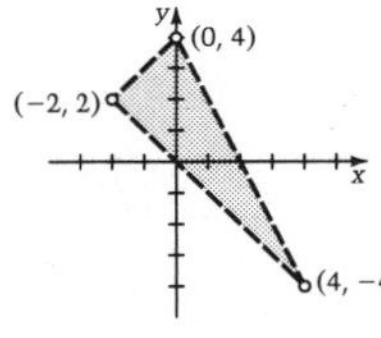

36. 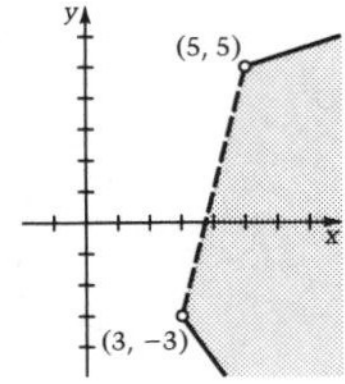

38. 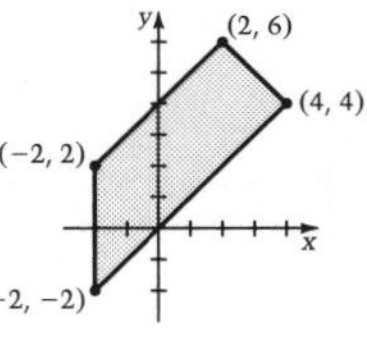

39. 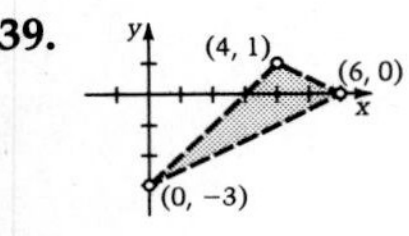

*41. 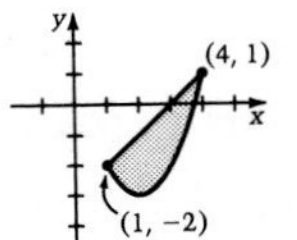

*43. 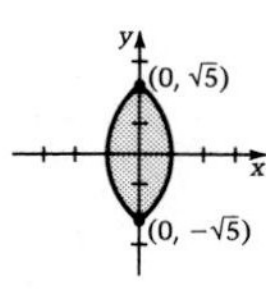

*45. 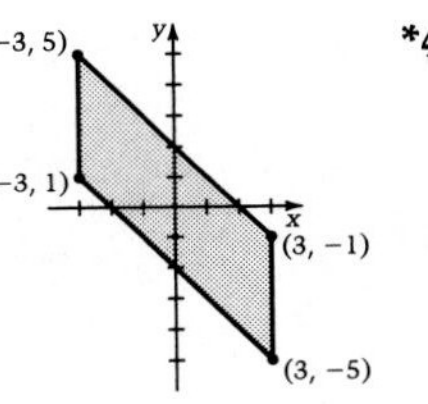

*47. 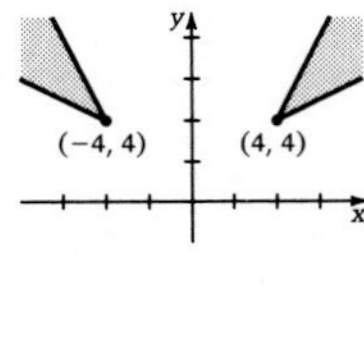

40. 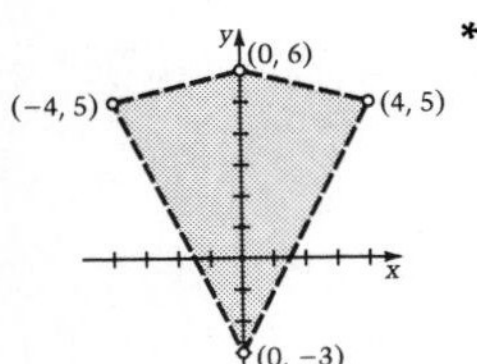

*42. 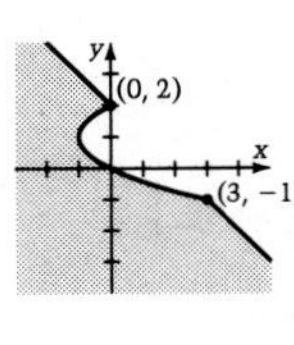

*44. 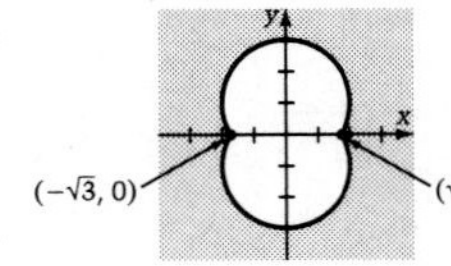

*46. 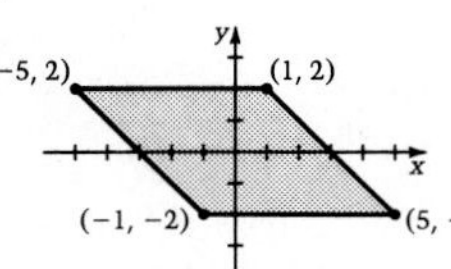

*48. 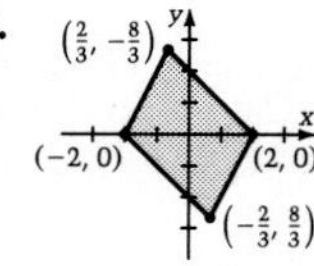

***49.** $y \le \frac{4}{3}x + 2$
$y > \frac{2}{3}x$
$y < 2$

***51.** $y < x + 2$
$y \ge x$
$x \ge -2$

***53.** $y < 3$
$x < 1$
$x \ge -2$
$y \ge -x - 2$

***55.** $y \ge -\frac{1}{2}x - \frac{1}{2}$
$y \ge 0$
$y \ge \frac{2}{3}x - \frac{2}{3}$
$y \le -\frac{1}{2}x + 4$

***50.** $y < \frac{1}{2}x + 1$
$y \ge -\frac{1}{2}x - 1$
$x \le 2$

***52.** $y \le x + 5$
$y \le 3$
$y < 4 - x$

***54.** $y < 4x$
$y \ge \frac{1}{3}x$
$y \ge x - 2$

***56.** $y \le x + 4$
$y \le 2 - x$
$y < 2 - \frac{5}{3}x$
$y \ge -\frac{3}{7}x - \frac{12}{7}$

---

*denotes Superset exercises

Section 9.3, p. 546

1. 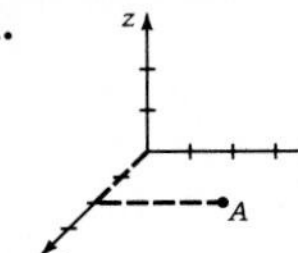

3. 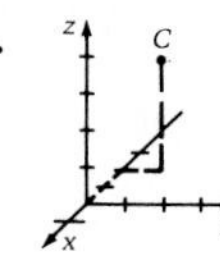

5. 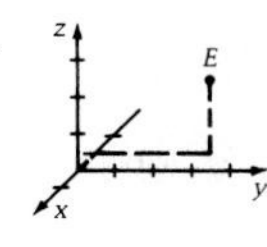

7. 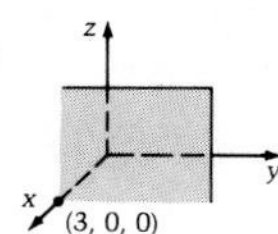

9. 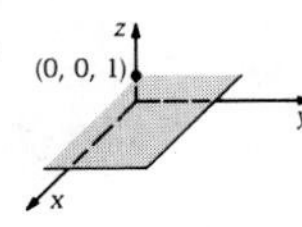

2. 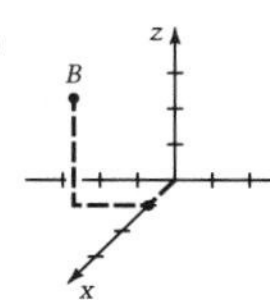

4. 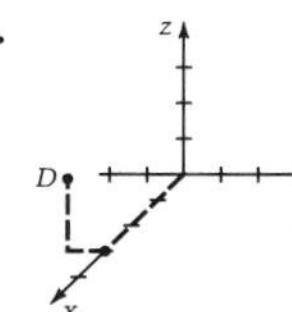

6. 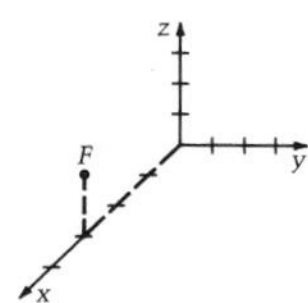

8. 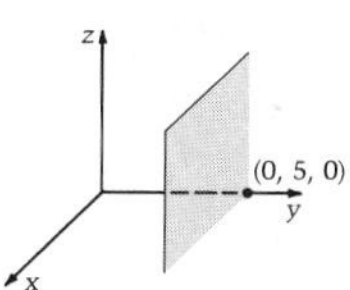

10. 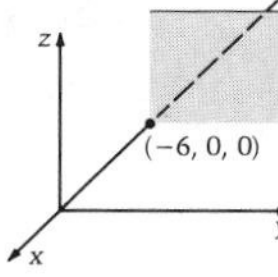

11. 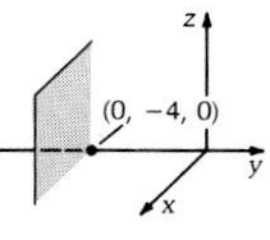

13. $(4, 2, 6)$

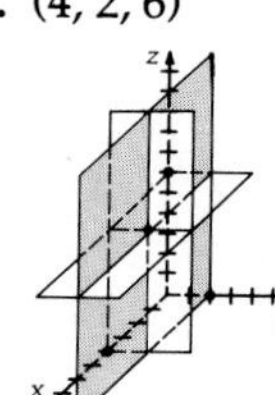

15. $(-3, 7, -1)$

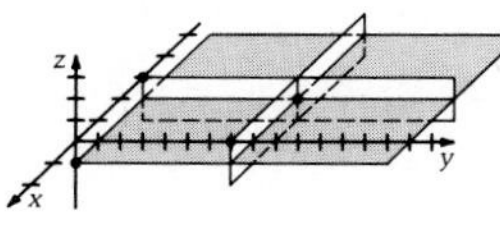

17. $(-2, -10, 0)$

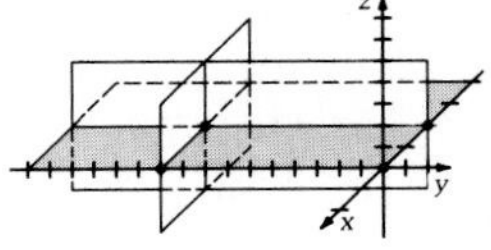

12. 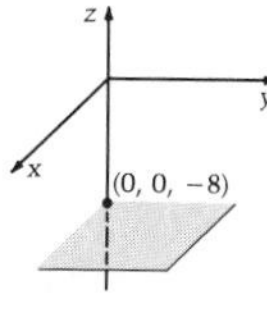

14. $(-1, 5, 4)$

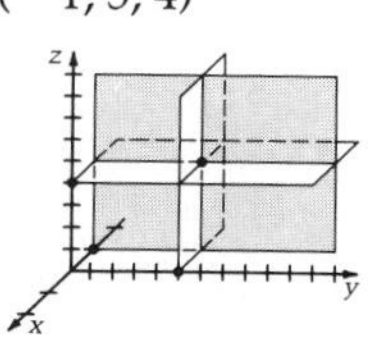

16. $(-6, -3, 5)$

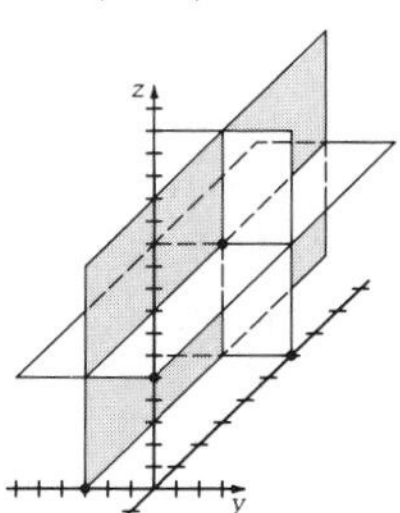

18. $(12, 0, -9)$

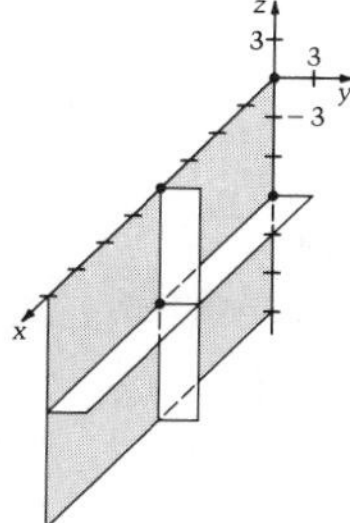

19. $(3, 1, -2)$ **21.** $\left(\frac{36}{17}, -\frac{9}{17}, \frac{10}{17}\right)$ **23.** $(1, 2, -4)$ **25.** $(1, 3, -2)$ **27.** no solutions

20. $(4, -5, 2)$ **22.** $(1, 0, -2)$ **24.** $\left(\frac{15}{7}, \frac{31}{21}, -\frac{23}{21}\right)$ **26.** no solutions **28.** $(-4, 7, -2)$

29. $\left(c, c + \frac{12}{7}, c - \frac{11}{7}\right)$, where c is any real number **31.** $\left(\frac{1}{2}, -\frac{2}{3}, \frac{5}{6}\right)$ **33.** $\left(\frac{5}{12}, -\frac{7}{16}, -\frac{15}{4}\right)$

30. $(-1, 0, 2)$ **32.** $\left(1 + \frac{2}{9}c, c, -1 + \frac{4}{3}c\right)$, where c is any real number **34.** $(-1, 4, 0)$

35. $\left(\frac{20}{9} - \frac{13}{9}c, c, \frac{1}{9} + \frac{7}{9}c\right)$, where c is any real number **37.** 9.80 **39.** 32.40

36. $\left(\frac{7}{18} - \frac{7}{18}c, c, \frac{1}{6} + \frac{5}{6}c\right)$, where c is any real number **38.** 30.72 **40.** 9.47

41. (a) $(-2.8, 0, 0)$ (b) $(0, -2.1, 0)$ (c) $(0, 0, 17)$ **42.** (a) $(4.9, 0, 0)$ (b) $(0, -78.5, 0)$ (c) $(0, 0, 2.0)$ **43.** (a) $(0.2, 0, 0)$ (b) $(0, 0.3, 0)$ (c) $(0, 0, -0.1)$ **44.** (a) $(-0.7, 0, 0)$ (b) $(0, -0.6, 0)$ (c) $(0, 0, 0.4)$ ***45.** 24 years old ***46.** \$1.74 ***47.** 223 ***48.** 330 ***49.** 7 years old ***50.** 78 **51.** \$450 ***52.** \$1.89

Section 9.4, p. 552

1. -1 **2.** -2 **3.** 0 **4.** 7

5. $\left[\begin{array}{cc|c} 3 & 7 & 8 \\ 2 & 9 & 1 \end{array}\right]$ **6.** $\left[\begin{array}{cc|c} 3 & 4 & 11 \\ 2 & -1 & 0 \end{array}\right]$

7. $\left[\begin{array}{ccc|c} 2 & -1 & 1 & 2 \\ 1 & 3 & 0 & 5 \\ 1 & 0 & -3 & 1 \end{array}\right]$ **8.** $\left[\begin{array}{ccc|c} -1 & 2 & -3 & 5 \\ 4 & 3 & 1 & 2 \\ 1 & -1 & -1 & 3 \end{array}\right]$

9. $\begin{cases} 12x + 9y = 6 \\ 8x + 6y = 4 \end{cases}$ **10.** $\begin{cases} 10x + 5y = 7 \\ -6x - 3y = -4 \end{cases}$

11. $\begin{cases} -x + y - 2z = 1 \\ x + y - z = 7 \\ 2x + z = 4 \end{cases}$ **12.** $\begin{cases} -5x + 3y + 2z = 2 \\ 3x + y - 4z = 6 \\ -x + 2y - z = 4 \end{cases}$

13. $\left[\begin{array}{cc|c} -2 & 7 & 8 \\ 3 & -1 & 7 \end{array}\right], \begin{bmatrix} -2 & 7 \\ 3 & -1 \end{bmatrix}, \begin{bmatrix} 8 \\ 7 \end{bmatrix}$ **14.** $\left[\begin{array}{cc|c} 2 & 3 & 6 \\ 5 & 7 & 13 \end{array}\right], \begin{bmatrix} 2 & 3 \\ 5 & 7 \end{bmatrix}, \begin{bmatrix} 6 \\ 13 \end{bmatrix}$

15. $\left[\begin{array}{cc|c} 8 & 4 & 5 \\ 6 & 3 & 7 \end{array}\right], \begin{bmatrix} 8 & 4 \\ 6 & 3 \end{bmatrix}, \begin{bmatrix} 5 \\ 7 \end{bmatrix}$ **16.** $\left[\begin{array}{cc|c} 2 & -3 & 5 \\ 1 & 2 & 3 \end{array}\right], \begin{bmatrix} 2 & -3 \\ 1 & 2 \end{bmatrix}, \begin{bmatrix} 5 \\ 3 \end{bmatrix}$

17. $\left[\begin{array}{cc|c} 3 & 5 & 1 \\ 4 & 3 & 5 \end{array}\right], \begin{bmatrix} 3 & 5 \\ 4 & 3 \end{bmatrix}, \begin{bmatrix} 1 \\ 5 \end{bmatrix}$ **18.** $\left[\begin{array}{cc|c} -4 & 10 & -2 \\ 2 & -5 & 1 \end{array}\right], \begin{bmatrix} -4 & 10 \\ 2 & -5 \end{bmatrix}, \begin{bmatrix} -2 \\ 1 \end{bmatrix}$

19. $\left[\begin{array}{ccc|c} 5 & -1 & 1 & -1 \\ 1 & 3 & 0 & 7 \\ 0 & 1 & 2 & -6 \end{array}\right], \begin{bmatrix} 5 & -1 & 1 \\ 1 & 3 & 0 \\ 0 & 1 & 2 \end{bmatrix}, \begin{bmatrix} -1 \\ 7 \\ -6 \end{bmatrix}$

20. $\left[\begin{array}{ccc|c} 1 & 2 & 1 & 4 \\ 2 & 0 & 3 & 1 \\ 0 & 5 & 2 & -1 \end{array}\right], \begin{bmatrix} 1 & 2 & 1 \\ 2 & 0 & 3 \\ 0 & 5 & 2 \end{bmatrix}, \begin{bmatrix} 4 \\ 1 \\ -1 \end{bmatrix}$

21. $\left[\begin{array}{ccc|c} 3 & 2 & -1 & 6 \\ -1 & -1 & 1 & 1 \\ 1 & 2 & 2 & 5 \end{array}\right], \begin{bmatrix} 3 & 2 & -1 \\ -1 & -1 & 1 \\ 1 & 2 & 2 \end{bmatrix}, \begin{bmatrix} 6 \\ 1 \\ 5 \end{bmatrix}$

22. $\left[\begin{array}{ccc|c} 3 & 1 & 2 & 1 \\ 4 & -1 & 4 & 4 \\ 1 & 1 & -2 & -5 \end{array}\right], \begin{bmatrix} 3 & 1 & 2 \\ 4 & -1 & 4 \\ 1 & 1 & -2 \end{bmatrix}, \begin{bmatrix} 1 \\ 4 \\ -5 \end{bmatrix}$

23. $\left[\begin{array}{ccc|c} 1 & -10 & 14 & 2 \\ -2 & -1 & 2 & 5 \\ -1 & 3 & -4 & 1 \end{array}\right], \begin{bmatrix} 1 & -10 & 14 \\ -2 & -1 & 2 \\ -1 & 3 & -4 \end{bmatrix}, \begin{bmatrix} 2 \\ 5 \\ 1 \end{bmatrix}$

24. $\left[\begin{array}{ccc|c} 2 & -1 & -3 & 3 \\ 1 & 1 & -1 & 1 \\ 4 & 2 & 1 & 8 \end{array}\right], \begin{bmatrix} 2 & -1 & -3 \\ 1 & 1 & -1 \\ 4 & 2 & 1 \end{bmatrix}, \begin{bmatrix} 3 \\ 1 \\ 8 \end{bmatrix}$

25. $\left[\begin{array}{ccc|c} 3 & -6 & 2 & -1 \\ 1 & 4 & 1 & 0 \\ 3 & 2 & -2 & 5 \end{array}\right], \begin{bmatrix} 3 & -6 & 2 \\ 1 & 4 & 1 \\ 3 & 2 & -2 \end{bmatrix}, \begin{bmatrix} -1 \\ 0 \\ 5 \end{bmatrix}$

26. $\left[\begin{array}{ccc|c} 5 & -3 & -2 & -2 \\ 1 & -2 & 1 & 0 \\ 3 & 1 & -4 & 3 \end{array}\right], \begin{bmatrix} 5 & -3 & -2 \\ 1 & -2 & 1 \\ 3 & 1 & -4 \end{bmatrix}, \begin{bmatrix} -2 \\ 0 \\ 3 \end{bmatrix}$

27. $(5, -1)$ **28.** $(1, 2)$ **29.** $\left(\frac{3}{2}, \frac{7}{6}, \frac{1}{6}\right)$ **30.** $(1, 0, -2)$ **31.** all the points on the line $x + \frac{3}{4}y = \frac{1}{2}$ **32.** no solutions **33.** no solutions **34.** $\left(c, c + \frac{10}{7}, c - \frac{8}{7}\right)$, where c is any real number

35. $(3, 2)$ **36.** $(-3, 4)$ **37.** no solutions **38.** $\left(\frac{19}{7}, \frac{1}{7}\right)$ **39.** $(2, -1)$ **40.** all the points on the line $x - \frac{5}{2}y = \frac{1}{2}$ **41.** $(1, 2, -4)$ **42.** $(5, 1, -3)$ **43.** $(5, -3, 3)$ **44.** $(-1, 0, 2)$

---

*denotes Superset exercises

45. $\left(c, 11 + 5c, 8 + \frac{7}{2}c\right)$, where c is any real number **47.** $\left(\frac{3}{4}, \frac{1}{8}, -\frac{5}{4}\right)$

46. $\left(\frac{36}{17}, -\frac{9}{17}, \frac{10}{17}\right)$ **48.** no solutions

***49.** all real numbers except 1 and 3 ***51.** none ***53.** none

***50.** 1, 3 ***52.** all real numbers except 3

Section 9.5, p. 563

1. 3 **3.** 6 **5.** 0 **7.** 2 **9.** $\left(\frac{2}{3}, \frac{1}{3}\right)$ **11.** no unique solution **13.** $\left(\frac{4}{9}, \frac{1}{9}\right)$ **15.** no unique solution

2. 4 **4.** -12 **6.** 0 **8.** 7 **10.** $\left(\frac{1}{4}, -\frac{3}{4}\right)$ **12.** $(4, -1)$ **14.** no unique solution **16.** no unique solution

17. $\left(-\frac{7}{5}, \frac{1}{5}\right)$ **19.** $(1, -2)$ **21.** $\frac{9}{2}$ **23.** no real number values **25.** $5, -3$ **27.** (a) 0 (b) 0 **29.** (a) 13 (b) 13

18. $(1, 2)$ **20.** $\left(\frac{2}{7}, \frac{3}{7}\right)$ **22.** $-\frac{3}{4}$ **24.** 0, 1 **26.** $\sqrt{3}, -\sqrt{3}$ **28.** (a) 5 (b) 5 **30.** (a) 0 (b) 0

31. (a) -18 (b) -18 **33.** 0 **35.** 13 **37.** -18 **39.** $(3, 1, -2)$ **41.** $\left(\frac{36}{17}, -\frac{9}{17}, \frac{10}{17}\right)$ **43.** $(1, 2, -4)$

32. (a) -3 (b) -3 **34.** 5 **36.** 0 **38.** -3 **40.** $(4, -5, 2)$ **42.** $(1, 0, -2)$ **44.** $\left(\frac{15}{7}, \frac{31}{21}, -\frac{23}{21}\right)$

45. $(1, 3, -2)$ **47.** no unique solution **49.** no unique solution **51.** $\left(\frac{1}{2}, -\frac{2}{3}, \frac{5}{6}\right)$ **53.** $\left(\frac{5}{12}, -\frac{7}{16}, -\frac{15}{4}\right)$

46. no unique solution **48.** $(-4, 7, -2)$ **50.** $(-1, 0, 2)$ **52.** no unique solution **54.** $(-1, 4, 0)$

55. no unique solution ***57.** 0 ***59.** -4 ***61.** Let $A = \begin{bmatrix} 0 & 0 & 0 \\ 1 & 2 & 3 \\ 4 & 5 & 6 \end{bmatrix}$; $\det A = 0$

56. no unique solution ***58.** -6 ***60.** 0 ***62.** Let $A = \begin{bmatrix} 1 & 2 & 3 \\ 1 & 2 & 3 \\ 4 & 5 & 6 \end{bmatrix}$; $\det A = 0$

***63.** $\det A = 9$; $\det B = \begin{vmatrix} 3 & 1 & -2 \\ -1 & 0 & 4 \\ 0 & 1 & 1 \end{vmatrix} = -9$ ***65.** $\det A = 9$; $\det B = \begin{vmatrix} -1 & 0 & 4 \\ 3 & 1 & -2 \\ -2 & 1 & 9 \end{vmatrix} = 9$

***64.** $\det A = 9$; $\det B = \begin{vmatrix} -1 & 0 & 4 \\ 3 & 1 & -2 \\ 0 & 2 & 2 \end{vmatrix} = 18$ ***66.** $\det A = 9$; $\det B = \begin{vmatrix} -10 & -3 & 10 \\ 3 & 1 & -2 \\ 0 & 1 & 1 \end{vmatrix} = 9$

***67.** Let $A = \begin{bmatrix} 0 & 1 & 2 \\ 0 & 3 & 4 \\ 0 & 5 & 6 \end{bmatrix}$; $\det A = 0$ ***69.** $\det A = 9$; $\det B = \begin{vmatrix} 0 & -1 & 4 \\ 1 & 3 & -2 \\ 1 & 0 & 1 \end{vmatrix} = -9$

***68.** Let $A = \begin{bmatrix} 1 & 1 & 2 \\ 3 & 3 & 4 \\ 5 & 5 & 6 \end{bmatrix}$; $\det A = 0$ ***70.** $\det A = 9$; $\det B = \begin{vmatrix} -1 & 0 & 8 \\ 3 & 1 & -4 \\ 0 & 1 & 2 \end{vmatrix} = 18$

***71.** $\det A = 9$; $\det B = \begin{vmatrix} -1 & 0 & 2 \\ 3 & 1 & 4 \\ 0 & 1 & 1 \end{vmatrix} = 9$

---

*denotes Superset exercises

***72.** $\det A = 9$; $\det B = \begin{vmatrix} -1 & 0 & 4 \\ 0 & 1 & -2 \\ -3 & 1 & 1 \end{vmatrix} = 9$ ***74.** $4x + 2y - 18 = 0$

***73.** Calculating the determinant and setting it equal to zero produces $y - y_0 = \dfrac{y_1 - y_0}{x_1 - x_0}(x - x_0)$.

Section 9.6, p. 573

1. $AB = \begin{bmatrix} 5 & 10 \\ 3 & 6 \end{bmatrix}$; $BA = \begin{bmatrix} 11 & 0 \\ 0 & 0 \end{bmatrix}$ **3.** $AB = \begin{bmatrix} 8 & 2 \\ -12 & -3 \end{bmatrix}$; $BA = [5]$ **5.** $AB = \begin{bmatrix} 0 & 3 \\ 6 & -1 \end{bmatrix}$; $BA = \begin{bmatrix} 0 & 3 & -4 \\ 4 & 3 & -2 \\ 0 & 3 & -4 \end{bmatrix}$

2. $AB = \begin{bmatrix} 0 & 0 \\ 14 & 0 \end{bmatrix}$; $BA = \begin{bmatrix} 6 & -9 \\ 4 & -6 \end{bmatrix}$ **4.** $AB = \begin{bmatrix} -3 & -1 \\ 6 & 2 \end{bmatrix}$; $BA = [-1]$ **6.** $AB = \begin{bmatrix} 0 & 2 \\ 0 & -3 \end{bmatrix}$; $BA = \begin{bmatrix} -2 & 4 & -1 \\ -1 & 2 & -2 \\ 0 & 0 & -3 \end{bmatrix}$

7. $AB = \begin{bmatrix} 3 & -1 & 0 & 3 \\ 1 & 3 & 0 & 0 \end{bmatrix}$; BA is not defined **9.** $AB = \begin{bmatrix} -2 & -1 & 3 \\ 1 & 4 & 1 \\ 0 & 4 & 2 \end{bmatrix}$; $BA = \begin{bmatrix} 1 & 6 & 4 \\ 2 & 4 & 2 \\ -1 & -1 & -1 \end{bmatrix}$ **11.** $\begin{bmatrix} -4 & 3 \\ 3 & -2 \end{bmatrix}$

8. $AB = \begin{bmatrix} 1 & 0 & 1 & 0 \\ -3 & 3 & -2 & 0 \end{bmatrix}$; BA is not defined **10.** $AB = \begin{bmatrix} 3 & -4 & 1 \\ 0 & 9 & -3 \\ 0 & 3 & -3 \end{bmatrix}$; $BA = \begin{bmatrix} -2 & 0 & 2 \\ -11 & 3 & -4 \\ 1 & 0 & 8 \end{bmatrix}$ **12.** $\begin{bmatrix} 1 & -2 \\ -\frac{1}{2} & \frac{3}{2} \end{bmatrix}$

13. $\begin{bmatrix} \frac{2}{3} & -\frac{5}{3} \\ -\frac{1}{3} & \frac{4}{3} \end{bmatrix}$ **15.** $\begin{bmatrix} 6 & -1 & -3 \\ -\frac{1}{2} & 0 & \frac{1}{2} \\ -4 & 1 & 2 \end{bmatrix}$ **17.** $\begin{bmatrix} -1 & 0 & 1 \\ \frac{7}{5} & \frac{1}{5} & -\frac{3}{5} \\ \frac{9}{5} & \frac{2}{5} & -\frac{6}{5} \end{bmatrix}$ **19.** $\begin{bmatrix} -\frac{1}{4} & \frac{7}{20} & \frac{1}{20} & -\frac{3}{20} \\ -1 & \frac{2}{5} & \frac{1}{5} & \frac{2}{5} \\ \frac{5}{4} & -\frac{7}{20} & -\frac{1}{20} & \frac{3}{20} \\ 2 & -1 & 0 & 0 \end{bmatrix}$ **21.** $(-15, 11)$

14. $\begin{bmatrix} -2 & 1 \\ \frac{5}{3} & -\frac{2}{3} \end{bmatrix}$ **16.** $\begin{bmatrix} \frac{8}{5} & -\frac{2}{5} & -\frac{3}{5} \\ -\frac{12}{5} & \frac{3}{5} & \frac{7}{5} \\ \frac{1}{5} & \frac{1}{5} & -\frac{1}{5} \end{bmatrix}$ **18.** $\begin{bmatrix} \frac{1}{7} & \frac{1}{7} & 0 \\ -\frac{2}{7} & \frac{5}{7} & 1 \\ \frac{8}{7} & -\frac{20}{7} & -3 \end{bmatrix}$ **20.** $\begin{bmatrix} -\frac{7}{20} & \frac{2}{5} & -\frac{27}{20} & -\frac{7}{5} \\ -\frac{1}{20} & \frac{1}{5} & -\frac{1}{20} & -\frac{1}{5} \\ -\frac{1}{4} & 0 & -\frac{1}{4} & 0 \\ 1 & 0 & 1 & 1 \end{bmatrix}$ **22.** $(5, -3)$

23. $\left(\frac{11}{3}, -\frac{7}{3}\right)$ **25.** $\left(4, \frac{1}{2}, -3\right)$ **27.** $\left(1, \frac{4}{5}, -\frac{2}{5}\right)$ ***29.** (a) $(-4, 10)$ (b) $\left(-\frac{5}{2}, \frac{15}{2}\right)$ (c) $(-1, 3)$

24. $\left(-7, \frac{17}{3}\right)$ **26.** $\left(\frac{9}{5}, -\frac{6}{5}, -\frac{2}{5}\right)$ **28.** $\left(\frac{1}{7}, \frac{12}{7}, -\frac{27}{7}\right)$ ***30.** (a) $\left(-\frac{10}{3}, \frac{4}{3}\right)$ (b) $\left(-\frac{16}{3}, \frac{7}{3}\right)$ (c) $\left(\frac{2}{3}, -\frac{2}{3}\right)$

***31.** (a) $(1, 1)$ (b) $(0, 0)$ (c) $(3, 3)$ ***33.** $\begin{bmatrix} \frac{1}{2} & 0 & 0 & -\frac{1}{4} \\ 0 & 1 & 0 & 0 \\ 0 & 0 & -1 & 0 \\ 0 & 0 & 0 & \frac{1}{2} \end{bmatrix}$ ***35.** $\begin{bmatrix} 1 & -1 & 0 & 0 \\ 0 & 1 & -1 & 0 \\ 0 & 0 & 1 & -1 \\ 0 & 0 & 0 & 1 \end{bmatrix}$

***32.** (a) $\left(-\frac{2}{3}, -\frac{1}{3}\right)$ (b) $(0, 0)$ (c) $(-3, -3)$ ***34.** $\begin{bmatrix} 1 & 0 & 0 & 0 \\ 0 & -1 & 0 & 0 \\ 0 & 0 & \frac{1}{3} & 0 \\ -1 & 0 & 0 & 1 \end{bmatrix}$ ***36.** $\begin{bmatrix} 1 & 0 & 0 & 0 \\ -1 & 1 & 0 & 0 \\ 0 & -1 & 1 & 0 \\ 0 & 0 & -1 & 1 \end{bmatrix}$

---

*denotes Superset exercises

***37.** $A^{-1} = \begin{bmatrix} \frac{1}{4} & -\frac{1}{4} \\ -\frac{1}{2t} & \frac{3}{2t} \end{bmatrix}, t \neq 0$ ***39.** $A^{-1} = \frac{-1}{t^2 - t - 6}\begin{bmatrix} 4 & -t-1 \\ 2-t & 1 \end{bmatrix}, t \neq 3 \text{ and } t \neq -2$

***38.** $A^{-1} = \frac{-1}{t^2 + 5t + 6}\begin{bmatrix} -2 & -t-5 \\ -t & 3 \end{bmatrix}, t \neq -2 \text{ and } t \neq -3$ ***40.** $A^{-1} = \frac{1}{2t^2 - 18}\begin{bmatrix} t & -3 \\ -6 & 2t \end{bmatrix}, t \neq 3 \text{ and } t \neq -3$

***41.** $a \neq 0$, $c \neq 0$, and b is any real number ***43.** Replace each diagonal entry by its multiplicative inverse.
***42.** The three nonzero entries are arranged so that precisely one of them appears in each of the three rows and in each of the three columns. ***44.** 8

Section 9.7, p. 583

1. 300 donuts, 300 brownies, \$81 **3.** 8000 45's, 3000 LP's, \$6320
2. 400 deluxe hubcaps, 300 standard hubcaps, \$10,100 **4.** 35 carts, 30 caddies, \$445
5. 20 multiple choice, 10 true/false
6. 4.5 oz of soup, 2.5 ounces of tuna salad
7. To minimize: 15 amateurs, 20 professionals; To maximize: 10 amateurs, 50 professionals
8. 480 cartons from manufacturer B and no cartons from manufacturer A, \$1032
9. 20.8 pounds of brand A, none of brand B **11.** 6 of Package I, 10 of Package II
10. 45 sports coats, 10 skirts **12.** 6 of Package I and 10 of Package II, or 14 of Package I and 4 of Package II
13. 98 acres of alfalfa, 30 acres of beans, \$27,400
14. 3 three-point exercises, 14 five-point exercises **16.** \$1.52
15. 20 gallons from F_1 to S_1, 20 gallons from F_1 to S_2, 5 gallons from F_2 to S_1, none from F_2 to S_2
***17.** All the points in the feasible regions that lie on the line $800 = 0.08b + 0.12s$ yield earnings of \$800. As E increases, there are fewer and fewer points in the feasible region that lie on the line. When $E > 1000$ or $E < 480$, the line does not intersect the feasible region. That is, the total earnings on the investments can range from \$480 to \$1000.
***18.** Yes, any of the functions that form the boundary of the feasible region.
***19.** Since $x^2 + y^2$ is the square of the distance from the origin to any point (x, y), the maximum value of $E = x^2 + y^2$ will occur at the point in the region that is farthest from the origin, and the minimum value of $E = x^2 + y^2$ will occur at the point in the region that is closest to the origin. Therefore, these points will lie on the boundary of the region.
***20.** The maximum value of $E = x^3$ would occur at the boundary point that is farthest from the y-axis. The minimum value of $E = x^3$ would occur at the boundary point that is closest to the y-axis. The maximum value of $E = y^3$ would occur at the boundary point that is farthest from the x-axis. The minimum value of $E = y^3$ would occur at the boundary point that is closest to the x-axis.
***21.** The minimum value would occur at the boundary point that is closest to the line $y = x$. ***23.** 3 chairs, 4 tables
***22.** The minimum value would occur at the boundary point that is closest to the line $y = \frac{2}{3}x$. ***24.** 1 chair, 5 tables

Chapter 9 Review Exercises, p. 595

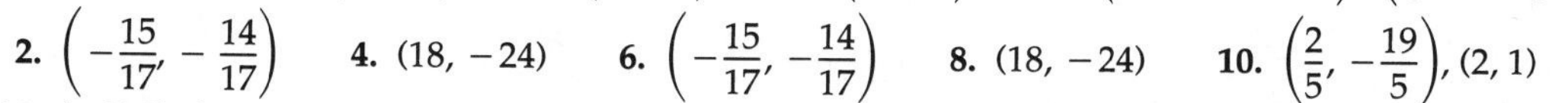

1. $\left(\frac{5}{19}, \frac{22}{19}\right)$ **3.** $\left(\frac{7}{13}, \frac{9}{13}\right)$ **5.** $\left(\frac{5}{19}, \frac{22}{19}\right)$ **7.** $\left(\frac{7}{13}, \frac{9}{13}\right)$ **9.** $\left(\frac{27}{8}, 1 + \frac{3}{8}\sqrt{15}\right), \left(\frac{27}{8}, 1 - \frac{3}{8}\sqrt{15}\right)$

2. $\left(-\frac{15}{17}, -\frac{14}{17}\right)$ **4.** $(18, -24)$ **6.** $\left(-\frac{15}{17}, -\frac{14}{17}\right)$ **8.** $(18, -24)$ **10.** $\left(\frac{2}{5}, -\frac{19}{5}\right), (2, 1)$

11. $(-7, 5), (1, -1)$ **13.** **15.** **17.** **19.**

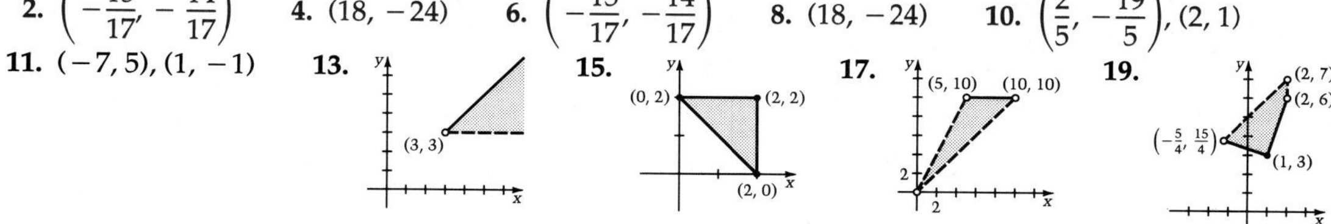

*denotes Superset exercises

12. $\left(-\frac{4}{5}, -\frac{2}{5}\right)$, (4, 2)

14.
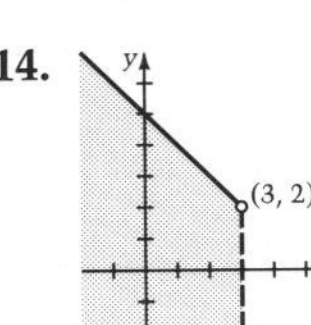

16.
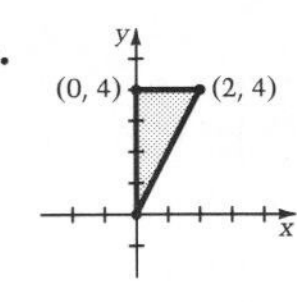

18.
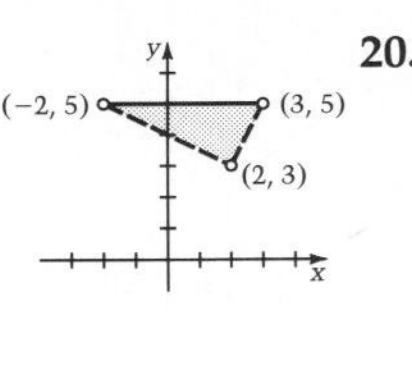

20.
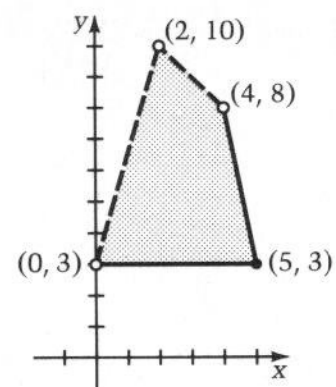

21. (3, −4, 2) **23.** $\left(\frac{19}{9}, \frac{2}{3}, -\frac{5}{9}\right)$ **25.** 1 **27.** −6 **29.** −11 **31.** 0 **33.** $\left(-\frac{5}{19}, \frac{17}{19}\right)$

22. (2, 1, −5) **24.** $\left(\frac{83}{38}, -\frac{1}{19}, -\frac{4}{19}\right)$ **26.** −10 **28.** 0 **30.** 0 **32.** 10 **34.** $\left(\frac{1}{22}, -\frac{35}{22}\right)$

35. no unique solution **37.** $\left(\frac{11}{8}, \frac{15}{16}, \frac{17}{8}\right)$ **39.** $AB = \begin{bmatrix} 7 & 9 \\ -3 & -1 \end{bmatrix}$; $BA = \begin{bmatrix} 2 & 6 \\ -2 & 4 \end{bmatrix}$

36. $\left(\frac{67}{7}, -\frac{52}{7}, \frac{33}{7}\right)$ **38.** no unique solution **40.** $AB = \begin{bmatrix} 8 & -3 & 5 \\ 5 & -1 & -3 \end{bmatrix}$; BA is not defined

41. $AB = \begin{bmatrix} 2 & 3 & -4 \\ 4 & 4 & 0 \end{bmatrix}$; BA is not defined **43.** $AB = \begin{bmatrix} -3 & -4 & 5 \\ 5 & -1 & 4 \\ 4 & 1 & -1 \end{bmatrix}$; $BA = \begin{bmatrix} -5 & -6 & 5 \\ 5 & -3 & 1 \\ 5 & -7 & 3 \end{bmatrix}$

42. $AB = \begin{bmatrix} 2 & 3 & 0 \\ 6 & -1 & 0 \\ 3 & -1 & 2 \end{bmatrix}$; $BA = \begin{bmatrix} -1 & 0 & 3 \\ 12 & 2 & -1 \\ 6 & 0 & 2 \end{bmatrix}$ **44.** $AB = \begin{bmatrix} -3 & 1 \\ 4 & -12 \end{bmatrix}$; $BA = \begin{bmatrix} -3 & 1 & -2 \\ 17 & -5 & 8 \\ -3 & 2 & -7 \end{bmatrix}$

45. $\begin{bmatrix} -2 & 3 \\ 3 & -4 \end{bmatrix}$ **47.** $\begin{bmatrix} 2 & 3 \\ \frac{3}{2} & \frac{5}{2} \end{bmatrix}$ **49.** $\begin{bmatrix} 10 & 3 & -4 \\ 5 & 1 & -2 \\ 7 & 2 & -3 \end{bmatrix}$ **51.** $\begin{bmatrix} 1 & \frac{1}{2} & \frac{1}{2} \\ 3 & 1 & 1 \\ 3 & \frac{1}{2} & \frac{3}{2} \end{bmatrix}$

46. $\begin{bmatrix} -\frac{2}{3} & \frac{5}{3} \\ 1 & -2 \end{bmatrix}$ **48.** $\begin{bmatrix} -3 & 2 \\ 8 & -5 \end{bmatrix}$ **50.** $\begin{bmatrix} -2 & 1 & 1 \\ 11 & -3 & -6 \\ 4 & -1 & -2 \end{bmatrix}$ **52.** $\begin{bmatrix} -2 & -6 & \frac{5}{2} \\ 2 & 5 & -2 \\ 1 & 3 & -1 \end{bmatrix}$

53. (a) (4, −5) (b) (−9, 13) **55.** (a) $\left(8, \frac{13}{2}\right)$ (b) (3, 2) **57.** (a) (−32, −16, −23) (b) (17, 9, 12)

54. (a) $\left(\frac{8}{3}, -3\right)$ (b) $\left(-\frac{11}{3}, 5\right)$ **56.** (a) (1, −2) (b) (−11, 29) **58.** (a) (7, −40, −14) (b) (−5, 25, 9)

59. (a) $\left(-\frac{1}{2}, -3, -\frac{3}{2}\right)$ (b) $\left(\frac{3}{2}, 5, \frac{11}{2}\right)$ **61.** 83, −5 **63.** 200 of item B and none of item A

60. (a) $\left(\frac{23}{2}, -10, -5\right)$ (b) (2, −1, −1) **62.** 54, 3 **64.** \$3500 in a savings account and \$3500 in stocks

65. (a) all real numbers except -2 (b) -2 (c) none

67.

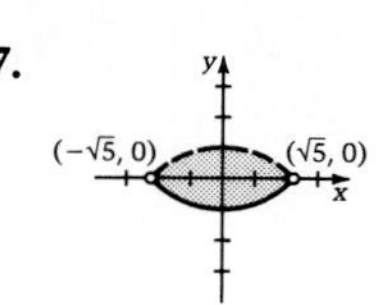

69. $\begin{cases} y \geq -\dfrac{9}{2}x - 13 \\ y < -\dfrac{9}{2}x + \dfrac{37}{2} \\ y < 5 \\ y \geq -4 \end{cases}$

66.

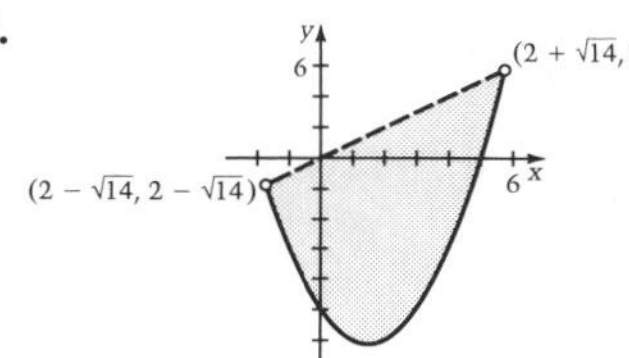

68.

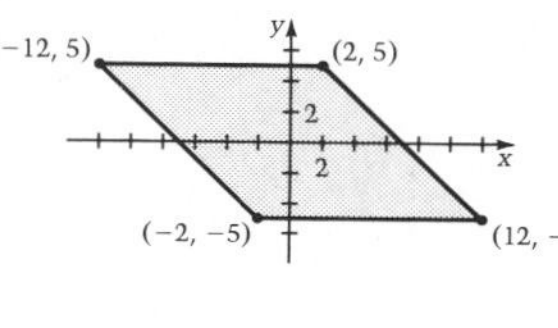

70. $\begin{cases} y < \dfrac{13}{5}x + \dfrac{56}{5} \\ y \leq -\dfrac{4}{7}x + \dfrac{34}{7} \\ y \leq -3x + 17 \\ y > \dfrac{3}{14}x - \dfrac{11}{2} \end{cases}$

Chapter 9 Test, p. 598

1. no **3.** (2, 0) **5.** no solutions **7.** $\left(\dfrac{9}{8}, \dfrac{3\sqrt{55}}{8}\right), \left(\dfrac{9}{8}, -\dfrac{3\sqrt{55}}{8}\right)$ **9.**

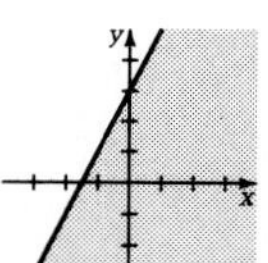

2. no **4.** all the points on the line $-25x + 10y = 5$ **6.** $\left(\dfrac{29}{26}, -\dfrac{24}{13}\right)$ **8.**

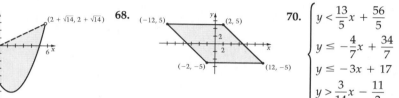

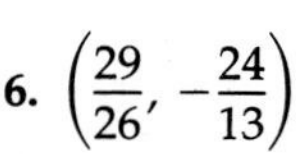

10.

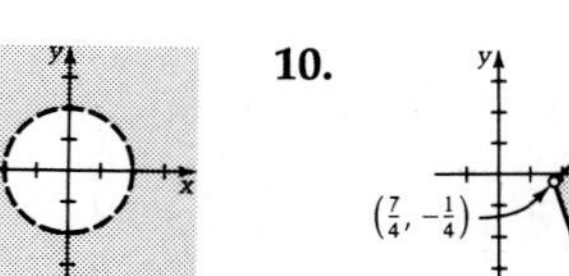

11.

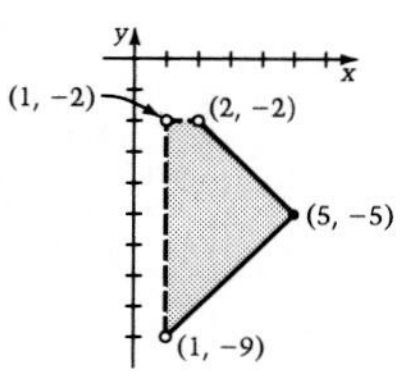

13.

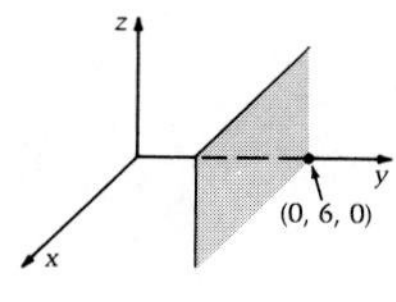

15. $(2, -1, 4)$ **17.** $\left[\begin{array}{ccc|c} 2 & 1 & -3 & -7 \\ 1 & -2 & 0 & 5 \\ 0 & 4 & 2 & 0 \end{array}\right]$, $\begin{bmatrix} 2 & 1 & -3 \\ 1 & -2 & 0 \\ 0 & 4 & 2 \end{bmatrix}$, $\begin{bmatrix} -7 \\ 5 \\ 0 \end{bmatrix}$

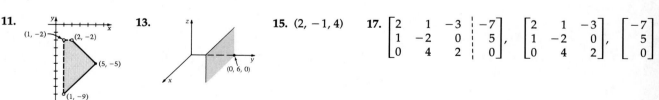

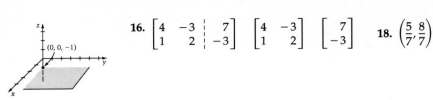

12. 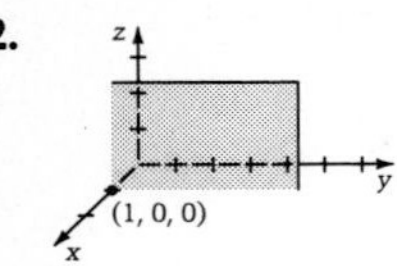

14. 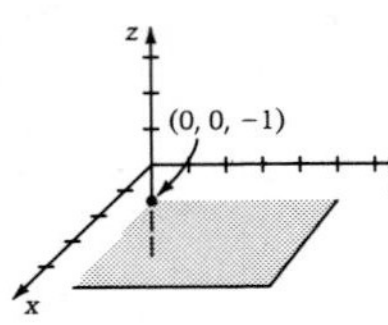

16. $\left[\begin{array}{cc|c} 4 & -3 & 7 \\ 1 & 2 & -3 \end{array}\right]$ $\begin{bmatrix} 4 & -3 \\ 1 & 2 \end{bmatrix}$ $\begin{bmatrix} 7 \\ -3 \end{bmatrix}$ **18.** $\left(\frac{5}{7}, \frac{8}{7}\right)$

19. $(1, 4, -2)$ **21.** (a) -3 (b) -3 **23.** $AB = \begin{bmatrix} 3 & -1 \\ 6 & -2 \end{bmatrix}$; $BA = [1]$ **25.** $\begin{bmatrix} \frac{2}{3} & -\frac{1}{3} \\ -\frac{5}{3} & \frac{4}{3} \end{bmatrix}$

20. -6 **22.** $\left(0, -\frac{3}{19}, \frac{10}{19}\right)$ **24.** $AB = \begin{bmatrix} 2 & 5 \\ 0 & -3 \end{bmatrix}$; $BA = \begin{bmatrix} 0 & 2 & 1 \\ 6 & -4 & 7 \\ 3 & -3 & 3 \end{bmatrix}$ **26.** $\begin{bmatrix} -2 & 3 & 1 \\ 2 & -2 & -1 \\ -\frac{3}{2} & 2 & \frac{1}{2} \end{bmatrix}$

27. 7000 hamburgers, 9000 hotdogs

Chapter 10

Section 10.1, p. 599

1. $-2, 1, 4, 7, 22, 55$ **3.** $\frac{1}{3}, \frac{2}{5}, \frac{3}{7}, \frac{4}{9}, \frac{9}{19}, \frac{20}{41}$ **5.** $0, \frac{1}{2}, \frac{2}{3}, \frac{3}{4}, \frac{8}{9}, \frac{19}{20}$ **7.** $4, 7, 12, 19, 84, 403$

2. $3, 2, 1, 0, -5, -16$ **4.** $\frac{3}{2}, 2, \frac{9}{4}, \frac{12}{5}, \frac{27}{10}, \frac{20}{7}$ **6.** $-1, \frac{1}{2}, -\frac{1}{3}, \frac{1}{4}, -\frac{1}{9}, \frac{1}{20}$ **8.** $6, 3, -2, -9, -74, -393$

9. $6, 2, 0, 0, 30, 272$ **11.** $0, 0, 6, 24, 504, 6840$ **13.** $\frac{3}{2}, \frac{5}{6}, \frac{7}{12}, \frac{9}{20}, \frac{19}{90}, \frac{41}{420}$

10. $12, 20, 30, 42, 132, 506$ **12.** $14, 48, 108, 200, 1350, 10920$ **14.** $\frac{1}{2}, \frac{1}{6}, \frac{1}{12}, \frac{1}{20}, \frac{1}{90}, \frac{1}{420}$

15. $\frac{1}{2}, \frac{1}{4}, \frac{1}{8}, \frac{1}{16}, \frac{1}{512}, \frac{1}{1{,}048{,}576}$ **17.** $3, 4, 5, 6, 7, 8$ **19.** $1, -1, -5, -13, -29, -61$

16. $-\frac{1}{2}, \frac{1}{4}, -\frac{1}{8}, \frac{1}{16}, -\frac{1}{512}, \frac{1}{1{,}048{,}576}$ **18.** $4, 3, 2, 1, 0, -1$ **20.** $5, 7, 11, 19, 35, 67$

21. $3, -4, -3, -10, 3, -4$ **23.** $1, 3, 2, -1, -3, -2$ **25.** $1, 3, 1, 3, 1, 3$ **27.** $4, 7, 10, 13, 16$

22. $1, 3, -1, 1, 1, 3$ **24.** $3, 1, 4, 5, 9, 14$ **26.** $1, 4, 3, -1, -4, -3$ **28.** $3, 9, 15, 21, 27$

29. $5, 3, 1, -1, -3$ **31.** $-2, 3, 8, 13, 18$ **33.** $\frac{1}{2}, \frac{3}{4}, 1, \frac{5}{4}, \frac{3}{2}$ **35.** $\frac{1}{4}, \frac{3}{8}, \frac{1}{2}, \frac{5}{8}, \frac{3}{4}$ **37.** yes, $d = 3$ **39.** no

30. $-5, -3, -1, 1, 3$ **32.** $2, -3, -8, -13, -18$ **34.** $\frac{1}{3}, \frac{1}{2}, \frac{2}{3}, \frac{5}{6}, 1$ **36.** $\frac{1}{6}, \frac{1}{2}, \frac{5}{6}, \frac{7}{6}, \frac{3}{2}$ **38.** no **40.** yes, $d = -5$

41. yes, $d = -8$ **43.** yes, $d = 0$ **45.** no **47.** yes, $d = \sqrt{3}$ **49.** $59, 119$ **51.** $-19, -49$ **53.** $77, 167$

42. no **44.** yes, $d = 1$ **46.** no **48.** no **50.** $-20, -50$ **52.** $71, 146$ **54.** $25, 55$

55. $-58, -118$ **57.** $\frac{31}{6}, \frac{61}{6}$ **59.** $78, 820$ **61.** $24, -200$ **63.** $48, 1000$ **65.** $-72, -800$ **67.** $8, \frac{220}{3}$

56. $33, 78$ **58.** $5, \frac{85}{8}$ **60.** $18, -220$ **62.** $81, 970$ **64.** $12, 320$ **66.** $-9, 390$ **68.** $\frac{33}{8}, \frac{265}{4}$

69. $191.8, 395.8, 19384$ **71.** $-15063, -37163, -1528400$ ***73.** $a_n = n^2$ ***75.** $a_n = 2^{n-1}$

70. $-23.07, -56.57, -2340.5$ **72.** $87336, 176436, 8822700$ ***74.** $a_n = 2^n - 1$ ***76.** $a_n = (-1)^{n+1}$

---

*denotes Superset exercises

***77.** $a_n = 2^{n-2}$, for $n \geq 2$ ***79.** 2.27 ***81.** 3, 7, 21 ***83.** 6, 2, −12 ***85.** \$375 ***87.** 666 ***89.** 8,062,500

***78.** $a_n = 2 - \frac{1}{2^{n-1}}$ ***80.** 0.00505 ***82.** 0, 4, 18 ***84.** 8, 8, 8 ***86.** 16 ***88.** $\frac{23}{4}, \frac{15}{2}, \frac{37}{4}$ ***90.** 8,051,750

Section 10.2, p. 607

1. no **3.** yes, $r = \frac{3}{2}$ **5.** no **7.** $32, 256, 2^{n-1}$ **9.** $64, 512, 2^n$ **11.** $128, 8192, \left(\frac{1}{8}\right)4^{n-1}$

2. no **4.** no **6.** yes, $r = -\frac{3}{2}$ **8.** $243, 6561, 3^{n-1}$ **10.** $-\frac{1}{4}, \frac{1}{32}, 8\left(-\frac{1}{2}\right)^{n-1}$ **12.** $-\frac{32}{9}, \frac{256}{243}, 27\left(-\frac{2}{3}\right)^{n-1}$

13. $\frac{1}{3}, -\frac{1}{81}, -81\left(-\frac{1}{3}\right)^{n-1}$ **15.** $81, 2187, 3^{n-2}$ **17.** $9\sqrt{6}, 81\sqrt{2}, \sqrt{2}(\sqrt{3})^{n-1}$ **19.** $-0.000001, 0.000000001, 0.1(-0.1)^{n-1}$

14. $-4, 32, \frac{1}{8}(-2)^{n-1}$ **16.** $8, 16\sqrt{2}, (\sqrt{2})^n$ **18.** $\frac{243}{2}, -\frac{6561}{16}, -16\left(-\frac{3}{2}\right)^{n-1}$ **20.** $\frac{1}{243}, \frac{1}{6561}, \left(\frac{1}{3}\right)^{n-1}$

21. $\frac{4}{125}, \frac{32}{15{,}625}, \frac{25}{8}\left(\frac{2}{5}\right)^{n-1}$ **23.** $2^n - 1, 127$ **25.** $-2(1 - 2^n), 254$ **27.** $\frac{1 - 4^n}{-24}, \frac{5461}{8}$

22. $0.000032, 0.000000256, 0.1(0.2)^{n-1}$ **24.** $\frac{1 - 3^n}{-2}, 1093$ **26.** $\frac{8[1 - (-1/2)^n]}{3}, \frac{43}{16}$ **28.** $\frac{81}{5}\left[1 - \left(-\frac{2}{3}\right)^n\right], \frac{463}{27}$

29. $\frac{-243[1 - (-1/3)^n]}{4}, -\frac{547}{9}$ **31.** $\frac{1 - 3^n}{-6}, \frac{1093}{3}$ **33.** $\frac{\sqrt{2} + \sqrt{6} - \sqrt{2}(\sqrt{3})^n - \sqrt{2}(\sqrt{3})^{n+1}}{-2}, 40\sqrt{2} + 13\sqrt{6}$

30. $\frac{1 - (-2)^n}{24}, \frac{43}{8}$ **32.** $(\sqrt{2})^{n+2} + (\sqrt{2})^{n+1} - \sqrt{2} - 2, 14 + 15\sqrt{2}$ **34.** $\frac{-32[1 - (-3/2)^n]}{5}, -\frac{463}{4}$

35. $\frac{[1 - (-0.1)^n]}{11}, 0.0909091$ **37.** $\frac{125[1 - (2/5)^n]}{24}, \frac{25{,}999}{5000}$

36. $\frac{3[1 - (1/3)^n]}{2}, \frac{1093}{729}$ **38.** $\frac{(1 - (0.2)^n)}{8}, 0.1249984$

***39.** Since $a_1, a_2, a_3, a_4, \ldots$ is a geometric sequence, we know that for $n \geq 1$, $a_{n+1}/a_n = r$. Thus, $a_{n+2}/a_n = r^2$.

***40.** Since $2^{a_1}, 2^{a_2}, 2^{a_3}, 2^{a_4}, \ldots$ is a geometric sequence, we know that for $n \geq 1$, $2^{a_{n+1}}/2^{a_n} = 2^{a_{n+1}-a_n} = r$. Thus, $a_{n+1} - a_n = \log_2 r$.

***41.** 9 ft, 3 ft, $\frac{1}{3}$ ft ***43.** 45 ft, 51 ft, $53\frac{2}{3}$ ft ***45.** \$80, \$86.40; \$469.33

***42.** 65.61% ***44.** \$19,660.80 ***46.** −1

***47.** Since $a \neq b$, we have $0 < (a - b)^2$ so that $4ab < a^2 + 2ab + b^2$ and then $2\sqrt{ab} < a + b$.

***48.** Apply the Pythagorean Theorem to each of the three triangles in the figure.

Section 10.3, p. 610

1. $1 + 4 + 7 + 10 + 13 = 35$ **3.** $13 + 8 + 1 = 22$ **5.** $3 + 3 + 3 + 3 = 12$ **7.** $0 + 1 - 2 + 3 - 4 = -2$

2. $7 + 5 + 3 + 1 - 1 - 3 = 12$ **4.** $12 + 12 + 10 = 34$ **6.** $4 + 3 + 2 + 1 = 10$ **8.** $24 + 64 + 160 + 384 = 632$

9. $\sum_{n=3}^{6} n$ **11.** $\sum_{n=1}^{5} n^2$ **13.** $\sum_{n=1}^{6} 2n$ **15.** $\sum_{n=1}^{4} (-1)^n$ **17.** 14, 91 **19.** 21, 1365 **21.** $\frac{7}{9}, \frac{182}{243}$ **23.** $\frac{95}{9}, \frac{3325}{243}$

10. $\sum_{n=7}^{10} n$ **12.** $\sum_{n=1}^{4} n^3$ **14.** $\sum_{n=1}^{5} 4n$ **16.** $\sum_{n=1}^{4} 5$ **18.** 20, −135 **20.** 43, 9331 **22.** $\frac{39}{25}, \frac{5187}{3125}$ **24.** $\frac{343}{25}, \frac{52{,}136}{3125}$

25. 8, 105 **27.** $\frac{125}{7}$ **29.** series has no sum **31.** $\frac{20}{7}$ **33.** series has no sum **35.** $\frac{13}{12}$ **37.** $-\frac{3}{2}$ **39.** 6.4

26. −10, 43 **28.** 256 **30.** series has no sum **32.** series has no sum **34.** $\frac{21}{2}$ **36.** $\frac{319}{420}$ **38.** $-\frac{11}{6}$ **40.** 1.6

41. 0.063 **43.** $\frac{242}{1125}$ **45.** $\frac{1082}{2475}$ **47.** $\frac{44}{333}$ **49.** $\frac{8}{2475}$ ***51.** $\sum_{n=1}^{\infty} 4(-1/3)^{n-1}$

---

*denotes Superset exercises

42. 0.233 **44.** $\frac{128}{225}$ **46.** $\frac{2339}{9900}$ **48.** $\frac{395}{999}$ **50.** $\frac{1}{11}$ ***52.** There are two such series: $\sum_{n=1}^{\infty} 6\left(\frac{1}{3}\right)^{n-1}$ and $\sum_{n=1}^{\infty} 3\left(\frac{2}{3}\right)^{n-1}$

***53.** $1, 2, 4, 8, 2^{n-1}$ ***55.** $3, 7, 11, 15, 4n-1$ ***57.** $-3, -1, 1, 3, 2n-5$ ***59.** $\sum_{n=1}^{\infty} (3n+1)$ ***61.** $\sum_{n=1}^{\infty} (n^2+1)$

***54.** $3, 5, 7, 9, 2n+1$ ***56.** $1, 3, 9, 27, 3^{n-1}$ ***58.** $1, 5, 9, 13, 4n-3$ ***60.** $\sum_{n=1}^{\infty} (2n+1)$ ***62.** $\sum_{n=1}^{\infty} (2^n+1)$

***63.** $\sum_{n=1}^{\infty} (3^n - 1)$ ***65.** $\frac{1}{2}$ ***67.** 54 feet ***69.** \$21,474,836.47 ***71.** \$20,807.68

***64.** $\sum_{n=1}^{\infty} (n^3 + 2)$ ***66.** $\frac{7}{24}$ ***70.** \$10,485.76 ***72.** 18th day

Section 10.4, p. 620

***17.** use: If $x^m x^k = x^{m+k}$, then $(x^m x^k)x = (x^{m+k})x$ so that $x^m(x^k \cdot x) = (x^{m+k}) \cdot x^1$ and thus $x^m x^{k+1} = x^{(m+k)+1} = x^{m+(k+1)}$
***18.** use: $(ab)^{b+1} = (ab)^1 (ab)^k = a^1 b^1 a^k b^k = a^k a^1 b^k b^1 = a^{k+1} b^{k+1}$
***19.** use: $2(k+1) = 2k + 2 \le 2^k + 2 \le 2^k + 2^k = 2^k + 1$
***20.** use: $1 + 2(k+1) = (1+2k) + 2 \le 3^k + 2 \le 3^k + 2 \cdot 3^k = 3^{k+1}$
***21.** from $k^2 + k > k^2$ we get $\sqrt{k}\sqrt{k+1} + 1 > k + 1$ and thus $\sqrt{k} + \frac{1}{\sqrt{k+1}} > \sqrt{k+1}$.
***25.** use: $x^{k+1} = x^k \cdot x > 1 \cdot x = x > 1$ ***27.** use: $(k+1)^2 + (k+1) = (k^2 + k) + 2(k+1)$
***26.** use: $0 < x^k \cdot x < 1 \cdot x = x < 1$ ***28.** use: $(k+1)^3 + 2(k+1) = (k^3 + 2k) + 3(k^2 + k + 1)$
***29.** use: $5^{k+1} - 1 = 5(5^k - 1) + (5 - 1)$
***30.** use: $6^{k+1} - 1 = 6(6^k - 1) + (6 - 1)$
***31–32.** Arbitrarily select one of the vertices of a $k+1$-sided convex polygon P and label it A. Label the two vertices adjacent to A as B and C. Since P is convex, diagonal BC is interior to P and subdivides P into a triangle (ABC) and a k-sided convex polygon P' having BC as one of its sides.

Section 10.5, p. 624

1. 6 **3.** 2 **5.** 1 **7.** 36 **9.** 6 **11.** 7 **13.** 28
2. 24 **4.** 40,320 **6.** 48 **8.** 1 **10.** 5 **12.** 20 **14.** 35
15. $x^6 + 6x^5y + 15x^4y^2 + 20x^3x^3 + 15x^2y^4 + 6xy^5 + y^6$
16. $x^8 + 8x^7y + 28x^6y^2 + 56x^5y^3 + 70x^4y^4 + 56x^3y^5 + 28x^2y^6 + 8xy^7 + y^8$
17. $a^7 - 7a^6b + 21a^5b^2 - 35a^4b^3 + 35a^3b^4 - 21a^2b^5 + 7ab^6 - b^7$ **19.** $81 + 216p + 216p^2 + 96p^3 + 16p^4$
18. $a^6 - 6a^5b + 15a^4b^2 - 20a^3b^3 + 15a^2b^4 - 6ab^5 + b^6$ **20.** $8 + 60s + 150s^2 + 125s^3$
21. $16 + 96p + 216p^2 + 216p^3 + 81p^4$ **23.** $\frac{t^3}{27} - \frac{t^2}{3} + t - 1$ **25.** $z^5 + 5z^3 + 10z + 10z^{-1} + 5z^{-3} + z^{-5}$
22. $125 + 150s + 60s^2 + 8s^3$ **24.** $\frac{a^4}{16} - \frac{a^3}{2} + \frac{3}{2}a^2 - 2a + 1$ **26.** $p^4 - 4p^2 + 6 - 4p^{-2} + p^{-4}$
27. $1 - 6\sqrt{x} + 15x - 20x\sqrt{x} + 15x^2 - 6x^2\sqrt{x} + x^3$ **29.** $s + 3s^{4/3} + 3s^{5/3} + s^2$ **31.** $1 + 12x + 66x^2 + 220x^3$
28. $1 + 5\sqrt{t} + 10t + 10t\sqrt{t} + 5t^2 + t^2\sqrt{t}$ **30.** $p + 4p^{3/2} + 6p^2 + 4p^{5/2} + p^3$ **32.** $1 - 10y + 45y^2 - 120y^3$
33. $256 - 1024s + 1792s^2 - 1792s^3$ **35.** $1 + 22p + 220p^2 + 1320p^3$ **37.** $1140t^3$ **39.** 70 **41.** $55y^{13/2}$
34. $19683 + 59049p + 78732p^2 + 61236p^3$ **36.** $1 + 36x + 594x^2 + 5940x^3$ **38.** $3060s^4$ **40.** $11520y^2$ **42.** 924
43. $12870p^{16}$ ***45.** $i + \frac{1}{2}\left(1 - \frac{1}{n}\right)i^2 + \frac{1}{6}\left(1 - \frac{3}{n} + \frac{2}{n^2}\right)i^3$ ***47.** follow the hint ***49.** 3^n
44. $28y$ ***46.** \$1061.83 ***48.** follow the hint with $x = -1$ ***50.** $\binom{n+k+1}{n+1}$

*denotes Superset exercises

***51.** Let k be an integer such that $1 \le k \le p-1$. Since $\binom{p}{k} = \frac{p!}{(p-k)!k!} = p \cdot \frac{(p-1)!}{(p-k)!k!}$ is an integer, the factors of the denominator $(p-k)!k!$ must divide evenly into the numerator $p!$. Since all the factors of the denominator are less than p and since p is a prime, the factors of $(p-k)!k!$ must in fact divide evenly into $(p-1)!$ That is, $\binom{p}{k} = pq$, where q is some integer.

If we let $p = 4$ and $k = 2$, $\binom{p}{k} = \binom{4}{2} = 6$ and 6 is not divisible by 4. ***53.** 1.21896, 1.21899442

***52.** 1.082856, 1.082856706 ***54.** 0.81704, 0.817072806 ***56.** 128.44867256, 128.4486726 ***58.** 1.030423662, 1.03
***55.** 0.922744, 0.922744694 ***57.** 19742.127793236, 19742.12779 ***59.** 1.072246668, 1.07
***60.** 1.2214, 1.221402758; 1.6484375, 1.648721271; 1.8214, 1.8221188; 2.1147461, 2.117000017; 2.2224, 2.225540928

Section 10.6, p. 631

1. (a) 1 (b) 3 (c) -1 **3.** (a) does not exist (b) does not exist (c) 0 **5.** (a) -3 (b) $\frac{3}{2}$
2. (a) -2 (b) 3 (c) 1 **4.** (a) -1 (b) 3 (c) does not exist **6.** (a) -2 (b) does not exist
7. (a) does not exist (b) does not exist **9.** $\frac{3}{2}$ **11.** 0 **13.** does not exist **15.** 0 ***17.** $\lim_{n\to\infty} \frac{2[1-(-1/2)^n]}{3}$
8. (a) does not exist (b) 0 **10.** 0 **12.** does not exist **14.** 7 **16.** 0 ***18.** $\lim_{n\to\infty} \frac{3[1-(1/3)^n]}{2}$
***19.** $\lim_{n\to\infty} 24\left[1-\left(\frac{3}{4}\right)^n\right]$ ***21.** $\lim_{n\to\infty}\left(\frac{1}{2}-\frac{1}{n+2}\right)$ ***23.** $\lim_{n\to\infty}\left(-\frac{13}{12}+\frac{1}{n+2}+\frac{1}{n+3}+\frac{1}{n+4}\right)$ ***25.** 0 ***27.** 6
***20.** $\lim_{n\to\infty} 5\left[1-\left(-\frac{2}{5}\right)^n\right]$ ***22.** $\lim_{n\to\infty}\left(\frac{3}{2}-\frac{1}{n+1}-\frac{1}{n+2}\right)$ ***24.** $\lim_{n\to\infty}\left(-\frac{1}{6}+\frac{1}{n+6}\right)$ ***26.** 2 ***28.** $2a$
***29.** 0 ***31.** 4.01, 4.001, 4.0001, 4 ***33.** 32.24, 32.024, 32.0024, 32
***30.** $3a^2$ ***32.** 12.06, 12.006, 12.0006, 12 ***34.** 80.8, 80.08, 80.008, 80
***35.** (a) approx: 1.2496, exact: 1.25, difference: 0.0004 (b) approx: 1.9375, exact: 2, difference: 0.0625 (c) approx: 2.3056, exact: 2.5, difference: 0.1944, approx: 3.0507812, exact: 4, difference: 0.9492188, approx: 3.3616, exact: 5, difference: 1.6384 (d) The five-term approximation is more accurate the closer x is to zero.
***36.** 0.18233067, 0.18232156, 0.00000911, 0.40729167, 0.40546511, 0.00182656, 0.475152, 0.47000363, 0.00514837, 0.57773438, 0.55961579, 0.01811859, 0.61380267, 0.58778667, 0.026016 The difference increases as x gets further from zero.
***37.** $0.1\overline{6}$, 0.25, 0.30, $0.3\overline{3}$, 0.357, 0.375, $0.3\overline{8}$, 0.4, $0.\overline{409}$, $0.41\overline{6}$
***38.** 0.3333, 0.4583, 0.5250, 0.5667, 0.5952, 0.6161, 0.6319, 0.6444, 0.6545, 0.6629

Section 10.7, p. 636

1. The tree with branches HHH, HHT, HTH, HTT, THH, THT, TTH, TTT.
2. The tree with branches TTTT, TTTF, TTFT, TTFF, TFTT, TFTF, TFFT, TFFF, FTTT, FTTF, FTFT, FTFF, FFTT, FFTF, FFFT, FFFF.
3. (a) The tree with branches 11, 12, 13, 14, 21, 22, 23, 24, 31, 32, 33, 34, 41, 42, 43, 44. (b) The tree with branches 12, 13, 14, 21, 23, 24, 31, 32, 34, 41, 42, 43.
4. The tree with branches MAT, MAH, MTA, MTH, MHA, MHT, AMT, AMH, ATM, ATH, AHM, AHT, TMA, TMH, TAM, TAH, THM, THA, HMA, HMT, HAM, HAT, HTM, HTA. **6.** 32 **8.** 2450
5. Let R, B, V, G, S, A, N, M, and F be abbreviations for red, blue, silver, beige, standard, automatic, no radio, AM radio, and AM / FM radio, respectively. Then the answer is the tree with branches RSN, RSM, RSF, RAN, RAM, RAF, BSN, BSM, BSF, BAN, BAM, BAF, VSN, VSM, VSF, VAN, VAM, VAF, GSN, GSM, GSF, GAN, GAM, GAF. **7.** 3125
9. 40,320 **11.** (a) 10^7 (b) 8×10^9 **13.** (a) 72 (b) 48 **15.** (a) yes (b) 10^7
10. (a) 11,232,000 (b) 17,576,000 (c) 15,818,400 **12.** 96 **14.** (a) 120 (b) 24 **16.** (a) 380 (b) 266 (c) 30
17. 840 **19.** 32,760 **21.** 1 **23.** 24 **25.** 2730 **27.** 406,535,035,200 **29.** 28 **31.** 715 **33.** 1

---

*denotes Superset exercises

18. 1320 **20.** 55,440 **22.** 24 **24.** 40,320 **26.** 104,652 **28.** 117,600 **30.** 84 **32.** 10 **34.** 1
35. 220 **37.** 3003 **39.** 120 **41.** (a) 60 (b) 125 (c) 12 **43.** (a) 210 (b) 140 (c) 21
36. 38,760 **38.** 2,598,960 **40.** 120 **42.** (a) 360 (b) 1296 (c) 180 **44.** (a) 364 (b) 364 (c) 4
***45.** The tree with branches B, WB, WWB, WWWB. ***47.** The tree with branches EG, GE, GG.
***46.** (a) 64 (b) 12 (c) 46 (d) 32 ***48.** (a) 4 (b) 2
***49.** (a) The tree with branches A, BA, BCA, CA, CBA. (b) If we permit replacement, then the tree would have an infinite number of branches that extend forever. ***51.** 5040
***50.** The tree with branches BR, BWR, BWWR, RB, RWB, RWWB, WBR, WBWR, WRB, WRWB, WWBR, WWRB. ***52.** 8
***53.** $\binom{n}{n} = \frac{n!}{(n-n)!n!} = \frac{n!}{0!n!} = \frac{n!}{n!} = 1, \binom{n}{0} = \frac{n!}{(n-0)!0!} = \frac{n!}{n!0!} = \frac{n!}{n!} = 1$ ***55.** 30 ***57.** 7 ***59.** 32
***54.** $\binom{n}{r} = \frac{n!}{(n-r)!r!} = \frac{n!}{r!(n-r)!} = \frac{n!}{[n-(n-r)]!(n-r)!} = \binom{n}{n-r}$ ***56.** 34,650 ***58.** (a) 5 (b) 10 (c) 10 (d) 5
***60.** 2^n

Section 10.8, p. 647

1. $\frac{1}{6}$ **3.** $\frac{11}{36}$ **5.** $\frac{1}{4}$ **7.** 0 **9.** $\frac{1}{9}$ **11.** $\frac{2}{5}$ **13.** $\frac{3}{5}$ **15.** 1 **17.** (a) $\frac{15}{28}$ (b) $\frac{3}{28}$ (c) $\frac{5}{14}$
2. $\frac{5}{12}$ **4.** $\frac{5}{9}$ **6.** $\frac{1}{9}$ **8.** 1 **10.** $\frac{1}{36}$ **12.** $\frac{3}{5}$ **14.** $\frac{2}{5}$ **16.** 0 **18.** (a) $\frac{14}{55}$ (b) $\frac{28}{55}$ (c) $\frac{1}{55}$ (d) $\frac{12}{55}$
19. (a) $\frac{1}{16}$ (b) $\frac{1}{16}$ (c) $\frac{1}{4}$ (d) $\frac{3}{8}$ (e) $\frac{1}{4}$ **21.** 0.98 **23.** $\frac{4}{5}$ **25.** $\frac{1}{5}$ **27.** $\frac{2}{5}$
20. (a) $\frac{11}{850}$ (b) $\frac{1}{5525}$ (c) $\frac{6}{5525}$ **22.** 0.003 **24.** $\frac{4}{5}$ **26.** 0 **28.** 0
29. (a) S = {TTTTT, HTTTT, THTTT, TTHTT, TTTHT, TTTTH, HHTTT, HTHTT, HTTHT, HTTTH, THHTT, THTHT, THTTH, TTHHT, TTHTH, TTTHH, HHHTT, HHTHT, HHTTH, HTHHT, HTHTH, HTTHH, THHHT, THHTH, THTHH, TTHHH, HHHHT, HHHTH, HHTHH, HTHHH, THHHH, HHHHH} (b) E_1 = {HHHTT, HHTHT, HHTTH, HTHHT, HTHTH, HTTHH, THHHT, THHTH, THTHH, TTHHH, HHHHT, HHHTH, HHTHH, HTHHH, THHHH, HHHHH} E_2 = {TTTTT, HTTTT, THTTT, TTHTT, TTTHT, TTTTH, HHTTT, HTHTT, HTTHT, HTTTH, THHTT, THTHT, THTTH, TTHHT, TTHTH, TTTHH} E_3 = {HTTTT, THTTT, TTHTT, TTTHT, TTTTH, HHHHT, HHHTH, HHTHH, HTHHH, THHHH} (c) $E_1 \cup E_2$: those outcomes that have at least three heads or at most two heads. $P(E_1 \cup E_2) = 1$ (d) $E_1 \cup E_3$: those outcomes that have at least three heads or exactly four tosses the same. $P(E_1 \cup E_3) = \frac{21}{32}$
30. (a) S = {AAA, BAA, ABA, AAB, BBA, BAB, ABB, BBB, CAA, ACA, AAC, CCA, CAC, ACC, CCC, CBB, BCB, BBC, CCB, CBC, BCC, ABC, ACB, BAC, BCA, CAB, CBA} (b) E_1 = {CCA, CAC, ACC, CCB, CBC, BCC}; E_2 = {AAA, BAA, ABA, AAB, BBA, BAB, ABB, CAA, ACA, AAC, CCA, CAC, ACC, ABC, ACB, BAC, BCA, CAB, CBA}; E_3 = {AAA, BAA, ABA, AAB, BBA, BAB, ABB, BBB} (c) $E_1 \cup E_3$: those outcomes having exactly two C's or no C's. $P(E_1 \cup E_3) = \frac{14}{27}$ (d) $E_2 \cup E_3$: those outcomes having at least one A or no C's. $P(E_2 \cup E_3) = \frac{20}{27}$
***31.** If we add the probability of tossing a head to the probability of tossing a tail, we obtain the value of 1.1.
***32.** If we add the probability that the car will break down to the probability that it will not break down, we obtain the value 1.1.
***33.** The probability of any event cannot be greater than 1.
***34.** The probability that you will pass or fail the next exam is 1.
***35.** Since the two events are mutually exclusive, the probability that either Tina or Bob will get a part is 0.3 + 0.25, or 0.55.
***36.** Since the two events are mutually exclusive, if their combined probability is 0.08 then the probability that the horse will come in second is 0.08 − 0.15 = −0.07.

*denotes Superset exercises

***37.** (a) The region outside E_1; 0.75 (b) The region outside E_2; 0.58 (c) The region inside E_1 or E_2; 0.67 (d) The sample space; 1 (e) The region outside the two sets; 0.33 (f) No region; 0 (g) This shaded diagram is the same as the one in part (e); 0.33 (h) This shaded diagram is the same as the one in part (d); 1

***38.** (a) The region outside E_1; 0.75 (b) The region outside E_2; 0.58 (c) The region inside E_1 or E_2; 0.57 (d) The region outside $E_1 \cap E_2$; 0.90 (e) The region outside $E_1 \cup E_2$; 0.43 (f) This shaded diagram is the same as the one in part (e); 0.43 (g) This shaded diagram is the same as the one in part (d); 0.90

Chapter 10 Review Exercises, p. 659

1. 0, 4, 0, 8, 0, 40 **3.** 8, 5, 0, −7, −72, −391 **5.** 2, −3, 7, −13, 27, −53 **7.** 1, 2, 2, 3, 4, 6
2. −4, −3, 0, 5, 60, 357 4. 13, 18, 5, 30, −67, 414 6. 1, 3, 6, 10, 15, 21 **8.** 1, 3, 5, 6, 6, 6
9. 2, 5, 8, 11, 14, 29, 610 **11.** 3, 5, 7, 9, 11, 21, 440 **13.** 243, 3^n, 1092
10. −2, 2, 6, 10, 14, 34, 720 **12.** 4, 2, 0, −2, −4, −14, −300 **14.** $-\frac{81}{2}, -128\left(-\frac{3}{4}\right)^{n-1}, -\frac{481}{8}$
15. $2, \frac{1}{8}(-2)^{n-1}, -\frac{21}{8}$ **17.** $\sum_{n=1}^{5}(5n-3)$ **19.** $\sum_{n=1}^{6} 3(-2)^{n-1}$ **21.** 3 **23.** $\frac{1}{6}$ **25.** series has no sum
16. $0.0048, 3(0.2)^{n-1}, 3.74976$ **18.** $\sum_{n=1}^{6}(-1)^n(3n-1)$ **20.** $\sum_{n=1}^{5} 20(0.1)^{n-1}$ **22.** 3 **24.** series has no sum **26.** −1
27. $-\frac{5}{6}$ **29.** $\frac{1031}{3300}$ **31.** $\frac{703}{2250}$ **35.** $81x^4 - 216x^3y + 216x^2y^2 - 96xy^3 + 16y^4$ **37.** $190x^{16}$
28. $-\frac{25}{12}$ **30.** $\frac{2033}{4500}$ **32.** $\frac{4513}{9990}$ **36.** $32x^5 - 80x^4y + 80x^3y^2 - 40x^2y^3 + 10xy^4 - y^5$ **38.** $1820y^8$
39. $-43008p^3$ **41.** (a) 0 (b) does not exist (c) 2 (d) 2 (e) 2 **43.** $\frac{2}{5}$ **45.** 0
40. $-196{,}830\,t^{19/2}$ **42.** (a) 2 (b) 2 (c) 0 (d) 1 (e) 0 **44.** does not exist **46.** $\frac{7}{3}$
47. The tree with branches AA, AS, AN, SA, SS, SN, NA, NS, NN. **49.** 120 **51.** 60 **53.** 30 **55.** 3,628,800
48. The tree with branches MC, MH, MP, SC, SH, SP. **50.** 1050 **52.** 504 **54.** 1 **56.** 6720
57. 1 **59.** 126 **61.** 5040 **63.** 96 **65.** 240 **67.** (a) $\frac{1}{126}$ (b) $\frac{20}{63}$ (c) $\frac{5}{126}$ **69.** $\frac{6}{7}$
58. 21 **60.** 1 **62.** 144 **64.** 240 **66.** 480 **68.** (a) $\frac{33}{66{,}640}$ (b) $\frac{1}{54{,}145}$ (c) $\frac{33}{54{,}145}$ **70.** 0

Chapter 10 Test, p. 661

1. 1, 2, 5, 12, 27, 503 **3.** 7, 2, −3, −8, −13 **5.** 222 **7.** $-\frac{16}{45}, \frac{128}{1215}, \frac{27}{10}\left(-\frac{2}{3}\right)^{n-1}$
2. (a) 2, 7, −2, 3, 2, 7 (b) 1, 3, 2, 6, 11, 19 **4.** 63, 123 **6.** (a) no (b) yes **8.** $6\left[1-\left(\frac{2}{3}\right)^n\right], \frac{422}{81}$
9. $-2 + 0 + 4 + 10 + 18 = 30$ **11.** $\frac{21}{4}, \frac{189}{32}$ **13.** $\frac{41}{110}$ **15.** $81 + 540t + 1350t^2 + 1500t^3 + 625t^4$ **17.** $-\frac{1}{5}$
10. $\sum_{n=2}^{6} n^2$ **12.** (a) 28 (b) $-\frac{1}{3}$ **16.** (a) does not exist (b) 2 (c) does not exist (d) 3 (e) 4 (f) −3 **18.** 0
19. does not exist **21.** 256 **23.** 75,287,520 **25.** (a) 84 (b) 7 (c) 21 **27.** (a) 1 (b) $\frac{1}{4}$ (c) $\frac{1}{2}$
20. The tree with branches GGG, GGB, GBG, GBB, BGG, BGB, BBG, BBB. **22.** (a) 1680 (b) 39,916,800 (c) 84 (d) 1
24. (a) 120 (b) 625 (c) 24 **26.** (a) $\frac{1}{8}$ (b) $\frac{3}{8}$ (c) $\frac{7}{8}$

---

*denotes Superset exercises

Index

Reciprocal Identities

$$\sin\theta = \frac{1}{\csc\theta} \qquad \cot\theta = \frac{1}{\tan\theta}$$

$$\cos\theta = \frac{1}{\sec\theta} \qquad \sec\theta = \frac{1}{\cos\theta}$$

$$\tan\theta = \frac{1}{\cot\theta} \qquad \csc\theta = \frac{1}{\sin\theta}$$

Quotient Identities

$$\tan\theta = \frac{\sin\theta}{\cos\theta} \qquad \cot\theta = \frac{\cos\theta}{\sin\theta}$$

Pythagorean Identities

$$\sin^2\theta + \cos^2\theta = 1$$

$$\tan^2\theta + 1 = \sec^2\theta$$

$$1 + \cot^2\theta = \csc^2\theta$$

Sum and Difference Identities

$$\cos(\alpha - \beta) = \cos\alpha\cos\beta + \sin\alpha\sin\beta$$

$$\cos(\alpha + \beta) = \cos\alpha\cos\beta - \sin\alpha\sin\beta$$

$$\sin(\alpha + \beta) = \sin\alpha\cos\beta + \cos\alpha\sin\beta$$

$$\sin(\alpha - \beta) = \sin\alpha\cos\beta - \cos\alpha\sin\beta$$

$$\tan(\alpha + \beta) = \frac{\tan\alpha + \tan\beta}{1 - \tan\alpha\tan\beta}$$

$$\tan(\alpha - \beta) = \frac{\tan\alpha - \tan\beta}{1 + \tan\alpha\tan\beta}$$

Double-Angle Identities

$$\sin 2\theta = 2\sin\theta\cos\theta$$

$$\cos 2\theta = \cos^2\theta - \sin^2\theta$$

$$\cos 2\theta = 1 - 2\sin^2\theta$$

$$\cos 2\theta = 2\cos^2\theta - 1$$

$$\tan 2\theta = \frac{2\tan\theta}{1 - \tan^2\theta}$$

Half-Angle Identities

$$\sin\frac{\theta}{2} = \pm\sqrt{\frac{1 - \cos\theta}{2}} \qquad \tan\frac{\theta}{2} = \frac{\sin\theta}{1 + \cos\theta}$$

$$\cos\frac{\theta}{2} = \pm\sqrt{\frac{1 + \cos\theta}{2}} \qquad \tan\frac{\theta}{2} = \frac{1 - \cos\theta}{\sin\theta}$$

$$\tan\frac{\theta}{2} = \pm\sqrt{\frac{1 - \cos\theta}{1 + \cos\theta}}$$

Conversion Identities

$$\sin\alpha\cos\beta = \frac{1}{2}[\sin(\alpha + \beta) + \sin(\alpha - \beta)]$$

$$\sin\alpha\sin\beta = \frac{1}{2}[\cos(\alpha - \beta) - \cos(\alpha + \beta)]$$

$$\cos\alpha\cos\beta = \frac{1}{2}[\cos(\alpha + \beta) + \cos(\alpha - \beta)]$$

$$\sin x + \sin y = 2\sin\frac{x + y}{2}\cos\frac{x - y}{2}$$

$$\sin x - \sin y = 2\cos\frac{x + y}{2}\sin\frac{x - y}{2}$$

$$\cos x + \cos y = 2\cos\frac{x + y}{2}\cos\frac{x - y}{2}$$

$$\cos x - \cos y = -2\sin\frac{x + y}{2}\sin\frac{x - y}{2}$$

Law of Sines

In any triangle with angles A, B, and C, and opposite sides a, b, and c, respectively,

$$\frac{\sin A}{a} = \frac{\sin B}{b} = \frac{\sin C}{c}.$$

Law of Cosines

In any triangle with angles A, B, and C, and opposite sides a, b, and c, respectively,

$$c^2 = a^2 + b^2 - 2ab\cos C$$

$$a^2 = b^2 + c^2 - 2ab\cos A$$

$$b^2 = a^2 + c^2 - 2ac\cos B$$